VIIIA

IIIA	IVA	VA	VIA	VIIA	2 **He** Helium 4.00260

IB	IIB	5 **B** Boron 10.81	6 **C** Carbon 12.011	7 **N** Nitrogen 14.0067	8 **O** Oxygen 15.9994	9 **F** Fluorine 18.998403	10 **Ne** Neon 20.179
		13 **Al** Aluminum 26.98154	14 **Si** Silicon 28.0855	15 **P** Phosphorus 30.97376	16 **S** Sulfur 32.06	17 **Cl** Chlorine 35.453	18 **Ar** Argon 39.948
29 **Cu** Copper 63.546	30 **Zn** Zinc 65.38	31 **Ga** Gallium 69.72	32 **Ge** Germanium 72.59	33 **As** Arsenic 74.9216	34 **Se** Selenium 78.96	35 **Br** Bromine 79.904	36 **Kr** Krypton 83.80
47 **Ag** Silver 107.868	48 **Cd** Cadmium 112.41	49 **In** Indium 114.82	50 **Sn** Tin 118.69	51 **Sb** Antimony 121.75	52 **Te** Tellurium 127.60	53 **I** Iodine 126.9045	54 **Xe** Xenon 131.30
79 **Au** Gold 196.9665	80 **Hg** Mercury 200.59	81 **Tl** Thallium 204.37	82 **Pb** Lead 207.2	83 **Bi** Bismuth 208.9804	84 **Po** Polonium (209)	85 **At** Astatine (210)	86 **Rn** Radon (222)

Left column partial: 28 **Ni** Nickel 58.70 · 46 **Pd** Palladium 106.4 · 78 **Pt** Platinum 195.09

64 **Gd** Gadolinium 157.25	65 **Tb** Terbium 158.9254	66 **Dy** Dysprosium 162.50	67 **Ho** Holmium 164.9304	68 **Er** Erbium 167.26	69 **Tm** Thulium 168.9342	70 **Yb** Ytterbium 173.04	71 **Lu** Lutetium 174.97
96 **Cm** Curium (247)	97 **Bk** Berkelium (247)	98 **Cf** Californium (251)	99 **Es** Einsteinium (254)	100 **Fm** Fermium (257)	101 **Md** Mendelevium (258)	102 **No** Nobelium (255)	103 **Lr** Lawrencium (260)

metals

nonmetals

metalloids

*Name not officially assigned

Values of atomic mass in parentheses are estimated and represent, in most cases, the mass number of the most stable isotope.

BASIC CHEMISTRY

Basic Chemistry

GENERAL, ORGANIC, BIOLOGICAL

DENIS M. CALLEWAERT

JULIEN GENYEA

DEPARTMENT OF CHEMISTRY

OAKLAND UNIVERSITY

WORTH PUBLISHERS, INC.

Worth Publishers, Inc.

444 Park Avenue South

New York, New York 10016

PREFACE

This textbook is the product of many years of experience teaching a two-semester sequence of courses in general, organic, and biological chemistry for students who have not necessarily had any prior exposure to chemistry. With this book, students in nursing, allied health, agriculture, home economics, and similar programs—where the study of chemistry is limited to two courses—can acquire a good understanding of general chemical concepts and fundamental biochemistry.

Anyone teaching an introductory science course must contend with the dilemma of how to treat material at an elementary level without introducing drastic oversimplifications that might result in misinterpretations. We have attempted to explain concepts in sufficient qualitative detail to avoid oversimplification. Based on our classroom experience, we have tried to provide especially careful explanations of difficult topics and to anticipate the most common student misconceptions. Students in our classes have found that the approach used here stimulates them intellectually without overtaxing their mathematical skills.

Basic Chemistry was written with the assumption that many students in the course have had no prior exposure to chemistry. It is our experience that the best introduction to chemistry, and the best preparation for any further study of this subject, must concentrate on explaining the essential chemical concepts, rather than providing a cursory treatment of a much larger number of topics. Also, it is very helpful for students to see where a particular topic fits into the whole picture, and to consider examples demonstrating the usefulness of the concepts being discussed. This textbook was written with these pedagogical principles constantly in mind.

Our selection of examples, exercises, problems, drawings, and photographs is intended to make the material relevant and interesting to all students. In addition, many of our examples and illustrations contain applications of chemical concepts that have been chosen to broaden students' awareness of how science affects their daily lives.

There is a unifying theme that is carried throughout this book—understanding the chemistry of the human body. The integration of organic and biochemical material into general chemistry, where appropriate, provides students with a holistic view of chemistry—encouraging them to see that the traditional branches of chemistry are parts of a unified science sharing common fundamental concepts.

Much of our discussion is qualitative in nature. We feel, however, that an adequate qualitative understanding of many chemical concepts can be achieved only if students work some simple associated quantitative problems. The mathematical skills required are arithmetic and occasionally simple algebra. Almost all of the problems are solved using a factor-unit framework that enables students to obtain numerical answers with ease and, at the same time, appreciate the chemical principles that are the basis for the factor-unit method of calculation. Our approach is designed to maximize students' understanding of what they are doing and why, when they work quantitative problems.

The essential features of a student study guide are contained as an integral part of the textbook. Chapters begin with an introduction that puts the material in the chapter into perspective, followed by a list of study objectives. These two brief sections are designed to help students focus on what they are expected to learn and why it is important. At the end of nearly every section in a chapter there are exercises designed to test whether the student has grasped the basic points covered in that section. Students are strongly encouraged to work these exercises before going on. Detailed solutions to them are presented at the end of the chapter so that students can test their comprehension. Each chapter ends with a concise summary list of the important concepts and relationships discussed in the chapter, followed by a set of problems. Answers to all of these problems are provided at the back of the book.

This text also includes self-contained units called Essential Skills. These units provide worked-out examples and exercises with complete solutions. Wherever an essential skill is needed in a chapter, students are reminded to turn to the appropriate Essential Skills unit in order to increase their facility with that skill.

The index uses a boldface number to identify the page on which a term is defined. Thus, students can easily find the definitions of terms in their correct context.

For the sake of flexibility, a few sections are clearly indicated as optional. These optional sections have been written so that they may be omitted without affecting the comprehension of subsequent topics.

Basic Chemistry is divided into five parts: General Chemistry, Organic Molecules, Biomolecules, Metabolism, and Essential Skills. The first four parts are divided into a total of 29 chapters. Each of these parts begins with an overview that outlines the material to be covered in the following group of chapters and places it in proper perspective.

General Chemistry

Beginning in Chapter 1, simple quantitative problems are included so that students who have difficulty with problem-solving can obtain or sharpen the necessary skills right at the beginning of the course. The backgrounds of students taking this course are quite varied, as are the skills they bring with them. Essential Skills 1 through 5—Exponential Numbers; Units (and Conversion of Units); Significant Figures: Rounding Off the Result of a Calculation; Direct Proportionality and the Factor-Unit Method; and Problem Solving—are meant to supplement Chapter 1, particularly for those students who are not well prepared. These topics are presented as separate units so as not to overburden Chapter 1, and to allow for maximum flexibility in the depth of coverage.

It is our experience that students have difficulty appreciating the significance of a measurement's uncertainty. In Chapter 1, we stress this point by demonstrating that how a measurement can be *used* depends on how *exact* it is.

We feel that the appropriate place for a discussion of nuclear reactions is in Chapter 2, along with a description of the basic features of atomic structure.

Treating nuclear reactions in Chapter 2 allows us to discuss an application of chemistry to human health early in the text. However, this discussion of nuclear reactions could easily be delayed until later in the course.

We describe some of the chemical and physical properties of several elements before discussing any model for the behavior of electrons in atoms. We also discuss the classification of compounds into ionic and covalent compounds based on their chemical and physical properties before presenting models for ionic and covalent bonds. We find this approach particularly useful for students who have had little prior exposure to chemistry.

We present in some qualitative detail the concept of free energy, since free-energy changes are an essential part of biochemical processes, and we use the concept of entropy to elucidate the important processes of diffusion and osmosis. We do not treat thermodynamic relationships quantitatively.

We have tried to help the many students who have difficulty with the mole concept: by carefully distinguishing between mass and number of atoms or molecules as measures of amount; by drawing an analogy between a mole and a dozen; and by explaining in detail why moles are a useful measure of amounts of chemical substances. Chemical stoichiometry is first discussed using mole amounts (the simplest way). Then it is shown that any problem involving mass amounts can easily be converted to an equivalent problem using moles. The practical questions of how to prepare solutions with specific concentrations and how to dilute solutions are treated in detail.

Rates of chemical reactions and chemical equilibria are treated qualitatively for the most part, but in sufficient depth that students will subsequently appreciate, for example, how a catalyst affects the rate of a chemical reaction, and how a buffer system controls the pH of a solution by shifting the position of equilibrium. An understanding of acid-base reactions is vital for any further study of chemistry. It is our experience that this is the topic students find most difficult. Our primarily qualitative treatment of acids and bases was written with the common student misconceptions in mind. Since the properties of buffer solutions are so important for an understanding of biochemical processes, we spend some time developing a clear picture of how a buffer system operates.

Our discussion of oxidation-reduction includes both inorganic and organic compounds as examples. In Chapter 12, we discuss reactions of electrolytes because at this point students have the background to consider all the major types of reactions of electrolytes in a unified manner.

These 12 chapters, together with the associated seven Essential Skills units, are also well suited for use in either a brief introductory chemistry course or a preparatory course for students who intend to take subsequently a standard sequence of two semesters of general chemistry and two semesters of organic chemistry. Therefore, these chapters and Essential Skills units are available separately under the title *Fundamentals of College Chemistry*.

Organic Molecules

In these seven chapters, the physical and chemical properties of simple organic molecules are discussed, with special emphasis on those classes of compounds whose functional groups are important components of biomolecules. Our treatment of organic chemistry is carefully limited to those topics that are fundamental to the understanding of basic biochemistry. Biochemical examples are often used in these chapters to illustrate the various types of organic reactions. Throughout our discussion of organic molecules, and later of biochemistry, we point out similarities and differences between related reactions as they occur in the human body and as they are carried out under laboratory conditions.

Biomolecules

These chapters treat the structures, properties, and reactivities of the major classes of biomolecules (with the exception of lipids, which are described later). Carbohydrates are covered in Chapter 20, and then three chapters are devoted to the crucial topic of proteins. The first discusses amino acids and the general structural features of proteins. The second describes the many functional roles of proteins in the human body, with separate sections devoted to transport proteins and to antibody structure and function. The next chapter presents the structure-function relationship of enzymes. The topics of enzyme kinetics, inhibition, and allosteric regulation are developed in sufficient qualitative detail to give students an appreciation for the precision of metabolic regulation and the molecular basis of drug action. The functions of coenzymes and their relationship to vitamins are discussed but detailed reaction mechanisms are not. The structural formulas for major coenzymes are presented at appropriate places in the text. These coenzymes, with their vitamin portions indicated, are also included in an appendix at the back of the book for reference. The final chapter in this unit treats nucleic acids and protein biosynthesis. This chapter has been written to give students a basic appreciation for human genetic processes, as opposed to the detailed molecular genetics of bacteria. Recent advances in our understanding of genetic organization, function, and regulation in higher organisms are included where appropriate.

Metabolism

In our presentation of the major metabolic pathways, we first present an overview of the process. Only then do we take up a step-by-step description of the individual reactions involved. Finally, we provide a summary of these reactions. We also discuss the integration and regulation of metabolic pathways in some detail, concentrating on the underlying control mechanisms.

It is our experience that many students find the study of metabolic pathways a hopeless exercise in the memorization of complex charts. For this reason we believe it is essential to relate each step in a pathway to the reaction of one or a few functional groups, to point out similarities in reactions, and to keep a constant eye on the overall process and its utility to the cell. We feel that the unified discussion of the interrelated topics of carbohydrate metabolism and bioenergetics in Chapter 25 is preferable to fragmenting these topics into separate chapters.

The structure and function of lipids are discussed in Chapter 26. The topic is placed here in order to aid students in their subsequent study of lipid metabolism (Chapter 27) and to provide some relief at this point from the study of metabolic pathways. The interrelationships of the various metabolic pathways are described in a final chapter, together with human nutrition and body fluids as these topics relate to metabolism. A discussion of the important topic of diabetes is included in this chapter to illustrate the necessity for proper metabolic regulation in the human body.

We have made every effort to keep our discussions as up-to-date as possible and have emphasized the dynamic nature of our knowledge of human biochemistry, including the fact that medical applications of this knowledge are actively changing and expanding.

Chapters 13 through 29, and the associated Essential Skills units, are also available as a separate textbook entitled *Fundamentals of Organic and Biological Chemistry*. This text is suitable for a one-semester course in organic and biological chemistry for students who have previously completed at least one semester of introductory chemistry.

Acknowledgments

We have been fortunate to have the assistance of many capable people during the preparation of these books. We thank Robert Stern for his help in getting this project started and for his assistance with early drafts. Steven Miller has been a continual source of constructive criticism through various drafts, including the preliminary edition and galley proofs. We also appreciate the contributions of the following chemists who reviewed drafts of the manuscript and/or the preliminary edition; they provided us with many valuable suggestions.

Wasi Ahmed, State University of New York at Binghamton
P. Wayne Ayers, East Carolina University
David Bak, Hartwick College
Jack Dalton, Boise State University
Leslie N. Davis, Community College of the Finger Lakes
Mary Delton, Oakland University
H. Ed Fiehler, Miami University
David Johnson, Wayne State University
Paul Ketchum, Oakland University
Gordon Lillis, San Joaquin Delta College
Tom Mines, St. Louis Community College at Florissant Valley
Clarence R. Perisho, Mankato State University
Susan Poulter, University of Utah
Wilmer Reed, DeAnza College
Joan Reeder, Eastern Kentucky University
William Schulz, Eastern Kentucky University
Cynthia Sevilla, Oakland University
Michael Sevilla, Oakland University
Peter Sheridan, State University of New York at Binghamton
Harry M. Smiley, Eastern Kentucky University
Carol B. Swezey, St. Elizabeth Hospital Medical Center,
    Lafayette, Indiana
L. G. Wade, Jr., Colorado State University
David Warr, Bristol Community College
Albert Zabady, Montclair State College

We thank the many students at Oakland University who used the preliminary editions of these texts, and who provided us with vitally important student feedback. We are indebted also to Geraldine Felton for her helpful advice.

We thank Susan Forgette for her skillful and rapid typing of many drafts, the preliminary edition, and the final manuscript. We are grateful to our families; they have been a source of strength for us. They have also contributed in more direct ways: We would like to thank Karen Callewaert for her help with initial drafts, and Carol Genyea for illustrating the preliminary edition.

Finally, we express our gratitude to Worth Publishers for their patience, guidance, and insistence upon the highest standards. In particular we wish to thank Ken Ekkens, whose attention to design and layout contributed so much to the "look" of the book, and Gordon Beckhorn, our editor, who served as a constant source of help and encouragement.

Denis M. Callewaert
Julien Genyea

Rochester, Michigan
March 1980

TO THE STUDENT

Basic Chemistry is intended to make your study of chemistry an interesting and enjoyable experience. Whether or not we are successful depends to a large extent on you, because learning chemistry will require your active intellectual participation. To learn chemistry you cannot merely memorize a collection of facts or a rote method for solving problems. You need to acquire sufficient understanding of general principles to be able to apply them. There are several features of this book that will assist you in this.

Each chapter begins with a brief **introduction,** which provides an overview of the topics to be discussed in the chapter, their relevance to our daily lives, and how they fit into the overall picture of chemistry.

Following the introduction there is a list of **study objectives** setting forth what you are expected to know and what types of problems you are expected to solve after you have learned the material in the chapter. These study objectives can be useful in reviewing for tests. For example, if one study objective is to be able to define a term, as part of your review you should write down an appropriate definition without looking at the chapter. Similarly, if a particular study objective is to be able to solve a certain type of problem, then as part of your review you should actually work out a few problems of that type.

There are **exercises** at the end of almost every section of every chapter; *these are your most important learning aid.* After you have studied the material in a particular section, it is imperative that you test your comprehension by working out the exercises before going on to the next section. At the end of each chapter, you will find worked-out solutions to all the exercises in that chapter. If you have any difficulty with an exercise, you should spend more time studying the material in that section.

Also, at the end of each chapter is a list of **summary** statements highlighting the major points in the chapter. These statements can be used as a brief outline for reviewing the chapter.

Finally, there is a set of **problems** at the end of every chapter, and it is important that you work all of them. The only way to gain an adequate understanding of a chemical principle is to work a number of problems that require use of that principle. As many students have learned the hard way, it is not sufficient merely to follow the worked-out examples presented in the text or in a lecture. Working problems yourself is the key to success in the study of chemistry.

The **answers** to all problems are given at the back of the book. A word of caution: Do not be satisfied with just getting the correct answer to a particular problem. If you had difficulty solving the problem, or if you are unclear about the method you used, then you should consider the problem further and perhaps ask your instructor for help.

There are a number of units called **Essential Skills** at the back of the book. In general, each Essential Skill is designed to provide you with a more intensive treatment of a basic topic, with extra drill exercises. For example, exponential notation is discussed in Chapter 1 and is used throughout the book. You are advised in Chapter 1 that Essential Skills 1 deals with exponential notation. If you are not familiar with this topic or need practice in handling numbers written in exponential form, you should read Essential Skills 1 and work the exercises.

To some extent, the study of chemistry is like learning a new language, in that you will encounter many new terms whose precise meanings you will be required to know. Since you will want on occasion to refresh your memory, the **index** at the back of the book uses boldface type to indicate the page on which each term is defined. Thus you can easily find the definition for every scientific term used in the text and the context in which it is used.

Chemistry is a fascinating subject that affects every aspect of our lives. Learning chemistry is not easy—it is a challenge. But it need not be a very difficult task. What is required is a moderate amount of time spent studying the basic principles of chemistry and working all the problems you are assigned. We believe that with a reasonable effort you will find the study of chemistry both interesting and rewarding.

CONTENTS

BASIC CHEMISTRY

GENERAL CHEMISTRY

Overview

Chemical Links Found Between Body Functions and Behavior
The New York Times

PCB Contamination Traced to 17 States
The Denver Post

Acid Rain: A Shower of Death
The Binghamton, N.Y., Press & Sun-Bulletin

Tampering With DNA: Sin or Salvation
The Detroit Free Press

These headlines, and many others that have appeared in newspapers and magazines in recent years, reflect the public's ever-growing concern about the effects of science on the quality of human life. Antibiotics, nuclear power, plastics, fertilizers, pesticides, and computers are but a few of the consequences of scientific research that have significantly altered the daily lives of people all over the world. The effects—positive or negative—of the further application of existing scientific knowledge, as well as the potential effects of the application of future scientific developments, are staggering. It is only in this latest period out of the whole span of human existence that people have been capable of making such drastic, widespread, and rapid changes in the natural world, and, potentially, even in the basic nature of living organisms. There has been, and will continue to be, serious public debate regarding whether or not certain scientific developments are being used wisely. It is vital that every one of us be able to participate knowledgeably in making judgments with regard to the application of scientific developments.

Your own participation in the decision-making process will be more meaningful, and the outcome will more likely be beneficial, if you have some basic understanding of science. Although you are probably beginning the study of chemistry because knowledge of the subject will be necessary in your chosen career, what you learn about science in general may well prove to be extremely valuable to all of us by enhancing your ability to contribute more effectively in the collective decisions of society.

Science is based on a procedure called the **scientific method.** The fundamental concept of the scientific method is that any idea or **hypothesis** used to explain an observation must be subjected to experimental tests. An **experimental test** consists of observing something under a controlled set of conditions, so that a clear connection can be established between cause and effect. To show what we mean, let us consider a specific example.

In a maternity ward of a hospital in Vienna in about 1845, a large proportion of the patients had been dying of an illness called childbed fever. A physician at the hospital, Ignaz Semmelweiss, noted that doctors and medical students often examined women right after dissecting cadavers in an adjacent autopsy room, usually without having previously washed their hands. Semmelweiss *hypothesized* that the childbed fever was caused by infectious material carried in from the autopsy room. He *tested* his hypothesis by requiring all medical personnel to wash their hands in a chemical solution that would kill infectious organisms before they examined maternity patients. When this was done, the incidence of childbed fever in the hospital decreased dramatically.

The result of Semmelweiss's experimental test was strong evidence that his hypothesis was correct, but it is possible that the observed result could be explained equally well by some different hypothesis. In general, if we perform experiments to test a hypothesis, and the results of these experiments agree with our expectations, this does not prove conclusively that our hypothesis is correct. But the more experimental tests we perform in which the results agree with the hypothesis, the more reason we have to believe that our hypothesis is valid. On the other hand, if we perform an experiment and obtain results that are not in agreement with our hypothesis, and we are confident that our measurements are reliable, then we should reject or modify our hypothesis. When scientists test a hypothesis, they usually perform the same experiment (or similar experiments) several times to ensure that their observations are **reproducible.** Scientists, like everyone else, can (and often do) have weird ideas, but scientists subject their scientific ideas to experimental tests, and then submit the validity of their observations to the critical judgment of other scientists.

Another aspect of the scientific method is that the results of many experiments are summarized into general statements called laws. A **scientific law** is a statement about some portion of the natural world that we believe to be true in all cases.

But scientists are not satisfied with merely generalizing their observations in the form of scientific laws. They want to understand why these generalizations hold. To do this they construct theories about the natural world. A **theory** provides an explanation for a large number of observed phenomena in terms of a relatively simple model. For example, all modern explanations of phenomena in chemistry, physics, and biology are based on the **atomic molecular theory.** The basis of the particular model used in the atomic molecular theory is that any macroscopic (large) sample of matter is composed of a vast number of small particles—atoms and molecules. Roughly speaking, a macroscopic sample of solid matter is one that is large enough to see and feel. In any experiment, we observe the properties of, and make measurements on, macroscopic samples. However, we explain the properties of these macroscopic samples in terms of the behavior of the atoms and molecules of which they are composed. In other words, the observable properties of any macroscopic sample of matter are assumed to result from the collective behavior of a large number of atoms or molecules. For example, one measurable property of a macroscopic sample of a gas is its pressure. We explain the pressure of a gas by saying that it results from the large number of atoms or molecules in the gas colliding with the walls of the container holding the gas. In the discussions in this book—and in all modern discussions of chemistry, physics, and biology—you will find a continual shift back and forth between descriptions of the behavior of atoms and molecules and descriptions of the properties of macroscopic samples of matter.

Chemistry is the science that deals specifically with the properties of matter and the changes that matter can undergo. Chemistry is loosely divided into several overlapping areas or fields. For example, the chemistry of substances that contain the element carbon has traditionally been considered a separate area known as **organic chemistry,** and the chemistry dealing with the reactions that take place in living systems is called **biochemistry.**

At one time people thought that certain properties of living systems could not be explained solely on the basis of the laws and principles of chemistry and physics. It was thought that biological phenomena involved additional factors called "vital forces." But we now believe that we shall eventually be able to explain all the characteristics of living organisms by using only the concepts of chemistry and physics. We know that living organisms are composed of atoms and molecules and that the basic interactions of atoms and molecules in living systems obey the same scientific laws as the atoms and molecules in nonliving matter. The molecules found in living matter are called *biomolecules,* and the major goal of biochemical research is to determine exactly how biomolecules interact with one another so as to generate and maintain life.

The fundamental principles of chemistry are common to all areas of chemistry; understanding these principles is essential before one can study any specific area of chemistry in greater detail. The aim of this book is to enable you to acquire an adequate understanding of these chemical principles.

Learning the principles of chemistry, or any other area of knowledge, cannot be a completely linear procedure, like building a brick wall in which each brick is laid down atop the bricks that have already been set in place. Rather, it is in some ways like constructing a spiral with interconnected loops.

We shall begin "at the beginning," with a discussion in Chapter 1 of some of the basic quantities we shall use, and then proceed in Chapter 2 to describe how matter is classified and to present the fundamentals of the atomic molecular theory. You will notice that we discuss some of these same concepts again in later chapters. There is a good reason for this. After you have built up some background knowledge of chemistry, you can gain a deeper understanding and appreciation of the fundamental ideas by reconsidering them.

And so, let us begin our study of chemistry.

CHAPTER 1

Basic Quantities: The Process of Measurement

1-1 INTRODUCTION

The science of chemistry is based on quantitative experimental measurements. In fact, modern chemistry began when the French chemist Antoine Lavoisier (1743–1794) realized the fundamental importance of measuring accurately the mass of each of the substances involved in a chemical reaction. Lavoisier met his end on the guillotine, but you should not let this fact deter you from studying chemistry.

Since quantitative measurements are an integral part of chemistry, we begin with a discussion of the process of measurement. We shall describe how mass, length, volume, and temperature are measured. We shall also see that the conclusions that can be drawn from the results of any measurement depend on the reliability of the measurement.

Our intention in this chapter is to treat, in a preliminary manner, some of the basic quantities and concepts that we shall be using throughout this text. Thus, in addition to mass, length, volume, and temperature, we shall discuss several other properties of matter, the differences between physical changes and chemical reactions, some of the fundamental features of energy and electrical forces, and the difference between heat and temperature. You will gain a deeper insight into the nature of these quantities and concepts and a greater appreciation of their importance as you encounter them in later chapters.

A most essential part of the basic material presented in this chapter is (1) the use of exponential numbers; (2) arithmetic operations with units and the conversion of units; (3) the solution of numerical problems involving two quantities that are directly proportional to one another; and (4) the use of significant figures and how to round off the result of a calculation. An extended discussion of these topics, along with corresponding drill exercises, is provided in Essential Skills 1, 2, 3, and 4, respectively, at the back of this book. You will be referred to these Essential Skills at appropriate places in this chapter; it is imperative that you know this material before proceeding with other chapters.

The energy released in this startlingly brilliant lightning burst can cause dramatic chemical and physical changes in our environment.

1-2   STUDY OBJECTIVES

After studying the material in this chapter and in Essential Skills 1, 2, 3, and 4, you should be able to:

1. Define the term mass, explain the difference between mass and weight, and state the law of conservation of mass.

2. Recognize common units for mass, length, and volume and know the SI units for these quantities.

3. Perform arithmetic operations with units and convert a given quantity expressed in terms of one unit into an equivalent quantity expressed in terms of some other unit.

4. Express a given number in standard scientific notation, write a given exponential number in decimal form, and perform the basic arithmetic operations of multiplication, division, addition, and subtraction with exponential numbers.

5. Define and use the prefixes kilo-, deci-, centi-, milli-, micro-, and nano-.

6. Define the terms density and specific gravity.

7. Calculate the density, mass, or volume of a sample of matter from given values of the other two quantities.

8. Define the terms precision, significant figures, systematic error, and accuracy.

9. Relate the precision and accuracy of a measurement to the extent that definite conclusions can be based on the measurement.

10. Round off the result of an arithmetic calculation so as to indicate correctly the uncertainty in the calculated value.

11. Define the terms physical property, physical change, chemical property, chemical reaction, reactant, and product, and be able to distinguish physical changes from chemical reactions.

12. Define the terms kinetic energy and potential energy.

13. State and use the law of conservation of energy.

14. State Coulomb's law and use it in a qualitative fashion.

15. Define the terms heat and temperature.

16. Express a temperature given on the Celsius, Fahrenheit, or Kelvin scale as an equivalent temperature on the other two scales.

17. (**Optional**) Define the term specific heat.

18. (**Optional**) Calculate any one of the following—heat input (or heat loss), specific heat, mass, or temperature increase (or temperature decrease) —from given values of the other three.

1-3   THE PROCESS OF MEASUREMENT: MASS AND ITS MEASUREMENT

Measurement involves the comparison of a quantity with a standard unit. We are all familiar with the process of measuring the size of objects. For example,

Figure 1-1   The length of a pencil can be determined by a comparison with the
standard length of 1 in.

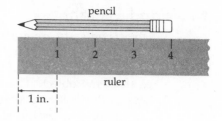

we can measure the length of a pencil by using a ruler (see Figure 1-1). The
standard unit in this example is one inch (1 in.). Thus, when we say that a
pencil is 4 in. long, we mean that it has a length equal to four standard inches. It
is essential to include the standard unit when stating the result of any measurement. If, for example, you say that a piece of wood has a length of 5, your statement has no meaning — it might be interpreted as 5 inches, 5 feet, even 5 yards.

Mass and Its Measurement

Mass is an intrinsic property of an object that expresses the amount of matter in
the object and the object's resistance to changes in its motion. Mass should not
be confused with weight (see below). The mass of an object does not depend on
its position, temperature, and so on, or on the state of its surroundings. Let us
see how the mass of an object can be measured.

The internationally accepted standard unit of mass is called the kilogram, abbreviated kg. **One kilogram** is defined as the mass of a particular platinum-
iridium metal cylinder, which is carefully preserved at the International Bureau
of Weights and Measures in Sèvres, France. The mass of a quart of milk is close
to 1 kg. If we want to measure the mass of an object, we need to compare its
mass with the standard kilogram at Sèvres. Rather than taking our object to
France, we compare it to a secondary standard mass that has been indirectly
compared to the international standard. In practice, sets of secondary standards,
with masses of 1 kg, 0.5 kg, 0.1 kg, and so on, are generally used.

One simple way to determine the mass of an object with a set of secondary
standard masses is to use an instrument called a **two-pan balance.** A two-pan
balance has a metal bar balanced on a knife edge in its middle with pans attached at each end (see Figure 1-2). The object is placed on one pan and standard masses are added to the other pan until the two pans balance (the metal
bar is level). When the beam is level there is an equal amount of mass in each pan.
If, for example, the pans balance when the amount of standard mass is 1.23 kg,
we say that the object has a mass of 1.23 kg. At present, instruments other than
the two-pan balance are more commonly used to measure mass, but all such instruments involve indirect comparison with the international standard of mass.

Figure 1-2   To determine the mass of a sample
of matter using a two-pan balance,
standard masses are added to the
pan on the right until the bar
at the top is level. When this bar
is level, there is an equal amount
of mass in each pan.

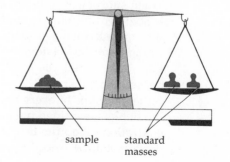

Mass and Weight

Objects on earth are attracted to the center of the earth by the force of gravity. This force is directly proportional to the mass of the object. The gravitational force pulling an object toward the center of the earth is called the **weight** on earth of the object. The farther from the center of the earth an object is, the smaller the pull of gravity, and hence, the less the weight. On the moon, the gravitational attraction of an object toward the center of the moon is less than the gravitational attraction of the same object on earth (see Figure 1-3). Thus, the weight of an object depends on its location, whereas the mass of the object does not.

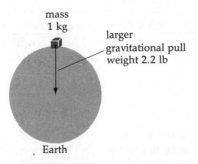

mass
1 kg

larger
gravitational pull
weight 2.2 lb

Earth

mass
1 kg   smaller
gravitational pull
weight 0.43 lb

Moon

Figure 1-3   Any object has the same mass on the earth's surface as it does on the surface of the moon. Its weight, however, is smaller on the moon than on earth because the gravita- tional attraction toward the center of the moon is weaker than the gravitational attraction toward the center of the earth.

When using a two-pan balance, the beam is level when the gravitational pull on each pan, or the weight on each pan, is the same. Since both sides of the balance are at the same distance from the center of the earth when the weights on each side are the same, the mass on each side is the same as well. Thus, for a two-pan balance or similar instrument, equality of weight indicates equality of mass. For this reason people say, for example, that an object with a mass of 5 kg weighs 5 kg on earth.

At this time in the United States, the most common unit of weight is the pound (abbreviated lb). At the earth's surface, a mass of 0.45 kg has a weight of 1 lb. We can therefore say that on earth a mass of 0.45 kg is equivalent to a weight of 1 lb. In the near future it is likely that the unit kilogram will replace the unit pound for everyday use in the United States.

Pharmacists and physicians often use the units grains, drams, and ounces, which are part of the older apothecary system of units, to specify the weight of a drug (see Appendix 1).

Law of Conservation of Mass

When substances undergo any type of change (except a nuclear reaction), there is no detectable change in the total mass of all of the substances present. This is why mass is such a fundamentally important quantity. The initial experimental evidence for this statement was obtained by Antoine Lavoisier. Lavoisier carefully determined the total mass of various substances in closed containers, burned the substances, and then measured the mass of each container and its contents again. In every case, he found that the total mass of the container and its contents remained unchanged. His experiments led to the fundamental generalization called the **law of conservation of mass.** All of the experimental evidence obtained since Lavoisier's time is in agreement with this law.

(1-1)   **Law of Conservation of Mass:** In any system that is enclosed so that matter cannot enter or leave, the total mass remains the same when a chemical reaction or any other process, with the exception of nuclear reactions, takes place.

In a nuclear reaction there is a detectable change in the total amount of mass (see Section 2-9).

Exercise 1-1

An object A, which weighs 1.0 lb, is placed on the left pan of a two-pan balance, and an object B, which has a mass of 0.45 kg, is placed on the right pan.
   (a) Will the beam tip down to the left, down to the right, or be level? Explain your answer.
   (b) If the balance with objects A and B in the pans were moved to the surface of the moon, what would be the position of the beam? Explain your answer.

1-4   LENGTH AND VOLUME

It is quite likely that people were measuring lengths and volumes even before they had a written language in which to record their observations. At different times and in different parts of the world, various standards of length and volume have been used. One of the early standards of length was the foot (see Figure 1-4).

Figure 1-4   A Roman tablet (written in Greek), which shows what the length of the standard foot was considered to be at about the time A.D. 300–500.

Length

The process of adopting an internationally accepted standard for length (and for other quantities as well) began in France at the time of the French Revolution and was developed into the *système métrique* (**metric system**) by the government of Napoleon. In the metric system, the standard of length is called the **meter,** abbreviated m. You are probably more familiar with the use of the units inch, foot, and yard as standards of length. A length of 1 meter is slightly greater than a length of 1 yard.

We often refer to lengths that are much smaller or larger than 1 m. For example, the length of a typical dust particle is about 0.00000005 m. It is quite inconvenient to write and keep track of all the zeros in a very small or very large number, so scientists most often write numbers such as these in an equivalent form called **exponential notation,** in which powers of 10 are used. For example, since 0.00000001 is equal to 1×10^{-8}, the length 0.00000005 m can be written as 5×10^{-8} m. Writing numbers in exponential form is a great asset when dealing with very large or small numbers and simplifies performing many numerical calculations. We shall be using numbers in exponential form throughout this book. If you are not familiar with exponential numbers or with performing arithmetic operations with numbers written in exponential form, it is absolutely essential that you study the material given in Essential Skills 1.

It is also convenient to define units of length that are larger and smaller than the basic unit, the meter. For example, a length of one foot (1 ft) is equal to a length of 12 in., and a length of 1 yard is equal to a length of 3 ft. In the metric system, larger or smaller multiples or submultiples of a basic unit are also defined, but these multiples are always simple powers of 10. Various powers of 10 are given names, and these names are used as prefixes in front of the basic unit. Several of these metric system prefixes are listed in Table 1-1.

Table 1-1 Metric System and SI Prefixes

Prefix	Multiplication Factor	Abbreviation	Example
kilo-	1000 or 10^3	k	1 kilogram = 1000 grams 1 kg = 10^3 g
deci-	0.1 or 10^{-1}	d	1 decimeter = 0.1 meter 1 dm = 10^{-1} m
centi-	0.01 or 10^{-2}	c	1 centimeter = 0.01 meter 1 cm = 10^{-2} m
milli-	0.001 or 10^{-3}	m	1 milliliter = 0.001 liter 1 ml = 10^{-3} liter
micro-	0.000001 or 10^{-6}	μ	1 microliter = 0.000001 liter 1 μl = 10^{-6} liter
nano-	0.000000001 or 10^{-9}	n	1 nanometer = 0.000000001 meter 1 nm = 10^{-9} m

The prefixes can be used with other quantities such as mass, volume, energy, and so on, in addition to length. Since these prefixes are used quite often in scientific work, you should become familiar with them. For example, the prefix centi- means 10^{-2}. Therefore, a length of 0.015 m can be expressed as either 1.5×10^{-2} m or 1.5 centimeters (1.5 cm).

The United States is gradually adopting the metric system for everyday measurements. Eventually, lengths or distances will commonly be expressed in meters, centimeters, and kilometers rather than in feet, inches, and miles. A length of 1 in., for example, is equal to a length of 2.54 cm (see Figure 1-5).

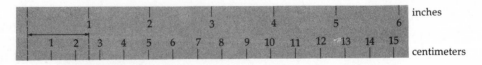

Figure 1-5   A comparison of the inch and centimeter units of length. A length of 1.00 in. is equal to a length of 2.54 cm.

This fact can be expressed as

(1-2)   1.00 in. = 2.54 cm

Equation 1-2 can be used to convert lengths expressed in units of inches to lengths expressed in units of centimeters, and likewise to convert lengths expressed in units of centimeters to lengths expressed in units of inches. You will often find it necessary to convert a quantity expressed in terms of one unit into an equivalent expression in terms of another unit, both in scientific work and everyday life. It is essential for you to be able to convert units. For example, we can convert the length 7.65 cm to a length expressed with the unit inch using the following three steps:

Step 1   Recognize that an *appropriate* conversion factor relates a length expressed in centimeters and the same length expressed in inches. Thus,

(length in cm) × (appropriate conversion factor) = (length in in.)

Step 2   Recognize that Eq. 1-2 can be used to obtain the conversion factors 1.00 in. per 2.54 cm (1.00 in./2.54 cm) and 2.54 cm per 1.00 in. (2.54 cm/1.00 in.).

Step 3   Use the conversion factor that gives the correct units. In this example, 7.65 cm multiplied by (1.00 in./2.54 cm) gives a length in inches, so this is the appropriate conversion factor to use. Thus,

$$(7.65 \ \text{cm}) \times \left(\frac{1.00 \ \text{in.}}{2.54 \ \text{cm}} \right) = 3.01 \ \text{in.}$$

A similar procedure can be used whenever you want to convert units. A more detailed discussion of conversion of units, together with several practice exercises and their solutions, can be found in Essential Skills 2.

Most scientists have traditionally used the metric system of units. In 1960, an international body, the General Conference of Weights and Measures, recommended the adoption of a somewhat modified form of the metric system called the **International System of Units** (Système International, abbreviated **SI**). SI units are the officially accepted units of the U.S. National Bureau of Standards.

SI units are gradually being used more and more by scientists in their work, and in the future they will probably be the only units used by the scientific community. Presently, however, most chemists still use the units with which they are most familiar, and in general, we shall use those familiar units in this textbook. We shall, however, describe the SI unit for the quantities we discuss, and there is a table of SI units in Appendix 2. The SI unit of length is the meter, and the SI unit of mass is the kilogram.

Signs similar to this one, from alongside a Canadian highway, are beginning to appear in the United States.

Volume

Volume is the term we use to refer to the amount of space occupied by an object. For example, the volume of a box is equal to the product of its length, width, and height. A cubic box with each side equal to the SI standard of one meter (1 m) has a volume of 1 m $\times$ 1 m $\times$ 1 m or 1 m³ (see Figure 1-6). The unit m³, called the cubic meter, is the SI unit of volume.

Figure 1-6   The SI unit of volume is the cubic meter, m³. A volume of 1.00 m³ is equal to a volume of 35.3 ft³ (or 35.3 cu. ft.). The storage space available in a 17.0-cu. ft. refrigerator is about 0.5 m³.

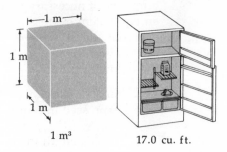

1 m³   17.0 cu. ft.

Chemists usually work with volumes that are much smaller than a cubic meter, so they use smaller units of volume such as the liter when reporting their measurements. **One liter** is defined to be 1/1000 of a cubic meter, or one cubic decimeter:

(1-3) $1 \text{ liter} = 10^{-3} \text{ m}^3 = 1 \text{ dm}^3$

Although commonly done, we shall not use the abbreviation l for the unit liter, since it is too easy to confuse the letter l with the number 1. (A script ℓ can be used without confusion as an abbreviation for liter.) As Figure 1-7 shows, the volume of a quart is slightly less than 1 liter (1 quart = 0.946 liter).

Figure 1-7   A comparison of the quart and liter units of volume. The volume of a quart milk carton is slightly less than 1 liter.

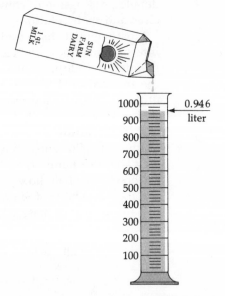

The prefixes listed in Table 1-1 can all be used with the unit liter. Perhaps the most common unit of volume that chemists use is the **milliliter** (ml). Since the prefix milli- means 10^{-3},

(1-4) $1 \text{ ml} = 10^{-3} \text{ liter}$

One milliliter (1 ml) and one cubic centimeter (1 cm³) refer to exactly the same volume. Using Eqs. 1-3 and 1-4, 1 ml is equal to 10^{-6} m³, which is also the volume of 1 cm³ (1 cm $= 10^{-2}$ m). Two abbreviations are used for the unit cubic centimeter, cm³ and cc. Therefore, 1 cm³, 1 cc, and 1 ml each refer to exactly the same volume (see Figure 1-8). Very small volumes are often expressed in terms of the unit microliter. One microliter (μl) is equivalent to 10^{-6} liter.

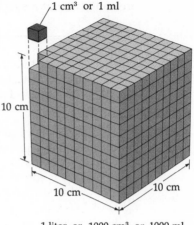

Figure 1-8 A milliliter and a cubic centimeter refer to exactly the same volume. There are 1000 cm³ or 1000 ml in 1 liter.

The apothecary units minim, fluid dram, and fluid ounce are often used to specify the volume of a liquid drug. The relationship of these apothecary units to the milliliter is given in Appendix 1.

In chemical laboratories and in medical facilities we need to measure volumes of liquids more frequently than volumes of solids and gases. A graduated cylinder or a hypodermic syringe can be used for this purpose (see Figure 1-9). Some other devices that chemists use to measure liquid volumes are volumetric flasks, pipets, and burets (see Figure 1-9). A volumetric flask is calibrated to contain a definite volume of liquid when it is filled to the calibration mark. A pipet is filled to a calibrated mark and then used to deliver a definite volume of liquid. A buret can be used to deliver various exact volumes of liquid by opening and closing the stopcock at the bottom of the calibrated tube.

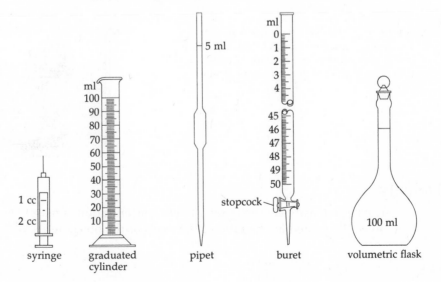

Figure 1-9 Some instruments that are used to measure liquid volumes.

The volume of an object, unlike its mass, changes with changes in the object's surroundings. Most solids and liquids expand when their temperatures are increased, whereas pressure has a negligible effect on the volume of a liquid or a solid. On the other hand, there is a very large change in the volume of a given mass of gas when either the temperature or the pressure is changed (see Chapter 5). The instruments we have described for measuring liquid volumes are calibrated at a specific temperature—usually 25°C.

Exercise 1-2
Consider the pencil in Figure 1-1. Suppose that we create a new unit of length called a "wee-inch" such that a length of one "wee-inch" is equal to a length of $\frac{1}{6}$ in. What is the length of the pencil in Figure 1-1 in units of "wee-inches"?

Exercise 1-3
    (a) 0.0025 liter is equivalent to _____ ml.
    (b) 1.35 nm is equivalent to _____ cm.
    (c) 0.0000035 g is equivalent to _____ μg.
    (d) 1500 cc is equivalent to _____ liters.
    (e) 7.25 liters = _____ qt.
    (f) 4.2 lb = _____ g.
    (g) 75 miles = _____ km.
    (h) 68 g = _____ oz.

If you have any difficulty with these exercises then you have not given sufficient attention to the material in Essential Skills 1 and 2.

1-5   DENSITY AND SPECIFIC GRAVITY

You are familiar with the fact that a chunk of iron (for example, an iron bolt) will sink in water (see Figure 1-10). This characteristic property of iron and water is often described by saying that "iron is heavier than water." Of course, this does not mean that all samples of iron have more mass than all samples of water. Rather, it means that for samples of iron and water that have the *same* volume, the mass of the iron sample is greater. We can say that the mass exists in a more compact or concentrated form in iron than in water.

Figure 1-10   The density of the iron bolt is greater than that of water, so it sinks. The density of cork is less than water, so a cork floats.

Density

The term density is used to refer to the concentration of mass. The **density** of an object is defined as the ratio of the mass of the object to its volume:

(1-5)
$$\text{Density} = \frac{\text{mass}}{\text{volume}}$$

The more mass there is in a given volume, the greater the density.

Neither the mass of an object nor its volume is a characteristic property that can be used to identify what kind of object it is. For example, a 10-g sample of matter might be a sample of water or iron or any of countless other substances. Density, however, is a characteristic property of a substance, and it can be used to distinguish one substance from another. For example, two samples of water might have different masses and different volumes but they, and all other samples of water, have the same density (see Figure 1-11).

Figure 1-11   Both beakers contain samples of water. The mass and the volume of each sample are different, but the density of the water in each beaker is the same.

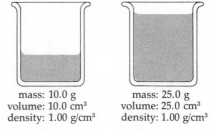

mass: 10.0 g
volume: 10.0 cm³
density: 1.00 g/cm³

mass: 25.0 g
volume: 25.0 cm³
density: 1.00 g/cm³

The density of water and iron are different. At room temperature the density of water is about 1.00 g/cm³ and the density of iron is about 7.86 g/cm³. Thus, a precise statement that compares iron and water is "the density of iron is greater than the density of water" (see Figure 1-12).

Figure 1-12   These samples of iron and water have the same volume, but the iron sample has a larger mass and thus a greater density.

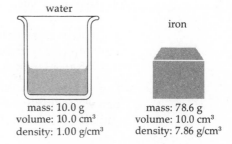

water

iron

mass: 10.0 g
volume: 10.0 cm³
density: 1.00 g/cm³

mass: 78.6 g
volume: 10.0 cm³
density: 7.86 g/cm³

One way to determine the density of a substance is to measure both the mass and the volume of a sample and then use Eq. 1-5 to calculate its density. For example, the mass of the iron sample in Figure 1-12 is 78.6 g, and its volume is 10.0 cm³. Therefore,

(1-6) $$\text{Density of iron} = \frac{\text{mass of iron sample}}{\text{volume of iron sample}} = \frac{78.6 \text{ g}}{10.0 \text{ cm}^3} = 7.86 \text{ g/cm}^3$$

Since the volume of a given mass of a solid or a liquid depends on the temperature (see Section 1-4), so does the density. Thus, to be precise when reporting the density of a solid or a liquid, we must indicate the temperature of the substance. The volume of a given mass of a gas depends on both the temperature and the pressure, so when stating the density of gases, both the temperature and the pressure must be specified. The densities of several common substances are listed in Table 1-2. As you can see from the examples in Table 1-2, solids and liquids have much higher densities than gases.

An extremely important application of the concept of density is to use Eq. 1-5 and the value of the density of a liquid to determine what volume of the liquid must be measured out in order to yield a given mass of the liquid. It is easier to measure the volume of a liquid and calculate its mass from Eq. 1-5 than it is to measure a liquid's mass directly, say, with a two-pan balance.

Table 1-2 Density of Common Substances

Substance	Density
Water (liquid), 25°C	0.997 g/cm³
Water (liquid), 4°C	1.000 g/cm³
Ice, 0°C	0.917 g/cm³
Ethyl alcohol, 25°C	0.785 g/cm³
Diethyl ether, 25°C	0.708 g/cm³
Mercury, 25°C	13.6 g/cm³
Iron, 25°C	7.86 g/cm³
Oxygen (gas), 25°C, 1 atm pressure	1.33×10^{-3} g/cm³; 1.33 g/liter
Nitrogen (gas), 25°C, 1 atm pressure	1.16×10^{-3} g/cm³; 1.16 g/liter
Salt (sodium chloride), 25°C	2.17 g/cm³
Whole blood	1.06 g/cm³
Urine—normal range	1.018–1.030 g/cm³

Consider the following problem: What volume of ethyl alcohol must be measured out at 25°C in order to have 5.0 g of ethyl alcohol?

We can solve this problem using the following three steps:

Step 1   Recognize that according to Eq. 1-5 the volume and mass of a sample are directly proportional to one another, so that

Volume of ethyl alcohol = (mass of ethyl alcohol) × (appropriate factor)

Step 2   Recognize that from the density of ethyl alcohol, 0.785 g/ml, we can obtain two factors: 0.785 g per 1.00 ml (0.785 g/1.00 ml), and 1.00 ml per 0.785 g (1.00 ml/0.785 g).

Step 3   Use the factor that gives us the correct units. In this problem when we multiply 5.0 g by the factor (1.00 ml/0.785 g), we get a volume in milliliters. Therefore,

(1-7)
$$\text{Volume of ethyl alcohol} = (5.0 \ \cancel{g}) \times \left(\frac{1.00 \ \text{ml}}{0.785 \ \cancel{g}}\right) = 6.4 \ \text{ml}$$

Two liquids that do not mix with each other separate into layers, with the denser liquid at the bottom. This chemist is using a separatory funnel to drain off the denser liquid.

Most of the problems you will encounter in introductory chemistry that require you to calculate the numerical value of a quantity involve a direct proportional relationship. In Essential Skills 3 there is an extensive discussion of direct proportionality, as well as other examples of the simple method for solving problems involving two quantities that are directly proportional to one another we used in the problem above. It is extremely important that you study this material and be able to solve problems of this type easily. As discussed in Essential Skills 3, for any two quantities that are directly proportional to one another, such as the volume and the mass of a sample, if you are given a known amount of one quantity and you want to find the amount of the other, then:

(1-8)   (Quantity to be determined) = (known quantity) × (appropriate factor)

As we saw in the problem above, in the case of the directly proportional quantities mass and volume of a sample, the appropriate factor to use is obtained from the density of the substance.

Exercise 1-4
The volume of air in a person's lungs is about 5.0 liters. Calculate the mass of 5.0 liters of inhaled air at 25°C and 1 atm pressure, where the density of air is 1.2 g/liter.

Exercise 1-5
    (a) Calculate the volume of 25.0 g of ethyl alcohol at 25°C.
    (b) At 25°C the density of diethyl ether is less than the density of ethyl alcohol. If 25.0 g of ethyl alcohol at 25°C completely fill a certain flask, can this flask contain 25.0 g of diethyl ether at 25°C?
    (c) At 25°C the density of diethyl ether is 0.708 g/cm³. Calculate the volume of 25.0 g of diethyl ether at 25°C.

Specific Gravity

Quite often the densities of substances are compared to the density of water as a reference. The ratio of the density of a substance to the density of water at the same temperature is called the **specific gravity** of the substance.

(1-9) $$\text{Specific gravity} = \frac{\text{density of substance}}{\text{density of water}}$$

For example, at 25°C the density of ethyl alcohol is 0.785 g/cm³ and its specific gravity is 0.785.

(1-10) $$\text{Specific gravity of ethyl alcohol} = \frac{0.785 \ \cancel{g/cm^3}}{1.00 \ \cancel{g/cm^3}} = 0.785$$

Note that the density of ethyl alcohol has units of grams per cubic centimeter, whereas the specific gravity is just a number, without units, since specific gravity is a ratio of two densities with the same units.

In medical facilities, the specific gravity of a person's urine is often measured and the result used for diagnostic purposes. For water solutions (such as urine), the big advantage of specific gravity over density is that there is a negligible change in specific gravity with a change in temperature. Thus, for example, the specific gravity of a person's urine *in vivo* (inside the body) at the normal body temperature of 37°C is essentially the same as the specific gravity of a urine sample taken from the person and measured at room temperature.

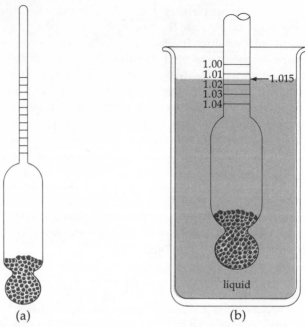

Figure 1-13   The specific gravity of a liquid can be measured with a hydrometer (a). The smaller the specific gravity of the liquid, the more the hydrometer will sink. Note the order of the scale markings on the neck of hydrometer (b). This liquid has a specific gravity of about 1.015.

The specific gravity of a liquid can be measured with a simple device called a **hydrometer.** Hydrometers that are used to measure the specific gravity of urine samples are sometimes called **urinometers.**

One type of hydrometer, illustrated in Figure 1-13, consists of a weighted glass bulb with a calibrated scale on its neck. The hydrometer is floated in the liquid whose specific gravity is to be determined. The smaller the specific gravity of the liquid, the farther the hydrometer will sink in the liquid. Thus the specific gravity of the liquid can be read from the calibrated scale on the neck of the hydrometer. The operation of a hydrometer is based on the principle that a floating object displaces a mass of liquid that is equal to the mass of the object. A given hydrometer with a fixed mass must displace the same mass of any liquid. Thus, the smaller the specific gravity of the liquid, the larger the volume of displaced liquid must be, and consequently the more the hydrometer will sink.

1-6   UNCERTAINTIES IN MEASUREMENT

Urine is the body fluid that contains the waste products of metabolism. The composition and specific gravity of urine are used as indicators of normal or abnormal metabolic and kidney function. The specific gravity of a person's urine also depends on such factors as the volume of liquid ingested, loss of water by evaporation (perspiration), and so on. The specific gravity of a urine specimen obtained from a healthy adult in the morning after a 12-hour period of rest and no fluid intake is normally in the range 1.018 to 1.030. Thus, when we measure the specific gravity of a person's urine, we want to be able to conclude from our measurement whether it falls within this normal range. However, we can draw a conclusion from our measurement only if it is sufficiently reliable.

For example, suppose that we measure the specific gravity of a urine sample with the crude urinometer, labeled A, shown in Figure 1-14. The scale in this

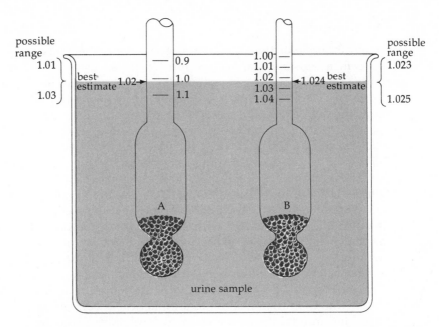

Figure 1-14   Using urinometer A, the best estimate of the specific gravity of the urine sample is 1.02, but it might be as low as 1.01 or as high as 1.03. A more precise determination of the specific gravity can be made with the second urinometer. Using urinometer B, the best estimate of the specific gravity is 1.024, and with reasonable certainty the specific gravity is no lower than 1.023 nor higher than 1.025.

urinometer is marked off in tenths, and when we read the level to which the urinometer sinks we can estimate the value in the hundredths place by reading between the scale lines. Thus, using urinometer A we estimate the level as 1.02. We might conclude that our urine sample, with a specific gravity of 1.02, is within the normal range, but upon closer examination we cannot be certain. For example, we, or someone else, might repeat the measurement and estimate a specific gravity of 1.01 or 1.03. Thus, even though our best estimate of the specific gravity is 1.02 using this crude hydrometer, it is reasonable that the actual specific gravity might be as high as 1.03 or as low as 1.01. A simple way of expressing this is to say that our measurement has an uncertainty or precision of ±0.01 and to report our value for the specific gravity as 1.02 ± 0.01.

Precision

The term **precision** is used to describe the extent to which repeated measurements are likely to differ. Measurements that are very precise are ones for which there are very small differences between repeated trials. On the other hand, if there are large differences between the values of repeated measurements, we say that the measurement has low precision.

When you make an inference based on a measured value, you must consider the precision of that measurement. In our example, the precision of the specific gravity measurement (±0.01) is too low to conclude with reasonable certainty that the specific gravity falls within the normal range.

Suppose, however, that we measure the specific gravity of our urine sample with the more precise urinometer, labeled B, shown in Figure 1-14. The scale divisions on this urinometer are in hundredths, and we can estimate a value in the thousandths place when making a measurement. With urinometer B, if we obtain a specific gravity of 1.024 ± 0.001, for example, we can be reasonably certain on the basis of this more precise measurement that the specific gravity of our urine sample falls within the normal range.

Significant Figures

As our example of the specific gravity of urine illustrates, to use a measured value in a sensible manner we must have an estimate of its precision. In common practice we do not include the precision of our measured value explicitly (as in 1.02 ± 0.01), but it is understood by scientists that there is an uncertainty of about one in the last digit on the right in the value we report. Thus, a specific gravity reported as 1.02 has an implied precision of ± 0.01, and a specific gravity of 1.024 has an implied precision of ± 0.001.

The numbers we obtain through a measurement are called **significant figures.** For example, the specific gravity measurement of 1.024 has four significant figures. The number 4 in the thousandths place is significant even though it is an estimated value. When we compare two measurements of the same physical quantity, the measurement with more significant figures is the more precise measurement. For example, for the specific gravity of our urine sample, the measured value 1.024, which has four significant figures, is a more precise measurement than the measured value 1.02, which has three significant figures.

A more detailed discussion of significant figures, as well as an explanation of how to round off the result of an arithmetical calculation so as to indicate correctly the uncertainty in the calculated value, is given in Essential Skills 4. Since calculations involving measured values appear throughout this textbook, and for that matter, in all scientific work, it is quite important that you understand the material in Essential Skills 4.

Accuracy

If we reconsider the process of measuring the specific gravity of our urine sample, we can see another factor that affects the reliability of our measure-

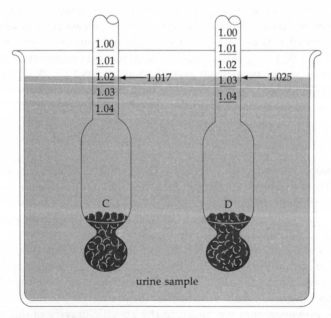

Figure 1-15   The only difference between urinometers C and D is the position of the scale markings. A specific gravity determined with urinometer C, which has a correctly marked scale, is an accurate measurement. But urinometer D, which has an incorrectly marked scale, yields an inaccurate measurement. The precision that can be obtained with either urinometer, however, is the same (± 0.001).

ments. Whether we used urinometer A or urinometer B, we had assumed that the scale markings were placed correctly. If the scales were incorrectly marked by the manufacturer of the urinometer, then our measured value of the specific gravity will differ from the true value. We cannot determine whether the scale in a particular urinometer is marked correctly or incorrectly simply by making repeated measurements of a urine sample with that urinometer and comparing the values we obtain. Since each measurement uses the same scale, if there is an error because a scale is marked incorrectly, the error will be the same for each measurement (see Figure 1-15). An error that is the same for repeated measures is called a **systematic error.** Systematic errors are possible in any measurement process. In our example, one way we could detect the presence or absence of a systematic error due to an incorrectly marked scale on the urinometer would be to place it in a liquid of known specific gravity and compare the value we obtain with this known value.

The term **accuracy** is used to refer to the amount of systematic error present in a measurement. Measurements with little systematic error are said to be very accurate, whereas those with larger systematic errors are clearly just inaccurate measurements.

We can rely on a measured value of a quantity only if we are reasonably certain that it is close to the true value. This reliability is possible only for precise and accurate measurements. The conclusion we reach based on a measured value is more definite the greater the precision and accuracy of that measurement.

Exercise 1-6

Substances A, B, C, D, and E are known substances with the following specific gravities:

Substance	Specific Gravity
A	1.120
B	1.549
C	1.145
D	1.798
E	1.087

It is known that a substance X is one of the substances A, B, C, D, and E. The specific gravity of X is measured, and a value of 1.1 is obtained. On the basis of this measurement (assuming no systematic error in the measurement):

(a) Can you confidently decide which substance X is?
(b) Can you confidently decide which substances X is not?
(c) Assume that the specific gravity of X is remeasured and a value of 1.118 ± 0.005 is obtained. In this case, what are the answers to parts (a) and (b)?

Exercise 1-7

Perform the following operations and round off your answer to indicate the proper uncertainty in your result.

(a) $21.26 \times 3.2 =$ (b) $(8.34 \times 10^5) \times (1.2 \times 10^{-2}) =$

(c) $\dfrac{1.23}{3476} =$ (d) $\dfrac{7.8 \times 10^{-5}}{4.239 \times 10^{-7}} =$

(e) $52.236 + 1.9 =$ (f) $(1.62 \times 10^3) + (2.456 \times 10^5) =$

If you have any difficulty with this exercise, then you have not given sufficient attention to the material in Essential Skills 1 and 4.

1-7   PHYSICAL AND CHEMICAL CHANGES AND PROPERTIES

Water and table salt are two familiar substances with different characteristic properties. Water is a liquid at room temperature, has a density of 0.997 g/cm³ at 25°C, and freezes at 0°C, whereas table salt (which has the chemical name sodium chloride) is a solid at room temperature, has a density of 2.17 g/cm³, and melts at 801°C. These characteristic properties of water and salt (solid, liquid, or gas at room temperature, density, and melting or freezing point) are examples of physical properties.

Physical Properties and Physical Changes

In general, a **physical property** of a substance is a consistent characteristic property that can be observed and measured without changing the substance into another substance. Specific heat (see Section 1-11), boiling point, heat of fusion and heat of vaporization (see Chapter 2), ability to conduct heat (thermal conductivity), and ability to conduct electricity (electrical conductivity) are some other examples of physical properties of substances. Different substances have different characteristic physical properties and usually they differ in their chemical composition as well (see Chapter 2).

A change in the physical properties of a substance is called a **physical change.** Water and most other substances can exist in either the solid, liquid, or gas form (see Chapter 2). A process, such as the melting of ice, in which a substance changes its form is called a **change of state.** Changes of state are examples of physical changes. The change in density of a substance that occurs with a change in temperature (see Section 1-5) is another example of a physical change.

Chemical Reactions and Chemical Properties

In a **chemical change** or **chemical reaction,** one or more substances are changed into other substances. The substances we start with in a chemical reaction are called **reactants,** and the new substances present after the reaction takes place are called **products.** For example, if a piece of iron is left outdoors, we can detect a slow change in its appearance as time elapses. The surface of the iron changes to a reddish color and has a different texture from the iron with which we started (see Figure 1-16a). If we scrape off the reddish material from the surface of the iron and measure its density, specific heat, melting point, and so on, we find properties that are significantly different from those of iron. This is conclusive evidence that the reddish material is not iron but a different substance and, consequently, that a chemical reaction has taken place. The common name for the product of this chemical reaction is "rust," and we commonly call the chemical reaction "rusting." If we investigated the process of rusting more carefully, we would find that besides iron it involves another substance as a reactant, namely, oxygen. We can represent the chemical reaction of rusting as

(1-11)

$$\text{Iron} + \text{oxygen} \longrightarrow \text{rust}$$
$$\text{Reactants} \longrightarrow \text{products}$$

Often when we observe a change in the physical properties of a system we suspect that a chemical reaction has taken place. But before we have proof that a new substance has been formed and that a chemical reaction has occurred, we must separate some material from our system and show that its characteristic properties, and usually its chemical composition (see Chapter 2), are different from those of the substances originally present.

Figure 1-16   Some common chemical reactions are (a) the rusting of iron, (b) the combustion of wood, and (c) the rising of bread dough before baking. In bread dough, the chemical action of yeast evolves carbon dioxide gas. These gas bubbles expand the dough to produce the light, airy consistency of the baked loaf of bread.

Let us look at a few more common examples of chemical reactions. Burning, or **combustion** (see Figure 1-16b), is a chemical reaction that proceeds much more rapidly than the rusting of iron and with the evolution of energy in the form of heat and light. In general, the release or absorption of energy when substances are put in contact with each other is an indication that a chemical reaction has taken place. The evolution of a gas and the formation of a solid in a liquid are other indications of a chemical reaction. Bread, for example, rises because of the carbon dioxide gas evolved as a product in a chemical reaction involving yeast (see Figure 1-16c). The separation of milk into a solid part (curds) and a watery part (whey) when it sours is a consequence of the chemical reaction involved in the souring process.

Most substances can participate in a great number of chemical reactions. A **chemical property** of a substance refers to a chemical reaction in which the substance is a reactant. For example, the fact that iron rusts is both a chemical property of iron and also a chemical property of oxygen, since both iron and oxygen are reactants in the rusting reaction (Eq. 1-11).

Exercise 1-8
Classify each of the following as a physical change or a chemical reaction:
  (a) The evaporation of ethyl alcohol
  (b) The digestion of sugar
  (c) The combustion of gasoline
  (d) The dissolving of sugar in water
  (e) The growth of a human child

Exercise 1-9
A 10.0-g piece of iron is completely changed into rust. The mass of the rust is 14.3 g. Using the law of conservation of mass, calculate the mass of oxygen that was used in this chemical reaction.

1-8   ENERGY

The two fundamental components of our universe are matter and energy. An object possesses **energy** if it is capable of causing a change in something else. Energy is divided into two basic forms: kinetic energy and potential energy.

Kinetic Energy

Energy associated with motion is called **kinetic energy.** A moving car, for example, possesses kinetic energy. If a moving car hits a parked car, it can alter the shape of the parked car (see Figure 1-17). It is a matter of common experience (unfortunately) that the amount of change in the shape of a parked car hit by a moving car is greater the larger the mass and the higher the speed, and thus the kinetic energy, of that moving car.

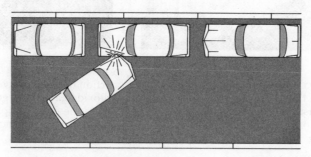

Figure 1-17   A moving car possesses kinetic energy that is greater the larger the mass and the higher the speed of the car. The kinetic energy of a moving car becomes clearly apparent when it hits a parked car. The resulting impact is greater the larger the mass and the higher the speed of the moving car.

Any moving object has kinetic energy, and this kinetic energy is directly proportional to the mass (m) of the object and increases with the square of the object's speed (s^2). The mathematical relationship between the kinetic energy of an object and its mass and speed is

(1-12)   Kinetic energy $= \frac{1}{2}ms^2$

Potential Energy

Potential energy, the other basic kind of energy, is stored energy that an object possesses because of its position in relation to another object or objects. For example, a rock located on a hillside possesses potential energy. The rock has the potential ability to alter the shape of an object at the bottom of the hill (see Figure 1-18). In this example, if the log holding the rock is removed, the rock will roll down the hill and strike the car parked below. The larger the mass of rock and the higher up the hill it was, the more damage the rock can do. Thus, the potential energy of the rock with respect to the car is directly proportional to the mass of the rock and its height above the position of the car.

(1-13)   Potential energy of rock $= K \times$ (mass) $\times$ (height above the car)

The force of gravity is expressed in the constant K in Eq. 1-13.

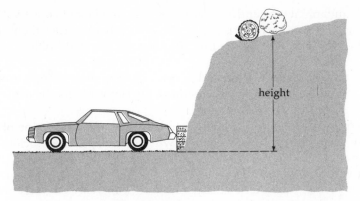

Figure 1-18 A rock on a hillside has the potential ability to alter the shape of a car parked at the bottom of the hill. Thus, the rock has potential energy. The larger the mass of the rock and the higher the rock is above the car, the greater the potential energy of the rock.

The Law of Conservation of Energy

The rock in Figure 1-18, if not constrained by the log, will move by itself down the hill, pulled by the earth's gravitational attraction. As the rock moves down the hill, its potential energy decreases. Note that when the rock is on top of the hill, the gravitational attraction between the rock and the earth is *less* than when the rock is at the bottom of the hill, but that the potential energy of the rock is *greater* when the rock is on top of the hill. If you want to move the rock up the hill, you must do work against the gravitational pull of the earth. This requires an input of energy and increases the rock's potential energy. If you carry the rock up the hill, your body supplies this energy. Chemical reactions that release energy take place within your body. Some of this released energy is used to increase the rock's potential energy and to increase your own potential energy as well, and the rest is given off as waste heat energy. The amount of energy input provided by the chemical reactions in your body is equal to the sum of the increase in potential energy of you and the rock and the amount of waste heat energy given off (see Figure 1-19). This last statement is an example of one of the

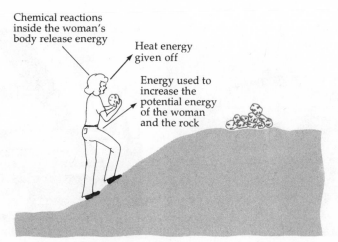

Chemical reactions inside the woman's body release energy

Heat energy given off

Energy used to increase the potential energy of the woman and the rock

Figure 1-19 As a woman walks up a slope carrying a rock, chemical reactions inside her body produce energy. Some of this energy is converted into an increase in potential energy of both the woman and the rock, and some is given off as heat.

most fundamental scientific generalizations, the law of conservation of energy. The original basis for this law was a large number of experiments conducted between 1840 and 1848 by the English scientist James P. Joule.

(1-14)   **Law of Conservation of Energy:** Energy may change from one form to another, but the total amount of energy remains the same.

The law of conservation of energy, like the law of conservation of mass, applies to any chemical or physical change (with the exception of nuclear reactions, see Chapter 2).

If the rock in Figure 1-18 rolls down the hill, it picks up speed and rotational motion. Thus, the kinetic energy of the rock increases and its potential energy decreases, but the total energy of the rock, the sum of its kinetic and potential energy, remains the same. (Actually, because of kinetic friction, there might be a small decrease in the total energy of the rock if some energy is lost by the rock and gained by the ground.) If the rolling rock hits a parked car, the rock stops moving and loses its kinetic energy. Some of this kinetic energy is used to change the shape of the parked car, some appears in the form of sound, and some is used to increase slightly the internal motion (i.e., molecular kinetic energy) of the molecules that make up the car and the rock.

Mechanical energy, sound energy, heat (thermal) energy, light energy, electrical energy, magnetic energy, gravitational energy, and chemical energy are all forms of kinetic and potential energy. Usually we become aware of the energy in a system when it changes from one form to another—for example, when the rock in Figure 1-18 hits the parked car. But according to the law of conservation of energy, the total amount of energy in a system remains constant no matter what changes occur in the form of the energy.

Units of Energy

As we have seen, there are many forms of energy, and scientists use several different units for each energy form. Chemists most often use the calorie (cal). One **calorie** is the amount of heat energy required to raise the temperature of 1 g of water 1° on the Celsius temperature scale (see Section 1.10). Although this definition of a calorie refers to heat energy, all other forms of energy, including mechanical energy, chemical energy, and electrical energy, can be expressed in units of calories.

A wide variety of devices are used to convert energy from one form to another. The water wheel changes kinetic energy of falling water into mechanical energy. Solar panels collect light energy from the sun and convert it to heat energy.

In California a large geothermal power project harnesses the heat energy of hot springs (at the left), sending high-pressure steam through insulated pipes to an electrical generating unit on the hill in the background.

The SI unit of energy is called the **joule** (J). (Recall that the law of conservation of energy was first proposed on the basis of Joule's work.) One calorie is equivalent to 4.184 joules:

(1-15)   1.000 cal = 4.184 J

Even though the international body that established the SI units in 1960 recommends that the joule be the only energy unit used, most scientists still use the energy units with which they are familiar in their routine work.

As with other units, the prefixes in Table 1-1 can be used in conjunction with the unit calorie. For example, one kilocalorie (1 kcal) is equivalent to 1000 cal. Nutritionists also use a unit called a "Calorie" or "big calorie" (note the capital letter C). The nutritionist's "big calorie" is equivalent to 1000 cal or 1 kcal.

Exercise 1-10
Water behind a dam is allowed to run downhill and turn a turbine. The turbine is connected to an electrical generator that produces electrical energy. Describe the changes in the form of energy that take place in this process.

Exercise 1-11
Energy is required to melt ice. It takes 79 cal to melt 1.0 g of ice, and this energy can be supplied in a variety of forms in addition to a direct input of heat energy. An experiment was performed in which 1.0000 g of ice at 0.00°C was held over a pail of water at 0.00°C and dropped into the water. After the ice hit the water and the ice and water stopped moving, the mass of ice in the water was determined to be 0.9990 g. The temperature of the ice and water in the pail was still found to be 0.00°C.
  (a) How much ice melted? How much energy was required to melt this amount of ice?
  (b) Where did the energy to melt the ice come from?
  (c) Using the law of conservation of energy, calculate the kinetic energy of the falling ice at the instant just before it hit the surface of the water and determine the change in the potential energy of the ice falling from its initial position above the pail to a position just at the surface of the water.

1-9   ELECTRICAL ATTRACTION AND REPULSION: COULOMB'S LAW

One of the most important features of the natural world is the presence of electrical charges. Perhaps the earliest record of electrical attraction dates from about 600 B.C., when the Greeks observed that if a piece of amber is rubbed, it attracts small, light objects such as leaves and feathers. The Greek word for amber is *elektron*. Much later, in the seventeenth century, it was noted that two glass rods that had been electrified by being rubbed with a silk cloth repelled one another, whereas an electrified glass rod and an ebony wood rod that was rubbed with fur attracted one another (see Figure 1-20). These observations suggested that there are two distinct kinds of electricity. Benjamin Franklin in 1747 proposed that electrification consists of the separation of equal amounts of positive and negative electrical charges. We now know that there is a fundamental particle of negative electrical charge called the **electron** and another fundamental particle, this one of positive electrical charge, the **proton,** and that the electron and proton have opposite charges of equal magnitude (see Chapter 2). When a glass rod is rubbed with silk, electrons are transferred from the glass to the silk, resulting in a negative electrical charge on the silk and a positive electrical charge on the glass rod.

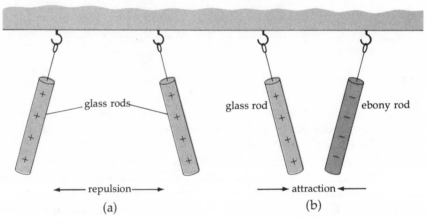

| glass rods | glass rod | ebony rod |

◄──── repulsion ────►   ────► attraction ◄────

(a)   (b)

Figure 1-20   (a) Each of the glass rods has an electrical charge. The rods move away from one another because of the repulsion between the positive charges on each rod. (b) The glass rod has a positive electrical charge, whereas the ebony rod is charged negatively. The glass rod and the ebony rod attract one another because of the attraction between positive and negative electrical charges.

Charles Coulomb (1736–1806) performed many quantitative experiments with electrically charged objects and established the fundamental relationship between the attraction and repulsion of charged objects. This relationship is now known as Coulomb's law:

(1-16)   **Coulomb's Law:**   (1) Like electrical charges repel one another and unlike electrical charges attract one another; (2) the force between any two electrical charges decreases as the distance between them increases (in fact, the force decreases as the square of this distance).

A positive and a negative electrical charge attract one another, and the closer they are together, the stronger this force of attraction. As a result, it takes energy to increase the distance between a positive and a negative charge. Thus a system of electrical charges possesses energy that depends on the relative positions of the charges—that is, the system possesses potential energy. The change in potential energy with distance for a positive and a negative charge that attract one another is similar to the variation of the potential energy with distance for two masses that attract one another. To increase the distance between a positive and a negative charge, energy must be supplied, and therefore the potential energy of the charged particles increases as they are separated (see Figure 1-21). Note that as the opposite electrical charges are separated, the force of attraction between them *decreases* but the potential energy of the charges *increases.* On the other hand, if a positive and negative charge come closer together, the force of attraction *increases* but their potential energy *decreases.*

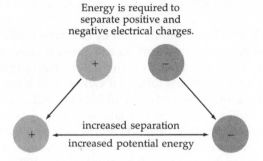

Energy is required to separate positive and negative electrical charges.

increased separation
increased potential energy

Figure 1-21   An energy input is required in order to increase the separation between positive and negative electrical charges. Because of this energy input, the charges have a larger potential energy when they are farther apart, even though the attractive force between them is smaller.

We shall be using these facts about electrical attraction and repulsion often. In particular, we shall see in Chapter 4 that electrical forces are responsible for holding atoms together in molecules.

Exercise 1-12
Two metal spheres 1 m apart each have a positive electrical charge.
(a) Is the force between these spheres attractive or repulsive?
(b) If the spheres are moved 2 m apart, does the force increase or decrease? Does the potential energy of the spheres increase or decrease?

1-10   HEAT AND TEMPERATURE

Heat and temperature are quite different, and we must be careful not to confuse them. **Heat** is the form of energy that flows from a hotter object to a colder object when they are in contact. **Temperature,** on the other hand, is not a form of energy; it is a quantitative measure of the degree of "hotness" or "coldness" of an object. "Hot" and "cold" are the terms we use to describe the temperature of an object qualitatively. However, to establish a quantitative temperature scale we must use some property of a standard material that varies with temperature. We cannot simply rely on our subjective sense of touch to measure the degree of "hotness" or "coldness" of an object.

Temperature Measurement and Temperature Scales

The most common temperature-measuring device is a **mercury thermometer.** A mercury thermometer can be constructed in the following manner. A glass bulb attached to a thin (capillary) tube is filled with mercury at a low temperature. If the temperature of the bulb is increased, the mercury in it expands, its volume increases, and a small amount of mercury moves into the capillary tube. The higher the temperature, the larger the volume of the mercury and the more mercury enters the capillary tube. Thus a temperature scale can be constructed on the basis of the length of the column of mercury in the capillary tube. Other liquids besides mercury can also be used to construct thermometers. Ethyl alcohol, for example, is used in thermometers to measure temperatures that are below the melting point of mercury ($-39°C$), since the melting point of ethyl alcohol is much lower ($-117°C$).

The temperature scale that is commonly used by chemists is called the **Celsius scale.** This scale is constructed by first putting a mercury thermometer in a mixture of ice and liquid water (water at its freezing point), marking the position of the mercury column on the capillary tube and calling this mark zero degrees Celsius ($0°C$). Then the thermometer is put in boiling water, and the higher level of the mercury column is marked and labeled $100°C$. (As we shall see in Chapter 5, the boiling point of water depends on the atmospheric pressure. The boiling point of water at a pressure of one atmosphere is $100°C$.) The distance between these two calibration marks is then divided into 100 equal parts (see Figure 1-22). For this reason the Celsius scale is sometimes referred to as the **centigrade scale.** (Recall that the prefix centi- means one-hundredth.)

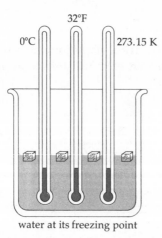
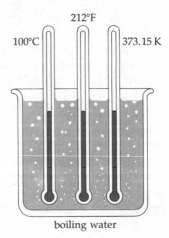

water at its freezing point   boiling water

Figure 1-22   Three identical mercury thermometers. When they are put in a mixture of ice and water, the mercury column in the capillary tube is at the same level in all three thermometers. This position is designated 0°C on the Celsius scale, 32°F on the Fahrenheit scale, and

273.15 K on the Kelvin scale. When these thermometers are put into boiling water, the common level of the mercury column is designated 100°C on the Celsius scale, 212°F on the Fahrenheit scale, and 373.15 K on the Kelvin scale.

The **Fahrenheit temperature scale** is the one that is in everyday use in the United States at this time, although it will soon be replaced by the Celsius scale. The Fahrenheit scale is constructed by labeling the mark on the thermometer corresponding to the freezing point of water as 32°F, and labeling the mark corresponding to the boiling point of water as 212°F. The relationship between a temperature on the Fahrenheit scale and the same temperature on the Celsius

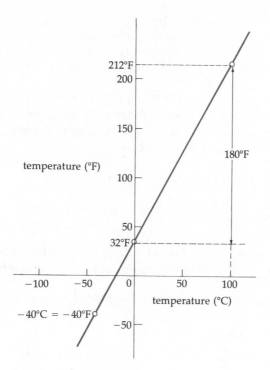

Figure 1-23   A linear relationship exists between a temperature measured on the Fahrenheit scale and a temperature measured on the Celsius scale. A change of 100°C is equivalent to a change of 180°F, and 0°C is equivalent to 32°F. Notice that −40°C is equivalent to −40°F.

scale is shown in Figure 1-23. On the Fahrenheit scale there are 180 divisions (or 180°F) between the freezing point and the boiling point of water, compared to 100 divisions (or 100°C) on the Celsius scale. Thus, a temperature change of 180°F is equivalent to a temperature change of 100°C:

(1-17)   Change of 180°F = change of 100°C

Equation 1-17 can be used to convert temperatures on the Celsius scale to those on the Fahrenheit scale and vice versa if we are careful to take into account that 0°C is equivalent to 32°F and not 0°F. Since a temperature of 0°C is not equivalent to 0°F, temperatures on the Celsius and Fahrenheit scales are *not* directly proportional to one another.

The following equation can be used to convert temperatures on the Celsius scale to temperatures on the Fahrenheit scale:

(1-18) $\text{Temperature in }°F = \dfrac{180°F}{100°C} \times (\text{temperature in }°C) + 32°F$

When you use Eq. 1-18, you must be careful and *first multiply* the temperature given in °C by the factor 180°F/100°C and *then add* 32°F. Note that a temperature of 0°C corresponds to 32°F and that a temperature of 100°C corresponds to a temperature of 212°F. It is helpful to use these facts to check that you have applied Eq. 1-18 correctly.

To convert temperatures on the Celsius scale to temperatures on the Fahrenheit scale, the following equation can be used:

(1-19) $\text{Temperature in }°C = \dfrac{100°C}{180°F} \times (\text{temperature in }°F - 32°F)$

When you use Eq. 1-19, you must be careful and *first subtract* 32°F from the temperature given in °F and *then multiply* the result by the factor 100°C/180°F. Note that a temperature of 32°F corresponds to 0°C and a temperature of 212°F corresponds to a temperature of 100°C. You can use these facts to check that you are applying Eq. 1-19 correctly.

In Eqs. 1-18 and 1-19 the factor 180°F/100°C or 100°C/180°F and the term 32°F are exact values, since they are based on defined quantities.

Normal body temperature is 98.6°F. To convert this to a temperature on the Celsius scale we can use Eq. 1-19 as follows:

(1-20)

$$\text{First subtract:} \quad 98.6°F - 32°F = 66.6°F$$

$$\text{Then multiply:} \quad \frac{100°C}{180°F} \times 66.6°F = 37.0°C$$

A common error in using Eq. 1-19 is to multiply first and then subtract. You can catch this type of error, like many others; simply <u>pay careful attention to units in your work</u>.

For instance, if you incorrectly first multiply 98.6°F by the factor 100°C/180°F, you will obtain

$$\frac{100°C}{180°F} \times (98.6°F) = 54.8°C$$

Since you cannot subtract 32°F from 54.8°C, you should realize that you have made a mistake.

The Fahrenheit temperature scale was constructed by Daniel Fahrenheit (1686–1736). Fahrenheit picked zero on his scale to correspond to the lowest temperature he could obtain, which was that of a mixture of the substance ammonium chloride and snow. As one story goes (it may or may not be true), Fahrenheit set up his scale so that 100 degrees would conform approximately to his own normal body temperature.

In any case, temperatures much lower than that of a mixture of ammonium chloride and snow (0°F or −18°C) can now be obtained. In fact, it is possible to obtain temperatures that are almost as low as −273.15°C. There are good theoretical reasons to believe that −273.15°C is the lowest temperature that could possibly be obtained (see Chapter 5). This temperature, **−273.15°C,** is therefore called **absolute zero,** and it is used to establish a temperature scale called the **absolute,** or **Kelvin, temperature scale.** The Kelvin scale is the SI temperature scale, and it is often used in scientific work (see Chapter 5). The Kelvin and Celsius scales are related by the following equations:

(1-21)

$$\text{Temperature in K} = \frac{1\text{ K}}{1°C} \times (\text{temperature in °C}) + 273.15\text{ K}$$

(1-22)

$$\text{Temperature in °C} = \frac{1°C}{1\text{ K}} \times (\text{temperature in K} - 273.15\text{ K})$$

Note that the SI convention for the unit of temperature on the Kelvin scale does not involve a degree symbol; it is K and not °K.

According to Eqs. 1-21 and 1-22, a temperature change of 1 K is equivalent to a temperature change of 1°C. Thus the only difference between the Kelvin and Celsius scales is the 273.15 difference in the zero points. A temperature of 0 K corresponds to −273.15°C, and 0°C corresponds to +273.15 K. The temperature of boiling water is 100°C or 373 K, and the normal body temperature for an adult human is 37.0°C or 310.2 K.

Exercise 1-13

Suppose that a new temperature scale called the "hypothetical" scale (°H) is constructed by labeling the mark on a mercury thermometer corresponding to the freezing point of water as 50°H and the mark corresponding to the boiling point of water as 250°H.

(a) A change of 100°C corresponds to a change of _____ °H.
(b) 0°C corresponds to a temperature of _____ °H.
(c) Normal body temperature, 37°C, corresponds to a temperature of _____ °H.

Exercise 1-14

(a) 50°C corresponds to _____ K and _____ °F.
(b) 100 K corresponds to _____ °C and _____ °F.
(c) −25°C corresponds to _____ K and _____ °F.
(d) −20°F corresponds to _____ °C and _____ K.

Optional

1-11 HEAT FLOW

We all know that we can raise the temperature of an object by heating it. The amount of heat energy (the heat input) that we must supply to an object in order to raise its temperature depends on three things:

1. The nature of the substance

2. The mass of the substance

3. The amount of the temperature increase

For a given substance the heat input is directly proportional to the mass of the substance and the temperature increase. For example, the temperature of 1 g of water can be raised 1°C by a heat input of 1 cal. Thus, to raise the temperature of 2 g of water by 1°C requires a heat input of 2 cal, and to increase the temperature of 2 g of water by 2°C requires 4 cal of heat input (see Figure 1-24).

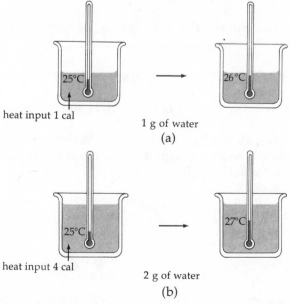

heat input 1 cal

1 g of water

(a)

heat input 4 cal

2 g of water

(b)

Figure 1-24   (a) An input of 1 cal of heat can raise the temperature of 1 g of water by 1°C. (b) To raise the temperature of 2 g of water by 2°C requires a heat input of 4 cal.

For a given substance the heat input required to raise 1 g of the substance by 1°C is called the **specific heat** of the substance. The specific heat of a substance is a characteristic physical property of that substance and its physical state (solid, liquid, or gas). Since for liquid water a heat input of 1.00 cal will raise the temperature of 1.00 g by 1.00°C,

(1-23) $$\text{Specific heat of water (liquid)} = 1.00 \frac{\text{cal}}{\text{g} \cdot {}^{\circ}\text{C}}$$

Equation 1-23 applies to water only in the liquid state, because solid water (ice) and gaseous water (steam) have different specific heats.

The specific heats of a few common substances are given in Table 1-3. The higher the specific heat of a substance, the more heat input is required to raise the temperature of 1 g of the substance by 1°C, and likewise, the more heat that is evolved when 1 g of the substance cools down by 1°C. Water has a relatively high specific heat, and this property of water is an advantage when water is used in hot water bottles, packs, baths, and solar heating systems.

Table 1-3 Specific Heat

Substance	Specific Heat (cal/g · °C)
Water (liquid)	1.00
Glass	0.09
Iron	0.11
Ethyl alcohol	0.60
Salt (sodium chloride)	0.21

We can express the relationship between the heat input required, the substance involved, the mass of the substance, and the temperature increase by the following equation:

(1-24)   Heat input = (specific heat) × (mass) × (temperature increase)

Similarly, the relationship of the heat lost by a hot body when it cools down and the temperature decrease is

(1-25)   Heat lost = (specific heat) × (mass) × (temperature decrease)

Let us consider two examples of how Eqs. 1-24 and 1-25 are used.

EXAMPLE 1 Compare the heat input needed to raise the temperature of 5.0 g of water from 20.0°C to 25.0°C with the heat input needed to obtain the same temperature rise for 5.0 g of glass.

For water: $\text{Heat input} = \left(1.00 \frac{\text{cal}}{\cancel{\text{g}} \cdot \cancel{{}^{\circ}\text{C}}}\right) \times (5.0 \, \cancel{\text{g}}) \times (5.0 {}^{\circ}\cancel{\text{C}}) = 25.0 \text{ cal}$

For glass: $\text{Heat input} = \left(0.09 \frac{\text{cal}}{\cancel{\text{g}} \cdot \cancel{{}^{\circ}\text{C}}}\right) \times (5.0 \, \cancel{\text{g}}) \times (5.0 {}^{\circ}\cancel{\text{C}}) = 2.3 \text{ cal}$

Note that because the specific heat of water is so much larger than that of glass, a great deal more heat input is required to raise the temperature of 5.0 g of water than 5.0 g of glass.

EXAMPLE 2 If 25.0 g of water at 25°C loses 100 cal of heat, what is the new temperature of the water?

Since the water is losing heat in this example, we can use Eq. 1-25 in the following rearranged form:

(1-26)
$$\text{Temperature decrease} = \frac{\text{(heat lost)}}{\text{(specific heat)} \times \text{(mass)}}$$

$$= \frac{(100 \ \cancel{cal})}{\left(1.00 \ \dfrac{\cancel{cal}}{\cancel{g} \cdot °C}\right) \times (25.0 \ \cancel{g})} = 4.0°C$$

Thus, since the starting temperature was 25.0°C, the new temperature after the water loses heat is 21.0°C.

See if you can do the following similar exercises. If you have any difficulty with them, the material in Essential Skills 5 should prove very helpful. Essential Skills 5 discusses the type of logical reasoning and mathematical techniques used to solve quantitative problems.

Exercise 1-15
An experiment was performed to determine the specific heat of ice. It was observed that it takes 90 cal of heat input to raise the temperature of 30 g of ice from −10.0°C to −4.0°C. From this data, calculate the specific heat of ice.

Exercise 1-16
The specific heat of gold is 3.1×10^{-2} cal/g · °C. When the temperature of a sample of gold changed from 37°C to 25°C, it was observed that 18 cal of heat were lost by the gold. Calculate the mass of the gold sample.

1-12 SUMMARY

1. Measurement involves comparison with a standard unit, and the standard unit must be included when using the result of any measurement.

2. The Système International, SI, is the internationally accepted system of units.

3. The SI standards of length, volume, and mass are the meter, the cubic meter (m^3), and the kilogram, respectively.

4. The prefixes kilo- (10^3), deci- (10^{-1}), centi- (10^{-2}), milli- (10^{-3}), micro- (10^{-6}), and nano- (10^{-9}) are often used in scientific work.

5. Chemists frequently use the volume units liter and milliliter (ml).

6. The law of conservation of mass states that the total amount of mass in any system remains the same when a chemical reaction or any other process (except a nuclear reaction) takes place in the system.

7. Density is a characteristic property of a substance and is defined by the equation: density = mass/volume.

8. The specific gravity of a substance is defined by the equation: specific gravity = density of substance/density of water.

9. The conclusions that can be reached based on a measured value are more definite the greater the precision and accuracy of the measurement.

10. A physical property of a substance is a characteristic property that can be observed and measured without changing the substance into another substance. Physical changes are changes in physical properties.

11. In a chemical reaction one or more substances are changed into other substances (reactants are changed into products).

12. The chemical properties of a substance are those chemical reactions in which the substance is a reactant.

13. Kinetic energy is energy associated with motion.

14. Potential energy is stored energy that an object has because of its relative position with respect to another object or objects.

15. The law of conservation of energy states that energy may change from one form to another, but when this occurs the total amount of energy (except in a nuclear reaction) remains the same.

16. Coulomb's law states that (1) like electrical charges repel one another and unlike electrical charges attract one another; and (2) the force between any two electrical charges decreases as the square of the distance between them increases.

17. One calorie is the amount of heat energy required to raise the temperature of 1 g of water by 1° on the Celsius temperature scale.

18. Heat is the form of energy that flows from a hotter to a colder object when they are in contact.

19. Temperature is a quantitative measure of the degree of "hotness" or "coldness" of an object.

20. The temperature of an object can be expressed in terms of the Celsius, Fahrenheit, or Kelvin temperature scale.

21. **(Optional)** The specific heat of a substance is the number of calories needed to raise the temperature of 1 g of the substance by 1° on the Celsius temperature scale.

PROBLEMS

1. Write the following numbers in standard scientific notation (see Essential Skills 1). `
   (a) 32500   (b) 0.00305   (c) 0.000068   (d) 20300000

2. Write the following exponential numbers in decimal form (see Essential Skills 1).
   (a) 7.3×10^{-2} (b) 8.5×10^{4} (c) 2.09×10^{-3} (d) 6.28×10^{4}

3. A 1.00-g sample of ice was partially melted by warming it. The mass of liquid water formed was 0.43 g. What was the mass of the ice left unmelted?

4. Express the answers to the following in standard scientific notation. Round off the result of each calculated value so as to indicate correctly the uncertainty in the calculated value (see Essential Skills 1 and 4).

   (a) $(4.56 \times 10^{-4}) \times (3.2 \times 10^{7}) =$

   (b) $\dfrac{4.20 \times 10^{-4}}{3.000 \times 10^{-9}} =$

   (c) $(205) \times (1237) =$

   (d) $(5.278 \times 10^{-5}) \times (2.32 \times 10^{-6}) =$

   (e) $(1.5873 \times 10^{-4}) - (1.6 \times 10^{-6}) =$

   (f) $\dfrac{3.634}{1.82} =$

   (g) $2589 + 22.35 =$

   (h) $\dfrac{9.875 \times 10^{-8}}{1.82 \times 10^{-10}} =$

5. Do the following conversions (see Essential Skills 3).
   (a) 12.3 mg = _____ g
   (b) 2.67 liters = _____ ml
   (c) 32 kg = _____ lb
   (d) 2.3 nm = _____ in.
   (e) 4.5 liters = _____ qt
   (f) 5.2×10^{-4} g = _____ μg
   (g) 24,000 cal = _____ kcal

6. The density of a person's urine is normally in the range 1.018 g/ml to 1.030 g/ml. If the density of Mary's urine is measured and a value of 1.1 g/ml is obtained, can we confidently conclude on the basis of this measurement that the density of Mary's urine falls outside the normal range?

7. The density of copper metal at 25°C is 8.92 g/cm³. Determine the volume of 5.3 g of copper at 25°C.

8. The density of nitrogen gas at 25°C and 1 atmosphere pressure is 1.16 g/liter. Determine the mass of 25 cm³ of nitrogen at 25°C and 1 atmosphere.

9. The density of ethyl alcohol decreases from 0.800 g/ml at 7°C to 0.775 g/ml at 37°C. Is the volume of 15 g of ethyl alcohol at 37°C larger than, smaller than, or the same as the volume at 7°C?

10. The density of human bone is about 1.8 g/ml. What is the volume of a human skeleton if the dry weight of the skeleton is 2.7 kg?

11. Calcium metal is a hard, brittle, gray solid. Calcium hydroxide is a milky white solid that is not soluble in water. Hydrogen is a gas. When calcium metal is put into water, a chemical reaction takes place in which calcium hydroxide and hydrogen are formed. If some calcium metal is put into water, list three things we could observe directly that would give evidence that a chemical reaction is taking place.

12. You are to administer a solution of a drug to a patient whose body weight is 150 lb. It is very important that the drug dosage not vary by more than 2% from exactly 5 ml of solution for each 100 lb of body weight. What volume of drug solution should you administer to the patient? What is the maximum and minimum volume you can administer and still be within the allowable 2% range? (Percentages are discussed in Essential Skills 6.)

13. The mass of an object is determined to be 5.12 g. What is the approximate uncertainty in this measurement?

14. A car moving on a level road slows down from 50 miles per hour (50 mph) to 30 mph. Its _____ (kinetic, potential) energy has _____ (decreased, increased).

15. When a positive and a negative electrical charge are brought closer together, the potential energy of the system _____ (increases, decreases, remains the same).

16. When a person shoots an arrow straight up into the air, as the arrow goes up, the potential energy of the arrow _____ (increases, decreases, remains the same) and the kinetic energy _____ (increases, decreases, remains the same).

17. The boiling point of liquid nitrogen is 77 K. What is the boiling point of liquid nitrogen on the Celsius scale and on the Fahrenheit scale?

18. The melting point of sodium chloride (salt) is 801°C. What is the melting point of sodium chloride on the Kelvin scale and on the Fahrenheit scale?

19. **(Optional)** Lithium metal has a specific heat of 0.85 cal/g · °C. Calculate the final temperature of 20.0 g of lithium at 24.2°C after 68 cal of heat have been added.

20. **(Optional)** A piece of cobalt metal at 98°C is put into a polystyrene cup that contains 25 g of water at a temperature of 22°C. The temperature of the water with the cobalt in it attains a constant temperature of 31°C. The specific heat of water is 1.0 cal/g · °C and the specific heat of cobalt is 0.11 cal/g · °C.
    (a) Calculate the heat absorbed by the water. (You can ignore the small amount of heat absorbed by the polystyrene cup.)
    (b) Determine the mass of the piece of cobalt metal.

SOLUTIONS TO EXERCISES

1-1 (a) Since A has a weight of 1 lb on the earth's surface, it has a mass of 0.45 kg. Therefore, since the masses are equal, the beam will be level.

(b) On the moon's surface the masses are still equal, and the beam is level there also. The gravitational pull on both masses is smaller on the moon by the same amount.

1-2 1.0 "wee-inch" = $\frac{1}{6}$ in.

$$\text{Length of pencil} = (4.0 \text{ in.}) \times \left(\frac{1.0 \text{ "wee-inch"}}{\frac{1}{6} \text{ in.}} \right) = 24 \text{ "wee-inches"}$$

1-3 (a) $(0.0025 \text{ liter}) \times \left(\frac{1000 \text{ ml}}{1.0 \text{ liter}} \right) = 2.5 \text{ ml}$

(b) $1.35 \text{ nm} = (1.35 \times 10^{-9} \text{ m}) \times \left(\frac{10^2 \text{ cm}}{1.00 \text{ m}} \right) = 1.35 \times 10^{-7} \text{ cm}$

(c) $0.0000035 \text{ g} = (3.5 \times 10^{-6} \text{ g}) \times \left(\frac{1.0 \text{ } \mu g}{10^{-6} \text{ g}} \right) = 3.5 \text{ } \mu g$

(d) $1500 \text{ cc} = 1500 \text{ ml} = 1500 \times 10^{-3} \text{ liter} = 1.5 \text{ liters}$

(e) $(7.25 \text{ liters}) \times \left(\frac{1.06 \text{ qt}}{1.00 \text{ liter}} \right) = 7.69 \text{ qt}$

(f) $(4.2 \text{ lb}) \times \left(\frac{454 \text{ g}}{1.0 \text{ lb}} \right) = 1.9 \times 10^3 \text{ g}$

(g) $(75 \text{ miles}) \times \left(\frac{1.6 \text{ km}}{1.0 \text{ mile}} \right) = 120 \text{ km}$

(h) $(68 \text{ g}) \times \left(\frac{1.00 \text{ oz}}{28.3 \text{ g}} \right) = 2.4 \text{ oz}$

1-4 $\text{Mass of air} = (5.0 \text{ liters}) \times \left(\frac{1.2 \text{ g}}{1.0 \text{ liter}} \right) = 6.0 \text{ g}$

1-5 (a) Mass and volume are directly proportional to one another. The density of ethyl alcohol is 0.785 g/cm^3 (Table 1-2).

$$\text{Volume of ethyl alcohol} = (25.0 \text{ g}) \times \left(\frac{1.00 \text{ cm}^3}{0.785 \text{ g}} \right) = 31.8 \text{ cm}^3$$

(b) Since the density of diethyl ether is less than the density of ethyl alcohol, 25 g of diethyl ether will occupy a larger volume than 25 g of ethyl alcohol. Thus, the flask cannot contain 25 g of diethyl ether.

(c) $\text{Volume of diethyl ether} = (25.0 \text{ g}) \times \left(\frac{1.00 \text{ cm}^3}{0.708 \text{ g}} \right) = 35.3 \text{ cm}^3$

1-6 (a) No. The implied uncertainty in the specific gravity of X is ± 0.1 (specific gravity of X = 1.1 ± 0.1). Therefore the specific gravity of X could reasonably be in the range from 1.0 to 1.2. Thus X could be A, C, or E.

(b) It is not reasonable that X is B or D, since the specific gravity of each of these substances is well outside the range 1.1 ± 0.1.

(c) The specific gravity of X could reasonably be in the range from 1.113 to 1.123. Only substance A has a specific gravity in this range.

1-7 For (a), (b), (c), and (d), use the rule for multiplication or division that the result is rounded off so that the number of significant figures in the result is equal to the number of significant figures in the factor that has the fewest number of significant figures (see Essential Skills 4).

(a) 68   (b) 1.0×10^4 (c) 3.54×10^{-4} (d) 1.8×10^2

For (e) and (f), determine which number has an uncertainty in the largest decimal place.

(e) Uncertainty in tenths place: 52.2 + 1.9 = 54.1

(f) $2.456 \times 10^5 = 245.6 \times 10^3$ (uncertainty in tenths place)

$(245.6 \times 10^3) + (1.62 \times 10^3) = 247.2 \times 10^3 = 2.472 \times 10^5$

1-8 (a) Physical change   (b) Chemical reaction   (c) Chemical reaction
     (d) Physical change   (e) Chemical reaction

1-9 Mass of iron + mass of oxygen = mass of rust
     Mass of oxygen = mass of rust − mass of iron = 14.3 g − 10.0 g = 4.3 g

1-10 As the water runs downhill, the potential energy of the water decreases and its kinetic energy increases. When the water turns the turbine, some of the kinetic energy of the water is converted to mechanical energy and the rest is converted to heat energy. When the turbine powers an electrical generator, some of the mechanical energy of the turbine is converted to electrical energy and the rest is converted to heat energy.

1-11 (a) 0.0010 g of ice melted. Since the energy required to melt 1.0 g of ice is 79 cal, the energy required to melt 0.0010 g of ice is

$$(0.0010 \text{ g}) \times \left(\frac{79 \text{ cal}}{1.0 \text{ g}}\right) = 0.079 \text{ cal}$$

(b) As the ice falls, the potential energy of the ice decreases and its kinetic energy increases. When the ice hits the water, it stops moving and the kinetic energy of the ice just before it hits the water is used to melt the ice.
(c) The kinetic energy of the ice just before it hits the water is equal to the energy required to melt 0.0010 g of ice or 0.079 cal. The change of potential energy of the ice must be equal to the kinetic energy of the ice just before it hits the water, or 0.079 cal.

1-12 (a) The charges are alike, so the force is repulsive.
(b) Force decreases as distance increases. Potential energy decreases. To bring two positively charged spheres closer together requires an energy input, so the potential energy increases. Therefore, if the distance between the spheres increases, the potential energy decreases.

1-13 (a) A change of 100°C corresponds to a change of 200°H.
(b) 0°C corresponds to 50°H.
(c) A change of 37°C is equivalent to a change of $(37°\cancel{C}) \times (200°H/100°\cancel{C}) = 74°H$. Therefore, since 0°C corresponds to 50°H, 37°C corresponds to 50°H + 74°H = 124°H.

1-14 (a) 50°C corresponds to 323 K and to 122°F.
(b) 100 K corresponds to −173°C and to −279°F.
(c) −25°C corresponds to 248 K and to −13°F.
(d) −20°F corresponds to −29°C and to 244 K.

1-15 (**Optional**) See Essential Skills 5.

Heat input = (specific heat) × (mass) × (temperature increase)

$$\text{Specific heat} = \frac{(\text{heat input})}{(\text{mass}) \times (\text{temp. increase})} = \frac{(90 \text{ cal})}{(30 \text{ g}) \times (6°C)} = 0.5 \frac{\text{cal}}{\text{g} \cdot °C}$$

1-16 (**Optional**)

Heat lost = (specific heat) × (mass) × (temperature decrease)

$$\text{Mass} = \frac{(\text{heat lost})}{(\text{specific heat}) \times (\text{temp. decrease})} = \frac{(18 \text{ cal})}{\left(3.1 \times 10^{-2} \frac{\text{cal}}{\text{g} \cdot °\cancel{C}}\right) \times (12°\cancel{C})} = 48 \text{ g}$$

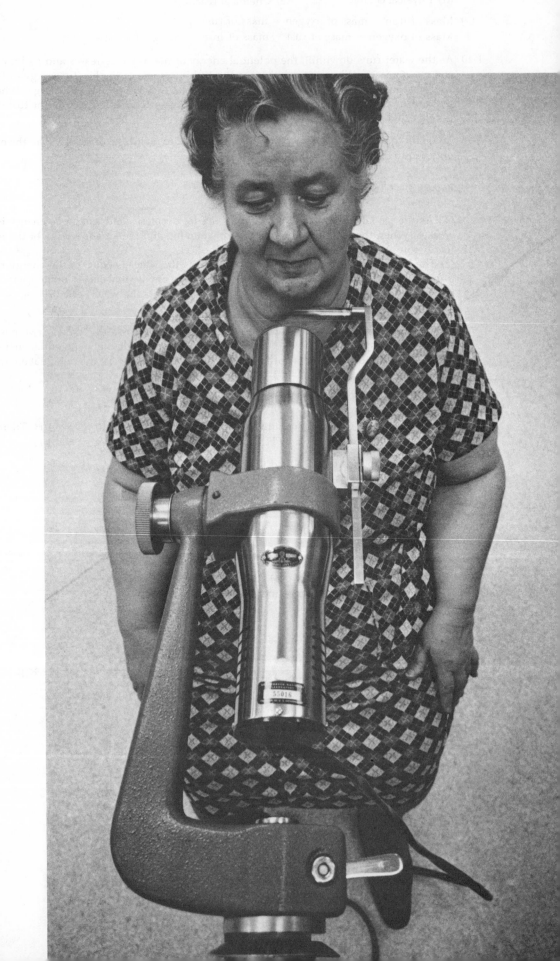

CHAPTER 2

Classification of Matter: Atomic Structure

2-1 INTRODUCTION

The alchemists of the Middle Ages tried in vain to convert common substances such as lead and iron into gold. Pursuit of the alchemists' dreams was abandoned in the early nineteenth century when Dalton's atomic theory of matter indicated that it is impossible to convert common metals into gold. In the nineteenth and twentieth centuries, however, using the atomic theory of matter as a guide, chemists have been able to accomplish a great deal in their sophisticated laboratories. Our daily lives are now filled with examples of synthetic products whose existence was made possible by the miracles of chemical research.

In fact, as a consequence of the discovery of the phenomenon of radioactivity in the early twentieth century, the modern-day "alchemist" can even produce radioactive gold in nuclear reactors. It is very expensive to make gold this way, and most of the gold that is produced lasts only a few days, but radioactive gold and many other radioactive substances are useful in medical diagnosis and in certain types of cancer therapy. Radioactive substances are also used in chemical research and have a wide variety of industrial applications, such as in the generation of electricity from the heat energy produced by a nuclear reactor. Although nuclear reactions can be used advantageously, we are all aware of the great dangers associated with them.

In this chapter we shall describe some of the basic macroscopic features of the solid, liquid, and gas states of matter. We shall discuss the fact that most samples of matter can be separated into simpler parts, and that there is a finite number of fundamental chemical substances called elements. We shall also describe Dalton's atomic model and see that chemists still use this model as a basic framework for the description of chemical reactions. We shall then discuss the essential features of the structure of atoms and describe the three fundamental atomic particles: electrons, protons, and neutrons. Finally, we shall study the phenomenon of radioactivity, and learn the basic features of nuclear reactions.

The activity of an isotope of iodine injected into a patient's bloodstream is measured in the vicinity of the thyroid gland. The rate of uptake of this radioactive tracer is a valuable diagnostic test of thyroid function.

2-2   STUDY OBJECTIVES

After studying the material in this chapter, you should be able to:

1. Describe the basic macroscopic and molecular features of solid, liquid, and gas states.

2. Define the terms melting point, boiling point, heat of fusion, and heat of vaporization.

3. Define the terms homogeneous and heterogeneous samples of matter, pure substance, solution, compound, and element.

4. Explain Dalton's atomic theory of matter.

5. Define the term chemical formula and use chemical formulas to write balanced chemical equations representing chemical reactions.

6. State the charge and mass in atomic mass units of the fundamental subatomic particles—electrons, protons, and neutrons—and give the basic features of the structure of an atom.

7. Define the terms atomic number, mass number, ion, cation, anion, and isotope.

8. Determine the number of electrons, protons, and neutrons in an atom or ion, given its mass number and a table relating chemical symbols and atomic numbers.

9. Define the terms radioactivity, nuclear reaction, nuclear decay, nuclear transmutation, nuclear fission, nuclear fusion, half-life, and activity.

10. Compare the basic features of nuclear reactions with those of chemical reactions.

11. Give the mass, charge, and symbol for alpha, beta, and gamma rays.

12. Represent nuclear reactions by balanced equations.

13. Relate the half-life of a radioactive isotope to the number of nuclei that decay in a given period of time.

14. Discuss the different effects of alpha, beta, and gamma rays on human body cells.

2-3   STATES OF MATTER

Matter is anything that has mass and occupies space. On the basis of our everyday experience, we know that there are three distinctly different forms, or states, of matter: solid, liquid, and gas. For example, water can exist as a solid (which we commonly call ice), as a liquid, or as a gas (which we call water vapor). A piece of ice, or any other solid form of matter, has both a fixed shape and a high density. It is also difficult to compress or expand ice and other solids. Liquids also have high densities, and they are also difficult to compress or expand, but the shape of a liquid is not definite. Liquids assume the shape of the container in which they are put. For almost all substances, the density of the solid form is slightly greater than that of the liquid state. Water is a rare exception. At 0°C the density of liquid water is slightly greater than that of ice; this is why ice floats. The densities of liquid and solid water at 0°C are 0.99987 g/ml and 0.9168 g/ml, respectively.

Gases are quite different from solids and liquids. A sample of gas will fill a container completely. It is relatively easy to compress and expand a gas, and under normal circumstances gases have very low densities. For example, at 100°C and a pressure of 1 atmosphere, the densities of water in the liquid and gas states are 0.958 g/ml and 5.964 × 10⁻⁴ g/ml, respectively. Therefore, under these particular conditions, the density of the liquid is about 1600 times the density of the gas, so that 1.0 ml of liquid water and 1600 ml of water vapor will have the same mass.

Whether a substance exists as a solid, a liquid, or a gas depends on the temperature and pressure. Consider water again. At standard pressure (1 atm), if we take an ice cube out of a freezer and put it in a room where the temperature is above 0°C, it will melt. Similarly, if we take a glass of liquid water outside on a cold day when the temperature is below 0°C, it will freeze. We therefore say that the stable form of water below 0°C is ice, and that above 0°C it is the liquid state that is stable.

What happens to water at 0°C? At this temperature, the solid and liquid states of water can coexist in a stable condition. The temperature at which both the liquid and solid states of a substance can coexist is called the **melting point** (abbreviated m.p.) of the substance. Therefore, the melting point of water is 0°C (see Figure 2-1).

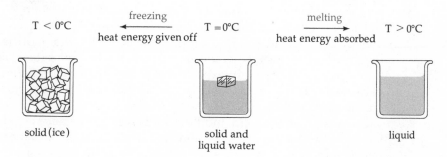

Figure 2-1   At 0°C both liquid water and solid water (ice) can coexist in a stable condition. If the temperature is lowered below 0°C, the liquid portion will freeze and all the water will be in the form of ice. In this process heat energy is given off to the surroundings. If the temperature is raised above 0°C, the solid portion will melt and all the water will be in the liquid form. Heat energy is absorbed from the surroundings in this process.

Energy is required to convert ice to liquid water. The most usual manner in which this energy is supplied is in the form of heat. To melt 1 g of ice requires 80 cal, and we say that the heat of fusion of water is 80 cal/g. For any substance, the **heat of fusion** is the energy required to convert 1 g of that substance from the solid to the liquid state. For any substance, the amount of heat required to melt a sample of the solid substance is directly proportional to the mass of the sample.

At some temperature above the melting point, an analogous situation occurs between the liquid and gas states of a substance. For water the liquid state is stable between 0°C and 100°C. Above 100°C, it is the gas state of water that is stable. If we heat some water in an open container, when the water reaches a certain temperature, bubbles form and rise to the surface, and steam is given off from the surface. When the pressure is 1 atm, the temperature at which this happens is 100°C, and we call 100°C the **boiling point** (b.p.) of water.

Vaporization is the process of going from the liquid state to the gas state, while the reverse process is called **condensation.** Energy is required to vaporize a liquid. The energy required to convert 1 g of a substance from the liquid to the gas state is called the **heat of vaporization** for that substance. For water, the heat of vaporization is 540 cal/g. For any substance, the amount of heat required to vaporize a sample of the liquid form of the substance is directly proportional to the mass of the sample.

Most substances can exist as a solid, liquid, or gas (depending on the temperature and pressure) and have a melting point, a boiling point, a heat of fusion, and a heat of vaporization that are characteristic properties of the substance.

We shall treat the properties of the solid, liquid, and gas states and the processes of fusion and vaporization in more detail in Chapter 5. However, on the basis of this preliminary discussion, we can already make some extremely important inferences about the molecular nature of matter.

A gas expands to fill completely the container in which it is put, but a solid and a liquid do not. Therefore, attractive forces must exist to hold solids and liquids together. These attractive forces are not present, or are very much weaker, in the gas state. If we postulate that any sample of matter of macroscopic size is composed of a large number of individual particles, then we can infer that there must be large attractive forces between individual particles when matter is in the solid or liquid state. The attractive forces between individual particles in the gas state must be much smaller or negligible. We can support this argument by comparing the densities of the solid, liquid, and gas states for a substance. Since the density of the gas state is much lower than the density of either the solid or liquid state, the individual particles must be farther apart on the average in the gas state than when the substance is in solid or liquid form. We therefore expect weaker attractive forces between individual particles in the gas state, since attractive forces between particles should decrease as the particles become farther apart.

The density of the solid and liquid forms of a substance are about the same, so we infer that in the solid and liquid states of a substance the distances between individual particles are roughly the same (see Figure 2-2).

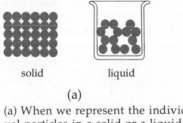

solid   liquid

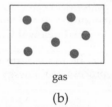

gas

(a)   (b)

Figure 2-2   (a) When we represent the individ-   The actual arrangement of particles
ual particles in a solid or a liquid   is quite different in the solid and the
as small balls, the small distance   liquid state. (b) In a gas, the average
between particles is indicated   distance between the individual
by the balls touching one another.   particles is large.

A simplified description of some of the changes that occur when a substance goes from one state to another follows: When a solid melts, the forces of attraction between individual particles are partially overcome; the energy needed to melt a solid is the heat of fusion. When a liquid vaporizes, the particles become separated by large distances and the forces of attraction between individual particles are extremely small; the energy required to vaporize a liquid is the heat of vaporization. Since there is a greater reduction in the forces of attraction between individual particles in vaporization than in melting, we might expect that the heat of vaporization of a substance is larger than its heat of fusion. For water and most other substances, this is indeed the case.

Exercise 2-1
The melting point of ethyl alcohol is −114°C, and its boiling point is 78°C. What state of this alcohol is stable at −10°C? What physical changes occur when the temperature of ethyl alcohol goes from −150°C to +150°C?

Exercise 2-2
The heat of fusion of ethyl alcohol is 25 cal/g. What mass of solid ethyl alcohol can be converted to the liquid state by 1.0 kcal?

2-4 SEPARATION OF MATTER

It is natural to try and separate samples of matter into simpler parts. Some samples of matter are **homogeneous;** that is, they have uniform chemical and physical properties throughout the sample. Any two parts of a homogeneous sample of matter have identical characteristic properties (melting point, boiling point, density, reactivity with other substances, and so on).

Other samples of matter have distinct regions with different characteristic properties. Such samples of matter are called **heterogeneous.** A simple example should clarify the distinction between homogeneous and heterogeneous samples of matter: If you add a very large amount of sugar to a cup of water, some of the sugar will dissolve in the water, and some will settle to the bottom of the cup. The material inside the cup is an obvious example of a heterogeneous sample of matter. It clearly has two distinct parts: solid and liquid. The two parts of this heterogeneous sample can be separated into homogeneous parts by the mechanical process of **filtration** (see Figure 2-3). In general, heterogeneous samples of matter can be separated into homogeneous parts by some simple mechanical means.

Figure 2-3   A heterogeneous mixture composed of a homogeneous solution of sugar and water and a homogeneous sample of excess solid sugar can be separated by the process of filtration. The homogeneous solid sugar is retained on the filter paper in the funnel, whereas the homogeneous liquid solution passes through the filter paper and is collected in the graduated cylinder.

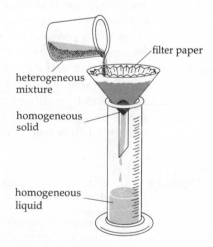

There is a fundamental difference, however, between the homogeneous solid and liquid parts of the heterogeneous sample in our example. The homogeneous liquid part can be further separated into two components by boiling off (vaporizing) the water and leaving solid sugar behind. Liquid water, without sugar in it, can be re-formed by cooling and condensing the hot water vapor (see Figure 2-4). In this process, called **distillation,** the liquid part of the original sample is separated into two homogeneous components by means of a change of state. Neither the sugar component nor the water component can be separated further in this manner. Sugar and water are examples of pure substances,

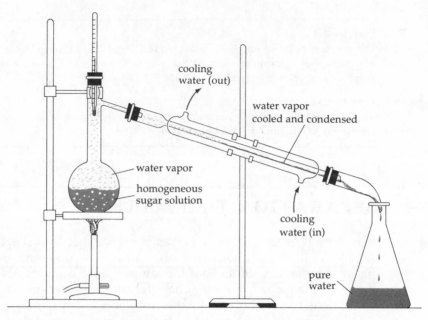

Figure 2-4   A homogeneous sugar solution can be separated into two pure substances by the process of distillation. When the sugar solution is heated, the water vaporizes. The water vapor condenses in the form of pure water. When all the water has boiled off, pure solid sugar remains in the flask.

and the homogeneous liquid mixture of the two is an example of a solution. We shall define the terms "solution" and "pure substance" shortly. First let us see if it is possible to decompose a pure substance into even simpler parts.

If we pass an electric current through a sample of liquid water, the water is decomposed into two gases called hydrogen and oxygen, which have different properties (see Figure 2-5). These gases *cannot* be decomposed into simpler substances by any ordinary chemical or physical means. The phrase "ordinary chemical or physical means" is used to exclude nuclear reactions, which involve enormous amounts of energy. Hydrogen and oxygen are examples of two of the simplest of chemical substances, which are called elements.

An **element** is defined as a substance that cannot be decomposed into simpler substances by ordinary chemical or physical means. At the present time 106 elements are known. Some of them do not occur naturally but can be generated only by nuclear reactions (see Section 2-9). Each element has a name and a one-

Figure 2-5   Water can be decomposed into its elements, hydrogen and oxygen, by passing an electric current through a sample of water in a process called electrolysis. Electrolysis is a complicated process, but the overall result of the electrolysis of water is that hydrogen gas is released at one electrode, while oxygen gas is released at the other electrode.

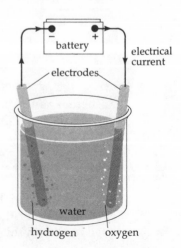

or a two-letter symbol to represent it. The names and symbols of the elements are given in the table printed on the inside back cover of this book. Not all of the 106 elements are important in the chemistry of the human body. Although a large number of elements play crucial roles in the chemistry of life, the most predominant ones in the human body are carbon (C), hydrogen (H), oxygen (O), and nitrogen (N), and to a lesser degree, phosphorus (P) and sulfur (S). Of the mass of the human body, 99% consists of the elements C, H, O, and N.

Pure substances are either elements or compounds. **Compounds,** such as water, can be decomposed into two or more elements, and compounds have a definite constant elemental composition. For example, every sample of pure water, whether in outer space, or in a muscle cell, is composed of the elements hydrogen and oxygen in the same mass ratio. The mass of any water sample is 11.1% hydrogen and 88.9% oxygen.

Now that we have defined the simplest chemical substances, elements, we can give a precise definition of the terms "pure substances" and "solutions." A **pure substance** (either a compound or an element) is a homogeneous sample of matter that cannot be separated into simpler substances by a change of state, and that has both a definite elemental composition and characteristic physical and chemical properties. A **solution** is a homogeneous mixture of two or more pure substances. The composition of a solution, such as a solution of sugar dissolved in water, is variable. We shall discuss solutions further in Chapter 8.

Our discussion thus far, of how samples of matter can be separated into simpler parts, is summarized in Figure 2-6. Of course, there are many questions about matter that we have yet to discuss, for example: What is the molecular nature of matter, and how are elements combined to form compounds?

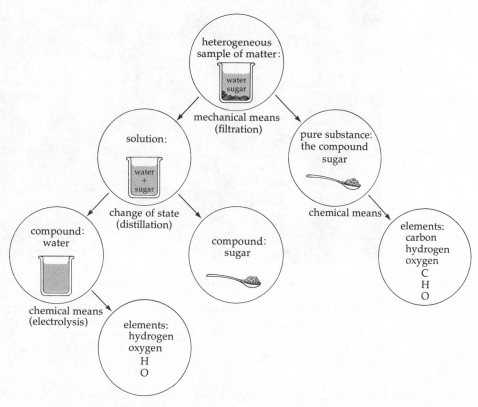

Figure 2-6   Heterogeneous samples of matter can be separated into homogeneous parts. Homogeneous solutions can be separated into pure substances. Compounds can be decomposed into their elements.

Exercise 2-3

You are given a mixture of common table salt and sand and asked to separate it into pure substances. Describe an experimental procedure that you could use to accomplish this type of separation.

2-5   THE ATOMIC MODEL

For a very long time people have speculated about the fundamental nature of matter. In 440 B.C. the ancient Greek philosopher Empedocles proposed that all matter is composed of four basic parts—earth, air, fire, and water. At about the same time, another Greek philosopher, Democritus, expounded quite a different belief—that matter consists of a large number of small unbreakable particles that cannot be seen. He called these particles atoms. The term "atom" comes from the Greek word *atomos,* which means indivisible.

Although the views expressed by these two Greek philosophers were radically different, they shared one common basic feature: Neither view was based on any quantitative experimental evidence. In fact, no significant experimental support for either model of matter was obtained until the latter part of the eighteenth century. A great deal of chemical work was done in the meantime, but almost all of it was qualitative in nature.

Early alchemists worked in primitive laboratories and without the benefit of the theoretical framework provided by Dalton's atomic model. Although these chemists failed to achieve their primary goal—the conversion of common metals to gold—they still uncovered a great deal of useful practical information and developed much of the basic laboratory equipment and techniques that made possible the evolution of modern chemistry.

In about 1780, however, a significant change took place in the study of chemical reactions. Chemists began to make precise and accurate quantitative measurements of the masses of the reactants and products in a chemical reaction. The French chemist Antoine Lavoisier (1743–1794) was perhaps the first

Antoine Lavoisier has been called the founder of modern chemistry because of the importance of his precise and accurate measurements of the masses of reactants and products in chemical reactions.

person to recognize the crucial importance of making this type of quantitative measurement, which is the cornerstone of modern chemistry.

Using quantitative data obtained by Lavoisier and other chemists, an English school teacher, John Dalton, proposed an **atomic theory of matter** in about 1808. The atomic models of Dalton and Democritus share many common features, but Dalton's hypotheses were supported by quantitative experimental evidence. In addition, Dalton's postulates suggested new chemical experiments. When these experiments were carried out, the results supported Dalton's model.

Some of the main features of Dalton's atomic model are the following:

1. Elements are composed of small indivisible particles called **atoms.**

2. Atoms of the same element are identical and therefore have identical properties.

3. Atoms of different elements are different and therefore have different properties.

4. Atoms can combine to form compounds in chemical reactions, and compounds can be rearranged to form other compounds.

5. Atoms of one element do not change into atoms of another element, and no atoms are lost or created in chemical reactions.

Dalton's postulates raised many questions about the nature of atoms. For example: What are the structure, size, and mass of atoms? What are the changes that occur when atoms combine to form compounds? Additional experimental evidence has also led to some modifications of Dalton's atomic model, which we shall discuss shortly. Dalton's original ideas, however, are still used by chemists as a framework to describe the most basic features of a chemical reaction.

2-6   MOLECULES AND CHEMICAL REACTIONS

An atom of a given element is represented by the symbol for that element. For example, the symbol O is used to represent an oxygen atom. Separate uncombined atoms are a rarity. In compounds, atoms of different elements are combined, and for many elements, atoms of the same element are combined with one another.

Molecules

When atoms of the same or different elements combine, the resulting particle is called a **molecule.** The atoms of a molecule are held together by forces called **chemical bonds.** Chemical bonds are discussed in Chapter 4.

The number and identity of the atoms in a molecule of a substance is determined by experiment. For example, experimental evidence shows that the element oxygen exists as molecules composed of two oxygen atoms bonded together, and not as individual uncombined atoms. Molecules of the compound water are composed of two atoms of hydrogen bonded to one atom of oxygen.

We represent molecules by giving the symbols for the elements present in the molecule. Numerical subscripts are used to indicate how many atoms of each element are in the molecule. For example, a molecule of the substance oxygen, which contains two oxygen atoms, is represented by the symbol O_2. Likewise, H_2O is the chemical symbol for a molecule of water, in which there are two hydrogen atoms and one oxygen atom. Chemical symbols for molecules are called **molecular formulas.**

It is very important to understand the difference between the symbols O_2 and $2O$. The symbol O_2 refers to a single oxygen molecule in which two oxygen atoms are held tightly, in close proximity, by a chemical bond. On the other hand, the symbol $2O$ represents two separate, uncombined oxygen atoms. The symbol $O + O$ is equivalent to the symbol $2O$ (see Figure 2-7).

two separate
oxygen atoms

oxygen
molecule

$2O$

O_2

Figure 2-7   Two separate oxygen atoms, which are represented by the symbol $2O$; and two oxygen atoms held close   together by a chemical bond in an oxygen molecule, which is represented by the symbol O_2.

We shall see in Chapter 4 that some chemical compounds, such as sodium chloride (salt), do not contain molecules in the solid state. Every substance, however, has a **chemical formula,** which indicates the relative number of atoms of each element in the substance. The chemical formula for a substance is obtained from experimental data. For sodium chloride, the chemical formula is NaCl. This chemical formula indicates that in the substance sodium chloride there is one atom of sodium for each atom of chlorine. If the experimental data for a substance indicates that the substance has molecular units, then we can speak of molecules of the substance and call the chemical formula a molecular formula. Chemical formulas are an essential part of the language of chemistry, and you must be able to use them correctly to refer to chemical substances.

Chemical Equations

Chemical reactions are represented by **chemical equations.** For example, the chemical equation for the reaction in which two oxygen atoms combine to form an oxygen molecule can be arrived at in the following manner:

(2-1)   Words: oxygen atoms $\longrightarrow$ oxygen molecule

(2-2)   Symbols: $O \longrightarrow O_2$

(2-3)   Balancing: $2O \longrightarrow O_2$

Chemical equations are written and read from left to right. Molecules on the left side of a chemical equation are called **reactants,** and those on the right are called **products.**

It is necessary to put a 2 in front of the symbol O so that the chemical equation representing the reaction is consistent with the experimental fact that oxygen atoms are not created or destroyed in the reaction. Going from statement 2-2 to statement 2-3 is referred to as **balancing** the chemical equation. A chemical equation must be balanced in order to represent a chemical reaction accurately.

As another example of writing and balancing a chemical equation, let us consider the decomposition of the compound water. This chemical reaction can be described by the words

(2-4)   Water $\longrightarrow$ hydrogen + oxygen

To convert statement 2-4 to a chemical equation, we need to know the chemical formulas for the substances involved in the reaction. Recall that chemical formulas are obtained from experimental data, and for the substances in this reaction they are water (H_2O), hydrogen (H_2), and oxygen (O_2). Note that when chemists speak of the substances hydrogen and oxygen, they are referring to molecules, not atoms. To refer to an atom of hydrogen, for example, the term "hydrogen atom" must be used.

Now, using chemical formulas, statement 2-4 can be rewritten as

(2-5) $H_2O \longrightarrow H_2 + O_2$

Equation 2-5 is not balanced, however, since one oxygen atom is indicated on the left side and there are two oxygen atoms on the right side.

How can we balance the number of oxygen atoms in Eq. 2-5? The equation would be balanced if the reactant were H_2O_2 instead of H_2O, but this is not permissible because H_2O_2 is *not* the chemical formula for the compound water. H_2O_2 is the molecular formula for the compound hydrogen peroxide, which is distinctly different from water. When you balance a chemical equation, *never* change the chemical formulas of the reactants or products.

The only way we can balance a chemical equation is to put numbers in front of the chemical formulas that appear in the equation. This procedure does not alter the substances being considered. The number in front of a molecular formula, for example, just indicates how many molecules with that formula are involved in the reaction.

The numbers in front of the chemical formulas in a balanced chemical equation are called **stoichiometric coefficients.** Now, one way to balance chemical equations is to guess the appropriate stoichiometric coefficients and then check to see if this guess is correct. Another method is to use an adaptation of the procedure illustrated below for the decomposition of water.

		H on Left	H on Right	O on Left	O on Right
Step 1	$H_2O \longrightarrow H_2 + O_2$	2	2	1	2

Step 2   Put a 2 in front of H_2O to balance the oxygen atoms.

		H on Left	H on Right	O on Left	O on Right
	$2H_2O \longrightarrow H_2 + O_2$	4	2	2	2

Note that the 2 in front of H_2O means two times the *whole* molecular formula H_2O. Now the hydrogen atoms are not balanced, so we need to put a 2 in front of the symbol H_2 to balance the number of hydrogen atoms.

		H on Left	H on Right	O on Left	O on Right
Step 3	$2H_2O \longrightarrow 2H_2 + O_2$	4	4	2	2

Therefore, the equation $2H_2O \rightarrow 2H_2 + O_2$ is a correctly balanced chemical equation symbolizing the decomposition of water to form hydrogen and oxygen (see Figure 2-8). Notice that the stoichiometric coefficient for O_2 is a 1. When the stoichiometric coefficient is 1, it is never written explicitly but rather is always implied.

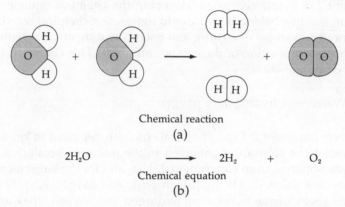

Chemical reaction
(a)

$2H_2O \longrightarrow 2H_2 + O_2$

Chemical equation
(b)

Figure 2-8   A schematic representation of the decomposition of the compound water into the elements hydrogen and oxygen. The chemical reaction (a) can be symbolized by the bal- anced chemical equation (b). Note that neither hydrogen atoms nor oxygen atoms are created or destroyed in this reaction.

The chemical equations

$$4H_2O \longrightarrow 4H_2 + 2O_2 \quad \text{and} \quad H_2O \longrightarrow H_2 + \tfrac{1}{2}O_2$$

are also balanced chemical equations for the decomposition of water. However, we usually write chemical equations using the *smallest* possible set of whole numbers.

Exercise 2-4
The compound ammonia has the chemical formula NH_3. Ammonia is produced commercially by the reaction of nitrogen, N_2, and hydrogen, H_2. Write a balanced chemical equation for this reaction.

Exercise 2-5
Ammonia, NH_3, is produced in the human body as a by-product of metabolism. Ammonia can react with carbon dioxide, CO_2, to form urea, CH_4N_2O, and water, H_2O, which are excreted in the urine. Write a balanced chemical equation for this reaction of ammonia and carbon dioxide.

2-7 ATOMIC STRUCTURE

Using Dalton's atomic theory as a model, chemists made great advances during the nineteenth century in their understanding of chemical reactions. However, little additional knowledge about the intrinsic nature of atoms was obtained until the very end of that century. Experimental results obtained in the late 1800s and early 1900s showed that atoms are not indivisible, as Dalton had proposed. It is now known that atoms themselves are composed of three kinds of **subatomic particles: electrons, protons,** and **neutrons.**

Electrons and protons are particles that have electrical charge, whereas the neutron is an uncharged particle. The charge of an electron is equal in magnitude but opposite in sign to the charge of a proton. In atomic charge units, the charge of an electron is -1 and the charge of a proton is $+1$. The electron is the fundamental particle of negative electrical charge, and the proton is the fundamental particle of positive charge.

Protons and neutrons have about the same mass; the mass of an electron is much less. A convenient scale for comparing the relative masses of different atoms is the atomic mass scale (see Section 2-8). In atomic mass units, abbreviated amu, the mass of a proton and the mass of a neutron are both about 1 amu, whereas the mass of an electron is only about 1/1840 amu. Table 2-1 summarizes the mass and charge properties of these basic subatomic particles.

Table 2-1 Fundamental Subatomic Particles

Particle	Electrical Charge	Mass (amu)	Mass (g)
Electron	-1	5.486×10^{-4}	9.110×10^{-28}
Proton	$+1$	1.007	1.673×10^{-24}
Neutron	0	1.009	1.675×10^{-24}

How are protons, neutrons, and electrons arranged in atoms? The basic features of the structure of an atom can be visualized as follows: The neutrons and the protons are packed together in a very small space, called the **nucleus,** in the center of the atom. The lighter electrons occupy a much larger volume around the nucleus. If we think of atoms and nuclei as spheres, then the diameters of atoms range from about 0.1 to 0.5 nm (1 nm = 1 nanometer = 10^{-9} m). In contrast, the diameter of a nucleus is only about 1/100,000 of the diameter of an atom—about 10^{-6} nm. You can get an idea of how extremely small a nucleus is compared to the size of an entire atom by visualizing the point of a pin in the center of a very large hot-air balloon. The pin point corresponds to the nucleus and the balloon to the entire atom. Figure 2-9 shows one artist's conception of the size of an atom and its nucleus.

Figure 2-9   A schematic representation of an atom. The nucleus, which contains the atom's protons and neutrons, is represented by a sphere with a diameter of 1.5 mm. On this scale, the diameter of the entire atom is 150 m or about 0.1 mile. The actual diameter of a nucleus is approximately 10^{-6} nm, whereas the diameter of an atom is about 10^{-1} nm.

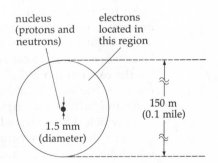

Atoms do not have a net electrical charge, so we know that the number of electrons must be equal to the number of protons in an atom. The number of protons in an atom is called the **atomic number** of that atom. Atoms of different elements have different numbers of protons (and electrons) and thus different atomic numbers. Thus an atom of an element can be represented by using the chemical symbol for that element or by using the atomic number of the atom. Atomic numbers of all the elements are included in a table printed on the inside back cover of this book. As we shall see in Chapter 4, the atomic number is the most significant factor in determining the chemical properties of an atom.

The electrons in an atom are kept reasonably close to the nucleus by an electrical force of attraction between the negatively charged electrons and the positively charged protons. The forces that hold the nucleus together are much more complex. In fact, a nucleus is really a very complicated entity consisting of more particles than just neutrons and protons. Fortunately, we need not be concerned with the detailed structure of the nucleus because in any chemical reaction the nucleus of every atom remains unchanged.

What does change in chemical reactions is the behavior of an atom's electrons. We shall treat this topic further in Chapter 4; however, on the basis of what we have said so far, we can give a modern interpretation of Dalton's postulate that atomic identity is preserved in a chemical reaction. For example, in the reaction $2H_2O \rightarrow 2H_2 + O_2$, the nuclei of the oxygen atoms are not changed. In the water molecule, however, the oxygen atom is not quite the same as that in the oxygen molecule because the behavior of the oxygen atom's electrons is somewhat different in the two molecules; but there is still enough of a similarity in the oxygen atoms of these molecules to warrant using the same name and symbol for them.

2-8   IONS AND ISOTOPES

As we mentioned previously, for any atom, the number of electrons is equal to the number of protons; therefore, atoms do not have a net electrical charge. The protons in an atom are held inside the nucleus by extremely strong nuclear forces. The number of protons in a nucleus can change only in a nuclear reaction involving relatively large amounts of energy. On the other hand, the number of electrons in an atom can be changed under much less drastic conditions.

Ions

In some chemical reactions, atoms lose or gain electrons and form particles called **ions,** which have net electrical charges. Since electrons have a negative charge, the loss of electrons results in the formation of a positively charged ion, called a **cation.** When an atom gains electrons, a negatively charged ion, called an **anion,** is produced.

An ion is represented by using the symbol for the atom from which it was produced, with the electrical charge of the ion written as a superscript on the upper right-hand corner of the symbol. For example, if a calcium atom, Ca, loses two electrons, a calcium cation with a charge of $+2$ is formed. The symbol for the calcium ion is Ca^{2+} (see Figure 2-10a). If a chlorine atom, Cl, gains one electron, the ion Cl^-, called the chloride anion, is formed (see Figure 2-10b).

An ion such as Ca^{2+} has very different chemical properties than the atom Ca from which it was formed. Therefore, one must always be sure to include the correct charge when writing the symbol for an ion. In Chapter 4 we shall discuss which ions, if any, an atom of a given element forms. Ions are extremely important components of the solutions in the human body.

electrical charge of the ion

$$Ca \longrightarrow Ca^{2+} + 2e^-$$

calcium atom   calcium ion   electrons

(a)

electrical charge of the ion

$$Cl + e^- \longrightarrow Cl^-$$

chlorine atom   electron   chloride ion

(b)

Figure 2-10   (a) When a calcium atom loses two electrons, a calcium cation with a charge of +2 is formed. The charge on the calcium ion is indicated by the superscript (2+) on the symbol Ca^{2+}. (b) When a chlorine atom gains one electron, a chloride anion with a charge of −1 is formed. The charge on the chloride ion is indicated by the superscript (−) on the symbol Cl^-. The number 1 is not used for ions with charges of +1 or −1.

Isotopes

Dalton postulated that all atoms of an element are exactly alike. Although they must all contain the same number of protons, we have just seen that they may gain or lose electrons to form ions. What about neutrons? It has been found that different atoms of the same element often have different numbers of neutrons. For example, the element chlorine has an atomic number of 17, and in a sample of chlorine, all of the chlorine atoms have 17 protons and 17 electrons, but some have 18 neutrons and some have 20. The only difference in these atoms is the mass of the nucleus, and this does not alter the chemical properties significantly. Atoms with the same atomic number but a different number of neutrons are called **isotopes** of that element. Some isotopes are radioactive and are extremely useful for a variety of medical diagnostic techniques (see Section 2-9).

The number of neutrons plus the number of protons in an atom is called the **mass number,** represented by the symbol A.

(2-6) A = number of neutrons + number of protons

When we need to identify isotopes of an element, we symbolize an isotope by using the mass number, A, as a superscript on the upper left-hand corner of the symbol for the element. For example, chlorine has the atomic number 17. Thus, the isotope of chlorine that has 20 neutrons has a mass number of $20 + 17 = 37$, and this isotope is represented by the symbol ^{37}Cl (see Figure 2-11). Since the mass of an atom depends almost entirely on the number of neutrons and the number of protons in the atom, and both neutrons and protons have a mass very close to 1 amu, the mass of an atom is just about equal to A amu. Thus, an atom of ^{37}Cl has a mass of 37 amu.

mass number $\longrightarrow$ ^{37}Cl $\longleftarrow$ symbol for the
element chlorine,
atomic number 17

Figure 2-11   A representation for the isotope of chlorine with 20 neutrons. The element symbol Cl indicates that the atomic number of this atom is 17. The superscript 37 is the mass number (the number of neutrons plus the number of protons). The symbol $^{37}_{17}Cl$, in which the subscript 17 indicates the atomic number explicitly, can also be used to represent this isotope.

Given the symbol for an isotope, it is easy to determine the number of electrons, protons, and neutrons in an atom of that isotope. Consider ^{11}B, for example; B is the chemical symbol for the element boron, which has an atomic number of 5. Therefore, an atom of ^{11}B has 5 electrons and 5 protons. Since the mass number for this atom is 11, the number of neutrons is $11 - 5 = 6$. Let us consider one other example: the ion $^{19}F^-$. The atomic number of fluorine, F, is 9, so this ion has 9 protons and $9 + 1 = 10$ electrons. Note that the $^{19}F^-$ anion is formed when a ^{19}F atom gains an additional electron. The mass number of ^{19}F is 19, so the number of neutrons in the $^{19}F^-$ ion is $19 - 9 = 10$.

Since isotopes have different masses, the atomic mass scale is based on one specific isotope of one particular element, carbon (C). The atomic mass scale is defined by arbitrarily assigning to an atom of the carbon isotope ^{12}C a mass of 12.00000 amu. When we get to Chapter 6 we shall discuss the relative masses of atoms in greater detail.

Exercise 2-6

Determine the number of electrons, protons, and neutrons in each of the following atoms or ions: ^{14}N, $^{16}O^{2-}$, $^{39}K^+$, and ^{131}I.

Exercise 2-7

How many ^1H atoms are needed so that a sample of ^1H atoms has the same mass as one ^{12}C atom?

2-9   RADIOACTIVITY: NUCLEAR REACTIONS

In 1896, the French scientist A. Henri Becquerel noticed that a covered photographic plate, which he had accidentally placed near a sample of uranium ore, had blackened. Becquerel proposed that the uranium ore spontaneously emitted high-energy radiation, which he called **radioactivity.** Two years later, Marie and Pierre Curie isolated two previously unknown elements, polonium and radium, from uranium-containing ores, and found both to be radioactive. We now know that 22 of the naturally occurring elements have one or more radioactive isotopes, and it is now possible by using complex techniques to produce radioactive isotopes of all of the elements. It has also been found that all of the isotopes of the elements with atomic number greater than 83 (that of bismuth) are radioactive.

Three different types of radiation account for the majority of radioactivity. They are called alpha, beta, and gamma rays after the first three letters of the Greek alphabet, α, β, and γ. Alpha rays are really streams of particles. An **alpha particle** consists of two protons and two neutrons, and has a charge of $+2$; it is the same as the nucleus of a helium atom. A beta ray or **beta particle** is a fast-moving electron. **Gamma rays** are not particles, but rather are a special form of electromagnetic radiation similar to x rays. A gamma ray has zero mass and zero electrical charge.

The alpha, beta, and gamma rays emitted by radioisotopes are produced by nuclear reactions. **Nuclear reactions** involve changes within atomic nuclei. The spontaneous nuclear reaction in which a radioactive isotope emits radiation is called **nuclear decay.** In most nuclear reactions, nuclei of one element are changed into nuclei of a different element. This is in contrast to chemical reactions, in which there are no changes within atomic nuclei. We shall also see that the energy changes involved when an atom undergoes a nuclear reaction are many times greater than those in chemical reactions.

Chemist and physicist Marie Sklodowska Curie shared the 1903 Nobel Prize in physics with A. Henri Becquerel and with her husband, Pierre Curie, for their work on radioactivity. In 1911, Madame Curie became the first scientist to be awarded a second Nobel Prize, this time in chemistry. The most commonly used unit of radioactivity, the curie, was named to honor her early work.

Representing Nuclear Reactions

When we represent chemical reactions we usually do not distinguish between isotopes of the same element, since they have essentially identical chemical properties. For nuclear reactions, however, we must always indicate the specific isotopes involved. For example, the isotope of carbon with a mass of 14 can be referred to as carbon-14, C-14, or ^{14}C. When representing nuclear reactions, the mass number of an isotope is added as a left-hand superscript on the symbol for the element, and it is also convenient to indicate the atomic number—the nuclear charge—by a left-hand subscript. The complete representation for the isotope radium-226, which is used in cancer therapy, is shown in Figure 2-12.

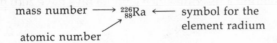

$$\text{mass number} \longrightarrow {}^{226}_{88}Ra \longleftarrow \text{symbol for the element radium}$$

$$\text{atomic number} \nearrow$$

Figure 2-12   The representation for the isotope radium-226. The superscript 226 is the mass number of the isotope, and the subscript 88 is the atomic number. When nuclear equations are written, the atomic numbers of the isotopes involved in the reaction always should be included as subscripts.

We now require symbols for the subatomic particles neutrons and electrons. Other subatomic particles are involved in some nuclear reactions, but we shall not consider them. The symbol for a neutron is $^{1}_{0}n$, since a neutron has zero electrical charge and a mass of 1 amu. The symbol for a beta particle is $^{0}_{-1}e$. Notice that we use the mass number zero for a beta particle, since its mass is only 1/1840 the mass of a neutron or a proton, or essentially zero. An alpha particle is an ion with a charge of +2, but for nuclear reactions we use the symbol $^{4}_{2}He$ and not $^{4}_{2}He^{2+}$. Similarly, we use the symbol $^{1}_{1}H$ for a proton involved in a nuclear reaction. Since nuclear reactions do not depend on the presence or absence of electrons around the nucleus, we need not represent the electrons (if any) around the nucleus. For example, radium ($^{226}_{88}$Ra) is converted to radon ($^{222}_{86}$Rn) by emitting an alpha particle. Radium-226 emits an alpha particle in the same nuclear reaction regardless of whether the radium is an atom or an ion or is bonded to another atom in a molecule. This spontaneous nuclear reaction is represented by

(2-7) $^{226}_{88}\text{Ra} \longrightarrow {}^{222}_{86}\text{Rn} + {}^{4}_{2}\text{He}$

Let us consider another example. Phosphorus-32 is an artificially produced radioisotope used to treat some forms of leukemia. It spontaneously emits a beta particle to produce sulfur-32 in the following reaction:

(2-8) $^{32}_{15}\text{P} \longrightarrow {}^{32}_{16}\text{S} + {}^{0}_{-1}e$

Notice the following about Eqs. 2-7 and 2-8.

1. The sum of the mass number(s) of the reactant(s) is equal to the sum of the mass numbers of the products (upper left superscripts).

2. The sum of the charge(s) of the reactant(s) is equal to the sum of the charges of the products (lower left subscripts).

The charges referred to here are **nuclear charges** of the isotopes and the charges of the subatomic particles (electrons and neutrons). Statements 1 and 2 are true of all nuclear reactions, and they can be used to determine the product nucleus formed when a radioactive isotope emits an alpha or a beta particle.

For example, consider the product when uranium-238 emits an alpha particle:

(2-9) $^{238}_{92}\text{U} \longrightarrow {}^{4}_{2}\text{He} + X$

The product nucleus, X, must therefore have a mass number of $238 - 4 = 234$ and a nuclear charge of $92 - 2 = 90$. The element with the atomic number 90 is thorium (Th). Thus, the equation representing the decay of U-238 is

(2-10) $^{238}_{92}\text{U} \longrightarrow {}^{4}_{2}\text{He} + {}^{234}_{90}\text{Th}$

Gamma rays are also emitted in nuclear reactions, but they are not usually included when representing a reaction since gamma rays have zero mass and zero charge.

Exercise 2-8

Selenium-75, which is used to study the function of the pancreas, emits a beta particle when it decays. Write the equation for the nuclear decay of Se-75.

Nuclear Energy

We are all aware of nuclear bombs and nuclear power plants. Nuclear reactions can obviously involve large energy changes. Energy is produced in these reactions at the expense of mass. The theory of relativity relates mass and energy by the **Einstein equation:**

(2-11) $\Delta E = \Delta m c^2$

where the symbol ΔE is the change in energy, Δm is the change in mass, and c is the velocity of light.

For example, cobalt-60 undergoes the nuclear reaction

(2-12) $^{60}_{27}Co \longrightarrow ^{0}_{-1}e + ^{60}_{28}Ni$

If 100 g of cobalt-60 react in this manner, the products weigh 5 mg less than the initial 100 g, and about 1×10^8 kcal of energy is released. To release this amount of energy by burning carbon, about 10,000,000 g of carbon would have to react with oxygen in the combustion reaction $C + O_2 \rightarrow CO_2$.

The Einstein equation also predicts mass and energy changes for chemical reactions, but the changes in mass in chemical reactions are too small to be detected and the resulting energy changes for the same amount of reactants are typically 10^5 times smaller than in nuclear reactions. It is for this reason that we are able to use the law of conservation of mass and the law of conservation of energy for chemical reactions but not for nuclear reactions.

Now let us look at a nuclear reaction used to generate energy in nuclear reactors and bombs. The uranium-235 nuclei, or some other nuclei such as U-233 or plutonium-239, split apart when struck by a neutron, producing a variety of smaller nuclei, more neutrons, and a lot of energy. This type of process is called **nuclear fission.** Two major ways in which U-235 splits are

(2-13) $^{1}_{0}n + ^{235}_{92}U \longrightarrow ^{142}_{56}Ba + ^{91}_{36}Kr + 3^{1}_{0}n$

and

(2-14) $^{1}_{0}n + ^{235}_{92}U \longrightarrow ^{137}_{52}Te + ^{97}_{40}Zr + 2^{1}_{0}n$

The additional neutrons produced in reaction 2-13 or in reaction 2-14 can then strike neighboring U-235 nuclei, thus beginning a chain reaction that is accompanied by a tremendous release of energy—a nuclear explosion. In a properly functioning nuclear reactor, the fission process is allowed to proceed in a controlled manner so that there is only a gradual release of energy and therefore no explosion.

The bombardment of nuclei with neutrons generated in a nuclear reactor is also used to produce artificially many radioactive isotopes, such as those used in modern medicine. For example, cobalt-60, used in cancer therapy, is produced in the nuclear reaction

(2-15) $^{59}_{27}Co + ^{1}_{0}n \longrightarrow ^{60}_{27}Co$

The production of cobalt-60 is an example of a nuclear transmutation. The term **nuclear transmutation** refers to any nuclear change that is caused by experimental techniques. Nuclear transmutations can also be accomplished by bombarding reactant nuclei with a variety of fast-moving ions and particles, such as protons and alpha particles, as well as with high-energy gamma rays. Most of the radioactive isotopes used in chemical research and medicine are produced by neutron bombardment.

Another type of nuclear reaction involves the merging of smaller nuclei to form larger ones and is called **nuclear fusion.** Nuclear fusion reactions are the source of the sun's energy and the energy released in hydrogen bombs.

Exercise 2-9
Sulfur-35, which does not occur naturally, can be made artificially by bombarding the naturally occurring isotope chlorine-35 with neutrons. Write the equation for this nuclear reaction.

Radioactive Decay

All radioactive isotopes decay, but they do so at vastly different rates. For example, a 10-g sample of iodine-131, which is used to study and treat diseases of the thyroid gland, contains about 4.6×10^{21} nuclei of I-131. After about 8 days half of these nuclei have decayed, leaving only 5 g of I-131 containing 2.3×10^{21} nuclei of I-131. After another 8 days only 2.5 g of I-131 containing 1.15×10^{21} nuclei would remain, and so on (see Figure 2-13). The time required for one-half of the nuclei in a sample of a particular radioactive isotope to decay is called the **half-life** of that isotope. The symbol used for half-life is $t_{1/2}$. Thus, $t_{1/2}$ for I-131 is 8 days. The half-lives of different radioisotopes range from fractions of a second to billions of years.

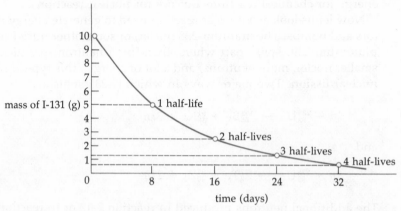

Figure 2-13   Iodine-131 decays by emitting beta particles and gamma rays in the nuclear reaction

$$^{131}_{53}\text{I} \longrightarrow {}^{131}_{54}\text{Xe} + {}^{0}_{-1}e$$

The half-life of I-131 is 8 days. In each 8-day period, the amount of I-131 is reduced by a factor of one-half. Note that as a sample of I-131 decays, it is the number of I-131 nuclei and the mass of I-131 that decreases. There is only a very small decrease in the total mass of the sample, since I-131 is converted into Xe-131 in the decay process.

The number of radioactive nuclei in a sample that decay in one second, or the rate of decay, is called the **activity** of the sample. The activity of a sample containing radioactive nuclei is directly proportional to the number of radioactive nuclei in the sample. Since the number of radioactive nuclei in a sample decreases with time, so does the sample's activity. A sample containing I-131, for example, will have only one-half of its original activity after 8 days, the $t_{1/2}$ for I-131. On the other hand, the activity of a sample of radium-226, which has a $t_{1/2}$ of 1602 years, will be essentially unchanged after 8 days. Only a minute fraction of the radium-226 will decay in 8 days, or in any other time period that is much less than 1602 years.

The **curie**, abbreviated Ci, is the commonly used unit of activity. An activity of 1 Ci corresponds to 3.7×10^{10} nuclear decays per second. One gram of pure radium-226, which contains 2.6×10^{21} nuclei of Ra-226, has an activity of 1 Ci. Typically, the activity of the samples of radioactive isotopes used for medical diagnosis have activities that range from several microcuries (μCi) to a few millicuries (mCi).

A variety of instruments, such as Geiger counters and scintillation counters, are available that can "count" the actual number of alpha, beta, or gamma rays emitted by a radioactive sample as it decays. These instruments allow us to measure even minute amounts of activity.

Exercise 2-10

The half-life of chromium-51, which is used for a variety of diagnostic purposes, is about 28 days. Suppose that a radioactive sample initially contains 200 μg of Cr-51.

(a) How much Cr-51 is left in the sample after 56 days?
(b) When is the mass of Cr-51 in the sample 12.5 μg?

Exercise 2-11

Another radioactive sample containing Cr-51 has an initial activity of 0.1 mCi. What is the activity of this sample after 84 days?

Radiation and Human Health

Because even minute amounts of a radioactive isotope can be detected and measured, and because a radioactive isotope of an element behaves chemically and physically (for the most part) just like the stable isotopes of that element, radioactive isotopes are used in several areas of research and medicine. Some researchers use radioisotopes to follow, or trace, the path and rate of a chemical reaction. A radioactive isotope of an element can also be used as a **tracer** to follow the metabolism of that element, or a particular compound containing that element, in the human body. Such radioisotopic tracers are being used increasingly in medical diagnosis. The thyroid gland, for example, is located in the neck and requires iodine to make thyroid hormone (see Chapter 28). The thyroid gland is the only organ in the human body that specifically requires iodine, which it absorbs from the bloodstream. Thus, the rate of uptake of iodine by a patient's thyroid can be measured by administering a very small amount of I-131 and monitoring the activity by placing a detector at the patient's neck.

In diagnostic applications, such as I-131 thyroid function tests, the amount of a radioisotope used is very small. On the other hand, large amounts of some radioisotopes can be used to kill cancer cells. The alpha, beta, or gamma rays emitted by radioisotopes have high energies and can destroy human cells. For example, a large amount of I-131 can be administered to destroy a tumor of the thyroid gland.

Specialized instruments, such as this full body radiation chamber, allow the diagnosis of mineral deficiencies and metabolic problems associated with the endocrine system. Minute amounts of radioactive isotopes administered to a patient can be traced to different parts of the body.

A very common treatment for cancer uses the radiation from cobalt-60 (see Eq. 2-12). In addition to beta particles, cobalt-60 emits very high-energy gamma rays when it decays. Gamma rays can pass right through a human body, leaving a trail of destruction in their wake. By focusing these gamma rays on the site of a tumor, therapists try to destroy a large number of cancer cells and a minimal number of normal cells. Because of the destructive power of gamma rays, personnel who work with isotopes such as cobalt-60 must protect themselves with lead shields and aprons to minimize their own exposure.

Alpha particles from an external source are stopped by dead skin cells and therefore cannot penetrate the human body. If a person ingests an isotope that is an emitter of alpha particles, however, the alpha particles will penetrate living cells and damage them extensively. For this reason alpha emitters are not used diagnostically.

Beta particles from an ingested beta emitter can also damage internal body cells. Beta rays emitted from an external source have sufficient energy to penetrate through the skin surface to a depth of a few millimeters and can damage living cells in this region.

2-10 SUMMARY

1. Most substances can exist as a solid, a liquid, or a gas.

2. The stable state of a substance is dependent upon the temperature and the pressure.

3. Energy must be supplied to melt or to vaporize a substance.

4. Heterogeneous samples of matter can be separated into homogeneous components by mechanical means.

5. Solutions can usually be separated into pure substances by changes of state.

6. Compounds can be broken down into elements by chemical reactions.

7. Elements are the simplest chemical substances. Pure substances can be either elements or compounds.

8. The chemical formula for a substance indicates the relative number of atoms of each element in the substance. Chemical formulas are obtained from experimental data.

9. Chemical reactions are symbolized by balanced chemical equations.

10. The atomic number of an atom is equal to the number of protons in that atom. For any atom, the number of protons is equal to the number of electrons.

11. The mass number, A, of an atom is equal to the number of protons plus the number of neutrons in that atom.

12. Positively or negatively charged ions are formed when atoms lose or gain electrons.

13. Isotopes are atoms of the same element that have different numbers of neutrons.

14. Nuclear reactions involve changes in atomic nuclei. For the same amount of reactant, a nuclear reaction involves a larger change in energy than does a chemical reaction.

15. Radioactive isotopes emit alpha, beta, and gamma rays in a process called nuclear decay. The emitted radiation can damage body cells.

16. Nuclear reactions are represented by equations in which mass numbers and electrical charge numbers of all the nuclei and subatomic particles involved are explicitly indicated and balanced.

17. The half-life, $t_{1/2}$, of a radioactive isotope is the time required for one-half of the nuclei to decay.

18. The activity of a radioactive sample is the number of nuclear decays per second. The unit of activity is the curie (Ci).

19. Radioactive isotopes are used as an aid in medical diagnosis and in cancer therapy.

PROBLEMS

1. The compound ammonia, NH_3, is a solid at 180 K, a liquid at 230 K, and a gas at 280 K. On this basis, the melting point of ammonia is between what two temperatures? The boiling point of ammonia is between what two temperatures?

2. The heat of vaporization of water is 540 cal/g. What amount of heat energy is required to convert 25 g of liquid water to the gas state?

3. When a sample of liquid water freezes, the energy of the water _____ (increases, decreases, remains the same).

4. Classify the following as an element, compound, solution, or heterogeneous mixture: (a) pure liquid water, H_2O; (b) pure gaseous oxygen, O_2; (c) clean air; (d) an apple; (e) a piece of copper wire.

5. Balance the following reactions:
   (a) _____ C_3H_6 + _____ $O_2 \longrightarrow$ _____ CO_2 + _____ H_2O
   (b) _____ Fe + _____ $O_2 \longrightarrow$ _____ Fe_2O_3
   (c) _____ NO_2 + _____ $H_2O \longrightarrow$ _____ HNO_3 + _____ NO
   (d) _____ P + _____ $O_2 \longrightarrow$ _____ P_2O_5
   (e) _____ NH_3 + _____ $O_2 \longrightarrow$ _____ NO + _____ H_2O

6. Fill in the blanks in the table below.

	No. of Protons	No. of Electrons	No. of Neutrons
^{33}S			
$^{51}V^{3+}$			
$^{18}O^{2-}$			
$^{87}Sr^{2+}$			
^{23}Na			
$^{35}Cl^-$			

7. What is the chemical symbol for the particle that contains 9 protons, 10 electrons, and 10 neutrons?

8. What is the chemical symbol for the particle that contains 12 protons, 10 electrons, and 12 neutrons?

9. What must be true about two atoms if they are isotopes of the same element?

10. Which of the following statements are *not* true?
    (a) Most of the mass of an atom is in its nucleus.
    (b) The nucleus of an atom occupies a small fraction of the volume of an atom.
    (c) An atomic nucleus contains protons and electrons.
    (d) A neutral atom has the same number of protons and electrons.

11. The first artificial conversion of one nucleus into another was performed in 1919. In this reaction a nitrogen-14 nucleus and an alpha particle reacted and an oxygen-17 nucleus and a proton were the products. Write a balanced equation for this nuclear reaction.

12. The radioactive isotope technetinium-99, which is used for a number of diagnostic purposes, has a half-life of only 6.0 hr. A sample containing technetinium-99 has an initial activity of 16 mCi. What is the activity of this sample after 18 hr? When will the activity of this sample be 1 mCi?

13. It was observed that a sample containing the radioactive isotope selenium-75 used in a pancreas scan had an initial activity of 300 μCi and 30 days later it had an activity of 75 μCi. What is the half-life of selenium-75?

14. Radioactive rubidium-90 emits a beta particle when it decays. Write the equation for the nuclear decay of Rb-90.

15. A typical activity of an I-131 sample used for thyroid function tests is 5 μCi. Would you expect the activity of an I-131 sample used to destroy a tumor of the thyroid to be larger, smaller, or equal to 5 μCi? Why?

SOLUTIONS TO EXERCISES

2-1 The liquid state of ethyl alcohol is stable between its melting point ($-114°C$) and its boiling point ($78°C$). Therefore, at $-10°C$ the stable state of ethyl alcohol is a liquid, while at $-150°C$ the alcohol is a solid. As the temperature is raised, ethyl alcohol melts at $-114°C$ and goes to the liquid state. At $78°C$ this liquid ethyl alcohol vaporizes and goes to the gas state.

2-2 The amount of heat required to melt a sample of a solid is directly proportional to the mass of the sample. To melt 1.0 g of ethyl alcohol requires 25 cal. Therefore, the amount of solid alcohol that can be melted is equal to

$$(1.0 \times 10^3 \text{ cal}) \times \left(\frac{1.0 \text{ g}}{25 \text{ cal}}\right) = 40 \text{ g}$$

2-3 You could dissolve the salt in water and separate this heterogeneous mixture into solid sand and a salt solution by filtration. Then you could obtain the solid salt by boiling off the water.

2-4 $N_2 + 3H_2 \longrightarrow 2NH_3$

2-5 $2NH_3 + CO_2 \longrightarrow CH_4N_2O + H_2O$

2-6 The number of protons is equal to the atomic number. For atoms, the number of electrons is equal to the number of protons. For ions, you must take into account the number of electrons lost or electrons gained when the ion is formed from the atom. Therefore, we have

number of neutrons = mass number − number of protons

	Protons	Electrons	Neutrons
^{14}N	7	7	7
$^{16}O^{2-}$	8	10	8
$^{39}K^+$	19	18	20
^{131}I	53	53	78

2-7 The mass of one ^1H is 1.0 amu, and the mass of one ^{12}C is 12.0 amu. Therefore, twelve ^1H atoms have the same mass as one ^{12}C atom.

2-8 $^{75}_{34}Se \longrightarrow ^{75}_{35}Br + ^{0}_{-1}e$

Selenium has a nuclear charge of 34. When it emits a beta particle, the product nucleus has a charge of 35. The element with a nuclear charge (atomic number) of 35 is Br.

2-9 Chlorine has a nuclear charge of 17 and sulfur a nuclear charge of 16. Therefore, in the reaction $^{35}_{17}Cl + ^1_0n \rightarrow ^{35}_{16}S + X$, the mass number of X must be 1 and the nuclear charge of X must also be 1. Thus, X is a proton, ^1_1H, so

$$^{35}_{17}Cl + ^1_0n \longrightarrow ^{35}_{16}S + ^1_1H$$

2-10 (a) After 28 days, one-half of the Cr-51 present initially is still present, or 100 μg. After another 28 days (56 days total), one-half of the 100 μg, or 50 μg, are still left.
(b) After 56 days there are 50 μg and after an additional 28 days (84 days total) there are 25 μg left. An additional 28 days later, after a total of 112 days, 12.5 μg of Cr-51 are left.

2-11 The activity of a sample is directly proportional to the number of radioactive nuclei in the sample. From Exercise 2-10, after 84 days the number of radioactive nuclei in any sample of Cr-51 will be $\frac{1}{2} \times \frac{1}{2} \times \frac{1}{2} = \frac{1}{8}$ of the number initially present. Therefore, after 84 days the activity of the sample is one-eighth of the initial activity, or

$$\frac{1}{8} \times (0.1 \text{ mCi}) = 0.0125 \text{ mCi}$$

CHAPTER 3

The Periodic Table: Electronic Structure of Atoms

3-1 INTRODUCTION

The world is a complicated place, and people have always searched for an understanding of the forces that affect their lives. One way of reducing the complexities of nature to a more manageable form is to find similarities among natural phenomena, similarities that bring order to apparent disarray. Thus, scientists classify plants, animals, chemical elements, and so on, into groups of similar entities. Grouping things that are similar makes their study easier.

Having perceived some order in the natural world, a scientist has a passionate desire to explain why this order arises. In this chapter we shall discuss a fascinating example of how science develops in this manner.

On the basis of a great deal of experimental work, chemists in the nineteenth century recognized that certain groups of elements have quite similar chemical and physical properties. In 1869, a Russian chemist, Dmitri Ivanovich Mendeleev, devised an extremely useful arrangement of the known chemical elements, which he called the periodic table. The basis for his arrangement was that elements with similar properties could be grouped together.

As we mentioned in Chapter 2, the nineteenth-century chemists worked without a model of the structure of the atom. Yet they argued that there must be some reason why certain elements are similar to one another. By the early part of the twentieth century, an atomic model capable of explaining the arrangement of the elements in the periodic table had been developed. In the course of the development of this model, scientists adopted a radically new way of viewing the behavior of atomic and subatomic particles. These new ideas were not accepted without a major controversy, which raged for many years within the scientific community.

We shall see in this chapter how the periodic table can be used as a valuable chemical tool. We shall also discuss some of the basic features of the model scientists now use to describe electrons in atoms, and some aspects of the development of this model. We shall see that this model provides an explanation for why certain elements have similar properties.

Our attempt to structure the complex world around us is constantly challenged by the seemingly limitless variety in the physical forms and combinations of substances to be studied.

3-2   STUDY OBJECTIVES

After studying the material in this chapter, you should be able to:

1. Describe the physical characteristics of metals and nonmetals, as well as the characteristic chemical properties of metals and nonmetals.

2. Define the terms group and period as they are used with reference to the periodic table of the elements.

3. Identify an element as a representative element, transition element, lanthanide, or actinide from its position in the periodic table.

4. Use the periodic table and information supplied about a compound to predict the properties of a related compound.

5. Predict the order of the representative elements, with regard to their ionization energy and atomic size, from their position in the periodic table.

6. Describe the basic features of the Bohr model of the hydrogen atom.

7. Define the terms orbital and energy-level diagram.

8. Describe both the similarities and the differences between Bohr's model of the hydrogen atom and the quantum mechanical description.

9. Define the terms inner-core electrons, valence electrons, and shell.

10. Describe the differences between two orbitals in different shells.

11. Determine the following for any representative element using the periodic table: (1) the number of valence electrons; (2) the shell in which the valence electrons are located.

12. Predict the ion that an atom of a given representative element will tend to form based on its position in the periodic table.

13. Write Lewis structures for atoms and ions.

14. **(Optional)** Determine the electron configuration for an atom given the ordering of the subshell energies for the atom.

15. **(Optional)** Describe the common characteristic valence-shell electron configuration for a group of representative elements.

3-3   METALS, NONMETALS, AND METALLOIDS

In the part of the earth's crust that we can sample, the abundance of the different chemical elements, either as uncombined "free elements" or more usually as parts of chemical compounds, varies tremendously. The 10 most abundant elements in order are oxygen, silicon, aluminum, iron, calcium, sodium, potassium, magnesium, hydrogen, and titanium. About 99% of the earth's crust is composed of these 10 elements, with the first two—oxygen and silicon—accounting for about 75%. The elemental composition of the human body is quite different. Four elements—oxygen, carbon, hydrogen, and nitrogen—make up about 99% of the mass of the human body.

The names and symbols for all the chemical elements are given on the inside back cover of this book. The vast majority of the elements in their pure state are solids at usual room temperature and pressure. Mercury and bromine are the only elements that are liquids below 25°C (cesium, gallium, and francium have melting points slightly above 25°C). A few elements are gases. Hydrogen, nitrogen, oxygen, fluorine, and chlorine are diatomic ("two-atom") gases with the molecular formulas H_2, N_2, O_2, and Cl_2, respectively. Helium, neon, argon, krypton, xenon, and radon are all monatomic ("one-atom") gases.

An early classification scheme divided the chemical elements into two broad groups—metals and nonmetals—on the basis of their physical properties. **Metals** are elements, such as silver, sodium, copper, and mercury, that (1) have a very good ability to conduct heat and electricity; (2) have a luster or shiny surface; and (3) usually can be pulled into long, thin wires (see Figure 3-1). Elements such as oxygen, nitrogen, sulfur, and bromine, which are not good conductors of heat and electricity and which do not have the other physical properties of metals, are called **nonmetals.**

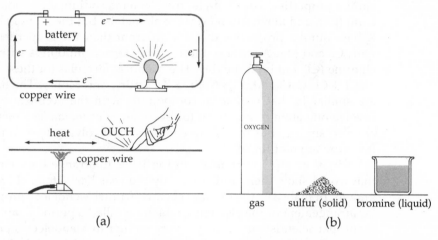

(a)   (b)

Figure 3-1   (a) Copper is a metal. Copper can   mine are nonmetals. These elements
be drawn into a thin wire, and   cannot be drawn into thin wires,
it conducts heat and electricity   and they are poor conductors of
easily. (b) Oxygen, sulfur, and bro-   heat and electricity.

Some elements, such as silicon, which have physical properties somewhere between those of metals and nonmetals, are called **metalloids.** Silicon is a semiconductor—the electrical conductivity of silicon is greater than that of typical nonmetals, but less than that of typical metallic elements.

The most characteristic chemical property of metals and nonmetals is that metals react with nonmetals to form compounds with distinctly different properties than the starting materials. For example, sodium metal is a soft, grayish solid, with a relatively low melting point. The nonmetal chlorine is a toxic, yellowish gas. When sodium and chlorine combine, they form the compound sodium chloride, which has the chemical formula NaCl. Sodium chloride is common table salt. It is a harmless white solid, with quite a high melting point. When NaCl dissolves in water, there is no direct sensory evidence of any further chemical reaction. However, the resulting solution conducts an electrical current because of the presence of separate Na^+ and Cl^- ions.

As we shall see in the next section, a very useful way of arranging the chemical elements is in a form called the periodic table. There is a periodic table on the inside front cover of this book. Different colors are used in this table to indicate the usual classification of elements into metals, nonmetals, and metalloids. Note that:

1. Most of the elements are metals.

2. Ten elements in the upper right-hand corner of the periodic table are nonmetals.

3. Eight elements along a diagonal line between the metals and nonmetals are classified as metalloids.

Exercise 3-1

Using the periodic table inside the front cover of this book, classify the elements phosphorus, barium, nickel, and germanium either as a metal, a nonmetal, or a metalloid.

3-4   THE PERIODIC TABLE

Early in the nineteenth century, chemists realized that the metals and nonmetals could be subdivided even further, that is, into even smaller groups with similar properties. For example, the elements sodium (Na), potassium (K), calcium (Ca), and magnesium (Mg) are all metals, but the pair of elements Na and K have similar properties that are different than the properties of the similar pair, Ca and Mg. All four of these metals form compounds with the nonmetals chlorine (Cl) and bromine (Br). The chemical formulas for these compounds are $NaCl$, KCl, $CaCl_2$, $MgCl_2$, $NaBr$, KBr, $CaBr_2$, and $MgBr_2$. The metals Na and K are similar in that they form compounds with the nonmetals Cl and Br that involve one atom of nonmetal for every atom of metal. The metals Ca and Mg are also similar, but they form compounds involving two atoms of nonmetal for every atom of metal.

Various attempts were made in the 1800s to arrange the elements in a table that would show their similarities. By the year 1869 a total of 63 elements were known. In that year, the Russian chemist Dmitri Mendeleev proposed a particular arrangement of the elements, which he called a periodic table. The arrangement that scientists use today is very similar to Mendeleev's periodic table. A modern version of the periodic table is shown in Figure 3-2. (Because it is used so frequently, a periodic table is also printed on the inside front cover of this book. This latter periodic table also includes the atomic weight of each element. The atomic weight, which is the relative weight of an average atom of an element, is discussed in Chapter 6.)

In the modern **periodic table,** the elements are ordered according to their atomic number and then are arranged in rows. The length of a row is adjusted so that those elements with similar properties fall under one another in vertical

In 1869, Dmitri Mendeleev, after a painstaking study of the properties of the 63 elements that were known at that time, arranged these elements into a periodic table. In a brilliant example of scientific reasoning, Mendeleev organized elements with similar physical and chemical properties into groups.

Groups of Representative Elements

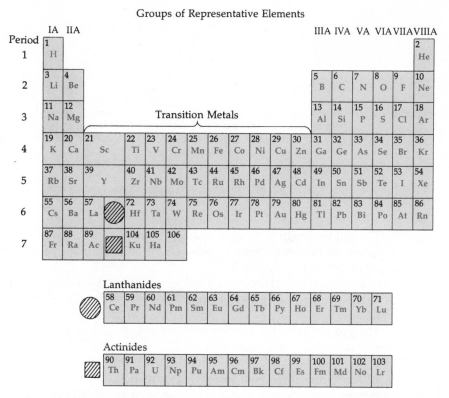

Figure 3-2   A periodic table of the elements.

columns. <u>Elements in the same vertical column in the periodic table form a</u>
group. For example, we saw that the elements sodium (atomic number 11) and
potassium (atomic number 19) are similar; hence, the length of the third row
is set at eight. This arrangement allows potassium to begin the fourth row and
fall under sodium in the first column so that sodium and potassium are in the
same group.

<u>The horizontal rows in the periodic table are called</u> **periods.** The term
"period" is used because when the elements are ordered on the basis of increasing atomic number, there is a *periodic* reoccurrence of elements with similar chemical properties. The empirical generalization that the chemical properties of the elements vary periodically with increasing atomic number is
sometimes called the **periodic law.**

Most elements can react with a number of other elements. For example,
sodium can react with hydrogen, oxygen, chlorine, sulfur, and so on. There are,
however, a group of very similar gaseous elements, namely, group VIIIA in the
periodic table, that are quite different from other elements. They have little or
no tendency to react with any of the other elements. This group of elements
is called the **noble gases.** There are no known naturally occurring compounds
involving the noble gas elements. A relatively small number of such compounds have been prepared in chemical laboratories; the first one was not made
until the early 1960s. The noble gas compound xenon difluoride, XeF_2, is one
example.

When the elements are ordered on the basis of their atomic number, perhaps
the most distinct and important example of the periodic variation of elements
with similar properties is the position of the noble gases. In the sequence of
atomic numbers, the noble gases are not close to one another, but occur periodically. The atomic numbers of the noble gas elements are 2, 10, 18, 36, 54, and 86.

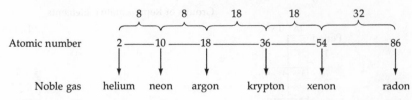

Figure 3-3   When the elements are ordered on the basis of atomic number, there is a periodic reoccurrence of the noble gases. Beginning with helium,   which has an atomic number of 2, the other noble gases occur after increases in atomic number of 8, 8, 18, 18, and 32.

Thus, if we begin with the noble gas helium (atomic number 2), the other noble gas elements occur after increases in atomic number of 8, 8, 18, 18, and 32, respectively (see Figure 3-3).

The most important feature of the periodic table is that elements in the same vertical column, or group, have similar properties. Elements in groups IA, IIA, IIIA, IVA, VA, VIA, VIIA, and VIIIA are called the **representative elements.** The properties of the members of one group of representative elements in general are significantly different from the properties of the elements in the other representative groups. For example, the properties of the elements in group IA differ significantly from those in groups IIA, IIIA, IVA, and so on. Within any representative group there is the same spacing between the atomic numbers of the elements as was seen in differences between the atomic numbers of the noble gas elements (see Figure 3-3). The element hydrogen is unique. In addition to being the element with atomic number 1, hydrogen does *not* fit in any one group in the periodic table. Some of the properties of hydrogen are similar to those of group IA elements and some are similar to those of the elements in group VIIA. We have chosen to place hydrogen with the group IA elements in our periodic table.

The representative elements are of most interest to us, so we shall present a brief description of a few of the chemical and physical properties of some members of each representative group.

Group IA

The group IA elements, also known as the **alkali metals,** are all soft, silvery metals. They all readily form compounds with oxygen and they also react vigorously with water to evolve hydrogen gas and compounds with the general chemical formula MOH, called alkali metal hydroxides. In the general chemical formula MOH, M represents any one of the alkali metals (as in NaOH and KOH). Alkali metals are stored under kerosene or in a vacuum to prevent their reacting with air and water vapor.

As we mentioned previously, all alkali metals react with the nonmetals in group VIIA to produce solid compounds such as NaCl, KF, LiBr, and so on. The general chemical formula for these compounds is MX, where M represents any group IA element and X represents any group VIIA element. When any solid compound of this type dissolves in water, the solid breaks apart and separate M^+ cations and X^- anions are present in the resulting solution. Sodium ions, Na^+, and potassium ions, K^+, are particularly important, since they are essential constituents of human body fluids.

Group IIA

The group IIA elements are also known as the **alkaline earth metals.** All of the alkaline earth metals react with oxygen to form compounds with the general chemical formula MO (for example, MgO and CaO). Under appropriate conditions all alkaline earth metals react with water to give hydrogen gas and alkaline earth hydroxides, with the general chemical formula $M(OH)_2$, in reactions similar to the reactions of alkali metals with water. All group IIA elements form solid compounds with all the group VIIA elements. These compounds, such as $CaCl_2$ and $MgBr_2$, have the general chemical formula MX_2. When any solid compound of this type dissolves in water, separate M^{2+} cations and X^- anions are present in the resulting solution. Ca^{2+} and Mg^{2+} ions are also essential components of human body fluids.

Group IIIA

The elements in group IIIA exhibit a range of properties. Boron (B) is a hard, brittle metalloid, whereas the other elements in group IIIA are metals. Aluminum (Al) is the third most abundant element on earth, and its use in the manufacture of a large number of items makes it the most familiar element in this group.

Rolled wire is only one of the many diverse forms into which the element aluminum can be shaped for our use.

Group IVA

The group IVA elements also exhibit a wide range of properties, ranging from the nonmetal carbon (C) to the metalloids silicon (Si) and germanium (Ge) and then to the metals tin (Sn) and lead (Pb) as the atomic number increases. Carbon is an extremely versatile element. There are a gigantic number of carbon-containing compounds, and one can say that carbon-containing compounds are the molecular basis for life. Carbon-containing compounds are the subject of Chapters 13 through 18.

Group VA

Group VA includes the nonmetals nitrogen (N) and phosphorus (P), the metalloids arsenic (As) and antimony (Sb), and the metal bismuth (Bi). Nitrogen and phosphorus are also essential for living organisms. Nitrogen is a major constituent of proteins (see Chapter 21) and nucleic acids (see Chapter 24). We shall study the chemistry of some nitrogen- and phosphorus-containing compounds in Chapters 17 and 18.

The application of nitrogen-rich fertilizers to a field before planting winter wheat should result in a larger harvest of grain. Nitrogen is one of the elements essential to the growth of plants. The farmer thus replaces nitrogen depleted from the soil by the previous crop.

Group VIA

The common group VIA elements oxygen (O) and sulfur (S), as well as selenium (Se), are nonmetals, whereas tellurium (Te) and radioactive polonium (Po) are metalloids. Oxygen is the most abundant element on earth; it will form compounds with almost every other chemical element. Molecular oxygen (O_2) in the atmosphere is vital to human existence, and all living organisms consist of many types of compounds that contain oxygen. The chemistry of some simple oxygen-containing compounds is discussed in Chapters 15, 16, and 17.

Group VIIA

All of the common group VIIA elements, except for radioactive astatine (At), are nonmetals. Group VIIA elements are also known as **halogens.** In the gas state, halogens are diatomic molecules (F_2, Cl_2, Br_2, I_2). The halogens are very reactive, and in nature they are found only in compounds combined with metallic elements. Halogens also form compounds with hydrogen—HF, HCl, HBr, and HI—which we shall discuss in Chapters 10 and 11. The chloride ion, Cl^-, is an essential component of human body fluids.

All the elements in the 10 columns between groups IIA and IIIA are called **transition elements.** The transition elements are all metals. In general, the differences between transition elements in different groups are not as large as the differences between representative elements in different groups. For example, copper (Cu) and zinc (Zn) are more alike than are carbon and nitrogen. Some of the transition metals, such as iron (Fe), copper (Cu), silver (Ag), and gold (Au), are familiar to us in our everyday lives. Some transition elements, such as manganese (Mn), iron (Fe), and zinc (Zn), are essential in trace amounts for living organisms.

In order to reduce the length of the periodic table, two series of 14 elements each are listed separately at the bottom of the table. The 14 elements with atomic numbers 58 through 71 are called the **lanthanide series** because they follow the element lanthanum (atomic number 57). There are very small differences in properties among the lanthanides. There are just a few industrial uses of lanthanides, and as far as we know, none of these elements is essential to any living organism. The other series of 14 elements, with atomic numbers 90 through 103, are called the **actinide series** (after actinium, atomic number 89). The actinides are also quite similar to one another, and all of them are radioactive. Uranium (U) and thorium (Th) are used as fuels in nuclear reactors.

Exercise 3-2
Write down the sequence of atomic numbers for the elements in group IVA. What are the spacings between the elements in this sequence? Compare these spacings for the group IVA elements with those for the noble gases.

Exercise 3-3
Classify the elements carbon, nickel, uranium, sulfur, barium, xenon, and phosphorus as either a representative element, a transition element, a lanthanide, or an actinide. Indicate in which group any representative element is found.

3-5 USING THE PERIODIC TABLE

The periodic table is an extremely useful tool, which you will use throughout this book. Recall that the fundamental feature of the periodic table is the arrangement of elements with similar properties in vertical columns called groups. Thus, if we know something about the chemistry of the element oxygen in group VIA, for example, we can often (but not always) successfully infer something about the chemistry of the element sulfur, which is also in group VIA. In this sense, as we proceed to learn some chemical facts, we can use the periodic table to expand this chemical knowledge.

We can often predict the chemical formula of a compound if we know the chemical formula of a related compound. For example, the chemical formula for the compound water is H_2O. On this basis, we might predict that a compound containing the elements hydrogen and sulfur, with the chemical formula H_2S, exists, since both oxygen and sulfur are elements in group VIA. The compound hydrogen sulfide, with the chemical formula H_2S, does in fact exist. Hydrogen sulfide is quite toxic and is also responsible for the smell of rotten eggs.

Let us consider another example of using the periodic table to predict chemical formulas. Phosphorus forms two different oxides, with the chemical formulas P_2O_5 and P_2O_3. What chemical formulas would you predict for the oxides of nitrogen? If your answer to this question is N_2O_5 and N_2O_3, on the grounds that both nitrogen and phosphorus are elements in group VA, then you are correct. Nitrogen does form oxides with molecular formulas N_2O_5 and N_2O_3.

The chemistry of the oxides of nitrogen (and chemistry in general), however, is a bit more complicated. Additional oxides of nitrogen exist, including the gaseous compounds NO_2 and NO, which are two of the major pollutants of our atmosphere. The existence of oxides of nitrogen with the chemical formulas NO_2 and NO could *not* be predicted from the information given here about the oxides of phosphorus. This example shows the utility of the periodic table, and also illustrates its limitations.

In addition to predicting chemical formulas, the periodic table can be used to infer many other chemical and physical properties of an element or a compound, given information about similar elements and compounds. Consider this example. Suppose that we know the following about the compound sodium chloride: It has the chemical formula NaCl; it is a solid at room temperature; it has a very high melting point; and when it dissolves in water, sodium chloride breaks apart, and separate Na^+ cations and Cl^- anions exist in solution. Can we use this information to predict anything about the compound potassium fluoride? Yes. We can make the following argument: Potassium, like sodium, is an element in group IA, and both fluorine and chlorine are elements in group VIIA. We therefore predict that potassium fluoride is a high-melting solid with the chemical formula KF, and that separate K^+ cations and F^- anions exist in solution when KF dissolves in water (see Table 3-1). All of these predictions agree with the results of experiments.

Table 3-1 Comparison of Sodium Chloride and Potassium Fluoride

Elements		Compounds	
Group IA	Group VIIA	Chemical Formula	Ions Present in Water
Na	Cl	NaCl	$Na^+ + Cl^-$
K	F	KF	$K^+ + F^-$

Exercise 3-4
Carbon forms a compound with oxygen, called carbon dioxide, which has the chemical formula CO_2. Carbon also forms a compound with sulfur. Predict the chemical formula for this second compound.

Exercise 3-5
When the solid compound calcium bromide, which has the chemical formula $CaBr_2$, dissolves in water, separate Ca^{2+} and Br^- ions exist in solution. What ions would you predict to be present when the compound magnesium iodide dissolves in water?

3-6   TRENDS IN THE PERIODIC TABLE

As we shall see shortly, the chemical behavior of an atom is intimately connected with that atom's ability to lose or gain electrons. Two properties of an atom that are closely related to an atom's tendency to lose or gain electrons in chemical reactions are the atom's ionization energy and its electron affinity. The **ionization energy** (IE) and the **electron affinity** (EA) for atoms of many different elements have been measured experimentally. In this section we shall define these properties and see how they vary from element to element. We shall also discuss how a measure of the size of an atom can be determined experimentally, and see that the atom's size can be correlated with its position in the periodic table. When we consider how the magnitude of a property of an atom varies from one atom to another, there is often a general trend going across a row (within a period) or down a column (within a group) in the periodic table. Since the representative elements are of most interest to us, we shall limit our discussion to those elements.

Ionization Energy

Atoms, except those that are radioactive, are stable entities. There must be a force of attraction holding the electrons of an atom near the nucleus. Energy is therefore required to move an electron from an atom to a distance far away from the nucleus. If we want to measure the amount of energy involved in this process, we must do it under circumstances where the atom is not interacting with other atoms. When is this process possible? Only when we have individual atoms in the gas state separated by large distances from other atoms.

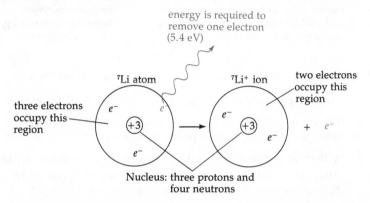

Figure 3-4   An electron can be removed from a lithium atom to produce a positively charged lithium ion. The energy required to remove one elec- tron from a lithium atom in the gas state—the first ionization energy of lithium—is 5.4 eV.

The **first ionization energy** of an atom is defined as the energy required to remove one electron from that atom when it is in the gas state (see Figure 3-4). Let us use the general symbol $X(g)$ to refer to an atom X in the gas state, the symbol $X(g)^+$ to refer to the cation formed when the atom X loses an electron, and the symbol $e^-(g)$ to represent an electron. Using these symbols, the first ionization energy of an atom X is the energy required for the following reaction:

(3-1) $X(g) \longrightarrow X(g)^+ + e^-(g)$

For example,

$Na(g) \longrightarrow Na^+(g) + e^-(g)$

It is convenient to express ionization energies in units called electron volts (abbreviated eV). **One electron volt** is the amount of energy that is equivalent to 3.83×10^{-20} cal. Thus 1.00 eV = 3.83×10^{-20} cal. The first ionization energies of the representative elements in the first four periods of the periodic table are given in Figure 3-5. Since we shall discuss only the representative elements, in this figure and in the subsequent figures in this section we shall use an abbreviated form of the periodic table in which only the representative elements are included.

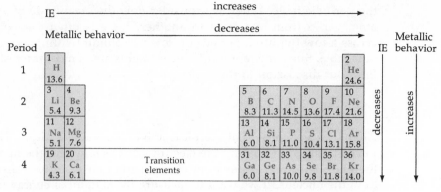

Figure 3-5   First ionization energies in electron volts: energy required for the reaction $X(g) \rightarrow X(g)^+ + e^-(g)$. Parallel trends exist in ionization energies and metallic behavior for the representative elements. Ionization energies increase and metallic behavior decreases going across a row in the periodic table, and ionization energies decrease and metallic behavior increases going down a column in the same group.

Look at Figure 3-5 and consider the following observations:

1. Group IA elements have relatively low ionization energies.

2. It is quite difficult (it requires a relatively large amount of energy) to remove an electron from a noble gas atom.

3. Within any period, it is relatively easy to remove an electron from a group IA atom, but in general this removal becomes increasingly more difficult as one goes across the row.

4. Within any group (going down a column in the periodic table), the ionization energy decreases as the atomic number increases.

We can summarize these observations by saying that <u>as a general trend, the first ionization energy increases on going across a row in the periodic table and decreases on going down a column within the same group.</u>

It is also possible to remove a second electron from an atom. The reaction involved in this process is

(3-2) $X(g)^+ \longrightarrow X(g)^{2+} + e^-(g)$ for example $Ca^+(g) \longrightarrow Ca(g)^{2+} + e^-(g)$

The energy required for reaction 3-2 is called the **second ionization energy.** For any atom, the second ionization energy is greater than the first IE, since the second ionization involves removing a negatively charged electron from a positively charged ion. For example, for calcium the first IE is 6.1 eV and the second IE is 11.9 eV. Three, or more, electrons can be removed from atoms, but each successive removal of an electron requires a larger amount of energy.

Electron Affinity

Many atoms have a tendency to gain an electron and form a negative ion. If there is a force of attraction between a neutral atom and an electron, then some energy is released when the atom binds an electron and forms a negative ion. The **electron affinity** (EA) of an atom X is defined as the energy released in the following reaction.

(3-3) $X(g) + e^-(g) \longrightarrow X(g)^-$ for example $F(g) + e^-(g) \longrightarrow F(g)^-$

Period

Group	IA	IIA		IIIA	IVA	VA	VIA	VIIA	VIIIA
1	1 H 0.7								2 He −0.6
2	3 Li 0.8	4 Be −0.2		5 B 0.3	6 C 1.2	7 N 0.05	8 O 1.5	9 F 3.6	10 Ne −0.2
3	11 Na 1.2	12 Mg 0.07		13 Al 0.5	14 Si 1.4	15 P 0.8	16 S 2.1	17 Cl 3.7	18 Ar −0.4

Figure 3-6

Electron affinities in electron volts: energy released when the reaction $X(g) + e(g)^- \rightarrow X(g)^-$ takes place. The group VIIA elements F and Cl have large positive electron affinities, indicating a strong tendency for these atoms to form negatively charged ions. On the other hand, it is energetically unfavorable for the noble gas elements He, Ne, and Ar to form negatively charged ions. The negative value of the electron affinity for these atoms indicates that energy is required in order for these noble gas elements to form negative ions.

Electron affinities are difficult to measure experimentally, and the EA has not been determined for many elements. Figure 3-6 shows the EA of the representative elements in the first three rows of the periodic table. The EA values of the elements beryllium, helium, neon, and argon are negative. A negative EA means that energy is *required* (not released) for reaction 3-3 to occur.

There does not appear to be a consistent general trend in the variation of the electron affinity from element to element. The following points, however, are worth mentioning: (1) The electron affinities of the group VIIA elements are very large. Thus atoms of group VIIA elements have a strong tendency to form negative ions with a charge of −1. Compare this with the small tendency for a group VIIA element to form a positive ion. For example, for the element chlorine, 13.0 eV of energy per atom is required to form a Cl^+ ion. On the other hand, 3.7 eV of energy is released when a Cl^- ion is formed. (2) The electron affinities of the noble gases that have been measured are negative. Recall that the ionization energies of the noble gases are very large. Thus, the noble gases have little tendency to form either positive or negative ions. This is an important fact, which we shall use when we discuss chemical bonding in Chapter 4.

Atomic Size

One important factor determining the chemical and physical properties of a compound is the size of the atoms in the compound. Recall the schematic picture of an atom that we presented in Chapter 2. The size of an atom is determined by the space occupied by the electrons. We cannot determine the size of an atom precisely for two reasons: (1) It is impossible to pinpoint the position of the electrons in an atom (see Section 3-7); (2) the behavior of some of an atom's electrons is altered when the atom combines with other atoms. Therefore, the size of an atom really depends on how the atom is bonded to other atoms (see Chapter 4). However, a usable measure of the approximate size of an atom can be determined by following three steps:

1. Find by experiment the distance D between the nucleus of one atom and the nucleus of its closest neighbor in a sample of the pure element.*

* For those elements that are solids, the distance D between the nuclei of two neighboring atoms can be determined by using a beam of x rays. In this experimental method, a beam of x rays strikes the solid sample and the direction and intensity of the scattered x rays are measured. The distance D can be determined from an analysis of the results of this type of experiment.

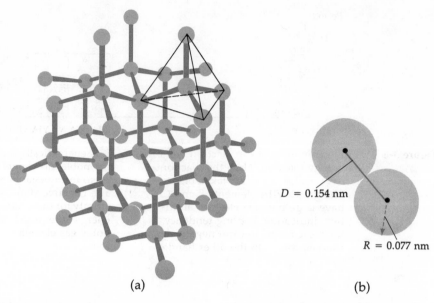

(a)   (b)

Figure 3-7   (a) Schematic model of the structure of carbon in the form of diamond. Each carbon atom is at the center of a regular tetrahedron. (b) Two of the nearest-neighbor carbon atoms represented by spheres that touch one another. The distance D between the centers of the spheres (the distance between the nuclei of the two carbon atoms) is 0.154 nm, as determined by experiment. On this basis the radius of each carbon atom sphere is 0.077 nm.

2. Assume that an atom is a spherical ball with a radius R, and assume that atoms which are closest neighbors touch one another.

3. Use this radius R as a measure of the size of the atom. Then the atom's radius R is equal to $D/2$.

An example of this approach for determining the atomic radius of carbon is illustrated in Figure 3-7. In the elemental form of carbon known as diamond, each carbon atom is at the center of a regular tetrahedron with four nearest neighbors. The distance D between two nearest neighbors is 0.154 nm (1 nm = 10^{-9} m). Therefore, the atomic radius of carbon is 0.077 nm.

Not all of the elements in the first four rows of the periodic table are solids, however. The elements hydrogen, nitrogen, oxygen, fluorine, and chlorine are gases under standard conditions. These gaseous elements consist of diatomic molecules with the molecular formulas H_2, N_2, O_2, F_2, and Cl_2, respectively. The stable state of bromine under standard conditions is a liquid. It is easy, however, to obtain bromine in the gas state as well as in the liquid state. In the gas state, bromine also consists of diatomic molecules with the molecular formula Br_2. Using other physical techniques that do not involve x rays, it is possible to measure the distance D between the nuclei of these diatomic molecules. Atomic radii determined in this manner for the representative elements in periods 1, 2, 3, and 4 are given in Figure 3-8. Note the general trend exhibited in Figure 3-8: Atomic radii decrease along a row in the periodic table and increase on going down a column.

The noble gases, neon, argon, and krypton, however, do not form diatomic molecules, so the method we described cannot be used to obtain atomic radii for these elements. For this reason, atomic radii for the noble gases are not included in Figure 3-8.

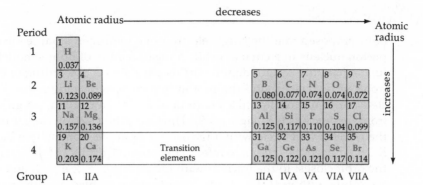

Figure 3-8   Atomic radii in nanometers.
Atomic radii for the representative
elements decrease going across a
row in the periodic table and they
increase going down a column
within the same group.

Exercise 3-6
Give an explanation for the fact that the second ionization energy for any atom is larger
than its first ionization energy.

Exercise 3-7
Predict which of the following elements has the lowest first ionization energy: carbon,
silicon, nitrogen, germanium, and phosphorus.

Exercise 3-8
Would you predict that the atomic radius of antimony (Sb) is larger or smaller than that
of arsenic (As)?

3-7   ELECTRONS IN ATOMS

A major problem for scientists in the early twentieth century was to devise a
model that would describe the behavior of electrons in atoms. In 1913, the
Danish physicist Niels Bohr developed a model for the hydrogen atom. He proposed that the electron revolves around the proton in a circular orbit. In this
model, the motion of the electron is like the movement of the earth around the
sun. Bohr's model was very successful in explaining some of the properties of
the hydrogen atom, but there were also some major difficulties and limitations
in his theory. Within 15 years after Bohr proposed his model, a new approach to
the treatment of all atoms and molecules, called quantum mechanics, was developed. As far as we know, quantum mechanical models are capable of giving a
completely satisfactory representation of the behavior of electrons in atoms and
molecules. The subject of quantum mechanics is complex, however, and mathematically quite involved.
   We shall not discuss quantum mechanics in any detail. What we shall do is
the following:

1. Describe the basic features of Bohr's model for the hydrogen atom. This
   discussion is worthwhile because many of the concepts developed by
   Bohr have been retained in a modified form in the quantum mechanical
   description.

2. Discuss the reasons why Bohr's model was unsatisfactory and needed
   modification.

3. Discuss in what way this modification was accomplished.

The Bohr Model

Bohr proposed that the single electron in a hydrogen atom revolves around the proton nucleus in a circular orbit. A major feature of Bohr's model was the postulate that only certain definite orbits are allowed. According to Bohr, the electron can revolve around the proton in a circle with radius R_1, close to the nucleus, or in one of the circles with radii R_2, R_3, and so on, progressively farther from the nucleus (see Figure 3-9). However, the electron *cannot* revolve around the nucleus in a circle with a radius, for example, somewhere between R_1 and R_2. Bohr used the letter n to refer to the possible orbits as follows: $n = 1$ refers to the orbit closest to the nucleus with radius R_1; $n = 2$, $n = 3$, and so on, refer to the orbits with radii R_2, R_3, and so on. Bohr derived a mathematical equation which he used to calculate the radii of the allowed circular orbits.

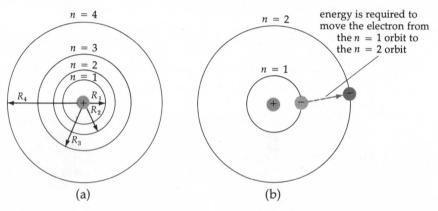

(a)   (b)

Figure 3-9   (a) Schematic representation of the $n = 4$. (b) The negatively charged
allowed orbits for the electron in a   electron is attracted to the positively
hydrogen atom according to the   charged nucleus. Therefore, energy
Bohr model. The nucleus, which is   is required to move the electron
a single proton with a charge of   from the $n = 1$ orbit close to the
$+1$, is at the center, and the orbits   nucleus to the $n = 2$ orbit farther
are labeled $n = 1$, $n = 2$, $n = 3$, and   away.

In developing his model, Bohr treated the motion of the electron around the proton by using the same approach that scientists had used since the time of Newton (1680), with one major exception: Only certain orbits for the electron were allowed. Bohr's postulate that only particular circular orbits are allowable paths for the electron was a radically new idea at the time. After all, there is no restriction of allowable paths in the motion of the large objects we normally encounter. For example, if you tie a piece of rope around a stone and then swing the stone around your head in a circle, the radius of this circle can be 2 ft, 5 ft or any distance in between. Scientists now agree that Bohr had the right idea, and **discrete** (or **quantized**) **behavior** is a part of all the models scientists now use to describe electrons in atoms and molecules.

The following comparison should help to clarify the meaning of this concept of discrete behavior: It is possible to enter many buildings by using either a ramp or a set of steps. Along a ramp there is a continuous change in elevation. In a series of steps, however, there are discrete changes in elevation.

There are two examples of discrete behavior in the Bohr model: (1) The electron cannot revolve around the nucleus in any arbitrary circle. Only certain definite circular orbits are allowable paths. (2) The electron cannot have any arbitrary energy. The energy of the electron can have only certain definite values. In quantum mechanics there are numerous examples of discrete or quantized behavior similar to these of the Bohr model.

Time-exposure photography reveals
the circular orbits of a carnival ride.

The most important property of an electron in an atom is its energy. Let us designate the energy of the orbit $n = 1$ by E_1, the energy of the orbit $n = 2$ by E_2, and so on. The phrase "energy of the orbit $n = 1$" is a shorthand way of referring to the energy of the hydrogen atom when the electron is in the orbit with $n = 1$. Bohr developed a mathematical equation from which he calculated exact values for the energies E_1, E_2, E_3, and so on. We shall not need to use this quantitative data, but it is extremely important that you have a qualitative understanding of how the energy of a Bohr orbit varies with its distance from the nucleus. The orbit with $n = 1$ has the lowest energy. Let us see why.

The nucleus of a hydrogen atom, a single proton, is positively charged, whereas the electron has a negative charge. Recall that there is an attractive force between positive and negative electrically charged particles, which decreases as the distance between the particles increases. Thus, it takes work or energy to overcome in part the force of attraction between the electron and the proton and move an electron from the circular orbit $n = 1$, closest to the proton, to the orbit with $n = 2$, which is farther away (see Figure 3-9). If we have to give the electron some additional energy to get it from the $n = 1$ orbit to the $n = 2$ orbit, then E_2, the energy of the electron when it is in the $n = 2$ orbit, must be higher than E_1, its energy in the $n = 1$ orbit. A similar argument shows that the energies E_1, E_2, E_3 of the allowed orbits continually increase as n increases. Since there is only a discrete set of allowed orbits in the Bohr model, there is only a discrete set of allowed energies for the electron. It is convenient to use a diagram, called an **energy-level diagram,** to illustrate the ordering of these energies. The energy-level diagram for four Bohr orbits is given in Figure 3-10.

Figure 3-10 The energy-level diagram for the first four Bohr orbits. The larger the value of n for the orbit, the greater the radius of the orbit and the higher the energy.

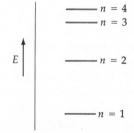

How was Bohr's model tested? The test of any scientific theory or model is whether or not predictions derived by the use of the model agree with the experimental data. In one way, Bohr's theory was spectacularly successful. The triumph of Bohr's model was its ability to predict the kind of light given off by hydrogen atoms that are heated in a flame. This success for the Bohr model convinced many people that there must be something to the radically new idea of discrete behavior that Bohr used in his model. Even though Bohr's model was so successful for the hydrogen atom, it could not, however, be extended to treat other atoms with more than one electron. When people tried to use Bohr's approach for other atoms, their predictions did not agree with the experimental data. Bohr's model needed modification, and a major change in scientific thinking had to take place before a model capable of explaining the properties of electrons in all atoms was developed.

Exercise 3-9

According to the Bohr model, which of the following requires more energy: taking the electron in the $n = 1$ orbit of a hydrogen atom and moving it very far away from the nucleus, or carrying out the same process starting with the electron in the $n = 2$ orbit?

The Quantum Mechanical Model

In the early twentieth century scientists concluded that the motion of small objects, such as electrons in atoms, had to be described in a fundamentally different way than the motion of large objects. In particular, they argued that Bohr's postulate of an orbiting electron in a hydrogen atom was an oversimplification and had to be abandoned. It can be shown that there is no possible way of measuring precisely the position of an electron in an atom without drastically altering its motion (see the optional Section 3-10). This statement is one consequence of a general principle of the quantum mechanical model, called the **uncertainty principle.** Thus, there is no way of determining the motion of an electron precisely enough to warrant talking about electrons in atoms being in definite orbits.

If electrons in atoms cannot be described by orbits, then how can they be described? The best that can be done is to specify the probability that any one electron in an atom is at a certain position in that atom. There is no way of saying definitely and exactly where the electron is. Describing an electron in an atom in terms of the probability of finding it at different positions is another of the fundamental concepts of quantum mechanics.

It does seem strange that the position of an electron in an atom cannot be determined precisely, but remember that we cannot directly observe electrons in atoms. Electrons are very small particles with very small mass, and it is really not unreasonable that the behavior of electrons should be quite different from the behavior of the large-size objects we normally encounter.

The probability of finding an electron at different positions in an atom is obtained from a mathematical expression called an **orbital.** For any atom, there are only certain discrete allowed orbitals.

Let us consider, as an example, the lowest-energy orbital in the hydrogen atom (i.e., the orbital in the hydrogen atom that describes the electron when it has the lowest possible energy). For this orbital, the probability of finding the electron at a certain position depends only on how far that position is from the nucleus. This probability decreases rapidly as the distance of the position from the nucleus increases. This is illustrated schematically in Figure 3-11. In Figure 3-11a, the density of the dots (the dot density) in a region is proportional to the probability of finding the electron in that region. A high dot density represents

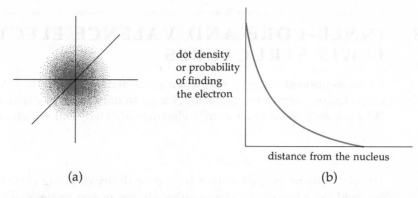

dot density
or probability
of finding
the electron

distance from the nucleus

(a)   (b)

Figure 3-11   (a) Schematic illustration of the shape of the lowest energy orbital for the hydrogen atom. The probability of finding the electron is high in a region where the density of dots is high, and the probability of finding the electron is low in a region where the density of dots is low. (b) The probability of finding the electron at a certain distance in any direction from the nucleus, as indicated by the density of dots in (a), decreases rapidly as the distance from the nucleus increases.

a region where there is a high probability of finding the electron. There is little probability of finding the electron in a region where the dot density is low. Figure 3-11b is a plot of the dot density, or the probability of finding the electron, as a function of distance from the nucleus. Note how rapidly the dot density decreases as the distance from the nucleus increases. The probability of finding the electron inside or outside an imaginary sphere with a certain radius about the nucleus can be calculated by using the quantum mechanical model. For example, there is a 90% probability that the electron will be found inside an imaginary sphere with a radius of 0.14 nm centered about the nucleus, and only a 10% probability that it will be found outside this sphere.

According to the quantum mechanical model, there are only certain discrete allowed orbitals for any atom. This is similar to Bohr's postulate that the electron in the hydrogen atom can travel only in certain definite circular orbits. Orbitals have energies in the same sense that orbits have energy in the Bohr model, and we use the phrase "energy of an orbital" as a shorthand way of stating the energy of an electron when it is described by that orbital. Since for any atom there is a set of allowed orbitals, there is a set of allowed energies for the electrons in the atom. We can illustrate the energies of these allowed orbitals with an energy-level diagram, as was done for the energies of the allowed orbits in the Bohr model.

The similarities and differences between the quantum mechanical description of an atom and Bohr's model of the hydrogen atom are summarized in Table 3-2.

Table 3-2 Bohr and Quantum Mechanical Models: Similarities and Differences

Bohr Model	Quantum Mechanical Model
1. Applicable to hydrogen atom	1. Applicable to all atoms
2. Electron moves in a circular orbit	2. Probability of finding the electron at a certain position in the atom described by an orbital
3. Only certain circular orbits allowed	3. Only certain orbitals allowed
4. Each circular orbit has a different energy	4. Each orbital has a definite energy, although some orbitals have the same energy

3-8   INNER-CORE AND VALENCE ELECTRONS: LEWIS STRUCTURES

As we mentioned in Chapter 2, when an atom bonds to another atom in a chemical reaction, there is a significant change in only some of the atom's electrons. Why are only some of an atom's electrons affected, and which are involved?

Inner-Core and Valence Electrons

The electrons in orbitals with a high probability of being close to the nucleus are held very tightly by the positive charge of the nucleus. These electrons, called **inner-core electrons,** are not significantly affected by a chemical reaction. On the other hand, electrons that have a high probability of being relatively far from the nucleus are bound less tightly to the nucleus than are the inner-core electrons. It is these electrons, called **valence electrons,** which are principally involved when an atom bonds to other atoms. Since the valence electrons are the electrons primarily involved in bonding atoms to one another, we need a description of the valence electrons for an atom.

A result of the quantum mechanical model is that the allowed orbitals which describe the electrons in an atom are grouped into various **shells.** The shells are labeled by a number n, called the **principal quantum number.** Thus, the designations $n = 1$, $n = 2$, $n = 3$, and so on, correspond to the first, second, and third shells, respectively.

The principal quantum number n (shell number) and the n-label for the Bohr orbits are similar in the following sense: (1) In general, the smaller the value of the shell number, the higher the probability of finding the electron close to the nucleus; and (2) the energy of an electron is determined primarily by what shell it is in. In general, the larger the shell number, the higher the energy of the electron in the atom and the smaller the amount of energy required to remove the electron from the atom.

For the representative elements (which are the elements we are primarily interested in) we can use the following simple general rule to determine the number of valence electrons and their shell: (1) The number of valence electrons is equal to the group number for that element, and (2) the shell that the valence electrons are in is equal to the period of the element (see Figure 3-12).

Group number of the element	IA	IIA	IIIA	IVA	VA	VIA	VIIA	VIIIA
Number of valence electrons	1	2	3	4	5	6	7	8

(a)

Period of the element	Shell of the valence electrons
1	1
2	2
3	3
4	4
5	5
6	6
7	7

(b)

Figure 3-12   (a) For a representative element, the number of valence electrons is equal to the group number for that element.   (b) For a representative element, the shell that the valence electrons are in is equal to the period of that element.

For example, sodium is in group IA, period 3, so it has one valence electron in the third shell. Carbon is in group IVA, period 2, so it has four valence electrons in the second shell.

It is of utmost importance that you remember the following two points about valence electrons for representative elements:

1. Atoms in the same group have the same number of valence electrons. This is the basic reason why elements in the same group have similar chemical properties.

2. As you go down a column within a group in the periodic table, the valence electrons occupy progressively higher shells.

It is also important that you understand the relationship between the trends in the periodic table, which we discussed in Section 3-6, and the properties of the valence electrons. First of all, the noble gas elements, group VIIIA, which are chemically quite unreactive and do not tend to form either positive or negative ions, have eight valence electrons, except for helium, which has two valence electrons. Thus, it appears that a particularly stable arrangement for electrons in an atom is to have eight electrons in the occupied shell with highest energy. This idea can be used to predict what ions an atom tends to form, and it is also important for understanding how one atom bonds to another (see Chapter 4).

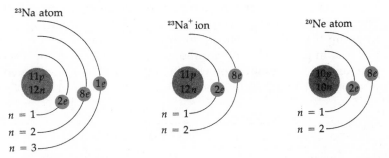

Figure 3-13   Schematic illustration of the arrangement of the electrons in shells for a sodium atom, ^{23}Na, a sodium ion, ^{23}Na$^+$, and a neon atom, ^{20}Ne. Both a sodium ion and a neon atom have two electrons in the first shell and eight electrons in the second shell.

Compare, for example, a neon atom (atomic number 10) with a sodium atom (atomic number 11). Since neon is in Group VIIIA, period 2, it has eight valence electrons in the second shell and the remaining two inner-core electrons are in the first shell (see Figure 3-13). Sodium, in group IA, period 3, has one valence electron in the third shell and 10 remaining inner-core electrons. The 10 inner-core electrons in a sodium atom are arranged in shells like the 10 electrons in neon (i.e., two electrons in the first shell and eight in the second shell) (see Figure 3-13). When a sodium atom loses its valence electron and a sodium ion, Na$^+$, is formed, both the Na$^+$ ion and a Ne atom have similar electron arrangements, with two electrons in the first shell and eight electrons in the second shell. Thus, we can say that a sodium atom tends to form a sodium ion because a sodium ion has a particularly stable electron arrangement similar to that of a noble gas. For any element in group IA, if the single valence electron is lost, a noble gas-type electron arrangement results. Similarly, if any atom in group IIA loses two electrons, or any atom in group IIIA loses three electrons, a noble gas-type electron arrangement is left. Thus group IA elements tend to form ions with a charge of +1, group IIA elements tend to form ions with a charge of +2, and some group IIIA elements form ions with a charge of +3.

On the other side of the periodic table, each of the nonmetal atoms in group VIIA has seven valence electrons. A group VIIA atom can achieve a noble gas-type arrangement, with eight electrons in the outermost shell, by gaining one electron. Atoms of elements in group VIIA tend to gain one electron and form ions with a charge of -1. Similarly, we can predict that each nonmetal atom in group VIA tends to gain two electrons and form an anion with a charge of -2. For example, if an oxygen atom gains two electrons, the O^{2-} ion with a total of eight electrons in the second shell results (see Figure 3-14).

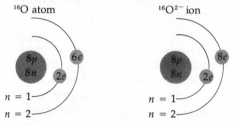

Figure 3-14   Schematic illustration of the arrangement of valence electrons in shells for an oxygen atom, ^{16}O, and an oxide ion, $^{16}O^{2-}$. Both an oxide ion (with two more electrons than an oxygen atom) and a neon atom have two electrons in the first shell and eight electrons in the second shell.

The situation is more complicated for the elements in groups IVA and VA. The nonmetals—carbon and silicon in group IVA and nitrogen and phosphorus in group VA—do not usually form ions. On the other hand, the metals—tin and lead in group IVA and antimony and bismuth in group VA—do form positive ions. The generalization we can make about the ions formed by elements in groups IA, IIA, IIIA, VIA, and VIIA is summarized in Table 3-3.

Table 3-3 Ions Formed from Atoms of Some Elements

Group	IA	IIA	IIIA	VIA	VIIA
Usual charge of the ion	$+1$	$+2$	$+3$	-2	-1
Example	Na^+	Ca^{2+}	Al^{3+}	O^{2-}	Cl^-

We can also use our model for the electrons in atoms to explain how the size of an atom depends on its position in the periodic table. Let us compare two elements that are next to one another in the same period—for example, sodium and magnesium. When we go from Na to Mg, there are two important changes that influence the size of the atom: (1) the change in the number of protons and (2) the change in the number of electrons. Going from Na to Mg, the number of protons increases by one and thus the positive charge on the nucleus also increases by one. Other things being equal, we would expect that this would result in a greater pull of the electrons toward the nucleus and a decrease in size. The number of electrons also increases by one going from Na to Mg. We would expect that, other things being equal, the more electrons in an atom, the bigger the size of the atom.

We can now speculate as to which of these effects is more important. As we go across a row in the periodic table, electrons go into the same shell. The increase in nuclear charge is more important than the increase in the number of electrons, and as a result the atomic radii decrease along a period (see Figure 3-8). However, when we go down a column in the periodic table, the valence electrons are going into shells with higher energy. For example, sodium has one

When electricity passes through the tubing of a neon sign, the atoms of this noble gas give off a bright glow of light.

valence electron in the third shell and 10 inner-core electrons, whereas potassium has one valence electron in the fourth shell and 18 inner-core electrons. The nuclear charge increases from $+11$ to $+19$ upon going from Na to K. But the valence electron is shielded from the nucleus by eight more inner-core electrons in K than in Na. The net result is that upon going down a column within a group, the fact that the electrons go into shells with progressively higher principal quantum numbers is more important than the increase in nuclear charge, so atomic radii increase.

We can use much the same argument to explain the general trend of ionization energies. If we go across a row in the periodic table, the size of the atom decreases. Thus the valence electrons are held more tightly to the nucleus and the energy necessary to remove one should increase. Upon going down a column, the size of the atom increases, the valence electrons are held less tightly by the nucleus, and the ionization energy decreases.

Exercise 3-10
Determine for N, B, S, I$^-$, and O^{2-} the number of valence electrons and in which shell they are found.

Exercise 3-11
Predict the ion that each of the following atoms will tend to form: Br, Mg, P, S, Kr.

Lewis Structures

It is convenient to represent an atom or an ion with its valence electrons by a symbol, called an **electron-dot** or **Lewis structure.** G. N. Lewis was an American chemist who first introduced these symbols. In a Lewis structure for an atom or an ion, the element symbol for an atom is used to represent the nucleus of the atom *and* its inner-core electrons. The valence electrons are indicated by dots around the element symbol. Some examples of Lewis structures are shown below:

$$\text{Na}\cdot \quad \cdot\overset{\cdot}{\underset{\cdot}{\text{C}}}\cdot \quad :\overset{\cdot\cdot}{\underset{\cdot\cdot}{\text{O}}} \quad :\overset{\cdot\cdot}{\underset{\cdot\cdot}{\text{Cl}}}:^- \quad \text{Na}^+ \quad :\overset{\cdot\cdot}{\underset{\cdot\cdot}{\text{O}}}:^{2-}$$

In Lewis structures the dots are just a convenient way of indicating the number of valence electrons. There is no relationship between the position of the dots and the actual position of the valence electrons in the atom.

Exercise 3-12
Write Lewis structures for the following: N, Ba, S, I$^-$.

Optional

3-9   ELECTRON CONFIGURATION

The quantum mechanical model of an atom yields a more detailed description of the allowed orbitals, and one that is capable of providing a deeper insight into the properties of atoms and molecules than the simple grouping into shells and division into inner-core and valence electrons that we discussed in the last section. Some of the features of this more detailed description that we shall present in this section are the following:

1. **Subshells:** Orbitals are grouped not only into shells, but also into sub-shells, designated by the letters s, p, d, f, g, and h. An orbital in an atom is classified according to its shell and subshell.

2. **Allowed subshells/orbitals:** Different shells have different numbers of allowed subshells, and there is a different number of allowed orbitals for each subshell.

3. **Orbital energy:** The energy of an orbital depends on the identity of the atom, as well as on the shell and subshell the orbital is in.

4. **Orbital occupancy:** A maximum of two electrons can occupy any particular orbital.

5. **Electron configuration:** Electrons in an atom occupy the allowed orbitals beginning with the orbital of lowest energy. The designation of the number of electrons in each subshell for an atom of an element is called the electron configuration for the atom.

Subshells

To refer to an orbital we use a symbol in which the principal quantum number for the orbital is given first (to specify the shell), followed by the letter s, p, d, f, g, or h (to specify the subshell). Thus, for example, $3d$ refers to an orbital in the d subshell of the third shell. Recall that the major difference between orbitals in different shells is that, in general, the smaller the shell number, the greater the probability of finding the electron close to the nucleus. Orbitals that are in different subshells have different shapes. For example, s orbitals are spherical. For an electron in an s orbital, the probability of finding the electron at a certain position depends *only* on how far that position is from the nucleus, and not on the direction. The lowest-energy orbital for a hydrogen atom, which we described earlier (see Figure 3-11a), is a $1s$ orbital. All s orbitals have a similar spherical shape, although the detailed description of the orbital depends on its shell and the identity of the atom. For example, when we compare the $1s$ orbitals for atoms of different elements, we find that the larger the atomic number, the greater the probability of finding an electron in the $1s$ orbital closer to the nucleus. (An atom with a larger atomic number has a larger nuclear charge, and this higher positive charge pulls an electron in a $1s$ orbital closer to the nucleus.) On the other hand, for the *same* atom, an electron in the $1s$ orbital, for example, will more likely be closer to the nucleus than an electron in the $2s$ orbital for the atom.

Only s orbitals are spherical. The nonspherical shape of a $2p$ orbital is illustrated in Figure 3-15, where we use the same type of schematic representation as was used for the $1s$ orbital in Figure 3-11a. The shapes of d, f, g, and h, orbitals are increasingly more complicated and need not be considered here.

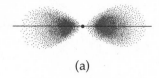

(a)

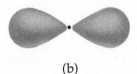

(b)

Figure 3-15 (a) Two-dimensional representation of the nonspherical shape of a $2p$ orbital. The probability of finding the electron is high in a region where the density of dots is high, and the probability of finding the electron is low in a region where the density of dots is low.

(b) Schematic three-dimensional representation of the region where an electron in a $2p$ orbital is likely to be found. The electron is almost certainly somewhere inside this region. The probability of finding the electron outside this region is negligibly small.

Allowed Subshells/Orbitals

The allowed subshells for the shells of interest to us are given in Table 3-4.

Table 3-4 Allowed Subshells

Shell	Allowed Subshells
1	$1s$
2	$2s$ and $2p$
3	$3s$, $3p$, and $3d$
4	$4s$, $4p$, $4d$, and $4f$
5	$5s$, $5p$, $5d$, $5f$, and $5g$
6	$6s$, $6p$, $6d$, $6f$, $6g$, and $6h$

The number of allowed orbitals for the subshells of interest to us are given in Table 3-5. Table 3-5 applies to the allowed subshells of *any* shell. For example, in any of the d subshells ($3d$, $4d$, $5d$, and so on) there are five allowed orbitals.

Table 3-5 Allowed Orbitals

Subshell	Allowed Orbitals
s	1
p	3
d	5
f	7

Orbital Energy

For an atom of a given element, the energy of an orbital depends on its shell and subshell. An energy-level diagram, similar to the one we used for the Bohr orbits (Figure 3-10), can be used to illustrate the order of the energies of the allowed orbitals. The energy-level diagram for an atom of a given element can be determined from a combination of quantum mechanical theory and experimental data. A portion of the energy-level diagram for a sodium atom is given in Figure 3-16. Note that a horizontal line is used to indicate the energy of *each* orbital. Thus, three horizontal lines are used for the $2p$ and $3p$ energy levels, since for each of these p subshells there are three allowed orbitals.

Figure 3-16   The energy-level diagram for a sodium atom. To save space, a break was made in the scale between the 1s and 2s levels. The 1s level has a much lower energy level than the 2s level. The energy separation between the 1s and 2s levels is actually about 12 times the separation between the 2s and 2p levels.

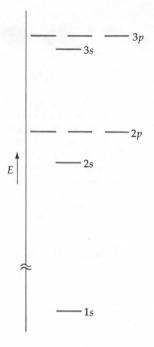

Each element has the same type of subshells and orbitals but, as we discussed earlier, the detailed description of these orbitals depends on the indentity of the atom. For example, for every atom, the 1s orbital has the lowest energy, but the actual numerical value of this energy is different for different atoms. An electron in the 1s orbital of a sodium atom, for example, is held more tightly to the nucleus, and therefore has a lower energy, than an electron in a 1s orbital of a hydrogen atom, because the sodium nucleus has a charge of $+11$ whereas the nuclear charge of a hydrogen atom is only $+1$.

Orbital Occupancy

A consequence of a general principle of quantum mechanics, called the **Pauli exclusion principle,** is that no more than two electrons can occupy any one orbital. The maximum number of electrons that are allowed in any subshell can be determined using this result and the information in Table 3-5 (see Table 3-6).

Table 3-6 Maximum Number of Electrons

Subshell	Electrons
s	2
p	6
d	10
f	14

Electron Configuration

For any atom, we can easily determine the number of electrons in each subshell (the electron configuration for the atom) if we have the energy-level diagram for the atom. To do this, we (1) determine the total number of electrons in the atom by using the atom's atomic number, and (2) put electrons in orbitals, beginning with the orbital of lowest energy, and making sure to put no more than two electrons in any one orbital.

Figure 3-17   The occupation of orbitals for a
sodium atom. The electrons are
represented by X's on the energy-
level diagram. The electrons in a
sodium atom occupy the orbitals, in
turn, starting with the orbital of
lowest energy and allowing no more
than two electrons in any one
orbital. The electron configuration
for a sodium atom is $1s^2 2s^2 2p^6 3s^1$.

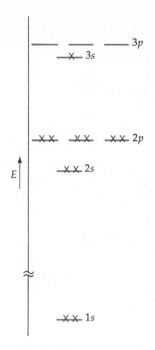

Let us do this for a sodium atom using the energy-level diagram for sodium, which we have already discussed (Figure 3-16). Sodium, with an atomic number of 11, has 11 electrons. We can represent an electron by an X, and use Figure 3-16 to obtain Figure 3-17, which indicates the orbitals occupied by the 11 electrons in a sodium atom.

A more convenient way to specify the electron configuration of an atom is to list the subshells that have electrons in them, and use a superscript for each subshell to indicate the number of electrons in that subshell. Thus, the electron configuration for a sodium atom is designated $1s^2 2s^2 2p^6 3s^1$.

The electron configuration $1s^2 2s^2 2p^6 3s^1$ that we have determined for a sodium atom is the electron configuration with the lowest possible energy; it is called the **ground state.** However, if the electron in the $3s$ subshell, for example, was removed and placed in the $3p$ subshell, the atom would have the electron configuration $1s^2 2s^2 2p^6 3p^1$. A sodium atom with this configuration would have a higher energy than the ground state, and a configuration of higher energy such as this is called an **excited state.** At room temperature, for any element, almost all the atoms have their ground-state electron configurations (for instance, $1s^2 2s^2 2p^6 3s^1$ for a Na atom). An atom can go from the ground state to an excited state when it absorbs energy in just the right amount. One way this change can be accomplished is for the atom to absorb light with the proper amount of energy.

We can now determine the electron configuration for an atom of any element, in a manner similar to the one we used for sodium, once we have the energy-level diagram for an atom of the element. Each element has its own energy-level diagram, but there is a general common ordering of the subshell energies that can be used to give the correct electron configuration for most elements. This common order is

(3-4) $1s, 2s, 2p, 3s, 3p, 4s, 3d, 4p, 5s, 4d, 5p, 6s, 4f, 5d, 6p, 7s, 5f, 6d$

Using what we have learned about the electron configurations, we can show that not only do the representative elements in the same group have the same number of valence electrons (Figure 3-12), but that these valence electrons are distributed between s and p subshells in the same manner for atoms in the same group. As an example, the electron configurations for the group IVA elements are given in Table 3-7. You should verify the results in Table 3-7 using Eq. 3-4 and Table 3-6.

Table 3-7 Electron Configurations for Group IVA Elements

Element	Period	Atomic No.	Electron Configuration	Valence-Shell Electron Configuration
C	2	6	$1s^22s^22p^2$	$2s^22p^2$
Si	3	14	$1s^22s^22p^63s^23p^2$	$3s^23p^2$
Ge	4	32	$1s^22s^22p^63s^23p^63d^{10}4s^24p^2$	$4s^24p^2$
Sn	5	50	$1s^22s^22p^63s^23p^63d^{10}4s^24p^64d^{10}5s^25p^2$	$5s^25p^2$
Pb	6	82	$1s^22s^22p^63s^23p^63d^{10}4s^24p^64d^{10}4f^{14}5s^25p^65d^{10}6s^26p^2$	$6s^26p^2$

Recall that the valence electrons are the electrons in the shell with highest energy. Note in Table 3-7 that all the group IVA elements have the similar valence-shell configuration, which can be represented by ns^2np^2, where n is the period of the element. In a similar manner we can obtain the results in Table 3-8 for the valence-shell electron configurations of the other representative elements.

Table 3-8 General Valence-Shell Electron Configurations

Group	Valence-Shell Electron Configuration
IA	ns^1
IIA	ns^2
IIIA	ns^2np^1
IVA	ns^2np^2
VA	ns^2np^3
VIA	ns^2np^4
VIIA	ns^2np^5
VIIIA	ns^2np^6

As we see clearly from Table 3-8, each group of representative elements has a common characteristic valence-shell electron configuration, which is in keeping with the fact that elements in the same group have similar chemical properties.

Exercise 3-13
Determine the electron configuration for all the group VIA elements and verify that they all have the similar valence-shell electron configuration ns^2np^4.

Optional

3-10 DO ELECTRONS IN AN ATOM MOVE IN ORBITS?

In order to decide whether or not a body is moving in an orbit, we have to measure precisely both the position and the velocity of the body. It is possible to do this for large objects, such as planets, cars, and tennis balls. There is no way, however, to measure precisely both the position and the velocity of small objects such as electrons in atoms. Let us consider how we can determine both the position and the velocity of a large object. Consider a car moving along a road.

To find the velocity of the car, we determine its position d_1 at some instant t_1, and then measure its position d_2 at some later time t_2. The velocity is determined by the distance traveled, $d_2 - d_1$, in the time interval, $t_2 - t_1$. Thus,

$$\text{velocity} = \frac{d_2 - d_1}{t_2 - t_1}$$

This description of how to measure a car's velocity seems straightforward, but we have tacitly assumed that we can do one critical thing. We have assumed that the measurement of the car's position at time t_1 did not significantly alter the motion of the car. Now if we use our eyes, for example, to estimate a car's velocity, we obtain information about the car's position from visible light that bounces off the car. Light hitting the car does not change the car's motion. Similarly, a state trooper might determine the velocity of a car more precisely by bouncing a radar beam off the car. Neither visible light nor the radar beam interferes with the motion of the car, because the mass of the car is very large compared with the small amount of energy in the impinging visible light or radar beam. (Of course the state trooper might alter the motion of the car by some other means if it were going too fast.) But what if the state trooper used a cannon instead of a radar beam to determine the position of the car? The car's position at time t_1 could be determined by hitting it with a cannon ball. But the impact would be likely to knock the car off the road. In this event, we have obviously affected the car's motion and we have no hope of determining the original velocity of the car.

Attempting to measure both the position and velocity of an electron in an atom is somewhat analogous to trying to measure the position and velocity of a car with a cannon ball. Even light bouncing off an electron in an atom has sufficient energy to disturb the electron significantly. Light affects electrons but not cars because electrons have such very small mass. As a consequence, we cannot say that atoms contain electrons orbiting about the nucleus. Rather, we must describe electrons in atoms with the more complex terminology associated with the concept of quantum mechanical orbitals.

3-11 SUMMARY

1. An element is classified as a metal, a nonmetal, or a metalloid on the basis of its physical properties.

2. Metals tend to lose electrons in chemical reactions, whereas nonmetals tend to gain electrons.

3. Elements in the same group of the periodic table have similar properties.

4. Ionization energies increase across a row and decrease down a column in the periodic table. Atomic size decreases across a row and increases down a column.

5. Bohr postulated that the electron in the hydrogen atom revolves around the nucleus in a circular orbit, and that only certain orbits are allowed.

6. It is impossible to measure precisely both the position and the velocity of an electron in an atom.

7. In the quantum mechanical description of atoms, electrons are described by orbitals.

8. An atom's electrons can be classified as inner-core or valence electrons. Valence electrons are those electrons in the shell with highest energy.

9. The valence electrons of an atom are the electrons that are primarily involved when that atom bonds to other atoms.

10. For a representative element, the number of valence electrons is equal to the group number of the element, and the shell they are in is equal to the period of the element.

11. For many elements, the ion that an atom of an element will tend to form can be predicted from the position of the element in the periodic table.

12. A Lewis structure for an atom or an ion is a convenient way of exhibiting the number of valence electrons for that atom or ion.

13. **(Optional)** The electron configuration for an atom is the designation of the number of electrons in each subshell.

PROBLEMS

1. Classify each of the following elements as a metal, a nonmetal, or a metalloid: (a) phosphorus, (b) beryllium, (c) titanium, (d) arsenic, (e) carbon.

2. A certain element is a reddish-brown liquid with a low electrical conductivity. Is this element a metal, a nonmetal, or a metalloid?

3. Arsenic forms a compound with sulfur with the chemical formula As_2S_5. On this basis, predict the formula for a compound formed between the elements antimony and oxygen.

4. Carbon and chlorine form a compound called carbon tetrachloride with the chemical formula CCl_4. On this basis, predict the chemical formula for the product of the following reaction: silicon + fluorine → _____.

5. When the solid compound cadmium chloride, with the chemical formula $CdCl_2$, dissolves in water, separate Cd^{2+} and Cl^- ions exist in solution. What ions would you predict to be present when the compound zinc bromide dissolves in water?

6. Which of the following atoms would you expect to have the largest size: O, F, S, Cl, Se?

7. Which of the following atoms would you expect to have the smallest size: P, S, O, N, C?

8. The chloride ion, Cl^-, has a different size than a chlorine atom, Cl. Would you expect the size of a Cl^- ion to be larger than, smaller than, or the same as the size of a Cl atom? Why?

9. The sodium ion, Na^+, has a different size than a sodium atom, Na. Would you expect the size of a Na^+ ion to be larger than, smaller than, or the same as the size of a Na atom? Why?

10. Which of the following atoms would you expect to have the largest first ionization energy: Si, P, S, As, Se?

11. Would you expect that the energy required to remove an electron from a helium atom, He, is larger than, smaller than, or the same as the energy required to remove an electron from a lithium ion, Li^+?

12. Using the Bohr model for the hydrogen atom, briefly describe the difference between the orbit with $n = 2$ and the orbit with $n = 3$. Would it require an energy input, or would energy be released, if the electron was taken from the $n = 2$ orbit and put in the $n = 3$ orbit?

13. Briefly describe the difference between a Bohr orbit and a quantum mechanical orbital.

14. Determine the number of valence electrons and the shell in which they are found for the following: (a) Sr, (b) P, (c) I^-, (d) K, (e) Ga, (f) B, (g) C.

15. Predict the ion that each of the following atoms will tend to form: (a) Ca, (b) C, (c) Se, (d) Li, (e) He, (f) H.

16. Write Lewis structures for the following: (a) N, (b) K, (c) Br^-, (d) Mg^{2+}, (e) Se.

17. **(Optional)** Determine the electron configuration for the following: (a) Ca, (b) Al, (c) Cl, (d) Rb.

SOLUTIONS TO EXERCISES

3-1 Metals: barium and nickel; nonmetal: phosphorus; metalloid: germanium.

3-2

	8	18	18	32
Atomic number	6⎯⎯14⎯⎯	32⎯⎯	50⎯⎯	82

$$\begin{array}{ccccc} & \downarrow & \downarrow & \downarrow & \downarrow & \downarrow \\ \text{Group IVA} & C & Si & Ge & Sn & Pb \end{array}$$

Group IVA elements and the noble gases have the same spacing.

3-3 Representative elements: carbon, group IVA; sulfur, group VIA; barium, group IIA; phosphorus, group VA; xenon, group VIIIA. Transition element: nickel. Actinide element: uranium.

3-4 Since sulfur and oxygen are both group VIA elements, we predict the chemical formula CS_2 in analogy to CO_2. (Carbon disulfide, CS_2, is a known compound.)

3-5 Since calcium and magnesium are both elements in group IIA and since bromine and iodine are both elements in group VIIA, we predict the chemical formula MgI_2 in analogy to $CaBr_2$, and we predict that when MgI_2 dissolves in water separate Mg^{2+} and I^- ions will be present.

3-6 Consider, for example, a calcium atom. For the second ionization energy an electron must be removed from a Ca^+ ion, which already has a positive charge. This requires more energy than when removing an electron from a neutral Ca atom (the first ionization energy).

3-7 Using the general trend that ionization energy increases along a row and decreases down a column in the periodic table, we predict that germanium has the lowest first ionization energy. (It does.)

3-8 Both antimony and arsenic are group VA elements. Using the general trend that atomic size increases as you go down a column in the periodic table, we predict that antimony is larger. (It is.)

3-9 According to the Bohr model, if the electron is in the $n = 1$ orbit it is closer to the nucleus and held more tightly to the nucleus than if it is farther away from the nucleus, as in the $n = 2$ orbit. Thus, to remove the electron from the $n = 1$ orbit requires more energy.

3-10 Use Figure 3-12 and the periodic table. Note that an I^- ion has one more electron than an I atom and that an O^{2-} ion has two more electrons than an O atom.
N: 5 valence electrons in second shell
B: 3 valence electrons in second shell
S: 6 valence electrons in third shell
I^-: 8 valence electrons in fifth shell
O^{2-}: 8 valence electrons in second shell

3-11 Use Table 3-3 and the periodic table.
Br^-, Mg^{2+}, and S^{2-}. P and Kr will not tend to form ions.

3-12 $:\!\overset{..}{N}\!\cdot$ Ba: $:\!\overset{..}{\underset{..}{S}}\!:$ $:\!\overset{..}{\underset{..}{I}}\!:^-$

3-13 **(Optional)**

Element	Period	Atomic No.	Electron Configuration	Valence-Shell Electron Configuration
O	2	8	$1s^22s^22p^4$	$2s^22p^4$
S	3	16	$1s^22s^22p^63s^23p^4$	$3s^23p^4$
Se	4	34	$1s^22s^22p^63s^23p^63d^{10}4s^24p^4$	$4s^24p^4$
Te	5	52	$1s^22s^22p^63s^23p^63d^{10}4s^24p^64d^{10}5s^25p^4$	$5s^25p^4$
Po	6	84	$1s^22s^22p^63s^23p^63d^{10}4s^24p^64d^{10}4f^{14}5s^25p^65d^{10}6s^26p^4$	$6s^26p^4$

CHAPTER 4

Chemical Compounds and Chemical Bonds

4-1 INTRODUCTION

The atoms in a molecule are held together by chemical bonds. But what is a chemical bond? We often use the word "bond" in quite a general fashion. We say that two pieces of paper are bonded together by glue; or that each brick in a wall is bonded by mortar to the adjacent bricks. We even consider people joined together by the bonds of matrimony. What, then, is the nature of a chemical bond? What "glue" makes individual atoms stick together to form molecules?

We shall see that the basis of a chemical bond is really quite simple. Atoms are held together in molecules by chemical bonds because there is an attractive force between positive and negative electrical charges. Chemists use a number of different models, however, to describe the properties of chemical bonds. Some of these models are qualitative descriptions of a chemical bond. Others, which are mathematically very involved, can be used to obtain quantitative values for bond properties.

We shall see that chemists classify chemical compounds into two broad classes—ionic compounds and covalent compounds—based on their chemical and physical properties. We shall also learn how to classify bonds into different types and why such a classification is useful. We shall discuss a simple method for predicting the chemical formulas of a large number of ionic compounds. For many covalent compounds, we shall see how to predict which atoms are bonded together in a molecule of the compound. In addition, we shall discuss several pictorial representations that chemists use to indicate bonding arrangements in molecules.

The chemical and physical properties of a molecule are determined not only by the bonding arrangement of the atoms in the molecule, but also by the molecular geometry, or shape, of the molecule. We shall investigate the relationship between the way the atoms in a molecule are bonded together and the shape of that molecule. Finally, in an optional section, we shall describe a method for predicting the shape of a simple molecule or polyatomic ion.

A Mexican couple join their lives with the bonds of matrimony. The atoms in a molecule are joined together by chemical bonds.

4-2   STUDY OBJECTIVES

After studying the material in this chapter, you should be able to:

1. Define the terms chemical bond, bond energy, and bond length.

2. List the general properties of ionic and covalent compounds.

3. Predict whether simple compounds are ionic or covalent.

4. Identify the polyatomic ions in Table 4-1 by name and chemical formula.

5. Write a chemical formula for an ionic compound or a binary covalent compound given its name.

6. Write the name of an ionic compound or a binary covalent compound given its chemical formula.

7. Describe a model used for ionic bonds.

8. Explain the nature of a bonding orbital, a shared pair of electrons, and a covalent bond.

9. Define the term electronegativity and recognize the general trend for electronegativity in the periodic table.

10. Define the terms polar covalent bond and nonpolar covalent bond.

11. Classify bonds as nonpolar covalent, polar covalent, or ionic using the difference in electronegativity for the atoms in the bond.

12. Define the terms structural formula, Lewis structure, octet rule, and resonance structure.

13. Determine the usual number of covalent bonds for elements in groups IVA, VA, VIA, and VIIA.

14. Draw structural formulas and Lewis structures for simple molecules and polyatomic ions given their molecular formulas.

15. Use resonance structures to draw a Lewis structure representation for a molecule or polyatomic ion when it is appropriate to do so.

16. Explain what is meant by the shape of a molecule.

17. **(Optional)** Predict the shape of a simple molecule or polyatomic ion given the chemical formula for the molecule or ion.

4-3   CHARACTERISTIC PROPERTIES OF CHEMICAL BONDS

In any molecule of a substance, the forces that hold the atoms close together are called **chemical bonds.** For example, in the hydrogen molecule, H_2, such a bond holds the two hydrogen atoms together. The distance between the nuclei of two bonded atoms in a molecule is called the **bond length.** Bond lengths can be determined by experiment. They are one of the characteristic properties of a chemical bond. The distance between the nuclei of the two hydrogen atoms in a H_2 molecule has been determined to be 0.074 nm.

Perhaps the most important property of a chemical bond is the energy needed to break it. For example, energy is required to pull apart a H_2 molecule, in which the two hydrogen atoms are 0.074 nm apart, into two hydrogen atoms separated by a very large distance (see Figure 4-1). We say that a chemical bond has been broken in this process.

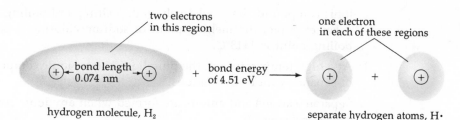

two electrons
in this region

one electron
in each of these regions

bond length
0.074 nm

+ bond energy
of 4.51 eV

hydrogen molecule, H_2

separate hydrogen atoms, H·

Figure 4-1   Schematic representation of the separation of a hydrogen molecule (H_2) into two separate hydrogen atoms. The energy required for this process—the bond energy of H_2—is 4.51 eV. The plus symbol $\oplus$ represents a proton—the nucleus of a hydrogen atom.

The energy required to break a chemical bond is a measure of the bond's strength. This energy is called the **bond energy.** Bond energies are determined from experimental data. The bond energy for the hydrogen-hydrogen bond in the H_2 molecule is 4.51 eV (1 eV = 3.83×10^{-20} cal) (Figure 4-1).

Exercise 4-1

The bond energies for the chlorine (Cl_2), oxygen (O_2), nitrogen (N_2), and hydrogen chloride (HCl) molecules are 2.51 eV, 5.15 eV, 9.83 eV, and 4.47 eV, respectively. In which of these molecules are the atoms bonded together more strongly than the hydrogen atoms in a H_2 molecule? The bond energy for a H_2 molecule is 4.51 eV.

4-4 CLASSIFICATION OF CHEMICAL COMPOUNDS AND CHEMICAL BONDS

Chemists divide chemical compounds into two broad classes—**ionic compounds** and **covalent compounds**—based on their chemical and physical properties. The chemical bonds in ionic compounds are called **ionic bonds,** and those in covalent compounds are called **covalent bonds.** Chemists use different models to describe ionic and covalent bonds. First let us see how different types of compounds are classified.

Every chemical compound has a name and a chemical formula. Some more familiar chemical compounds have both a systematic or formal name and a common name. For example, the compound $NaHCO_3$ has the systematic name sodium hydrogen carbonate; its common name is baking soda. The system for naming ionic compounds and binary covalent compounds is discussed in Essential Skills 7. It is *very important* that you be able to name a compound correctly if you are given its chemical formula, or to infer the chemical formula for a compound if you are given its name. The names of chemical compounds are part of the language of chemistry, and if we are to be able to communicate information about chemical compounds we must be able to use the language of chemistry correctly.

Ionic Compounds

Sodium chloride (NaCl, common table salt), potassium bromide (KBr), calcium fluoride (CaF_2), sodium sulfide (Na_2S), calcium hydrogen carbonate [$Ca(HCO_3)_2$], ammonium chloride (NH_4Cl), and many other compounds are called **ionic compounds** and have the following distinctive general properties:

1. Ionic compounds have relatively high melting and boiling points.
    For example, the melting point of sodium chloride is 801°C, and its boiling point is 1413°C.

2. At a high temperature in the pure liquid state, ionic compounds consist of separate cations and anions.

3. Separate cations and anions are formed when any ionic compound dissolves in water.
    The primary experimental evidence for the existence of separate ions in molten compounds and in solutions in water is that they can conduct an electrical current. In order for something to carry an electrical current, separate electrically charged particles that are capable of movement must be present. For example, molten sodium chloride, or a solution of sodium chloride in water, contains separate Na^+ cations and Cl^- anions (see Figure 4-2). A compound that produces a large concentration of ions when it is dissolved in water is often referred to as an **electrolyte.** Most electrolytes are ionic compounds that are relatively soluble in water like NaCl. A small number of covalent compounds are also electrolytes.

Figure 4-2   Schematic illustration of the separate sodium ions, Na^+, and chloride ions, Cl^-, present in solution when solid sodium chloride, NaCl, dissolves in water.

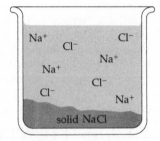

4. At the microscopic level, solid ionic compounds are made up of an extended, ordered three-dimensional array of cations and anions.
    For example, experimental evidence indicates that there is an ordered three-dimensional array of Na^+ cations and Cl^- anions in solid sodium chloride (see Figure 4-3). Notice that each Na^+ ion has six Cl^- ions as nearest neighbors, and that each Cl^- ion is surrounded by six Na^+ ions. Because of the extended arrangement of ions we cannot assign a particu-

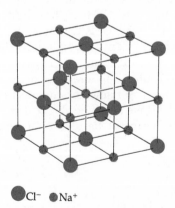

●Cl^- ●Na^+

Figure 4-3   In solid NaCl there is an ordered three-dimensional array of Na^+ cations and Cl^- anions. A micro- / scopic view of a grain of table salt offers evidence of such an ordered arrangement of ions.

lar Cl⁻ ion to a particular Na⁺ ion in solid sodium chloride. Consequently, we cannot identify a unique molecular unit in solid NaCl that involves one Na⁺ ion and one Cl⁻ ion. Thus, we must call NaCl a chemical formula for solid sodium chloride. We should *not* refer to NaCl as a molecular formula. The chemical formula NaCl for solid sodium chloride just indicates that in any sample of sodium chloride there are an equal number of Na⁺ ions and Cl⁻ ions. There is some type of ordered three-dimensional array of cations and anions in all solid ionic compounds, so we cannot identify a small molecular unit in any solid ionic compound.

Some ionic compounds consist of just two elements; these are called **binary compounds.** On the basis of experimental observations we know that binary ionic compounds consist of a metallic and a nonmetallic element. For example, the binary ionic compound sodium chloride consists of the metallic element sodium and the nonmetallic element chlorine. When a binary ionic compound, such as NaCl, dissolves in water, monatomic cations and monatomic anions are present in the resulting solution. Sodium chloride forms Na⁺ cations and Cl⁻ anions when it is dissolved in water.

Other ionic compounds—for example, sodium nitrate, $NaNO_3$—consist of more than two elements. When this type of ionic compound dissolves in water, the resulting solution contains a polyatomic ion. A **polyatomic ion** is an ion that contains more than one atom. For example, when $NaNO_3$ dissolves in water, the Na⁺ cation and the polyatomic anion NO_3^- are present in the solution. In a polyatomic ion, a number of atoms are bonded together by covalent bonds and the whole polyatomic unit has a net positive or negative electrical charge. When an ionic compound containing a polyatomic ion dissolves in water, the polyatomic ion does not dissociate, but stays together as a single, charged entity. There is no simple way of determining which elements will combine to form polyatomic ions. The names, formulas, and charges of several common polyatomic ions are given in Table 4-1. You should memorize the formulas and names of these polyatomic ions and learn to think of the group of atoms and its associated charge as a unit in its own right. For example, when you see the chemical formula $NaHCO_3$, you should recognize this as an ionic compound containing the Na⁺ cation and the HCO_3^- polyatomic anion. There are many common polyatomic anions, but the ammonium ion, NH_4^+, is the only common polyatomic cation. It is *quite important* that you remember to include the charge as a superscript when writing the formulas for polyatomic ions and that you realize that the electrical charge is associated with the whole unit, and not with any particular atom in it. For example, NO_2^- is the nitrite *ion*, whereas NO_2 is the nitrogen dioxide *molecule*. Leaving out the charge when it is supposed to be there completely changes the chemical meaning.

Table 4-1 Some Common Polyatomic Ions

Formula	Name	Charge	Formula	Name	Charge
OH⁻	Hydroxide	−1	$H_2PO_4^-$	Dihydrogen phosphate	−1
CO_3^{2-}	Carbonate	−2			
HCO_3^-	Hydrogen carbonate (bicarbonate)	−1	NO_3^-	Nitrate	−1
			NO_2^-	Nitrite	−1
			SO_4^{2-}	Sulfate	−2
PO_4^{3-}	Phosphate	−3	CN⁻	Cyanide	−1
HPO_4^{2-}	Monohydrogen phosphate	−2	NH_4^+	Ammonium	+1

Covalent Compounds

Hydrogen (H_2), chlorine (Cl_2), methane (CH_4), and carbon dioxide (CO_2) are a few examples of compounds that have characteristic properties distinctly different from those of ionic compounds. Compounds of this type are classified as **covalent compounds.** In contrast to ionic compounds, most (though not all) covalent compounds have the following properties:

1. They have relatively low melting and boiling points.

2. Separate ions are not formed in the pure liquid state or when the compound is dissolved in water.

3. There is an identifiable molecular unit involving a small number of atoms in the solid state.

    For example, below $-101°C$, chlorine (Cl_2) is a solid. Experimental evidence shows that in the solid state each chlorine atom has only one nearest-neighbor chlorine atom, which is 0.202 nm away. This is just about the same distance as the bond length (0.200 nm) of a Cl_2 molecule in the gas state. Because these two measurements are so close, we can say that there are Cl_2 molecules in solid chlorine as well as in gaseous chlorine. The Cl_2 molecular units in the solid are arranged in parallel layers that are about 0.37 nm apart. Within each layer, the distance between Cl_2 molecules is about 0.33 nm.

Most, but not quite all, covalent compounds contain only nonmetallic elements. By far the largest, and from our point of view the most important, group of covalent compounds contains the nonmetallic element carbon. As we shall see, carbon-containing compounds are the basis of life.

Some compounds consist of molecules in the solid and gaseous state even though they form ions when dissolved in water. Compounds of this type are also called covalent. Many important compounds classified as acids or bases, such as the acid hydrogen chloride (HCl), and the base ammonia (NH_3), have these properties (see Chapters 10 and 11). HCl and NH_3 are examples of covalent compounds that are also electrolytes.

Exercise 4-2

Predict whether the following compounds are ionic or covalent: sulfur trioxide (SO_3); magnesium bromide ($MgBr_2$); sodium hydrogen carbonate ($NaHCO_3$); nitrogen monoxide (NO); potassium cyanide (KCN); carbon dioxide (CO_2). For the ionic compounds, predict the ions that are present when the compound is dissolved in water.

4-5   CHEMICAL FORMULAS OF IONIC COMPOUNDS

In the previous section we said that binary ionic compounds involve a metal and a nonmetal. This is just what we should expect. If we consider the trends in ionization energy and electron affinity discussed in Chapter 3, we can say that metals are more likely to give up electrons than are nonmetals and that nonmetals are more willing to accept additional electrons than are metals. We can think of the formation of an ionic compound as involving an exchange of electrons between metal atoms and nonmetal atoms. The electrons lost by the metal atoms are gained by the nonmetal atoms. Therefore, it is not surprising that metals and nonmetals can combine and form ionic compounds.

Note that there are different relative numbers of cations and anions in different ionic compounds. For example, the relative number of cations and anions in NaCl is 1:1, whereas it is 1:2 in calcium fluoride (CaF_2) and 2:1 in sodium sulfide (Na_2S). There is a simple explanation for the number of atoms in each of these chemical formulas.

We saw in Section 3-8 that atoms of elements in groups IA, IIA, and IIIA in the periodic table can lose one, two, and three valence electrons to form cations with charges of $+1$, $+2$, and $+3$, respectively. The resulting cations have the electron arrangement of a noble gas atom. We also saw that the atoms of elements in groups VIIA and VIA in the periodic table can gain one and two valence electrons to form anions with charges of -1 and -2, respectively. These anions also have an electron arrangement similar to a noble gas atom. Recall that cations and anions with these charges are the ones that tend to form, since the electron arrangement of a noble gas atom is particularly stable. Thus, <u>atoms of group IA, IIA, and IIIA elements combine with atoms of group VIA and VIIA elements to form ionic compounds</u>. For example, sodium is an element in group IA, whereas chlorine is in group VIIA. Sodium and chlorine combine to form the compound sodium chloride. The compound itself must have zero net electrical charge. To form a sodium cation, a sodium atom must lose one electron, and to form a chloride anion, a chlorine atom must gain one electron. Therefore, in sodium chloride there must be one Na^+ cation for each Cl^- anion. This explains why the formula for sodium chloride is NaCl. Similarly, in calcium fluoride, for every Ca^{2+} cation, with a charge of $+2$, there must be two F^- anions, each with a charge of -1; thus the chemical formula for calcium fluoride is CaF_2. A Ca^{2+} cation is formed when a calcium atom loses two electrons. So two fluorine atoms are needed to accept these electrons and in the process they become two F^- anions.

We can predict <u>the chemical formula for any ionic compound if we know the charge of the cation and the charge of the anion contained in the compound.</u> We simply determine what number of cations and what number of anions will give a collection of ions with a zero net electrical charge. The smallest whole numbers that give a collection of ions with a zero net electrical charge are used in the chemical formula for the ionic compound. The following examples illustrate how this is done.

EXAMPLE 1   For a binary ionic compound involving the representative elements, the charge on the cation and the charge of the anion are determined from their positions in the periodic table. Thus to determine the formula for aluminum oxide we proceed as follows:

Step 1   Aluminum is in group IIIA, so the aluminum cation (Al^{3+}) has a charge of $+3$; oxygen is in group VIA, so the oxide anion (O^{2-}) has a charge of -2.

Step 2   The collection of one Al^{3+} cation and one O^{2-} anion has a net charge of $+1$, and the collection of one Al^{3+} cation and two O^{2-} anions has a net charge of -1. Since neither of these collections with one Al^{3+} has a net electrical charge of zero, we next try to use a collection with two Al^{3+} cations. Two Al^{3+} cations have a combined charge of $+6$, three O^{2-} anions have a combined charge of -6, and a collection of two Al^{3+} cations and three O^{2-} anions does have a net charge of zero. Thus, the chemical formula of aluminum oxide is Al_2O_3. Other collections of these ions—such as four Al^{3+} cations and six O^{2-} anions, for example—also have a net electrical charge of zero. In the chemical formula, however, the smallest whole numbers (2 and 3 in this case) that give a collection with zero net charge are used.

EXAMPLE 2 To determine the chemical formula for an ionic compound involving a polyatomic ion, we need to know (or look up in a table of polyatomic ions) the charge of the polyatomic ion involved. Thus, to determine the chemical formula for calcium hydrogen carbonate (also called calcium bicarbonate) we proceed as follows:

Step 1 The charge on the calcium cation (Ca^{2+}) is +2, and the charge on the hydrogen carbonate anion (HCO_3^-) is -1 (see Table 4-1).

Step 2 A collection of one Ca^{2+} cation and two HCO_3^- anions has a net charge of zero. The chemical formula for calcium hydrogen carbonate (calcium bicarbonate) is thus $Ca(HCO_3)_2$. Note that the parentheses and the subscript 2 in the chemical formula $Ca(HCO_3)_2$ are used to indicate that the compound contains two HCO_3^- anions for each Ca^{2+} cation. You should be able to recognize from the chemical formula $Ca(HCO_3)_2$ that when calcium hydrogen carbonate dissolves in water, separate Ca^{2+} and HCO_3^- ions are formed and the resulting solution contains two HCO_3^- anions for every Ca^{2+} cation.

When writing formulas for ionic compounds, we do not include the charges of the cation and the anion as superscripts. Thus the chemical formula for sodium chloride is written NaCl, not Na^+Cl^-. It is of utmost importance, however, to indicate electrical charges by superscripts when we want to represent separate ions. For instance, when NaCl dissolves in water, separate Na^+ cations and Cl^- anions are formed.

It is also *essential* to use the proper names and formulas of polyatomic ions. Notice, for example, that the name for NH_4^+ is "the ammonium ion," whereas NH_3 refers to the chemical compound with the name "ammonia." Also, the chemical formula for the monohydrogen phosphate ion is HPO_4^{2-} with a charge of -2, whereas the dihydrogen phosphate ion has the chemical formula $H_2PO_4^-$ and a charge of -1.

Exercise 4-3
Write the chemical formula for each of the following ionic compounds: calcium oxide, potassium carbonate, sodium monohydrogen phosphate, ammonium sulfate, aluminum nitrate.

4-6   IONIC BONDS

The forces that hold ions together in ionic compounds are called **ionic bonds.** What model do chemists use to describe ionic bonds?

Let us consider as a specific example the model for the ionic bond in sodium chloride. As we discussed in Section 4-4, solid sodium chloride consists of an extended three-dimensional array of Na^+ cations and Cl^- anions. The distance between a Na^+ cation and any of its six nearest-neighbor Cl^- anions is 0.281 nm. There are no NaCl molecules in the solid, but at very high temperatures when sodium chloride is a gas, NaCl molecules do exist. The bond length of a gaseous NaCl molecule is 0.236 nm, which is close to the separation of Na^+ and Cl^- nearest neighbors in solid sodium chloride. The bond energy of a NaCl molecule is 4.25 eV.

Thus, a reasonable model for the ionic bond in a NaCl molecule is to consider that the single valence electron on the sodium atom, Na·, is transferred completely to the chlorine atom, ·C̈l:. Recall that Na· and ·C̈l: are Lewis structures for a sodium atom and a chlorine atom, with the dots representing valence electrons. The resulting Na^+ cation and $:\overset{..}{\underset{..}{Cl}}:^-$ anion are then held close together by the electrical force of attraction between the positive charge of the cation and the negative charge of the anion (see Figure 4-4).

$$Na\overset{\curvearrowright}{} + \cdot\overset{..}{\underset{..}{Cl}}:$$

separate atoms

$$Na^+ \xrightarrow{\text{attraction}} \xleftarrow{} \ :\overset{..}{\underset{..}{Cl}}:^-$$

NaCl molecule in gas

Figure 4-4   We can think of a NaCl molecule in the gas state as being formed when the valence electron in a sodium atom is transferred to a chlorine atom. The ionic bond in a gaseous   NaCl molecule results from the electrical force of attraction between a Na^+ cation and a Cl^- anion that are 0.236 nm apart.

In a similar manner, chemists consider that there are positively charged cations and negatively charged anions in any ionic compound and that the ionic bond is the electrical attraction between these opposite charges.

Exercise 4-4
Describe the ionic bonds in potassium bromide and calcium fluoride.

4-7   COVALENT BONDS

The forces that hold atoms together in covalent compounds are called **covalent bonds.** Most covalent compounds involve only nonmetal elements. Thus, we do not expect electrons in covalent compounds to transfer completely from one atom to another and form ions. We therefore need a different model for the bonds in covalent compounds. Let us construct one.

Sharing of Electrons

Remember that the nucleus of an atom contains protons and neutrons. Thus all nuclei have a positive electrical charge. Also recall that two positive (or two negative) electrical charges repel one another, and that there is an attractive force between positive and negative electrical charges. Therefore, when we consider the nature of a covalent bond between two atoms, the question "what causes two positively charged nuclei to stay close together in spite of the repulsive force between them?" must be answered. Since atoms contain only nuclei and electrons, it must be the electrons that are responsible for holding the atomic nuclei together in a molecule. Let us consider how this is accomplished in the simplest covalent molecule, the hydrogen molecule.

As we discussed in Chapter 3, we cannot pinpoint the exact location of an electron in an atom. This is also true for electrons in molecules and ions. We can only give the *probability* of finding the electron at any particular point in space. Let us focus our attention on just one of the electrons in a H_2 molecule and suppose for the moment that this electron is at the position indicated in Figure 4-5a. There is an attractive force between the electron and each of the two protons. When the electron is in the region between the two protons, this attractive force counteracts the repulsive force between the two protons. We might think of this situation as analogous to an adult holding together two children who want to get away from one another.

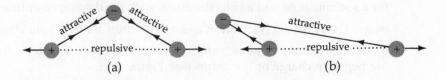

Figure 4-5   (a) When one of the electrons in a hydrogen molecule is in the region between the two positive nuclei, the attractive forces between the electron and each nucleus are opposed to and greater than the repulsive force between the nuclei.

(b) When an electron in a hydrogen molecule is outside of the region between the two nuclei, the repulsive force between the two nuclei is greater than the attractive forces.

Now recall that both attractive and repulsive forces between two electrically charged particles are smaller the greater the distance between the particles (Chapter 1). A calculation of the magnitude of the forces shows that the attractive force is greater than the repulsive force if the electron is at the position shown in Figure 4-5a, or at any other position in the region between the nuclei.

On the other hand, if the electron is located at the position indicated in Figure 4-5b, or at any other position outside the region between the two nuclei, then the repulsive force between the nuclei will be larger than the attractive forces. Similar considerations apply to the other electron in a H_2 molecule.

Thus if a H_2 molecule is going to hold together, the probability of finding the electrons in the region between the two nuclei must be greater than the probability of finding the electrons outside this region. We can describe this situation, in which the electrons are more likely to be found in the region between the two nuclei, by saying that the electrons are **shared** by the two nuclei. The shared electrons in H_2 are the "glue" that holds H_2 together. In a H_2 molecule, there is also a repulsive force between the two negatively charged electrons (see Figure 4-6). However, even with this additional repulsive force, the shared electrons can keep the positively charged nuclei in a H_2 molecule together.

Figure 4-6   Schematic illustration of the forces in a hydrogen molecule. There are attractive forces between each electron and each nucleus, repulsive forces between the two electrons, and repulsive forces between the two nuclei. The attractive forces are large enough to keep the two atoms bonded together in a hydrogen molecule.

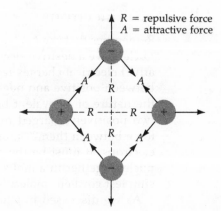

Orbitals in Molecules

According to the quantum mechanical model (Chapter 3), electrons in molecules, as well as in atoms, are described by orbitals. Recall that an orbital is a mathematical expression that gives the probability of finding an electron at each point in space. We did not describe orbitals in atoms in any detail, and we shall not do so for molecules either. But it is quite instructive to consider the qualitative changes that occur in orbitals when two atoms bond together.

Suppose that we consider how a H_2 molecule might be formed: Two hydrogen atoms, which are initially at a very large distance apart, approach one another, and finally come to rest 0.074 nm apart—the bond length in the H_2 molecule (see Figure 4-1). As we discussed in Section 3-7, the electron in an isolated hydrogen atom does not "know" one direction from another. For example, in Figure 4-7a the electron in an isolated hydrogen atom is just as likely to be found 0.05 nm from the nucleus at position (1) as 0.05 nm in the opposite direction, at position (2). Thus the orbital for the electron in an isolated hydrogen atom is "nondirectional." But when the two hydrogen atoms approach one another, this is no longer true. In order for the electrons to bind the nuclei together in a hydrogen molecule, the probability of finding the electron at position (1) must be greater than the probability at position (2) (Figure 4-7b). Thus the orbital that describes the electrons in a hydrogen molecule must result in an increased probability of finding the electrons in the bonding region between the two atoms. An orbital of this type is called a **bonding orbital**.

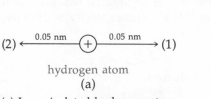

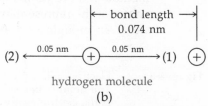

hydrogen atom

(a)

hydrogen molecule

(b)

Figure 4-7   (a) In an isolated hydrogen atom the electron is just as likely to be to the right of the nucleus (plus sign), at (1), as to the left, at (2). (b) In a hydrogen molecule, the electrons occupy a bonding orbital, in which an electron is more likely to be found 0.05 nm from a given nucleus in the direction of the other nucleus, at (1), than in the opposite direction, at (2).

Now, remember that in Section 3-8 we described one general principle of quantum mechanics, the Pauli exclusion principle, one result of which is that a maximum of two electrons can occupy any one orbital. The two electrons in a H_2 molecule occupy the same bonding orbital. Two electrons occupying a bonding orbital (or, equivalently, a shared pair of electrons) describes what chemists usually mean by a **covalent bond**.

It is important to remember that when one atom bonds to another, it is the *valence electrons* that are principally involved in the bonding (Section 3-8). Each atom in a molecule holds on to its inner-core electrons so tightly that these electrons play only a very minor role in molecular bonding.

Exercise 4-5
Describe the covalent bond in a chlorine molecule, Cl_2.

4-8   ELECTRONEGATIVITY

It is quite useful to have a measure of the relative ability of an atom to attract the shared pair of electrons in a covalent bond. Chemists call this relative ability of an atom to attract electrons the atom's **electronegativity,** abbreviated EN. For any two bonded atoms, the atom with the higher electronegativity value pulls the shared pair of electrons more toward itself. One method of assigning electronegativity values to atoms makes use of the atom's ionization energy and its electron affinity (Section 3-6). For a bonded atom to have a large ability to attract the shared pair of electrons in the bond, the atom must:

1. Hold onto its own electrons very tightly, so that the other atom to which it is bonded cannot pull them away—the atom must have a large ionization energy; and

2. Have a large attraction for one or more additional electrons—the atom must have a large electron affinity.

For these reasons, one method of assigning an electronegativity value to an atom is to let the atom's electronegativity be directly proportional to the sum of the atom's ionization energy (IE) and its electron affinity (EA); that is, EN is directly proportional to IE + EA. On this basis, atoms with a large IE and a large EA have large electronegativity values. Chemists use this method and several others to assign electronegativity values to atoms. For any particular atom, there is only a slight difference between the EN value assigned using one method and the value assigned using any other method. The electronegativity values of the representative elements in the first five rows of the periodic table are given in Figure 4-8. Note that electronegativity is dimensionless.

Figure 4-8   Electronegativity values. The number beneath each element is the electronegativity of that element, and the number above is the atomic number. Electronegativity values increase going across a row in the periodic table and tend to decrease going down a column within the same group.

						1 H 2.1

3 Li 1.0	4 Be 1.5	5 B 2.0	6 C 2.5	7 N 3.0	8 O 3.5	9 F 4.0
11 Na 0.9	12 Mg 1.2	13 Al 1.5	14 Si 1.8	15 P 2.1	16 S 2.5	17 Cl 3.0
19 K 0.8	20 Ca 1.0	31 Ga 1.6	32 Ge 1.8	33 As 2.0	34 Se 2.4	35 Br 2.8
37 Rb 0.8	38 Sr 1.0	49 In 1.7	50 Sn 1.8	51 Sb 1.9	52 Te 2.1	53 I 2.5

Notice that nonmetals have large electronegativity values, whereas metals have small electronegativity values. In particular, note the large electronegativity values of the elements nitrogen, oxygen, and fluorine. Also, there is a general trend in the electronegativity of the elements in the periodic table: Electronegativity values generally increase along a row, and decrease down a column. You should recognize that this general trend is consistent with the general trend of the ionization energy of the elements and the position of metals and nonmetals in the periodic table (Section 3-6).

In a bond between two atoms, A and B, the degree to which the sharing of the electrons is unequal is related directly to the difference in electronegativity values, EN(B) − EN(A), of atoms A and B. This is why electronegativity values are so useful. Recall that the Greek letter delta, Δ, is often used to represent a difference. Thus we can represent the difference in electronegativity values, EN(B) − EN(A), by the symbol ΔEN. For any two atoms, A and B, the *larger* the ΔEN, the *more unequal* is the sharing of electrons in a bond between them (see Figure 4-9). Also, in the bond between atoms A and B, the shared electrons will be pulled toward the atom with the larger electronegativity value. The

slight shift of
bonding electrons

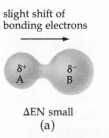

ΔEN small
(a)

large shift of
bonding electrons

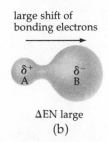

ΔEN large
(b)

Figure 4-9

(a) Schematic illustration of the slight shift in the bonding electrons toward the more electronegative element B when the difference in electronegativities, ΔEN, is small. This slight shift produces a small partial negative charge, δ⁻, on atom B and an equally small partial

positive charge, δ⁺, on atom A. (b) When ΔEN is large, the large shift in the bonding electrons toward the more electronegative element B produces a large partial negative charge on B and an equally large partial positive charge on A.

bonding electrons are shared completely equally when the atoms in the bond are identical. In this case ΔEN = 0. Whenever the bonded atoms are different there is some degree of unequal sharing, and we can use the computed value of ΔEN as a measure of this inequality. Let us consider a few examples.

What is the ΔEN for the ionic bond in a binary ionic compound, such as sodium chloride? The electronegativity of sodium is 0.9 and that of chlorine is 3.0 (see Figure 4-8). Thus ΔEN for the bond in a sodium chloride molecule in the gas state is 2.1. Now, this value of 2.1 is quite large. If you compare the electronegativity values in Figure 4-8, you will see that the largest possible value for ΔEN is 3.2. Thus a ΔEN value of 2.1 for a sodium chloride bond is consistent with a bond in which the electrons are shared very unequally. In fact, as we have discussed before (page 108), the sharing is so unequal that we can consider the shared electron to be completely pulled away from the sodium atom and completely acquired by the chlorine atom.

How large must ΔEN be in order for us to consider that the bond between atoms A and B is ionic? In other words, how large is ΔEN for those bonds where the bonding electrons are pulled completely away from the atom of lower electronegativity and attached to the atom with the far larger electronegativity? The answer to this question is somewhat arbitrary. For most binary ionic compounds (those binary compounds with distinctly ionic properties), ΔEN is larger than 2.0. Therefore, as a rule of thumb, chemists consider that a bond is ionic if ΔEN is 2.0 or larger, and covalent if ΔEN is less than 2.0.

As another example, consider the molecule

H—Cl

Chlorine is more electronegative than hydrogen, and ΔEN is 0.9 for this bond. The shared pair of electrons is pulled more toward the chlorine atom and away from the hydrogen atom, but the inequality of the sharing is not extreme so we do not consider this bond to be ionic. There is some shift of shared electrons, however, toward the chlorine atom, which results in a partial negative charge on the chlorine atom and a partial positive charge on the hydrogen atom. If we represent a partial negative charge by δ⁻ and a partial positive charge by δ⁺, this shift of shared electrons in a H—Cl molecule can be indicated by

δ⁺ δ⁻
H—Cl

Bonds in which there is a significant degree of unequal sharing, but not enough to be called ionic (ΔEN between 0.4 and 2.0), are called **polar covalent bonds.** The bond in HCl is an example. Figure 4-10 indicates how chemists classify a bond as nonpolar covalent, polar covalent, or ionic on the basis of the difference in the electronegativity values, ΔEN, for the atoms in the bond.

Bond Type	Examples		
Ionic	Na^+Cl^-	K^+Br^-	$Ca^{2+}O^{2-}$
Polar covalent	$\overset{\delta^+\ \ \delta^-}{H—Cl}$	$\overset{\delta^+\ \ \delta^-}{N—O}$	$\overset{\delta^+\ \ \delta^-}{H—O}$
Nonpolar covalent	H—H	F—F	C—H

Figure 4-10   A bond is classified as nonpolar covalent if ΔEN for the atoms in the bond is between 0 and 0.4; the bond is classified as polar covalent if ΔEN is greater than 0.4 but less than 2.0; and the bond is classified as ionic if ΔEN is 2.0 or larger.

Electronegativity is a very useful concept when discussing the properties of covalent bonds, and we shall use it often in this text. One word of caution, however: Remember that the electronegativity is a measure of the *relative* ability of an atom to attract and hold onto electrons. Thus it is the *difference* in electronegativity values, ΔEN, and not the electronegativity of the atoms in a bond, that determines how the electrons in the bond are shared. Fluorine, for example, is the element with the highest electronegativity. But this does not mean that all bonds involving fluorine are ionic. In the fluorine molecule, F_2, for example, the electrons are shared equally between the two fluorine atoms, ΔEN is zero, and the bond is nonpolar and covalent.

Exercise 4-6
Consider the following bonds: C—H, C—O, S—H, N—O, Na—F, N—C. (a) Using electronegativities, arrange these bonds in a series going from most equal sharing to most unequal sharing of bonding electrons. (b) For each bond, determine which atom pulls the bonding electrons more toward itself. (c) Classify each bond as being ionic, polar covalent, or nonpolar covalent.

4-9   STRUCTURAL FORMULAS AND THE OCTET RULE

The most basic feature of a molecule, besides its molecular formula, is the way in which the atoms in the molecule are bonded together. The position of the centers of the atoms (the nuclei of the atoms) in a molecule can be determined experimentally. When the nuclei of two atoms in a molecule are very close together, there must be a bond between these two atoms.

Structural Formulas

Chemists use pictures called **structural formulas** to indicate the atoms in a molecule that are bonded together by covalent bonds. In a structural formula for a molecule, an atom of an element is represented by the chemical symbol for the element, and a line between the symbols for two elements represents a covalent bond between the two atoms. Structural formulas based on experimental evidence for some simple covalent compounds are given in Figure 4-11. All of the bonding arrangements in Figure 4-11a agree with the following generalization: In covalent compounds, atoms of hydrogen, fluorine, oxygen, nitrogen, and carbon form one, one, two, three, and four covalent bonds, respectively. Now, fluorine, oxygen, nitrogen, and carbon are all elements in the second row of the periodic table. Since we know that elements in the same group have similar chemical properties, we might expect that elements in the same groups as these four elements will bond similarly. Thus, in the third row of the periodic table, we expect to find that chlorine forms one covalent bond, sulfur two bonds, phosphorus three bonds, and silicon four bonds. The structural formulas for the covalent compounds given in Figure 4-11b, as well as for many other compounds, are consistent with this expectation.

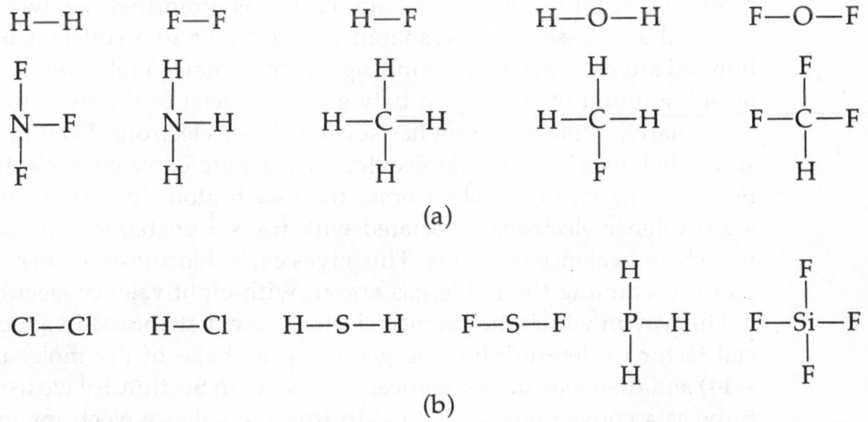

(a)

(b)

Figure 4-11 (a) Structural formulas for some compounds involving hydrogen and the second-row elements fluorine, oxygen, nitrogen, and carbon. (b) Structural formulas for some compounds involving hydrogen, fluorine, and the third-row elements chlorine, sulfur, phosphorus, and silicon.

We can generalize our discussion in the following manner: Usually in covalent compounds (but not always) group VIIA atoms form one covalent bond, group VIA atoms form two covalent bonds, group VA atoms form three covalent bonds, and group IVA atoms form four covalent bonds (see Table 4-2). Most covalent compounds involve elements in groups IVA, VA, VIA, and VIIA. Compounds containing elements in groups IA, IIA, and IIIA are almost always ionic. A few covalent compounds, which we shall not discuss, involve group IA, IIA, or IIIA elements.

Table 4-2 Usual Number of Covalent Bonds in Covalent Compounds

Group	IVA	VA	VIA	VIIA
Covalent bonds	4	3	2	1

The information in Table 4-2 is very useful. In the next section we shall see that it can often be used to predict which atoms in a molecule are bonded together. Note, however, that Table 4-2 gives only the *usual* number of bonds formed. There are numerous exceptions to the generalizations given in this table. For example, sulfur forms three different compounds with fluorine. These compounds have the molecular formulas SF_2, SF_4, and SF_6. In the compound SF_2 the sulfur atom does form two bonds, which is consistent with the generalizations in Table 4-2. But in the compounds SF_4 and SF_6 the sulfur atom forms four and six bonds, respectively.

Lewis Structures and the Octet Rule

There is a simple explanation for the generalizations in Table 4-2. Recall that noble gas atoms have the most stable arrangement of electrons (Chapter 3). How can an atom in a molecule acquire an electron arrangement resembling that of a noble gas atom without forming an ion? One way is by forming covalent bonds and sharing valence electrons with other atoms in the molecule. For example, a hydrogen atom has one electron, but a hydrogen molecule, H_2, has a shared electron pair associated simultaneously with both atoms. Therefore, each hydrogen atom in a hydrogen molecule has an electron arrangement that resembles an atom of helium, the noble gas atom that has two electrons. It is reasonable to associate the shared electron pair in a covalent bond with both bonded atoms, since these bonding electrons have a relatively large probability of being found in the region between the nuclei of the two atoms.

Similarly, a chlorine atom has seven valence electrons (Section 3-8). If we consider that in a chlorine molecule, Cl_2, a pair of valence electrons is shared between the two chlorine atoms, then each atom in a chlorine molecule has eight valence electrons associated with it—six unshared valence electrons and two shared valence electrons. This gives each chlorine atom an electron arrangement resembling the noble gas argon, with eight valence electrons.

The way in which the valence electrons are distributed in a molecule is a crucial factor in determining the geometrical shape of the molecule (see Section 4-14) and the molecule's chemical reactivity. In Section 3-9 we used Lewis structures as a convenient way of exhibiting the valence electrons in atoms. Lewis structures are also used to show the approximate arrangement of the valence electrons in molecules. Lewis structures and structural formulas are very similar. Besides indicating which atoms are bonded together, the dots in a Lewis structure for a molecule also indicate approximately the regions in space where there is a high probability of finding valence electrons. Thus, in a Lewis structure for a molecule, a pair of dots next to an element symbol is used to represent an unshared pair of electrons, whereas a pair of dots or a line between two element symbols is used to represent a shared electron pair, or bond, between two atoms (see Figure 4-12).

Recall that for an atom of a representative element, the number of valence electrons is equal to the group number of that element (Section 3-8). Consider carbon: Since carbon is an element in group IVA, a carbon atom has four valence electrons. Thus the electron arrangement of a carbon atom approaches that

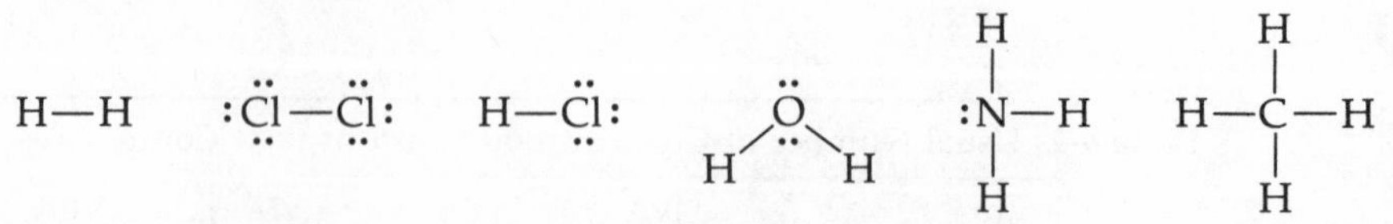

$$H-H \qquad :\ddot{Cl}-\ddot{Cl}: \qquad H-\ddot{Cl}: \qquad \overset{\displaystyle \ddot{O}}{\underset{H \qquad H}{}} \qquad :\overset{\displaystyle H}{\underset{H}{N}}-H \qquad H-\overset{\displaystyle H}{\underset{H}{C}}-H$$

Figure 4-12   Lewis structures for H_2, Cl_2, HCl, H_2O, NH_3, and CH_4.

of the noble gas atom neon when a carbon atom acquires four more electrons by forming four covalent bonds with other atoms. A carbon atom with four covalent bonds has a total of eight shared electrons around it, one pair of shared electrons in each of the four bonds. For example, in the compound methane, which has the molecular formula CH_4, the carbon atom forms a bond with each of the four hydrogen atoms. In this way each atom acquires an electron arrangement similar to a noble gas atom. This is indicated by the Lewis structure for methane (Figure 4-12). In general, a carbon atom in a molecule will form four bonds. Likewise, since atoms of all of the other group IVA elements have four valence electrons, atoms of all group IVA elements usually form four bonds.

Using the same line of reasoning, you should now be able to explain why atoms in group VA usually form three covalent bonds, atoms in group VIA usually form two covalent bonds, whereas atoms in group VIIA usually form only one covalent bond.

The Lewis structures for NH_3 and H_2O are also given in Figure 4-12. Notice that, like the carbon atom in methane and the chlorine atom in Cl_2, the nitrogen atom in NH_3 and the oxygen atom in H_2O also have eight shared and unshared electrons around them. In general, we can consider that the covalent bonding arrangement in molecules involving the representative elements in groups IVA, VA, VIA, and VIIA is a consequence of each atom's tendency to attain a noble gas arrangement of electrons by sharing valence electrons with other atoms.

Since all noble gas atoms except helium have eight valence electrons, it is convenient to refer to the observation that "an atom tends to attain a noble gas arrangement by forming covalent bonds with other atoms" by calling this the **octet rule.** Thus, for most molecules involving the representative elements in groups IVA, VA, VIA, and VIIA, a Lewis structure that satisfies the octet rule provides a good qualitative picture of how the valence electrons are distributed in the molecule. Of course, when we use the octet rule, we must remember that a covalently bonded hydrogen atom has only two and not eight shared electrons around it, and that the octet rule is a useful generalization but one with numerous exceptions.

Covalent bonds hold the atoms in polyatomic ions together, and a simple Lewis structure that is consistent with the octet rule can be drawn for many polyatomic ions. For example, the SO_4^{2-} polyatomic ion has a total of 32 valence electrons (six from each of the four oxygen atoms, six from the sulfur atom, and two additional electrons, since the charge on this ion is -2). You can readily verify that the Lewis structure for SO_4^{2-} in Figure 4-13 is consistent with the octet rule. Note, however, that in this Lewis structure the sulfur and oxygen atoms do not have their usual number of two covalent bonds.

Experimental results have shown that the sulfur atom in SO_4^{2-} is at the center of this polyatomic ion, and for this reason the sulfur atom rather than an oxygen atom is put at the center of the Lewis structure for SO_4^{2-}. This bonding arrangement in SO_4^{2-} is an example of the following generalization: <u>When a molecule or polyatomic ion contains just two elements, A and B, with one atom of element A and more than one atom of element B, the A atom is usually at the center of the molecule or polyatomic ion and all the B atoms are bonded to it.</u>

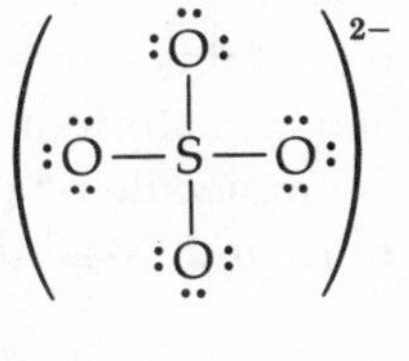

Figure 4-13   Lewis structure for the polyatomic ion SO_4^{2-}.

Exercise 4-7

Which of the following Lewis structures are consistent with the octet rule?

(a)
$$\begin{array}{c} H \\ | \\ :N-H \\ | \\ H \end{array}$$

(b)
$$\begin{array}{c} :\ddot{C}l: \\ | \\ :\ddot{C}l-C-\ddot{C}l: \\ | \\ :\ddot{C}l: \end{array}$$

(c)
$$\begin{array}{c} :\ddot{F}: \quad :\ddot{F}: \\ B \\ | \\ :\ddot{F}: \end{array}$$

(d)
$$\begin{array}{c} \ddot{S} \\ / \quad \backslash \\ :\ddot{F} \qquad \ddot{F}: \end{array}$$

(e)
$$\left[:\ddot{C}l-\ddot{O}:\right]^{-}$$

4-10 PREDICTING STRUCTURAL FORMULAS AND LEWIS STRUCTURES

Given the chemical formula for a covalent compound or a polyatomic ion, we can usually use the generalizations about bonding arrangements that we have discussed to predict the correct structural formula for a molecule of the compound or the ion. We can also use the octet rule to draw a Lewis structure that conveys the appropriate features of the distribution of the valence electrons in the molecule or ion.

Let us consider the example of a water molecule, whose structural formula we already know (Figure 4-11a). By itself, the molecular formula H_2O does not convey the information that the oxygen atom is bonded to each of the hydrogen atoms in a water molecule. The molecular formula, H_2O, is consistent with the incorrect structural formula, H—H—O, as well as the correct structural formula, H—O—H. If we know, however, that a hydrogen atom forms only one covalent bond and an oxygen atom usually forms two covalent bonds, then the only possible bonding arrangement for water that is consistent with the usual number of bonds formed by hydrogen and oxygen is indicated by the structural formula H—O—H.

Since we have the structural formula for H_2O, we can construct a Lewis structure as follows: In a H_2O molecule there are a total of eight valence electrons (one for each hydrogen atom and six for the oxygen atom). In the structural formula H—O—H, four valence electrons are accounted for by the two covalent bonds. We satisfy the octet rule by indicating the remaining four valence electrons as two unshared pairs about the oxygen atom (see Figure 4-12).

Given their molecular formulas, we can use a similar line of reasoning to predict structural formulas and draw Lewis structures for other simple covalent compounds and polyatomic ions. We proceed as follows:

1. We draw a structural formula for the molecule or ion using the following generalizations as a guide.

    (a) In covalent compounds, atoms of the representative elements in groups IVA, VA, VIA, and VIIA usually form four, three, two, and one covalent bonds, respectively; and a hydrogen atom forms one covalent bond.

    (b) When a molecule or polyatomic ion contains two elements, A and B, with one atom of A and more than one atom of B, then usually (1) the A atom is at the center of the molecule or polyatomic ion; and (2) all the B atoms are bonded to it; and (3) all the A—B bonds have identical properties.

2. We then determine the total number of valence electrons in the molecule.

3. Finally, we draw a Lewis structure by distributing these valence electrons consistent with the octet rule.

Let us consider a few examples. The molecular formula for the compound ethane is C_2H_6. As step 1, we can draw the structural formula for ethane (see Table 4-3). There are a total of 14 valence electrons in C_2H_6, and these are all accounted for by the seven covalent bonds. Hence the structural formula for ethane is also a Lewis structure.

Table 4-3 Molecular Formulas, Lewis Structures, and Condensed Structural Formulas for Ethane, Ethylene, and Acetylene

	Ethane	Ethylene	Acetylene
Molecular formula	C_2H_6	C_2H_4	C_2H_2
Number of valence electrons (molecular formula)	$\left(2C \times \dfrac{4e^-}{C}\right)$ $+ \left(6H \times \dfrac{1e^-}{H}\right) = 14$	$\left(2C \times \dfrac{4e^-}{C}\right)$ $+ \left(4H \times \dfrac{1e^-}{H}\right) = 12$	$\left(2C \times \dfrac{4e^-}{C}\right)$ $+ \left(2H \times \dfrac{1e^-}{H}\right) = 10$
Structural formula and Lewis structure	H—C—C—H (with H above and below each C)	C=C (with two H on each C)	H—C≡C—H
Number of valence electrons (Lewis structure)	7 bonds $\times \dfrac{2e^-}{bond} = 14$	6 bonds $\times \dfrac{2e^-}{bond} = 12$	5 bonds $\times \dfrac{2e^-}{bond} = 10$
Condensed structural formula	CH_3CH_3	$CH_2{=}CH_2$	$CH{\equiv}CH$

The compound ethylene, which has the molecular formula C_2H_4, also contains two carbon atoms, but only four hydrogen atoms. Thus in ethylene there are not enough hydrogen atoms available for each carbon atom to form four single bonds. We can, however, draw a structural formula for ethylene in which each carbon atom forms four bonds if there are two bonds between the two carbon atoms (see Table 4-3). Two bonds between a pair of atoms is called a **double bond**. The 12 valence electrons in C_2H_4 are accounted for by the six covalent bonds, and therefore, our structural formula is also a Lewis structure for ethylene.

We can use a similar line of reasoning to draw the structural formula and Lewis structure for the compound acetylene, with the molecular formula C_2H_2. In the structural formula shown in Table 4-3, each carbon atom forms four bonds, one bond to a hydrogen atom and a **triple bond** between the two carbon atoms. As we shall see in Chapter 13, a great many carbon-containing compounds are known, and most have some carbon-hydrogen single bonds. For carbon-containing compounds we often use **condensed structural formulas,** in which the lines indicating carbon-hydrogen bonds are omitted. Examples of condensed structural formulas, in this case for ethane, ethylene, and acetylene, are given in Table 4-3.

As another example, consider carbon dioxide, for which the molecular formula is CO_2. On the basis of step 1 of our general procedure, we can draw a structural formula for CO_2 in which each oxygen atom forms a double bond with the carbon atom (see Table 4-4). In CO_2 there are a total of 16 valence electrons (step 2). Using step 3, since there are eight valence electrons indicated in four bonds of our structural formula, eight valence electrons must be distributed as unshared pairs. There are already eight shared electrons around the carbon atom in our structural formula, so we do not want any more electrons about this atom. We can, however, get an octet of electrons about both oxygen atoms if we put two unshared pairs on each of them. Note that in the resulting Lewis structure for CO_2, shown in Table 4-4, the total number of valence electrons is equal to 16, which is the correct number of valence electrons for CO_2.

Table 4-4 Molecular Formulas, Structural Formulas, and Lewis Structures for Carbon Dioxide and Formaldehyde

	Carbon Dioxide	Formaldehyde
Molecular formula	CO_2	CH_2O
Number of valence electrons (molecular formula)	$\left(1C \times \dfrac{4e^-}{C}\right) +$ $\left(2O \times \dfrac{6e^-}{O}\right) = 16$	$\left(1C \times \dfrac{4e^-}{C}\right) +$ $\left(2H \times \dfrac{1e^-}{H}\right) + \left(1O \times \dfrac{6e^-}{O}\right) = 12$
Structural formula	$O{=}C{=}O$	$\begin{array}{c} O \\ \parallel \\ H-C-H \end{array}$
Number of valence electrons (structural formula)	$4 \text{ bonds} \times \dfrac{2e^-}{\text{bond}} = 8$	$4 \text{ bonds} \times \dfrac{2e^-}{\text{bond}} = 8$
Lewis structure	$\ddot{O}{=}C{=}\ddot{O}$	$\begin{array}{c} :\!\ddot{O}\!: \\ \parallel \\ H-C-H \end{array}$
Number of valence electrons (Lewis structure)	$\left(4 \text{ bonds} \times \dfrac{2e^-}{\text{bond}}\right) +$ $\left(4 \text{ unshared pairs} \times \dfrac{2e^-}{\text{unshared pair}}\right) = 16$	$\left(4 \text{ bonds} \times \dfrac{2e^-}{\text{bond}}\right) +$ $\left(2 \text{ unshared pairs} \times \dfrac{2e^-}{\text{unshared pair}}\right) = 12$

Let us consider one more example. The molecular formula for the compound formaldehyde is CH_2O. If the carbon atom in a formaldehyde molecule is to have four bonds, the oxygen atom two bonds, and the hydrogen atoms each one bond, then the structural formula for formaldehyde must have the carbon atom bonded to each hydrogen atom with a single bond as well as to the oxygen atom with a double bond (see Table 4-4). This structural formula indicates eight valence electrons, and there are 12 valence electrons in CH_2O. Thus, if two unshared pairs are put on the oxygen atom, the resulting Lewis structure, shown in Table 4-4, has 12 valence electrons and each atom satisfies the octet rule.

Exercise 4-8
Draw Lewis structures for the following:
(a) CH_4 (b) CH_2Cl_2 (c) CH_5N (d) CO (e) PO_4^{3-}

4-11 BOND PROPERTIES AND STRUCTURAL FORMULAS

The structural formulas we use to represent bonding arrangements in molecules should be consistent with the actual properties of bonds as determined by experiment. Remember that the most important properties of a bond are (1) its bond energy and (2) its bond length (see Section 4-3). A stronger bond has a larger bond energy and a shorter bond length. Are the structural formulas we have written thus far consistent with the experimental data? Let us see.

In the structural formulas for ethane, ethylene, and acetylene (Table 4-3), carbon-hydrogen single bonds are indicated and these bonds are represented by the same notation, C—H. This is justified because the bond energy and the bond lengths of the carbon-hydrogen bonds in all three compounds are nearly the same. In general, in all molecules containing a particular type of bond, the bond energy and bond length of this type of bond are nearly the same.

In addition to C—H single bonds, the structural formula of ethane has a carbon-carbon single bond, C—C, whereas the structural formulas for ethylene and acetylene have a carbon-carbon double bond, C=C, and a carbon-carbon triple bond, C≡C, respectively. From experimental data, it has been determined that the carbon-carbon bond energies increase on going from ethane to ethylene to acetylene, whereas the bond lengths decrease (see Table 4-5).

Table 4-5 Carbon-Carbon Bond Energies and Bond Lengths

Bond	Bond Energy (eV)	Bond Length (nm)
C—C	3.58	0.154
C=C	6.32	0.134
C≡C	8.65	0.120

Therefore, the structural formulas that we use for the carbon-carbon bond in ethane, C—C, ethylene, C=C, and acetylene, C≡C, are consistent with the relative strengths of these bonds as determined by experiment. The bond energies and bond lengths given in Table 4-5 represent average values of these quantities. Bond energies and bond lengths for the same type of bond do vary by a small amount from one compound to another.

Exercise 4-9

Consider the following Lewis structure for the compound acetic acid, $C_2H_4O_2$, with the carbon-oxygen bonds labeled (1) and (2).

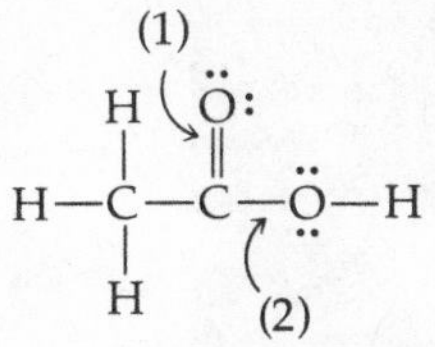

(a) Are bonds (1) and (2) equivalent?
(b) Which bond, (1) or (2), is stronger?
(c) Which bond, (1) or (2), is longer?

4-12   RESONANCE STRUCTURES

For many molecules and polyatomic ions one cannot draw a single Lewis structure that is consistent with both the octet rule and the known experimental facts about the molecule or ion. For example, experimental results show that in a molecule of SO_2 the sulfur atom is at the center of the molecule and that the bonds between the sulfur atom and each oxygen atom have exactly the same bond energy, bond length, and other properties. The Lewis structure for SO_2 in Figure 4-14a is consistent with the octet rule, but this picture implies that the sulfur-oxygen bond on the right is stronger than the bond on the left. This implication conflicts with the experimental facts. The American chemist and Nobel Prize laureate, Linus Pauling, developed a method of extending the use of Lewis structures so that they could be used for SO_2 and other molecules for which a single Lewis structure is inadequate. For SO_2 this extension uses a composite picture involving two Lewis structures, as shown in Figure 4-14b.

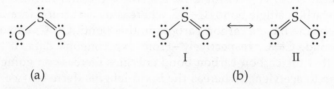

(a)   (b)

Figure 4-14   (a) A single Lewis structure implies that the S—O bond on the right is stronger than the S—O bond on the left, which is in conflict with the experimental facts.

(b) A composite picture involves two resonance structures and implies that the S—O bonds are equivalent, which is in agreement with the experimental facts.

The double arrow between the Lewis structures I and II indicates that the actual arrangement of valence electrons in SO_2 is intermediate between the arrangement implied by structure I and the arrangement implied by structure II. In this sense, Figure 4-14b conveys the information (1) that the sulfur-oxygen bonds in SO_2 have the same properties and (2) that each bond is intermediate between a single bond and a double bond. This composite picture is not meant to imply that some of the valence electrons are actually hopping back and forth continually from one side of the SO_2 molecule to the other.

Chemists use the term **resonance structures** to describe a picture like Figure 4-14b in which more than one Lewis structure is used to represent the approximate arrangement of valence electrons in a molecule or ion. Resonance struc-

Linus Pauling, recipient of the Nobel Prize in chemistry in 1954, also received the Nobel Peace Prize in 1962 for his efforts to halt atmospheric testing of nuclear weapons.

tures involving two, three, four, or even more individual Lewis structures are used for a molecule or ion when a single Lewis structure is not adequate.

Benzene, C_6H_6, is another compound for which resonance structures are used. The six carbon atoms in a benzene molecule are bonded together and form a regular hexagonal (six-sided) ring. A hexagonal benzene ring unit is part of many biologically and industrially important molecules. All the carbon-carbon bonds in benzene have identical properties, as do all the carbon-hydrogen bonds. A Lewis structure representation consistent with these experimental facts and involving two resonance structures is given in Figure 4-15.

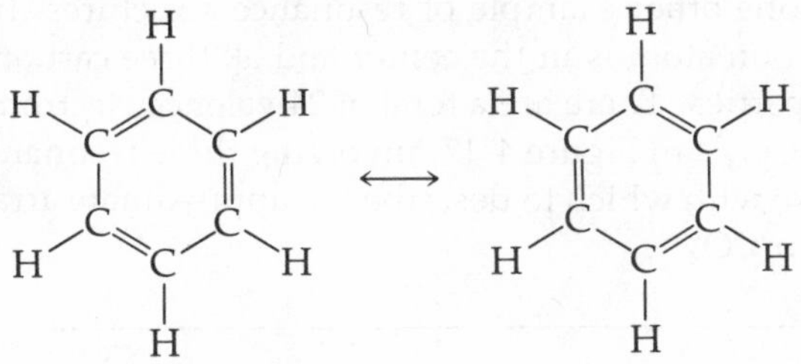

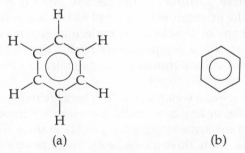

I	II

Figure 4-15   A Lewis structure representation for benzene, involving two resonance structures. This composite picture implies that all six carbon-carbon bonds are equivalent.

Note that both resonance structures I and II in Figure 4-15 are consistent with the octet rule and that each carbon atom has its usual number of four covalent bonds and each hydrogen atom has its usual one bond. These resonance structures imply that all the carbon-carbon bonds are equivalent (in agreement with the experimental facts) and that each carbon-carbon bond in benzene is intermediate between a carbon-carbon single bond and a carbon-carbon double bond. Experiments have shown that in benzene (1) the carbon-carbon bond length is 0.139 nm, which is intermediate between the length of a carbon-carbon single bond (0.154 nm) and a carbon-carbon double bond (0.134 nm) (Table 4.5); and (2) the bond energy of the carbon-carbon bond in benzene is 5.2 eV, which is intermediate between the bond energy of a carbon-carbon single bond (3.58 eV) and that of a carbon-carbon double bond (6.32 eV).

In the composite picture for benzene in Figure 4-15, three bonds are in different positions in the two resonance structures. Chemists often use the picture in Figure 4-16a as an equivalent representation for the resonance structure of benzene. The circle inside the hexagonal ring signifies that there are three bonds involving all six carbon atoms. For simplicity, the skeleton symbol in Figure 4-16b is also used to represent a benzene molecule.

(a)	(b)

Figure 4-16   (a) A simplified representation for benzene. The circle inside the hexagonal ring signifies that there are three additional bonds involving all six carbon atoms. (b) Skeleton representation for benzene. Visualize a carbon atom at the intersection of each of the six straight lines and that each carbon atom has a hydrogen atom bonded to it.

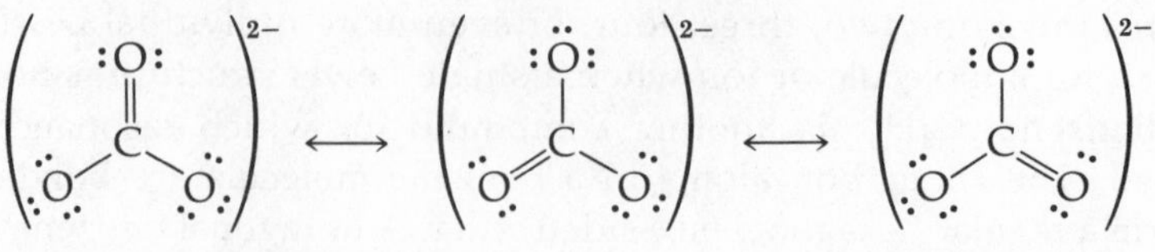

Figure 4-17   Three resonance structures for the carbonate ion, $CO_3{}^{2-}$. The composite picture involving these Lewis structures implies that all three carbon-oxygen bonds are equivalent, which is in agreement with the experimental facts.

Let us consider one other example of resonance structures. In the carbonate ion, $CO_3{}^{2-}$, the carbon atom is in the center and all three carbon-oxygen bonds have identical properties. There are a total of 24 valence electrons in $CO_3{}^{2-}$. The representation for $CO_3{}^{2-}$ in Figure 4-17, involving three resonance structures, is a reasonable picture with which to describe the approximate arrangement of the valence electrons in $CO_3{}^{2-}$.

Exercise 4-10

Draw Lewis structures involving resonance for: (a) SO_3; (b) $NO_2{}^-$; (c) $NO_3{}^-$; and (d) $HCO_3{}^-$, where the hydrogen atom is bonded to one of the oxygen atoms.

4-13   THE REPRESENTATION AND SHAPE OF MOLECULES AND POLYATOMIC IONS

We have used various symbols and pictures to represent molecules, and it is timely to summarize the information that each of these various representations is designed to convey.

As we discussed in Section 4-10, a **molecular formula** merely indicates the number of atoms of each element in the molecule. A **structural formula** for a molecule shows which atoms in the molecule are bonded together, and a **Lewis structure** is a structural formula that gives a crude picture of the position of the valence electrons as well. Once we have learned something about a compound, we should naturally bear in mind what we already know about the compound when we look at any particular picture or symbol for it. For example, when we now see the molecular formula H_2O for the compound water, we should also think of the structural formula H—O—H, in which the oxygen atom is bonded to each of the two hydrogen atoms.

An extremely important feature of a molecule is its shape. The **shape** of a molecule refers to the relative positions of the centers (nuclei) of all the atoms; equivalently, it means the orientation in space of all the bonds in the molecule. Another term for the shape of a molecule is **molecular geometry.** We shall see that the chemical properties of a compound are related directly to the shape of its molecules. This is particularly true of the large molecules involved in most chemical reactions of biological interest.

Let us consider the shape of a water molecule. There are two possible kinds of shape for a H_2O molecule, or any other molecule with only three atoms: either (1) all three nuclei lie on the same straight line, or (2) all three nuclei do not lie on the same straight line. If the three atoms lie on the same straight line we say that the molecule is **linear;** if the three atoms do not lie on the same straight line we say that the molecule is **bent.**

Experimental observations have shown that the shape of a H_2O molecule is bent. We can represent this bent shape of a H_2O molecule by

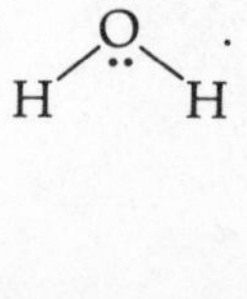

The angle between the two H—O bonds in a H_2O molecule is 105°, and this angle is referred to as the H—O—H bond angle (see Figure 4-18a). In general, whenever one atom in a molecule forms bonds with two other atoms, the angle between the two bonds is called a **bond angle** (see Figure 4-18b).

Figure 4-18   (a) The angle between the hydrogen-oxygen bonds in a water molecule is 105°. (b) The angle between the A—B and B—C bonds is called the A—B—C bond angle.

(a)   (b)

Molecules of most compounds contain more than three atoms, and consequently they have more complicated shapes than that of a water molecule. In particular, for molecules of most compounds, all the atomic nuclei do not lie in the same plane. As a result we need perspective drawings, or better still, three-dimensional models to indicate the shape of most molecules. The shape of most molecules is *not* conveyed by their structural formulas alone. Thus, when you see a structural formula or a Lewis structure, you should not think that these are representations of a molecule's shape.

Exercise 4-11
In ammonia, NH_3, the hydrogen atoms are bonded to the central nitrogen atom and all three N—H bonds are equivalent. What are the possible geometric shapes for a molecule of ammonia?

Optional

4-14   PREDICTING THE SHAPE OF A MOLECULE OR POLYATOMIC ION

In general, the shape of a molecule is determined from experimental observations. It is possible, however, to predict the shape of many simple molecules by using the following principle: The shared electron pairs (bonds) and unshared electron pairs surrounding an atom in a molecule are oriented in space so as to be as far apart from one another as possible. This principle is reasonable because the repulsion between different electron pairs is smallest when the pairs are as far away from one another as possible. It is easy to predict the shape of a molecule using this principle if we have an adequate Lewis structure for the molecule. In this section we shall discuss only molecules with a central atom. We shall begin by considering the shape of a few molecules in which there are no unshared pairs about the central atom.

In a molecule of beryllium dichloride, $BeCl_2$, the central Be atom is bonded to each Cl atom (see Figure 4-19). The Be—Cl bonds are farthest apart when the nuclei of all three atoms are on the same line, with an angle of 180° between the two Be—Cl bonds. Thus, on the basis of our principle we predict that $BeCl_2$ has a **linear shape.** Our prediction agrees with the experimental evidence.

Figure 4-19   The two Be—Cl bonds are as far apart as possible in the linear shape of $BeCl_2$. Note that there is not an octet of electrons on the central Be atom.

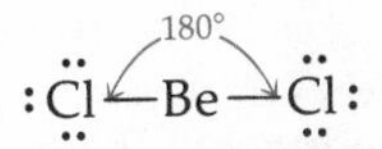

Boron trifluoride, BF_3, is a molecule with a central B atom bonded to each of three F atoms (see Figure 4-20). The three B—F bonds will be farthest apart if (1) all three bonds and all four atoms are in the same plane, and (2) the bonds are directed toward the corners of an equilateral triangle. In this **triangular planar shape** the bond angle between any pair of B—F bonds is 120°. We therefore predict that BF_3 has a triangular planar shape; our prediction agrees with the experimental observations.

Figure 4-20   The three B—F bonds are as far apart as possible in the triangular planar shape of boron trifluoride, BF_3. Note that there is not an octet of electrons on the central B atom.

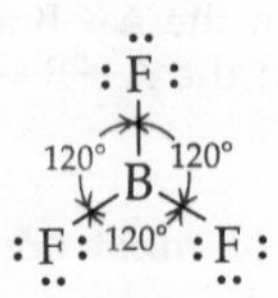

Methane, CH_4, is an example of a molecule in which the central atom forms four bonds (see Figure 4-21). It is a geometric fact that four lines radiating from a central point are farthest apart when the central point is at the center of a regular tetrahedron and the four lines are oriented toward the four corners of the tetrahedron. The angle between any two lines in such a figure is 109.5°. On this basis we expect that CH_4 has a **tetrahedral shape,** with the carbon atom in the center, and that the bond angle between any two C—H bonds is 109.5°. These predictions agree with the experimental data for methane.

Figure 4-21   The four C—H bonds in methane, CH_4, are as far apart as possible in a tetrahedral shape.

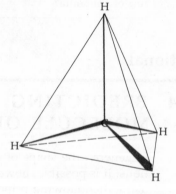

We can predict that for any molecule that has a central atom with no unshared pairs on it, the shape will be linear, triangular planar, or tetrahedral, depending on whether the central atom is bonded to two, three, or four other atoms.

When we apply our shape-predicting principle to molecules for which the Lewis structure contains a double or triple bond about the central atom, we consider that all the electrons in a multiple bond are oriented in the same direction in space. This is reasonable because every bond in the double or triple bond is directed toward the same atom. For example, we predict that carbon dioxide, CO_2, with two double bonds about the central carbon atom, is linear (see Figure 4-22) and that the carbonate ion, CO_3^{2-}, with two single bonds and one double bond about the central carbon atom in each of its three resonance structures, has a triangular planar shape (see Figure 4-23). The experimentally demonstrated shape of CO_2 is linear and that of CO_3^{2-} is triangular planar.

Figure 4-22   Carbon dioxide, with two double bonds about the central carbon atom, has a linear shape.

$$\overset{..}{O} = C = \overset{..}{O}$$

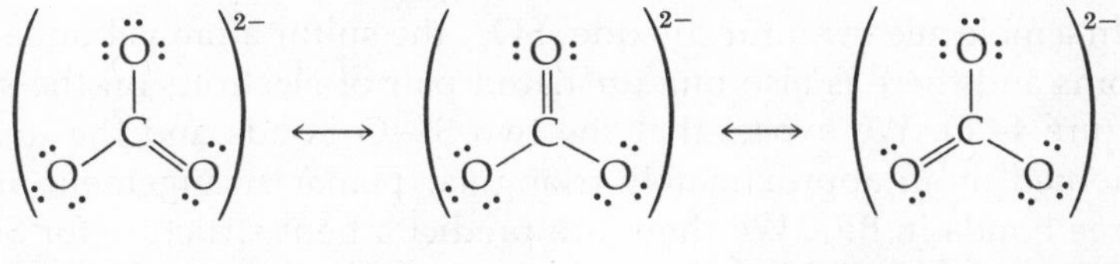

$$\left(\begin{matrix} :\ddot{O}: \\ :\ddot{O} \overset{..}{\underset{..}{C}} \overset{..}{O}: \end{matrix}\right)^{2-} \longleftrightarrow \left(\begin{matrix} :O: \\ \| \\ :\ddot{O} \overset{..}{C} \overset{..}{O}: \end{matrix}\right)^{2-} \longleftrightarrow \left(\begin{matrix} :\ddot{O}: \\ \ddot{O} \overset{..}{\underset{..}{C}} \overset{..}{O}: \end{matrix}\right)^{2-}$$

Figure 4-23   The carbonate ion, $CO_3{}^{2-}$, with   bonds about the central carbon
three equivalent carbon-oxygen   atom, has a triangular planar shape.

Recall that the pair of electrons in a bond is described by an orbital that gives the electrons a relatively high probability of being in the region between the two bonded atoms. An unshared electron pair on an atom is described by an orbital that gives the unshared pair a relatively high probability of being in a region of space directed away from the bonding orbitals of the atom. In this sense we speak of the **orientation in space** of an unshared pair of electrons. Thus, for example, we expect that the three bonds and one unshared pair of electrons about the nitrogen atom in ammonia, NH_3, are oriented approximately toward the corners of a regular tetrahedron in a manner analogous to the orientation of the four bonds about the carbon atom in methane (see Figure 4-24). However, the shape of NH_3 is *not* tetrahedral. Recall that the shape of a molecule is determined by the position of the nuclei of the atoms in the molecule. Thus, according to Figure 4-24, the geometric figure formed by the N atom and the three H atoms is a pyramid. We therefore predict that the NH_3 has a **pyramidal shape.** From experimental evidence, NH_3 does have a pyramidal shape, with a bond angle of 107° between any pair of N—H bonds.

Figure 4-24   A molecule of ammonia, NH_3, with three nitrogen-hydrogen bonds and one unshared pair of electrons about the central nitrogen atom, has a pyramidal shape.

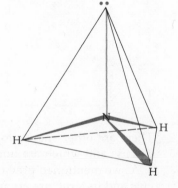

If a molecule has a central atom bonded to two other atoms and there are two unshared electron pairs on the central atom, such as in H_2O (see Figure 4-25), we expect the two bonds and two unshared pairs to be oriented tetrahedrally and that the molecule has a **bent shape.**

Figure 4-25 H_2O, with two hydrogen-oxygen bonds and two unshared pairs of electrons on the central oxygen atom, has a bent shape.

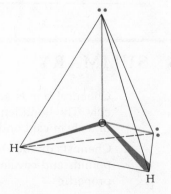

In a molecule of sulfur dioxide, SO_2, the sulfur atom is bonded to the oxygen atoms and there is also one unshared pair of electrons on the sulfur atom (see Figure 4-26). We expect that the two S—O bonds and the unshared pair are oriented in an approximately triangular planar arrangement analogous to the three bonds in BF_3. We therefore predict a bent structure for SO_2; our prediction and the experimental data agree.

Figure 4-26 SO_2, with two equivalent sulfur-oxygen bonds and one unshared pair of electrons on the central sulfur atom, has a bent shape.

Table 4-6 summarizes the results of our discussion. We can predict the shape of many simple molecules by using this table as follows:

1. Draw a Lewis structure for the molecule.
2. Determine the number of bonded atoms and the number of unshared electron pairs around the central atom.
3. Use Table 4-6 to predict the shape of the molecule.

Table 4-6 Predicting the Shape of a Molecule or Polyatomic Ion

Number of Atoms Bonded to the Central Atom	Number of Unshared Electron Pairs Around the Central Atom	Shape	Examples
4	0	Tetrahedral	CH_4, $SiCl_4$, SO_4^{2-}
3	0	Triangular planar	BF_3, CO_3^{2-}, SO_3
3	1	Pyramidal	NH_3, PCl_3, BrO_3^-
2	0	Linear	$BeCl_2$, CO_2
2	1	Bent	SO_2, NO_2^-
2	2	Bent	H_2O, ClO_2^-

There are other shapes for molecules with a central atom which we shall not discuss and which are not listed in Table 4-6.

As we mentioned previously, the chemical and physical properties of a molecule and its structure are intimately linked. We shall discuss the shapes of more complicated molecules in appropriate places throughout this text.

Exercise 4-12
Predict the shape of the following: (a) CCl_4 (b) NO_3^- (c) H_2S (d) SO_3^{2-}

4-15 SUMMARY

1. Chemical bonds are the forces that hold two atoms together in a molecule. The bond length and bond energy are the characteristic properties of a chemical bond.

2. Chemists divide chemical compounds into two broad classes, ionic compounds and covalent compounds, based on their chemical and physical properties.

3. An ionic bond involves cations and anions held close together by the electrical force of attraction between opposite charges.

4. A covalent bond involves two electrons occupying a bonding orbital—a shared electron pair.

5. The relative ability of an atom to attract and hold onto electrons is called the atom's electronegativity.

6. Chemists classify a bond between two atoms as ionic, polar covalent, or nonpolar covalent on the basis of the difference in electronegativity, ΔEN, for the atoms in the bond.

7. A structural formula for a molecule indicates the bonding arrangement in the molecule. A Lewis structure, in addition to showing the bonding arrangement, indicates the approximate region in space where there is a high probability of finding the unshared valence electrons.

8. The generalization that the covalent bonding arrangement in many molecules is a consequence of each atom's tendency to attain a noble gas arrangement of valence electrons by sharing valence electrons with other atoms is called the octet rule.

9. Resonance structures are a composite picture involving two or more Lewis structures.

10. The shape of a molecule, or molecular geometry, refers to the position of the centers of all the atoms in the molecule.

11. **(Optional)** Molecular shapes are determined experimentally. It is often possible to predict the shape of a simple molecule or a polyatomic ion.

PROBLEMS

1. If a compound is an electrolyte, which of the following can you conclude?
   (a) The compound is definitely an ionic compound.
   (b) The compound is definitely a covalent compound.
   (c) The compound is most likely ionic, but it may be a covalent compound.
   (d) The compound is most likely covalent, but it may be an ionic compound.

2. Predict whether the following compounds are ionic or covalent. For the ionic compounds, predict the ions that are present when the compound is dissolved in water: (a) K_2CO_3; (b) N_2O_3; (c) H_2S; (d) $Ba(OH)_2$; (e) Na_2S; (f) PCl_5; (g) $(NH_4)_2SO_4$.

3. Name the following compounds: (a) N_2O_3; (b) PbS; (c) $CuBr_2$; (d) As_2S_5; (e) $Fe_2(SO_4)_3$; (f) P_2I_4; (g) $Cu(HCO_3)_2$; (h) $Mg_3(PO_4)_2$.

4. Write chemical formulas for (a) strontium bromide; (b) ammonium nitrate; (c) tetraphosphorus triselenide; (d) iron(II) iodide; (e) lead(II) carbonate; (f) oxygen difluoride; (g) potassium monohydrogen phosphate.

5. Which of the following atoms has the largest electronegativity: O, N, C, S, P?

6. Classify the following bonds as ionic, polar covalent, or nonpolar covalent: (a) C—P; (b) O—H; (c) Ba—Cl; (d) Cl—Cl; (e) N—H; (f) P—O; (g) Li—F.

7. Draw Lewis structures for the following: (a) HCN; (b) OH^-; (c) C_2Cl_4; (d) BrO_3^-; (e) N_2H_4; (f) H_2CO_3 (the hydrogens are bonded to oxygen atoms); (g) ClO_4^-.

8. Draw a Lewis structure representation involving resonance structures for the following: (a) ozone, O_3; (b) BF_3; (c) $H_2PO_3^-$ (one of the hydrogen atoms is bonded to the phosphorus atom and the other is bonded to one of the oxygen atoms).

9. Which of the following diatomic molecules has the strongest bond: O_2, N_2, F_2, H_2, Br_2?

10. The Lewis structures for ethyl alcohol and formaldehyde are given below. On this basis, predict whether the carbon-oxygen bond energy in ethyl alcohol is larger than, smaller than, or the same, as the carbon-oxygen bond energy in formaldehyde and also predict whether the carbon-oxygen bond length in ethyl alcohol is larger than, smaller than, or the same as the carbon-oxygen bond length in formaldehyde.

$$
\begin{array}{cc}
\text{H} \quad \text{H} & \qquad :\!\overset{..}{O}\!: \\
| \quad | & \qquad \| \\
\text{H—C—C—}\overset{..}{\underset{..}{O}}\text{—H} & \quad \text{H—C—H} \\
| \quad | & \\
\text{H} \quad \text{H} & \\
\text{ethyl alcohol} & \text{formaldehyde}
\end{array}
$$

11. **(Optional)** Predict the shape of the following: (a) PH_3; (b) O_3; (c) SiO_4^{4-}; (d) $SnCl_3^-$; (e) ClO_4^-; (f) ClO_3^-; (g) HCN.

SOLUTIONS TO EXERCISES

4-1 The bond energy is a measure of the bond strength. The bond in O_2 and in N_2 is stronger than the bond in H_2.

4-2 Ionic: $MgBr_2 \rightarrow Mg^{2+} + 2Br^-$ $\qquad$ $NaHCO_3 \rightarrow Na^+ + HCO_3^-$
Covalent: SO_3, NO, CO_2

4-3 Calcium oxide, CaO; potassium carbonate, K_2CO_3; sodium monohydrogen phosphate, Na_2HPO_4; ammonium sulfate, $(NH_4)_2SO_4$; aluminum nitrate, $Al(NO_3)_3$.

4-4 Potassium bromide: $\qquad$ K^+ $\quad :\overset{..}{\underset{..}{Br}}:^-$ $\left.\rule{0cm}{0.8cm}\right\}$ electrical force of attraction between

Calcium fluoride: $\quad :\overset{..}{\underset{..}{F}}:^-$ Ca^{2+} $\quad :\overset{..}{\underset{..}{F}}:^-$ oppositely charged ions

4-5 The covalent bond in Cl_2 involves the sharing of a pair of valence electrons between the two Cl atoms.

4-6 The degree of unequal sharing depends on ΔEN. The atom with the largest electronegativity in the bond pulls the bonding electrons toward itself. (See Figure 4-10.)

Bond	ΔEN	More Electronegative Atom	Type
C—H	0.4	C	Nonpolar covalent
S—H	0.4	S	Nonpolar covalent
N—C	0.5	N	Polar covalent
N—O	0.5	O	Polar covalent
C—O	1.0	O	Polar covalent
Na—F	3.1	F	Ionic

4-7 (a), (b), (d), (e) are consistent; (c) is inconsistent (there is no octet around the B atom).

4-8 (a) $\quad \begin{array}{c} \text{H} \\ | \\ \text{H—C—H} \\ | \\ \text{H} \end{array}$ $\qquad$ (b) $\begin{array}{c} \text{H} \\ | \\ \text{H—C—}\overset{..}{\underset{..}{Cl}}: \\ | \\ :\overset{..}{\underset{..}{Cl}}: \end{array}$

(c) To agree with the usual number of covalent bonds, the carbon and nitrogen atoms must be bonded together and the carbon bonded to three H atoms and the nitrogen to two H atoms:

$$
\begin{array}{c}
\text{H} \quad \text{H} \\
| \quad | \\
\text{H—C—N}: \\
| \quad | \\
\text{H} \quad \text{H}
\end{array}
$$

(d) :C≡O: With a triple bond, both the C and O atoms obey the octet rule.

(e) The different atom P is the central atom with all the O atoms bonded to it. The total number of valence electrons is 32 (there are three additional valence electrons because of the charge -3):

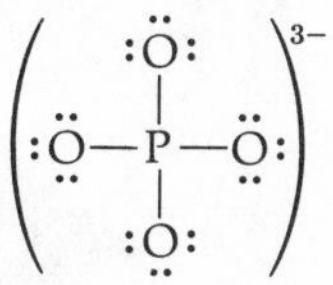

4-9 Bonds (1) and (2) are not equivalent. The C═O double bond (1) is stronger and shorter than the C—O single bond (2).

4-10

(a)

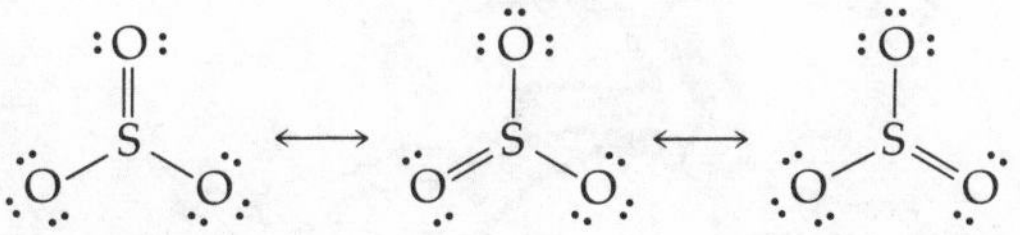

(b)

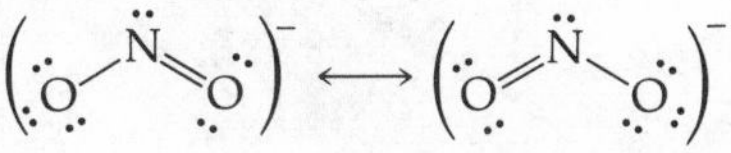

(c)

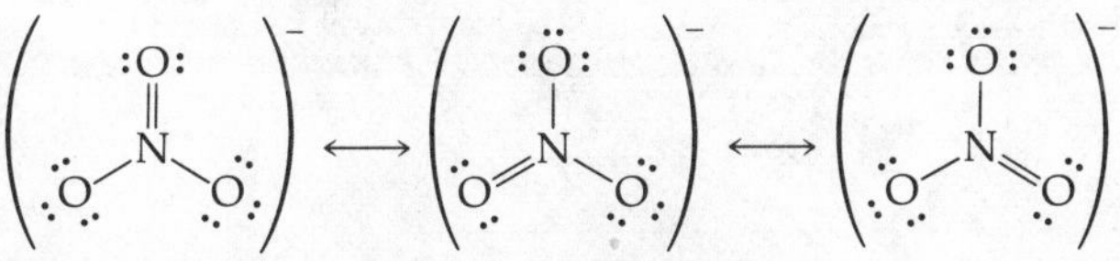

(d) Compare with Figure 4-17 for CO_3^{2-}.

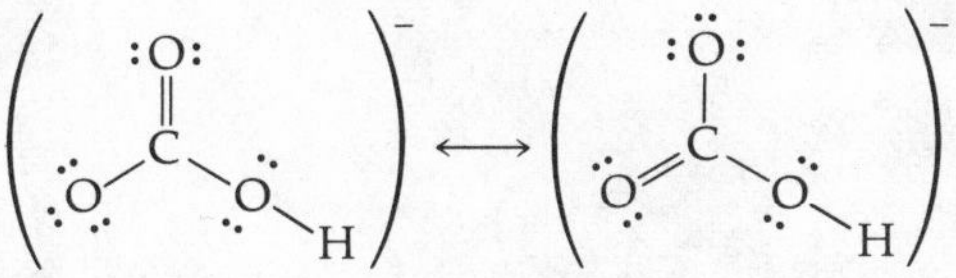

4-11 There are two possible shapes: (1) a triangular planar arrangement with all three bonds and all four atoms in the same plane and the bonds directed toward the corners of an equilateral triangle; or (2) a pyramidal shape with the three hydrogen atoms in a plane and the nitrogen atom above the plane of the hydrogen atoms at the apex of a pyramid. (The actual shape of an ammonia molecule is pyramidal.)

4-12 **(Optional)**

(a)

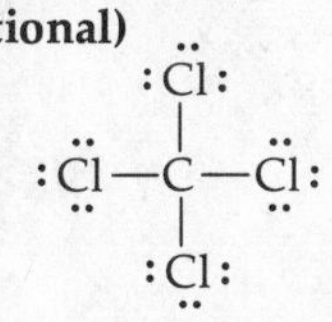

Four bonds and no unshared pairs on central carbon atom. Shape: tetrahedral.

(b) See Exercise 4-10c. Three bonded atoms and no unshared pairs on central nitrogen atom. Shape: triangular planar.

(c)

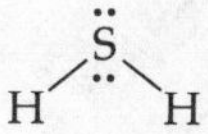

Two bonded atoms and two unshared pairs on central sulfur atom. Shape: bent.

(d)

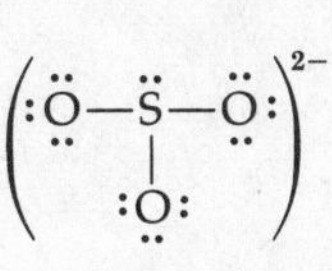

Three bonded atoms and one unshared pair on the central sulfur atom. Shape: pyramidal.

CHAPTER 5

Gases, Solids, and Liquids

5-1 INTRODUCTION

Did you ever wonder why sweating cools you off, or why ice melts above 0°C, but not below, or why the air pressure in car tires increases when the car is driven? In this chapter we shall provide a framework within which questions such as these can be answered. We shall look at the molecular behavior of gases, liquids, and solids and learn in qualitative terms how the observable properties of a macroscopic sample of a gas, liquid, or solid are consequences of the activity going on at the molecular level. We shall see that a set of assumptions about the molecular behavior of gases—the kinetic molecular theory—can be used to explain the observable macroscopic properties of gases and to give a simple interpretation of the absolute zero of temperature. In addition, we shall treat some of the quantitative relationships among the pressure, volume, temperature, and amount of a gas.

The observable properties of many of the macroscopic systems we encounter do not change with time. Such systems are said to be in equilibrium. We shall see that for many equilibrium systems, such as a mixture of ice and water at 0°C, changes are actually taking place at the molecular level, but that these different changes merely compensate for one another. Such a system is said to be in a state of dynamic equilibrium.

Our discussion of the molecular behavior of solids, liquids, and gases will reveal that a fundamental aspect of any molecular system is how disordered or random the system is. Scientists use the term entropy to describe the amount of disorder in a molecular system. We shall see that the direction of an observable change in a macroscopic system can be explained on the basis of a natural tendency for any system to go to a state where the potential energy, or enthalpy, is lower, and the molecular disorder, or entropy, is higher. We shall also see that we can predict the direction of change for a macroscopic system by considering a quantity known as the free energy, which is simply related to both the enthalpy and entropy of the system. The free energy of a system is a useful tool with which we can judge when the system is in a state of equilibrium.

Water can often be found as a gas (water vapor), as a solid (ice), and as a liquid in the same picturesque winter setting.

5-2   STUDY OBJECTIVES

After studying the material in this chapter, you should be able to:

1. Describe the assumptions of the kinetic molecular theory.

2. Give a kinetic molecular definition of the absolute zero of temperature.

3. Solve quantitative problems involving changes of pressure, volume, and temperature for gases.

4. Define the term partial pressure and use Dalton's law of partial pressures.

5. Relate the macroscopic properties of solids, liquids, and gases to the molecular structure and molecular motion of these states of matter.

6. Describe the different classes of solid compounds.

7. Define the terms equilibrium, dynamic equilibrium, vapor pressure, boiling point, and normal boiling point.

8. Define the terms entropy, change in enthalpy, and change in free energy.

9. Describe how the direction of change for a nonequilibrium system can be explained in terms of the natural tendency for the system to go to a state of lower enthalpy and higher entropy.

10. Describe the relationship between the direction of change for a nonequilibrium system and whether the change in free energy is positive, negative, or zero.

5-3   GASES

In Section 2-3 we described some of the properties of gases. Recall that a gas fills a container completely, has a low density, and is relatively easy to compress and expand. In Chapter 2 we also inferred that (1) the atoms or molecules in a gas must be, on the average, quite far apart from one another, and (2) the attractive forces between individual particles in a gas must be extremely weak. These two inferences about the molecular nature of a gas are part of a model of the gas state called the kinetic molecular theory.

Kinetic Molecular Theory

The **kinetic molecular theory** consists of the following set of simple assumptions about the behavior of gases:

1. Most of the volume of a gas is empty space.

2. The molecules of a gas are continually moving rapidly and randomly.

3. There are no attractive or repulsive forces between gas molecules except when they collide.

4. Individual molecules of a gas move at different speeds and therefore have different kinetic energies.

5. The average kinetic energy of the molecules in a gas is directly proportional to the absolute temperature. (Recall that absolute temperature refers to temperature on the Kelvin scale, as discussed in Section 1-10.)

The observable properties of most gases under usual conditions are consistent with these assumptions.

Let us discuss briefly the assumptions of the kinetic molecular theory. The first assumption is that most of the volume of a gas is empty space. Consider the following example. The air around us is about 20% oxygen and 80% nitrogen, with small amounts of other gases. If we think of an oxygen or nitrogen molecule as a small sphere with a molecular diameter D, then the average distance between two gas molecules in the air we breathe is about 12 times the diameter of a molecule, or $12D$. (See Figure 5-1). Thus, there is a lot of empty space between gas molecules. In the air around us, this empty space accounts for more than 99.9% of the total volume.

Figure 5-1   Most of the air we breathe is just empty space. When an oxygen or a nitrogen molecule is pictured as a small sphere with a diameter D, then the average distance between molecules is $12D$.

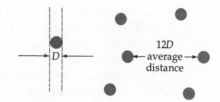

The key idea in the second assumption of the kinetic molecular theory is a concept of **random motion.** If we could watch a single gas molecule, we would not see it moving always with the same speed and in the same direction, like a car traveling along a straight and level road. Rather, we would see that the gas molecule frequently changes both its speed and its direction as a result of collisions with other molecules and with the walls of the container. In Figure 5-2 we represent an attempt to describe the chaotic path of an individual gas molecule. Because of the random motion of gas molecules and large volume of empty space in a gas, gases tend to spread out (diffuse) rapidly in all directions. In everyday life we often become aware of the diffusion of gases that can be detected by their odor. In fact, you may want to try a simple experiment. Pour some household ammonia (a solution of ammonia gas in water) into a warm glass and set it in a corner of a room where there are no air currents. Observe the time it takes for you to detect the odor of ammonia in some other part of the room. This is the time required for some of the ammonia molecules to diffuse to where you are. (Ammonia can be quite toxic. Do not breathe even barely detectable amounts of ammonia for very long.)

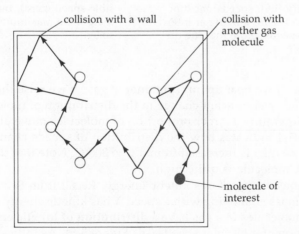

collision with a wall

collision with another gas molecule

molecule of interest

Figure 5-2   Gas molecules exhibit continuous random motion. A given gas molecule changes its direction and its speed very frequently as it collides with other gas molecules and the walls of the container.

The third assumption of the kinetic molecular theory states that there are no attractive or repulsive forces between gas molecules. This assumption is almost, but not quite, entirely true for real gases. Actually, there are very weak attractive forces between gas molecules, which become important at low temperatures or high pressures. However, under normal conditions we can safely assume that there are no significant forces between the molecules of a gas.

Although gas molecules move about rapidly (assumption 2), they do not all move at the same speed (assumption 4). Consider the speed of an individual gas molecule. When this molecule collides with other gas molecules, or with the walls of the container holding the gas, an exchange of energy takes place. The speed of the particular molecule we are considering is likely to increase or decrease as a result of these collisions. Now, if we could measure the speed of each individual gas molecule in a container at the same instant, we would find a few molecules moving very slowly and a few molecules moving extremely fast, but the majority of the gas molecules would be traveling at speeds between these two extremes. We could then express our findings in the form of a graph, in which we indicate the number of gas molecules traveling at each speed. Such a graph is called a **distribution of molecular speeds,** and an example is shown in Figure 5-3. If we could measure the speed of each gas molecule at a later time, we would find that some of the molecules are moving faster while others are moving slower. We would find, however, that the distribution of molecular speeds does not change with time.

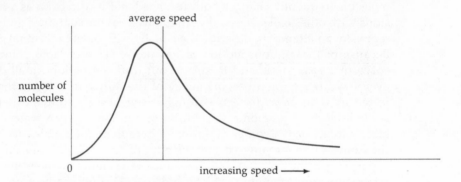

Figure 5-3   Distribution of molecular speeds. At any instant, relatively few gas molecules are moving at either very low or very high speeds; most of the molecules are moving at intermediate speeds. Note that the distribution of molecular speeds is not symmetric about the average speed—there is a minimum possible speed (zero), but theoretically there is no maximum possible speed.

What happens if we heat up our container of gas? If we raise the temperature of the gas, there is a significant change in the distribution of molecular speeds. At a higher temperature a larger proportion of molecules move about at higher speeds. Figure 5-4 indicates how the distribution of speeds changes when the temperature of oxygen is increased from 0°C to 200°C. Note that the speed of an average oxygen molecule is quite high.

A moving molecule possesses kinetic energy. Recall from Section 1-8 that a particle with a mass m, moving with a speed s, has kinetic energy equal to $\frac{1}{2}ms^2$. Therefore, the molecules in a gas have a **distribution of kinetic energies** related to the distribution of molecular speeds (see Figure 5-5). As indicated in Figures 5-4 and 5-5, the average speed and the average kinetic energy increase as the temperature increases.

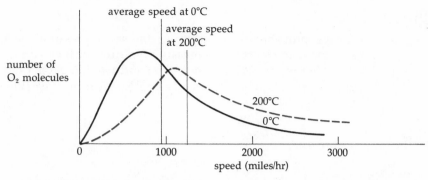

Figure 5-4   Distribution of molecular speeds for oxygen at 0°C and 200°C. At higher temperatures, there are proportionately more molecules moving about at higher speeds. At 0°C the average speed of an oxygen molecule is about 950 miles/hr, whereas at 200°C the average speed increases to about 1250 miles/hr.

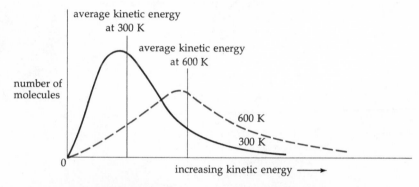

Figure 5-5   Distribution of kinetic energies. At any instant, the molecules in a gas have different kinetic energies related to their different molecular speeds. At higher temperatures, proportionately more molecules have higher kinetic energies. The average kinetic energy for a gas at 600 K is twice the average kinetic energy at 300 K.

Assumption 5 states that this average kinetic energy is directly proportional to the absolute temperature. For example, if the absolute temperature is changed from 300 K to 600 K (a doubling of the absolute temperature), then the average kinetic energy of the gas molecules also doubles. Thus, the absolute temperature of a gas is a measure of the average kinetic energy associated with the random motion of the gas molecules. We shall see in Chapter 9 that the proportion of molecules that move with high speeds, and that therefore possess a large amount of kinetic energy, is an extremely important factor in determining the rate of a chemical reaction.

Exercise 5-1

Using the assumptions of the kinetic molecular theory, explain the fact that gases can be compressed quite easily.

Exercise 5-2

A tiny solid particle (for example, a smoke particle) suspended in air and viewed under a microscope moves about randomly. When the temperature increases, this random motion increases. What causes the solid particle to move, and to move faster (on the average) at higher temperatures?

Pressure

You have probably heard or used the term pressure when talking about gases—for example, the "air pressure" in tires, or the "atmospheric pressure" in a weather report. What is pressure, and why do gases exert pressure?

Pressure results when a force is applied over an area. **Pressure** is defined mathematically by the equation

(5-1)

$$\text{pressure} = \frac{\text{force}}{\text{area}}$$

You can easily feel the difference between force and pressure by pushing on a balloon with the palms of your hands and then with two fingers (see Figure 5-6).

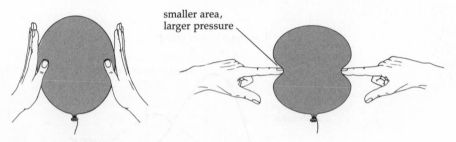

smaller area, larger pressure

Figure 5-6   There is a greater indentation in a balloon when you push in on it using two fingers than when you exert the same force using your flat hands. The pressure is greater using your fingers because the area of a single finger is less than the area of a whole hand.

Gases exert pressure because the molecules of a gas are continually moving and colliding with the walls of the container holding the gas. Any area of the wall experiences the force of many impinging gas molecules at any particular time (see Figure 5-7).

The more molecules that strike a given area at any instant of time, and the harder they hit, the larger the gas pressure. As a consequence, the pressure of a gas depends on

1. The number of gas molecules
2. The volume of the container the gas is in
3. The temperature

It is easy to predict, at least qualitatively, how the pressure of a gas depends on these three factors. For example, if we increase the amount of gas in a container of fixed volume and at a fixed temperature, the pressure will increase. There are now more molecules moving about in the container and hence more collisions with the walls at any particular instant (see Figure 5-7b).

On the other hand, if we increase the volume of a container holding a constant amount of gas at a fixed temperature, the pressure will decrease. In this case, the same number of molecules are moving around with the same average

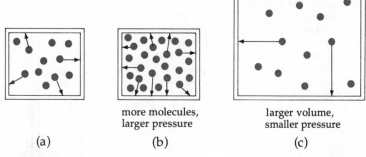

more molecules,
larger pressure

larger volume,
smaller pressure

(a)　　　　　　　(b)　　　　　　　　　(c)

Figure 5-7　　At a constant temperature, the
pressure of a gas depends on the
number of molecules and the volume. The volume is the same in (a)
and (b), but there are more mole-

cules in (b) and consequently a
larger pressure. The number of
molecules is the same in (a) and (c),
but the volume is larger in (c) and
consequently the pressure is lower.

speed, but in a larger space. Since the molecules now have more room to move
around, there will be fewer collisions with the walls of the container at any
given instant (see Figure 5-7c).

Now, if we increase the temperature of a given amount of gas in a container
of fixed volume, the pressure will increase. There are two reasons for this. Recall
that as the temperature increases, the average speed of the gas molecules increases. The faster the molecules are moving, the more collisions they will make
with the walls of the container at any particular instant. Also, a faster moving
molecule has a larger kinetic energy and exerts a greater force when it hits a
wall of the container.

Measuring Pressure

Weather forecasters have found that measuring the atmospheric pressure, or
barometric pressure, is useful in predicting weather conditions. Atmospheric
pressure is measured with a device known as a **mercury barometer.** One such
atmospheric pressure measurement might be 740 millimeters of mercury
(740 mmHg). The units used here, mmHg, may seem rather peculiar for a pressure. A millimeter is a unit of length and may not seem appropriate for a pressure. To understand why pressure can be measured in terms of millimeters of
mercury, we must consider how a mercury barometer works.

A mercury barometer can be constructed by completely filling a glass tube
(which is closed at one end) with mercury and inverting the open end under a
pool of mercury, as shown in Figure 5-8. When this is done, mercury will leave
the tube until the pressure inside the barometer, resulting from the weight of
the column of mercury, is equal to the air pressure on the mercury outside. The
higher the atmospheric pressure, the higher will be the column of mercury in
the barometer. Thus, the atmospheric pressure is directly proportional to the
height of the mercury column.

Mercury, like most liquids and solids, expands as the temperature increases,
so the height of a mercury column depends on the temperature. Therefore,
when reporting atmospheric pressures, the height of the mercury column at the
temperature of the air is converted to a height at a standard temperature, which
is chosen as 0°C. In addition, the weight of a column of mercury depends on the
effect of gravity, which in turn depends on the altitude. Therefore, the height of
a column of mercury in a barometer is converted to a height at some standard altitude, which is chosen to be sea level.

Figure 5-8   A mercury barometer. At any given level, the pressure is the same throughout a liquid. Thus, for a mercury barometer, the atmospheric pressure is equal to the pressure due to the weight of the mercury column. Atmospheric pressure = the height of the mercury column (mmHg).

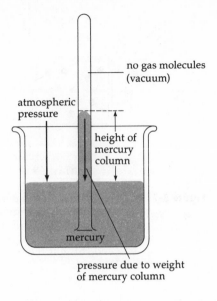

One millimeter of mercury at 0°C and sea level is referred to as one torr. The pressure unit torr is named after the Italian physicist, Evangelista Torricelli (1608–1647), who invented the barometer.

Atmospheric pressure varies with meteorological conditions and geographical location. A sudden drop in atmospheric pressure often occurs just prior to a storm, and a gradually rising atmospheric pressure is an indicator of approaching fair weather.

Pressures are also often described by using a unit known as the **atmosphere,** which is abbreviated atm. One atmosphere is equivalent to 760 torr or 760 mm of mercury. The pressure 1.000 atm is referred to as **standard pressure.** You must be careful to distinguish between the pressure unit *atmosphere* and *atmospheric pressure*. Atmospheric pressure is very rarely exactly equal to the standard pressure 1.000 atm.

Exercise 5-3
In the accompanying diagrams, the mercury in the bent tube (called a *manometer*) is used to indicate the pressure inside the round container. The pressure at any given level is the same throughout the liquid. (a) Which gas has the lowest pressure? Which gas (or gases) has a pressure greater than atmospheric pressure? (b) If the atmospheric pressure is 745 mmHg, what is the pressure inside container A, inside container B, and inside container C?

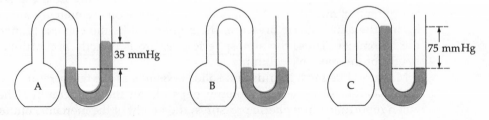

Exercise 5-4
The device commonly used to measure blood pressure is called a *sphygmomanometer.* In this device, a hollow cuff is wrapped around a person's arm and air is pumped into the cuff. Explain how and why the pressure in the cuff changes as air is pumped in.

Boyle's Law

It is quite easy to take a fixed amount of gas, keep its temperature constant, and measure its volume at different pressures. Robert Boyle (1627–1691), an English scientist, did experiments of this sort in the seventeenth century.

In his experiments, Boyle added mercury to a glass tube with a U-shaped bend, as shown in Figure 5-9. This trapped some air in the closed end. The volume of air, V, is equal to the product of the length, L, and the cross-sectional area of the U-tube. As Boyle added mercury to the open end, the trapped air was compressed and its volume decreased.

Figure 5-9   Boyle determined the pressure and volume of a sample of trapped air in an experimental apparatus. As he added mercury to the open end, the volume of the trapped air decreased and the pressure of the trapped air increased.

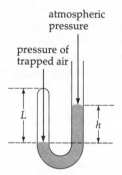

The pressure of the trapped air could be determined simply by measuring the atmospheric pressure with a barometer and the height h. In a liquid the pressure at any given level is the same throughout the liquid, so

(5-2)   Pressure of trapped air (P) = (pressure due to mercury column
of height h) + (atmospheric pressure)

(Do you see why the atmospheric pressure is included in Eq. 5-2?) Therefore, if the height h is 100 mm and the atmospheric pressure is 750 mmHg, then the pressure of the trapped air is 850 mmHg.

Boyle could thus determine both the pressure and volume of the trapped air under different conditions by measuring the height, h, and the length, L, as mercury was added to the tube.

The data in Table 5-1 are similar to the data that Boyle obtained in his experiments. Notice that for each set of conditions in Table 5-1, the product of the pressure times the volume has the same value even though the pressures and volumes vary considerably. Many other experiments of this type also gave results in which the product of the pressure and the volume turned out to be a constant.

Table 5-1 Boyle's Law Experiment

Atmospheric pressure = 750 mmHg
Cross-sectional area of U-tube = 1.00 cm²

h (mm)	L (cm)	P (mmHg)	V (cm³)	$P \times V$ (mmHg × cm³)
0	20.0	750	20.0	1.50×10^4
100	17.6	850	17.6	1.50×10^4
250	15.0	1000	15.0	1.50×10^4
500	12.0	1250	12.0	1.50×10^4
750	10.0	1500	10.0	1.50×10^4

On the basis of the results of these experiments, one of the earliest quantitative scientific laws was proposed. This law is now known as Boyle's law.

(5-3)   **Boyle's Law:** The product, pressure times volume, of a given amount of gas at a fixed temperature does not vary.

We can express Boyle's law in the following mathematical form:

(5-4) $P \times V = C$ for a fixed temperature and a fixed amount of gas

The symbol C in Eq. 5-4 stands for a constant that always has the same value as long as the temperature and the amount of gas do not change. According to Boyle's law, if the pressure increases the volume must decrease, and if the pressure decreases the volume must increase. This relationship is graphed in Figure 5-10. Thus Boyle's law is consistent with our previous qualitative conclusion that an increase in the volume of a fixed amount of gas at a fixed temperature will result in a decrease in pressure.

Figure 5-10   Boyle's law. For a fixed temperature and fixed amount of gas, the product $P \times V$ has a constant value. Going from A to B to C, the volume decreases, the pressure increases, but $P \times V$ has the same value of, in this case, 4 liters · atm.

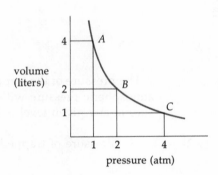

It is convenient to put Boyle's law in an alternative but equivalent form for calculating the changes that occur when a gas that is initially at a pressure P_1 and a volume V_1 changes to a new pressure P_2 and a new volume V_2. If the amount and temperature of the gas remain the same, then from Boyle's law, the product $P \times V$ has the same value for both the initial and new conditions.

(5-5) $P_1 \times V_1 = P_2 \times V_2$ for a fixed temperature and fixed amount of gas

EXAMPLE 1   Gases, such as the oxygen used in medical facilities, are normally compressed under high pressure into strong metal tanks for storage. The pressure of compressed oxygen in a 100-liter tank might be 120 atm. What volume would this amount of oxygen occupy at 1.00 atm?

SOLUTION   We should note that since the new pressure is *less* than the initial pressure, the new volume must be *larger* than the initial volume. In this example, $P_1 = 120$ atm, $V_1 = 100$ liters, and $P_2 = 1.00$ atm. We can rearrange Eq. 5-5 and put in the values of P_1, V_1, and P_2, to obtain

(5-6) $V_2 = V_1 \times \dfrac{P_1}{P_2} = 100 \text{ liters} \times \dfrac{120 \text{ atm}}{1.00 \text{ atm}} = 1.20 \times 10^4 \text{ liters}$

Note that the calculated value of the new volume V_2 is larger than the initial volume V_1, in agreement with our prior qualitative considerations. Since the volume of air entering a person's lungs in a single breath is about 0.5 liter, one tank of compressed oxygen contains roughly 24,000 "lungsful."

If we consider our example and Eq. 5-6 from a slightly different point of view, we can obtain a very simple practical method of solving problems involving Boyle's law. This method involves a good qualitative understanding of the content of Boyle's law but does not require an algebraic rearrangement of Eq. 5-5. A similar approach can be used for other gas law problems that we shall discuss shortly. Let us see how this method works.

In Example 1 we knew that V_2 must be larger than V_1 before we did the calculation because at a fixed temperature gases expand when the pressure is reduced. Therefore we could have solved this problem by using the following logical reasoning:

1. Boyle's law applies in this problem, so the new volume V_2 is equal to the initial volume V_1 multiplied by a *pressure-ratio factor*.

2. Since P_2 is less than P_1, V_2 must be bigger than V_1.

3. The pressure-ratio factor that is required to make V_2 larger than V_1 is 120 atm/1.00 atm.

4. Thus, $V_2 = 100$ liters $\times \dfrac{120 \text{ atm}}{1.00 \text{ atm}} = 1.20 \times 10^4$ liters.

A similar line of reasoning can be used to obtain the new pressure in a problem in which the volume is changed but the temperature is kept constant.

EXAMPLE 2 Five grams of oxygen gas at a pressure of 1.2 atm in a 2.5-liter container are compressed to a volume of 0.5 liter keeping the temperature constant. What is the pressure of the compressed O_2 gas?

SOLUTION Boyle's law applies, and thus the new pressure is equal to the initial pressure multiplied by a *volume-ratio factor*. Since the new volume is smaller than the initial volume, the new pressure must be higher than the initial pressure. Thus the volume-ratio factor must be 2.50 liters/0.50 liter. Therefore the new pressure P_2, the pressure of the compressed O_2 gas, is given by

(5-7) $P_2 = 1.2 \text{ atm} \times \dfrac{2.5 \text{ liters}}{0.5 \text{ liter}} = 6.0 \text{ atm}$

Effect of Temperature on the Pressure of a Fixed Volume of Gas

Boyle's law applies to a fixed amount of gas at constant temperature. What happens if the temperature of the gas changes? We have already discussed the fact that the pressure of a given amount of gas in a container of fixed size increases as the temperature increases. As an example, Figure 5-11 shows a graph of some measurements of the pressure of 1.0 g of oxygen gas in a 700-ml container as the temperature is varied. The graph in Figure 5-11 shows that $-70°C$ is the lowest temperature for which data was obtained. Below this temperature the dashed line represents *extrapolated values*, that is, the pressure values that would be *expected* if additional measurements had been made at lower temperatures. Notice that according to the extrapolated line, the pressure of the O_2 gas should be zero at $-273°C$. If we had used different gases with different amounts and volumes, we would still have obtained a similar extrapolated value. For all gases, the extrapolated pressure value reaches zero at about $-273°C$.

Recall that a gas exerts a pressure because the gas molecules are moving. Thus in order for a gas to exert zero pressure, the molecules must not move at all. At a temperature of $-273°C$, all molecular motion would cease. This is the lowest temperature that can possibly be obtained, and it is called **absolute zero.** On the

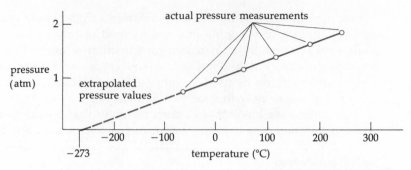

Figure 5-11   Pressure versus temperature for a given amount of oxygen gas at a fixed volume. Pressure can be seen to vary linearly with temperature in the range from −70°C to 250°C, temperatures for which experimental data are available. An extrapolation of this linear relationship to zero pressure yields a temperature of −273°C.

basis of more precise measurements, absolute zero has been found to be −273.15°C. The idea that all molecular motion would stop at absolute zero is consistent with the fifth assumption of the kinetic molecular theory—that the average kinetic energy is directly proportional to the absolute temperature. For according to kinetic molecular theory, if the absolute temperature is zero, the average kinetic energy is also zero, and this could happen only if all the gas molecules were completely stationary.

Recall from the discussion in Section 1-10 that absolute zero is the starting point for temperatures on the absolute or Kelvin scale and that the relationship between the Kelvin and Centigrade scales is

(5-8)   Temperature on Kelvin scale (K) = (temperature in °C) + 273.15

The data in Figure 5-11 have been replotted in Figure 5-12 using the Kelvin scale. The results in Figure 5-12 can be described by saying that the pressure of a given amount of gas in a fixed volume is directly proportional to the absolute temperature. (Direct proportional relationships are discussed in Essential Skills 3.) This statement can be written in the form of a mathematical equation:

(5-9) $P = C \times T$ or $\dfrac{P}{T} = C$ for a fixed amount of gas in a fixed volume

In Eq. 5-9, the symbol T refers to the absolute temperature and C is a constant that does not change as long as the amount and volume of gas remain the same.

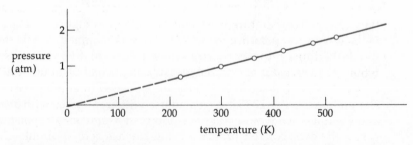

Figure 5-12   Pressure versus temperature for a given amount of gas at a fixed volume. The pressure is directly proportional to the temperature when temperatures on the Kelvin scale are used.

When performing a calculation, it is convenient to put Eq. 5-9 in the following equivalent form, similar to the one we used for Boyle's law:

(5-10)
$$\frac{P_1}{T_1} = \frac{P_2}{T_2}$$

One use of Eq. 5-10 is to determine the increase in gas pressure when a container of gas is heated. Consider the following example.

EXAMPLE 3   The temperature of a gas in a 10-liter container at 1.0 atm is 25°C. What is the pressure of the gas after the temperature is raised to 100°C?

SOLUTION   It is easy to solve this problem with the same reasoning that we used for problems involving Boyle's law. In this case the new pressure is equal to the initial pressure multiplied by a *temperature-ratio factor*. Since the new temperature is higher than the initial temperature, the new pressure will be greater than the initial pressure, so the temperature-ratio factor is 373 K/298 K.) (We use temperatures on the Kelvin scale.) Thus, the new pressure P_2 is

(5-11)
$$P_2 = 1.0 \text{ atm} \times \frac{373 \text{ K}}{298 \text{ K}} = 1.24 \text{ atm}$$

Charles' Law

What happens when a sample of gas is heated at constant pressure? The volume increases. When the volume of a fixed amount of gas at constant pressure is measured at different temperatures, the volume of the gas is found to be directly proportional to the absolute temperature of the gas. This relationship, shown in Figure 5-13, is called Charles' law.

(5-12)
Charles' Law: The volume of a fixed amount of gas at constant pressure is directly proportional to the absolute temperature.

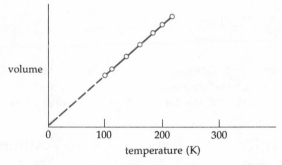

Figure 5-13   The volume of a fixed amount of a gas at a constant pressure is directly proportional to the absolute temperature. Experimental data are available for this gas down to 100 K. Below this temperature, the dashed line signifies that the values are extrapolated.

Jacques Alexandre César Charles (1746–1823) was a French scientist who first proposed this relationship on the basis of experiments conducted in the late eighteenth century. Charles' law can be expressed in three equivalent forms:

(5-13)
$$V = C \times T \quad \text{or} \quad \frac{V}{T} = C \quad \text{or} \quad \frac{V_1}{T_1} = \frac{V_2}{T_2}$$

Problems that involve determining temperature-related changes in the volume of a fixed amount of gas at a constant pressure (or determining how the temperature changes if the volume is changed under these conditions), by using Charles' law, can be solved by the same type of reasoning we used for those problems involving Boyle's law and Eq. 5-9.

Combined Gas Law

Boyle's law, Eq. 5-9, and Charles' law can be collected together in an expression called the **combined gas law:**

(5-14) $$\frac{PV}{T} = C \quad \text{or} \quad \frac{P_1 V_1}{T_1} = \frac{P_2 V_2}{T_2} \qquad \text{for a fixed amount of gas}$$

Eq. 5-14 is useful when we have a problem where we want to determine (1) how the pressure of a sample of a gas changes if *both* the volume and temperature are changed; or (2) how the volume of a sample of a gas changes if *both* the pressure and temperature are changed; or (3) how the temperature of a sample of a gas changes if *both* the pressure and volume are changed. We can solve any of these types of problems by the same type of reasoning that we used previously for other types of gas law problems. Let us consider an example.

EXAMPLE 4   At a pressure of 1.5 atm in a 2.3-liter tank at 100°C, 3.6 g of O_2 is compressed to a volume of 1.1 liters and also cooled to 0°C. What is the new pressure of the gas?

SOLUTION   In this example, the new pressure is equal to the initial pressure multiplied by *both* a volume-ratio factor and a temperature-ratio factor. Since the new volume is less than the original volume, the volume-ratio factor must tend to *increase* the pressure and it must therefore be 2.3 liters/1.1 liters. Since the new temperature is less than the original temperature, the temperature-ratio factor must tend to *decrease* the pressure, so it must be 273 K/373 K. (Note again that we always use temperatures on the Kelvin scale when solving gas law problems.) Therefore, the new pressure P_2 is

(5-15) $$P_2 = 1.5 \text{ atm} \times \frac{2.3 \text{ liters}}{1.1 \text{ liters}} \times \frac{273 \text{ K}}{373 \text{ K}} = 2.3 \text{ atm}$$

Real gases do not exactly obey Charles' law, Boyle's law, or Eq. 5-9. But for most common gases these relationships are quite accurate if the pressure is not much greater than 1 atm and the temperature is not much below 0°C.

Effect of Amount on the Pressure of a Fixed Volume of Gas at a Fixed Temperature

As we discussed earlier in this section, the pressure of a gas in a given container at a fixed temperature is directly proportional to the number n of gas molecules in the container. We can represent this statement by these equations:

(5-16) $$P = C \times n \quad \text{or} \quad \frac{P_1}{n_1} = \frac{P_2}{n_2} \qquad \text{for a fixed volume and a fixed temperature}$$

Chemists very often express amounts of chemical compounds using the unit mole. We shall consider how the properties of a gas sample depend on the amount of gas in the sample in a bit more detail after we have discussed the mole concept in the next chapter.

Exercise 5-5
The total volume of a person's lungs is about 5.5 liters. A person at rest moves about 0.5 liter of air in and out of the lungs with each breath. Suppose your lungs were *not* open to the atmosphere, so that the amount of air in them remained constant when your lung volume increased from 5.5 liters to 6.0 liters. Calculate the pressure of the air in your lungs when the volume is 6.0 liters if it is 1.0 atm when the volume is 5.5 liters.

Exercise 5-6
A balloon has a volume of 1.20 liters inside a room where the temperature is 25°C. What is the volume of this balloon when it is taken outside, where the temperature is −10°C?

Exercise 5-7
What volume will a sample of gas occupy at a pressure of 0.9 atm and 37°C if the same sample of gas occupies 1.40 liters at 1.10 atm and 20°C?

If you had difficulty with Exercises 5-5, 5-6, or 5-7, then you should find the suggestions about a general approach to problem solving in Essential Skills 5 to be helpful.

Partial Pressures

So far we have discussed only pure gases, but often we are interested in a mixture of gases. For example, the gas in your lungs is a mixture of oxygen, carbon dioxide, water vapor, and nitrogen, as well as very small amounts of other gases that we shall ignore. When such a mixture of gases is put into a container, the molecules of each gas strike the walls of the container independently, and each gas contributes to the total pressure in direct proportion to the number of molecules of that gas (see Figure 5-14). That part of the total pressure of a gas mixture due to the pressure of a particular gas is called the **partial pressure** of that particular gas. The total pressure of a gas mixture is just the sum of the partial pressures of all the gases in the mixture. This statement is called Dalton's law of partial pressures, and is named after John Dalton, who first proposed it in 1807. John Dalton was the same individual who proposed the modern atomic theory of matter (Chapter 2). Subscripts are used to indicate the partial pressures of different gases in a mixture. The letters $a, b, c, \ldots$ are used as general subscripts. The molecular formula of a gas is used as a subscript when referring to a specific gas; thus, P_{O_2} indicates the partial pressure of oxygen in a mixture. **Dalton's law of partial pressures** can be written in this general form:

(5-17) $$P_{total} = P_a + P_b + P_c + \cdots$$

Figure 5-14   For a mixture of gases, the total pressure is the sum of the partial pressures for all the gases in the mixture. The partial pressure of each gas is directly proportional to the number of gas molecules for that gas. There are twice as many nitrogen molecules as oxygen molecules in this gas mixture, and the partial pressure of nitrogen is twice as large as the partial pressure of oxygen.

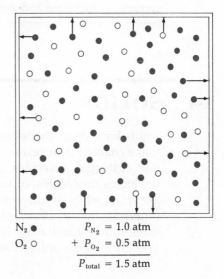

N_2 ● P_{N_2} = 1.0 atm
O_2 ○   + P_{O_2} = 0.5 atm
                  ───────────────
            P_{total} = 1.5 atm

The total pressure exerted by the mixture of gases in your lungs is equal to the atmospheric pressure. For this mixture, Eq. 5-17 is

(5-18) $$P_{\text{atmospheric}} = P_{O_2} + P_{CO_2} + P_{H_2O} + P_{N_2}$$

The composition of the air we inhale is different from that of the air we exhale. Inspired air is richer in oxygen and has relatively less carbon dioxide and water vapor than expired air. However, the air that exchanges with blood is neither inspired or expired air. This air is contained in a myriad of tiny hollow sacs in the lungs, called alveoli. The total pressure of alveolar air goes up and down rhythmically by about 1 mmHg during a respiratory cycle, but the average pressure is equal to the atmospheric pressure. Usually the composition of alveolar air is approximately constant throughout the respiratory cycle. Expired air and alveolar air are saturated with water vapor at normal body temperature, 37°C, whereas the water content of inspired air depends on the relative humidity. (We shall discuss relative humidity and water vapor saturation in the next section.) The partial pressures of the gases in inspired, expired, and alveolar air are given in Table 5-2 for a normal person under typical conditions.

Table 5-2 Component Gases in Human Respiration*

Gas	Partial pressure (mmHg)		
	Inspired Air (26°C)	Expired Air (37°C)	Alveolar Air (37°C)
O_2	158.2	116.2	101.2
CO_2	0.3	28.5	40.2
H_2O	6.3	47.0	47.0
N_2	595.2	568.3	571.6

* At atmospheric pressure of 760 mmHg and relative humidity of 25%.

Note that the atmospheric pressure is 760 mmHg for the data in Table 5-2 and that the sum of the partial pressures of O_2, CO_2, H_2O, and N_2 is 760 mmHg for inspired air, as well as for expired air and for alveolar air.

Exercise 5-8
Atmospheric pressure decreases as altitude increases. At the top of Mt. Everest the atmospheric pressure is 245 mmHg. At this altitude, the partial pressure of nitrogen is 196 mmHg. Calculate the partial pressure of O_2 on top of Mt. Everest, assuming that air is composed of nitrogen and oxygen only. Compare this value with the value in Table 5-2.

5-4   SOLIDS

Solids have fixed shapes and much larger densities than gases. We have seen that a characteristic feature of gases on the molecular level is rapid random motion of the gas molecules. Solids are quite different. An important property of the atoms, molecules, or ions in a solid is that they occupy fixed positions and are arranged in an ordered, regular pattern. For example, recall from Chapter 4 (Figure 4-3) that solid ionic compounds are described in terms of an ordered three-dimensional array of cations and anions, and that most covalent compounds have an ordered arrangement of molecules in the solid state.

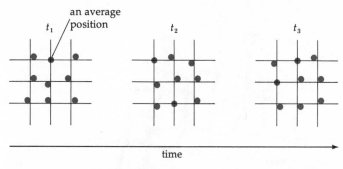

Figure 5-15   Schematic illustration of the vibrational motion in a solid. The atoms, molecules, or ions vibrate to a small extent about their average positions.

However, the atoms, molecules, or ions in a solid are *not completely* fixed (stationary). Each vibrates to a small extent about an average position (see Figure 5-15). The intensity of this vibrational motion increases as the temperature increases. At a sufficiently high temperature, the vibrational motion is so large that the forces of attraction between the molecules or ions are not strong enough to maintain the ordered arrangement of a solid. At this characteristic temperature (the melting point), the solid melts and forms a liquid.

Table 5-3 Classes of Solid Substances

Type	Description	Examples
Ionic	Extended array of anions and cations; strong electrostatic forces	$NaCl$, K_2SO_4, $Ca(NO_3)_2$
Molecular	Small molecular units; weaker intermolecular forces	Cl_2, H_2O, NH_3
Extended covalent	Extended array of atoms; bonded covalently	Diamond (C), carborundum (SiC)
Metallic	Positive ions in a "sea" of electrons	Na, Fe

It is useful to classify solid substances on the basis of the type of attractive forces that hold the component atoms, molecules, or ions together, as in Table 5-3. In all **solid ionic compounds,** there is an ordered arrangement of cations and anions held together by the electrical force of attraction between positive and negative charges. Since these attractive forces are quite strong, the melting points of ionic compounds are quite high (see Chapter 4, Section 4-4).

There are two types of solid covalent substances. A **molecular solid,** in which molecular units exist in the solid state, is one type. Solid chlorine (which we also discussed in Section 4-4) is a simple example of a molecular solid. The attractive forces between molecules in molecular solids are relatively weak, and therefore molecular solids have relatively low melting points. We shall describe the nature of these intermolecular forces in Chapter 7. The other type of covalent solid does not contain small molecular units but rather consists of an extended three-dimensional array of atoms with strong covalent bonds between the atoms. We shall call this type of solid an **extended covalent solid.** Diamond, one of several forms of solid carbon, is an example of an extended covalent solid. In diamond, each carbon atom shares a pair of valence electrons with four other carbon atoms (see Figure 5-16). Extended covalent substances are

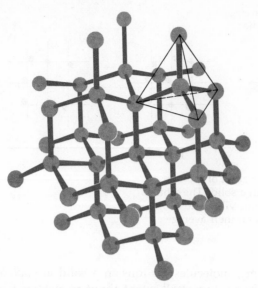

Figure 5-16   The structure of diamond, which   atoms, resulting in an extended
                   is an extended covalent solid. Each   three-dimensional network of
                   carbon atom forms four covalent   bonded carbon atoms.
                   bonds with four other carbon

similar to ionic compounds in that there is no single molecular unit consisting
of a small number of atoms. In a sense, the entire solid is one giant molecule.
Because strong covalent bonds link all the atoms together, diamond—and all
other extended covalent substances—have relatively high melting points. The
melting point of diamond, for instance, is 3500°C.

Metals constitute a fourth general class of solids. A characteristic physical
property of metals is their ability to conduct an electrical current easily. In order
for a metal to have a high electrical conductivity, there must be electrons in the
metal that are relatively free to move. An approximate model for a solid metal
assumes that each metal atom loses one or more of its readily ionized valence
electrons. The resulting positive ions are arranged in an ordered three-
dimensional pattern. All the electrons lost by the metal atoms are in some sense
dispersed throughout the whole metal sample. For example, solid sodium metal
can be crudely described as an ordered arrangement of Na^+ cations in a "sea" of
electrons (see Figure 5-17). The negatively charged electrons act like a "glue"
holding the positively charged Na^+ cations together. The forces of attraction
and the melting points of metallic solids vary over a wide range. Sodium has a
low melting point (98°C), but the melting point of iron is quite high (1530°C).

Figure 5-17   Schematic illustration of the struc-
                   ture of sodium metal. An ordered
                   three-dimensional arrangement of
                   sodium ions (Na^+) is embedded in
                   a "sea" of electrons. The electrons
                   are relatively free to move about
                   the metal. Since there are equal
                   numbers of Na^+ cations and elec-
                   trons, the metallic solid has no net
                   electrical charge.

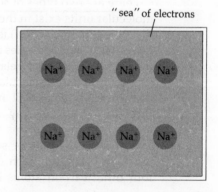

Exercise 5-9
Which of the following can be expected to form ionic solids: K_2CO_3, HCl, Ar, LiCl, Cu?

Exercise 5-10
Given the following experimental facts, predict the type of solids formed by SiO_2 and CO_2: (1) Both SiO_2 and CO_2 are poor conductors of electricity. (2) The melting point of SiO_2 is 1713°C, and CO_2 will melt at -56.5°C under appropriate conditions.

5-5   LIQUIDS

Liquids, like solids, have high densities and are very difficult to compress. Unlike solids, however, liquids do not have fixed shapes. **Diffusion** is a property that liquids have in common with gases. You can observe the diffusion of liquids simply by putting a drop of ink into a glass of water. For a short time the ink remains in a small, fairly well-defined region, but gradually the ink droplet loses its shape and its distinct color as it diffuses into the water (Figure 5-18).

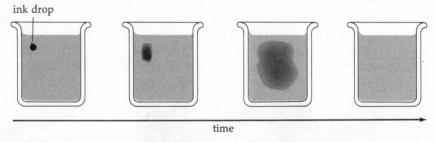

ink drop

time

Figure 5-18   When a drop of ink is put into some water, the ink diffuses gradually and becomes uniformly distributed throughout the liquid.

Diffusion occurs in a liquid because the molecules are fairly free to move. The molecules in a liquid, however, are much closer together than those in a gas. As a consequence, diffusion in a liquid occurs at a much slower rate than diffusion in a gas. Diffusion can also take place in solids, but it is usually an *extremely* slow process.

Another property of liquids with which we are all familiar is the **evaporation** of a liquid from an open container. For example, if we leave a glass of water in a room, the amount of water in the glass will gradually decrease (see Figure 5-19).

Figure 5-19   Evaporation of water. At any instant, some of the water molecules in the liquid have sufficiently high energy to overcome the attractive forces due to other molecules in the liquid and they escape into the gas state. Heat absorbed from the surroundings supplies the energy required for this process. The molecules have a higher potential energy in the gas state than in the liquid state.

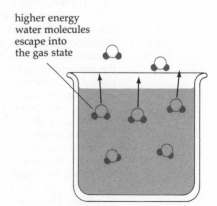

higher energy water molecules escape into the gas state

Evaporation of water occurs because some of the liquid water molecules have enough energy, at any instant of time, to escape into the gas state. In a liquid, there is a distribution of kinetic energy somewhat similar to the distribution in a gas (see Figure 5-5). At any particular instant, some of the water molecules are moving relatively fast and their kinetic energy is large, while other molecules are moving slower and have less kinetic energy. A water molecule with sufficient kinetic energy to overcome the intermolecular attractive forces that hold the liquid together can escape into the gas state. Thus a molecule that leaves the liquid state and goes into the gas state must have more than the average kinetic energy. When such a molecule leaves the liquid state, it carries energy away with it. In the gas the attractive forces between the molecules are much less than in the liquid, and the potential energy of the molecules in the gas state is larger than in the liquid state. If a small amount of liquid is in thermal contact with surroundings of large size (as in a glass of water in a room), then the loss of energy from the liquid due to evaporating molecules is compensated for by a flow of heat energy from the surroundings. As a result, the temperature of an evaporating liquid does not decrease, but remains the same as the temperature of its surroundings.

The evaporation of water is used naturally by humans (and other mammals) as a cooling mechanism. How does this work? When your body produces an excess amount of heat (for example, when you exercise vigorously), you sweat. Water molecules evaporating from the skin (or in the case of dogs, from the tongue) carry off energy. That energy is supplied by the body. The net result is the loss of excess heat energy. For the same reason, alcohol rubs are used to help reduce the body temperature of a person with a high fever.

Several strong demands are made on the body of a competitor in a race. Two of these are evident in these photographs. The contestants in a women's 800-meter race run at the Air Force Academy (see above) suffered added difficulty in meeting the oxygen demands of their muscle cells because of the high altitude in Colorado, where both the atmospheric pressure and the partial pressure of oxygen in the air are lower than at sea level (see Exercise 5-8). Excess body heat generated by the muscle cells of marathon runners is dissipated to a large extent by the evaporation of sweat. Water or fruit juices are consumed by marathoners to replace the body fluids lost in sweating. However, upon finishing a marathon, the body temperature of a runner may cool to below normal—a dangerous condition referred to as hypothermia. To protect against hypothermia, some of the runners shown on the right have wrapped themselves in insulating Mylar blankets.

At the molecular level, liquids are intermediate between gases and solids. The molecules of a liquid exhibit neither the large random motions of a gas nor the regular ordered arrangements characteristic of solids. As a consequence, it is difficult to give a simple qualitative picture of a liquid. At any instant most of the molecules in a liquid are grouped into clusters. Within each cluster, molecules are arranged in a somewhat orderly fashion—akin to the ordered arrangement in a solid. But there is much more freedom of motion in a liquid than in a solid. The clusters of molecules in a liquid are continually breaking apart and re-forming. In this process, a particular molecule does not remain in the same cluster but moves from one cluster to another (see Figure 5-20). Thus liquids are more disordered than solids but more ordered than gases. The phrase **flickering clusters** has been used to describe the molecular structure and molecular motion of liquids.

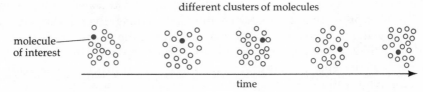

Figure 5-20   Schematic illustration of the molecular motion in a liquid. At any instant a given molecule is grouped in a cluster with several other molecules. These clusters are continually breaking and re-forming with different molecules in each cluster. As a consequence of this process, a given molecule can move from one region of the liquid to another region some distance away.

Vapor Pressure

Liquids evaporate from open containers. But what happens to a liquid in a closed container? A liquid in a closed container is an example of a general type of system known as a closed system. A **closed system** is one in which there is neither an input nor an output of matter. For example, suppose that we put 10 g of liquid water in a closed 1.0-liter container at 25°C. What happens? At first, some water molecules will evaporate from the liquid. But after a period of time the observable evaporation of water will cease, with most of the water remaining in the liquid state. When this occurs we say that the system has reached a state of equilibrium. A closed system is at **equilibrium** when no changes take place in the observable properties of the system (see Figure 5-21).

Figure 5-21   Dynamic equilibrium between the liquid and gas states of water. In a closed container, molecules are constantly evaporating from the liquid; however, an equal number of water molecules are condensing back to the liquid from the gas. The amount of water in either the gas or liquid state thus remains constant.

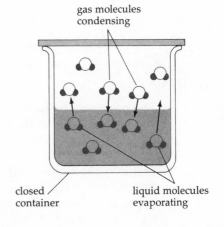

gas molecules condensing

closed container

liquid molecules evaporating

In our example, equilibrium is reached when the amounts of water in the gas and liquid no longer change. But equilibrium does *not* mean the absence of all motion. When the water in our example is at equilibrium, water molecules are still continually evaporating from the liquid, but the molecules that leave the liquid are compensated for by an equal number of water molecules condensing back to the liquid from the gas. Thus the *observable* amounts of water in the gas and liquid remain constant. We can describe this situation more graphically by calling it a state of dynamic equilibrium. A system can be said to be in a state of **dynamic equilibrium** when there are no changes in the observable properties of the system, even though changes are taking place at the molecular level.

When a liquid and a gas are in equilibrium at a certain temperature, the pressure of the vapor is the same no matter what size container the liquid-gas system is in. This equilibrium pressure is called the **vapor pressure** of the liquid. The magnitude of the vapor pressure depends on the liquid and on the temperature. The vapor pressure of water at 25°C is 23.8 mmHg. Thus at 25°C in any size container, as long as liquid water and water vapor are in equilibrium, the pressure of the water vapor is 23.8 mmHg. Even if the gas above the liquid water is a mixture of water vapor and air, the partial pressure of water, P_{H_2O}, is still 23.8 mmHg when liquid and water vapor are in equilibrium at 25°C.

The vapor pressure of water and other liquids increases with an increase in temperature. A graph of this relationship is shown in Figure 5-22. The higher the temperature, the greater the average kinetic energy of the molecules in the liquid. This results in an increased tendency for the molecules to evaporate into the gas state.

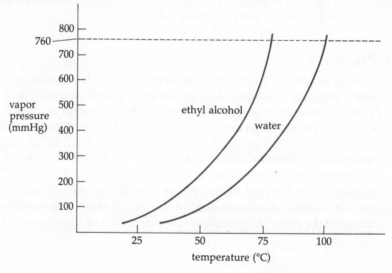

Figure 5-22   Variation with temperature of the vapor pressure for ethyl alcohol and water. The vapor pressure of a liquid increases rapidly with an increase in temperature.

Boiling

The property of a liquid that we commonly refer to as boiling is related to the vapor pressure of the liquid. As we heat a liquid such as water in an open container, the temperature and thus the magnitude of the vapor pressure both rise. At some temperature the vapor pressure of water is equal to the atmospheric pressure (see Figure 5-22). At this temperature, large bubbles form in the water, and the rate at which water vapor is produced becomes quite rapid. When this happens we say that the water boils, and we call this characteristic temperature the **boiling point** of water (see Figure 5-23).

Figure 5-23   Boiling water. A liquid boils in an open container at the temperature at which the vapor pressure of the liquid is equal to the atmospheric pressure. The vapor pressure of water at 99.6°C is 749 mmHg. Thus, when the atmospheric pressure is 749 mmHg, water boils at 99.6°C.

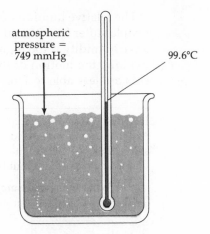

Since atmospheric pressure varies from day to day and place to place, the temperature at which water boils is not always the same. For example, if the atmospheric pressure is 760 torr (1.000 atm), then the boiling point of water is 100°C. But if the atmospheric pressure is 733 torr, then the temperature at which water boils is 99°C. The **normal boiling point** of a liquid is defined as the temperature at which the vapor pressure of the liquid is equal to the standard pressure of 1.000 atm. Thus the normal boiling point of water is 100°C. Usually, when the properties of a liquid are given, the term "boiling point" is used without the word "normal." When this is done, the standard pressure of 1.000 atm is implied even though it is not stated explicitly.

Relative Humidity

Water vapor is always present in the air around us, and our well-being depends on this fact. When we described the evaporation of water from an open container or from a person's skin, we actually assumed that the surrounding air was not already saturated with water vapor. This is usually the situation found inside buildings. But when the surrounding air is saturated, no evaporation will occur.

Humidity is the term we use to describe the amount of water vapor in air. The **relative humidity** of air is a measure of how close the air is to being saturated with water vapor. The partial pressure of water in air is a measure of the amount of water vapor in the air. Air that is saturated with water vapor at a given temperature has a partial pressure of water equal to the vapor pressure of water at that temperature. Thus the relative humidity is defined by the equation

(5-19) $$\text{Relative humidity} = \left(\frac{\text{observed partial pressure of water vapor}}{\text{vapor pressure of water at that temperature}} \right) \times 100\%$$

The closer the relative humidity is to 100%, the wetter the air is. Air saturated with water vapor has a relative humidity of 100%. Dry air has a very low relative humidity.

As we have seen, humans use the evaporation of water as a means of maintaining a constant body temperature. On the other hand, the loss of too much water can be quite serious. The rate of evaporation of water from the human body depends on the relative humidity and temperature of the surrounding air. A person exposed in hot, dry air can rapidly become dangerously dehydrated. Therefore, to remain alive outdoors in a desert, a person must slow down the rate of evaporation by wearing garments that cover most of the body.

The relative humidity of air is important for a number of other reasons. Many people suffer from the effects of sinus or bronchial discomfort whenever the relative humidity is too high or too low. In addition, when the relative humidity is too low, the motion of the cilia in the mucous membranes decreases, and the cilia are less able to filter out airborne bacteria.

Exercise 5-11

At 25°C liquid water and water vapor are in equilibrium in the left-hand container shown below. The volume of this container is decreased while the temperature is maintained at 25°C. After equilibrium has been reestablished:

  (a) What is the pressure of the gas?
  (b) What change (if any) has taken place in the amount of liquid?

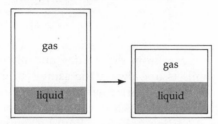

Exercise 5-12

What is the partial pressure of water vapor in air at 25°C if the relative humidity is 50%? The vapor pressure of water at 25°C is 23.8 torr.

5-6   ENTHALPY, ENTROPY, FREE ENERGY, AND THE DIRECTION OF CHANGE

Notice the ball above the rock in Figure 5-24. If the rock is removed, the ball will roll down the hill. It will not remain still or roll up the hill. Recall from our discussion in Chapter 1 that the potential energy of a mass is lower the closer it is to the center of the earth, where the attractive force is greater. Thus, the ball on the hill is a simple example of a general natural tendency for systems to go to a state of lower potential energy. There are many common examples, however, where this tendency to go to lower potential energy *cannot* explain what change will actually take place in a system.

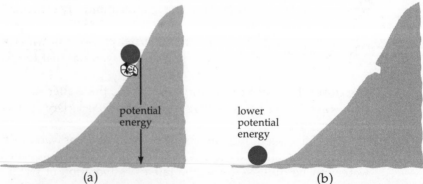

(a)   (b)

Figure 5-24   If the rock is removed, the ball will roll down the hill. The ball has a lower potential energy at the

bottom of the hill in (b) than it does higher up the hill in (a).

Consider the diffusion of ink, which we discussed in Section 5-5 (see Figure 5-18). The ink has essentially the same potential energy whether it is concentrated into a small region or has diffused and spread uniformly throughout the whole liquid volume. Yet we all know what happens when a drop of ink is put into some water. We also know that once the ink has spread throughout the volume of water, it will not "diffuse back" and re-form a concentrated ink drop in one part of the liquid.

Consider another example. In a dry room, water will evaporate from your skin. Recall that evaporation of water from your skin extracts heat energy from your body, and that water molecules in the gas state have a higher potential energy than water molecules in the liquid state. When a system at a constant pressure absorbs heat energy, we say that its heat content or **enthalpy** increases. Thus, water molecules in the gas state have a larger potential energy and a higher enthalpy than water molecules in the liquid state. Why does liquid water evaporate and go to the gas state where water has a higher potential energy or higher enthalpy? If we consider the fundamental difference between a rolling ball and molecules that are diffusing or evaporating, we can begin to get an answer to this question.

Let us return once again to our ink droplet. Ink molecules that are spread throughout the entire volume of water constitute a more disordered or random arrangement of molecules than ink molecules concentrated in a small droplike region. Similarly, as we discussed previously, there is more disorder or randomness to water molecules in the gas state than to water molecules in the liquid. Therefore, in both diffusion and evaporation the direction of change is toward a state with less order or greater randomness.

Scientists use the term **entropy** to describe the amount of randomness in a system. The larger the entropy of a system, the less order or more randomness the system has. We could say that the direction of change in diffusion or evaporation is toward a state of higher entropy. The "push" behind diffusion and evaporation is the natural tendency for a system to go to a state of higher entropy (see Figure 5-25).

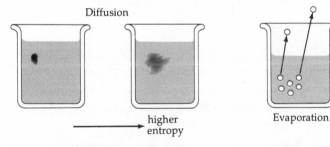

Figure 5-25   The "push" behind both diffusion and evaporation is the natural tendency for a system to go to a state of more disorder or larger entropy.

Our discussion of enthalpy and entropy effects is actually quite general. If we are interested in predicting the possibility of a system undergoing a chemical or physical change, we have to consider how both the enthalpy and entropy would change. We can say that there is a natural tendency for any system to make a spontaneous change to some state which has (1) a lower potential energy or lower enthalpy and (2) a greater randomness or higher entropy. Since a **spontaneous change** is one that can take place by itself, without work having to be done on the system, diffusion, evaporation, and a ball rolling down a hill are all examples of spontaneous changes.

A ball on a hill will go down the hill to a state of lower potential energy if it is not prevented from doing so. In this case the "push" behind the spontaneous process is the tendency to go to a state of lower potential energy. The molecules inside the ball behave the same way and thus have the same entropy at the top or bottom of the hill.

Ink molecules in water diffuse away from a more concentrated region. The "push" behind this spontaneous process is the tendency to go to a state of higher entropy (see Figure 5-25). There is no significant difference in potential energy or enthalpy whether the ink is in a concentrated region or spread throughout the water.

But what happens in the case of the evaporation of a liquid or the melting of a solid, where there is both an enthalpy change and an entropy change? In these situations, the tendency of a system to go *both* to a state of lower enthalpy and to a state of higher entropy can be taken into account with a quantity that is called **free energy.** Recall that Δ is the general symbol we use to represent a change. The change in free energy is related to the change in enthalpy and entropy by the following equation:

(5-20) Δ in free energy = (Δ in enthalpy) $- T(\Delta$ in entropy)

In Eq. 5-20, T is the temperature on the Kelvin scale.

We use the letter H to represent enthalpy; therefore, ΔH refers to a change in enthalpy. The letter S is used to represent entropy, so that ΔS refers to a change in entropy. Finally, the letter G is used to represent free energy and ΔG represents a change in free energy. Therefore, Eq. 5-20 can be rewritten as

(5-21) $\Delta G = \Delta H - T(\Delta S)$

We can predict whether or not a system can undergo a given change spontaneously using the following general principle:

(5-22)   Under conditions of constant temperature and pressure, a system can undergo a spontaneous change if and only if the ΔG is negative.

Note that:

1. For a change to a state of *lower enthalpy*, ΔH is negative and this will tend to make ΔG negative and favor a spontaneous process.

2. For a change to a state of *higher entropy*, ΔS and $T(\Delta S)$ are positive, but $-T(\Delta S)$ is negative. Thus a positive ΔS tends to make ΔG negative and favors a spontaneous process.

3. For a change where the lower enthalpy "push" and the higher entropy "push" are in opposite directions, ΔG can be used to determine whether or not this change can occur spontaneously.

Let us apply these ideas to one example—the change of water from the liquid to the solid state or from the solid to the liquid state.

(5-23) $H_2O(solid) \rightleftharpoons H_2O(liquid)$

Double arrows are used in Eq. 5-23 to indicate that the reaction can go in either direction. At any temperature liquid H_2O has both a higher enthalpy and a higher entropy than solid H_2O (see Figure 5-26). Remember that the liquid state has less order or more randomness than the solid state (Section 5-5), and that heat is absorbed when ice melts, raising the enthalpy of water. Thus the natural tendency for a system to go to the state of lower potential energy or

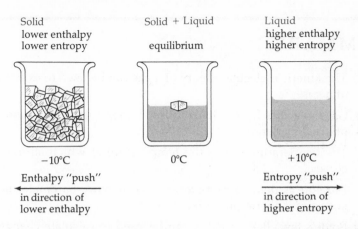

Figure 5-26   The stable state of the substance water is a solid (ice) for temperatures below 0°C and a liquid for temperatures above 0°C. At 0°C both solid and liquid water can coexist in equilibrium.

lower enthalpy generates a "push" in the direction of solid H_2O, but the natural tendency for a system to go to a state of higher entropy generates a "push" in the direction of the liquid state. We might say that the enthalpy "push" favors the solid and that the entropy "push" favors the liquid. Which one wins? We know that the answer to this question depends on the temperature:

Below 0°C: If the temperature is below 0°C, a sample of H_2O(liquid) will freeze spontaneously. Thus, below 0°C, the enthalpy "push," favoring the solid, is larger than the entropy "push" in the direction of the liquid, and ΔG for the change H_2O(liquid) $\rightarrow$ H_2O(solid) is negative.

Above 0°C: At temperatures above 0°C, ice melts spontaneously. At these temperatures, the entropy "push" favoring the liquid outweighs the enthalpy "push" favoring the solid, and ΔG for the change H_2O(solid) $\rightarrow$ H_2O(liquid) is negative. Note that ΔS in Eq. 5-21 is multiplied by the temperature T. Thus at higher temperatures the term $T(\Delta S)$ is larger and more important in determining the sign of ΔG and the direction of spontaneous change.

At 0°C: Solid and liquid water can coexist in equilibrium at 0°C. At this temperature the entropy "push" and the enthalpy "push" just balance one another, and ΔG for either the change H_2O(liquid) $\rightarrow$ H_2O(solid) or the change H_2O(solid) $\rightarrow$ H_2O(liquid) is zero.

The approach we have used to discuss liquid and solid water can be used quite generally. All chemical and physical changes can be analyzed in terms of a "push" toward a state of lower enthalpy (a negative ΔH), and a "push" toward a state of higher entropy (a positive ΔS). Quite often these "pushes" are in opposite directions, and then the direction in which a spontaneous change takes place is determined by which "push" is larger (in which direction ΔG is negative). When the "pushes" in opposite directions have the same magnitude, $\Delta G = 0$, and an equilibrium state is possible.

Exercise 5-13
The melting point of benzene, C_6H_6, is 5.5°C.
   (a) For the change benzene (solid) $\rightarrow$ benzene(liquid), the ΔH is _____ (positive, negative, zero) and ΔS is _____ (positive, negative, zero).
   (b) At what temperatures will benzene(solid) spontaneously melt? What is the sign of ΔG for the spontaneous change benzene(solid) $\rightarrow$ benzene(liquid)?
   (c) At what temperatures will benzene(liquid) spontaneously freeze? What is the sign of ΔG for the spontaneous change benzene(liquid) $\rightarrow$ benzene(solid)?
   (d) At what temperature can solid and liquid benzene coexist in equilibrium? At this temperature what is ΔG for the change benzene(solid) $\rightarrow$ benzene(liquid)?

5-7   SUMMARY

1. The kinetic molecular theory of a gas can be used to explain the behavior of a gas.

2. Gas molecules have a distribution of molecular speeds and a distribution of kinetic energies.

3. The temperature 0 K is the temperature at which all molecular motion ceases.

4. Pressure is defined as force/area. A mercury barometer is used to measure atmospheric pressure.

5. Boyle's law, $P_1V_1 = P_2V_2$, can be used to calculate changes in pressure and volume for a fixed amount of gas at a fixed temperature.

6. The relationship $P_1/T_1 = P_2/T_2$ can be used to calculate changes in pressure and temperature for a fixed amount of gas at a fixed volume.

7. Charles' law, $V_1/T_1 = V_2/T_2$, can be used to calculate changes in volume and temperature for a fixed amount of gas at a fixed pressure.

8. The combined gas law, $P_1V_1/T_1 = P_2V_2/T_2$, can be used to calculate changes in pressure, volume, and temperature for a fixed amount of gas.

9. The partial pressure of one gas in a gas mixture is that part of the total pressure due to that particular gas.

10. Dalton's law, $P_{total} = P_a + P_b + P_c + \cdots$, states that the total pressure of a gas mixture is equal to the sum of the partial pressures of the component gases.

11. The atoms, molecules, or ions in a solid occupy fixed positions in an ordered regular pattern.

12. Solid substances can be classified on the basis of the type of attractive force between the atoms, molecules, or ions into (a) ionic solids, (b) molecular solids, (c) extended covalent solids, and (d) metals.

13. A system in which opposing molecular processes compensate for each other, and where no observable change can be seen, is said to be in a state of dynamic equilibrium.

14. When the liquid and gas states of a substance are in dynamic equilibrium, the pressure of the substance in the gas state is called the vapor pressure of the substance.

15. The vapor pressure of a substance increases as the temperature increases.

16. The boiling point of a substance is the temperature at which the vapor pressure is equal to the atmospheric pressure. The normal boiling point of a substance is the temperature at which the vapor pressure is equal to the standard pressure of 1.000 atm.

17. The entropy of a system is a measure of the amount of randomness or disorder of the system at the molecular level. The higher the entropy, the less ordered or more random the system is. At the molecular level, liquids are more disordered than solids but more ordered than gases.

18. There is a natural tendency for a nonequilibrium system to go to a state of lower enthalpy and higher entropy.

19. A system can undergo a spontaneous change (under conditions of constant temperature and pressure) if and only if the change in free energy, ΔG, for the change is negative.

PROBLEMS

1. Using the assumptions of the kinetic molecular theory, explain why the pressure of a fixed amount of gas in a rigid container decreases when the temperature decreases.

2. Why does a weather balloon filled with helium expand as it rises in the atmosphere?

3. What is the level of mercury in the column on the right of the accompanying diagram if the atmospheric pressure is 730 torr and the pressure of the gas is 500 torr?

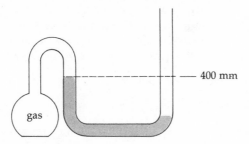

4. Some nitrogen in a 2.50-liter container at 100°C has a pressure of 3.00 atm. The nitrogen is cooled until the temperature is 50°C. Calculate the new pressure of the gas.

5. With no temperature change, 4.0 liters of oxygen is compressed to 0.30 liter. The new pressure of the gas is 7.0 atm. Find the pressure of the gas before it was compressed.

6. Some nitrogen in a 4.2-liter container has a pressure of 2.35 atm. All of this nitrogen is transferred to a larger container with a volume of 12.3 liters. What is the new pressure of the gas?

7. Some SO_2 gas in a container at 27°C has a pressure of 3.2 atm. The temperature of the gas is changed until the new pressure is 5.1 atm. Calculate the new temperature of the gas.

8. Some CO_2 gas in a 5.0-liter container at 20°C has a pressure of 1.3 atm. All of this CO_2 gas is transferred to a new container with a volume of 3.0 liters. The temperature of the gas in the new container is 0°C. What is the pressure of the CO_2 in the new container?

9. Some O_2 gas at 1000 torr in a 10.0-liter container is transferred to a 24.6-liter container in which the temperature is 27°C. The pressure of the gas in the 24.6-liter container is 1.00 atm. Calculate the temperature of the gas when it was in the 10.0-liter container.

10. A sample of air at 37°C (assumed to contain O_2, CO_2, N_2, and water vapor) has a total pressure of 740 torr. If the partial pressure of O_2 is 110 torr, the partial pressure of N_2 is 565 torr, and the partial pressure of CO_2 is 30 torr, what is the partial pressure of water vapor in this sample of air?

11. Which of the following would you expect to form an ionic solid: NH_3, CaF_2, SO_2, Mg, $NaNO_3$, H_2S?

12. Using the assumptions of the kinetic molecular theory, explain why the vapor pressure of water increases as the temperature increases.

13. The vapor pressure of water at 25°C is 25 torr. The partial pressure of $H_2O(gas)$ must be 25 torr for which of the following:
    (a) 10.0 g of $H_2O(gas)$ in a closed container at 25°C
    (b) 10.0 g of $H_2O(liquid)$ in an open container at 25°C
    (c) 10.0 g of $H_2O(liquid)$ and 10.0 g of $H_2O(solid)$ in an open container at 25°C
    (d) 10.0 g of $H_2O(liquid)$ and 10.0 g of $H_2O(gas)$ in a closed container at 25°C?

14. The vapor pressure of water at 25°C is 25 torr. For the change $H_2O(liquid) \rightleftharpoons H_2O(gas)$ in a closed container at 25°C and a partial pressure of H_2O of 25 torr: (a) Is ΔH positive, negative, or zero? (b) Is ΔS positive, negative or zero? (c) Is ΔG positive, negative, or zero?

SOLUTIONS TO EXERCISES

5-1 Gases can be compressed quite easily according to the assumptions of the kinetic molecular theory because most of the volume of a gas is empty space and there are no forces between gas molecules.

5-2 The suspended solid particle moves randomly because it is hit by air molecules. At a higher temperature air molecules are moving faster on the average and hit the suspended solid particle more often and with more force.

5-3 (a) Consider the gas pushing down on the mercury in one side of the bent tube and the atmospheric pressure pushing down on the other side. Thus, gas C has the lowest pressure. Gas A has a pressure greater than the atmospheric pressure. (b) At the levels indicated by the dashed lines, the pressure is the same on both sides of the bent tube.

$$\text{Pressure of gas A} = 35 \text{ mmHg} + 745 \text{ mmHg} = 780 \text{ mmHg}$$

$$\text{Pressure of gas B} = 745 \text{ mmHg}$$

Therefore, Pressure of gas C + 75 mmHg = 745 mmHg, so that

$$\text{Pressure of gas C} = 670 \text{ mmHg}$$

5-4 As air is pumped in, the pressure rises. The increase in the number of air molecules in the cuff causes a larger pressure. This effect is larger than the increase in volume of the cuff.

5-5 The temperature and amount of the gas remain constant, so Boyle's law applies. The new volume is larger than the initial volume, so the pressure decreases. The volume-ratio factor is then 5.5 liters/6.0 liters. Thus, the new pressure is

$$1.0 \text{ atm} \times \frac{5.5 \text{ liters}}{6.0 \text{ liters}} = 0.92 \text{ atm}$$

5-6 The pressure and amount of the gas remain constant, so Charles' law applies. The new temperature is less than the initial temperature, so the volume will decrease. Therefore the temperature-ratio factor is 263 K/298 K ($-10°C = 263$ K; $25°C = 298$ K). Thus, the new volume is

$$1.20 \text{ liters} \times \frac{263 \text{ K}}{298 \text{ K}} = 1.06 \text{ liters}$$

5-7 The amount of gas is fixed, so the combined gas law applies. Since the new pressure is less than the original pressure, the pressure-ratio factor must tend to increase the volume, so it must be 1.10 atm/0.90 atm. Since the new temperature ($37°C = 310$ K) is larger than the original temperature ($20°C = 293$ K), the temperature-ratio factor must tend to increase the volume, so it must be 310 K/293 K. Thus, the new volume is

$$1.40 \text{ liters} \times \frac{1.10 \text{ atm}}{0.90 \text{ atm}} \times \frac{310 \text{ K}}{293 \text{ K}} = 1.8 \text{ liters}$$

5-8 $P_{\text{atmospheric}} = P_{O_2} + P_{N_2}$
245 mmHg = P_{O_2} + 196 mmHg
Therefore, P_{O_2} = 49 mmHg on top of Mt. Everest, compared to 158.2 mmHg when the atmospheric pressure is 760 mmHg.

5-9 Ionic solids: K_2CO_3 (contains K^+ and CO_3^{2-} ions); LiCl (contains Li^+ and Cl^- ions).

5-10 SiO_2 and CO_2 are covalent compounds. The high melting point of SiO_2 is indicative of an extended covalent solid, and the low melting point of CO_2 indicates that CO_2 is a molecular solid.

5-11 (a) As long as liquid and vapor are in equilibrium at 25°C, the pressure of the gas will be the vapor pressure of water at 25°C, which is 23.8 mmHg.

(b) As the volume is decreased, some of the gas condenses and forms an additional amount of liquid. The decrease in the number of water molecules in the gas offsets the decrease in volume and the pressure remains constant.

5-12 If the relative humidity is 50%, then the partial pressure of water vapor is one-half the vapor pressure at 25°C, or $\frac{1}{2} \times 23.8$ mmHg = 11.9 mmHg (Eq. 5-19).

5-13 (a) Both ΔH and ΔS are positive.

(b) Above 5.5°C, where ΔG for the change benzene(solid) → benzene(liquid) is found to be negative.

(c) Below 5.5°C, where ΔG for the change benzene(liquid) → benzene(solid) is found to be negative.

(d) At 5.5°C, where ΔG for the change benzene(solid) → benzene(liquid) or the change benzene(liquid) → benzene(solid) is zero.

CHAPTER 6

Chemical Reactions: Stoichiometry and Energy Changes

6-1 INTRODUCTION

When you go to a hardware store to buy nails, you usually need a certain number of nails to do a job. However, nails are generally sold by the pound, because it is more convenient to weigh a large number of nails than to count them.

In a similar way, we measure amounts of chemical substances by determining their masses, even though we are interested primarily in the numbers of atoms or molecules. Not only is it inconvenient, it is actually impossible to count out large numbers of atoms or molecules. On the other hand, it is relatively easy to determine the mass of a substance. For example, when hydrogen and oxygen react to form water, two molecules of H_2 and one molecule of O_2 react to form two molecules of water ($2H_2 + O_2 \rightarrow 2H_2O$). In this reaction there is a simple relationship between the number of hydrogen molecules that react, the number of oxygen molecules that react, and the number of water molecules that are formed. This is a characteristic feature of all chemical reactions. To determine the number of molecules or atoms in a sample of a substance, we need only know the relationship between the mass of the substance, which we can measure, and the number of molecules or atoms per unit of mass.

Consider our analogy to nails. A carpenter would know from experience, for example, that 1 lb of a certain size and type of nails contains 200 nails. If a second type of nails are twice as heavy, then 1 lb of these nails would contain only 100 nails. In this chapter we shall see how to relate a given mass of a chemical substance to the number of molecules the substance contains.

Consider another analogy. Eggs and doughnuts are often packaged and sold by the dozen. We shall see that it is also convenient to group atoms or molecules into "packages" called *moles*. The only difference between a dozen and a mole is that there are 12 objects in a dozen, whereas a mole contains an enormous number—6.022×10^{23}—of atoms or molecules.

In addition to the relationship between mass and number of atoms or molecules, we shall discuss the relationship between the quantities of reactants and products in a chemical reaction, as well as the energy changes involved in chemical reactions.

Vendors of fruit and vegetables frequently sell their wares on the basis of weight rather than number.

6-2   STUDY OBJECTIVES

After studying the material in this chapter, you should be able to:

1. Use a table of atomic weights to determine (a) the relative masses of atoms of different elements; (b) the molecular weights or formula weights of compounds; and (c) the relative masses of molecules of different compounds, given the molecular framework of the compounds.

2. Define the terms gram atomic weight, gram molecular weight, gram formula weight, Avogadro's number, and mole.

3. Find (a) the mass, (b) the number of moles, or (c) the number of atoms, molecules, or formula units of a substance, given its chemical formula and any one of these measures.

4. Do the following types of problems when the amounts of reactants and products are expressed as either moles or mass, using a balanced equation for the reaction:
   (a) Given the amount of one reactant, determine the amounts of the other reactants and the amounts of the products formed.
   (b) Given the amount of product formed, determine the amounts of the reactants.
   (c) Given the amounts of all the reactants, determine the maximum amount of products that can be formed and the amount of reactants left unused when a maximum amount of product is formed.

5. Explain why energy is released when some chemical reactions occur, whereas energy is required in order for other chemical reactions to take place; define the terms exothermic reaction and endothermic reaction.

6. Define the terms free energy, spontaneous reaction, nonspontaneous reaction, and exergonic reaction, and know the relationship between the change in free energy for a reaction and the maximum amount of useful work that can be accomplished when the reaction takes place.

7. Calculate the maximum amount of useful work that can be obtained from a specified amount of a reactant, given the value of the free energy change per mole for that reaction.

6-3   ATOMIC AND MOLECULAR WEIGHTS

As we mentioned in Section 6-1, we measure amounts of chemical substances in terms of *mass*, but in chemical reactions it is the *number* of atoms or molecules of the reactant substances that is of fundamental importance. For example, suppose that we want to compare the effectiveness of two drugs, A and B, in treating a certain disease. We wish to give one group of patients drug A, while a second group is to receive drug B. For a useful comparison we also want to give both groups of patients the same number of drug molecules. Suppose we know that a molecule of B is twice as heavy as a molecule of A. Then, if we give each patient in one group 30 mg of drug A, we must give each patient in the other group 60 mg of drug B in order for the patients in both groups to get the same number of drug molecules.

Notice that in this example we do not have to know the actual mass of either a molecule of A or a molecule of B in order to determine the ratio of the mass of drug B to the mass of drug A. It was sufficient to know the relative masses of A and B molecules. We know that the ratio of the mass of a molecule of B to the mass of a molecule of A is 2 to 1 (or 2/1). We can summarize our discussion in the form of a simple equation. For samples of drug A and drug B to contain the same number of molecules, the following ratios must be equal:

(6-1)
$$\frac{\text{mass of drug A}}{\text{mass of drug B}} = \frac{\text{mass of a molecule of A}}{\text{mass of a molecule of B}}$$

It is quite important that you understand the difference between the actual mass of an object and the relative masses of two objects. Usually, chemical problems concerned with amounts of substances are similar to the example involving drugs A and B in that the actual masses of the atoms or molecules are not needed. It is sufficient to know the *relative masses* of the atoms or molecules involved. For this reason, a scale of relative masses, called the atomic weight scale, is used a great deal in chemistry.

Atomic Weight

In the nineteenth century, chemists were forced to use a scale of relative masses because they could not determine the actual mass of any atom of any element. They established a scale of atomic weights by assigning a value of 1.00 to a hydrogen atom and determining the atomic weights of other elements in relation to this value. A carbon atom, for example, which is about 12 times heavier than a hydrogen atom, has an atomic weight of 12 on this scale; see Figure 6-1.

Figure 6-1   Schematic illustration of the relative masses (atomic weights) of carbon and hydrogen atoms. Twelve hydrogen atoms and one carbon atom have the same mass.

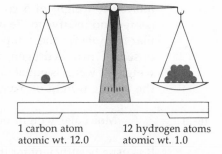

1 carbon atom   12 hydrogen atoms
atomic wt. 12.0   atomic wt. 1.0

After the existence of isotopes was discovered, it was more reasonable to base an atomic weight scale on an atom of a single isotope. Recall from Chapter 2 that isotopes of an element have different masses. In 1961, scientists throughout the world agreed to use an atomic weight scale based on assigning the value 12.000 to an atom of the isotope carbon-12. Atomic weights of the elements relative to carbon-12 are given in the table on the inside back cover of this text and also in the periodic table on the inside front cover. The difference between the atomic weight for any element based on carbon-12 and the atomic weight for that element used by chemists in the nineteenth century is extremely small.

Recall the atomic mass scale, which we introduced in Section 2-8. In that section we said that because (1) the mass of an atom depends almost entirely on the number of neutrons and protons in the atom and (2) both neutrons and protons have a mass very close to 1 on the atomic mass scale, (3) the mass of any atom is close to the integer A, where A is the mass number of the atom (A = number of neutrons + number of protons). However, *atomic weight* values for many elements are not close to integer values. For example, the atomic weight of the element chlorine is 35.453. This is because all naturally occurring substances containing the element chlorine contain a mixture of the isotopes ^{35}Cl and ^{37}Cl; about 75% of the chlorine atoms are ^{35}Cl and about 25% are ^{37}Cl. The atomic weight for the element chlorine represents an average value for this mixture of isotopes (see Figure 6-2). Similarly, the **atomic weight** value for any element is an average value for the naturally occurring mixture of isotopes of that element relative to the mass of an atom of carbon-12. For example, using a table of atomic weights, the mass of an average oxygen atom compared to the mass of an average hydrogen atom is equal to 15.9994/1.0079 or 15.874. Thus, an average oxygen atom is about 16 times heavier than an average hydrogen atom.

Figure 6-2   Any naturally occurring chlorine sample has a mixture of the isotopes ^{35}Cl and ^{37}Cl in the proportions of about 75% ^{35}Cl and 25% ^{37}Cl. The average value for this mixture is the atomic weight of chlorine: $(0.75 \times 35) + (0.25 + 37) = 35.5$.

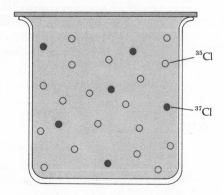

Molecular Weight and Formula Weight

If we know the molecular formula for a compound, it is quite simple to use a table of atomic weights to obtain the molecular weight of that compound. The molecular weight of a compound is the mass of an average molecule of that compound relative to the mass of an atom of carbon-12. For example, the molecular formula for the compound water is H_2O, so that a water molecule is composed of two hydrogen atoms and one oxygen atom. Hence the molecular weight of water is equal to the sum of twice the atomic weight of hydrogen plus the atomic weight of oxygen.

(6-2)   Molecular weight of $H_2O = (2 \times 1.0079) + 15.9994 = 18.0152$

It is usually sufficient for the problems in this text to calculate molecular weights to three significant figures. Using three significant figures the molecular weight of a water molecule is 18.0. This means that a water molecule is about 18 times heavier than a hydrogen atom (see Figure 6-3). The **molecular weight** of a compound is equal to the sum of the atomic weights for all the atoms in a molecule of that compound. For example, the molecular weight of hydrogen (H_2) is about 2.0, whereas that of methane (CH_4) and glucose ($C_6H_{12}O_6$) are about 16 and 180, respectively.

Figure 6-3   Schematic illustration of the relative masses of a water molecule and hydrogen atoms. Eighteen hydrogen atoms and one water molecule have the same mass.

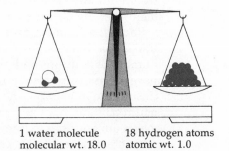

1 water molecule   18 hydrogen atoms
molecular wt. 18.0   atomic wt. 1.0

As we discussed in Section 4-3, ionic compounds in the solid state do not contain molecular units with a small number of atoms (or ions) in each unit. Consequently, we cannot, for example, call the chemical formula NaCl for the ionic compound sodium chloride a molecular formula. The relative weight of a NaCl *formula unit* involving one sodium atom and one chlorine atom is 58.5 (the sum of the atomic weight of sodium, 23.0, and the atomic weight of chlorine, 35.5). We call 58.5 the **formula weight** for NaCl.

We can summarize our discussion as follows:

(6-3)   Molecular weight or formula weight = sum of the atomic weights for all the atoms in the molecule or formula unit

---

Exercise 6-1

Using 16 for the atomic weight of oxygen, 12 for the atomic weight of carbon, and 18 for the molecular weight of water, fill in the blanks below.

(a) The mass of three oxygen atoms is _____ times the mass of one carbon atom.
(b) The mass of three oxygen atoms is equal to the mass of _____ carbon atoms.
(c) The mass of three carbon atoms is equal to the mass of _____ water molecules.

Exercise 6-2

Calculate the molecular weight or formula weight for the following compounds:

(a) H_2CO_3 (b) NaH_2PO_4 (c) $CH_3CH_2CH_2CH_3$
(d) $Ca(OH)_2$ (e) $C_2H_4O_2$

6-4   AVOGADRO'S NUMBER AND MOLES

Using atomic weights we know that an atom of oxygen is about 16 times heavier than an atom of hydrogen. Therefore, for a sample of the element oxygen to contain the *same number* of atoms as a sample of the element hydrogen, the mass of the oxygen sample must be about 16 times greater than the mass of the hydrogen sample. For example, a 1.0079-g sample of the element hydrogen and a 15.9994-g sample of the element oxygen contain the *same number* of atoms. (Compare this with the example of drugs A and B in Section 6-3.) A sample of hydrogen with a mass of 1.0079 g is a convenient size sample because it has a mass in grams that is numerically equal to the atomic weight of hydrogen. This size sample of hydrogen is given a special name. It is called one gram atomic weight of hydrogen. Similarly, one gram atomic weight of the element oxygen has a mass of 15.9994 g. Thus the number of atoms of hydrogen in 1 gram atomic weight of hydrogen is equal to the number of atoms of oxygen in 1 gram atomic weight of oxygen.

For any element, a sample of that element with a mass in grams numerically equal to the atomic weight is called one **gram atomic weight, GAW.** One gram atomic weight of any element contains the same number of atoms. The number of atoms in 1 GAW of any element is called **Avogadro's number,**[*] and it is given the symbol N. Therefore, there are N atoms of carbon in 12.0111 g of carbon, N atoms of hydrogen in 1.0079 g of hydrogen, N atoms of oxygen in 15.9994 g of oxygen, and so on.

---

[*] Lorenzo Romarro Amadeo Carlo Avogadro di Quaregua e di Cerreto (Amadeo Avogadro), 1776–1856, was an Italian scientist whose work was of fundamental importance in the development of the atomic molecular model of matter.

Similarly, for any covalent compound, a sample of that compound with a mass in grams numerically equal to the molecular weight is called one **gram molecular weight, GMW.** There are Avogadro's number of molecules in 1 GMW of a covalent compound. For ionic compounds, one **gram formula weight, GFW,** contains Avogadro's number of formula units. For example, there are N molecules of glucose, $C_6H_{12}O_6$, in 180 g of glucose, N formula units of NaCl in 58.5 g of sodium chloride, and N atoms of carbon in 12 g of carbon (see Figure 6-4).

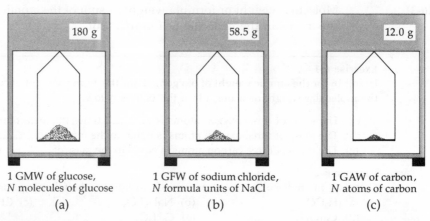

1 GMW of glucose, N molecules of glucose	1 GFW of sodium chloride, N formula units of NaCl	1 GAW of carbon, N atoms of carbon
(a)	(b)	(c)

Figure 6-4   Schematic illustration of a type of chemical balance that indicates the mass of the sample directly. (a) The molecular weight of glucose is 180. The quantity 180 g, which is 1 GMW of glucose, contains N (Avogadro's number) glucose molecules. (b) The formula weight of sodium chloride is 58.5. The quantity 58.5 g is 1 GFW of NaCl, and it contains N NaCl formula units. (c) The atomic weight of carbon is 12.0. The quantity 12.0 g is 1 GAW of carbon, and it contains N carbon atoms.

The value of Avogadro's number, N, the number of atoms in 1 GAW or the number of molecules in 1 GMW or the number of formula units in 1 GFW, has been determined from experimental data. N has the value 6.022×10^{23}. This is an extremely large number!

Atoms and molecules are very small, so there are a very large number of atoms in 1 GAW of an element or 1 GMW or 1 GFW of a compound. Similarly, there are a very large number of atoms and molecules in any size sample of any substance we actually use. For example, even a small sample of carbon, with a mass equal to the mass of a penny, contains about 0.5 GAW of carbon or some 3×10^{23} carbon atoms.

The Mole

Because the samples of elements and compounds we use contain very large numbers of atoms or molecules, chemists find it convenient to group a collection of identical atoms or molecules into sets containing Avogadro's number, or 6.022×10^{23}, atoms or molecules. We accomplish a similar thing when we group a collection of eggs into sets of 12. We call 12 objects 1 dozen and, for example, we refer to a collection of 24 eggs as 2 dozen eggs. Avogadro's number, or 6.022×10^{23} objects, is called **1 mole.** Moles are used to refer to the number of identical atoms, molecules, or formula units in a sample in the same manner as dozens are used when referring to numbers of eggs or doughnuts. For example, we refer to the collection of 12.044×10^{23} water molecules as 2.0000 moles of water, and when we use the phrase 3.0 moles of water we are speaking about a collection of 18×10^{23} water molecules.

For a collection of any objects, there is a simple relationship between the number of moles and the number of objects, which is similar to the relationship between the number of dozens and the number of objects. Compare the following relationships:

(6-4)   1.0 dozen = 12 objects

(6-5)   1.000 mole = 6.022×10^{23} objects

Note that in either Eq. 6-4 or Eq. 6-5 the objects could be eggs, oranges, atoms, molecules, or anything else. We can use Eq. 6-5 and the general principles for conversion of units (see Essential Skills 3) to convert from number of objects to moles or from moles to number of objects. Recall that when converting units, an equation such as Eq. 6-5 can be used to obtain two conversion factors. In this case the conversion factors are

(6-6) $$\frac{6.022 \times 10^{23} \text{ objects}}{1.000 \text{ mole}} \quad \text{and} \quad \frac{1.000 \text{ mole}}{6.022 \times 10^{23} \text{ objects}}$$

Recall also that when we convert units we use whichever of the two conversion factors gives us the correct units for our answer. Let us consider an example.

EXAMPLE 1   A sample of methane, CH_4, contains 39×10^{23} molecules. How many moles of methane are there in this sample?

In this problem we want to convert number of molecules to moles. The conversion factor that will give us the unit mole is the second factor in Eq. 6-6. Therefore,

(6-7) $$\text{moles of methane} = (39 \times 10^{23} \text{ molecules}) \times \frac{1.000 \text{ mole}}{6.022 \times 10^{23} \text{ molecules}}$$
$$= 6.5 \text{ moles}$$

Mass and Moles

One gram atomic weight of any element contains 6.022×10^{23} atoms of that element or 1 mole of atoms. Another way of expressing this fact is to say that the mass of 1 mole of atoms of an element is equal to the gram atomic weight for that element. For example, the mass of 1 mole of hydrogen atoms is 1.0079 g, the mass of 1 mole of oxygen atoms is 15.9994 g, and so on. Similarly, the mass of 1 mole of any covalent compound is equal to the gram molecular weight for that compound and the mass of 1 mole of any ionic compound is equal to the gram formula weight for that compound. For example, the mass of 1 mole or the mass per mole (mass/mole) of hydrogen molecules, H_2, is 2.0158 g/mole; the mass/mole for water molecules, H_2O, is about 18 g/mole; and the mass/mole for NaCl is about 58.5 g/mole. Notice that the mass of 1 mole of hydrogen molecules is twice the mass of 1 mole of hydrogen atoms because each hydrogen molecule contains two hydrogen atoms.

The relationship between the number of moles in a sample of a chemical substance and the mass of the sample is very simple. The mass of a sample is directly proportional to the number of moles.

Recall from Essential Skills 3 that (1) the factor-unit method for solving numerical problems is applicable whenever two quantities are directly proportional to one another; and (2) the amount of the quantity we want to determine is related to the amount of the known quantity by

(6-8)   Amount of quantity we want to determine
= (amount of known quantity) × (appropriate factor)

The appropriate factor relating mass and moles for a given substance is obtained from the atomic weight, molecular weight, or formula weight for the substance. Let us consider a few examples.

EXAMPLE 2   What is the mass of 2.5 moles of the compound carbon dioxide, which has the molecular formula CO_2?

The appropriate factor relating mass and number of moles for CO_2 is obtained from the molecular weight of CO_2. The molecular weight of carbon dioxide is $12 + 16 + 16 = 44$, which means that the mass of 1 mole of CO_2 is equal to 44 g. Therefore, for CO_2 we have the two factors: 44 g/1.0 mole and 1.0 mole/44 g.

In this problem, since we want to determine the mass of 2.5 moles, we want to multiply 2.5 moles by an appropriate factor that will give us *grams* as the units in our answer. The appropriate factor is 44 g/1.0 mole, and we obtain

(6-9) $$\text{Mass of 2.5 moles of } CO_2 = (2.5 \text{ moles}) \times \frac{44 \text{ g}}{1.0 \text{ mole}} = 110 \text{ g}$$

EXAMPLE 3   How many moles and how many molecules are there in a 120-g sample of benzene, which has the molecular formula C_6H_6?

The molecular weight of benzene is $(6 \times 12) + (6 \times 1) = 78$. Thus, for benzene the factors relating mass and moles are 78 g/1.0 mole and 1.0 mole/78 g. When we multiply the given mass of benzene, 120 g, by the appropriate factor 1.0 mole/78 g, we get an answer with the units *moles*. Therefore

(6-10) $$\text{No. of moles of benzene (in 120 g)} = (120 \text{ g}) \times \frac{1.0 \text{ mole}}{78 \text{ g}} = 1.5 \text{ moles}$$

We can obtain the *number of molecules* if we multiply 1.5 moles by the conversion factor 6.022×10^{23} molecules/1.000 mole. Thus,

(6-11) $$1.5 \text{ moles of benzene} = (1.5 \text{ moles}) \times \frac{6.022 \times 10^{23} \text{ molecules}}{1.000 \text{ mole}}$$

$$= 9.0 \times 10^{23} \text{ molecules of benzene}$$

Let us summarize our discussion so far. We have considered three different ways of referring to the amount of a substance: (1) the mass of the substance; (2) the number of moles of the substance; and (3) the number of atoms, molecules, or formula units of the substance. For example, we can refer to the *same amount* of the element oxygen in any of the following three ways: (1) 32 g of oxygen; (2) 1.0 mole of oxygen; or (3) 6.022×10^{23} molecules of oxygen.

Given any one of these three ways of referring to the amount of a substance, and given its chemical formula, we can determine the other two. The factors that relate the mass of the substance and the number of moles of that substance can be obtained from the atomic weight, molecular weight, or formula weight for the substance. Avogadro's number, N (6.022×10^{23} objects/mole), is the factor that relates the number of moles and the number of atoms, molecules, or formula units. These relationships are illustrated schematically as follows.

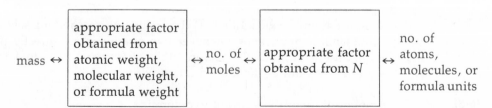

Exercise 6-3

(a) How many moles of glucose, $C_6H_{12}O_6$, are there in 1.0 g of glucose?   (b) How many molecules of glucose are there in 1.0 g of glucose?

Exercise 6-4

Calculate the mass of 3.5×10^{24} molecules of carbon monoxide, CO.

Molar Volume of a Gas

It is easier to measure the volume of a gas sample than it is to measure its mass. The volume of a gas sample, however, varies to a large extent with the temperature and pressure. We can use the volume of a gas sample as a measure of the amount of the gas, provided that we specify the temperature and pressure. The temperature and pressure most often specified are 0°C (273 K) and 1.00 atm, conditions called **standard temperature and pressure,** and abbreviated STP.

The volume of 1 mole of any gas at STP is 22.4 liters. The volume of 1 mole of gas is called the **molar volume.** Thus, for any gas at STP the molar volume is 22.4 liters/mole.

At STP, the volume of a gas sample is directly proportional to the number of moles of gas in the sample. Therefore, we use the factors 22.4 liters/1.00 mole and 1.00 mole/22.4 liters to relate the volume of a gas sample at STP and the number of moles of gas in the sample. Let us consider a few examples.

EXAMPLE 4   What is the volume of 0.32 mole of oxygen, O_2, at STP?

The appropriate factor relating volume of oxygen and number of moles of O_2 is 22.4 liters/1.00 mole. Therefore,

$$\text{(6-12)} \qquad \text{Volume of } O_2 \text{ at STP} = (0.32 \text{ mole}) \times \frac{22.4 \text{ liters}}{1.00 \text{ mole}} = 7.2 \text{ liters}$$

EXAMPLE 5   What is the volume of 15.0 g of oxygen at STP?

It is instructive and convenient to do this problem in two steps: (1) determine the number of moles of oxygen; and then (2) determine the volume at STP.

Step 1   The molecular weight of O_2 is 32.0, so 1 mole of O_2 has a mass of 32.0 g and

$$\text{(6-13)} \qquad \text{Moles of } O_2 = (15.0 \text{ g}) \times \frac{1.00 \text{ mole}}{32.0 \text{ g}} = 0.469 \text{ mole}$$

Step 2

$$\text{(6-14)} \qquad \text{Volume of } O_2 \text{ at STP} = (0.469 \text{ mole}) \times \frac{22.4 \text{ liters}}{1.00 \text{ mole}} = 10.5 \text{ liters}$$

The molar volume of a gas is 22.4 liters *only* at STP. If you need to know the molar volume of a gas at a temperature and pressure that are different from 273 K and 1.00 atm, you can obtain this value by using the molar volume at STP and the combined gas law (Eq. 5-14).

Exercise 6-5

A sample of nitrogen occupies a volume of 2.5 liters at STP. How many moles of nitrogen are in this sample?

Exercise 6-6

What is the volume of 82.5 g of CO_2 at STP?

6-5   CHEMICAL REACTIONS: MOLE AMOUNTS

When we consider a chemical reaction, we are often interested in such questions as (1) what amounts of reactants we need to form a certain amount of product or (2) what amount of product can possibly be formed from given amounts of reactants. In this section we shall see how to answer such questions when the amounts of the substances involved in the chemical reaction are expressed in terms of moles, and in the following section we shall consider amounts of substances expressed in terms of mass.

We know that atoms and molecules react with one another in a simple way on the basis of number. Consider, for example, the balanced chemical equation for the formation of water from hydrogen and oxygen:

(6-15)
$$2H_2 + O_2 \longrightarrow 2H_2O$$

Equation 6-15 tells us that 2 molecules of hydrogen and 1 molecule of oxygen must react to form every 2 molecules of water. So, for example, to form six molecules of water, six molecules of H_2 and three molecules of O_2 must react. Several other examples of the information contained implicitly in Eq. 6-15 are given in Table 6-1.

Table 6-1 $2H_2 + O_2 \rightarrow 2H_2O$

Amount of H_2 That Reacts	Amount of O_2 That Reacts	Amount of H_2O Formed
2 molecules	1 molecule	2 molecules
10 molecules	5 molecules	10 molecules
2 moles	1 mole	2 moles
6 moles	3 moles	6 moles
1 mole	$\frac{1}{2}$ mole	1 mole
3.2 moles	1.6 moles	3.2 moles
0.48 mole	0.24 mole	0.48 mole

As Table 6-1 shows, Eq. 6-15 tells us that (1) the amount of hydrogen that reacts and the amount of oxygen that reacts are directly proportional to one another; and (2) the amount of water formed is directly proportional to the amount of hydrogen or the amount of oxygen that reacts. We can use these facts to find useful relations between the reactants and products in the formation of water.

For each 2 moles of hydrogen that react, 1 mole of oxygen must also react, so these two quantities are related by the factors

(6-16)
$$\frac{2 \text{ moles of } H_2 \text{ (that react)}}{1 \text{ mole of } O_2 \text{ (that reacts)}} \quad or \quad \frac{1 \text{ mole of } O_2 \text{ (that reacts)}}{2 \text{ moles of } H_2 \text{ (that react)}}$$

Similarly, for 2 moles of water to be formed, 2 moles of hydrogen must react, so these two quantities are related by the factors

(6-17)
$$\frac{2 \text{ moles of } H_2O \text{ (formed)}}{2 \text{ moles of } H_2 \text{(that react)}} \quad or \quad \frac{2 \text{ moles of } H_2 \text{ (that react)}}{2 \text{ moles of } H_2O \text{ (formed)}}$$

Also, for 2 moles of water to be formed, 1 mole of oxygen must react, so these two quantities are related by the factors

(6-18)
$$\frac{2 \text{ moles of } H_2O \text{ (formed)}}{1 \text{ mole of } O_2 \text{ (that reacts)}} \quad or \quad \frac{1 \text{ mole of } O_2 \text{ (that reacts)}}{2 \text{ moles of } H_2O \text{ (formed)}}$$

Recall that the numbers in front of chemical substances in a balanced chemical equation are called *stoichiometric coefficients* (see Section 2-6). The factors spelled out in relations 6-16, 6-17, and 6-18 can be obtained by simply using the appropriate stoichiometric coefficients in the balanced chemical equation for the reaction $2H_2 + O_2 \rightarrow 2H_2O$. A similar procedure can be used for *any* balanced chemical equation. Let us see how we can use these factors in a few numerical problems.

EXAMPLE 6 How many moles of hydrogen are needed to react with 2.5 moles of oxygen to form water?

The factors relating the amounts of hydrogen and oxygen that react are given in relation 6-16. If we multiply 2.5 moles of O_2 by 2 moles H_2/1 mole O_2, our answer will have the units "moles of H_2." Therefore,

(6-19)
$$\text{No. of moles of } H_2 \text{ (that react)} = (2.5 \text{ moles of } O_2) \times \frac{2 \text{ moles of } H_2}{1 \text{ mole of } O_2}$$

$$= 5.0 \text{ moles of } H_2$$

Note that because stoichiometric coefficients are exact numbers, our answer has the correct number of significant figures.

Of course, putting materials together does not guarantee that they will react. All we have determined is that if 2.5 moles of oxygen do react to form water, then 5.0 moles of hydrogen must react as well. Thus, if we have a 2.5-mole sample of oxygen in a flask, and we want all of it to react with hydrogen and form water, we must introduce at least 5.0 moles of hydrogen into the flask. We may put more than 5.0 moles of hydrogen into the flask, but if we put in any less, it is not possible for all 2.5 moles of oxygen to react.

EXAMPLE 7 If 2.5 moles of oxygen react with 5.0 moles of hydrogen and form water, how many moles of water are formed?

The factors relating moles of water formed and moles of hydrogen that react are given in relation 6-17. In this case 2 moles of H_2O/2 moles of H_2 is the appropriate factor.

(6-20)
$$\text{No. of moles of } H_2O \text{ (formed)} = (5.0 \text{ moles of } H_2) \times \frac{2 \text{ moles of } H_2O}{2 \text{ moles of } H_2}$$

$$= 5.0 \text{ moles of } H_2O$$

We can obtain the same answer by considering the number of moles of oxygen that react and the appropriate factor from relation 6-18.

(6-21)
$$\text{No. of moles of } H_2O \text{ (formed)} = (2.5 \text{ moles of } O_2) \times \frac{2 \text{ moles of } H_2O}{1 \text{ mole of } O_2}$$

$$= 5.0 \text{ moles of } H_2O$$

Our answer to this problem is based on the information given that the oxygen and hydrogen react and form water. However, if we mix 5.0 moles of hydrogen and 2.5 moles of oxygen and not all the hydrogen and oxygen react, then less than 5.0 moles of water will be formed. Similarly, if some of the hydrogen and oxygen react, but form a compound other than water, say, hydrogen peroxide (H_2O_2), then less than 5.0 moles of water will be produced. Thus, if 5.0 moles of hydrogen and 2.5 moles of oxygen are mixed together, 5.0 moles of water will form if, and only if, all the hydrogen and all the oxygen react to form water.

EXAMPLE 8   If 7.5 moles of H_2 and 3.0 moles of oxygen are put into a container and react to form water, what is the *maximum* number of moles of water that can form?

In this example we have to be a bit careful. We can use the appropriate factor (relation 6-17) to find the number of moles of water formed if we know the number of moles of hydrogen that react. It may not be possible, however, for all 7.5 moles of hydrogen to react. The 3.0 moles of oxygen that we have may not be sufficient to react with 7.5 moles of hydrogen. Let us see.

We can determine the number of moles of oxygen that must react if 7.5 moles of hydrogen react using the appropriate factor in relation 6-16.

(6-22) $$\text{No. of moles oxygen} = (7.5 \cancel{\text{ moles of } H_2}) \times \frac{1 \text{ mole of } O_2}{2 \cancel{\text{ moles of } H_2}}$$

$$= 3.75 \text{ moles of } O_2$$

According to Eq. 6-22, if 7.5 moles of hydrogen react, then 3.75 moles of oxygen must also react. This is not possible here, because we have only 3.0 moles of oxygen. So not all the hydrogen can react. On the other hand, there is enough hydrogen for all the oxygen to react with some of the hydrogen. If we assume that all 3.0 moles of oxygen react, and use the factor 2 moles of H_2/1 mole of O_2 from relation 6-16, then 6.0 moles of hydrogen must react as well.

Now that we know the maximum numbers of moles of hydrogen and oxygen that can react, we can use the appropriate factor from relation 6-17 or 6-18 and find that 6.0 moles of water can be formed. Thus, of the initial 7.5 moles of hydrogen, 6.0 moles react and 1.5 moles remain unused (see Table 6-2).

Table 6-2 Example 8: $2H_2 + O_2 \rightarrow 2H_2O$

Before Reaction	Reacting Amounts	After Reaction
7.5 moles of H_2	6.0 moles of H_2	1.5 moles of H_2
3.0 moles of O_2	3.0 moles of O_2	0.0 moles of O_2
0.0 moles of H_2O		6.0 moles of H_2O

In this example, since the formation of the product water is limited by the amount of oxygen, we call oxygen the **limiting reagent.** Also, since there is more than enough hydrogen to react with all of the oxygen, we say that the hydrogen is in *excess*.

Note that if we had mistakenly assumed that because we had 7.5 moles of hydrogen, 7.5 moles of hydrogen could react, we would have erroneously concluded that 7.5 moles of water could be formed. Thus, in a problem in which we are given the amounts of the reactants, we must first determine the limiting reagent before we can calculate the amount of product formed.

EXAMPLE 9   An important reaction for living organisms is the reaction of the sugar glucose, which has the molecular formula $C_6H_{12}O_6$, with oxygen to form carbon dioxide and water. If 2.2 moles of glucose react with a sufficient amount of oxygen, how many moles of carbon dioxide and how many moles of water are formed?

We begin with the balanced chemical equation for this reaction,

(6-23) $$C_6H_{12}O_6 + 6O_2 \longrightarrow 6CO_2 + 6H_2O$$

According to Eq. 6-23, when 1 mole of glucose reacts, 6 moles of carbon dioxide are formed. Thus the number of moles of glucose that react and the number of moles of carbon dioxide that are formed are related by the factors

(6-24) $$\frac{6 \text{ moles of } CO_2 \text{ (formed)}}{1 \text{ mole of glucose (reacts)}} \quad \text{or} \quad \frac{1 \text{ mole of glucose (reacts)}}{6 \text{ moles of } CO_2 \text{ (formed)}}$$

If we multiply 2.2 moles of glucose by the factor 6 moles of CO_2/1 mole of glucose, our answer will have the units "moles of CO_2." Therefore,

(6-25) $$\text{Moles of } CO_2 \text{ (formed)} = (2.2 \text{ moles of glucose}) \times \frac{6 \text{ moles of } CO_2}{1 \text{ mole of glucose}}$$

$$= 13.2 \text{ moles of } CO_2$$

According to Eq. 6-23, an equal number of moles of water and carbon dioxide are formed when glucose reacts with oxygen. Thus 13.2 moles of water will also be formed.

Exercise 6-7

In the human body, the breakdown of proteins results in the formation of ammonia, NH_3. Ammonia is converted to urea, CH_4N_2O, which is excreted in the urine. The balanced equation for the latter reaction is

$$2NH_3 + CO_2 \longrightarrow CH_4N_2O + H_2O$$

If 5.2 moles of urea are formed, how many moles of carbon dioxide reacted and how many moles of ammonia reacted?

Exercise 6-8

If 4.3 moles of glucose react according to Eq. 6-23, how many moles of oxygen must also react?

Exercise 6-9

If 2.70 moles of glucose and 15.2 moles of oxygen are mixed together and react according to Eq. 6-23, what is the maximum number of moles of carbon dioxide that can form?

6-6   CHEMICAL REACTIONS: MASS AMOUNTS

In the previous section we described amounts of reactants and products in terms of moles. In experimental practice, however, the amount of a reactant or product is determined by measuring its mass. Chemical balances give the number of grams of a sample, but they cannot count molecules. How can we determine, for example, the mass of a product formed if we are given the mass of a reactant? Using what we have just learned, it is easy. If we convert the mass of the reactant into the number of moles of reactant, we can find the number of moles of product using the procedures discussed in the last section. We can then determine the mass of that number of moles of product. Similarly, any problem involving masses of reactants and products can be converted to a problem involving moles. Let us consider a couple of examples.

EXAMPLE 10   If 100 g of glucose react with oxygen to produce carbon dioxide and water, what mass of oxygen must also react, and what are the masses of carbon dioxide and water produced?

It is instructive to do this problem, and in a similar manner other problems of this type, in a series of steps: (1) write the balanced chemical equation for the reaction; (2) determine the number of moles of glucose from the given mass of glucose; (3) determine the number of moles of oxygen that react, and the number of moles of carbon dioxide and water formed, using the balanced chemical equation; and then (4) determine the mass of oxygen that reacts and the masses of carbon dioxide and water formed; (5) finally, check the answer using the law of conservation of mass. The central role of the balanced chemical reaction in problems of this type is illustrated in the following diagram.

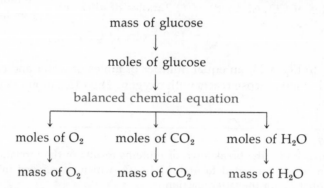

Step 1   The balanced chemical equation for the reaction is (Eq. 6-23)

$$C_6H_{12}O_6 + 6O_2 \longrightarrow 6CO_2 + 6H_2O$$

Step 2   From $(6 \times 12) + (12 \times 1) + (6 \times 16) = 180$, which is the molecular weight of glucose, we obtain the two factors, 180 g/1.00 mole and 1.00 mole/180 g, relating mass and number of moles for glucose. Thus,

$$\text{No. of moles of glucose} = (100 \text{ g}) \times \frac{1.00 \text{ mole}}{180 \text{ g}} = 0.556 \text{ mole}$$

Step 3   Using the balanced chemical equation, we obtain
(a) No. of moles of O_2 (that react)

$$= (0.556 \text{ moles of glucose}) \times \frac{6 \text{ moles of } O_2}{1 \text{ mole of glucose}} = 3.33 \text{ moles}$$

(b) No. of moles of CO_2 formed = 3.33 moles, since the number of moles of CO_2 formed is equal to the number of moles of O_2 that react.

(c) No. of moles of H_2O formed = 3.33 moles, since the number of moles of CO_2 formed and the number of moles of H_2O formed are equal.

Step 4
(a) Using the molecular weight of O_2, which is $2 \times 16 = 32.0$,

$$\text{Mass of } O_2 \text{ (that reacts)} = 3.33 \text{ moles} \times \frac{32.0 \text{ g}}{1.00 \text{ mole}} = 106 \text{ g}$$

(b) Using the molecular weight of CO_2, which is $12 + (2 \times 16) = 44.0$,

$$\text{Mass of } CO_2 \text{ (formed)} = 3.33 \text{ moles} \times \frac{44.0 \text{ g}}{1.00 \text{ mole}} = 146 \text{ g}$$

(c) Using the molecular weight of H_2O, which is $(2 \times 1) + 16 = 18.0$,

$$\text{Mass of } H_2O \text{ (formed)} = 3.33 \text{ moles} \times \frac{18.0 \text{ g}}{1.00 \text{ mole}} = 60 \text{ g}$$

Step 5   We have determined that 100 g of glucose and 106 g of oxygen react and form 146 g of carbon dioxide and 60 g of water. The total mass of the reactants, glucose and oxygen, is 206 g, which is equal to the total mass of the products, carbon dioxide and water. This equality is in agreement with the law of conservation of mass.

Engraving by Albrecht Dürer, 1514.
Perhaps the angel is contemplating
the measure of it all.

EXAMPLE 11   If 25.0 g of oxygen and 3.00 g of hydrogen are put into a container and react to form water, what is the maximum mass of water that can be formed?

In this problem we do not know whether the oxygen or the hydrogen is in excess. Therefore, we must first determine the limiting reagent as in Example 8. We can determine the maximum amount of water that can be formed by using the method of Example 10. Thus, we (1) write the balanced chemical equation for the reaction; (2) determine the moles of hydrogen and oxygen present initially; (3) determine the limiting reagent; (4) determine the maximum number of moles of water that can be formed using the number of moles of the limiting reagent; (5) determine the maximum mass of water that can be formed; and (6) check the answer using the law of conservation of mass

Step 1   The balanced chemical equation is $2H_2 + O_2 \longrightarrow 2H_2O$.

Step 2   No. of moles of O_2 = $(25.0 \text{ g}) \times \dfrac{1.00 \text{ mole}}{32.0 \text{ g}}$ = 0.781 mole

No. of moles of H_2 = $(3.00 \text{ g}) \times \dfrac{1.00 \text{ mole}}{2.00 \text{ g}}$ = 1.50 moles

Step 3   According to the balanced chemical equation, if 1 mole of oxygen reacts, 2 moles of hydrogen must react. Therefore, if 0.781 mole of oxygen reacts, the number of moles of hydrogen that react with the oxygen is

$$(0.781 \text{ mole of } O_2) \times \frac{2 \text{ moles of } H_2}{1 \text{ mole of } O_2} = 1.562 \text{ moles of } H_2$$

Since we have only 1.50 moles of hydrogen available, hydrogen is the limiting reagent and oxygen is in excess. *All* of the hydrogen can react with *some* of the oxygen.

Step 4   According to the balanced chemical equation, the number of moles of water formed is equal to the number of moles of hydrogen that react. Therefore, if 1.50 moles of hydrogen react, 1.50 moles of water are formed.

Step 5   The molecular weight of H_2O is 18.0. Therefore, the maximum mass of water that can be formed is

$$(1.50 \text{ moles}) \times \frac{18.0 \text{ g}}{1.00 \text{ mole}} = 27.0 \text{ g}$$

Step 6   When the total mass of the two reactants, hydrogen and oxygen, is 3.00 g + 25.0 g = 28.0 g, the mass of water formed is 27.0 g. If the reaction takes place, does a mass of 1.0 g disappear? No! Remember that oxygen is in excess. If 1.50 moles of hydrogen react, then 0.750 moles of oxygen must also react (for every 2 moles of hydrogen that react, 1 mole of oxygen must also react). Hence, of the 0.781 mole of oxygen present initially, 0.750 mole of oxygen reacts and 0.031 mole of oxygen remains unreacted. The mass of unreacted oxygen is then

$$(0.031 \text{ mole}) \times \frac{32.0 \text{ g}}{1.00 \text{ mole}} = 1.0 \text{ g}$$

Thus our calculations agree with the law of conservation of mass.

If we start with 25.0 g of oxygen and 3.0 g of hydrogen, or a total of 28.0 g, and if the maximum amount of water is formed, the total mass is still 28.0 g—27.0 g of water and 1.0 g of oxygen (see Table 6-3). The law of conservation of mass says that in a chemical reaction the total mass must remain unchanged. But when a chemical reaction takes place, the total number of moles usually does change. In this example, we started with 0.781 mole of oxygen and 1.50 moles of hydrogen, or a total of 2.281 moles. If the maximum possible amount of reaction takes place, we wind up with 1.50 moles of water and 0.031 mole of unreacted oxygen, or a total of 1.531 moles. Thus the total number of moles decreases by 0.75 mole. The reason for this decrease becomes clear when we look at the chemical equation for the reaction. For each 2 moles of hydrogen and 1 mole of oxygen that react, 2 moles of water are formed. Thus the total number of moles decreases by 1 mole for each 2 moles of water produced.

Table 6-3 Example 11: $2H_2 + O_2 \rightarrow 2H_2O$

Before Reaction	Reactants	After Reaction
25.0 g of O_2	24.0 g of O_2 (reacting)	27.0 g of H_2O
3.0 g of H_2	3.0 g of H_2 (reacting)	1.0 g of O_2
28.0 g = total mass	1.0 g of O_2 (nonreacting)	28.0 g = total mass
	28.0 g = total mass	

Of course, in a chemical reaction the number of *atoms* of each element does not change. In this example we started with 1.50 moles of H_2, or 3.0 moles of hydrogen atoms in the form of hydrogen molecules. If the maximum possible amount of reaction takes place, 1.50 moles of H_2O are formed. This amount of water contains 3.0 moles of hydrogen atoms. Similarly, we started with 0.781 mole of O_2, or 1.562 moles of oxygen atoms. After the maximum possible amount of reaction takes place, we have 1.50 moles of oxygen atoms contained in water molecules and 0.062 mole of oxygen atoms in the form of 0.031 mole of unreacted oxygen molecules. Thus the total number of moles of oxygen atoms after the maximum possible amount of reaction takes place is 1.562 moles—the same number of oxygen atoms as we started with.

Exercise 6-10
Ammonia, NH_3, is produced commercially by the direct reaction of nitrogen and hydrogen: $N_2 + 3H_2 \rightarrow 2NH_3$. If 25.0 g of hydrogen react with sufficient nitrogen, what mass of ammonia is formed? What mass of nitrogen reacted?

Exercise 6-11
If 25.0 g of nitrogen and 5.00 g of hydrogen are put into a container and react to form ammonia, what is the maximum mass of ammonia that can be formed?

6-7   CHEMICAL REACTIONS: ENERGY CHANGES

Recall that potential energy is stored energy that an object possesses because of its position in relation to another object or objects (Section 1-8). Most chemical reactions involve a change in the relative position of atoms because the atoms are bonded together differently in the reactants and the products. Thus, in general, <u>a chemical reaction is accompanied by a change in potential energy.</u>

For example, for the reaction

(6-26) $2H_2(gas) + O_2(gas) \longrightarrow 2H_2O(liquid)$

the reactants involve two hydrogen atoms bound close together in a hydrogen molecule and two oxygen atoms bound close together in an oxygen molecule, whereas the product involves two hydrogen atoms and one oxygen atom bound close together in a water molecule. For reaction 6-26 the potential energy of the product is less than the potential energy of the reactants. Therefore when reaction 6-26 takes place the chemical system loses potential energy (see Figure 6-5).

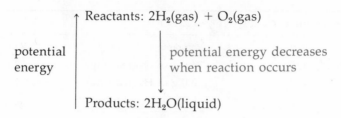

Figure 6-5   There is a decrease in the potential   when the following reaction occurs:
energy of the chemical system $2H_2(gas) + O_2(gas) \rightarrow 2H_2O(liquid)$

Recall that the total amount of energy in any nonnuclear process is constant (law of conservation of energy) but that energy can be converted from one form to another (Section 1-8). When reaction 6-26 occurs, the chemical system loses energy and this energy appears in another form or forms depending on the manner in which the reaction takes place. For example, the energy lost by the chemical system when reaction 6-26 takes place can all appear as heat that is absorbed by the surroundings. When a system at a constant pressure evolves heat, we say that its heat content or enthalpy decreases (Section 5-6). Thus, hydrogen and oxygen atoms bound together in water molecules have a lower potential energy and a lower enthalpy than the same number of hydrogen and oxygen atoms in the form of hydrogen molecules and oxygen molecules (see Figure 6-6).

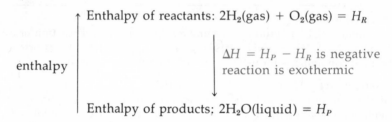

Figure 6-6   The enthalpy change,   When this reaction takes place, the
$\Delta H = H_P - H_R$,   chemical system loses enthalpy and
is negative for the reaction   an equivalent amount of heat is
$2H_2(gas) + O_2(gas) \rightarrow 2H_2O(liquid)$.   given off to the surroundings.

We represent enthalpy by the letter H, and when we consider chemical reactions we represent the enthalpy of the reactants by H_R, the enthalpy of the products by H_P, and the change in enthalpy by ΔH, where $\Delta H = H_P - H_R$. For reaction 6-26, H_P is less than H_R and thus ΔH is negative. Chemical reactions or physical changes for which ΔH is negative are called **exothermic.**

For many chemical reactions or physical changes the products have a higher potential energy and a higher enthalpy than the reactants and ΔH is positive. Chemical reactions or physical changes for which ΔH is positive are called **endothermic.**

For example, you should recall that water molecules in the gas state have a higher enthalpy than water molecules in the liquid state (Section 5-6). Thus, for the vaporization of water,

(6-27) $H_2O(liquid) \longrightarrow H_2O(gas)$

ΔH is positive and the change is endothermic.

It is quite important that you understand the following facts about the energy changes that accompany chemical reactions. For a given exothermic reaction involving a fixed quantity of reactants:

1. The chemical system loses a definite amount of energy no matter how the reaction takes place.

2. An equal amount of energy appears in the form of heat, or as a combination of heat and some form of work, depending on the specific reaction conditions.

For example, the sugar glucose, $C_6H_{12}O_6$, can react with oxygen to form carbon dioxide and water:

(6-28) $C_6H_{12}O_6 + 6O_2 \longrightarrow 6CO_2 + 6H_2O$

For reaction 6-28 (the oxidation of glucose), ΔH is -670 kcal/mole of glucose. This means that when 1 mole of glucose reacts according to Eq. 6-28, the chemical system loses 670 kcal of energy. If the oxidation of 1 mole of glucose is carried out inside a closed metal container, then all 670 kcal is evolved in the form of heat and no work is done (see Figure 6-7a). However, the oxidation of glucose also takes place inside human body cells in quite a different manner which provides the cells with a very important source of energy to do work. Cells in living organisms require energy to provide, among other things, the following essential kinds of work:

1. The chemical work needed to synthesize larger molecules, for example, the synthesis of proteins from amino acids (Chapter 24), the synthesis of nucleic acids from nucleotides (Chapter 24), and the synthesis of polysaccharides from monosaccharides (Chapter 25)

2. The work needed to transport some substances across cell membranes (Chapter 22)

3. The mechanical work involved in muscle contractions

When the oxidation of glucose takes place in our body cells, approximately 300 kcal/mole of work is accomplished and the remaining 370 kcal/mole is lost in the form of heat (see Figure 6-7b).

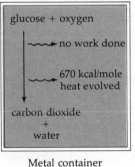
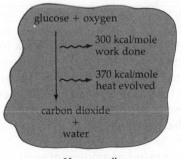

Metal container
(a)

Human cell
(b)

Figure 6-7   Reaction of glucose with oxygen to form carbon dioxide and water is an exothermic reaction. When this reaction takes place, a definite amount of energy (per mole of glucose that reacts) is converted into other forms depending on the reaction conditions. When the reaction occurs in a closed metal container (a), all the chemical energy that is released is converted into heat. When the reaction occurs in a human cell (b), some of the chemical energy released is used to accomplish work and the rest is converted into heat.

Green plants, such as the rice being cultivated here, use energy from the sun to carry out the endothermic reactions of photosynthesis. In this process, solar energy is converted to chemical energy, which is then stored by the plant as starches and sugars. When we consume the grains of rice and digest the stored compounds, exothermic reactions take place in our bodies to release this chemical energy, which is the driving force that sustains life.

Do human cells operate efficiently? Could the oxidation of glucose under other conditions produce more than 300 kcal/mole of work? It is important to know the maximum amount of work that can possibly be obtained from chemical reactions such as the oxidation of glucose.

Work can be obtained only from spontaneous chemical reactions. Recall that:

1. A spontaneous change is a change that can occur by itself without an energy input.

2. The change in free energy, ΔG, is negative for a spontaneous change.

Chemical reactions with a negative ΔG, are called **exergonic**. Thus, work can be obtained only from exergonic reactions.

The maximum amount of work (other than the small amount involved in any change of volume) that can possibly be obtained from an exergonic reaction is equal to the magnitude of the free-energy change, ΔG, for that reaction.

For the oxidation of glucose (reaction 6-28), ΔG is equal to -688 kcal/mole of glucose at 25°C and 1.0 atm. Note that the negative sign of ΔG agrees with the fact that the oxidation of glucose is a spontaneous reaction capable of providing energy to do work.

The maximum amount of work that can possibly be obtained when 1 mole of glucose undergoes reaction 6-28 is equal to the magnitude of ΔG (688 kcal). Since about 300 kcal/mole of work is actually obtained in human cells, approximately [(300 kcal/mole)/(688 kcal/mole)] × 100 or 44% of the maximum possible amount is actually obtained. Human cells are quite efficient compared to machines. For example, the amount of work actually obtained in the operation of an automobile engine is less than 10% of the maximum possible amount that could be obtained from the combustion of gasoline.

It is very important that you understand the following facts about the ΔG for a reaction:

1. The sign of ΔG tells whether or not a reaction is spontaneous, that is, if it can be a source of energy to do work. A reaction with a negative ΔG is spontaneous, but this does not mean that the reaction will *necessarily* take place. In general, spontaneous reactions will take place *only* under the right set of conditions. Reaction 6-28 will take place inside human cells at 37°C (98.6°F), but will not occur when a sample of glucose in a beaker is exposed to the oxygen in the air and warmed to 37°C.

2. The magnitude of ΔG for a reaction tells you the maximum work that can *possibly* be obtained from the reaction under any conditions. The amount of work *actually* obtained from the chemical reaction might be zero, or anything up to the maximum possible amount, depending on the specific set of conditions under which the reaction takes place.

Values of ΔG for chemical reactions are usually tabulated for 1 mole of reactant (or 1 mole of product). The free-energy change that *actually* occurs when a chemical reaction takes place is directly proportional to the number of moles of reactant (or the number of moles of product). Since we usually measure the amount of a chemical substance by its mass, we can (1) determine the number of moles of reactant from the mass of the reacting substance; and then (2) determine the free-energy change for this number of moles of reacting substance. Let us consider an example.

EXAMPLE 12 What is the maximum amount of work that can be obtained from the oxidation of 1.00 g of glucose? For this reaction $\Delta G = -686$ kcal/mole of glucose.

Step 1 The molecular weight of glucose, $C_6H_{12}O_6$, is 180. Therefore, the number of moles of glucose is

$$(1.00 \, \cancel{g}) \times \frac{1.00 \text{ mole}}{180 \, \cancel{g}} = 5.55 \times 10^{-3} \text{ mole}$$

Step 2 For 1 mole of glucose the maximum amount of work that can be obtained is 686 kcal, so the maximum amount of work that can be obtained from 1.0 g of glucose is

$$(5.55 \times 10^{-3} \, \cancel{\text{mole}}) \times \frac{686 \text{ kcal}}{1.00 \, \cancel{\text{mole}}} = 3.81 \text{ kcal}$$

Exercise 6-12

Palmitic acid, $C_{16}H_{32}O_2$, is a component of many fats. The overall equation for the oxidation of palmitic acid in human cells is

$$C_{16}H_{32}O_2 + 23O_2 \longrightarrow 16CO_2 + 16H_2O$$

ΔG for this reaction is -2340 kcal/mole of palmitic acid. Is the oxidation of palmitic acid a possible source of work? What is the maximum amount of work that can be done when 2.00 g of palmitic acid react in this manner?

Exercise 6-13

Compare the maximum amount of work that can be done when 1.0 g of the fatty acid palmitic acid reacts with oxygen with the maximum amount of work that can be done when 1.0 g of the sugar glucose reacts with oxygen.

6-8   SUMMARY

1. The atomic weight scale is based on the assignment of the value 12.000 for the mass of an atom of carbon-12. The atomic weight of an element is the mass of an average atom of that element relative to the mass of an atom of carbon-12.

2. The molecular weight or formula weight of a compound is equal to the sum of the atomic weights for all the atoms in a molecule or formula unit of the compound.

3. A mole contains Avogadro's number of objects: 1.000 mole $= 6.022 \times 10^{23}$ objects.

4. The mass of any sample of a substance is directly proportional to the number of moles. The appropriate factors relating mass and moles for a given substance are obtained from the atomic weight, or molecular weight, or formula weight for the substance.

5. STP refers to the standard temperature of 0°C and standard pressure of 1 atm. At STP the volume of a gas is directly proportional to the number of moles. The volume of 1 mole of any gas at STP is 22.4 liters.

6. The factors relating the number of moles of one reactant to the number of moles of another reactant, or the factors relating the number of moles of a product to the number of moles of a reactant, can be obtained directly from the stoichiometric coefficients in the balanced chemical equation for that reaction.

7. Any problem involving masses of reactants and products in a chemical reaction can be converted to a problem involving moles by using the relationships between mass and moles for the substances in the chemical reaction.

8. In general, a chemical reaction is accompanied by a change in potential energy.

9. Chemical reactions or physical changes for which ΔH is negative are called exothermic, and if ΔH is positive they are called endothermic.

10. Only chemical reactions with a negative ΔG, called exergonic reactions, are possible sources of work.

11. The maximum possible amount of work that can be obtained from a spontaneous chemical reaction is equal to the magnitude of the change in free energy for that reaction.

PROBLEMS

1. The mass of one molecule of SO_2 is equal to the mass of _____ molecules of O_2, _____ molecules of CH_4, and _____ molecules of H_2. Use 1, 12, 16, and 32 for the atomic weights of H, C, O, and S, respectively.

2. Determine the molecular weight or formula weight for the following compounds:
   (a) H_2SO_4 (b) $(NH_4)_2CO_3$ (c) Al_2O_3 (d) CH_4N_2O

   (e)
$$
\begin{array}{c}
\ \ \ \ \text{H}\ \ \ \ \text{O}\ \ \ \ \text{H} \\
\ \ \ \ |\ \ \ \ \ ||\ \ \ \ \ | \\
\text{H}-\text{C}-\text{C}-\text{C}-\text{H} \\
\ \ \ \ |\ \ \ \ \ \ \ \ \ \ \ | \\
\ \ \ \ \text{H}\ \ \ \ \ \ \ \ \ \ \ \text{H}
\end{array}
$$

   (f)
$$
\begin{array}{c}
\ \ \ \ \text{H}\ \ \ \ \text{O}\ \ \ \ \ \ \ \ \ \text{H}\ \ \ \ \text{H} \\
\ \ \ \ |\ \ \ \ \ ||\ \ \ \ \ \ \ \ \ \ |\ \ \ \ | \\
\text{H}-\text{C}-\text{C}-\text{O}-\text{C}-\text{C}-\text{H} \\
\ \ \ \ |\ \ \ \ \ \ \ \ \ \ \ \ \ \ \ \ |\ \ \ \ | \\
\ \ \ \ \text{H}\ \ \ \ \ \ \ \ \ \ \ \ \ \ \ \text{H}\ \ \ \ \text{H}
\end{array}
$$

3. Determine the mass of:
   (a) 3.2 moles of H_2O
   (b) 5.4×10^{-2} mole of glucose ($C_6H_{12}O_6$)
   (c) 7.3×10^{24} molecules of ammonia (NH_3)
   (d) 9.6×10^{21} molecules of carbon monoxide (CO)
   (e) 6.4×10^{-4} mole of urea (CH_4N_2O)

4. Determine the number of moles in:
   (a) 7.2×10^{-2} g of SO_2
   (b) 3.4×10^{22} molecules of N_2
   (c) 295 g of glucose ($C_6H_{12}O_6$)
   (d) 1.1 mg of methane (CH_4)
   (e) 8.4×10^{24} molecules of HCl

5. For each of the pairs given below, determine which member contains the larger number of molecules:
   (a) 6.2 g of H_2O or   6.2 moles of H_2O
   (b) 6.2 g of H_2O or   6.2 g of NH_3
   (c) 4 g of CH_4 or   16 g of SO_2
   (d) 200 g of glucose ($C_6H_{12}O_6$)   or   100 g of urea (CH_4N_2O)

6. Determine the mass of:
   (a) one molecule of glucose ($C_6H_{12}O_6$)
   (b) one molecule of SO_2

7. Consider the reaction: $N_2 + 3H_2 \rightarrow 2NH_3$.
   (a) If 3.2 moles of N_2 react, then _____ moles of H_2 must also react and _____ moles of NH_3 are formed.
   (b) If 4.8 moles of H_2 react with sufficient N_2, then _____ moles of NH_3 are formed.
   (c) To form 6.5 moles of NH_3, _____ moles of N_2 and _____ moles of H_2 must react.
   (d) If 2.6 moles of H_2 and 7.5 moles of N_2 are mixed and react, the maximum amount of NH_3 that can form is _____ moles.

8. Consider the reaction for the combustion (burning) of acetylene (C_2H_2):
   $2C_2H_2 + 5O_2 \rightarrow 4CO_2 + 2H_2O$.
   (a) If 10.0 moles of CO_2 are formed, then _____ moles of C_2H_2 and _____ moles of O_2 must have reacted.
   (b) If 3.1 moles of C_2H_2 react, then _____ moles of O_2 must also react and _____ moles of CO_2 and _____ moles of H_2O are formed.
   (c) _____ moles of O_2 are needed to react with 0.86 mole of C_2H_2.
   (d) If 2.5 moles of C_2H_2 and 6.0 moles of O_2 are mixed and react, the maximum amount of CO_2 that can form is _____ moles.

9. Consider the reaction for the formation of urea (CH_4N_2O) from NH_3 and CO_2:
   $2NH_3 + CO_2 \rightarrow CH_4N_2O + H_2O$.
   (a) To form 100 g of urea, _____ g of NH_3 and _____ g of CO_2 must react.
   (b) If 8.3 g of CO_2 react, then _____ g of NH_3 must also react, which results in the formation of _____ g of urea and _____ g of H_2O.
   (c) If 38 g of NH_3 and 45 g of CO_2 are mixed and react, the maximum mass of urea that can form is _____ g.

10. Consider the reaction for the combustion of methane:
    $CH_4 + 2O_2 \rightarrow CO_2 + 2H_2O$.
    (a) If 100 g of methane react with sufficient oxygen, _____ g of CO_2 and _____ g of H_2O are formed.
    (b) If 72 g of H_2O are formed, then _____ g of CO_2 must also be formed and _____ g of O_2 must have reacted.
    (c) If 8.00 g of CH_4 and 30.0 g of oxygen are mixed and react, the maximum amount of CO_2 that can form is _____ g.

11. When a substance burns, the substance combines with oxygen and energy is released in the form of heat. The energy released in such a reaction is called the heat of combustion. In the combustion of methane ($CH_4 + 2O_2 \rightarrow CO_2 + 2H_2O$), 213 kcal of heat are liberated per mole of methane. In the combustion of acetylene ($C_2H_2 + \frac{5}{2}O_2 \rightarrow 2CO_2 + H_2O$), 310 kcal of heat are liberated per mole of acetylene.
    (a) How much heat is evolved in the combustion of 1.0 g of methane and in the combustion of 1.0 g of acetylene?
    (b) What mass of methane and what mass of acetylene must react with oxygen (burn) to give 1000 kcal of heat?

12. In the human body a process, called glycolysis, takes place that involves the formation of lactic acid ($C_3H_6O_3$) from glucose ($C_6H_{12}O_6$). The balanced chemical equation is $C_6H_{12}O_6 \rightarrow 2C_3H_6O_3$. For this overall reaction $\Delta G = -47.0$ kcal/mole.
    (a) Can the glycolysis process in the human body be a possible source of work?
    (b) Determine the maximum amount of work that can be obtained when 2.5 g of glucose is converted to lactic acid in the glycolysis process.
    (c) Compare the maximum amount of work in glycolysis with the maximum amount of work that can be obtained when glucose reacts with O_2 and forms CO_2 and H_2O.

13. Consider the following reaction for the combustion of H_2(gas):

    $$2H_2(gas) + O_2(gas) \rightarrow 2H_2O(liquid) \qquad \Delta G = -113 \text{ kcal}$$

    Would it be possible to run an engine using the combustion of H_2 as the energy source? Explain your answer.

14. Aluminum metal is produced from the ore bauxite (Al_2O_3) by passing an electrical current through molten Al_2O_3. The overall reaction that takes place is

    $$2Al_2O_3(solid) \rightarrow 4Al(solid) + 3O_2(gas) \qquad \Delta G = +754 \text{ kcal}$$

    Does the production of aluminum metal by this method require the utilization of electrical energy, or is electrical energy generated in the process? Explain your answer.

15. What is the volume of 98 g of SO_2 gas at STP?

16. A sample of oxygen at STP has a volume of 1.0 liter. What is the mass of oxygen in this sample?

SOLUTIONS TO EXERCISES

6-1 (a) four   (b) four   (c) two

6-2 Molecular weight or formula weight is equal to the sum of the atomic weights for all the atoms in the chemical formula.
    (a) H_2CO_3: $(2 \times 1) + 12 + (3 \times 16) = 62$
    (b) NaH_2PO_4: $23 + (2 \times 1) + 31 + (4 \times 16) = 120$
    (c) $CH_3CH_2CH_2CH_3$ or C_4H_{10}: $(4 \times 12) + (10 \times 1) = 58$
    (d) $Ca(OH)_2$: $40 + (2 \times 16) + (2 \times 1) = 74$
    (e) $C_2H_4O_2$: $(2 \times 12) + (4 \times 1) + (2 \times 16) = 60$

6-3 The molecular weight of glucose, $C_6H_{12}O_6$, is 180.

$$\text{Moles of glucose} = (1.0 \text{ g}) \times \frac{1.00 \text{ mole}}{180 \text{ g}} = 5.6 \times 10^{-3} \text{ mole}$$

$$\text{Molecules of glucose} = (5.6 \times 10^{-3} \text{ mole}) \times \frac{6.022 \times 10^{23} \text{ molecules}}{1.000 \text{ mole}}$$
$$= 3.4 \times 10^{21} \text{ molecules}$$

6-4 Calculate the moles and then the mass of CO (molecular weight 28).

$$\text{Moles of CO} = (3.5 \times 10^{24} \text{ molecules}) \times \frac{1.000 \text{ mole}}{6.022 \times 10^{23} \text{ molecules}} = 5.8 \text{ moles}$$

$$\text{Mass of CO} = (5.8 \text{ moles}) \times \left(\frac{28 \text{ g}}{1.0 \text{ mole}}\right) = 160 \text{ g}$$

6-5 At STP the volume of 1 mole of any gas is 22.4 liters.

$$\text{No. of moles of nitrogen} = (2.5 \text{ liters}) \times \frac{1.00 \text{ mole}}{22.4 \text{ liters}} = 0.11 \text{ mole}$$

6-6 Determine number of moles of CO_2 (molecular weight = 44) and then the volume.

$$\text{No. of moles of } CO_2 = (82.5 \text{ g}) \times \frac{1.00 \text{ mole}}{44.0 \text{ g}} = 1.88 \text{ moles}$$

$$\text{Volume of } CO_2 = (1.88 \text{ moles}) \times \frac{22.4 \text{ liters}}{1.00 \text{ mole}} = 42.1 \text{ liters}$$

6-7 Use the balanced chemical equation to obtain the factors relating moles of urea formed to moles of carbon dioxide and moles of ammonia that react.

$$\text{No. of moles of } CO_2 \text{ that react} = (5.2 \text{ moles of urea}) \times \frac{1 \text{ mole of } CO_2}{1 \text{ mole of urea}} = 5.2 \text{ moles}$$

$$\text{No. of moles of } NH_3 \text{ that react} = (5.2 \text{ moles of urea}) \times \frac{2 \text{ moles of } NH_3}{1 \text{ mole of urea}} = 10.4 \text{ moles}$$

6-8 $\text{Moles of } O_2 \text{ that react} = (4.3 \text{ moles of glucose}) \times \dfrac{6 \text{ moles of } O_2}{1 \text{ mole of glucose}} = 26 \text{ moles}$

6-9 First determine the limiting reagent. Then determine the maximum number of moles of CO_2 that can form using the number of moles of the limiting reagent. If 2.7 moles of glucose react, then the number of moles of O_2 that must also react is

$$(2.70 \text{ moles of glucose}) \times \frac{6 \text{ moles of } O_2}{1 \text{ mole of glucose}} = 16.2 \text{ moles}$$

Since only 15.2 moles of oxygen are available, O_2 is the limiting reagent and glucose is in excess. All 15.2 moles of O_2 will react with some of the glucose. Since the number of moles of CO_2 formed is equal to the number of moles of O_2 that react, 15.2 moles of CO_2 is the maximum number of moles of CO_2 that can form.

6-10 Method: (1) determine the number of moles of H_2; (2) determine the number of moles of NH_3 and of N_2; (3) determine the mass of NH_3 and of N_2; (4) check using the law of conservation of mass.

(1) $\text{No. of moles of } H_2 = (2.50 \text{ g}) \times \dfrac{1.00 \text{ mole}}{2.00 \text{ g}} = 12.5 \text{ moles}$

(2) $\text{Moles of } NH_3 \text{ formed} = (12.5 \text{ moles of } H_2) \times \dfrac{2 \text{ moles of } NH_3}{3 \text{ moles of } H_2} = 8.33 \text{ moles}$

$\text{No. of moles } N_2 \text{ that react} = (12.5 \text{ moles of } H_2) \times \dfrac{1 \text{ mole of } N_2}{3 \text{ moles of } H_2} = 4.17 \text{ moles}$

(3) $\text{Mass of } NH_3 \text{ formed} = (8.33 \text{ moles}) \times \dfrac{17.0 \text{ g}}{1.00 \text{ mole}} = 142 \text{ g}$

$\text{Mass of } N_2 \text{ that reacts} = (4.17 \text{ moles}) \times \dfrac{28.0 \text{ g}}{1.00 \text{ mole}} = 117 \text{ g}$

(4) Total mass of reactants = 25 g + 117 g = 142 g
Mass of products = 142 g

6-11 Method: (1) determine the number of moles of N_2 and of H_2 present initially; (2) determine the limiting reagent; (3) determine the maximum number of moles of NH_3 that can form using the number of moles of the limiting reagent; (4) determine the mass of NH_3; (5) check using the law of conservation of mass.

(1) No. of moles of N_2 = $(25.0\,g) \times \dfrac{1.00\ \text{mole}}{28.0\,g}$ = 0.893 mole

No. of moles of H_2 = $(5.00\,g) \times \dfrac{1.00\ \text{mole}}{2.00\,g}$ = 2.50 moles

(2) If 2.50 moles of H_2 react, then the number of moles of N_2 that must react is

$(2.50\ \text{moles of } H_2) \times \dfrac{1\ \text{mole of } N_2}{3\ \text{moles of } H_2}$ = 0.833 mole

We have 0.893 mole of N_2 initially, therefore N_2 is in excess and H_2 is the limiting reagent.

(3) Moles of NH_3 that can form = $(2.50\ \text{moles of } H_2) \times \dfrac{2\ \text{moles of } NH_3}{3\ \text{moles of } H_2}$ = 1.67 moles

(4) Mass of NH_3 that can form = $(1.67\ \text{moles}) \times \dfrac{17.0\ g}{1.00\ \text{mole}}$ = 28.3 g

(5) If 2.50 moles of H_2 react, then the number of moles of N_2 that react is 0.833 mole, and 0.893 mole $-$ 0.833 mole = 0.060 mole of N_2 remain unreacted. Therefore the mass of unreacted N_2 is

$(0.060\ \text{mole}) \times \dfrac{28\ g}{1.0\ \text{mole}}$ = 1.7 g

Total mass initially = 25.0 g of N_2 + 5.00 g of H_2 = 30.0 g

Total mass after reaction = 28.3 g of NH_3 + 1.7 g of N_2 = 30.0 g

6-12 Since ΔG is negative, the oxidation of palmitic acid is a spontaneous reaction and a possible source of work. The molecular weight of palmitic acid is 256. Therefore,

No. of moles of palmitic acid = $(2.00\,g) \times \dfrac{1.0\ \text{mole}}{256\,g}$ = 7.81×10^{-3} mole

Thus the maximum amount of work (for 2.00 g) is

$(7.81 \times 10^{-3}\ \text{mole}) \times \dfrac{2340\ \text{kcal}}{1.00\ \text{mole}}$ = 18.3 kcal

6-13 From Exercise 6-12, the maximum amount of work that can be obtained from 1.0 g of palmitic acid is 9.2 kcal. When glucose reacts with oxygen, $C_6H_{12}O_6 + 6O_2 \rightarrow 6CO_2 + 6H_2O$, $\Delta G = -686$ kcal/mole. Therefore, since the molecular weight of glucose is 180,

No. of moles of glucose = $(1.0\,g) \times \dfrac{1.0\ \text{mole}}{180\,g}$ = 5.6×10^{-3} mole

and the maximum amount of work from 1.0 g of glucose is

$(5.6 \times 10^{-3}\ \text{mole}) \times \dfrac{686\ \text{kcal}}{1.00\ \text{mole}}$ = 3.8 kcal

Thus the maximum amount of work from 1.0 g of palmitic acid is about 9.2/3.8 = 2.5 times as much as from 1.0 g of glucose.

CHAPTER 7
Intermolecular Attractions

7-1   INTRODUCTION

We all know that water exists as a liquid at room temperature; and we have learned that a relatively large amount of energy is required to boil water, that is, to change water from the liquid state to the gas state. However, water simply should not behave the way it does! Water has a relatively low molecular weight (MW = 18), and most substances with low molecular weights exist as gases at room temperature—nitrogen (MW = 28), oxygen (MW = 32), even chlorine (MW = 71). Why then is water a liquid?

In this chapter we shall discuss the reason for the special properties of water, which result principally from the existence of particularly strong attractive forces between molecules. In general, attractive forces between molecules are called intermolecular attractions. These attractions between molecules, such as between water molecules, are weaker than covalent bonds, but they are quite important in determining the chemical and physical properties of a substance. Intermolecular attractions are critical to the existence of life. Try to imagine the disasterous consequences if water were a gas at room temperature!

Intermolecular attractions are responsible for the ability of molecular substances to exist as liquids or solids, since they hold the individual molecules of a substance together in these compressed states. There are also attractive forces between molecules of different compounds and between different parts of large molecules, attractions that are vital to the existence and function of living organisms. The interactions between water and other molecules are particularly important, since most chemical reactions in living organisms (as well as most of those carried out in laboratories) take place in an aqueous environment. In this chapter we shall therefore discuss intermolecular attractions in aqueous environments. We shall see that the attractive forces between molecules can be divided into three types: (1) London dispersion forces, which are the major attractive interactions between most molecules; (2) dipolar forces, which contribute to a small degree to the intermolecular attractions between polar molecules; and (3) a specialized attractive force, called a hydrogen bond, which is extremely important in the structure and function of biological molecules.

An oil-coated seabird is but one grim piece of evidence of an oil spill. Because of differences in intermolecular attractions, oil does not dissolve in water.

7-2   STUDY OBJECTIVES

After studying the material in this chapter, you should be able to:

1. Define and give examples of intermolecular attractions, dipole moments, polar molecules, nonpolar molecules, London dispersion forces, and hydrogen bonds.

2. Correlate the heat of vaporization and the boiling point of a liquid with the strength of the intermolecular attractions present.

3. Predict, by using electronegativities, whether a bond is polar (has a large dipole moment) or nonpolar (has a zero or very small dipole moment).

4. Correlate the strength of London forces with molecular weight.

5. Classify molecules (or the parts of large molecules or large ions) as either hydrophilic or hydrophobic on the basis of the structural formula.

6. Determine if a hydrogen bond can form between two molecules, and if so, indicate which atoms are involved in the hydrogen bond, given structural formulas for the molecules.

7. Define and give examples of hydration of an ion.

7-3   DIPOLAR FORCES AND LONDON DISPERSION FORCES

The attractive forces *between* molecules are called **intermolecular attractions.** Energy is required in order to disrupt intermolecular attractions. The energy needed to melt ice and vaporize liquid water are examples of this. However, since water molecules remain intact when ice melts or liquid water vaporizes, the energy needed to break apart a water molecule must be quite a bit greater than the energy needed to overcome the intermolecular attractions. For example, the heat of vaporization of water is 9.72 kcal/mole, whereas more than

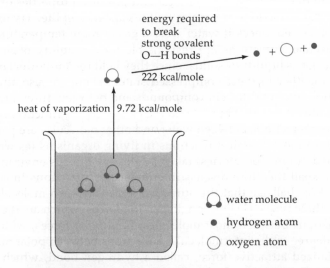

Figure 7-1   A total of 9.72 kcal of energy is needed to vaporize 1 mole of water. A much larger amount of energy, 222 kcal, is needed to break 1 mole of gaseous water into separate hydrogen and oxygen atoms.

20 times this energy is required to break the O—H bonds of water (see Figure 7-1). From these simple considerations we conclude that in water, and in other similar substances, the intermolecular attractions are much weaker than the chemical bonds that hold the atoms in a molecule together.

In general, both the heat of vaporization and the boiling point of a substance are rough measures of the intermolecular attractions in the liquid state of that substance. The larger the heat of vaporization, or the higher the boiling point, the stronger the intermolecular attractions.

Table 7-1 Boiling Points of Cl_2, Br_2, ICl, and I_2

Compound	MW	Boiling Point (K)
Cl_2	71	238
Br_2	160	332
ICl	162	370
I_2	254	457

Consider Table 7-1, which shows the boiling points of four similar compounds—Cl_2, Br_2, ICl, and I_2—which are composed of group VIIA elements and exist as diatomic molecules in the gas state. By examining the physical basis for the differences in the boiling points of these substances, we can better understand some of the important aspects of intermolecular attractions. As you can see, the boiling points of Cl_2, Br_2, ICl, and I_2 increase with increasing molecular weight. This suggests that it might be useful to plot the boiling point of these substances versus their molecular weights in order to investigate this relationship more closely. We do this in Figure 7-2, where we see that (1) there is an approximately linear relationship between boiling point and molecular weight, but (2) the boiling point of ICl is somewhat larger than the value we would predict on the basis of the boiling points of Cl_2, Br_2, and I_2. By focusing attention on an apparent exception to a general relationship, we can often obtain clues to help us better understand natural phenomena. This approach is frequently used in science. Thus, let us see how ICl differs from Cl_2, Br_2, and I_2.

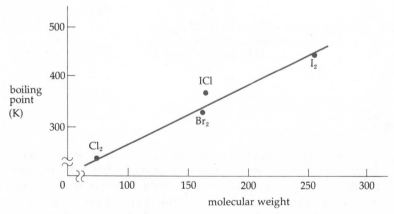

Figure 7-2   The boiling points of Cl_2, Br_2, ICl, and I_2 are plotted versus molecular weight. The boiling point of ICl is somewhat larger than one would predict on the basis of a linear relationship between boiling point and molecular weight.

Dipolar Forces

The Lewis structures for Cl_2, Br_2, I_2, and ICl, shown in Figure 7-3, indicate that there is a single covalent bond between the atoms in each of these four molecules. The pair of bonding electrons in Cl_2, Br_2, and I_2 is shared equally between the identical atoms. This is not the case, however, for ICl. Since the electronegativity of iodine (EN = 2.5) is less than that of chlorine (EN = 3.0), the bonding electrons in ICl are pulled slightly toward the more electronegative Cl atom.

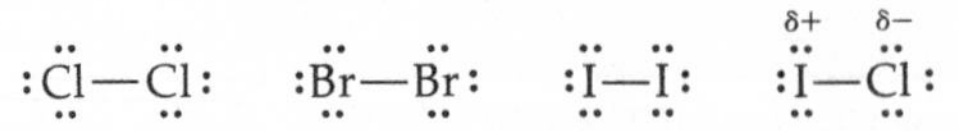

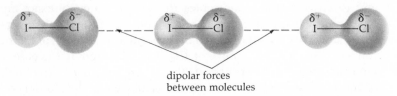

Figure 7-3   Lewis structures for Cl_2, Br_2, I_2, and ICl. The bonding electrons in ICl are pulled toward the more electronegative chlorine atom, which results in a partial negative charge on the chlorine atom and an equal partial positive charge on the iodine atom.

Molecules of ICl will tend to align themselves so that the negative end of one molecule is close to the positive end of another molecule, as shown in Figure 7-4. In this position, there is a net attractive force between each of the molecules. These are examples of dipolar forces.

dipolar forces
between molecules

Figure 7-4   Dipolar forces in ICl. Molecules of ICl tend to align themselves so that the negative end of one molecule and the positive end of another molecule are close together.

Diatomic molecules such as ICl, in which there is a separation of positive and negative charge, are said to have a **dipole moment.** A dipole moment has a direction, which can be indicated by an arrow going from the partial positive charge to the partial negative charge, as shown in Figure 7-5.

Figure 7-5   The dipole moment in ICl is indicated by the arrow going from the partial positive charge to the partial negative charge.

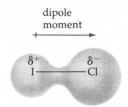

dipole moment

In molecules with more than two atoms, each polar bond (where there is a separation of positive and negative charges) has a **bond dipole moment.** The dipole moment of a molecule as a whole depends on the dipole moments of all the bonds in the molecule and on the shape of the molecule. In general, molecules that have dipole moments are called **polar molecules,** and molecules in which the dipole moment is zero or extremely small (essentially zero) are called **nonpolar molecules.** The compound ICl has a nonzero dipole moment and is thus a polar molecule, whereas Cl_2, Br_2, and I_2 are nonpolar.

In general, the attractive force between polar molecules due to their dipole moments is called a **dipolar force.** At the same distance apart, the dipolar force between two ICl molecules, for example, is considerably weaker than the attractive force between a positive ion and a negative ion. There are two reasons for this: (1) ICl molecules (and other polar molecules) have only a partial positive charge and a partial negative charge; and (2) the positive ends of the ICl molecules repel each other, as do the negative ends of the molecules.

Cl_2, Br_2, and I_2 are nonpolar, so attractive dipolar forces are not present. This is the reason that ICl has a somewhat higher boiling point than we might expect on the basis of the boiling points of Cl_2, Br_2, and I_2. We have now explained why ICl is an exception to the general relationship in Figure 7-2. We shall discuss the physical basis for this general relationship between molecular weight and boiling point shortly.

Generally, there are attractive dipolar forces between any two polar molecules but not between nonpolar molecules. Molecules that contain only nonpolar bonds or bonds with a very small polarity have a zero or extremely small dipole moment and are essentially nonpolar molecules. For example, the substance ethane is composed of carbon and hydrogen, which have almost the same electronegativity. Thus, the carbon-carbon bonds and carbon-hydrogen bonds in ethane, and in similar compounds, are nonpolar bonds.

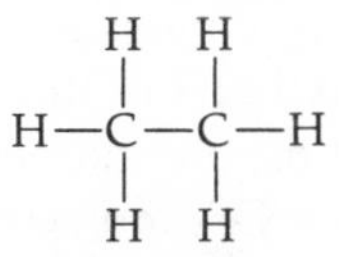

$$\begin{array}{cc} \text{H} & \text{H} \\ | & | \\ \text{H}-\text{C}-\text{C}-\text{H} \\ | & | \\ \text{H} & \text{H} \end{array}$$

ethane

Given the structural formula for a larger molecule or ion, we can divide it up into polar and nonpolar parts depending on whether the bonds in a given region of the molecule or ion are polar or nonpolar. An example of a larger molecule that contains a portion composed exclusively of carbon-carbon and carbon-hydrogen bonds, as well as a polar portion, is 1-aminohexane, shown in Figure 7-6. We shall see many more examples later.

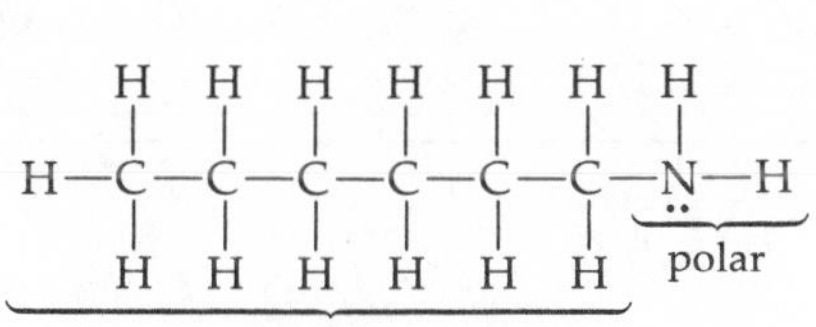

$$\begin{array}{ccccccc} \text{H} & \text{H} & \text{H} & \text{H} & \text{H} & \text{H} & \text{H} \\ | & | & | & | & | & | & | \\ \text{H}-\text{C}-\text{C}-\text{C}-\text{C}-\text{C}-\text{C}-\text{N}-\text{H} \\ | & | & | & | & | & | \\ \text{H} & \text{H} & \text{H} & \text{H} & \text{H} & \text{H} \quad \text{polar} \end{array}$$

nonpolar

Figure 7-6   A molecule of 1-aminohexane, with a polar part and a nonpolar part. The end of the molecule containing the polar nitrogen-hydrogen bonds and nitrogen-carbon bond is the polar portion.

As we mentioned previously, both the dipole moment and the polarity of a molecule as a whole depend on the shape of the molecule, as well as on the dipole moments of all the bonds in the molecule. Some small molecules with quite polar bonds are nonpolar because they have a very symmetrical shape. For example, recall that carbon dioxide, CO_2, is a linear molecule with the carbon atom in the center (see Figure 7-7a). Each carbon-oxygen bond in a molecule of carbon dioxide is quite polar, with the electrons pulled toward the more electronegative oxygen atom.

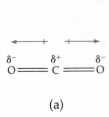

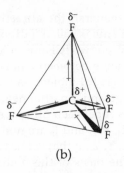

(a)   (b)

Figure 7-7   Both carbon dioxide (a) and carbon   bond dipole moments cancel one
tetrafluoride (b) are nonpolar mole-   another because of the symmetrical
cules with polar bonds. In each of   shape, and thus the overall dipole
these molecules, the individual   moment of the molecule is zero.

Because of its linear shape, however, the dipole moment in one carbon-oxygen bond in CO_2 is exactly compensated for by the dipole moment in the other carbon-oxygen bond in the opposite direction. As a result, the dipole moment of a CO_2 molecule as a whole is zero, and CO_2 is nonpolar. For a similar reason, the molecule carbon tetrafluoride, CF_4, which has four very polar carbon-fluorine bonds, is nonpolar because it is symmetrical, with the carbon atom in the center of a regular tetrahedron and the fluorine atoms at the corners (see Figure 7-7b). Here, too, the individual bond dipole moments compensate for one another, so the overall dipole moment of a CF_4 molecule is zero.

Exercise 7-1
The boiling point of N_2 (MW = 28) is 77 K; the boiling point of NO (MW = 30) is 121 K; and the boiling point of O_2 (MW = 32) is 90 K. Explain why NO has the highest boiling point of these three compounds.

Exercise 7-2
Using the electronegativity values in Figure 4-8, which of the following bonds would you expect to be very polar and thus have a significantly large bond dipole moment?
    (a) N—H   (b) C—H   (c) P—O
    (d) S—O   (e) C—N

London Dispersion Forces

Although there are no attractive dipolar forces between nonpolar molecules, the concept of a dipole was extended by F. London in 1930 and used to explain the intermolecular attraction between nonpolar molecules. London calculated the magnitude of these intermolecular attractions using complicated equations. We can construct a simple but useful picture of the origin of London dispersion forces if we allow ourselves to talk about electrons moving about in molecules and ignore for the moment the difficulties associated with this type of description for the behavior of electrons in molecules (Section 3-7).

Consider, for example, a nonpolar Cl_2 molecule, in which there is no permanent separation of charge. If the electrons are moving about, however, at any one instant there might be slightly more negative charge on one side of the Cl_2 molecule and a corresponding decrease of negative charge (a positive charge) on the other side. Thus a Cl_2 molecule might have a temporary dipole with more negative charge to the right at one instant, and at another instant a temporary dipole with more negative charge to the left (see Figure 7-8).

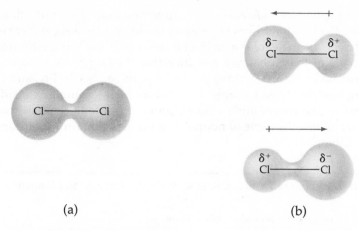

(a)   (b)

Figure 7-8   (a) Schematic illustration of the average symmetric distribution of electrons in an isolated chlorine molecule with no permanent dipole moment. (b) Schematic illustration of the temporary dipole moments in a Cl_2 molecule produced by electrons shifting from one side of the molecule to the other. The arrows indicate temporary dipole moments.

Now let us consider two separate Cl_2 molecules that are as close together as the Cl_2 molecules in solid or liquid chlorine. There will be a slight preference for the arrangements of the temporary dipoles illustrated in Figure 7-9 over other possible arrangements because in these preferred arrangements the negative charge on one molecule is close to the positive charge on the other molecule. When two Cl_2 molecules are close to one another, *related* temporary charge separations or temporary dipoles are induced in each molecule. The attractive forces between induced temporary dipoles in two molecules that are close together are called **London dispersion forces** or simply London forces. London forces result from the interaction of induced temporary charge separations, whereas dipolar forces occur only in polar molecules in which there is a permanent separation of charge. London forces are the *only* type of intermolecular attraction in nonpolar molecules, whereas *both* London forces and dipolar forces contribute to the intermolecular attractions in polar molecules.

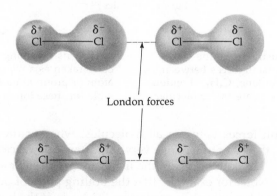

London forces

Figure 7-9   Schematic illustration of the temporary charge separations and dipole moments responsible for the London dispersion forces between two chlorine molecules. In these preferred arrangements, the temporary positive end of one molecule is close to the temporary negative end of another molecule, so there is a temporary dipolar force between the molecules.

For a group of molecules with approximately the same shape and types of atoms, the London forces increase as the molecular weight increases. In general, the larger the molecular weight of a substance, the more electrons there are in a molecule of that substance and therefore the larger the induced temporary dipoles. The increase in boiling point along the series Cl_2, Br_2, and I_2 is an example of this general trend (Table 7-1). As another example, consider the boiling points of the compounds ethane, propane, butane, and pentane, which is a similar series, this time of nonpolar molecules containing carbon and hydrogen (see Table 7-2).

Table 7-2 Boiling Points of Ethane, Propane, Butane, and Pentane

Compound	Formula	MW	Boiling Point (K)
Ethane	C_2H_6	30	185
Propane	C_3H_8	44	231
Butane	C_4H_{10}	58	273
Pentane	C_5H_{12}	72	309

For most substances, the magnitude of the London forces is determined primarily by the molecular weight. The London force between two large molecules is an interaction between each molecule taken as a whole and not between any specific atoms in the molecule (see Figure 7-10).

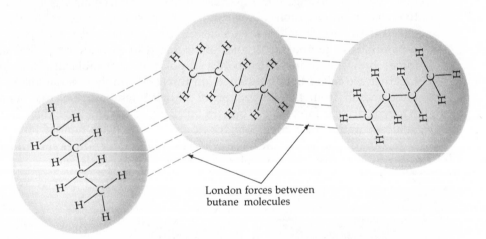

London forces between butane molecules

Figure 7-10 Schematic illustration of the London dispersion forces between molecules of butane, C_4H_{10}. London dispersion forces are intermolecular attractions involving entire molecules taken as a whole. No specific atom or group of atoms is responsible for these forces.

There are London forces between all molecules, whether they are polar or nonpolar, and with the exception of one very important special case—which we shall discuss next—London forces are the major intermolecular attraction even in polar molecules. Note, for example, that the boiling point of polar ICl is only slightly higher than what we might predict on the basis of the boiling points of the nonpolar substances Cl_2, Br_2, and I_2 (Figure 7-2).

Exercise 7-3
Using Table 7-2, predict the boiling points of methane, CH_4, and hexane, C_6H_{14}.

7-4   HYDROGEN BONDS

The one special case in which the total intermolecular attractions are very much stronger than the London forces alone involves the element hydrogen. Consider the boiling points of the hydrides of group IVA, VA, VIA, and VIIA elements shown in Figure 7-11. Note that:

1. The boiling points of the group IVA hydrides increase regularly with increasing molecular weight, as we would predict on the basis of increasing London forces with increasing molecular weight.

2. The boiling points of NH_3, H_2O, and HF, which are the lightest compounds in groups VA, VIA, and VIIA, respectively, are abnormally high.

This second observation suggests that other strong intermolecular attractions, in addition to London forces, are present in NH_3, H_2O, and HF.

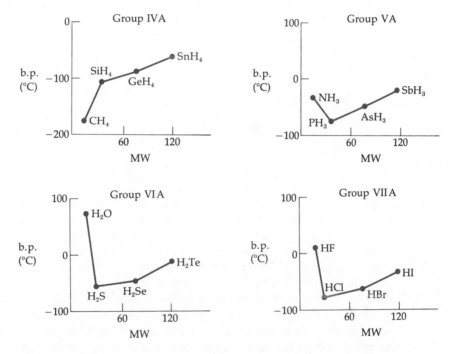

Figure 7-11   Boiling points of the group IVA, VA, VIA, and VIIA hydrides. The boiling points of NH_3 in group VA, H_2O in group VIA, and HF in group VIIA are abnormally high.

The special feature of NH_3, H_2O, and HF responsible for these strong intermolecular attractions is that these substances all involve one or more hydrogen atoms bonded to another atom that has a very large electronegativity and a relatively small size. Because the electronegativities of N, O, and F are so large compared to H, the shared electrons are pulled away from the hydrogen atom in the nitrogen-hydrogen bonds in NH_3, the oxygen-hydrogen bonds in H_2O, and the fluorine-hydrogen bond in HF. This unequal sharing results in N—H, O—H, and F—H bonds that are very polar, with a large partial positive charge on the hydrogen atoms and a large partial negative charge on the N, O, and F atoms. In addition, the small size of a hydrogen atom and the relatively small size of a N, O, or F atom allows NH_3 molecules or H_2O molecules or HF molecules to get quite close to one another. As a consequence, there is a particularly strong attractive interaction between the large partial positive charge on a hydrogen atom in one molecule of NH_3, H_2O, or HF and the large partial negative charge on the N, O, or F atom of a neighboring molecule (see Figure 7-12).

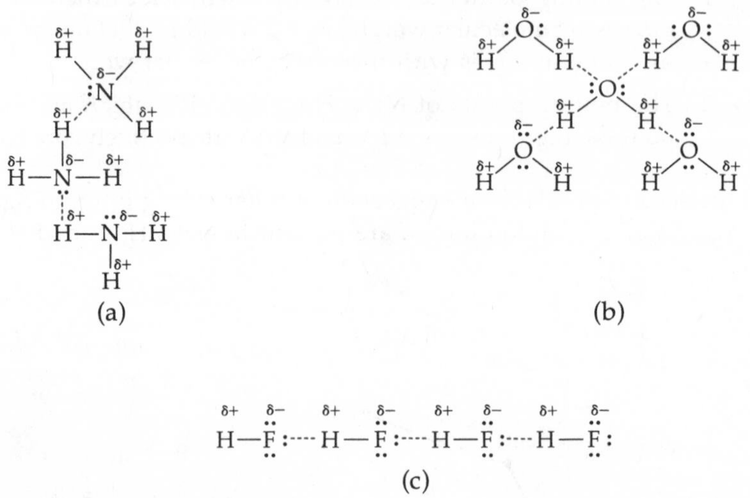

Figure 7-12   Hydrogen bonds in (a) NH_3, (b) H_2O, and (c) HF are shown   as dashed blue lines in these schematic illustrations.

This type of strong intermolecular attraction is called a hydrogen bond because a hydrogen atom is the bridge that links two interacting molecules together. In general, a **hydrogen bond** is a particularly strong intermolecular attraction between (1) a hydrogen atom in one molecule (called the *donor* molecule) bonded covalently to a small atom with a large electronegativity (principally N, O, or F) and (2) a small atom with a large electronegativity (principally N, O, or F) on another molecule (called the *acceptor* molecule).

Although hydrogen bonds are particularly strong in comparison to other intermolecular attractions, they are still considerably weaker than the covalent bonds that hold the atoms together in a molecule. The strength of a hydrogen bond depends on the atoms and molecules involved; a typical hydrogen bond is about one-tenth as strong as a typical covalent bond. You should recognize that a hydrogen bond between two molecules, in contrast to a London force, is a *localized* interaction involving specific atoms in each molecule.

Strong hydrogen bonds involve only N, O, and F atoms (in addition to hydrogen) because these are the only atoms that combine the requisite properties of large electronegativity and small size. Recall the following general trends in the periodic table: (1) electronegativity increases along a row and decreases down a column, and (2) atomic size decreases along a row and increases down a column. For this reason only the second-row hydrides in groups VA, VIA, and VIIA, namely, NH_3, H_2O, and HF, exhibit abnormally high boiling points that can be attributed to hydrogen bonding between the molecules.

As we have noted, carbon-hydrogen bonds are essentially nonpolar, since the electronegativities of carbon (2.5) and hydrogen (2.1) are almost the same. Because of this, hydrogen atoms bonded to carbon atoms do not participate in hydrogen bonds. Note that CH_4, the second-row group IVA hydride, does not have an abnormally high boiling point (Figure 7-11).

Hydrogen bonding plays an extremely important role in determining the chemical and physical properties of one particularly vital substance, water. The especially high boiling point of water indicates that there is a large amount of hydrogen bonding in water when it is in the liquid state. As indicated in Figure 7-12, a single water molecule can form hydrogen bonds with several other water molecules at the same time.

Hydrogen bonds can also form between donor and acceptor molecules that are different. For example, hydrogen bonds can form between the molecules of dimethyl ether and water, as shown in Figure 7-13a. Because chemical reactions in water are so important, the ability (or lack of ability) of molecules to form hydrogen bonds with water is an important property of a substance. In the hydrogen bonds between dimethyl ether molecules and water molecules, the water molecules are the donor molecules and the dimethyl ether molecules are the acceptors.

Ethyl alcohol (ethanol) is another substance that can form hydrogen bonds with water (see Figure 7-13b). In this case, a hydrogen bond can form with an ethyl alcohol molecule as the acceptor and a water molecule as a donor or with an ethyl alcohol molecule as a donor and a water molecule as an acceptor.

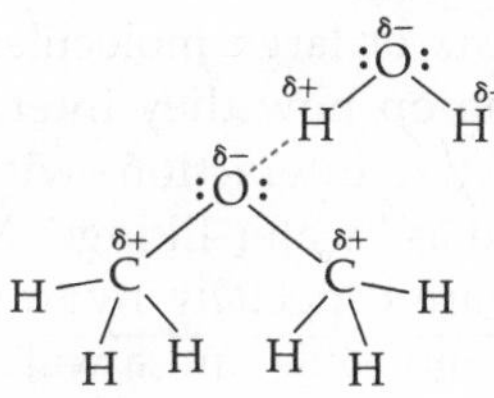

Figure 7-13   Dashed blue lines represent hydrogen bonds (a) between dimethyl ether and water and (b) between ethanol and water. In the latter case, a hydrogen bond can form either with ethanol as the acceptor and water as the donor or with ethanol as the donor and water as the acceptor.

In larger molecules, hydrogen bonds can also form between different parts of the *same* molecule. We shall see later that this type of hydrogen bonding is a critical factor in determining the structure and function of proteins and nucleic acids.

Exercise 7-4
Hydrogen bonds can form between which of the following pairs of molecules? Draw structural formulas indicating the hydrogen bonds that can form.

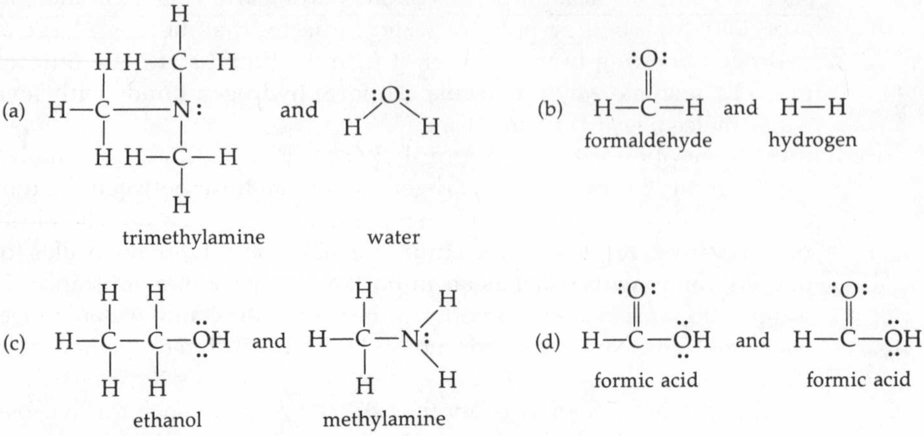

(a)
trimethylamine   water

(b) formaldehyde   hydrogen

(c) ethanol   methylamine

(d) formic acid   formic acid

7-5   INTERACTIONS WITH WATER

Hydrophilic and Hydrophobic Interactions

Molecules (or parts of large molecules or large ions) can be divided into two groups depending on how they interact with water molecules. Molecules that form strong attractive interactions with water molecules are called **hydrophilic,** which can be read as "water-liking." Molecules that form hydrogen bonds with water molecules are especially hydrophilic. The force of attraction between a very hydrophilic molecule and a water molecule is comparable to the forces of attraction between water molecules. Polar molecules, and the polar parts of molecules that do not form strong hydrogen bonds with water molecules, are hydrophilic to a lesser extent. For example, ammonia, NH_3, is a hydrophilic molecule, and the portion of a molecule of 1-hexanol involving the —OH group is also hydrophilic (see Figure 7-14a).

Molecules (or parts of large molecules) that do not form strong attractive interactions with water molecules are called **hydrophobic,** meaning "water-fearing." Nonpolar molecules are hydrophobic. Hydrophobic molecules do not repel water molecules. Rather, the force of attraction between a hydrophobic molecule and a water molecule is much weaker than the force of attraction between two water molecules. Thus methane, CH_4, is hydrophobic, and so is the part of the 1-hexanol molecule involving the nonpolar C—C and C—H

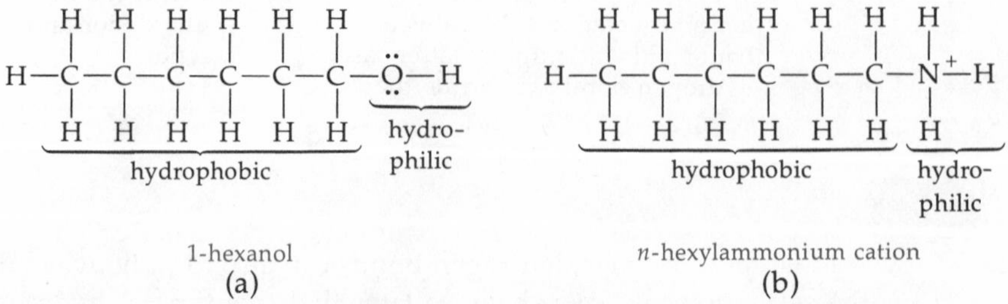

1-hexanol
(a)

n-hexylammonium cation
(b)

Figure 7-14   A molecule of 1-hexanol (a) and the n-hexylammonium cation (b)   each contain hydrophobic and hydrophilic parts.

bonds (see Figure 7-14a). When water is present, hydrophobic molecules, and the hydrophobic parts of larger molecules, tend to cluster together.

Large ions can also have hydrophobic and hydrophilic parts. For example, in the ion shown in Figure 7-14b, the positive charge is localized on the nitrogen atom. This charged part of the ion attracts water molecules around it and is hydrophilic, whereas the rest of the ion is nonpolar and hydrophobic. We shall see that hydrophilic and hydrophobic interactions are extremely important in determining the extent to which a substance will dissolve in water (Chapter 8), as well as in determining the structure and function of proteins and nucleic acids.

Exercise 7-5

Indicate the hydrophilic and hydrophobic parts of the following molecules:

(a) 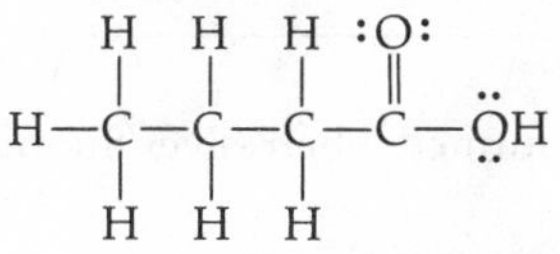 butyric acid

(b) 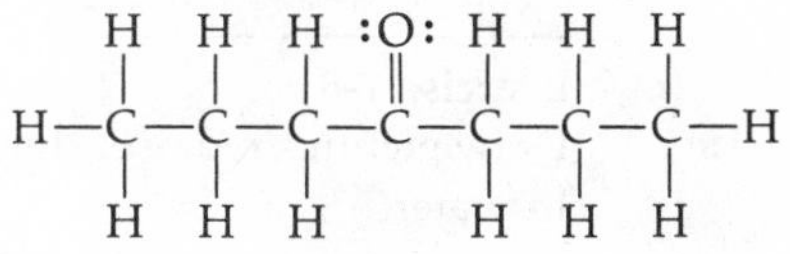 dipropyl ether

(c) 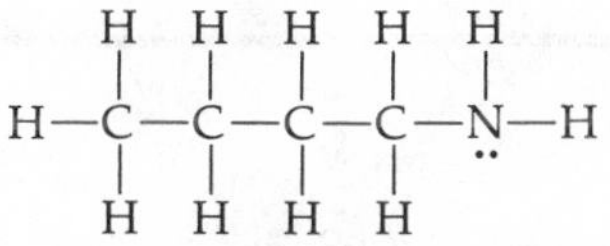 n-butylamine

Hydration of Ions

Because they possess a full electrical charge (or charges), ions interact particularly strongly with water molecules. For example, when the ionic compound NaCl dissolves in water, separate Na^+ and Cl^- ions are present. There is a strong attraction between the positively charged Na^+ cations and the partial negative charge on the oxygen atom of a water molecule. The Na^+ ion pulls several water molecules close to it, oriented so that the oxygen atoms of the water molecules are closest (see Figure 7-15a). Similarly, the attractive force between the negatively charged Cl^- ion and partial positive charge on the hydrogen atoms in a water molecule results in a cluster of water molecules forming about a Cl^- ion, but oriented so that the hydrogen atoms are closest (Figure 7-15b).

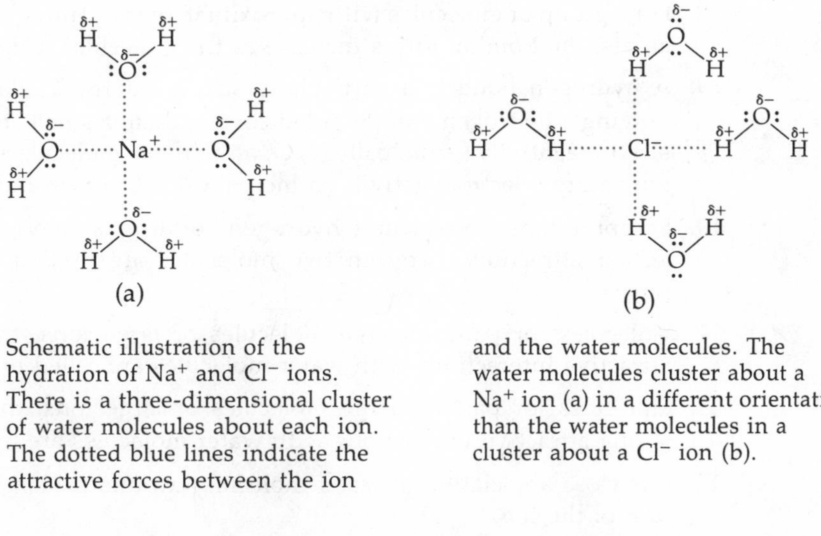

(a)   (b)

Figure 7-15   Schematic illustration of the hydration of Na^+ and Cl^- ions. There is a three-dimensional cluster of water molecules about each ion. The dotted blue lines indicate the attractive forces between the ion and the water molecules. The water molecules cluster about a Na^+ ion (a) in a different orientation than the water molecules in a cluster about a Cl^- ion (b).

The close association of water molecules around an ion is called **hydration** of the ion. The force of attraction between a hydrated ion and its cluster of associated water molecules is quite large. A measure of the strength of this attraction is the energy it would take to pull an ion out of water—the **hydration energy.** For example, in a large amount of liquid water containing 1 mole of dissolved NaCl, it would take about 185 kcal to pull out the Na^+ and Cl^- ions and form separate ions in the gas state.

The hydration energy of an ion depends on the charge and the size of the ion. The larger the amount of charge on the ion (either positive or negative) and the smaller the size of the ion, the more strongly held is that cluster of water molecules around the ion and the larger the hydration energy of the ion. We shall see in Chapter 8 that the hydration energy of ions is one of the major factors determining the extent to which an ionic substance dissolves in water.

Exercise 7-6

Draw pictures representing the hydrated ion formed when silver fluoride, AgF, dissolves in water.

7-6   SUMMARY

1. Attractive forces between molecules are called intermolecular attractions.

2. The magnitude of the heat of vaporization and the boiling point of a substance are rough measures of the intermolecular attractions in the liquid state of that substance.

3. A bond in which there is a separation of positive and negative charge has a bond dipole moment.

4. The dipole moment of a molecule as a whole depends on the dipole moments of all the bonds in the molecule and on the shape of the molecule.

5. Molecules with dipole moments are called polar molecules, and those with zero or extremely small dipole moments are nonpolar molecules.

6. Dipolar forces are the attractive forces between polar molecules due to their dipole moments.

7. The attractive forces between induced temporary dipoles in two molecules that are close together are called London forces.

8. For a group of molecules with approximately the same shape and types of atoms, the London forces increase as the molecular weight increases.

9. A hydrogen bond is a particularly strong intermolecular attraction involving a hydrogen atom bonded covalently to a small atom with a large electronegativity (principally N, O, or F) in one molecule and a small atom with a large electronegativity (principally N, O, or F) in another molecule.

10. Except in those cases where hydrogen bonding is involved, the intermolecular attractions between two molecules are predominately London forces.

11. Molecules, or parts of large molecules or large ions, that form strong attractive interactions with water molecules are called hydrophilic.

12. Molecules, or parts of large molecules or large ions, that do not form strong attractive interactions with water molecules are hydrophobic.

13. The close association of water molecules around an ion is called hydration of the ion.

PROBLEMS

1. The boiling point of CO is 83 K, and the boiling point of N_2 is 77 K. Suggest a reason why the boiling point of CO is slightly higher than the boiling point of N_2 even though they have the same molecular weight.

2. Order the following compounds with respect to their boiling points: CCl_4, CF_4, CBr_4.

3. Considering the bond dipole moments and the shape of the molecule, which of the following would you expect to be polar (i.e., have a nonzero dipole moment): (a) SO_2 (bent shape); (b) CCl_4 (tetrahedral shape); (c) H_2S (bent shape); (d) HCN (linear shape)?

4. Considering the factors that influence the strength of hydrogen bonds, explain why the data in Figure 7-11 indicate strong hydrogen bonding for NH_3 but weak or negligible hydrogen bonding for HCl, even though chlorine and nitrogen both have an electronegativity of 3.0.

5. Can the compound dimethyl ether,

$$
\begin{array}{ccc}
\text{H} & & \text{H} \\
| & \ddot{} & | \\
\text{H---C---O---C---H} \\
| & \ddot{} & | \\
\text{H} & & \text{H}
\end{array}
$$

form hydrogen bonds in the pure state? Can dimethyl ether form hydrogen bonds with water molecules?

6. A hydrogen bond can form between which of the following pairs of molecules? Draw structural formulas indicating the hydrogen bond for those cases where a hydrogen bond can form.

(a)

$$
\begin{array}{ccccccc}
\text{H} & :\text{O}: & \text{H} & & \text{H} & & \text{H} \\
| & \| & | & & | & \ddot{} & | \\
\text{H---C---C---N---H} & \text{and} & \text{H---C---O---C---H} \\
| & & \ddot{} & & | & \ddot{} & | \\
\text{H} & & & & \text{H} & & \text{H}
\end{array}
$$

(b)

$$
\begin{array}{ccccc}
& \text{H} & & & \text{H} \\
& | & & & | \\
\text{H} & \text{H---C---H} & & \text{H} & \text{H---C---H} \\
| & | & & | & | \\
\text{H---C-------N:} & \text{and} & \text{H---C-------N:} \\
| & | & & | & | \\
\text{H} & \text{H---C---H} & & \text{H} & \text{H---C---H} \\
& | & & & | \\
& \text{H} & & & \text{H}
\end{array}
$$

(c) HF and H_2O

(d)

$$
\begin{array}{ccccc}
\text{H} & & & & \text{H} \\
| & & & & | \\
\text{H---C---O---H} & \text{and} & \text{H---C---O---H} \\
| & \ddot{} & & & | & \ddot{} \\
\text{H} & & & & \text{H}
\end{array}
$$

7. Both the substances 1-hexanol (see Figure 7-14) and 1 butanol,

$$
\begin{array}{ccccc}
\text{H} & \text{H} & \text{H} & \text{H} \\
| & | & | & | \\
\text{H---C---C---C---C---O---H} \\
| & | & | & | & \ddot{} \\
\text{H} & \text{H} & \text{H} & \text{H}
\end{array}
$$

have hydrophilic and hydrophobic parts. Can you suggest a reason why 1-hexanol dissolves in water to a much lesser extent than 1-butanol?

8. Explain how water can hydrate both positive and negative ions.

SOLUTIONS TO EXERCISES

7-1 N_2 and O_2 are nonpolar. The bond in NO is polar: NO has a dipole moment and it is a polar molecule. Thus, there are attractive dipolar forces between NO molecules and no such attractive forces between nonpolar N_2 or O_2 molecules.

7-2 (a) N—H, $\Delta EN = 0.9$, very polar
     (b) C—H, $\Delta EN = 0.4$, essentially nonpolar
     (c) P—O, $\Delta EN = 1.4$, extremely polar
     (d) S—O, $\Delta EN = 1.0$, very polar
     (e) C—N, $\Delta EN = 0.5$, slightly polar

7-3 As we go along the series in Table 7-2, each substance has a MW that is 14 larger than the previous one. The corresponding increases in b.p. are 56 K (between ethane and propane), 42 K (between the butane and propane), and 36 K (between pentane and butane). We therefore predict that the b.p. of methane (MW = 16) is about 65 K less than the b.p. of ethane, or 120 K, and that the b.p. of hexane (MW = 86) is about 32 K higher than the b.p. of pentane, or 341 K. The experimentally determined b.p. of methane is 112 K, and for hexane it is 342 K.

7-4

(a) 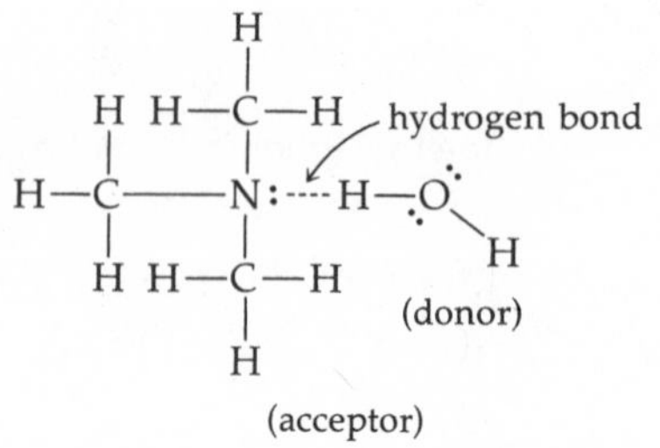 [structure: hydrogen bond between an amine and water, N: ---- H—O ; labeled (donor) for H—O and (acceptor) for N]

(b) No hydrogen bond between formaldehyde and hydrogen.

(c) 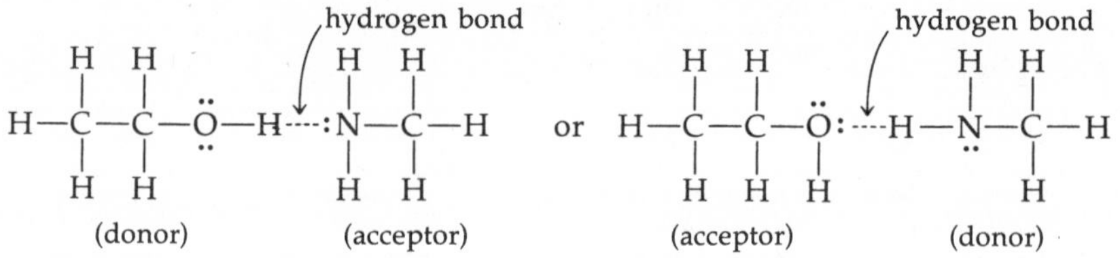 [structures showing hydrogen bonds between ethanol (donor) and an amine (acceptor), or with ethanol O as acceptor and amine N—H as donor]

(d) 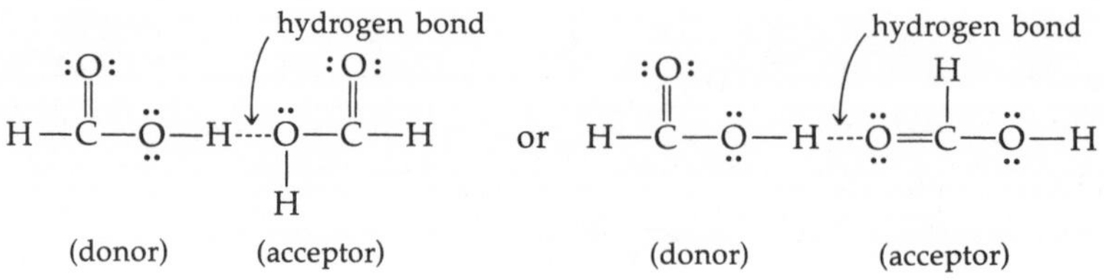 [structures showing hydrogen bonds between acetic acid molecules, donor and acceptor]

7-5

(a) 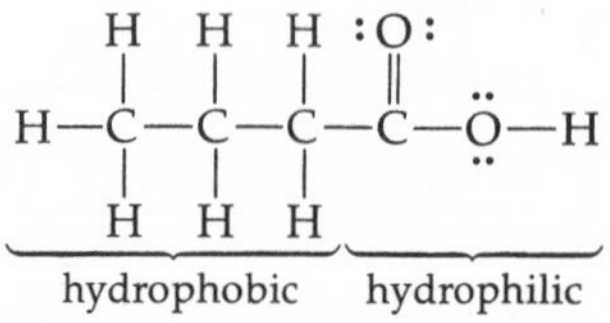 [structure H—C—C—C—C—O—H with hydrophobic and hydrophilic portions labeled]

(b) [structure with hydrophilic :O: region and two hydrophobic regions labeled]

(c) [structure H—C—C—C—C—N: with hydrophobic and hydrophilic portions labeled]

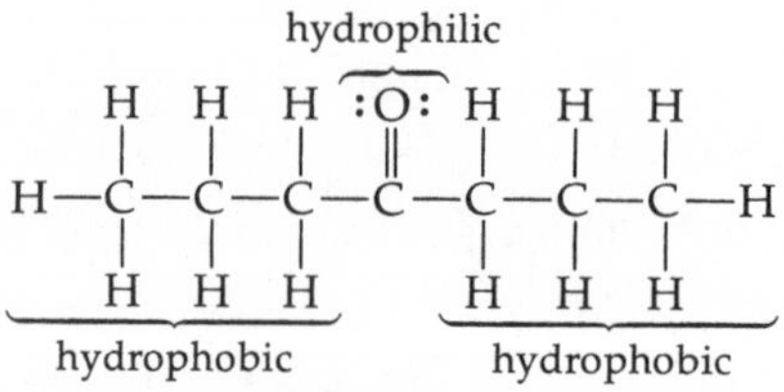

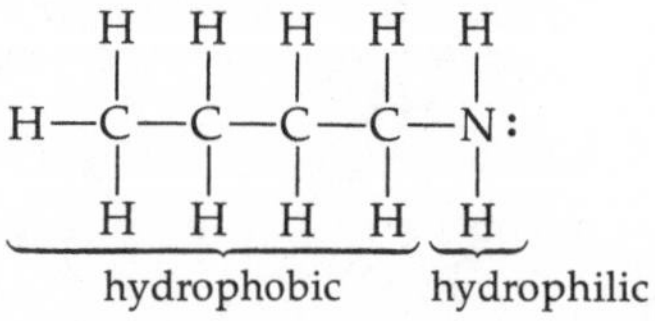

7-6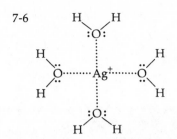

CHAPTER 8

Solutions and Colloids

8-1   INTRODUCTION

Air, soda water, salt water, gasoline, vodka, household ammonia, vinegar, dental fillings, and sterling silver are familiar examples of solutions. In Section 2-4 we defined a solution as a homogeneous mixture of two or more substances. But recognizing when to call a mixture a solution can be a problem. There is not always a clear distinction between a homogeneous mixture and a heterogeneous one. Cigarette smoke, fog, milk, and blood plasma are examples of mixtures with properties somewhat intermediate between those of clearly homogeneous and clearly heterogeneous mixtures. Mixtures of this intermediate type are called colloids, and they are quite important in biological systems.

A typical solution, such as salt water, or a glucose (sugar) solution used for intravenous feeding, contains a small amount of one substance, called the solute, dissolved in a large amount of another substance, called the solvent. Solutions in which the solvent is water are called aqueous solutions. We shall discuss aqueous solutions almost exclusively, since the solutions in the human body are aqueous solutions and they are the most common solutions found in most chemical laboratories as well.

The properties of a solution, such as the sweetness of a sugar solution, are dependent on the concentration of the solution, that is, on the amount of solute compared to the amount of solvent. We shall describe how the concentration of a solution is defined quantitatively and discuss several types of problems involving solution concentration that have important clinical applications. We shall also consider briefly the factors that determine the amount of a solute that will dissolve in a given amount of water, or the solubility of the solute in water. Finally, we shall investigate the behavior of solutions that are separated by a membrane. For example, the shriveling of a cucumber in brine to form a pickle, the absorption of water through the roots of plants, and the function of natural and artificial kidneys are all processes involving the selective passage of material from one solution to another through a membrane.

Honey is a familiar example of what is essentially a very concentrated solution of several sugars dissolved in water.

8-2   STUDY OBJECTIVES

After studying the material in this chapter, you should be able to:

1. Define the terms solution, solvent, solute, concentration, dilute solution, and concentrated solution.

2. Describe the preparation of a solution with a given concentration, determine the amount of a solution (with a known concentration) that contains a given amount of solute, and describe the preparation of a given amount of solution with a smaller concentration from a sufficient amount of solution with a higher concentration (i.e., dilute a solution), using the units of concentration g/ml, (W/V)%, or molarity.

3. Determine the normality of an ionic solution given its molarity and find the molarity given a solution's normality.

4. Define the terms saturated solution, solubility, soluble substance, insoluble substance, miscible liquid, and immiscible liquid.

5. Use the relationship between the solubility of a substance and the mass of solute in a saturated solution.

6. Use the fact that the solubility of a gas in a liquid is directly proportional to the partial pressure of the gas.

7. Use the qualitative relationship between a substance's solubility in water and its hydrophilic and/or hydrophobic characteristics.

8. Define the terms colloid, dispersed substance, dispersing medium, aerosol, foam, emulsion, sol, Brownian motion, sedimentation, and Tyndall effect.

9. Compare the properties of an aqueous colloidal dispersion with those of an aqueous solution.

10. Define the terms semipermeable membrane, osmosis, osmotic pressure, isotonic, hypotonic, hypertonic, dialysis, hemodialysis, hemolysis, and crenation.

11. Use the fact that the osmotic pressure of a solution is directly proportional to the sum of the molarities of all the dissolved or dispersed particles.

8-3   SOLUTIONS AND THEIR PREPARATION

A **solution** is a homogeneous mixture of two or more pure substances. Let us consider, for example, a solution made by dissolving some solid sugar in water. The sugar molecules are dispersed individually and randomly throughout the water and there are no clumps of sugar visible to the eye, even with the use of a microscope (see Figure 8-1). In addition, a visible drop of the sugar solution contains a vast number of sugar molecules, and each drop is just like any other drop of the solution with the same size. It is in this sense that we say that the sugar solution is homogeneous, even though it is not uniform at the molecular level—there are obvious differences between sugar and water molecules.

We usually think of forming a solution by dissolving a small amount of one substance, such as sugar, in a larger amount of another substance, such as water. In a solution such as this, the substance present in the larger amount is called the **solvent,** and the substance present in the smaller amount is called the **solute** (see Figure 8-2). For a mixture to be homogeneous, and therefore be a so-

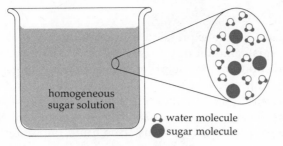

water molecule
sugar molecule

Figure 8-1   Schematic illustration of a homogeneous sugar solution. No clumps
of sugar are visible in the container
(left), and the sugar molecules can
be viewed as being individually
and randomly dispersed at the
molecular level (right).

lution, the solute particles must be sufficiently small—roughly less than 2 nm
(2×10^{-9} m) in their largest dimension—and they must not clump together to
form larger aggregates in the solvent. The dimensions of most molecules and
ions are quite a bit less than 2 nm. Biological systems, however, contain some
huge molecules, or macromolecules, such as proteins and nucleic acids, whose
dimensions are much larger than 2 nm. These macromolecules are present in
the human body in an aqueous environment but not as a solution. We shall
discuss shortly the properties of these and other mixtures containing larger-
size particles.

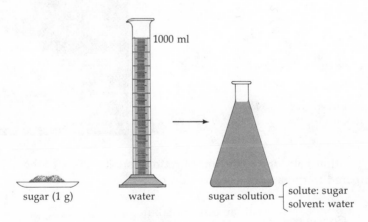

1000 ml

sugar (1 g)   water   sugar solution — { solute: sugar
                                                      { solvent: water

Figure 8-2   When 1 g of sugar is dissolved in
1000 ml of water, the result is a
sugar solution in which sugar is
the solute and water is the solvent.

We are concerned primarily with aqueous solutions because of their importance in the human body, but there are also other types of solutions. Air
(without any pollutant particles suspended in it) is a gaseous solution composed of a mixture of several gases—mostly oxygen (~20%) and nitrogen
(~80%). Soda water and household ammonia are solutions formed by dissolving in water the gases carbon dioxide and ammonia, respectively. Vodka is
a solution of two liquids—ethyl alcohol and water. Dental fillings and sterling
silver are examples of solid solutions. Dental fillings contain liquid mercury dissolved in solid silver, and sterling silver has copper dissolved in silver. We shall
limit our discussion of solutions to aqueous solutions.

Concentration

We are all familiar with the expression **dilute solution,** referring to a solution with a relatively small amount or low concentration of solute, and the expression **concentrated solution,** for a solution with a relatively large amount or high concentration of solute (see Figure 8-3). The terms "dilute" and "concentrated" describe the concentration of a solution in a rough, qualitative manner, but we shall need a more precise measure of solution concentration. Many drugs are prepared and administered as solutions. It is quite important that you know how to:

1. Prepare a solution with a given concentration.

2. Determine the amount of a solution (with a known concentration) that contains a given amount of solute.

3. Prepare a given amount of solution with a smaller concentration from a solution with a higher concentration (i.e., dilute a solution).

Figure 8-3   When a small amount of sugar is dissolved in a glass of water, the result is a dilute solution. If several cubes of sugar were to be dissolved in the same volume of water, the result would be a more concentrated solution.

Quantitatively, the term **concentration** usually refers to the amount of solute compared to the amount of solution.

(8-1)
$$\text{Concentration} = \frac{\text{amount of solute}}{\text{amount of solution}}$$

The concentration of a specific solution can be expressed in a number of different ways depending on how the amount of solute and the amount of solution are measured. In all cases, however, concentration involves a ratio of amounts. It is convenient to express concentration differently in different situations. But once you understand how to work with concentrations expressed in one manner, it is easy to use any other form.

The mass of a solid is easy to measure, and thus for solid solutes, concentration is often expressed as the mass of solute per unit volume of solution, with the mass of solute in grams and the volume of solution in milliliters.

(8-2)
$$\text{Concentration} = \frac{\text{mass of solute (g)}}{\text{volume of solution (ml)}}$$

For example, if 500 ml of a glucose solution contains 20 g of dissolved glucose, then for this solution

(8-2a) $$\text{Concentration of glucose} = \frac{20\text{ g}}{500\text{ ml}} = 0.040\text{ g/ml}$$

It is helpful to consider the parallel between the *concentration* of a solution and the *density* of a substance, which we can consider is a measure of the concentration of mass in the substance (density = mass/volume). Recall that for a sample of a substance such as water, the larger the volume of the sample the larger its mass, but the ratio of the mass compared to the volume (the density) remains the same. Similarly, for a specific solution, such as a glucose solution with a concentration of 0.040 g/ml, the larger the volume of this solution the larger the mass of glucose in it. However, although all samples of pure water have the same density (0.997 g/ml) at 25°C, we can prepare glucose solutions with different concentrations (0.040 g/ml, 0.050 g/ml, 0.0126 g/ml, etc.).

An equivalent way of expressing the relationship between the mass of solute and the volume of a solution with a specific concentration is to say that <u>the mass of solute in a solution and the volume of the solution are directly proportional to one another</u>. The appropriate factor relating the mass of solute and the volume of solution is obtained from the concentration of the solution. For example, 500 ml of a glucose solution with a concentration of 0.040 g/ml contains

$$(500 \text{ m\!l}) \times \frac{0.040\text{ g}}{1.0\text{ m\!l}} = 20\text{ g of glucose}$$

whereas 250 ml of this solution contains

$$(250 \text{ m\!l}) \times \frac{0.040\text{ g}}{1.0\text{ m\!l}} = 10\text{ g of glucose}$$

or half as much glucose (see Figure 8-4).

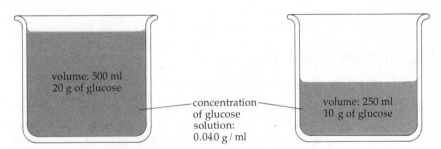

Figure 8-4   The solutions on the left and right contain different masses of glucose and have different volumes, but they both have the same glucose concentration, 0.040 g/ml.

The concentration of a solid solute is also frequently expressed as the mass of solute per 100 ml of solution. This quantity is commonly referred to as the weight-volume percent (W/V) %.

(8-3) $$\text{Concentration in (W/V) \%} = \frac{\text{mass of solute (g)}}{\text{volume of solution (ml)}} \times 100\%$$

Thus, the glucose solution we have used as an example has a concentration of 0.040 g/ml, which is 4.0 g/100 ml or 4.0 (W/V) %. The designation (W/V) is usually omitted, and this solution is simply referred to as a 4.0% glucose solution. Note that, since 1 ml of water has a mass of about 1 g, it is common practice to refer to the quantity 4.0 g/100 ml as a percent even though this quantity has units g/ml and a percent is a number without units. There is a discussion of percentages in Essential Skills 6.

For solutions with a liquid solute, concentrations are expressed as volume-volume percent, (V/V) %, since it is easy to measure the volume of a liquid.

(8-4) $$\text{Concentration in (V/V) \%} = \frac{\text{volume of solute (ml)}}{\text{volume of solution (ml)}} \times 100\%$$

Thus a solution prepared by using 15 ml of ethanol (ethyl alcohol) and enough water to give a total solution volume of 100 ml is a 15 (V/V) % ethanol solution, or simply a 15% ethanol solution.

Although it is not a good idea to do so, the designation (W/V) or (V/V) is frequently omitted when the concentration of a solution is specified. If this is done, you must know if the pure solute is a solid or a liquid in order to know whether the amount of solute was determined by its mass or its volume.

Note that all of the concentration quantities we have considered involve the volume of *solution*. In general, the volume of a solution formed with a solid solute is somewhat different from the volume of the pure solvent. For example, if you dissolve 10 g of NaCl in 100 ml of water, the volume of the resulting NaCl solution is 103 ml and not 100 ml. Also, in general, the volume of a solution formed by mixing two liquids is somewhat different from the sum of the volumes of the separate pure liquid components. For example, if you mix 30 ml of ethyl alcohol and 70 ml of water, the volume of the resulting solution is 97 ml and not 100 ml.

Exercise 8-1

Determine the concentration of a glucose solution that contains 2.4 g of glucose and has a volume of 300 ml. Express the concentration in terms of (a) g/ml; and (b) (W/V) %.

Preparation of a Solution

How can you prepare 500 ml of a 5.0% NaCl solution?

Step 1 Determine the required mass of NaCl as follows: NaCl is a solid, therefore 5.0% means that 100 ml of solution contains 5.0 g of NaCl (5.0 g/100 ml). Since the mass of solute and the volume of solution are directly proportional to one another,

(8-5) $$\text{Mass of NaCl} = (500 \text{ ml}) \times \frac{5.0 \text{ g}}{100 \text{ ml}} = 25 \text{ g}$$

Step 2 Prepare the actual solution. Use a chemical balance to measure 25 g of NaCl and put this amount in a 500-ml volumetric flask (see Figure 8-5). Fill the flask about half full with water and dissolve the NaCl. Then add sufficient water to reach exactly the mark on the volumetric flask that indicates 500 ml of solution. Note that we do *not* have to determine the amount of water to use. We want the volume of *solution* to be 500 ml, and the volume of solution will be 500 ml using this method.

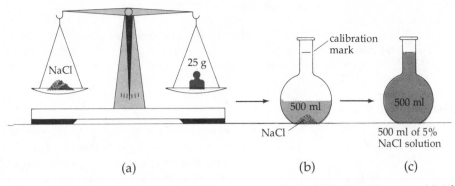

(a)   (b)   (c)

Figure 8-5   Preparation of 500 ml of a 5.0% NaCl solution. (a) 25 g of NaCl is weighed out. (b) The salt is put into a 500-ml flask, which is filled about halfway with water. (c) After the NaCl has dissolved, just enough additional water is added to reach the 500-ml calibration mark.

Exercise 8-2

Determine the mass of glucose needed to prepare the following glucose solutions:
    (a) 300 ml of a 2.8% solution   (b) 600 ml of a 2.8% solution
    (c) 600 ml of a 1.4% solution

Determination of the Volume of a Solution That Contains a Desired Amount of Solute

Chemical laboratories usually have stock solutions of several common reagents, and hospitals keep stock solutions of various drugs. These stock solutions are labeled with the name of the reagent or drug and its concentration, for example, "Demerol hydrochloride; 50 mg/1.0 ml." What volume of this stock solution would contain 37 mg of the drug Demerol hydrochloride?

The designation 50 mg/1.0 ml means that 1.0 ml of solution contains 50 mg of Demerol hydrochloride. Therefore,

(8-6) $$\text{Volume of stock solution} = (37 \text{ mg}) \times \frac{1.0 \text{ ml}}{50 \text{ mg}} = 0.74 \text{ ml}$$

To obtain this volume you could withdraw 0.74 ml of the drug solution from the stock solution bottle with a syringe, as shown in Figure 8-6.

Figure 8-6   This Demerol hydrochloride stock solution has a concentration of 50 mg/1.0 ml. Thus, a volume of 0.74 ml contains 37 mg of Demerol hydrochloride.

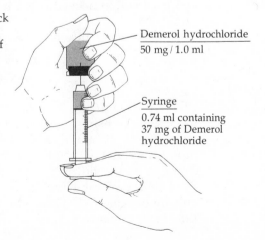

Demerol hydrochloride
50 mg / 1.0 ml

Syringe
0.74 ml containing
37 mg of Demerol
hydrochloride

Exercise 8-3
Determine the volume of a 1.8% NaCl solution that contains
  (a) 5.0 g of NaCl   (b) 2.5 g of NaCl   (c) 5.0 mg of NaCl

Dilution of a Solution

Chemical reagents or drugs are often stored as stock solutions with a relatively high concentration to save space and for ease in preparation. You must be able to determine the volume of a concentrated solution needed to prepare a given volume of a more dilute solution. For example, suppose that you need to prepare 500 ml of a 2.0% NaCl solution from a 15% stock solution. The key concept in a problem of this type is that when you take a sample of a stock solution and add water to it in order to dilute it, the volume and concentration change, *but* the amount of solute remains the same. We can express this fact by the following equation:

(8-7)   Mass of solute in volume of stock solution taken
                = mass of solute in final diluted solution

For any solution, the mass of solute and the volume of solution are related by

(8-8)   Mass of solute = (volume of solution) × (concentration)

We can use Eq. 8-8 to relate the mass of solute to the volume of solution and its concentration for *both* the original stock solution and for the final diluted solution. Thus, using Eqs. 8-7 and 8-8, we obtain

(8-9)   (Volume of stock solution) × (concentration of stock solution)
                = (volume of diluted solution) × (concentration of diluted solution)

When we divide both sides of Eq. 8-9 by the term (concentration of stock solution), we obtain

(8-10)   Volume of stock solution required

                = (volume of diluted solution) × $\left(\dfrac{\text{concentration of diluted solution}}{\text{concentration of stock solution}}\right)$

In our example, the concentration of the stock solution is 15% or 15 g/100 ml, the concentration of diluted solution is 2.0% or 2.0 g/100 ml, and the volume of diluted solution needed is 500 ml. Therefore,

(8-11)   Volume of 15% NaCl solution required

                $= (500 \text{ ml}) \times \dfrac{2.0 \text{ g/100 ml}}{15 \text{ g/100 ml}} = 67 \text{ ml}$

To actually prepare the dilute solution, we measure out 67 ml from the NaCl stock solution using a graduated cylinder and transfer this amount to a 500-ml volumetric flask. We then fill the flask halfway with pure water, mix the two liquids thoroughly, and then add water up to the 500-ml mark and mix again. These steps are shown in Figure 8-7. Note that when diluting a solution, as in the case of preparing a solution (Figure 8-5), you do *not* have to determine the amount of water to use. You need only determine the volume of the stock solution required and know what volume of the final diluted solution you need.

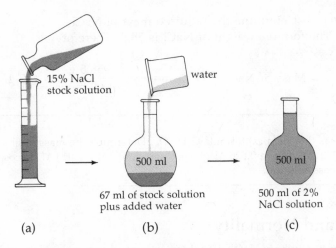

15% NaCl
stock solution

water

500 ml

500 ml

67 ml of stock solution
plus added water

500 ml of 2%
NaCl solution

(a)   (b)   (c)

Figure 8-7   Diluting a solution. To prepare   about halfway with water (b). After
500 ml of a 2% NaCl solution from   the contents of the flask have been
a 15% NaCl stock solution, 67 ml of   thoroughly mixed, additional water
stock solution is measured out (a),   is added to just reach the 500-ml
put in a 500-ml flask, and filled   calibration mark (c).

Exercise 8-4
A stock bottle of 3.0% hydrogen peroxide solution is available. How would you prepare
25 ml of 0.40% hydrogen peroxide solution?

Molarity

There are a variety of ways of expressing the concentration of a solution. When
we considered chemical reactions we saw that the number of moles of a substance is a more fundamental measure of the amount of the substance than is
the mass of the substance. For this reason, the most common way chemists
express the concentration of a solution is to use *moles* for the amount of solute
and volume in *liters* for the amount of solution. This particular way of expressing concentration is called molarity. The **molarity** of a solution is the
number of moles of solute per liter of solution. Thus,

(8-12) $$\text{Molarity} = \frac{\text{moles of solute}}{\text{volume of solution (liters)}}$$

The units of molarity are moles per liter. This combination of units is often
represented by the word **molar** and by the symbol M. Therefore, the expression
"3.42 M NaCl" represents a 3.42-molar sodium chloride solution. Such a solution contains 3.42 moles of NaCl per liter of solution.

Recall that it is a simple matter to go back and forth between the mass of a
substance and the number of moles by using the substance's molecular weight
or formula weight. For example, to determine the mass of NaCl required to
prepare 500 ml of a 3.42 M NaCl solution, we can proceed as follows:

Step 1   Determine the number of moles of NaCl required.
A 3.42 M NaCl solution contains 3.42 moles per 1.00 liter. Therefore,
since 500 ml = 0.500 liter,

(8-13) $$\text{Moles of NaCl} = (0.500 \text{ liter}) \times \frac{3.42 \text{ moles}}{1.00 \text{ liter}} = 1.71 \text{ moles}$$

Step 2   Determine the required mass of NaCl.
The formula weight of NaCl is 58.5. Therefore,

(8-14) $\text{Mass of NaCl} = (1.71 \; \cancel{\text{moles}}) \times \dfrac{58.5 \text{ g}}{1.00 \; \cancel{\text{mole}}} = 100 \text{ g}$

Exercise 8-5
Glucose has the chemical formula $C_6H_{12}O_6$. Determine the mass of glucose required to prepare 300 ml of (a) a 0.020 *M* glucose solution; (b) a 0.010 *M* glucose solution; (c) a 2.0% glucose solution.

Equivalents and Normality

As we know, when NaCl dissolves in water, separate Na^+ and Cl^- ions are present. Similarly, *any* ionic substance that dissolves in water has separate anions and cations. In a NaCl solution, there are the same number of Na^+ and Cl^- ions; however, this is not the case for all ionic substances. For example, in a solution of calcium nitrate, $Ca(NO_3)_2$, for each calcium ion with a charge of +2, there are two nitrate ions with a charge of −1. Since *one* Ca^{2+} ion has the same amount of positive charge as the negative charge of *two* NO_3^- ions, one Ca^{2+} ion is *equivalent* to two NO_3^- ions in terms of electrical charge.

Sometimes, as in the case of electrolytes in human body fluids, this concept of equivalence is used to specify the number of ions and their concentrations. In general, the **number of equivalents** of an ion is defined by the relationship

(8-15)   Number of equivalents
= (number of moles) × (number of electrical charges per ion)

Thus, if 0.1 mole of $Ca(NO_3)_2$ is dissolved in water, the resulting solution contains 0.1 mole of Ca^{2+} ions and 0.2 mole of NO_3^- ions, or 0.2 equivalent of Ca^{2+} and 0.2 equivalent of NO_3^- (see Figure 8-8). Notice that in this solution the number of moles of NO_3^- is twice the number of moles of Ca^{2+}, but the solution contains an *equal* number of equivalents of Ca^{2+} and NO_3^- ions.

The concentration of ions can be expressed as equivalents per liter. The concentration of ions in human body fluids is fairly low, and is often expressed as milliequivalents per liter (meq/liter). For example, in human blood plasma the concentration of calcium ion is normally about 5 meq/liter.

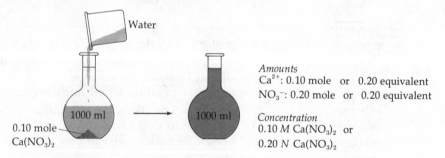

Amounts
Ca^{2+}: 0.10 mole   or   0.20 equivalent
NO_3^-: 0.20 mole   or   0.20 equivalent

Concentration
0.10 *M* $Ca(NO_3)_2$ or
0.20 *N* $Ca(NO_3)_2$

Figure 8-8   When 0.10 mole of $Ca(NO_3)_2$ is dissolved in water to give a solution with a volume of 1000 ml, the concentration of the resulting solution is 0.10 *M* $Ca(NO_3)_2$ or 0.20 *N* $Ca(NO_3)_2$. Note that even though the concentration of NO_3^- is 0.20 *M* and the concentration of Ca^{2+} is 0.10 *M*, this solution is referred to as a 0.10 *M* $Ca(NO_3)_2$ because it was formed by dissolving 0.10 mole of $Ca(NO_3)_2$ in 1 liter of water.

The concentration unit of equivalents per liter is also referred to as the **normality** (with the symbol N). Thus, for ionic solutions,

(8-15a)
$$\text{Normality} = \frac{\text{number of equivalents}}{\text{volume of solution (liters)}}$$

For example, we can refer to the solution with 0.1 mole of $Ca(NO_3)_2$ and a volume of 1.0 liter as either a 0.1 M $Ca(NO_3)_2$ solution or as a 0.2 N $Ca(NO_3)_2$ solution (see Figure 8-8).

As another example of the use of normality, consider a solution that contains 0.20 mole of Na_2HPO_4 in 1.0 liter. There are 0.40 mole or 0.40 equivalent of Na^+, and 0.20 mole or 0.40 equivalent of HPO_4^{2-} in this solution. Thus, this solution can be referred to as either a 0.20 M Na_2HPO_4 solution or as a 0.40 N Na_2HPO_4 solution.

Normalities are used in other contexts in addition to specifying the concentration of ionic solutions. In several of these other cases, the use of normalities has been ambiguous. Because of this, equivalents and normalities are not permitted in the SI system of units. Many chemists, however, still continue to use these units, although we hope this practice will soon cease.

Exercise 8-6
Determine the molarity of the following ions:

(a) Bicarbonate, 28 meq/liter   (b) Chloride, 105 meq/liter
(c) Phosphate, 2.0 meq/liter   (d) Magnesium, 1.8 meq/liter
(e) Potassium, 4.5 meq/liter   (f) Sodium, 140 meq/liter

8-4   SOLUBILITY

When you were a child, did you ever keep pouring salt or sugar into a glass of water? If you did, you found out that there was a limit to the amount of salt or sugar that would dissolve in a given amount of water.

In general, if we keep on adding a solid solute to a given amount of water, we find that eventually some solid solute remains on the bottom of the container no matter how much we stir the solution or how long we wait. At this point the solution and the solid solute have reached a state of *dynamic equilibrium* in which some solute molecules are dissolving and going into solution but other solute molecules in solution are crystallizing out of solution and forming solid solute at the same time (see Figure 8-9). A solution that is in dynamic equilibrium with solid solute is called a **saturated** solution. We can separate a saturated

Figure 8-9   Schematic illustration of the state of dynamic equilibrium in a saturated solution. At any instant, the number of solute molecules that crystallize out of solution are compensated for by an equal number of molecules dissolving and going into solution from the solid solute.

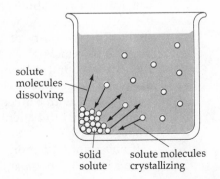

solute molecules dissolving

solid solute

solute molecules crystallizing

solution from the excess solid by pouring off the liquid (decanting), and the solution will remain saturated. The concentration of solute in a saturated solution is called the **solubility** of the solute substance in that solvent.

Except under some very special circumstances, the solubility of a substance at a given temperature in a given solvent represents the highest attainable concentration for a solution of that substance. At 25°C the solubility of NaCl is 6.15 M, and we can easily prepare salt solutions with concentrations ranging from just above 0 M to 6.15 M. At 25°C, any NaCl solution with a concentration of less than 6.15 M is called an **unsaturated** solution.

The solubilities of most substances vary over a wide range depending on the nature of the solute and the solvent. As noted above, the solubility of a solid substance also depends on the temperature. The solubility of most, but not all, solid substances increases as the temperature increases. As you probably know, more sugar will dissolve in a given amount of hot tea than in iced tea. In our further discussion we shall assume that the solvent is water and thus that the term solubility refers to the concentration of a saturated aqueous solution. The solubilities of a few common substances are given in Table 8-1. The terms soluble and insoluble are often used to describe substances with relatively high and low solubilities, respectively. Roughly speaking, if the solubility of a substance is greater than 0.1 M, we say that it is **soluble;** and if its solubility is much less than 0.1 M, we say that it is **insoluble.**

Table 8-1   Solubilities of Some Common Substances at 25°C

Substance	Solubility (g/100 ml)
Table salt (NaCl)	36
Table sugar (sucrose)	210
Urea	120
Magnesium hydroxide	1×10^{-3}
Gasoline	4×10^{-2}

Some liquids can also form saturated aqueous solutions. Bromine, for example, is a liquid at room temperature, and the solubility of bromine in water is about 3.5 g/100 ml at 25°C. Other liquids, such as ethyl alcohol, can form solutions with water in any proportion between 0% and 100% ethyl alcohol. Two liquids that can form solutions in any proportion are said to be **miscible.**

The solubility of a gas in a liquid depends on the partial pressure of the gas over the solution and on the temperature. The solubility of many gases at a given temperature is directly proportional to the partial pressure of the gas.

The solubility of oxygen in water when the partial pressure of O_2 is 1.0 atm is about 4.0×10^{-2} g/liter (at 25°C). In atmospheric air with a total pressure of 1.0 atm, the partial pressure of O_2 is about 1/5 atm. The solubility of oxygen in water exposed to atmospheric air is thus about $(1/5) \times (4.0 \times 10^{-2}$ g/liter) or 0.8×10^{-2} g/liter. As you can see, the solubility of oxygen in water is very slight. The solubility of oxygen in blood, which is an aqueous medium, is likewise extremely low. Almost all the oxygen carried in the bloodstream is chemically bound to hemoglobin molecules and does not exist as physically dissolved oxygen molecules. In a variety of clinical situations, pure oxygen, or a gas mixture with a much higher partial pressure of oxygen than in normal air, is administered to patients in order to increase the amount of oxygen in their blood.

If you want to demonstrate the effect of pressure on the solubility of a gas, open a bottle of champagne. Champagnes (and other carbonated beverages) are bottled under a carbon dioxide pressure slightly greater than 1 atm. When you open the bottle, CO_2 bubbles out of solution because the partial pressure of CO_2 in air is extremely low.

Once you have opened the champagne, don't forget to drink it fairly soon. If you leave it around at room temperature for any length of time it will go "flat." One other reason for this is that the solubility of carbon dioxide, and most other gases, decreases as the temperature increases. The decreased solubility of oxygen in water at increased temperatures is one of the major aspects of thermal pollution. As a by-product of many industrial operations, the temperature of the water in nearby lakes and rivers is raised. The decreased amount of oxygen at the higher temperature can be disastrous for aquatic organisms.

Exercise 8-7

The solubility of urea, a waste product of human metabolism excreted in urine, is about 120 g/100 ml at 25°C. Determine the mass of urea contained in 350 ml of a saturated urea solution.

Exercise 8-8

The solubility of carbon dioxide at a partial pressure of 1.00 atm is about 0.15 g/liter at 25°C. The partial pressure of CO_2 in the alveoli in a person's lungs is about 40 torr. What is the solubility of CO_2 at this partial pressure and 25°C?

8-5   DETERMINING SOLUBILITY

We can make a few useful generalizations about the solubility of substances in water, but there is no set of simple rules that will allow us to predict how soluble a substance will be. The solubility of a substance in water depends in a complex way on a number of factors. One of the most important of these factors is how the solute molecules or ions affect the extensive hydrogen bonding that occurs between water molecules.

Solubility of Molecular Substances

Interspersing a solute molecule among water molecules disturbs the hydrogen bonding between water molecules. In general, this can occur if the attraction between solute molecules and water molecules is sufficiently strong. Thus, many substances composed of polar molecules that can form strong hydrogen bonds with water molecules (hydrophilic molecules) are quite soluble in water. For example, urea (Figure 8-10a) is a by-product of human metabolism that is excreted in urine. Urea molecules can hydrogen bond with water molecules,

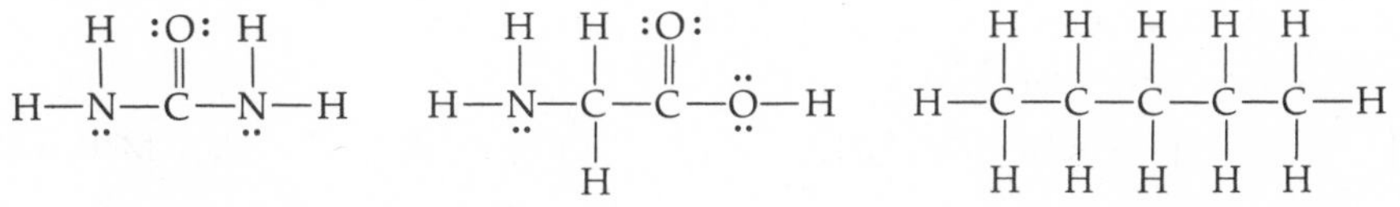

Urea (hydrophilic)
(a)

Glycine (hydrophilic)
(b)

Pentane (hydrophobic)
(c)

Figure 8-10   Molecules of urea and glycine can hydrogen bond with water molecules, and these hydrophilic compounds have a high solubility in water. Pentane is a nonpolar hydrophobic compound that has a very low solubility in water.

and the solubility of urea is about 120 g/100 ml at 25°C. Glycine (Figure 8-10b), which is a component of proteins, is another substance whose molecules can hydrogen bond with water molecules. The solubility of glycine is about 25 g/100 ml at 25°C. On the other hand, nonpolar hydrophobic substances have extremely low solubilities (or in many cases are virtually insoluble) in water. The solubility of pentane (Figure 8-10c) is only about 0.04 g/100 ml at 25°C.

In general, when we compare the solubilities of a set of similar substances whose molecules contain both a hydrophilic and a hydrophobic part, we find that the solubility decreases the larger the hydrophobic portion of the molecule. For example, the solubility of 1-pentanol is about 3 g/100 ml at 25°C, whereas, the solubility of 1-hexanol, which has a somewhat larger hydrophobic part, is about 0.6 g/100 ml at 25°C (see Figure 8-11).

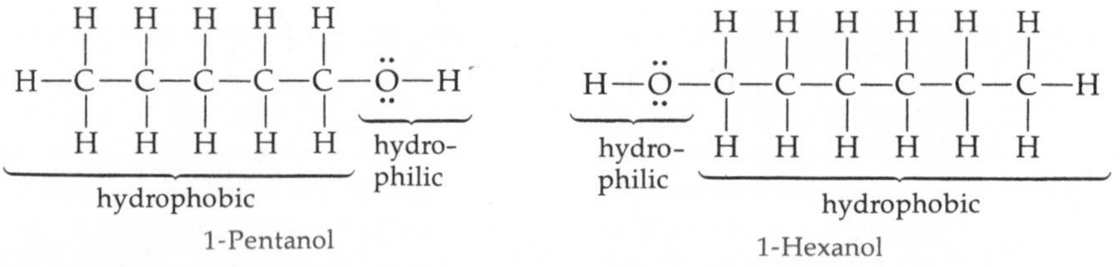

Figure 8-11   The respective hydrophilic parts of 1-pentanol and 1-hexanol are the same. However, the hydrophobic part of 1-hexanol is larger than that of 1-pentanol; hence, 1-hexanol has a lower solubility in water.

Gasoline and oil are mixtures of molecules that are all nonpolar. If gasoline or oil is added to water, the nonpolar gasoline or oil forms a separate layer on top of the water. The hydrophobic molecules of the gasoline or oil cluster together because the attraction between these hydrophobic molecules and the water molecules is not large enough to compensate for the disturbance of the hydrogen-bonded water molecules that would occur if gasoline or oil molecules were dispersed in the water. In effect, water molecules resist the intrusion of hydrophobic molecules, and this causes the hydrophobic molecules to cluster together. This clustering together of hydrophobic molecules, or hydrophobic parts of molecules, in an aqueous environment is the essential feature of any hydrophobic interaction. As we shall see later, hydrophobic interactions play an important role in determining the properties of many systems, such as biological membranes, proteins, nucleic acids, and soaps and detergents.

Exercise 8-9

Which of the following substances would you expect to be soluble and which insoluble in water?

(a) Ammonia, NH_3 (b) Nitrogen, N_2 (c) Methanol, $H-\overset{\displaystyle H}{\underset{\displaystyle H}{C}}-\overset{..}{\underset{..}{O}}-H$

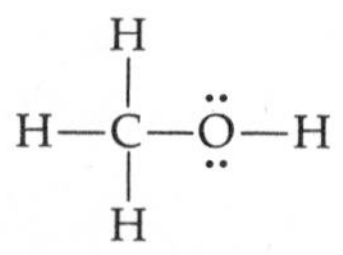

(d) Ethane, C_2H_6 (e) Methylamine, $H-\overset{\displaystyle H}{\underset{\displaystyle H}{C}}-\overset{\displaystyle H}{\underset{..}{N}}-H$

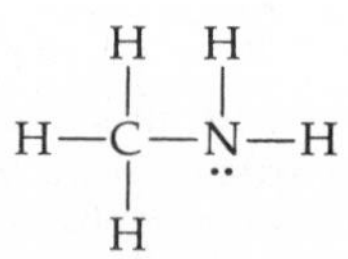

Solubility of Ionic Substances

Solubilities of ionic substances vary over an enormous range. The solubility of NaCl is quite large—6.15 M at 25°C—but the solubility of CaF_2 is extremely small—about 3.7×10^{-4} M at 25°C. There is no simple explanation for why some ionic substances are very soluble in water whereas others are virtually insoluble. The process of dissolving an ionic substance in water can be visualized as occurring in two steps: (1) pulling the solid apart and forming separate ions in the gas state; and (2) hydrating the ions (see Section 7-5). A large amount of energy is needed to accomplish step 1, but as we discussed previously, when an ion is hydrated a large amount of energy is released. Thus it is hard to predict for a given ionic substance whether the overall energy change is negative, thus favoring solution of the substance, or positive, and thus favoring insolubility.

Recall that you must also consider the "push" toward a state of higher entropy (more disorder) when trying to predict the direction of change. There is a large entropy *increase* when a solid breaks apart and separate ions are formed in the gas state (step 1), but there is also a large entropy *decrease* when an ion is hydrated (step 2), and the overall entropy change may be positive or negative. A hydrated ion holds a cluster of water molecules tightly, which decreases the ability of these water molecules to move around. Thus ions in water alter the structure of water to a more ordered arrangement with a lower entropy.

As you can see, several factors influence the extent to which an ionic substance dissolves in water. As a result, there is no simple way to predict the solubility of a particular ionic compound. Generalizations about the solubility of ionic compounds based on experimental facts are discussed in Section 12-8.

8-6   COLLOIDS

As we mentioned earlier, it is not always possible to make a clear distinction between homogeneous mixtures and heterogeneous mixtures. Let us see why.

When we dissolve some sugar in water and form a sugar solution, we cannot see the individual sugar molecules, they do not settle out under the influence of gravity, and the sugar cannot be separated from the water by filtration. On the other hand, a slurry of very fine sand particles suspended in water is definitely a heterogeneous mixture. We can see the individual sand particles, gravity will cause them to settle out, and they can be separated from the water by filtration. Blood plasma, however, is a mixture that contains giant protein molecules, which are much larger than sugar molecules but much smaller than sand particles, so it cannot be classified as either a solution or a heterogeneous mixture. Blood plasma is an example of a colloidal dispersion and has properties somewhere in between those of a solution and a heterogeneous mixture.

A colloidal dispersion, or more simply, a colloid, like a solution, can be a gas, a liquid, or a solid. We can think of a colloid as involving one substance in smaller amount, the **dispersed substance,** suspended in a larger amount of a second substance, the **dispersing medium.** In a colloid, the dispersed substance is analogous to the solute in a solution, and the dispersing medium is analogous to the solvent. In general, a mixture is called a **colloid** if the largest dimension of the dispersed particles is roughly between 2 nm and 200 nm. In blood plasma, for example, the largest proportion of the colloidally dispersed protein molecules are albumins. An albumin macromolecule is about 14 nm by 5 nm by 5 nm. Compare the dimensions of an albumin macromolecule with the largest dimension of a glucose (sugar) molecule, 0.7 nm.

The dispersed substance in a colloid may exist as individual macromolecules, but more often it consists of aggregates of many smaller particles. Aggregates of colloidal size can be formed either by breaking up larger pieces of a substance or by causing a large number of individual, small molecules to clump together.

Colloids are classified according to the physical state of the dispersed substance and the dispersing medium, and various types of colloids are given special names. An **aerosol** is a colloid in which the dispersing medium is a gas and the dispersed substance is either a solid or a liquid. Smoke is an aerosol of solid particles dispersed in air, and fog contains droplets of liquid water dispersed in air. When any two gases are mixed they never form a colloid or a heterogeneous mixture, but rather, they always form a homogeneous solution.

When the dispersing medium is a liquid, the resulting colloid is called a **foam,** an **emulsion,** or a **sol,** depending on whether the dispersed substance is a *gas,* a *liquid,* or a *solid,* respectively. Soap suds and whipped cream are foams, and milk is an emulsion containing liquid globules of butter fat dispersed in an aqueous medium. Milk of magnesia is a sol that has solid particles of magnesium hydroxide dispersed in water. Colloids in which the dispersing medium is a solid are called solid foams, solid emulsions, or solid sols.

The most important colloidal systems are those in which the dispersing medium is water and the dispersed substance is a solid or a liquid (sols and emulsions). Let us briefly compare some of the properties of these type of colloids with those of an aqueous solution.

In a solution, the small solute molecules or ions do not settle out under the influence of gravity because they are moving about randomly in much the same manner as the water molecules. In a colloid, the larger dispersed particles also move about because they are continually being struck by the moving water molecules. This buffeting by the water molecules causes the colloidal particles to move in an erratic manner akin to the random motion of the water molecules but at a slower rate. The erratic motion of a colloidally dispersed particle is called **Brownian motion** (see Figure 8-12). In 1827, Robert Brown, a Scottish botanist, noticed that pollen grains suspended in water moved continually in a haphazard manner. Brownian motion keeps particles of colloidal size from settling out but will not prevent the settling of larger particles.

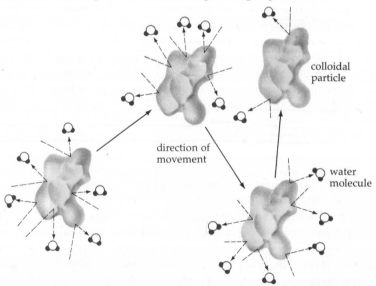

colloidal particle

direction of movement

water molecule

Figure 8-12   Schematic illustration of Brownian motion. A colloidal particle moves about haphazardly because it is continually struck by water molecules. At any given time, more water molecules might strike one side of the particle, and this unequal distribution of collisions will move the particle in a particular direction. At a later time, more collisions on another side might move it in a different direction.

Collodial dispersions (such as protein macromolecules in water) will never settle out of their own accord under the force of gravity. An instrument called an **ultracentrifuge,** which spins samples at a rate of 50,000 revolutions per minute or more, can produce an artificial gravitational force that is many thousands of times larger than the gravitational force of the earth; this device can be used to separate colloidal dispersions. **Sedimentation** is what the settling out of particles is called. Ultracentrifuge sedimentation can also provide accurate information about the molecular weight, size, and shape of colloidal particles. This is one of the techniques that scientists use to obtain valuable information about proteins and other biologically important macromolecules.

A great deal of useful information about the size and shape of macromolecules can be obtained by carefully analyzing the character of the light scattered from a colloidal dispersion of these molecules when a light beam is passed through the colloid. Looking from the side, you cannot see a beam of light as it passes through a solution. However, when a beam of light is passed through a colloidal dispersion, the beam is visible from the side (see Figure 8-13). This characteristic property of a colloidal dispersion is called the **Tyndall effect,** after J. Tyndall, who studied the phenomenon of light scattering by suspended particles around 1869. The larger-size particles in a colloidal dispersion can scatter a light beam, whereas the smaller solute particles in a solution cannot.

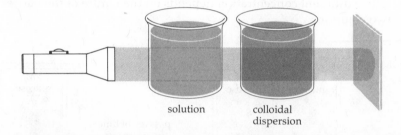

solution   colloidal
                  dispersion

Figure 8-13   The Tyndall effect. An aqueous solution does not scatter a light beam, but the suspended particles in an aqueous colloidal dispersion do. When viewed from the side, a beam of light from a flashlight, or from a spotlight in a darkened theater, is also visible because of the Tyndall effect. In this case, the light is scattered by suspended particles of dust or smoke in the aerosol that we breathe.

Exercise 8-10

A mixture of a particular substance in water does not separate when standing and exhibits the Tyndall effect. What can you conclude about the nature of this mixture?

8-7   OSMOSIS AND DIALYSIS

Recall that diffusion is a natural process that occurs in liquids. Suppose that we have a container with two sides, A and B, which are separated by a thin wall, and we put a sugar solution in side A and a sugar solution with a *different* concentration in side B. If we then punch a hole in the wall separating side A from side B, sugar molecules and water molecules will move back and forth between the two sides of the container. However, there will be a *net movement* of sugar molecules from the more concentrated to the more dilute solution. After a sufficient period of time has elapsed, a state of dynamic equilibrium will be reached in which there is the *same* sugar concentration on both sides of the container (see Figure 8-14). Also, recall that the "push" behind diffusion is the general natural tendency for any system to go to a more disordered state—to a state with higher entropy. In our example, the final equilibrium state with a uniform sugar concentration has a higher entropy than the initial state with unequal sugar concentrations in sides A and B. In general, whenever two solutions with different concentrations are in contact, there will be a "push" toward equalizing the two concentrations. The same consideration applies to two colloidal dispersions with different concentrations of a dispersed substance. However, what (if anything) will *actually* diffuse from one solution to another solution with a different concentration depends on the nature of the barrier between the two solutions.

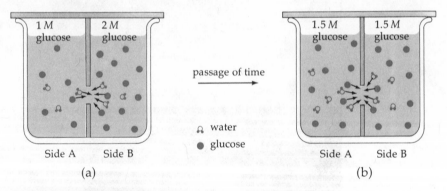

Figure 8-14   (a) If a hole is punched in a thin wall separating two glucose solutions with different concentrations, sugar and water molecules will diffuse back and forth between the two sides, with a net movement of sugar molecules from high to low concentration. (b) After sufficient time has passed, the glucose concentrations on both sides are equal, and there is no movement of sugar molecules in either direction.

The existence of life depends on cell membranes that *selectively* allow for the diffusion of some substances and not others. Membranes that are selective in what will diffuse through them are called **semipermeable membranes.** We shall discuss some features of cell membranes shortly, but first let us consider a simpler situation.

Osmosis

If two solutions with different concentrations are separated by a semipermeable membrane that will permit *only* the passage of water, then there will be a spontaneous flow of water from the more dilute to the more concentrated solution, as shown in Figure 8-15. The diffusion of water across a semipermeable mem-

Figure 8-15   Schematic illustration of osmosis. Dilute and concentrated glucose solutions are separated by a semi-permeable membrane that allows only the passage of water. There is a net movement of water through the membrane from the dilute to the concentrated solution.

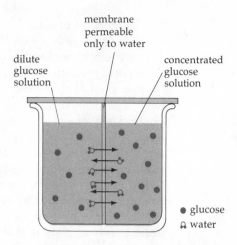

brane is called **osmosis** (Greek for "push"). The drive behind osmosis and diffusion is the same—namely, an attempt to equalize the concentration of the two solutions. In osmosis, however, the semipermeable membrane prevents the solute from diffusing from the more concentrated to the more dilute solution, but water can cross the semipermeable membrane, so water diffuses in the opposite direction, from the more dilute to the more concentrated solution.

The flow of water from the more dilute to the more concentrated solution will not continue indefinitely. As water flows from left to right in Figure 8-15, the amount of the water on the right side increases. The resulting increased water pressure on the right side causes a push in the opposite direction. Eventually a state of equilibrium is reached when the thrust to equalize concentrations, which produces diffusion of water from left to right, is balanced by the excess pressure on the right side (see Figure 8-16a). A state of equilibrium can also be obtained with equal liquid levels on each side, if the correct extra pressure is applied to the more concentrated solution (see Figure 8-16b). When a solution is separated from pure water by a semipermeable membrane, the extra pressure that must be applied to the solution in order to prevent the flow of water into the solution is called the **osmotic pressure** of the solution. The osmotic pressure

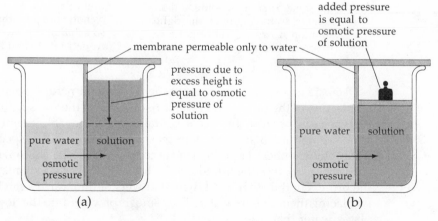

Figure 8-16   Schematic illustration of osmotic pressure. When pure water and a solution are separated by a membrane permeable only to water, there is a net flow of water into the solution (a). At equilibrium, the liquid levels will not be equal and the pressure exerted by the extra column of water on the right of the membrane will be equal to the osmotic pressure of the solution. The two liquid levels can be maintained at the same height if an extra pressure equal to the osmotic pressure of the solution is applied to the solution on the right of the membrane (b).

of a solution is not actually a pressure produced by the solution itself. Osmotic pressure arises only when the solution is separated from water by a semipermeable membrane. In this case, if the solution or colloidal dispersion is dilute, its osmotic pressure is directly proportional to the sum of the molarities of all the dissolved or dispersed particles.

Any particle that *cannot* diffuse across the semipermeable membrane, whether it is a molecule, an ion, or a macromolecule, contributes to the osmotic pressure of solution or a colloidal dispersion. Thus a 0.02 M sugar solution has twice the osmotic pressure of a 0.01 M sugar solution. However, a 0.01 M NaCl solution and a 0.02 M sugar solution have the same osmotic pressure, since 0.01 M NaCl has a total concentration of 0.02 M (0.1 M Na^+ + 0.1 M Cl^-). It is possible to determine the molecular weight of the dissolved solute or dispersed substance in a solution or colloidal dispersion from a measurement of the osmotic pressure. This is one of the techniques scientists use to determine the molecular weight of proteins and other large molecules.

Two solutions, A and B, that have the same osmotic pressure are said to be **isotonic.** If a solution A has an osmotic pressure less than a solution B, then solution A is said to be **hypotonic** with respect to solution B. A dilute solution is hypotonic with respect to a more concentrated solution. A concentrated solution with a higher osmotic pressure is said to be **hypertonic** with respect to a dilute solution with a lower osmotic pressure (see Figure 8-17).

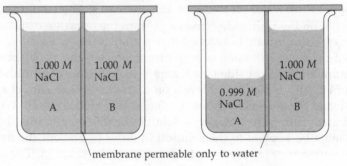

membrane permeable only to water

Figure 8-17   The two solutions on the left have the same osmotic pressure and are *isotonic.* In the figure on the right, solution A has a lower osmotic pressure than solution B. Solution A is *hypotonic* with respect to solution B; and solution B has a higher osmotic pressure than solution A, so solution B is *hypertonic* with respect to solution A.

Osmosis is responsible for a number of familiar natural phenomena. For example, one of the reasons that water rises through the roots to the tops of plants and trees is because the roots contain semipermeable membranes and the aqueous medium inside the plant cells is hypertonic with respect to the ground water. A cucumber placed in brine (a concentrated salt solution) loses water and shrivels up because the liquid inside the cucumber is hypotonic with respect to the brine. On the other hand, wilted flowers or limp carrots can be "freshened" by placing them in pure water. The osmosis of water into the flower or carrot replaces water that was previously lost.

Dialysis

Small ions and molecules, but not larger particles of colloidal size, can diffuse through some natural membranes, such as animal bladders, or through some synthetic membranes, such as cellophane. The selective diffusion of small ions and molecules, but not colloidal particles, through a membrane is called

dialysis. Diffusion due to a difference in concentration is the driving force behind both dialysis and osmosis. The distinction between the two processes is due merely to a difference in what can diffuse through the semipermeable membrane.

As illustrated in Figure 8-18, dialysis is used to separate large molecules such as proteins from smaller ions and molecules. Suppose that an aqueous mixture containing some protein and smaller ions and molecules is placed inside a bag made out of a substance through which small ions and molecules but not colloidal particles can diffuse—a dialysis bag. Then if the dialysis bag is continually bathed with pure water, the smaller ions and molecules will diffuse out of the bag and are removed by the external flow of water. After a sufficient amount of time only protein and water will be left inside the dialysis bag.

Figure 8-18   Dialysis is accomplished by use of a bag made of material that is permeable to small ions and small molecules but not large protein molecules. Small ions, such as Na^+ and Cl^-, and small molecules, such as sugar, can diffuse out of the bag, but protein molecules cannot. The small ions and small molecules are removed by the flow of water past the bag. After a period of time, only protein molecules and water will remain inside the dialysis bag.

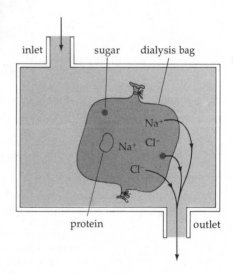

One of the functions of a normal kidney is to allow small molecules that are the waste products of metabolism, such as urea, to diffuse from blood across the semipermeable membranes of the kidney tubules into the urine. In a process called **hemodialysis,** illustrated in Figure 8-19, an artificial kidney machine mimics this natural kidney function. Blood taken from one of the patient's arteries is pumped through tubing, usually made of cellophane, and returned to one of the patient's veins. The cellophane tubing is bathed in a solution that contains the ions found in blood plasma at their normal concentration. The cellophane serves as a dialysis membrane since it is permeable to small ions and molecules but not to macromolecular proteins. Urea and other small waste-product molecules diffuse from the patient's blood through the cellophane membrane and into the bath solution. In addition, if the concentration of any ion in the patient's blood is lower than normal, some of that ion will diffuse from the bath solution into the blood; diffusion of the ion will take place in the opposite direction if the concentration of the ion is higher than normal. Thus an artificial kidney machine is used not only to eliminate waste product molecules but also to restore any ionic imbalance in the blood. A typical hemodialysis treatment lasts about six hours, and the bath solution is usually changed every two hours.

Exercise 8-11
Why is the bath solution changed regularly during a hemodialysis treatment?

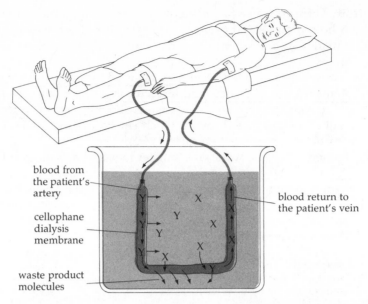

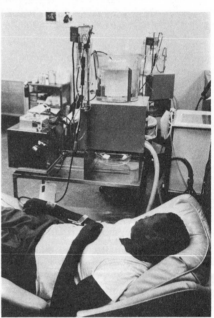

blood from
the patient's
artery

cellophane
dialysis
membrane

waste product
molecules

blood return to
the patient's vein

Figure 8-19   Schematic illustration of the process of hemodialysis with an artificial kidney machine. The concentration of ions and small molecules in the bath solution is the same as in normal blood plasma, except that no waste product molecules are present initially. The X's represent ions or small molecules whose concentration is below normal in the patient's blood plasma, whereas the Y's are those ions or small molecules whose concentration is above normal. The photograph shows a patient undergoing a hemodialysis treatment.

Cell Membranes

The membranes enclosing the cells in the human body are quite complex entities. They serve as semipermeable membranes and allow some substances, but not others, to diffuse spontaneously across the cell membrane from a region of high concentration to a region of low concentration. At the same time, however, some substances move across a cell membrane from a region of low concentration to one of high concentration. This movement is opposite from the direction of simple spontaneous diffusion and requires the expenditure of energy by the cell. The process is called active transport (see Chapter 22).

Water will move across a cell membrane if there is a difference between the osmotic pressure of the fluid inside of and external to the cell. All the smaller ions and molecules as well as the larger protein macromolecules contribute to the osmotic pressure of the fluid inside a cell.

The flow of water into or out of a red blood cell can be easily observed under a microscope. If red blood cells are put into a hypotonic environment, such as pure water, then water will diffuse into the blood cells and cause them to burst. This process is called hypotonic **hemolysis** of blood cells. On the other hand, if red blood cells are put into a hypertonic solution, water will diffuse out of the cells and the cells will shrink. This process is called **crenation.** Therefore, when intravenous fluids are administered to a patient, they must be isotonic with respect to the intracellular fluid in order to avoid hemolysis or crenation of the red blood cells and similar damage to other types of body cells. For example, a 0.9% sodium chloride solution is isotonic with respect to the fluid inside red blood cells. Thus, sodium chloride solutions with this concentration are called physiological saline solutions. Neither hemolysis nor crenation occurs when red blood cells are placed in a physiological saline solution (see Figure 8-20).

normal red blood cells

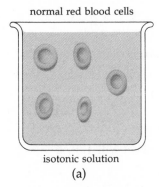

isotonic solution

(a)

crenation

hypertonic solution

(b)

Figure 8-20   (a) There is no net diffusion of water molecules either into or out of red blood cells when they are in an isotonic solution. (b) In a hypertonic solution, there is a net diffusion of water out of the cells, which causes them to shrink. (c) In a hypotonic solution, there is a net diffusion of water into the cells, which causes the cells to swell and burst.

hemolysis

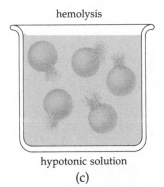

hypotonic solution

(c)

Exercise 8-12

If a 0.01 M glucose solution and a 0.01 M NaCl solution are put into the left- and right-hand sides, respectively, of a container divided by a membrane permeable only to water, would there be (a) a net flow of water from left to right; (b) a net flow of water from right to left; or (c) no net flow of water?

Exercise 8-13

Is a 0.9% glucose ($C_6H_{12}O_6$) solution isotonic, hypotonic, or hypertonic with respect to a 0.9% NaCl (physiological saline) solution?

Exercise 8-14

Why is the bath solution in an artificial kidney machine made isotonic with respect to the blood of a patient receiving a hemodialysis treatment?

8-8   SUMMARY

1. A solution is defined to be a homogeneous mixture of two or more pure substances.

2. Concentration usually refers to the amount of solute compared to the amount of solution; g/ml, (W/V) %, (V/V) %, and molarity are common units of concentration.

3. Ion concentrations are often expressed in units of equivalents per liter or normality (N).

4. A solution that is in dynamic equilibrium with solid solute is said to be saturated, and the concentration of solute in a saturated solution is called the solubility of the solute. The solubility of a solute depends on the temperature as well as other factors.

5. At a given temperature, the solubility of a gas in a liquid is directly proportional to the partial pressure of the gas.

6. The solubility of most gases decreases as the temperature increases.

7. Small hydrophilic molecules are soluble in water, whereas hydrophobic molecules are not.

8. A colloid consists of dispersed particles, with a largest dimension between 2 nm and 200 nm, suspended in a dispersing medium.

9. Membranes that are selective in what will diffuse through them are called semipermeable membranes.

10. The diffusion of water across a semipermeable membrane that allows only the passage of water is called osmosis.

11. The osmotic pressure of a solution is directly proportional to the sum of the molarities of all the dissolved or dispersed particles.

12. Two solutions with the same osmotic pressure are called isotonic. If a given solution A has an osmotic pressure less than a second solution B, then solution A is hypotonic with respect to B; and if the osmotic pressure of A is greater than that of B, then solution A is hypertonic with respect to solution B.

13. The selective diffusion of small ions and molecules, but not colloidal particles, through a membrane is called dialysis.

PROBLEMS

1. Determine the concentration for each of the following solutions as g/ml, (W/V) %, and molarity.
   (a) Mass of glucose = 2.5 g; volume of solution = 850 ml
   (b) Mass of NaCl = 4.2 g; volume of solution = 2.5 liters
   (c) Number of moles of $NaHCO_3$ = 0.15 mole; volume of solution = 420 ml
   (d) Number of moles of KCl = 0.28 mole; volume of solution = 1.5 liters

2. In each of the following, determine the mass and the number of moles of solute required to prepare the given solution:
   (a) 250 ml of a 0.80% NaCl solution
   (b) 2.2 liters of a 0.15 M $NaHCO_3$ solution
   (c) 350 ml of a glucose solution with a concentration of 5.0 g/100 ml
   (d) 650 ml of 0.40% KCl solution.

3. In each of the following, determine the volume of solution that contains the designated amount of solute.
   (a) A 2.5% glucose solution that contains 50 g of glucose
   (b) A 1.8% NaCl solution that contains 0.12 mole of NaCl
   (c) A 2.0 M KCl solution that contains 12 g of KCl
   (d) A 1.5 M NaHCO$_3$ solution that contains 20 g of NaHCO$_3$

4. In each of the following, determine the volume of stock solution required to prepare the given solution:
   (a) 200 ml of 0.25 M NaCl. Stock solution is 4.0 M NaCl.
   (b) 100 ml of 1.5% glucose solution. Stock solution is 5.0% glucose.
   (c) 800 ml of 0.20 M NaHCO$_3$. Stock solution is 1.5 M NaHCO$_3$.
   (d) 150 ml of 0.025 M KCl. Stock solution is 0.20 M KCl.

5. Determine the concentration of the ions in the following solutions both as moles per liter and as normality.
   (a) 0.28 M Na$_2$CO$_3$ (b) 0.52 M MgCl$_2$ (c) 0.12 M KHCO$_3$ (d) 1.5% Na$_3$PO$_4$

6. The solubility of NaCl is 36 g/100 ml at 25°C. What mass, and what number of moles, of NaCl will 20 ml of a saturated sodium chloride solution contain?

7. The solubility of sucrose is 210 g/100 ml at 25°C. If the water is boiled off from 250 ml of a saturated sucrose solution, what mass of sucrose remains?

8. The solubility of urea, CH$_4$N$_2$O, is 120 g/100 ml at 25°C. Can 50 ml of a urea solution at 25°C contain 2.2 moles of urea?

9. The solubility of oxygen in water when the partial pressure of O$_2$ is 1.0 atm is found to be 4.0×10^{-2} g/liter at 25°C. The partial pressure of oxygen in alveolar air is about 100 torr. What is the solubility of oxygen in water at 25°C when the partial pressure of oxygen is 100 torr?

10. You are given an aqueous mixture in a beaker. How would you decide whether this mixture was a solution, a collodial dispersion, or a heterogeneous mixture?

11. A 0.9% NaCl solution is isotonic with respect to the fluid contents of red blood cells. If red blood cells are placed in a 1.1% NaCl solution, state whether the red blood cells will experience hemolysis, crenation, or neither.

12. What concentration of glucose solution is isotonic with respect to a 0.9% NaCl solution?

13. Two solutions, A and B, which are separated by a membrane permeable only to water, are at equilibrium, as indicated in the diagram below. Which solution has the higher osmotic pressure? Is solution B hypertonic, hypotonic, or isotonic with respect to solution A?

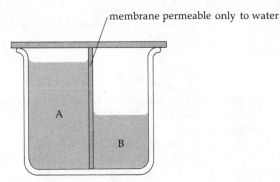

14. For the following pairs, determine which solution has the higher osmotic pressure:
    (a) 0.1 M NaCl   or   0.1 M CaCl$_2$
    (b) 0.2 M NaCl   or   0.1 M Na$_3$PO$_4$
    (c) 1.0 M NaCl   or   1% NaCl

15. When a flexible dialysis bag containing some protein is placed in water, it swells in size. Explain why this happens.

SOLUTIONS TO EXERCISES

8-1 (a) Concentration $= \dfrac{2.4 \text{ g}}{300 \text{ ml}} = 8.0 \times 10^{-3} \dfrac{\text{g}}{\text{ml}}$

(b) Concentration $= \left(8.0 \times 10^{-3} \dfrac{\text{g}}{\text{ml}}\right) \times 100\% = 0.80 \left(\dfrac{W}{V}\right)\%$

8-2 (a) 2.8% means 2.8 g/100 ml.

Mass of glucose $= (300 \text{ ml}) \times \dfrac{2.8 \text{ g}}{100 \text{ ml}} = 8.4$ g

(b) Mass of glucose $= (600 \text{ ml}) \times \dfrac{2.8 \text{ g}}{100 \text{ ml}} = 16.8$ g

(c) Mass of glucose $= (600 \text{ ml}) \times \dfrac{1.4 \text{ g}}{100 \text{ ml}} = 8.4$ g

8-3 1.8% means 1.8 g/100 ml.

(a) Volume $= (5.0 \text{ g}) \times \dfrac{100 \text{ ml}}{1.8 \text{ g}} = 280$ ml

(b) Volume $= (2.5 \text{ g}) \times \dfrac{100 \text{ ml}}{1.8 \text{ g}} = 140$ ml

(c) Volume $= (5.0 \times 10^{-3} \text{ g}) \times \dfrac{100 \text{ ml}}{1.8 \text{ g}} = 0.28$ ml

8-4 Volume of stock solution (required)

$= \text{(volume of diluted solution)} \times \left(\dfrac{\text{concentration of diluted solution}}{\text{concentration of stock solution}}\right)$

Concentration of stock solution = 3.0%
Concentration of diluted solution = 0.40%.

Therefore, volume of stock solution (required) $= (25 \text{ ml}) \times \dfrac{0.40\%}{3.0\%} = 3.3$ ml.

To prepare the dilute solution, take 3.3 ml of stock solution and add water until the final volume is 25 ml.

8-5 300 ml = 0.300 liter.

(a) Determine the number of moles of glucose and then the mass.

Number of moles $= (0.300 \text{ liter}) \times \dfrac{0.020 \text{ mole}}{1.0 \text{ liter}} = 6.0 \times 10^{-3}$ mole

Molecular weight of glucose = 180

Mass of glucose $= (6.0 \times 10^{-3} \text{ mole}) \times \dfrac{180 \text{ g}}{1.0 \text{ mole}} = 1.1$ g

(b) Since the volume is the same but the concentration is one-half that in part (a), the mass required is one-half that in part (a), or 0.55 g.

(c) 2.0% means 2.0 g/100 ml.

Mass of glucose $= (300 \text{ ml}) \times \dfrac{2.0 \text{ g}}{100 \text{ ml}} = 6.0$ g

8-6 (a) Bicarbonate, HCO_3^-; one charge per ion. No. of equivalents = no. of moles; therefore 28 meq/liter is 28 millimoles/liter = 28 mM.

(b) Chloride, Cl^-; one charge per ion, 105 mM.

(c) Phosphate, PO_4^{3-}; three charges per ion. No. of equivalents = 3 × no. of moles or no. of moles = $\frac{1}{3}$ × no. of equivalents. Therefore 2.0 meq/liter = 0.67 mM.

(d) Magnesium, Mg^{2+}; two charges per ion, 0.9 mM.

(e) Potassium, K^+; one charge per ion, 4.5 mM.

(f) Sodium, Na^+; one charge per ion, 140 mM.

8-7 Since the solution is saturated, the concentration of urea is equal to the solubility, 120 g/100 ml.

Mass of urea $= (350 \text{ ml}) \times \dfrac{120 \text{ g}}{100 \text{ ml}} = 420$ g

8-8 Solubility is directly proportional to partial pressure. 1.00 atm = 760 torr

Therefore, solubility at 40 torr = $\left(0.15\dfrac{g}{liter}\right) \times \dfrac{40\ \text{torr}}{760\ \text{torr}} = 7.9 \times 10^{-3}\ \dfrac{g}{liter}$

8-9 (a) Soluble. (b) Insoluble. (c) Soluble. (d) Insoluble. (e) Soluble.

8-10 Since the aqueous mixture does not separate on standing, it is either a solution or a colloidal dispersion. It exhibits the Tyndall effect, so it is an aqueous collodial dispersion.

8-11 If the bath solution is not changed regularly, the concentration of urea and other small waste-product molecules would gradually increase, and the desired diffusion of these molecules through the cellophane tubing into the bath solution would slow down and eventually stop.

8-12 The total concentration of particles and the osmotic pressure of the NaCl solution on the right will be higher than that of the glucose solution on the left, since a 0.01 M NaCl solution is 0.01 M Na$^+$ and 0.01 M Cl$^-$. Therefore, the water will flow from the more dilute to the more concentrated solution, or from left to right.

8-13 Determine the concentration on a mole basis (the molarity). 0.9% means 0.9 g/100 ml. The molecular weight of glucose is 180. The formula weight of NaCl is 58.5.
For the glucose solution,
$$\left(\dfrac{0.9\ g}{100\ ml}\right) \times \left(\dfrac{1.0\ mole}{180\ g}\right) = \dfrac{5 \times 10^{-3}\ mole}{100\ ml} = 5 \times 10^{-2}\ M$$
For the NaCl solution,
$$\left(\dfrac{0.9\ g}{100\ ml}\right) \times \left(\dfrac{1.0\ mole}{58.5\ g}\right) = \dfrac{1.5 \times 10^{-2}\ mole}{100\ ml} = 1.5 \times 10^{-1}\ M$$
The NaCl solution is 0.15 M Na$^+$ and 0.15 M Cl$^-$. A 0.9% glucose solution is therefore hypotonic with respect to a 0.9% NaCl solution.

8-14 The bath solution is made isotonic with blood to prevent hemolysis and/or crenation of red blood cells.

CHAPTER 9

Rates of Chemical Reactions and Chemical Equilibrium

9-1   INTRODUCTION

Some chemical reactions, such as the formation of rust on a car, are very slow. Other chemical reactions, such as the explosion of dynamite, are very rapid. In this chapter we shall discuss the rates at which chemical reactions occur and examine some of the factors that influence these rates.

Most chemical reactions in the human body are very rapid. Many of these reactions can proceed outside the human body, but at a much slower rate. We shall see that cells in the human body use enzymes to speed up the rate of reactions. Enzymes are examples of catalysts, substances that speed up the rate of reaction but are themselves unaltered by the overall reaction. We shall see that the way in which a given overall reaction takes place, its path, depends upon the reaction conditions. In general, a catalyst provides an alternative pathway for a reaction which allows the reaction to proceed at a faster rate than it does in the absence of the catalyst.

We shall also see that many chemical reactions are reversible; that is, the product(s) can react to re-form the original reactants. We shall see that reversible reactions can attain a state of dynamic equilibrium, where the macroscopic properties of the system do not change with time and where the forward rate is equal to the reverse rate. We shall also discuss how a reversible reaction in a state of dynamic equilibrium adjusts to an applied stress.

These topics—the rate of a chemical reaction and chemical equilibrium—have very important practical applications. For example, suppose we want to make a particular compound that has some medical or industrial uses, and that this compound is a product of a particular reversible reaction. The reaction is a reasonable one to use *only* if the reaction (and the temperature and pressure used) are such that a sufficiently large amount of product is present when the state of equilibrium is attained. In addition, the reaction rate *must* be fast enough so that the equilibrium state is attained within a reasonable length of time. Thus, a practical method of producing a desired compound must involve a reaction and a set of reaction conditions that provide *both* a sufficiently rapid reaction rate and a favorable equilibrium situation.

A modern highway and the meandering old road at its right are alternative paths through these mountains. The same chemical reaction can also occur via alternative pathways at vastly different rates.

9-2   STUDY OBJECTIVES

After studying the material in this chapter, you should be able to:

1. Define the terms reaction rate, reaction mechanism, reaction mechanism step, reaction intermediate, energy-profile diagram, activated complex, and activation energy.

2. Construct an energy-profile diagram for a given reaction of the type $AB + CD \rightarrow AC + BD$, which is assumed to proceed by AB and CD colliding and forming an activated complex, and identify the position of the reactants, products, activated complex, activation energy, E_a, and overall energy change, ΔE, on the diagram.

3. Use the model and energy-profile diagram for $AB + CD \rightarrow AC + BD$ to explain why the reaction rate for this reaction is directly proportional to both [AB] and [CD], and describe how the reaction rate depends on the orientation with which AB and CD collide, the kinetic energy of the collision between AB and CD, and the temperature.

4. Define the term catalyst and describe how the presence of a catalyst affects the reaction mechanism and the reaction rate.

5. Identify the reaction intermediates and/or catalyst when given a reaction mechanism.

6. Define the terms reversible reaction, dynamic equilibrium, equilibrium constant expression, and equilibrium constant.

7. Write the equilibrium constant expression for a reversible reaction when given the reaction equation.

8. Relate qualitatively the magnitude of K_{eq} with the position of equilibrium.

9. Describe how K_{eq} and the position of equilibrium change with a temperature increase or decrease given the sign of ΔH for a reaction.

10. Describe Le Châtelier's principle and use it to predict the effect on the position of equilibrium of an increase or a decrease in any one of the chemical species involved in a reversible reaction.

9-3   RATES OF CHEMICAL REACTIONS

The rate of a chemical reaction is analogous to the rate at which manufactured items come off an assembly line. The rate at which manufactured items are produced could be expressed as "the number of items produced per minute" or "the number of items produced per hour":

(9-1)
$$\text{Production rate} = \frac{\text{no. of items produced}}{\text{time interval}}$$

For a chemical reaction we can say that the **reaction rate** is the amount of chemical change per unit time. Since we are usually interested in the products, or more specifically, the concentration of the products, for a chemical reaction, we give the reaction rate as "the increase in product concentration per minute" or "the increase in product concentration per second":

(9-2)
$$\text{Reaction rate} = \frac{\text{increase in product concentration}}{\text{time interval}}$$

An assembly line can run at different speeds, and likewise chemical reactions can proceed at different rates. When chemists study the rate of a particular reaction they determine how various reaction conditions, such as different concentrations or different temperatures, affect the reaction rate.

In general, the rate of a chemical reaction increases as the reactant concentration increases, and for most chemical reactions, there is a large increase in the reaction rate for even a small increase in temperature. For example, the rate of decomposition of hydrogen peroxide ($2H_2O_2 \rightarrow 2H_2O + O_2$) increases by a factor of about 2.5 when the temperature is changed by only 10°C from 25°C to 35°C (for the same initial concentration of H_2O_2).

When we compare the rates of different reactions under identical conditions of concentration and temperature, we can find an enormous variation in reaction rates. A convenient way to compare the rates of different reactions is to compare the time it takes for half of the initial amount of reactants to be transformed into products, that is, the half-life (Section 2-9). Some chemical reactions are extremely slow, with half-lives of years, and are inconvenient to study. The half-lives of many other reactions are in the range of 1 minute to several hours, and the rates of these reactions can be conveniently measured by simply mixing the reactants, recording the time with a stopwatch, and determining the concentration of products at various times. Many other important reactions, such as the acid-base reactions that we shall discuss in Chapter 10 and the biologically important reactions involving enzyme catalysts to be discussed in Section 9-4, are very fast, with half-lives ranging from 10 sec to 10^{-6} sec. Special methods are needed to study the rates of these fast reactions; however, it is now possible to study the rates of certain reactions that have half-lives as short as 10^{-12} sec!

Let us consider a simple chemical reaction that has been studied extensively. At high temperatures the diatomic gases H_2 and I_2 will react to form HI:

(9-3) $$H_2 + I_2 \longrightarrow 2HI$$

When we put some H_2 and I_2 in a container and keep it at a constant temperature, the concentration of HI increases with time, as shown in Figure 9-1. Note that we use the symbol [HI] to represent "the concentration of HI in units of moles per liter." In addition, notice that the increase in the [HI] per unit time (i.e., the reaction rate) is largest at the beginning of the reaction and that the rate decreases as the reaction proceeds. The rate at the beginning of the reaction is called the **initial rate**.

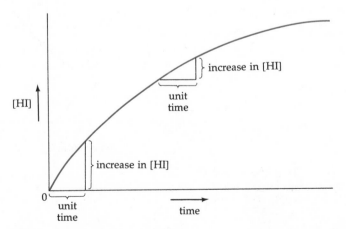

Figure 9-1 A plot of [HI] versus time for the reaction $H_2 + I_2 \rightarrow 2HI$. At any time, the reaction rate is the increase in [HI] per unit time. Note that the reaction rate is largest at the beginning of the reaction, but that it decreases with the passage of time.

Table 9-1 $H_2 + I_2 \rightarrow 2HI$ (at 700 K)

	[H_2]	[I_2]	Initial Rate (moles/liter · sec)
Experiment I	0.010 M	0.010 M	6.4×10^{-6}
Experiment II	0.020 M	0.010 M	12.8×10^{-6}
Experiment III	0.010 M	0.020 M	12.8×10^{-6}

The rate of a reaction depends on the concentration of the reactants. For example, the initial rate of formation of HI at 700 K was measured in three separate experiments for three different sets of H_2 and I_2 concentrations. The results of these experiments are given in Table 9-1. When we compare the results of Experiments I and II we see that, for the same [I_2], doubling the [H_2] doubles the initial rate. Thus, the rate of formation of HI is directly proportional to [H_2]. Similarly, when we compare the results of Experiments I and III we see that, for the same [H_2], doubling the [I_2] doubles the rate. Thus, the rate of formation of HI is *directly proportional* to *both* the concentration of H_2 (I versus II) *and* the concentration of I_2 (I versus III). Therefore, we can predict that if we did a fourth experiment in which the concentrations of both H_2 and I_2 were 0.020 M, the initial rate of formation of HI would be quadrupled, or $4 \times (6.4 \times 10^{-6})$ or 25.6×10^{-6} mole/liter · sec. Reaction 9-3 is typical in that the rate of formation of the product increases as the concentrations of the reactants increase. For most reactions, however, the rate of formation of products is *not* directly proportional to the concentration of reactants, as it is in this case.

The reaction rate does increase markedly as the temperature increases for almost all reactions. Consider, for example, the initial rate of formation of HI (reaction 9-3) over the temperature range 550 K to 600 K when the concentrations of H_2 and I_2 are both 0.010 M (see Figure 9-2). Note that the initial rate at 600 K is about 18 times the initial rate at 550 K!

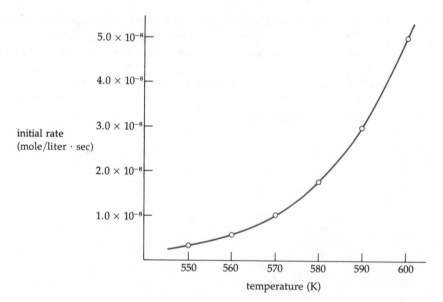

Figure 9-2 A plot of initial rate versus temperature for the reaction $H_2 + I_2 \rightarrow 2HI$. The initial concentrations of H_2 and I_2 are both 0.010 M.

Exercise 9-1

For reaction 9-3, use Figure 9-2 to predict the initial rate of formation of HI when [H_2] = 0.030 M, [I_2] = 0.020 M, and the temperature is 600 K.

Reaction Mechanisms

The most common representation for an overall chemical reaction, such as $H_2 + I_2 \rightarrow 2HI$, gives the reactants, the products, and the relative number of moles of reactants and products, but it does *not* convey any information about *how* the reaction takes place. A model describing the molecular changes that occur during a chemical reaction is called a **reaction mechanism.** For illustrative purposes we shall discuss one possible reaction mechanism for the formation of HI. Our discussion will reveal some general concepts that are quite helpful in explaining how chemical reactions take place.

The following is one simple reaction mechanism for the formation of HI.

(9-4)   **Postulated Reaction Mechanism:** H_2 and I_2 collide, and when they collide with sufficient kinetic energy and with a favorable orientation, molecular changes occur that result in the formation of HI.

According to this model, the rate of formation of HI is directly proportional to the rate at which H_2 and I_2 molecules collide. (The more collisions, the faster the rate.) If we double $[H_2]$ while keeping $[I_2]$ the same in a given container at a constant temperature, the rate at which H_2 and I_2 collide will double. Similarly, doubling $[I_2]$ while keeping $[H_2]$ the same doubles the rate of collisions, and doubling both $[H_2]$ and $[I_2]$ results in four times the rate of collisions of H_2 and I_2 molecules. Thus, the rate at which H_2 and I_2 molecules collide is directly proportional to both $[H_2]$ and $[I_2]$. Therefore, according to our model, the rate of formation of HI is also directly proportional to both $[H_2]$ and $[I_2]$. This is in agreement with the experimental data given in Table 9-1.

Our model also includes the reasonable assumption that, for a reaction to take place, molecular changes sufficient to form the product HI must occur when H_2 and I_2 molecules collide. The molecular changes that must occur are a stretching, and consequently, a weakening, of both the H—H and I—I bonds, together with bond formation between hydrogen and iodine atoms. Figure 9-3 is a schematic representation of this process.

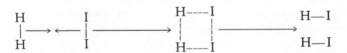

Figure 9-3   Schematic illustration of the molecular changes that occur upon collision and formation of products for the reaction $H_2 + I_2 \rightarrow 2HI$. In this process the H—H and I—I bonds are stretched and weakened (dashed black) as the H—I bonds are formed (dashed blue).

These molecular changes involve changes in potential energy. Since it requires energy to stretch a bond, a stretched bond has a higher potential energy than a bond that is not stretched. Thus, according to our model, as the H—H and I—I bonds stretch, the potential energy increases. A special diagram, the **potential energy-profile diagram,** is used to show the relationship between the molecular changes during a collision and the changes in potential energy. Figure 9-4 is a potential energy-profile diagram for the reaction $H_2 + I_2 \rightarrow 2HI$. Note that the potential energy increases as the H—H and I—I bonds are stretched. As this stretching occurs, a hydrogen atom and an iodine atom are also coming closer together, forming a H—I bond. This latter process tends to reduce the potential energy.

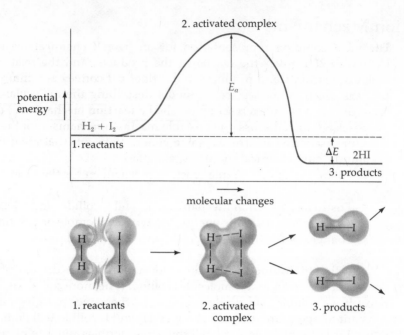

Figure 9-4   An energy-profile diagram for the reaction $H_2 + I_2 \rightarrow 2HI$. Potential energy is plotted versus molecular changes. State 1 represents the separate reactants, H_2 and I_2. State 2 represents the activated complex, a configuration involving all four atoms and having the greatest potential energy. State 3 represents the products, 2HI. The activation energy, E_a, is the potential energy increase that must occur to form the activated complex—about 41 kcal/mole for this reaction. The change in energy, ΔE, is about -4 kcal for this reaction and is equal to $E(\text{products}) - E(\text{reactants})$.

The particular configuration of all four atoms that has the greatest potential energy is called the activated complex. For any reaction mechanism similar to the one we have been discussing, an **activated complex** is the configuration of atoms that has the greatest potential energy. An activated complex is *not* a stable chemical species. Rather it is a transitory configuration of atoms that is used in our model to describe the process of bond breaking and bond formation during a collision. Once the activated complex is formed, the formation of products requires no further increase in energy.

The potential energy increase that must occur in order to form the activated complex in a reaction mechanism is called the **activation energy, E_a**. The activation energy for the reaction $H_2 + I_2 \rightarrow 2HI$, which can be determined from a quantitative treatment of how the reaction rate varies with temperature (see Figure 9-2), is about 41 kcal/mole of activated complex. The energy difference, ΔE, for this reaction, that is, E (products) $- E$ (reactants), is about -4 kcal.

Potential energy-profile diagrams, such as Figure 9-4, are very useful when discussing chemical reaction rates. The changes in potential energy that occur in chemical reactions are analogous to the changes in potential energy that occur when you walk over a hill. Different reactions have different activation energies, just as different hills have different heights. Also, when you walk over a hill, you might wind up at a height higher or lower than the height you started at. Similarly, ΔE for some reactions is positive and for others it is negative (ΔE is positive if more energy is required to break the bonds in the reactants than is released when the new bonds in the product are formed, whereas ΔE is negative if the reverse is true). The ΔE for the reaction $H_2 + I_2 \rightarrow 2HI$ is negative. This means that energy is released in the overall reaction. Figure 9-5 shows an energy-profile diagram for a reaction in which ΔE is positive. Since the activation energy for a reaction is the energy needed to distort the reactant molecules and form the activated complex, activation energies are always positive.

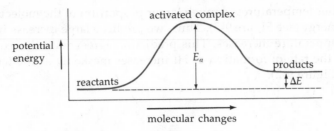

Figure 9-5   An energy-profile diagram for a reaction with a positive ΔE. The positive ΔE indicates that, overall, energy is absorbed when this reaction takes place.

The energy that is required to form the activated complex (see Figure 9-4) in the reaction $H_2 + I_2 \rightarrow 2HI$ can come only from the kinetic energy that the H_2 and I_2 molecules have when they collide. When molecules of H_2 and I_2 collide, their kinetic energy is used to stretch the bonds and increase the potential energy of the system. Conservation of energy requires that the kinetic energy decrease by the same amount that the potential energy increases. Therefore, H_2 and I_2 can react to form HI only when they collide with kinetic energy equal to or greater than the energy needed to form the activated complex, that is, the activation energy. If the colliding H_2 and I_2 molecules do not have sufficient kinetic energy, the complex cannot become fully activated and products cannot be formed. Only a small fraction of actual collisions have enough kinetic energy to form a fully activated complex.

In Chapter 5 we saw that there is a distribution of kinetic energies in a sample of a gas: At any instant different molecules are moving with different speeds and as a result there is a similar distribution of kinetic energies for gas molecule collisions (see Figure 9-6). Our model predicts that of those collisions only the fraction with kinetic energy greater than the activation energy can actually result in products. Furthermore, the larger the activation energy, the smaller the fraction of collisions that can result in products being formed.

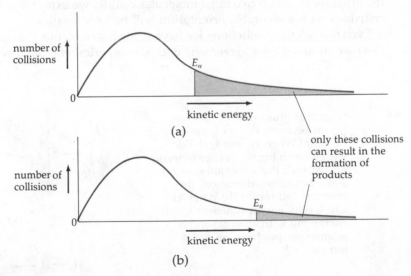

Figure 9-6   The number of collisions between molecules is plotted versus the kinetic energy of collision. Only collisions represented by the shaded areas, those with energies greater than the activation energies, can result in product formation.

The activation energy, E_a, for reaction (b) is greater than the activation for reaction (a). Therefore, a smaller fraction of the total number of collisions can result in product formation in reaction (b).

At higher temperatures, a much larger proportion of the molecules have high kinetic energy (see Figure 9-7). Thus, we predict a large increase in reaction rate as the temperature increases. This prediction agrees with the experimental result that the rate of formation of HI increases markedly as the temperature increases (Figure 9-2).

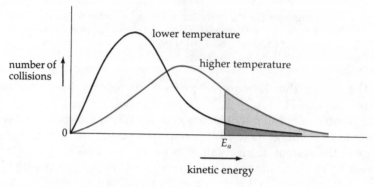

Figure 9-7   The number of collisions between molecules is plotted versus the kinetic energy of collision at two different temperatures. For a given reaction, the fraction of the collisions with kinetic energy greater than the activation energy is much larger at the higher temperature than at the lower temperature.

Our model also states that in addition to having sufficient kinetic energy, the H_2 and I_2 molecules must collide with a **favorable orientation** in order for a bond to form between a hydrogen atom and an iodine atom (Figure 9-8a). Figure 9-8b shows a collision orientation for which formation of HI is highly unlikely. For simple diatomic molecules like those in our example, we would predict that a relatively large fraction of the collisions would have a favorable orientation. On the other hand, when two large molecules collide, we expect that the fraction of collisions with a favorable orientation will be very small.

Even though the predictions we have made based on our reaction mechanism (9-4) are in qualitative agreement with the experimental data, we cannot con-

Figure 9-8   Schematic illustration of two collisions between H_2 and I_2 molecules. (a) When H_2 and I_2 collide with enough kinetic energy (energy greater than the activation energy) and a favorable orientation, the products 2HI result. (b) When H_2 and I_2 collide with sufficient kinetic energy but with an unfavorable orientation, products are not formed.

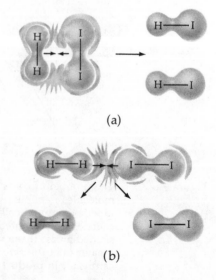

(a)

(b)

clude that the formation of HI actually takes place in accordance with this model. Although reaction rates for the reaction $H_2 + I_2 \rightarrow 2HI$ have been studied since the 1890s, chemists continue to argue about the nature of the actual reaction mechanism.

Exercise 9-2

Draw an energy-profile diagram for the reaction

$$CO(gas) + NO_2(gas) \rightarrow CO_2(gas) + NO(gas)$$

assuming that the reaction mechanism involves CO and NO_2 colliding and forming an activated complex with an energy of activation of 32 kcal/mole. Identify the position of the reactants, products, activated complex, activation energy, and overall energy change. ΔE is about -54 kcal/mole for this reaction.

Exercise 9-3

The reaction $(A + B \rightarrow C + D)$ takes place by means of A and B colliding and forming an activated complex. A second reaction $(E + F \rightarrow G + H)$ proceeds by a similar mechanism. At the same temperature and concentrations of reactants, the rate of the first reaction is much higher than the rate of the second reaction. Assuming that the fraction of collisions with a favorable orientation is the same for both reactions, which reaction has the larger activation energy?

9-4 CATALYSIS

It is possible to alter the reaction path for a reaction without changing either the reactants or the products. This is analogous to the fact that you can drive from San Francisco, California, to Boulder, Colorado, by many different routes, some more mountainous than others. A reaction path can be altered by adding a catalyst to the reaction mixture. A **catalyst** is a substance that alters the rate of a reaction but that has the same chemical form and is present in the same amount before and after the reaction takes place. A catalyst is chemically modified during the course of the reaction—it is involved in the reaction mechanism; but it experiences no net overall change—it is neither a reactant nor a product. In general, a catalyst increases the rate of a chemical reaction.

All living organisms depend on a group of catalysts called enzymes for their very existence (see Chapter 23). Without enzyme catalysts, many vital chemical reactions in living organisms would proceed too slowly to maintain life. For example, an enzyme known as "catalase" acts to increase the rate of decomposition of hydrogen peroxide to form water and oxygen in the human body.

(9-5) $$2H_2O_2 \longrightarrow 2H_2O + O_2$$

The bubbling action when hydrogen peroxide is used on cuts is evidence that reaction 9-5 is taking place. The rate of reaction 9-5 at 22°C is 0.35 M/sec if $[H_2O_2] = 10^{-3}$ M and [catalase] $= 10^{-5}$ M. In the absence of catalase, or some other catalyst, the rate of reaction 9-5 at 22°C is 10^{-13} M/sec. Thus, under these conditions, the enzyme catalase enhances the decomposition of hydrogen peroxide by a factor of 3.5×10^{12}. It is not unusual for enzymes and other catalysts to have such a profound effect as this on the rate of a reaction. Catalysts are also widely used in industry, and much chemical research is devoted to the search for more effective and less expensive industrial catalysts.

Let us illustrate the basic features of catalysis with the general reaction

(9-6) $AB + CD \xrightarrow{\text{X}} AC + BD$

where X represents the catalyst. Suppose that the reaction mechanism for reaction 9-6 involves two elementary reaction steps:

(9-7)   Step I $AB + X \longrightarrow \underset{\underset{X}{\diagdown\diagup}}{A{-}B}$

(9-8)   Step II $\underset{\underset{X}{\diagdown\diagup}}{A{-}B} + CD \longrightarrow AC + BD + X$,

According to this postulated reaction mechanism, the products, AC and BD, are formed only when *both* steps I and II take place *successively*. The molecule ABX whose formation is postulated in step I is an example of a reaction intermediate. **Reaction intermediates** are stable chemical species (they do not break apart spontaneously) that are formed in one step of a reaction mechanism but are subsequently used as reactants in another step. Thus, reaction intermediates exist (or are postulated to exist) during the course of the reaction but are not present initially or after the reaction ceases—they are neither reactants nor products. A reaction intermediate is quite different from an activated complex. If the concentration of a reaction intermediate is high enough during the course of a reaction, its presence can be detected and some of its properties can even be measured. This is not true of an activated complex. An activated complex is not a stable entity but rather the transitory configuration of atoms with the greatest potential energy that is postulated to occur during the course of a collision.

Notice that the catalyst X is used to form the reaction intermediate $\underset{\underset{X}{\diagdown\diagup}}{A{-}B}$ in step I, but that an equal amount of X is regenerated in step II. To form the reaction intermediate $\underset{\underset{X}{\diagdown\diagup}}{A{-}B}$, step I must proceed through the activated complex $\underset{\underset{X}{\diagdown\diagup}}{A{-}{-}{-}B}$ (see Figure 9-9), and the reactants must overcome an activation energy $E_a(\text{I})$.

Suppose that step II proceeds through the activated complex $\begin{smallmatrix} A{-}{-}{-}B \\ |\diagdown\diagup| \\ |X| \\ |\diagup\diagdown| \\ C{-}{-}{-}D \end{smallmatrix}$ and, that an activation energy $E_a(\text{II})$ must be overcome in this step (Figure 9-9). In the two-step process with the catalyst X, two activation energies, $E_a(\text{I})$ and $E_a(\text{II})$, must be surmounted. However, if X has the appropriate chemical properties, both $E_a(\text{I})$ and $E_a(\text{II})$ are considerably lower than the activation energy for the uncatalyzed reaction, which is also shown in Figure 9-9. For example, in step II, the fact that B is bonded to X as well as to A in the intermediate $\underset{\underset{X}{\diagdown\diagup}}{A{-}B}$ might weaken the A—B bond and make it easier (require less energy) to form the activated complex in step II than to form the activated complex in the uncatalyzed reaction. Recall that at the same temperature a reaction with a lower activation energy proceeds at a much faster rate than one with a higher activation energy. Therefore, our catalyst X speeds up the reaction $AB + CD \rightarrow AC + BD$ by providing an alternative path (reaction mechanism) with a lower overall activation energy. In general, any catalyst that speeds up a reaction functions in this manner. Note that a catalyst does not merely reduce the activation energy of the uncatalyzed reaction.

Let us return for a moment to our analogy of driving from San Francisco to Boulder. You could take one route which involves going over Pike's Peak, but a route that avoids all large mountains would probably require less energy and be faster. (Note, however, that the difference in altitude between San Francisco and Boulder is the same no matter what route you take.) Similarly, for the overall chemical reaction there is a definite ΔE that depends solely on the nature of the reactants and the products and not on the reaction mechanism. The reactants in each step in our reaction mechanism must collide with a favorable orientation. Besides providing a reaction path with a lower activation energy, many enzymes and industrial catalysts also provide a surface on which it is easier for the reactant molecules to interact with a favorable orientation.

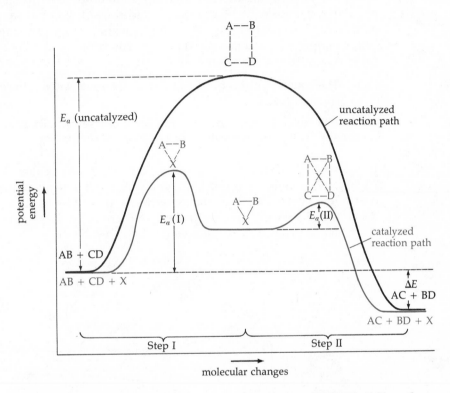

Figure 9-9   Energy-profile diagram for the reaction $AB + CD \rightarrow AC + BD$. When the catalyst X is not present, the reaction takes place in a single step with a very large activation energy, E_a (uncatalyzed). With the catalyst X present, the reaction mechanism involves two steps and a reaction intermediate. The activation energy for step I, $E_a(I)$, and the activation energy for step II, $E_a(II)$ are both much smaller than the activation energy for the uncatalyzed reaction. The overall energy change, ΔE, for the reaction $AB + CD \rightarrow AC + BD$ is the same for both reaction mechanisms.

Exercise 9-4

The following is a reaction mechanism for a certain overall reaction:

Step I $NO + \frac{1}{2}O_2 \rightarrow NO_2$
Step II $NO_2 + SO_2 \rightarrow NO + SO_3$

(a) What is the overall reaction?
(b) Which species (if any) are reaction intermediates?
(c) Which species (if any) are catalysts?

9-5   CHEMICAL EQUILIBRIUM

Reversible Reactions—Dynamic Equilibrium

It is a matter of experimental fact that in many chemical reactions the reactants are not completely converted to products even if the reaction proceeds for a very long time. For example, at high temperatures carbon monoxide and water react to form carbon dioxide and hydrogen.

(9-9) $$CO(g) + H_2O(g) \longrightarrow CO_2(g) + H_2(g)$$

The letter in parentheses after each compound specifies its *state* (gas, liquid, or solid) for these particular reaction conditions. If you put some carbon monoxide and water (not necessarily in equal amounts) into an empty container at a high temperature and measured the concentrations of CO, H_2O, CO_2, and H_2 at various later times, you would find that:

1. The concentrations of CO and H_2O decrease initially but then level off to constant values.

2. The concentrations of CO_2 and H_2 increase initially from zero and then approach constant values as well (see Figure 9-10).

If you were to measure the concentration of any one of the substances in the reaction after time t in Figure 9-10, you would not detect any change in its concentration. We therefore say that after time t the system has reached *equilibrium*. Recall that a system is in equilibrium if there is no observable change in the macroscopic properties of the system.

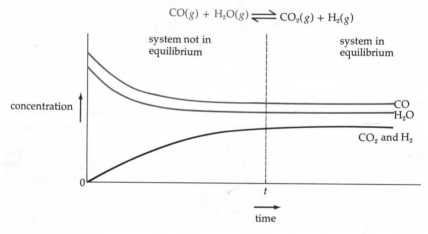

$$CO(g) + H_2O(g) \rightleftharpoons CO_2(g) + H_2(g)$$

Figure 9-10   When some CO and some H_2O are placed in an empty container, they react to form CO_2 and H_2. The concentrations of reactants and products are plotted versus time for the reaction. After time t, there are no detectable changes in the concentrations of CO, H_2O, CO_2, and H_2; that is, the system is in equilibrium.

Now, if you put some of the products of reaction 9-9, CO_2 and H_2, into an empty container at a high temperature, you would find that:

1. They react to form CO and H_2O.

2. An equilibrium state is also reached in which some CO and H_2O as well as CO_2 and H_2 are present (see Figure 9-11).

From these experimental results we infer that CO and H_2O can react to form CO_2 and H_2, and that these products CO_2 and H_2 can *also* react to form CO and

$$CO_2(g) + H_2(g) \rightleftharpoons CO(g) + H_2O(g)$$

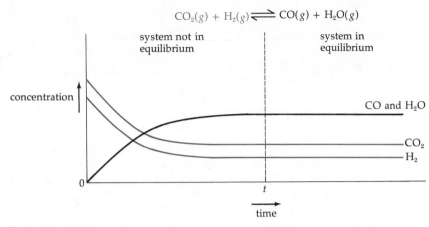

Figure 9-11   When some CO_2 and some H_2 are placed in an empty container, they react to form CO and H_2O. The concentrations of reactants and products are plotted versus time for the reaction. After time t, there are no detectable changes in the concentrations of CO_2, H_2, CO, and H_2O; that is, the system is in equilibrium.

H_2O. We therefore conclude that reaction 9-9 can proceed in either direction. We call a reaction that can proceed in either direction a **reversible reaction,** and when writing its equation a double arrow is used to indicate reversibility. Thus, Eq. 9-9 becomes

(9-10) $$CO(g) + H_2O(g) \rightleftharpoons CO_2(g) + H_2(g)$$

In our first experiment, we started with only CO and H_2O present, and reaction 9-10 initially proceeded from left to right; but as soon as some CO_2 and H_2 were formed, they reacted and henceforth reaction 9-10 proceeded in both directions. After a sufficient time had elapsed, a state of dynamic equilibrium resulted in which reaction 9-10 proceeded forward (from left to right) and also in reverse (from right to left) at the same rate (Figure 9-12). A state of dynamic equilibrium with equality of rates in the forward and reverse directions was

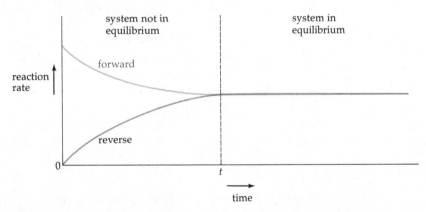

Figure 9-12   Rate is plotted versus time for a reversible reaction. Initially, only the species on the left of the reaction are present. The forward rate is thus very fast, and the reverse rate is zero. As the forward reaction proceeds, the rate of the forward reaction slows down, while the rate of the reverse reaction increases. After time t, the forward rate and the reverse rate are equal, and the system is in a state of dynamic equilibrium.

also reached when we started with only CO_2 and H_2O. In general, a **state of dynamic equilibrium,** <u>characterized by equal rates in the forward and reverse directions, is reached by any reversible reaction after a sufficient period of time, no matter what the composition of the starting mixture.</u> Thus, when a state of dynamic equilibrium is reached, we have

(9-11)   Rate forward = rate reverse

We shall see shortly that the state of dynamic equilibrium reached by a chemical reaction is characterized by a quantity that has a fixed value at any fixed temperature, and that this quantity changes to a new fixed value if the temperature is changed.

The Equilibrium Constant Expression

Let us return to reaction 9-10 and consider the quantitative results of some possible experiments. Suppose that in one such experiment (Experiment I) we put enough CO and H_2O in an empty container so that the initial concentrations of CO and H_2O are each 10^{-2} M. We then keep the temperature at 1000°C and at later times we withdraw samples of the resulting gas mixture, analyze these samples, and determine the concentrations of CO, H_2O, CO_2, and H_2. When equilibrium is reached we find that the concentrations of these gases have the values indicated in Table 9-2. We do several other experiments of this type and tabulate all our data. Note that the equilibrium concentrations are the same in Experiments I and II. In these two cases we reach the same equilibrium point whether we start from the left or the right in reaction 9-10. By inspecting the data for Experiments I through V (Table 9-2), it is *not* apparent that there is a quantity involving the equilibrium concentrations that has the same value for each of the experiments. However, as you can easily verify for yourself, the quantity $\left\{ \dfrac{[CO_2]_{eq}[H_2]_{eq}}{[CO]_{eq}[H_2O]_{eq}} \right\}$ has essentially the same numerical value for all five experiments (see Table 9-2, column C).

Table 9-2   Initial and Equilibrium Concentrations, $CO + H_2O \rightleftharpoons CO_2 + H_2$ (at 1000°C)

	A				B				C
	Initial Concentration ($M \times 10^3$)				Equilibrium Concentration ($M \times 10^3$)				$\left\{ \dfrac{[CO_2]_{eq}[H_2]_{eq}}{[CO]_{eq}[H_2O]_{eq}} \right\}$
Experiment	[CO]	[H$_2$O]	[CO$_2$]	[H$_2$]	[CO]$_{eq}$	[H$_2$O]$_{eq}$	[CO$_2$]$_{eq}$	[H$_2$]$_{eq}$	
I	10.0	10.0	0	0	5.65	5.65	4.35	4.35	0.59
II	0	0	10.0	10.0	5.65	5.65	4.35	4.35	0.59
III	20.0	20.0	0	0	11.25	11.25	8.75	8.75	0.60
IV	10.0	20.0	0	0	4.10	14.10	5.90	5.90	0.60
V	5.0	10.0	5.0	20.0	7.10	12.10	2.90	17.90	0.60

If we performed any number of additional experiments with mixtures of CO, H_2O, CO_2, and H_2 at 1000°C, we would find the same relationship among the equilibrium concentrations. We can therefore say that for any set of equilibrium concentrations at 1000°C,

(9-12) $\dfrac{[CO_2][H_2]}{[CO][H_2O]} = 0.60$ (for equilibrium concentrations at 1000°C)

There are an infinite number of different sets of equilibrium concentrations similar to the five sets given in Table 9-2 that satisfy Eq. 9-12, but this equation is valid only for a set of *equilibrium* concentrations and *only* if the temperature of the system is 1000°C. At a different temperature the expression on the left-hand side of Eq. 9-12 is also equal to a constant, but the value of the constant is different. We can express these facts through the following equation:

(9-13)
$$\frac{[CO_2][H_2]}{[CO][H_2O]} = K_{eq}$$

Equation 9-13 is an example of an **equilibrium constant expression,** and K_{eq} is called an **equilibrium constant.** At a given temperature K_{eq} has a constant value that does not depend on the concentrations, but the value of K_{eq} is different at different temperatures. For Eq. 9-13, K_{eq} has the value 0.60 at 1000°C, but this constant is 0.35 at 1200°C and 1.25 at 800°C.

There is an equilibrium constant expression that relates equilibrium concentrations for any reversible reaction. For example, for the reaction

(9-14)
$$SO_2(g) + NO_2(g) \rightleftharpoons SO_3(g) + NO(g)$$

the equilibrium constant expression is

(9-15)
$$\frac{[SO_3][NO]}{[SO_2][NO_2]} = K_{eq}$$

In Eq. 9-15 and similar expressions, it is understood that the concentrations are equilibrium concentrations.

Let us symbolize a general reversible reaction as

(9-16)
$$aA + bB + cC \cdots \rightleftharpoons dD + eE + fF \cdots$$

In Eq. 9-16 the capital letters A, B, C, . . . and D, E, F, . . . represent chemical substances, and the lowercase letters represent stoichiometric coefficients.

The equilibrium constant expression for the general reaction 9-16 is

(9-17)
$$K_{eq} = \frac{[D]^d[E]^e[F]^f}{[A]^a[B]^b[C]^c}$$

Let us consider three examples of how Eqs. 9-16 and 9-17 can be used.

EXAMPLE 1   Nitric oxide (NO) is a pollutant in the atmosphere. It is formed at a high temperature by the reversible reaction

(9-18)
$$N_2(g) + O_2(g) \rightleftharpoons 2NO(g)$$

When N_2, O_2, and NO attain a state of dynamic equilibrium, the equilibrium constant expression is

(9-19)
$$K_{eq} = \frac{[NO]^2}{[N_2][O_2]}$$

EXAMPLE 2   A reversible reaction that is very important industrially involves the direct synthesis of ammonia, NH_3, from its elements:

(9-20)
$$N_2 + 3H_2 \rightleftharpoons 2NH_3$$

Using our general relationship 9-17, the equilibrium constant expression for reaction 9-20 is

(9-21)
$$K_{eq} = \frac{[NH_3]^2}{[N_2][H_2]^3}$$

Equation 9-17 applies to *any* reversible reaction. It is applicable to reversible reactions in aqueous solutions, as well as to reversible reactions in the gas state. We shall use Eq. 9-17 extensively in our discussion of acid-base reactions in aqueous solutions in Chapters 10 and 11. Equation 9-17 is also applicable to reversible reactions involving both a solid and a gas, or both a liquid and a gas, and so on. For example, iron(II) oxide, FeO, and hydrogen, H_2, in a closed container at a high temperature attain a state of dynamic equilibrium with elemental iron, Fe, and water vapor, H_2O:

(9-22)
$$FeO(s) + H_2(g) \rightleftharpoons Fe(s) + H_2O(g)$$

According to Eq. 9-17, the right-hand side of the equilibrium constant expression for reaction 9-22 should be

(9-23)
$$\frac{[H_2O][Fe]}{[H_2][FeO]}$$

However, since FeO and Fe are pure solids, they have a *fixed* concentration at any given temperature and pressure. By convention, the fixed values of [FeO] and [Fe] are omitted from expression 9-23, so the equilibrium constant expression for reaction 9-22 is written in the following simpler form:

(9-24)
$$K_{eq} = \frac{[H_2O]}{[H_2]}$$

The equilibrium constant expression for any reversible reaction involving a pure solid or a pure liquid is written in a similar manner, with the fixed concentrations of the pure substances omitted from the right-hand side of the equilibrium constant expression. For example, for the reaction

(9-25)
$$CH_4(g) + 2O_2(g) \rightleftharpoons CO_2(g) + 2H_2O(l)$$

the equilibrium constant expression is

(9-26)
$$K_{eq} = \frac{[CO_2]}{[CH_4][O_2]^2}$$

Since the concentration of pure liquid H_2O has a fixed value, it is omitted from the right-hand side of Eq. 9-26.

Exercise 9-5

Write the appropriate equilibrium constant expression for the following reactions.
(a) $2SO_2(g) + O_2(g) \rightleftharpoons 2SO_3(g)$
(b) $2H_2O_2(g) \rightleftharpoons 2H_2O(g) + O_2(g)$
(c) $H_2(g) + I_2(g) \rightleftharpoons 2HI(g)$
(d) $FeO(s) + CO(g) \rightleftharpoons Fe(s) + CO_2(g)$

The Position of Equilibrium and the Magnitude of K_{eq}

The **position of equilibrium** for a reversible reaction refers to the concentrations of the substances on the right-hand side compared to the concentration of the substances on the left-hand side at equilibrium. Whether the position of equilibrium lies more to the right or more to the left depends on the magnitude of the equilibrium constant, K_{eq}, for the reversible reaction.

Consider, for example, reaction 9-10, for which K_{eq} is 0.60 at 1000°C. In Experiment I (Table 9-2) the equilibrium concentrations of the substances on the left are greater than the equilibrium concentrations of the substances on the right, and we say that the position of equilibrium lies to the left.

On the other hand, for reaction 9-14, K_{eq} is 3.00 at a certain temperature T, and the position of equilibrium lies to the right. At temperature T, when we start with initial concentrations of SO_2 and NO_2 of 10^{-2} M, which are similar to the initial concentrations of CO and H_2O in Experiment I (see Table 9-2), the equilibrium concentrations of SO_3 and NO (the substances on the right in Eq. 9-14) are 6.35×10^{-3} M, and the equilibrium concentrations of SO_2 and NO_2 (the substances on the left in Eq. 9-14) are 3.65×10^{-3} M. Note that the equilibrium constant for reaction 9-14 at temperature T ($K_{eq} = 3.00$) is larger than the equilibrium constant for reaction 9-10 at 1000°C ($K_{eq} = 0.60$), and that, for similar initial concentrations, the position of equilibrium is farther to the right for reaction 9-14 compared to reaction 9-10 (compare $[SO_3]_{eq}$ and $[NO]_{eq} = 6.35 \times 10^{-3} M$, with the $[CO_2]_{eq}$ and $[H_2]_{eq}$ for Experiment I, 4.35×10^{-3} M.) In general, for similar initial concentrations, the larger the equilibrium constant the more the position of equilibrium will lie to the right (see Figure 9-13).

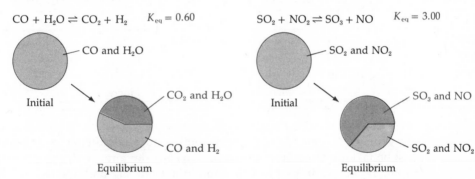

$$CO + H_2O \rightleftharpoons CO_2 + H_2 \qquad K_{eq} = 0.60$$

$$SO_2 + NO_2 \rightleftharpoons SO_3 + NO \qquad K_{eq} = 3.00$$

Figure 9-13   Schematic illustration of the relationship between the position of equilibrium and the magnitude of K_{eq}. For each reaction, the initial state (represented by the upper circles) consists only of the compounds on the left of the reversible reaction—at concentrations of 10^{-2} M. The equilibrium states are represented by the lower circles, in which the sizes of the blue and gray areas indicate the respective relative concentrations of the compounds on the right and left of the reversible reactions. Note that the reaction with the larger K_{eq} has the larger blue area, indicating that the position of equilibrium is farther to the right for this reaction.

Exercise 9-6

The equilibrium constant for the reaction $FeO(s) + H_2(g) \rightleftharpoons Fe(s) + H_2O(g)$ at 700°C is about 0.43, and for the reaction $FeO(s) + CO(g) \rightleftharpoons Fe(s) + CO_2(g)$ at 700°C the equilibrium constant is about 0.68. Suppose that 0.1 mole of FeO and 0.1 mole of H_2 are placed in a 1.0-liter container A, and 0.1 mole of FeO and 0.1 mole of CO are placed in another 1.0-liter container B. When equilibrium has been reached in both cases, which container contains the larger mass of Fe?

Effect of Temperature on K_{eq} and the Position of Equilibrium

As we have seen, for reaction 9-10, $CO(g) + H_2O(g) \rightleftharpoons CO_2(g) + H_2(g)$, the equilibrium constant decreases as the temperature increases. For this reaction, K_{eq} decreases from 1.25 at 800°C to 0.60 at 1000°C and to 0.35 at 1200°C.

For other reversible reactions, K_{eq} increases as the temperature increases. At high temperatures N_2 and O_2 undergo a reversible reaction to form NO:

(9-27) $$N_2(g) + O_2(g) \rightleftharpoons 2NO(g)$$

For reaction 9-27, K_{eq} increases from 8×10^{-4} at 1800°C to 2×10^{-3} at 2000°C and then to 4×10^{-3} at 2200°C. In general, for a reversible reaction, K_{eq} increases as the temperature increases for an endothermic reaction (ΔH positive), and K_{eq} decreases as the temperature increases for an exothermic reaction (ΔH negative). Recall that ΔH is equal to the sum of the enthalpies of the substances on the right minus the sum of the enthalpies of the substances on the left (see Section 6-7).

For reaction 9-27, ΔH is positive (+43.2 kcal), and as we just saw, K_{eq} increases as the temperature increases. On the other hand, for reaction 9-10, ΔH is negative (-9.9 kcal), and we saw that K_{eq} decreases as the temperature increases for this reaction. Our discussion is summarized in Table 9-3.

Table 9-3 Variation of K_{eq} with temperature

Endothermic reaction $N_2 + O_2 \rightleftharpoons 2\,NO$; $\Delta H = 43.2$ kcal			Exothermic reaction $CO + H_2O \rightleftharpoons CO_2 + H_2$; $\Delta H = -9.9$ kcal		
T (°C)	K_{eq}		T (°C)	K_{eq}	
1800	8×10^{-4}		800	1.25	
2000	2×10^{-3}		1000	0.60	
2200	4×10^{-3}		1200	0.35	
	increase	increase		increase	decrease

In general, the change of K_{eq} with temperature is quite large, and this effect must be taken into account whenever we consider a reversible reaction. For example, if we want to prepare ammonia from N_2 and H_2 by the reaction $N_2 + 3H_2 \rightarrow 2NH_3$, we want the reaction to proceed at a reasonable rate. Since the rates of almost all chemical reactions increase as the temperature increases, we want to run our reaction at a high temperature. However, the reaction between N_2 and H_2 to form NH_3 (Eq. 9-20) is a reversible reaction with a negative ΔH. Since ΔH is negative, the higher the temperature, the smaller K_{eq} is and the farther the position of equilibrium lies to the left. Thus, if we run the reaction at too high a temperature, equilibrium is reached very quickly but there is a negligible amount of NH_3 in the equilibrium mixture. We cannot run the reaction at too low a temperature either, even though K_{eq} is large at a low temperature and the equilibrium mixture would contain a large amount of NH_3, because at too low a temperature the reaction rate is too slow. We must use an intermediate temperature to obtain a reasonable amount of NH_3 from N_2 and H_2 within a reasonable time.

Exercise 9-7

For the reaction $SO_2(g) + NO_2(g) \rightleftharpoons SO_3(g) + NO(g)$, the value of ΔH is negative. How does a decrease in temperature affect K_{eq} and the position of equilibrium?

Position of Equilibrium and Rate of Reaction

It is important to remember the following experimental fact: The position of equilibrium for any reversible reaction depends on neither the reaction mechanism nor on the time it takes to reach equilibrium. Thus, for example, one can drastically change the rate of a chemical reaction by adding a suitable catalyst to the reaction mixture, but doing this will *not* alter the position of equilibrium. For any reversible reaction, the position of equilibrium depends solely on the equilibrium constant expression, the value of the equilibrium constant, and the initial concentrations, and not on the time it takes to get to that equilibrium position. Knowing the value of the equilibrium constant for a particular reversible reaction tells absolutely nothing about how long it might take to reach equilibrium starting from some nonequilibrium concentrations. For many reversible reactions the rate of approach to equilibrium is extremely slow in the absence of a catalyst, and quite rapid when a suitable catalyst is present.

9-6   LE CHÂTELIER'S PRINCIPLE

Systems that are in a state of dynamic equilibrium are in a sense quite sluggish. The equilibrium situation is a preferred state, and resistance to any disturbance can be observed experimentally.

Let us consider a simple example. If you put 1 mole of pure liquid water in an empty 1-liter closed container at 25°C, some of the water vaporizes and the following dynamic equilibrium is established.

(9-28) $H_2O(liquid) \rightleftharpoons H_2O(gas)$

The equilibrium gas pressure in this situation is about 24 torr, which is the vapor pressure of water at 25°C.

As we know, the vapor pressure of water and other liquids depends only on the temperature and not on the size of the container the liquid is in. Thus, if we keep the temperature constant and increase the size of the container to 2 liters, the pressure initially drops, but more liquid evaporates and when equilibrium has been reestablished the pressure is still 24 torr. If the temperature is fixed at 25°C, as the volume is continually increased the pressure will remain constant at 24 torr as long as there is a dynamic equilibrium between liquid and vapor (see Figure 9-14). Increasing the volume is a disturbance that momentarily reduces the pressure, so the evaporation rate exceeds the condensation rate and the proportion of water molecules in the gas state increases. After a short period of time, the system returns to a state of dynamic equilibrium—the rate of evaporation is equal to the rate of condensation and the gas pressure returns to its previous equilibrium value of 24 torr. The net overall effect of the increase in volume is a shift of water molecules to the right in reaction 9-28. Of course this process cannot continue indefinitely. At some large volume, all the liquid water will have evaporated. When there is no longer a dynamic equilibrium between liquid water and vapor, the pressure will drop below 24 torr with any further increase in volume because there is no more $H_2O(liquid)$ in reaction 9-28 to form $H_2O(gas)$.

The situation we have just described is an example of how a system in dynamic equilibrium behaves when it is disturbed. It is an example of a very useful general principle called Le Châtelier's principle.

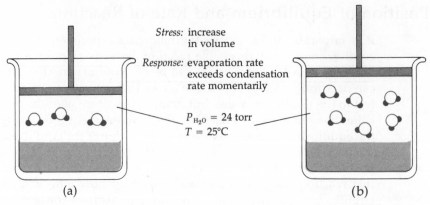

Figure 9-14   Le Châtelier's principle. (a) Liquid and gaseous water are in dynamic equilibrium at 25°C. If the volume is increased at 25°C, the evaporation rate exceeds the condensation rate momentarily. (b) When equilibrium has been reestablished, there are more molecules in the gas state. Note that for both (a) and (b) the pressure of H₂O is equal to the vapor pressure of water at 25°C, since liquid and gaseous water are in dynamic equilibrium at 25°C.

Henry Louis Le Châtelier was a French scientist who studied the behavior of equilibrium systems in the late nineteenth century. **Le Châtelier's principle** states that when a stress is applied to a system in dynamic equilibrium, the equilibrium system readjusts so as to reduce the effect of the stress as much as possible. In our previous example, the stress was an increase in volume, which would have resulted in a decrease in pressure if it were not for the fact that the equilibrium reaction 9-28 readjusts; more molecules go into the gas state, so that the equilibrium pressure remains at 24 torr.

Le Châtelier's principle also applies to reversible chemical reactions when they are in a state of dynamic equilibrium. A stress in this case might be an increase or decrease in temperature, pressure, or volume; but for chemical reactions the most important type of stress is an increase or decrease in the concentration of one or more of the substances involved in the reaction. As a specific example, let us again consider the reaction,

(9-29) $$CO(g) + H_2O(g) \rightleftharpoons CO_2(g) + H_2(g)$$

with the equilibrium constant expression

(9-30) $$\frac{[CO_2][H_2]}{[CO][H_2O]} = 0.60 \quad \text{(at 1000°C)}$$

In Table 9-2 (Experiment IV) we saw that one set of equilibrium concentrations was $[CO] = 4.10 \times 10^{-3}\,M$, $[H_2O] = 14.10 \times 10^{-3}\,M$, $[CO_2] = 5.90 \times 10^{-3}\,M$, and $[H_2] = 5.90 \times 10^{-3}\,M$. Suppose that we add to this equilibrium mixture additional H₂O, say, 10^{-2} mole. How will this system behave, and what will be its *new* equilibrium state?

Le Châtelier's principle enables us to predict whether the new equilibrium concentration of any particular substance in the reaction will be larger or smaller than it was before *without* performing any numerical calculations. When we add more H₂O to our equilibrium system, [H₂O] is increased and as a result of this applied stress the system is momentarily no longer at equilibrium. Our dynamic system, however, can readjust itself and reduce this stress if some of the additional H₂O reacts with an equal amount of CO and forms more CO₂ and more H_2. It is possible to check our qualitative predictions by actually calculating values for the new equilibrium concentrations. We shall not discuss the method for doing this type of calculation, but using Eq. 9-30 we calculate that when 10^{-2} mole of H_2O is added to the original equilibrium mixture, the new equilibrium concentrations are $[CO] = 3.25 \times 10^{-3}\,M$, $[H_2O] = 23.25 \times 10^{-3}\,M$, and $[CO_2] = [H_2] = 6.75 \times 10^{-3}\,M$, as shown in Table 9-4.

Table 9-4 Shift in Position of Equilibrium: $CO + H_2O \rightleftharpoons CO_2 + H_2$

	Concentration ($M \times 10^3$)				$\left\{ \dfrac{[CO_2]\,[H_2]}{[CO]\,[H_2O]} \right\}$*
	CO	H_2O	CO_2	H_2	
Original equilibrium	4.10	14.10	5.90	5.90	$0.60 = K_{eq}$
Just after addition of 10^{-2} mole of H_2O	4.10	24.10	5.90	5.90	$0.35 \neq K_{eq}$
New equilibrium	3.25	23.25	6.75	6.75	$0.60 = K_{eq}$

* Note that the value of this quantity is equal to 0.60, the value of the equilibrium constant, for both the original and the new equilibrium concentrations, but that this quantity does not have the value 0.60 for the nonequilibrium concentrations.

Thus, the quantitative calculation of the direction in which the equilibrium concentrations change is in complete agreement with our qualitative predictions on the basis of Le Châtelier's principle. The equilibrium readjusts with some of the added 10^{-2} mole of H_2O—specifically, 0.85×10^{-3} mole of H_2O—reacting with an equal amount of CO and forming an additional 0.85×10^{-3} mole of both CO_2 and H_2O. Since the equilibrium concentrations of CO_2 and H_2 increase, we can describe these changes in the equilibrium concentrations by saying that <u>the stress of adding more H_2O causes the system to readjust and shift the position of equilibrium more to the right</u>, as shown in Figure 9-15. Of course, when we added the H_2O we did not add any additional CO, so the new equilibrium concentration of CO is less than what it was previously.

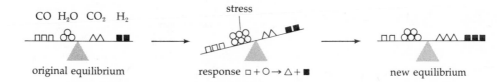

CO H_2O CO_2 H_2

original equilibrium   response $\square + \circ \rightarrow \triangle + \blacksquare$ new equilibrium

stress

Figure 9-15   Le Châtelier's principle illustrated schematically for the chemical reaction $CO + H_2O \rightleftharpoons CO_2 + H_2$. Initially, CO and H_2O are in dynamic equilibrium with CO_2 and H_2 (left). Some additional H_2O is added (center), causing a momen-tary nonequilibrium situation. The system's response to this stress is for some of the additional H_2O to react with CO, forming more CO_2 and H_2, until a new equilibrium position is established (right).

Similarly, we predict on the basis of Le Châtelier's principle that <u>if we remove some H_2O from our original equilibrium mixture, the system readjusts to reduce this stress as much as possible by shifting the position of equilibrium to the left</u>. In this case, the new equilibrium concentration of CO is greater than it was originally, and the new equilibrium concentrations of CO_2 and H_2 are less than their previous equilibrium values. When we apply Le Châtelier's principle in a similar manner to an increase or decrease in the concentrations of CO, CO_2, or H_2O, we obtain the results listed in Table 9-5.

Table 9-5 Le Châtelier's Principle: $CO + H_2O \rightleftharpoons CO_2 + H_2$

| Stress | Changes in Equilibrium Concentrations | | | |
	[CO]	[H_2O]	[CO_2]	[H_2]
Addition of CO	↑	↓	↑	↑
Removal of some CO	↓	↑	↓	↓
Addition of H_2O	↓	↑	↑	↑
Removal of some H_2O	↑	↓	↓	↓
Addition of CO_2	↑	↑	↑	↓
Removal of some CO_2	↓	↓	↓	↑
Addition of H_2	↑	↑	↓	↑
Removal of some H_2	↓	↓	↑	↓
Addition of a catalyst	No change			

Le Châtelier's principle is likewise applicable to *any* reversible reaction in a state of dynamic equilibrium. We shall see that the ability of reversible reactions in dynamic equilibrium to readjust the position of equilibrium so as to compensate partially for an increase or decrease in concentration of one of the substances involved in the reaction is vital to the proper functioning of living systems.

Exercise 9-8
The reaction $CH_4(g) + 2O_2(g) \rightleftharpoons CO_2(g) + 2H_2O(g)$ is a reversible reaction. A gaseous mixture of CH_4, O_2, CO_2, and H_2O is in a state of dynamic equilibrium. Predict the effect on the position of equilibrium of (a) adding more CH_4; (b) removing some H_2O; (c) removing some O_2; (d) adding more CO_2.

9-7   SUMMARY

1. The rate of a chemical reaction is defined as the increase of the product concentration per unit time.

2. In general, rates of chemical reactions increase if the concentrations of reactants are increased, and they increase markedly if the temperature is increased.

3. A model describing the molecular changes that occur during a chemical reaction is called a reaction mechanism. Many reaction mechanisms involve a series of sequential reaction steps.

4. A useful reaction mechanism for the reaction $AB + CD \rightarrow AC + BD$ involves a single step in which AB and CD collide and form products if they collide with a favorable orientation and sufficient kinetic energy.

5. An energy-profile diagram describes the relationship between the molecular changes during a collision and the changes in potential energy.

6. The configuration of atoms with the greatest potential energy in an energy-profile diagram is called an activated complex.

7. The increase in potential energy that occurs when going from the reactants to the activated complex is called the activation energy, E_a.

8. Reaction intermediates are stable chemical species formed in one step of a reaction mechanism and subsequently used as a reactant in a later step.

9. A catalyst is a substance that alters the rate of a reaction but has the same chemical form and is present in the same amount before and after the reaction takes place.

10. A reversible reaction is a reaction that can proceed in either direction.

11. Reversible reactions can attain a state of dynamic equilibrium with equality of rates in the forward and reverse directions.

12. The general reversible reaction $aA + bB + cC \cdots \rightleftharpoons dD + eE + fF \cdots$ has the equilibrium constant expression

$$K_{eq} = \frac{[D]^d[E]^e[F]^f}{[A]^a[B]^b[C]^c} \qquad \text{where } K_{eq} \text{ is the equilibrium constant.}$$

13. In general, the larger the K_{eq}, the more the position of equilibrium will lie toward the right of a reversible reaction; the smaller the K_{eq}, the more the position of equilibrium will lie toward the left of the reaction.

14. The value of K_{eq} does not depend on the reaction mechanism, and is therefore unaffected by a catalyst.

15. If $\Delta H > 0$ (an endothermic reaction), then K_{eq} increases as the temperature increases; and if $\Delta H < 0$ (an exothermic reaction), then K_{eq} decreases as the temperature increases.

16. Le Châtelier's principle states that when a stress is applied to a system in dynamic equilibrium, the equilibrium system readjusts so as to reduce the effect of the stress as much as possible.

PROBLEMS

1. At 700 K, the initial rate of formation of HI is 6.4×10^{-6} mole/liter · sec when 0.01 mole of H_2 and 0.01 mole of I_2 are put in a 1-liter box. If 0.01 mole of H_2 and 0.01 mole of I_2 are put in a 2-liter box at 700 K, what is the initial rate of formation of HI?

2. At 700 K, the initial rate of formation of HI is 6.4×10^{-6} mole/liter · sec when 0.01 mole of H_2 and 0.01 mole of I_2 are put in a 1-liter box. (a) Assuming that the reaction rate remains constant at its initial value, how many moles of HI will be formed in 1.0 min? (b) What is the percentage decrease in the amount of H_2 and I_2 after 1.0 min?

3. Consider the reversible reaction

$$A + B \underset{\text{reverse}}{\overset{\text{forward}}{\rightleftharpoons}} C + D$$

Assume that the reaction mechanism in both the forward and reverse directions is a simple one-step mechanism that can be described by an energy-profile diagram, and that ΔE for the reaction is positive.
(a) Is the activation energy in the forward direction greater than or less than the activation energy in the reverse direction? Make a sketch of the energy-profile diagram.
(b) If equal amounts of A, B, C, and D are put in a container, will the rate in the forward direction be greater than or less than the rate in the reverse direction? (Assume that the fraction of collisions with a favorable orientation is the same for the forward and reverse reactions.)

4. It takes longer for milk to sour in a refrigerator than it does at room temperature. Give a possible explanation for this fact.

5. The following is a reaction mechanism for a certain overall reaction:

Step I	$H_2 + NO \longrightarrow H_2O + N$
Step II	$N + NO \longrightarrow N_2 + O$
Step III	$O + H_2 \longrightarrow H_2O$

(a) What is the overall reaction?
(b) Which species (if any) are reaction intermediates?
(c) Which species (if any) are catalysts?

6. Write the equilibrium constant expression for each of the following reactions:
(a) $O_3(g) + NO(g) \rightleftharpoons O_2(g) + NO_2(g)$
(b) $2N_2O(g) + O_2(g) \rightleftharpoons 4NO(g)$
(c) $CaCO_3(s) \rightleftharpoons CaO(s) + CO_2(g)$
(d) $2NO_2(g) \rightleftharpoons 2NO(g) + O_2(g)$

7. Consider the reaction $C(s) + CO_2(g) \rightleftharpoons 2CO(g)$. When the system is at equilibrium at a certain temperature T, 0.10 mole of $C(s)$ is present. When the temperature is decreased and equilibrium is reestablished at the lower temperature, the amount of $C(s)$ is larger than it was at the higher temperature. Is ΔH for the reversible reaction positive, negative, or zero?

8. Consider the following experiment: 1×10^{-2} mole of $H_2S(g)$ is put in an empty 1-liter container A and the equilibrium $H_2S(g) \rightleftharpoons H_2(g) + S(s)$ is established. 1×10^{-2} mole of $SO_2(g)$ is then put in another empty 1-liter container B and the equilibrium $SO_2(g) \rightleftharpoons O_2(g) + S(s)$ is established. The amount of $S(s)$ in container B is less than that in container A.
(a) Which reversible reaction has the larger equilibrium constant?
(b) In which container will the time needed to reach equilibrium be shorter?

9. Consider the equilibrium $N_2(g) + 3H_2(g) \rightleftharpoons 2NH_3(g)$. Predict the effect on the position of equilibrium at a constant temperature of (a) adding more NH_3; (b) adding more N_2; (c) removing some H_2; (d) adding a catalyst.

SOLUTIONS TO EXERCISES

9-1 If $[H_2] = [I_2] = 0.010\ M$ at 600 K, then (from Figure 9-2) the initial rate is found to be 5.0×10^{-8} mole/liter $\cdot$ sec. Since the initial rate is directly proportional to both $[H_2]$ and $[I_2]$, tripling $[H_2]$ (from 0.010 M to 0.030 M) will increase the initial rate by a factor of 3, and doubling $[I_2]$ (from 0.01 M to 0.020 M) will increase the initial rate by a factor of 2. Therefore, the initial rate when $[H_2] = 0.030\ M$ and $[I_2] = 0.02\ M$ is 6 times 5.0×10^{-8} mole/liter $\cdot$ sec or 3.0×10^{-7} mole/liter $\cdot$ sec.

9-2

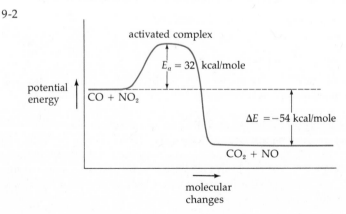

9-3 With all other factors the same, the larger the activation energy the slower the rate of reaction. Therefore, the second reaction, with the slower rate, has the larger activation energy.

9-4 (a) Add steps I and II to get the overall reaction, $SO_2 + \frac{1}{2}O_2 \rightarrow SO_3$.
(b) NO_2 is a reaction intermediate.
(c) NO is a catalyst.

9-5 Use Eq. 9-17.

(a) $K_{eq} = \dfrac{[SO_3]^2}{[SO_2]^2[O_2]}$

(b) $K_{eq} = \dfrac{[H_2O]^2[O_2]}{[H_2O_2]^2}$

(c) $K_{eq} = \dfrac{[HI]^2}{[H_2][I_2]}$

(d) $K_{eq} = \dfrac{[CO_2]}{[CO]}$ (concentrations of pure solids are omitted from the right-hand side of the equilibrium constant expression)

9-6 The reactions have similar initial conditions, but K_{eq} for the reaction in container B is larger. Thus, for the reaction in container B, the position of equilibrium will be farther to the right. There will be more Fe(s) in container B at equilibrium than in container A.

9-7 Since ΔH is negative, a decrease in temperature will increase K_{eq} and thus shift the position of equilibrium farther to the right.

9-8 Use Le Châtelier's principle.

	Changes in Equilibrium Concentrations			
Stress	$[CH_4]$	$[O_2]$	$[CO_2]$	$[H_2O]$
(a) Adding more CH_4	↑	↓	↑	↑
(b) Removing some H_2O	↓	↓	↑	↓
(c) Removing some O_2	↑	↓	↓	↓
(d) Adding more CO_2	↑	↑	↑	↓

CHAPTER 10

Acids and Bases

10-1 INTRODUCTION

Will vinegar react with baking soda? How does a chemist go about answering a question like this? One way, of course, is to do an experiment. You may want to try this one yourself. Another method, which is extremely helpful as a predictive tool, involves a classification scheme in which chemists divide chemical substances and chemical reactions into several broad groups. For example, acids and bases are two important classes of chemical substances, and there is a general type of reaction between any acid and any base. Thus if we can identify vinegar as an acidic solution, and baking soda as basic, we can predict what reaction will take place between them.

Many of the reactions vital for our health involve the reaction of an acid with a base. In addition, in order for the human organism to function properly, a large number of chemical reactions must take place and proceed at just the right rate. The rates of many of these crucial reactions depend on the presence of certain biological catalysts called enzymes. Most enzymes have acidic and basic characteristics, and their effect on the rate of a reaction is often critically dependent on the acidity of the reaction medium. If the acidity of the medium is either too high or too low, the rate of the reaction involving the enzyme can be disastrously altered. Because of this, there are elaborate control mechanisms that operate in the human body to ensure that the acidity of the various body fluids is maintained at the proper level within very narrow limits. Serious physical disability and even death can result if these regulatory mechanisms fail.

In this chapter we sketch briefly the development of the acid-base concept, describe how Arrhenius defined an acid and a base, and then discuss in some detail the more modern Brønsted-Lowry model for acid-base reactions, since this latter formulation is applicable to a wide range of situations. We discuss the distinction between strong and weak acids and between strong and weak bases, and describe a quantitative measure of acid and base strength. We also discuss how and why net ionic equations are used to describe chemical reactions. The last part of this chapter focuses on the connection between the nature of the bonds in a substance and that substance's acid-base behavior.

The air above the city of Leeds, England, shows "clear" evidence of the effects of heavy industry and a dense population. Sulfur dioxide and nitrogen oxides, major industrial pollutants, are converted to acids in the atmosphere.

10-2   STUDY OBJECTIVES

After studying the material in this chapter, you should be able to:

1. Define an Arrhenius acid and an Arrhenius base and be able to write a chemical equation for the neutralization reaction between such an acid and base.

2. Describe why a proton is strongly hydrated in water and know the symbols used for a hydrated proton.

3. Define a Brønsted-Lowry acid and a Brønsted-Lowry base and be able to write a chemical equation for the proton-transfer reaction between such an acid and base.

4. Identify conjugate acid-base pairs and, given the chemical formula of an acid, write the chemical formula for its conjugate base, and vice versa.

5. Define the terms acid strength, base strength, strong acid, strong base, weak acid, and weak base.

6. Recognize the common strong acids and bases as such.

7. Compare qualitatively the relative strengths of two acids given their equilibrium constants K_a, and compare qualitatively the relative strengths of two bases given their equilibrium constants K_b.

8. Compare the relative strengths of two bases given the relative strengths of their conjugate acids, and vice versa.

9. Write a net ionic equation for the reaction of an acid with a base.

10. Relate the acid-base properties of a substance with the general formula

$$-\overset{\displaystyle |}{\underset{\displaystyle |}{X}}-OH$$ to the electronegativity of X and the atoms bonded to X.

11. Identify those atoms in a molecule with unshared pairs of electrons as proton acceptors (bases) given a Lewis structure for the molecule.

12. **(Optional)** Predict the relative strengths of simple acids and bases.

10-3   DEVELOPMENT OF THE ACID-BASE CONCEPT

The terms "acid" and "base" have been used in a variety of ways. The way in which the meaning of these terms has changed is an example of how scientific concepts continually grow and develop.

Acid and *base* were first used in a rather loose way to group substances that exhibit certain directly observable properties. For example, many foods have a sour taste because they contain acids (lactic acid in sour milk, citric acid in lemons or oranges, or acetic acid in vinegar) and many bases, such as sodium hydroxide (lye) or sodium carbonate (washing soda), feel soapy. In 1887, Svanté Arrhenius, a young Swedish scientist, formulated a more precise definition of the terms acid and base using a molecular model of how substances behave in water. As the body of chemical knowledge increased, chemists began to see parallels between reactions in water (classified as acid-base reactions according to the Arrhenius definition) and other reactions that do not take place in water and therefore could not be considered acid-base reactions according to the Arrhenius model. In addition, chemists obtained experimental data that did not completely agree with Arrhenius's molecular model. In 1923, Johannes N. Brønsted and Thomas M. Lowry, working independently, revised and expanded the definition of the terms acid and base. Since 1923, there have been several other important and useful extensions of the acid-base concept.

For the chemical reactions that are of greatest interest to us, the Brønsted-Lowry description of acids and bases is most useful. In common usage, when scientists discuss aqueous solutions, the Arrhenius framework is often employed, so familiarity with it is also helpful. Another acid-base concept, developed by G. N. Lewis, is used extensively in treating mechanisms of organic reactions.

10-4 ACIDS AND BASES ACCORDING TO ARRHENIUS

Arrhenius defined an **acid** as a substance that dissociates (breaks apart) to produce hydrogen ions (or protons), H^+, when put in water, and he defined a **base** as a substance that dissociates to give hydroxide ions, OH^-, in water. According to Arrhenius, HCl is an acid which will dissociate in water to give H^+ and Cl^- ions:

(10-1) $$HCl \longrightarrow H^+ + Cl^- \quad \text{(in water)}$$

NaOH is an Arrhenius base which dissociates in water to produce Na^+ and OH^- ions:

(10-2) $$NaOH \longrightarrow Na^+ + OH^- \quad \text{(in water)}$$

An Arrhenius acid can react with an Arrhenius base and can form water and an ionic compound, or salt, in a reaction called a neutralization. The general form for a **neutralization reaction** is

(10-3) $$\text{acid} + \text{base} \longrightarrow \text{water} + \text{salt}$$

Hydrochloric acid (HCl) and sodium hydroxide (NaOH) undergo the following neutralization reaction:

(10-4) $$HCl + NaOH \longrightarrow H_2O + NaCl$$

Hydrobromic acid (HBr) is an acid and potassium hydroxide (KOH) is a base, and the neutralization reaction between them is

(10-5) $$HBr + KOH \longrightarrow H_2O + KBr$$

In any neutralization reaction, H^+ from the acid combines with OH^- from the base to form H_2O. For example, for reaction 10-4

$$H^+ + Cl^- + Na^+ + OH^- \longrightarrow \overset{H \qquad H}{:\!O\!:} + NaCl$$

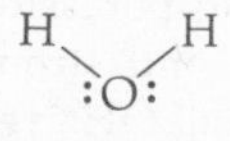

Of course, many substances contain the element hydrogen but do not dissociate to form H^+ ions in water. Glucose, $C_6H_{12}O_6$, for example, is very soluble in water, but it does not dissociate to produce hydrogen ions and therefore it is not an acid. Acetic acid, $C_2H_4O_2$, is an example of a substance in which only one of the hydrogen atoms can dissociate in water. Hydrogen atoms that can dissociate and form H^+ ions when a substance is put in water are often called **acidic hydrogens.** In acetic acid *only* the hydrogen atom that is bonded to the oxygen atom is capable of dissociating.

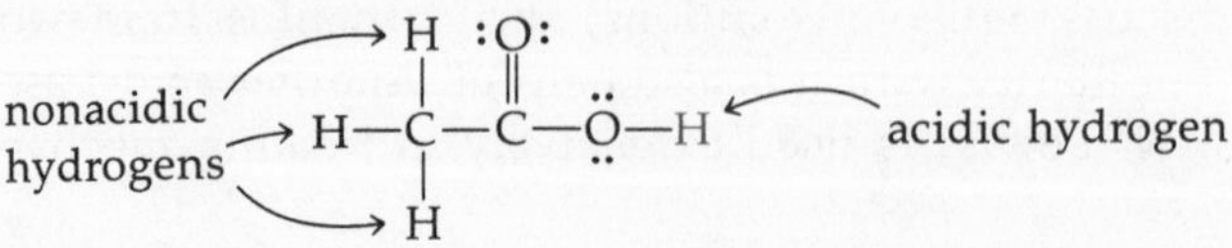

In Section 10-9 we shall discuss why only certain hydrogen atoms in some substances are acidic.

Some acids can dissociate and produce more than one H^+ per molecule. One molecule of sulfuric acid, H_2SO_4, can yield two H^+ ions. If all the acidic hydrogens react, the neutralization reaction between sulfuric acid and sodium hydroxide is

(10-6) $$H_2SO_4 + 2NaOH \longrightarrow 2H_2O + Na_2SO_4$$

Metal hydroxides, such as NaOH, KOH, and $Ca(OH)_2$ (remember that metals are elements on the left-hand side of the periodic table), which are soluble in water, are Arrhenius bases, because they dissociate to give OH^- ions. One mole of NaOH or 1 mole of KOH produces 1 mole of OH^- when it dissolves in water, but if 1 mole of calcium hydroxide, $Ca(OH)_2$, dissolves in water, 2 moles of OH^- ions are produced:

(10-7) $$Ca(OH)_2 \longrightarrow Ca^{2+} + 2OH^-   \text{(in water)}$$

Many other substances that contain the hydroxyl group, $-OH$, in their structural formulas do not dissociate in water to produce hydroxide ions, OH^-, and are not Arrhenius bases. Ethanol (or ethyl alcohol), C_2H_5OH, has the structural formula H—C—C—O—H. When ethanol is dissolved in water, no OH^- ions are produced, so ethanol is not an Arrhenius base. Whether or not a substance that contains the hydroxyl group will dissociate and give hydroxide ions in water is discussed in Section 10-9.

Exercise 10-1
Write the formulas for the ions that are present when the following salts dissolve in water (if you need to, consult Table 4-1):

(a) $NaHCO_3$
(b) K_3PO_4
(c) $(NH_4)_2CO_3$
(d) Na_2HPO_4

Exercise 10-2
Write an equation for the neutralization of:
(a) Nitric acid, HNO_3, and potassium hydroxide, KOH
(b) Sulfuric acid, H_2SO_4, and strontium hydroxide, $Sr(OH)_2$ (assume that all acidic hydrogens and all OH^- react)

10-5 THE HYDRATED PROTON

Arrhenius said that an acid, such as HCl, produces H^+ ions in water. We now know that ions in water are hydrated. An H^+ ion (or proton) is quite small, and therefore its electrical charge is concentrated in a very small volume. It interacts so strongly with water molecules that a bond is formed between H^+ and H_2O. When one H^+ ion and one H_2O molecule bond, the species H_3O^+, called the **hydronium ion,** is formed (see Figure 10-1a). In the hydronium ion, the three hydrogens and the three hydrogen-oxygen covalent bonds are all equivalent.

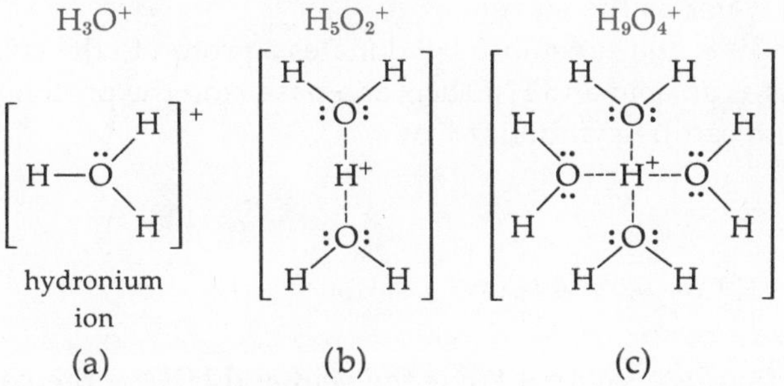

Figure 10-1 Forms for the hydrated proton.
(a) In a hydronium ion, the proton is covalently bonded to the oxygen atom of a water molecule, forming three equivalent hydrogen-oxygen covalent bonds. (b) In the $H_5O_2^+$ ion, the proton is strongly associated with two water molecules. (c) In the $H_9O_4^+$ ion, the proton is strongly associated with four water molecules.

Hydronium ions are not the only form in which a proton exists in water. Experimental evidence indicates that protons also form $H_5O_2^+$ ions with two water molecules (see Figure 10-1b), and $H_9O_4^+$ ions with four water molecules (see Figure 10-1c). Thus, no single symbol can adequately represent the complicated way in which protons are hydrated in water. For simplicity, we shall use the symbol H_3O^+ for the hydrated proton. The symbol $H^+(aq)$, where (aq) refers to an aqueous solution, is also used to represent the hydrated proton. Thus, the symbols H_3O^+ and $H^+(aq)$ can be used interchangeably, since they refer to the same physical entity.

10-6 THE BRØNSTED-LOWRY ACID-BASE MODEL

Some 35 years after Arrhenius formulated his acid-base model, Brønsted and Lowry developed a broader concept of acids and bases when they noticed that quite a number of chemical reactions had a common feature. Can you identify what the following reactions have in common?

(10-8) $HF + H_2O \longrightarrow F^- + H_3O^+$

(10-9) $H_2O + NH_3 \longrightarrow OH^- + NH_4^+$

(10-10) $HCN + OH^- \longrightarrow CN^- + H_2O$

If you are good at recognizing patterns, perhaps you have picked out that in each case the first reactant on the left loses a H^+ ion and the other reactant gains it. This transfer of a proton (H^+) from one substance to another is the basis of the Brønsted-Lowry acid-base model.

According to Brønsted and Lowry, an **acid-base reaction** is a **proton-transfer reaction** in which one substance donates a proton and another substance accepts it. An **acid** is defined as a substance that *donates* a proton. A **base** is defined as a substance that *accepts* a proton. From now on, we shall refer to Brønsted-Lowry acid-base reactions as proton-transfer reactions. We indicated in Section 10-3 that proton-transfer reactions take place in solvents other than water, but since water is the most common solvent in the laboratory, and since body fluids are also aqueous, we shall consider only proton-transfer reactions that involve water as the solvent.

In reaction 10-8, the substance HF donates a proton to the substance H_2O. HF therefore acts as an acid and H_2O acts as a base, and the proton-transfer reaction between them can be symbolized by

(10-11)
$$\overset{\overset{\textstyle\widehat{H^+}}{\displaystyle\frown}}{HF} \ + \ H_2O \longrightarrow F^- + H_3O^+$$

For this reaction, we say that HF is the acid and H_2O is the base. However, in many other acid-base reactions, such as reaction 10-9, H_2O acts as an acid.

Exercise 10-3
Identify the acid and the base, and show the proton-transfer reactions, for reactions 10-9 and 10-10.

Comparison of the Arrhenius and Brønsted-Lowry Models

Before we continue our discussion of proton-transfer reactions, it is worthwhile to compare the Arrhenius and Brønsted-Lowry definitions of the terms "acid" and "base." Arrhenius and Brønsted-Lowry would classify the same substances as acids when they are present in aqueous solutions. For example, the reaction of HCl with water is an experimental fact that does not depend on the model we use to describe the reaction. Arrhenius would say that the acid HCl dissociates in water, and he would represent this dissociation reaction by

(10-12) $HCl \longrightarrow H^+ + Cl^-$

Brønsted and Lowry would describe the reaction of HCl with water as a proton-transfer reaction between the acid HCl and the base water. They would represent the reaction by

(10-13) $HCl + H_2O \longrightarrow H_3O^+ + Cl^-$

In either description, HCl is classified as an acid.

The Brønsted-Lowry model, however, contains a more general and inclusive definition of the term "base." According to Brønsted-Lowry, a base is *any* substance that can act as a proton acceptor, but an Arrhenius base must contain a hydroxyl group, —OH, in its structural formula, and it must dissociate and form OH^- ions in water. Arrhenius would not classify NH_3, for example, as a base, but in reaction 10-9, NH_3 accepts a proton and therefore acts as a Brønsted-Lowry base.

Conjugate Acid-Base Pairs

Let us consider reaction 10-11 again and look at the products F^- and H_3O^+. Is it possible for a proton-transfer reaction to occur between them? Yes, F^- is capable of accepting a proton (acting as a base), and H_3O^+ can donate a proton (act as an acid). The proton-transfer reaction

(10-14) $F^- + H_3O^+ \longrightarrow HF + H_2O$

which is the reverse of reaction 10-11, can take place.

If we compare reactions 10-11 and 10-14, we see that when the acid HF loses a proton, the base F^- is formed. Two chemical species, such as HF and F^-, whose only difference is a single H^+ ion, are called a **conjugate acid-base pair.** F^- is the conjugate base of the acid HF, and HF is the conjugate acid of the base F^-. Similarly, in reaction 10-11, the base H_2O accepts a proton and its conjugate acid H_3O^+ is formed. Thus, reactions 10-11 and 10-14 can be rewritten to illustrate conjugate acid-base pairs and the fact that Brønsted-Lowry acid-base reactions are generally reversible reactions (see Figure 10-2).

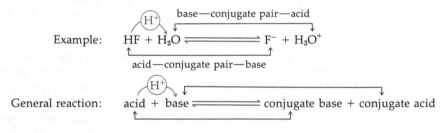

Example:

$$HF + H_2O \rightleftharpoons F^- + H_3O^+$$

General reaction:

$$acid + base \rightleftharpoons conjugate\ base + conjugate\ acid$$

Figure 10-2   A Brønsted-Lowry acid-base reaction is a proton-transfer reaction between an acid (proton donor) and a base (proton acceptor). In a conjugate acid-base pair, the species differ by only a single H^+. The species with the additional H^+ is the acid, whereas the other species is the base.

Any proton-transfer reaction is analogous to reaction 10-11. The reactant acid loses a proton and its conjugate base is one of the products. The reactant base gains a proton, and its conjugate acid is the other product. Figure 10-2 also shows the general form of all proton-transfer reactions.

It is really quite simple to determine whether or not two substances form a conjugate acid-base pair, and which is which. Look at the chemical formulas for the substances: If the only difference in the formulas is a single H^+, then they are a conjugate acid-base pair, and of the two species, the one that has the additional H^+ is the acid, whereas the other is the base. For example, HF is an acid and F^- is its conjugate base.

It is worthwhile to emphasize two aspects of proton-transfer reactions:

1. One substance can act as an acid only if another substance acts as a base. In other words, in order for one substance to act as a proton donor, another substance must act as a proton acceptor.

2. When an acid loses a proton, the species that remains is capable of accepting a proton, and is therefore a base. Similarly, when a base gains a proton, the species that is formed is then capable of donating a proton and is therefore an acid.

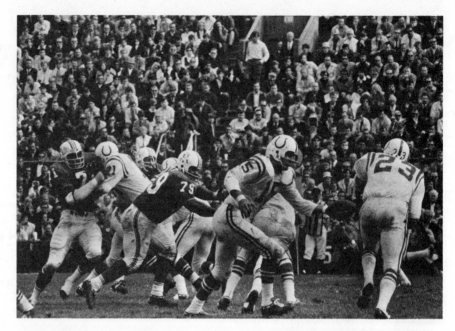

A football play can be viewed as analogous to an acid-base reaction. When the quarterback (the acid) passes off (donates) the football (the proton, H^+), the running back (the base) accepts it.

We expect that a proton-transfer reaction will take place to some extent between any substance that has been identified as a proton donor and any other substance that has been identified as a proton acceptor when the substances are put in water. In addition, we can predict the products of the expected proton-transfer reaction no matter how complicated the substances are. A proton-transfer reaction just involves an H^+ ion leaving the donor (acid) and bonding to the acceptor (base). For example, the ion $CH_3NH_3^+$ is an acid, and we have already seen that F^- is a base. Even though you probably have never seen the formula $CH_3NH_3^+$ before, once you know that it is an acid you can use the general principles you have just learned to predict that the following reaction will occur to some degree:

(10-15)
$$CH_3NH_3^+ \;+\; F^- \rightleftharpoons CH_3NH_2 + HF$$

Exercise 10-4
Identify the conjugate acid-base pairs in reactions 10-9 and 10-10.

The Same Substance Acting as a Proton Donor and as a Proton Acceptor

Let us return to Eqs. 10-8 and 10-9 for a moment and use these reactions to illustrate a very important point about Brønsted-Lowry acids and bases. In reaction 10-8 ($HF + H_2O \rightarrow F^- + H_3O^+$), since H_2O accepts a proton from HF, it is acting as a base. In reaction 10-9 ($H_2O + NH_3 \rightarrow OH^- + NH_4^+$), however, since H_2O donates a proton to NH_3, it is acting as an acid. Thus water can act both as an acid and as a base.

It is quite reasonable that some substances (water among them) have the ability both to donate and also to accept protons. In the presence of NH_3, which has a tendency to accept a proton and a negligible tendency to donate a proton, water exhibits its ability to donate a proton and acts as an acid. On the other hand, in the presence of HF, which has a tendency to donate a proton and a negligible tendency to accept a proton, water exhibits its proton-accepting ability and acts as a base. According to the Brønsted-Lowry acid-base model, water and other substances capable of both accepting and donating a proton are not "acids" or "bases"; rather, they are substances that *act* as an acid or a base, depending on the other chemical species present.

Exercise 10-5
Carbonic acid, H_2CO_3, and the monohydrogen carbonate (bicarbonate) ion, HCO_3^-, are extremely important components of human blood. In water, H_2CO_3 can act as an acid, HCO_3^- can act as an acid and a base, and CO_3^{2-}, the carbonate ion, can act only as a base.
  (a) Write equations for the proton-transfer reactions between H_2CO_3 and H_2O, between HCO_3^- and H_2O, and between CO_3^{2-} and H_2O.
  (b) Which of the following constitute conjugate acid-base pairs? H_2CO_3, HCO_3^-; H_2CO_3, CO_3^{2-}; HCO_3^-, CO_3^{2-}; H_2O, OH^-; H_2O, H_3O^+; H_3O^+, OH^-. For each pair, identify the acid and the base.

Exercise 10-6
The ion HPO_4^{2-} can act as either an acid or a base. Predict the reaction that will occur between HPO_4^{2-} and the hydronium ion, H_3O^+.

10-7 STRENGTHS OF ACIDS AND BASES

If 1 mole of HCl is dissolved in 1 liter of water, no evidence can be found of HCl molecules in the resulting solution. The solution contains 1 mole of hydronium ions, H_3O^+, and 1 mole of Cl^- ions (reaction 10-13). We summarize this situation by saying that a 1 M HCl solution is 100% dissociated. (Note: We call the solution that results when 1 mole of HCl is dissolved in 1 liter of water a "1 M HCl solution" even though there are essentially no HCl molecules in it.)

Strengths of Acids

Chemists call HCl, and all other acids for which a 1 M solution is virtually 100% dissociated, **strong acids.** HI, HBr, and HNO_3 are other examples of strong acids. The term "strong acid" is used for acids whose 1 M solutions are 100% dissociated because the characteristic acidic properties of a solution are associated with a high hydronium ion concentration.

Not all acids are strong acids. In fact, 1 M solutions of most acids are significantly less than 100% dissociated, and these acids are called **weak acids.** Acetic acid, $HC_2H_3O_2$, for example, is a weak acid. A 1.00 M solution of acetic acid is 0.42% dissociated, since as a matter of experimental fact, the concentration of acetate ions, $C_2H_3O_2^-$, the conjugate base of acetic acid, formed by the reaction

(10-16) $$HC_2H_3O_2 + H_2O \rightleftharpoons H_3O^+ + C_2H_3O_2^-$$

is 4.2×10^{-3} M. In a 1.00 M acetic acid solution about 4 of every 1000 acetic acid molecules are dissociated, and thus there are only about 4 acetate anions for every 996 acetic acid molecules.

Classifying an acid as weak or strong is a qualitative measure of the acid's strength. We shall define the **strength of an acid** quantitatively as

(10-17)   Acid strength = percent dissociation of a 1 M solution of the acid
                   = percentage of the conjugate base in a 1 M solution

Lactic acid, $HC_3H_5O_3$, is another example of a weak acid:

(10-18) $HC_3H_5O_3 + H_2O \rightleftharpoons H_3O^+ + C_3H_5O_3^-$

As a matter of experimental fact, the concentration of the lactate ion, $C_3H_5O_3^-$, in a 1.00 M solution of lactic acid is 1.18×10^{-2} M. Thus, a 1.00 M solution of lactic acid is 1.18% dissociated. Since a 1.00 M solution of acetic acid is only 0.42% dissociated, acetic acid is a weaker acid than lactic acid. The relative strengths of the strong acid HCl and the weak acids, lactic acid and acetic acid, are compared in Figure 10-3.

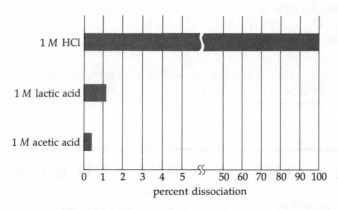

Figure 10-3   A comparison of the strengths of the strong acid hydrochloric acid, which is 100% dissociated, with the weak acid lactic acid, which is 1.18% dissociated, and with the weaker acid acetic acid, which is only 0.42% dissociated.

K_{eq} for the Dissociation of an Acid

When acetic acid is dissolved in water, a state of dynamic equilibrium is quickly established for the reversible reaction 10-16, and the following equilibrium constant relationship applies:

(10-19) $$K_{eq} = \frac{[H_3O^+][C_2H_3O_2^-]}{[HC_2H_3O_2][H_2O]}$$

Recall that $[H_3O^+]$ refers to the equilibrium concentration of H_3O^+ in units of moles per liter (M) and similarly for the other species in Eq. 10-19.

The term $[H_2O]$ in Eq. 10-19 is the concentration of water. In all the situations we shall discuss, water is the solvent and $[H_2O]$ is very large. In 1.00 liter of pure water, the number of moles of water is 55.5, and thus $[H_2O]$ is 55.5 M. If a proton-transfer reaction, such as reaction 10-16, takes place in a large amount of water, the percentage change in $[H_2O]$ is very small. Thus, $[H_2O]$ is essentially fixed at 55.5 M. In this case, as in the examples of reversible reactions involving a pure liquid or a pure solid that we discussed in Section 9-5, we omit the fixed value of $[H_2O]$ from the right-hand side of the equilibrium constant expression.

Equation 10-19 now has the simpler form

(10-20) $$K_a = \frac{[H_3O^+][C_2H_3O_2^-]}{[HC_2H_3O_2]}$$

The subscript a in Eq. 10-20 indicates that the equilibrium involves the dissociation of a weak acid.

Any weak acid is involved in a reversible reaction similar to that given for acetic acid (reaction 10-16). If we represent any acid by the formula HA, the general reversible reaction of HA with water is

(10-21) $$HA + H_2O \rightleftharpoons H_3O^+ + A^-$$

and the general equilibrium constant expression is

(10-22) $$K_a = \frac{[H_3O^+][A^-]}{[HA]}$$

Recall the general relationship between the magnitude of the equilibrium constant for a reversible reaction and the relative concentration of species on the left and right of the reversible reaction—the larger the equilibrium constant the more the position of equilibrium lies to the right in the reversible reaction. Thus, for the general reaction 10-21, the larger K_a, the larger the concentration of A^- (the larger the percentage dissociation for a 1.00 M solution) and, consequently, the stronger the acid.

For example, we previously mentioned the experimental facts that a 1.00 M acetic acid solution is 0.42% dissociated, and that a 1.00 M lactic acid solution is 1.18% dissociated. We therefore expect that K_a for lactic acid, the stronger acid, to be larger than K_a for acetic acid, the weaker of the two acids. We can test our qualitative predictions by calculating the numerical values of K_a for each acid in the following manner:

Step 1   From the experimental value for the percent dissociation, calculate the equilibrium concentrations of H_3O^+, A^-, and HA.

Step 2   Then substitute these values into the general equilibrium constant expression 10-22 to obtain the value of K_a.

Thus, for acetic acid ($HC_2H_3O_2 + H_2O \rightleftharpoons H_3O^+ + C_2H_3O_2^-$),

Step 1   A 1.000 M solution of acetic acid is 0.42% dissociated, so $[H_3O^+] = [C_2H_3O_2^-] = 0.0042\ M = 4.2 \times 10^{-3}\ M$, and $[HC_2H_3O_2]$ is found by subtraction: $1.0000\ M - 0.0042\ M = 0.9958\ M$.

Step 2 $$K_a = \frac{[H_3O^+][C_2H_3O_2^-]}{[HC_2H_3O_2]} = \frac{(4.2 \times 10^{-3}\ M) \times (4.2 \times 10^{-3}\ M)}{(0.9958\ M)}$$

$$= 1.77 \times 10^{-5}\ \text{mole/liter}$$

Using the same method of calculation for lactic acid, which is 1.18% dissociated, the value of K_a(lactic acid) is 1.41×10^{-4} mole/liter.

Thus, the K_a values we have calculated agree with our qualitative predictions: K_a for the stronger acid, lactic acid, 1.41×10^{-4} mole/liter, is larger than K_a for the weaker acid, acetic acid, 1.77×10^{-5} mole/liter. In general, for any weak acid, the larger K_a, the stronger the acid. Therefore, a convenient way of comparing the strengths of two weak acids is to compare their K_a values.

The K_a values for several acids are given in Table 10-1. Notice that the values for K_a decrease going down the column of acids. Thus the strongest acid in the table is HSO_4^-, and the weakest is HPO_4^{2-}.

Table 10-1   Strengths of Acids and Bases: K_a and K_b at 25°C

Acid		K_a (mole/liter)
HSO_4^-	Monohydrogen sulfate ion	2.0×10^{-2}
H_3PO_4	Phosphoric acid	7.5×10^{-3}
HF	Hydrofluoric acid	3.5×10^{-4}
$HC_3H_5O_3$	Lactic acid	1.4×10^{-4}
$HC_2H_3O_2$	Acetic acid	1.8×10^{-5}
H_2CO_3	Carbonic acid	4.3×10^{-7}
$H_2PO_4^-$	Dihydrogen phosphate ion	6.2×10^{-8}
NH_4^+	Ammonium ion	5.6×10^{-10}
HCN	Hydrogen cyanide	4.0×10^{-10}
HCO_3^-	Monohydrogen carbonate ion	5.6×10^{-11}
$CH_3NH_3^+$	Methylammonium ion	2.7×10^{-11}
HPO_4^{2-}	Monohydrogen phosphate ion	2.2×10^{-13}

decreasing
acid strength

Some strong acids: HCl, HI, HBr, HNO_3, H_2SO_4
All dissociate completely to give H_3O^+

There are a number of acids in Table 10-1 that are ions. One example is the monohydrogen carbonate ion, HCO_3^-, an important component of human blood. Of course, it is not possible to obtain a sample of the HCO_3^- ion by itself, since negative ions are always accompanied by an equivalent number of positive ions in an aqueous solution. A small concentration of HCO_3^- and H_3O^+ is generated when H_2CO_3 dissociates in water. A large concentration of HCO_3^- can be obtained from the dissociation in water of a salt such as $NaHCO_3$ (sodium monohydrogen carbonate), which contains the monohydrogen carbonate ion. A large concentration of HCO_3^- can also be produced in water by proton-transfer reactions in which H_2CO_3 or CO_3^{2-} is one of the reactants (see Section 11-5).

The general symbol, HA, can be used to represent an acid that is an ion, and reaction 10-21 and Eq. 10-22 also apply to this type of acid. When HA is used to represent the acid HCO_3^-, the symbol A^- represents the base CO_3^{2-}, and reaction 10-21 becomes

(10-23) $$HCO_3^- + H_2O \rightleftharpoons H_3O^+ + CO_3^{2-}$$

Also included in Table 10-1 are some common strong acids that are 100% dissociated in water. Since all strong acids are 100% dissociated in water, and we have defined acid strength on the basis of the percent dissociation in water, all strong acids have the same strength. However, two acids that are 100% dissociated in water might be less than 100% dissociated in some other solvent. Therefore, in that solvent one could say that one acid was stronger than the other on the basis of percent dissociation.

Sulfuric acid, H_2SO_4, is an interesting strong acid that is 100% dissociated in water according to the reaction

(10-24) $$H_2SO_4 + H_2O \longrightarrow HSO_4^- + H_3O^+$$

Base		K_b (mole/liter)
SO_4^{2-}	Sulfate ion	5.0×10^{-13}
$H_2PO_4^-$	Dihydrogen phosphate ion	1.3×10^{-12}
F^-	Fluoride ion	2.8×10^{-11}
$C_3H_5O_3^-$	Lactate ion	7.1×10^{-11}
$C_2H_3O_2^-$	Acetate ion	5.7×10^{-10}
HCO_3^-	Monohydrogen carbonate ion	2.3×10^{-8}
HPO_4^{2-}	Monohydrogen phosphate ion	1.6×10^{-7}
NH_3	Ammonia	1.8×10^{-5}
CN^-	Cyanide ion	2.5×10^{-5}
CO_3^{2-}	Carbonate ion	1.8×10^{-4}
CH_3NH_2	Methylamine	3.5×10^{-4}
PO_4^{3-}	Phosphate ion	4.5×10^{-2}

increasing
base strength

Some strong bases: $NaOH$, KOH, $Ca(OH)_2$
All dissociate completely to give OH^-
Na_2S, BaO, KNH_2
All react completely with water to give OH^-

However, the monohydrogen sulfate ion, HSO_4^-, is a weak acid, and it is involved in the reversible reaction

(10-25) $$HSO_4^- + H_2O \rightleftharpoons SO_4^{2-} + H_3O^+$$

K_a for reaction 10-25 is 2.0×10^{-2} mole/liter. Thus HSO_4^- is quite a "strong" weak acid.

Exercise 10-7

The value of K_a for ascorbic acid (vitamin C) is 8.1×10^{-5} mole/liter. Compare the strength of ascorbic acid with the strength of acetic acid and lactic acid (see Table 10-1).

Strength of Bases

In water, the acetate ion, $C_2H_3O_2^-$, acts as a base in the reaction

(10-26) $$C_2H_3O_2^- + H_2O \rightleftharpoons HC_2H_3O_2 + OH^-$$

A large concentration of $C_2H_3O_2^-$ can be obtained by dissolving a salt containing the acetate ion, such as sodium acetate, in water. We refer to a solution that has a volume of 1 liter and contains 1 mole of sodium acetate as a 1 M sodium acetate solution, even though reaction 10-26 takes place to a small extent and $[C_2H_3O_2^-]$ in this solution is slightly less than 1 M. Reaction 10-26 is one particular example of the typical behavior of weak bases in water. We can use the symbol A^- to refer to a weak base. The general reversible reaction of A^- with water is

(10-27) $A^- + H_2O \rightleftharpoons HA + OH^-$

and the associated general equilibrium constant expression is

(10-28) $K_b = \dfrac{[HA][OH^-]}{[A^-]}$

The **strength of a base** is defined by

(10-29)   Base strength = the percentage of HA in a 1 M solution of A^-

Note that base strength and acid strength are defined in an analogous manner in terms of a percentage of reaction of A^- or HA with water.

The larger K_b, the farther the position of the equilibrium lies to the right in reaction 10-27 and the stronger the base. The K_b values for the conjugate bases of the acids given in Table 10-1 are also listed in that table. Note that the values for K_b *increase* going down the column of bases. Reaction 10-27 and Eq. 10-28 also apply to ammonia, which is uncharged, and to bases that have more than one negative charge. For ammonia, we associate NH_3 with A^- and NH_4^+ with HA. For CO_3^{2-}, we associate CO_3^{2-} with A^- and HCO_3^- with HA.

A strong base is a substance that is soluble in water and that dissociates completely in water or reacts completely with water to give OH^- ions. The strong bases NaOH, KOH, $Ca(OH)_2$, Na_2S, BaO, and KNH_2 are included in Table 10-1. The soluble ionic compounds Na_2S, BaO, and KNH_2 are strong bases because S^{2-}, O^{2-}, and NH_2^- react 100% with water to produce OH^-.

(10-30) $S^{2-} + H_2O \longrightarrow HS^- + OH^-$

(10-31) $O^{2-} + H_2O \longrightarrow OH^- + OH^-$

(10-32) $NH_2^- + H_2O \longrightarrow NH_3 + OH^-$

Sulfides and oxides, such as FeS and CuO, which are insoluble in water, are not strong bases.

Amphoteric Substances

An **amphoteric substance** is a substance that can act as either an acid or a base. We have already mentioned the most important amphoteric substance—water. There are other examples of amphoteric substances in Table 10-1, such as $H_2PO_4^-$, HPO_4^{2-}, and HCO_3^-. For each of these, both a K_a and a K_b are listed.

Water is such an important substance that it warrants special attention. In any aqueous solution, one molecule of water can act as an acid and another molecule of water can act as a base, and the following fundamental and important reversible reaction always takes place:

(10-33) $H_2O + H_2O \rightleftharpoons H_3O^+ + OH^-$
                    acid   base

Reaction 10-33 is referred to as the **self-ionization of water,** since the substance H_2O is the only reactant, and the products are the ions H_3O^+ and OH^-. The equilibrium constant expression for reaction 10-33 is

(10-34) $K_w = [H_3O^+][OH^-]$

In Eq. 10-34, as in the equilibrium constant expressions 10-22 and 10-28, the fixed value of $[H_2O]$ is omitted from the right-hand side of the equilibrium constant expression, and the equilibrium constant is designated K_w. The value of K_w is 1.0×10^{-14} mole^2/liter^2 at 25°C. The magnitude of K_w is so small because the tendency of reaction 10-33 to proceed from left to right is small. In pure water, $[H_3O^+]$ must equal $[OH^-]$, and both have the value 10^{-7} mole/liter. We shall discuss Eq. 10-34 further in Section 11-3.

Relationship Between the Strength of an Acid and the Strength of Its Conjugate Base

For any conjugate acid-base pair, the stronger the acid the weaker the base, and the weaker the acid the stronger the base. In fact, K_a for any given acid and K_b for its conjugate base are simply related by the equation

(10-35) $$K_a \times K_b = K_w$$

You can easily verify that Eq. 10-35 is true for the conjugate acid-base pairs in Table 10-1. For example, K_a for acetic acid is 1.7×10^{-5} mole/liter at 25°C, and K_b for its conjugate base, the acetate anion, is 5.7×10^{-10} mole/liter; therefore,

$$K_a \times K_b = (1.7 \times 10^{-5}\,\text{mole/liter}) \times (5.7 \times 10^{-10}\,\text{mole/liter})$$
$$= 1.0 \times 10^{-14}\,\text{mole}^2/\text{liter}^2$$

Lactic acid, with a K_a of 1.41×10^{-4} mole/liter at 25°C, is a stronger acid than acetic acid. The conjugate base of lactic acid, the lactate anion, which has a K_b of 7.1×10^{-11} mole/liter, is a weaker base than the conjugate base of acetic acid, the acetate anion, which has a K_b of 5.7×10^{-10} mole/liter.

Exercise 10-8
Formic acid, HCO_2H, is one of the many irritants in ant or bee stings. K_b for the formate ion, HCO_2^-, is 5.6×10^{-11} mole/liter. Is formic acid a stronger or a weaker acid than acetic acid?

10-8 SYMBOLIZING CHEMICAL REACTIONS: NET IONIC EQUATIONS

At the end of the nineteenth century, Arrhenius symbolized the neutralization reaction between HCl and NaOH by the equation

(10-36) $$\text{HCl} + \text{NaOH} \longrightarrow H_2O + \text{NaCl}$$

Today chemists would usually symbolize this reaction in a somewhat different form called the **net ionic equation.** When writing a net ionic equation, chemists use chemical formulas that correspond as much as possible to the actual molecular or ionic nature of the substances involved in the overall reaction. Also, substances that are present in solution but do not take part in the chemical reaction do not appear in a net ionic equation. When net ionic equations are used for several similar reactions, the similarities among the reactions become more apparent.

How do we put Eq. 10-36 in net ionic form? In water, the strong acid HCl dissociates 100% to form H_3O^+ and Cl^- ions. The strong base NaOH and the salt NaCl are also 100% dissociated. What about the self-ionization of water? The dissociation of water itself is very small, so the overwhelming proportion of the substance water exists as molecules of H_2O. If we use appropriate symbols to indicate the ionic or molecular nature of the substances in reaction 10-36, it can be rewritten as

(10-37) $$H_3O^+ + Cl^- + Na^+ + OH^- \longrightarrow Na^+ + Cl^- + 2H_2O$$

In order to balance Eq. 10-37, two molecules of H_2O are needed on the product side of the equation. Note that when a H^+ leaves the H_3O^+ ion, one molecule of H_2O is left, and when this H^+ ion combines with a OH^- ion, another molecule of H_2O is formed.

Now look at reaction 10-37 and Figure 10-4. What substances actually react? Only H_3O^+ and OH^-. Thus the net ionic equation, which emphasizes what is really taking place chemically in the reaction between HCl and NaOH, is

(10-38) $$H_3O^+ + OH^- \longrightarrow H_2O + H_2O$$

Figure 10-4   A solution of sodium hydroxide contains separate Na^+ and OH^- ions, and a solution of hydrochloric acid contains separate H_3O^+ and Cl^- ions. When these solutions are mixed, the reaction $H_3O^+ + OH^- \rightarrow H_2O + H_2O$ takes place. Note that the Na^+ and Cl^- ions are not involved in this chemical reaction, and they are still present as separate ions after the reaction occurs.

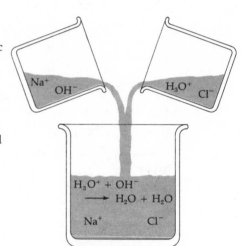

Exercise 10-9
Write the net ionic equation for the reaction of HNO_3 with KOH.

Since HNO_3 is a strong acid and KOH is a strong base, the net ionic equation for the neutralization reaction between HNO_3 and KOH is the same as the neutralization reaction between HCl and NaOH, as shown in reaction 10-38. Thus the chemical reaction between any strong acid and any strong base is the same. This fact becomes readily apparent when net ionic equations are used.

The net ionic equation for reactions involving weak acids and weak bases presents a different picture. Suppose, for example, that we want to write a net ionic equation for the reaction of acetic acid, $HC_2H_3O_2$, and sodium hydroxide in water. Before this reaction takes place, $HC_2H_3O_2$ dissociates in water to form some H_3O^+ and some $C_2H_3O_2^-$ ions. Acetic acid is a weak acid, however, and its percent dissociation is very small. Therefore, since acetic acid exists predominately as undissociated molecules in water, we represent this substance by the symbol $HC_2H_3O_2$. The net ionic equation for the reaction is then

(10-39) $$HC_2H_3O_2 + OH^- \rightleftharpoons H_2O + C_2H_3O_2^-$$

Since the percent dissociation of all weak acids is small, they are all symbolized by their undissociated form when they appear as reactants in a net ionic equation.

How do we write a net ionic equation for a reaction involving a weak base—for example, the reaction of ammonia and HCl in water? Before the reaction, in pure water, ammonia molecules and ammonium ions are in equilibrium, $NH_3 + H_2O \rightleftharpoons NH_4^+ + OH^-$, but since the vast proportion of this substance is present as ammonia molecules, we represent it by the symbol NH_3 when writing an equation for its reaction with HCl. The net ionic equation is therefore

(10-40) $$NH_3 + H_3O^+ \rightleftharpoons NH_4^+ + H_2O$$

Can you write a net ionic equation for the reaction of sodium monohydrogen carbonate and sodium hydroxide? Like any salt, sodium monohydrogen carbonate, $NaHCO_3$, dissociates completely. The monohydrogen carbonate anion, HCO_3^-, can act as both a weak acid and as a weak base. It will act as a proton donor (acid) in the presence of the strong base OH^-. The net ionic equation for the reaction is then

(10-41) $$HCO_3^- + OH^- \rightleftharpoons H_2O + CO_3^{2-}$$

Exercise 10-10
Write a net ionic equation for the reaction of lactic acid, $HC_3H_5O_3$, with potassium hydroxide, KOH.

Exercise 10-11
Write a net ionic equation for the reaction of the salt sodium dihydrogen phosphate, NaH_2PO_4, with HCl.

10-9 ACID, BASE, OR NEITHER: HOW DOES ONE PREDICT?

Why can a given substance act as an acid, or as a base, or as neither an acid nor a base? Why can some substances act as both an acid and as a base? We must consider the nature of the chemical bonds in the substance if we hope to answer these questions.

Let us consider first whether a given substance will dissociate to produce H_3O^+ ions (act as an acid) or OH^- ions (act as a strong base). Sodium hydroxide, NaOH, is a strong base. Nitric acid, HNO_3, has the structural formula O—N—O—H and is a strong acid. There is a common part of the bonding
∥
O
arrangement in both NaOH and HNO_3, involving in part the hydroxyl group, even though these substances act quite differently when dissolved in water. To emphasize this common feature, we represent the bonding arrangement in both

|
of these substances by the symbol —X—O—H. For NaOH, X represents the
|

Na atom. For HNO_3, X represents the N atom with two oxygen atoms bonded to it. Many (but not all) strong bases, and many (but not all) strong and weak acids have this type of bonding arrangement.

Sodium hydroxide is an ionic compound with an ionic bond between the Na^+ cation and the polyatomic anion OH^-. The Lewis structure for NaOH is $Na^+ :\ddot{O}: H^-$. Recall that separate ions are present whenever any ionic compound dissolves in water. Thus when NaOH dissolves in water, separate Na^+ and OH^- ions are present. In general, any soluble ionic compound containing OH^- ions is a strong base (Figure 10-5a). Thus KOH, $Ba(OH)_2$, and so on, are strong bases. For example, $Mg(OH)_2$ is an ionic compound that contains OH^- but it is not a strong base because it is insoluble in water.

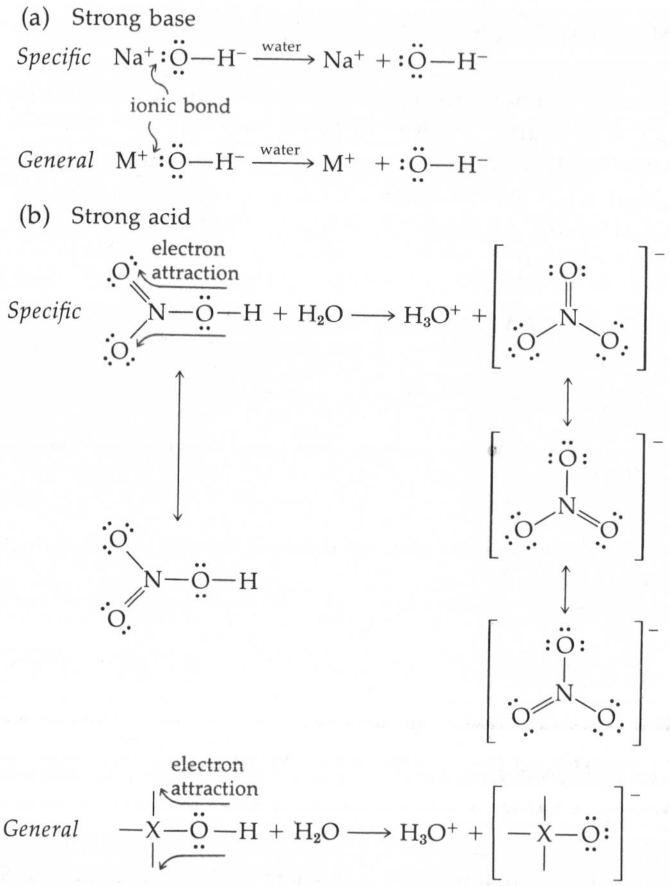

(a) Strong base

Specific $\quad Na^+ :\ddot{O}-H \xrightarrow{\text{water}} Na^+ + :\ddot{O}-H^-$

ionic bond

General $\quad M^+ :\ddot{O}-H \xrightarrow{\text{water}} M^+ + :\ddot{O}-H^-$

(b) Strong acid

Specific

electron attraction

General

electron attraction

$\quad -X-\ddot{O}-H + H_2O \longrightarrow H_3O^+ + \left[-X-\ddot{O}: \right]^-$

Figure 10-5 (a) Any soluble ionic compound that contains OH^- ions, such as sodium hydroxide, is a strong base. (b) A substance with the bonding arrangement $-X-O-H$, such as nitric acid, can act as a strong acid if the atom X, in combination with the atoms bonded to it, exerts a sufficiently strong attraction for electrons so that a H^+ ion can break away. Note the resonance structures used for HNO_3 and NO_3^-.

Let us compare the nitrogen-oxygen bond in HNO_3 with the ionic sodium-oxygen bond in NaOH. In order to form OH^- ions in water, the N—OH bond would have to break so that *both* bonding electrons go with the oxygen atom. This is not reasonable, since both nitrogen and oxygen are nonmetals with large electronegativities. There is not a large inequality in the sharing of the bonding electrons in the nitrogen-oxygen bond. This bond is not ionic; it is a polar covalent bond.

On the other hand, since a molecule of HNO_3 dissolved in water does produce a H^+ ion, it must be the O—H bond that breaks. What is it about HNO_3 that favors this process?

The nitrogen atom, and the other oxygen atoms that are bonded to the nitrogen atom in the molecule, exert a strong attraction for electrons, which pulls the bonding electrons in the O—H bond away from the hydrogen atom, making it easier for a H^+ ion to break away (see Figure 10-5b).

We can generalize this argument for any substance that has the form

$$-\overset{|}{\underset{|}{X}}-O-H.$$ If the atom X, by itself, or the atom X in combination with the

atoms bonded to it, exerts a strong attraction for electrons, we predict that the XO—H bond will break when the substance is put in water and thus we expect that such a substance can act as an acid (Figure 10-5b).

Ethanol, or ethyl alcohol, C_2H_5OH, which has the structural formula

$$H-\overset{\overset{\displaystyle H}{|}}{\underset{\underset{\displaystyle H}{|}}{C}}-\overset{\overset{\displaystyle H}{|}}{\underset{\underset{\displaystyle H}{|}}{C}}-OH,$$ is a substance of the form $-\overset{|}{\underset{|}{X}}-O-H$, but since the elec-

tronegativity of carbon is not very large, and the carbon atom in the group

$$-\overset{|}{\underset{|}{C}}-O-H$$ is not bonded to other atoms that strongly attract electrons, we

expect that when ethanol dissolves in water there is little tendency for the CO—H bond in the molecule to break and for a proton to be transferred to a water molecule. In aqueous solutions, ethanol does not act as an acid. Ethanol is just one example of a large class of similar organic compounds called alcohols, which we discuss in Chapter 15; many alcohols are physiologically and biochemically important.

Acetic acid, $CH_3-\overset{\overset{\displaystyle O}{\|}}{C}-O-H$, is one example of another class of organic compounds called carboxylic acids, which we discuss in Chapter 17. As their name implies, carboxylic acids are acids. As a matter of fact, almost all of them are weak acids. A carboxylic acid and an alcohol both have the arrangement C—OH. In a carboxylic acid, however, the carbon atom of this group is also bound to a second oxygen atom—an atom that attracts electrons strongly. This

combination, which we can represent schematically by $-\overset{\overset{\displaystyle O}{\|}}{C}-O-H$, is responsible for the acidic behavior of carboxylic acids. Many carboxylic acids are involved in the process of human metabolism; lactic acid (see Table 10-1) is a very important carboxylic acid.

Exercise 10-12
In lactic acid, $HC_3H_5O_3$, the central carbon atom has an —OH group bonded to it. Draw the structural formula for lactic acid.

Now let us see if we can explain why some substances are proton acceptors, or Brønsted-Lowry bases. Water and ammonia are two common weak bases. If we look at the Lewis structures for these substances, shown in Figure 10-6, we see that there is an unshared pair of electrons on the nitrogen atom in ammonia, and two unshared pairs on the oxygen atom of the water molecule.

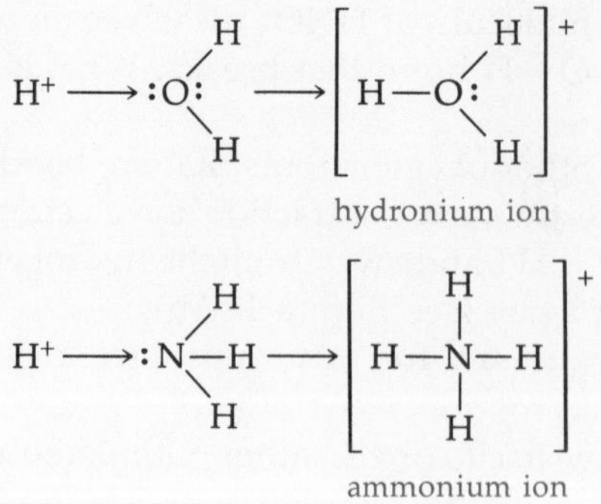

$$H^+ \longrightarrow \ddot{O}: \quad \longrightarrow \quad \left[H-\overset{H}{\underset{H}{\ddot{O}}} \right]^+$$

hydronium ion

$$H^+ \longrightarrow \ddot{N}-H \quad \longrightarrow \quad \left[H-\overset{H}{\underset{H}{N}}-H \right]^+$$

ammonium ion

Figure 10-6   The oxygen atom in a molecule of water has two unshared pairs of electrons; it can act as a proton acceptor. When one of these unshared pairs is used to form a covalent bond with a proton, the product is a hydronium ion. The nitrogen atom in an ammonia molecule has an unshared pair of electrons, so it too can act as a proton acceptor. When this unshared pair is used to form a covalent bond with a proton, the product is the ammonium ion.

When an atom has an unshared pair of electrons, it is capable of attracting a proton, H^+, which has a positive charge. An atom with an unshared pair of electrons can share both of these electrons with a H^+ ion, which has no electrons. When this occurs, a covalent bond is formed between the H^+ atom and the atom that had the unshared pair (Figure 10-6). <u>All Brønsted-Lowry bases have atoms with unshared pairs of electrons capable of forming a covalent bond with a proton</u>.

For example, methylamine, $H-\overset{H}{\underset{H}{C}}-\overset{H}{\underset{H}{N}}:$, is an example of a class of organic compounds called amines (discussed in Chapter 18). Both methylamine and ammonia contain a nitrogen atom with an unshared pair of electrons, and both are weak Brønsted-Lowry bases. When methylamine accepts a proton, the methylammonium cation is formed.

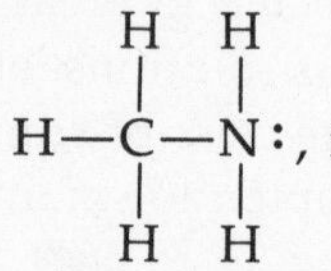

Recall that an amphoteric substance can act both as a proton acceptor (base) and as a proton donor (acid). A very important class of amphoteric substances are called amino acids. Proteins are composed of a large number of amino acids (see Chapter 21). Glycine, $:\overset{H}{\underset{H}{N}}-\overset{H}{\underset{H}{C}}-\overset{:O:}{C}-\ddot{O}-H$, is an example of an amino acid.

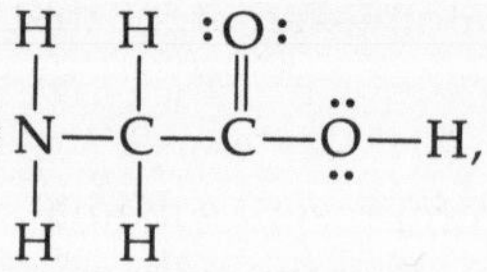

Exercise 10-13
Draw the Lewis structure for the methylammonium cation.

Exercise 10-14
Identify the acidic and basic parts of glycine. When glycine acts as a base, what is the structural formula for the product? What ion is formed when glycine behaves as an acid?

Exercise 10-15

A Lewis structure representation for $SO_2(OH)_2$ is $H-\ddot{O}-\overset{:O:}{\underset{:O:}{S}}-\ddot{O}-H$. Would you expect this substance to act as an acid, as a base, as both, or as neither an acid nor a base?

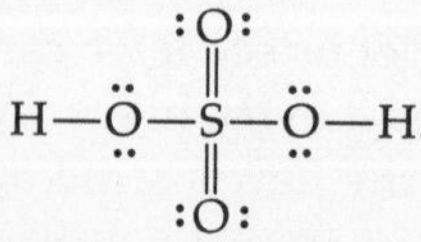

Optional

10-10 PREDICTING THE RELATIVE STRENGTHS OF ACIDS AND BASES

We mentioned in the previous section that all Brønsted-Lowry bases have an unshared pair of electrons. A base will have a larger tendency to accept a H^+ ion (be a stronger base) the more the unshared electron pair is localized in a small region of space. The same number of electrons concentrated in a small volume will attract a H^+ ion to a greater extent than if the electrons are spread out over a large volume. This is a very general principle that can often be used to predict the relative strength of bases. It can also be used to predict the relative strengths of acids, as we shall see shortly.

First let us compare the strength of two very simple bases: the F^- ion and the Cl^- ion. Both fluorine and chlorine are group VIIA elements, with the same negative charge, but in terms of size Cl^- is quite a bit larger than F^-. Recall the general trend: Within a family in the periodic table, size increases as the atomic number increases (Chapter 3). Thus the concentration of electron charge is greater in the F^- ion than it is in the Cl^- ion (see Figure 10-7). If we use the general principle we have just described, we predict that F^- is a stronger base than Cl^-. What is the experimental evidence? F^- is a weak base, but Cl^- is such a weak base that, practically speaking, it is not a base at all. On a relative basis, F^- is indeed the stronger base.

Figure 10-7   The relative sizes of a fluoride ion versus a chloride ion. The F^- ion is smaller than a Cl^- ion, so the concentration of electrical charge is greater in the F^- ion than it is in the Cl^- ion.

The argument we just used to predict the relative strengths of the bases F^- and Cl^- is also sufficient to explain why HCl is a stronger acid than HF. Recall the general relationship between the strength of a base and the strength of its conjugate acid. Between Cl^- and F^-, Cl^- is the weaker base. Therefore, HCl, the conjugate acid of Cl^-, is a stronger acid than HF, the conjugate acid of F^-.

Let us consider one other example. We discussed the acidic properties of nitric and acetic acids in Section 10-9. Nitric acid is the stronger acid. One way of explaining this observation is to consider the relative strengths of the conjugate bases: the nitrate ion, NO_3^-, and the acetate ion, $C_2H_3O_2^-$. We can describe these bases by means of the Lewis structures shown in Figure 10-8.

An oxygen atom is the proton-attracting site in each ion. In the NO_3^- ion the additional negative charge is spread out, or delocalized, over the three equivalent oxygen atoms, whereas in the $C_2H_3O_2^-$ ion the same amount of negative charge is delocalized over only two oxygen atoms. Therefore, the electron charge is more concentrated, or localized, on an oxygen atom in the acetate ion than it is on an oxygen atom in the nitrate ion. Hence, we predict that the acetate ion is the stronger base. If this is true, then acetic acid is the weaker acid.

The relative strengths of many other acids and bases can be compared using the general principles we have illustrated in this section.

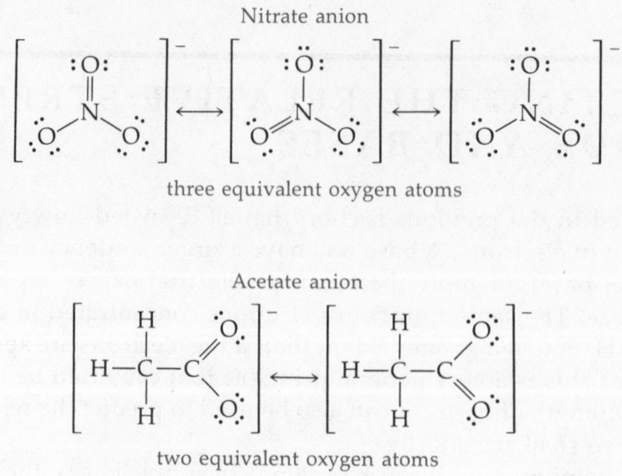

Nitrate anion

three equivalent oxygen atoms

Acetate anion

two equivalent oxygen atoms

Figure 10-8   Lewis structures for the nitrate and the acetate anions. In the NO_3^- ion, the additional negative charge is spread out over the three equiv-alent oxygen atoms, whereas in the $C_2H_3O_2^-$ ion, the additional nega-tive charge is spread out over the two equivalent oxygen atoms.

Exercise 10-16

Predict which of the following is a stronger acid: HNO_2 or HNO_3.

10-11 SUMMARY

1. The classification of chemical substances and chemical reactions into general types is used extensively by chemists. Acid-base reactions are an important class of chemical reactions.

2. There has been a continual process of change in the acid-base concept.

3. Arrhenius defined an acid as a substance that dissociates in water to pro-duce H^+ ions, and a base as a substance that gives OH^- ions upon disso-ciation in water. The general form of the Arrhenius neutralization reaction is acid + base → water + salt.

4. H^+ ions in water are bound to water molecules. The hydronium ion, H_3O^+, consists of one H^+ ion and one H_2O molecule bonded together.

5. In the Brønsted-Lowry model an acid is a proton donor, a base is a proton acceptor, and an acid-base reaction is a proton-transfer reaction whose general form is

$$\text{acid} + \text{base} \rightleftharpoons \text{conjugate base} + \text{conjugate acid}$$

6. Two chemical species whose only difference is a single H^+ ion constitute a conjugate acid-base pair. The acid is the one with the additional H^+.

7. The strength of an acid is defined as the percent dissociation of a 1 M so-lution of the acid or, equivalently, as the percentage of the conjugate base in a 1 M solution. Strong acids are virtually 100% dissociated in water.

For weak acids the percent dissociation is small. For a weak acid, HA, the larger the equilibrium constant, K_a, for the general reversible reaction, $HA + H_2O \rightleftharpoons H_3O^+ + A^-$, the stronger the acid.

8. OH^- is a strong base, as is any soluble ionic substance that dissociates completely in water or reacts completely with water to produce OH^- ions.

9. The strength of a base, A^-, is defined as the percentage of the conjugate acid HA in a 1 M solution of the base. For a weak base, A^-, the larger the equilibrium constant, K_b, for the reaction $A^- + H_2O \rightleftharpoons HA + OH^-$, the stronger the base.

10. For a conjugate acid-base pair, the stronger the acid, the weaker the base, and vice versa.

11. Water is both a proton donor (an acid) and proton acceptor (a base). The self-ionization of water,

$$H_2O + H_2O \rightleftharpoons H_3O^+ + OH^-$$

is the fundamental equilibrium that applies to any aqueous solution. The equilibrium constant expression for the self-ionization of water is $K_w = [H_3O^+][OH^-]$.

12. Net ionic equations are the best simple representation for chemical reactions in water.

13. Whether a substance acts as an acid, as a base, as both an acid and a base, or as neither an acid nor a base, can often be predicted by considering the structural formula for the substance and the nature of the bonds in it.

14. **(Optional)** A base will be weaker, and its conjugate acid stronger, the more the electron charge is delocalized from the proton-accepting atom in the base.

PROBLEMS

1. Write an equation for the neutralization reaction of
   (a) HI and NaOH   (b) HNO_3 and $Ca(OH)_2$ (c) H_2SO_4 and KOH

2. The monohydrogen sulfite ion, HSO_3^-, can act as an acid or as a base.
   (a) Write the equation representing HSO_3^- acting as an acid and water acting as a base. Identify the conjugate acid-base pairs.
   (b) Write the equation representing HSO_3^- acting as a base and water acting as an acid. Identify the conjugate acid-base pairs.
   (c) Write the net ionic equation for the reaction of $NaHSO_3$ and KOH in water.

3. (a) When $H_2PO_4^-$ acts as an acid, what is the chemical formula for its conjugate base?
   (b) When $H_2PO_4^-$ acts as a base, what is the chemical formula for its conjugate acid?

4. In the reaction $H_2BO_3^- + H_3O^+ \rightleftharpoons H_3BO_3 + H_2O$ identify the conjugate acid-base pairs and identify which is the acid and which is the base for each pair.

5. Boric acid, H_3BO_3, has a K_a of 5.8×10^{-10} mole/liter.
   (a) Is boric acid a stronger acid than HCN?
   (b) Is boric acid a stronger acid than HCO_3^-?
   (c) Which is a stronger base, $H_2BO_3^-$ or CO_3^{2-}?
   (d) Which is a stronger base, $H_2BO_3^-$ or CN^-?

6. K_a for hydrogen sulfide, H_2S, is 1.0×10^{-7} mole/liter. Is the percent dissociation of a $0.1\ M\ H_2S$ solution less than, greater than, or equal to the percent dissociation of a $0.1\ M$ acetic acid solution?

7. A $1.0\ M$ solution of acetic acid is 0.42% dissociated. A $1.0\ M$ solution of HNO_2 is 2.1% dissociated. Is K_a for HNO_2 larger than or smaller than K_a for acetic acid?

8. A solution of the ionic compound sodium benzoate contains the benzoate ion $C_7H_5O_2{}^-$, which is a weak base. K_b for the benzoate ion is 1.5×10^{-10} mole/liter.
   (a) Write the reaction between the benzoate ion and water.
   (b) Is benzoic acid a stronger or a weaker acid than acetic acid?

9. For each of the following, write the net ionic reaction for the reaction that occurs (if any) when solutions of the two substances are mixed:
   (a) $NaHSO_4$ and $NaOH$
   (b) K_2CO_3 and HCl
   (c) KCl and HNO_3
   (d) CH_3NH_3Cl and $NaOH$
   (e) Na_2HPO_4 and HBr

10. **(Optional)** (a) Draw a structural formula for boric acid, H_3BO_3.
    (b) Would you predict H_3BO_3 to be a strong or weak acid?
    (c) Predict whether H_3BO_3 is stronger, weaker, or about the same strength as HNO_2.

SOLUTIONS TO EXERCISES

10-1 (a) $NaHCO_3 \rightarrow Na^+ + HCO_3{}^-$
      (b) $K_3PO_4 \rightarrow 3K^+ + PO_4{}^{3-}$
      (c) $(NH_4)_2CO_3 \rightarrow 2NH_4{}^+ + CO_3{}^{2-}$
      (d) $Na_2HPO_4 \rightarrow 2Na^+ + HPO_4{}^{2-}$

10-2 (a) $HNO_3 + KOH \rightarrow KNO_3 + H_2O$
      (b) $H_2SO_4 + Sr(OH)_2 \rightarrow SrSO_4 + 2H_2O$

10-3 $H_2O + NH_3 \longrightarrow OH^- + NH_4{}^+$ $HCN + OH^- \longrightarrow CN^- + H_2O$
      acid   base   acid   base

10-4 $H_2O + NH_3 \rightleftharpoons OH^- + NH_4{}^+$ $HCN + OH^- \rightleftharpoons CN^- + H_2O$
         ⌐—conjugate pair—⌐   ⌐—conjugate pair—⌐
         ⌐—conjugate pair—⌐   ⌐—conjugate pair—⌐

10-5 (a) $H_2CO_3 + H_2O \rightleftharpoons HCO_3{}^- + H_3O^+$
          $HCO_3{}^- + H_2O \rightleftharpoons H_2CO_3 + OH^-$
          $HCO_3{}^- + H_2O \rightleftharpoons CO_3{}^{2-} + H_3O^+$
          $CO_3{}^{2-} + H_2O \rightleftharpoons HCO_3{}^- + OH^-$
      (b) Conjugate acid-base pairs:

Acid	Base
H_2CO_3	$HCO_3{}^-$
$HCO_3{}^-$	$CO_3{}^{2-}$
H_2O	OH^-
H_3O^+	H_2O

10-6 H_3O^+ is a strong acid. Therefore $HPO_4{}^{2-}$ will act as a base.
      $H_3O^+ + HPO_4{}^{2-} \rightleftharpoons H_2PO_4{}^- + H_2O$

10-7 The larger K_a, the stronger the acid. Therefore ascorbic acid ($K_a = 8.1 \times 10^{-5}$ mole/liter) is a stronger acid than acetic acid ($K_a = 1.77 \times 10^{-5}$ mole/liter) but a weaker acid than lactic acid ($K_a = 1.41 \times 10^{-4}$ mole/liter).

10-8 The larger K_b, the stronger the base and the weaker the conjugate acid. K_b for the acetate ion (5.7×10^{-10} mole/liter) is larger than K_b for the formate ion (5.6×10^{-11} mole/liter). Therefore K_a for acetic acid is less than K_a for formic acid, and formic acid is a stronger acid than acetic acid.

10-9 $H_3O^+ + OH^- \rightarrow 2H_2O$ (nitric acid strong acid, KOH strong base)

10-10 $HC_3H_5O_3 + OH^- \rightarrow C_3H_5O_3^- + H_2O$ (lactic acid weak acid, KOH strong base)

10-11 $H_2PO_4^- + H_3O^+ \rightarrow H_3PO_4 + H_2O$ ($H_2PO_4^-$ can act as an acid or a base; with H_3O^+ present it will act as a base)

10-12.
$$
\begin{array}{c}
\quad\; H \;\; H \;\; O \\
\quad\; | \quad\; | \quad\; \| \\
H-C-C-C-O-H \\
\quad\; | \quad\; | \\
\quad\; H \;\; OH
\end{array}
$$

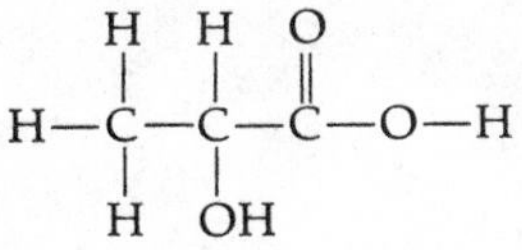

10-13
$$
\left[
\begin{array}{c}
\;\; H \;\; H \\
\;\; | \quad\; | \\
H-C-N-H \\
\;\; | \quad\; | \\
\;\; H \;\; H
\end{array}
\right]^+
$$

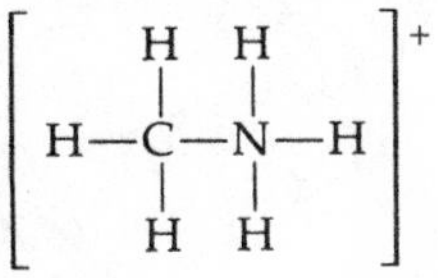

10-14 Glycine has the $\overset{\displaystyle O}{\overset{\displaystyle \|}{-C}}-O-H$ group of atoms common to carboxylic acids and a nitrogen atom with an unshared pair of electrons, which is common to amine bases.

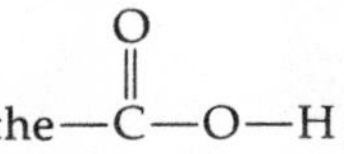

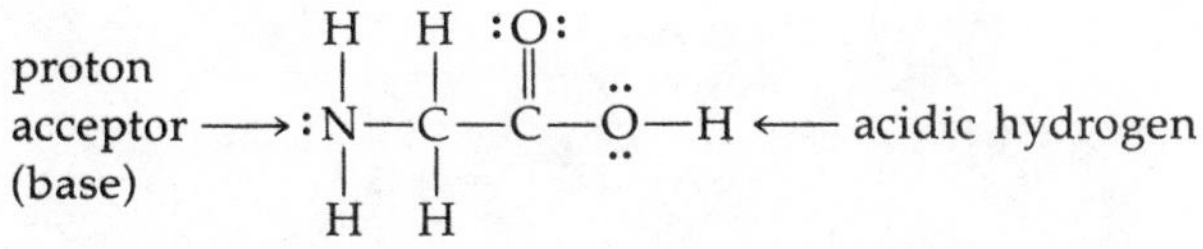

proton
acceptor $\longrightarrow$: N—C—C—Ö—H $\longleftarrow$ acidic hydrogen
(base)

When glycine acts as a base, H^+—N—C—C—Ö—H is produced.

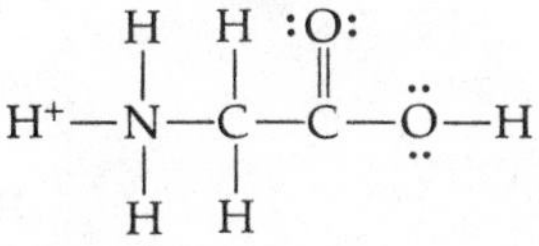

When glycine acts as an acid,   : N—C—C—Ö : $^-$ is produced.

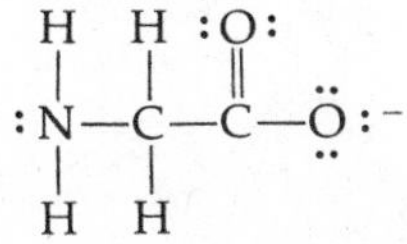

10-15 Sulfur is a nonmetal with a large electronegativity, and it is bound to two oxygen atoms that attract electrons. Therefore it can be predicted that $SO_2(OH)_2$ can act as an acid. $SO_2(OH)_2$ is sulfuric acid, H_2SO_4.

10-16 **(Optional)** Compare the Lewis structure representation for the nitrate ion, NO_3^- (Figure 10-8) with the Lewis structure representation for the nitrite ion, NO_2^-:

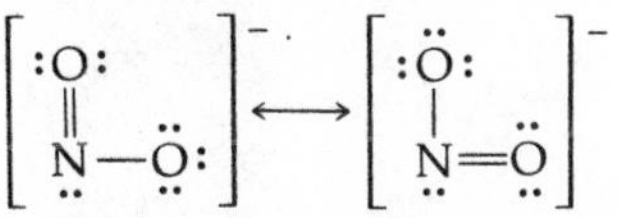

$$
\left[
\begin{array}{c}
:\!O\!: \\
\| \\
N-\ddot{O}\!:
\end{array}
\right]^{-} \longleftrightarrow
\left[
\begin{array}{c}
:\ddot{O}\!: \\
| \\
N=\ddot{O}
\end{array}
\right]^{-}
$$

In NO_3^- the negative charge is spread out over three oxygen atoms, whereas it is spread out over only two oxygen atoms in NO_2^-. Therefore NO_3^- is the weaker base and HNO_3 is the stronger acid.

CHAPTER 11

Measurement and Control of Acidity

11-1   INTRODUCTION

In this chapter, we define what is meant by the acidity or basicity of a solution and introduce the pH scale as a way of expressing the degree to which a solution is acidic or basic. We consider what is meant by the "extent" of a proton-transfer reaction, and how the concentration of an acidic or basic solution can be determined by an experimental method called titration. The properties of buffer solutions are also discussed, and these ideas are applied to the buffer systems in human blood.

Acid-base considerations are fundamental (one might even say, basic!) to understanding many of the chemical reactions that occur not only in living things but also in our environment. Let us take a moment to look at a worldwide environmental problem that has only recently come to be recognized, that of acid rain.

The water that eventually becomes rain or snow originates by evaporation and transpiration (water vapor lost by plants) and is essentially pure. In the atmosphere, this water vapor comes into contact with atmospheric gases, particularly carbon dioxide (CO_2), together with small amounts of dust and debris swept up by the wind. The result is that normal rainfall is often slightly acidic.

Today, however, the rain and snow falling east of Indiana and north of Florida are 50 times more acidic than they were a generation ago. This acid precipitation is disrupting the ecosystems of lakes and damaging plant and animal species as well. In the most extreme example yet recorded—a storm in Scotland in 1974—the rain was as acidic as vinegar.

The main reason for this trend is the rise in the emission of sulfur and nitrogen oxides from the burning of fossil fuels, particularly by steel mills, power plants, petroleum refineries, and metal smelters. In the atmosphere, sulfur dioxide and nitrogen oxides react with water vapor and within two to four days

This famous statue, one of a pair of once-magnificent lions beside the entrance to the New York Public Library, shows graphically how the acidity of rainfall affects our monuments and buildings. Note the loss of detail in the lion's face, which is missing its whiskers!

are converted to sulfuric acid and nitric acid. Ironically, an antipollution measure has contributed to the problem: in recent years, taller and taller smokestacks have been built at industrial plants to relieve local pollution problems. The result is that global wind currents now pick up the pollutants and carry them aloft. By the time they have turned to acids, they may have drifted 100 to 1000 miles from their source.

The scope and magnitude of the adverse effects caused by acid rain are not at all clear—especially on a long-term basis. There is, however, significant evidence that the problem is serious. Many aquatic life forms can survive and reproduce only if the acidity of their aquatic habitat is within a fairly narrow range. Researchers estimate that acid rain has killed the salmon and trout in nearly 15,000 lakes in Sweden. Fish can no longer survive in half of the lakes found at altitudes of more than 2000 feet above sea level in the Adirondack Mountains of New York.

A potentially more serious problem is that acid rain leaches out vital plant nutrients (calcium, magnesium, potassium, and so on) from plant leaves and soil, reducing, for example, forest productivity. This drop in productivity is more pronounced in areas where the soil is naturally acid and thus already somewhat deficient in plant nutrients. Acid rain also promotes the decomposition of stone and concrete, as well as the rusting of iron and steel, thus accelerating the deterioration of buildings and other structures.

Much more research on the effects of acid rain is needed. The United States and Canada are now discharging over 41 million tons of nitrogen and sulfur compounds into the sky each year. As oil becomes scarce and we turn to high-sulfur coal as a fuel source, the sulfur pollution problem is almost certain to become more severe. What goes up the chimney flue must return to earth in one form or another, so methods for dealing responsibly with the waste products of our industrial activity must be found.

11-2 STUDY OBJECTIVES

After studying the material in this chapter, you should be able to:

1. Define the terms acidic, basic, and neutral solution.

2. Calculate $[OH^-]$ given $[H_3O^+]$, and vice versa, for a solution.

3. Predict whether a salt solution will be acidic, basic, or neutral by considering the acid-base properties of the salt's anion and cation.

4. Calculate, for any aqueous solution, approximate values for the pH, pOH, $[H_3O^+]$, and $[OH^-]$, given any one of these four values.

5. Predict (qualitatively) the extent of a proton-transfer reaction by considering the strengths of the acid and base involved.

6. Describe the titration procedure, and calculate the concentration of an acid solution given experimental data obtained from the titration of that acid with a strong base. Perform a similar calculation to determine the concentration of a base solution.

7. Describe how a buffer system operates, calculate $[H_3O^+]$ for a buffer solution, and determine the change in $[H_3O^+]$ if some strong acid or strong base is added to the solution.

8. Describe the basic features of the blood buffer systems and the causes of respiratory acidosis and alkalosis.

11-3 ACIDIC, BASIC, OR NEUTRAL?

Recall that in any aqueous solution the reversible reaction

(11-1) $H_2O + H_2O \rightleftharpoons H_3O^+ + OH^-$

always takes place, and that the related equilibrium constant expression is

(11-2) $K_w = [H_3O^+][OH^-]$

where $K_w = 1.0 \times 10^{-14}$ mole^2/liter^2 at 25°C. In any aqueous solution, no matter what other chemical substances are present, there is always some of the acid H_3O^+ and some of the base OH^-, and the concentrations of these ions are related by Eq. 11-2. Thus, for aqueous solutions, H_3O^+ is the fundamental acid, and OH^- is the fundamental base. That is why the acidity or basicity of an aqueous solution depends on the *relative* concentration of H_3O^+ compared to the concentration of OH^-.

An aqueous **acidic** solution is one in which $[H_3O^+]$ is greater than $[OH^-]$. In an aqueous **basic** solution $[H_3O^+]$ is less than $[OH^-]$. An aqueous solution is called **neutral** if $[H_3O^+]$ and $[OH^-]$ are equal (see Figure 11-1).

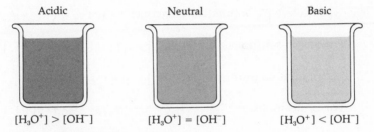

Acidic	Neutral	Basic
$[H_3O^+] > [OH^-]$	$[H_3O^+] = [OH^-]$	$[H_3O^+] < [OH^-]$

Figure 11-1   Representation of acidic, neutral, and basic aqueous solutions.

In pure water, for every H_3O^+ ion produced by reaction 11-1, an OH^- ion is also produced. Thus in pure water $[H_3O^+]$ and $[OH^-]$ are equal, and pure water is an example of a neutral solution. For a neutral solution at room temperature $[H_3O^+]$ is 10^{-7} M and $[OH^-]$ is 10^{-7} M.

When an acid is added to pure water, the resulting solution is, of course, acidic. For example, if 0.1 mole of HCl is dissolved in water to form 1.0 liter of solution, 0.1 mole of H_3O^+ is formed by the complete dissociation of HCl. The value of $[H_3O^+]$ thus increases to 10^{-1} M.

Since H_3O^+ is on the right in the reversible reaction 11-1, we predict, using Le Châtelier's principle, that when HCl is added to pure water the position of equilibrium will shift to the left in an attempt to compensate partially for the increase in $[H_3O^+]$. Therefore $[OH^-]$ will decrease below 10^{-7} M.

We can use Eq. 11-2 to obtain a numerical value for $[OH^-]$ in the solution, since we know $[H_3O^+]$. Rearranging Eq. 11-2 yields

(11-3) $[OH^-] = \dfrac{K_w}{[H_3O^+]} = \dfrac{1.0 \times 10^{-14} \text{ mole}^2/\text{liter}^2}{1.0 \times 10^{-1} \text{ mole/liter}} = 1.0 \times 10^{-13}$ mole/liter

Thus, for our example, $[H_3O^+]$ is 10^{-1} M, and $[OH^-]$ is 10^{-13} M. The value of $[H_3O^+]$ is *greater than* that of $[OH^-]$ in our acidic solution.

Naturally, when a base is added to pure water, the resulting solution is basic. A strong base, such as NaOH, dissociates completely to produce OH^- ions. When 0.1 mole of NaOH is dissolved in water to form 1.0 liter of solution, $[OH^-]$ is 0.1 M and $[H_3O^+]$ is 10^{-13} M.

The product $[H_3O^+] \times [OH^-]$ must always equal 10^{-14} mole^2/liter^2 for any aqueous solution according to Eq. 11-2. Thus, if $[H_3O^+]$ for a solution increases, $[OH^-]$ of that solution must decrease, and vice versa (see Table 11-1). Another way of saying that $[H_3O^+]$ increases is to say that the acidity of the solution increases. Similarly, to say that $[OH^-]$ increases is the same as saying that the solution becomes more basic.

Table 11-1 $[H_3O^+]$ and $[OH^-]$ of Some Aqueous Solutions

Solution	$[H_3O^+]$	$[OH^-]$	$[H_3O^+][OH^-]$ (mole^2/liter^2)
0.1 M HCl	10^{-1} M	10^{-13} M	10^{-14}
0.01 M HCl	10^{-2} M	10^{-12} M	10^{-14}
Pure water	10^{-7} M	10^{-7} M	10^{-14}
0.01 M NaOH	10^{-12} M	10^{-2} M	10^{-14}
0.1 M NaOH	10^{-13} M	10^{-1} M	10^{-14}

It is worthwhile to summarize our conclusions:

(11-4)   For acidic solutions: $[H_3O^+] > 10^{-7}$ M and $[OH^-] < 10^{-7}$ M

(11-5)   For neutral solutions: $[H_3O^+] = [OH^-] = 10^{-7}$ M

(11-6)   For basic solutions: $[H_3O^+] < 10^{-7}$ M and $[OH^-] > 10^{-7}$ M

Exercise 11-1
In a certain solution, $[OH^-]$ is 2×10^{-5} M. What is $[H_3O^+]$ in this solution? Is the solution acidic, basic, or neutral?

Hydrolysis of Salts

Whenever any acid (HNO_3, HCN, $HC_2H_3O_2$, etc.) is dissolved in pure water, the resulting solution is acidic, but solutions of many salts, such as a solution of ammonium chloride, NH_4Cl, are also acidic. Solutions of many other salts are basic.

Any soluble salt dissociates completely in water. When NH_4Cl dissociates in water, the NH_4^+ ion, which is a weak acid (see Table 10-1), is present; the proton-transfer reaction

(11-7) $NH_4^+ + H_2O \rightleftharpoons H_3O^+ + NH_3$

takes place, and so an aqueous solution of NH_4Cl is slightly acidic.

Whenever a salt dissociates in water and one or both of the separate ions present undergo a proton-transfer reaction with H_2O, the process is called hydrolysis of the salt. **Hydrolysis** is a general term meaning reaction with water. We say that NH_4Cl hydrolyzes in water to give a slightly acidic solution.

Not all salts hydrolyze. One, or both, of the ions coming from the salt must act as an acid or a base in order for hydrolysis to occur. Neither Na^+ nor Cl^- hydrolyzes in water, and NaCl solutions are neutral. Solutions of KBr, $CaCl_2$, and $NaNO_3$ are also neutral. Solutions of NaCN undergo hydrolysis and are slightly basic because the CN^- ion present when NaCN is dissolved in water acts as a base.

(11-8) $CN^- + H_2O \rightleftharpoons HCN + OH^-$

In general, to predict whether a salt solution is acidic, basic, or neutral:

1. Determine what ions are present when the salt dissolves in water.

2. Consider separately the possibility of hydrolysis for the cation and anion.

To do this you need to know whether the anion or cation can act as an acid, as a base, as both an acid and a base, or as neither.
    Consider the following examples.

EXAMPLE 1 Is a solution of sodium phosphate (Na_3PO_4) acidic, basic, or neutral?

   Step 1 A solution of Na_3PO_4 contains separate Na^+ and PO_4^{3-} ions.

   Step 2 Na^+ does not hydrolyze. However, PO_4^{3-} can act as a weak base (see Table 10-1), and thus it undergoes the hydrolysis reaction

(11-9) $PO_4^{3-} + H_2O \rightleftharpoons HPO_4^{2-} + OH^-$

Therefore, a solution of Na_3PO_4 is basic because reaction 11-9 takes place.

EXAMPLE 2 Is a solution of potassium monohydrogen carbonate ($KHCO_3$) acidic, basic, or neutral?

   Step 1 A $KHCO_3$ solution contains separate K^+ and HCO_3^- ions.

   Step 2 K^+ does not hydrolyze. However, HCO_3^- can act as either an acid or a base (see Table 10-1) and it does hydrolyze.

(11-10a) $HCO_3^- + H_2O \rightleftharpoons CO_3^{2-} + H_3O^+$

(11-10b) $HCO_3^- + H_2O \rightleftharpoons H_2CO_3 + OH^-$

The equilibrium constant for reaction 11-10a, $K_a(HCO_3^-)$, is 5.6×10^{-11} mole/liter, whereas the equilibrium constant for reaction 11-10b, $K_b(HCO_3^-)$, is 2.3×10^{-8} mole/liter (see Table 10-1).
    Recall from Chapter 9 that for a reversible reaction, the larger K_{eq} the farther the position of equilibrium lies to the right. Thus, since $K_b(HCO_3^-)$ is larger than $K_a(HCO_3^-)$, the position of equilibrium lies farther to the right for reaction 11-10b than for reaction 11-10a. Since OH^- is on the right in 11-10b whereas H_3O^+ is on the right in reaction 11-10a, $[OH^-]$ is greater than $[H_3O^+]$ and the $KHCO_3$ solution is basic.

Exercise 11-2
Separate solutions of the following salts are prepared: NaF, CH_3NH_3Cl, KNO_3, and Na_2CO_3. Which of these solutions will be basic, which acidic, and which neutral?

11-4 pH AND pOH

In aqueous solutions $[H_3O^+]$ can vary over a wide range of values (Table 11-1). There is also a very large variation in $[H_3O^+]$ for solutions of many common substances, as is seen in Table 11-2.

Table 11-2 Approximate Acidity of Some Common Substances

	$[H_3O^+]$	pH
Gastric juice	$7 \times 10^{-2}\ M$ to $10^{-3}\ M$	1.2 to 3.0
Soft drinks and wine	$10^{-2}\ M$ to $10^{-4}\ M$	2.0 to 4.0
Urine	$10^{-5}\ M$ to $10^{-8}\ M$	5.0 to 8.0
Milk	$4 \times 10^{-7}\ M$ to $2 \times 10^{-7}\ M$	6.3 to 6.7
Pure water	$10^{-7}\ M$	7.0
Blood plasma	$4 \times 10^{-8}\ M$	7.4
Eggs	$2 \times 10^{-8}\ M$ to $10^{-8}\ M$	7.6 to 8.0
Baking soda solution ($10^{-2}\ M$)	$6.5 \times 10^{-10}\ M$	9.2
Milk of magnesia	$10^{-10}\ M$	~10.0
Household ammonia	$10^{-12}\ M$	~12.0

When a quantity can range from very large to very small numbers, it is often convenient to use an alternative compressed scale for that quantity. The compressed scale used for $[H_3O^+]$ is called the **pH scale** and is shown in Figure 11-2.

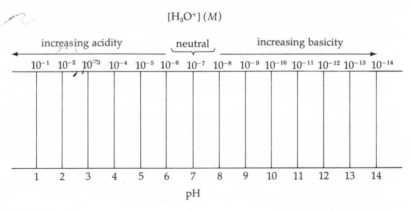

Figure 11-2   The pH scale. The more acidic the solution, the larger $[H_3O^+]$ and the smaller the pH. The more basic the solution, the smaller $[H_3O^+]$ and the larger the pH.

Giving the pH of a solution is an alternative way of expressing $[H_3O^+]$ for that solution. For example, using Figure 11-2, a solution with $[H_3O^+] = 10^{-1}\ M$ has a pH of 1. If $[H_3O^+]$ decreases by a factor of 1000, so that it is $10^{-4}\ M$, the pH changes only from 1 to 4. This example illustrates why we call the pH scale a compressed scale.

Look at Figure 11-2 carefully, and check the following important facts about the relationship between pH and $[H_3O^+]$:

1. In neutral solutions, $[H_3O^+] = 10^{-7}\,M$ and the pH = 7.
   In any acidic solution, $[H_3O^+] > 10^{-7}\,M$ and the pH < 7.
   In any basic solution, $[H_3O^+] < 10^{-7}\,M$ and the pH > 7.

2. The *larger* $[H_3O^+]$ is (moving from right to left in Figure 11-2), the more acidic the solution and the *lower* the pH.

3. The *smaller* $[H_3O^+]$ is (moving from left to right in Figure 11-2), the more basic the solution and the *higher* the pH.

For example, a very acidic solution has a very high $[H_3O^+]$ but a very low pH.

Determination of Approximate pH

If the numerical value of $[H_3O^+]$ is exactly a power of 10 ($10^{-2}\,M$, $10^{-10}\,M$, etc.), we can use Figure 11-2 directly to find the pH, and vice versa.

How do we determine the pH of a solution with $[H_3O^+]$, for example, equal to $5.3 \times 10^{-5}\,M$? It is usually sufficient to obtain an approximate value for the pH. This is easy to do in the following manner: 5.3×10^{-5} is certainly greater than 1×10^{-5} and less than 10×10^{-5}, or 1×10^{-4}. We know that a solution with $[H_3O^+] = 1 \times 10^{-5}\,M$ has a pH of 5, and one with $[H_3O^+] = 10^{-4}\,M$ has a pH of 4. Therefore the pH of a solution with $[H_3O^+] = 5.3 \times 10^{-5}\,M$ must be somewhere between 4 and 5.

We can use the same approach to obtain an approximate value for $[H_3O^+]$ for a solution if we know the pH. For example, in a solution with a pH of 9.8, $[H_3O^+]$ is between $10^{-9}\,M$ and $10^{-10}\,M$.

There is an alternative compressed scale for $[OH^-]$ as well. It is called the pOH scale, and it is used occasionally. We do not need another figure, similar to Figure 11-2, for this scale, because there is a very simple relationship between the pH and the pOH for a given solution. This relationship is

(11-11) $$pH + pOH = 14$$

Equation 11-11 follows from Eq. 11-2, the relationship between $[H_3O^+]$ and $[OH^-]$ for any aqueous solution, and the method by which the pH scale is constructed.

Thus four quantities—$[H_3O^+]$, $[OH^-]$, pH, and pOH—can be used to specify the acidity or basicity of a solution, but any one of them is sufficient. If we know the value of any one of $[H_3O^+]$, $[OH^-]$, pH, and pOH for a solution, we can use Eq. 11-2, Eq. 11-11, and Figure 11-2 to determine approximate values for the other three. For example, if $[H_3O^+] = 10^{-3}\,M$, the pH is 3 (Figure 11-2), the pOH is 11 (Eq. 11-11), and $[OH^-]$ is $10^{-11}\,M$ (Eq. 11-2).

Exercise 11-3
For a certain solution $[H_3O^+]$ is $7.6 \times 10^{-8}\,M$. What is an approximate value for the pH of this solution?

Exercise 11-4
The pH of a solution is 5.3. What is an approximate value of $[OH^-]$ in this solution?

Precise Determination of pH

The complete pH scale is actually constructed using the logarithmic function. pH is defined by the mathematical equation

(11-12) $$pH = -\log[H_3O^+]$$

A fundamental property of logarithms is that, for any number n, $\log 10^n = n$. Thus, for example, if $[H_3O^+]$ is 10^{-2} M, the pH is 2. If $[H_3O^+]$ is not an exact power of 10—for example, 5.3×10^{-5} M—and if it is necessary to obtain a precise value for the pH, Eq. 11-12 can be used to accomplish this with the aid of a calculator or a table of logarithms. Using either one of these devices, the pH of a 5.3×10^{-5} M solution is determined to be 4.28. This value agrees with our qualitative estimate that the pH of a solution with $[H_3O^+] = 5.3 \times 10^{-5}$ M is somewhere between 4 and 5.

We can also obtain a precise value of $[H_3O^+]$ in a solution, given the pH, by using Eq. 11-12 and a calculator or a table of logarithms. For example, using either of these devices, $[H_3O^+]$ for a solution with a pH of 9.8 is 1.6×10^{-10} M. This value agrees with our qualitative estimate that $[H_3O^+]$ for a solution with a pH of 9.8 is between 10^{-9} M and 10^{-10} M.

11-5   EXTENT OF PROTON-TRANSFER REACTIONS

When an equal number of moles of an acid and a base react in water, the solution will contain some mixture of reactants and products after the reaction occurs and equilibrium is established. For an acid-base reaction, we use the phrase **extent of reaction** to mean the relative proportion of products compared to reactants in the equilibrium mixture attained after the acid and the base are mixed in equal molar amounts. The larger the proportion of products, the greater the extent of the reaction.

There are four general types of proton-transfer reactions, since the acid or base can be either weak or strong. We shall consider each type separately.

Strong Acid and Strong Base

Recall that the net ionic equation for the reaction of any strong acid with any strong base is

(11-13) $$H_3O^+ + OH^- \rightleftharpoons H_2O + H_2O$$

Since H_3O^+ has a very large tendency to donate a proton, and an OH^- ion has a very large tendency to accept a proton, the extent of reaction 11-13 is extremely large. For example, when 1 liter of 1 M HCl and 1 liter of 1 M NaOH are mixed, all of the H_3O^+ from the dissociation of the HCl reacts with all of the OH^- from the dissociation of the NaOH. After the reaction takes place, $[H_3O^+]$ and $[OH^-]$ for the solution have the same extremely small values as for pure water. The solution is neutral, with a pH of 7. We can describe this situation by saying that reaction 11-13 goes to **completion.** After the reaction takes place, the solution contains 1 mole of NaCl and has a volume of 2 liters. It is a 0.5 M NaCl solution.

Strong Acid and Weak Base

The net ionic equation for the reaction of a weak base, A^-, with a strong acid is

(11-14) $$H_3O^+ + A^- \rightleftharpoons HA + H_2O$$

The acetate ion, $C_2H_3O_2^-$, is a weak base; in this case Eq. 11-14 becomes

(11-15) $$H_3O^+ + C_2H_3O_2^- \rightleftharpoons HC_2H_3O_2 + H_2O$$

The extent of reaction 11-15, and of the general reaction 11-14, is quite large, but the reaction does not quite go to completion for a very simple reason. H_3O^+ is a very good proton donor, but a weak base such as $C_2H_3O_2^-$ is not as good a proton acceptor as the strong base OH^-. If, for example, 1 mole of sodium acetate is dissolved in 1 liter of 1 M HCl, almost all the $C_2H_3O_2^-$ is converted to $HC_2H_3O_2$. There is, however, a very small but detectable amount of $C_2H_3O_2^-$ present in the solution after equilibrium is established. A numerical calculation shows that this amount of $C_2H_3O_2^-$ is 4.2×10^{-3} mole. Thus we are justified in saying that reaction 11-15 goes nearly to completion. In general, the extent of reaction between any weak base and any strong acid is very large and goes nearly to completion.

Now consider reaction 11-15 again. If, before the reaction takes place, the number of moles of H_3O^+ and the number of moles of $C_2H_3O_2^-$ are equal, then they are also equal in the equilibrium solution. Thus $[H_3O^+]$ after the reaction takes place is 4.2×10^{-3} M, and the solution is acidic. In general, when equal numbers of moles of a strong acid and a weak base are put in water, react, and attain equilibrium, the resulting solution is acidic.

Strong Base and Weak Acid

Acetic acid is a weak acid. Let us estimate the extent of reaction between sodium hydroxide and acetic acid. The appropriate net ionic equation is

(11-16) $$HC_2H_3O_2 + OH^- \rightleftharpoons H_2O + C_2H_3O_2^-$$

Now, OH^- is a very good proton acceptor, but acetic acid is not as good a proton donor as the strong acid H_3O^+. We therefore predict that the extent of reaction 11-16 is very large, but that it does not quite go to completion.

The general net ionic equation for the reaction of a weak acid with a strong base is

(11-17) $$HA + OH^- \rightleftharpoons H_2O + A^-$$

In general, the extent of reaction between any weak acid and any strong base is very large and goes nearly to completion. If we start with equal numbers of moles of HA and OH^- and reaction 11-17 takes place, the equilibrium solution will be slightly basic, since not quite all of the OH^- will have reacted with HA.

Weak Acid and Weak Base

There is no general principle that we can use to predict the extent of reaction between a weak acid and a weak base. In this case, the extent of reaction may be very large, very small, or anywhere between these two extremes, depending on the *specific* weak acid and the *specific* weak base involved. However, we can say that the extent of reaction between any weak acid and water (acting as a base) is small, and the extent of reaction between any weak base and water (acting as an acid) is likewise small.

An important reaction between a weak acid and a weak base, in which the extent of reaction is very large, is the reaction between lactic acid and bicarbonate ion (acting as a base), which takes place in the human body. Lactic acid, $HC_3H_5O_3$, is produced in human tissues during strenuous exercise and is one of the causes of muscle soreness. It can be neutralized by the HCO_3^- ion in the blood, because of the large extent of the reaction

(11-18) $$HC_3H_5O_3 + HCO_3^- \rightleftharpoons C_3H_5O_3^- + H_2CO_3$$

Our discussion of the extent of proton-transfer reactions is summarized in Figure 11-3.

$$\text{Strong acid + strong base} \longrightarrow \text{complete}$$

$$H_3O^+ + OH^- \longrightarrow H_2O + H_2O$$

$$\text{Strong acid + weak base} \longrightarrow \text{nearly complete}$$

General: $H_3O^+ + A^- \;\rightleftharpoons\; HA \;+ H_2O$

Example: $H_3O^+ + CN^- \rightleftharpoons HCN + H_2O$

$$\text{Strong base + weak acid} \longrightarrow \text{nearly complete}$$

General: $OH^- + HA \rightleftharpoons A^- + H_2O$

Example: $OH^- + HF \rightleftharpoons F^- + H_2O$

Figure 11-3   Net ionic equations illustrating the extent of proton-transfer reactions. The extent of the proton-transfer reaction between a weak acid and a weak acid and a specific weak base involved. Note that for reversible reactions that go nearly to completion, a longer arrow to the right is used.

Let us apply the ideas we have just discussed about the extent of proton-transfer reactions to one important biochemical example. We indicated in Section 10-1 that the biochemical activity of many substances depends on the pH of the solution they are in. We can get a clue as to why this is so if we consider the behavior of one specific amino acid.

In Section 10-9 we indicated that the amino acid glycine is both an acid and a base (see Figure 11-4). Pure glycine and other amino acids exist as solids up to a relatively high temperature, and they are more soluble in water than in non-polar solvents. These properties are characteristic of ionic compounds.

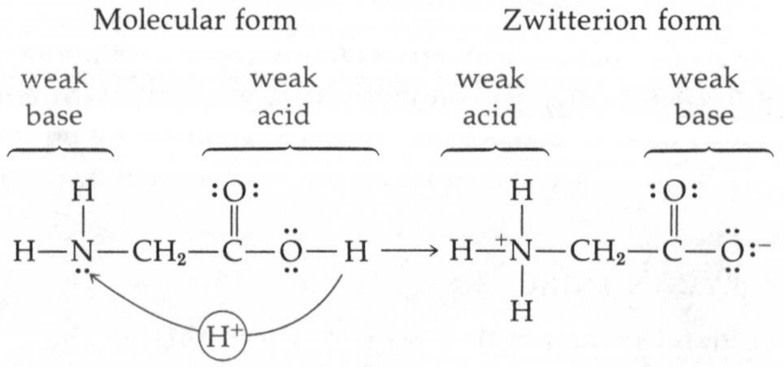

Figure 11-4   The amino acid glycine in its molecular form (left) and zwitterion form (right). The parts of glycine that can act as a weak acid or a weak base are indicated. The for- mation of the zwitterion from the molecular form is represented as an intramolecular proton-transfer reaction.

How can glycine have saltlike properties? Glycine and other amino acids exist in the solid state as species with both a positively charged and a negatively charged part (Figure 11-4). A species with both positively and negatively charged parts, but with zero overall electrical charge, is called a **zwitterion.** (*Zwitter* is a German word meaning hermaphrodite—an organism with both male and female sexual organs.) We can think of the zwitterion of glycine as being formed by the internal acid-base reaction indicated in Figure 11-4.

In the solid state, glycine exists in the zwitterion form, which accounts for its saltlike properties. When glycine is dissolved in pure water, almost all of the dissolved glycine also exists as a zwitterion. If, however, the pH of the solution is very low ($[H_3O^+]$ is very large), the following acid-base reaction takes place to a large extent:

(11-19)
$$^+NH_3-CH_2-\overset{\overset{\textstyle O}{\|}}{C}-O^- + H_3O^+ \rightleftharpoons {}^+NH_3-CH_2-\overset{\overset{\textstyle O}{\|}}{C}-OH + H_2O$$

Thus in a low-pH solution nearly all of the dissolved glycine exists as a positive cation.

On the other hand, if the pH of the solution is very high ($[OH^-]$ is very large), nearly all of the dissolved glycine exists as a negative anion because the following acid-base reaction takes place to a large extent:

(11-20)
$$^+NH_3-CH_2-\overset{\overset{\textstyle O}{\|}}{C}-O^- + OH^- \rightleftharpoons NH_2-CH_2-\overset{\overset{\textstyle O}{\|}}{C}-O^- + H_2O$$

Thus, in solution, glycine exists predominately as a cation at very low pH, as a zwitterion with no net electrical charge in neutral solution, and as an anion at very high pH (see Figure 11-5).

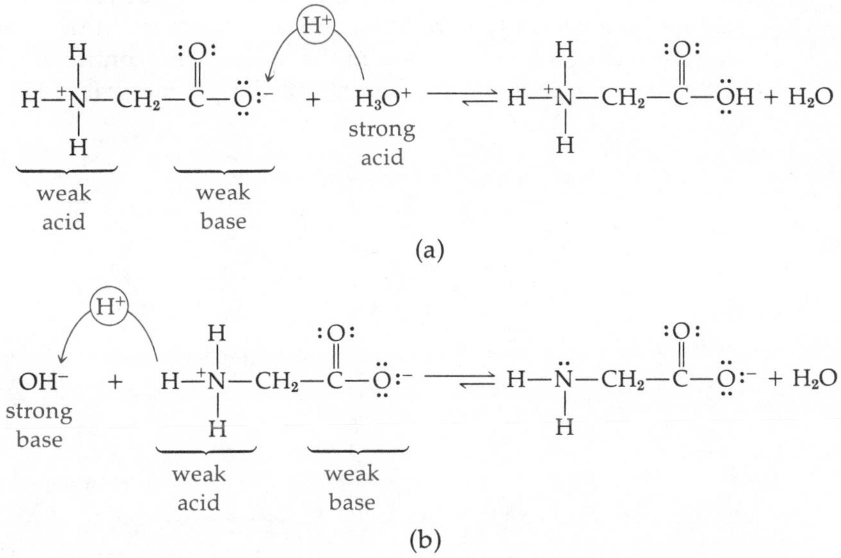

(a)

(b)

Figure 11-5   (a) In a solution with a low pH (high $[H_3O^+]$), the zwitterion form of glycine acts as a weak base. (b) In a solution with a high pH (low $[H_3O^+]$ and high $[OH^-]$, the zwitterion form of glycine acts as a weak acid.

Exercise 11-5

A large concentration of the monohydrogen carbonate ion, HCO_3^-, can be produced in water by a proton-transfer reaction in which either H_2CO_3 or CO_3^{2-} is one of the reactants. For each case, identify the other reactant and write the net ionic equation for the reaction.

Exercise 11-6

Predict the extent of reaction between the following:

(a) HCN and NaOH   (b) KOH and HBr   (c) NH_4Cl and NaOH

11-6   TITRATION

The concentration of an acid or a base solution can be determined quite precisely by an experimental method called titration. A **titration** involves a chemical reaction between solutions of two different substances: one of unknown concentration, whose concentration is to be determined, and a second, whose concentration is known. The solution of known concentration is referred to as a **standard solution.**

We shall first consider how we can determine by titration the concentration of a solution of a **monoprotic acid,** which has one acidic proton per molecule, such as HCl, HNO_3, acetic acid, or lactic acid. We shall then discuss the more complicated case of the titration of a solution of a **diprotic acid,** which has two acidic protons per molecule, such as H_2SO_4 and H_2CO_3. Finally, we shall see that titration procedures can also be used to determine the concentration of a base solution.

Titration of a Monoprotic Acid

To determine the concentration of an acid solution using a titration procedure, we precisely measure out a definite volume of the acid solution using a pipet, and transfer this amount of acid solution to a beaker. We also fill a buret with a standard NaOH solution. We then proceed to add small increments of the standard NaOH solution from the buret to the acid solution in the beaker, while we continually stir the solution in the beaker, and continually measure the pH of this solution with an instrument called a pH meter (see Figure 11-6).

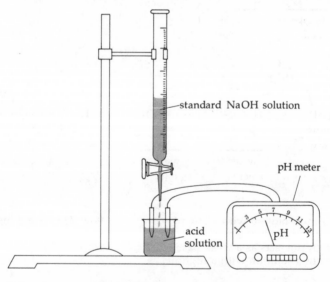

Figure 11-6   Laboratory apparatus used to perform titrations.

As we add OH^- solution from the buret, an acid-base reaction takes place and $[H_3O^+]$ or the pH of the solution, changes in a characteristic manner. The exact way in which the pH of the solution changes as OH^- is added depends on whether the acid is strong or weak, and for a weak acid, on which weak acid is used. Figure 11-7 is a schematic representation of the typical way pH changes during the titration of a weak monoprotic acid with OH^-.

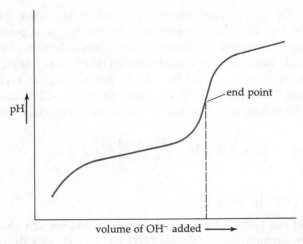

Figure 11-7 pH changes during the titration of a typical weak monoprotic acid with OH⁻. The end point is the point at which the pH increases most rapidly.

Before we add any OH⁻ to the acid solution, $[H_3O^+]$ is large and the pH is small. As we add small amounts of OH⁻, essentially all the added OH⁻ will react with some of the acid in the solution. Recall that the strong base OH⁻ reacts completely with a strong acid and essentially completely with a weak acid. On addition of OH⁻, $[H_3O^+]$ decreases slowly and the pH rises slowly. At one point in the titration, however, just enough OH⁻ has been added to react with all the acid in solution. After this point has been reached, even a very small addition of OH⁻ will cause a very large increase in pH, since now there is essentially no more acid in solution. The point in the titration where the pH rises most rapidly is called the **end point** of the titration. The volume of OH⁻ solution needed to reach the end point can be determined from the curve of pH versus added OH⁻ in Figure 11-7. The end point of a titration can also be determined by using an appropriate **indicator,** a substance that abruptly changes color at the point where the pH rises rapidly. At the end point, the amount of OH⁻ that has been added is just sufficient to react with all the monoprotic acid present initially. Since a monoprotic acid and OH⁻ react in equimolar amounts, at the end point of a titration we have

(11-21)   Number of moles of monoprotic acid present initially
          = number of moles of OH⁻ needed to reach the end point

It is very easy to use Eq. 11-21 to obtain the concentration of a monoprotic acid solution. If we let M_a represent the molarity of the monoprotic acid solution, and V_a its volume in liters, then

(11-21a)   Number of moles of acid present initially = $M_a \times V_a$

Similarly, if we let M_b stand for the molarity of the standard OH⁻ solution, and V_b stand for the volume in liters of the OH⁻ solution needed to reach the end point, then

(11-21b)   Number of moles of OH⁻ needed to reach the end point = $M_b \times V_b$

Substitution of Eqs. 11-21a and 11-21b into Eq. 11-21 gives

(11-22) $M_a \times V_a = M_b \times V_b$ (for a monoprotic acid titrated with OH⁻)

We can use Eq. 11-22 to determine the value of M_a, since the values of the other terms in Eq. 11-22 are determined by the titration procedure. For example, suppose that we want to determine the concentration of a solution of the monoprotic acid, lactic acid. We start with 25.00 ml of the lactic acid solution, carry out the titration, and find that 37.75 ml of 0.525 M OH⁻ solution are needed to reach the end point. If we rearrange Eq. 11-22 and substitute these data, we get the following value for M_a (the concentration of lactic acid):

(11-23) $$M_a = M_b \times \frac{V_b}{V_a} = 0.525\ M \times \frac{37.75\ \text{ml}}{25.00\ \text{ml}} = 0.793\ M$$

Titration of a Diprotic Acid

A diprotic acid has two acidic protons. For example, we saw in the last section that the cation formed from glycine, $^+NH_3CH_2CO_2H$, is a diprotic acid. This cation can be obtained in solution by dissolving the solid ionic compound $ClNH_3CH_2CO_2H$ in water. When an $^+NH_3CH_2CO_2H$ solution is titrated with OH⁻, the following two reactions take place in a stepwise fashion:

(11-24)
$$^+NH_3CH_2\overset{\displaystyle O}{\overset{\|}{C}}{-}OH + OH^- \longrightarrow\ ^+NH_3CH_2\overset{\displaystyle O}{\overset{\|}{C}}{-}O^- + H_2O$$

and

(11-25)
$$^+NH_3CH_2\overset{\displaystyle O}{\overset{\|}{C}}{-}O^- + OH^- \longrightarrow NH_2CH_2\overset{\displaystyle O}{\overset{\|}{C}}{-}O^- + H_2O$$

As OH⁻ is slowly added, essentially all of the $^+NH_3CH_2CO_2H$ is converted to $^+NH_3CH_2CO_2^-$ by reaction 11-24 before any of the $^+NH_3CH_2CO_2^-$ formed in reaction 11-24 can be converted further to $NH_2CH_2CO_2^-$ by reaction 11-25. In the curve of pH versus added OH⁻ for this titration (Figure 11-8), there are two points at which the pH rises rapidly; that is, there are two end points. At the first end point the amount of OH⁻ that has been added is just sufficient to react

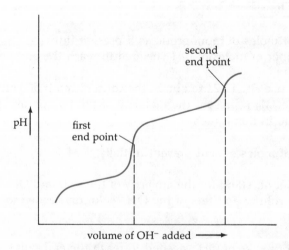

Figure 11-8   pH change during the titration of a $^+NH_3CH_2CO_2H$ solution as OH⁻ is added. At the first end point, the reaction, $^+NH_3CH_2CO_2H + OH^- \rightarrow$ $^+NH_3CH_2CO_2^- + H_2O$, has gone essentially to completion. At the second end point, the further reaction, $^+NH_3CH_2CO_2^- + OH^- \rightarrow$ $NH_2CH_2CO_2^- + H_2O$, has gone essentially to completion.

with all of the $^+NH_3CH_2CO_2H$ present initially. Thus we have

(11-26)   Number of moles of OH^- needed to reach the first end point
                  = number of moles of $^+NH_3CH_2CO_2H$ present initially

Equation 11-26 can be used (in a manner analogous to the way we used Eq. 11-21 for a monoprotic acid) to obtain the concentration of the $^+NH_3CH_2CO_2H$ solution from the titration data.

When the second end point in the titration of $^+NH_3CH_2CO_2H$ is reached, the amount of OH^- added is just sufficient to convert all the $^+NH_3CH_2CO_2H$ initially present to $NH_2CH_2CO_2^-$. This corresponds to the overall reaction

(11-27) $$^+NH_3CH_2\overset{\overset{\textstyle O}{\|}}{C}-OH + 2OH^- \longrightarrow NH_2CH_2\overset{\overset{\textstyle O}{\|}}{C}-O^- + 2H_2O$$

which is the sum of reactions 11-24 and 11-25.

According to reaction 11-27,

(11-28)   Number of moles of OH^- needed to reach the second end point
                  = 2 × (number of moles of $^+NH_3CH_2CO_2H$ present initially)

(Note: Since the number of moles of OH^- needed to reach the second end point is twice the number of moles of OH^- needed to reach the first end point, the volume of OH^- added to reach the second end point is twice the volume of OH^- added to reach the first end point; see Figure 11-8).

Since 1 mole of the diprotic acid $^+NH_3CH_2CO_2H$ can react with 2 moles of OH^-, 1 mole of this diprotic acid is equivalent to 2 moles of a monoprotic acid in its overall capacity to react with OH^- (see reaction 11-27).

The following equilibria apply for any diprotic acid, using the general formula H_2A for the diprotic acid:

(11-29) $H_2A + H_2O \rightleftharpoons H_3O^+ + HA^-$ $K_{a_1} = \dfrac{[H_3O^+][HA^-]}{[H_2A]}$

and

(11-30) $HA^- + H_2O \rightleftharpoons H_3O^+ + A^{2-}$ $K_{a_2} = \dfrac{[H_3O^+][A^{2-}]}{[HA^-]}$

K_{a_1} is associated with the reaction of H_2A with H_2O, and K_{a_2} is associated with the reaction of HA^- with water.

Any diprotic acid for which K_{a_1} is very much larger than K_{a_2} will exhibit a pH versus OH^- titration curve similar to Figure 11-8 for $^+NH_3CH_2CO_2H$, where K_{a_1} is 4.5×10^{-3} mole/liter and K_{a_2} is 2.5×10^{-10} mole/liter.

Diprotic acids such as H_2SO_4, for which K_{a_1} is not much larger than K_{a_2}, have a pH versus OH^- titration curve with only a single end point corresponding to the overall reaction:

(11-31) $H_2A + 2OH^- \longrightarrow A^{2-} + 2H_2O$

For example, when a solution of H_2SO_4 is titrated with OH^-, there is a single end point corresponding to the overall reaction

(11-32) $H_2SO_4 + 2OH^- \longrightarrow 2H_2O + SO_4^{2-}$

Note that according to reaction 11-32, 1 mole of H_2SO_4 is equivalent to 2 moles of the monoprotic acid HCl in its capacity to react with OH^-.

Titration of a Base

The concentration of a base solution can be determined using titration procedures analogous to those we have described for acid solutions. In the case of a base, a definite volume of base solution is titrated with a standard acid solution. For example, to determine the concentration of a NaOH solution, we would titrate a definite volume of the NaOH solution with a standard acid solution, such as a 0.100 M HCl solution. In this titration, the initial pH is very high and there is a rapid decrease in the pH at the end point, similar to that shown in Figure 11-9. At the end point of this titration,

(11-33)   Number of moles of NaOH present initially
                  = number of moles of HCl added

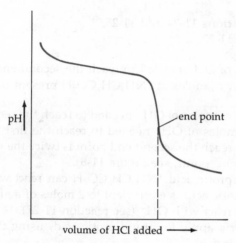

pH

volume of HCl added ⟶

end point

Figure 11-9   pH changes during the titration of a typical weak base with HCl. The   end point is the point at which the pH decreases most rapidly.

Exercise 11-7
A volume of 25.00 ml of a monoprotic acid solution is titrated with 0.348 M OH^-. It takes 22.45 ml of the OH^- solution to reach the end point. Calculate the concentration of the acid solution. Why don't we need to know what specific acid is involved?

11-7 BUFFERS

In a healthy individual, the pH of most body fluids is kept within a very narrow range. We have already indicated some reasons for this in Section 10-1. The human body has elaborate mechanisms to control the pH level of the various body fluids. A solution that can maintain the pH at nearly a constant level even when additional strong acid, H_3O^+, or additional strong base, OH^-, is added, is called a **buffer solution.**

Pure water is not a buffer solution. If one adds 0.1 mole of HCl to 1 liter of water, the value of $[H_3O^+]$ changes from 10^{-7} M to 10^{-1} M, and the pH changes from 7 to 1—a millionfold increase in $[H_3O^+]$! Similarly, if one adds 0.1 mole of NaOH to 1 liter of water, the pH changes from 7 to 13.

The reason for these large pH changes is clear. When we add H_3O^+ to pure water, there is no base present for it to react with, nor is there any acid present in pure water capable of reacting with added OH^-. If we want a solution that can resist changes in pH, we must have substances present that are capable of reacting with both H_3O^+ and OH^-. We must therefore have a solution with both a relatively large concentration of an acid, to react with added OH^-, and a relatively large concentration of a base, to react with added H_3O^+.

What kinds of acid and base can we use? A strong acid and a strong base will not work, because these two species react completely with each other. For the same reason, it is not possible to have a solution with large concentrations of a weak acid and a strong base, or a weak base and a strong acid. We are left with only one other possibility, a solution with large concentrations of both a weak acid and a weak base. In general, a solution with large concentrations of both a weak acid and its conjugate base can act as a buffer solution.

Some examples of buffer solutions are a solution with [acetic acid] = 1 M and [acetate ion] = 1 M; a solution with $[H_2PO_4^-]$ = 0.5 M and $[HPO_4^{2-}]$ = 0.3 M, and a solution with $[NH_4^+]$ = 0.2 M and $[NH_3]$ = 0.4 M.

Since buffer solutions are so important, let us analyze one in some detail: the acetic acid-acetate ion buffer system. The reversible reaction between acetic acid and acetate ion is

$$(11\text{-}34) \qquad HC_2H_3O_2 + H_2O \rightleftharpoons H_3O^+ + C_2H_3O_2^-$$

and the associated equilibrium constant expression is

$$(11\text{-}35) \qquad K_a = \frac{[H_3O^+][C_2H_3O_2^-]}{[HC_2H_3O_2]} \qquad K_a = 1.77 \times 10^{-5} \text{ mole/liter}$$

Equation 11-35 can be rearranged algebraically to the following equivalent form:

$$(11\text{-}36) \qquad [H_3O^+] = \frac{[HC_2H_3O_2]}{[C_2H_3O_2^-]} \times K_a = \frac{[HC_2H_3O_2]}{[C_2H_3O_2^-]} \times (1.77 \times 10^{-5})$$

The form of Eq. 11-36 reveals an important fact: $[H_3O^+]$ is directly proportional to the ratio $[HC_2H_3O_2]/[C_2H_3O_2^-]$.

Let us consider a specific solution prepared by dissolving 0.5 mole of acetic acid and 0.5 mole of sodium acetate in sufficient water to form 1.00 liter of solution. If reaction 11-34 does not take place in either direction to any significant extent, $[HC_2H_3O_2]$ is 0.5 M and $[C_2H_3O_2^-]$ is also 0.5 M. Thus, if our assumption of no significant reaction is valid, the ratio $[HC_2H_3O_2]/[C_2H_3O_2^-]$ = 1, and $[H_3O^+]$ = 1.77 $\times$ 10^{-5} M according to Eq. 11-36.

Our assumption is in fact correct. A solution that is 0.5 M in acetic acid and 0.5 M in sodium acetate does have $[H_3O^+]$ = 1.77 $\times$ 10^{-5} M. It is instructive to use what we know about reversible acid-base reactions to explain this fact.

Acetic acid is a weak acid, so if we put acetic acid in pure water, the extent of reaction 11-34 from left to right is very small. When a large concentration of $C_2H_3O_2^-$ is also present in the solution, the extent of this reaction will, according to Le Châtelier's principle, be smaller still. Similarly, if we dissolve sodium acetate in pure water, in which $[H_3O^+]$ is very small (10^{-7} M), the extent of reaction 11-34 from right to left is very small; thus, using Le Châtelier's principle, we expect that when a large concentration of $HC_2H_3O_2$ is also present, the extent of reaction will be even smaller. The net result is that when a large amount of acetic acid and a large amount of sodium acetate are put into the same solution, reaction 11-34 does not take place in either direction to any significant extent.

Since the solution we have been considering is supposed to be a buffer solution, let us refer to it as solution B. Thus, for solution B, $[HC_2H_3O_2]$ is 0.5 M, $[C_2H_3O_2^-]$ is 0.5 M, and $[H_3O^+]$ is 1.77×10^{-5} M. Let us now consider what happens to the $[H_3O^+]$ of solution B if we add strong acid or strong base. The results of our analysis are given in Figure 11-10. The precise pH values were obtained using Eq. 11-12.

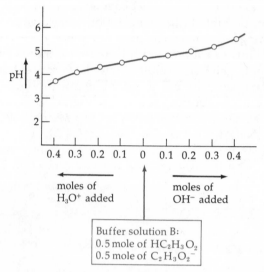

Figure 11-10 Buffering ability of an acetic acid–acetate ion solution. When a large amount of H_3O^+ is added to buffer solution B, there is only a slight decrease in pH because the reaction $C_2H_3O_2^- + H_3O^+ \rightarrow HC_2H_3O_2 + H_2O$ takes place. When a large amount of OH^- is added to buffer solution B, there is only a slight increase in pH because the reaction $HC_2H_3O_2 + OH^- \rightarrow C_2H_3O_2^- + H_2O$ takes place.

If we add H_3O^+ (from HCl, for example) to solution B, almost all of the added H_3O^+ will react with some of the $C_2H_3O_2^-$ in solution according to

(11-37) $$H_3O^+ + C_2H_3O_2^- \longrightarrow HC_2H_3O_2 + H_2O$$

Recall that the reaction of a strong acid (HCl) and a weak base ($C_2H_3O_2^-$) goes nearly to completion. After this reaction takes place, the amount of $HC_2H_3O_2$ will be larger, the amount of $C_2H_3O_2^-$ will be smaller, and the ratio $[HC_2H_3O_2]/[C_2H_3O_2^-]$ will increase. Thus, by Eq. 11-36, $[H_3O^+]$ will increase. The magnitude of this increase, however, will be very small, provided that we had a large amount of both $HC_2H_3O_2$ and $C_2H_3O_2^-$ present before we added the H_3O^+ and that the amount of H_3O^+ added is not too large. How much change is there in $[H_3O^+]$? Let us consider quantitatively a specific example. The calculation is easy to do, and makes explicit how a buffer works.

Suppose that we add 0.1 mole of H_3O^+ to solution B (Table 11-3). Reaction 11-37 will take place almost completely, and thus the amount of $C_2H_3O_2^-$ will decrease by 0.1 mole and the amount of $HC_2H_3O_2$ will increase by 0.1 mole. Since there is now 0.6 mole of $HC_2H_3O_2$ and 0.4 mole of $C_2H_3O_2^-$, the ratio $[HC_2H_3O_2]/[C_2H_3O_2^-]$ is 1.5. Using Eq. 11-36, the new $[H_3O^+]$ is 2.63×10^{-5} M, which is a pH of 4.58. The change in pH is 0.16, and this change is very much smaller than the change in pH of 6 (from 7 to 1) that occurs if we add 0.1 mole of H_3O^+ to pure water. Thus, upon addition of H_3O^+ to solution B, a nearly

Table 11-3 Acetic Acid-Acetate Ion Buffer Solution

	Moles of $HC_2H_3O_2$	Moles of $C_2H_3O_2^-$	$\dfrac{[HC_2H_3O_2]}{[C_2H_3O_2^-]}$	$[H_3O^+]$*	pH
1.00 liter of buffer solution B	0.500	0.500	1.00	1.77×10^{-5} M	4.76
Solution B + 0.1 mole of H_3O^+	0.600	0.400	1.50	2.63×10^{-5} M	4.58
Solution B + 0.2 mole of H_3O^+	0.700	0.300	2.33	4.08×10^{-5} M	4.39
Solution B + 0.3 mole of H_3O^+	0.800	0.200	4.00	7.00×10^{-5} M	4.16
Solution B + 0.1 mole of OH^-	0.400	0.600	0.667	1.17×10^{-5} M	4.93
Solution B + 0.2 mole of OH^-	0.300	0.700	0.429	7.50×10^{-5} M	5.13
Solution B + 0.3 mole of OH^-	0.200	0.800	0.250	4.38×10^{-6} M	5.36

* Note that $[H_3O^+]$ is 4.38×10^{-6} M when 0.3 mole of OH^- is added to 1 liter of buffer solution B, but $[H_3O^+]$ is 7.00×10^{-5} M, or 16 times larger, when 0.3 mole of H_3O^+ is added to 1 liter of buffer solution B. On the other hand, when 0.3 mole of OH^- is added to 1 liter of pure water, $[H_3O^+]$ is 3.3×10^{-14} M; whereas when 0.3 mole of H_3O^+ is added to 1 liter of pure water, $[H_3O^+]$ is 3×10^{-1} M, a whopping 1.1×10^{13} times larger than when 0.3 mole of OH^- is added to pure water.

constant pH is maintained. We are therefore justified in calling solution B a buffer solution, provided that the pH change is also very small when OH^- is added to the original buffer solution B.

If we add 0.1 mole of NaOH to solution B, essentially 0.1 mole of OH^- will react with some of the $HC_2H_3O_2$ in solution according to

(11-38) $$OH^- + HC_2H_3O_2 \longrightarrow H_2O + C_2H_3O_2^-$$

Using the same method of calculation we employed for the addition of 0.1 mole of H_3O^+, we find that the new $[H_3O^+]$ is 1.17×10^{-5} M and the new pH is 4.93. Thus addition of OH^-, like addition of H_3O^+, does not cause a large change in the pH of solution B. We can now conclude that solution B is a buffer solution.

Now refer to Table 11-3 and Figure 11-10 while you consider the following properties of buffer solutions. A buffer solution cannot keep the pH exactly constant if H_3O^+ or OH^- is added. If we add H_3O^+ to a buffer solution, the pH decreases, and the more H_3O^+ we add, the larger the decrease. The change in pH, however, is relatively small—much smaller than if we add H_3O^+ to pure water. Similarly, if we add OH^- to a buffer, the pH increases somewhat and the increase is larger the more OH^- we add. The change in pH, however, is relatively small—much smaller than if we add OH^- to pure water.

Obviously, we cannot keep on adding H_3O^+ or OH^- to a buffer solution without destroying its properties as a buffer at some point. For example, if we start with solution B and add 0.5 mole of H_3O^+, we use up essentially all the $C_2H_3O_2^-$ and destroy the buffering capacity of the solution.

The amount of H_3O^+ or OH^- a buffer can absorb without a significant change in the pH is called the **buffering capacity** of the buffer. When does the change in pH become significant? That depends on how closely we want the buffer solution to control the pH. For example, in one case we might want to control the pH to ±0.1. For this solution the buffering capacity is the amount of H_3O^+ it can absorb before its pH drops by 0.1 or the amount of OH^- it can absorb before its pH rises by 0.1. In another situation, even with the same buffer solution, it might be sufficient to control the pH to ±0.5. The same volume of the same buffer solution could absorb more H_3O^+ or OH^- and stay within the ±0.5 range than if the tolerable range were ±0.1. Of course, the larger the volume of the buffer solution, the greater its buffering capacity.

We have been discussing buffer solution B, which was prepared with equimolar amounts of $HC_2H_3O_2$ and $C_2H_3O_2^-$. Will a solution that does not contain equimolar amounts still be a buffer solution? Yes, the ratio $[HC_2H_3O_2]/[C_2H_3O_2^-]$

can be somewhat larger or smaller than 1 and the solution will still be a buffer solution, provided that both $[HC_2H_3O_2]$ and $[C_2H_3O_2^-]$ are not too low. If $[HC_2H_3O_2]$ is very low, not enough weak acid will be present in solution to react with added OH^-, and if $[C_2H_3O_2^-]$ is very low there is a similar difficulty with added H_3O^+.

For such a buffer solution, $[H_3O^+]$ will be directly proportional to the ratio $[HC_2H_3O_2]/[C_2H_3O_2^-]$ (see Eq. 11-36). For example, in a buffer solution that is prepared with 0.60 mole of $HC_2H_3O_2$ and 0.40 mole of $C_2H_3O_2^-$, the ratio $[HC_2H_3O_2]/[C_2H_3O_2^-]$ is $0.60/0.40 = 1.5$ and $[H_3O^+]$ is 2.63×10^{-5} M (compare this with the result in Table 11-3).

Everything we have considered about the acetic acid/acetate ion buffer system holds in general for a buffer involving a weak acid-weak base conjugate pair. Let us generalize our results as a means of summarizing what we have learned about buffers.

If HA is a weak acid with dissociation constant K_a, and A^- is its conjugate base, the equilibrium constant expression for the general reversible reaction $HA + H_2O \rightleftharpoons H_3O^+ + A^-$ can be put in the form

(11-39) $$[H_3O^+] = \frac{[HA]}{[A^-]} \times K_a(HA)$$

A solution with a sufficiently large concentration of both HA and A^- is a buffer solution. This solution stabilizes the pH because there is a large amount of weak acid present which can react with added OH^-, and a large amount of weak base which can react with added H_3O^+. For a solution that contains HA and A^-, $[H_3O^+]$ is directly proportional to the ratio $[HA]/[A^-]$. Solutions with a $[HA]/[A^-]$ ratio that is not very much larger or very much smaller than 1 are buffer solutions.

Consider the following example.

EXAMPLE 3

(a) What is the $[H_3O^+]$ of a 1.0-liter buffer solution that contains 1.0 mole of HCN and 0.80 mole of NaCN?

Step 1 Write the equation for the reaction under consideration and its associated equilibrium constant expression in the form of Eq. 11-39:

(11-40) $$HCN + H_2O \rightleftharpoons H_3O^+ + CN^- \qquad [H_3O^+] = \frac{[HCN]}{[CN^-]} \times K_a(HCN)$$

Step 2 Substitute the values for the concentration of weak acid, weak base, and K_a into the relationship in step 1. From the given data, $[HCN] = 1.00$ M and $[CN^-] = 0.80$ M; and from Table 10-1, the value of $K_a(HCN) = 4.0 \times 10^{-10}$ mole/liter. Therefore,

(11-41) $$[H_3O^+] = \frac{1.0\ M}{0.80\ M} \times (4.0 \times 10^{-10}\ \text{mole/liter}) = 5.0 \times 10^{-10}\ \text{mole/liter}$$

(b) What is the new $[H_3O^+]$ when 0.20 mole of HCl is added to the buffer solution described in part (a)?

Step 3 Write the equation for the reaction that takes place:

(11-42) $$H_3O^+ + CN^- \rightleftharpoons HCN + H_2O$$

Step 4   Determine the new concentrations of the weak acid and the weak base. According to Eq. 11-42, when 0.20 mole of H_3O^+ reacts with 0.20 mole of CN^-, an additional 0.20 mole of HCN is formed and the number of moles of CN^- is decreased by 0.20 mole. Therefore, after the reaction takes place,

(11-43)   Number of moles of HCN = 1.00 + 0.20 = 1.20 moles
$$[HCN] = 1.20\ M$$

(11-44)   Number of moles of CN^- = 0.80 − 0.20 = 0.60 mole
$$[CN^-] = 0.60\ M$$

Step 5   Substitute the new concentrations of weak acid and weak base in the relationship in step 1 to obtain the new $[H_3O^+]$:

(11-45) $$[H_3O^+] = \frac{1.20\ M}{0.60\ M} \times (4.0 \times 10^{-10}\ \text{mole/liter}) = 8.0 \times 10^{-10}\ \text{mole/liter}$$

(c) What is the new $[H_3O^+]$ when 0.20 mole of NaOH is added to the buffer solution described in part (a)?
Repeat steps 3, 4, and 5.

Step 3

(11-46) $$OH^- + HCN \rightleftharpoons H_2O + CN^-$$

Step 4

(11-47)   Number of moles of HCN = 1.00 − 0.20 = 0.80 mole
$$[HCN] = 0.80\ M$$

(11-48)   Number of moles of CN^- = 0.80 + 0.20 = 1.00 mole
$$[CN^-] = 1.00\ M$$

Step 5

(11-49) $$[H_3O^+] = \frac{0.80\ M}{1.00\ M} \times (4.0 \times 10^{-10}\ \text{mole/liter}) = 3.2 \times 10^{-10}\ \text{mole/liter}$$

Body fluids contain several conjugate acid-base pairs that act as buffer systems. We shall discuss a very important one, the blood buffer system, in the next section.

Exercise 11-8
An important buffer system in the human body involves the $H_2PO_4^-/HPO_4^{2-}$ conjugate acid-base pair (see Table 10-1).
   (a) What is the equilibrium reaction between these ions?
   (b) If 1.0 mole of NaH_2PO_4 and 1.5 mole of Na_2HPO_4 are dissolved in water to form 1.0 liter of buffer solution, what is $[H_3O^+]$ for this buffer?
   (c) Write an equation for the reaction that takes place when 0.20 mole of HCl is added to half the solution in part (b) and calculate $[H_3O^+]$ in the resulting solution.
   (d) Write an equation for the reaction that takes place when 0.20 mole of NaOH is added to the other half of the solution in part (b) and calculate $[H_3O^+]$ for the resulting solution.

11-8 BUFFERS IN BLOOD

Three interrelated mechanisms are used to maintain the proper pH of the body fluids: (1) kidney function; (2) respiration rate; and (3) buffer action. In this section we shall discuss some of the basic features of the buffers in blood. We shall treat the whole subject of regulation and control of body fluids in Chapter 29.

In blood there are three major buffer systems: (1) the H_2CO_3/HCO_3^- buffer, (2) the $H_2PO_4^-/HPO_4^{2-}$ buffer, and (3) the buffering action of various proteins. The major buffering action is accomplished via the H_2CO_3/HCO_3^- conjugate acid-base pair. Carbonic acid, H_2CO_3, is generated in the blood plasma (which is an aqueous medium) from dissolved CO_2 gas. Carbon dioxide can enter or leave the blood from the lungs. If we represent the CO_2 gas in the lung alveoli by $CO_2(g)$ and the CO_2 dissolved in blood plasma by $CO_2(aq)$, then the following equilibrium is established in the lungs:

$$(11\text{-}50) \qquad CO_2(g) \rightleftharpoons CO_2(aq)$$

In water, one molecule of dissolved CO_2 can react with one molecule of H_2O to form carbonic acid, which has the molecular formula H_2CO_3. Water, $CO_2(aq)$, and H_2CO_3 can be in equilibrium according to the reaction,

$$(11\text{-}51) \qquad CO_2(aq) + H_2O \rightleftharpoons H_2CO_3$$

The erythrocytes (red blood cells) contain an enzyme called carbonic anhydrase, which catalyzes reaction 11-51 in both the forward and reverse directions. If the equilibrium between $CO_2(aq)$, H_2O, and H_2CO_3 in blood is disturbed, carbonic anhydrase catalyzes reaction 11-51 from either left to right or from right to left so that equilibrium is rapidly reestablished.

Carbonic acid is also in equilibrium with the bicarbonate anion:

$$(11\text{-}52) \qquad H_2CO_3 + H_2O \rightleftharpoons H_3O^+ + HCO_3^-$$

The effective K_a for H_2CO_3 in blood is about 8.0×10^{-7} mole/liter, which is slightly different from the value for dilute aqueous solutions given in Table 10-1. If we apply the general equation 11-39 to reaction 11-52, we obtain

$$(11\text{-}53) \qquad [H_3O^+] = \frac{[H_2CO_3]}{[HCO_3^-]} \times (8.0 \times 10^{-7} \text{ mole/liter})$$

The normal pH of human blood is 7.4, which corresponds to a $[H_3O^+]$ of about 4.0×10^{-8} M. If we substitute this value for $[H_3O^+]$ into Eq. 11-43, we find that the ratio $[H_2CO_3]/[HCO_3^-]$ is about 1/20 in blood. Therefore, in normal blood, $[H_2CO_3]$ must be much less than $[HCO_3^-]$.

From our general discussion of buffers in Section 11-7, it might seem surprising that a solution with a $[H_2CO_3]/[HCO_3^-]$ ratio as low as 1/20 can provide effective buffering action. But in human blood this buffer system is coupled to the large supply of $CO_2(g)$ present in the lungs (Eqs. 11-50 and 11-51). Therefore, even though at any moment $[H_2CO_3]$ in the blood is low, H_2CO_3 is still capable of reacting with excess base (thereby stabilizing the pH) because additional H_2CO_3 can be generated by absorption of $CO_2(g)$. In short, the large supply of $CO_2(g)$ in the lungs makes the effective buffering capacity of the H_2CO_3/HCO_3^- buffer very large.

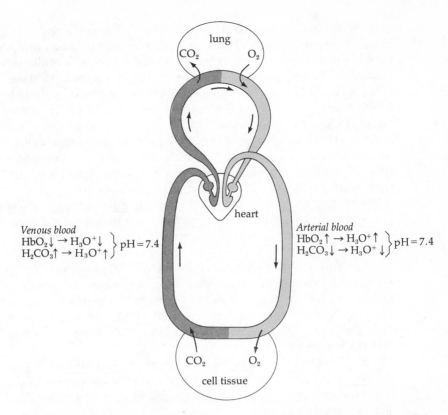

Venous blood
$HbO_2\downarrow \rightarrow H_3O^+\downarrow$
$H_2CO_3\uparrow \rightarrow H_3O^+\uparrow$ $\Big\}$ pH = 7.4

Arterial blood
$HbO_2\uparrow \rightarrow H_3O^+\uparrow$
$H_2CO_3\downarrow \rightarrow H_3O^+\downarrow$ $\Big\}$ pH = 7.4

Figure 11-11   Schematic illustration of the control of pH in arterial and venous blood. In arterial blood, the relatively smaller $[CO_2]$ results in a relatively smaller $[H_2CO_3]$, which tends to reduce $[H_3O^+]$. However, this tendency is largely compensated for by a relatively higher concentration of the weak acid oxyhemoglobin, HbO_2, and so the pH is stabilized at about 7.4. In venous blood, the relatively larger $[CO_2]$ results in a relatively larger $[H_2CO_3]$, which tends to increase $[H_3O^+]$. However, this tendency is largely compensated for by a relatively lower concentration of oxyhemoglobin, and the pH is again stabilized at about 7.4.

Blood leaving the lungs and going to the body tissues—arterial blood—is relatively rich in oxygen and poor in CO_2 compared to venous blood, which circulates back from the body tissues to the lungs (see Figure 11-11). Since arterial blood contains less CO_2 than venous blood, it would appear that the position of equilibrium for reaction 11-51 would be farther to the left for arterial blood, and consequently, that arterial blood would contain less H_2CO_3 and be less acidic than venous blood. If this were to occur to any significant extent, there would be disastrous consequences for the human organism. If the pH of your blood goes as high as 7.8 or as low as 7.0 for very long, you will die. Needless to say, some other mechanism(s) must operate to keep the pH of venous and arterial blood essentially the same, and very close to a pH of 7.4 at all times. Refer to Figure 11-11 as you consider the following control mechanism.

Oxygen in the blood exists mostly bound to a protein called hemoglobin. This complex is called **oxyhemoglobin,** and we shall represent it symbolically by the formula HbO_2, where Hb stands for hemoglobin. Since arterial blood contains more O_2 than venous blood, arterial blood is richer in HbO_2 and poorer in Hb than venous blood. Now, both HbO_2 and Hb are weak acids, but HbO_2 is the stronger acid of the two. Since arterial blood contains more of the stronger acid, HbO_2, we would expect that arterial blood would be more acidic if the relative amounts of HbO_2 and Hb were the only factors determining the blood pH.

We thus have two mechanisms that tend to operate in opposite directions. By itself, the relatively smaller amount of CO_2 in arterial blood compared to that in venous blood would tend to make arterial blood less acidic, but the greater amount of the stronger acid HbO_2 in arterial blood tends to make arterial blood more acidic. When the human body is operating properly, these two tendencies just balance one another. The net result is that the pH of arterial and venous blood are just about the same, 7.4.

Any appreciable change in the blood pH from the normal value of 7.4 is a serious condition. A condition in which the blood is more acidic than normal (pH < 7.4) is called **acidosis,** whereas one in which the blood pH is more basic than normal (pH > 7.4) is referred to as **alkalosis.**

Respiratory acidosis can result if your rate of breathing is too slow, that is, if you **hypoventilate.** Hypoventilation can occur in many conditions, such as drug overdose, pneumonia, coronary attack, and so on. Why does hypoventilation lead to acidosis?

When you hypoventilate, you do not expel CO_2 from your lungs fast enough. The partial pressure of CO_2 in your lungs increases and equilibrium reactions 11-50 and 11-51 are both shifted to the right. Thus $[H_2CO_3]$ in your blood goes up; $[HCO_3^-]$ also increases, by reaction 11-52, but not as much as $[H_2CO_3]$. Thus the ratio $[H_2CO_3]/[HCO_3^-]$ increases, $[H_3O^+]$ rises according to Eq. 11-53, and the pH drops.

If you breathe too rapidly, that is, if you **hyperventilate,** respiratory alkalosis can result. When you hyperventilate, the partial pressure of CO_2 in the lungs is lower than normal, the $[H_2CO_3]$ in the blood goes down, the $[H_3O^+]$ decreases, and the pH rises.

The action of the kidneys can compensate for a rise or drop in blood pH, and restore the pH balance if the condition that caused the pH to deviate from the normal value of 7.4 is not too severe.

Acidosis, and less commonly alkalosis, can also result from certain metabolic disturbances. The topic of metabolic acidosis and metabolic alkalosis will be discussed in Chapter 29.

The other major blood buffer system involves the $H_2PO_4^-/HPO_4^{2-}$ conjugate acid-base pair. The equilibrium between these species is

$$(11\text{-}54) \qquad H_2PO_4^- + H_2O \rightleftharpoons HPO_4^{2-} + H_3O^+$$

If we use the value of K_a for $H_2PO_4^-$ from Table 10-1, and if we apply the general equation 11-39 to this equilibrium, we obtain

$$(11\text{-}55) \qquad [H_3O^+] = \frac{[H_2PO_4^-]}{[HPO_4^{2-}]} \times (6.2 \times 10^{-8} \text{ mole/liter})$$

At a pH of 7.4, the normal blood pH, $[H_3O^+]$ is 4.0×10^{-8} M. If we substitute this value into Eq. 11-55, the ratio $[H_2PO_4^-]/[HPO_4^{2-}]$ is found to be 0.64. Thus blood plasma at the normal pH of 7.4 contains comparable amounts of $H_2PO_4^-$ and HPO_4^{2-}, and this buffer system is also effective in maintaining a constant blood pH.

Exercise 11-9

A person is extremely anxious and afraid and is breathing rapidly. A friend encourages him to breathe into and out of a paper bag. Can you explain why this is a good idea?

11-9   SUMMARY

1. For an acidic solution $[H_3O^+] > 10^{-7}$ M and $[OH^-] < 10^{-7}$ M; for a neutral solution $[H_3O^+] = [OH^-] = 10^{-7}$ M; and for a basic solution $[H_3O^+] < 10^{-7}$ M and $[OH^-] > 10^{-7}$ M.

2. Some salts hydrolyze in water to give acidic or basic solutions because the anion and/or the cation of the salt acts as an acid or as a base.

3. The pH scale is an alternative way of expressing $[H_3O^+]$ for a solution. For an acidic solution, pH < 7; for a neutral solution, pH $= 7$; and for a basic solution, pH > 7.

4. The reaction of a strong acid and a strong base goes to completion. The reactions of a strong acid and a weak base, and a strong base and a weak acid, go nearly to completion. The extent of the proton-transfer reaction between a weak acid and a weak base depends on the specific acid and specific base involved.

5. The concentration of an acid or a base solution can be determined by the experimental method called titration.

6. A solution that can maintain the pH at nearly a constant level when additional H_3O^+ or additional OH^- is added is called a buffer solution.

7. A solution with a relatively large concentration of a weak acid and a relatively large concentration of its conjugate base is a buffer solution.

8. Human blood contains three major buffer systems: H_2CO_3/HCO_3^-; $H_2PO_4^-/HPO_4^{2-}$; and various proteins. Normally the pH of blood is very close to 7.4. Acidosis is a condition in which the blood pH is less than 7.4, and alkalosis is one in which the blood pH is greater than 7.4.

PROBLEMS

1. Determine $[H_3O^+]$ for each of the following solutions:
   (a) $[OH^-] = 2.3 \times 10^{-8}$ M
   (b) $[OH^-] = 5.8 \times 10^{-12}$ M
   (c) $[OH^-] = 7.2 \times 10^{-2}$ M

2. Determine $[OH^-]$ for each of the following solutions:
   (a) $[H_3O^+] = 4.5 \times 10^{-7}$ M
   (b) $[H_3O^+] = 6.8 \times 10^{-10}$ M
   (c) $[H_3O^+] = 2.1 \times 10^{-3}$ M

3. For each of the following solutions, predict whether the solution will be acidic, basic, or neutral (see Table 10-1):
   (a) 1 M NH$_4$Br   (b) 1 M NaI
   (c) 1 M Na$_3$PO$_4$ (d) 1 M NaHCO$_3$

4. Determine an approximate value for the pH of each of the following solutions:
   (a) $[H_3O^+] = 2.2 \times 10^{-3}$ M
   (b) $[H_3O^+] = 8.3 \times 10^{-10}$ M
   (c) $[OH^-] = 4.5 \times 10^{-6}$ M

5. Determine an approximate value of $[H_3O^+]$ for each of the following solutions:
   (a) pH $= 6.2$ (b) pH $= 1.5$
   (c) pH $= 8.9$ (d) pH $= 11.3$

6. Write net ionic equations for the reactions that occur and predict the extent of reaction:
   (a) $NaHCO_3$ + NaOH
   (b) HCl + $Ca(OH)_2$
   (c) NaCN + HCl
   (d) NH_4Cl + NaOH

7. 25.00 ml of a NaOH solution is titrated with 0.156 M HCl. To reach the end point requires 18.52 ml of the HCl solution. Calculate the concentration of the NaOH solution.

8. Some acetic acid was dissolved in 25.00 ml of water and the resulting solution was titrated with a standard 0.242 M NaOH solution. To reach the end point required 22.53 ml of the NaOH solution. Determine the mass of acetic acid that was dissolved to form the acetic acid solution.

9. A buffer solution is obtained when 1.0 mole of HCN and 0.30 mole of NaOH react in water to produce some NaCN.
   (a) If the volume of this buffer solution is 1.0 liter, calculate $[H_3O^+]$ for this buffer.
   (b) 0.05 mole of NaOH are added to half the buffer solution. Calculate $[H_3O^+]$ in the resulting solution.
   (c) 0.05 mole of HCl are added to the other half of the original buffer solution. Calculate $[H_3O^+]$ in the resulting solution.
   (d) What will $[H_3O^+]$ be if the solutions in parts (b) and (c) are mixed?

SOLUTIONS TO EXERCISES

11-1 $[H_3O^+] = \dfrac{K_w}{[OH^-]} = \dfrac{1 \times 10^{-14}}{2 \times 10^{-5}} = 5 \times 10^{-10}\ M$ The solution is basic.

11-2   NaF (basic): $F^- + H_2O \rightleftharpoons HF + OH^-$
      CH_3NH_3Cl (acidic): $CH_3NH_3^+ + H_2O \rightleftharpoons H_3O^+ + CH_3NH_2$
      KNO_3 (neutral):   no reaction
      Na_2CO_3 (basic): $CO_3^{2-} + H_2O \rightleftharpoons HCO_3^- + OH^-$

11-3 $[H_3O^+]$ is between $10^{-7}\ M$ and $10^{-8}\ M$, so the pH is between 7 and 8.

11-4   The pH is between 5 and 6, so $[H_3O^+]$ is between $10^{-5}\ M$ and $10^{-6}\ M$. Thus, $[OH^-]$ is between $10^{-9}\ M$ and $10^{-8}\ M$.

11-5   Using H_2CO_3 and NaOH: $H_2CO_3 + OH^- \rightarrow HCO_3^- + H_2O$
      Using Na_2CO_3 and HCl: $CO_3^{2-} + H_3O^+ \rightarrow HCO_3^- + H_2O$

11-6   (a) HCN is a weak acid, NaOH is a strong base.
        The reaction $HCN + OH^- \rightarrow H_2O + CN^-$ is nearly complete.
     (b) KOH is a strong base, HBr is a strong acid.
        The reaction $H_3O^+ + OH^- \rightarrow H_2O + H_2O$ is complete.
     (c) NH_4^+ is a weak acid, NaOH is a strong base.
        The reaction $NH_4^+ + OH^- \rightarrow NH_3 + H_2O$ is nearly complete.

11-7   Use Eq. 11-22, which applies for the titration of a monoprotic acid with OH^-,

$$M_a = M_b \times \frac{V_b}{V_a} = 0.348\ M \times \frac{22.45\ \text{ml}}{25.00\ \text{ml}} = 0.313\ M$$

We do not need to know which specific acid is involved, since Eq. 11-22 applies for any monoprotic acid.

11-8   (a) $H_2PO_4^- + H_2O \rightleftharpoons H_3O^+ + HPO_4^{2-}$
        $K_a = 6.2 \times 10^{-8}$ mole/liter   (from Table 10.1)

     (b) $[H_3O^+] = \dfrac{[H_2PO_4^-]}{[HPO_4^{2-}]} \times K_a = \dfrac{1.0\ M}{1.5\ M} \times (6.2 \times 10^{-8}\ \text{mole/liter}) = 4.1 \times 10^{-8}$ mole/liter

(c) In half the solution there is 0.50 mole of $H_2PO_4^-$ and 0.75 mole of HPO_4^{2-}. When 0.20 mole of HCl is added, 0.20 mole of H_3O^+ reacts with 0.20 mole of HPO_4^{2-} according to the reaction $H_3O^+ + HPO_4^{2-} \rightarrow H_2PO_4^- + H_2O$. Thus, after the reaction takes place, $0.50 + 0.20 = 0.70$ mole of $H_2PO_4^-$ and $0.75 - 0.20 = 0.55$ mole of HPO_4^{2-} are present.

$$\text{New } [H_3O^+] = \frac{0.70 \text{ mole}}{0.55 \text{ mole}} \times (6.2 \times 10^{-8} \text{ mole/liter}) = 7.8 \times 10^{-8} \text{ mole/liter}$$

(d) 0.20 mole of OH^- from NaOH will react with 0.20 mole of $H_2PO_4^-$ according to the reaction $OH^- + H_2PO_4 \rightarrow HPO_4^{2-} + H_2O$. There will be $0.75 + 0.20 = 0.95$ mole of HPO_4^{2-} after the reaction and $0.50 - 0.20 = 0.30$ mole of $H_2PO_4^-$ after the reaction.

$$\text{New } [H_3O^+] = \frac{0.30 \text{ mole}}{0.95 \text{ mole}} \times (6.2 \times 10^{-8} \text{ mole/liter}) = 2.0 \times 10^{-8} \text{ mole/liter}$$

11-9 Anxiety and fear often produce hyperventillation and thus a lower-than-normal partial pressure of CO_2 in the lungs. Breathing into and out of a paper bag increases the partial pressure of CO_2 in the lungs.

CHAPTER 12

Oxidation-Reduction Reactions and Reactions of Electrolytes in Aqueous Solutions

12-1 INTRODUCTION

The rusting of iron, the manufacture of margarine from liquid vegetable oils, the electrolysis of molten sodium chloride, and the degradation of fats and carbohydrates in the human body are all chemical reactions with one common feature. They are all examples of a large class of reactions collectively called oxidation-reduction reactions.

There are several parallels between oxidation-reduction reactions and the acid-base reactions we discussed in Chapters 10 and 11. First, both are transfer reactions. According to the Brønsted-Lowry definition of acid and base, acid-base reactions involve the transfer of *protons*; oxidation-reduction reactions, on the other hand, involve the transfer of *electrons*. Second, both types of reactions require two substances. In acid-base reactions, a substance can act as a proton acceptor (base) only if another substance acts as a proton donor (acid). Similarly, in an oxidation-reduction reaction, a substance can be oxidized (losing electrons) only if some other substance is reduced (gaining electrons). Third, both the concept of acid-base and the concept of oxidation-reduction have changed over the years to include an ever-widening variety of reactions.

We shall begin our study of oxidation-reduction reactions by considering some simple electron-transfer reactions. We shall then develop the concept of oxidation-reduction in a more general way. We shall also look at some of the characteristic features of typical oxidation-reduction reactions that occur in living cells.

After studying oxidation-reduction reactions in this chapter, and acid-base reactions in the previous two chapters, we shall have the necessary background to discuss the types of reactions that electrolytes undergo in aqueous solutions. We shall consider the possibilities that can occur when two different electrolyte solutions are mixed—that a solid precipitate is formed, that an acid-base reaction or an oxidation-reduction reaction takes place, or that a complex ion is formed.

A dramatic example of an oxidation-reduction reaction.

12-2   STUDY OBJECTIVES

After studying the material in this chapter, you should be able to:

1. Define oxidation and reduction in terms of the loss and gain of electrons.

2. Identify the substance oxidized and the substance reduced in reactions in which binary ionic compounds are formed from their elements.

3. Define the term oxidation number, and assign oxidation numbers to atoms in molecules and ions.

4. Define oxidation and reduction in terms of increase and decrease in the oxidation number.

5. Define the terms oxidizing agent and reducing agent.

6. Determine if a balanced chemical reaction is an oxidation-reduction reaction, and if it is, determine the substance oxidized, the substance reduced, the oxidizing agent, and the reducing agent.

7. Identify a reaction involving the breaking of carbon-hydrogen bonds or the formation of carbon-oxygen bonds as an oxidation, and identify a reaction involving the formation of carbon-hydrogen bonds or the breaking of carbon-oxygen bonds as a reduction, given structural formulas for the reactants and products.

8. **(Optional)** Predict oxidation-reduction reactions that will occur, given a table of half-reactions with oxidizing and reducing agents ordered according to their strength.

9. Determine what precipitate, if any, will form when 0.1 M solutions of two electrolytes are mixed, given a table of solubility generalizations.

12-3   LOSS AND GAIN OF ELECTRONS

If an electric current is passed through molten sodium chloride at a high temperature, sodium cations, Na^+, and chloride anions, Cl^-, are converted into elemental sodium and elemental chlorine, respectively. (Recall that separate cations and anions are a characteristic feature of any ionic compound in the liquid state, as discussed in Chapter 4.) This type of process is called **electrolysis,** and is illustrated in Figure 12-1. The two electrically conducting, chemically inert rods inserted into the liquid and connected to the battery in Figure 12-1 are called **electrodes.**

In the electrolysis of NaCl, electrons are pulled away from chloride anions at one electrode, called the **anode,** and chlorine atoms are formed.

(12-1)   Anode reaction $:\ddot{Cl}:^- \longrightarrow :\ddot{Cl}\cdot + e^-$

Two chlorine atoms combine to form one chlorine molecule, $2Cl \longrightarrow Cl_2(g)$. The electrons lost by the chloride anions at the anode are pushed around to the other electrode, called the **cathode.** At the cathode, sodium cations combine with electrons and form sodium atoms.

(12-2)   Cathode reaction $Na^+ + e^- \longrightarrow Na\cdot$

Energy is required for electrolysis to take place. The energy in this instance is provided by a battery.

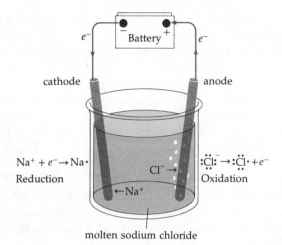

Figure 12-1   The electrolysis of molten sodium chloride. A battery pushes electrons to the cathode, where they react with sodium ions to form elemental sodium. At the anode, the battery pulls electrons away from Cl^- ions to form Cl atoms, which combine to form Cl_2 molecules.

The overall reaction for the electrolysis of molten sodium chloride is the sum of the two electrode reactions 12-1 and 12-2.

(12-3) $Na^+ + :\overset{..}{\underset{..}{Cl}}:^- \longrightarrow Na\cdot + :\overset{..}{\underset{..}{Cl}}\cdot \longrightarrow Na + \tfrac{1}{2}Cl_2(g)$

Thus in the electrolysis of molten sodium chloride, Cl^- anions each lose one electron and Na^+ cations each gain one electron.

Loss of electrons is called **oxidation** and gain of electrons is called **reduction**. Thus reaction 12-1 represents the oxidation of Cl^- anions and reaction 12-2 represents the reduction of Na^+ cations. The overall reaction 12-3 for the electrolysis process is an example of the simplest type of **oxidation-reduction reaction:** the transfer of electrons from one substance to another. In reaction 12-3, an electron is transferred from a Cl^- anion to a Na^+ cation.

Although separate Na^+ and Cl^- ions are actually present in molten sodium chloride, recall that our model for the ionic bond in a NaCl molecule in the gas state involves a Na^+ cation and a Cl^- anion held together by the attractive electrical force between their opposing charges (Chapter 4). Let us consider the formation of a NaCl molecule in the gas state from sodium and chlorine atoms:

(12-4) $Na\cdot(g) + :\overset{..}{\underset{..}{Cl}}\cdot(g) \longrightarrow Na^+ :\overset{..}{\underset{..}{Cl}}:^-(g)$

In reaction 12-4, the Na atom loses an electron and is oxidized, while the Cl atom gains an electron and is reduced. Thus reaction 12-4 also involves the transfer of an electron and is an oxidation-reduction reaction (see Figure 12-2).

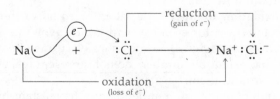

Figure 12-2   Schematic representation of the formation of a sodium chloride molecule in the gas state, involving the transfer of an electron from a sodium atom to a chlorine atom.

In fact, the formation of NaCl (reaction 12-4) is very similar to the reverse of reaction 12-3. Remember, however, that a Na^+ and a Cl^- ion are held close together in a gaseous NaCl molecule, whereas separate ions are present in molten NaCl.

The formation of sodium chloride from sodium and chlorine is a typical example of the formation of a binary ionic compound from a metallic element and a nonmetallic element. These reactions are oxidation-reduction reactions in which the metallic element loses electrons and is oxidized and the nonmetallic element gains electrons and is reduced.

Historically, the term oxidation originally referred to the combination of an element with oxygen, as in the rusting of iron ($4Fe + 3O_2 \rightarrow 2Fe_2O_3$). Reduction was a term first used in metallurgy to refer to a process whereby a metallic element is produced from a compound of the metal. For example, metallic tin, Sn, is produced from the ore cassiterite, SnO_2, by a reduction process involving carbon ($SnO_2 + C \rightarrow Sn + CO_2$).

Exercise 12-1
Identify the direction of electron transfer, the substance oxidized, and the substance reduced in each of the following reactions:

(a) $2K + Br_2 \rightarrow 2KBr$ (b) $O_2 + 2Ba \rightarrow 2BaO$ (c) $Sr + Cl_2 \rightarrow SrCl_2$

12-4   OXIDATION NUMBER

It is quite useful to extend the oxidation-reduction concept to certain reactions involving covalent compounds. To see how this is done, let us consider the formation of HCl from hydrogen and chlorine atoms:

(12-5) $H\cdot + :\overset{..}{\underset{..}{Cl}}\cdot \longrightarrow HCl$

The bond in HCl is not ionic; it is a polar covalent bond. Since chlorine is more electronegative than hydrogen, the shared pair of electrons in the HCl molecule is pulled toward the chlorine atom and away from the hydrogen. Recall that we can represent this shift of the shared pair of electrons by the symbol $\overset{\delta^+}{H}—\overset{\delta^-}{Cl}$, where δ^+ and δ^- represent the partial positive and negative charges, respectively. This is the description we used in Chapter 4 for the bond in HCl, and it is a reasonable qualitative picture of that bond.

Suppose, however, that we assign both of the shared electrons in the hydrogen-chlorine bond in HCl to the more electronegative chlorine atom. If we do this, the chlorine atom now has eight valence electrons associated with it and thus acquires a hypothetical *assigned* "charge" of (-1). The hydrogen atom in HCl, in turn, now has no valence electrons associated with it, and acquires a hypothetical *assigned* "charge" of $(+1)$. The (-1) and $(+1)$ "charges" are examples of oxidation numbers.

In general, the **oxidation number** of an atom is equal to the hypothetical "charge" on the atom when, for each covalent bond involving that atom, the shared pair of electrons is assigned to the more electronegative element. To emphasize the fact that oxidation numbers do not represent real electric charge, we use parentheses around them.

Using our definition of oxidation numbers, both the reactant hydrogen atom and the reactant chlorine atom in reaction 12-5 are assigned the oxidation number (0), since as reactants they are each separate, unbounded atoms. Thus in reaction 12-5 the oxidation number of hydrogen changes from (0) to (+1) and the oxidation number of chlorine changes from (0) to (-1). We can use such changes in oxidation numbers to broaden our concept of oxidation-reduction in a very useful way.

We define **oxidation** as either a loss of electrons or an increase in oxidation number, and we define **reduction** as either a gain of electrons or a decrease in oxidation number. Many reactions can be conveniently grouped together as oxidation-reduction reactions by using this extended definition of oxidation-reduction. In reaction 12-5, for example, since the oxidation number of hydrogen increases, hydrogen is oxidized, and since the oxidation number of chlorine decreases, chlorine is reduced (see Figure 12-3). Reaction 12-5 is thus an oxidation-reduction reaction.

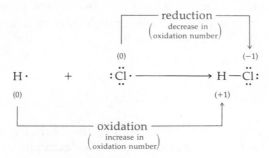

Figure 12-3   Schematic representation of the oxidation of hydrogen atoms, and the reduction of chlorine atoms, when they react and form HCl molecules. The numbers in parentheses are oxidation numbers.

It is quite important to note that in reaction 12-5, and in any reaction in which one substance is oxidized, another substance must be reduced, and vice versa. Oxidation and reduction go hand in hand. A reaction cannot involve one without the other. Reaction 12-3 involves a transfer of negatively charged electrons. We can view reaction 12-5 in a similar manner as involving the transfer of *negative* assigned "charge" from the hydrogen atom to the chlorine atom. Thus, we can consider that the *increase* in oxidation number of the hydrogen atom in reaction 12-5 is a consequence of that atom *losing* negative assigned "charge", and that the *decrease* in oxidation number of the chlorine atom is a consequence of that atom *gaining* negative assigned charge. From this point of view, any oxidation-reduction reaction involves the transfer of either real negatively charged electrons, as in reaction 12-3, or negative assigned "charge", as in reaction 12-5.

We could assign oxidation numbers to atoms in any molecule or ion using a procedure similar to the one we used for HCl. It is simpler, however, and almost always sufficient, to assign oxidation numbers using the following set of rules:

Rule 1   The oxidation number of an atom of an element in its elemental form is zero. Thus, hydrogen in H_2, oxygen in O_2, and carbon in C all have an oxidation number of (0).

Rule 2   The oxidation number of a monatomic ion is equal to the actual charge on the ion. Thus, for example, the oxidation number for H^+ is $(+1)$, for Fe^{3+} it is $(+3)$, and for Cl^- it is (-1).

Rule 3   In almost all compounds containing oxygen, the oxidation number of oxygen is (-2). An oxygen atom has six valence electrons, forms two covalent bonds, and has a large electronegativity. Therefore, in almost all compounds, oxygen tends to pull two additional electrons toward itself. When these two electrons are assigned to the oxygen atom, the oxidation number (-2) results. For example, the oxidation number of oxygen in H—O—H is (-2).

Rule 4   In most compounds, hydrogen has an oxidation number of $(+1)$ since in most compounds a hydrogen atom forms a bond with an atom more electronegative than itself. For example, in H—O—H, if we assign the electron for each hydrogen atom to the oxygen atom, then each of the hydrogen atoms acquires a "charge" of $(+1)$.

Rule 5   The assignment of oxidation numbers must be consistent with the principle of charge conservation. We cannot create or destroy electrical charge by assigning oxidation numbers. Thus:

(a) For neutral molecules (which have no electrical charge), the sum of the oxidation numbers for all atoms must be zero. Consider a water molecule again. When each of the two hydrogen atoms has an oxidation number of $(+1)$, and the oxygen atom has an oxidation number of (-2), the sum of the oxidation numbers is equal to zero: $(+1) + (+1) + (-2) = 0$.

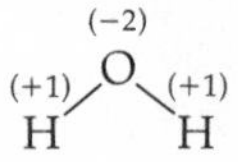

$$\overset{(-2)}{\underset{H \qquad H}{\overset{(+1)}{\diagdown}O\overset{(+1)}{\diagup}}}$$

(b) For a polyatomic ion, the sum of the oxidation numbers must equal the actual charge on the ion. For example, in the hypochlorite ion, ClO^- (the active ingredient in common laundry bleaches), the oxidation number of oxygen is (-2), so that the oxidation number of chlorine must be $(+1)$ in order that the sum of these two oxidation numbers equal the actual charge on the ion, -1: $(-2) + (+1) = -1$ (charge of the ClO^- ion).

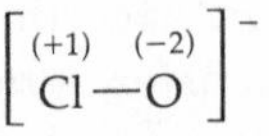

$$\left[\overset{(+1)\quad(-2)}{Cl—O} \right]^-$$

Exercise 12-2

Determine the oxidation number of all the atoms in each of the following:

(a) CO (b) SO_2
(c) MnO_4^- (d) Cu_2O
(e) Fe^{2+} (f) ClO_3^-

12-5   OXIDIZING AND REDUCING AGENTS

In oxidation-reduction reactions, the substance responsible for oxidizing something else is called an **oxidizing agent.** It is not surprising that in many reactions elemental oxygen is an oxidizing agent. Consider the reaction

(12-6) $C + O_2 \longrightarrow CO_2$

In this reaction, in which C represents elemental carbon in the form of graphite, the oxidation number of carbon increases from (0) to $(+4)$ and the oxidation number of each oxygen atom decreases from (0) to (-2). Therefore, carbon is oxidized and oxygen is the oxidizing agent in this reaction.

Oxygen is not the only oxidizing agent. There are many others. Do you see that in the reaction

(12-7) $$ClO^- + NO_2^- \longrightarrow Cl^- + NO_3^-$$

the hypochlorite ion, ClO^-, is an oxidizing agent (Figure 12-4)? The ClO^- ion is the active ingredient in laundry bleaches because of its ability to act as an oxidizing agent.

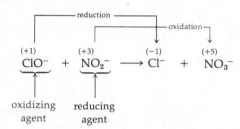

Figure 12-4   In this reaction ClO^- oxidizes NO_2^- and is itself reduced, while NO_2^- reduces ClO^- and is itself oxidized.   The numbers in parentheses are oxidation numbers for chlorine and nitrogen.

Similarly, the substance responsible for reducing something else is called a **reducing agent.** In the reaction

(12-8) $$C + 2H_2 \longrightarrow CH_4$$

the oxidation number of carbon decreases from (0) to (−4) while the oxidation number of each hydrogen atom increases from (0) to (+1). Hydrogen thus reduces carbon in this reaction and is therefore a reducing agent.

We expect that an oxidation-reduction reaction can occur when a sufficiently strong oxidizing agent and a sufficiently strong reducing agent make contact. For example, under an appropriate set of conditions, hydrogen and oxygen can react to form water ($2H_2 + O_2 \rightarrow 2H_2O$). In this reaction, hydrogen acts as a reducing agent and reduces oxygen, while oxygen acts as an oxidizing agent and oxidizes hydrogen. Once substances have been classified as oxidizing agents or reducing agents, and have been ordered according to strength, the possibility of a great number of reactions can be predicted. We shall see how this can be done in the optional Section 12-7. Recall that classifying substances as acids or bases is also an extremely valuable tool in predicting the results of many chemical reactions.

Exercise 12-3

Identify the substance oxidized, the substance reduced, the oxidizing agent, and the reducing agent in each of the following reactions:
  (a) $CH_4 + 2O_2 \rightarrow CO_2 + 2H_2O$
       Methane, CH_4, is the major component in natural gas.
  (b) $Zn + HgO \rightarrow ZnO + Hg$
       This reaction takes place in the so-called mercury battery.
  (c) $2Cu_2O + C \rightarrow 4Cu + CO_2$
       This reaction is involved in the production of copper from copper oxide ore.

12-6   COMPOUNDS CONTAINING CARBON

Although there are many applications of the oxidation-reduction concept in the study of chemical reactions, the oxidation-reduction reactions that take place inside the cells of living organisms are predominately of one type. Virtually all of the chemical compounds in living cells contain the element carbon.

We can determine if a reaction involving carbon-containing compounds is an oxidation-reduction reaction by assigning oxidation numbers to the carbon and the other atoms in the reactants and products. However, there is a simpler alternative procedure. We merely consider the bonds that carbon atoms form in the reactant molecules and what *changes*, if any, occur in these bonds as a result of the chemical reaction.

Recall that a carbon atom has four valence electrons and forms four covalent bonds (Chapter 4). Carbon-carbon bonds, carbon-hydrogen bonds, and carbon-oxygen bonds are the most important bonds when considering oxidation-reduction reactions involving carbon. Let us consider each of these types of bonds separately.

In the compound methane, $H\!-\!\overset{\displaystyle H}{\underset{\displaystyle H}{\overset{|}{\underset{|}{C}}}}\!-\!H$, the oxidation number of carbon is (-4), and it is $(+1)$ for each hydrogen. If we consider each carbon-hydrogen bond separately, then for each bond the oxidation number of carbon is (-1) and that of hydrogen is $(+1)$.

Considering the bonds for a carbon atom separately is very convenient for reactions in which some of the bonds involving the carbon atom remain the same. For example, the compound ethylene, C_2H_4, can react with hydrogen to form the compound ethane, C_2H_6:

(12-9)
$$H\!-\!\overset{\displaystyle H}{\overset{|}{C}}\!=\!\overset{\displaystyle H}{\overset{|}{C}}\!-\!H + H_2 \longrightarrow H\!-\!\overset{\displaystyle H}{\underset{\displaystyle H}{\overset{|}{\underset{|}{C}}}}\!-\!\overset{\displaystyle H}{\underset{\displaystyle H}{\overset{|}{\underset{|}{C}}}}\!-\!H$$

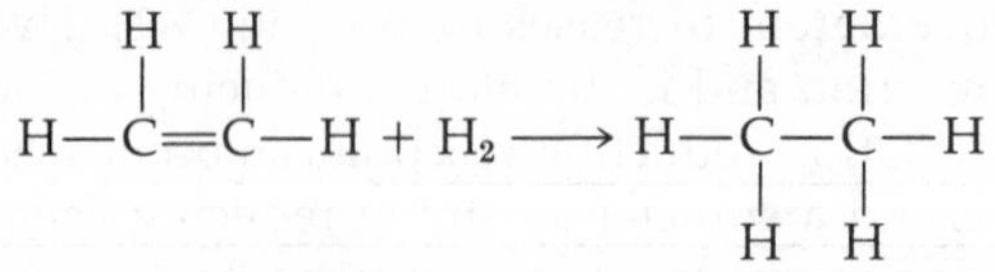

In this reaction we need consider only the bonds that *change*. We can ignore the bonds that remain the same. For each carbon atom in ethylene, a carbon-carbon bond is broken and a carbon-hydrogen bond is formed, as illustrated in Figure 12-5. Each carbon atom in ethylene is also bonded to two hydrogen atoms, but these bonds are present in the product ethane as well, so we can ignore them. Thus, considering the carbon-carbon bond in ethylene only, the oxidation number of each carbon atom is zero, since the bonded atoms are identical. When we ignore the carbon-hydrogen bonds that do not change, the oxidation number of each carbon atom in ethane is (-1). We therefore conclude that

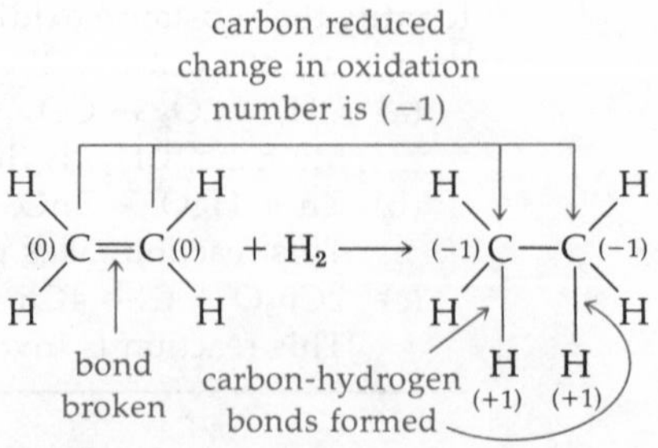

Figure 12-5   In the reaction of ethylene (C_2H_4) with hydrogen (H_2) to form ethane (C_2H_6), one carbon-carbon bond is broken and two carbon-hydrogen bonds are formed. The oxidation numbers are determined by ignoring the four carbon-hydrogen bonds that do not change during the reaction.

in a reaction such as 12-9, in which a carbon atom loses a carbon-carbon bond and forms a carbon-hydrogen bond, the change in oxidation number for that carbon atom is (-1); in other words the carbon atom involved is reduced. The key thing to remember when we are interested in determining the change in oxidation number for any carbon atom is that we can ignore the bonds that remain the same.

In general, increasing the number of carbon-hydrogen bonds represents a reduction of carbon. For reaction 12-9, since the carbon atoms in ethylene are reduced, we say that the reaction involves the reduction of ethylene.

If the formation of carbon-hydrogen bonds represents a reduction of carbon, then the loss of carbon-hydrogen bonds represents an oxidation of carbon. For example, the reverse of reaction 12-9 can occur under suitable conditions:

(12-10) $$C_2H_6 \longrightarrow C_2H_4 + H_2$$

In reaction 12-10 the reactant ethane loses carbon-hydrogen bonds and the carbon atoms in ethane are oxidized. The hydrogen atoms that leave the ethane are reduced. Since our main interest here is in carbon-containing compounds, we say that reaction 12-10 represents an oxidation of ethane. Thus, increasing the number of carbon-hydrogen bonds in a compound involves a reduction of the compound, and decreasing the number of carbon-hydrogen bonds represents an oxidation.

As we shall see, many chemical reactions that occur in the human body involve a compound losing or gaining carbon-hydrogen bonds. In the human body, however, the substance hydrogen, H_2, is not a reactant or a product in these reactions. Other substances are the donors or acceptors of hydrogen atoms.

The oxidation of the carbon-containing compounds we eat provides the energy necessary to sustain life.

A slightly different way of viewing reaction 12-10 is to say that it involves ethane losing electrons. Let us see why. A hydrogen atom contains an electron. Thus, when a compound loses a hydrogen atom, it also loses an electron. Recall that the basic definition of oxidation is a loss of electrons, so considering loss of hydrogen as an oxidation is consistent with oxidation referring to a loss of electrons. Note the fundamental difference between a hydrogen atom, $H\cdot$, and a hydrogen ion, H^+. The H^+ ion does not contain an electron. Loss or gain of H^+ ions is an acid-base reaction, not an oxidation-reduction reaction.

When we use a line of reasoning for carbon-oxygen bonds similar to the one we used for carbon-hydrogen bonds, we see that increasing the number of carbon-oxygen bonds in a compound represents an oxidation of the compound and decreasing the number of carbon-oxygen bonds represents a reduction. For example, in reaction 12-6, $C + O_2 \rightarrow CO_2$, we saw that the carbon atom is oxidized. The structural formula for the product CO_2 in this reaction is $O{=}C{=}O$. Thus four carbon-oxygen bonds are formed in reaction 12-6, and for each carbon-oxygen bond formed the oxidation number of the carbon atom increases from (0) to ($+1$).

Let us now consider an example of an oxidation-reduction reaction that takes place in the human body. In one reaction of an energy-production process in human cells, the compound malic acid is converted to oxaloacetic acid as shown in Figure 12-6. Consider the carbon atom of malic acid that is marked with an asterisk. As you can see, only some of the bonds involving this carbon atom are altered: a carbon-hydrogen bond is lost and a carbon-oxygen bond is formed in its place. Both of these changes represent an oxidation of this carbon atom. Thus we say that malic acid is *oxidized* to form oxaloacetic acid. When we ignore the bonds that remain the same in the oxidation of malic acid, the change in oxidation number of the marked carbon atom is ($+2$). This increase in oxidation number is a consequence of the loss of two bonded hydrogen atoms, each of which contains one negatively charged electron.

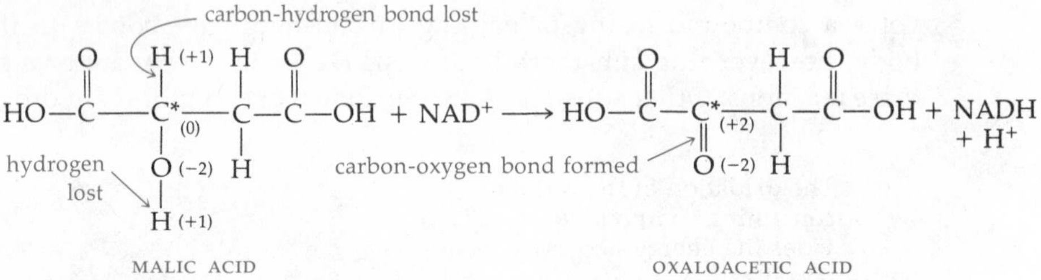

Figure 12-6   The oxidation of malic acid to oxaloacetic acid. The numbers in parentheses are oxidation numbers, which are determined by ignoring the bonds that do not change in the reaction. The C* in malic acid loses one carbon-hydrogen bond and gains one carbon-oxygen bond. The change in the oxidation number of this carbon atom is ($+2$); thus, it is oxidized. NAD$^+$ is the acronym for the coenzyme that is reduced in this reaction.

The oxidation of malic acid to oxaloacetic acid is catalyzed by an enzyme. A substance associated with the enzyme, called a coenzyme, is reduced. We shall discuss coenzymes in more detail in Chapter 23. Coenzymes have complex structures and long names, so we refer to them by acronyms, or abbreviations made up of the initial letters of parts of their chemical names. The acronym for the coenzyme involved in the conversion of malic acid to oxaloacetic acid is NAD$^+$, which is an ion with a single positive charge. When malic acid loses two hydrogen atoms and is oxidized, a hydrogen atom with an additional electron, called a hydride ion, H:$^-$, combines with NAD$^+$ to form a substance called NADH

(12-11) $NAD^+ + H{:}^- \longrightarrow NADH$

The remaining hydrogen, which does not have an electron, is left as a separate proton, H$^+$. Note that two hydrogen atoms, (H· + H·), contain two electrons, and so does the combination of one hydride ion and one proton, (H:$^-$ + H$^+$).

Of course, not all the reactions involving hydrogen and/or oxygen are oxidation-reduction reactions. For example, many reactions that take place in the human body involve a carbon-containing compound that gains or loses water. Malic acid, for instance, is generated in the human body when the compound fumaric acid gains a water molecule (see Figure 12-7). The conversion of fumaric acid to malic acid is *not* an oxidation-reduction reaction. Why? In this reaction, one of the marked carbon atoms in fumaric acid gains a carbon-hydrogen bond and is reduced, but the other marked carbon atom gains a carbon-oxygen bond and is oxidized. Thus, overall, fumaric acid is neither oxidized nor reduced. Likewise, any reaction in which a compound simply gains or loses water is not an oxidation-reduction reaction.

$$HO-\underset{\text{O}}{\overset{\text{O}}{\parallel}}C-\underset{(0)}{\overset{\text{H}}{\overset{|}{C^*}}}=\underset{(0)}{\overset{\text{H}}{\overset{|}{C^*}}}-\underset{\text{O}}{\overset{\text{O}}{\parallel}}C-OH + HOH \longrightarrow HO-\overset{\text{O}}{\overset{\parallel}{C}}-\overset{\text{H}}{\overset{|}{\underset{(+1)}{C^*}}}-\underset{(-1)}{\overset{\text{H}}{\overset{|}{C^*}}}-\overset{\text{O}}{\overset{\parallel}{C}}-OH$$

carbon-oxygen bond formed — O(−2) H(+1) — carbon-hydrogen bond formed

H(+1)

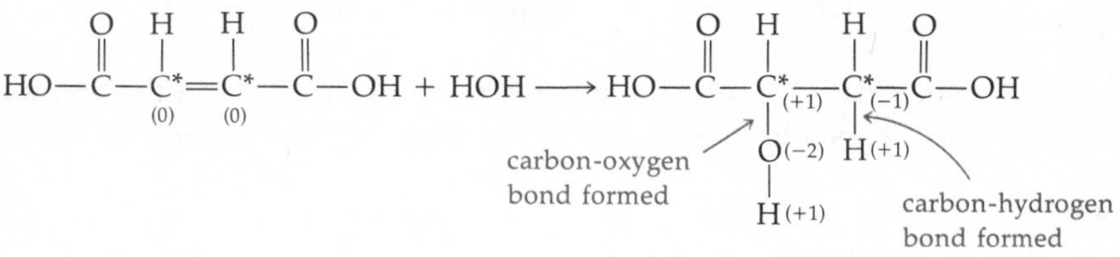

FUMARIC ACID   MALIC ACID

Figure 12-7   The reaction of fumaric acid with water to form malic acid is not an oxidation-reduction reaction. The numbers in parentheses are oxidation numbers, which are determined by ignoring the bonds that do not change in the reaction.

Note that the change in oxidation number for one of the marked carbons is (+1), whereas it is (−1) for the other marked carbon. Thus, the total change in oxidation number for the carbon atoms involved in the reaction is zero.

Exercise 12-4

Identify the bonds formed and the bonds lost by the carbon atoms in the following reactions, and classify the reactions as an oxidation of carbon, a reduction of carbon, or neither.

(a) 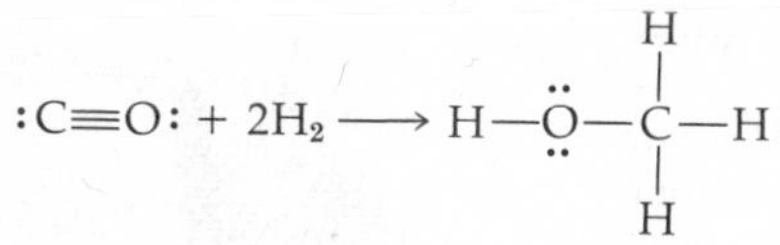 $:C{\equiv}O: + 2H_2 \longrightarrow H-\overset{\text{H}}{\underset{\text{H}}{\overset{|}{\underset{|}{\overset{..}{\underset{..}{O}}-C}}}}-H$

In this reaction, carbon monoxide, CO, is converted to methyl alcohol, CH_3OH.

(b) 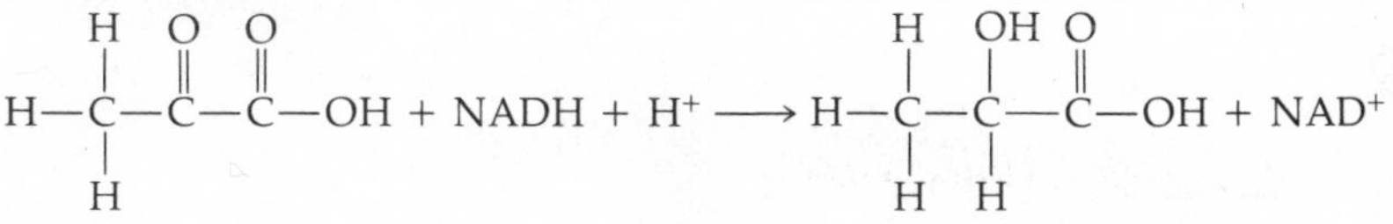 $H-\overset{\text{H}}{\underset{\text{H}}{\overset{|}{\underset{|}{C}}}}-\overset{\text{O}}{\overset{\parallel}{C}}-\overset{\text{O}}{\overset{\parallel}{C}}-OH + NADH + H^+ \longrightarrow H-\overset{\text{H}}{\underset{\text{H}}{\overset{|}{\underset{|}{C}}}}-\overset{\text{OH}}{\underset{\text{H}}{\overset{|}{\underset{|}{C}}}}-\overset{\text{O}}{\overset{\parallel}{C}}-OH + NAD^+$

pyruvic acid   lactic acid

This reaction is involved in energy production in some human cells.

(c) 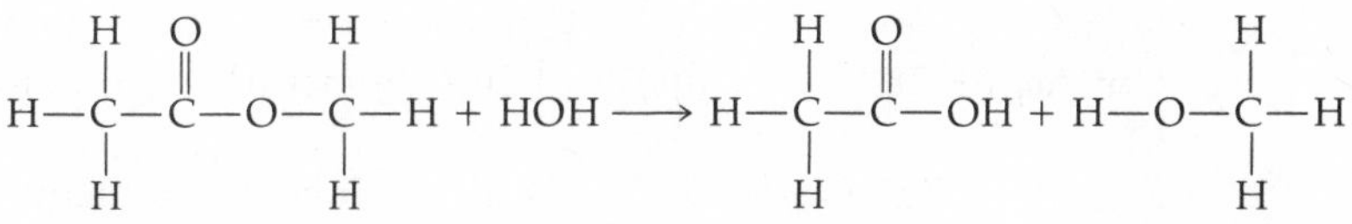 $H-\overset{\text{H}}{\underset{\text{H}}{\overset{|}{\underset{|}{C}}}}-\overset{\text{O}}{\overset{\parallel}{C}}-O-\overset{\text{H}}{\underset{\text{H}}{\overset{|}{\underset{|}{C}}}}-H + HOH \longrightarrow H-\overset{\text{H}}{\underset{\text{H}}{\overset{|}{\underset{|}{C}}}}-\overset{\text{O}}{\overset{\parallel}{C}}-OH + H-O-\overset{\text{H}}{\underset{\text{H}}{\overset{|}{\underset{|}{C}}}}-H$

methyl acetate   acetic acid   methyl alcohol

Optional

12-7   PREDICTING OXIDATION-REDUCTION REACTIONS

Recall (Section 12-3) that the overall reaction for the electrolysis of molten sodium chloride is composed of two physically separate electrode reactions, usually referred to as **half-reactions.** Thus, the electrolysis of NaCl can be represented as

(12-12)

Reduction half-reaction	$Na^+ + e^- \longrightarrow Na$
Oxidation half-reaction	$Cl^- \longrightarrow \frac{1}{2}Cl_2(g) + e^-$

Overall reaction = sum of half-reactions $Na^+ + Cl^- \longrightarrow Na + \frac{1}{2}Cl_2(g)$

Note that electrons are not reactants or products in the overall reaction for the electrolysis of molten NaCl, but that they are present in each half-reaction.

Half-reactions

It is quite useful to consider any overall oxidation-reduction reaction, not just electrolysis reactions, as the sum of two half-reactions. Consider, for example, the following experiment: A piece of Zn metal is put into a 1 M solution of $Cu(NO_3)_2$; some of the Zn goes into solution and forms Zn^{2+} ions, while some of the Cu^{2+} ions come out of solution and form a thin plate of Cu metal on the surface of the remaining Zn metal (see Figure 12-8). The overall chemical reaction that takes place in this experiment is

(12-13) $Zn(s) + Cu^{2+}(aq) \longrightarrow Zn^{2+}(aq) + Cu(s)$

Figure 12-8   If a piece of Zn metal is put into 1 M $Cu(NO_3)_2$ the reaction $Zn(s) + Cu^{2+}(aq) \rightarrow Zn^{2+}(aq) + Cu(s)$ takes place. Some of the Zn goes into solution and forms Zn^{2+} ions, while some of the Cu^{2+} ions come out of solution and form a copper coating on the zinc.

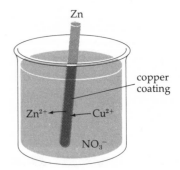

Reaction 12-13 does not take place as two physically separate half-reactions in our experiment. However, it is still very convenient to think of the overall reaction as the sum of the following two half-reactions:

(12-14) $Cu^{2+}(aq) + 2e^- \longrightarrow Cu(s)$ [$Cu^{2+}(aq)$ acting as an oxidizing agent]

(12-15) $Zn(s) \longrightarrow Zn^{2+}(aq) + 2e^-$ [$Zn(s)$ acting as a reducing agent]

When we put both a strip of Zn metal and a strip of Cu metal in a solution con-

Figure 12-9   When both a strip of Zn metal
and a strip of Cu metal are put in a
solution that is both 1 M $Cu(NO_3)_2$
and 1 M $Zn(NO_3)_2$, some of the Zn
goes into solution and forms Zn^{2+}
ions, while some of the Cu^{2+} ions
come out of solution and form a
copper coating on the zinc. However, no reaction takes place that
involves the Cu metal.

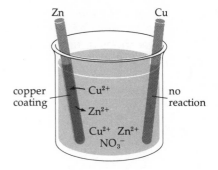

taining both 1 M Cu^{2+} and 1 M Zn^{2+} (see Figure 12-9), reaction 12-13 takes
place as before, but none of the Cu(s) goes into solution and no Zn(s) is plated
out onto the copper strip. The reaction

(12-16) $$Zn^{2+}(aq) + Cu(s) \longrightarrow Zn(s) + Cu^{2+}(aq)$$

in which $Zn^{2+}(aq)$ acts as an oxidizing agent and Cu(s) as a reducing agent, does
not occur. We therefore conclude that $Cu^{2+}(aq)$ is a stronger oxidizing agent
than $Zn^{2+}(aq)$ and that Zn(s) is a stronger reducing agent than Cu(s).

By performing other experiments we can compare the relative strengths of a
large number of oxidizing and reducing agents. It is very convenient to tabulate
the results of these experiments as half-reactions, such as 12-14 and 12-15, in a
standard form in which:

1. For each half-reaction the oxidizing agent is on the left and the reducing
   agent is on the right.

2. The oxidizing agents increase in strength going up the table on the left
   and the reducing agents increase in strength going down the table on the
   right.

The half-reactions 12-14 and 12-15 are shown in this standard form in Table
12-1. The double arrows in Table 12-1 indicate that each half-reaction has a
species capable of acting as an oxidizing agent [such as $Cu^{2+}(aq)$] and a species
capable of acting as a reducing agent [such as Cu(s)].

Table 12-1 The Standard Form for Half-reactions

Increasing strength ↑ as oxidizing agents	$Cu^{2+}(aq) + 2e^- \rightleftharpoons Cu(s)$	Increasing strength ↓ as reducing agents
	$Zn^{2+}(aq) + 2e^- \rightleftharpoons Zn(s)$	

Predicting Oxidation-Reduction Reactions

Tables like 12-1 are of great value in predicting overall oxidation-reduction reactions since when we consider the coupling of two half-reactions, <u>the overall
oxidation-reduction that can occur is the one involving the *stronger* oxidizing
agent and the *stronger* reducing agent as reactants</u>. Thus, using Table 12-1 we
would predict that the overall oxidation-reduction reaction Zn(s) + $Cu^{2+}(aq)$ →
$Zn^{2+}(aq)$ + Cu(s) can occur, since $Cu^{2+}(aq)$ is a stronger oxidizing agent than
$Zn^{2+}(aq)$ and Zn(s) is a stronger reducing agent than Cu(s). Our prediction is
in agreement with the experimental facts.

Table 12-2 is a more extensive tabulation of half-reactions in standard form. Note that the atoms and the electrical charges are balanced in all of the half-reactions in this table. The strength of an ion as an oxidizing or reducing agent depends on its concentration. For the half-reactions in Table 12-2, all the ions have a concentration of 1 M. The strength of a gas as an oxidizing or reducing agent also depends on its partial pressure. For the half-reactions in Table 12-2, all the gases have a partial pressure of 1 atm. Let us use Table 12-2 to predict some overall oxidation-reduction reactions.

Table 12-2 Half-reactions

Increasing strength
as oxidizing agents

1	$F_2(g) + 2e^- \rightleftharpoons 2F^-(aq)$
2	$H_2O_2(aq) + 2H^+(aq) + 2e^- \rightleftharpoons 2H_2O$
3	$MnO_4^-(aq) + 8H^+(aq) + 5e^- \rightleftharpoons Mn^{2+}(aq) + 4H_2O$
4	$Cl_2(g) + 2e^- \rightleftharpoons 2Cl^-(aq)$
5	$O_2(g) + 4H^+(aq) + 4e^- \rightleftharpoons 2H_2O$
6	$Br_2(l) + 2e^- \rightleftharpoons 2Br^-(aq)$
7	$NO_3^-(aq) + 4H^+(aq) + 3e^- \rightleftharpoons NO(g) + 2H_2O$
8	$Ag^+(aq) + e^- \rightleftharpoons Ag(s)$
9	$MnO_4^-(aq) + 2H_2O + 3e^- \rightleftharpoons MnO_2(s) + 4OH^-(aq)$
10	$I_2(s) + 2e^- \rightleftharpoons 2I^-(aq)$
11	$O_2(g) + 2H_2O + 4e^- \rightleftharpoons 4OH^-(aq)$
12	$Cu^{2+}(aq) + 2e^- \rightleftharpoons Cu(s)$
13	$2H^+(aq) + 2e^- \rightleftharpoons H_2(g)$
14	$Pb^{2+}(aq) + 2e^- \rightleftharpoons Pb(s)$
15	$Fe^{2+}(aq) + 2e^- \rightleftharpoons Fe(s)$
16	$Zn^{2+}(aq) + 2e^- \rightleftharpoons Zn(s)$
17	$2H_2O + 2e^- \rightleftharpoons H_2(g) + 2OH^-(aq)$
18	$Al^{3+}(aq) + 3e^- \rightleftharpoons Al(s)$
19	$Na^+(aq) + e^- \rightleftharpoons Na(s)$
20	$Ca^{2+}(aq) + 2e^- \rightleftharpoons Ca(s)$
21	$K^+(aq) + e^- \rightleftharpoons K(s)$

Increasing strength
as reducing agents

EXAMPLE 1 According to Table 12-2, the nonmetals F_2, Cl_2, Br_2, and I_2 are all relatively strong oxidizing agents. This fact agrees with our view of nonmetals as elements that have relatively large tendencies to form negative ions by grabbing electrons from other substances. Also, the relative strength of F_2, Cl_2, Br_2, and I_2 as oxidizing agents is in agreement with the electronegativities of these elements (F_2 is the strongest oxidizing agent, and it has the largest electronegativity.) Consider half-reactions 4 and 6 in Table 12-2. $Cl_2(g)$ is a stronger oxidizing agent than $Br_2(l)$, and $Br^-(aq)$ is a stronger reducing agent than $Cl^-(aq)$. Therefore we predict that the oxidation-reduction reaction

(12-17) $$Cl_2(g) + 2Br^-(aq) \longrightarrow Br_2(l) + 2Cl^-(aq)$$

will occur if $Cl_2(g)$ is bubbled through an aqueous solution containing 1 M Br^- (see Figure 12-10).

Figure 12-10 When $Cl_2(g)$ is bubbled through an aqueous 1 M Br⁻ solution, the reaction

$$Cl_2(g) + 2Br^-(aq) \rightarrow Br_2(l) + 2Cl^-(aq)$$

takes place. This reaction is used as a common test for Br⁻ ion in a solution, since $Br_2(l)$ has a characteristic dark red color.

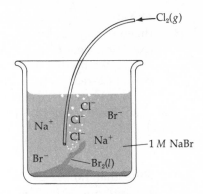

EXAMPLE 2 When we consider half-reactions 8 and 12 in Table 12-2, we predict that an oxidation-reduction reaction can occur, with Ag^+ as the oxidizing agent and Cu as the reducing agent. When we combine any two half-reactions to obtain an equation for an overall oxidation-reduction reaction, both half-reactions must have the *same* number of electrons, since electrons are neither reactants nor products in the overall reaction. In this example we must therefore multiply half-reaction 8 by 2 and then combine it with half-reaction 12 to obtain the following overall oxidation-reduction reaction equation:

(12-18)
$$2Ag^+(aq) + 2e^- \longrightarrow 2Ag(s)$$
$$\underline{Cu(s) \longrightarrow Cu^{2+}(aq) + 2e^-}$$
$$2Ag^+(aq) + Cu(s) \longrightarrow Cu^{2+}(aq) + 2Ag(s) \qquad \text{overall reaction}$$

The following experimental facts are consistent with our prediction (Eq. 12-18). Copper metal placed in a 1 M $AgNO_3$ solution is rapidly coated with metallic silver. No reaction occurs when silver metal is added to a solution of $Cu(NO_3)_2$.

EXAMPLE 3 Based on the relative position of half-reactions 17 and 19 in Table 12-2, we predict, in agreement with the experimental facts, that Na(s) can react with H_2O, liberating $H_2(g)$.

(12-19)
$$2Na(s) + H_2O \longrightarrow H_2(g) + 2Na^+(aq) + 2OH^-(aq)$$

Similarly, we predict that K, Ca, and Al can react with H_2O, liberating $H_2(g)$. K and Ca do react with H_2O. The rates for Na, K, and Ca reactions with H_2O are quite different. No information about *reaction rates* is contained in Table 12-2.

When a strip of Al is put in water, no reaction takes place. This is not in agreement with our predictions. However, any piece of Al that has been exposed to air is coated with a thin layer of aluminum oxide, Al_2O_3. This Al_2O_3 layer prevents the metal Al and the water from coming into contact with one another.

EXAMPLE 4 Based on the position of the half-reactions 14, 15, and 16 with respect to half-reactions 17 and 13, we predict that Pb, Fe, and Zn will not react with H_2O to produce $H_2(g)$, but that $H_2(g)$ will be liberated when any one of these metals is put in acid solution. For example, we predict that

(12-20a)
$$Pb(s) + H_2O \longrightarrow \text{no reaction}$$

(12-20b)
$$Pb(s) + 2H^+(aq) \longrightarrow Pb^{2+}(aq) + 2H_2(g)$$

Again our predictions are in agreement with the experimental facts.

EXAMPLE 5 Using Table 12-2, we predict (1) that Ag and Cu cannot be oxidized by H_2O or by H^+, but (2) that Ag and Cu can be oxidized by the stronger oxidizing agent NO_3^- in an acidic solution (half-reaction 7). Thus, for example, we predict (in agreement with the experimental facts) that the following oxidation-reduction reaction will occur when Cu metal is added to an aqueous nitric acid solution. (Recall that nitric acid, HNO_3, is a strong acid which dissociates completely in water to give H^+ and NO_3^- ions.)

(12-21) $$3Cu(s) + 2NO_3^-(aq) + 8H^+(aq) \longrightarrow 3Cu^{2+}(aq) + 2NO(g) + 4H_2O$$

Note that Eq. 12-21 was obtained by multiplying half-reaction 7 by 2 and half-reaction 12 by 3 (so that both half-reactions involve 6 electrons) before adding the half-reactions.

Exercise 12-5

Using Table 12-2, predict in each case if the substances can react. In the cases where a reaction can occur, write a balanced equation for the overall oxidation-reduction reaction. Assume that the concentration of all ions is 1 M.

(a) $MnO_4^-(aq)$ and I^- (in acidic solution)   (b) $Al^{3+}(aq)$ and $Pb(s)$
(c) $Fe(s)$ and $Cu^{2+}(aq)$ (d) $HNO_3(aq)$ and $Ag(s)$

12-8 REACTIONS OF ELECTROLYTES IN AQUEOUS SOLUTIONS

A very important class of reactions are those involving electrolytes in aqueous solutions. Recall that (1) an electrolyte is a compound that produces a large concentration of ions when it is dissolved in water; (2) electrolytes are soluble ionic compounds (Chapter 4) or strong acids (since they dissociate completely); and (3) a substance is soluble if its solubility is greater than 0.1 M, and a substance is insoluble if its solubility is much less than 0.1 M (Chapter 8). Also recall from Section 8-5 that there is no simple way to explain the fact that some ionic compounds are very soluble in water, whereas other ionic compounds are virtually insoluble. We can, however, make some useful generalizations about the solubility of some ionic compounds based on experimental facts; see Table 12-3.

We now have the background to consider some general types of reactions that can occur when two solutions of different electrolytes are mixed: formation of a solid precipitate; acid-base reactions; oxidation-reduction reactions; and formation of a complex ion.

Table 12-3 Solubility Generalizations for Ionic Compounds

1. All **nitrates** [$NaNO_3$, $Cu(NO_3)_2$, etc.] are soluble.
2. All **chlorides** ($NaCl$, $CuCl_2$, etc.) are soluble, except $AgCl$, $PbCl_2$, and Hg_2Cl_2 (Hg_2Cl_2 contains the polyatomic cation Hg_2^{2+}).
3. All **sulfates** (Na_2SO_4, $CuSO_4$, etc.) are soluble, except $CaSO_4$, $SrSO_4$, $BaSO_4$, Hg_2SO_4, $HgSO_4$, $PbSO_4$, and Ag_2SO_4.
4. All **hydroxides** [$Cu(OH)_2$, $Pb(OH)_2$, etc.] are insoluble, except those of the group IA elements and $Sr(OH)_2$ and $Ba(OH)_2$. $Ca(OH)_2$ is slightly soluble.
5. All **carbonates** ($CuCO_3$, $PbCO_3$, etc.) are insoluble, except those of the group IA elements and $(NH_4)_2CO_3$.

Formation of a Solid Precipitate

A very common type of reaction that can occur when two different electrolyte solutions are mixed is the formation of an insoluble ionic compound. The insoluble ionic compound precipitates out of solution as a solid. For example, when we mix a 0.1 M AgNO$_3$ solution and a 0.1 M NaCl solution, white solid AgCl precipitates out of solution (see Figure 12-11).

(12-22) $Ag^+(aq) + Cl^-(aq) \longrightarrow AgCl(s)$

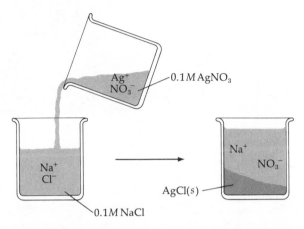

Figure 12-11   When a 0.1 M AgNO$_3$ solution is added to an equal volume of 0.1 M NaCl solution (left), the reaction $Ag^+(aq) + Cl^-(aq) \rightarrow AgCl(s)$ takes place. After the reaction, the silver chloride formed precipitates out of the mixture as a white solid and the solution contains Na$^+$ and NO$_3^-$ ions (right). The concentrations of Ag$^+$ and Cl$^-$ in the solution after the reaction takes place are extremely small, since the solubility of AgCl(s) is extremely low.

We could have predicted that reaction 12-22 would occur when 0.1 M AgNO$_3$ and 0.1 M NaCl are mixed, by using the solubility generalizations in Table 12-3. A AgNO$_3$ solution contains the ions Ag$^+$ and NO$_3^-$ and a NaCl solution contains the ions Na$^+$ and Cl$^-$. When these two solutions are mixed, the *possible* precipitates are NaNO$_3$(s) and AgCl(s). NaNO$_3$(s) is soluble, according to Table 12-3, whereas AgCl(s) is insoluble.

Let us consider two other examples of how we can use the information in Table 12-3 to predict whether or not a precipitation reaction will occur.

EXAMPLE 1 Will a precipitation reaction occur when a 0.1 M Ba(OH)$_2$ solution and a 0.1 M Cu(NO$_3$)$_2$ solution are mixed?

A solution of Ba(OH)$_2$ contains Ba^{2+} and OH$^-$ ions, whereas a solution of Cu(NO$_3$)$_2$ contains Cu^{2+} and NO$_3^-$ ions. When these solutions are mixed, the possible precipitates are Ba(NO$_3$)$_2$(s) and Cu(OH)$_2$(s). Ba(NO$_3$)$_2$(s) is soluble, whereas Cu(OH)$_2$(s) is insoluble. We therefore predict, in agreement with the experimental facts, that when 0.1 M Ba(OH)$_2$ and 0.1 M Cu(NO$_3$)$_2$ are mixed, the following precipitation reaction will occur.

(12-23) $Cu^{2+}(aq) + 2OH^-(aq) \longrightarrow Cu(OH)_2(s)$

Early on a snowy morning in a city parking lot: Imagine that the car tracks represent the movement of ions in an aqueous solution, where the ions may eventually come together and form an insoluble precipitate (cars).

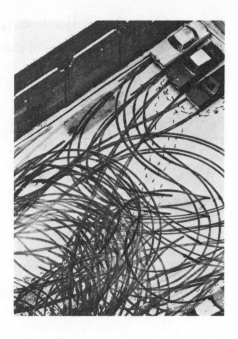

EXAMPLE 2   Will a precipitation reaction occur when a 0.1 M KCl solution and a 0.1 M $(NH_4)_2SO_4$ solution are mixed?

A KCl solution contains K^+ and Cl^-, and a $(NH_4)_2SO_4$ solution contains NH_4^+ and SO_4^{2-} ions. The possible precipitates are $K_2SO_4(s)$ and $NH_4Cl(s)$. According to Table 12-3, both of these compounds are soluble, so we predict that no precipitation reaction will occur. This again agrees with the experimental facts.

Exercise 12-6

In each case, predict whether a precipitate will form when 0.1 M solutions of the two substances are mixed, and write a net ionic reaction for any precipitation reaction. Use Table 12-3.

(a) KOH and $ZnCl_2$ (b) $(NH_4)_2CO_3$ and $Cu(NO_3)_2$

Acid-Base Reactions

Acid-base reactions were discussed extensively in Chapters 10 and 11. Recall that most acid-base reactions that take place to a large extent are either (1) the reaction of a strong acid with a strong base, (2) the reaction of a strong acid with a weak base, or (3) the reaction of a strong base with a weak acid. For example, when a 0.1 M Na_2CO_3 solution (which contains Na^+ ions and the weak base CO_3^{2-}) is mixed with an equal amount of 0.1 M HCl (which contains the ions H_3O^+ and Cl^- since HCl is a strong acid), the following acid-base reaction occurs:

(12-24) $$CO_3^{2-}(aq) + H_3O^+(aq) \longrightarrow HCO_3^-(aq) + H_2O$$

Oxidation-Reduction Reactions

As we discussed in the previous section, Table 12-2 can be used to predict oxidation-reduction reactions. We expect that an oxidation-reduction reaction will occur when an electrolyte solution containing a strong oxidizing agent is mixed with another electrolyte solution containing a strong reducing agent.

For example, we can mix a 0.1 M $KMnO_4$ solution (which contains the ions K^+ and MnO_4^-) with a 0.1 M solution of the strong acid HBr (which contains the ions H^+ and Br^-). We predict, on the basis of the relative positions of half-reactions 3 and 6 in Table 12-2, that the following oxidation-reduction reaction will occur, with MnO_4^- acting as the oxidizing agent and Br^- acting as the reducing agent:

(12-25) $$2MnO_4^-(aq) + 10Br^-(aq) + 16H^+(aq) \longrightarrow 2Mn^{2+}(aq) + 5Br_2(l) + 8H_2O$$

Note that in order to obtain the balanced equation 12-25 we multiplied half-reaction 3 by 2 and half-reaction 6 by 5 (so that both half-reactions have 10 electrons) and then combined them.

Formation of a Complex Ion

A **complex ion** is an ion that contains a central metal cation with two or more anions or molecules bound to the central metal cation. $Fe(CN)_6^{4-}$ and $Ag(NH_3)_2^+$ are two examples of complex ions. Complex ion formation is another type of reaction that can occur when solutions of electrolytes are mixed.

For example, when a 0.1 M $Fe(NO_3)_2$ solution (which contains Fe^{2+} and NO_3^- ions) is mixed with a 0.1 M NaCN solution (which contains Na^+ and CN^- ions), the complex ion $Fe(CN)_6^{4-}$ is formed:

(12-26) $$Fe^{2+}(aq) + 6CN^-(aq) \longrightarrow Fe(CN)_6^{4-}(aq)$$

A large number of complex ions are known, particularly complex ions involving transition metal cations. Complex ions are often formed when solutions of two electrolytes are mixed. Many biologically important substances, including chlorophyll, vitamin B_{12}, and hemoglobin, contain complex ions as essential parts.

12-9   SUMMARY

1. Oxidation is defined as a loss of electrons or an increase in oxidation number. Reduction is defined as a gain of electrons or a decrease in oxidation number.

2. The oxidation number of an atom is equal to the hypothetical "charge" on that atom when, for each covalent bond involving that atom, the shared electrons are assigned to the more electronegative element. Oxidation numbers do not represent actual electrical charges.

3. Oxidation numbers are assigned to atoms according to set rules.

4. In an oxidation-reduction reaction, an oxidizing agent oxidizes another substance while itself being reduced, and a reducing agent reduces another substance while itself being oxidized.

5. For carbon-containing compounds, oxidation generally involves the loss of carbon-hydrogen bonds or the gain of carbon-oxygen bonds, whereas reduction involves the gain of carbon-hydrogen bonds or the loss of carbon-oxygen bonds.

6. **(Optional)** It is possible to predict oxidation-reduction reactions that will occur using a table of half-reactions in standard form.

7. Precipitation reactions, acid-base reactions, oxidation-reduction reactions, and complex-ion formation are four general types of reactions that electrolytes in aqueous solutions undergo.

PROBLEMS

1. Determine the oxidation number of all the atoms in the following:
   (a) HNO_3 (b) CO_3^{2-}
   (c) NH_4^+ (d) Fe_2O_3
   (e) HPO_4^{2-} (f) SO_3

2. Determine which of the following are oxidation-reduction reactions. For the oxidation-reduction reactions, identify the substance oxidized, the substance reduced, the oxidizing agent, and the reducing agent.
   (a) $2Fe(s) + 3Cl_2(g) \rightarrow 2Fe^{3+}(aq) + 6Cl^-(aq)$
   (b) $3Ag_2S(s) + 2NO_3^-(aq) + 8H^+(aq) \rightarrow 6Ag^+(aq) + 3S(s) + 2NO(g) + 4H_2O$
   (c) $HCO_3^-(aq) + OH^-(aq) \rightarrow H_2O + CO_3^{2-}(aq)$
   (d) $3Br_2(l) + 6OH^-(aq) \rightarrow 5Br^-(aq) + BrO_3^-(aq) + 3H_2O$

3. Consider the reaction

$$H-\underset{\underset{H}{|}}{\overset{\overset{H}{|}}{C}}-\overset{\overset{O}{\|}}{C}-H + X + H_2O \longrightarrow H-\underset{\underset{H}{|}}{\overset{\overset{H}{|}}{C}}-\overset{\overset{O}{\|}}{C}-OH + XH_2$$

   (a) Does this reaction represent an oxidation of carbon, a reduction of carbon, or neither?
   (b) Is the substance X an oxidizing agent, a reducing agent, or neither?

4. **(Optional)** Answer the following questions using Table 12-2.
   (a) Can $I_2(s)$ oxidize $Cu(s)$?
   (b) Can $Zn^{2+}(aq)$ oxidize $Fe(s)$?
   (c) Can $MnO_4^-(aq)$ oxidize $NO(g)$ in acid solution?
   (d) Can $Cl^-(aq)$ reduce $H_2O_2(aq)$ in acid solution?

5. **(Optional)** Write balanced equations for the oxidation-reduction reactions in Problem 4.

6. For each of the following, predict whether a precipitate will form when 0.1 M solutions of the two substances are mixed. Use Table 12-3.
   (a) $Pb(NO_3)_2$ and Na_2SO_4 (b) $CuCl_2$ and NH_4NO_3
   (c) $Ba(OH)_2$ and $Fe(NO_3)_3$ (d) Na_2CO_3 and $Zn(NO_3)_2$

7. Write a net ionic reaction for each precipitation reaction in Problem 6.

SOLUTIONS TO EXERCISES

12-1 (a)
$$2K + Br_2 \longrightarrow 2K^+Br^-$$
with "2e⁻" transfer noted, "Br₂ reduced" and "K oxidized"

(b)
$$O_2 + 2Ba \longrightarrow 2Ba^{2+}O^{2-}$$
with "4e⁻" transfer noted, "O₂ reduced" and "Ba oxidized"

(c)
$$Sr + Cl_2 \longrightarrow Sr^{2+}(Cl^-)_2$$
with "2e⁻" transfer noted, "Cl₂ reduced" and "Sr oxidized"

12-2   (a)   CO: oxygen (-2), carbon $(+2)$
      (b) SO_2: oxygen (-2), sulfur $(+4)$
      (c) MnO_4^-: oxygen (-2), manganese $(+7)$
      (d) Cu_2O: oxygen (-2), copper $(+1)$
      (e) Fe^{2+}: iron $(+2)$
      (f) ClO_3^-: oxygen (-2), chlorine $(+5)$

12-3   (a) CH_4 is oxidized, since the oxidation number of carbon goes from (-4) in CH_4 to $(+4)$ in CO_2. O_2 is reduced, since the oxidation number of oxygen goes from (0) in O_2 to (-2) in H_2O. O_2 is the oxidizing agent, and CH_4 is the reducing agent.
      (b) Zn is oxidized, since the oxidation number of zinc goes from (0) in Zn to $(+2)$ in ZnO. HgO is reduced, since the oxidation number of mercury goes from $(+2)$ in HgO to (0) in Hg. HgO is the oxidizing agent, and Zn is the reducing agent.
      (c) C is oxidized, since the oxidation number of carbon goes from (0) in C to $(+4)$ in CO_2. Cu_2O is reduced, since the oxidation number of copper goes from $(+1)$ in Cu_2O to (0) in Cu. Cu_2O is the oxidizing agent, and C is the reducing agent.

12-4   (a) The carbon atom in CO loses two carbon-oxygen bonds and forms three carbon-hydrogen bonds. Therefore this reaction involves a reduction of carbon.
      (b) The central carbon atom in pyruvic acid loses one carbon-oxygen bond and forms one carbon-hydrogen bond. Therefore this reaction involves a reduction of carbon.
      (c) There is no change in either the total number of carbon-oxygen or the total number of carbon-hydrogen bonds, so this reaction is neither an oxidation nor a reduction of carbon.

12-5   **(Optional)**
      (a) Considering the relative positions of half-reactions 3 and 10, we predict that the reaction $2MnO_4^-(aq) + 10I^-(aq) + 16H^+(aq) \rightarrow 2Mn^{2+}(aq) + 5I_2(s) + 8H_2O$ can occur.
      (b) Considering the relative positions of half-reactions 14 and 18, we predict that no reaction can occur.
      (c) Considering the relative positions of half-reactions 12 and 15, we predict that the reaction $Cu^{2+}(aq) + Fe(s) \rightarrow Fe^{2+}(aq) + Cu(s)$ can occur.
      (d) Considering the relative positions of half-reactions 7 and 8, we predict that the reaction $NO_3^-(aq) + 3Ag(s) + 4H^+(aq) \rightarrow NO(g) + 3Ag^+(aq) + 2H_2O$ can occur.

12-6   (a) A solution of KOH contains K^+ and OH^- ions, and a solution of $ZnCl_2$ contains Zn^{2+} and Cl^- ions. When these solutions are mixed, the possible precipitates are $Zn(OH)_2(s)$ and KCl(s). KCl(s) is soluble, whereas $Zn(OH)_2(s)$ is insoluble. We therefore predict that the precipitation reaction, $Zn^{2+}(aq) + 2OH^- \rightarrow Zn(OH)_2(s)$, will occur.
      (b) A solution of $(NH_4)_2CO_3$ contains NH_4^+ and CO_3^{2-} ions, and a solution of $Cu(NO_3)_2$ contains Cu^{2+} and NO_3^- ions. When these solutions are mixed, the possible precipitates are $NH_4NO_3(s)$ and $CuCO_3(s)$. $NH_4NO_3(s)$ is soluble, whereas $CuCO_3(s)$ is insoluble. We therefore predict that the precipitation reaction, $Cu^{2+}(aq) + CO_3^{2-}(aq) \rightarrow CuCO_3(s)$, will occur.

ORGANIC MOLECULES

Overview

An interesting episode of the science fiction television program, *Star Trek*, involved a visit to a planet where the only life form is a creature who eats rocks. This being obtains its energy from chemical reactions involving the element silicon, a major constituent of the rocks it munches. Thus the life of this fictional creature depends on silicon-containing compounds. Now, as far as we know, no such form of life actually exists, or even could exist. All the living systems we know about—human beings, plants, animals, even single-celled organisms—use compounds containing carbon, rather than silicon, as the source of both their energy and their substance. Carbon-containing compounds are the basis of life as we know it.

Carbon atoms are unique in their ability to bond with one another and with other elements. Each carbon atom in virtually every molecule forms a total of four bonds to other atoms. Because of carbon's versatile bonding properties, more than 3 million carbon-containing compounds are known to exist. Chemists are continually synthesizing new ones, and identifying previously unknown carbon-containing compounds derived from natural sources. Oil, natural gas, and coal are natural sources of many carbon-containing compounds, such as those used in the synthesis of nylon, plastics, and the majority of modern medicines.

As a consequence of the importance and the diversity of carbon-containing compounds, chemists have separated the study of these compounds into a special field, called organic chemistry. Chemists call most compounds that contain the element carbon organic compounds.

Proteins, carbohydrates, vitamins, DNA, RNA, and most other substances unique to living systems are very large, complicated organic molecules, often called bio-organic molecules, or simply biomolecules. In the next seven chapters, we shall study the chemistry of some of the simpler organic molecules. Once we have become familiar with some of their chemical properties, we can apply what we have learned to the study of biomolecules.

How can we possibly study the 3 million and more organic compounds? Obviously we cannot investigate them one by one. We must group together compounds with similar chemical properties and study the characteristics that are common to each group.

The classification of organic compounds begins with their separation into three broad groups, based on the elements they contain: (1) those that contain the elements carbon and hydrogen only; (2) those that contain the element oxygen in addition to carbon and hydrogen; and (3) those that contain nitrogen, sulfur, or other atoms in addition to carbon, hydrogen, and possibly oxygen.

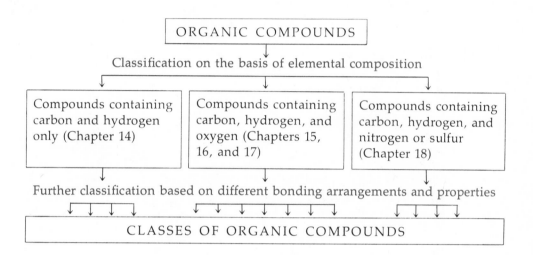

Further classification of organic compounds beyond these three broad groups is based on the different bonding arrangements that are found in compounds composed of the same elements, and on the properties associated with these bonding arrangements. For example, both ethyl alcohol (in alcoholic beverages) and acetic acid (in vinegar) contain the elements carbon, hydrogen, and oxygen. However, the bonding arrangements and the properties of ethyl alcohol and acetic acid are quite different. Ethyl alcohol and acetic acid are therefore placed in different classes of organic compounds. Note the arrows in the diagram, which indicate that we shall be concerned with four classes of organic compounds made up of only carbon and hydrogen, seven classes that have carbon, hydrogen, and oxygen as their constituent elements, and four classes of compounds containing nitrogen or sulfur, in addition to carbon and hydrogen.

In Chapter 13 we shall describe several classes of organic compounds and the basis for naming organic compounds. Then, in Chapters 14, 15, 16, 17, and 18 we shall study the structure and properties of several classes of organic compounds. In Chapter 19 we shall discuss stereoisomers—compounds with the same bonding arrangement but with different shapes, and therefore different properties.

After our study of relatively simple organic compounds, we shall be ready to study the structure and properties of more complex biomolecules.

CHAPTER 13

Organic Chemistry

13-1   INTRODUCTION

Several thousand of the organic compounds that occur in nature have been isolated and studied by chemists. An even larger number of organic compounds have been synthesized in the laboratory. Even a casual glance at the widespread use of synthetic chemicals in modern society attests to the success of the organic chemist in synthesizing new compounds that can benefit society.

Both the isolation of naturally occurring organic compounds and the synthesis of new organic compounds require a detailed understanding of the physical and chemical properties of a relatively small number of structural features that are common to large numbers of organic compounds.

For example, one structural feature found in many organic compounds consists of a nitrogen atom bonded to a carbon atom and to two hydrogen atoms,

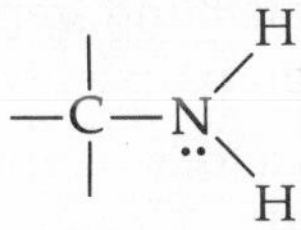

$$-\overset{\textstyle |}{\underset{\textstyle |}{C}}-\overset{\textstyle H}{\underset{\textstyle H}{N}}$$. This group of bonded atoms is called an amine group. The nitrogen

atom of an amine group can act as a Brønsted-Lowry base (Chapter 10) and both the nitrogen atom and the hydrogen atoms of the amine group can participate in hydrogen bonds with water molecules (Chapter 7). These properties of amine groups can be used to separate compounds containing amine groups from compounds with other structural features. Many synthetic drugs contain the amine group, and since the ammonium salt of an amine is usually more soluble in aqueous solutions than the corresponding amine, chemists frequently convert drugs containing amines to the corresponding ammonium salts in order to increase their solubility in body fluids.

In this chapter we shall discuss how chemists classify and name organic compounds. The vast number of organic compounds requires a systematic way of naming them. Learning these names is akin to learning a new language. It is very important that you learn this new language well because, as in any language, slight differences in a word can impart a very different meaning. Finally, we shall discuss the shapes of organic molecules. We will then be ready to study individual classes of organic compounds in subsequent chapters.

A very large number of naturally occurring organic compounds have been isolated from green plants. Petroleum is the starting material for many synthetic organic compounds.

13-2   STUDY OBJECTIVES

After studying the material in this chapter, you should be able to:

1. Define the terms functional group and class of organic compounds.

2. Define the terms isomer, structural formula, structural isomer, functional group isomer, and positional isomer.

3. Tell whether two compounds are, or are not, structural isomers, given structural formulas for the compounds.

4. Draw the structural formula for a simple alkane, given its common or IUPAC name, and for a complex alkane, given its IUPAC name.

5. Write the IUPAC name for an alkane, given its structural formula.

6. Define the term alkyl group and know the structural formulas and names for all of the alkyl groups with one to four carbon atoms.

7. Explain the meaning of the phrase, shape of a molecule.

8. Describe how the bonds of a carbon atom are oriented in space for a molecule in which the carbon atom forms four single bonds.

13-3   FUNCTIONAL GROUPS: CLASSES OF ORGANIC COMPOUNDS

The chemical properties of a compound depend on the elements it contains, on which atoms in a molecule are bonded together, and on whether these bonds are single, double, or triple bonds. Recall from Chapter 4 that:

1. The representation of a molecule that indicates its bonding arrangement is called a structural formula.

2. In structural formulas we use a line between two atoms to represent a bond that consists of a shared electron pair. A double bond, represented by two parallel lines, consists of two shared electron pairs.

3. Each carbon atom in virtually every molecule forms a total of four bonds, whereas every hydrogen atom forms only one bond.

4. A nitrogen atom tends to form a total of three bonds, and an oxygen atom tends to form a total of two bonds.

Consider the structural formulas for the compounds ethene and 2-butene:

$$
\begin{array}{cc}
\overset{\displaystyle H\quad H}{\underset{\displaystyle}{\,|\quad\,|\,}} & \overset{\displaystyle H\quad H\quad H\quad H}{\underset{\displaystyle H\quad\quad H}{\,|\quad\,|\quad\,|\quad\,|\,}} \\
H-C=C-H & H-C-C=C-C-H \\
\text{ethene} & \text{2-butene}
\end{array}
$$

Notice that a carbon-carbon double bond is a structural feature that both ethene and 2-butene have in common. When molecules of two different compounds have structural formulas with a common feature, the two compounds usually have a number of similar chemical properties that are characteristic of this common structural feature. For example, both ethene and 2-butene react with water, in the presence of a suitable catalyst, to give product compounds in which an —OH group is bonded to a carbon atom:

$$H-\overset{\overset{\displaystyle H}{|}}{C}=\overset{\overset{\displaystyle H}{|}}{C}-H + H_2O \xrightarrow{\text{catalyst}} H-\overset{\overset{\displaystyle H}{|}}{\underset{\underset{\displaystyle H}{|}}{C}}-\overset{\overset{\displaystyle H}{|}}{\underset{\underset{\displaystyle OH}{|}}{C}}-H$$

(13-1)

$$H-\overset{\overset{\displaystyle H}{|}}{\underset{\underset{\displaystyle H}{|}}{C}}-\overset{\overset{\displaystyle H}{|}}{C}=\overset{\overset{\displaystyle H}{|}}{C}-\overset{\overset{\displaystyle H}{|}}{\underset{\underset{\displaystyle H}{|}}{C}}-H + H_2O \xrightarrow{\text{catalyst}} H-\overset{\overset{\displaystyle H}{|}}{\underset{\underset{\displaystyle H}{|}}{C}}-\overset{\overset{\displaystyle H}{|}}{\underset{\underset{\displaystyle H}{|}}{C}}-\overset{\overset{\displaystyle H}{|}}{\underset{\underset{\displaystyle OH}{|}}{C}}-\overset{\overset{\displaystyle H}{|}}{\underset{\underset{\displaystyle H}{|}}{C}}-H$$

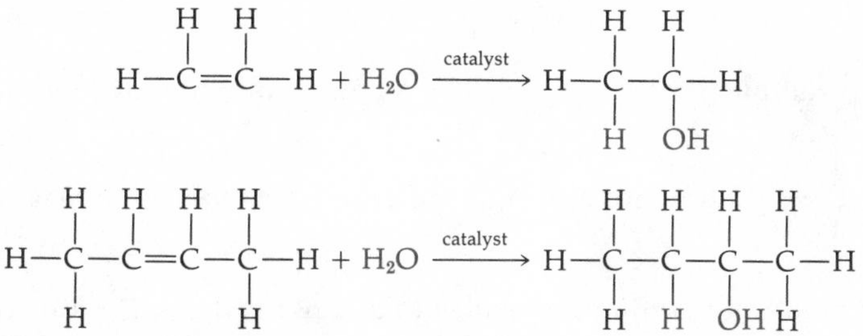

Experiments with many other compounds show that, in general, compounds with a carbon-carbon double bond react with water in a manner similar to reactions 13-1. We can say that the addition of water is a characteristic chemical property of molecules with a carbon-carbon double bond.

Another characteristic chemical property of molecules with a carbon-carbon double bond is their reaction with hydrogen in the presence of a suitable catalyst. The reaction of 2-butene with hydrogen is an example:

(13-2)

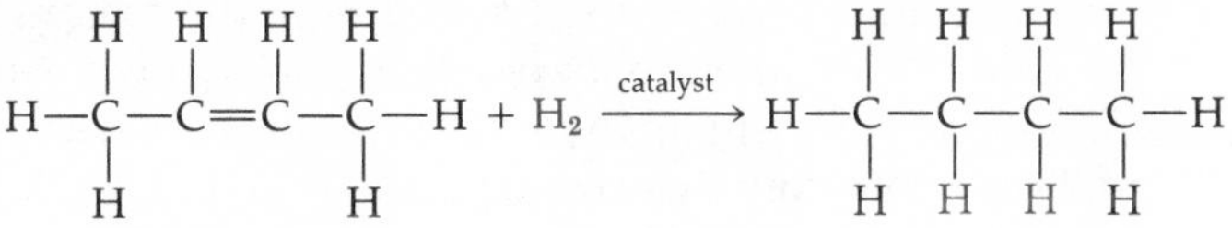

$$H-\overset{\overset{\displaystyle H}{|}}{\underset{\underset{\displaystyle H}{|}}{C}}-\overset{\overset{\displaystyle H}{|}}{C}=\overset{\overset{\displaystyle H}{|}}{C}-\overset{\overset{\displaystyle H}{|}}{\underset{\underset{\displaystyle H}{|}}{C}}-H + H_2 \xrightarrow{\text{catalyst}} H-\overset{\overset{\displaystyle H}{|}}{\underset{\underset{\displaystyle H}{|}}{C}}-\overset{\overset{\displaystyle H}{|}}{\underset{\underset{\displaystyle H}{|}}{C}}-\overset{\overset{\displaystyle H}{|}}{\underset{\underset{\displaystyle H}{|}}{C}}-\overset{\overset{\displaystyle H}{|}}{\underset{\underset{\displaystyle H}{|}}{C}}-H$$

Chemists call a group of atoms and bonds that behave in a similar manner in many different compounds a **functional group.** Functional groups are used to classify organic compounds. A carbon-carbon double bond is one example of a functional group. Organic compounds that contain a carbon-carbon double bond and no other functional group are called **alkenes.** Examples of some of the classes of organic compounds and their corresponding functional groups are given in Table 13-1.

Table 13-1 Some Classes of Organic Compounds and Their Functional Groups

Class	Functional Group	Example Formula	Name
Alkene	$\underset{\diagup}{\overset{\diagdown}{}}C=C\overset{\diagup}{\underset{\diagdown}{}}$	$\underset{H}{\overset{H}{\diagdown}}C=C\underset{\diagdown H}{\overset{\diagup H}{}}$	Ethylene or ethene
Alcohol	$-\overset{|}{\underset{|}{C}}-O-H$	$H-\overset{\overset{H}{|}}{\underset{\underset{H}{|}}{C}}-\overset{\overset{H}{|}}{\underset{\underset{H}{|}}{C}}-O-H$	Ethyl alcohol or ethanol
Carboxylic acid	$-\overset{\overset{O}{\|}}{C}-O-H$	$H-\overset{\overset{H}{|}}{\underset{\underset{H}{|}}{C}}-\overset{\overset{O}{\|}}{C}-O-H$	Acetic acid or ethanoic acid
Ester	$-\overset{\overset{O}{\|}}{C}-O-\overset{|}{\underset{|}{C}}-$	$H-\overset{\overset{O}{\|}}{C}-O-\overset{\overset{H}{|}}{\underset{\underset{H}{|}}{C}}-H$	Methyl formate
Amine	$-\overset{|}{\underset{|}{C}}-N\overset{\diagup}{\underset{\diagdown}{}}$	$H-\overset{\overset{H}{|}}{\underset{\underset{H}{|}}{C}}-\overset{\overset{H}{|}}{\underset{\underset{H}{|}}{C}}-N\underset{\diagdown H}{\overset{\diagup H}{}}$	Ethylamine

For example,

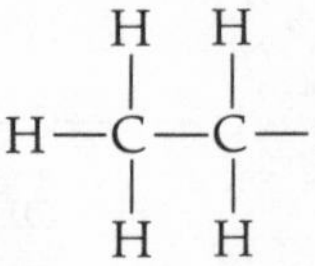

$$H-\underset{\underset{H}{|}}{\overset{\overset{H}{|}}{C}}-\underset{\underset{H}{|}}{\overset{\overset{H}{|}}{C}}-OH,$$

ethanol, and similar molecules that contain the

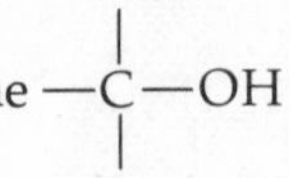

$$-\underset{|}{\overset{|}{C}}-OH$$

functional group are called **alcohols,** whereas

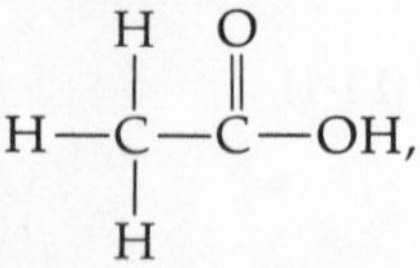

$$H-\underset{\underset{O}{\|}}{\overset{\overset{H}{|}}{C}}-\overset{\overset{O}{\|}}{\underset{H}{\underset{|}{C}}}-OH,$$

acetic acid, and similar molecules that contain the

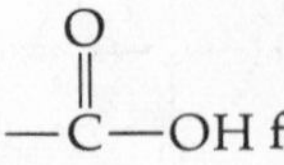

$$-\overset{O}{\overset{\|}{C}}-OH$$

functional group are called **carboxylic acids.**

The simplest class of organic molecules, the alkanes, contain only carbon-carbon and carbon-hydrogen single bonds. Alkanes are reactants in only a few types of reactions (see Chapter 14), so we do not consider alkanes to contain a functional group. Most organic compounds in other classes contain one or more portions that are alkanelike in that they contain only C—C and C—H single bonds. The alkanelike portions of organic compounds are also relatively nonreactive.

In the next few chapters we shall discuss the characteristic physical and chemical properties and the rules used for naming the major classes of simple organic compounds that contain only one functional group. More complicated organic compounds may contain two, three, four, or even more functional groups. A molecule that contains two or more functional groups can usually, but not always, undergo all of the characteristic reactions of each of the functional groups. For example, the compound β-hydroxybutyric acid, which is present in excessive amounts in the bloodstream and urine of diabetic individuals, has both an alcohol functional group and a carboxylic acid functional group (see Figure 13-1). The compound β-hydroxybutyric acid can undergo the chemical reactions characteristic of *both* alcohols and carboxylic acids.

Figure 13-1   The compound β-hydroxybutyric acid has both an alcohol and a carboxylic acid functional group.

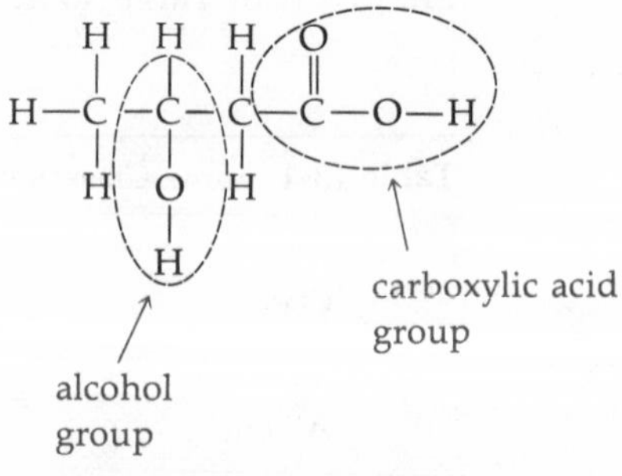

In general, we can think of a complicated organic molecule or biomolecule as being composed of several functional groups and capable of reacting in a variety of ways depending on the nature of the functional groups in the molecule.

13-4   ISOMERISM: STRUCTURAL ISOMERS

There are many different classes of carbon-containing compounds. The basic reason for carbon's versatility, as we mentioned before, lies in the carbon atom's unique ability to bond to other carbon atoms and to a variety of other atoms. Another consequence of carbon's bonding ability is the existence of different carbon-containing compounds with the same molecular formula. Different compounds that have the same molecular formula are called **isomers.** For compounds to be different, (1) there must be some difference in the chemical and physical properties of the compounds, and (2) it must be possible to separate one compound from another. The compounds ethyl alcohol and dimethyl ether, shown in Figure 13-2, are examples of isomers. Ethyl alcohol and dimethyl ether are compounds with very different chemical and physical properties, but they both have the same molecular formula, C_2H_6O.

Figure 13-2   Ethyl alcohol and dimethyl ether are structural isomers.

ethyl alcohol   dimethyl ether

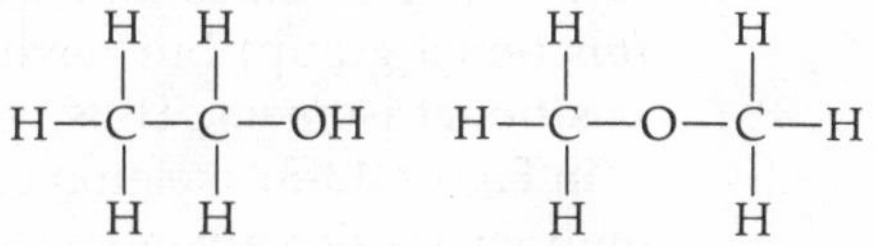

There are two major types of isomers: **structural isomers** and **stereoisomers.** We shall discuss structural isomers in this section. Stereoisomers will be discussed later, in Chapters 14 and 19.

Recall that a structural formula for a molecule indicates which atoms in the molecule are bonded together. We can say that a structural formula is a picture representing the bonding arrangement in the molecule. We can see from the structural formulas given in Figure 13-2 that ethyl alcohol and dimethyl ether have different bonding arrangements. For example, the oxygen atom in ethyl alcohol is bonded to a carbon atom and a hydrogen atom, whereas in dimethyl ether the oxygen atom forms bonds with two different carbon atoms. Ethyl alcohol and dimethyl ether are examples of structural isomers. <u>**Structural isomers** are isomers that have different structural formulas; in other words, they are different compounds that have the same molecular formula but differ in the way the atoms are bonded together</u>. Ethyl alcohol and dimethyl ether are structural isomers that belong to different classes of organic compounds. Structural isomers that have different functional groups, and thus belong to different classes of organic compounds, are referred to as **functional-group isomers.**

There are also many examples of structural isomers that belong to the same class of organic compounds. For example, there are two different alkane compounds with the molecular formula C_4H_{10}. One compound is called normal butane or n-butane, and the other is isobutane. (How alkanes are named is discussed in the next section.) The physical properties of n-butane and isobutane are different (see Figure 13-3), and there are some ways in which the chemical properties of these compounds differ as well.

n-butane	isobutane

Structural formula: Structural formula:

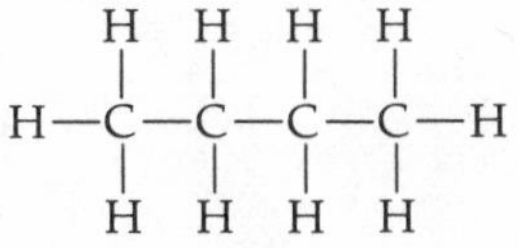

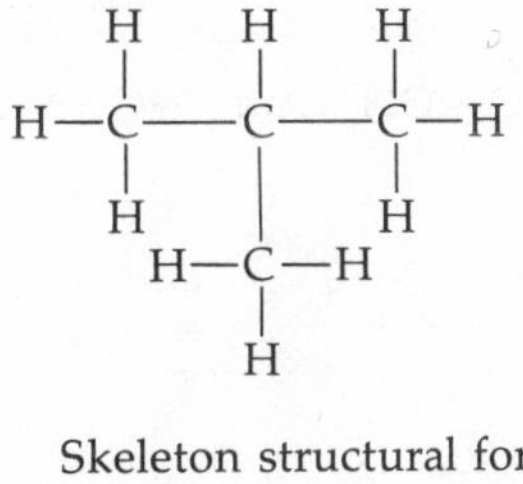

Skeleton structural formula: Skeleton structural formula:

C—C—C—C   C—C—C
                                 |
                                 C

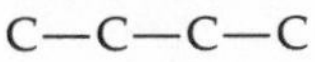

Boiling point: 0°C   Boiling point: −12°C
Melting point: −138°C   Melting point: −159°C
Density (25°C) 0.622 g/ml   Density (25°C) 0.604 g/ml

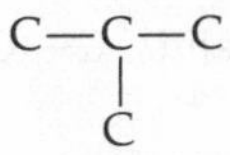

Figure 13-3   The structural isomers n-butane and isobutane have different physi-   cal properties, and some different chemical properties as well.

Notice that the bonding arrangement in n-butane is different than the bonding arrangement in isobutane. Each carbon atom in n-butane is bonded to either one or two other carbon atoms. In isobutane, on the other hand, each carbon atom is bonded to either one or three other carbon atoms. Thus n-butane and isobutane have different structural formulas and are examples of structural isomers. Structural isomers that contain *identical* functional groups (for the alkanes there is no functional group) but have *different* bonding arrangements are referred to as **positional isomers.** Thus, n-butane and isobutane are positional isomers.

In Figure 13-3, skeleton structural formulas in which the hydrogen atoms are omitted are also given for n-butane and isobutane. Skeleton structural formulas are often useful when we are concerned primarily with carbon-carbon bonding. When looking at a skeleton structural formula, we just have to keep in mind that each carbon atom in a molecule forms four bonds and that the bonds that are omitted in a skeleton structural formula are carbon-hydrogen bonds.

We can draw pictures that look different but have the same bonding arrangement. Notice that there is the same bonding arrangement in each of the representations in Figure 13-4, which are all equivalent skeleton structural formulas for isobutane.

Figure 13-4   All of these skeleton structural formulas represent isobutane.

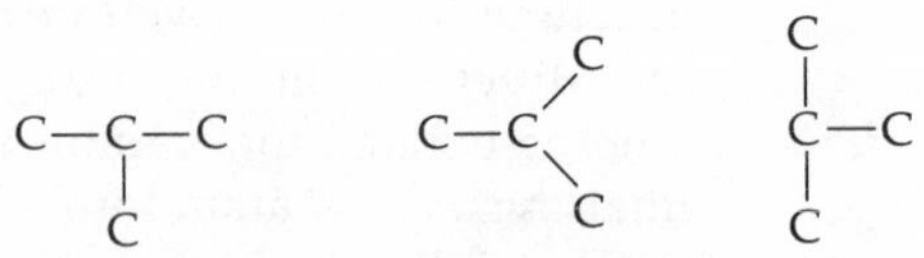

$$C-C-C \qquad C-C{<}{\overset{C}{\underset{C}{}}} \qquad \overset{C}{\underset{C}{C-C}}$$

Another abbreviated way of representing the compounds n-butane and isobutane is: $CH_3{-}CH_2{-}CH_2{-}CH_3$ (n-butane) and $CH_3{-}CH{-}CH_3$ (isobutane).

$$\underset{CH_3}{|}$$

In these representations, called **condensed structural formulas,** subscripts are used to indicate the number of hydrogen atoms bonded to each carbon atom. This is often more convenient than drawing all of the C—H bonds.

Exercise 13-1

Which of the following pairs are structural formulas for structural isomers? Indicate which structural isomers are functional-group isomers and which are positional isomers.

(a) $CH_3{-}CH_2{-}CH_2{-}\overset{O}{\overset{\|}{C}}{-}OH$ and $CH_3{-}\overset{O}{\overset{\|}{C}}{-}O{-}CH_2{-}CH_3$

(b) $CH_2{=}CHCH_2CH_3$ and $CH_3CH{=}CHCH_3$

(c) $CH_3\overset{O}{\overset{\|}{C}}{-}OH$ and $H\overset{O}{\overset{\|}{C}}{-}OCH_2CH_3$

(d) 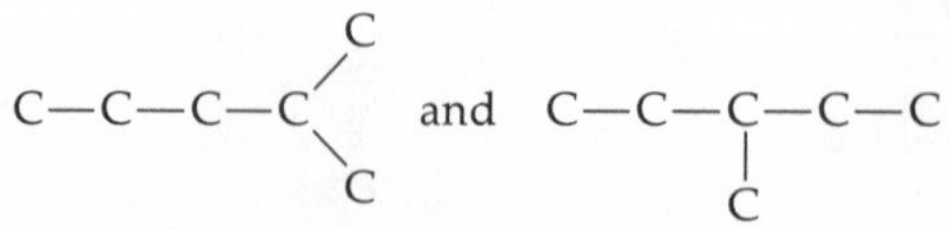 $C{-}C{-}C{-}C{<}{\overset{C}{\underset{C}{}}}$ and $C{-}C{-}\underset{\underset{C}{|}}{C}{-}C{-}C$

(e) 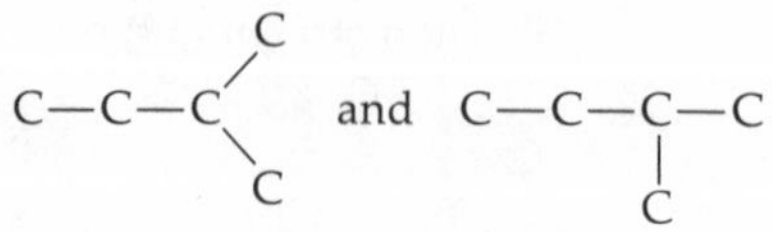 $C{-}C{-}C{<}{\overset{C}{\underset{C}{}}}$ and $C{-}C{-}\underset{\underset{C}{|}}{C}{-}C$

Exercise 13-2

Draw skeleton structural formulas for all of the positional isomers that have the molecular formula C_5H_{12}.

13-5 NAMING ORGANIC MOLECULES: THE ALKANES

The first organic compounds to be isolated and studied were named in a non-systematic fashion. For example, the compound

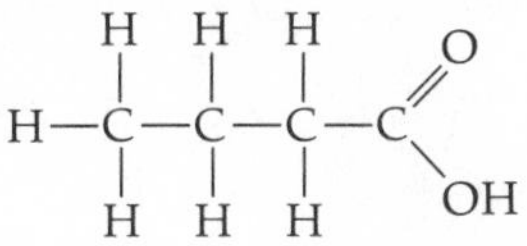

$$
H-\underset{\underset{H}{|}}{\overset{\overset{H}{|}}{C}}-\underset{\underset{H}{|}}{\overset{\overset{H}{|}}{C}}-\underset{\underset{H}{|}}{\overset{\overset{H}{|}}{C}}-C\overset{\nearrow O}{\underset{\searrow OH}{}}
$$

is commonly called butyric acid because it was originally found in rancid butter (the Latin word for butter is *butyrum*). As more and more organic compounds were discovered, attempts were made to develop a systematic method for naming them. At the present time the system devised by the International Union of Pure and Applied Chemistry (IUPAC) is universally accepted by chemists. The systematic IUPAC names for many organic compounds are, however, very long and cumbersome. For this reason we often use the less formal common names for many compounds.

Naming Alkanes

The alkanes comprise the simplest class of organic compounds; they contain only C—C and C—H single bonds. The condensed structural formulas and common names for the unbranched alkanes containing one to 10 carbon atoms are given in Table 13-2. An unbranched alkane is an alkane whose structural formula can be written as a single linear chain of carbon atoms. The common names (and also the IUPAC names) for all alkanes end in *-ane*. Notice that the names for the unbranched alkanes with five or more carbon atoms begin with the Greek word for the number of carbon atoms. Also notice that alkanes with four or more carbon atoms have positional isomers. In order to distinguish one positional isomer from another, we can attach prefixes to their names. The prefix *n-* (for normal) is used for an unbranched alkane.

Table 13-2 Unbranched Alkanes

Common Name	Structural Formula	No. of Carbon Atoms	No. of Positional Isomers
Methane	CH_4	1	1
Ethane	$CH_3—CH_3$	2	1
Propane	$CH_3—CH_2—CH_3$	3	1
n-Butane	$CH_3—CH_2—CH_2—CH_3$	4	2
n-Pentane	$CH_3—CH_2—CH_2—CH_2—CH_3$	5	3
n-Hexane	$CH_3—CH_2—CH_2—CH_2—CH_2—CH_3$	6	5
n-Heptane	$CH_3—CH_2—CH_2—CH_2—CH_2—CH_2—CH_3$	7	9
n-Octane	$CH_3—CH_2—CH_2—CH_2—CH_2—CH_2—CH_2—CH_3$	8	18
n-Nonane	$CH_3—CH_2—CH_2—CH_2—CH_2—CH_2—CH_2—CH_2—CH_3$	9	35
n-Decane	$CH_3—CH_2—CH_2—CH_2—CH_2—CH_2—CH_2—CH_2—CH_2—CH_3$	10	75

As we saw in the previous section, *n*-butane has one positional isomer, which is called isobutane (Figure 13-3). There are three alkanes with five carbon atoms. The names of these positional isomers are *n*-pentane, isopentane, and neopentane (see Figure 13-5). Alkanes with more carbon atoms have larger and larger numbers of positional isomers.

Figure 13-5   Positional isomers of pentane.

n-Pentane $CH_3—CH_2—CH_2—CH_2—CH_3$

Isopentane
$$CH_3—\overset{\displaystyle |}{\underset{\displaystyle CH_3}{CH}}—CH_2—CH_3$$

Neopentane
$$CH_3—\overset{\displaystyle CH_3}{\underset{\displaystyle CH_3}{\overset{|}{\underset{|}{C}}}}—CH_3$$

IUPAC Nomenclature

To identify each of the 75 alkanes with 10 carbon atoms by a different prefix would be ridiculous. The use of IUPAC nomenclature makes this unnecessary. The IUPAC name for an organic compound consists of three component parts:

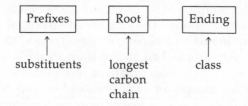

1. The **root** of the IUPAC name. <u>The root specifies the longest continuous chain of carbon atoms in a molecule of the compound</u>. It is derived from the name for the unbranched alkane with that number of carbon atoms. For this reason the names for the unbranched alkanes in Table 13-2 should be committed to memory.

2. The **ending** of an IUPAC name specifies either the class of compounds to which the compound belongs or the major functional group in a molecule of the compound. The ending *-ane* specifies an alkane. (Some other IUPAC endings include *-ene* for alkenes, with a $\overset{\diagdown}{\diagup}C{=}C\overset{\diagup}{\diagdown}$ functional group, and *-ol* alcohols, with a $—\overset{|}{\underset{|}{C}}—OH$ functional group.)

3. **Prefixes** are used to specify the identity and location of atoms or groups of atoms (other than hydrogen), called **substituents,** which are attached to the longest carbon chain. The order, numbering, and punctuation used for prefixes follow a set of rules that we shall discuss shortly.

Let us consider a few examples of how to draw a structural formula for a compound given its IUPAC name.

EXAMPLE 1   Draw a structural formula for pentane.

Begin by identifying the root and ending of this IUPAC name:

pentane

root   ending

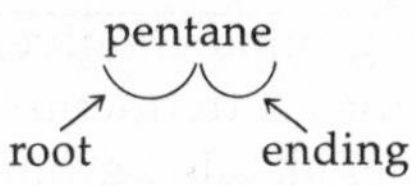

The root of this IUPAC name specifies that the longest continuous chain of carbon atoms is five carbons long. The ending *-ane* specifies that the compound is an alkane. The structural formula for this compound is therefore drawn by first constructing a chain of five carbon atoms,

C—C—C—C—C

then adding hydrogen atoms for a total of four bonds for each carbon atom:

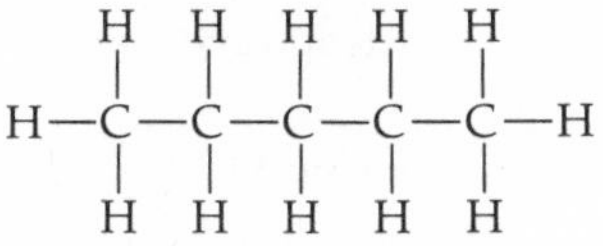

$$H-\overset{\displaystyle H}{\underset{\displaystyle H}{C}}-\overset{\displaystyle H}{\underset{\displaystyle H}{C}}-\overset{\displaystyle H}{\underset{\displaystyle H}{C}}-\overset{\displaystyle H}{\underset{\displaystyle H}{C}}-\overset{\displaystyle H}{\underset{\displaystyle H}{C}}-H$$

The condensed structural formula for a molecule of pentane is

$H_3C-CH_2-CH_2-CH_2-CH_3$

The prefix *n-*, which is used in the common name for this compound, is not needed, because the root of the IUPAC name specifies a single continuous carbon chain.

EXAMPLE 2   Draw a structural formula for methylbutane.

First, identify the components of this IUPAC name:

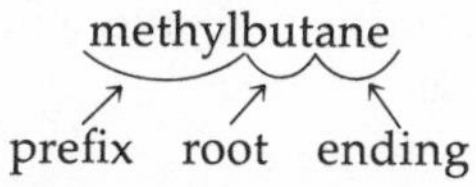

methylbutane

prefix   root   ending

The root and ending of this IUPAC name specify a continuous chain or backbone of four carbon atoms,

C—C—C—C

and that the compound is an alkane. The prefix, methyl-, specifies that a methyl substituent, $-CH_3$, is bonded to this chain. The condensed structural formula for a molecule of methylbutane is therefore

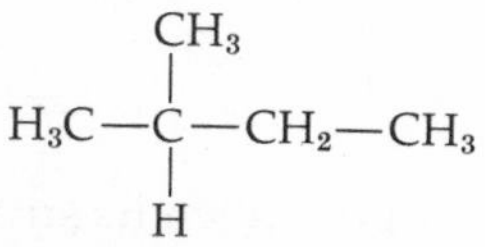

$$H_3C-\overset{\displaystyle CH_3}{\underset{\displaystyle H}{C}}-CH_2-CH_3$$

The common name for this compound is isopentane. The IUPAC name uses the root + ending *butane* because the longest continuous chain of carbon atoms in a molecule of this compound is 4. On the other hand, common names for alkanes are based on the *total* number of carbon atoms in a molecule of the compound (five in this case).

The methyl group refers to the group of atoms that would result if one hydrogen were removed from methane. In general, an **alkyl group** refers to a group of atoms that would result if one hydrogen atom were removed from an alkane. The names for the most common alkyl groups are given in Table 13-3. The names of these common alkyl groups are used frequently and should be committed to memory. Notice that the names of all alkyl groups end in -*yl*.

Table 13-3 Common Alkyl Groups

Structural Formula	Name
CH_3-	Methyl
CH_3-CH_2-	Ethyl
$CH_3-CH_2-CH_2-$	*n*-Propyl
$CH_3-CH-CH_3$ $\quad\quad\mid$	Isopropyl
$CH_3-CH_2-CH_2-CH_2-$	*n*-Butyl
$CH_3-CH_2-CH-CH_3$ $\quad\quad\quad\quad\mid$	*sec*-Butyl (secondary butyl)
CH_3 $\mid$ $CH_3-CH-CH_2-$	Isobutyl
CH_3 $\mid$ CH_3-C-CH_3 $\mid$	*tert*-Butyl or *t*-butyl (tertiary butyl)

EXAMPLE 3   Draw the structural formula for 3,5-diethyl-4-methyloctane.
The root + ending of this IUPAC name specify an alkane whose longest continuous chain has eight carbon atoms, that is,

$$C_1-C_2-C_3-C_4-C_5-C_6-C_7-C_8$$

The prefixes specify ethyl substituents bonded to the third and fifth carbon atoms, and a methyl group bonded to the fourth carbon atom. The structural formula for a molecule of 3,5-diethyl-4-methyloctane is thus

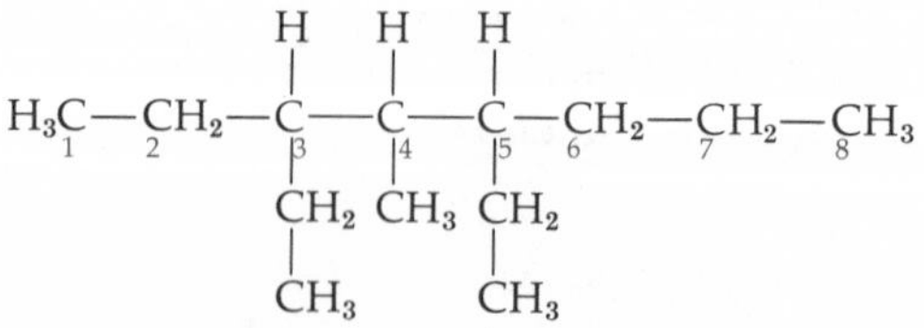

$$H_3C-CH_2-\underset{3}{C}-\underset{4}{C}-\underset{5}{C}-CH_2-CH_2-CH_3$$

with H on C3, C4, C5; CH_2 (then CH_3) on C3, CH_3 on C4, CH_2 (then CH_3) on C5.

In order to write the IUPAC name of a compound given its structural formula, you must learn a few additional rules. The vast majority of alkanes can be named using the following rules:

1. The root of the IUPAC name must specify the longest continuous chain of carbon atoms in the molecule, even though the structural formula for a compound may be drawn so that the longest carbon chain is not a straight line.

For example, in the skeleton structural formula

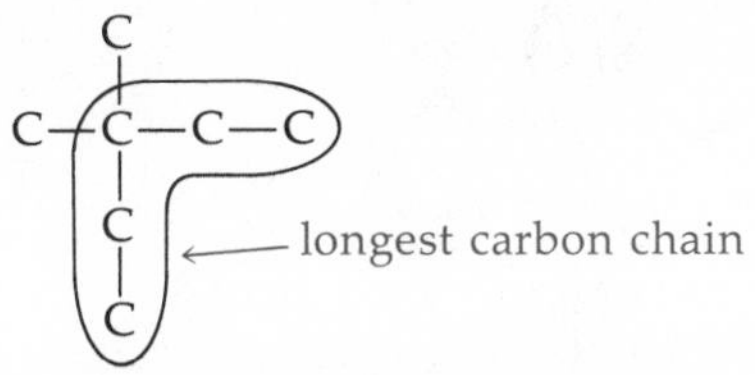

$$
\begin{array}{c}
C \\
| \\
C\!-\!\!\!\boxed{\begin{array}{ccc} C\!-\!C\!-\!C \\ | \\ C \\ | \\ C \end{array}} \quad \longleftarrow \text{longest carbon chain}
\end{array}
$$

the longest continuous chain consists of five carbon atoms.

2. Substituents are identified by name and by a number that indicates the carbon atom of the longest chain to which they are attached. The longest continuous chain must be numbered so that the positions of the substituents will have the lowest possible numbers. The prefixes *di-*, *tri-*, and *tetra-* before the name of a substituent indicate two, three, or four of that substituent in the molecule. For example, *diethyl* indicates two ethyl groups in a molecule.

3. When a molecule contains more than one substituent, the substituents are arranged alphabetically. Notice that di*ethyl* comes before *methyl* in Example 3; the prefixes *di-*, *tri-*, and so on, are not counted when arranging substituents alphabetically.

4. IUPAC names are written as a single word with numbers separated from one another by commas and numbers separated from letters by hyphens (see Example 3). No punctuation or space is used between the name of a substituent and the root name.

The following examples illustrate how these rules are used.

EXAMPLE 4 Write the IUPAC name for the compound with the condensed structural formula

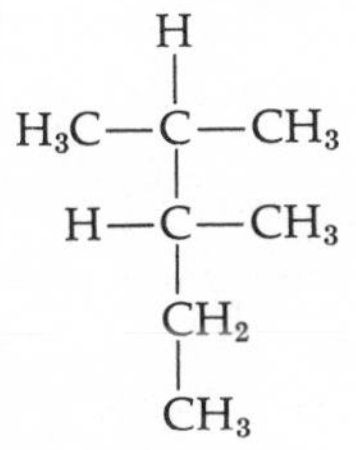

$$
\begin{array}{c}
H \\
| \\
H_3C\!-\!C\!-\!CH_3 \\
| \\
H\!-\!C\!-\!CH_3 \\
| \\
CH_2 \\
| \\
CH_3
\end{array}
$$

Proceed as follows: (a) Identify the longest continuous chain of carbon atoms and thus determine the root of the name. (b) Identify the class of compound and thereby obtain the ending for the IUPAC name. (c) Identify the substituents and number the longest carbon chain so that the substituents have the lowest possible numbers. (d) Name the substituents as prefixes of the IUPAC name. Thus:

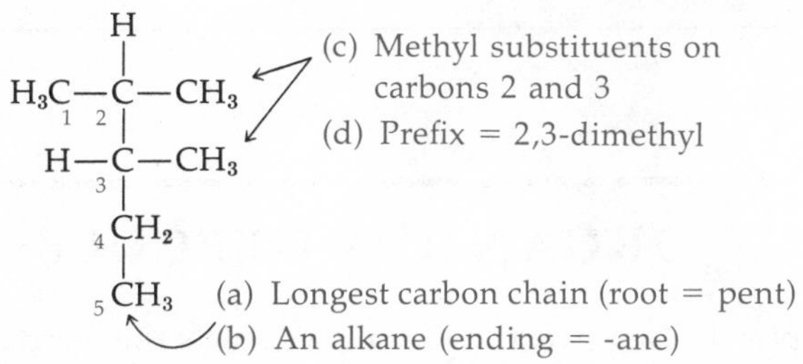

$$
\begin{array}{c}
H \\
| \\
H_3C\!-\!\underset{1\ \ 2}{C}\!-\!CH_3 \\
| \\
H\!-\!\underset{3}{C}\!-\!CH_3 \\
| \\
_4CH_2 \\
| \\
_5CH_3
\end{array}
$$

(c) Methyl substituents on carbons 2 and 3
(d) Prefix = 2,3-dimethyl
(a) Longest carbon chain (root = pent)
(b) An alkane (ending = -ane)

The IUPAC name for this compound is therefore 2,3-dimethylpentane.

EXAMPLE 5 Write the IUPAC name for the compound with the condensed structural formula

$$CH_3$$
$$|$$
$$H_3C—CH \quad CH_3$$
$$HC—CH$$
$$CH_3—CH_2—CH_2 \quad CH_3$$

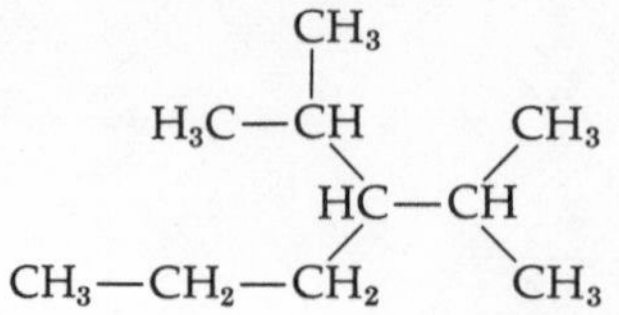

Proceeding as we did in Example 4, we determine the following:

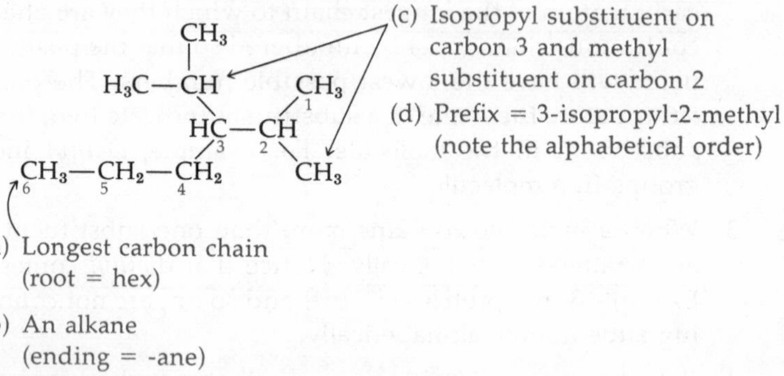

$$CH_3$$
$$|$$
$$H_3C—CH \quad CH_3$$
$$\quad\quad 1$$
$$HC—CH$$
$$3 \quad 2$$
$$CH_3—CH_2—CH_2 \quad CH_3$$
$$6 \quad 5 \quad 4$$

(a) Longest carbon chain
   (root = hex)

(b) An alkane
   (ending = -ane)

(c) Isopropyl substituent on carbon 3 and methyl substituent on carbon 2

(d) Prefix = 3-isopropyl-2-methyl (note the alphabetical order)

Thus, the IUPAC name for this compound is 3-isopropyl-2-methylhexane. (Notice that in this example the carbon chain could be numbered differently, but the name of this compound would be the same.)

Exercise 13-3
Draw condensed structural formulas for the following compounds:
   (a)   3-Ethylpentane
   (b)   3-Ethyl-2-methyloctane
   (c)   2,2,4-Trimethylpentane

Exercise 13-4
Write the IUPAC name for each of the following compounds:

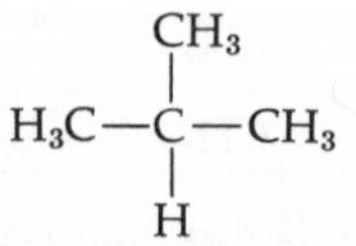

(a) 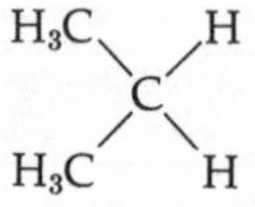
$$H_3C \quad\quad H$$
$$\quad C$$ 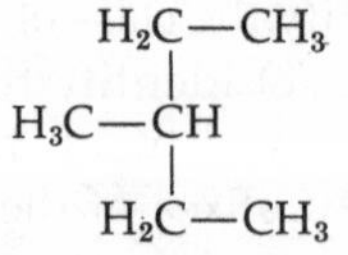
$$H_3C \quad\quad H$$

(b)
$$CH_3$$
$$|$$
$$H_3C—C—CH_3$$
$$|$$
$$H$$

(c)
$$H_2C—CH_3$$
$$|$$
$$H_3C—CH$$
$$|$$
$$H_2C—CH_3$$

(d) 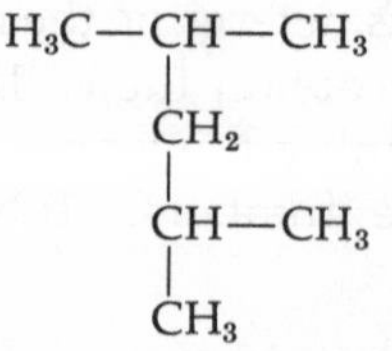
$$H_3C—CH—CH_3$$
$$|$$
$$CH_2$$
$$|$$
$$CH—CH_3$$
$$|$$
$$CH_3$$

13-6   SHAPES OF ORGANIC MOLECULES

Recall from Chapter 4 that a structural formula depicts the bonding arrangement in a molecule, but it is *not* a picture of the shape of a molecule. The **shape** of a molecule is specified by the positions of the centers of all atoms in the molecule, or the orientation in space of the bonds in the molecule. It is important to consider this orientation, since the chemical and physical properties of a compound depend critically on the shape of the molecules of the compound.

The molecular shape is a particularly crucial factor determining the chemical properties of large biomolecules. We shall use the phrase "structure of a molecule" to refer to a combination of both the bonding arrangement in the molecule and the shape of the molecule. Thus a structural formula for a molecule gives only a partial description of the structure of that molecule. Other pictures must be used to indicate the shape of the molecule.

Molecules of different compounds sometimes have the same structural formula but different shapes. This leads to a more subtle, but extremely important, kind of isomerism called stereoisomerism. Two different compounds that have the same molecular formula and the same structural formula but different shapes are called **stereoisomers.** One type of stereoisomerism is discussed in Chapter 14 and another type is discussed in Chapter 19. We shall consider the shape of a few simple molecules here. We shall discuss the shape of more complicated molecules and see numerous examples of the intimate relationship between molecular shape and chemical behavior throughout the remainder of this textbook.

The simplest organic compound is methane, which has the molecular formula CH_4. Recall that experimental evidence shows that the carbon atom in methane is at the center of a regular tetrahedron with a hydrogen atom at each of the four corners (Section 4-13). We can think of the tetrahedral shape of a methane molecule as being determined by four equivalent single bonds on the carbon atom oriented toward the corners of a regular tetrahedron. Chemists use various types of perspective drawings, such as that in Figure 13-6, to picture the shape of methane and other molecules. Most compounds have molecules whose atoms do not all lie in the same plane. Therefore, the best way to visualize the shape of a molecule, and to make correct inferences about those properties of the molecule that depend on its shape, is to construct a three-dimensional model of the molecule. Chemists often use wooden balls and sticks, or plastic tetrahedra and tubing, for this purpose.

Figure 13-6   Methane molecules are tetrahedral in shape.

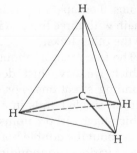

Each bond angle in a molecule of methane with its tetrahedral shape—that is, the angle between any two carbon-hydrogen bonds—is 109.5°. This tetrahedral orientation of the four single bonds about the carbon atom is the typical orientation of bonds about any carbon atom that forms four single bonds. This is true, for example, for each carbon atom in ethane and ethyl alcohol (Figure 13-7). We can think of the shape of ethane or ethyl alcohol as consisting of two tetrahedra joined together by a carbon-carbon single bond. In order to have a full understanding of the shape of ethane or ethyl alcohol, however, we have to know how the two tetrahedra in these molecules are oriented with respect to one another. Is one orientation of the two tetrahedra preferred over all other orientations?

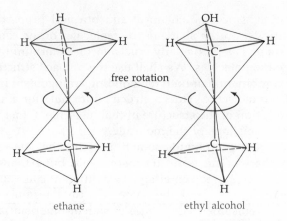

ethane   ethyl alcohol

Figure 13-7   In ethane and ethyl alcohol   rotation about the carbon-carbon
molecules, there is essentially free   single bond.

On the basis of experimental evidence, it has been found that there is very little energy difference between different orientations of the two tetrahedra in ethane or ethyl alcohol. Thus, for example, ethane is a gas at room temperature, and we can think of the ethane molecules as moving about with the two tetrahedra essentially rotating freely about the carbon-carbon single bond. Because of this unrestricted rotation of the two tetrahedra about the carbon-carbon single bond, it is not possible to separate two (or more) kinds of ethane—one form corresponding to one orientation of the two tetrahedra with respect to one another, and the other corresponding to a different relative orientation of the two tetrahedra. Thus ethane does *not* have any stereoisomers.

What we have said about the carbon-carbon bond in ethane is generally true of carbon-carbon single bonds in simple molecules. At room temperature, there is essentially free rotation about any carbon-carbon single bond in a simple molecule. Thus we should not think of molecules containing carbon-carbon single bonds as rigid entities. Rather, we must consider them as having a large degree of flexibility. For example, when we see the skeleton structural formula for butane, C—C—C—C, we should not think of a rigid linear arrangement of atoms. Perhaps a good way to visualize the shape of butane is to think of a chain with three links, with each link capable of moving randomly with respect to the other links.

The free rotation about the carbon-carbon single bond in simple molecules, which we have just described, is not always found in some complicated bonding situations. For example, there is no free rotation about the carbon-carbon single bonds in a class of compounds called cycloalkanes (Chapter 14), which have carbon atoms joined in a ring. There is also an absence of free rotation about the double bond in alkene molecules. In addition, as we shall discuss later, various nonbonding interactions between parts of large biomolecules, such as proteins and nucleic acids, cause these molecules to adopt very definite shapes. The shape of a biomolecule is a primary factor in determining its chemical behavior.

13-7   SYMBOLS, PICTURES, AND REPRESENTATIONS OF MOLECULES

As you know, we use various symbols and pictures to represent molecules. It is essential that you understand what information is contained in a given representation of a molecule. Moreover, it is equally important that you understand what information is *not* conveyed by a given representation. This is particularly important to keep in mind when you consider isomers. Remember that a molecular formula for a compound, such as C_4H_{10}, does not imply a particular bonding arrangement for molecules of that compound. In fact, we have seen that there are two different compounds, the isomers *n*-butane and isobutane, which have the same molecular formula, C_4H_{10}.

Likewise, a structural formula is not a picture of the shape of a molecule. Thus, for example, we must not draw the erroneous conclusion that the carbon atom in a methane molecule is at the center of a square by merely looking at the structural formula

Once we have learned something about the shapes of molecules, however, we should be able to use this information to infer the correct shape of a molecule from its structural formula. For example, when we see the condensed structural formula CH_3CH_3 for ethane, we should be able to infer that the shape of ethane can be described as two tetrahedra joined together by a carbon-carbon single bond (Figure 13-7). In this sense, a structural formula for a molecule can be considered an abbreviation for the structure of the molecule.

13-8   SUMMARY

1. A group of atoms and bonds that behaves in a similar manner in many different compounds is called a functional group. Organic compounds are classified on the basis of their functional groups.

2. The simplest class of organic molecules, the alkanes, contain only C—C and C—H single bonds, are generally not reactive, and are therefore not considered to possess a functional group.

3. A molecule that possesses two or more functional groups can usually undergo all of the reactions characteristic of each of its component functional groups.

4. Isomers are different compounds with the same molecular formula; structural isomers have different structural formulas, whereas stereoisomers have the same structural formulas but different shapes. There are two types of structural isomers: functional-group isomers and positional isomers.

5. The IUPAC method for naming organic compounds is universally accepted by chemists. It consists of a set of rules whereby names are constructed of a root to which an ending and prefixes are attached. The root specifies the longest carbon chain in molecules of that compound. The ending (-*ane* for alkanes) specifies the classification of the compound or the major functional group in the molecule, and the prefixes specify the identity and location of substituents bonded to the longest carbon chain.

6. If one hydrogen is removed from an alkane, the remaining group of atoms is called an alkyl group.

7. The shape of a molecule is specified by the positions of the centers of all of the atoms in the molecule. The typical arrangement of atoms about a carbon atom with four single bonds is tetrahedral.

8. There is essentially free rotation about C—C single bonds, except those C—C single bonds that are part of a ring structure.

PROBLEMS

1. Draw structural formulas for and name all of the alkyl groups with the formula C_4H_9—.

2. Draw structural formulas for each of the following alkanes:
   (a) Isobutane
   (b) *n*-Octane
   (c) Octane
   (d) 2,2-Dimethylnonane
   (e) 3-Ethylpentane

3. Determine which pairs of the following compounds are structural isomers:
   (a) 2,2-Dimethylbutane   and   2-methylpentane
   (b) *n*-Octane   and   4-ethyloctane

   (c) 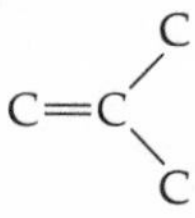 $C=C$ and $C=C$ (skeleton structures)

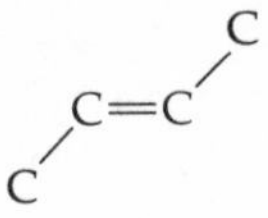

   (d) 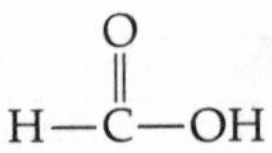 $H-\overset{\overset{\displaystyle O}{\|}}{C}-OH$ and $H-\overset{\overset{\displaystyle O}{\|}}{C}-H$

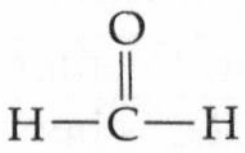

   For those pairs that are structural isomers, indicate whether they are functional-group or positional isomers.

4. Write the IUPAC names for the following alkanes:

   (a) 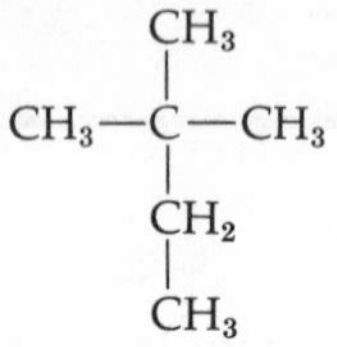 $CH_3-\overset{\overset{\displaystyle CH_3}{|}}{\underset{\underset{\displaystyle CH_3}{|}}{\underset{\overset{\displaystyle CH_2}{|}}{C}}}-CH_3$

   (b) 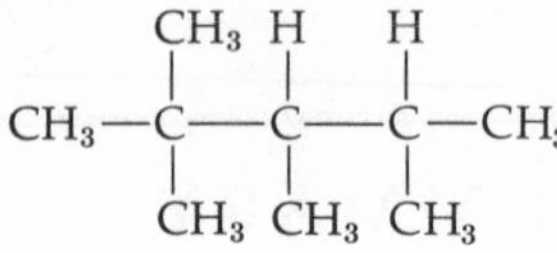 $CH_3-\overset{\overset{\displaystyle CH_3}{|}}{\underset{\underset{\displaystyle CH_3}{|}}{C}}-\overset{\overset{\displaystyle H}{|}}{\underset{\underset{\displaystyle CH_3}{|}}{C}}-\overset{\overset{\displaystyle H}{|}}{\underset{\underset{\displaystyle CH_3}{|}}{C}}-CH_3$

   (c) $CH_3-CH_2-CH_2-CH_3$

   (d) 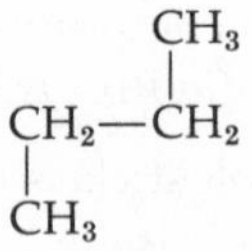 $CH_2-\overset{\overset{\displaystyle CH_3}{|}}{CH_2}$ $\underset{\displaystyle CH_3}{|}$

   (e) 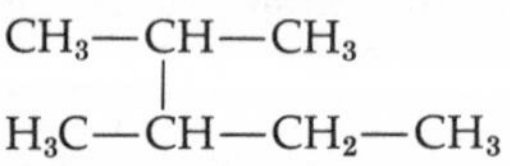 $CH_3-CH-CH_3$
        $H_3C-CH-CH_2-CH_3$

   (f) 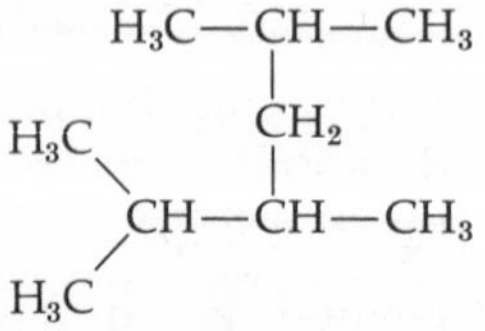 H_3C $CH-CH-CH_3$ with $H_3C-CH-CH_3$, CH_2, H_3C

SOLUTIONS TO EXERCISES

13-1 (a) Both of these molecules have the same molecular formula, $C_4H_8O_2$. However, they have different functional groups and are thus functional-group isomers.
   (b) These are positional isomers, because the $C=C$ double bond is located in different positions in these two molecules.
   (c) These have different molecular formulas and are therefore not isomers.
   (d) These are skeleton structural formulas for two positional isomers with the molecular formula C_6H_{14}.
   (e) These two skeleton structural formulas both represent the same compound.

13-2 (a) C—C—C—C—C

   (b) 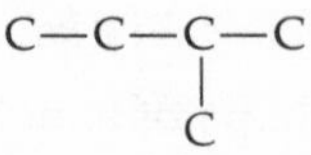 C—C—C—C
             |
             C

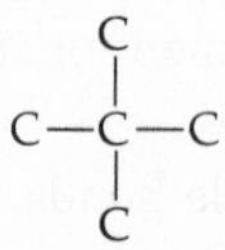

   (c) C—C—C
          |
          C

13-3 (a)

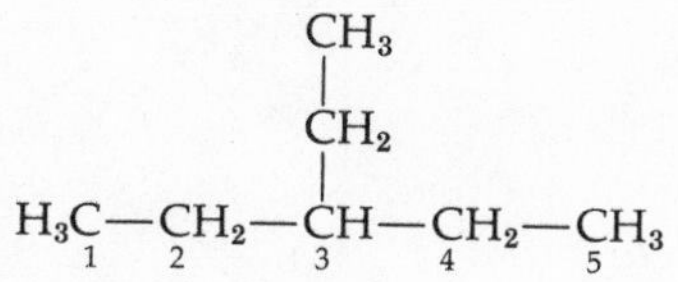

$$
\begin{array}{c}
CH_3 \\
| \\
CH_2 \\
| \\
H_3C-CH_2-CH-CH_2-CH_3 \\
\;\;1\quad\;\;2\quad\;\;3\quad\;\;4\quad\;\;5
\end{array}
$$

Root = pent, so the longest chain has five carbon atoms

Ending = -ane, an alkane

Prefix = 3-ethyl; an ethyl substituent is located on carbon 3 of the longest chain

(b)

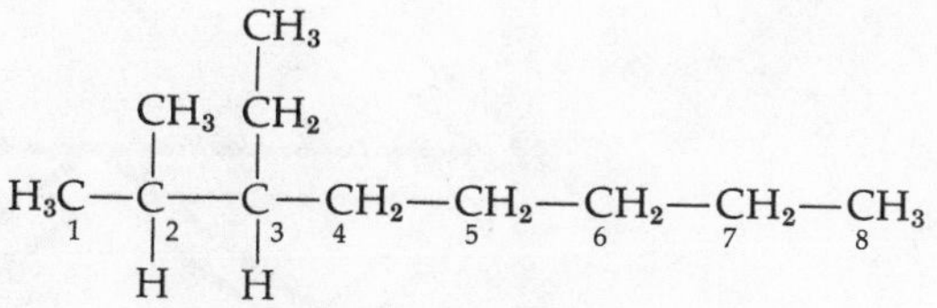

$$
\begin{array}{c}
CH_3 \\
| \\
CH_3\;\;CH_2 \\
|\quad\; | \\
H_3C-C--C-CH_2-CH_2-CH_2-CH_2-CH_3 \\
\;\;1\;\;\;|2\;\;\;|3\quad\;4\quad\;\;\;5\quad\;\;\;6\quad\;\;\;7\quad\;\;8 \\
\;\;\;\;\;H\;\;\;H
\end{array}
$$

(c)

$$
\begin{array}{c}
CH_3\quad\quad CH_3 \\
|\quad\quad\quad\; | \\
H_3C-C-CH_2-CH-CH_3 \\
\;\;1\;\;\;|2\quad3\quad\;4\quad\;5 \\
\;\;\;\;CH_3
\end{array}
$$
Prefix = 2,2,4-trimethyl

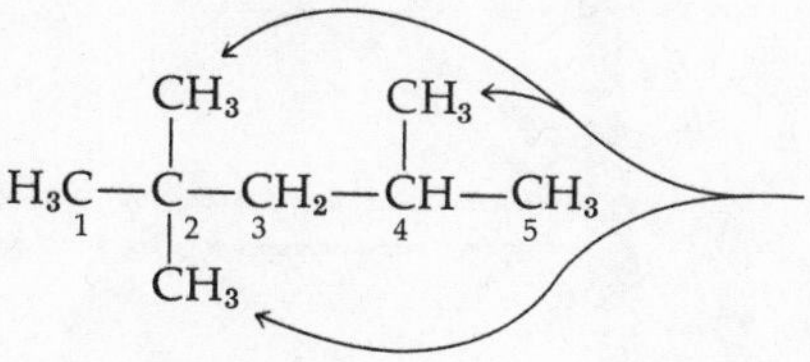

13-4 (a) Longest chain = three carbons; root = prop.
An alkane; ending = -ane.
No substituents.
Thus the IUPAC name is propane.
(b) Root + ending = propane.
A methyl substituent is located on the second carbon but need not be numbered, because if it were on one of the other carbon atoms the longest chain would have four carbons.
Thus the IUPAC name is methylpropane.
(c) Longest carbon chain = 5; root = pent.
An alkane; ending = -ane.
A methyl substituent is located on carbon 3.
The IUPAC name is thus 3-methylpentane.
(d) Root + ending = pentane.
Two methyl substituents located on carbons 2 and 4; prefix = 2,4-dimethyl.
Thus the IUPAC name is 2,4-dimethylpentane.

CHAPTER 14

Hydrocarbons

14-1 INTRODUCTION

The simplest broad group of organic compounds are those that contain only the elements carbon and hydrogen—the hydrocarbons. Hydrocarbons have a number of common physical and chemical properties. For example, they are all nonpolar and thus quite insoluble in water, and they can all burn, a process in which they react vigorously with oxygen.

None of the organic compounds found in the human body are hydrocarbons; these compounds all contain other elements, especially oxygen and nitrogen. But molecules of most of the chemical compounds in human cells do contain hydrocarbonlike parts consisting solely of carbon and hydrogen. Thus, in order to understand the chemical properties of more complex biomolecules, we must have some understanding of the structure, physical properties, and chemical reactivity of hydrocarbons.

There is another important and timely reason to become familiar with the properties and reactivity of hydrocarbons. The hydrocarbons found in natural gas and petroleum play a crucial role in the workings of modern industrial society. We depend on naturally occurring hydrocarbons for fuel and as a source of raw materials for the manufacture of plastics, synthetic rubber, fertilizers, and hundreds of compounds that we use in our daily lives. A large number of synthetic drugs are also manufactured from the hydrocarbons found in petroleum. Since the world's supply of naturally occurring petroleum is limited, a great deal of chemical research is currently aimed at discovering economical alternative sources of hydrocarbons.

Coal, shale oil, and even garbage are potential sources of vast amounts of hydrocarbons. There are major problems, however, with each of these potential sources. For example, although there is a huge amount of coal in the world, with our existing technology most of it is very costly to mine. In addition, much of the coal in nature contains some sulfur. When coal with a high sulfur content is burned, large amounts of atmospheric pollutants are produced. These pollutants are a major source of environmental damage.

The large number of derricks that pump oil from beneath the ocean floor off the coast of Venezuela are a graphic example of our dependence on limited reserves of naturally occurring petroleum.

14-2   STUDY OBJECTIVES

After careful study of the material in this chapter, you should be able to:

1. Classify a hydrocarbon as an alkane, alkene, alkyne, aromatic hydro-carbon, cycloalkane, or cycloalkene, given its structural formula.

2. Explain the use of the symbols R and R'.

3. Draw the structural formula for a hydrocarbon molecule, given its IUPAC name, and vice versa.

4. Draw structural formulas for structural and/or geometric isomers of a hydrocarbon molecule, given its molecular formula.

5. Describe the polarity and solubility of each class of hydrocarbons.

6. Identify the hydrocarbonlike portions of a molecule, given its structural formula.

7. Explain why cyclohexane does not preferentially exist in a planar shape, and why some disubstituted cycloalkanes possess stereoisomers.

8. Describe the shape of an ethene molecule, and explain why some disub-stituted alkenes possess geometric isomers.

9. Draw the structural formulas and give the names of the products of re-actions in which alkenes are (a) reduced, (b) hydrated, or (c) halogenated.

10. Draw the structural formulas for the products obtained in reactions that form carbon-carbon double bonds.

11. Describe the chemical reactivity of aromatic molecules.

12. Define the term polymer, and describe the structure of polyolefins.

14-3   CLASSES OF HYDROCARBONS

Hydrocarbons are subdivided into four classes that have markedly different properties. Recall that alkanes which contain only C—C and C—H single bonds participate in relatively few chemical reactions and do not have a charac-teristic functional group. Alkanes are combustible, however, and are used extensively as a primary source of energy. For example, methane is the major component of natural gas. Although there are no alkanes in human cells, most biomolecules do contain alkanelike portions.

Hydrocarbons that contain one or more C=C double bond functional groups belong to the class of molecules called **alkenes.** In living cells there are many molecules that contain C=C double bonds. Thus, after studying the properties of alkenes, we shall be able to predict those properties of more complex mole-cules that depend on the presence of C=C double bonds. **Alkynes** are a class of hydrocarbons that contain C≡C triple bonds. There are only a handful of com-pounds in nature that contain the C≡C functional group.

Some hydrocarbons contain rings of carbon atoms. **Cycloalkanes** are a subclass of alkanes that contain rings of carbon atoms and C—C and C—H single bonds only. Cycloalkanes are interesting because a number of biomole-cules contain rings of carbon atoms joined by single bonds. Alkenes containing rings of carbon atoms with one or more C=C functional groups also exist; they are called **cycloalkenes.** Compounds that contain a benzene ring belong to a class of hydrocarbons called **aromatic hydrocarbons.**

Exercise 14-1
For each class of hydrocarbons, name the characteristic functional group.

14-4 ALKANES

In the last chapter we discussed the structure and nomenclature of alkanes. Recall that in an alkane each carbon atom forms four covalent bonds either with hydrogen atoms or with other carbon atoms. Since alkanes contain only C—C and C—H single bonds, and each carbon atom in the skeleton formula forms the maximum number of C—H bonds, we say they are **saturated.** You should also recall that the IUPAC name for an alkane consists of: (1) a root derived from the longest continuous chain of carbon atoms, (2) the ending -*ane*, and (3) prefixes that indicate the nature and position of substituent alkyl groups that are bonded to the longest continuous chain of carbon atoms. The rules for naming all classes of organic compounds are reviewed in Essential Skills 8.

Properties of Alkanes

Alkanes are relatively unreactive. The most important exception is that alkanes and all other classes of hydrocarbons can react with oxygen to produce CO_2 and H_2O. The process is known as **combustion** (burning). For example, propane reacts with oxygen as follows:

(14-1)
$$CH_3—CH_2—CH_3 + 5O_2 \longrightarrow 3CO_2 + 4H_2O$$

Reaction 14-1 is highly exothermic (heat releasing), $\Delta H = -531$ kcal/mole, and this is one reason for the use of bottled propane gas as a fuel for heating homes and campers. Larger alkanes, such as octane and its isomers, are also used as fuels (for example, in gasoline). The human body cannot use the combustion of alkanes as a source of energy. In fact, most alkanes are poisonous.

Alkyl groups and alkanelike portions of large molecules are also relatively nonreactive. There are, however, a few important enzyme-catalyzed reactions in the human body in which C—C single bonds are oxidized to produce a C=C double bond. In the majority of these reactions, however, the C—C single bond is very close to a functional group that strongly enhances its reactivity. A typical example is the oxidation of the —CH_2—CH_2— portion of succinic acid:

(14-2)

$$
\begin{array}{ccc}
\underset{\text{succinic acid}}{
\begin{array}{c}
\text{O} \\
\| \\
\text{C—OH} \\
| \\
\text{CH}_2 \\
| \\
\text{CH}_2 \\
| \\
\text{C—OH} \\
\| \\
\text{O}
\end{array}}
& + \text{FAD} \longrightarrow &
\underset{\text{fumaric acid}}{
\begin{array}{c}
\text{O} \\
\| \\
\text{C—OH} \\
| \\
\text{CH} \\
\| \\
\text{CH} \\
| \\
\text{C—OH} \\
\| \\
\text{O}
\end{array}} + \text{FADH}_2
\end{array}
$$

Reaction 14-2 is necessary for energy production in the human body (Chapter 25). In this reaction, succinic acid is oxidized (loses hydrogen) and the complex molecule represented by the symbol FAD (Chapter 23) is reduced (gains hydrogen).

Natural gas, a mixture of simple gaseous alkanes, is burned off at this well in a Libyan oil field.

Transport of natural gas to countries in need of fuel is often not economical.

Not only are alkanes generally unreactive, but they are also nonpolar. Recall that C—C and C—H are nonpolar covalent bonds (Chapter 4). Thus, alkanes cannot form hydrogen bonds with water and are insoluble in it. Modern society is very aware of this fact, as a result of the unfortunate incidences of oil spills in the oceans. Recall that we use the word **hydrophobic** to refer to molecules or parts of molecules that tend to be insoluble in water (Chapter 7). The alkanelike portions of biomolecules are also nonpolar and hydrophobic, and thus biomolecules with large alkanelike portions are not soluble in water.

For example, fatty acids with a large alkanelike portion, such as palmitic acid, which has the structural formula

$$\begin{array}{c}
\;\;\;\;\;\text{H}\;\;\;\text{H}\;\;\;\text{H}\;\;\;\text{H}\;\;\;\text{H}\;\;\;\text{H}\;\;\;\text{H}\;\;\;\text{H}\;\;\;\text{H}\;\;\;\text{H}\;\;\;\text{H}\;\;\;\text{H}\;\;\;\text{H}\;\;\;\text{H}\;\;\;\text{H}\;\;\;\text{O} \\
\;\;\;\;\;|\;\;\;\;\;|\;\;\;\;\;|\;\;\;\;\;|\;\;\;\;\;|\;\;\;\;\;|\;\;\;\;\;|\;\;\;\;\;|\;\;\;\;\;|\;\;\;\;\;|\;\;\;\;\;|\;\;\;\;\;|\;\;\;\;\;|\;\;\;\;\;|\;\;\;\;\;|\;\;\;\;\;\parallel \\
\text{H}-\text{C}-\text{C}-\text{C}-\text{C}-\text{C}-\text{C}-\text{C}-\text{C}-\text{C}-\text{C}-\text{C}-\text{C}-\text{C}-\text{C}-\text{C}-\text{C}-\text{OH} \\
\;\;\;\;\;|\;\;\;\;\;|\;\;\;\;\;|\;\;\;\;\;|\;\;\;\;\;|\;\;\;\;\;|\;\;\;\;\;|\;\;\;\;\;|\;\;\;\;\;|\;\;\;\;\;|\;\;\;\;\;|\;\;\;\;\;|\;\;\;\;\;|\;\;\;\;\;| \\
\;\;\;\;\;\text{H}\;\;\;\text{H}\;\;\;\text{H}\;\;\;\text{H}\;\;\;\text{H}\;\;\;\text{H}\;\;\;\text{H}\;\;\;\text{H}\;\;\;\text{H}\;\;\;\text{H}\;\;\;\text{H}\;\;\;\text{H}\;\;\;\text{H}\;\;\;\text{H}
\end{array}$$

are not water soluble, despite the presence of the polar carboxylic acid functional group on these molecules. We shall see that this feature of fatty acids is important in their function as components of cell membranes (Chapter 26).

Other molecules in the human body, including enzymes, also contain hydrophobic alkyl groups. We shall see that the hydrophobic nature of alkyl groups plays an important role in the structure and properties of enzymes and other proteins (Chapter 21).

Exercise 14-2
Draw structural formulas for the following molecules:

  (a) 3-Methylheptane
  (b) 3-Ethyl-4-*tert*-butyloctane
  (c) 3-Isopropylpentane

Exercise 14-3
Draw the structural formulas and write the IUPAC names for all of the positional isomers of hexane.

Exercise 14-4

Identify the alkanelike portions of the following molecules:

(a) 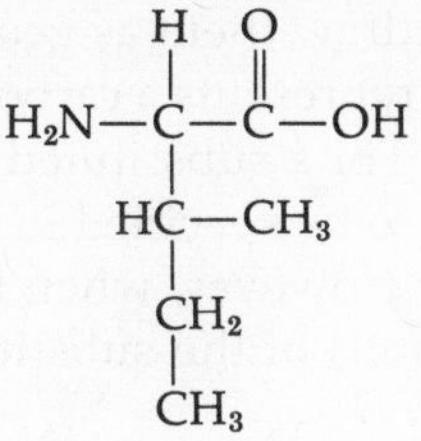
$$H_2N-\overset{\overset{\displaystyle H}{|}}{C}-\overset{\overset{\displaystyle O}{||}}{C}-OH$$
with $\overset{|}{HC}-CH_3$, $\overset{|}{CH_2}$, $\overset{|}{CH_3}$

isoleucine
(an amino acid)

(b) 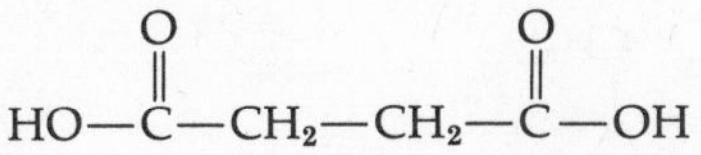
$$HO-\overset{\overset{\displaystyle O}{||}}{C}-CH_2-CH_2-\overset{\overset{\displaystyle O}{||}}{C}-OH$$
succinic acid

(c)
$$H_3C-(CH_2)_{12}-\overset{\overset{\displaystyle H}{|}}{C}=\overset{\overset{\displaystyle}{}}{C}-\overset{\overset{\displaystyle H}{|}}{\underset{\underset{\displaystyle OH}{|}}{C}}-\overset{\overset{\displaystyle H}{|}}{\underset{\underset{\displaystyle NH_2}{|}}{C}}-CH_2OH$$
with H below first C

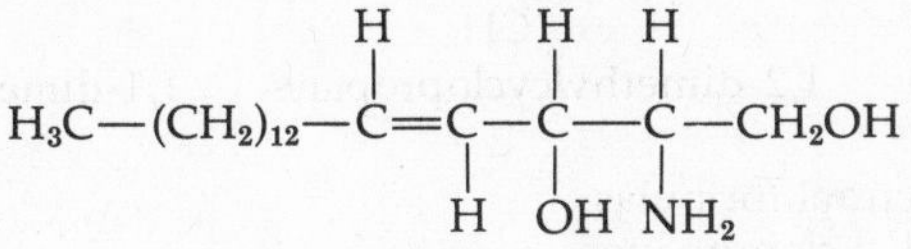

sphingosine
(a component of
cell membranes)

(d) 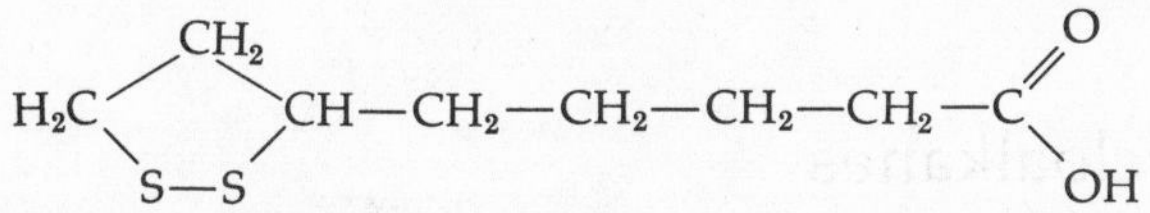
$$H_2C\overset{\diagup}{\underset{S-S}{\diagdown}}CH-CH_2-CH_2-CH_2-CH_2-C\overset{\diagup O}{\diagdown OH}$$
lipoic acid
(a coenzyme)

14-5 CYCLOALKANES

We have just discussed alkanes, which consist of linear chains of carbon atoms. Alkanes with rings of carbon atoms also exist. They are called **cycloalkanes.** The C—C and C—H bonds in cycloalkanes are generally similar to those in alkanes, and like the alkanes, cycloalkanes are not components of human cells. However, several important molecules in human cells contain rings of five or six atoms. The study of cycloalkanes will aid our understanding of these complex molecules.

Structure and Nomenclature

Cycloalkanes are named by placing the prefix *cyclo-* before the name of the corresponding alkane. The structures and names of some simple cycloalkanes are given in Table 14-1. Since an additional C—C single bond is used to close the

Table 14-1 Some Cycloalkanes

Name	Structural Formula	Simplified Representation
Cyclopropane	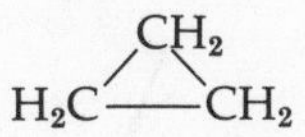 $H_2C\overset{\diagup CH_2 \diagdown}{-\!-\!-}CH_2$	△
Cyclobutane	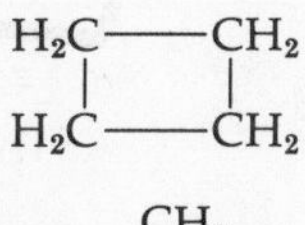 $\begin{array}{l} H_2C\text{---}CH_2 \\ \;\;\mid\qquad\;\mid \\ H_2C\text{---}CH_2 \end{array}$	 ▢
Cyclopentane	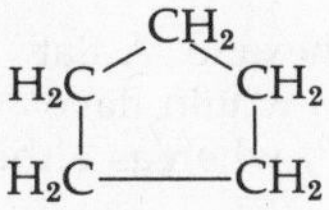 $\begin{array}{c} CH_2 \\ H_2C\diagup\;\;\diagdown CH_2 \\ \mid\qquad\qquad\mid \\ H_2C\text{---}CH_2 \end{array}$	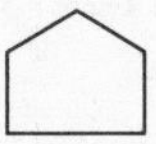 ⬠
Cyclohexane	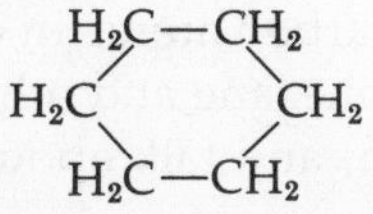 $\begin{array}{c} H_2C\text{---}CH_2 \\ H_2C\diagup\qquad\diagdown CH_2 \\ H_2C\text{---}CH_2 \end{array}$	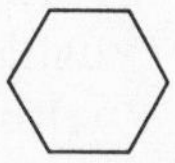 ⬡

ring, a cycloalkane contains two fewer hydrogen atoms than the open-chain alkane with the same number of carbon atoms. Chemists often abbreviate the
structural formulas for cycloalkanes, and just draw them as geometrical figures
(triangles, squares, etc.) in which each corner represents a carbon atom, and the
hydrogen atoms are omitted (see Table 14-1). For a substituted cycloalkane the
position of a single substituent does not need to be specified in the name, since
all of the positions in the ring are equivalent. However, when there are two or
more substituents in a cycloalkane, the positions of the substituents are numbered as for alkanes (see Figure 14-1).

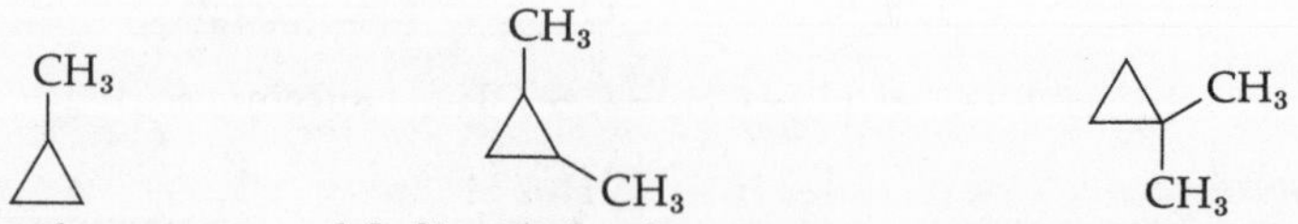

methylcyclopropane   1,2-dimethylcyclopropane   1,1-dimethylcyclopropane

Figure 14-1   The names and structural formulas
for some substituted cyclopropanes.

The Shape of Cycloalkanes

As we saw in Section 13-6, a tetrahedral orientation of bonds with bond angles
of 109.5° is typical for carbon atoms that form four single bonds. The tetrahedral
arrangement of four bonds around a carbon atom is preferred whenever possible because this arrangement is the one with the lowest energy. For some
cycloalkanes, however, a tetrahedral arrangement for all carbon atoms is not
possible. The $\overset{C \frown C}{\underset{C}{}}$ bond angles in cyclopropane must be 60°, and they are
close to 90° in cyclobutane.

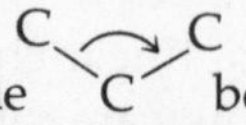

All of the carbon atoms in larger cycloalkanes such as cyclohexane can possess
a tetrahedral arrangement of bonds when all of the carbon atoms do not lie in
the same plane. Cyclohexane with the carbon atoms at the corners of a flat hexagon would have bond angles of 120° (see Figure 14-2a). However, if the carbon
atoms in a molecule of cyclohexane are not all in the same plane, the bond
angles can be the 109.5° required for tetrahedral bonding. Cyclohexane preferentially exists in a nonplanar shape, called the **chair conformation,** where all
of the bonds angles are 109.5° (see Figure 14-2b). The chair conformation of
cyclohexane has the lowest energy. However, cyclohexane is not rigid and can
assume other shapes. Another shape where all of the bond angles are 109.5° is
called the **boat conformation** (see Figure 14-2c). Since there is very little energy
difference between these and other conformations for cyclohexane, it is not
possible to isolate cyclohexane molecules with different shapes. Hence, these
conformations of cyclohexane are *not* stereoisomers.

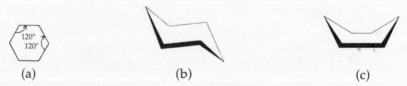

(a)   (b)   (c)

Figure 14-2   The shape of cyclohexane. A flat   the chair conformation (b) or the
hexagonal shape (a) would have   boat conformation (c) has bond
bond angles of 120°, whereas either   angles of 109.5°.

Although all of the carbon atoms in cyclohexane are not in the same plane, we
usually represent cyclohexane and other molecules containing a six-carbon ring
with a planar hexagon, and talk about bonds above and below this plane.

Geometric Isomers of Cycloalkanes

Although there is essentially free rotation around C—C single bonds in alkanes (Section 13-6), this is not possible for cycloalkanes. The ring structure prevents free rotation (see Figure 14-3). For example, rotation of one carbon atom 180° around another carbon atom would require a C—C single bond to be broken, which requires a large amount of energy.

Figure 14-3   Free rotation occurs around C—C single bonds in open-chain alkanes (left) but is not possible around the C—C single bonds in a cycloalkane (right).

free rotation   free rotation not possible

Because of the lack of free rotation around C—C bonds in cycloalkanes, disubstituted cycloalkanes with the same structural formula may have different shapes; that is, they may be stereoisomers. For example, consider a molecule of 1,2-dimethylcyclohexane (Figure 14-4a). The two methyl groups in this molecule can be either on the same side of the ring or on opposite sides. Since the methyl groups cannot rotate from one side to the other, molecules of the two compounds depicted in Figure 14-4a are distinct. These two compounds have different properties and can thus be separated from each other. Since these compounds have the same structural formula but different shapes, they are stereoisomers (Section 13-6). Stereoisomers that differ from each other in the spatial arrangement (the geometry) of substituents are called **geometric** or **cis-trans isomers.** We use the prefix *cis-* to denote the isomer in which both substituents are on the same side of the ring and *trans-* to denote the isomer in which they are on opposite sides. Not all disubstituted cycloalkanes, however, possess cis-trans isomers. For example, there are no geometric isomers of 1,1-dimethylcyclohexane (Figure 14-4b).

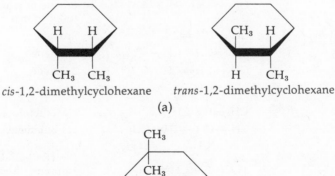

cis-1,2-dimethylcyclohexane   *trans*-1,2-dimethylcyclohexane

(a)

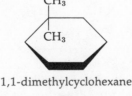

1,1-dimethylcyclohexane

(b)

Figure 14-4   There are two geometric (cis-trans) isomers of 1,2-dimethylcyclohexane (a), but there are no geometric isomers of 1,1-dimethylcyclohexane (b).

Exercise 14-5
The molecular formula for cyclohexane is C_6H_{12}. Why are there two fewer hydrogen atoms in cyclohexane than there are in hexane?

Exercise 14-6
Why is the shape of cyclohexane molecules not simply a planar hexagon?

Exercise 14-7
Draw the structural formula of *trans*-1-ethyl-2-isopropylcyclohexane.

14-6 ALKENES

Hydrocarbons that contain the $C=C$ double bond functional group are called alkenes. Molecules containing $C=C$ double bonds abound in nature and are very important to the human body.

Structure and Nomenclature

In the IUPAC system of nomenclature, an alkene is named after the alkane with the same number of carbon atoms by changing the ending from *-ane* to *-ene*. A numerical prefix is used to indicate the position of the $C=C$ double bond. Hence, the compound $H_2C=CH_2$ is named ethene. The common name for ethene is ethylene. The names and structural formulas of a few alkenes are given in Table 14-2. Notice that there are three different positional isomers that contain four carbon atoms and one $C=C$ double bond. According to IUPAC rules, the positional isomer with the $C=C$ double bond between the first and second carbon atoms is called 1-butene. The other two are 2-butene and 2-methyl-propene. Note that the IUPAC name for $CH_3-CH_2-CH=CH_2$ is 1-butene, and not 3-butene. The IUPAC rules require that the lowest possible numbers be used to specify the location or locations of all double bonds or substituents.

Table 14-2 Some Alkenes

No. of C Atoms	Structural Formula	Name
2	$H_2C=CH_2$	Ethene or ethylene (used to make polyethylene)
3	$CH_3-CH=CH_2$	Propene
4	$CH_2=CH-CH_2-CH_3$	1-Butene
4	$CH_3-CH=CH-CH_3$	2-Butene
4	$CH_3-\overset{\overset{\displaystyle CH_3}{\vert}}{C}=CH_2$	2-Methylpropene
4	$CH_2=CH-CH=CH_2$	1,3-Butadiene
5	$H_2C=\overset{\overset{\displaystyle CH_3}{\vert}}{C}-CH=CH_2$	2-Methyl-1,3-butadiene or isoprene (used to make rubber)

Alkenes that contain two or more double bonds also exist. For these alkenes the ending *-ene* is changed to *-diene* (or *-adiene*) for alkenes with two double bonds, *-triene* (or *-atriene*) for alkenes with three double bonds, and so on. A numerical prefix is also used to specify the position of each double bond. For example, the IUPAC name for $H_2C=CH-CH=CH_2$ is 1,3-butadiene. The compound 2-methyl-1,3-butadiene is commonly called isoprene (Table 14-2). An isoprene-type unit is a common structural feature of some biomolecules, including vitamins A and K (Chapter 26), as well as natural and synthetic rubber (Section 14-9).

Shapes of Alkenes

It has been determined experimentally that all six atoms in the ethene molecule lie in the same plane, with angles of 120° between the three bonds formed by each carbon atom (see Figure 14-5). We can view the shape of ethene as two CH_2

Figure 14-5 The shape of ethene molecules. All six atoms in ethene lie in the same plane, with bond angles of 120°.

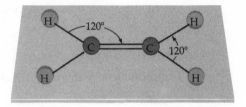

triangles joined together by the C=C double bond. It has also been determined experimentally that a great deal of energy is required to rotate one CH_2 triangle with respect to the other. Thus, the CH_2 triangles are kept in a rigid coplanar arrangement and we say that there is not free rotation about the C=C double bond. As a consequence of this fact, substituted alkenes may have geometric isomers. The simplest example is 2-butene:

$$\underset{cis\text{-2-butene}}{\overset{\displaystyle H_3C \diagdown \qquad \diagup CH_3}{\underset{\displaystyle H \diagup \qquad \diagdown H}{C=C}}} \qquad\qquad \underset{trans\text{-2-butene}}{\overset{\displaystyle H_3C \diagdown \qquad \diagup H}{\underset{\displaystyle H \diagup \qquad \diagdown CH_3}{C=C}}}$$

Two geometric isomers of 2-butene exist: *cis*-2-butene, in which both methyl groups are located on the same side of the C=C double bond; and *trans*-2-butene, in which the methyl groups are on opposite sides. These geometric isomers have somewhat different physical properties and chemical reactivity (see Table 14-3), so they can be separated from each other. Thus there are a total of four isomers (positional and geometric) of butene.

Table 14-3 Properties of Butene Isomers

Name	Structural Formula	Boiling Point (°C)
1-Butene	$H_2C=CH-CH_2-CH_3$	−6
cis-2-Butene	$\overset{H_3C \diagdown \qquad \diagup CH_3}{\underset{H \diagup \qquad \diagdown H}{C=C}}$	+1
trans-2-Butene	$\overset{H_3C \diagdown \qquad \diagup H}{\underset{H \diagup \qquad \diagdown CH_3}{C=C}}$	+4
2-Methylpropene	$H_2C=C\overset{\diagup CH_3}{\diagdown CH_3}$	−7

Properties of Alkenes

The physical properties of alkenes are generally similar to those of the corresponding alkanes. They are nonpolar, insoluble in water, and less dense than water. However, the chemical reactivity of alkenes is quite different than that of alkanes. The $C=C$ double bond functional group can participate in several reactions. Three important reactions of alkenes are reduction, halogenation, and hydration. All of these reactions involve breaking one of the two bonds between the doubly bonded carbon atoms, leaving a $C-C$ single bond, and the formation of an additional single bond between each carbon atom and a hydrogen, oxygen, or halogen atom. These three types of reactions are referred to as **addition reactions,** since the product molecule has an additional atom (or groups of atoms) bonded to each of the carbon atoms of the double bond in the reactant.

CHEMICAL REACTIVITY

The **reduction** of the $C=C$ double bond in an alkene produces an alkane in the general reaction:

(14-3) $$H_2 + R'_1-CH=CH-R'_2 \xrightarrow{\text{catalyst}} R'_1-CH_2-CH_2-R'_2$$

Chemists frequently use the symbol R for an alkyl group. Thus, R can represent $-CH_3$ (methyl), $-CH_2-CH_3$ (ethyl), and so on. The symbol R' denotes an alkyl group or a hydrogen atom. The subscripts 1 and 2 indicate that the R groups may not be identical. Thus, if R'_1 is the ethyl alkyl group and R'_2 is a hydrogen atom, reaction 14-3 becomes

(14-3a) $$H_2 + CH_3-CH_2-CH=CH_2 \xrightarrow{\text{catalyst}} CH_3-CH_2-CH_2-CH_3$$

The reduction of a $C=C$ double bond is also referred to as the **hydrogenation** of a $C=C$ double bond. Hydrogenation is commercially important in the production of gasoline and other fuels, shortenings, and some detergents. Hydrogenation reactions in industry and chemical laboratories require high pressures, high temperatures, and a catalyst such as platinum. On the other hand, the metabolism of fats and carbohydrates in the human body requires the enzyme-catalyzed reduction of $C=C$ double bonds under much milder conditions. For example, one step in the biosynthesis of a fatty acid molecule is the reduction of a double bond in the reaction

(14-4) $$R-CH_2-CH_2-CH=CH-\overset{\overset{\displaystyle O}{\|}}{C}-S-CoA + NADPH + H^+ \xrightarrow{\text{enzyme}}$$

$$R-CH_2-CH_2-\overset{\overset{\displaystyle H}{|}}{\underset{\underset{\displaystyle H}{|}}{C}}-\overset{\overset{\displaystyle H}{|}}{\underset{\underset{\displaystyle H}{|}}{C}}-\overset{\overset{\displaystyle O}{\|}}{C}-S-CoA + NADP^+$$

In reaction 14-4, NADPH represents a coenzyme that donates hydrogen and is oxidized to $NADP^+$. A coenzyme portion of the other reactant is represented as CoA. Another coenzyme, FAD, is frequently used as a hydrogen acceptor in reactions in which CH_2-CH_2 single bonds are oxidized to $HC=CH$ double bonds in the human body. See reaction 14-2 for an example.

Another addition reaction of $C=C$ double bonds is **halogenation.** The reaction between an alkene and Br_2 or Cl_2 results in the addition of two halogen atoms, forming a dibromo- or a dichloro- derivative of the alkene. For example, 2-methylpropene will react with Br_2 to produce 1,2-dibromo-2-methylpropane:

(14-5)

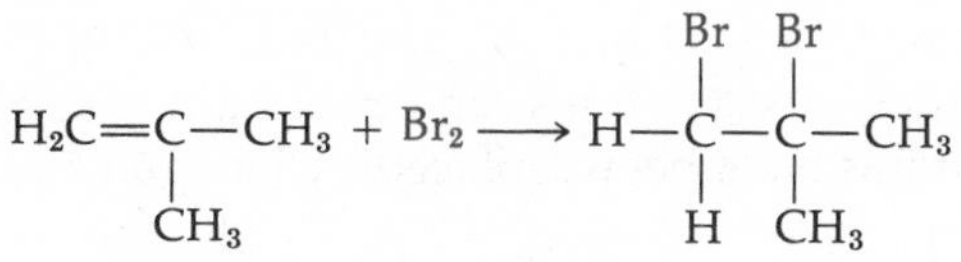

$$H_2C=\overset{\underset{\displaystyle CH_3}{|}}{C}-CH_3 + Br_2 \longrightarrow H-\overset{\overset{\displaystyle Br}{|}}{\underset{\underset{\displaystyle H}{|}}{C}}-\overset{\overset{\displaystyle Br}{|}}{\underset{\underset{\displaystyle CH_3}{|}}{C}}-CH_3$$

Notice in reaction 14-5 that one bromine atom adds to each of the two carbon atoms involved in the C=C double bond. When Br_2 or Cl_2 reacts with any C=C double bond, one halogen atom will bond to each of the two C atoms that formed the double bond.

Alkenes will also participate in addition reactions with hydrogen bromide or hydrogen chloride to produce derivatives that contain a single halogen atom. For example,

(14-6)

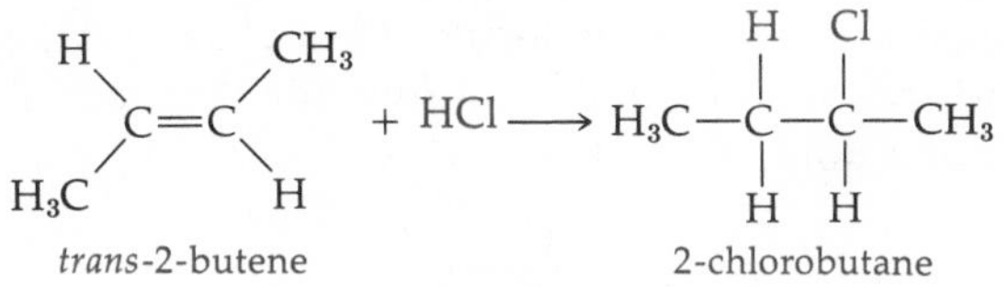

trans-2-butene   2-chlorobutane

and

(14-7)

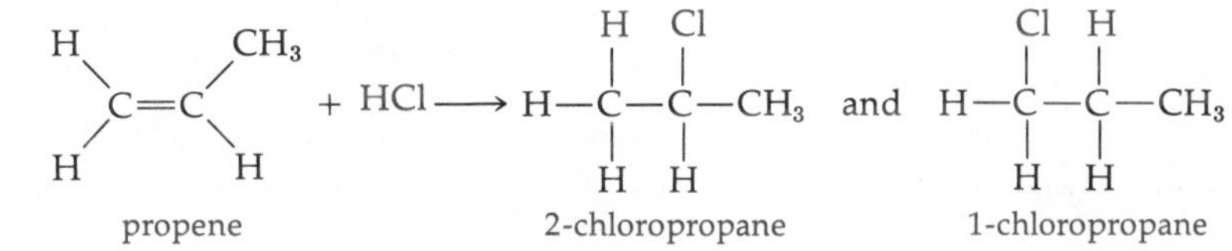

propene   2-chloropropane   1-chloropropane

In reaction 14-6 the only product formed is 2-chlorobutane, since the reactant, *trans*-2-butene, is a symmetric molecule. Propene, on the other hand, is not a symmetric molecule, and the products of reaction 14-7 are a mixture of the positional isomers 2-chloropropane and 1-chloropropane.

It has been determined experimentally that the product of reaction 14-7 is almost entirely 2-chloropropane. Similarly, when any nonsymmetric alkene reacts with HCl, HBr, or certain other compounds such as H_2O, the hydrogen atom usually adds to the carbon atom of the C=C double bond that *already* has the most hydrogen atoms. This generalization is known as **Markovnikov's rule,** after the Russian chemist Vladimir Markovnikov, who studied the addition reactions of alkenes.

For some nonsymmetric alkenes the products of an addition reaction are roughly a 50–50 mixture of positional isomers. Consider the reaction of HBr with *trans*-2-pentene:

(14-8)

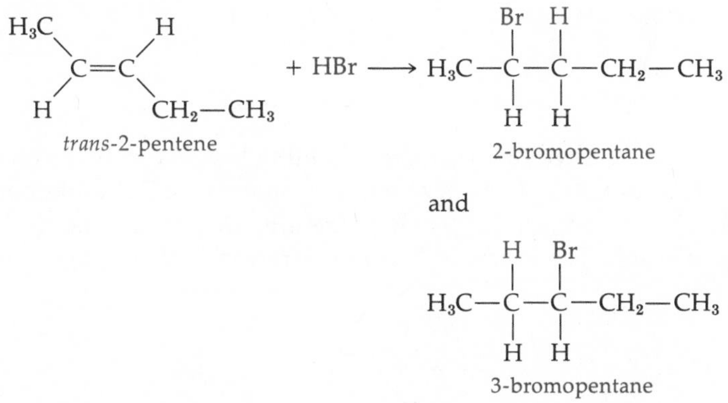

trans-2-pentene   2-bromopentane

and

3-bromopentane

Although *trans*-2-pentene is nonsymmetric, each of the carbon atoms in the C=C double bond has the *same* number of hydrogen atoms (one) bonded to it. Therefore, both 2-bromopentane and 3-bromopentane are formed in roughly equal amounts.

The products of reactions 14-5, 14-6, 14-7, and 14-8 belong to a class of compounds called **alkyl halides.** Alkyl halides can be thought of as alkanes in which one or more hydrogen atoms have been replaced by halogen atoms (Table 14-4).

Table 14-4 Some Important Alkyl Halides

Formula	Common Name	Use
CCl_4	Carbon tetrachloride	Solvent; dry cleaning
$CHCl_3$	Chloroform	Solvent for fats, oils, and rubber
$CH_3—CH_2Cl$	Ethyl chloride	Local anesthetic

A very important addition reaction of $C{=}C$ double bonds is **hydration,** where the components of a water molecule, H— and —OH, are added to the $C{=}C$ double bond, producing an alcohol functional group. The general reaction for hydration of a double bond is

(14-9)

$$\begin{array}{c} R_1' \\ \diagdown \\ C{=}C \\ \diagup \quad \diagdown \\ H \qquad R_2' \end{array} \quad \begin{array}{c} H \\ \diagup \\ \\ \diagdown \\ H \end{array} \quad + H_2O \xrightarrow{\text{catalyst}} \begin{array}{c} H \quad OH \\ | \quad | \\ R_1'{-}C{-}C{-}R_2' \\ | \quad | \\ H \quad H \end{array}$$

The hydration of a nonsymmetric alkene also follows Markovnikov's rule. For example, the hydration of propene yields predominately 2-propanol:

(14-10)

$$\begin{array}{c} H \\ \diagdown \\ C{=}C \\ \diagup \quad \diagdown \\ H \qquad CH_3 \end{array} \quad \begin{array}{c} H \\ \diagup \\ \\ \diagdown \\ \end{array} \quad + H_2O \xrightarrow{\text{catalyst}} \begin{array}{c} H \quad OH \\ | \quad | \\ H{-}C{-}C{-}CH_3 \\ | \quad | \\ H \quad H \end{array}$$

propene   2-propanol

The naming of alcohols is discussed in Chapter 15. Several enzyme-catalyzed reactions in the human body involve hydration of $C{=}C$ double bonds. One example is the hydration of fumaric acid to produce malic acid:

(14-11)

$$\begin{array}{c} O \\ \| \\ C{-}OH \\ | \\ H{-}C \\ \| \\ C{-}H \\ | \\ C{-}OH \\ \| \\ O \end{array} \quad + H_2O \xrightarrow{\text{an enzyme}} \begin{array}{c} O \\ \| \\ C{-}OH \\ | \\ H{-}C{-}OH \\ | \\ H{-}C{-}H \\ | \\ C{-}OH \\ \| \\ O \end{array}$$

fumaric acid   malic acid

In Section 14-4 we saw that $C{=}C$ double bonds can be formed by the oxidation of the saturated parts of some molecules. $C{=}C$ double bonds can also be produced from an alcohol by **dehydration,** that is, by the removal of the elements of water. For example, the dehydration of 2-propanol yields propene:

(14-12)

$$\begin{array}{c} H \quad OH \\ | \quad | \\ H{-}C{-}C{-}CH_3 \\ | \quad | \\ H \quad H \end{array} \xrightarrow{\text{catalyst}} \begin{array}{c} H \\ \diagdown \\ C{=}C \\ \diagup \quad \diagdown \\ H \qquad H \end{array} \begin{array}{c} CH_3 \\ \diagup \\ \\ \diagdown \\ \end{array} + H_2O$$

2-propanol   propene

We shall see several more examples of reactions involving $C{=}C$ double bonds when we study human metabolism.

Exercise 14-8

Draw the structural formulas for the major products of the following reactions:

(a) 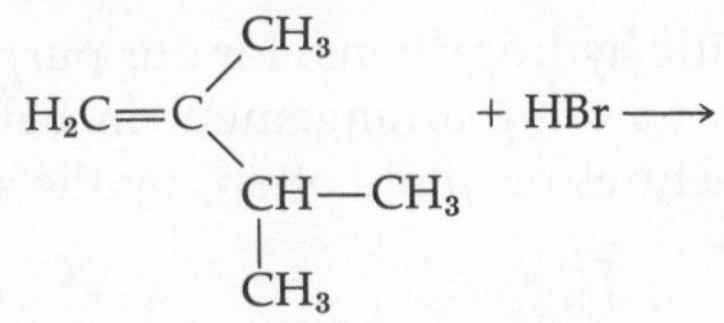 $H_2C=C$ with CH_3 up, $CH-CH_3$ down (with CH_3 below), $+ HBr \longrightarrow$

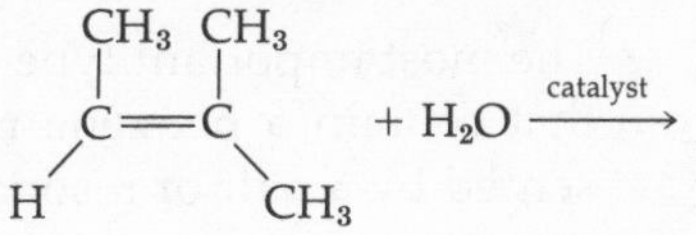

(b) $C=C$ with CH_3 and CH_3 up on left and right carbons, H lower left, CH_3 lower right, $+ H_2O \xrightarrow{catalyst}$

(c) $H_2 + H_2C=C$ with CH_2-CH_3 up and CH_2-CH_3 down $\xrightarrow{catalyst}$

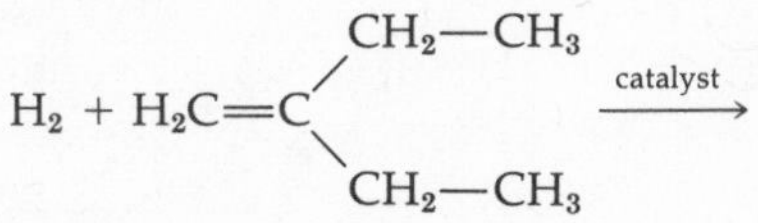

(d) 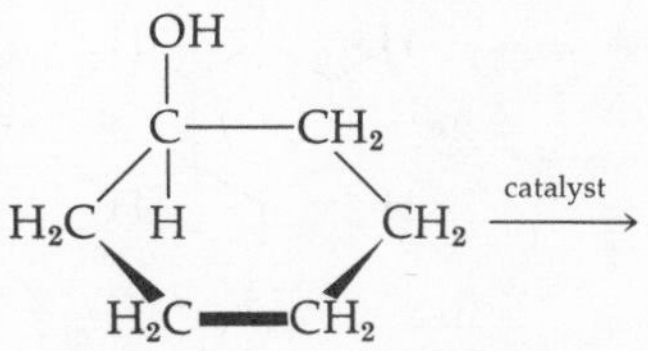 a six-membered ring: top carbon bearing OH, with H_2C, H, CH_2, CH_2, H_2C, CH_2 $\xrightarrow{catalyst}$

If you have difficulty with this type of problem, refer to Essential Skills 9 for some helpful hints.

Exercise 14-9

Draw structural formulas for:

  (a) *trans*-4-Ethyl-3-methyl-3-heptene
  (b) *cis*-3,4-Dimethyl-3-hexene
  (c) *trans*-2-Pentene

14-7 ALKYNES

Hydrocarbons that contain one or more C≡C triple bonds are called **alkynes**. The IUPAC ending for an alkyne is *-yne*. Very few alkynes are found in nature. The simplest alkyne is ethyne, H—C≡C—H, commonly called acetylene, which is used as a fuel for welding torches and other processes that require very high temperatures.

Acetylene-fueled torches are used in a variety of industrial welding operations, but in this instance an artist uses an acetylene torch to construct a metal sculpture.

14-8   AROMATIC HYDROCARBONS

The most important type of aromatic hydrocarbons, for our purposes, are those that contain a benzene ring. The bonding arrangement in benzene is represented by a pair of resonance structures or, more often, by the simple symbols

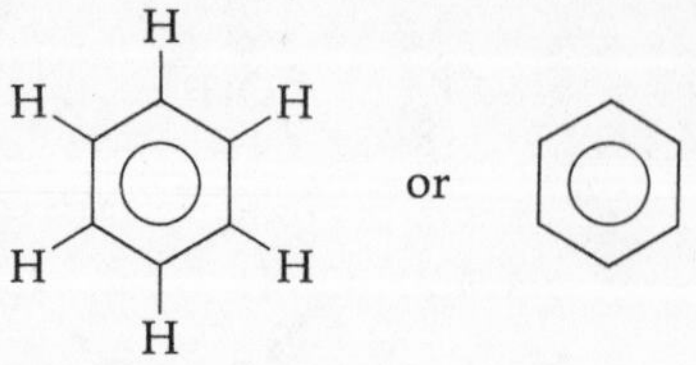

Nomenclature and Structure

Most molecules that contain benzene rings are referred to by their common names, rather than by their IUPAC names. For example, toluene is the name given to the aromatic hydrocarbon that results when a methyl group is substituted for a hydrogen atom in benzene. The names and structures of some substituted aromatic hydrocarbons are given in Table 14-5. Notice that there are only three positional isomers of xylene or dimethyl benzene. Unlike cyclohexane, aromatic molecules are planar. All of the six carbon and six hydrogen atoms in benzene lie in the same plane (see Figure 14-6). Therefore, substituted aromatic compounds such as xylene do not exhibit cis-trans isomerism. When there are two substituent groups on a benzene ring, their positions are

Table 14-5 Some Aromatic Hydrocarbons

Structural Formula	Name
	Benzene
	Toluene (methyl benzene)
	Dimethyl benzenes: o-Xylene (1,2-dimethylbenzene)
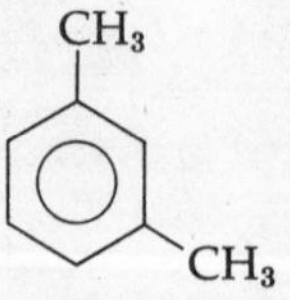	m-Xylene (1,3-dimethylbenzene)
	p-Xylene (1,4-dimethylbenzene)

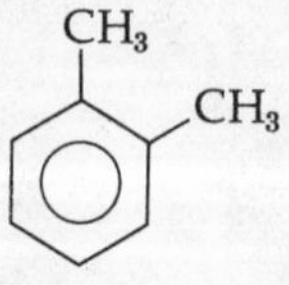

Positional isomers

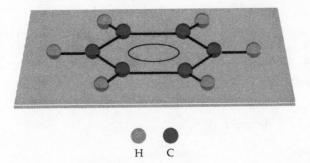

H   C

Figure 14-6   The shape of benzene molecules. All six carbon atoms and all six hydrogen atoms in benzene lie in the same plane.

indicated by the prefixes *o*- (for ortho), *m*- (for meta), and *p*- (for para) or by numbers (see Table 14-5). For a benzene ring with three or more substituents, numbers are used to indicate the positions of the substituents.

The group of atoms obtained by removing one hydrogen from benzene is called a **phenyl** group. One compound that contains a phenyl group is the amino acid phenylalanine (see Figure 14-7).

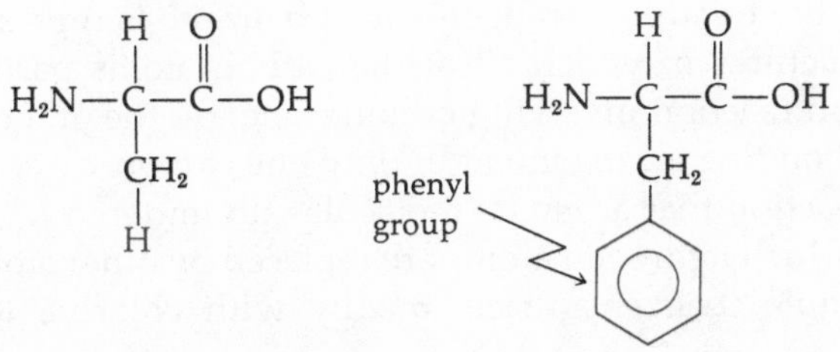

$$H_2N-\underset{\underset{H}{|}}{\overset{\overset{H}{|}}{C}}-\overset{\overset{O}{||}}{C}-OH$$

$$CH_2$$

alanine

$$H_2N-\underset{\underset{CH_2}{|}}{\overset{\overset{H}{|}}{C}}-\overset{\overset{O}{||}}{C}-OH$$

phenyl group

phenylalanine

Figure 14-7   The difference between the amino acids alanine (on the left) and phenylalanine (on the right) is in the presence of a phenyl group in phenylalanine instead of one of the hydrogens in alanine.

Properties of Aromatic Compounds

PHYSICAL PROPERTIES

The physical properties of benzene and other aromatic hydrocarbons are similar to those of alkanes and alkenes. They are nonpolar and thus insoluble in water. The components of proteins and nucleic acids contain aromatic portions that, because of their hydrophobic nature, are excluded from the aqueous environment of cells. For example, the amino acid phenylalanine contains a benzene ring that associates with other hydrophobic amino acids in proteins to form a hydrophobic interior, which is partially responsible for the unique structure of these large molecules (see Figure 14-8).

Figure 14-8   Hydrophobic groups, such as aromatic rings, form the interior or hydrophobic core in protein molecules, whereas the hydrophilic portions of the protein are found on the exterior, exposed to the aqueous environment.

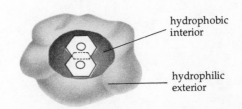

hydrophobic interior

hydrophilic exterior

CHEMICAL REACTIVITY

Benzene and substituted benzenes do not readily participate in the same type of addition reactions as alkenes. For example, cyclohexene rapidly reacts with hydrogen gas in the presence of nickel to give cyclohexane at 25°C and a pressure of about 1.5 atm:

(14-13)

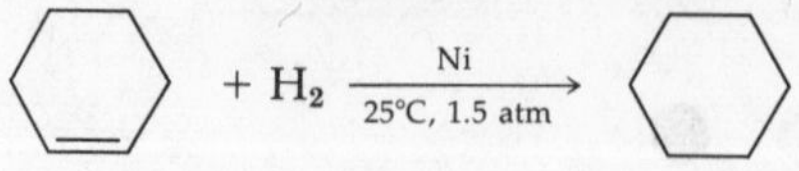

cyclohexene   cyclohexane

Benzene also reacts with hydrogen in the presence of nickel to give cyclohexane, but even at 200°C and a pressure of 100 atm, the reaction proceeds slowly:

(14-14)

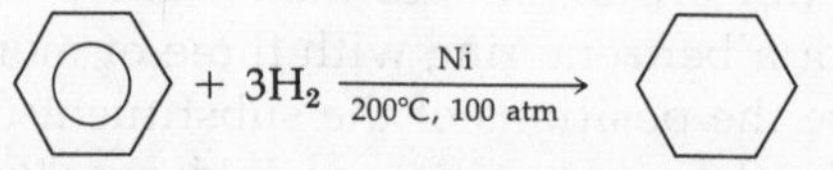

benzene   cyclohexane

An explanation for this difference in reactivity is that an addition reaction involves a change in the bonding arrangement of all of the carbon atoms in benzene. Recall that the bonding arrangement in benzene is represented by a pair of resonance structures in which *all* of the carbon atoms participate (Section 4-12). An addition reaction involving cyclohexene, on the other hand, requires a change in the bonding arrangement of only one carbon-carbon double bond.

One type of reaction that aromatic molecules do undergo relatively easily is **substitution,** in which hydrogen atoms are replaced by other atoms or groups of atoms. For example, benzene reacts readily with chlorine to form chlorobenzene:

(14-15)

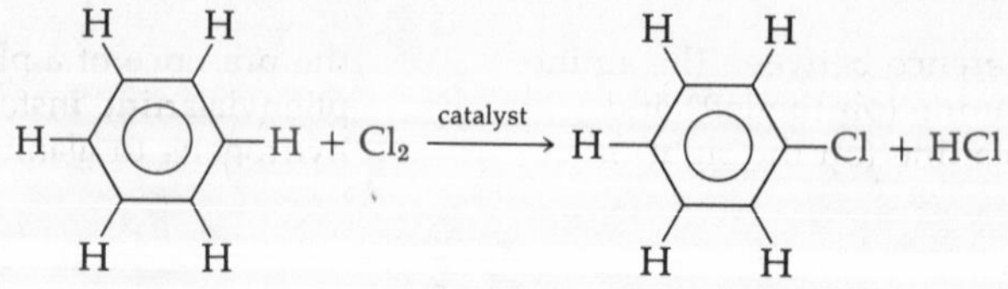

benzene   chlorobenzene

A number of products that can be obtained in such substitution reactions are listed in Table 14-6. One aromatic substitution reaction that occurs in the human body is the substitution of a hydroxyl group for one of the hydrogen atoms in the amino acid phenylalanine:

(14-16)

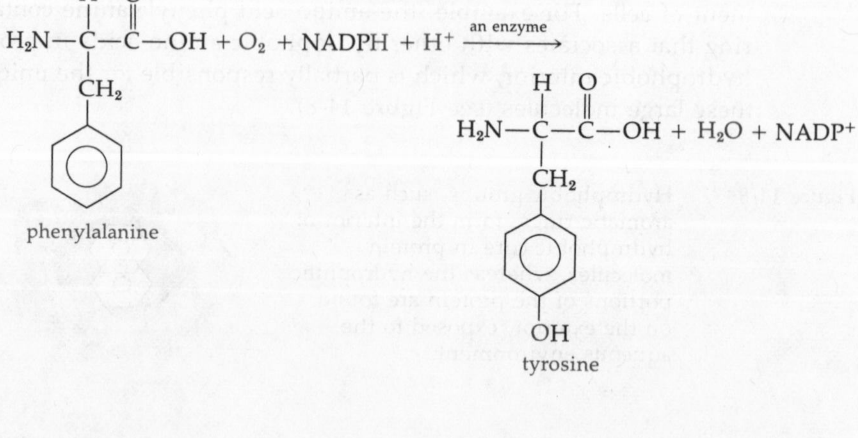

phenylalanine

tyrosine

Table 14-6 Some Aromatic Hydrocarbons Used Commercially

Structural Formula	Name	Some Uses
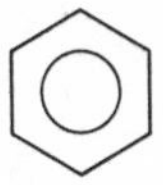 (benzene ring)	Benzene	Manufacture of leather, linoleum, varnishes, etc. (benzene is a carcinogen)
 CH_3 (on benzene ring)	Toluene	Manufacture of explosives, dyes, etc., and to extract chemicals from plants
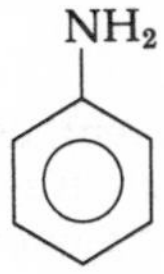 NH_2 (on benzene ring)	Aniline	Manufacture of dyes, medicines, perfume, and as a solvent
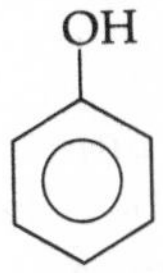 OH (on benzene ring)	Phenol	Used as a disinfectant and for manufacture of a variety of materials
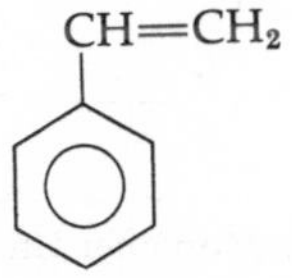 $CH{=}CH_2$ (on benzene ring)	Styrene	Used to make polystyrene and other plastics and synthetic rubbers

Tyrosine is also an amino acid required for protein synthesis. Interestingly, some tyrosine is also converted to hormones and neurotransmitters, which carry chemical messages from one cell to another. These conversions involve additional substitution reactions in which other hydroxyl groups (or in one case, iodine atoms) replace hydrogens on the benzene ring (Chapter 28).

Exercise 14-10
Draw structural formulas for:

    (a) *m*-Ethylphenol
    (b) 2,4-Dichlorotoluene
    (c) 1,2,4-Trimethylbenzene

Exercise 14-11
Describe the shape of a toluene molecule.

14-9 HYDROCARBON POLYMERS

A **polymer** is a very large molecule composed of simpler units or building blocks, called **monomers,** linked together in a long chain. Many naturally occurring compounds (e.g., starch and proteins) are polymers, whereas a large number of other polymers (e.g., nylon and polystyrene) are synthesized commercially. In order for a simple molecule to be a monomer unit, it must be capable of reacting with at least two other monomers. The reaction that links monomers together is called **polymerization** (see Figure 14-9).

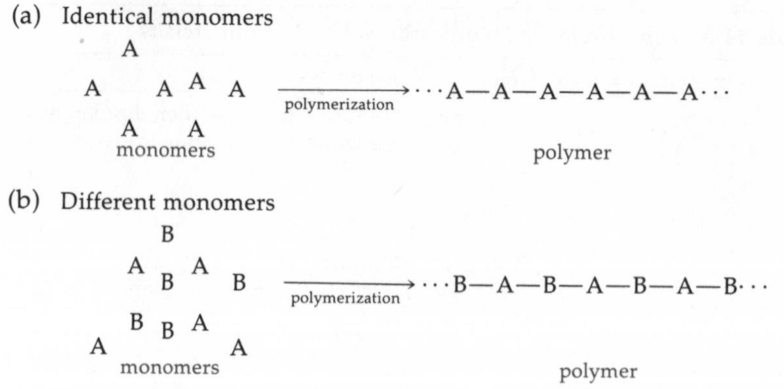

(a) Identical monomers

(b) Different monomers

Figure 14-9   Thousands of monomers can join together to form long polymer chains. The monomers may all be identical in a polymer (a), or they may be different, as in (b), where a polymerization reaction involving two different monomers is shown.

Some polymers are composed of identical monomer units, whereas others are composed of at least two different monomers. Table 14-7 lists the repeating monomer units found in some natural and synthetic polymers. Notice in Table

Table 14-7 Some Important Polymers

Name and Source	Monomer(s)	Repeating Monomer Unit (n Very Large)
Polyethylene (synthetic)	$CH_2{=}CH_2$ ethylene	$+(CH_2-CH_2)_n$
Rubber (natural and synthetic)	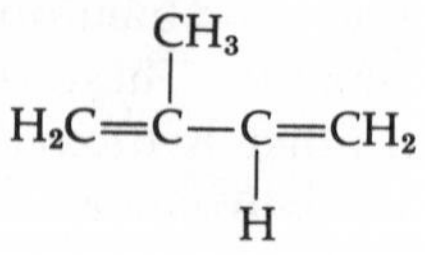 isoprene	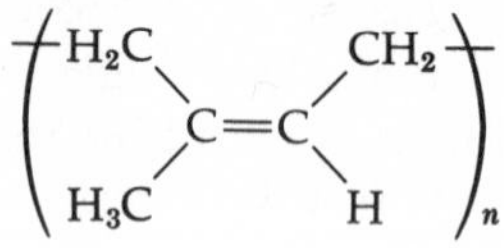
Nylon (synthetic)	a diamine and 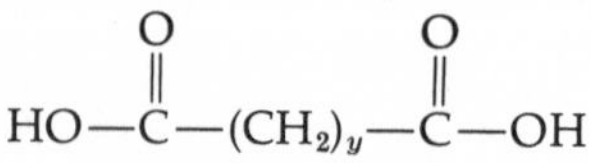 a dicarboxylic acid	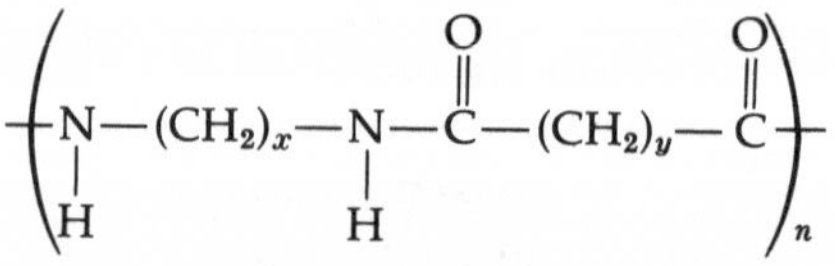 where x and y refer to the lengths of alkanelike carbon chains
Proteins (naturally occurring)	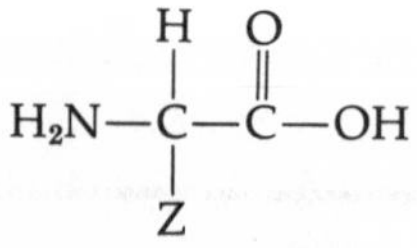 amino acids	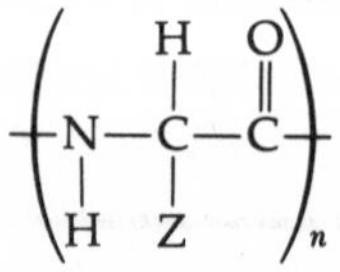 where Z is a group of atoms specific for each amino acid monomer
Starch (naturally occurring)	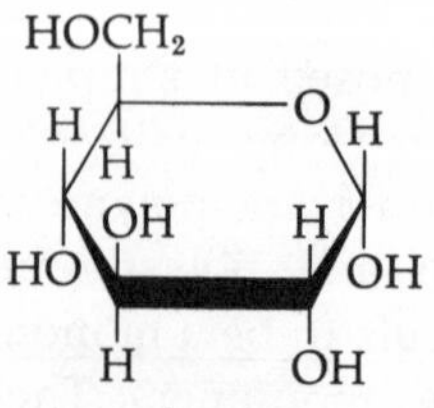 glucose	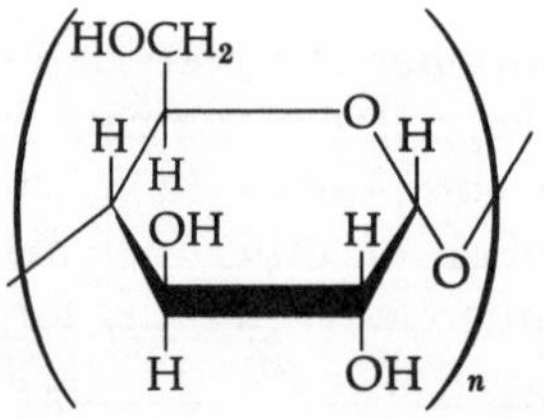

14-7 that the repeating monomer unit of a polymer has a bonding arrangement that is different from that in the individual monomers, since some bonds are broken and other bonds are formed during polymerization. Also notice that polymerization reactions can involve a variety of functional groups, including $C=C$ double bonds. Compounds containing $C=C$ double bonds (i.e., alkenes) were originally called olefins. Hence, polymers formed from alkene monomers are called **polyolefins.** Polymers formed from monomers containing other types of functional groups will be discussed later in this text.

The simplest polyolefin is polyethylene, which is formed by the polymerization of the monomer ethylene:

(14-17)
$$n(CH_2{=}CH_2) \xrightarrow[\text{polymerization}]{\text{a catalyst}} \cdots CH_2{-}CH_2{-}CH_2{-}CH_2{-}CH_2 \cdots$$
$$\text{ethylene} \qquad\qquad\qquad\qquad \text{polyethylene}$$

It is important to note that polymers such as polyethylene do not form automatically when their monomers are mixed together. Ethylene will react with itself to form polyethylene only under certain special conditions and in the presence of a catalyst. The particular polyethylene product obtained is somewhat dependent on the specific set of reaction conditions and the specific catalyst used. Many commonly used items are manufactured from polyethylene, including packaging materials, buckets, bottles, and syringes. Finished products made from synthetic polymers are called **plastics.**

A number of other commercial hydrocarbon polymers are similar to polyethylene in that their monomer unit is a substituted ethylene molecule. Compare the monomer units in polypropylene, polystyrene, polyvinyl chloride, and in Teflon with the monomer unit of polyethylene (see Table 14-8). Polypropylene, for example, has a methyl group attached to every other carbon atom, which gives it somewhat different properties from polyethylene.

Table 14-8 Polyolefins Related to Polyethylene

Name	Structure	Monomer	Uses
Polypropylene	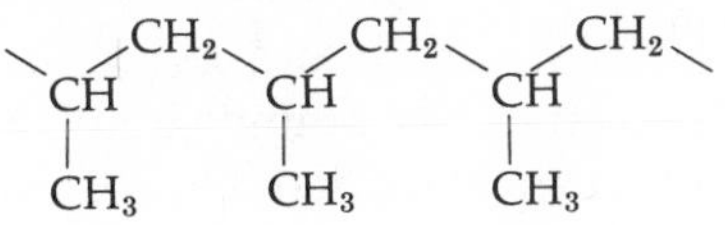	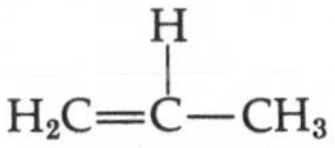 $H_2C{=}\overset{\displaystyle H}{\underset{\displaystyle }{C}}{-}CH_3$ propylene	Many—as for polyethylene
Polystyrene	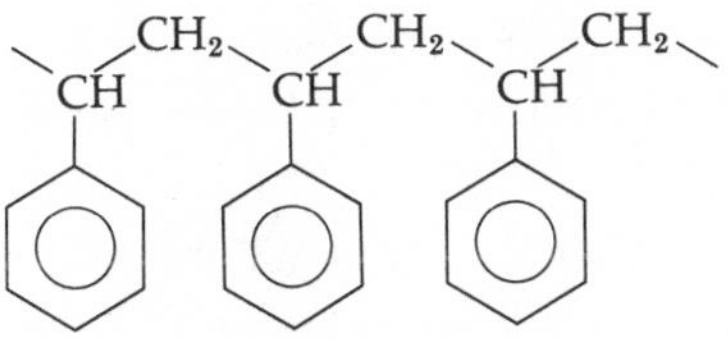	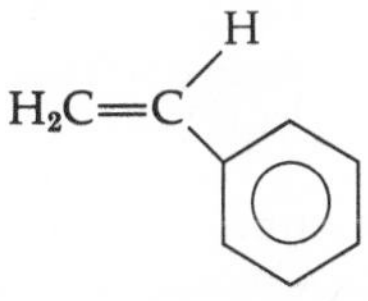 $H_2C{=}C$ (styrene) styrene	Polystyrene foam used for insulation
Polyvinyl chloride (PVC)	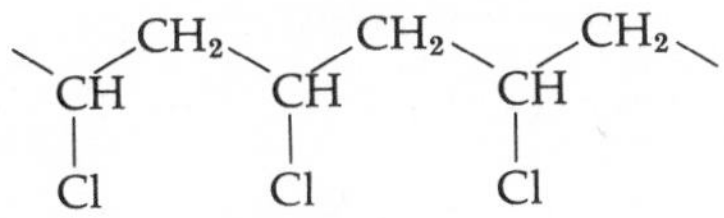	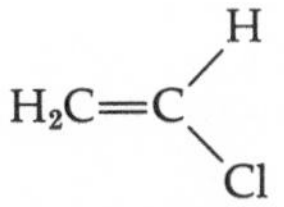 $H_2C{=}C{-}Cl$ vinyl chloride	Clear plastic bottles and tubing (vinyl chloride is a carcinogen)
Teflon	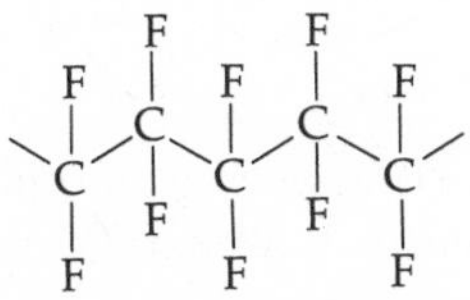	$F_2C{=}CF_2$ tetrafluoroethylene	Chemically inert slippery coatings

Synthetic rubber is formed by the polymerization of isoprene monomer units (Table 14-7), which contain two C=C double bonds. One of these bonds is broken during polymerization, while the other is shifted to another carbon atom (see Figure 14-10). Natural rubber is identical in structure to one of the synthetic rubber polymers but is polymerized in a different manner.

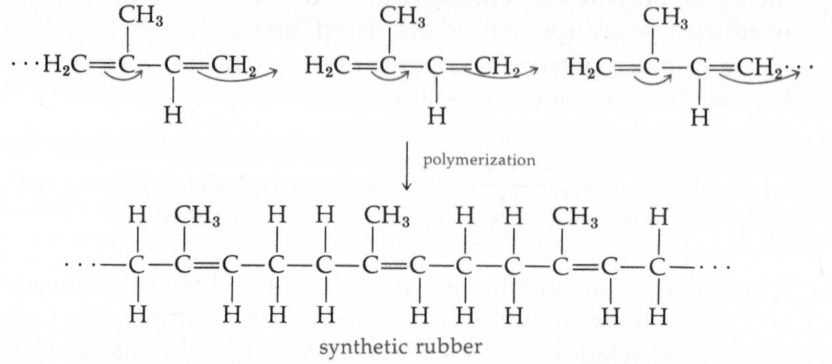

$$\cdots H_2C=\underset{\underset{H}{|}}{C}-C=CH_2, \quad H_2C=\underset{\underset{H}{|}}{C}-C=CH_2, \quad H_2C=\underset{\underset{H}{|}}{C}-C=CH_2; \cdots$$

polymerization

$$\cdots -\underset{\underset{H}{|}}{\overset{\overset{H}{|}}{C}}-\underset{\underset{H}{|}}{\overset{\overset{CH_3}{|}}{C}}=C-\underset{\underset{H}{|}}{\overset{\overset{H}{|}}{C}}-\underset{\underset{H}{|}}{\overset{\overset{H}{|}}{C}}-\underset{\underset{H}{|}}{\overset{\overset{CH_3}{|}}{C}}=C-\underset{\underset{H}{|}}{\overset{\overset{H}{|}}{C}}-\underset{\underset{H}{|}}{\overset{\overset{H}{|}}{C}}-\underset{\underset{H}{|}}{\overset{\overset{CH_3}{|}}{C}}=C-\underset{\underset{H}{|}}{\overset{\overset{H}{|}}{C}}- \cdots$$

synthetic rubber

Figure 14-10   The polymerization of isoprene monomers to form synthetic rubber can be represented by curved arrows to indicate the changes in the bonding arrangement and the linking together of monomer units.

14-10 SUMMARY

1. Hydrocarbons are a group of organic molecules that contain only carbon and hydrogen. Hydrocarbons that contain only C—C and C—H single bonds are called alkanes. The classes of alkene and alkyne compounds contain C=C double bond and C≡C triple bond functional groups, respectively. Hydrocarbons that contain a benzene ring are called aromatic hydrocarbons.

2. Hydrocarbons are nonpolar, and hydrocarbonlike portions of larger molecules are also nonpolar.

3. Alkanes are relatively unreactive compounds, but the —H$_2$C—CH$_2$— portions of certain molecules can be oxidized to HC=CH double bonds.

4. Alkenes readily participate in addition reactions, including reduction to form alkanes, hydration to form alcohols, and halogenation. Addition reactions involving nonsymmetric compounds usually follow Markovnikov's rule.

5. The symbol R represents an alkyl group, whereas R' represents an alkyl group or a hydrogen atom.

6. Aromatic hydrocarbons can undergo substitution reactions, in which hydrogen atoms are replaced by substituents. The phenyl group is

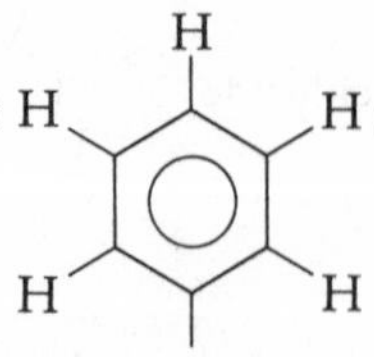

called a phenyl group.

7. A polymer is a very large molecule composed of monomer units that are polymerized to form a long chain. Polymers formed from alkene monomers are called polyolefins.

PROBLEMS

1. To which class of compounds do molecules with the following structural formulas belong?

(a) 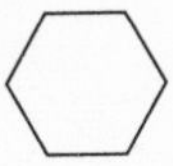 [hexagon structure]

(b) 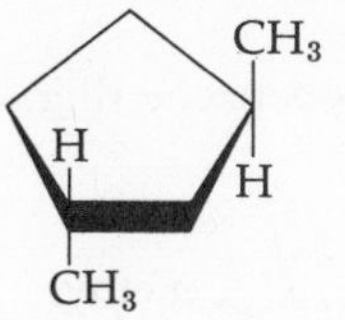 [cyclopentane structure with CH₃, H, H, CH₃ groups]

(c) 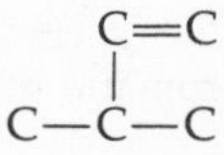 $C=C$ / $C-C-C$ structure

(d) 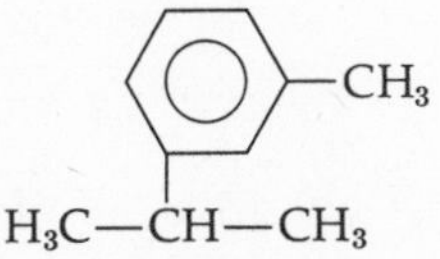 [benzene ring with $-CH_3$ group] $H_3C-CH-CH_3$

(e) 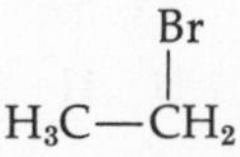 [Br above] H_3C-CH_2

2. Draw the structural formulas for each of the following:
   (a) cis-2-Pentene
   (b) trans-1,3-Dimethylcyclobutane
   (c) 2,3-Dimethyl-2-butene
   (d) 1,4-Diethylbenzene
   (e) trans-1,2-Diethylcyclopentane

3. Write IUPAC names for each of the following:

(a) 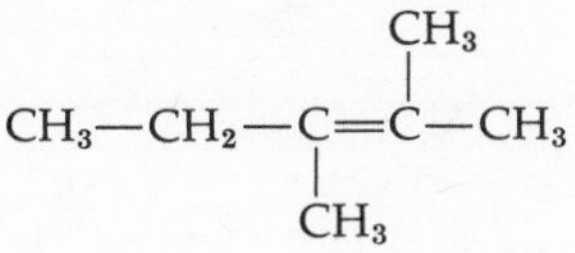 $CH_3-CH_2-C=C-CH_3$ with CH_3 above and CH_3 below the double-bonded carbons

(b) H_3C and CH_2-CH_3 / $C=C$ / H and H

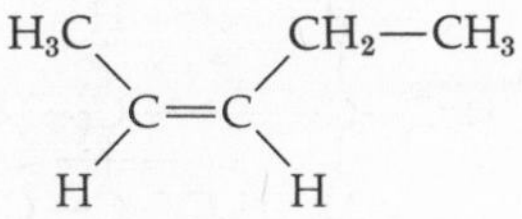

(c) [cyclohexane ring with $-CH_3$]

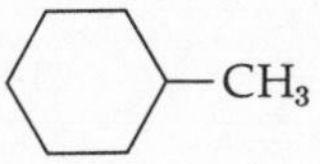

(d) 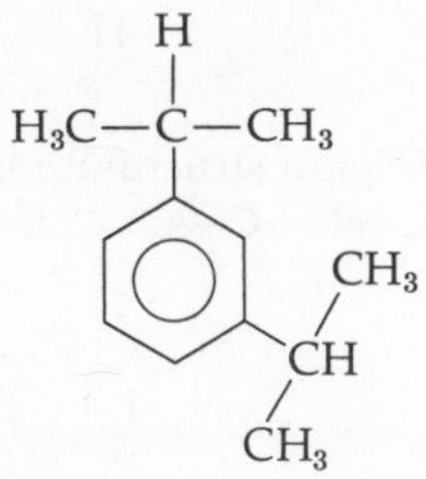 [benzene ring with $H_3C-C-CH_3$ (H above) group and CH_3/CH/CH_3 group]

(e) 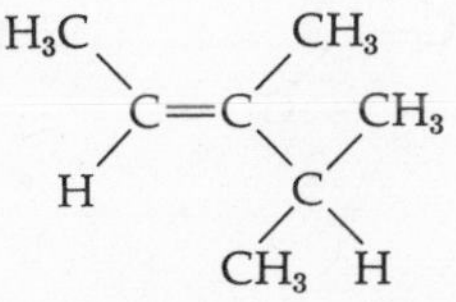 H_3C and CH_3 / $C=C$ / H and CH_3 / C / CH_3 H

4. Draw structural formulas for all of the hydrocarbons with the formula C_4H_8. Which of these are geometric isomers, and which are structural isomers?

5. Draw structural formulas for the products of the following reactions:

(a) 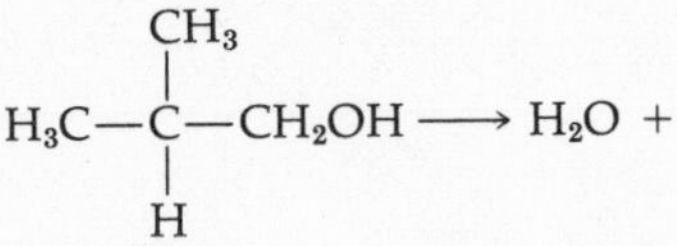 $H_3C-C-CH_2OH \longrightarrow H_2O +$ (with CH_3 above and H below the central C)

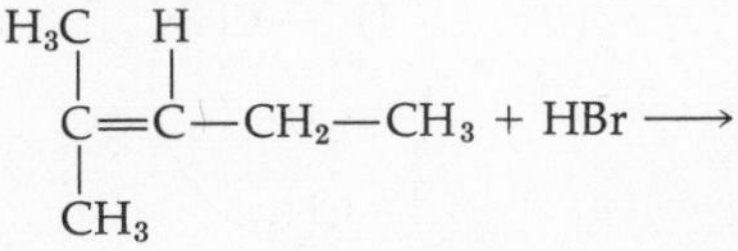

(b) H_3C H / $C=C-CH_2-CH_3 + HBr \longrightarrow$ (with CH_3 below)

(c) [cyclohexene ring with H and CH_3] $+ H_2O \longrightarrow$

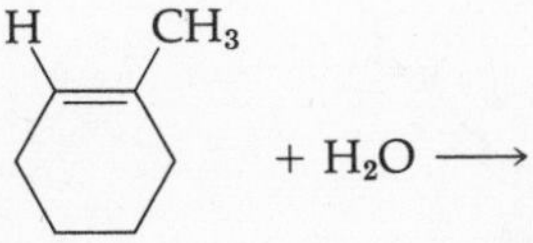

(d) Propene $+ H_2 \longrightarrow$

6. The polyolefin Saran, which is used in packaging, is polymerized from the monomer vinylidene chloride, $CH_2=CCl_2$. Draw a structural formula for a short segment of Saran.

SOLUTIONS TO EXERCISES

14-1 Alkanes; no characteristic functional group.
Alkenes; carbon-carbon double bond, C=C.
Alkynes; carbon-carbon triple bond, C≡C.

Aromatic hydrocarbons; a benzene ring, ⬡ .

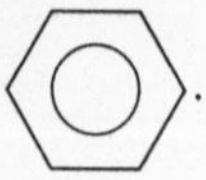

14-2 (a)

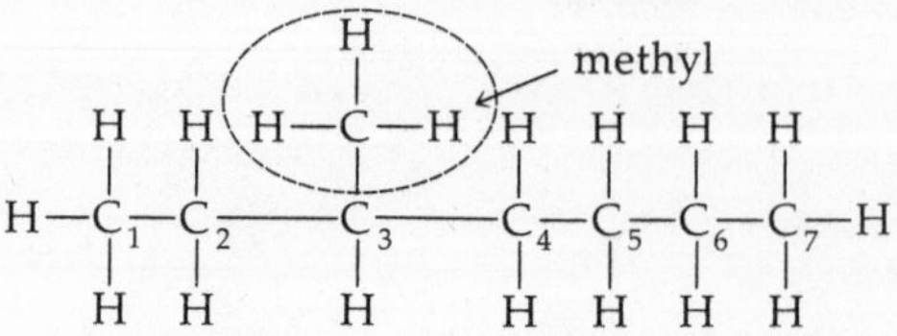

```
               H
               |
       H  H  H—C—H  H   H   H   H
       |  |  |      |   |   |   |
  H—C——C——C——————C——C——C——C—H
     1 | 2 | 3   4 | 5 | 6 | 7 |
       H  H  H      H   H   H   H
```
methyl

(b)

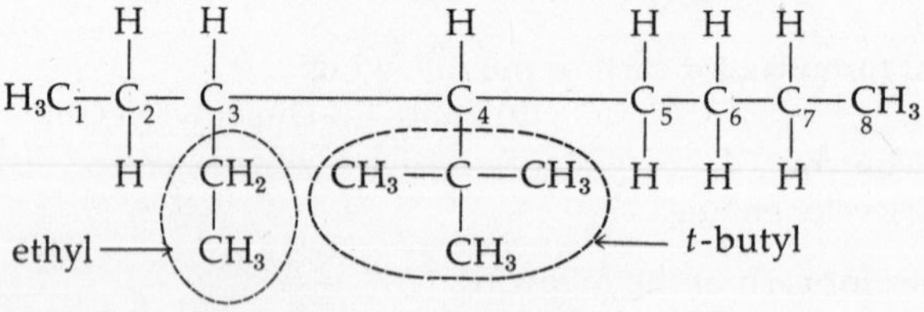

```
        H   H              H           H   H   H
        |   |              |           |   |   |
  H₃C——C——C——————————————C——————————C——C——C——CH₃
     1  | 2 | 3            4         5 | 6 | 7 | 8
        H  CH₂          CH₃—C—CH₃      H   H   H
           |               |
  ethyl——→ CH₃             CH₃    ←—— t-butyl
```

(c)

```
        H           H           H
        |           |           |
  H₃C——C——————————C——————————C——CH₃
     1  | 2       3           4 | 5
        H    H₃C—C—CH₃          H
                 |
                 H    ←—— isopropyl
```

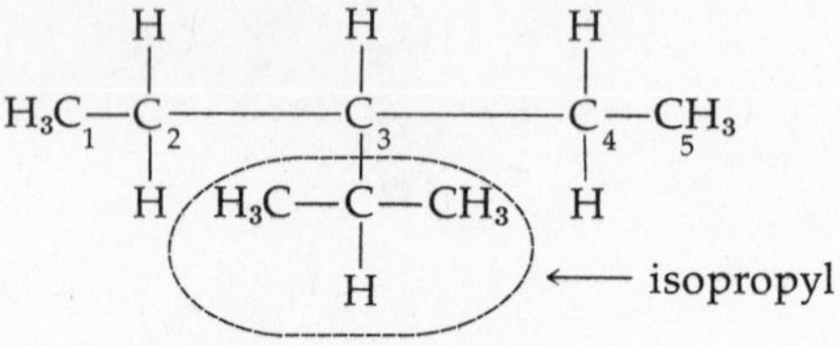

14-3 Using skeleton structural formulas:

C—C—C—C—C—C   hexane

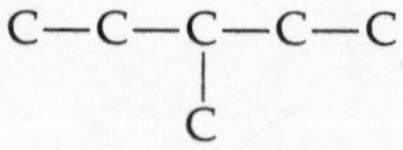

```
C—C—C—C—C        3-methylpentane
      |
      C
```

```
    C
    |
C—C—C—C    2,2-dimethylbutane
    |
    C
```

```
            C
           /
C—C—C—C        2-methylpentane
           \
            C
```

```
C—C—C—C    2,3-dimethylbutane
  | |
  C C
```

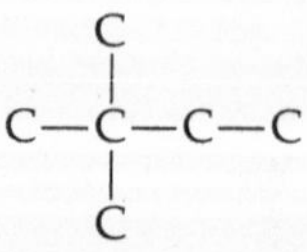

14-4 (a)

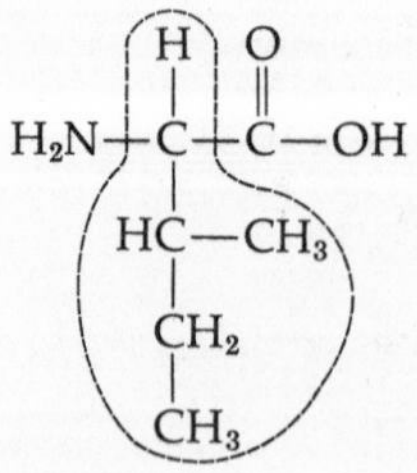

```
        H   O
        |   ||
  H₂N——C——C—OH
        |
       HC—CH₃
        |
       CH₂
        |
       CH₃
```

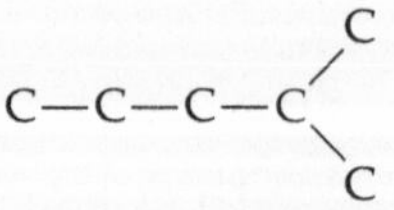

(b)

```
       O                    O
       ||                   ||
  HO—C—(CH₂—CH₂)—C—OH
```

(c)

```
                   H   H   H
                   |   |   |
  (H₃C—(CH₂)₁₂)—C=C——C———C—CH₂OH
                   |   |   |
                   H   OH  NH₂
```

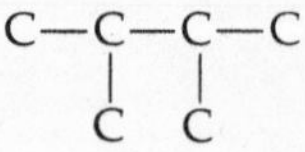

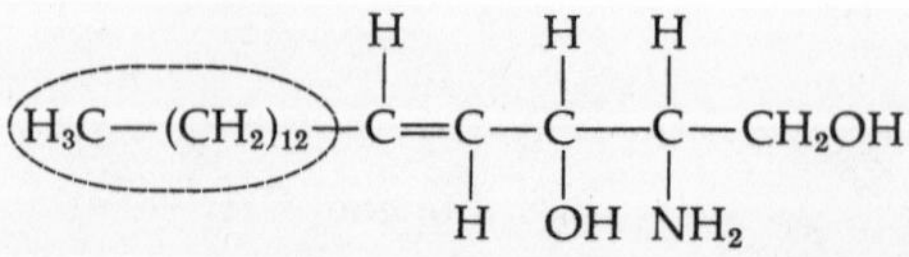

(d)

```
        CH₂
       /    \
  H₂C      CH—CH₂—CH₂—CH₂—CH₂—C≡O
      \    /                      
       S—S                       OH
```

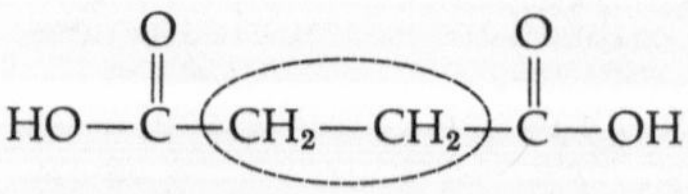

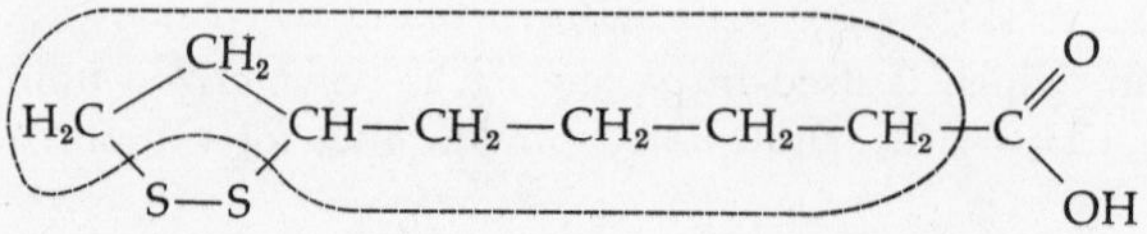

14-5 Cycloalkane rings, such as cyclohexane, involve one more carbon-carbon single bond than alkanes with the same number of carbon atoms. The carbon atoms involved in the additional C—C single bond can each bond to one fewer H atom than they could if this C—C bond were not present.

14-6 A planar hexagon would have bond angles of 120°, as opposed to the tetrahedral (109.5°) bond angles preferred by carbon atoms involved in four single bonds.

14-7

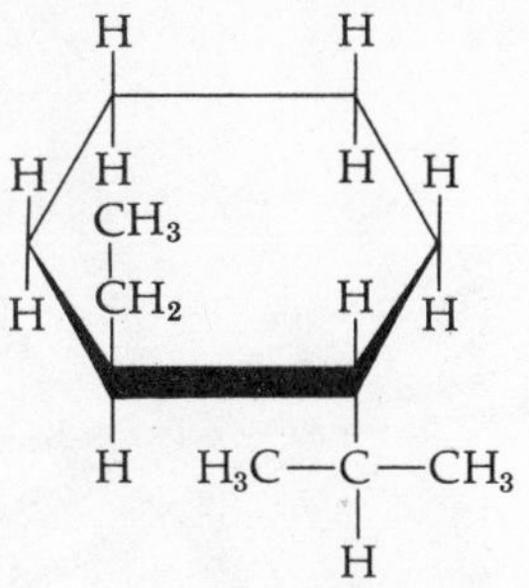

14-8 (a)

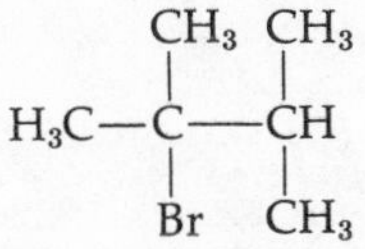

2-bromo-2,3-dimethylbutane

(b) 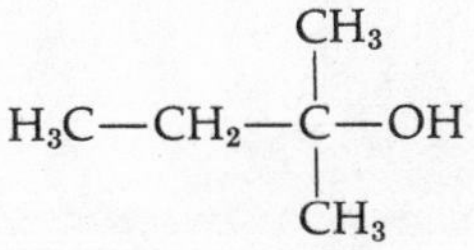 $H_3C—CH_2—\overset{\displaystyle CH_3}{\underset{\displaystyle CH_3}{C}}—OH$

2-methyl-2-butanol

(c)

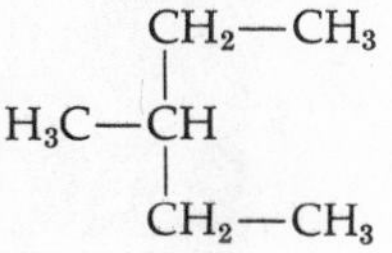

3-methylpentane

(d)

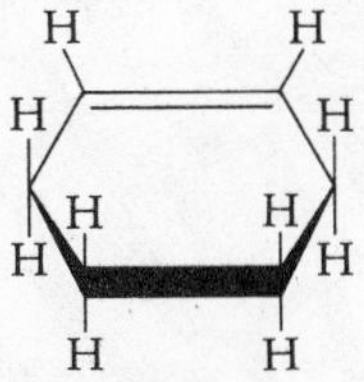

cyclohexene

Note that there are no geometric isomers of cyclohexene.

14-9 (a) 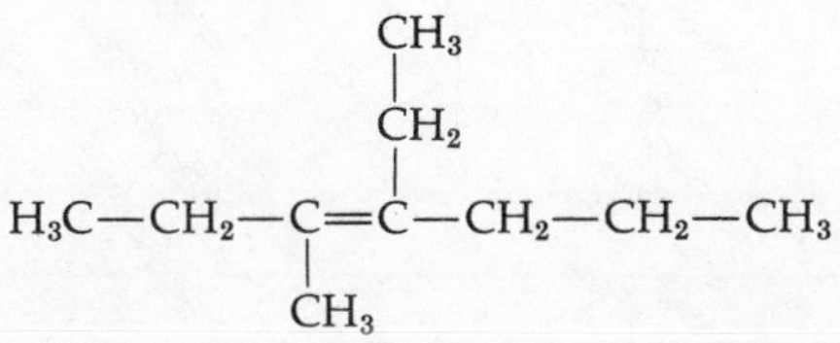 $H_3C—CH_2—\underset{\displaystyle CH_3}{\overset{\displaystyle \overset{\textstyle CH_3}{\underset{\textstyle |}{CH_2}}}{C}}{=}C—CH_2—CH_2—CH_3$

(b) $H_3C—CH_2—\underset{\displaystyle CH_3}{C}{=}\underset{\displaystyle CH_3}{C}—CH_2—CH_3$

(c) $H_3C—\underset{\displaystyle H}{\overset{\displaystyle H}{C}}{=}C—CH_2—CH_3$

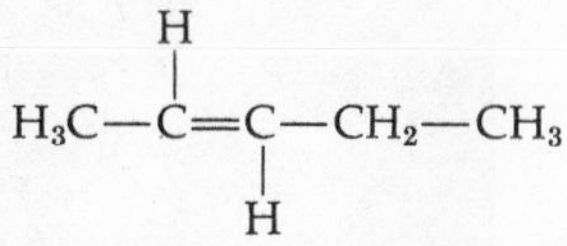

14-10 (a)

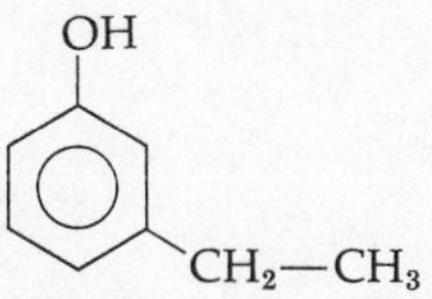

(b)

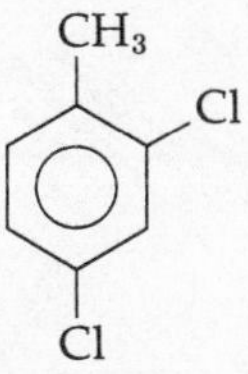

(c)

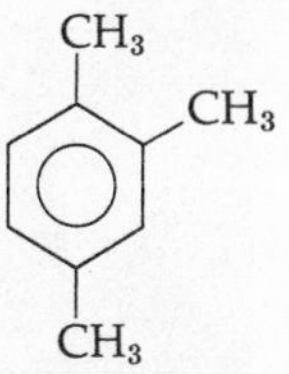

14-11 The methyl carbon atom is located at the center of a tetrahedron. At three corners of this tetrahedron are hydrogen atoms. At the fourth corner of this tetrahedron is one carbon atom of the phenyl group. The entire phenyl group is located in one plane. Thus, the shape of toluene can be represented as

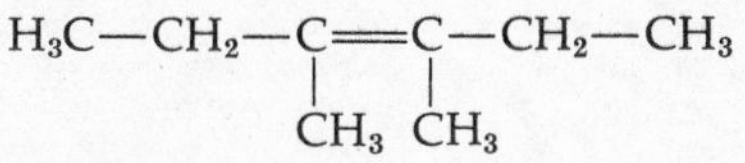

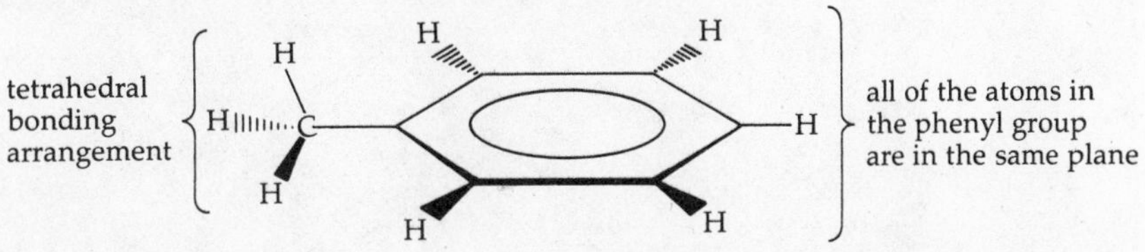

tetrahedral bonding arrangement

all of the atoms in the phenyl group are in the same plane

CHAPTER 15

Alcohols, Phenols, and Ethers

15-1 INTRODUCTION

Several billion years ago, certain early inhabitants of our planet developed the ability to make O_2 from H_2O by photosynthesis. Their closest relatives among organisms existing today are blue-green algae, and we humans are literally indebted to their ancestors for the air we breathe! Today there is a considerable supply of O_2 in the Earth's atmosphere, and the O_2 that is used by humans and other living organisms is continually replenished by the photosynthetic activity of plants and algae.

Now, in addition to requiring molecular oxygen for survival, humans make and degrade a large number of oxygen-containing organic molecules. In fact, just about every molecule in the human body contains the element oxygen.

The large group of oxygen-containing organic compounds is subdivided into several classes on the basis of their component functional groups. Three classes of organic compounds—alcohols, ethers, and phenols—all contain a C—O single bond. We shall discuss the structure, nomenclature, physical properties, and chemical reactivity of simple alcohols, phenols, and ethers in this chapter. In the next two chapters, we shall discuss those classes of oxygen-containing organic compounds that contain a C=O double bond: aldehydes and ketones (Chapter 16), and carboxylic acids, esters, and anhydrides (Chapter 17).

After studying Chapters 15, 16, and 17, you will not only know much about the chemistry of oxygen-containing compounds with one functional group, but you will also know much of the chemistry of compounds containing two or more functional groups. For example, the compound β-hydroxybutyric acid,

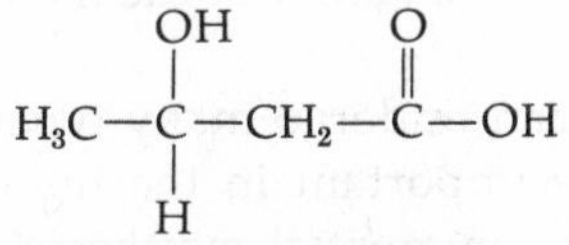

$$H_3C-\underset{\underset{H}{|}}{\overset{\overset{OH}{|}}{C}}-CH_2-\overset{\overset{O}{\|}}{C}-OH$$

contains both an alcohol and a carboxylic acid functional group, and it undergoes chemical reactions characteristic of both of these functional groups.

Wine is stored in huge vats for aging. The anaerobic (absence of free O_2) aging process involves several chemical reactions catalyzed by enzymes present in yeast cells and culminates in the reduction of acetaldehyde to ethanol. When the concentration of alcohol in wine reaches about 12%, the yeast cells die and further production of ethanol ceases.

15-2   STUDY OBJECTIVES

After careful study of this chapter, you should be able to:

1. Identify alcohol, ether, and phenol functional groups, given the structural formula for a compound, and draw the general structural formula for each of these classes of compounds.

2. Determine whether a specific alcohol, ether, or phenol can form hydrogen bonds in the pure state and/or with water.

3. Describe the acid-base properties of a specific alcohol, phenol, or ether.

4. Explain how hydrogen bonding affects the boiling point and solubility of alcohols, ethers, or phenols.

5. Draw the structural formulas for simple alcohols, phenols, and ethers, given their names, and vice versa.

6. Draw structural formulas for the products of reactions in which alcohols are oxidized or dehydrated.

7. Describe the major differences between an enzyme-catalyzed reaction in a living cell and the same reaction performed in the laboratory using nonenzyme catalysts.

15-3   GENERAL PROPERTIES OF ORGANIC COMPOUNDS CONTAINING OXYGEN

The major classes of oxygen-containing organic compounds are listed in Table 15-1. Alcohols, ethers, and phenols, the subjects of this chapter, all contain a C—O single bond. Aldehydes and ketones (Chapter 16) contain the carbonyl group, whereas carboxylic acids, esters, and anhydrides (Chapter 17) contain the carboxyl group.

The physical properties and the chemical reactivity of all of these oxygen-containing compounds are quite different from those of hydrocarbons. Recall that an oxygen atom has six valence electrons and a large electronegativity, and that in covalent compounds an oxygen atom usually forms two covalent bonds. In a carbon-oxygen bond or in an oxygen-hydrogen bond, the bonding electrons are pulled closer to the more electronegative oxygen atom, making these bonds polar, with a partial negative charge on the oxygen atom. Thus, organic functional groups that contain oxygen are polar groups and can participate in hydrogen bonding with water (Chapter 7). Therefore, oxygen-containing functional groups are hydrophilic, and the presence of these functional groups in a molecule *tends* to make it water soluble. Of course, the solubility of a molecule is determined by *all* of the atoms in the molecule and how they are bonded together.

Oxygen-containing organic molecules can undergo many types of reactions. We shall discuss those reactions that are important in the human body, and some others that are important for the commercial synthesis of drugs and plastics.

The kinds of reactions that occur inside living cells can also be carried out in the laboratory, but there are several important differences. Generally speaking, organic compounds that are capable of reacting do so only very slowly in the laboratory. A catalyst must usually be added in order for reactions to take place

Table 15-1 Classes of Oxygen-Containing Organic Compounds

Characteristic Functional Group		Class
C—O Single bond	—C—OH	Alcohols
	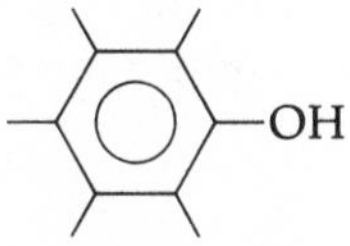 ⬡—OH	Phenols
	—C—O—C— or 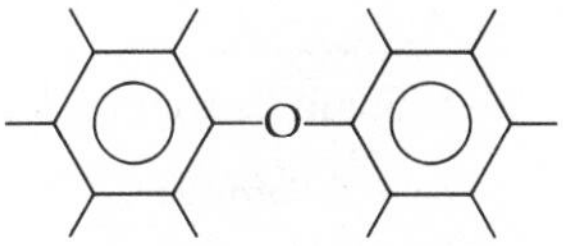 ⬡—O—⬡	Ethers
O‖—C— Carbonyl group	—C—H (O)	Aldehydes
	—C—C—C— or ⬡—C—⬡ (O)	Ketones
O‖—C—O— Carboxyl group	—C—OH (O)	Carboxylic acids
	—C—O—C— or —C—O—⬡ (O)	Esters
	—C—O—C— (O O)	Anhydrides

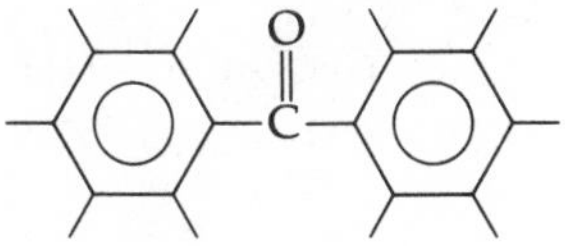

in a reasonable period of time. In living cells, highly efficient catalysts (called enzymes) are naturally present. Enzymes enable reactions to proceed at rates that are much faster than those that can be achieved using nonenzyme catalysts. In addition, enzymes are highly *specific* catalysts, assisting in the formation of only a single set of products. Nonenzyme catalysts, on the other hand, catalyze other reactions, in addition to the one desired, and so a mixture of products is often formed. Thus in the laboratory or industrial situation one has the additional task of separating the desired product from undesired ones. These differences between enzyme- and nonenzyme-catalyzed reactions should be kept in mind when studying specific reactions of organic compounds.

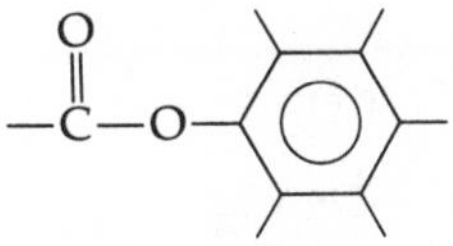

15-4 ALCOHOLS

Alcohols are organic derivatives of water, in which one of the hydrogen atoms of water is replaced by an alkyl group. Thus, an alcohol can be viewed as an alkyl group, R, bonded to a **hydroxyl** group, —OH, so that an alcohol has the general formula R—OH. Note that the alcohol functional group is —C—OH, whereas the group of two atoms, —OH, is called the hydroxyl group.

Structure and Nomenclature

The common names of alcohols are formed by adding the word "alcohol" to the name of the alkyl group to which the hydroxyl group is attached. Thus,

$$H_3C—\overset{\overset{\displaystyle OH}{|}}{C}H—CH_3$$

is called isopropyl alcohol, which is frequently used as rubbing alcohol. The structural formulas and names of some common alcohols are given in Table 15-2.

Table 15-2 Some Common Alcohols: General formula = R—OH

Structural Formula	Common Name	IUPAC Name
$CH_3—OH$	Methyl alcohol (wood alcohol)	Methanol
$CH_3—CH_2—OH$	Ethyl alcohol (grain alcohol)	Ethanol
$CH_3—CH_2—CH_2—OH$	n-Propyl alcohol	1-Propanol
$CH_3—\underset{\underset{\displaystyle OH}{\vert}}{C}H—CH_3$	Isopropyl alcohol	2-Propanol
$CH_3—CH_2—CH_2—CH_2—OH$	n-Butyl alcohol	1-Butanol
$CH_3—CH_2—\underset{\underset{\displaystyle OH}{\vert}}{C}H—CH_3$	sec-Butyl alcohol	2-Butanol
$CH_3—\overset{\overset{\displaystyle CH_3}{\vert}}{\underset{\underset{\displaystyle OH}{\vert}}{C}}—CH_3$	t-Butyl alcohol	2-Methyl-2-propanol
$\underset{\underset{\displaystyle OH}{\vert}}{H_2C}——\underset{\underset{\displaystyle OH}{\vert}}{CH_2}$	Ethylene glycol	1,2-Ethanediol
$\underset{\underset{\displaystyle OH}{\vert}}{H_2C}——\underset{\underset{\displaystyle OH}{\vert}}{CH}—\underset{\underset{\displaystyle OH}{\vert}}{CH_2}$	Glycerol (glycerine)	1,2,3-Propanetriol

The IUPAC names for simple alcohols are formed by adding the ending -ol to the name of the parent alkane (the e is dropped). Thus, the IUPAC name for methyl alcohol is methanol. The position of the hydroxyl group on the carbon chain of propanol and larger alcohols is indicated by a numerical prefix. Thus

$$H_3C—\overset{\overset{\displaystyle OH}{|}}{C}H—CH_3$$

the IUPAC name for $H_3C—CH—CH_3$ is 2-propanol. When a compound contains two, three, or more alcohol functional groups, the IUPAC ending is -diol, -triol, and so on (the e is not dropped in this case).

Ethanol, also called grain alcohol, is the only simple alcohol that is tolerated by the human body. In small doses it often produces pleasant sensations. However, ethanol acts as a physiological depressant. Many drivers who drink have become painfully aware of ethanol's ability to slow down their response time and decrease their muscle coordination. In very large doses, ethanol is lethal.

Methanol, also called wood alcohol, is frequently used as a solvent and as a fuel for cooking. Ethylene glycol is the major component of the antifreeze used in car radiators, and glycerol is a component of many oils and creams. We shall see in Chapter 26 that glycerol, also called glycerine, is a major component of the fats stored in the human body.

Hydrogen Bonding

The large electronegativity of the oxygen atom makes the hydroxyl group quite polar. Thus, an organic molecule that contains a hydroxyl group can act as either a hydrogen bond donor or a hydrogen bond acceptor. For example, alcohol molecules can form hydrogen bonds with water and other alcohol molecules (see Figure 15-1).

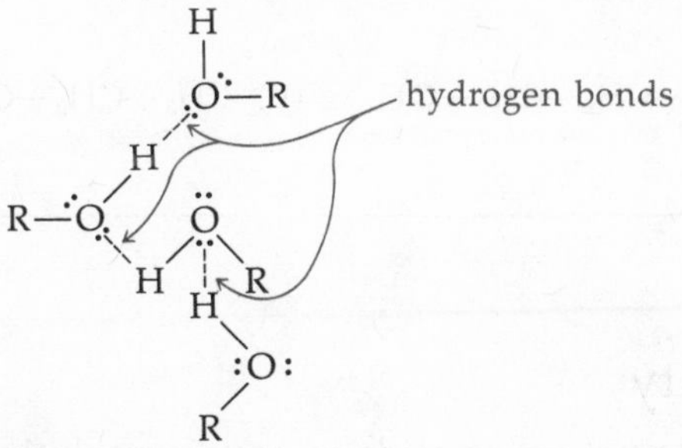

Figure 15-1 Alcohols form hydrogen bonds with water molecules and, in the pure state, with other alcohol molecules.

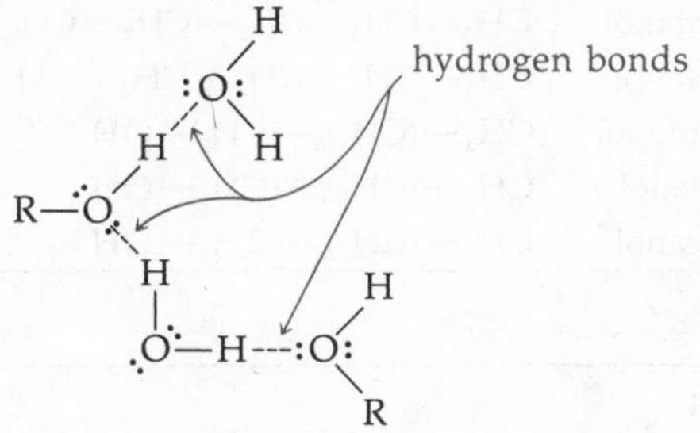

Because of their ability to form hydrogen bonds between identical molecules, pure alcohols have higher boiling points than molecules of similar molecular weight that do not form hydrogen bonds in the pure state, such as the hydrocarbons (see Table 15-3).

Table 15-3 Boiling Points of Some Hydrocarbons and Alcohols with Similar Molecular Weights

Hydrocarbons			Alcohols		
Name and Structural Formula	MW	Boiling Point (°C)	Name and Structural Formula	MW	Boiling Point (°C)
Ethane H_3C-CH_3	30	−89	Methanol H_3C-OH	32	+65
Propane $H_3C-CH_2-CH_3$	44	−42	Ethanol H_3C-CH_2-OH	46	+78
Butane $H_3C-CH_2-CH_2-CH_3$	58	0	1-Propanol $H_3C-CH_2-CH_2-OH$	60	+97

The formation of hydrogen bonds between the hydroxyl group of alcohol and water molecules also tends to make alcohols soluble in water, whereas hydrocarbon molecules are hydrophobic. However, alcohols that contain long alkyl chains are relatively insoluble in water because of the nonpolar hydrophobic character of the alkyl group. Notice that, in the series of alcohols in Table 15-4, the larger the nonpolar portion, the lower the solubility.

Table 15-4 Solubility of Alcohols

Name	Structural Formula	Solubility (g/100 ml of water)
Methanol	CH_3—OH	
Ethanol	CH_3—CH_2—OH	Completely miscible
1-Propanol	CH_3—CH_2—CH_2—OH	
1-Butanol	CH_3—CH_2—CH_2—CH_2—OH	7.9
1-Pentanol	CH_3—CH_2—CH_2—CH_2—CH_2—OH	2.3
1-Hexanol	CH_3—CH_2—CH_2—CH_2—CH_2—CH_2—OH	0.6
1-Heptanol	CH_3—$(CH_2)_5$—CH_2—OH	0.2
1-Octanol	CH_3—$(CH_2)_6$—CH_2—OH	0.05
1-Decanol	CH_3—$(CH_2)_8$—CH_2—OH	~0

Exercise 15-1

Arrange the following compounds in order of increasing solubility in water:

(a) CH_3—CH_2—CH_2—OH   (b) CH_3—CH_2—CH_3 (c) $CH_3-\underset{\underset{CH_3}{|}}{\overset{\overset{CH_3}{|}}{C}}-(CH_2)_3-OH$

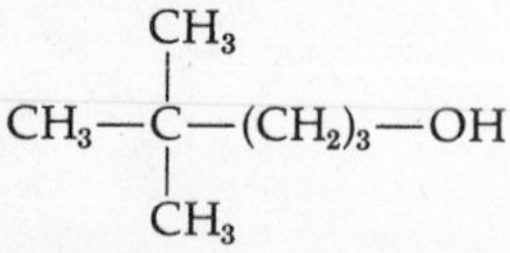

Acidity-Basicity

Alcohols do not act as acids in aqueous solutions. Let us consider the alcohol functional group, $-\overset{|}{\underset{|}{C}}-O-H$. The hydroxyl proton found in this group would be acidic if electrons were "pulled" away from the O—H bond, facilitating the loss of a proton (H^+). This is not the case, however. The electronegativity of the carbon atom in the alcohol functional group is not very large, and the carbon atom is not bonded to other atoms that strongly attract electrons. On the other hand, phenols are weak acids, as we shall see shortly.

Alcohols (and phenols and ethers) are extremely weak (weaker than water) Brønsted-Lowry bases. The oxygen atom in an alcohol, phenol, or ether has two unshared pairs of electrons, but the ability of these unshared electrons to accept a proton is quite small.

Chemical Reactivity

DEHYDRATION

In Section 14-6 we saw that an alkene could be formed by the dehydration of an alcohol in the reaction

(15-1)

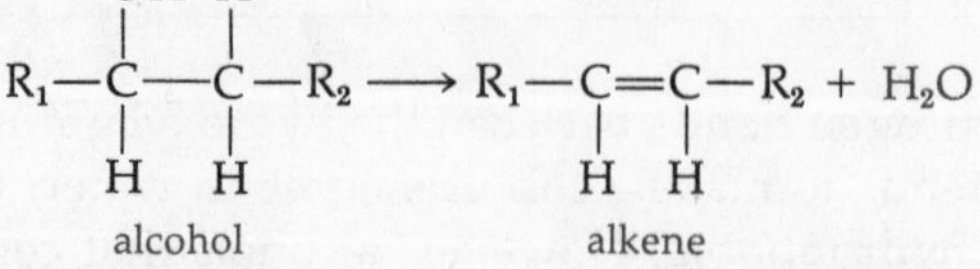

$$R_1-\underset{\underset{H}{|}}{\overset{\overset{OH}{|}}{C}}-\underset{\underset{H}{|}}{\overset{\overset{H}{|}}{C}}-R_2 \longrightarrow R_1-\underset{\underset{H}{|}}{\overset{}{C}}=\underset{\underset{H}{|}}{\overset{}{C}}-R_2 + H_2O$$

　　　　alcohol　　　　　　　alkene

Consider the following specific example. When a chemist dehydrates 2-butanol,

$CH_3—CH_2—CH—CH_3$, a mixture of three products is obtained—the alkenes
                    |
                    OH

1-butene, $CH_3—CH_2—CH=CH_2$, and *cis*-2-butene and *trans*-2-butene, $CH_3—CH=CH—CH_3$ (*trans*-2-butene is the major product). In later chapters we shall see several important reactions in human cells where the dehydration of the alcohol functional group to form a carbon-carbon double bond takes place. In the human body, however, a mixture of products is usually *not* obtained, since enzyme-catalyzed reactions yield only a single set of products.

Ethers can also be formed by the dehydration of alcohols under appropriate conditions. In this dehydration reaction, the elements of water are split off from two reacting alcohol molecules:

(15-2) $R—OH + R—OH \longrightarrow R—O—R + H_2O$
                  two alcohol molecules   an ether

OXIDATION

Another type of reaction that alcohols can undergo is oxidation. Recall from Chapter 12 that for organic molecules we can say that oxidation involves the loss of hydrogen or the gain of oxygen, whereas reduction involves the gain of hydrogen or the loss of oxygen. Hydration and dehydration reactions, on the other hand, involve neither oxidation nor reduction.

An alcohol functional group, $—\overset{\displaystyle |}{\underset{\displaystyle |}{C}}—OH$, can lose hydrogen and be oxidized to

a carbonyl group, $—\overset{O}{\overset{\|}{C}}—$. The ability of a given alcohol to be oxidized (and the product obtained) depends on the nature of the alkyl portion of the alcohol molecule. In an alcohol the carbon atom to which the hydroxyl group is attached is called the hydroxylic carbon atom (see Figure 15-2). If the hydroxylic carbon atom is bonded to none or one other carbon atom, it is called a *primary* carbon atom and the alcohol is called a **primary alcohol.** If the hydroxylic carbon atom is bonded to two or three other carbon atoms, then this carbon atom is called a *secondary* or *tertiary* carbon atom, respectively, and the alcohol is called a **secondary** or **tertiary alcohol.**

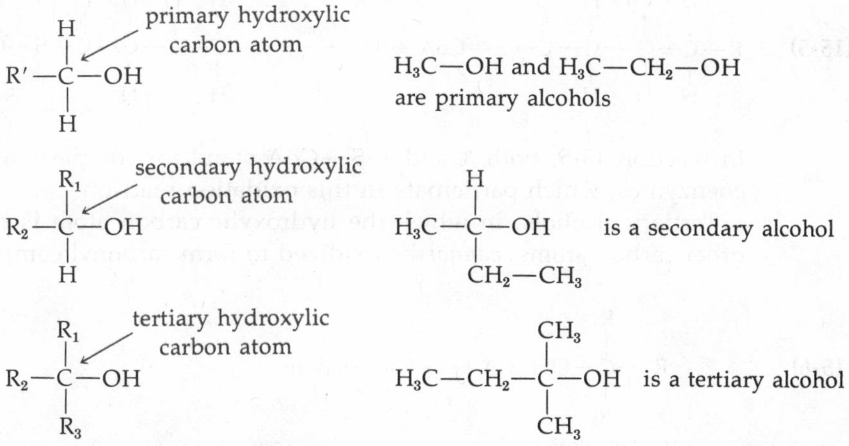

Figure 15-2   Alcohols are classified as primary, secondary, or tertiary depending on the bonding arrangement of the hydroxylic carbon atom.

The oxidation of a primary alcohol produces an **aldehyde**:

(15-3)
$$R'-\underset{\underset{H}{|}}{\overset{\overset{H}{|}}{C}}-OH + X \longrightarrow R'-\overset{\overset{O}{\|}}{C}-H + XH_2$$

primary   aldehyde
alcohol

Reaction 15-3 represents the general reaction for the oxidation of a primary alcohol. In this representation, X symbolizes the oxidizing agent. Thus, for example, the primary alcohols methyl alcohol and ethyl alcohol can be oxidized to the aldehydes $H_2C{=}O$ (formaldehyde) and $CH_3-\overset{\overset{O}{\|}}{C}-H$ (acetaldehyde).

The laboratory synthesis of aldehydes from primary alcohols is quite difficult, since the aldehyde product is generally further oxidized to a carboxylic acid. The enzyme-catalyzed oxidation of a primary alcohol to an aldehyde, on the other hand, readily produces only the aldehyde. We shall discuss how aldehydes are named in the next chapter.

The oxidation of a secondary alcohol produces a **ketone**:

(15-4)
$$R_1-\underset{\underset{H}{|}}{\overset{\overset{OH}{|}}{C}}-R_2 + X \longrightarrow R_1-\overset{\overset{O}{\|}}{C}-R_2 + XH_2$$

secondary   ketone
alcohol

Reaction 15-4 represents the general reaction for the oxidation of any secondary alcohol. Again, X symbolizes the oxidizing agent. For example, the secondary alcohol isopropyl alcohol can be oxidized to the ketone $H_3C-\overset{\overset{O}{\|}}{C}-CH_3$ (acetone).

Another example of the oxidation of a secondary alcohol to a ketone takes place in human cells as part of the oxidation of fatty acid molecules (Chapter 27):

(15-5)
$$R-\underset{\underset{H}{|}}{\overset{\overset{H}{|}}{C}}-\underset{\underset{H}{|}}{\overset{\overset{OH}{|}}{C}}-\underset{\underset{H}{|}}{\overset{\overset{H}{|}}{C}}-\overset{\overset{O}{\|}}{C}-S-CoA + X \xrightarrow{\text{enzyme}} R-\underset{\underset{H}{|}}{\overset{\overset{H}{|}}{C}}-\overset{\overset{O}{\|}}{C}-\underset{\underset{H}{|}}{\overset{\overset{H}{|}}{C}}-\overset{\overset{O}{\|}}{C}-S-CoA + XH_2$$

In reaction 15-5, both X and —S—CoA stand for complex substances, called coenzymes, which participate in this oxidation reaction.

Tertiary alcohols, in which the hydroxylic carbon atom is bonded to three other carbon atoms, *cannot* be oxidized to form carbonyl compounds:

(15-6)
$$R_1-\underset{\underset{R_3}{|}}{\overset{\overset{R_2}{|}}{C}}-OH + X \longrightarrow \text{No reaction}$$

tertiary alcohol

As we shall see in the following chapters, alcohols can be formed by reducing aldehydes and ketones in reactions that are similar to the reverse of reactions 15-3 and 15-4. Alcohols will also react with aldehydes, ketones, and carboxylic acids. The reactions of alcohols are summarized in Essential Skills 9.

Exercise 15-2

Why does pure methanol, with a molecular weight of 32, exist as a liquid at room temperature, whereas methane (MW = 16), ethane (MW = 30), and propane (MW = 44) are gases at room temperature?

Exercise 15-3

Classify as primary, secondary, or tertiary the hydroxylic carbon atoms in glycerol,

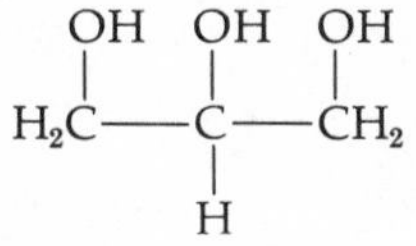

$$
\begin{array}{ccc}
\text{OH} & \text{OH} & \text{OH} \\
| & | & | \\
\text{H}_2\text{C} \!\!-\!\!\!-\!\! & \text{C} \!\!-\!\!\!-\!\! & \text{CH}_2 \\
& | & \\
& \text{H} &
\end{array}
$$

Exercise 15-4

Write the IUPAC names for each of the following alcohols, and draw structural formulas for the products of the given reactions (X represents an oxidizing agent):

(a) $2(\text{CH}_3\text{—CH}_2\text{—OH}) \rightarrow \text{H}_2\text{O} +$ (b) $\text{H}_3\text{C—CH}_2\text{—OH} \rightarrow \text{H}_2\text{O} +$

(c) 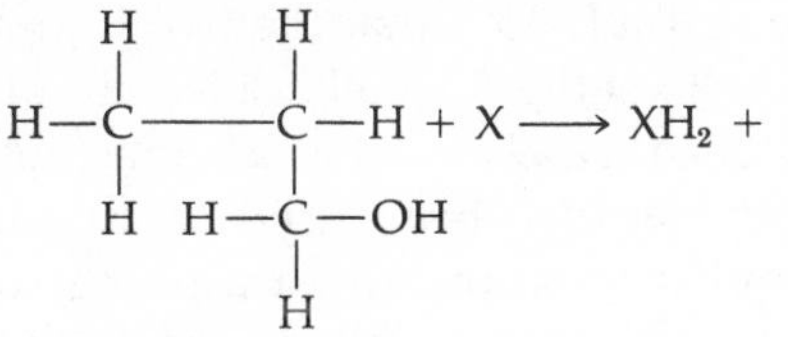
$$
\begin{array}{cc}
\text{H} & \text{H} \\
| & | \\
\text{H—C} \!\!-\!\!\!-\!\! & \text{C—H} + \text{X} \longrightarrow \text{XH}_2 + \\
| & | \\
\text{H} & \text{H—C—OH} \\
& | \\
& \text{H}
\end{array}
$$

(d) 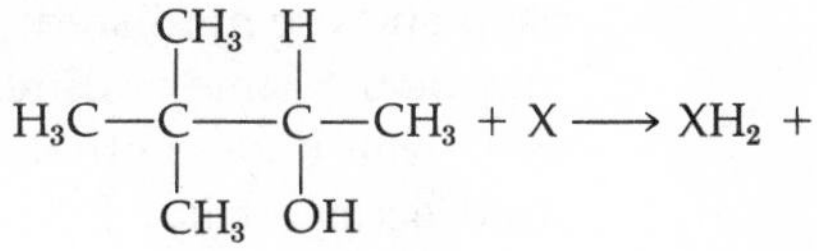
$$
\begin{array}{cc}
\text{CH}_3 & \text{H} \\
| & | \\
\text{H}_3\text{C—C} \!\!-\!\!\!-\!\! & \text{C—CH}_3 + \text{X} \longrightarrow \text{XH}_2 + \\
| & | \\
\text{CH}_3 & \text{OH}
\end{array}
$$

If you have difficulty working this type of problem, consult Essential Skills 9.

15-5 PHENOLS

Phenols are compounds in which a hydroxyl group is bonded to a benzene ring or to a substituted benzene ring. The simplest member of this class of compounds is phenol,

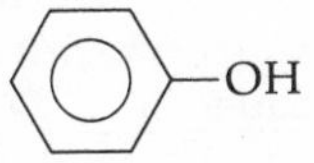

which is sometimes called carbolic acid. Note that phenol (feenol) is pronounced differently than phenyl (fenil). Phenol was the first antiseptic used in modern medicine. Phenols are similar to alcohols in many ways. They can form hydrogen bonds with phenols, with alcohols, and with water.

As we mentioned previously, phenols, in contrast to alcohols, are weak acids in aqueous solution. The phenyl group tends to pull electrons away from the atoms to which it is bonded. Thus, in a phenol, the phenyl group pulls electrons away from the —OH group. As a consequence, the hydrogen present in this —OH group is weakly acidic (Figure 15-3). We shall not discuss the complex nomenclature and reactions of phenols. Some phenols are used in photographic developing processes, whereas others are used in the manufacture of drugs, plastics, and other complex compounds.

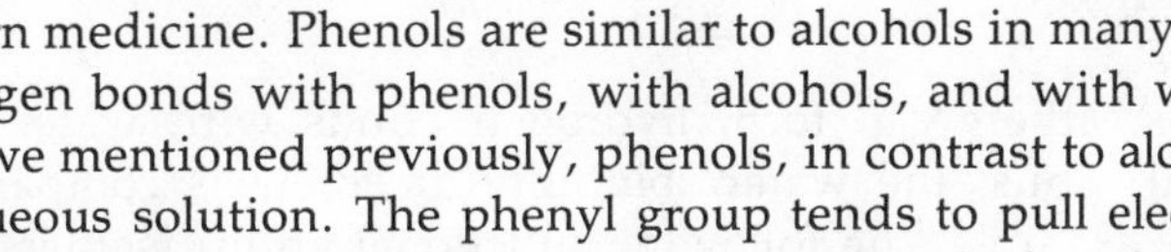

Figure 15-3   Phenol is a weak acid. The phenyl group pulls electrons away from the —OH group, thus facilitating the transfer of a proton to a water molecule

15-6 ETHERS

Ethers can also be considered organic derivatives of water in which *both* of the hydrogen atoms in a water molecule, H—O—H, are replaced by alkyl or phenyl groups, that is, R_1—O—R_2, or

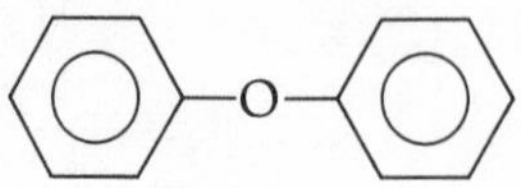

Structure and Nomenclature

The common names of ethers are formed by adding the word "ether" to the names of the R groups. The structural formulas and names of some common ethers are shown in Table 15-5. The IUPAC names for simple ethers are rarely used and are not listed. Notice in Table 15-5 that the two R groups on an ether may be the same or different. Diethyl ether, usually called simply "ether," is the most well-known ether. It was once widely used as an anesthetic and is still so used occasionally. Divinyl ether has also been used as an anesthetic. Notice that ethylene oxide is an example of an ether with a ring structure. Ethylene oxide is a gas that is used to sterilize plastic syringes and other materials that cannot be sterilized by high temperatures.

Table 15-5 Some Common Ethers

Structural Formula	Common Name
CH_3—O—CH_3	Dimethyl ether
CH_3—CH_2—O—CH_3	Methyl ethyl ether
CH_3—CH_2—O—CH_2—CH_3	Diethyl ether
CH_2=CH—O—CH=CH_2	Divinyl ether
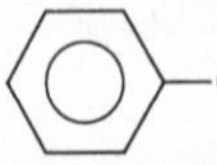 ⬡—O—CH_3	Methyl phenyl ether (anisole)
H_2C——CH_2 \\ O /	Ethylene oxide

Properties

Like alcohols, ethers can form hydrogen bonds with water molecules (see Figure 15-4a). Thus, the water solubility of ethers is comparable to that of alcohols. For example, the following functional group isomers, diethyl ether, CH_3—CH_2—O—CH_2—CH_3, and 1-butanol, CH_3—CH_2—CH_2—CH_2—OH, have similar solubilities, namely, 7.5 g/100 g of H_2O and 9 g/100 g of H_2O. Unlike alcohols, however, ethers *cannot* form hydrogen bonds in the pure state because they do not contain any hydrogen atoms bonded to highly electronegative atoms (Figure 15-4b).

As a result of their inability to form hydrogen bonds in the pure state, ethers have much lower boiling points than alcohols with similar molecular weights. For example, the boiling point of diethyl ether (MW = 74) is 35°C, which is much closer to that of the alkane pentane (MW = 72; b.p. = 36°C) than it is to 1-butanol (MW = 74; b.p. = 117°C). Because of its low boiling point and high vapor pressure, diethyl ether is easily inhaled, which is helpful when it is used as an anesthetic. However, ether vapor is extremely flammable. Flames and even static electrical sparks must be avoided when it is used. Operating room

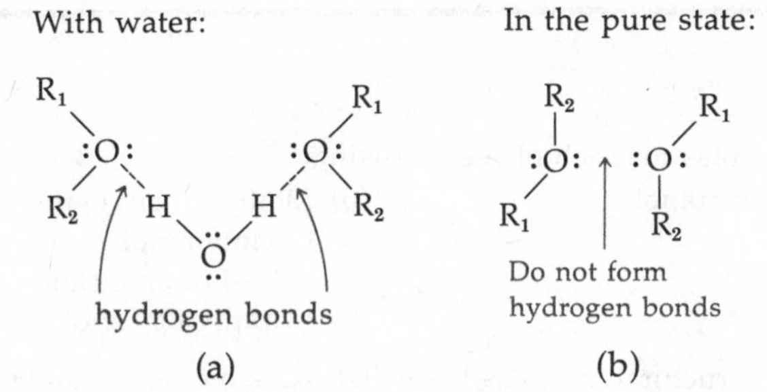

With water: In the pure state:

hydrogen bonds   Do not form
(a)   hydrogen bonds
                               (b)

Figure 15-4   Ethers form hydrogen bonds with   hydrogen bonds with other ether
water molecules (a) but cannot form   molecules in the pure state (b).

personnel wear conductive shoes and take other special precautions to prevent static electrical sparks, which could have disastrous effects if ether or other highly flammable materials are present.

The only reaction of ethers that we need to consider is their formation from alcohols; see reaction 15-2: $R—OH + R—OH \rightarrow R—O—R + H_2O$.

Exercise 15-5
Draw structural formulas for each of the following compounds:

(a) 3-Methyl-2-pentanol   (b) 2,3-Hexanediol   (c) Diisopropyl ether

Exercise 15-6
Identify the alcohol, ether, or phenol functional groups in the following structural formulas:

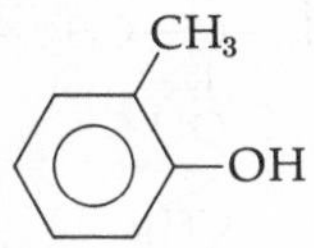

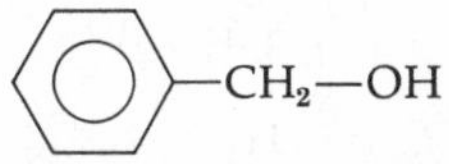

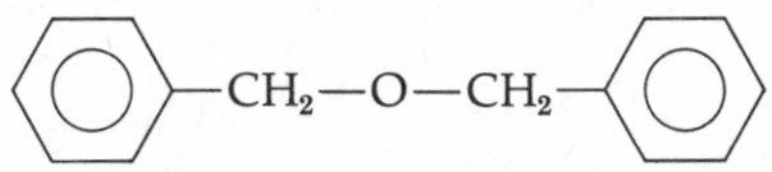

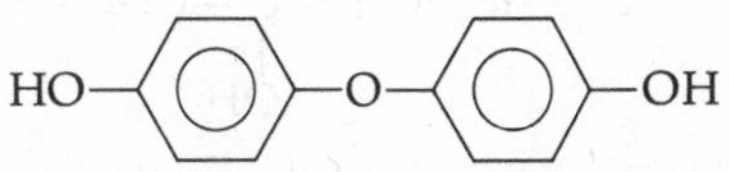

15-7 SUMMARY

1. The oxygen-containing portions of organic compounds are polar and hydrophilic.

2. The $-\overset{|}{\underset{|}{C}}-OH$ group is the alcohol functional group, and the $-OH$ group of alcohols and phenols is called the hydroxyl group.

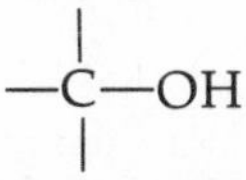

3. In aqueous solution, alcohols and ethers do not act as acids, and act as only extremely weak bases. Phenols act as weak acids and extremely weak bases.

4. Alcohols and ethers may be considered as organic derivatives of water. The general formula for an alcohol is $R—OH$; the general formula for an ether is $R_1—O—R_2$. The IUPAC ending for alcohols is -*ol*.

5. Alcohols can be dehydrated to form alkenes or ethers. Primary alcohols can be oxidized to form aldehydes. Secondary alcohols can be oxidized to ketones. Tertiary alcohols cannot be oxidized under similar conditions.

PROBLEMS

1. Draw structural formulas for each of the following:
   (a) 2,2-Dimethyl-1-pentanol
   (b) Methyl phenyl ether
   (c) Methyl ethyl ether
   (d) Cyclobutanol
   (e) 2,3-Heptanediol
   (f) 1,2,3-Propanetriol
   (g) Phenol
   (h) Isopropyl alcohol

2. Given the following structural formulas, write IUPAC names for (a) and (b), and write the common name for (c).

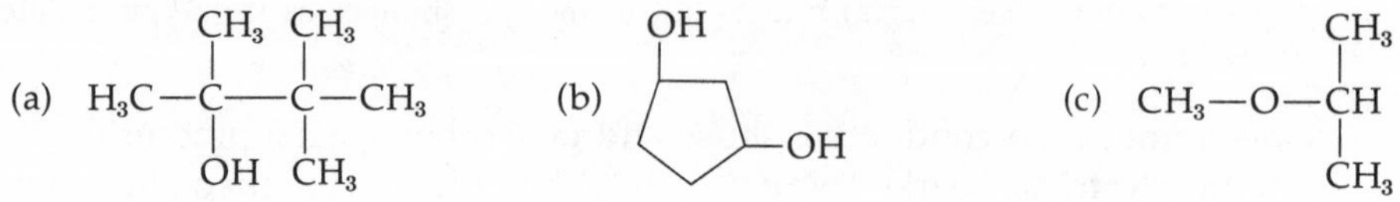

$$\text{(a)} \quad \overset{\displaystyle CH_3}{\underset{\displaystyle OH}{H_3C-\underset{|}{\overset{|}{C}}-}}\overset{\displaystyle CH_3}{\underset{\displaystyle CH_3}{\underset{|}{\overset{|}{C}}-CH_3}} \qquad\qquad \text{(b)} \qquad\qquad \text{(c)} \quad CH_3-O-\overset{\displaystyle CH_3}{\underset{\displaystyle CH_3}{\underset{|}{\overset{|}{CH}}}}$$

3. Arrange the compounds in Problem 2 in order of increasing solubility in water and predict which one has the lowest boiling point.

4. Draw the general structural formulas for
   (a) A secondary alcohol
   (b) A phenol
   (c) A nonaromatic ether

5. Draw structural formulas for the products of the following reactions (X symbolizes an oxidizing agent, and XH_2 symbolizes a reducing agent):

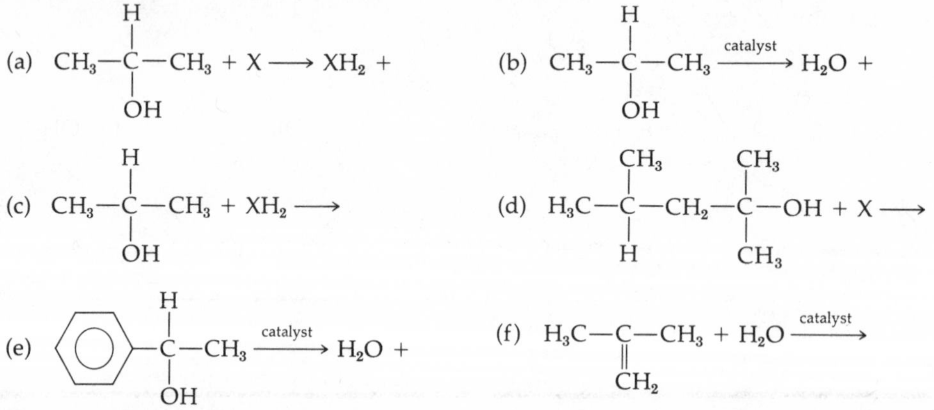

$$\text{(a)} \quad CH_3-\overset{\displaystyle H}{\underset{\displaystyle OH}{\underset{|}{\overset{|}{C}}}}-CH_3 + X \longrightarrow XH_2 +$$

$$\text{(b)} \quad CH_3-\overset{\displaystyle H}{\underset{\displaystyle OH}{\underset{|}{\overset{|}{C}}}}-CH_3 \xrightarrow{\text{catalyst}} H_2O +$$

$$\text{(c)} \quad CH_3-\overset{\displaystyle H}{\underset{\displaystyle OH}{\underset{|}{\overset{|}{C}}}}-CH_3 + XH_2 \longrightarrow$$

$$\text{(d)} \quad H_3C-\overset{\displaystyle CH_3}{\underset{\displaystyle H}{\underset{|}{\overset{|}{C}}}}-CH_2-\overset{\displaystyle CH_3}{\underset{\displaystyle CH_3}{\underset{|}{\overset{|}{C}}}}-OH + X \longrightarrow$$

$$\text{(e)} \quad \overset{\displaystyle H}{\underset{\displaystyle OH}{\underset{|}{\overset{|}{C}}}}-CH_3 \xrightarrow{\text{catalyst}} H_2O +$$

$$\text{(f)} \quad H_3C-\underset{\displaystyle CH_2}{\overset{\|}{C}}-CH_3 + H_2O \xrightarrow{\text{catalyst}}$$

6. Draw structural formulas for all of the structural isomers with the following molecular formula: C_3H_8O.

SOLUTIONS TO EXERCISES

15-1 $CH_2-CH_2-CH_3 < H_3C-\overset{\displaystyle CH_3}{\underset{\displaystyle CH_3}{\underset{|}{\overset{|}{C}}}}-(CH_2)_3-OH < CH_3-CH_2-CH_2OH$

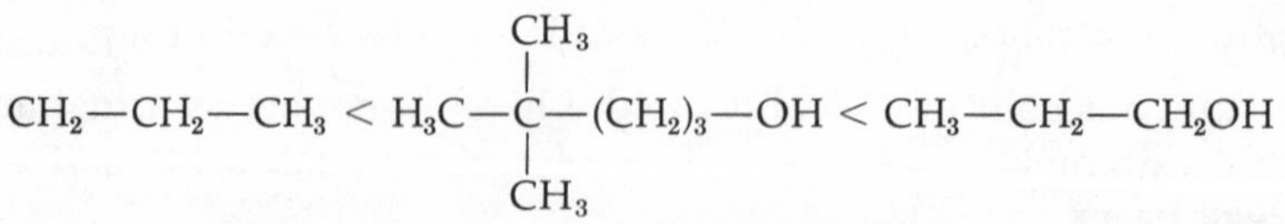

       insoluble alkane   soluble alcohols

Because of its smaller nonpolar portion, 1-propanol is most soluble.

15-2 Methanol is an alcohol and can act as both a hydrogen bond acceptor and donor, forming hydrogen bonds in the pure state. The other compounds are nonpolar alkanes, which cannot form hydrogen bonds.

15-3

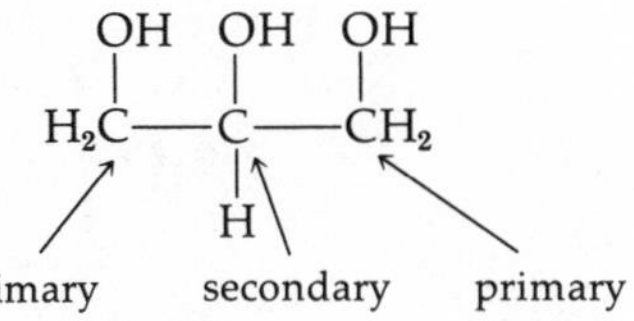

$$\underset{\text{primary}}{\text{H}_2\text{C}}-\underset{\overset{|}{\text{H}}}{\underset{\text{secondary}}{\text{C}}}-\underset{\text{primary}}{\text{CH}_2}$$

with OH OH OH above the carbons.

15-4 Name   Products

(a) Ethanol $\text{CH}_3-\text{CH}_2-\text{O}-\text{CH}_2-\text{CH}_3$

(b) Ethanol $\text{H}_2\text{C}{=}\text{CH}_2$

(c) 1-Propanol $\text{CH}_3-\text{CH}_2-\overset{\displaystyle \overset{\text{O}}{\|}}{\text{C}}-\text{H}$

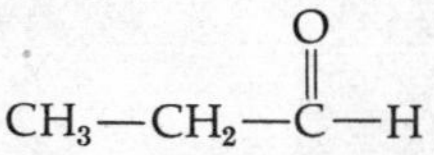

(d) 3,3-Dimethyl-2-butanol $\text{CH}_3-\overset{\overset{\text{CH}_3}{|}}{\underset{\underset{\text{CH}_3}{|}}{\text{C}}}-\overset{\overset{\text{O}}{\|}}{\text{C}}-\text{CH}_3$

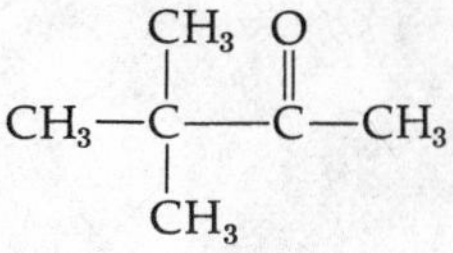

15-5 (a) 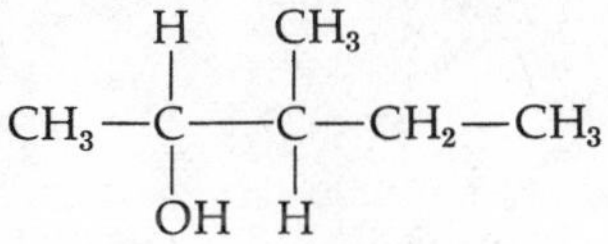 $\text{CH}_3-\overset{\overset{\text{H}}{|}}{\underset{\underset{\text{OH}}{|}}{\text{C}}}-\overset{\overset{\text{CH}_3}{|}}{\underset{\underset{\text{H}}{|}}{\text{C}}}-\text{CH}_2-\text{CH}_3$

(b) $\text{CH}_3-\overset{\overset{\text{OH}}{|}}{\text{CH}}-\overset{\overset{\text{OH}}{|}}{\text{CH}}-\text{CH}_2-\text{CH}_2-\text{CH}_3$

(c) $\underset{\underset{\text{CH}_3}{|}}{\overset{\overset{\text{CH}_3}{|}}{\text{HC}}}-\text{O}-\underset{\underset{\text{CH}_3}{|}}{\overset{\overset{\text{CH}_3}{|}}{\text{CH}}}$

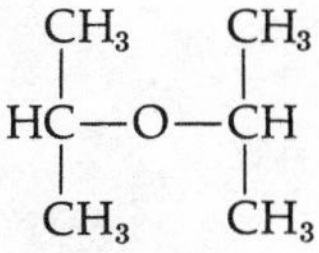

15-6 (a) 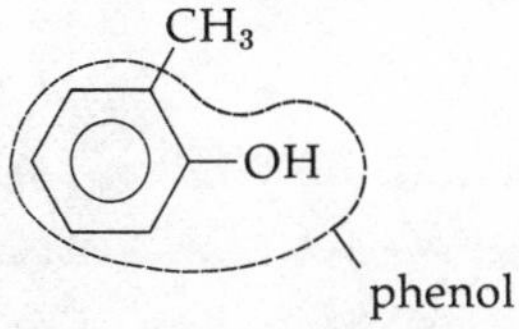 phenol ring with CH$_3$ and —OH, circled "phenol"

(b) 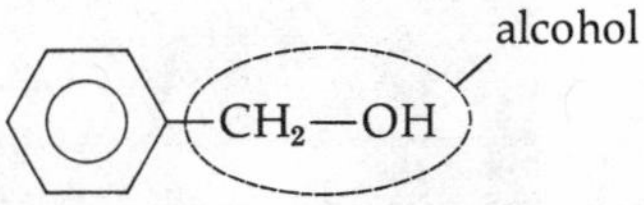 ring —CH$_2$—OH, labeled "alcohol"

(c) 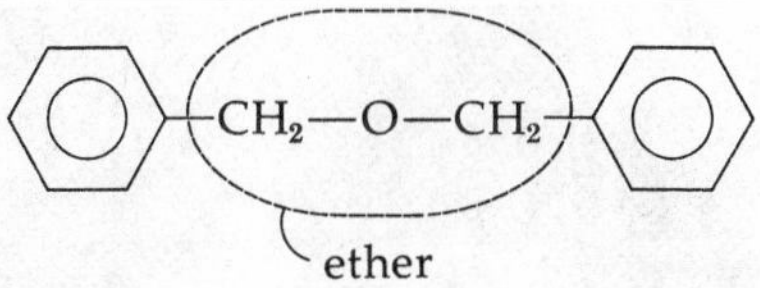 ring —CH$_2$—O—CH$_2$— ring, labeled "ether"

(d) 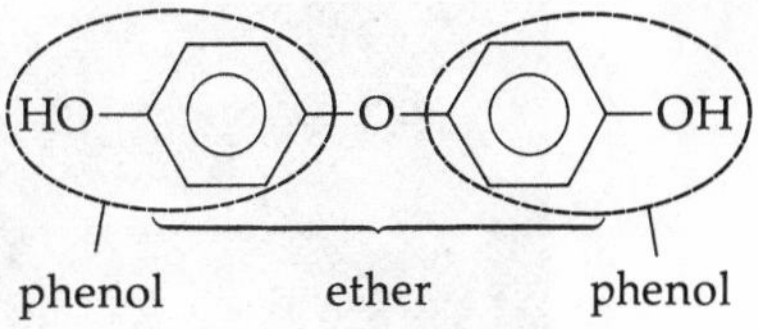 HO— ring —O— ring —OH, labeled "phenol   ether   phenol"

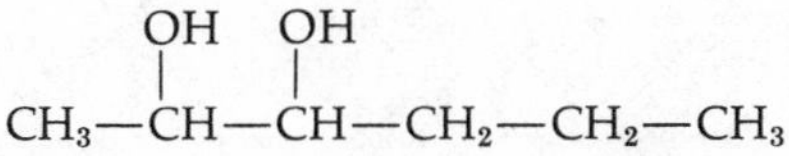

CHAPTER 16

Aldehydes and Ketones

16-1   INTRODUCTION

The compounds we studied in Chapter 15 contained an oxygen atom bonded to a carbon atom with a single bond. All of the other classes of oxygen-containing organic compounds that we shall discuss contain a carbon-oxygen double

bond, $-\overset{\overset{\displaystyle O}{\|}}{C}-$, which is called a **carbonyl group.** Most of the organic compounds found in nature, including DNA, proteins, and simple carbohydrates, contain carbonyl groups. Many synthetic organic compounds, including a number of polymers, also contain carbonyl groups.

The simplest organic compound that contains a carbonyl group is formalde-

hyde, $H-\overset{\overset{\displaystyle O}{\|}}{C}-H$, the only compound with a carbonyl group that is a gas at room temperature. It is a widespread atmospheric pollutant, formed in the incomplete combustion of such fuels as coal and wood, and it tends to accumulate over large urban areas. Formaldehyde is quite soluble in water; formalin, a concentrated solution of formaldehyde in water (usually 40 grams of formaldehyde per 100 ml of solution), is widely used in modern society as a disinfectant and germicide, in photographic processing, and as one of the starting materials in the manufacture of polymers, dyes, explosives, and other complex synthetic organic chemicals. Formaldehyde solutions are also used as embalming fluids.

Aldehydes and ketones, the subject of this chapter, have only one oxygen atom bonded to the carbon atom of the carbonyl group. The properties of both of these classes of compounds are due primarily to the carbonyl group. On the other hand, carboxylic acids and their derivatives have an additional oxygen atom bonded to the carbonyl carbon atom, giving the carboxyl group,

$-\overset{\overset{\displaystyle O}{\|}}{C}-O-$. Many of the properties of carboxylic acids, which will be discussed in the next chapter, are quite different from those of aldehydes and ketones.

Formaldehyde has several uses. It is used as the starting material for the industrial synthesis of a number of complex chemicals. It is also the active ingredient in laboratory preservatives and embalming fluids.

16-2   STUDY OBJECTIVES

After careful study of this chapter, you should be able to:

1. Identify carbonyl groups and distinguish an aldehyde or a ketone from other compounds that contain a carbonyl group, given the structural formula for a compound.

2. Explain the role of hydrogen bonding in the water solubility of aldehydes and ketones.

3. Draw the structural formula for a simple aldehyde or ketone, given its name, or vice versa.

4. Draw structural formulas for the products of reactions in which: (a) aldehydes or ketones are oxidized or reduced; (b) aldehydes undergo an aldol condensation; (c) hemiacetals, acetals, hemiketals, or ketals are formed or are hydrolyzed, given the structural formulas of the reactants.

16-3   ALDEHYDES

Aldehydes have the general formula $R'-\overset{\overset{\displaystyle O}{\|}}{C}-H$, where R' can be a hydrogen atom, an alkyl group, an aromatic group, or some other group of atoms.

Structure and Nomenclature

The IUPAC names for aldehydes are derived from the name of the parent alkane, with the ending -*al* to indicate the presence of the aldehyde functional

group. Thus, the IUPAC names of the aldehydes $H-\overset{\overset{\displaystyle O}{\|}}{C}-H$ and $CH_3-\overset{\overset{\displaystyle O}{\|}}{C}-H$ are methanal and ethanal, respectively. Note that the final -*e* of the alkane name is dropped when forming the name of the aldehyde. The rules for naming aldehydes and other organic compounds are summarized in Essential Skills 8.

Most aldehydes are usually referred to by their common names. The common name of an aldehyde is derived from the common name of the carboxylic acid that is produced when the aldehyde is oxidized. For example, the common name for methanal is formaldehyde, since it can be oxidized to produce formic

acid, $H-\overset{\overset{\displaystyle O}{\|}}{C}-OH$. The structural formulas and names of some simple aldehydes are given in Table 16-1.

The aromatic aldehyde benzaldehyde is used in the manufacture of dyes and for almond flavoring. In high concentrations benzaldehyde is a narcotic, and in very high concentrations it is lethal.

When common names of aldehydes are used, the Greek letters α, β, γ, δ, and so on, are used to indicate the positions of substituents on the hydrocarbon chain. For example, the structural formula of β-hydroxybutyraldehyde is

$$\underset{\gamma}{CH_3}-\underset{\underset{\displaystyle\beta}{\overset{\overset{\displaystyle OH}{|}}{CH}}}-\underset{\alpha}{CH_2}-\overset{\overset{\displaystyle O}{\|}}{C}-H$$

In the IUPAC name for this compound, the position of the hydroxyl group is indicated by a numerical prefix. Its IUPAC name is 3-hydroxybutanal. Note that in the common names for aldehydes the α position refers to the carbon atom *next* to the carbonyl group carbon atom, whereas in the IUPAC name this is carbon atom 2 (the carbonyl group carbon atom is carbon atom 1). The components of IUPAC and common names must never be mixed. For example, β-hydroxybutanal and 3-hydroxybutyraldehyde are *not* acceptable names for β-hydroxybutyraldehyde (3-hydroxybutanal).

Table 16-1 Some Simple Aldehydes

IUPAC Name	Common Name	Structural Formula
Methanal	Formaldehyde	$H-\overset{\overset{\displaystyle O}{\|\|}}{C}-H$
Ethanal	Acetaldehyde	$CH_3-\overset{\overset{\displaystyle O}{\|\|}}{C}-H$
Propanal	Propionaldehyde	$CH_3-CH_2-\overset{\overset{\displaystyle O}{\|\|}}{C}-H$
	Benzaldehyde	$\langle\bigcirc\rangle-\overset{\overset{\displaystyle O}{\|\|}}{C}-H$

Properties of Aldehydes

As we discussed in Chapter 15, the carbonyl group is polar as a result of the large electronegativity of the oxygen atom. Thus, low-molecular-weight aldehydes are soluble in water because hydrogen bonds form between the aldehyde molecules and water molecules (see Figure 16-1a). Aldehydes cannot, however, form hydrogen bonds in the pure state (see Figure 16-1b). The hydrogen atom

bonded to the carbonyl carbon atom, $-\overset{\overset{\displaystyle O}{\|\|}}{C}-H$, does not participate in hydrogen bonds. Thus, the boiling points of aldehydes are lower than those of hydrogen bond-forming alcohols with about the same molecular weight (see Table 16-2).

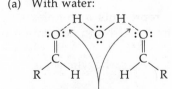

(a) With water: (b) In the pure state:

hydrogen bonds   no hydrogen bonds

Figure 16-1   Aldehydes form hydrogen bonds with water (a), but they cannot form hydrogen bonds with each other in the pure state (b).

Note, however, that because of the polar nature of the carbonyl group, the intermolecular attractions are larger and thus the boiling points of aldehydes are higher than those of nonpolar alkanes of similar molecular weight.

In aqueous solution, aldehydes do not act as acids and only as extremely weak Brønsted-Lowry bases.

Table 16-2 Boiling Points of Aldehydes and Other Compounds with Similar Molecular Weights

Name	Structural Formula	MW	Boiling Point (°C)
Propane	$CH_3-CH_2-CH_3$	44	−42
Acetaldehyde	$CH_3-\overset{\displaystyle O}{\overset{\|}{C}}-H$	44	20
Ethanol	CH_3-CH_2-OH	46	78
Butane	$CH_3-CH_2-CH_2-CH_3$	58	0
Propionaldehyde	$CH_3-CH_2-\overset{\displaystyle O}{\overset{\|}{C}}-H$	58	49
1-Propanol	$CH_3-CH_2-CH_2-OH$	60	97

Reactions of Aldehydes

OXIDATION-REDUCTION

In the last chapter we saw that an aldehyde can be produced by the oxidation of a primary alcohol. The reverse reaction, formation of a primary alcohol by the reduction of an aldehyde, can also occur under appropriate conditions:

(16-1)
$$R'-\overset{\displaystyle O}{\overset{\|}{C}}-H + XH_2 \xrightarrow{\text{catalyst}} R'-\overset{\displaystyle OH}{\underset{\displaystyle H}{\overset{\|}{C}}}-H + X$$

aldehyde   primary alcohol

Reaction 16-1 represents the general reaction for the reduction of an aldehyde. In this representation, the reducing agent is symbolized by XH_2. For example, yeasts produce the primary alcohol ethanol in the enzyme-catalyzed reduction of acetaldehyde:

(16-2)
$$H_3C-\overset{\displaystyle O}{\overset{\|}{C}}-H + NADH + H^+ \xrightarrow{\text{a specific enzyme}} CH_3-CH_2-OH + NAD^+$$

acetaldehyde   ethanol

In this reaction, the symbol NAD^+ represents a complex substance called a coenzyme (Chapter 23), which is produced by the oxidation of the reducing agent NADH.

Aldehydes can also be oxidized to form carboxylic acids:

(16-3)
$$R'-\overset{\displaystyle O}{\overset{\|}{C}}-H + X + H_2O \longrightarrow R'-\overset{\displaystyle O}{\overset{\|}{C}}-OH + XH_2$$

aldehyde   carboxylic acid

Reaction 16-3 represents the general reaction for the oxidation of an aldehyde. In this representation, X symbolizes the oxidizing agent, and the additional oxygen atom in the carboxylic acid product is supplied by a water molecule. The reactant aldehyde loses a C—H bond and gains a C—O bond, so reaction 16-3 does indeed represent the oxidation of the aldehyde. The oxidation of ethyl alcohol to acetaldehyde and then to acetic acid occurs when wine is exposed to air. Vinegar (an aqueous solution of acetic acid) is the final product.

FORMATION OF HEMIACETALS AND ACETALS

An aldehyde can also react with an alcohol to form another compound, called a **hemiacetal,** in the general reaction

(16-4)

$$\underset{\text{aldehyde}}{R'-\overset{\overset{\textstyle O}{\|}}{C}-H} + \underset{\text{alcohol}}{H-OR} \rightleftharpoons \underset{\text{hemiacetal}}{R'-\overset{\overset{\textstyle OH}{|}}{\underset{\underset{\textstyle H}{|}}{C}}-OR}$$

Hemiacetal formation is an addition reaction in which the components of an alcohol, —OR and —H, add to the C=O bond. The addition of an alcohol to a C=O bond is similar to the hydration reaction in which the components of water are added to a C=C double bond (see Figure 16-2).

Figure 16-2   The addition of an alcohol to the C=O bond of an aldehyde forms a hemiacetal. This is similar to the addition of water to a C=C bond in which an alcohol is formed.

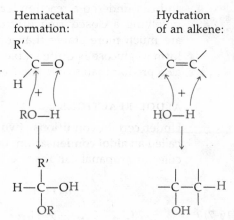

Reaction 16-4 is carried out in the laboratory by dissolving an aldehyde in an alcohol. In this alcoholic solution the equilibrium (reaction 16-4) is established, but most hemiacetals that do not involve a closed ring of atoms are not very stable and cannot be isolated from the alcoholic solution.

If an acid, such as HCl, is added to the equilibrium mixture in reaction 16-4, the hemiacetal can react further with another molecule of the alcohol to produce a compound called an **acetal:**

(16-5)

$$\underset{\text{hemiacetal}}{R'-\overset{\overset{\textstyle OH}{|}}{\underset{\underset{\textstyle H}{|}}{C}}-OR} + \underset{\text{alcohol}}{ROH} \xrightarrow{\text{H}^+} \underset{\text{acetal}}{R'-\overset{\overset{\textstyle OR}{|}}{\underset{\underset{\textstyle H}{|}}{C}}-OR} + H_2O$$

The H^+ in reaction 16-5 acts as a catalyst. Acetals, in contrast to most open-chain hemiacetals, can be isolated by evaporating off the excess alcohol from the reaction mixture. On the other hand, if an acetal is dissolved in water and acid is added, a hydrolysis reaction equivalent to the reverse of reactions 16-4 and 16-5 takes place. The general reaction for the hydrolysis of an acetal is

(16-6)

$$\underset{\text{acetal}}{R'-\overset{\overset{\textstyle OR}{|}}{\underset{\underset{\textstyle OR}{|}}{C}}-H} + H_2O \xrightarrow{\text{H}^+} \underset{\text{aldehyde}}{R'-\overset{\overset{\textstyle O}{\|}}{C}-H} + 2ROH$$

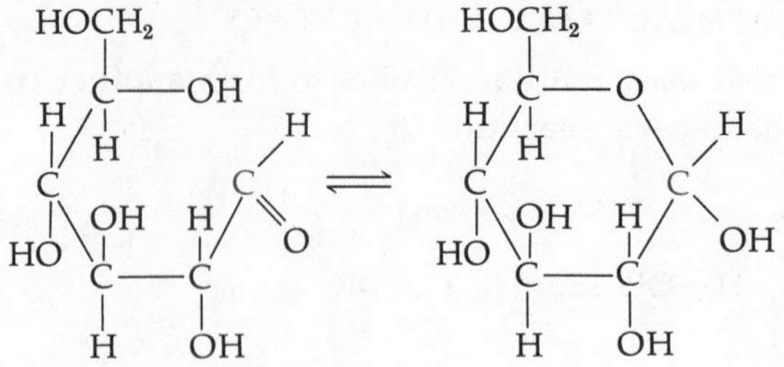

Figure 16-3   Internal hemiacetal formation by glucose. The free aldehyde form of glucose (left) reacts to form the hemiacetal (right). More than 99% of the dissolved glucose found in the human body is in the hemiacetal form, and solid glucose exists only as the hemiacetal.

Sugars, such as glucose, contain both aldehyde and alcohol functional groups and can undergo a reaction leading to the formation of an **internal hemiacetal** involving a closed ring of atoms (see Figure 16-3). Closed-ring hemiacetals are much more stable than open-chain hemiacetals. The internal hemiacetal form of glucose is quite stable and can be isolated from solution. In solution it also predominates over the free aldehyde form.

ALDOL REACTIONS

Under certain conditions, two aldehyde molecules can combine in a reaction called an **aldol condensation,** or aldol addition reaction. For example, two molecules of propanal can react in an aqueous solution of sodium hydroxide:

(16-7)
$$2(CH_3-CH_2-\overset{\overset{O}{\|}}{C}-H) \xrightarrow{OH^-} CH_3-CH_2-\overset{\overset{OH}{|}}{CH}-\underset{\underset{CH_3}{|}}{CH}-\overset{\overset{O}{\|}}{C}-H$$

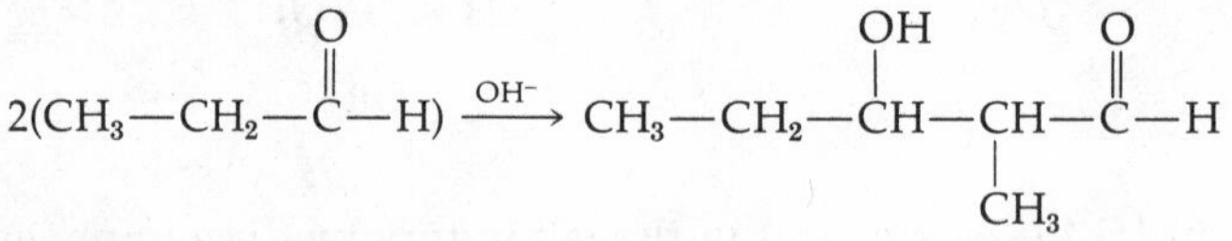

The product of reaction 16-7, 3-hydroxy-2-methylpentanal, has both **ald**ehyde and alcoh**ol** functional groups. This is where the name **aldol** comes from.

Aldol condensations, such as reaction 16-7, take place by means of a mechanism in which the base, OH^-, pulls off one of the weakly acidic α-hydrogens (Figure 16-4a). The resulting negatively charged ion combines with a second propanal molecule (Figure 16-4b). The oxygen atom of this second propanal then pulls a H^+ from a water molecule (Figure 16-4c). Note that the overall reaction for the mechanism shown in Figure 16-4 is reaction 16-7, and that the catalyst OH^- is regenerated.

Reaction 16-7 can be viewed as the α-carbon and an α-hydrogen of one aldehyde adding to the carbonyl group of a second aldehyde to give a β-hydroxy-aldehyde product. The general aldol reaction for aldehydes that have an α-hydrogen can be represented as

(16-8)
$$2\left(R-\overset{\overset{R'}{|}}{CH}-\overset{\overset{O}{\|}}{C}-H\right) \xrightarrow{OH^-} R-\underset{\underset{H}{|}}{\overset{\overset{R'}{|}}{C}}-\underset{\underset{H}{|}}{\overset{\overset{OH}{|}}{C}}-\underset{\underset{R}{|}}{\overset{\overset{R'}{|}}{C}}-\overset{\overset{O}{\|}}{C}-H$$

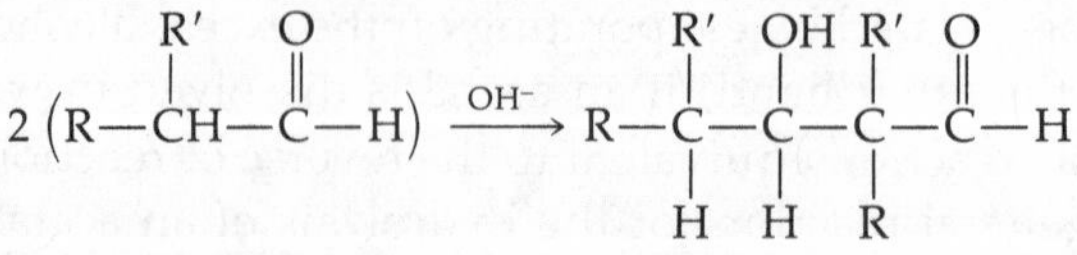

Recall that R′ represents R or H. In order for this general reaction to take place, the reactant aldehyde must have an α-hydrogen (see Figure 16-4a). Thus, the α-carbon must be a primary or secondary carbon atom; aldehydes with a tertiary α-carbon atom do not undergo an aldol condensation.

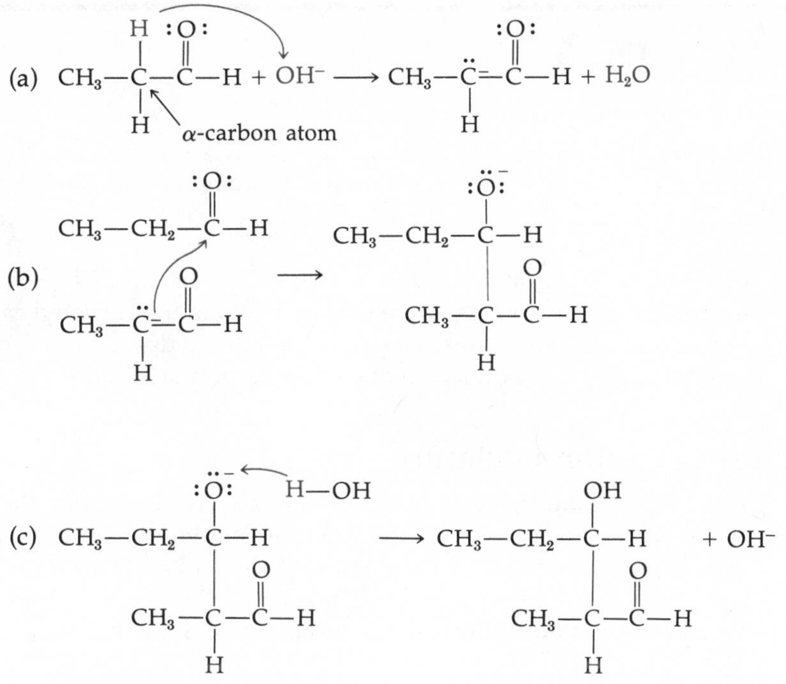

(a) $CH_3-\overset{\overset{\displaystyle H}{|}}{\underset{\underset{\displaystyle H}{|}}{C}}-\overset{\overset{\displaystyle :O:}{\|}}{C}-H + OH^- \longrightarrow CH_3-\overset{..}{C}=\overset{\overset{\displaystyle :O:}{\|}}{\underset{\underset{\displaystyle H}{}}{C}}-H + H_2O$

α-carbon atom

(b) $CH_3-CH_2-\overset{\overset{\displaystyle :O:}{\|}}{C}-H$ $CH_3-CH_2-\overset{\overset{\displaystyle :\overset{..}{O}:^-}{|}}{C}-H$

$CH_3-\overset{..}{C}=\overset{\overset{\displaystyle O}{\|}}{C}-H$ $\longrightarrow$ $CH_3-\overset{\overset{\displaystyle O}{\|}}{C}-C-H$

(c) $CH_3-CH_2-\overset{\overset{\displaystyle :\overset{..}{O}:^-}{|}}{C}-H$ $H-OH$ $\longrightarrow$ $CH_3-CH_2-\overset{\overset{\displaystyle OH}{|}}{C}-H$ $+ OH^-$

$CH_3-\overset{\overset{\displaystyle O}{\|}}{C}-\overset{\underset{\underset{\displaystyle H}{|}}{}}{C}-H$ $CH_3-\overset{\overset{\displaystyle O}{\|}}{C}-\overset{\underset{\underset{\displaystyle H}{|}}{}}{C}-H$

Figure 16-4 Aldol condensation reactions take place by a three-step mechanism. (a) The α-hydrogen atoms in aldehydes are weakly acidic because of the pull on their electrons by the neighboring oxygen atom. A base, OH$^-$ (or some other catalyst), first removes one of the α-hydrogens as a proton, leaving an unshared pair of electrons and a negative charge on the α-carbon atom. (b) The resulting negatively charged α-carbon atom attacks the carbonyl carbon atom of another aldehyde molecule, which has a partial positive charge. (c) The negatively charged oxygen atom of the product in (b) extracts a proton from a water molecule.

Aldol reactions are important in the commercial synthesis of many chemicals. Aldol-type reactions are also vital to the metabolism of carbohydrates in human cells (Chapter 25). In human cells, however, the catalysts for aldol reactions are specific enzymes.

Exercise 16-1
Draw structural formulas for the products of each of the following reactions (X symbolizes an oxidizing agent and XH$_2$ a reducing agent):

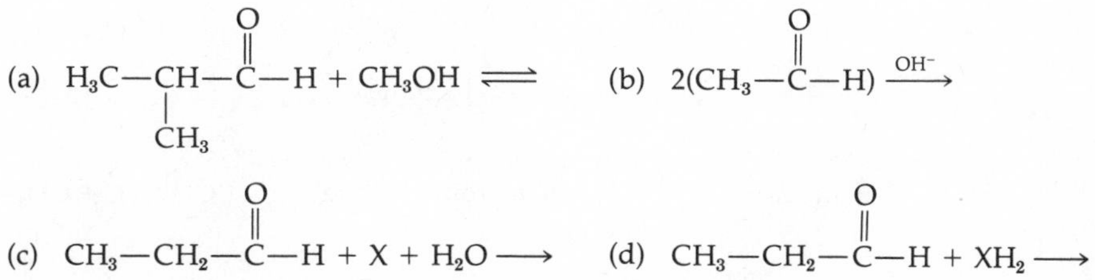

(a) $H_3C-\underset{\underset{\displaystyle CH_3}{|}}{CH}-\overset{\overset{\displaystyle O}{\|}}{C}-H + CH_3OH \rightleftharpoons$

(b) $2(CH_3-\overset{\overset{\displaystyle O}{\|}}{C}-H) \xrightarrow{OH^-}$

(c) $CH_3-CH_2-\overset{\overset{\displaystyle O}{\|}}{C}-H + X + H_2O \longrightarrow$

(d) $CH_3-CH_2-\overset{\overset{\displaystyle O}{\|}}{C}-H + XH_2 \longrightarrow$

If you have difficulty with this type of problem, review the material in Essential Skills 9.

Exercise 16-2
Write the IUPAC names for the reactants in (a) and (b) of Exercise 16-1.

16-4   KETONES

Ketones have the general formula R_1—$\overset{\displaystyle O}{\overset{\|}{C}}$—$R_2$, where R_1 and R_2 represent alkyl, aromatic, or other groups of atoms. Ketones and aldehydes with the *same* number of carbon atoms, such as CH_3—$\overset{\displaystyle O}{\overset{\|}{C}}$—$CH_3$ and CH_3—CH_2—$\overset{\displaystyle O}{\overset{\|}{C}}$—H, are functional group isomers. Note that in a ketone the carbonyl carbon is bonded to two other carbon atoms, whereas in all aldehydes, except formaldehyde, the carbonyl carbon is bonded to one other carbon atom and a hydrogen atom.

Structure and Nomenclature

The IUPAC names for simple ketones are derived from the parent alkane by use of the ending -*one* to indicate the ketone functional group. The ending -*one* is pronounced as in bone without the *b*. When a ketone has a chain of more than four carbon atoms, the position of the carbonyl group is indicated by adding a prefix number. For example, the IUPAC name for the ketone

CH_3—CH_2—$\overset{\displaystyle O}{\overset{\|}{C}}$—$CH_2$—$CH_3$ is 3-pentanone. Common, rather than IUPAC, names are often used for simple ketones. The common names for many ketones are formed by adding the name *ketone* to the names for the alkyl groups, R_1 and R_2, bonded to the carbonyl group. Thus, the common name for 3-pentanone is diethyl ketone. The structural formulas and names for some simple ketones are given in Table 16-3. Of the ketones listed in Table 16-3, acetone is the best known. It is an excellent solvent for many organic compounds. Acetone is also normally produced in small quantities in the human body. In diabetics, large quantities of acetone and other ketones may be produced. These "ketone bodies" are responsible for many of the serious effects of diabetes (Chapter 27). Acetophenone has a pleasant odor, like orange blossoms, and because of this property, it is used in some perfumes. This ketone is also used for the industrial synthesis of a number of compounds.

Table 16-3 Some Simple Ketones

IUPAC Name	Common Name	Structural Formula
Propanone	Acetone (dimethyl ketone)	CH_3—$\overset{\displaystyle O}{\overset{\|}{C}}$—$CH_3$
Butanone	Methyl ethyl ketone	CH_3—$\overset{\displaystyle O}{\overset{\|}{C}}$—$CH_2$—$CH_3$
2-Pentanone	Methyl *n*-propyl ketone	CH_3—$\overset{\displaystyle O}{\overset{\|}{C}}$—$CH_2$—$CH_2$—$CH_3$
3-Pentanone	Diethyl ketone	CH_3—CH_2—$\overset{\displaystyle O}{\overset{\|}{C}}$—$CH_2$—$CH_3$
	Acetophenone	CH_3—$\overset{\displaystyle O}{\overset{\|}{C}}$—⬡

Properties of Ketones

Low-molecular-weight ketones, like low-molecular-weight aldehydes, are water soluble because hydrogen bonds form between water molecules and the carbonyl oxygen atom (see Figure 16-5a). Ketones, like aldehydes, do not form hydrogen bonds in the pure state (see Figure 16-5b) and hence have boiling points comparable to their aldehyde isomers. For example, the boiling point of butanal is 76°C, whereas that of butanone is 80°C.

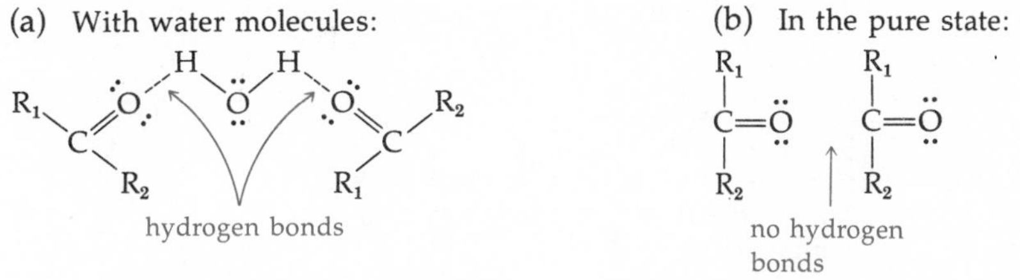

(a) With water molecules:

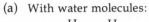

hydrogen bonds

(b) In the pure state:

$$\underset{R_2}{\overset{R_1}{\mid}} C = \ddot{O} \qquad \underset{R_2}{\overset{R_1}{\mid}} C = \ddot{O}$$

no hydrogen bonds

Figure 16-5 Ketones form hydrogen bonds with water (a), but they cannot form hydrogen bonds with each other in the pure state (b).

Reactions of Ketones

OXIDATION-REDUCTION

Recall that the oxidation of a secondary alcohol produces a ketone (Chapter 15). In the reverse reaction a secondary alcohol can be formed by the reduction of a ketone:

(16-9)

$$\underset{\text{ketone}}{R_1 - \overset{O}{\overset{\|}{C}} - R_2} + XH_2 \longrightarrow \underset{\substack{\text{secondary} \\ \text{alcohol}}}{R_1 - \overset{OH}{\underset{H}{\overset{\mid}{C}}} - R_2} + X$$

Reaction 16-9 represents the general reaction for the reduction of a ketone, where the reducing agent is symbolized by XH_2.

A biologically important example is the reduction of the ketone functional group in the compound pyruvic acid to the hydroxyl group in the compound lactic acid:

(16-10)

$$\underset{\text{pyruvic acid}}{HO - \overset{O}{\overset{\|}{C}} - \overset{O}{\overset{\|}{C}} - CH_3} + NADH + H^+ \xrightarrow[\text{enzyme}]{\text{a specific}} \underset{\text{lactic acid}}{HO - \overset{O}{\overset{\|}{C}} - \overset{OH}{\underset{H}{\overset{\mid}{C}}} - CH_3} + NAD^+$$

Reaction 16-10 occurs in active muscle cells. In this reaction the reducing agent is the complex coenzyme symbolized by NADH.

Unlike aldehydes, ketones *cannot* be oxidized to carboxylic acids.

FORMATION OF HEMIKETALS AND KETALS

Certain ketones, in reactions similar to those of aldehydes, can also react with some alcohols to produce compounds called **hemiketals** and **ketals.** Ketals can be hydrolyzed to yield an alcohol and a ketone in a reaction similar to the hydrolysis of an acetal. In Chapter 20 we shall see that sugars, such as fructose, that possess both alcohol and ketone functional groups can undergo a reaction leading to the formation of an **internal hemiketal** involving a closed ring of atoms. The formation of an internal hemiketal by fructose (see Figure 16-6) is similar to the formation of an internal hemiacetal by the aldehyde sugar glucose (Figure 16-3).

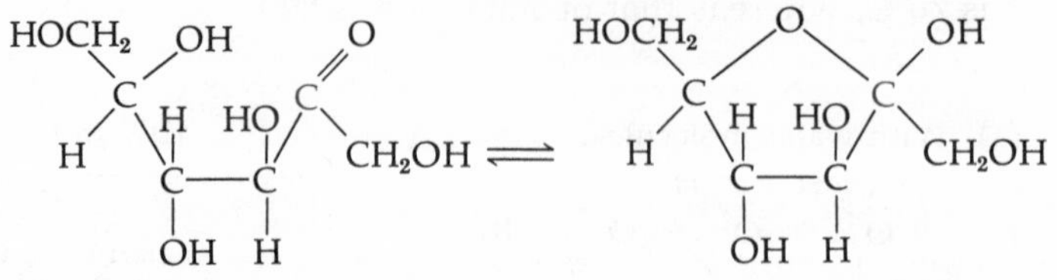

Figure 16-6   The free ketone form of fructose (left) reacts to form the internal hemiketal (right).

ALDOL CONDENSATION

Ketones that contain at least one primary or secondary α-carbon atom can undergo an aldol condensation reaction to give a β-hydroxyketone. For example,

(16-11)

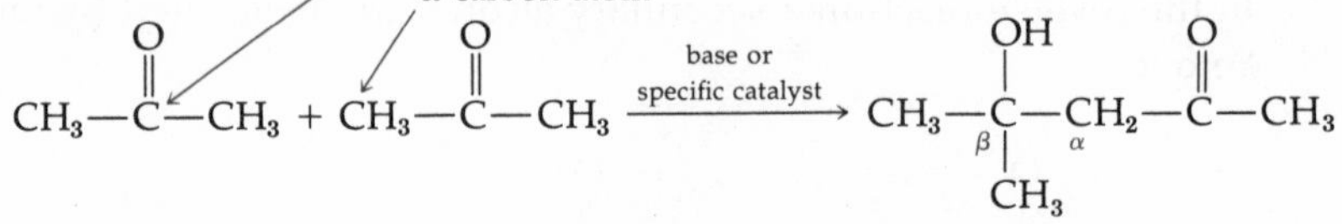

Note in the product of reaction 16-11 that the carbonyl carbon atom of one reactant ketone is bonded to an α-carbon atom of the other reactant ketone.

Exercise 16-3
Write the IUPAC name of the product of reaction 16-11.

Exercise 16-4
Draw structural formulas for the products of the following reactions (if no reaction occurs, write N.R.). Note that X is the symbol for an oxidizing agent and XH_2 represents a reducing agent.

(a) 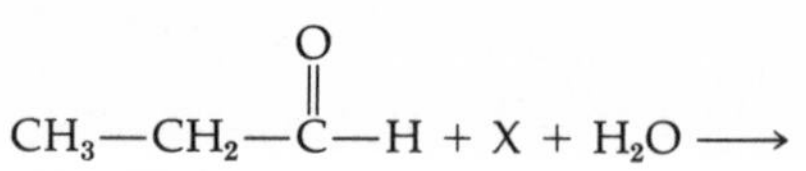 $CH_3—CH_2—\overset{\overset{\displaystyle O}{\|}}{C}—H + X + H_2O \longrightarrow$

(b) $H—\overset{\overset{\displaystyle CH_3}{|}}{\underset{\underset{\displaystyle CH_3}{|}}{C}}—\overset{\overset{\displaystyle O}{\|}}{C}—CH_3 + XH_2 \longrightarrow$

(c) $H_3C—\overset{\overset{\displaystyle CH_3}{|}}{\underset{\underset{\displaystyle CH_2—CH_3}{|}}{C}}—OH \quad + X \longrightarrow$

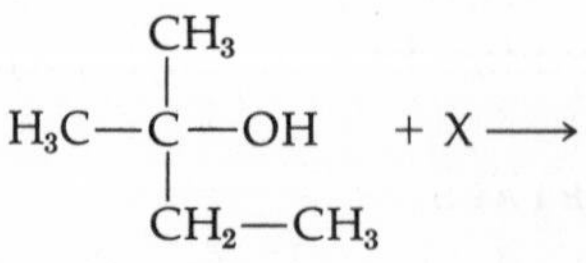

(d) 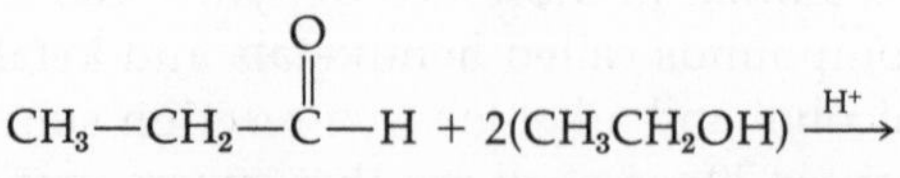 $CH_3—CH_2—\overset{\overset{\displaystyle O}{\|}}{C}—H + 2(CH_3CH_2OH) \overset{H^+}{\longrightarrow}$

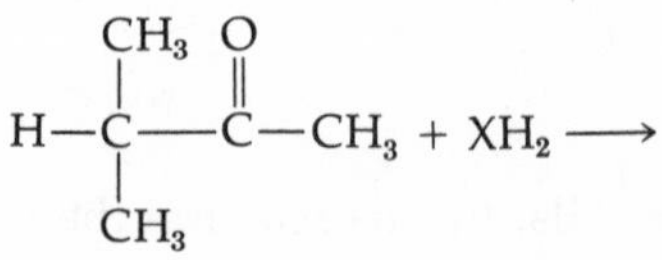

16-5   SUMMARY

1. The $-\overset{\displaystyle O}{\overset{\|}{C}}-$ group is called the carbonyl group.

2. Aldehydes have the general structural formula $R'-\overset{\displaystyle O}{\overset{\|}{C}}-H$, and their IUPAC names end in *-al*.

3. Ketones have the general structural formula $R_1-\overset{\displaystyle O}{\overset{\|}{C}}-R_2$, and their IUPAC names end in *-one*.

4. Aldehydes and ketones are polar and form hydrogen bonds with water. They do not form hydrogen bonds in the pure state.

5. Aldehydes and ketones can be formed by the oxidation of primary and secondary alcohols, respectively. In reverse reactions, primary and secondary alcohols can be formed by the reduction of aldehydes and ketones, respectively.

6. Aldehydes can be oxidized to form carboxylic acids; ketones cannot.

7. Aldehydes can react with alcohols to form hemiacetals and acetals. Some ketones can similarly form hemiketals and ketals. Acetals and ketals can be hydrolyzed to produce an aldehyde or ketone plus an alcohol.

8. Aldehydes and ketones that possess primary or secondary α-carbon atoms can undergo a type of reaction called aldol condensation to form β-hydroxyaldehydes or β-hydroxyketones, respectively.

PROBLEMS

1. Draw structural formulas for each of the following:
   (a) 2,2-Dimethylpentanal
   (b) 2,2-Dimethyl-3-pentanone
   (c) 4-Hydroxybutanal
   (d) α-Hydroxypropionaldehyde

2. Write the IUPAC names for the following compounds:

   (a) $CH_3-\underset{\underset{\displaystyle CH_3}{|}}{CH}-\overset{\displaystyle O}{\overset{\|}{C}}-H$

   (b) $\underset{\underset{\displaystyle CH_3}{|}}{\overset{\overset{\displaystyle CH_3}{|}}{HC}}-\overset{\displaystyle O}{\overset{\|}{C}}-\underset{\underset{\displaystyle CH_3}{|}}{\overset{\overset{\displaystyle CH_3}{|}}{CH}}$

   (c) $CH_3-\underset{\underset{\displaystyle H}{|}}{\overset{\overset{\displaystyle OH}{|}}{C}}-\underset{\underset{\displaystyle O}{\|}}{C}-CH_3$

   (d)

   (e) $CH_3-\overset{\displaystyle O}{\overset{\|}{C}}-CH_2-CH_3$

3. Which compounds in Problem 2 are isomers, and what kind of isomers are they?

4. Is compound (a) or (c) in Problem 2 more soluble in water? Which has the higher boiling point?

5. Draw the structural formulas for the products of the following reactions. If no reaction occurs, write N.R. (Note: X is the symbol for an oxidizing agent and XH_2 represents a reducing agent.)

(a) $CH_3-CH_2-\overset{\overset{\displaystyle O}{\|}}{C}-CH_2-CH_3 + X \longrightarrow$

(b) $CH_3-CH_2-\overset{\overset{\displaystyle O}{\|}}{C}-CH_2-CH_3 + XH_2 \longrightarrow$

(c) $2(H_3C-\overset{\overset{\displaystyle O}{\|}}{C}-H) \xrightarrow{OH^-}$

(d) $CH_3-\overset{\overset{\displaystyle CH_3}{|}}{\underset{\underset{\displaystyle CH_3}{|}}{C}}-\overset{\overset{\displaystyle O}{\|}}{CH} + H_2O + X \longrightarrow$

(e) $CH_3-CH_2-\overset{\overset{\displaystyle O}{\|}}{C}-H + XH_2 \longrightarrow$

(f) $CH_3-\overset{\overset{\displaystyle O}{\|}}{C}-H + 2(CH_3OH) \xrightarrow{H^+} H_2O +$

(g) $H_2O + CH_3-\overset{\overset{\displaystyle OCH_3}{|}}{\underset{\underset{\displaystyle CH_3}{|}}{C}}-OCH_3 \xrightarrow{H^+}$

(h) $CH_3-CH_2-CH_2-CH_2-\overset{\overset{\displaystyle O}{\|}}{C}-CH_2-CH_3 + X \longrightarrow$

(i) $CH_3-CH_2-CH_2-CH_2-\overset{\overset{\displaystyle O}{\|}}{C}-CH_2-CH_3 + XH_2 \longrightarrow$

6. Draw the structural formula for each of the following compounds:
   (a) 2-Pentanol
   (b) Propionaldehyde
   (c) Acetone
   (d) Dimethyl ketone
   (e) 3-Methyl-2-butanone
   (f) Methyl isopropyl ketone

7. Draw structural formulas for all of the structural isomers with the molecular formula C_3H_6O.

SOLUTIONS TO EXERCISES

16-1 (a) $CH_3-\overset{\overset{\displaystyle OH}{|}}{\underset{\underset{\displaystyle CH_3}{|}}{CH}}-\overset{\underset{\underset{\displaystyle OCH_3}{|}}{}}{C}-H$

(b) $CH_3-\overset{\overset{\displaystyle OH}{|}}{\underset{\underset{\displaystyle H}{|}}{C}}-CH_2-\overset{\overset{\displaystyle O}{\|}}{C}-H$

(c) $CH_3-CH_2-\overset{\overset{\displaystyle O}{\|}}{C}-OH + XH_2$

(d) $CH_3-CH_2-CH_2OH + X$

16-2 (a) 2-Methylpropanal and methanol
(b) Ethanal

16-3 4-Hydroxy-4-methyl-2-pentanone

16-4 (a) $CH_3-CH_2-\overset{\overset{\displaystyle O}{\|}}{C}-OH + XH_2$

(b) $\underset{\displaystyle CH_3}{\overset{\displaystyle CH_3}{HC}}-\underset{\displaystyle H}{\overset{\displaystyle OH}{C}}-CH_3 + X$

(c) N.R.

(d) $CH_3-CH_2-\underset{\displaystyle OCH_2CH_3}{\overset{\displaystyle OCH_2CH_3}{CH}} + H_2O$

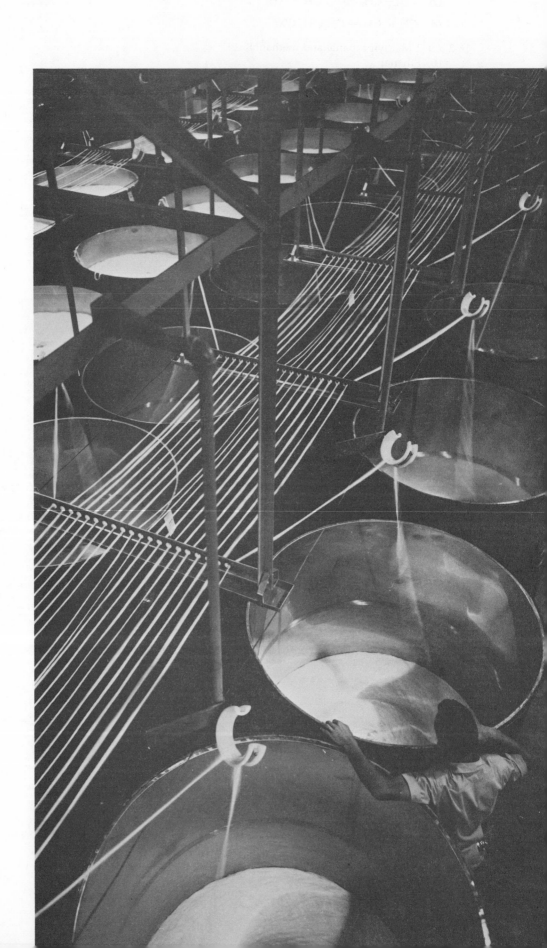

CHAPTER 17

Carboxylic Acids and Their Derivatives

17-1 INTRODUCTION

The carboxyl functional group, $-\overset{\overset{\displaystyle O}{\displaystyle \|}}{C}-OH$, contains two highly electronegative oxygen atoms. Molecules that contain this functional group, that is, carboxylic acids, behave quite differently from other oxygen-containing organic molecules.

In this chapter we shall discuss the unique properties of carboxylic acids that enable them to participate in large numbers of chemical reactions. Carboxylic acids found throughout nature include the amino acids used to make proteins (Chapter 21); lactic acid, which is produced by active muscle cells (Chapter 25); and citric acid, which is an important intermediate in cellular energy production (Chapter 25). Other naturally occurring carboxylic acids are found in bee venom, vinegar, soap, butter, cheese, and many other substances. Carboxylic acids are also extremely important commercially. Some are used to manufacture the synthetic fabrics Dacron and nylon. In the United States alone, more than 2000 tons of one carboxylic acid, acetylsalicylic acid (aspirin), are made yearly.

When a carboxylic acid reacts with an alcohol, a compound called an ester is formed. Similar reactions between two carboxylic acid molecules produce compounds called anhydrides. In this chapter we shall also study the structure, nomenclature, properties, and reactivity of these classes of oxygen-containing organic compounds. In later chapters we shall study several esters that are vital to the workings of cells in the human body. Simple anhydrides formed from carboxylic acids are not widespread in nature, although anhydrides are extremely important reactants in a number of industrial processes, including the production of some polymers.

After we have studied carboxylic acids, esters, and anhydrides, we shall briefly discuss the structure and properties of phosphoric acid and its derivatives. We shall see that phosphate esters and anhydrides are in many ways similar to the esters and anhydrides formed from carboxylic acids, and are involved in some of the most important chemical reactions that occur within human cells.

The manufacture of the synthetic fiber Dacron is an important application of the chemistry of carboxylic acids. Here continuous strands of Dacron polyester are being drawn from large vats into a stretching machine.

17-2 STUDY OBJECTIVES

After careful study of this chapter, you should be able to:

1. Draw the general structural formula for a carboxylic acid, an ester, or an anhydride.

2. Draw structural formulas for simple carboxylic acids, esters, and anhydrides, given their names, and vice versa.

3. Explain how hydrogen bonding affects the boiling point and solubility of carboxylic acids, esters, and anhydrides.

4. Explain why carboxylic acids are almost completely dissociated in the aqueous environment of human cells.

5. Draw structural formulas for the products of reactions in which carboxylic acids (a) react with bases, (b) are reduced, or (c) are decarboxylated, and for reactions in which anhydrides or esters are formed or hydrolyzed, given the structural formulas of the reactants.

6. Relate the properties of phosphoesters and phosphoanhydrides to esters and anhydrides of carboxylic acids.

17-3 CARBOXYLIC ACIDS

Structure and Nomenclature

The structural formulas of several carboxylic acids, together with their common and IUPAC names, are given in Table 17-1. The common names of some carboxylic acids are quite old and indicate some of the places where they are found in nature. The name formic acid, for instance, comes from the Latin word for ant (L. *formica* = *ant*). Acetic acid is named after vinegar (L. *acetum* = *vinegar*), of which it is the active ingredient. Butyric acid is named after butter (L. *butyrum* = *butter*) and is found in rancid butter. Common names of carboxylic acids end in *-ic* followed by the word "acid." The IUPAC names of carboxylic acids are formed from the parent alkane by using the suffix *-oic* to indicate the presence of the carboxylic acid functional group. The common names of carboxylic acids are frequently used. When common names of carboxylic acids are used, the Greek letters α, β, γ, δ, and so on, are used to indicate the positions of other substituents on the hydrocarbon chain in the same manner as for aldehydes (Chapter 16). For example, the formula of β-hydroxybutyric acid is

$$\underset{\gamma}{CH_3}-\underset{\beta}{\overset{\overset{\displaystyle OH}{|}}{CH}}-\underset{\alpha}{CH_2}-\overset{\overset{\displaystyle O}{||}}{C}-OH$$

Carboxylic acids with long hydrocarbon chains, generally 12 to 20 carbon atoms, such as palmitic acid and stearic acid (Table 17-1) are called **fatty acids** and are used in several ways by the human body (Chapters 26 and 27). Fatty acids are not water soluble because of their long hydrophobic hydrocarbonlike chains. Some **dicarboxylic acids,** with two carboxyl functional groups (for example, succinic acid), and the **tricarboxylic acid** citric acid are also important in human biochemistry (Chapter 25).

Table 17-1 Some Carboxylic Acids

Structural Formula	Common Name	IUPAC Name
$$H-\overset{\overset{\displaystyle O}{\|\|}}{C}-OH$$	Formic acid	Methanoic acid
$$CH_3-\overset{\overset{\displaystyle O}{\|\|}}{C}-OH$$	Acetic acid	Ethanoic acid
$$CH_3-CH_2-\overset{\overset{\displaystyle O}{\|\|}}{C}-OH$$	Propionic acid	Propanoic acid
$$CH_3-CH_2-CH_2-\overset{\overset{\displaystyle O}{\|\|}}{C}-OH$$	Butyric acid	Butanoic acid
$$CH_3-\overset{\overset{\displaystyle CH_3}{\|}}{CH}-\overset{\overset{\displaystyle O}{\|\|}}{C}-OH$$	α-Methylpropionic acid	2-Methylpropanoic acid
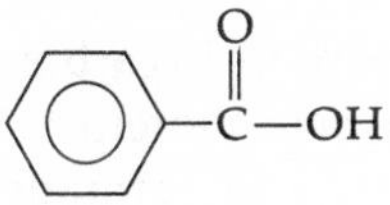 Benzene ring $-\overset{\overset{\displaystyle O}{\|\|}}{C}-OH$	Benzoic acid	
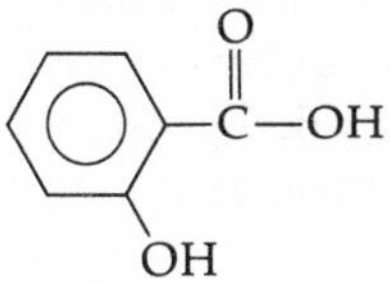 Benzene ring $-\overset{\overset{\displaystyle O}{\|\|}}{C}-OH$ with OH	Salicylic acid	
$$CH_3-(CH_2)_{14}-\overset{\overset{\displaystyle O}{\|\|}}{C}-OH$$	Palmitic acid, a fatty acid	
$$CH_3-(CH_2)_{16}-\overset{\overset{\displaystyle O}{\|\|}}{C}-OH$$	Stearic acid, a fatty acid	
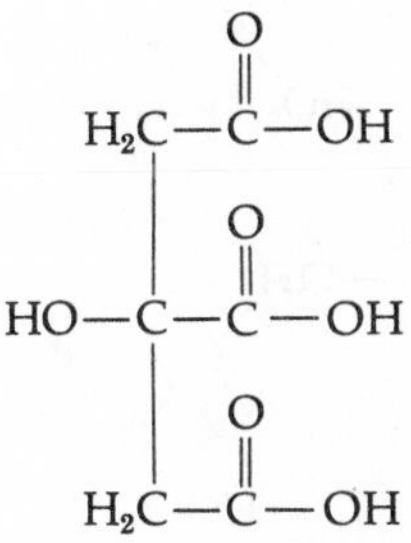 $$H_2C-\overset{\overset{\displaystyle O}{\|\|}}{C}-OH$$ $$HO-\overset{}{C}-\overset{\overset{\displaystyle O}{\|\|}}{C}-OH$$ $$H_2C-\overset{\overset{\displaystyle O}{\|\|}}{C}-OH$$	Citric acid, a tricarboxylic acid	

Properties of Carboxylic Acids

The carboxyl functional group, $-\overset{\overset{\displaystyle O}{\|\|}}{C}-OH$, contains the $-\overset{\overset{\displaystyle O}{\|\|}}{C}-$ group present in aldehydes and ketones and the $-\overset{}{\underset{}{C}}-OH$ group present in alcohols. The properties of carboxylic acids, however, are enormously different from the properties of aldehydes, ketones, or alcohols, because in carboxylic acids both oxygen atoms are bonded to the *same* carbon atom. There is no carbon-carbon single bond "insulation" between the $-\overset{\overset{O}{\|}}{C}-$ and $-\overset{|}{\underset{|}{C}}-OH$ groups. Therefore, the carboxyl group functions as a single functional group.

A molecule such as β-hydroxybutyraldehyde, $H_3C-\overset{\overset{OH}{|}}{\underset{\underset{H}{|}}{C}}-CH_2-\overset{\overset{O}{\|}}{C}-H$, on the other hand, has enough carbon-carbon single bond "insulation" between the $-\overset{\overset{O}{\|}}{C}-$ and $-\overset{|}{\underset{|}{C}}-OH$ groups, and it exhibits the separate characteristic properties of an aldehyde and an alcohol, and not those of a carboxylic acid.

HYDROGEN BONDING

The presence of two very electronegative oxygen atoms in carboxylic acids makes these molecules very polar. Carboxylic acids can therefore form hydrogen bonds with water molecules (see Figure 17-1). The formation of hydrogen bonds with water molecules makes low-molecular-weight carboxylic acids soluble in aqueous solution. Fatty acids (Table 17-1), however, are not water soluble because of the hydrophobic nature of their long alkyl groups. Carboxylic acids can also form hydrogen bonds with other carboxylic acid molecules (see Figure 17-1), which results in pure carboxylic acids having much higher boiling points than some other molecules of comparable molecular weight (see Table 17-2).

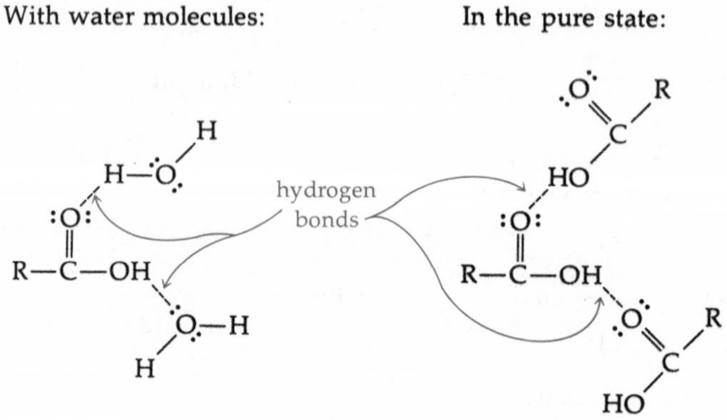

Figure 17-1   Carboxylic acids form hydrogen bonds with water (left), and in the pure state, with other carboxylic acid molecules (right).

ACIDIC PROPERTIES

We have already seen that the carboxylic acid functional group can serve as a proton donor, and thus that molecules containing this group are weak acids (Section 10-9). The doubly bonded oxygen in the carboxyl group pulls electrons toward itself and away from the O—H bond, thus facilitating the loss of this hydrogen as a proton (H^+). When a carboxylic acid is added to pure water, only a very small fraction of the carboxylic acid molecules lose a proton. Consider the following example: When 1 mole of acetic acid is dissolved in 1 liter of pure water (see reaction 17-1), only about 0.004 mole of acetate ions and 0.004 mole of hydronium ions are formed. Most of the acetic acid, 0.996 mole, is present

Table 17-2 Boiling Points of Compounds with Comparable Molecular Weights

Name	Structural Formula	MW	Boiling Point (°C)
Formic acid	$\overset{\displaystyle O}{\overset{\|}{H-C-OH}}$	46	101
Ethanol	CH_3-CH_2-OH	46	78
Acetaldehyde	$\overset{\displaystyle O}{\overset{\|}{CH_3-C-H}}$	44	20
Propane	$CH_3-CH_2-CH_3$	44	-42

as undissociated acetic acid molecules. The pH of this solution is about 2.4.

(17-1)
$$\underset{\text{acetic acid}}{\overset{\displaystyle O}{\overset{\|}{CH_3-C-OH}}} + H_2O \rightleftharpoons \underset{\text{acetate ion}}{\overset{\displaystyle O}{\overset{\|}{CH_3-C-O^-}}} + H_3O^+$$

What happens, however, when some acetic acid is dissolved in a buffer that maintains the pH at about 7.4? Recall that pH 7.4 is the pH of normal body fluids. The pH does not drop significantly upon addition of the acetic acid, because most of the added acetic acid reacts with the base present in the buffer and is converted to acetate ions. We can determine the ratio of acetate ions to undissociated acetic acid molecules by using the equilibrium constant expression for reaction 17-1:

(17-2)
$$K_a = \frac{[H_3O^+][A^-]}{[HA]} = \frac{[H_3O^+][\text{acetate ion}]}{[\text{acetic acid molecules}]}$$

Equation 17-2 can be rearranged algebraically to give Eq. 17-3:

(17-3)
$$\frac{K_a}{[H_3O^+]} = \frac{[\text{acetate ion}]}{[\text{acetic acid molecules}]}$$

K_a is the dissociation constant for acetic acid and is about 1.8×10^{-5} mole/liter, and a pH of 7.4 corresponds to a $[H_3O^+]$ of about 4×10^{-8} M. Using these values in Eq. 17-3, the ratio of acetate ions to undissociated acetic acid molecules in a solution at pH 7.4 is

(17-4)
$$\frac{[\text{acetate ion}]}{[\text{acetic acid molecules}]} = \frac{1.8 \times 10^{-5}}{4 \times 10^{-8}} \simeq \frac{400}{1}$$

In other words, in an aqueous solution with the pH maintained at 7.4, virtually all of the acetic acid is dissociated and exists in the form of acetate ions. Similar considerations apply to all other carboxylic acids. Thus, in the human body, where buffers maintain the pH at about 7.4, virtually all carboxylic acid functional groups are ionized. The ionized form of a carboxyl group is called a

carboxylate ion $(\overset{\displaystyle O}{\overset{\|}{R-C-O^-}})$. The name for a carboxylate ion is derived from the name of the carboxylic acid from which it is derived, by changing the -ic ending to -ate. Carboxylate ions are negatively charged and thus are attracted to positively charged parts of molecules. These ionic interactions play an important role in our body cells, as we shall see later.

Reactions of Carboxylic Acids

Some important reactions of carboxylic acids are summarized in Table 17-3. Reaction of a carboxylic acid in water with an equivalent amount of a strong base, such as NaOH, leads to the formation of a salt when water is evaporated:

$$\text{(17-5)} \qquad R'\!-\!\overset{\displaystyle O}{\overset{\|}{C}}\!-\!OH + NaOH \longrightarrow R'\!-\!\overset{\displaystyle O}{\overset{\|}{C}}\!-\!O^-Na^+ + H_2O$$

The sodium salt of stearic acid, for example, is sodium stearate, which is commonly used as a soap.

$$CH_3\!-\!(CH_2)_{16}\!-\!\overset{\displaystyle O}{\overset{\|}{C}}\!-\!O^-Na^+$$
sodium stearate

We have already seen that carboxylic acids can be formed by the oxidation of aldehydes (reaction 16-3). Alternatively, the reduction of a carboxylic acid produces an aldehyde:

$$\text{(17-6)} \qquad R'\!-\!\overset{\displaystyle O}{\overset{\|}{C}}\!-\!OH + XH_2 \longrightarrow R'\!-\!\overset{\displaystyle O}{\overset{\|}{C}}\!-\!H + X + H_2O$$

where XH_2 represents the reducing agent. Under laboratory conditions, the aldehyde produced in reaction 17-6 will also react with the reducing agent, XH_2, to give the corresponding primary alcohol. It is therefore difficult to obtain significant amounts of an aldehyde using this reaction. In the human body, however, the reduction of a carboxylic acid to an aldehyde is catalyzed by a specific enzyme which produces only the aldehyde and not the corresponding alcohol.

Table 17-3 Reactions of Carboxylic Acids

Ionization	$R'\!-\!\overset{O}{\overset{\|}{C}}\!-\!OH + H_2O \rightleftharpoons R'\!-\!\overset{O}{\overset{\|}{C}}\!-\!O^- + H_3O^+$ (a carboxylate ion)
Salt formation	$R'\!-\!\overset{O}{\overset{\|}{C}}\!-\!OH + NaOH \rightleftharpoons R'\!-\!\overset{O}{\overset{\|}{C}}\!-\!O^-Na^+ + H_2O$ (a sodium salt)
Reduction	$R'\!-\!\overset{O}{\overset{\|}{C}}\!-\!OH + XH_2 \rightleftharpoons R'\!-\!\overset{O}{\overset{\|}{C}}\!-\!H + X + H_2O$ (an aldehyde)
Decarboxylation	$R\!-\!\overset{O}{\overset{\|}{C}}\!-\!OH \longrightarrow RH + CO_2$
Anhydride formation	$R_1\!-\!\overset{O}{\overset{\|}{C}}\!-\!OH + HO\!-\!\overset{O}{\overset{\|}{C}}\!-\!R_2 \rightleftharpoons R_1\!-\!\overset{O}{\overset{\|}{C}}\!-\!O\!-\!\overset{O}{\overset{\|}{C}}\!-\!R_2 + H_2O$ (an anhydride)
Ester formation	$R'\!-\!\overset{O}{\overset{\|}{C}}\!-\!OH + HO\!-\!R \rightleftharpoons R'\!-\!\overset{O}{\overset{\|}{C}}\!-\!O\!-\!R + H_2O$ (an ester)

Another reaction that molecules containing carboxyl functional groups can undergo is **decarboxylation.** In a decarboxylation reaction, the carboxyl functional group is lost and a molecule of CO_2 is formed. Most simple carboxylic acids do not readily undergo decarboxylation reactions. However, carboxylic acids that have a ketone functional group located β to the carboxyl group (see reaction 17-7) are readily decarboxylated, both in the laboratory or by specific enzymes in the human body. Decarboxylation reactions reduce the length of a carbon chain by one carbon atom and are important in several normal biological processes. The decarboxylation of acetoacetic acid produces acetone in persons suffering from severe diabetes:

(17-7)

$$\underset{\text{acetoacetic acid}}{CH_3-\overset{O}{\overset{\|}{C}}-CH_2-\overset{O}{\overset{\|}{C}}-OH} \xrightarrow[\text{enzyme}]{\text{a specific}} \underset{\text{acetone}}{CH_3-\overset{O}{\overset{\|}{C}}-CH_3} + CO_2$$

β-ketone

Carboxylic acids also participate in a number of condensation reactions. A **condensation reaction** is a reaction in which two molecules condense or join together. We have already seen examples of condensation reactions, such as hemiacetal formation and aldol condensations, in Chapter 16. Carboxylic acids can participate in condensation reactions (1) with other carboxylic acid molecules to form anhydrides; (2) with alcohol molecules to form esters; (3) with phosphoric acid (see Section 17-5) to form phosphoesters; and (4) with other compounds, including some with nitrogen-containing or sulfur-containing functional groups (see Chapter 18).

When two carboxylic acid molecules react, an **anhydride** is formed in a condensation reaction accompanied by the elimination of a molecule of water:

(17-8)

$$\underset{\text{two carboxylic acids}}{R_1-\overset{O}{\overset{\|}{C}}-OH + HO-\overset{O}{\overset{\|}{C}}-R_2} \longrightarrow \underset{\text{an anhydride}}{R_1-\overset{O}{\overset{\|}{C}}-O-\overset{O}{\overset{\|}{C}}-R_2} + H_2O$$

Some anhydrides are formed just by heating the reactants, whereas the formation of other anhydrides requires the presence of a catalyst. Anhydrides can be hydrolyzed to carboxylic acids in the reverse of reaction 17-8.

When a carboxylic acid reacts with an alcohol, an **ester** is formed and a molecule of water is eliminated. The general condensation reaction for the formation of an ester is

(17-9)

$$\underset{\text{carboxylic acid}}{R'-\overset{O}{\overset{\|}{C}}-OH} + \underset{\text{alcohol}}{HO-R} \longrightarrow \underset{\text{an ester}}{R'-\overset{O}{\overset{\|}{C}}-O-R} + H_2O$$

Esters can also be hydrolyzed to produce a carboxylic acid and an alcohol:

(17-10)

$$R'-\overset{O}{\overset{\|}{C}}-O-R + H_2O \longrightarrow R'-\overset{O}{\overset{\|}{C}}-OH + HO-R$$

In the laboratory, reactions 17-9 and 17-10 are catalyzed by H^+ and usually result in a mixture of reactants and products in equilibrium. The desired product must then be separated from the other compounds in the mixture. The synthetic fabric Dacron is a polymer that is formed from dialcohol monomers and dicarboxylic acid monomers, which are joined together with ester bonds in an alternating arrangement (see Figure 17-2).

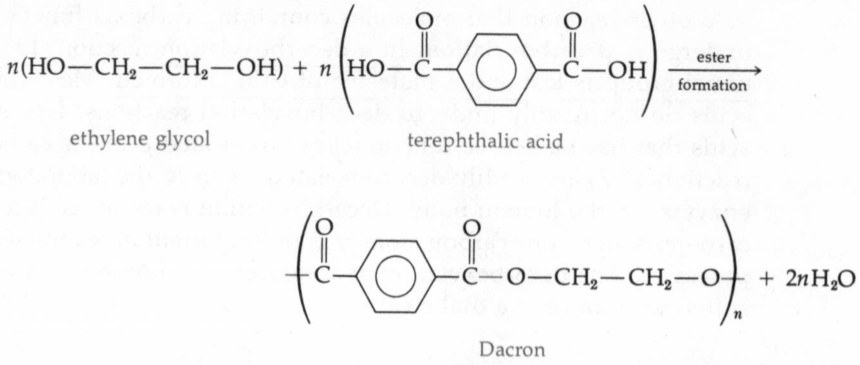

$$n(HO-CH_2-CH_2-OH) + n\left(HO-\overset{O}{\overset{\|}{C}}-\langle\bigcirc\rangle-\overset{O}{\overset{\|}{C}}-OH\right) \xrightarrow[\text{formation}]{\text{ester}}$$

ethylene glycol   terephthalic acid

$$\left(\overset{O}{\overset{\|}{C}}-\langle\bigcirc\rangle-\overset{O}{\overset{\|}{C}}-O-CH_2-CH_2-O\right)_n + 2nH_2O$$

Dacron

Figure 17-2   The polymerization of Dacron from   monomers by the formation of ester
                    ethylene glycol and terephthalic acid   bonds (shown in color).

The hydrolysis of esters can also be catalyzed by OH^-. Commercially, esters
of fatty acids are hydrolyzed by addition of a strong base, such as NaOH. This
alkaline hydrolysis of fatty acid esters is called **saponification,** since the sodium
salt of the acid that is formed is a soap. Hydrolysis and formation of esters are
catalyzed by specific enzymes in the human body.

Exercise 17-1

Draw structural formulas for the products of the following reactions. (X represents an
oxidizing agent, and XH_2 a reducing agent. If no reaction occurs, write N.R.)

(a) $CH_3-\overset{O}{\overset{\|}{C}}-OH + X \longrightarrow$ (b) $CH_3-\overset{O}{\overset{\|}{C}}-OH + XH_2 \longrightarrow$

(c) $2(CH_3-\overset{O}{\overset{\|}{C}}-OH) \xrightarrow[\text{catalyst}]{\text{heat}} H_2O +$

(d) $CH_3-CH_2-\overset{O}{\overset{\|}{C}}-OH + CH_3-CH_2OH \xrightarrow{\text{catalyst}} H_2O +$

If you have difficulty with this exercise, review Essential Skills 9.

17-4   ESTERS AND ANHYDRIDES

Structure and Nomenclature

Esters and anhydrides can be considered as derivatives of carboxylic acids. The
structural formulas and common names of some esters are given in Table 17-4.
Esters are named in a manner similar to ionic salts. For example, the ester

$CH_3-\overset{O}{\overset{\|}{C}}-O-CH_3$ is named methyl acetate, whereas $CH_3-\overset{O}{\overset{\|}{C}}-O^-Na^+$ is
called sodium acetate. The *acetate* portion of these names comes from the

carboxylate ion of acetic acid, that is, $CH_3-\overset{O}{\overset{\|}{C}}-O^-$, the acetate ion. Recall that
the names of carboxylate ions are formed by changing the suffix of the name for
the carboxylic acid to -*ate*. The *methyl* part of the name methyl acetate comes from
the name of the alkyl group that is bonded to the carboxylate part of the ester.

Table 17-4 Some Common Esters

Structural Formula	Common Name
$CH_3-\overset{\overset{\displaystyle O}{\|\|}}{C}-O-CH_3$	Methyl acetate
$CH_3-\overset{\overset{\displaystyle O}{\|\|}}{C}-O-CH_2-CH_3$	Ethyl acetate
$CH_3-CH_2-\overset{\overset{\displaystyle O}{\|\|}}{C}-O-CH_2-CH_3$	Ethyl propionate
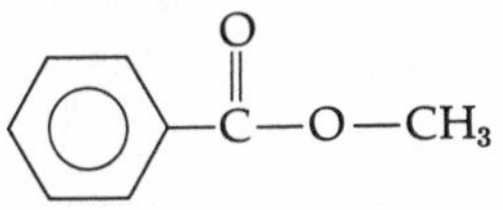 $\text{C}_6\text{H}_5-\overset{\overset{\displaystyle O}{\|\|}}{C}-O-CH_3$	Methyl benzoate
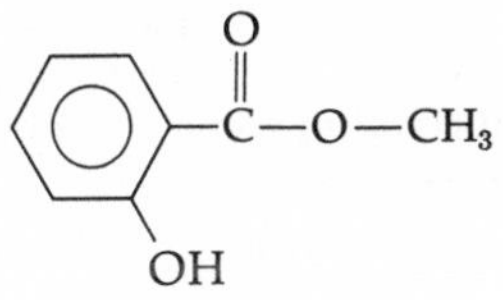 $\text{C}_6\text{H}_4(\text{OH})-\overset{\overset{\displaystyle O}{\|\|}}{C}-O-CH_3$	Methyl salicylate
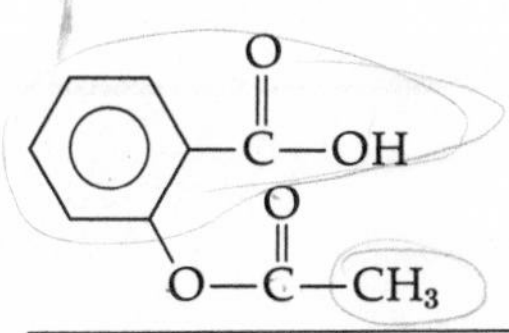 aspirin structure; $R-\overset{\overset{\displaystyle O}{\|\|}}{C}-O-R'$	Aspirin

Many esters have pleasant odors. Methyl butyrate is the ester responsible for the smell of pineapples. Esters of the triol glycerol, $HOCH_2-\underset{\underset{\displaystyle OH}{\|}}{CH}-CH_2OH$,

with three fatty acids, are called triglycerides. They are used for food storage in humans and are often referred to as just plain FAT!

Anhydrides are derivatives of two carboxylic acid molecules. The structural formulas and names of some simple anhydrides are given in Table 17-5. For symmetric anhydrides, that is, those derived from two molecules of the same carboxylic acid, the word *anhydride* is added to the name of the parent

carboxylic acid, after dropping the word *acid*. Thus, $CH_3-\overset{\overset{\displaystyle O}{\|\|}}{C}-O-\overset{\overset{\displaystyle O}{\|\|}}{C}-CH_3$ is called acetic anhydride.

Table 17-5 Some Simple Anhydrides

Structural Formula	Common Name
$CH_3-\overset{\overset{\displaystyle O}{\|\|}}{C}-O-\overset{\overset{\displaystyle O}{\|\|}}{C}-CH_3$	Acetic anhydride
$CH_3-CH_2-\overset{\overset{\displaystyle O}{\|\|}}{C}-O-\overset{\overset{\displaystyle O}{\|\|}}{C}-CH_2-CH_3$	Propionic anhydride
$CH_3-CH_2-CH_2-\overset{\overset{\displaystyle O}{\|\|}}{C}-O-\overset{\overset{\displaystyle O}{\|\|}}{C}-CH_2-CH_2-CH_3$	Butyric anhydride

Properties of Esters and Anhydrides

The ester $\overset{\displaystyle O}{\overset{\|}{-C}}-O-\overset{|}{\underset{|}{C}}-$ and anhydride $\overset{\displaystyle O}{\overset{\|}{-C}}-O-\overset{\displaystyle O}{\overset{\|}{C}}-$ functional groups are polar and can form hydrogen bonds with water. Thus, molecules containing these functional groups tend to be water soluble. Note, however, that esters and anhydrides do not contain hydrogen-oxygen bonds, so they do not form hydrogen bonds in the pure state. Recall that ethers, aldehydes, and ketones also do not form hydrogen bonds in the pure state.

The formation and hydrolysis reactions of esters and anhydrides have already been discussed (Section 17-3). Esters and anhydrides are not acidic, and are extremely weak (weaker than water) Brønsted-Lowry bases.

Exercise 17-2

Would you expect acetic anhydride to have a lower, higher, or similar boiling point when compared to (a) hexane, (b) octanoic acid, (c) methyl acetate? Explain your answers.

17-5 PHOSPHOESTERS AND PHOSPHOANHYDRIDES

The structure of phosphoric acid is somewhat similar to that of a carboxylic acid (see Figure 17-3).

Figure 17-3   A carboxylic acid and phosphoric acid are somewhat similar in structure, but phosphoric acid has two additional acidic hydrogens.

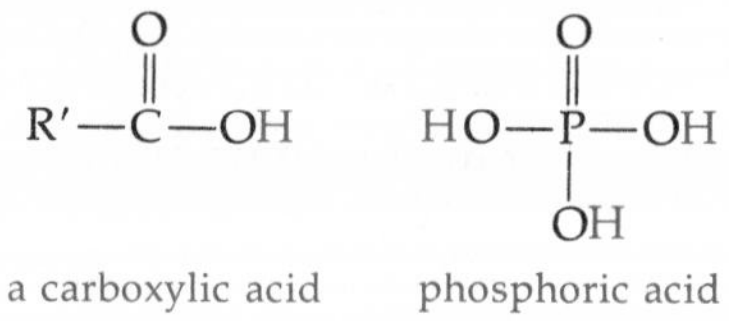

$$R'-\overset{\displaystyle O}{\overset{\|}{C}}-OH \qquad HO-\overset{\displaystyle O}{\underset{\underset{\displaystyle OH}{|}}{\overset{\|}{P}}}-OH$$

a carboxylic acid   phosphoric acid

Like carboxylic acids, phosphoric acid can form esters with alcohols:

(17-11)
$$R-OH + HO-\overset{\displaystyle O}{\underset{\underset{\displaystyle OH}{|}}{\overset{\|}{P}}}-OH \longrightarrow R-O-\overset{\displaystyle O}{\underset{\underset{\displaystyle OH}{|}}{\overset{\|}{P}}}-OH + H_2O$$

alcohol   phosphoric acid   organic phosphate ester

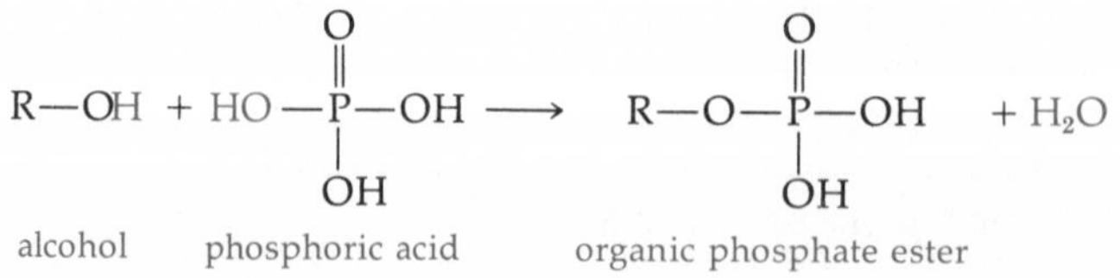

The resulting organic phosphate esters, or **phosphoesters,** are extremely important to living organisms, as we shall see in Chapter 25.

Phosphoric acid can also form anhydrides with itself,

(17-12)
$$HO-\overset{\displaystyle O}{\underset{\underset{\displaystyle OH}{|}}{\overset{\|}{P}}}-OH + HO-\overset{\displaystyle O}{\underset{\underset{\displaystyle OH}{|}}{\overset{\|}{P}}}-OH \longrightarrow HO-\overset{\displaystyle O}{\underset{\underset{\displaystyle OH}{|}}{\overset{\|}{P}}}-O-\overset{\displaystyle O}{\underset{\underset{\displaystyle OH}{|}}{\overset{\|}{P}}}-OH + H_2O$$

pyrophosphoric acid

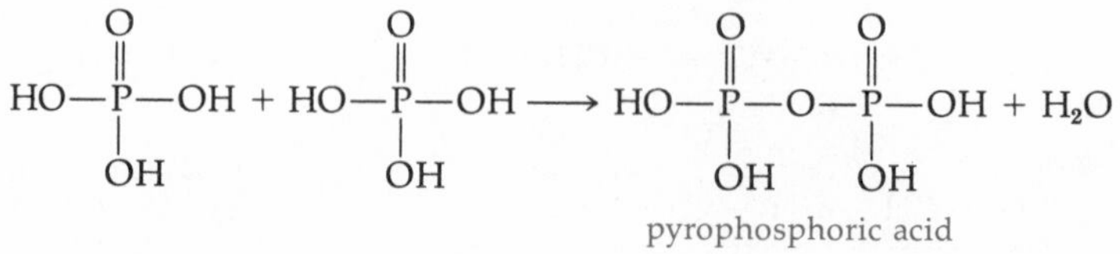

as well as with a carboxylic acid,

(17-13)
$$R'-\overset{\overset{\displaystyle O}{\|}}{C}-OH + HO-\overset{\overset{\displaystyle O}{\|}}{\underset{\underset{\displaystyle OH}{|}}{P}}-OH \longrightarrow R'-\overset{\overset{\displaystyle O}{\|}}{C}-O-\overset{\overset{\displaystyle O}{\|}}{\underset{\underset{\displaystyle OH}{|}}{P}}-OH + H_2O$$

      a carboxylic   phosphoric
        acid   acid

Since phosphoric acid has *three* acidic hydrogens, a phosphoester or a phospho-anhydride can form further anhydrides and esters. For example, the acid part of a phosphoester molecule can react with a molecule of phosphoric acid and form a molecule that is both an ester and an anhydride:

(17-14)
$$R-O-\overset{\overset{\displaystyle O}{\|}}{\underset{\underset{\displaystyle OH}{|}}{P}}-OH + HO-\overset{\overset{\displaystyle O}{\|}}{\underset{\underset{\displaystyle OH}{|}}{P}}-OH \longrightarrow R-O-\overset{\overset{\displaystyle O}{\|}}{\underset{\underset{\displaystyle OH}{|}}{P}}-O-\overset{\overset{\displaystyle O}{\|}}{\underset{\underset{\displaystyle OH}{|}}{P}}-OH + H_2O$$

                                          a diphosphate ester

Similarly,

(17-15)
$$R-O-\overset{\overset{\displaystyle O}{\|}}{\underset{\underset{\displaystyle OH}{|}}{P}}-O-\overset{\overset{\displaystyle O}{\|}}{\underset{\underset{\displaystyle OH}{|}}{P}}-OH + HO-\overset{\overset{\displaystyle O}{\|}}{\underset{\underset{\displaystyle OH}{|}}{P}}-OH \longrightarrow$$

$$R-O-\overset{\overset{\displaystyle O}{\|}}{\underset{\underset{\displaystyle OH}{|}}{P}}-O-\overset{\overset{\displaystyle O}{\|}}{\underset{\underset{\displaystyle OH}{|}}{P}}-O-\overset{\overset{\displaystyle O}{\|}}{\underset{\underset{\displaystyle OH}{|}}{P}}-OH + H_2O$$

                                      a triphosphate ester

A very important property of phosphoanhydrides is the large amount of energy released upon hydrolysis. Consider the reaction

(17-16)
$$H_2O + R-O-\overset{\overset{\displaystyle O}{\|}}{\underset{\underset{\displaystyle OH}{|}}{P}}-O-\overset{\overset{\displaystyle O}{\|}}{\underset{\underset{\displaystyle OH}{|}}{P}}-O-\overset{\overset{\displaystyle O}{\|}}{\underset{\underset{\displaystyle OH}{|}}{P}}-OH \longrightarrow$$

          a triphosphate ester

$$R-O-\overset{\overset{\displaystyle O}{\|}}{\underset{\underset{\displaystyle OH}{|}}{P}}-O-\overset{\overset{\displaystyle O}{\|}}{\underset{\underset{\displaystyle OH}{|}}{P}}-OH + HO-\overset{\overset{\displaystyle O}{\|}}{\underset{\underset{\displaystyle OH}{|}}{P}}-OH$$

           a diphosphate ester   phosphoric acid

The hydrolysis reaction 17-16 is quite exergonic ($\Delta G \approx -8$ kcal/mole). Therefore, this reaction is a potential source of a large amount of work (Section 6-7). In the human body, hydrolysis reactions in which a phosphoanhydride bond in a triphosphate or a diphosphate is broken are an important source of energy (Chapter 25).

Exercise 17-3
Draw structural formulas for (a) ethyl phosphate and (b) glycerol 3-phosphate.

17-6 SUMMARY

1. Carboxylic acids, esters, and anhydrides are polar molecules and can form hydrogen bonds with water molecules. Carboxylic acids also form hydrogen bonds in the pure state; esters and anhydrides of carboxylic acids do not.

2. Carboxylic acids are weak acids with the general formula $R'-\overset{\displaystyle O}{\overset{\|}{C}}-OH$. The IUPAC suffix for carboxylic acids is -*oic*. In the human body, where the pH is normally 7.4, virtually all carboxylic acid molecules are ionized to carboxylate ions.

3. The general formulas for esters and anhydrides are $R'-\overset{\displaystyle O}{\overset{\|}{C}}-O-R$ and $R_1-\overset{\displaystyle O}{\overset{\|}{C}}-O-\overset{\displaystyle O}{\overset{\|}{C}}-R_2$, respectively.

4. Carboxylic acids can undergo several reactions, including: (a) ionization, (b) decarboxylation, (c) reduction to form an aldehyde, (d) condensation with another carboxylic acid molecule to form an anhydride, and (e) condensation with an alcohol to form an ester.

5. Anhydrides can be hydrolyzed to form carboxylic acids, and esters can be hydrolyzed to form carboxylic acids and alcohols. The alkaline hydrolysis of a fatty acid ester is called saponification.

6. Phosphoric acid behaves somewhat like a carboxylic acid. Phosphoesters and phosphoanhydrides are important components of human cells.

PROBLEMS

1. Which of the following has the highest boiling point? Which has the highest solubility in water? Explain.

   (a) $CH_3-O-\overset{\displaystyle O}{\overset{\|}{C}}-CH_2-CH_3$ (b) $CH_3-CH_2-CH_2-\overset{\displaystyle O}{\overset{\|}{C}}-OH$

2. Are the compounds in Problem 1 isomers? If so, what kind of isomers are they? Identify their functional groups.

3. Draw structural formulas for each of the following:
   (a) 2-Methylpropanoic acid
   (b) Methyl propionate
   (c) α-Methylpropionic acid
   (d) 3-Hydroxy-3-methylbutanoic acid
   (e) Butyric anhydride

4. Give the common names for each of the following:

   (a) 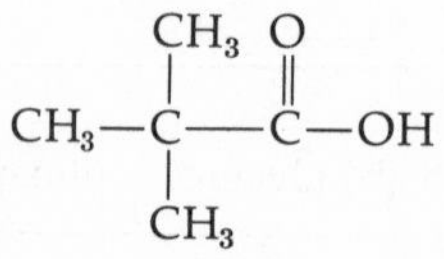 $CH_3-\overset{\displaystyle CH_3}{\underset{\displaystyle CH_3}{\overset{|}{\underset{|}{C}}}}-\overset{\displaystyle O}{\overset{\|}{C}}-OH$ (b) 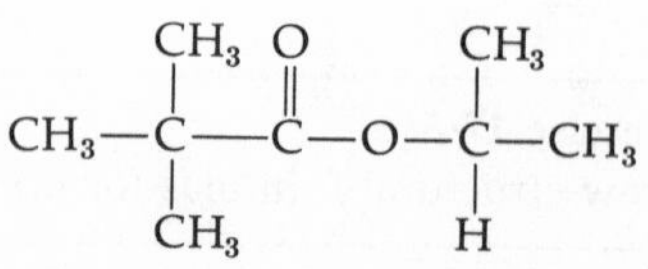 $CH_3-\overset{\displaystyle CH_3}{\underset{\displaystyle CH_3}{\overset{|}{\underset{|}{C}}}}-\overset{\displaystyle O}{\overset{\|}{C}}-O-\overset{\displaystyle CH_3}{\underset{\displaystyle H}{\overset{|}{\underset{|}{C}}}}-CH_3$

(c) $CH_3-CH_2-\overset{\overset{\displaystyle O}{\|}}{C}-O-CH_2-CH_3$

(d) 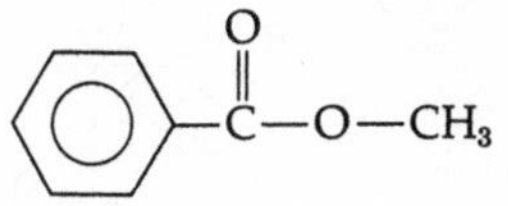 $\langle\bigcirc\rangle-\overset{\overset{\displaystyle O}{\|}}{C}-O-CH_3$

(e) $CH_3-CH_2-O-\overset{\overset{\displaystyle O}{\|}}{\underset{\underset{\displaystyle OH}{|}}{P}}-OH$

5. Draw structural formulas for the products of the following reactions. (X represents an oxidizing agent, and XH_2 a reducing agent. If no reaction occurs, write N.R.)

(a) $H_2O + CH_3-\overset{\overset{\displaystyle O}{\|}}{C}-O-CH_3 \xrightarrow{OH^-}$

(b) $OH^- + CH_3-CH_2-\overset{\overset{\displaystyle O}{\|}}{C}-OH \longrightarrow$

(c) $H_2O + CH_3-CH_2-\overset{\overset{\displaystyle O}{\|}}{C}-OH \xrightarrow{H^+}$

(d) $CH_3-\overset{\overset{\displaystyle O}{\|}}{C}-CH_3 + X + H_2O \longrightarrow$

(e) $CH_3-CH_2-\overset{\overset{\displaystyle O}{\|}}{C}-OH \longrightarrow CO_2 +$

(f) $CH_3-\overset{\overset{\displaystyle O}{\|}}{C}-O-\overset{\overset{\displaystyle O}{\|}}{C}-CH_3 + H_2O \xrightarrow{H^+}$

6. Are any of the reactants in Problem 5 functional group isomers? If so, which ones?

7. If you mix vodka (H_2O + ethanol) and vinegar (H_2O + acetic acid) over ice, what happens? What if you add a strong acid to this mixture and let it sit on the table for a day or so?

SOLUTIONS TO EXERCISES

17-1 (a) N.R.

(b) $CH_3-\overset{\overset{\displaystyle O}{\|}}{C}-H + X + H_2O$

(c) $CH_3-\overset{\overset{\displaystyle O}{\|}}{C}-O-\overset{\overset{\displaystyle O}{\|}}{C}-CH_3$

(d) $CH_3-CH_2-\overset{\overset{\displaystyle O}{\|}}{C}-O-CH_2-CH_3$

17-2 (a) Higher, since hexane is a nonpolar alkane with a lower molecular weight.
(b) Lower, since octanoic acid has a higher molecular weight and can form hydrogen bonds in the pure state.
(c) Higher, since methyl acetate has a lower molecular weight.

17-3 (a) $CH_3-CH_2-O-\overset{\overset{\displaystyle OH}{|}}{\underset{\underset{\displaystyle OH}{|}}{P}}=O$

(b) H_2C-OH
    $HC-OH$
    $H_2C-O-\overset{\overset{\displaystyle O}{\|}}{\underset{\underset{\displaystyle OH}{|}}{P}}-OH$

Note that according to our rules, the name for this compound should be glycerol 1-phosphate. The name glycerol 3-phosphate is used by biochemists and is derived from its reactions in body cells.

CHAPTER 18

Organic Compounds Containing Nitrogen or Sulfur

18-1   INTRODUCTION

The elements nitrogen and sulfur are important constituents of the human body. For example, 16% of the weight of a typical protein is nitrogen and 0.4% is sulfur. Nitrogen is also an important component of DNA.

How does the human body obtain the nitrogen and sulfur that it requires? Although about 80% of the air is molecular nitrogen, N_2, the human body cannot use nitrogen in the form of N_2. Our primary source of nitrogen is dietary protein. Large protein molecules in our diet are broken down during digestion into small organic molecules with nitrogen-containing functional groups. These small molecules, called amino acids, are then used in the synthesis of new proteins or to make other nitrogen-containing molecules. Dietary protein is also our primary source for the sulfur-containing functional groups needed to make proteins and other molecules. In some parts of the world there is tragic evidence of the lack of sufficient dietary protein. Newborn children in these regions thrive when they ingest their mother's milk, which usually contains an ample supply of proteins and other nutrients. However, these children often develop a disease called kwashiorkor when they stop breast-feeding. This disease results from a diet that is low in protein content, often one consisting primarily of starch. Early symptoms of kwashiorkor are diarrhea and loss of appetite; left untreated, the disease often leads to death. Kwashiorkor can be treated and prevented by providing children with ample amounts of dietary protein.

In preparation for our study of proteins, DNA, and other molecules with functional groups containing nitrogen or sulfur, we shall examine the structures and properties of simple organic molecules that contain these elements. As in the past few chapters, we shall look first at those classes of organic compounds that are determined by the presence of functional groups containing nitrogen or sulfur atoms. Then we shall see how these compounds react with some of the compounds we have studied previously.

The nitrogen and sulfur requirements of the human body are satisfied by dietary protein, which in many cultures is derived from meat. Before the advent of refrigeration, Native Americans preserved meat by drying it in the sun.

18-2   STUDY OBJECTIVES

After careful study of this chapter, you should be able to:

1. Identify amine, amide, sulfhydryl, thioester, and disulfide functional groups, given the structural formula of a compound.

2. Distinguish among primary, secondary, and tertiary amines and quaternary ammonium ions.

3. Write the structural formula for an amine, amide, thiol, or thioester, given its name, and vice versa.

4. Explain the solubility of amines and amides in water in terms of their ability to form hydrogen bonds.

5. Describe the acid-base properties of amines and amides, and how the predominant form of an amine in solution depends on the pH.

6. Draw the structural formula for an organic ammonium salt, given its name, and describe its acid-base properties and solubility.

7. Write structural formulas for the products of reactions in which (a) amines react with acids to form ammonium salts, (b) amines react with carboxylic acids to form amides, and (c) amides are hydrolyzed, given the structural formulas of the reactants.

8. Explain the difference between the boiling points of a thiol and an alcohol of similar molecular weight.

9. Draw structural formulas for the products of reactions in which (a) a thioester is formed, (b) a thioester is hydrolyzed, (c) a disulfide is formed, or (d) a disulfide is reduced, given the structural formulas of the reactants.

18-3   CLASSES OF COMPOUNDS CONTAINING NITROGEN OR SULFUR ATOMS

The names of the major classes of organic compounds that contain nitrogen or sulfur atoms are shown in Table 18-1. Notice that **amines** can be considered as organic derivatives of ammonia, NH_3. Primary, secondary, and tertiary amines can be considered as ammonia derivatives in which one, two, or three of the hydrogen atoms, respectively, have been replaced by alkyl groups. Like the base ammonia, which can accept an additional proton to form the positively charged ammonion ion, NH_4^+, primary, secondary, and tertiary amines are bases and can accept an additional proton. In addition, ions exist that contain four alkyl groups and have a charge of +1. These ions are called **quaternary ammonium ions.** Some of these ions are, in fact, components of cell membranes, as we shall see in Chapter 26.

An amine, like an alcohol, can react with a carboxylic acid in a dehydration reaction. The resulting compound is called an amide (see Table 18-1). **Amides**

contain the functional group $-\overset{\overset{\displaystyle O}{\|}}{C}-\underset{\underset{\displaystyle H}{|}}{N}-$. The amide functional group is respon-

sible for joining amino acids together in proteins (Chapter 21). Monomer units in the synthetic polymer nylon are also bonded together by amide bonds.

Table 18-1 Classes of Organic Compounds Containing Nitrogen or Sulfur

Name	Structure	Example
Primary amine	$R\!-\!NH_2$	$CH_3\!-\!NH_2$ methylamine
Secondary amine	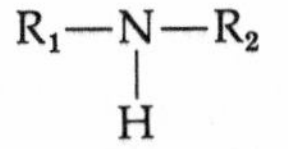 $R_1\!-\!\underset{\underset{H}{\vert}}{N}\!-\!R_2$	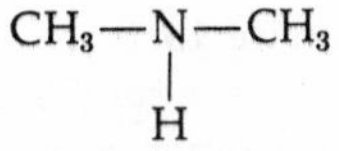 $CH_3\!-\!\underset{\underset{H}{\vert}}{N}\!-\!CH_3$ dimethylamine
Tertiary amine	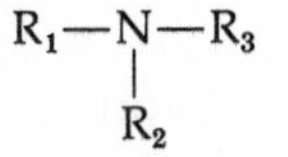 $R_1\!-\!\underset{\underset{R_2}{\vert}}{N}\!-\!R_3$	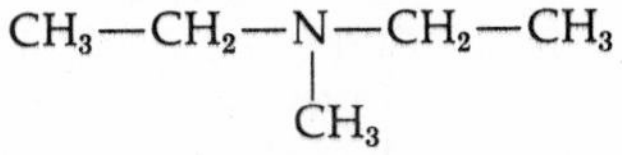 $CH_3\!-\!CH_2\!-\!\underset{\underset{CH_3}{\vert}}{N}\!-\!CH_2\!-\!CH_3$ diethylmethylamine
Quaternary ammonium salt	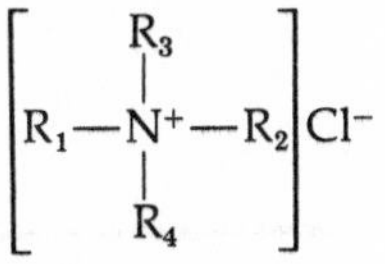 $\left[R_1\!-\!\overset{\overset{R_3}{\vert}}{\underset{\underset{R_4}{\vert}}{N^+}}\!-\!R_2\right]Cl^-$	$[(CH_3)_4N^+]Cl^-$ tetramethylammonium chloride
Simple amide	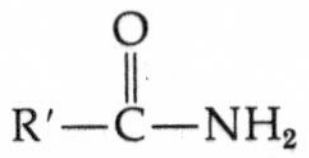 $R'\!-\!\overset{\overset{O}{\|}}{C}\!-\!NH_2$	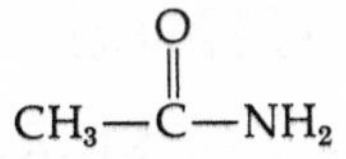 $CH_3\!-\!\overset{\overset{O}{\|}}{C}\!-\!NH_2$ acetamide
N-substituted amide	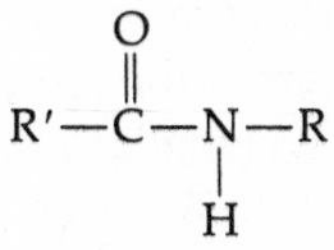 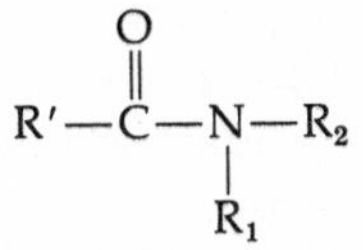 $R'\!-\!\overset{\overset{O}{\|}}{C}\!-\!\underset{\underset{H}{\vert}}{N}\!-\!R$ or $R'\!-\!\overset{\overset{O}{\|}}{C}\!-\!\underset{\underset{R_1}{\vert}}{N}\!-\!R_2$	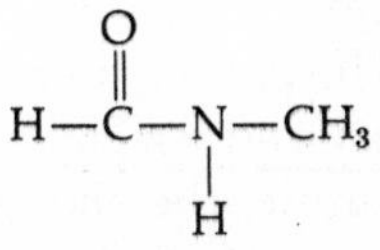 $H\!-\!\overset{\overset{O}{\|}}{C}\!-\!\underset{\underset{H}{\vert}}{N}\!-\!CH_3$ N-methylformamide
Thiol (mercaptan)	$R\!-\!SH$	$CH_3\!-\!CH_2\!-\!SH$ ethanethiol
Thioether	$R_1\!-\!S\!-\!R_2$	$CH_3\!-\!CH_2\!-\!S\!-\!CH_3$ methyl ethyl thioether
Thioester	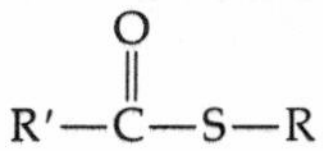 $R'\!-\!\overset{\overset{O}{\|}}{C}\!-\!S\!-\!R$	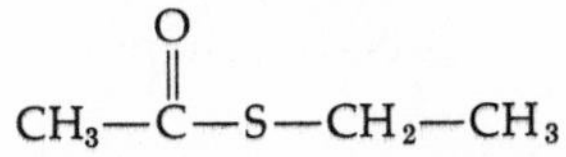 $CH_3\!-\!\overset{\overset{O}{\|}}{C}\!-\!S\!-\!CH_2\!-\!CH_3$ ethyl thioacetate
Disulfide	$R_1\!-\!S\!-\!S\!-\!R_2$	$CH_3\!-\!S\!-\!S\!-\!CH_3$ methyl disulfide

Since sulfur and oxygen are both group VI elements, many sulfur-containing compounds have a functional group that can be considered as the sulfur analog of an oxygen-containing functional group. These analogous sulfur- and oxygen-containing functional groups have somewhat similar properties. Thus, the —SH functional group, called the sulfhydryl group, is somewhat similar to the —OH group in an alcohol. Compounds containing the —SH group are therefore called thioalcohols or **thiols**. The prefix *thio-* indicates that a sulfur atom has replaced an oxygen atom in the characteristic functional group of these compounds. Thiols are also referred to as mercaptans (Table 18-1). We shall see that many reactions of thiols are similar to those of alcohols. For example, thiols can react with carboxylic acids to form **thioesters** (Table 18-1).

Compounds containing the functional group —S—S— are called **disulfides** (Table 18-1). Disulfides are important structural components of protein molecules.

Exercise 18-1

Draw structural formulas for all of the amines with the molecular formula C_3H_9N. Label each isomer as a primary, secondary, or tertiary amine.

Exercise 18-2

When ethanethiol reacts with acetic acid, a thioester is formed. Without looking ahead in the text, see if you can draw the structural formula for this thioester based on your knowledge of the reaction of alcohols and acids to form esters.

18-4   AMINES

Structure and Nomenclature

Table 18-2 shows the structural formulas and common names of several amines. The IUPAC names of amines are rather complicated, especially those for secondary and tertiary amines, and we shall not use them. In the common name for an amine, the alkyl groups attached to the nitrogen atoms are generally indicated by prefixes before the ending *-amine*. For example, ethyldimethylamine has two methyl groups and one ethyl group attached to the nitrogen atom. It is therefore a tertiary amine. The common names of some amines, such as aniline and alanine, are very old and do not indicate structural features. Notice in Table 18-2 that the name for the secondary amine, N-ethylaniline, contains the prefix N- to indicate that the ethyl group is bonded to the nitrogen and not to one of the carbon atoms in this molecule. This prefix is likewise used in the names of other amines where it is necessary to specify that an R group is bonded to the nitrogen and not another atom. The amine alanine also contains a carboxyl group. Compounds with both an amine and a carboxyl group are called **amino acids.** Amino acids that contain α-carboxyl groups, such as alanine (see Figure 18-1), are especially interesting because they are used to make proteins in the human body (Chapter 21).

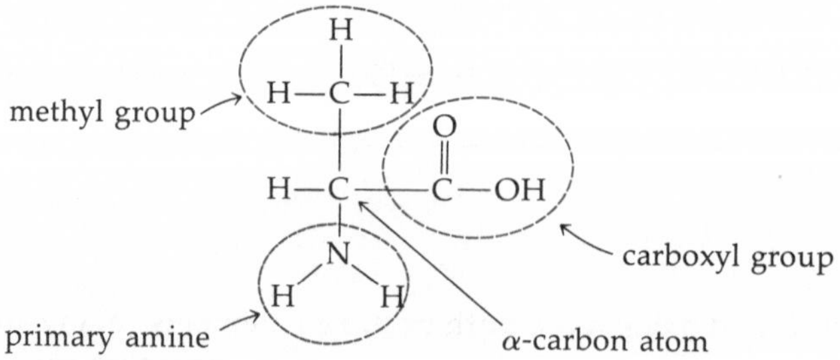

Figure 18-1   The α-amino acid alanine has a methyl group bonded to the α-carbon atom. Other α-amino acids have different groups of atoms, called side chains, bonded to the α-carbon atom.

Quaternary ammonium ions are also very important in the human body. Compounds containing quaternary ammonium cations and anions such as Cl^-, OH^-, SO_4^{2-}, and so on, are ionic compounds.

Table 18-2 Some Common Amines

Structural Formula	Common Name
Primary amines:	
NH_2 $\|$ CH_3-C-CH_3 $\|$ H	Isopropylamine
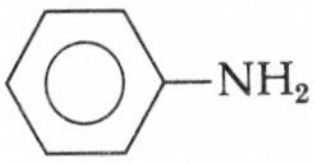 ◯$-NH_2$	Aniline (an aromatic amine)
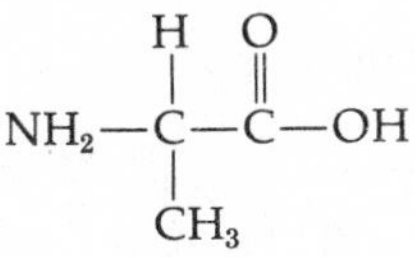 $H\ \ O$ $\|\ \ \ \|\|$ $NH_2-C-C-OH$ $\|$ CH_3	Alanine (an amino acid)
Secondary amines:	
H $\|$ $CH_3-CH_2-CH_2-N-CH_3$	Methyl-n-propylamine
◯$-N-CH_2-CH_3$ $\|$ H	N-ethylaniline (phenylethylamine)
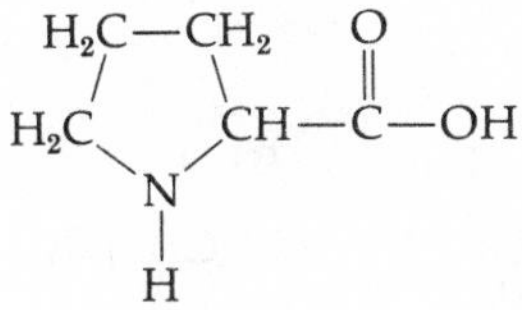 $H_2C-CH_2\ \ \ O$ $\diagup\qquad\diagdown\quad \|\|$ $H_2C\qquad CH-C-OH$ $\diagdown\quad\diagup$ N $\|$ H	Proline (an amino acid)
Tertiary amine:	
CH_3-N-CH_3 $\|$ CH_3	Trimethylamine
Quaternary ammonium compounds:	
$\begin{bmatrix} CH_3 \\ \| \\ CH_3-CH_2-N^+-CH_2-CH_3 \\ \| \\ CH_3 \end{bmatrix} Br^-$	Diethyldimethylammonium bromide
$\begin{bmatrix} CH_3 \\ \| \\ CH_3-N^+-◯ \\ \| \\ CH_3 \end{bmatrix} OH^-$	Trimethylphenylammonium hydroxide

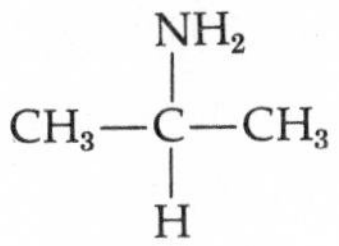

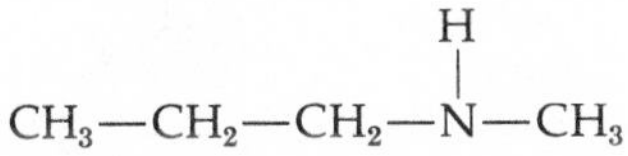

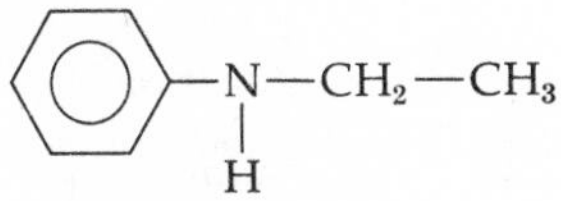

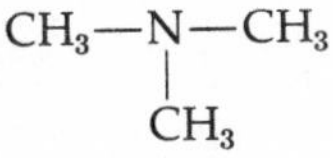

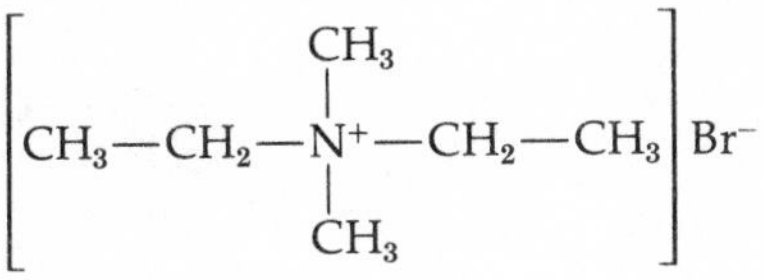

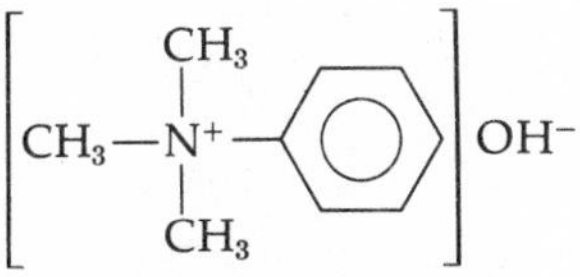

A number of important natural biomolecules as well as several drugs contain amine functional groups, often as part of a complicated structure that includes rings of atoms. The structural formulas of some of these complex amines are given in Table 18-3.

Table 18-3 Some Complex Amines

Drugs	Naturally Occurring in Humans
Morphine 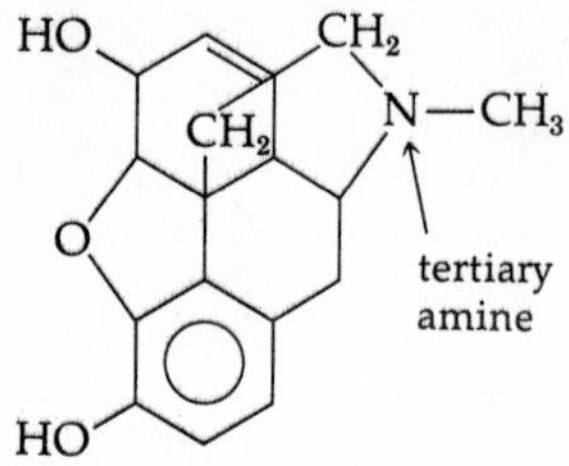	Adenine (a component of nucleic acids) 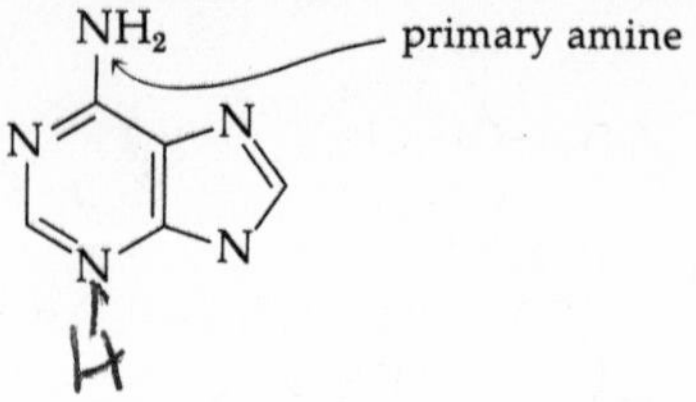

Morphine

tertiary amine

Adenine (a component of nucleic acids)

NH_2 — primary amine

Codeine

N—CH$_3$

tertiary amine

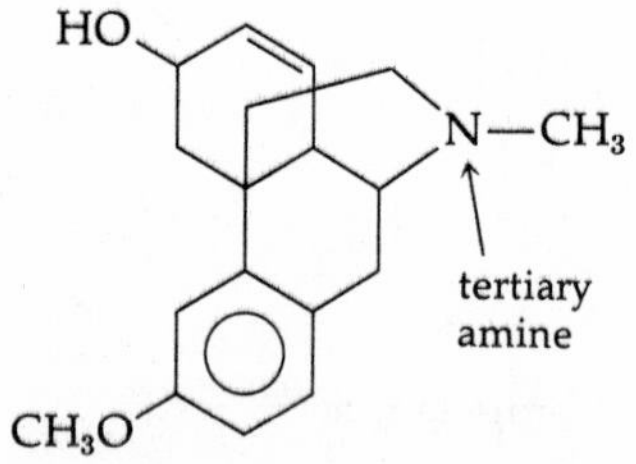

Tryptophan (an amino acid)

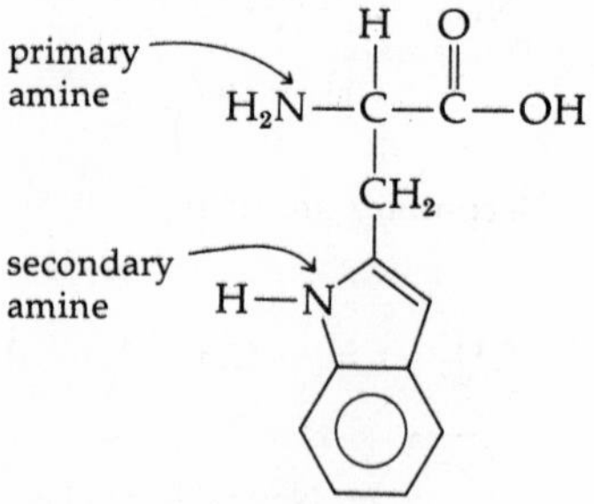

primary amine

H_2N—C—C—OH

secondary amine

H—N

Heroin

CH_3—C—O

N—CH$_3$

tertiary amine

CH_3—C—O

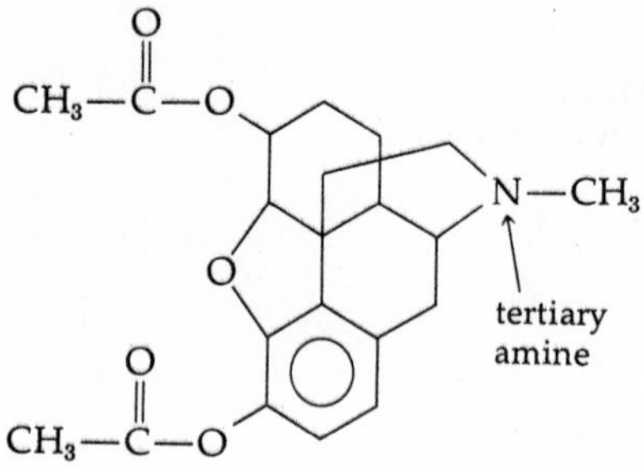

Epinephrine (adrenaline)

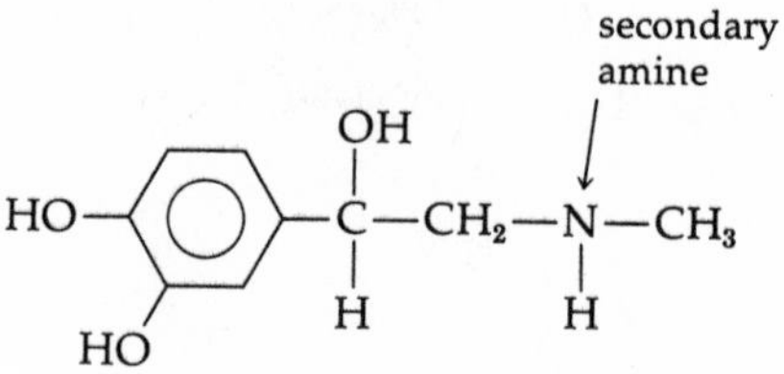

secondary amine

HO—

C—CH$_2$—N—CH$_3$

HO—

Amphetamine

CH$_2$

CH

NH$_2$ CH$_3$

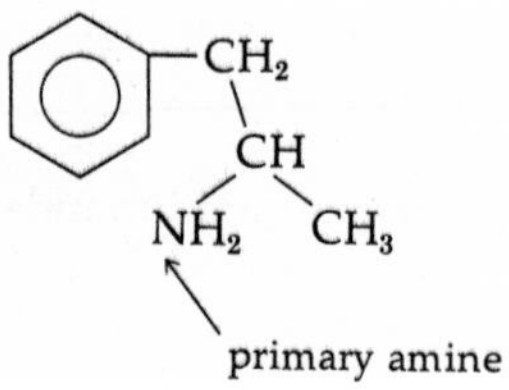

primary amine

Histamine

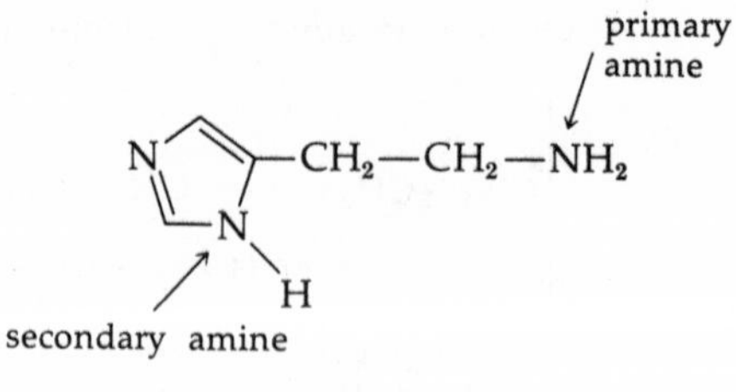

primary amine

CH$_2$—CH$_2$—NH$_2$

secondary amine

Nicotine (a poison)

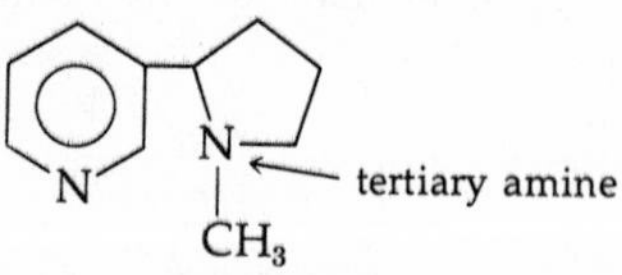

N— tertiary amine

CH$_3$

Procaine (Novocaine)

H_2N—

C—O—CH$_2$—CH$_2$—N

CH$_2$—CH$_3$

CH$_2$—CH$_3$

primary amine

tertiary amine

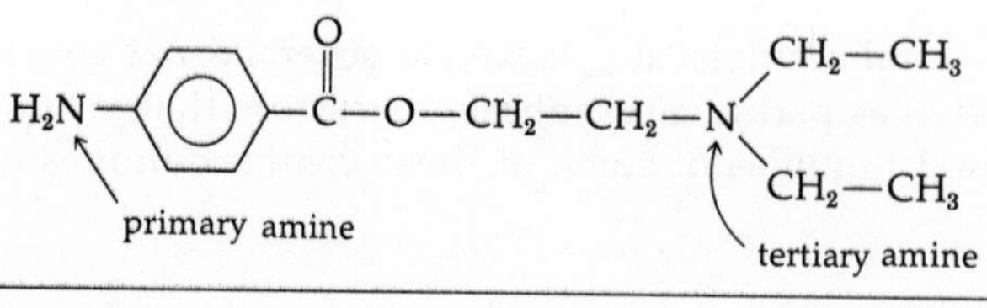

Physical Properties

An **amine** functional group is a very polar group that can hydrogen bond with water molecules. The nitrogen atom in a primary, a secondary, or a tertiary amine can act as a hydrogen bond acceptor; hence these amines can all form hydrogen bonds with water and other molecules (see Figure 18-2).

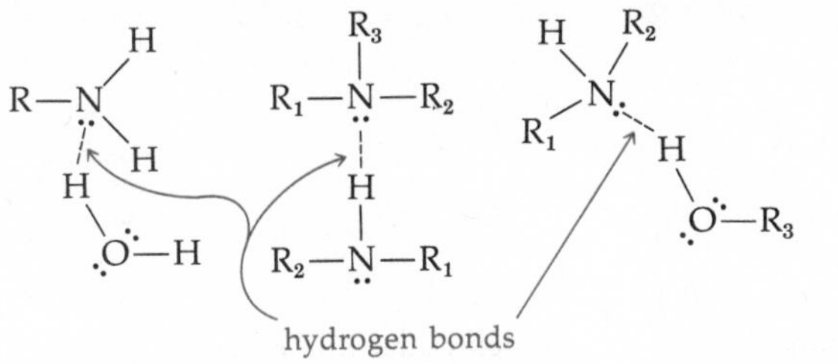

Figure 18-2   The amine nitrogen is a hydrogen atom acceptor in hydrogen bonds between an amine and water (left), between different amine molecules (center), and between an amine and an alcohol (right).

In addition, the hydrogen atom (or atoms) bonded to the nitrogen atom in primary or secondary amines can act as the hydrogen donor in the formation of a hydrogen bond with the oxygen atom in a water molecule or a very electronegative atom in another molecule (see Figure 18-3). Low-molecular-weight amines are therefore quite soluble in water. However, high-molecular-weight amines with large nonpolar parts are quite insoluble in water.

Figure 18-3   Primary (left) and secondary (right) amines are hydrogen atom donors in hydrogen bonds with molecules containing very electronegative atoms, such as the oxygen atom in water.

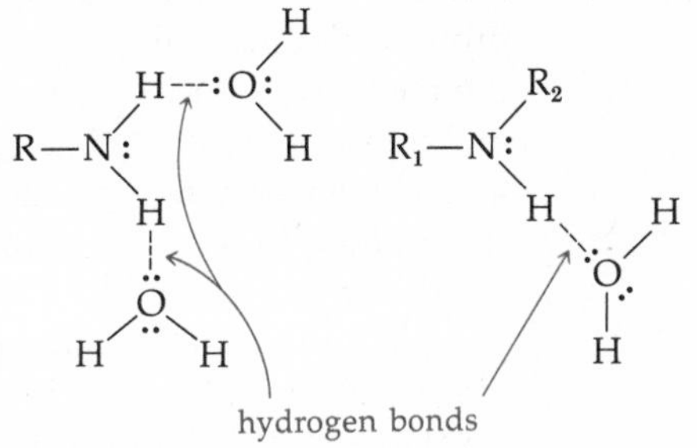

Chemical Reactivity

Ammonia can act as a proton acceptor, and therefore is a weak Brønsted-Lowry base. Likewise, primary, secondary, and tertiary amines can accept protons and are also weak bases. Thus, a solution of dimethylamine in water, for example, will be alkaline as a result of the acid-base reaction between water and the amine, producing the dimethylammonium ion and a hydroxide ion:

(18-1)

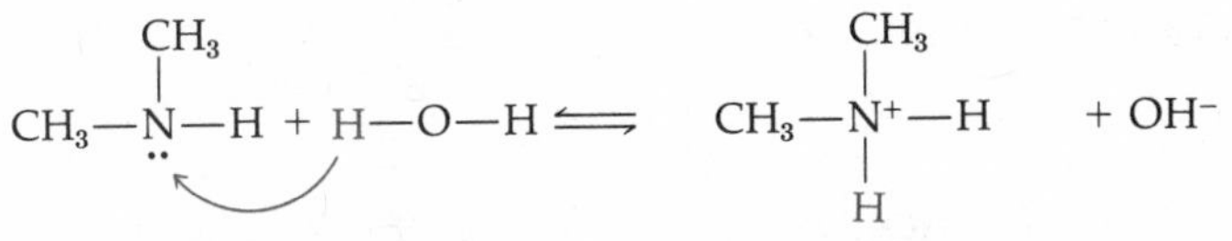

$$CH_3-\overset{\overset{\displaystyle CH_3}{|}}{\underset{\cdot\cdot}{N}}-H + H-O-H \rightleftharpoons CH_3-\overset{\overset{\displaystyle CH_3}{|}}{\underset{\underset{\displaystyle H}{|}}{N^+}}-H + OH^-$$

dimethylamine   dimethylammonium ion

Since amines are weak bases, they can react with strong acids and form an ammonium salt. For example,

(18-2)

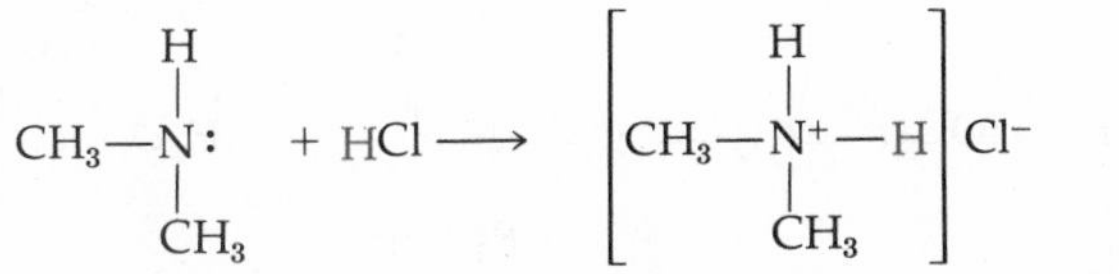

$$CH_3-\overset{\overset{\displaystyle H}{|}}{\underset{\underset{\displaystyle CH_3}{|}}{N}}: + HCl \longrightarrow \left[CH_3-\overset{\overset{\displaystyle H}{|}}{\underset{\underset{\displaystyle CH_3}{|}}{N^+}}-H \right] Cl^-$$

dimethylamine   dimethylammonium chloride

The salt formed in reaction 18-2, dimethylammonium chloride, is an ionic compound and is very soluble in water. It is often advantageous to convert an insoluble amine to the more soluble ammonium salt. For example, the solubility of morphine in water is 1 g/5000 ml, whereas 1 g of the ammonium sulfate salt of morphine will dissolve in only 15.5 ml of water.

Whether an amine will exist predominantly as a neutral molecule or as a protonated ion in an aqueous medium depends on the pH. Consider the equilibrium:

$$
(18\text{-}3) \qquad \underset{\substack{| \\ H \\ \text{dimethylamine}}}{CH_3 - \overset{..}{N} - CH_3} + H^+ \;\rightleftharpoons\; \underset{\substack{| \\ H \\ \text{dimethylammonium ion}}}{\overset{\overset{\displaystyle H}{|}}{CH_3 - \overset{+}{N} - CH_3}}
$$

Dimethylamine, like all amines, is a weak base, and in basic solutions, where $[H^+]$ is very small, the equilibrium is shifted to the left and the predominant form is the neutral molecule $(CH_3)_2NH$. On the other hand, in neutral or acidic solutions, $[H^+]$ is large enough so that the equilibrium is shifted to the right and the predominant form is the protonated ammonium ion $(CH_3)_2\overset{+}{N}H_2$. Since the predominant form of an amine depends on the pH, the solubility of an amine can be drastically altered by changing the pH. Amines with large nonpolar groups are virtually insoluble in solutions of high pH because they exist predominantly in the form of neutral insoluble molecules. But in a neutral or acidic solution, the neutral amine molecule is converted to the much more soluble protonated ammonium ion form. Note that when an amine is dissolved in water, the resulting solution is not neutral—it is basic. The $[H^+]$ in such a solution is much less than $10^{-7}\ M$, and the equilibrium in reaction 18-3 is shifted to the left. However, when an amine is dissolved in a buffer solution where $[H^+]$ is maintained at $10^{-7}\ M$ (neutral pH), the $[H^+]$ is sufficiently large to shift the equilibrium in reaction 18-3 to the right. As we shall see in Chapters 21, 22, and 23, amine functional groups are important parts of protein molecules, and their chemical behavior is important to the biological activity of proteins.

Quaternary ammonium salts are ionic compounds (see Table 18-2). Several substances containing quaternary (four-bonded) nitrogen ions have special functions in the human body. For example, we have frequently referred to a coenzyme called NAD^+. The positive charge of this coenzyme results from the presence of a quaternary nitrogen ion. Quaternary nitrogens have a strong tendency to attract electrons, which allows NAD^+ to function as a coenzyme in many oxidation-reduction reactions in the human body (see Figure 18-4).

The most important chemical property of primary and secondary amines is their ability to react with carboxylic acids to form amides, with the elimination of a molecule of water:

$$
(18\text{-}4) \qquad \underset{\substack{| \\ H \\ \text{a primary} \\ \text{amine}}}{R - N - H} + \underset{\substack{\text{a carboxylic} \\ \text{acid}}}{HO - \overset{\overset{\displaystyle O}{\|}}{C} - R'} \longrightarrow \underset{\substack{| \\ H \\ \text{an amide}}}{R - N - \overset{\overset{\displaystyle O}{\|}}{C} - R'} + H_2O
$$

$$
(18\text{-}5) \qquad \underset{\substack{| \\ R_2 \\ \text{a secondary} \\ \text{amine}}}{R_1 - N - H} + \underset{\substack{\text{a carboxylic} \\ \text{acid}}}{HO - \overset{\overset{\displaystyle O}{\|}}{C} - R'} \longrightarrow \underset{\substack{| \\ R_2 \\ \text{an amide}}}{R_1 - N - \overset{\overset{\displaystyle O}{\|}}{C} - R'} + H_2O
$$

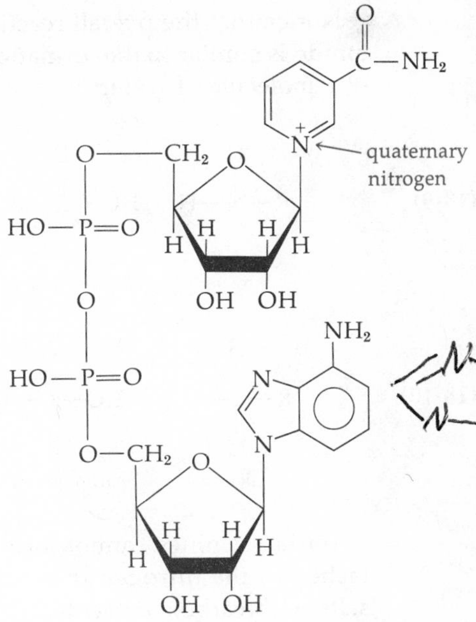

Figure 18-4 The coenzyme nicotinamide adenine dinucleotide (NAD$^+$).

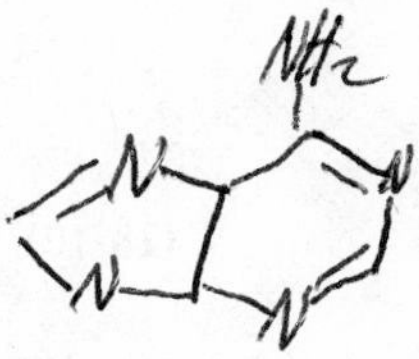

The reaction of an amine and a carboxylic acid can be viewed as an acid-base reaction to form a salt, followed by the loss of a molecule of water to form an amide. In the laboratory this is accomplished by heating the mixture to boil off the water:

(18-6)
$$R-\underset{\underset{H}{|}}{\overset{\overset{H}{|}}{N}}: + HO-\overset{\overset{O}{\|}}{C}-R' \longrightarrow \left[R-\overset{+}{N}H_3\right]\left[{}^-O-\overset{\overset{O}{\|}}{C}-R'\right]$$
salt

(18-7)
$$\left[R-\overset{+}{N}H_3\right]\left[{}^-O-\overset{\overset{O}{\|}}{C}-R'\right] \xrightarrow{\text{heat}} R-\underset{\underset{H}{|}}{N}-\overset{\overset{O}{\|}}{C}-R' + H_2O$$
salt   amide

In the industrial manufacture of amides, high temperature is used to drive off the water and produce the amide product. The synthetic polymer nylon (Section 14-9) is made by a polymerization reaction that involves the formation of amide bonds between alternating dicarboxylic acid and diamine monomer units:

(18-8)
$$n\left(HO-\overset{\overset{O}{\|}}{C}-(CH_2)_x-\overset{\overset{O}{\|}}{C}-OH\right) + n(H_2N-(CH_2)_y-NH_2) \longrightarrow$$
a dicarboxylic acid   a diamine

$$\left(-\underset{\underset{H}{|}}{N}-(CH_2)_y-\underset{\underset{H}{|}}{N}-\overset{\overset{O}{\|}}{C}-(CH_2)_x-\overset{\overset{O}{\|}}{C}-\right)_n + 2nH_2O$$
nylon

where x and y refer to the number of carbon atoms between the functional groups in these monomers. Nylon 66, produced from the six-carbon-atom dicarboxylic acid ($x = 4$) and the six-carbon-atom diamine ($y = 6$), is a typical commercial nylon polymer.

Notice that the overall reaction of an amine and a carboxylic acid to form an amide is similar to the formation of an ester in that both involve the elimination of a molecule of water:

$$
\begin{array}{cccc}
& \text{H} & \text{O} & \text{O} \\
& | & \| & \| \\
(18\text{-}9) & \text{R—N—H} + \text{HO—C—R}' \longrightarrow & \text{R—N—C—R}' + \text{H}_2\text{O} \\
& & & | \\
& & & \text{H}
\end{array}
$$

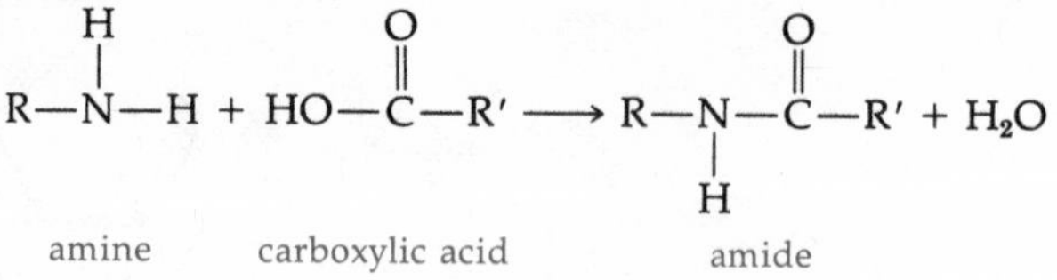

$$
\quad\quad\text{amine}\quad\quad\text{carboxylic acid}\quad\quad\quad\text{amide}
$$

$$
\begin{array}{cccc}
& \text{H} & \text{O} & \text{H}\quad\text{O} \\
& | & \| & |\quad\| \\
(18\text{-}10) & \text{R—C—OH} + \text{HO—C—R}' \longrightarrow & \text{R—C—O—C—R}' + \text{H}_2\text{O} \\
& | & & | \\
& \text{H} & & \text{H}
\end{array}
$$

$$
\quad\quad\text{alcohol}\quad\quad\text{carboxylic acid}\quad\quad\quad\text{ester}
$$

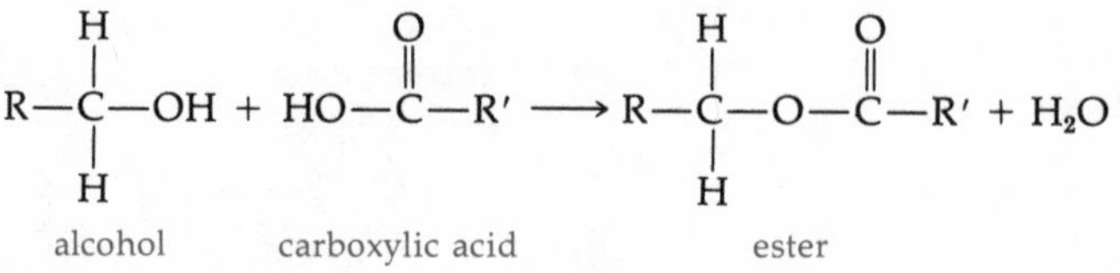

Tertiary amines cannot form amides, because there is no hydrogen atom attached to the nitrogen in a tertiary amine. However, tertiary amines do form salts with carboxylic acids.

Amides can also be formed by the reaction of an ester and an amine:

$$
\begin{array}{cccc}
\text{H} & \text{H}\quad\text{O} & \text{H}\;\;\text{O} & \text{H} \\
| & |\quad\| & |\;\;\| & | \\
(18\text{-}11)\quad\text{R—N—H} + \text{R}_2\text{—C—O—C—R}_1 \xrightarrow{\text{catalyst}} \text{R—N—C—R}_1 + \text{R}_2\text{—C—OH} \\
& | & & | \\
& \text{H} & & \text{H}
\end{array}
$$

$$
\quad\text{amine}\quad\quad\quad\text{ester}\quad\quad\quad\quad\text{amide}\quad\quad\quad\text{alcohol}
$$

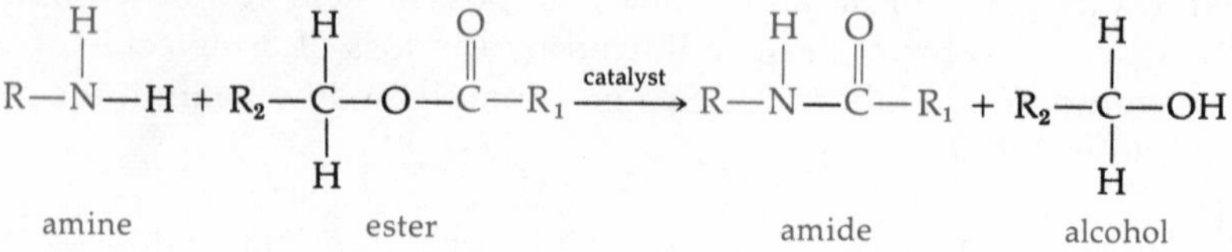

All of the enzymes and the other proteins in the human body are composed of amino acids held together by amide bonds (Chapter 21). In Chapter 24 we shall see that these amide bonds in proteins are formed from amines and esters in enzyme-catalyzed reactions similar to reaction 18-11.

Exercise 18-3

Indicate which of the following (1) can form a hydrogen bond by donating a hydrogen, (2) can form a hydrogen bond by accepting a hydrogen, (3) is a base, (4) is an acid.

(a) $CH_3—NH_2$ (b) $CH_3—\overset{+}{N}H_3$ (c) $CH_3—CH_2—\overset{\displaystyle |}{\underset{\displaystyle CH_3}{N}}—H$

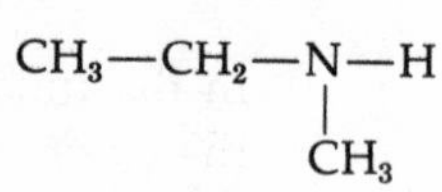

(d) $(CH_3)_3N$ (e) $(CH_3)_4N^+$

Exercise 18-4

Draw structural formulas for the products of the following reactions:

(a) $CH_3—\overset{\displaystyle H}{\overset{\displaystyle |}{N}}—CH_3 + CH_3—CH_2—\overset{\displaystyle O}{\overset{\displaystyle \|}{C}}—OH \longrightarrow H_2O +$

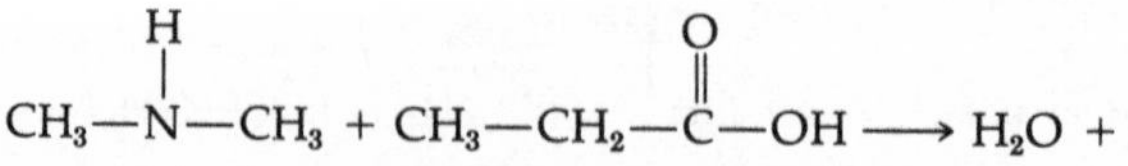

(b) $(CH_3)_3N + CH_3—CH_2—\overset{\displaystyle O}{\overset{\displaystyle \|}{C}}—OH \longrightarrow$

(c) $CH_3—\overset{\displaystyle NH_2}{\overset{\displaystyle |}{\underset{\displaystyle \underset{\displaystyle CH_3}{|}}{C}}}—H + H—\overset{\displaystyle O}{\overset{\displaystyle \|}{C}}—OH \longrightarrow H_2O +$

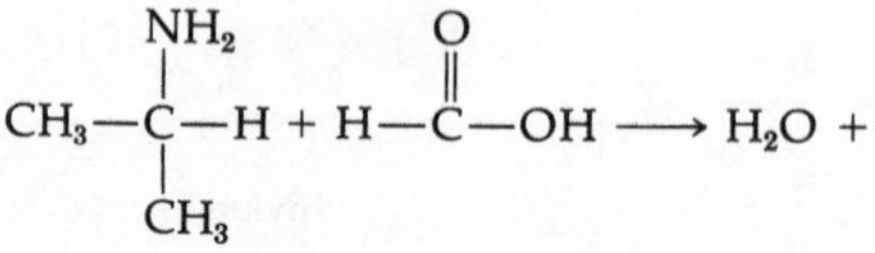

(d) $NH_3 + CH_3—\overset{\displaystyle O}{\overset{\displaystyle \|}{C}}—OH \longrightarrow H_2O +$

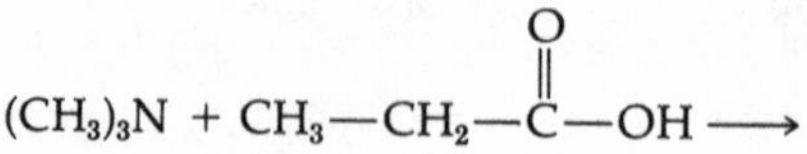

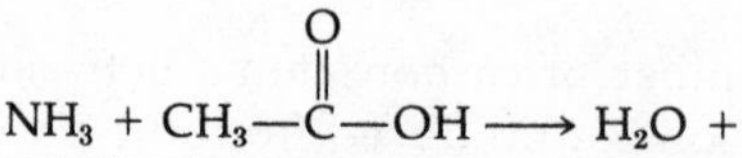

18-5   AMIDES

Structure and Nomenclature

The $\overset{\overset{\text{O}}{\|}}{-\text{C}-\text{N}-}$ group is called the **amide** functional group. When the nitrogen

atom of the amide group is bonded to two hydrogen atoms, the molecule is

called a **simple amide,** with the general formula $\text{R}'-\overset{\overset{\text{O}}{\|}}{\text{C}}-\text{NH}_2$.

Table 18-4 Some Simple Amides

Structural Formula	Common Name
$\text{H}-\overset{\overset{\text{O}}{\|}}{\text{C}}-\text{NH}_2$	Formamide
$\text{CH}_3-\overset{\overset{\text{O}}{\|}}{\text{C}}-\text{NH}_2$	Acetamide
$\text{CH}_3-\text{CH}_2-\overset{\overset{\text{O}}{\|}}{\text{C}}-\text{NH}_2$	Propionamide
$\overset{\overset{\text{OH}}{\|}}{\underset{\underset{\text{NH}_2}{\|}}{\text{H}-\text{C}-}}\overset{\text{C}=\text{O}}{}\text{CH}_2-\overset{\overset{\text{O}}{\|}}{\text{C}}-\text{NH}_2$	Asparagine (an amino acid)
$\overset{\overset{\text{OH}}{\|}}{\underset{\underset{\text{NH}_2}{\|}}{\text{H}-\text{C}-}}\overset{\text{C}=\text{O}}{}\text{CH}_2-\text{CH}_2-\overset{\overset{\text{O}}{\|}}{\text{C}}-\text{NH}_2$	Glutamine (an amino acid)
Ph$-\overset{\overset{\text{O}}{\|}}{\text{C}}-\text{NH}_2$	Benzamide
pyridinyl$-\overset{\overset{\text{O}}{\|}}{\text{C}}-\text{NH}_2$	Nicotinamide (a B vitamin)

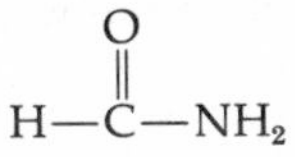

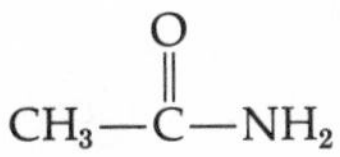

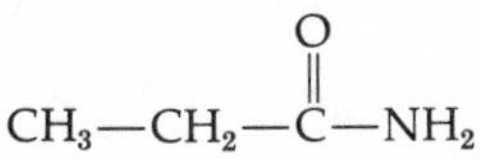

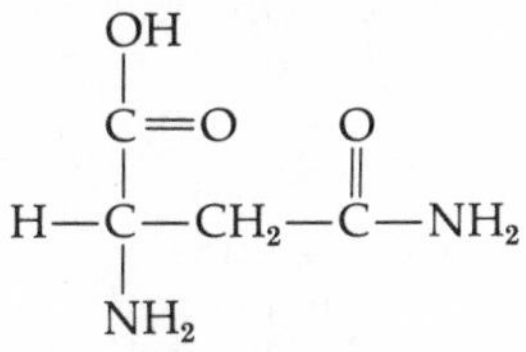

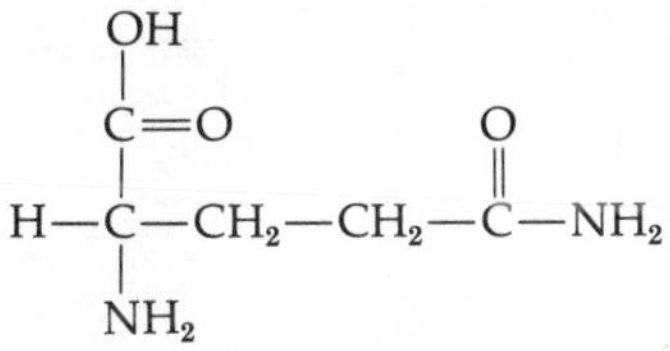

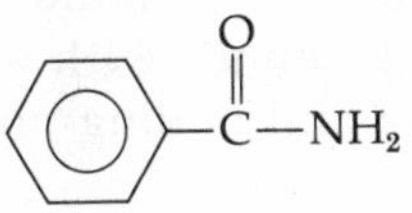

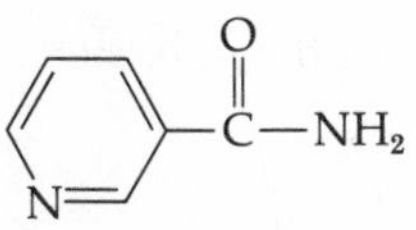

The structural formulas and common names of several simple amides are given in Table 18-4. The common name for a simple amide is derived from the common name of the parent carboxylic acid by dropping the suffix *-ic* (which denotes an acid) and adding the suffix *-amide*. Glutamine and asparagine are two amino acids that contain simple amide groups. Another important simple amide is nicotinamide, which is the active form of the B vitamin niacin, and is used in the formation of the coenzyme NAD^+ (see Figure 18-4).

N-substituted amides contain one or two R groups attached to the N atom of the amide group. The structural formulas of a few N-substituted amides are given in Table 18-5. Notice in Table 18-5 that the prefix N- is used to indicate that an R group is attached to the nitrogen atom in an N-substituted amide. We shall study N-substituted amides in more detail in Chapter 21, when we consider the amide bonds that join amino acids together to form proteins.

Table 18-5 Some N-Substituted Amides

Structural Formula	Name
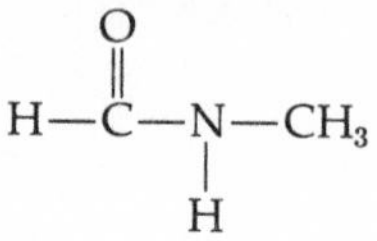 $H-\overset{\overset{\textstyle O}{\|\|}}{C}-\underset{\underset{\textstyle H}{\|}}{N}-CH_3$	N-methylformamide
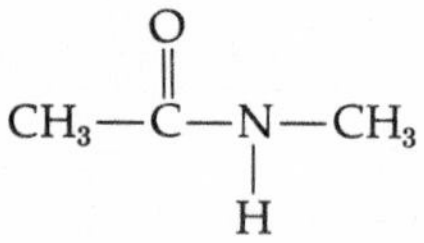 $CH_3-\overset{\overset{\textstyle O}{\|\|}}{C}-\underset{\underset{\textstyle H}{\|}}{N}-CH_3$	N-methylacetamide
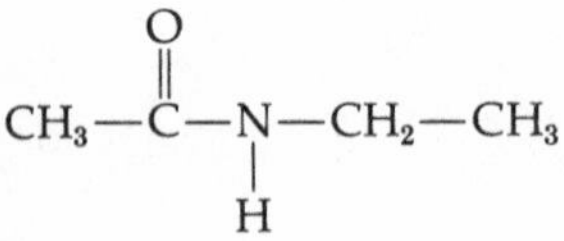 $CH_3-\overset{\overset{\textstyle O}{\|\|}}{C}-\underset{\underset{\textstyle H}{\|}}{N}-CH_2-CH_3$	N-ethylacetamide
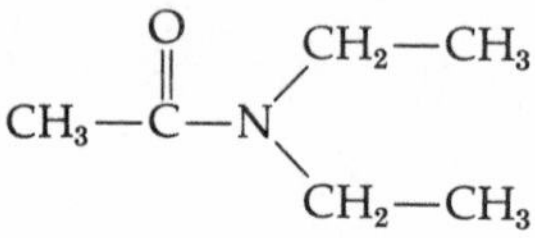 $CH_3-\overset{\overset{\textstyle O}{\|\|}}{C}-N\overset{\diagup CH_2-CH_3}{\diagdown CH_2-CH_3}$	N,N-diethylacetamide
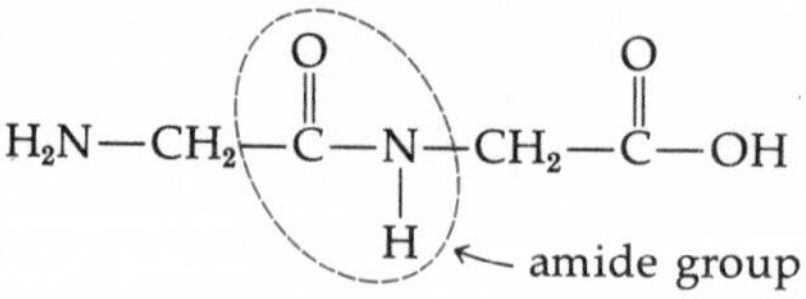 $H_2N-CH_2-\overset{\overset{\textstyle O}{\|\|}}{C}-\underset{\underset{\textstyle H}{\|}}{N}-CH_2-\overset{\overset{\textstyle O}{\|\|}}{C}-OH$ ← amide group	Glycylglycine (two amino acids bonded by an amide linkage)

Properties of Simple Amides

Simple amides are very polar, and low-molecular-weight amides are quite soluble in water because they readily form hydrogen bonds with water (see Figure 18-5). Simple amides also have higher melting and boiling points than many

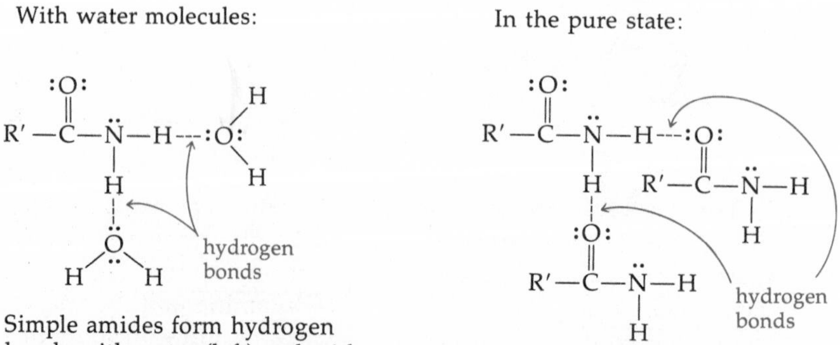

With water molecules:

In the pure state:

Figure 18-5   Simple amides form hydrogen bonds with water (left) and with other amide molecules in the pure state (right).

other compounds of similar molecular weight, because of their ability to form hydrogen bonds in the pure state (see Figure 18-5). However, whereas amines are weak bases, amides are not, even though the amide group, like an amine, contains a nitrogen atom. It has been determined experimentally that the carbon-nitrogen bond in amides is shorter than a typical carbon-nitrogen single bond, and that there is no free rotation about the amide carbon-nitrogen bond. One representation of the amide bond that accounts for these observations is a composite picture of two resonance structures, one of which involves a C=N double bond (Figure 18-6). Note that in the structure on the right side of this figure, there are no unshared electrons on the nitrogen atom. The electronegative oxygen atom tends to pull these electrons away from the nitrogen atom, making it unable to serve as a proton acceptor (base).

Figure 18-6   Resonance structures for an amide. This composite representation accounts for the experimental observations that (1) an amide carbon-nitrogen bond is shorter than a typical carbon-nitrogen single bond; (2) the amide nitrogen atom is not a proton acceptor (a base); and (3) there is no free rotation around the carbon-nitrogen bond in amides.

$$\underset{H}{\overset{:O:}{\underset{|}{\overset{\|}{R'-C-\overset{\cdot\cdot}{N}-R}}}} \longleftrightarrow \underset{H}{\overset{:\overset{\cdot\cdot}{O}:^{-}}{\underset{|}{\overset{|}{R'-C=\overset{+}{N}-R}}}}$$

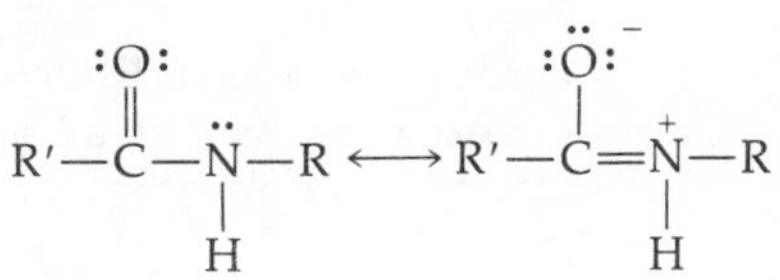

We have already seen how N-substituted amides are formed by the reaction of a carboxylic acid with a primary or a secondary amine (reactions 18-4 and 18-5). Simple amides can likewise be formed by the reaction of a carboxylic acid and ammonia:

(18-12)
$$R'-\overset{\overset{O}{\|}}{C}-OH + NH_3 \xrightarrow{\text{heat}} R'-\overset{\overset{O}{\|}}{C}-NH_2 + H_2O$$

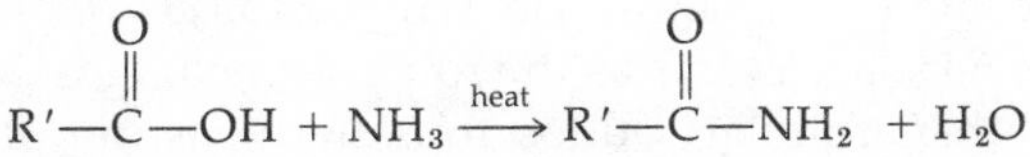

Note that the formation of both a simple and an N-substituted amide involves dehydration—removal of a molecule of water.

For our purposes, the most important reaction of amides is their hydrolysis to form amines (or ammonia) and carboxylic acids. Hydrolysis of an amide is exactly opposite to amide formation. For example,

(18-13)   Hydrolysis of a simple amide:

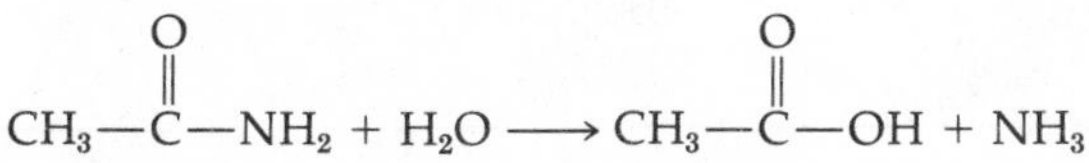

$$CH_3-\overset{\overset{O}{\|}}{C}-NH_2 + H_2O \longrightarrow CH_3-\overset{\overset{O}{\|}}{C}-OH + NH_3$$

(18-14)   Hydrolysis of an N-substituted amide:

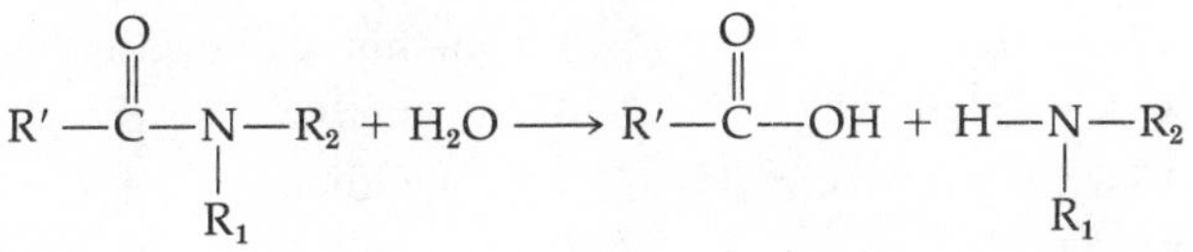

$$R'-\overset{\overset{O}{\|}}{\underset{\underset{R_1}{|}}{C}}-N-R_2 + H_2O \longrightarrow R'-\overset{\overset{O}{\|}}{C}-OH + \underset{\underset{R_1}{|}}{H}-N-R_2$$

Hydrolysis of the amide bonds in proteins is the major reaction that occurs during the digestion of proteins in the human body (Chapter 28).

Exercise 18-5

Draw structural formulas for the following compounds: (a) α-Methylpropionamide; (b) N,N-dimethylpropionamide; (c) N-cyclopentylformamide.

Exercise 18-6

Draw structural formulas for the products of each of the following reactions:

(a) 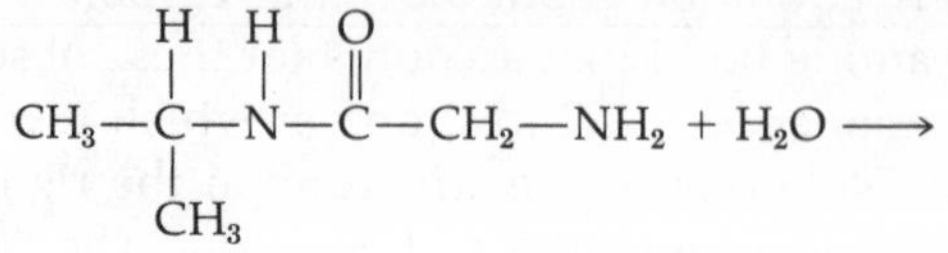
$$
\underset{\underset{CH_3}{|}}{CH_3-\overset{\overset{H}{|}}{C}-\overset{\overset{H}{|}}{N}-\overset{\overset{O}{\|}}{C}-CH_2-NH_2} + H_2O \longrightarrow
$$

(b) $NH_3 + H_2O \longrightarrow$

(c) 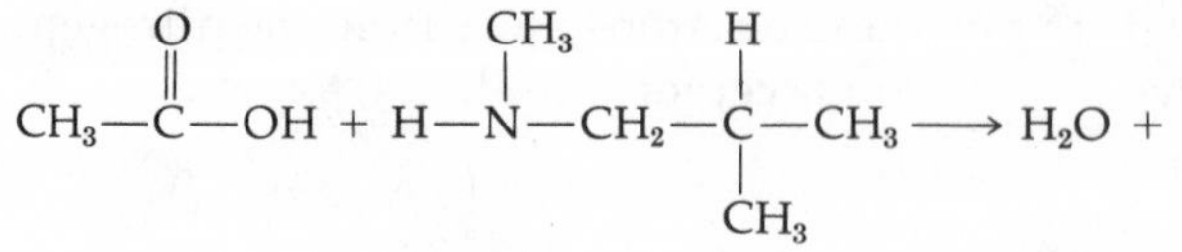
$$
CH_3-\overset{\overset{O}{\|}}{C}-OH + H-\overset{\overset{CH_3}{|}}{N}-CH_2-\underset{\underset{CH_3}{|}}{\overset{\overset{H}{|}}{C}}-CH_3 \longrightarrow H_2O +
$$

Exercise 18-7

For the reactions in Exercise 18-6, indicate which of the organic reactants and products can act as an acid, as a base, and as both an acid and a base.

18-6 THIOLS, DISULFIDES, AND THIOESTERS

The —SH functional group is called the **sulfhydryl** group, and is similar in many ways to the —OH hydroxyl group of alcohols. The class of organic molecules that contain only the sulfhydryl group are called **thiols**. The IUPAC name for a thiol is formed in the same manner as that of the corresponding alcohol, except the suffix -*ethiol* is used instead of the alcohol suffix -*ol*. The common name **mercaptan** is sometimes used. The names and structural formulas of a few thiols are given in Table 18-6. Cysteine is one of the amino acids used to make proteins. Ethyl mercaptan (ethanethiol) has the dubious distinction of being listed in the *Guinness Book of Records* as the foulest-smelling compound known! Minute amounts of ethanethiol are added to natural gas in order for gas leaks to be detectable (pure natural gas is odorless). 1-Butanethiol was long thought to be the malodorous molecule produced by skunks, but this was disproven in 1974. The odor of skunks is actually due to a mixture of sulfur-containing compounds, including 1-propanethiol.

Although the sulfhydryl group, —SH, and the alcohol group, —OH, are similar, remember that sulfur atoms are much less electronegative than oxygen atoms. Consequently, thiols do not form hydrogen bonds as readily as alcohols, have lower boiling points than alcohols of comparable weight (see Table 18-7), and are less soluble in water than alcohols.

Table 18-6 Some Thiols (Mercaptans)

Structural Formula	IUPAC Name	Common Name
CH_3-SH	Methanethiol	Methyl mercaptan
CH_3-CH_2-SH	Ethanethiol	Ethyl mercaptan
$CH_3-CH_2-CH_2-SH$	1-Propanethiol	n-Propyl mercaptan
$CH_3-CH_2-CH_2-CH_2-SH$	1-Butanethiol	n-Butyl mercaptan
$HO-\overset{\overset{O}{\|}}{C}-\underset{\underset{NH_2}{|}}{\overset{\overset{H}{|}}{C}}-CH_2-SH$		Cysteine (an amino acid)

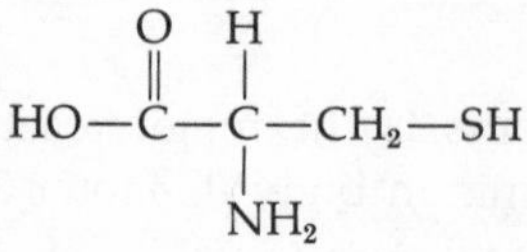

Table 18-7 Boiling Points of Some Alcohols and Thiols

Alcohol	Boiling Point (°C)	Thiol	Boiling Point (°C)
Methanol	65	Methanethiol	6
Ethanol	78	Ethanethiol	36
1-Butanol	117	1-Butanethiol	98

Two reactions involving sulfhydryl groups are of particular interest because of their importance in the human body. First, thiols can be easily oxidized to form disulfides in the reaction

(18-15) $\quad R_1—SH + HS—R_2 + X \longrightarrow R_1—S—S—R_2 + XH_2$
   a disulfide

In reaction 18-15, the two thiols each lose a hydrogen atom and couple together to form a disulfide. Since the thiols lose hydrogen, they are oxidized and the compound (X) that accepts the hydrogen atoms is reduced. Disulfides can be reduced back to the thiols in the reverse of reaction 18-15. Disulfide bonds between the sulfhydryl groups on cysteine amino acids are very important in determining the structure of proteins (Chapter 21).

The other important reaction of thiols is their combination with carboxylic acids to form **thioesters:**

(18-16)

$$\underset{\substack{\text{carboxylic}\\\text{acid}}}{R'—\overset{\overset{\displaystyle O}{\|}}{C}—OH} + \underset{\text{thiol}}{HS—R} \longrightarrow \underset{\text{thioester}}{R'—\overset{\overset{\displaystyle O}{\|}}{C}—S—R} + H_2O$$

The $—\overset{\overset{\displaystyle O}{\|}}{C}—S—$ group is called a **thioester** group, and reaction 18-16 is similar to the reaction of alcohols and carboxylic acids to form esters.

Thioesters formed between a substance called coenzyme A (see Figure 18-7) and carboxylic acids are vital to energy production in the human body. We shall

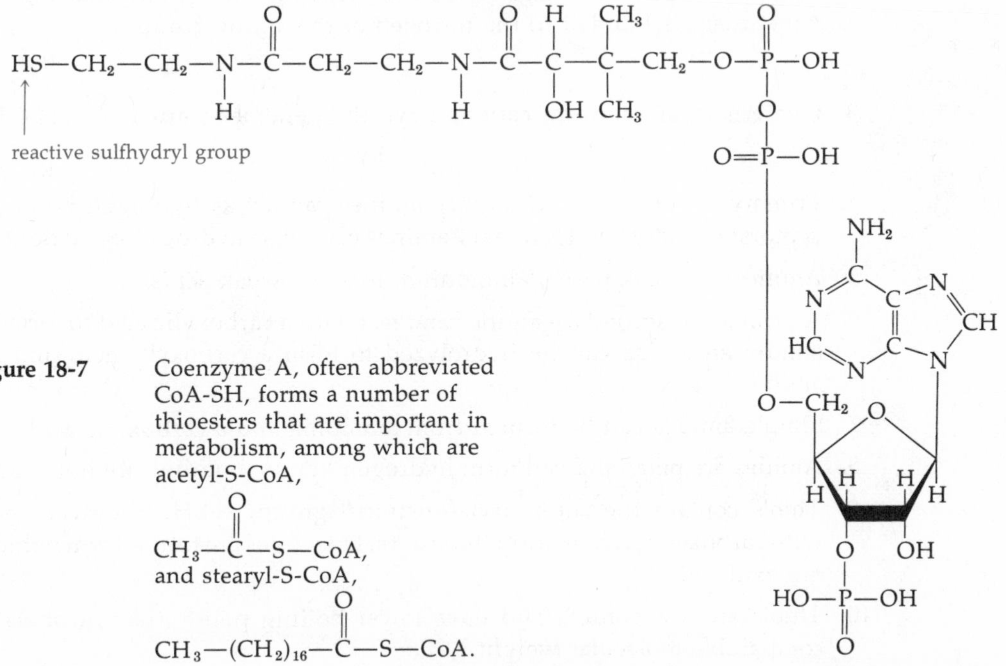

$$HS—CH_2—CH_2—\underset{\underset{H}{|}}{N}—\overset{\overset{\displaystyle O}{\|}}{C}—CH_2—CH_2—\underset{\underset{H}{|}}{N}—\overset{\overset{\displaystyle O}{\|}}{C}—\overset{\overset{H}{|}}{\underset{\underset{OH}{|}}{C}}—\overset{\overset{CH_3}{|}}{\underset{\underset{CH_3}{|}}{C}}—CH_2—O—\overset{\overset{\displaystyle O}{\|}}{\underset{\underset{O}{|}}{P}}—OH$$

reactive sulfhydryl group

$O{=}P{-}OH$

Figure 18-7 Coenzyme A, often abbreviated CoA-SH, forms a number of thioesters that are important in metabolism, among which are acetyl-S-CoA,

$$CH_3—\overset{\overset{\displaystyle O}{\|}}{C}—S—CoA,$$

and stearyl-S-CoA,

$$CH_3—(CH_2)_{16}—\overset{\overset{\displaystyle O}{\|}}{C}—S—CoA.$$

see that the thioester acetyl-S-coenzyme A is of central importance in metabo-

lism (Section 27-3). The $CH_3\!-\!\overset{\displaystyle O}{\overset{\|}{C}}\!-$ portion of acetyl-S-coenzyme A is called an
acetyl group. The general term **acyl** group refers to the atoms that remain when a
hydroxyl group is removed from a carboxylic acid.

Hydrolysis of thioesters is similar to hydrolysis of esters:

(18-17)
$$R'\!-\!\overset{\displaystyle O}{\overset{\|}{C}}\!-\!S\!-\!R + H_2O \longrightarrow R'\!-\!\overset{\displaystyle O}{\overset{\|}{C}}\!-\!OH + HS\!-\!R$$

Thioester hydrolysis is highly exergonic. In fact, hydrolysis of acetyl-S-CoA and
other thioesters provides the energy needed for some endergonic reactions in the
human body.

Exercise 18-8

Draw structural formulas for the products of each of the following reactions:

(a) Methanethiol + ethanethiol + X (where X is an oxidizing agent) →

(b) Acetic acid + 1-butanethiol → H_2O +

(c) Methylthiopropionate + H_2O →

18-7 SUMMARY

1. Amines can be viewed as organic derivatives of ammonia. Primary
   amines, secondary amines, and tertiary amines have these respective
   general formulas: $R\!-\!NH_2$, $R_1\!-\!\underset{\underset{\textstyle H}{|}}{N}\!-\!R_2$, and $R_1\!-\!\underset{\underset{\textstyle R_2}{|}}{N}\!-\!R_3$.

2. The common names for amines use the names of the alkyl group attached to the nitrogen atom as prefixes followed by the ending *-amine*.
   The prefix N- is also used when it is necessary to specify that a given
   substituent is bonded to the nitrogen of the amine group.

3. Quaternary ammonium cations have the general formula $R_1\!-\!\overset{\overset{\textstyle R_2}{|}}{\underset{\underset{\textstyle R_4}{|}}{\overset{+}{N}}}\!-\!R_3$.

4. Primary, secondary, and tertiary amines can act as hydrogen bond acceptors; primary and secondary amines can act as hydrogen bond donors.

5. Amines are weak bases; ammonium ions are weak acids.

6. A primary or secondary amine can react with a carboxylic acid to form an
   amide; an amide can be hydrolyzed to form a carboxylic acid and an
   amine.

7. Simple amides can be formed from ammonia and a carboxylic acid.

8. Amides are polar and can form hydrogen bonds, but they are not bases.

9. Thiols contain the sulfhydryl functional group, —SH. They can react
   with carboxylic acids to form thioesters. Disulfides are formed when thiols
   are oxidized.

10. Thiols are less soluble and have lower boiling points than alcohols of
    comparable molecular weight.

PROBLEMS

1. Identify and name each functional group in the following structural formulas:

   (a) 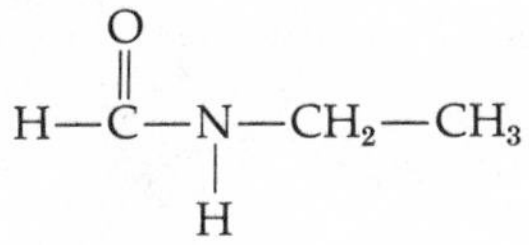 $H-\overset{\overset{\displaystyle O}{\|}}{C}-\underset{\underset{\displaystyle H}{|}}{N}-CH_2-CH_3$

   (b) $H_2N-CH_2-\overset{\overset{\displaystyle O}{\|}}{C}-CH_3$

   (c) $CH_3-\underset{\underset{\displaystyle H}{|}}{N}-\overset{\overset{\displaystyle O}{\|}}{C}-CH_3$

   (d) $CH_3-S-S-CH_3$

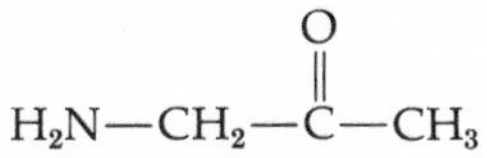

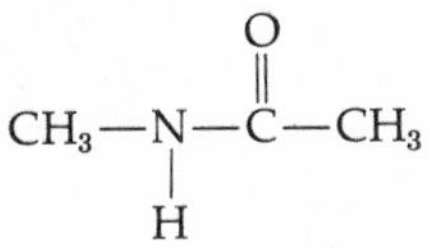

   (e) 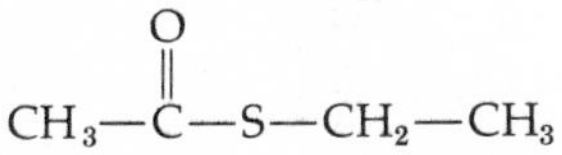 $CH_3-\overset{\overset{\displaystyle O}{\|}}{C}-S-CH_2-CH_3$

   (f) 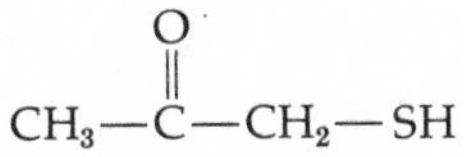 $CH_3-\overset{\overset{\displaystyle O}{\|}}{C}-CH_2-SH$

   (g) 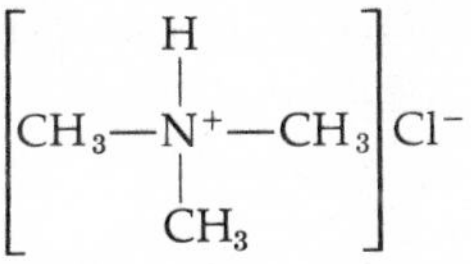 $\left[CH_3-\underset{\underset{\displaystyle CH_3}{|}}{\overset{\overset{\displaystyle H}{|}}{N^+}}-CH_3 \right] Cl^-$

2. Write the names for the compounds with the structural formulas (a), (c), (d), (e), and (g) in Problem 1.

3. Draw structural formulas for each of the following:
   (a) Tetraethylammonium bromide
   (b) *n*-Propylthioacetate
   (c) N-methylbutyramide
   (d) 2-Propanethiol
   (e) N,N-diethylbutyramide

4. Would you expect compound (a) or (e) in Problem 3 to be more soluble in aqueous solution? Explain.

5. Write structural formulas for the products of each of the following reactions:

   (a) 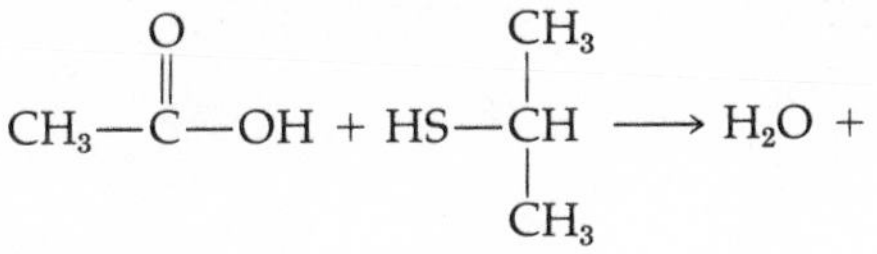 $CH_3-\overset{\overset{\displaystyle O}{\|}}{C}-OH + HS-\underset{\underset{\displaystyle CH_3}{|}}{\overset{\overset{\displaystyle CH_3}{|}}{CH}} \longrightarrow H_2O +$

   (b) 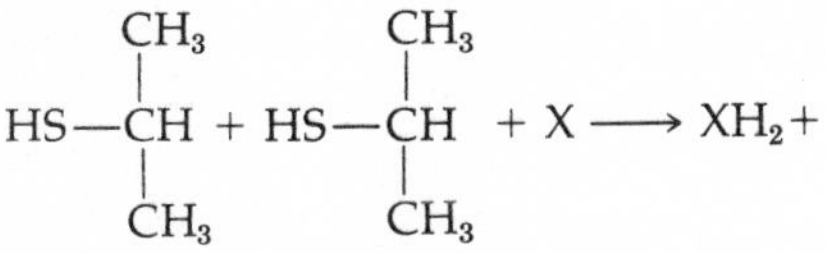 $HS-\underset{\underset{\displaystyle CH_3}{|}}{\overset{\overset{\displaystyle CH_3}{|}}{CH}} + HS-\underset{\underset{\displaystyle CH_3}{|}}{\overset{\overset{\displaystyle CH_3}{|}}{CH}} + X \longrightarrow XH_2 +$

   (c) $H_2N-CH_2-CH_3 + H_2O \rightarrow$

   (d) 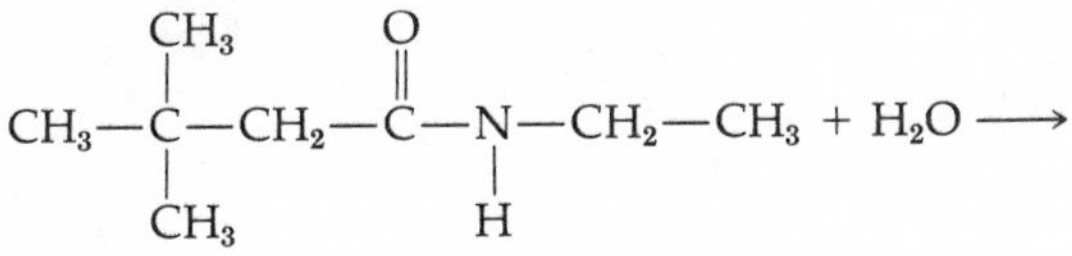 $CH_3-\underset{\underset{\displaystyle CH_3}{|}}{\overset{\overset{\displaystyle CH_3}{|}}{C}}-CH_2-\overset{\overset{\displaystyle O}{\|}}{C}-\underset{\underset{\displaystyle H}{|}}{N}-CH_2-CH_3 + H_2O \longrightarrow$

   (e) 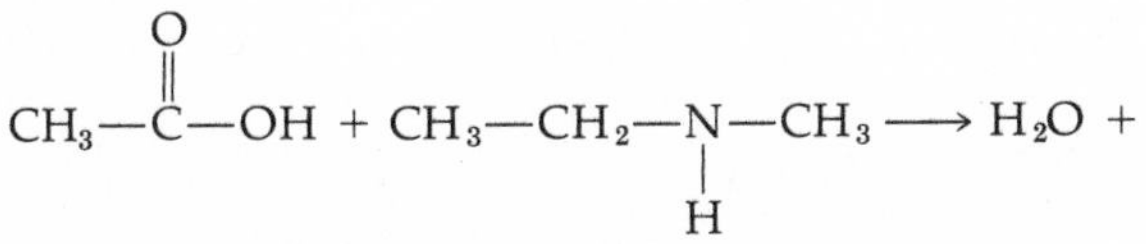 $CH_3-\overset{\overset{\displaystyle O}{\|}}{C}-OH + CH_3-CH_2-\underset{\underset{\displaystyle H}{|}}{N}-CH_3 \longrightarrow H_2O +$

   (f) $CH_3-S-S-CH_3 + XH_2 \longrightarrow X +$

6. Which has the lower boiling point, ethanol or ethanethiol? Explain your answer.

SOLUTIONS TO EXERCISES

18-1 CH_3—CH_2—CH_2—NH_2 (*n*-propylamine), primary amine.

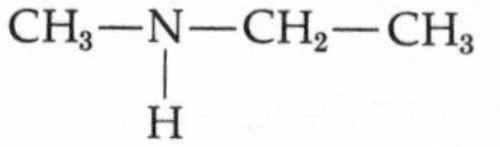

CH_3—$\underset{\underset{H}{|}}{N}$—$CH_2$—$CH_3$ (methylethylamine), secondary amine.

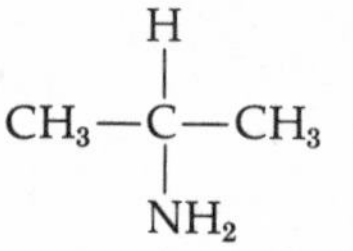

CH_3—$\underset{\underset{NH_2}{|}}{\overset{\overset{H}{|}}{C}}$—$CH_3$ (isopropylamine), primary amine.

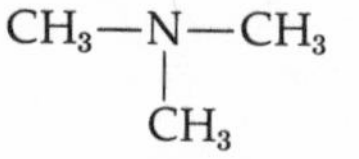

CH_3—$\underset{\underset{CH_3}{|}}{N}$—$CH_3$ (trimethylamine), tertiary amine.

18-2 CH_3—$\overset{\overset{O}{\|}}{C}$—$S$—$CH_2$—$CH_3$ is the thioester formed. Compare this reaction to the formation of ethyl acetate:

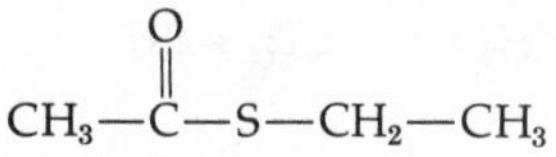

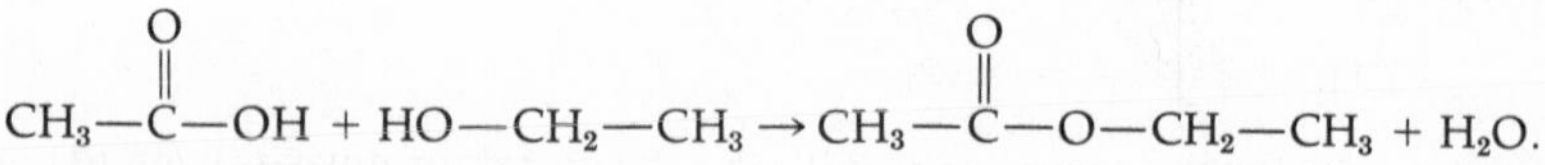

$$CH_3-\overset{\overset{O}{\|}}{C}-OH + HO-CH_2-CH_3 \rightarrow CH_3-\overset{\overset{O}{\|}}{C}-O-CH_2-CH_3 + H_2O.$$

18-3 (a) 1, 2, 3
(b) 1, 4
(c) 1, 2, 3
(d) 2, 3
(e) None of these

18-4 (a) 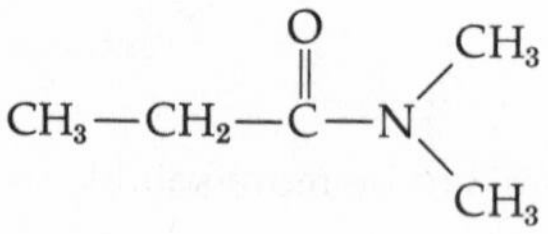 CH_3—CH_2—$\overset{\overset{O}{\|}}{C}$—$N\overset{\diagup CH_3}{\diagdown CH_3}$

(b) $\left[CH_3-CH_2-\overset{\overset{O}{\|}}{C}-O^- \right]\left[H-\overset{\overset{CH_3}{|}}{\underset{\underset{CH_3}{|}}{N^+}}-CH_3 \right]$

(c) H—$\overset{\overset{O}{\|}}{C}$—$\underset{\underset{H}{|}}{N}$—$\underset{\underset{H}{|}}{\overset{\overset{CH_3}{|}}{C}}$—$CH_3$

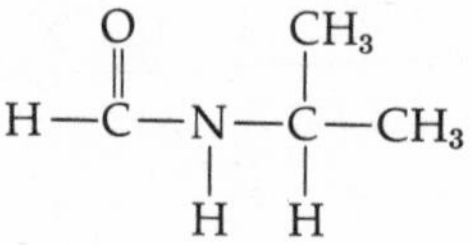

(d) 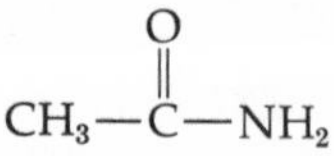 CH_3—$\overset{\overset{O}{\|}}{C}$—$NH_2$

18-5 (a) 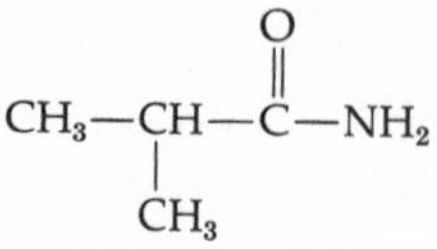 CH_3—$\underset{\underset{CH_3}{|}}{CH}$—$\overset{\overset{O}{\|}}{C}$—$NH_2$

(b) CH_3—CH_2—$\overset{\overset{O}{\|}}{C}$—$\underset{\underset{CH_3}{|}}{N}$—$CH_3$

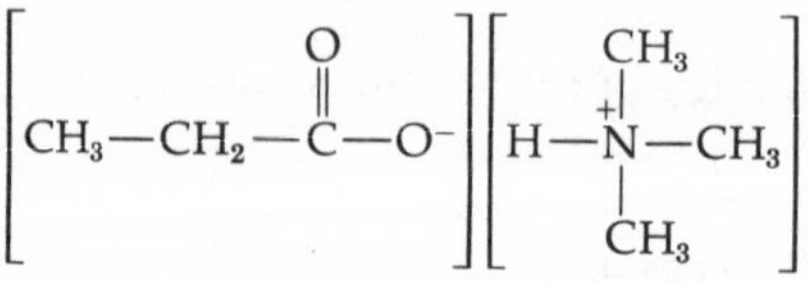

(c) H—$\overset{\overset{O}{\|}}{C}$—$\underset{\underset{H}{|}}{N}$—⬠

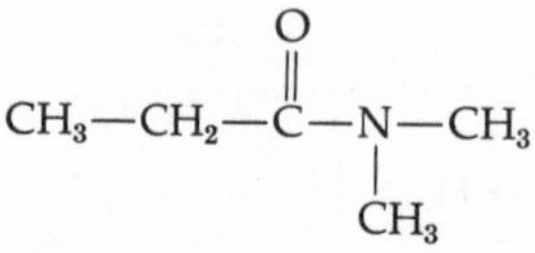

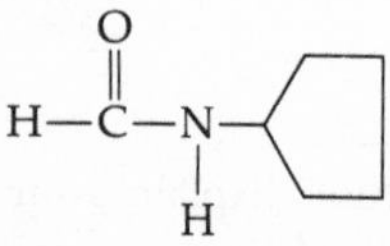

18-6 and 18-7

(a) 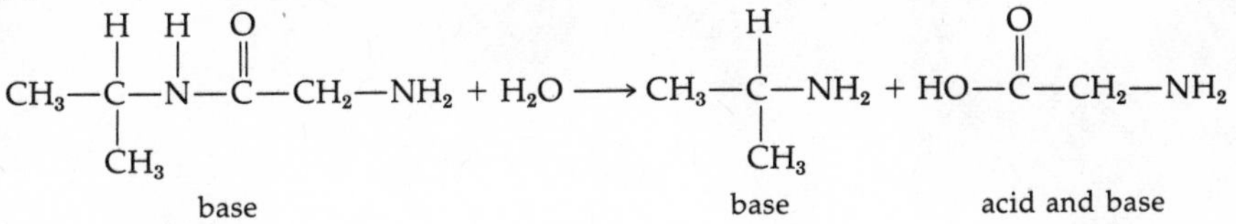
$$CH_3-\underset{\underset{CH_3}{|}}{\overset{\overset{H}{|}}{C}}-\underset{\overset{H}{|}}{N}-\overset{\overset{O}{||}}{C}-CH_2-NH_2 + H_2O \longrightarrow CH_3-\underset{\underset{CH_3}{|}}{\overset{\overset{H}{|}}{C}}-NH_2 + HO-\overset{\overset{O}{||}}{C}-CH_2-NH_2$$

base   base   acid and base

(b) $NH_3 + H_2O \longrightarrow NH_4^+ + OH^-$

(c) 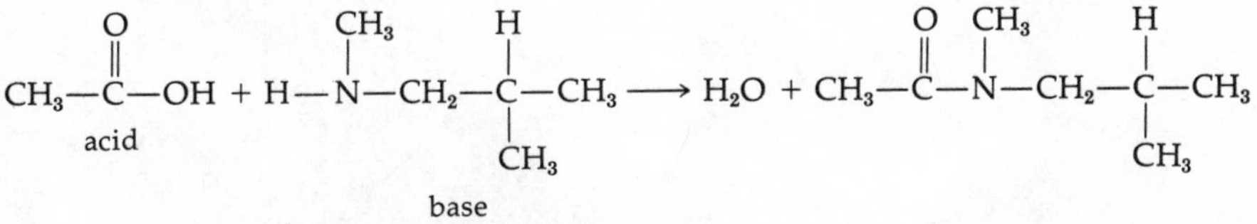
$$CH_3-\overset{\overset{O}{||}}{C}-OH + H-\underset{\underset{\underset{CH_3}{|}}{\overset{\overset{CH_3}{|}}{C}}-CH_3}{N}-CH_2 \longrightarrow H_2O + CH_3-\overset{\overset{O}{||}}{C}-\underset{\overset{CH_3}{|}}{N}-CH_2-\underset{\underset{CH_3}{|}}{\overset{\overset{H}{|}}{C}}-CH_3$$

acid

base

18-8 (a) $XH_2 + CH_3-S-S-CH_2-CH_3$

(b) 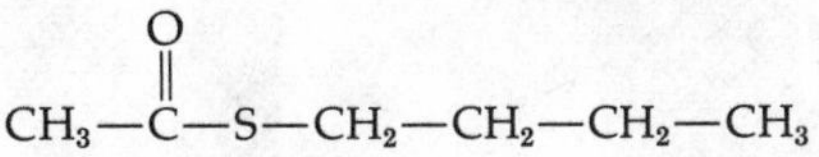 $CH_3-\overset{\overset{O}{||}}{C}-S-CH_2-CH_2-CH_2-CH_3$

(c) $CH_3-CH_2-\overset{\overset{O}{||}}{C}-OH + CH_3-SH$

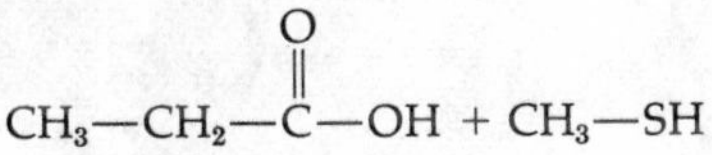

CHAPTER 19

Stereochemistry

19-1 INTRODUCTION

Did you ever try to put your right foot into your left shoe? If you did, you probably found that your right foot fits a lot better in your right shoe, and that walking is a great deal easier when your shoes are on the correct feet.

Stereochemistry deals with the shapes of molecules. A molecule's shape is a significant factor in determining how that molecule can react with other molecules. We shall see in this chapter that stereoisomers can be subdivided into two classes—enantiomers and diastereomers. Enantiomers come in pairs, like shoes and feet. The difference in shape between a pair of enantiomers is analogous to the difference in shape between your right and left shoes or your right and left feet. Enantiomers are mirror images of one another. We are all familiar with objects that differ in shape but are mirror images, such as hands, feet, gloves, shoes, right-handed and left-handed threads on nuts and bolts, and so on.

For many compounds that can exist as a pair of enantiomers, only one of the enantiomers is found naturally. When enantiomers of two different compounds interact, there can be a large difference in the rate of reaction, depending on which specific enantiomer of each compound is involved in the reaction. The difference in the chemical properties of enantiomers is an essential feature of the enzyme-catalyzed reactions that take place in living cells. The reactants for most of these reactions are compounds that can exist as a pair of enantiomers. We shall see that, in most cases, the enzymes produced by a cell are capable of catalyzing reactions involving only one enantiomer of a pair.

In this chapter we also develop some useful ways to represent the shapes of molecules. We shall see how to describe the difference in shape between two enantiomers. It is interesting to note that while this difference in shape may result in quite different chemical properties, the only difference in the physical properties of enanatiomers is the manner in which they interact with polarized light.

A ballet dancer uses her reflection to fine-tune her movements. Bear in mind the difference between the original and its mirror image as you study the structures and properties of enantiomer pairs.

19-2   STUDY OBJECTIVES

After studying the material in this chapter, you should be able to:

1. Define the terms isomer, structural isomer, stereoisomer, enantiomer, diastereomer, superimposable, nonsuperimposable, chiral molecule, and asymmetric carbon atom.

2. Determine whether two given perspective drawings or Fischer projections represent (a) the same molecule, (b) enantiomers, (c) diastereomers, or (d) different molecules that are not isomers.

3. Draw perspective drawings or Fischer projections for all the stereoisomers of a compound, given a structural formula for the compound, and label which stereoisomers are enantiomers and which are diastereomers.

4. Explain why the two enantiomers of a chiral molecule can react differently with a single enantiomer of another chiral molecule, but will react identically with a nonchiral molecule.

5. Define the terms linearly polarized light, unpolarized light, transmission axis, and rotation.

6. Describe how the rotation of linearly polarized light by a solution can be measured with a polarimeter.

7. Explain what the designations (+) and (−) for enantiomers of a chiral molecule mean, and the distinction between the labels (+) and (−) and the labels D- and L- for a pair of enantiomers.

19-3   CLASSIFICATION OF STEREOISOMERS

As you read this chapter it is important to keep in mind the following points, which we have discussed previously:

1. The **shape,** or molecular geometry, of a molecule refers to the relative positions of all the atomic nuclei in the molecule.

2. A **structural formula** is a representation of the bonding arrangement in the molecule that indicates which atoms in the molecule are bonded together.

3. **Isomers** are different compounds that have the same molecular formula.

4. For two compounds to be different, there must be some differences in their chemical and physical properties that will allow their separation from a mixture.

5. Isomers are classified as either structural isomers or stereoisomers.

Structural isomers are compounds that have the same molecular formula but a different bonding arrangement (they have different structural formulas), such as the functional group isomers, ethyl alcohol and dimethyl ether, or the positional isomers, *n*-butane and isobutane. **Stereoisomers,** on the other hand, are isomers that have the same structural formula but differ in the way the bonds in the molecule are oriented in space. We previously described one class of stereoisomers, geometric or cis-trans stereoisomers. 1,2-Dimethylcyclohexane (Section 14-5) and 2-butene (Section 14-6) are examples of compounds that have cis-trans stereoisomeric forms.

Cis-trans stereoisomerism, however, is not the only type of stereoisomerism. For example, glyceraldehyde (see Figure 19-1) is a compound that exists in two stereoisomeric forms that are not of the cis-trans type. The two stereoisomers of glyceraldehyde are labeled D-glyceraldehyde and L-glyceraldehyde. In Section 19-7 we shall see why we label them D and L. The differences between the stereoisomers D- and L-glyceraldehyde are more subtle than the differences between, say, the cis-trans stereoisomers of 2-butene. In contrast to *cis*- and *trans*-2-butene, D- and L-glyceraldehyde have the same melting point, boiling point, solubility in water and other solvents, and so on. In fact, the physical properties of D- and L-glyceraldehyde are identical except for the way in which they interact with linearly polarized light (Section 19-6). There are, however, significant and important differences in the chemical properties of D- and L-glyceraldehyde, which we shall discuss shortly (Section 19-5).

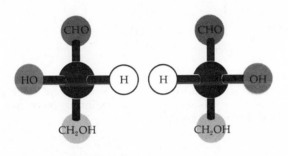

L-glyceraldehyde   D-glyceraldehyde

Figure 19-1   Ball-and-stick schematic models of the two stereoisomers of glyceraldehyde. Note that the H atom and the —OH group point toward you, and that the —CHO and —CH$_2$OH groups point away from you.

Look at the representations of D- and L-glyceraldehyde in Figure 19-1 closely. Do you see that they are **mirror images** of one another? Your left hand and your right hand function differently in certain circumstances because they have different shapes. Your right hand, for example, can only be used to shake hands properly with another right hand. In an analogous manner, D- and L-glyceraldehyde are examples of a class of stereoisomers called enantiomers. **Enantiomers are stereoisomers that are mirror images of one another.**

An actual three-dimensional model is the best way to visualize the shape of a molecule, to show correctly the spatial relationship between the atoms in the molecule, and to test whether or not two molecules are enantiomers. A variety of commercially available kits can be used to construct molecular models.

Several types of pictorial representations, such as those in Figure 19-1, are used to indicate molecular shapes, but none of these is as good as an actual model. Perspective drawings are a simpler way of indicating molecular shape than the ball-and-stick pictures in Figure 19-1. Since we are concerned primarily with the shapes of molecules that contain carbon atoms, we shall describe the way perspective drawings of carbon-containing compounds are interpreted, using D- and L-glyceraldehyde as examples.

Recall that for any carbon atom with four single bonds, the four bonds are oriented approximately toward the corners of a tetrahedron with the carbon atom in the center of the tetrahedron. The central carbon atom in glyceraldehyde has four single bonds and, as shown in Figure 19-2:

1. The central carbon atom is represented by a circle lying in the plane of the paper.

2. The H atom and the —OH group bonded to this carbon atom are represented as being closer to you, *in front* of the plane of the paper that contains the central carbon atom.

3. The —CH$_2$OH group and the aldehyde group, —$\overset{\overset{\displaystyle O}{\|}}{C}$—H (shown as —CHO), that are bonded to the central carbon atom are represented as being *behind* the plane of the paper.

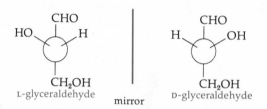

$$
\begin{array}{cc}
\text{CHO} & \text{CHO} \\
\text{HO} \diagdown | \diagup \text{H} & \text{H} \diagdown | \diagup \text{OH} \\
\bigcirc & \bigcirc \\
| & | \\
\text{CH}_2\text{OH} & \text{CH}_2\text{OH} \\
\text{L-glyceraldehyde} & \text{D-glyceraldehyde}
\end{array}
$$

mirror

Figure 19-2   Perspective drawings of D- and L-glyceraldehyde, which are mirror images of one another. Visualize the H atom and the —OH group as pointing toward you, and the —CHO and —CH$_2$OH groups as pointing away from you.

Look at the perspective drawings of D- and L-glyceraldehyde in Figure 19-2 closely. It should be clear to you that they represent mirror images.

Figure 19-3 shows perspective drawings of two additional examples of enantiomers: 2-butanol and *cis*-[Co(H$_2$N—CH$_2$—CH$_2$—NH$_2$)$_2$Cl$_2$]$^+$, which is a transition metal complex ion containing cobalt.

2-Butanol   *cis*-[Co(H$_2$N—CH$_2$—CH$_2$—NH$_2$)$_2$Cl$_2$]$^+$

$$
\begin{array}{cc}
\text{CH}_3 & \text{CH}_3 \\
\text{HO} \diagdown | \diagup \text{H} & \text{H} \diagdown | \diagup \text{OH} \\
\bigcirc & \bigcirc \\
| & | \\
\text{CH}_2\text{CH}_3 & \text{CH}_2\text{CH}_3
\end{array}
$$

mirror

mirror

Figure 19-3   Perspective drawings of the enantiomers of 2-butanol and *cis*[Co(H$_2$N—CH$_2$—CH$_2$—NH$_2$)$_2$Cl$_2$]$^+$

$$
\begin{array}{c}
\quad\quad\quad\quad \text{H} \quad \text{H} \quad \text{H} \quad \text{H} \\
\quad\quad\quad\quad | \quad\ | \quad\ | \quad\ | \\
\overset{\frown}{\text{N} \quad \text{N}} = :\text{N}-\text{C}-\text{C}-\text{N}: \\
\quad\quad\quad\quad | \quad\ | \quad\ | \quad\ | \\
\quad\quad\quad\quad \text{H} \quad \text{H} \quad \text{H} \quad \text{H}
\end{array}
$$

There is a fairly simply way of telling if a given compound can exist in two enantiomeric forms using both a model for a molecule of the compound and a model for its mirror image: When the molecule and its mirror image have the same shape, *all* the chemical and physical properties of the molecule and its mirror image are identical and there are no enantiomeric forms for the compound. If, and only if, the molecule and its mirror image have different shapes can a pair of enantiomeric stereoisomers exist.

We can test whether an object and its mirror image have the same or different shape by a *superposition procedure* in which we perform the following steps:

1. Think of the object and its mirror image as capable of penetrating one another without distorting their shapes.

2. Carry out this mental interpenetration procedure.

3. If all of the atoms in the resulting combination of the object and its penetrated mirror image are in positions identical to the original object, we say that the object and the mirror image are **superimposable.** If the resulting combination is not identical to the original object, we say that the object and its mirror image are **nonsuperimposable.**

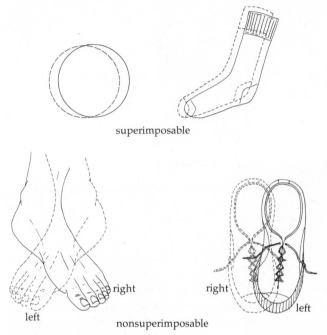

Figure 19-4   A ball and its mirror image, and a sock and its mirror image, are superimposable, whereas right and left feet, or right and left shoes, are nonsuperimposable.

Figure 19-4 illustrates several common examples of superimposable and nonsuperimposable objects. D-Glyceraldehyde and L-glyceraldehyde are nonsuperimposable mirror images of one another (see Figure 19-5). When a molecule and its mirror image are nonsuperimposable, the molecule is called a **chiral** molecule. Enantiomers can exist only for those compounds whose molecules are chiral.

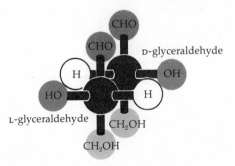

Figure 19-5   Ball-and-stick schematic models of D- and L-glyceraldehyde illustrate that these enantiomers are nonsuperimposable.

The term chiral comes from the Greek word *cheir*, meaning "hand." Glyceraldehyde, 2-butanol, and the *cis*-[Co(H$_2$N—CH$_2$—CH$_2$—NH$_2$)$_2$Cl$_2$]$^+$ ion are examples of chemical species that are chiral, and these species have enantiomeric forms. Methane, dichloromethane, chlorodifluoromethane, and 2-chloropropane (see Figure 19-6) are examples of compounds whose molecules are not chiral, and these compounds do not have enantiomers.

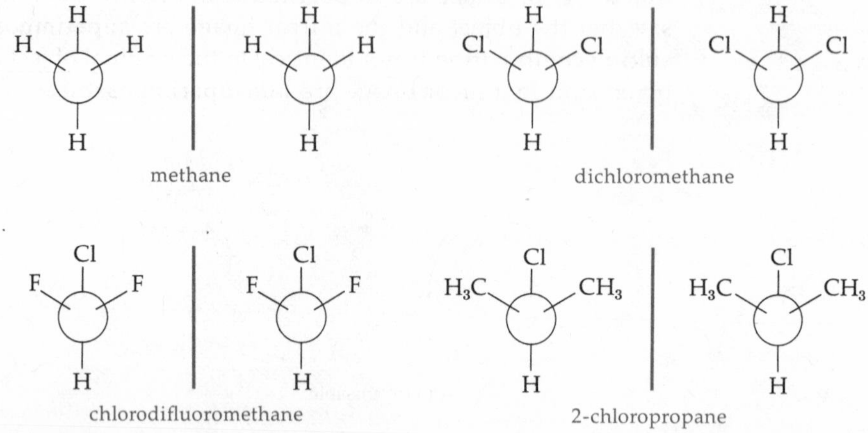

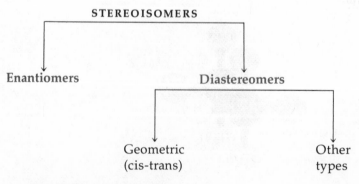

methane   dichloromethane

chlorodifluoromethane   2-chloropropane

Figure 19-6   Perspective drawings of nonchiral molecules. Note that there are four similar atoms attached to the carbon atom in methane, two pairs of similar atoms attached to the carbon atom in dichloromethane, one pair of similar atoms attached to the carbon atom in chlorodifluoromethane, and one pair of similar methyl groups attached to the central carbon atom in 2-chloropropane.

An equivalent way of describing stereoisomers that are enantiomers is to call them **mirror-image stereoisomers.** As we have mentioned previously, not all stereoisomers are mirror images of each other. Stereoisomers that are *not* mirror images of one another are called **diastereomers.** Thus, there are two classes of stereoisomers (see Figure 19-7):

1. Mirror-image stereoisomers or enantiomers
2. Nonmirror-image stereoisomers or diastereomers

Geometric isomers, such as *cis*- and *trans*-2-butene, are only one of several types of diastereomers. We shall discuss other types presently.

STEREOISOMERS

Enantiomers   Diastereomers

Geometric   Other
(cis-trans)   types

Figure 19-7   Enantiomers and diastereomers are the two major classes of stereoisomers.

Many different types of compounds can exist as enantiomers, but since we are interested primarily in organic compounds, in our further discussion we shall consider only enantiomers that are carbon-containing compounds.

Exercise 19-1

Which of the following molecules are chiral? For each chiral molecule, draw a perspective drawing of the other enantiomer. The symbol —COOH is an alternative way of repre-

senting the carboxyl group, $-\overset{\overset{O}{\|}}{C}-OH$.

(a) The amino acid alanine,

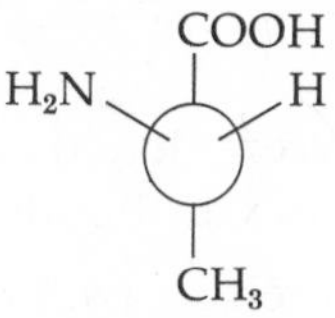

(b) Chloroform,

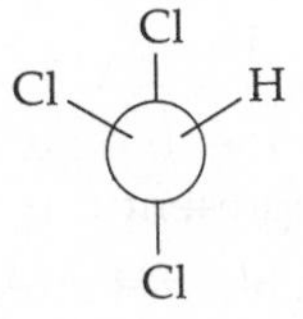

(c) Lactic acid,

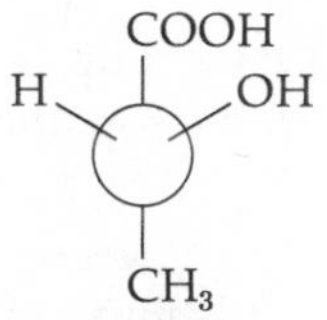

(d) 1,2-Dichloroethene

19-4 ENANTIOMERS AND DIASTEREOMERS OF CARBON-CONTAINING COMPOUNDS

Notice that in the chiral molecules glyceraldehyde and 2-butanol (Figures 19-2 and 19-3) there is a carbon atom that is bonded to four different atoms or groups of atoms. This is not true of the compounds in Figure 19-6, which are not chiral.

The central carbon atom in glyceraldehyde is bonded to a H atom, an —OH group, a —CH₂OH group, and a —CHO group, and the number 2 carbon in 2-butanol is bonded to a H atom, an —OH group, a —CH₃ group, and a —CH₂CH₃ group. Even though the number 2 carbon atom in 2-butanol is bonded to a carbon atom in a methyl group and another carbon atom in an ethyl group, the methyl group and the ethyl group are different.

Asymmetric Carbon

A carbon atom that is bonded to four different groups of atoms is called a **chiral carbon** or an **asymmetric carbon.** Usually (but not always) a molecule that has one or more asymmetric carbons is a chiral molecule. As our examples of glyceraldehyde and 2-butanol illustrate, in most cases carbon-containing molecules are chiral if they contain an asymmetric carbon; and molecules such as methane, dichloromethane, chlorodifluoromethane, and 2-chloropropane, which do not contain an asymmetric carbon, are not chiral. There are, however, exceptions to these generalizations: (1) Some carbon-containing compounds that are chiral do not have an asymmetric carbon; and (2) some carbon-containing compounds that have asymmetric carbons are not chiral. Thus, although the presence or absence of asymmetric carbons is a good indication of whether or not a molecule is chiral, the proof for chirality is nonsuperimposable mirror images.

We shall discuss only chiral molecules that contain one or more asymmetric carbons. Sugars and most amino acids possess asymmetric carbons. We shall not need to discuss chiral molecules that do not have an asymmetric carbon.

For larger molecules, with more than one asymmetric carbon, the perspective drawings we have used so far are tedious to draw and to use. It is more convenient to use planar representations called **Fischer projections.** Let us first illustrate how Fischer projections are drawn and used with a familiar example—glyceraldehyde.

Fischer Projections

To draw the Fischer projections for D- and L-glyceraldehyde we proceed as follows (see Figure 19-8a):

1. Consider the perspective drawing for D-glyceraldehyde and draw a cross to represent the asymmetric carbon with its four bonds.

2. Attach the H atom and the —OH group, which are coming toward you in the perspective drawing, to the left and right horizontal lines of the cross, respectively.

3. Attach the —CHO group and the —CH₂OH group, which are going away from you in the perspective drawing, to the top and bottom vertical lines of the cross, respectively.

Note that when you see a planar Fischer projection, it is very important that you visualize the horizontal lines as bonds coming toward you, and the vertical lines as bonds going away from you. As illustrated in Figure 19-8b, the Fischer projection for L-glyceraldehyde can be obtained by reflecting the Fischer projection for D-glyceraldehyde in a mirror.

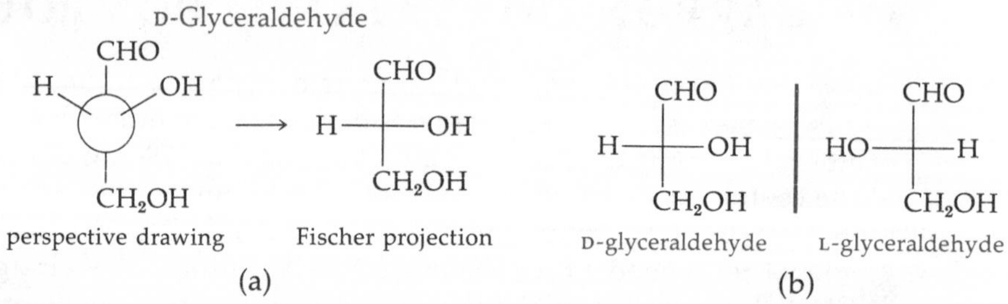

Figure 19-8   (a) Perspective drawing and Fischer projection for D-glyceraldehyde. Note that you should view the Fischer projection as if the H atom and the —OH group point

toward you, and the —CHO and —CH₂OH groups point away from you. (b) The Fischer projections of D- and L-glyceraldehyde are mirror images of one another.

We know that D- and L-glyceraldehyde are nonsuperimposable mirror images, and this fact can be inferred from the Fischer projections, provided that the Fischer projections are used appropriately. Notice that the Fischer projection for L-glyceraldehyde *cannot* be superimposed upon the Fischer projection for D-glyceraldehyde by moving the L-projection to the left and then rotating the L-projection while keeping it in the plane of the paper (see Figure 19-9). We could superimpose the Fischer projections for D- and L-glyceraldehyde if we lifted the L-projection out of the plane of the paper, flipped it over, and put it on top of the D-projection, but *this operation is not permissible.* Fischer projections are two-dimensional representations; they can be used to test for superimposability only if you keep them in the projection plane and do not raise them from the surface of this plane and try to flip them over.

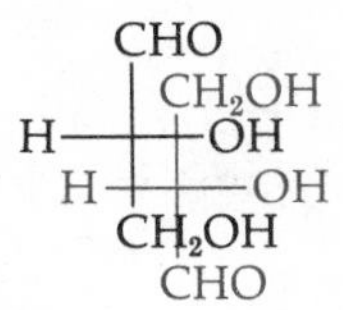

Figure 19-9   The Fischer projections for
D-glyceraldehyde (black) and
L-glyceraldehyde (blue) are
nonsuperimposable.

$$
\begin{array}{c}
\text{CHO} \\
\text{CH}_2\text{OH} \\
\text{H} - \overset{|}{\underset{|}{\text{C}}} - \text{OH} \\
\text{H} - \overset{|}{\underset{|}{\text{C}}} - \text{OH} \\
\text{CH}_2\text{OH} \\
\text{CHO}
\end{array}
$$

Let us now construct Fischer projections for the stereoisomers of a compound with two asymmetric carbons, for example, 2-bromo-3-chlorobutane. In Figure 19-10, the two asymmetric carbons are labeled with an asterisk (*). The perspective drawing and corresponding Fischer projection for one of the possible shapes for 2-bromo-3-chlorobutane is given in Figure 19-10a. The Fischer projection is labeled I and its mirror image is labeled II. To determine if projections I and II represent a pair of enantiomers, we have to test if I and II are superimposable. We cannot superimpose I and II (remember to keep them in the same plane), and therefore I and II do correspond to a pair of stereoisomers that are enantiomers.

These enantiomers are not the only stereoisomers of 2-bromo-3-chlorobutane. Another possible shape for this compound is represented in the perspective drawing and corresponding Fischer projection III given in Figure 19-10b. Fischer projection III is not superimposable on either projection I or II, and therefore Fischer projection III represents the shape of a third stereoisomer of our compound. The stereoisomer represented by Fischer projection III is not the mirror image of the shapes represented by Fischer projection I or II. Therefore, I and III represent diastereomers, as do II and III. When we consider Fischer projection IV, the mirror image of III, we find that III and IV are not superimposable. Therefore, III and IV correspond to a second pair of enantiomers. Thus there are two pairs of enantiomers, for a total of four stereoisomeric forms of 2-bromo-3-chlorobutane.

2-Bromo-3-chlorobutane:

$$
\text{CH}_3 - \overset{\overset{\text{H}}{|}}{\underset{\underset{\text{Br}}{|}}{\text{C}}} - \overset{\overset{\text{H}}{|}}{\underset{\underset{\text{Cl}}{|}}{\text{C}}} - \text{CH}_3
$$

Perspective drawing   Fischer projection I   Perspective drawing   Fischer projection III

I   II   III   IV

I and II are an enantiomer pair   III and IV are an enantiomer pair

(a)   (b)

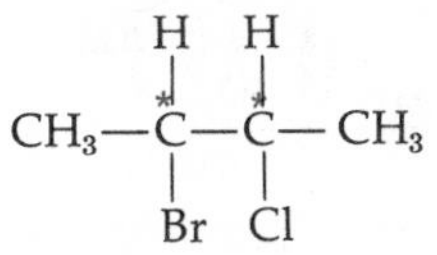

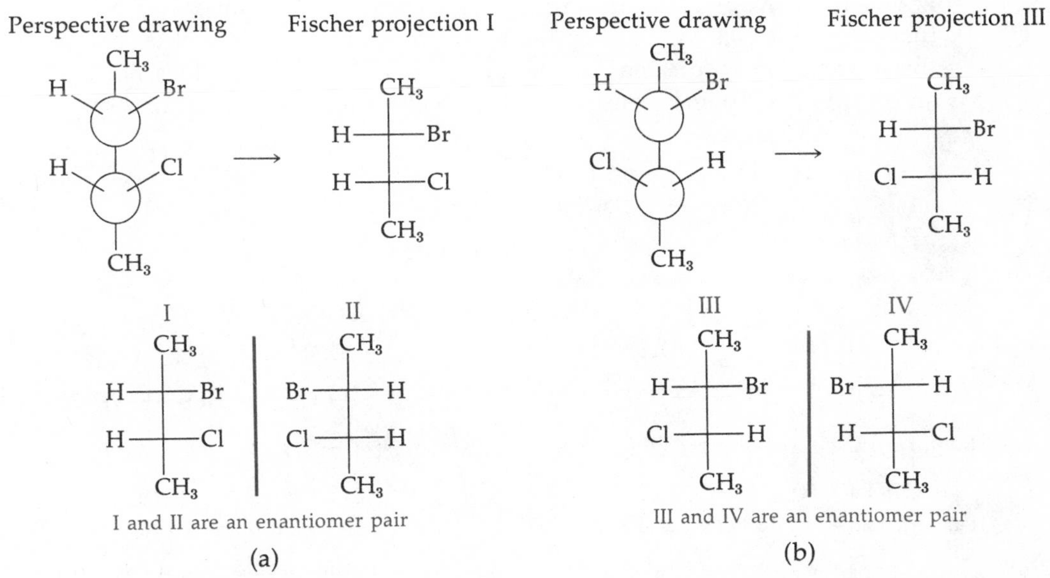

Figure 19-10   2-Bromo-3-chlorobutane has two asymmetric carbon atoms and four stereoisomers. The Fischer projections I and II correspond to one enantiomer pair (a), and the Fischer projections III and IV correspond to the other enantiomer pair (b).

Notice that there are two stereoisomers for glyceraldehyde, which has one asymmetric carbon, and four stereoisomers for 2-bromo-3-chlorobutane, which has two asymmetric carbons. The compound

$$HO-CH_2-\overset{*}{C}H-\overset{*}{C}H-\overset{*}{C}H-\overset{O}{\overset{\|}{C}}-H$$
$$\qquad\qquad\ \ OH\ \ \ OH\ \ \ OH$$

has three asymmetric carbons, and there are a total of eight stereoisomers of this compound. <u>As a general rule, for a compound with n asymmetric carbon atoms, the number of stereoisomers is no larger than 2^n ($2 \times 2 \times 2 \times \cdots n$ times).</u> For most compounds, such as the ones we have just described, the actual number of stereoisomers is equal to 2^n. For some compounds, however, the number of stereoisomers is less than 2^n. Tartaric acid, for example,

$$HO-\overset{O}{\overset{\|}{C}}-\overset{*}{C}H-\overset{*}{C}H-\overset{O}{\overset{\|}{C}}-OH$$
$$\qquad\qquad OH\ \ \ OH$$

has two asymmetric carbons but only three stereoisomers. We can use Fischer projections to see why this is so (see Figure 19-11). Fischer projections A and B are nonsuperimposable mirror images, so they correspond to an enantiomeric pair, called L-tartaric acid (A) and D-tartaric acid (B). The first pair of enantiomers to be discovered involved a salt of tartaric acid. This discovery of enantiomers, which marked the beginning of the study of molecular shapes, was made by Louis Pasteur in 1848. Fischer projection C is not superimposable on either Fischer projection A or B, so Fischer projection C is a representation for another stereoisomer of tartaric acid, called *meso*-tartaric acid. *meso*-Tartaric acid and either D- or L-tartaric acid are diastereomers. Fischer projections C and D are superimposable, and therefore they do *not* represent an enantiomeric pair. Fischer projections C and D are thus *equivalent* representations for *meso*-tartaric acid, which has the shape indicated by the perspective drawing in Figure 19-12. Notice that the shape of *meso*-tartaric acid consists of two halves that are mirror images of one another—one half above and the other half below the mirror plane indicated in the figure. *meso*-Tartaric acid does not exist as an enantiomeric pair because of this *internal* mirror plane of symmetry.

$$Tartaric\ acid:\ \ HO-\overset{O}{\overset{\|}{C}}-\overset{*}{C}H-\overset{*}{C}H-\overset{O}{\overset{\|}{C}}-OH$$
$$\qquad\qquad\qquad\qquad OH\ \ \ OH$$

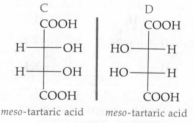

A	B	C	D
COOH	COOH	COOH	COOH
H——OH	HO——H	H——OH	HO——H
HO——H	H——OH	H——OH	HO——H
COOH	COOH	COOH	COOH
L-tartaric acid	D-tartaric acid	*meso*-tartaric acid	*meso*-tartaric acid

Figure 19-11   Fischer projections A and B are nonsuperimposable, and correspond to the enantiomer pair L- and D-tartaric acid, respectively. Fischer projections C and D are superimposable, and do *not* correspond to an enantiomer pair. C and D are equivalent representations for *meso*-tartaric acid.

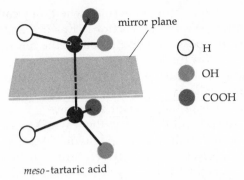

meso-tartaric acid

Figure 19-12   Perspective drawing of *meso*-tartaric acid.

Exercise 19-2
Draw structural formulas for each of the following, and indicate with an asterisk (*) any asymmetric carbons.

    (a) 2-Methyl-1-butanol   (b) 2,3-Dichlorobutane
    (c) 2-Methyl-2-butanol   (d) 1,2-Dibromopropane

Exercise 19-3
Draw Fischer projections for all the stereoisomers in Exercise 19-2 and indicate which are enantiomers and which are diastereomers.

19-5   REACTIONS OF STEREOISOMERS

We have seen that chiral molecules have different shapes from their mirror images. Because of these differences in shape, a chiral molecule and its mirror image can have significantly different chemical reactivity.

A chemical reaction involving a *single* enantiomer of a chiral molecule X (which we shall label D-X) and a *pair* of enantiomers, such as D- and L-glyceralde-hyde, is analogous to trying to put your left foot into your right and left shoes. The fit between the enantiomer D-X and D-glyceraldehyde can be very different from the fit between D-X and L-glyceraldehyde (see Figure 19-13). Recall that

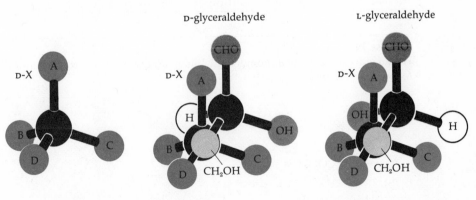

Figure 19-13   Ball-and-stick model of the chiral molecule D-X (left). The fit between D-X and D-glyceraldehyde (center) is different than the fit between D-X and L-glyceraldehyde (right).

the relative orientation of two reacting molecules when they collide (the fit) is a crucial factor in determining the rate of reaction. Thus if X and glyceraldehyde are molecules that react, it is possible that the rate of the reaction between D-X and D-glyceraldehyde is so slow that effectively no reaction takes place, whereas the rate of reaction between D-X and L-glyceraldehyde is very rapid. As we shall see, for most biologically important enzyme-catalyzed reactions, the enzyme fits with only one enantiomer of the reactant substance.

As we have just seen, there is a difference in the way D-X, a single enantiomer of the chiral molecule X, fits with D-glyceraldehyde or L-glyceraldehyde. On the other hand, there is an *identical* fit between any nonchiral molecule and D-glyceraldehyde or L-glyceraldehyde. This fact is illustrated in Figure 19-14, where Y is used as a label for the nonchiral molecule. A sock is analogous to a molecule that is not chiral, whereas your feet are analogous to a pair of enantiomers. The identical fit between the molecule Y, which is not chiral, and D- and L-glyceraldehyde is analogous to the identical fit between a sock and your left foot and the same sock and your right foot.

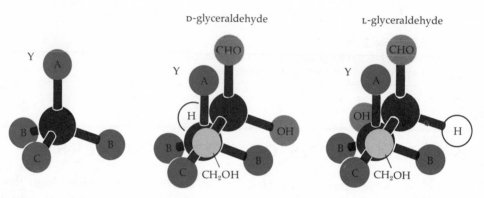

Figure 19-14   Model of the nonchiral molecule Y (left). The fit between Y and D-glyceraldehyde (center) is the same as that between Y and L-glyceraldehyde (right).

It is generally true that (1) a single enantiomer of a chiral molecule will fit differently, and react differently, with the two enantiomers of another chiral molecule; and that (2) a molecule that is not chiral will have an identical fit, and react identically, with the two enantiomers of a chiral molecule.

Exercise 19-4

Would you expect the enantiomers of 2-butanol (Figure 19-3) to react identically with the following molecules?

(a)

$$
\begin{array}{c}
OH \\
H_3C \diagdown \overset{|}{\diagup} H \\
| \\
COOH
\end{array}
$$

(b)

$$
\begin{array}{c}
H \\
H_3C \diagdown \overset{|}{\diagup} H \\
| \\
COOH
\end{array}
$$

19-6 OPTICAL ACTIVITY

The physical properties of two enantiomers of a chiral molecule are identical except for the way in which they interact with polarized light. Let us see what is meant by polarized light.

Light is an electromagnetic wave with an electric field and a magnetic field perpendicular to the direction of wave propagation. The magnitudes of both the electric and magnetic fields of any electromagnetic wave increase and decrease in a regular manner. For the electromagnetic wave in Figure 19-15, the direction of propagation is the y direction, the electric field is in the z direction, and the magnetic field (not shown) is in the x direction. The electric field direction is called the **direction of polarization.** We say that the electromagnetic wave in Figure 19-15 is **linearly polarized** in the z direction.

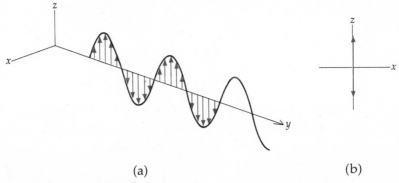

(a)   (b)

Figure 19-15   (a) Schematic representation of a linearly polarized electromagnetic wave propagating in the y direction, with the electric field (represented by arrows) oscillating in the z direction, perpendicular to the direction of propagation. There is also a magnetic field (not shown) oscillating in the x direction. The electric field, viewed with the y axis coming directly toward you, is shown in (b).

Other electromagnetic waves, with different directions of polarization, can propagate in the y direction. We can think of these waves as being obtained from the wave pictured in Figure 19-15 by rotating that wave by varying amounts about the y axis. Thus, the electric field of an electromagnetic wave propagating in the y direction can be in the z direction, or in the x direction, or in any direction in the xz plane (see Figure 19-16). Light coming from the sun, an electric light bulb, or any usual light source, is a combination of a very large number of electromagnetic waves with the electric field of the waves in all directions in the xz plane; light of this nature is called **unpolarized light.**

Figure 19-16   Representation of unpolarized light propagating in the y direction. The blue arrows indicate the electric fields of the component electromagnetic waves.

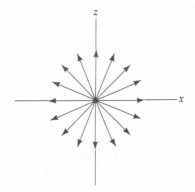

Certain substances transmit only those electromagnetic waves with the electric field in a single direction called the **transmission axis.** One of these substances is the commercial material Polaroid, which is used in sunglasses. When unpolarized light is passed through a piece of Polaroid, only those electromagnetic waves with the electric field in a single direction emerge, and this transmitted light is **linearly polarized light** (see Figure 19-17). If this linearly polarized light emerging from one piece of Polaroid is passed through a second piece of Polaroid, the light emerging from the second piece of Polaroid has a maximum intensity when the two pieces of Polaroid are aligned so that their respective transmission axes are in the same direction (Figure 19-18a); and the light has a minimum intensity when the two pieces of Polaroid are aligned so that their respective transmission axes are perpendicular to one another (Figure 19-18b). In an arrangement such as this, the first piece of Polaroid is called a polarizer, and the second piece of Polaroid is called an analyzer. An instrument with both a polarizer and an analyzer is called a **polarimeter.**

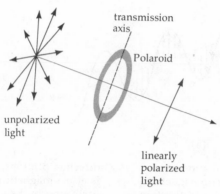

| Figure 19-17 | Schematic illustration of one way by which linearly polarized light can be generated. Unpolarized light is passed through a piece of Polaroid, which has a unique | transmission axis. The emergent light is linearly polarized with its electric field in the same direction as the transmission axis of the Polaroid. |

A solution consisting of one enantiomer of a chiral molecule in a nonchiral solvent rotates the direction of linearly polarized light. For example, if an aqueous solution of D-tartaric acid (Figure 19-11) is placed between the polarizer and the analyzer of a polarimeter, the direction of polarization of the linearly polarized light emerging from the polarizer is changed by an angle θ (see Figure 19-19). The angle θ is called the **rotation** of the solution. Compounds that rotate linearly polarized light are said to be **optically active.**

The rotation of the solution is easy to measure, since the light emerging from the analyzer will be at a maximum when the analyzer is rotated by the angle θ. When this is done, the direction of polarization of the light entering the analyzer is parallel to the transmission axis of the analyzer. When the analyzer is rotated to the right by an angle θ, we designate the rotation of the solution by $(+)\theta$, and when the analyzer is rotated to the left by an angle θ, we designate the rotation of the solution by $(-)\theta$.

The magnitude of the rotation of a D-tartaric acid solution (the size of the angle θ) is directly proportional to the number of D-tartaric acid molecules that interact with the light as it passes through the solution. Thus, the higher the concentration of the D-tartaric acid solution and the longer the length of the tube in the polarimeter containing the solution, the larger the rotation.

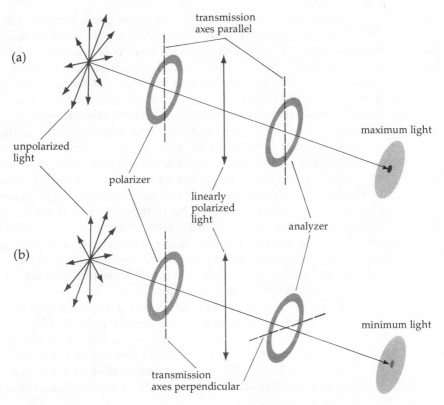

Figure 19-18   (a) When unpolarized light passes through two pieces of Polaroid (a polarizer and an analyzer) that have parallel transmission axes, the intensity of the transmitted light is at a maximum. (b) The intensity of the transmitted light is at a minimum when the transmission axes of the polarizer and the analyzer are perpendicular.

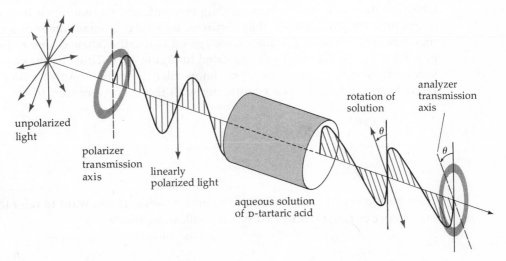

Figure 19-19   When linearly polarized light is passed through a solution of D-tartaric acid, the direction of polarization is rotated by an angle θ to the left. Thus, the light emerging from the analyzer has a maximum intensity when the analyzer is also rotated θ to the left, so that the transmission axis of the analyzer and the direction of polarization of the light emerging from the D-tartaric acid solution are parallel.

A solution of D-tartaric acid with a given concentration will rotate linearly polarized light a certain amount in one direction. A solution of L-tartaric acid with the same concentration put in the same polarimeter tube will rotate linearly polarized light by the *same amount* as the D-tartaric acid solution, *but in the opposite direction*. It is an experimental fact that a D-tartaric acid solution rotates linearly polarized light to the left (−), and that an L-tartaric acid solution rotates linearly polarized light to the right (+). When linearly polarized light interacts with a single D-tartaric acid molecule, the light is rotated to the left by a tiny amount. A single L-tartaric acid molecule, on the other hand, rotates linearly polarized light to the right by the same small amount. The measurable rotation of a solution containing a mixture of the *same* number of D-tartaric acid and L-tartaric acid molecules is zero. As linearly polarized light passes through such a solution, the light is rotated a bit to the left when it encounters a D-tartaric acid molecule and a bit to the right when it encounters an L-tartaric acid molecule, but since the number of D- and L-tartaric acid molecules is the same, the net overall rotation is zero. In general, a mixture of equal amounts of the two enantiomers of a chiral molecule is called a **racemic mixture,** and solutions of racemic mixtures have a net rotation of zero.

For any chiral molecule, one enantiomer rotates linearly polarized light to the right and the other enantiomer rotates linearly polarized light to the left by the same amount. The amount of this rotation is different for different chiral molecules. For example, the rotation of a 1 M solution of D-tartaric acid in a 10-cm polarimeter tube is (−)1.80°, whereas the rotation of a 1 M solution of D-lactic acid in a 10-cm polarimeter tube is (−)0.34°. Lactic acid is a chiral molecule with

the formula CH_3—C—C—OH.

19-7   WHICH ENANTIOMER IS WHICH?

Chemists have adopted certain labeling conventions, so that when they communicate with one another, they can refer to a specific enantiomer of a chiral molecule. For example, chemists have agreed to use the labels D- and L- for the enantiomers of glyceraldehyde indicated in Figure 19-1. Thus, if you know the conventions that chemists use, you know that the name "D-glyceraldehyde" refers to the glyceraldehyde enantiomer with the shape represented by

in which the —OH group is on the right. Likewise, if you want to refer to the enantiomer of tartaric acid that has the following shape:

with the bottom —OH group on the left, you would use the label "L-tartaric acid."

The D- and L-convention is one way chemists designate enantiomers. In Chapter 20, we shall discuss the D- and L-labeling conventions for the enantiomers of glucose and other sugars.

In the early twentieth century, chemists were aware that there were two enantiomeric forms of glyceraldehyde, that one enantiomer rotated linearly polarized light to the right (+), and that the other enantiomer rotated linearly polarized light to the left (−), but they did not know which enantiomer was (+) and which was (−). The (+) enantiomer was arbitrarily assigned the shape D-glyceraldehyde, even though based on the information known at the time, the (+) enantiomer was equally likely to have the shape of L-glyceraldehyde. In 1949, however, experiments were performed which showed that this arbitrary choice was in fact correct. The (+) enantiomer of glyceraldehyde *does* have the shape shown above for D-glyceraldehyde.

It is important that you realize that it is a matter of *experimental fact* that the enantiomer with this shape rotates linearly polarized light to the right (+), whereas it is a *convention* to designate this enantiomer D-glyceraldehyde. At the present time, chemists have determined experimentally which enantiomer is (+) and which is (−) for many other chiral molecules.

Whether a given enantiomer of a chiral molecule is (+) or (−) depends on the specific atoms in the molecule and how they are bonded together. There is no general way of predicting which enantiomer of the chiral molecule is (+) and which is (−) by looking, for example, at the Fischer projection for that enantiomer. Thus, unless you have memorized the appropriate experimental fact, you cannot look at the Fischer projection

$$
\begin{array}{c}
\text{COOH} \\
\text{H}\!-\!\!|\!-\!\text{OH} \\
\text{HO}\!-\!\!|\!-\!\text{H} \\
\text{COOH}
\end{array}
$$

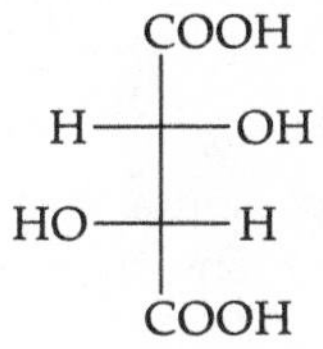

for an enantiomer of tartaric acid and call it (+)-tartaric acid. More important, even if you know this experimental fact, were you to refer to (+)-tartaric acid in your conversation with someone else, they would not know which enantiomer of tartaric acid you were talking about unless they also knew the same experimental fact. On the other hand, if you and another person both use the same labeling conventions, you could (1) look at the Fischer projection above and (2) notice that the bottom —OH group is on the left, and therefore call it L-tartaric acid; (3) refer to L-tartaric acid in conversation; and (4) expect the other person to know which enantiomer of tartaric acid you were talking about.

There is no relationship between the labels D- and L- assigned on the basis of certain conventions and the direction in which an enantiomer rotates linearly polarized light, (+) or (−). For example, we previously stated the experimental fact that L-tartaric acid is (+). On the other hand, D-glyceraldehyde is also (+).

Exercise 19-5

The Fischer projection for the amino acid L-alanine is

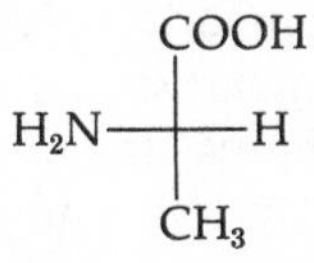

$$
\begin{array}{c}
\text{COOH} \\
\text{H}_2\text{N}\!-\!\!|\!-\!\text{H} \\
\text{CH}_3
\end{array}
$$

(a) Can you tell from the information given whether L-alanine is (+) or (−)?

(b) If you cannot tell whether L-alanine is (+) or (−) on the basis of the given information, what information is needed and how can it be obtained?

19-8 SUMMARY

1. Isomers are different compounds that have the same molecular formula. Structural isomers have different structural formulas. Stereoisomers have the same structural formula but differ in the way the bonds in the molecule are oriented in space.

2. Enantiomers are stereoisomers that are mirror images of one another, and diastereomers are stereoisomers that are not mirror images of one another. Geometric, or cis-trans stereoisomers, are one type of diastereomers.

3. Whether an object and its mirror image have the same shape or different shapes can be tested by a mental procedure called superposition. If an object and its mirror image have the same shape, then they are superimposable; if the object and its mirror image have different shapes, they are nonsuperimposable.

4. Molecules that are nonsuperimposable on their mirror images are called chiral molecules.

5. Enantiomers can exist only for those compounds whose molecules are chiral.

6. A carbon atom that is bonded to four different groups of atoms is called an asymmetric carbon.

7. Generally, but not always, carbon-containing compounds are chiral if they contain an asymmetric carbon, and molecules that do not contain an asymmetric carbon are not chiral.

8. Fischer projections are two-dimensional representations for the shapes of carbon-containing molecules that possess one or more asymmetric carbons.

9. For most, but not all, compounds with n asymmetric carbons, the total number of stereoisomers is 2^n.

10. In general, a single enantiomer of a chiral molecule will react differently with the two enantiomers of another chiral molecule, and a molecule that is not chiral will react identically with the two enantiomers of a chiral molecule.

11. An electromagnetic wave that is propagating in the y direction and that has the electric field in the z direction is said to be linearly polarized in the z direction.

12. Light coming from usual light sources is unpolarized.

13. Polaroid and certain other materials can convert unpolarized light to linearly polarized light.

14. A solution containing a single enantiomer of a chiral molecule can rotate the direction of linearly polarized light.

15. The physical properties of the two enantiomers of a chiral molecule are identical except for the way that they rotate linearly polarized light. For any chiral molecule, one enantiomer rotates linearly polarized light to the right and is known as the (+) enantiomer, whereas the other enantiomer rotates linearly polarized light to the left and is known as the (−) enantiomer.

16. Which enantiomer of a chiral molecule is (+) and which enantiomer is (−) is determined experimentally.

PROBLEMS

1. Describe the differences between structural isomers and stereoisomers. Draw structural formulas for pairs of molecules that are examples of each type of isomer.

2. Two substances, A and B, have the same molecular formula and the same structural formula but have slightly different melting points, boiling points, and solubilities in water. A and B are which of the following:
   (a) Structural isomers
   (b) Diastereomers
   (c) Enantiomers
   (d) Not isomers

3. Which of the following Fischer projections represent diastereomers? Which represent enantiomers?

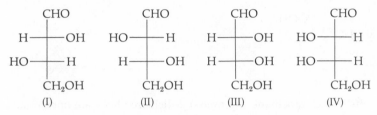

4. Which of these Fischer projections represent diastereomers? Which represent enantiomers?

5. Which of the following solutions would rotate linearly polarized light?
   (a) An equal mixture of III and IV
   (b) An equal mixture of I and II
   (c) Only I
   (d) Only III
   (e) None of these solutions

6. If substances C and D are enantiomers, which of the following statements are true?
   (a) The mirror image of C is identical with D.
   (b) C and D rotate linearly polarized light the same amount but in opposite directions.
   (c) If X is *any* substance that reacts with C, then X will react with D to a different degree.
   (d) C and D have slightly different melting points, boiling points, and solubility in water.

7. Draw structural formulas for each of the following and indicate with an asterisk (*) any asymmetric carbon atoms. Draw Fischer projections for all the stereoisomers (if any) of each, and indicate which are enantiomers and which are diastereomers.
   (a) α-Hydroxypropionamide
   (b) 2,2-Dichloro-3-methylbutane
   (c) 2,3-Hexanediol
   (d) α-Aminobutyric acid

8. There are four dimethylcyclopropane isomers (structural and stereoisomers). (a) Draw structural formulas for these four isomers. (b) Indicate which are structural isomers and which are stereoisomers. (c) For the stereoisomers, indicate which are enantiomers and which are diastereomers.

9. Describe the difference between linearly polarized light and unpolarized light.

10. Suppose that you have two bottles, one of which contains D-glyceraldehyde and the other contains L-glyceraldehyde. What experimental test could you perform and what information about the isomers could you utilize that would enable you to determine which isomer is in which bottle?

SOLUTIONS TO EXERCISES

19-1   (a)   Chiral,

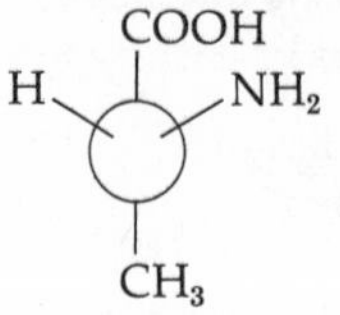

$$H \diagdown \underset{\displaystyle CH_3}{\overset{\displaystyle COOH}{\bigcirc}} \diagup NH_2$$

(b)   Nonchiral. (Note that although the molecule and its mirror image may at first seem different, when you carry out the mental superposition procedure, the mirror image can be turned so that it is superimposable. Thus, the mirror image is a representation of the same molecule.)

(c)   Chiral,

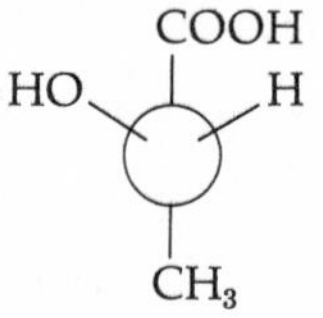

$$HO \diagdown \underset{\displaystyle CH_3}{\overset{\displaystyle COOH}{\bigcirc}} \diagup H$$

(d)   Both *cis*-1,2-dichloroethene and *trans*-1,2-dichloroethene are nonchiral.

19-2   (a)

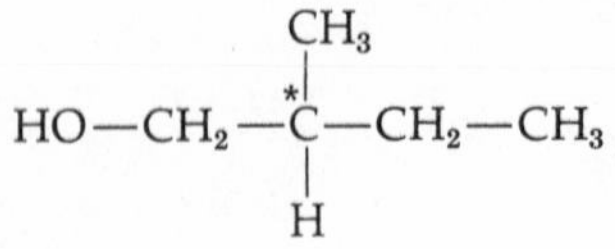

$$HO-CH_2-\overset{\displaystyle CH_3}{\underset{\displaystyle H}{\overset{|}{\underset{|}{C}}}}{}^*-CH_2-CH_3$$

(b)

$$CH_3-\overset{\displaystyle H}{\underset{\displaystyle Cl}{\overset{|}{\underset{|}{C}}}}{}^*-\overset{\displaystyle H}{\underset{\displaystyle Cl}{\overset{|}{\underset{|}{C}}}}{}^*-CH_3$$

(c)

$$CH_3-\overset{\displaystyle OH}{\underset{\displaystyle CH_3}{\overset{|}{\underset{|}{C}}}}-CH_2-CH_3$$

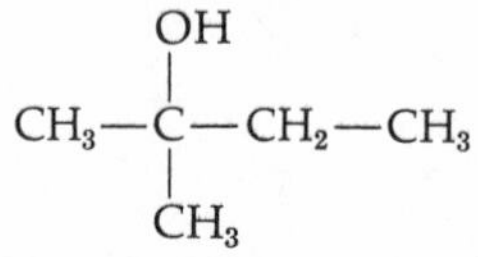

(no asymmetric carbon)

(d)

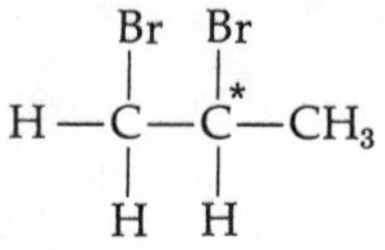

$$H-\overset{\displaystyle Br}{\underset{\displaystyle H}{\overset{|}{\underset{|}{C}}}}-\overset{\displaystyle Br}{\underset{\displaystyle H}{\overset{|}{\underset{|}{C}}}}{}^*-CH_3$$

19-3   (a)

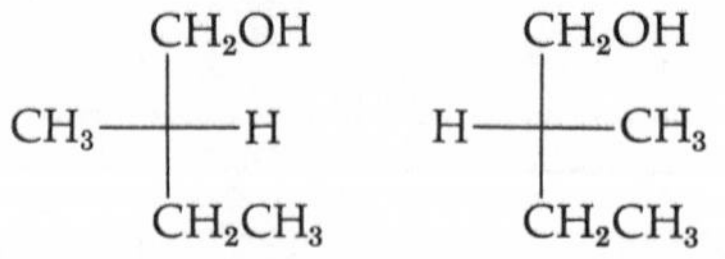

CH₃——H   H——CH₃

(with CH₂OH on top, CH₂CH₃ on bottom for both)

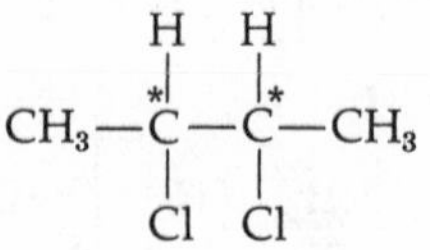

enantiomer pair

(b)

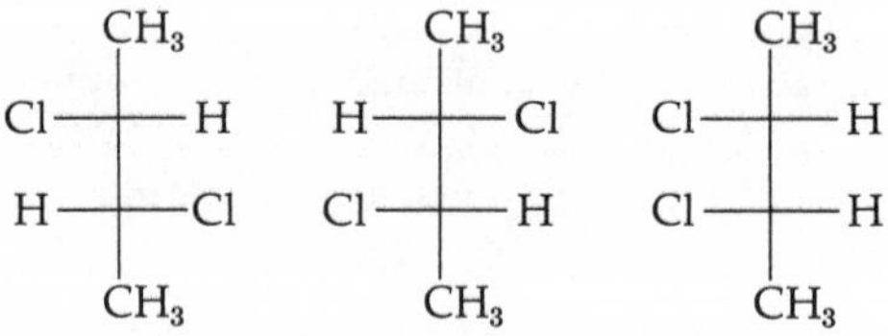

(1)   (2)   (3)

(each with CH₃ on top and CH₃ on bottom)

(1): Cl——H / H——Cl

(2): H——Cl / Cl——H

(3): Cl——H / Cl——H

(1) and (2) are an enantiomer pair
(1) and (3) are diastereomers, as are (2) and (3)
(3) is a *meso*-type compound

(d)

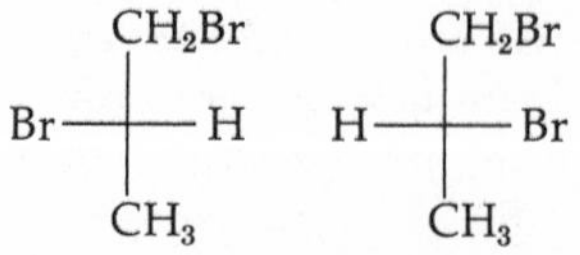

Br——H   H——Br

(each with CH₂Br on top and CH₃ on bottom)

enantiomer pair

19-4   (a) This is a chiral molecule, so it should react differently with the two enantiomers of 2-butanol.

(b) This is not a chiral molecule, so it should react identically with both enantiomers of 2-butanol.

19-5   (a) No. There is no relationship between the labels D- and L- and the direction in which an enantiomer rotates linearly polarized light.
(b) To determine whether L-alanine is (+) or (−), you can measure the rotation of an L-alanine solution with a polarimeter.

BIOLOGICAL MOLECULES

Overview

You have seen how the basic principles of chemistry can be applied to the study of simple molecules. Now you are ready to consider the much more complicated **biomolecules** found in living cells. In order to understand the function of biomolecules in healthy human cells, you must recognize that they are composed of simple parts whose properties you have already studied.

We can divide biomolecules into four categories: (1) carbohydrates, (2) amino acids and proteins, (3) nucleic acids, and (4) lipids. The structures and functions of the first three types of biomolecules will be presented in this part of the book. We shall begin with a study of carbohydrates, proceeding from small carbohydrates such as glucose to some very large carbohydrates such as starch and cellulose. We shall see that both starch and cellulose are composed of long chains of glucose building blocks bonded together in a particular arrangement.

Although proteins and nucleic acids are large and complex, they are also composed of smaller building blocks. We shall study these building blocks and how they are assembled in much the same way as you might attempt to understand how a radio operates by first learning about its component parts. Then we can consider the function of a protein molecule, keeping in mind that the function of any molecule, no matter how big, depends on its structure.

The relationship between the structure and the function of biomolecules will be heavily emphasized in our discussion. There is no magic involved—their function simply depends on the nature and orientation of their functional groups. We have already studied virtually all of the functional groups we shall encounter, as well as the types of chemical reactions that these functional groups can undergo.

One major property of all biomolecules is that they are found in living cells. Cells are very complicated, but we believe that an entire cell is no more than the sum of its parts. **Biochemistry** is the study of the component parts of cells and the interactions that take place between them. It is not enough to understand the structure of a protein if you do not understand its role in the cell. Yet there are thousands of proteins in each cell, and many other kinds of molecules as well. How can we possibly keep track of all of these molecules and what they are doing? To help out, we shall make extensive use of an analogy between a living cell and a factory. Factories have walls and cells have membranes. Factories have skilled workers and cells have enzymes. Factories have assembly lines and cells have metabolic pathways (a series of enzyme-catalyzed reactions leading to the synthesis or breakdown of a particular molecule).

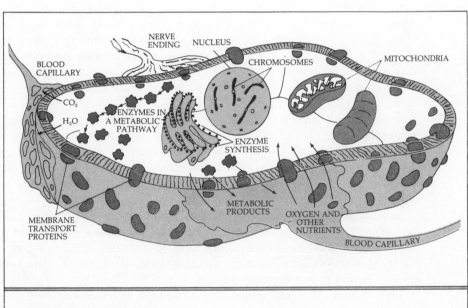

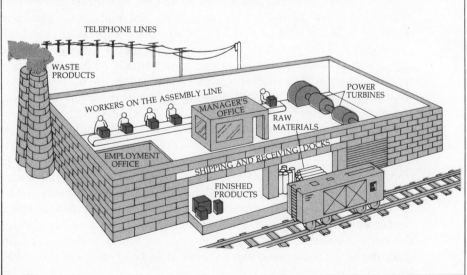

In the human body, cells with different functions are grouped together into tissues and organs. The entire human body is a highly integrated network of these tissues and organs. In order to visualize the interrelationships among molecules, cells, tissues, organs, and the entire human body, we can carry our cell-factory analogy one step further. We can think of the human body as being analogous to an industrial society. An industrial society must include mining, transportation, production, communication, and other industries. The human body also needs these kinds of services. Mining, or the intake of raw materials, is handled by cells in the lungs (for oxygen) and cells in the intestines (for food and water). Transportation is handled by the circulatory system. In fact, we might go so far as to compare transport proteins and red blood cells to trucks. Communication is handled by nerve cells (the telephone company) and hormones (Western Union).

Biochemists use several other analogies to describe the functions and interrelationships of biomolecules. In Chapter 24 we shall use an analogy in which the genetic information in DNA is compared to a coded message. When we study very complicated biomolecules and their functions, analogies add some fun as well as clarity to our discussions.

One aspect of biomolecules that must be kept in mind is that we do not know everything about them. Because some biological molecules are so complex; and because there are a lot of very specialized ones (especially proteins); and because many of the techniques for studying them are quite new—often we can only guess at how a particular biomolecule actually accomplishes its functional task. But these will be *educated* guesses, since there are no new rules, no magic in biochemistry. No biomolecule has yet been discovered that violates the chemical principles we have studied in the past several chapters.

Not having all the answers to our questions about biomolecules is sometimes hard to live with, especially when answers to these questions could be helpful to the practice of modern medicine. A great many of the answers to past questions have resulted in the saving of lives. For example, a better understanding of how enzymes work has allowed chemists to design and then develop drugs to combat certain specific diseases. Eventually, we shall also answer questions concerning the causes of diabetes, cancer, and other diseases. Understanding the answers to these questions, when they are found, will require a firm understanding of the basic principles we have already learned and those we are about to study.

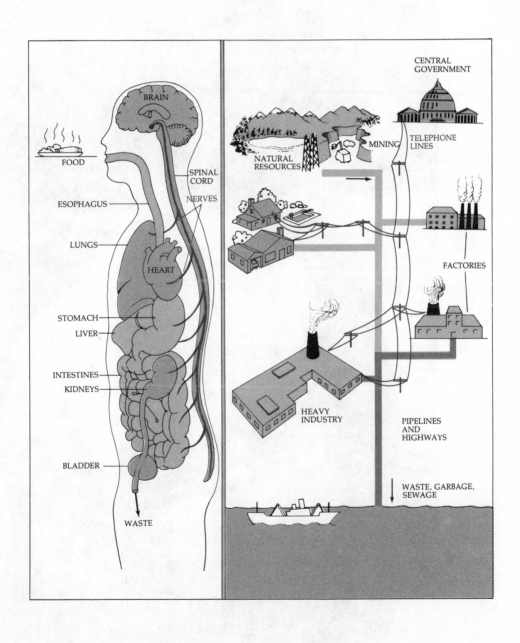

CHAPTER 20

Carbohydrates

20-1 INTRODUCTION

Compounds that contain an aldehyde or ketone functional group and two or more alcohol functional groups (i.e., polyhydroxy aldehydes or ketones), as well as compounds that can be hydrolyzed to give polyhydroxy aldehydes or ketones, are called carbohydrates. Carbohydrates are the most abundant organic compounds in nature, and they have been adapted for a wide variety of uses. Table sugar, the starch in potatoes and other vegetables, and the cellulose in wood are all carbohydrates.

Starch and cellulose are both large polymers that are assembled by plants from the same building block, D-glucose, a simple carbohydrate with the molecular formula $C_6H_{12}O_6$. The primary difference between starch and cellulose, as we shall see in this chapter, is simply a difference in the orientation of the bonds that join together the component D-glucose units.

In cellulose the D-glucose units form linear chains that are held together by hydrogen bonds to form long bundles. These bundles of cellulose molecules provide the structural form for all green plants, from blades of grass to huge trees. The availability and structural strength of trees make wood an important building material. In addition, wood pulp is the primary raw material in the manufacture of paper products. Cellulose-containing parts of other plants also have widespread applications; for example, cotton, which is about 90% cellulose, is a cornerstone of the textile industry. Synthetically prepared cellulose derivatives are also components of a large variety of commercial products. For example, nitrocellulose is used to make some lacquers and explosives, and cellulose acetate is used to make photographic films and phonograph records.

In starch the component D-glucose units are joined together differently than in cellulose. As a result, starch molecules have a spiral structure, rather than the linear structure of cellulose. Starch made by plants for energy storage is, in turn, used by humans and many other organisms for food. Most organisms possess enzymes that catalyze the breakdown of starch into its component glucose units, whereas relatively few organisms possess enzymes that catalyze the breakdown of cellulose.

For centuries, people throughout the world have used cotton to make clothing. Even today, despite the availability of a wide variety of synthetic fabrics, cotton remains a major raw material for the textile industry.

Like cellulose, starch is insoluble in water, but in boiling water it does form a colloidal dispersion that gels upon cooling. This property of starch is used in the preparation of jellies. Likewise, cornstarch is frequently used to thicken gravy, and clothing is occasionally starched to make it more resistant to wrinkling. Interestingly, bread generally becomes hard, not because it has lost water, but because a large number of starch chains have associated by forming additional hydrogen bonds. Hard, stale bread can be softened by warming it because the increased temperature breaks some of the hydrogen bonds. Starch is also used in the commercial manufacture of pastes and adhesives, as well as in the formation of drug tablets.

The human body produces several important carbohydrates. Glycogen, a polymer very similar to starch, is one of the compounds used for energy storage. An individual's blood type is determined by a specific carbohydrate on the surface of his or her red blood cells. An important component of the genetic material DNA is also a carbohydrate. In later chapters, especially Chapter 25, we shall study the specific roles of carbohydrates in the human body. In this chapter we shall explore the structure, nomenclature, and chemical reactions of carbohydrates in general, and study in detail those carbohydrates that are most important for humans.

20-2   STUDY OBJECTIVES

After studying the material in this chapter, you should be able to:

1. Define the terms carbohydrate, monosaccharide, disaccharide, polysaccharide, aldose and ketose, anomeric carbon, and glycosidic linkage.

2. Identify a monosaccharide as an aldose or ketose, and classify it to indicate the number of carbon atoms, given its structural formula.

3. Draw a Haworth projection for the hemiacetal or hemiketal form of a monosaccharide, given the Fischer projection for the free aldehyde or ketone form.

4. Identify the anomeric carbon atom in the hemiacetal or hemiketal form of a monosaccharide, tell if it is D or L, and specify whether the sugar is α or β, given its Haworth projection.

5. Draw Haworth projections for the monosaccharides glucose, galactose, ribose, and fructose.

6. Draw structural formulas for the products of reactions in which monosaccharides form internal hemiacetals or hemiketals, acetals, disaccharides, and esters, and for reactions in which acetals, disaccharides, and esters are hydrolyzed.

7. Classify the glycosidic bond in a disaccharide as α or β, given its Haworth projection.

8. Draw Haworth projections for the disaccharides maltose, lactose, and sucrose.

9. Determine if a disaccharide is a reducing sugar, given its Haworth projection.

10. Describe the structural differences between the polysaccharides starch (amylose and amylopectin), glycogen, and cellulose, and know which the human body is capable of digesting.

20-3 STRUCTURE AND CLASSIFICATION OF CARBOHYDRATES

We have defined **carbohydrates** as polyhydroxy aldehydes or ketones, or compounds that can be hydrolyzed to produce polyhydroxy aldehydes or ketones. Many carbohydrates have the general formula $C_n(H_2O)_n$, hence the name "carbo(n)-hydrate." Carbohydrates are also called sugars or saccharides. Many of the carbohydrates found in nature are very large molecules called polysaccharides. Starch and cellulose are examples of polysaccharides. These large polysaccharides are polymers. In Chapter 14 we saw that polymers are large molecules composed of simpler units called monomers. The simple units that form polysaccharides are called monosaccharides (see Figure 20-1). **Monosaccharides** are carbohydrates that cannot be hydrolyzed into simpler compounds. Polysaccharides can be hydrolyzed by acids or specific enzymes into monosaccharide units. For example, acid-catalyzed hydrolysis of starch or cellulose yields the monosaccharide glucose.

$$\text{polysaccharides} + H_2O \xrightarrow[\text{hydrolysis}]{\text{catalyst}} \text{monosaccharides}$$

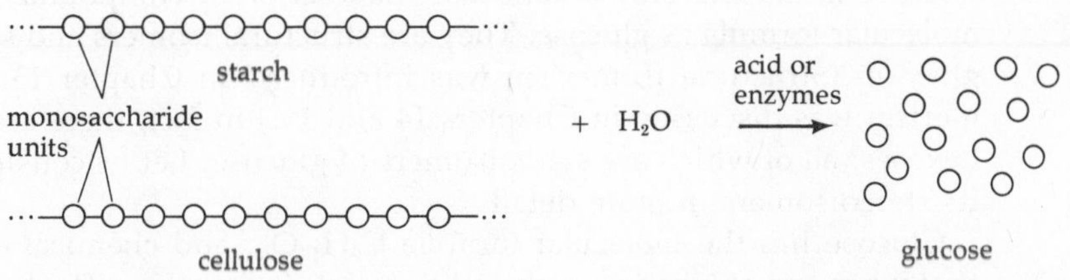

starch

monosaccharide units

cellulose

glucose

Figure 20-1   Polysaccharides can be hydrolyzed into their component monosaccharide units in reactions catalyzed by acids or specific enzymes.

The hydrolysis of either starch or cellulose yields the monosaccharide glucose.

Starch and cellulose are both polymers of glucose; however, the human body can digest starch but not cellulose. What is the basic difference between these two glucose polymers? In order to answer this question, and to explain other properties of polysaccharides, we must first study the structures and properties of monosaccharides. Glucose, the monosaccharide unit in cellulose and starch, is the most abundant carbohydrate in nature. In addition, glucose, which is also known as dextrose or "blood sugar," is vital to the human body. We shall therefore frequently use glucose as our example when discussing the chemistry of monosaccharides.

Classification of Monosaccharides

The general formula for many simple monosaccharides is $C_n(H_2O)_n$, where n is 3 or larger. For example, $n = 6$ for glucose. The molecular formula for glucose is thus $C_6H_{12}O_6$. The ending -ose is often used in the names of carbohydrates. Monosaccharides are classified in two ways: first, as aldehydes or ketones; and second, by the number of carbon atoms (n). A monosaccharide that contains the aldehyde functional group is called an **aldose,** whereas a monosaccharide that contains the ketone functional group is called a **ketose.**

Table 20-1   Classification of Monosaccharides
by Number of Carbon Atoms (n)

General Formula	n	Name
$C_3H_6O_3$	3	Triose
$C_4H_8O_4$	4	Tetrose
$C_5H_{10}O_5$	5	Pentose
$C_6H_{12}O_6$	6	Hexose
$C_7H_{14}O_7$	7	Heptose

Glucose is an aldose. Table 20-1 shows the names used to indicate the number of carbon atoms in a monosaccharide. For example, glucose has six carbon atoms and is therefore a hexose. In order to specify both the number of carbon atoms and the nature of the carbonyl-containing functional group, the names in Table 20-1 are combined with the name aldose or ketose. For example, glucose is both an aldose and a hexose; hence, glucose is an **aldohexose.** Other sugars important in the human body are aldopentoses, ketohexoses, aldotrioses, and so on.

Stereoisomers of Monosaccharides

Glucose is not the only aldohexose. Several other compounds have the same molecular formula as glucose. They are structural isomers and stereoisomers of glucose. (Structural isomerism was introduced in Chapter 13, and stereoisomerism was discussed in Chapters 14 and 19.) In fact, there are 31 other aldohexoses, all of which are stereoisomers of glucose. Let us consider glucose and its stereoisomers in more detail.

Glucose has the molecular formula $C_6H_{12}O_6$, and chemical evidence shows that it has one aldehyde group and five hydroxyl groups. The following structural formula is consistent with this experimental data:

$$H-\underset{\underset{OH}{|}}{\overset{\overset{H}{|}}{C}}-\underset{\underset{OH}{|}}{\overset{\overset{H}{|}}{C}}{}^*-\underset{\underset{OH}{|}}{\overset{\overset{H}{|}}{C}}{}^*-\underset{\underset{OH}{|}}{\overset{\overset{H}{|}}{C}}{}^*-\underset{\underset{OH}{|}}{\overset{\overset{H}{|}}{C}}{}^*-\overset{\overset{\displaystyle O}{\diagup}}{\underset{\diagdown}{C}}{}_H$$

This structural formula has four asymmetric carbon atoms, which are indicated by asterisks. We therefore expect 16 stereoisomers or 8 pairs of enantiomers. Each pair of enantiomers has a different name, and the prefixes D- and L- are used to distinguish between an enantiomer pair (see Chapter 19). The Fischer projections for one of these enantiomer pairs, D-glucose and L-glucose, are shown in Figure 20-2. The D- and L-enantiomers of a monosaccharide are designated as follows: Draw the Fischer projection as a vertical carbon atom chain with the carbonyl group at the top of the projection, and then look at the *lowest* asymmetric carbon atom. If the hydroxyl group is to the right in the projection, the sugar is the D-enantiomer. If it is to the left, it is the L-enantiomer. Cells in the human body can use only sugars in the D-configuration, such as D-glucose. D-Galactose is another D-aldohexose that is important in the human body. Recall that the symbol D- indicates the configuration around only one particular carbon atom of a monosaccharide, not the rotation of polarized light by that compound. If the rotation of light needs to be specified, the symbols (+) and (−) are used to designate rotation to the right and left, respectively. What about D-glucose? D-Glucose happens to be dextrorotatory; it rotates polarized light to the right. Thus, D-(+)-glucose is often called **dextrose.**

Figure 20-2 Fischer projections for the enantiomers L-glucose and D-glucose. The four asymmetric carbon atoms in each enantiomer are indicated by asterisks.

L-glucose   D-glucose

Formation of Internal Hemiacetals: Haworth Projections

The structure of D-(+)-glucose is not as simple as is indicated by the Fischer projection in Figure 20-2. This representation for aldohexoses shows only four asymmetric carbon atoms, and accounts for 16 stereoisomers. But we have stated that there are 32 stereoisomeric aldohexoses, which would indicate the presence of five asymmetric carbon atoms.

Recall from Chapter 16 that aldehyde and alcohol groups on the same molecule can react to form internal hemiacetals, and that internal hemiacetals with rings of six atoms are favored. D-Glucose can form *two* different hemiacetals, each containing a ring of six atoms, by a reaction of the aldehyde group and the alcohol group involving carbon atom number 5. The same is true for all the other 16 aldohexoses. Thus there are a total of 32 aldohexose stereoisomers. Fischer projections for the two different internal hemiacetals of D-glucose are given in Figure 20-3. Notice that each of these internal hemiacetals of D-glucose has a total of five asymmetric carbon atoms, because the carbon atom from the aldehyde group, carbon 1, is also asymmetric in the hemiacetal form. The additional carbon atom in an internal hemiacetal form is called an **anomeric** carbon atom. These two forms of D-(+)-glucose can be obtained as pure solids with different properties (see Table 20-2) and are called the α- and β-anomers of glucose. In general, **anomers** are diastereomers that differ only in the bonding arrangement of the anomeric carbon atom. The α-D-(+)-glucose anomer is the naturally occurring solid form of glucose. In addition to glucose and its stereoisomers, most other monosaccharides have α- and β-anomeric forms.

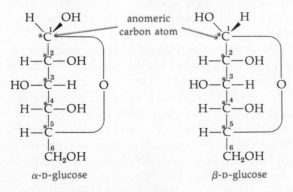

α-D-glucose β-D-glucose

Figure 20-3 Fischer projections for the two internal hemiacetal forms of D-glucose, the α- and β-anomers of D-glucose.

Table 20-2 Physical Properties of α- and β-D-Glucose Anomers

	Melting Point (°C)	Specific Rotation
α-D-Glucose	146	+112°
β-D-Glucose	150	+19°

The Fischer projections for α- and β-D-glucose, shown in Figure 20-3, are not very good representations of these molecules. Structural differences between the α- and β-forms, for example, are not easy to see from Fischer projections. Models would be much better for this purpose.

The structure of α-D-glucose, β-D-glucose, and the other stereoisomers of glucose resemble the structure of cyclohexane. Recall from Chapter 14 that cyclohexane, and other compounds containing rings of six carbon atoms, exist preferentially in a shape called the chair form, which allows all of the bond angles in the ring to be 109.5°. Thus the chair form is the best representation for the structures of the α- and β-forms of glucose (see Figure 20-4).

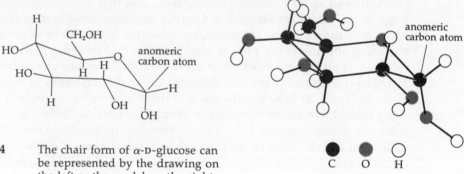

Figure 20-4   The chair form of α-D-glucose can be represented by the drawing on the left or the model on the right.

However, drawing the chair form of a monosaccharide is tedious, so a compromise method, called a **Haworth projection,** was developed in which all of the atoms in the ring are represented as being in one plane. Although Haworth projections are not the best representations of the shapes of monosaccharides, they are superior to Fischer projections for monosaccharides that exist in the internal hemiacetal or hemiketal form.

By applying a few simple rules, it is possible to draw Haworth projections for the α- and β-anomers of a monosaccharide, given the Fischer projection for the open-chain form. For all of the pentoses and hexoses we shall discuss, the hemiacetal or hemiketal group involves the carbonyl carbon atom and the lowest asymmetric carbon in the Fischer projection. Take glucose, for example. Figure 20-5 shows the Fischer projections for D- and L-glucose and the corresponding Haworth projections for α-D-glucose, β-D-glucose, α-L-glucose, and β-L-glucose. Note that the ring of atoms in a Haworth projection is to be visualized as projecting out from the plane of the paper. To emphasize this we usually use darker lines to represent the three bonds in the ring that are visualized as closer to the reader. Hydrogen atoms and hydroxyl groups are visualized as projecting up or down from the carbon atoms in the horizontal ring.

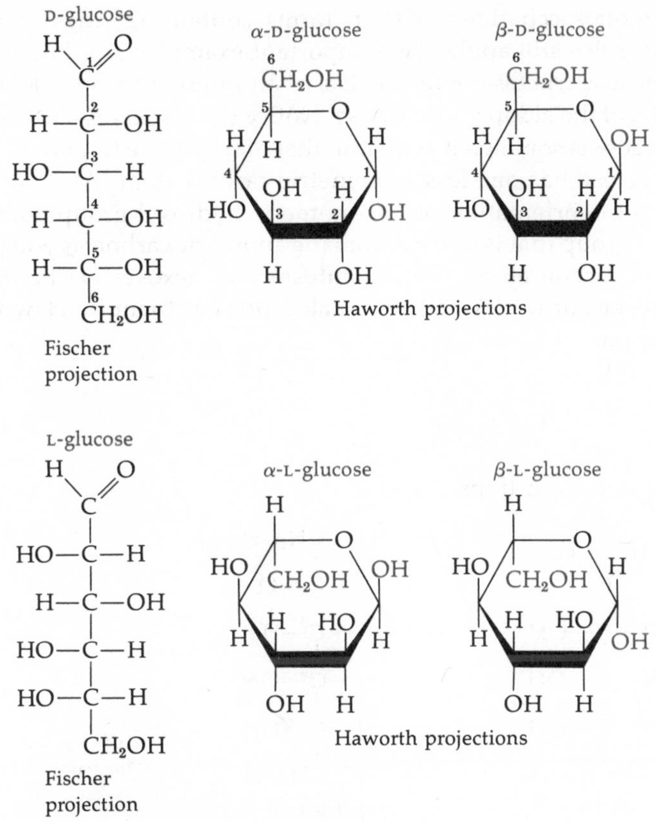

Figure 20-5   Fischer projections for D-glucose and L-glucose and the Haworth projections for their corresponding α- and β-anomeric forms.

The rules for drawing a Haworth projection are as follows:

1. Draw the anomeric carbon to the right and continue to draw the remaining atoms in the ring in a clockwise direction, with the ring oxygen at the top of the drawing.

2. Hydroxyl groups located on the *right* of the chain in the Fischer projection are drawn *down* in the Haworth projection. Hydroxyl groups drawn to the *left* in the Fischer projection are drawn *up* in the Haworth projection.

3. The last —CH₂OH group is drawn *above* the ring (up) in the Haworth projection of a D-sugar and *below* the ring (down) for an L-sugar.

4. The hydroxyl group attached to the anomeric carbon atom (the anomeric hydroxyl group) is drawn on the *same* side of the ring as the last —CH₂OH group for the β-anomer, and the *opposite* side of the ring for the α-anomer. When the α- or β-anomer is not specified, as in the designation D-glucose, the Haworth projection for the anomeric hydroxyl group is written as

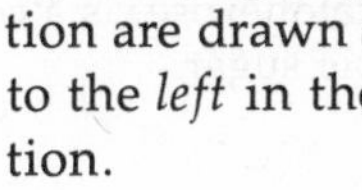

to indicate that the sugar is either α or β.

Some monosaccharides exist in forms containing rings of five atoms, for which our rules still apply. Two important examples are shown in Figure 20-6: the ketohexose fructose, which exists as an internal hemiketal, and the internal hemiacetal of the aldopentose ribose. Notice the Fischer projection for D-fructose in Figure 20-6 is somewhat similar to that of D-glucose (Figure 20-5), but that the ketose fructose has one less asymmetric carbon atom. Also notice that α- and β- refer to the orientation of the anomeric hydroxyl group with respect to the —CH₂OH group that is *farthest* from the anomeric carbon as you proceed around the ring of carbon atoms. Since pentoses and hexoses exist almost exclusively as hemiacetals or hemiketals, we shall represent them by Haworth projections from now on.

Fischer projections:

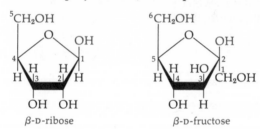

Haworth projections (for the predominant anomers):

β-D-ribose   β-D-fructose

Figure 20-6   Fischer and Haworth projections for two important monosaccharides that exist in forms containing rings of five atoms. β-D-Ribose, an aldopentose (left), is a component of   DNA and other complex biomolecules, whereas β-D-fructose (right), a ketohexose, is a component of table sugar.

When naturally occurring α-D-glucose is dissolved in water and the rotation of polarized light is measured immediately, the specific rotation is +112°. The specific rotation is the rotation for a 1.0 g/1.0 ml solution in a 1.0-decimeter polarimeter tube. However, over a period of time the specific rotation decreases until it finally reaches +52.7°. On the other hand, when β-D-glucose is dissolved in water, the specific rotation increases from +19° to +52.7°. These data indicate that, in solution, glucose exists as an equilibrium mixture of the α-form and the β-form (see Figure 20-7). At equilibrium, 36% of the D-glucose exists in the α-form, 64% in the β-form, and only 0.02% exists in the free aldehyde or open-chain form.

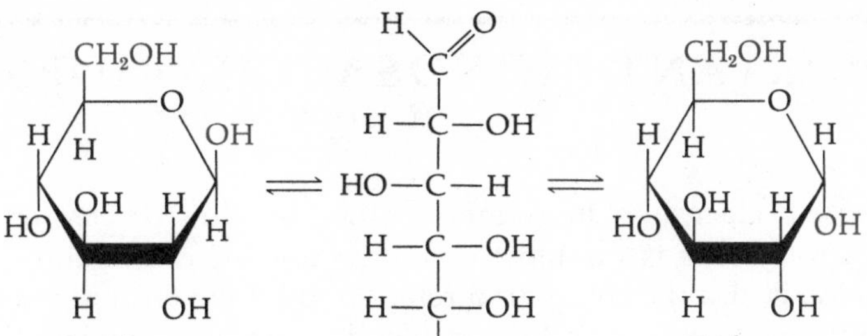

β-D-glucose (64%) ⇌ D-glucose open-chain form (0.02%) ⇌ α-D-glucose (36%)

Figure 20-7   In aqueous solution almost all D-glucose molecules exist in the β- or α-anomer forms. However, these anomers are in equilibrium with a small amount of the open-chain aldehyde form.

Exercise 20-1

Draw Fischer projections for all 8 of the D-aldohexoses in the open-chain form.

Exercise 20-2

The Fischer projection for D-galactose is given below. Draw the Haworth projection for β-D-galactose.

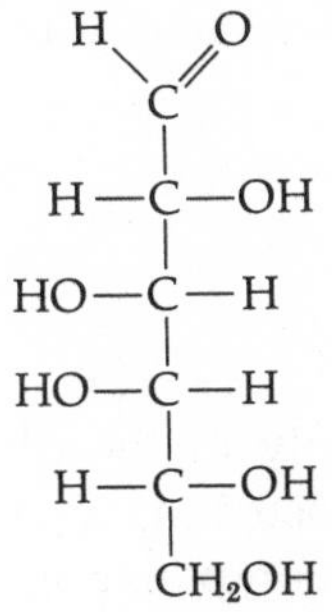

Exercise 20-3

The Haworth projections for three monosaccharides are given below. (a) Name these monosaccharides to indicate whether they are aldehydes or ketones and the number of carbon atoms (e.g., D-glucose is an aldohexose). (b) Indicate which carbon atom is the anomeric carbon atom, and determine if these are D- or L- and α- or β-anomers.

(1)

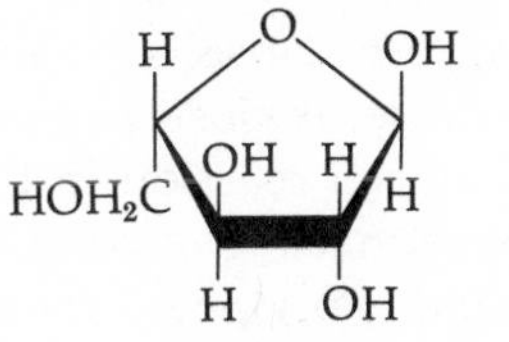

(2)

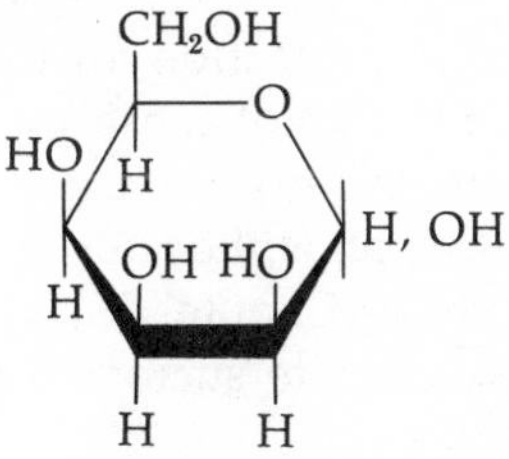

(3)

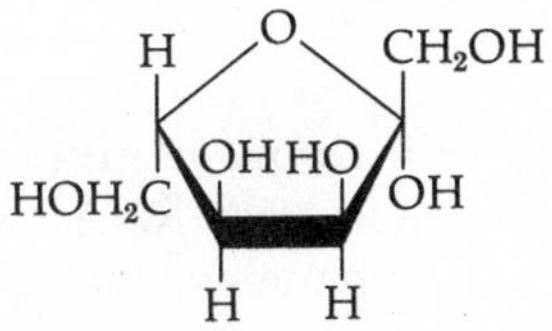

20-4 IMPORTANT MONOSACCHARIDES

D-Glucose

We chose D-glucose as an example in the preceding discussion for a very good reason. D-Glucose is the most abundant organic compound found in nature. D-Glucose is also the compound referred to by the term **blood sugar.** The concentration of D-glucose in human blood is carefully regulated, as we shall see later (Chapter 25). D-Glucose is used by cells in the human body as a source of nourishment. Too little glucose, and cells, especially those in the brain, starve.

D-Glucose is the monosaccharide component from which the polysaccharides cellulose, starch, and glycogen are built (see Section 20-7), and it is a component of several more complex polysaccharides. It is also a component of the important disaccharides sucrose, maltose, and lactose (see Section 20-6).

D-Galactose

The Haworth projection for the β-anomer of D-galactose, a stereoisomer of D-glucose, is

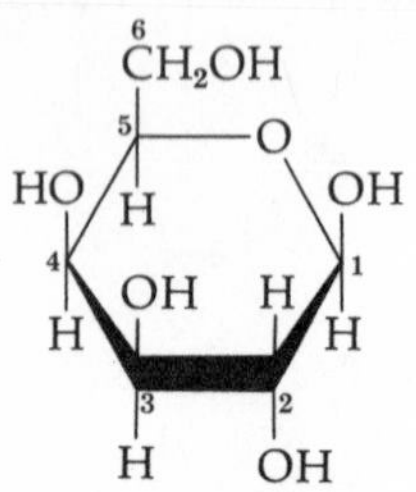

Notice that D-galactose differs from D-glucose only in the orientation of the groups bonded to carbon atom number 4. D-Galactose is a component of the disaccharide lactose found in milk, and of some complex polysaccharides. Ingested D-galactose is normally converted to D-glucose in the human body. The inability to perform this isomerization (conversion of one isomer to another) results in the disease galactosemia (Chapter 25).

D-Fructose

The Haworth projection for the β-anomer of D-fructose is given in Figure 20-6. D-Fructose is a ketohexose which, like galactose, is closely related to D-glucose in structure. A very important enzyme-catalyzed reaction in the human body converts the phosphoester D-glucose 6-phosphate to D-fructose 6-phosphate and vice versa (Chapter 25). β-D-Fructose and α-D-glucose are the components of the disaccharide sucrose (see Section 20-6), which is common table sugar.

D-Ribose

The Haworth projection for the β-anomer of D-ribose is given in Figure 20-6. This aldopentose is a component of ribonucleic acid, RNA (see Chapter 24). β-D-Ribose is also a component of several coenzymes, including NAD$^+$ (see Figure 18-4) and the compound ATP (Figure 25-4), which supplies energy for reactions in the human body.

The sugar cane plant is one of our major sources of sucrose (common table sugar), a disaccharide composed of the monosaccharide units β-D-fructose and α-D-glucose.

Many other monosaccharides, ranging in size from trioses to heptoses, are also found in nature. We shall encounter some of these in later chapters when we discuss human metabolism.

Exercise 20-4

D-Galactose and D-fructose can be converted to D-glucose by isomerization reactions. Can D-ribose be converted to D-glucose by such a reaction? Explain.

20-5 PROPERTIES OF MONOSACCHARIDES

Physical Properties

Pure monosaccharides are solids at room temperature. Monosaccharides contain alcohol groups and either a carbonyl group or a hemiacetal or hemiketal group. Like alcohols, they readily form hydrogen bonds with water or other molecules containing hydrophilic functional groups. Because a monosaccharide molecule such as glucose contains several alcohol groups, it can hydrogen bond with many water molecules and thus is extremely soluble in pure water and in aqueous solutions such as blood serum.

Reactions of Monosaccharides

In addition to the formation of internal hemiacetals or hemiketals, monosaccharides can participate in other reactions typical of compounds containing both alcohol and carbonyl groups. The most important specific reactions involving monosaccharides are those that occur in the human body. We shall discuss these reactions in detail in Chapter 25.

Let us consider some of the general reactions of monosaccharides, one of which is an oxidation reaction that can be used to determine the amount of a sugar present in a solution. Again, let us use glucose as our example. The α- and β-anomers of glucose in aqueous solution are in equilibrium with the free aldehyde form of glucose (see Figure 20-7). This equilibrium in aqueous solution between a free aldehyde and internal hemiacetal(s) is typical of all aldoses and other compounds that form internal hemiacetals. In solution, an internal hemiketal, such as fructose, is also in equilibrium with its free ketone form. Recall from Chapters 15 and 16 that aldehydes can be oxidized to form carboxylic acids, and that alcohols can also be oxidized to form aldehydes or ketones.

Benedict's, Fehling's, and Tollen's reagents are specific oxidizing agents that can oxidize aldoses with a free aldehyde group. Consider glucose. When one of these reagents reacts with the free aldehyde form of glucose, the oxidizing agent is reduced and glucose is converted to a complex mixture of products. Both Benedict's and Fehling's reagents contain Cu^{2+}, which is reduced, forming red Cu_2O. In solution, very little glucose (less than 1%) exists as a free aldehyde. The free aldehyde, however, is in equilibrium with the internal hemiacetal forms. As the free aldehyde form of glucose reacts with one of these reagents, more and more of the cyclic hemiacetal form is converted to the free aldehyde form and oxidized, until finally all of the glucose has been oxidized (see Figure 20-8). By measuring the amount of color change upon reaction with Benedict's or Fehling's reagents, one can determine the amount of aldoses such as glucose in a solution.

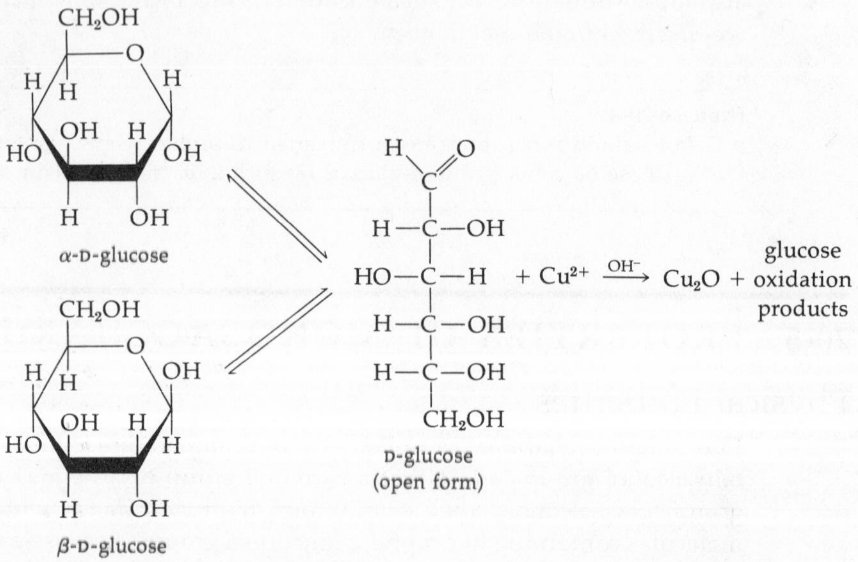

Figure 20-8   As the open-chain aldehyde form of glucose is oxidized, the two equilibria between this form and the α- and β-anomers shift, generating more of the open-chain aldehyde. In this manner all of the glucose in a solution can be oxidized.

Sugars that can reduce Fehling's, Benedict's, or Tollen's reagents are called **reducing sugars.** All aldoses are reducing sugars. Ketoses are also reducing sugars because they have a ketone functional group immediately adjacent to an alcohol functional group. The neighboring ketone group gives these adjacent alcohol groups slightly different properties than most alcohols. In particular, they can be more easily oxidized, and consequently D-fructose and other ketoses will also reduce Benedict's, Fehling's, or Tollen's reagents. We shall see that polysaccharides and some disaccharides are not reducing sugars.

Glucose is the only reducing sugar that is normally present in any appreciable concentration in human blood. Thus, Benedict's or Fehling's reagent can be used to measure "blood sugar." In patients with diabetes, these reagents can also be used to test for glucose in the urine. However, Benedict's and Fehling's reagents are not specific for glucose; diseases in which other reducing sugars accumulate in blood or urine can be misdiagnosed as diabetes when these reagents are used. Because of this possible difficulty, most clinical laboratories now use more specific reagents, usually enzymes, to determine the amount of glucose in a solution.

Another important reaction of carbohydrates is ester formation. Recall from Chapter 17 that alcohols can react with carboxylic acids or phosphoric acid to produce esters. There are many enzymes in the human body that catalyze the formation of phosphoesters from monosaccharides. For example, although glucose is present in fairly high concentration in the bloodstream, almost no free glucose exists within cells. In the cells of the human body, glucose is rapidly converted to the phosphoester called glucose 6-phosphate, where the phosphate group is bonded to carbon atom number 6 of glucose:

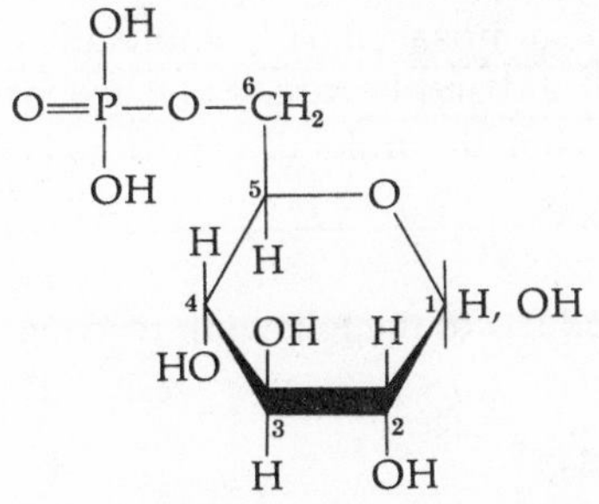

Glucose 6-phosphate is then oxidized in a series of reactions to provide energy for cellular functions. We shall encounter several other phosphoesters of monosaccharides when we study the metabolism of carbohydrates.

In Chapter 16 we saw that hemiacetals and hemiketals can react with alcohols to yield acetals and ketals, respectively. Monosaccharides that exist primarily as internal hemiacetals, such as glucose and ribose, or as internal hemiketals, such as fructose, also form acetals or ketals. When sugars react with alcohols to form acetals or ketals, the bond formed between the anomeric carbon atom of the sugar and the oxygen atom from the alcohol is called a **glycosidic bond,** and the term **glycoside** is used to refer to an acetal derivative of a carbohydrate. Glycosides of glucose are given the more specific name of **glucosides.** For example, under suitable conditions, α-D-glucose can react with methyl alcohol to produce the acetal called methyl-α-D-glucoside:

(20-1)

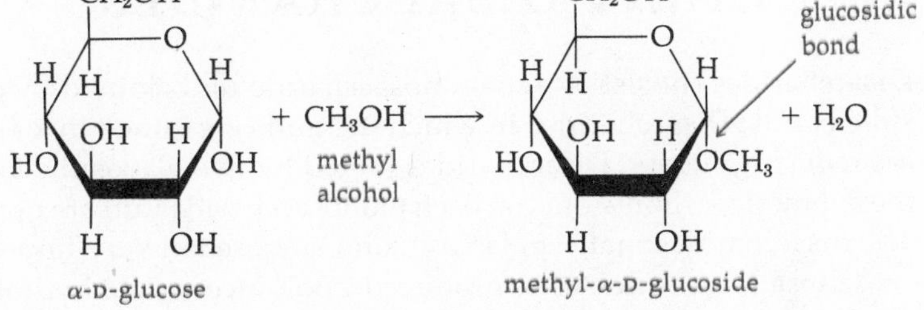

Notice that a glycosidic bond is a bond between the anomeric carbon and an oxygen atom that is not in the ring. Glycosidic linkages may be α or β. When β-D-glucose and methyl alcohol react, methyl-β-D-glucoside is formed.

Recall that acetals are much more stable in solution than are hemiacetals. The glycoside derivatives of monosaccharides, such as methyl-α-D-glucoside, do not exist as equilibrium mixtures of aldehydes and alcohols, and the carbon atoms in glycosidic bonds cannot be oxidized by Benedict's, Fehling's, or Tollen's reagents. However, glycosides can be hydrolyzed and the glycosidic bond broken in reactions catalyzed by an acid or by specific enzymes. An example is the acid-catalyzed hydrolysis of ethyl-β-D-riboside:

(20-2)

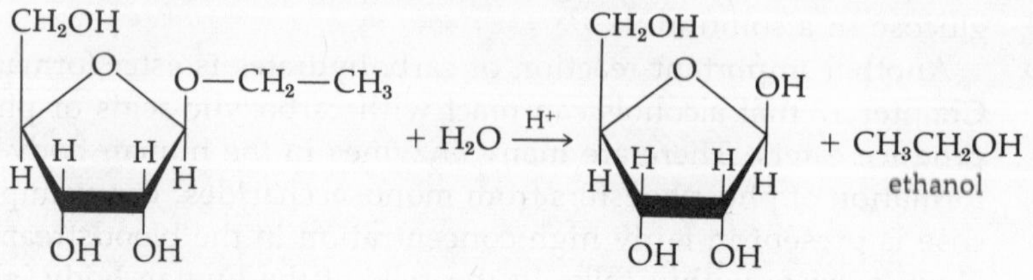

Glycosidic bonds can likewise be formed when the anomeric hydroxyl group of one monosaccharide and an alcohol functional group of another monosaccharide react. The joining of two monosaccharide molecules by a glycosidic bond(s) results in the formation of a disaccharide. Polysaccharides are formed by joining three or more monosaccharide molecules with glycosidic bonds.

Exercise 20-5
Draw Haworth projections for the products of the following reactions.

(a)

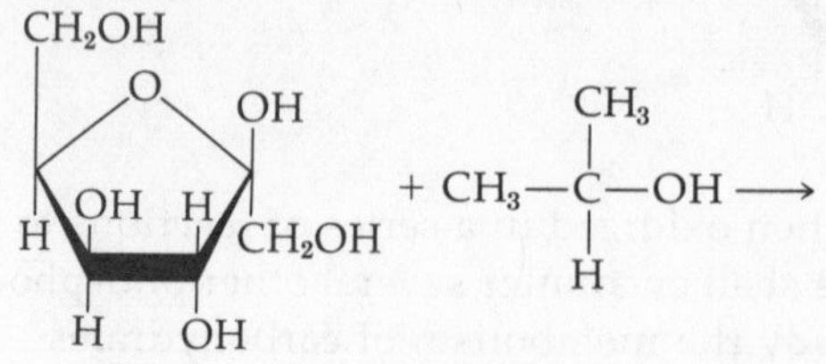

(b)

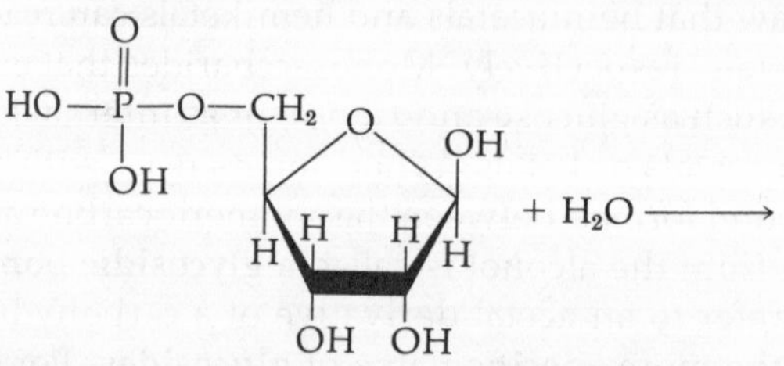

20-6 IMPORTANT DISACCHARIDES

Disaccharides consist of two monosaccharide units joined by one or two glycosidic bonds. Disaccharides in which the units are joined by one glycosidic bond are reducing sugars. Disaccharides joined by two glycosidic bonds that involve the anomeric carbon atoms on each monosaccharide unit are not reducing sugars. The disaccharides maltose, lactose, and sucrose are very important to humans.

Maltose is a disaccharide produced upon incomplete hydrolysis of the polysaccharide starch. It is found in germinating seeds. It is also produced commercially for use in the production of beer. Maltose is composed of two D-glucose units joined by an α-glucosidic bond between the anomeric carbon of one glucose unit and the number 4 carbon of the other glucose unit. This specific bond, termed an **α-1,4-glucosidic bond,** is also found in starch and glycogen:

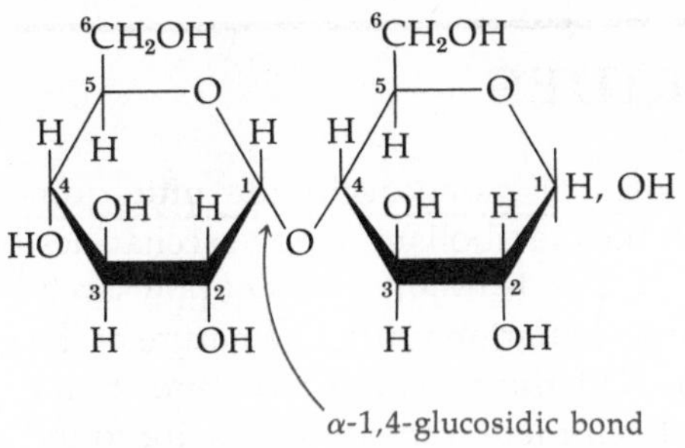

α-1,4-glucosidic bond

Note that the anomeric hydroxyl group of one of the glucose units participates in the glucosidic bond and therefore cannot be easily oxidized. However, the anomeric hydroxyl of the other glucose unit is not so occupied, and this glucose unit exists in equilibrium with the free aldehyde in solution. Thus maltose is oxidized by Benedict's, Fehling's, or Tollen's reagent and is a reducing sugar.

Lactose constitutes some 3% to 5% of the milk of mammals, including cows and humans. This disaccharide is composed of one galactose unit and one glucose unit joined by a glycosidic bond between the β-anomer of galactose and the number 4 carbon of glucose, a β-1,4-glycosidic bond:

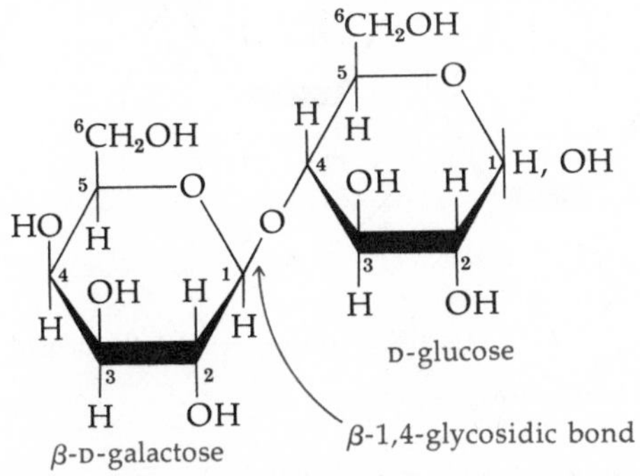

The glucose unit of lactose still exists as an equilibrium mixture of the α- and β-anomers and the free aldehyde in solution. Lactose is thus a reducing sugar.

Sucrose is found in honey, fruits, and vegetables. Sugar cane and sugar beets are the commercial sources for sucrose used as table sugar. Sucrose is composed of one glucose and one fructose unit, joined by two glycosidic bonds involving the anomeric carbons of both α-D-glucose and β-D-fructose:

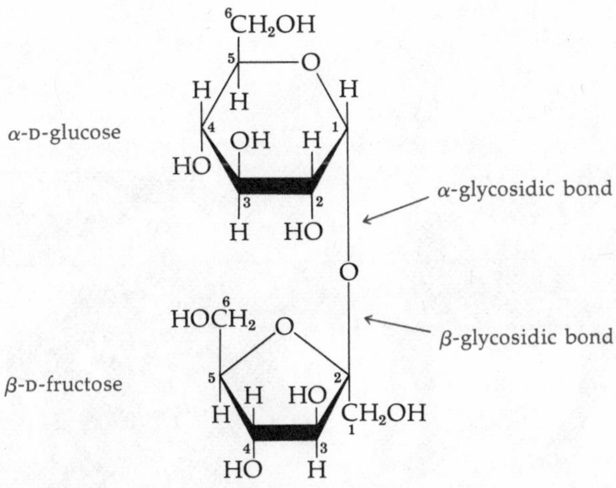

Sucrose is not a reducing sugar, since both anomeric carbons participate in the glycosidic bonds and thus no free aldehyde or ketone exists in solution. The metabolism of lactose and sucrose will be discussed in Chapter 25.

20-7 POLYSACCHARIDES

Polysaccharides are polymers of monosaccharide units joined by glycosidic bonds, and the first step in the metabolism of polysaccharides by humans is the hydrolysis of their glycosidic bonds. Many types of polysaccharides exist in nature, differing in their monosaccharide units, the nature of the glycosidic bonds that join the sugar units, and their overall structure. Some polysaccharides function as components of cell membranes. Others are found attached to proteins. Three of the most important polysaccharides found in nature are cellulose, starch, and glycogen. All three of these polysaccharides are built up from a single monosaccharide component, D-glucose. The difference between them is in the type of glucosidic bonds and in their overall structures.

Cellulose is a very large polymer of glucose units joined by β-1,4-glucosidic bonds in long chains:

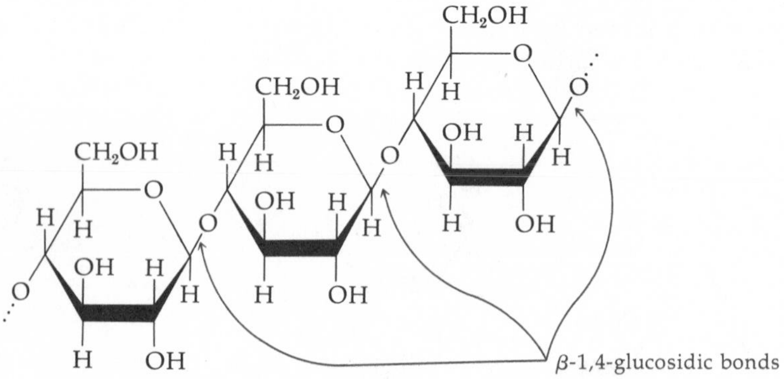

cellulose

β-1,4-glucosidic bonds

One cellulose molecule may contain up to 10,000 glucose units. Because of the very large molecular weight of cellulose molecules, cellulose is insoluble in water. Cellulose is used by plants to form rigid cell walls. As we have already seen, there are many commercial uses for plant cellulose.

The cellulose found in grasses can be digested by herbivorous animals such as this horse, but the men and their dog show no appetite for this vegetation, which they cannot digest.

Humans cannot digest cellulose. We do not have any enzymes capable of hydrolyzing a β-1,4-glucosidic bond between two glucose units. Thus, the cellulose in our diet passes through the digestive tract intact. Note, however, that humans do possess an enzyme capable of hydrolyzing the β-1,4-glycosidic bond between the galactose and glucose units of lactose.

Starch is also a very large polymer of glucose units. However, the bonds that hold the glucose units together are α-1,4- and not β-1,4- as in cellulose. Starch is used by plants for food storage. There are two basic types of starch, amylose and amylopectin (see Figure 20-9). **Amylose** is a linear polymer of α-1,4-linked glucose units, and is water soluble. **Amylopectin** is a branched polymer. The main chains of amylopectin are joined by α-1,4-glucosidic bonds, as in amylose. However, about every 20 to 30 glucose units there are branches joined by α-1,6-glucosidic bonds. Amylopectin is not soluble in water. About 80% to 90% of the starch obtained from plants is amylopectin. These values, 20 to 30 glucose units and 80% to 90% amylopectin, are averages from a wide variety of plants. Different specific starches are made by different plants.

Humans possess the enzyme **amylase,** which catalyzes the hydrolysis of the α-1,4-glucosidic bonds in starch, and another enzyme that can break the α-1,6-bonds. As a consequence of the action of these enzymes, starch is completely hydrolyzed to individual glucose units during digestion. The salivary glands and the pancreas produce amylase, releasing it into the mouth and intestines, respectively. The free glucose is then absorbed by cells in the small intestine and transferred to the bloodstream.

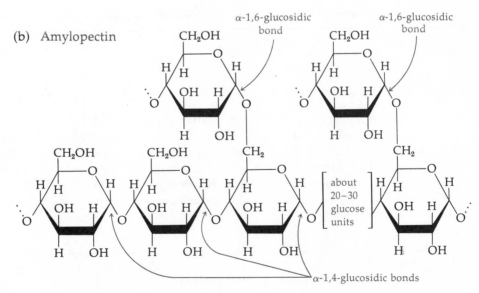

Figure 20-9   Amylose (a) and amylopectin (b) are two different forms of starch. Both have linear chains of glucose units joined by α-1,4-glucosidic bonds. However, amylopectin also has branch chains that are joined to the main chain by α-1,6-glucosidic bonds.

Glycogen is very similar to amylopectin in structure. It is a polymer of glucose units joined by α-1,4-bonds, with α-1,6-bonded branches:

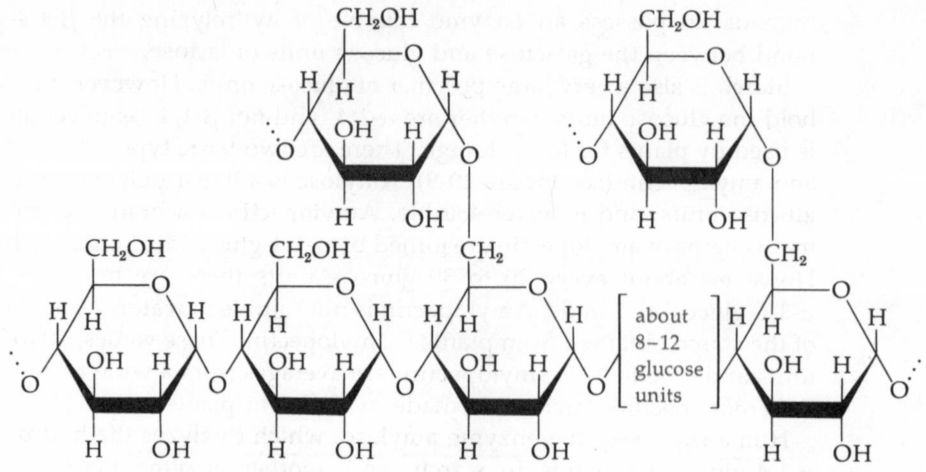

glycogen

However, it is more highly branched than amylopectin, with one branch point about every 8 to 12 glucose units. Glycogen is used for food storage in animal cells. The normal human body has about a one-day supply of stored glycogen. We shall study the metabolism of glycogen in Chapter 25.

Complex polysaccharides are carbohydrate polymers composed of more than one type of monosaccharide unit joined by glycosidic bonds. Some of the individual monosaccharide units that make up complex polysaccharides can also be fairly complex. For example, glucosamine and N-acetylgalactosamine are monosaccharides found in several complex polysaccharides:

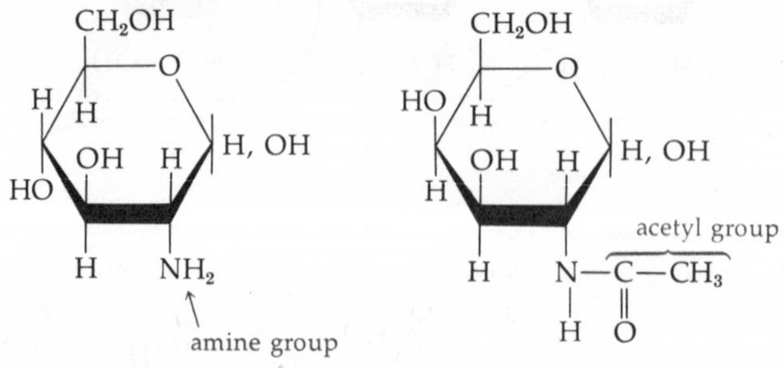

glucosamine   N-acetylgalactosamine

Although several complex polysaccharides are found in the human body, their exact functions are not clearly understood. Some complex polysaccharides are found attached to proteins, including some enzymes and antibodies (Chapter 22). Other complex polysaccharides are found on cell membranes, probably also attached to proteins. For example, the ABO and Rh blood-group antigens found on red blood cell membranes are known to be specific complex polysaccharides.

Exercise 20-6
How do cellulose and glycogen differ in structure?

Exercise 20-7
Would you expect amylopectin to reduce Benedict's reagent?

20-8 SUMMARY

1. Carbohydrates (also called saccharides or sugars) are polyhydroxy aldehydes or ketones or compounds that can be hydrolyzed to give polyhydroxy aldehydes or ketones.

2. Carbohydrates that cannot be hydrolyzed into simpler compounds are called monosaccharides.

3. Aldehyde monosaccharides are called aldoses, whereas ketone monosaccharides are called ketoses.

4. Many monosaccharides have the general formula $C_n(H_2O)_n$, where n is 3 or larger; in general, there are many stereoisomers of a particular monosaccharide.

5. The prefixes D- and L- are used to distinguish between the enantiomers of an enantiomeric pair of monosaccharides. Humans can metabolize only D-sugars.

6. Pentoses and hexoses form cyclic internal hemiacetals or hemiketals whose shapes can be represented by Haworth projections.

7. When a monosaccharide forms an internal hemiacetal or hemiketal, two different diastereomers, α- and β-anomers, can form. The carbonyl carbon used to form the hemiacetal or hemiketal becomes asymmetric and is called the anomeric carbon.

8. A monosaccharide exists in solution as an equilibrium mixture of the α-anomer and the β-anomer, with a small percentage in the open-chain form.

9. The monosaccharide D-glucose is the most abundant organic compound in nature. It is also called dextrose or blood sugar. D-Glucose is the monomer unit (building block) for the polysaccharides cellulose, starch, and glycogen, and is a component of the disaccharides maltose, sucrose, and lactose.

10. D-Galactose, D-fructose, and D-ribose are other monosaccharides vital to living organisms.

11. Monosaccharides are hydrophilic and are very soluble in water.

12. Monosaccharides and disaccharides that can be oxidized by Benedict's, Fehling's, or Tollen's reagent are called reducing sugars.

13. Monosaccharides can form esters with carboxylic or phosphoric acids. Phosphate esters of monosaccharides are very important in the human body.

14. Monosaccharides can react with alcohols to produce acetals or ketals that are called glycosides. Glycosides of glucose are called glucosides. Glycosidic bonds join monosaccharide units together to form disaccharides and polysaccharides.

15. Three important disaccharides are maltose (two α-1,4-bonded glucose units), lactose (a galactose joined to the number 4 carbon of glucose by a β-1,4-glycosidic bond), and sucrose (α-glucose joined to β-fructose).

16. Cellulose, starch, and glycogen are three important polysaccharides. Cellulose is a polymer of glucose units joined by β-1,4-glucosidic bonds. The glucosidic bonds in the polymers starch and glycogen are primarily α-1,4-bonds. In addition, amylopectin and glycogen contain α-1,6-branch points.

PROBLEMS

1. Draw Haworth projections for the indicated anomers of the following sugars:

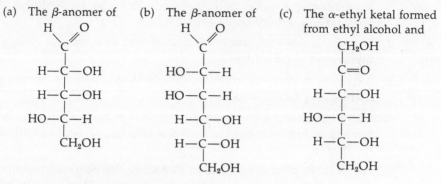

(a) The β-anomer of

$$\begin{array}{c} H\diagdown \quad O \\ \,\,\,\,\,\,\,\,\, C \\ H-C-OH \\ H-C-OH \\ HO-C-H \\ CH_2OH \end{array}$$

(b) The β-anomer of

$$\begin{array}{c} H\diagdown \quad O \\ \,\,\,\,\,\,\,\,\, C \\ HO-C-H \\ HO-C-H \\ H-C-OH \\ H-C-OH \\ CH_2OH \end{array}$$

(c) The α-ethyl ketal formed from ethyl alcohol and

$$\begin{array}{c} CH_2OH \\ C=O \\ H-C-OH \\ HO-C-H \\ H-C-OH \\ CH_2OH \end{array}$$

2. Draw Haworth projections for the following:
   (a) β-D-Fructose 1,6-diphosphate   (b) Methyl-α-D-riboside

3. Draw Fischer projections for all of the tetroses.

4. Draw Haworth projections for a short (three or four glucose units) segment of
   (a) Amylose   (b) Cellulose

5. What is an anomeric carbon atom?

6. Draw Haworth projections for maltose and sucrose. Identify the glycosidic bonds and indicate if these substances are reducing sugars.

7. Draw structural formulas or Haworth projections for the products of the following reactions. If no reaction occurs, write N.R.

(a)

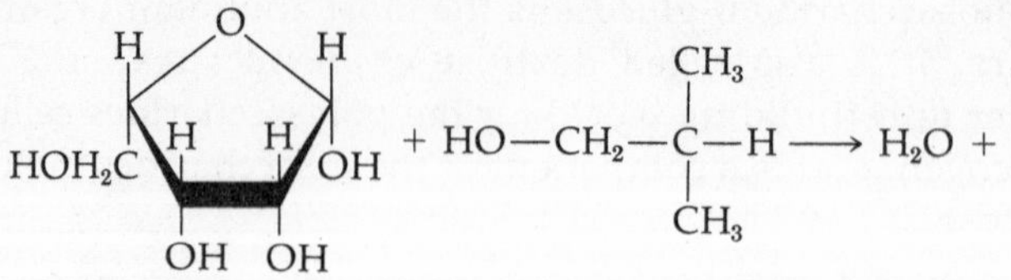

$$HOH_2C \quad \begin{array}{c} H \quad O \quad H \\ H \quad H \\ OH \\ OH \quad OH \end{array} \quad + \; HO-CH_2-\overset{\displaystyle CH_3}{\underset{\displaystyle CH_3}{C}}-H \longrightarrow H_2O \; +$$

(b)

$$\begin{array}{c} H\diagdown \quad O \\ \,\,\,\,\,\,\,\,\, C \\ H-C-OH \\ CH_2OH \end{array} \quad + \; H_2O \longrightarrow$$

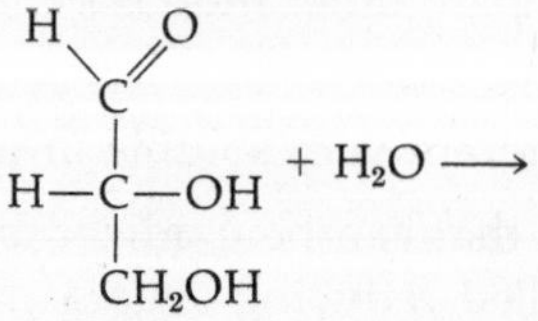

8. Draw Haworth projections for all of the D-aldopentoses.

9. Draw Fischer projections for all of the monosaccharide isomers of D-glyceraldehyde.

SOLUTIONS TO EXERCISES

20-1
$$\begin{array}{c} H\diagdown \quad O \\ \,\,\,\,\,\,\,\,\, C \\ H-C-OH \\ H-C-OH \\ H-C-OH \\ H-C-OH \\ CH_2OH \end{array}$$

$$\begin{array}{c} H\diagdown \quad O \\ \,\,\,\,\,\,\,\,\, C \\ H-C-OH \\ H-C-OH \\ HO-C-H \\ H-C-OH \\ CH_2OH \end{array}$$

$$\begin{array}{c} H\diagdown \quad O \\ \,\,\,\,\,\,\,\,\, C \\ H-C-OH \\ HO-C-H \\ HO-C-H \\ H-C-OH \\ CH_2OH \end{array}$$

$$\begin{array}{c} H\diagdown \quad O \\ \,\,\,\,\,\,\,\,\, C \\ HO-C-H \\ HO-C-H \\ HO-C-H \\ H-C-OH \\ CH_2OH \end{array}$$

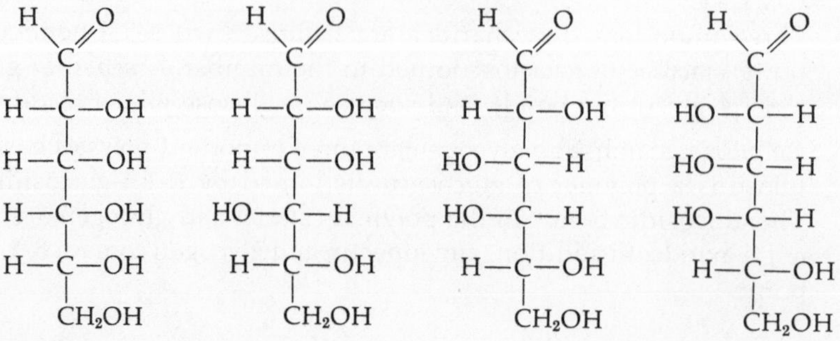

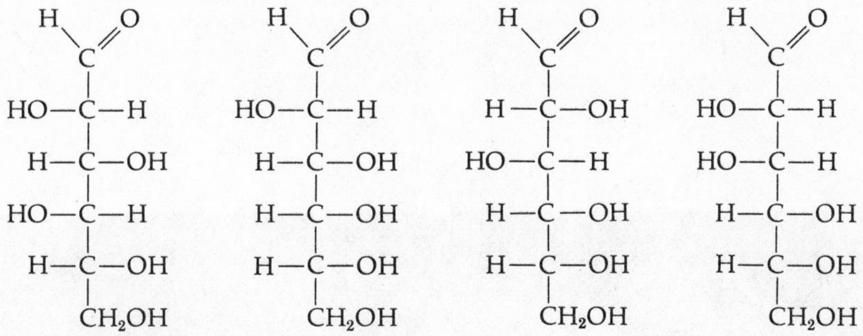

H   O
 C
HO—C—H
H—C—OH
HO—C—H
H—C—OH
CH₂OH

H   O
 C
HO—C—H
H—C—OH
H—C—OH
H—C—OH
CH₂OH

H   O
 C
H—C—OH
HO—C—H
H—C—OH
H—C—OH
CH₂OH

H   O
 C
HO—C—H
HO—C—H
H—C—OH
H—C—OH
CH₂OH

20-2

CH₂OH
O
HO H   OH
OH H
H   H
H   OH

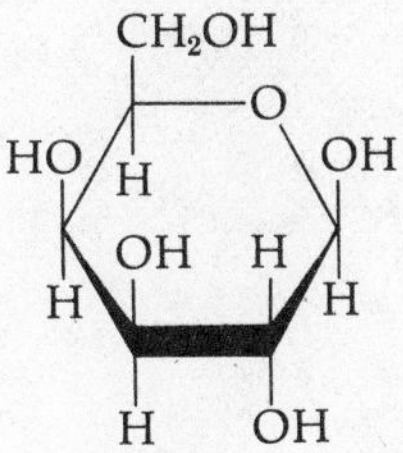

20-3 (a) (1) Aldopentose (2) Aldohexose (3) Ketohexose

(b) (1)

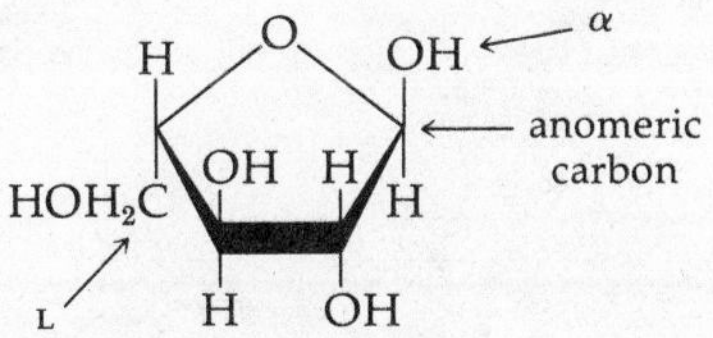

H   O   OH ← α
      ← anomeric carbon
OH H
HOH₂C   H
L   H   OH

(2)

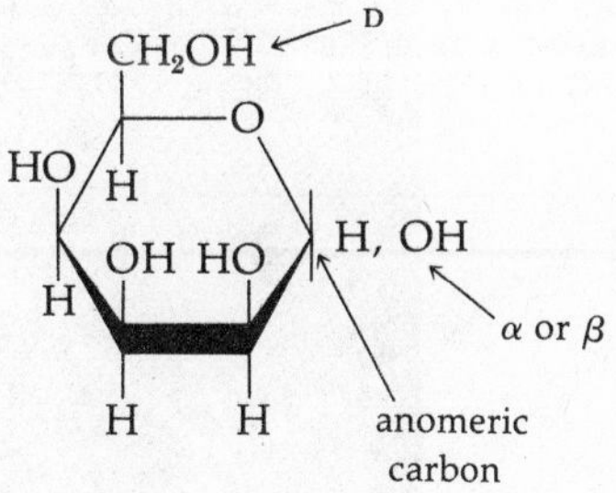

CH₂OH ← D
O
HO H
OH HO   H, OH
H   ← α or β
H   H   anomeric carbon

(3)

H   O   CH₂OH
      ← anomeric carbon
OH HO
HOH₂C   OH
L   H   H   β

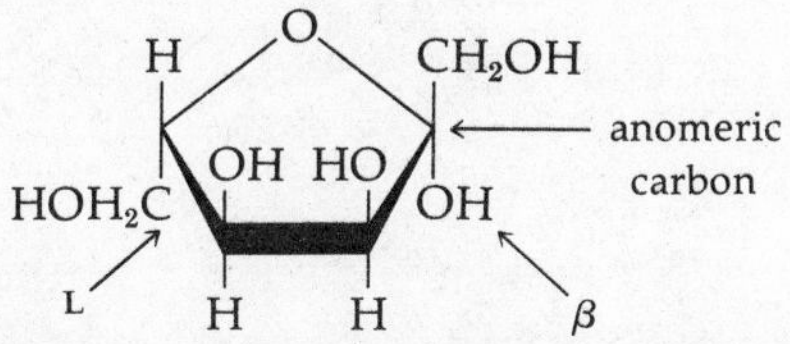

20-4 No, because D-ribose is an aldopentose, not an aldohexose; it is therefore not an isomer of glucose.

20-5 (a)

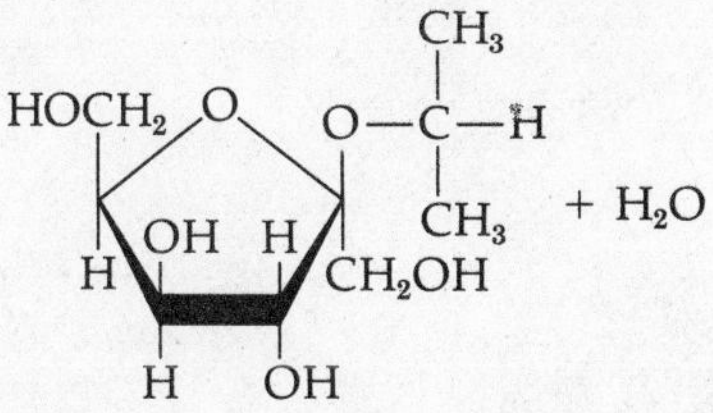

CH₃
HOCH₂ O   O—C—H
OH H   CH₃   + H₂O
H   CH₂OH
H   OH

(b)

HOCH₂ O   OH
H   H
H   H   + HO—P—OH
OH OH   OH
O

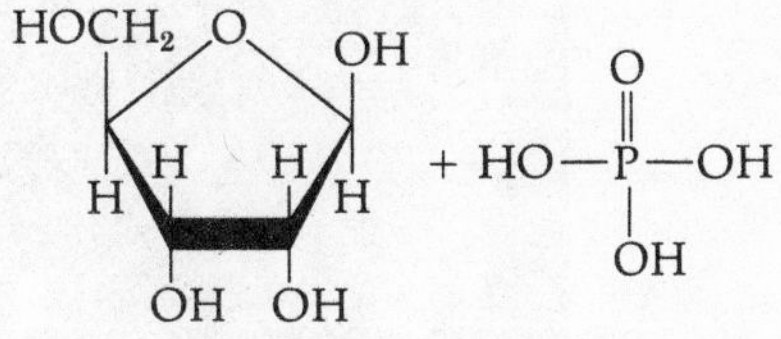

20-6 Cellulose contains only β-1,4-glucosidic bonds. Glycogen contains α-1,4-glucosidic bonds with α-1,6-branch points.

20-7 One molecule of amylopectin is composed of a very large number of glucose units. However, only one of them has a free anomeric hydroxyl group. Therefore amylopectin is not a reducing sugar, since the sole glucose unit that can be oxidized is negligible compared to the entire molecule.

CHAPTER 21

Amino Acids and Protein Structure

21-1 INTRODUCTION

The human body is largely made up of proteins. In fact, second only to water, proteins are the most plentiful chemical compounds in the human body. Protein molecules are complex—much more complex than water—and the body contains thousands of different ones. Hair, connective tissue, fingernails, and tendons are composed of proteins. Enzymes, the body's catalysts, are proteins. Antibodies, the molecules of immunity, are also proteins, as are a very large number of other important molecules. Proteins are vital to the functioning of cellular factories and to the integrated operation of all body organs.

It is not feasible to examine the detailed structure and function of each of the thousands of proteins in the human body. In fact, little is known about the structure of many proteins. In this chapter we shall examine the common structural features and properties of proteins. This will provide us with some insight into the relationship between protein structure and function.

Proteins are polymers. In the last chapter we saw that polysaccharides such as starch and cellulose are polymers composed of monosaccharide building blocks. The building blocks for proteins are called amino acids. Starch is composed of only a single building block, glucose, but there are 20 different amino acids used to assemble proteins. For example, hair consists predominantly of a type of protein called keratin. The properties of the keratin in hair are due to the unique arrangement, or sequence, of its component amino acids, which is different from the sequence of amino acids in every other protein in the body.

It is natural to begin our discussion of proteins by studying their amino acid building blocks. We shall then examine the way in which these amino acid building blocks are bonded together to form proteins. Only then can we study the forces that are responsible for the unique three-dimensional structure and properties of various proteins. We shall then be able to see why hair forms long strands, why it curls, and even the chemistry of permanent waving. More important, we shall have a basis for understanding the vital functions of enzymes, hemoglobin, protein hormones, and other proteins in the body.

The spatial relationship of the dewdrops on this spiderweb, intricate as it is, can only begin to convey the complexity of the arrangement of amino acid components in a protein. The structure and function of proteins are critically dependent on the precise arrangement of their component amino acids.

21-2 STUDY OBJECTIVES

After careful study of the material in this chapter, you should be able to:

1. Write the general formula for an α-amino acid.

2. Draw a peptide group and identify a peptide bond.

3. Define the terms peptide, protein, amino acid sequence, essential amino acid, zwitterion, N-terminal, C-terminal, hydrophobic core, oligomer, subunit, and isoenzyme.

4. Classify an amino acid as acidic, basic, neutral hydrophilic, or neutral hydrophobic, given the structural formula of its side-chain group.

5. Discuss the acid-base properties of a given amino acid and determine which form will predominate at low, neutral, or high pH.

6. Draw the structural formula for a peptide, given its abbreviated representation and the structural formulas for its component amino acids.

7. Define the terms primary, secondary, tertiary, and quaternary protein structure.

8. Describe the involvement of hydrogen bonds, disulfide bonds, hydrophilic and hydrophobic interactions, and salt bridges in maintaining protein shape.

9. Draw and describe an α-helix and a β-pleated sheet.

10. Discuss the process of protein denaturation and differentiate between reversible denaturation and coagulation of a protein.

21-3 PROTEIN POLYMERS

Proteins are polymers composed of amino acid building blocks that are joined together by amide bonds. The building blocks (monomers) for proteins are **α-amino acids,** so-called because each monomer possesses an amine functional group located on the carbon atom immediately adjacent (α) to a carboxyl functional group (see Figure 21-1). All 20 of the α-amino acids used to make proteins have the same general structure, but each has a different R' group. The R' group in an amino acid is often referred to as its **side-chain group.**

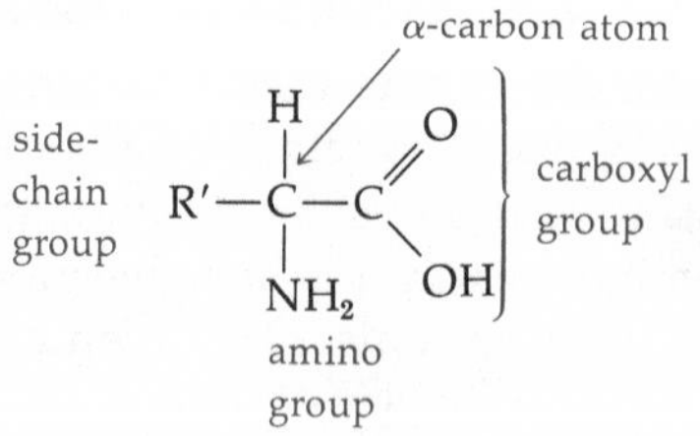

Figure 21-1   The general structural formula for an α-amino acid. The —NH$_2$ portion of a primary amine functional group is called an amino group.

In a protein, the carboxyl group of one amino acid is coupled to the amine group of the next amino acid by an **amide bond.** In Chapter 18 we saw that amides are formed by the reaction of a carboxylic acid and an amine:

(21-1)

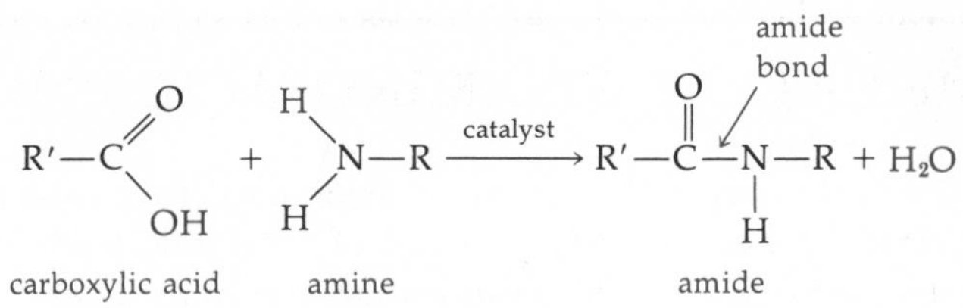

We also saw in Chapter 18 that amide bonds join the diamine and dicarboxylic acid monomer units together in the synthetic polymer nylon. Nylon has an ordered sequence of monomer units and is made simply by catalyzing the formation of amide bonds in a vat filled with a diamine and a dicarboxylic acid (reaction 18-8). Recall that the monomer units in nylon can link up in only one way, that is, as alternate dicarboxylic acid and diamine units. Proteins, however, are quite different. Since each amino acid possesses *both* of the functional groups needed to form an amide bond, and 20 different amino acids can be used, an enormous number of different proteins—each with a different sequence of amino acids—is possible.

In a protein, the amide bond between the α-carboxyl group of one amino acid and the α-amino group of another amino acid is usually called a **peptide bond.** Figure 21-2 shows a short string of four amino acids joined by peptide bonds. Short strings of amino acids such as this are called *peptides* to distinguish them from the very long strings of amino acids called *proteins*. Peptides are sometimes classified according to the number of amino acids they contain. A peptide composed of two amino acids that are joined by a single peptide bond is called a *dipeptide*. A string of three amino acids joined by two peptide bonds is called a *tripeptide*. Figure 21-2 represents a *tetrapeptide*. Usually strings of fewer than 100 amino acids are called **peptides,** whereas longer strings are called **proteins.** The line separating peptides and proteins is somewhat arbitrary, but very long strings of amino acids are always called proteins.

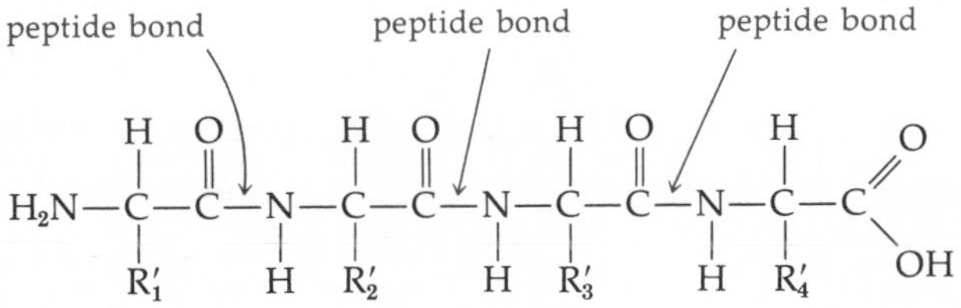

Figure 21-2   The general structural formula for a tetrapeptide.

The sequence of amino acids in a protein (or in a peptide) can be specified simply by indicating the order of the R' groups in the protein (or peptide). For example, R_1'—R_2'—R_3'—R_4' refers to the sequence of amino acids in the tetrapeptide shown in Figure 21-2. These 4 different amino acids can form a total of 24 different tetrapeptides with different amino acid sequences (that is, R_2'—R_1'—R_4'—R_3', R_4'—R_3'—R_2'—R_1', etc.). We shall see that the exact sequence of amino acids in a peptide or a protein is crucial to its overall shape and therefore to its function. Because the sequence is so critical, the manufacture of each protein in the human body is carefully controlled so that the amino acids can be assembled in just the right sequence (see Chapter 24).

Exercise 21-1

A tripeptide composed of three different amino acids has the sequence R_1'—R_2'—R_3'. Write the sequence for all the other tripeptides containing R_1', R_2', and R_3'.

21-4 CLASSIFICATION OF AMINO ACIDS

It is very useful to divide the 20 amino acids used for protein synthesis into groups or classes based on the properties of their side-chain (R′) groups. We shall divide amino acids into four classes: neutral hydrophobic, neutral hydrophilic, acidic, and basic. Although other schemes are also used to classify these amino acids, this one is simple and will help us to understand how the amino acid side chains of a protein interact with each other and with other molecules. The name, abbreviation, structural formula, and classification of each of the 20 amino acids are given in Table 21-1.

Neutral Amino Acids

Fifteen of the 20 amino acids are classified as **neutral** because their side chains have no charge at the pH of body cells (which is about 7). These neutral amino acids are further subdivided according to how the side-chain group interacts with water molecules.

Table 21-1 The Amino Acids

A. Neutral Hydrophobic Amino Acids

R′ groups

Alanine
ala

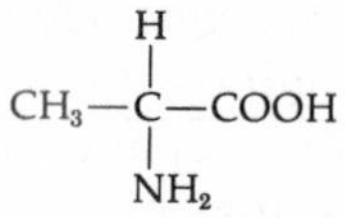

Valine
val

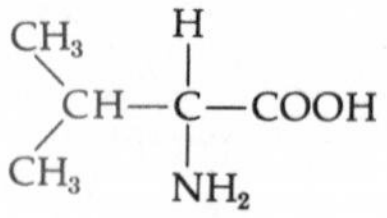

Leucine
leu

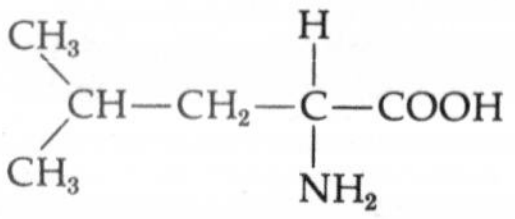

Isoleucine
ile

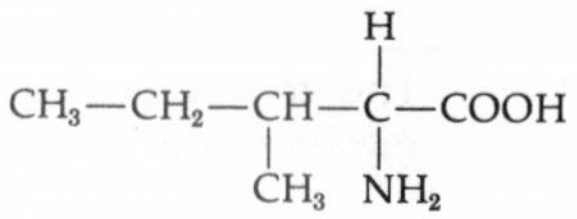

Proline
pro

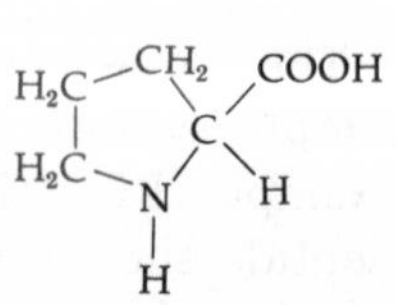

Phenylalanine
phe

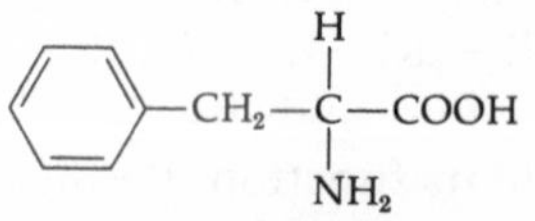

Tryptophan
trp

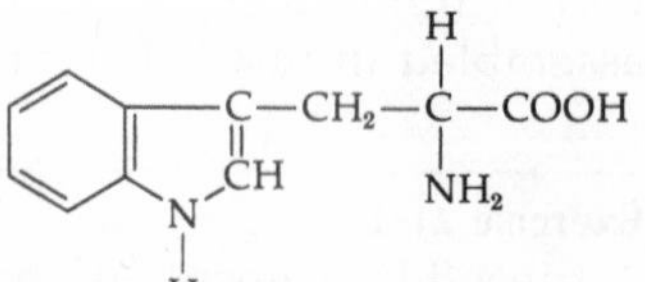

NEUTRAL HYDROPHOBIC AMINO ACIDS

Seven neutral amino acids have side chains (R') that are nonpolar or hydrophobic (Table 21-1A). These hydrophobic groups are either alkyl or aromatic in nature. The side chain of alanine (ala) is a methyl group, that of valine (val) is an isopropyl group, leucine (leu) has an isobutyl group, and isoleucine (ile) has a sec-butyl group. The amino acid proline (pro) is unusual in that the alkyl side chain is bonded to both the α-carbon and the α-amino group, forming a five-atom ring. The side chain of the amino acid phenylalanine (phe) is the aromatic benzyl group. The amino acid tryptophan (trp) has a rather complex aromatic side chain involving two rings.

NEUTRAL HYDROPHILIC AMINO ACIDS

Eight of the neutral amino acids are classified as hydrophilic (Table 21-1B). In general, these amino acids are more soluble in water than hydrophobic amino acids. The side chain of glycine (gly) is just a hydrogen atom. The other seven neutral hydrophilic amino acids have side chains that can form either strong or weak hydrogen bonds with water. Three have a hydroxyl group in their side chains: serine (ser), threonine (thr), and tyrosine (tyr). Two contain an amide functional group: asparagine (asn) and glutamine (gln). The remaining two contain a sulfur atom: cysteine (cys) and methionine (met).

B. Neutral Hydrophilic Amino Acids

R' groups

Glycine gly	$H-\overset{\overset{\displaystyle H}{	}}{\underset{\underset{\displaystyle NH_2}{	}}{C}}-COOH$
Serine ser	$HO-CH_2-\overset{\overset{\displaystyle H}{	}}{\underset{\underset{\displaystyle NH_2}{	}}{C}}-COOH$
Threonine thr	$CH_3-\overset{\overset{\displaystyle OH}{|}}{\underset{\underset{\displaystyle H}{|}}{C}}-\overset{\overset{\displaystyle H}{|}}{\underset{\underset{\displaystyle NH_2}{|}}{C}}-COOH$
Tyrosine tyr	$HO-\langle\bigcirc\rangle-CH_2-\overset{\overset{\displaystyle H}{	}}{\underset{\underset{\displaystyle NH_2}{	}}{C}}-COOH$
Asparagine asn	$\underset{O}{\overset{NH_2}{>}}C-CH_2-\overset{\overset{\displaystyle H}{	}}{\underset{\underset{\displaystyle NH_2}{	}}{C}}-COOH$
Glutamine gln	$\underset{O}{\overset{NH_2}{>}}C-CH_2-CH_2-\overset{\overset{\displaystyle H}{	}}{\underset{\underset{\displaystyle NH_2}{	}}{C}}-COOH$
Cysteine cys	$HS-CH_2-\overset{\overset{\displaystyle H}{	}}{\underset{\underset{\displaystyle NH_2}{	}}{C}}-COOH$
Methionine met	$CH_3-S-CH_2-CH_2-\overset{\overset{\displaystyle H}{	}}{\underset{\underset{\displaystyle NH_2}{	}}{C}}-COOH$

Acidic Amino Acids

Acidic amino acids have side chains that contain a second carboxyl group (Table 21-1C). The side chain of aspartic acid (asp) is $-CH_2-COOH$ and that of glutamic acid (glu) is $-CH_2-CH_2-COOH$. At the pH of cells in the body, these carboxyl groups exist primarily as negatively charged carboxylate ions and thus interact strongly with water molecules.

Table 21-1 (continued)

C. Acidic Amino Acids

R' groups

Aspartic acid asp	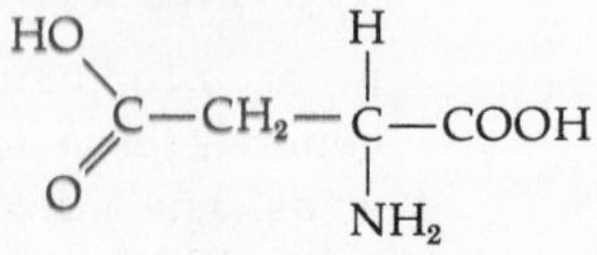 $\underset{\displaystyle O}{\overset{\displaystyle HO}{>}}C-CH_2-\underset{NH_2}{\overset{H}{\underset{\mid}{\overset{\mid}{C}}}}-COOH$
Glutamic acid glu	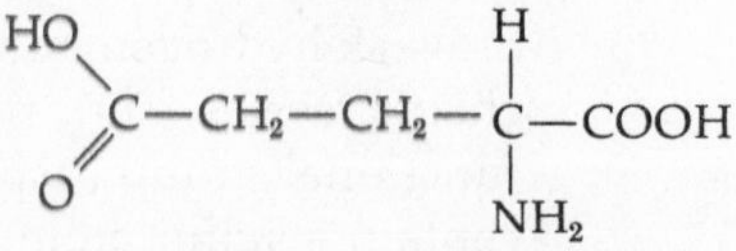 $\underset{\displaystyle O}{\overset{\displaystyle HO}{>}}C-CH_2-CH_2-\underset{NH_2}{\overset{H}{\underset{\mid}{\overset{\mid}{C}}}}-COOH$

Basic Amino Acids

Three of the amino acids contain a side chain that can act as a proton acceptor or base; they are thus classified as basic amino acids (Table 21-1D). The basic amino acids are lysine (lys), arginine (arg), and histidine (his). The basic properties of these three amino acids will be discussed in Section 21-5.

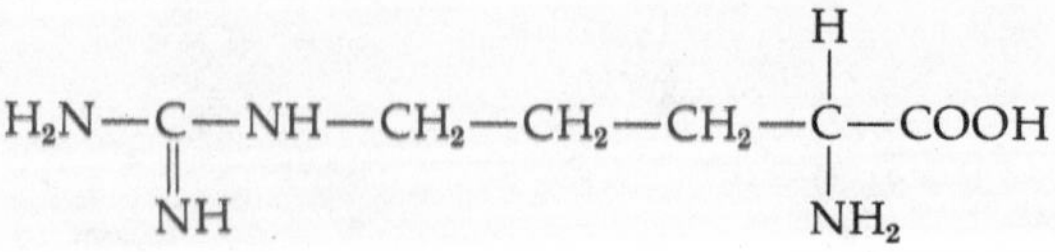

Table 21-1 (continued)

D. Basic Amino Acids

R' groups

Lysine lys	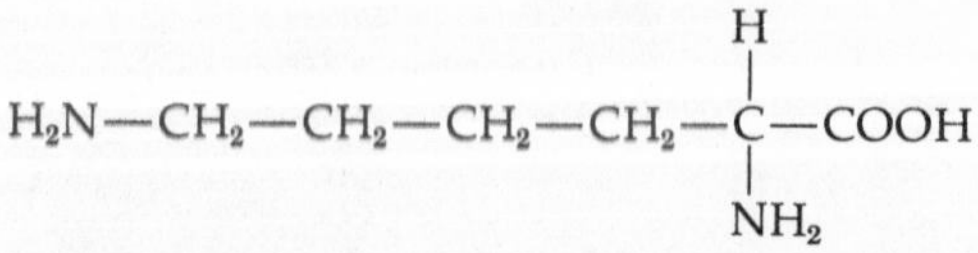 $H_2N-CH_2-CH_2-CH_2-CH_2-\underset{NH_2}{\overset{H}{\underset{\mid}{\overset{\mid}{C}}}}-COOH$
Arginine arg	$H_2N-\underset{NH}{\overset{\parallel}{C}}-NH-CH_2-CH_2-CH_2-\underset{NH_2}{\overset{H}{\underset{\mid}{\overset{\mid}{C}}}}-COOH$
Histidine his	$HC\!=\!\!=\!\!C-CH_2-\underset{NH_2}{\overset{H}{\underset{\mid}{\overset{\mid}{C}}}}-COOH$

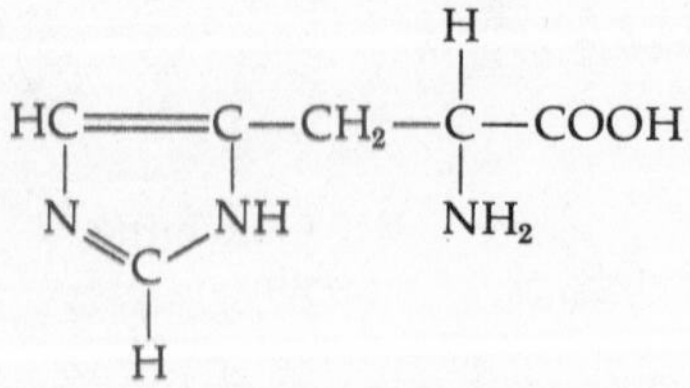

Derivatives of Amino Acids

The amino acids listed in Table 21-1 are the *only* amino acids used by living cells to manufacture proteins. Once certain proteins have been made, however, the side chains of a few of their component amino acids are sometimes modified by further chemical reactions.

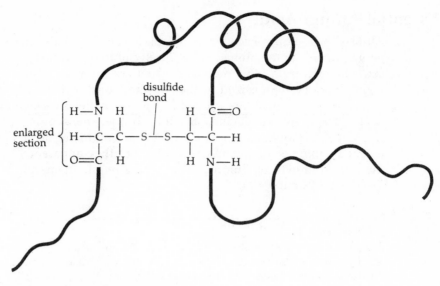

Figure 21-3   Schematic representation of a protein containing a disulfide bond between two component cysteines.

Recall that two thiols can be easily oxidized to form a disulfide bond (see Chapter 18). Frequently the thiol groups in the side chains of two cysteines (R' = —CH$_2$—SH) in the same protein combine to form a covalent disulfide bond between the two sulfur atoms (see Figure 21-3). If two cysteine molecules that are not part of a protein link up by forming a disulfide bond, the product is called cystine:

cystine

In some connective tissue proteins, the side chains of the amino acids lysine and proline are modified by specific enzymes to produce hydroxylysine and hydroxyproline:

hydroxylysine   hydroxyproline

The side-chain functional groups of some amino acids, especially the carboxyl groups of aspartic acid and glutamic acid and the hydroxyl groups of serine and threonine, can also react with a variety of nonprotein molecules, including carbohydrates and phosphoric acid.

A number of the α-amino acids used for protein synthesis are also used as starting materials for the manufacture of a variety of small molecules with very interesting functions, such as the hormone adrenaline (see Chapter 28).

Essential Amino Acids

All 20 of the α-amino acids listed in Table 21-1 are needed to make the many different proteins in the human body. Twelve of these amino acids can be synthesized by cells from other substances that are present in the body. The other eight cannot be synthesized by the body and must be included in the diet. The amino acids that must be included in a person's diet are called **essential amino acids.** They are isoleucine, leucine, lysine, methionine, phenylalanine, threonine, tryptophan, and valine. Arginine and histidine do not appear to be essential amino acids for human adults, but they apparently are essential for the normal growth of children. The dietary requirements for amino acids are discussed in Chapter 29.

Exercise 21-2

Draw the structural formulas for: (a) two acidic amino acids; (b) two basic amino acids; and (c) two neutral amino acids. In each case, indicate the functional group responsible for its classification as acidic, basic, or neutral.

Exercise 21-3

Classify each of the following amino acids as either neutral hydrophilic or neutral hydrophobic.

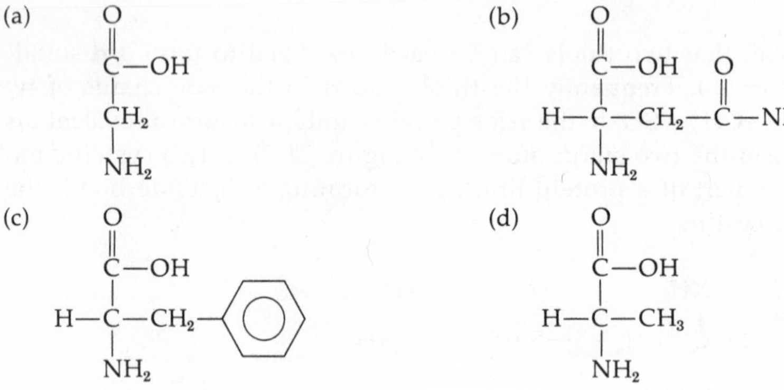

(a)
$$\begin{array}{c} O \\ \| \\ C-OH \\ | \\ CH_2 \\ | \\ NH_2 \end{array}$$

(b)
$$\begin{array}{cc} O & \\ \| & \\ C-OH & O \\ | & \| \\ H-C-CH_2-C-NH_2 \\ | & \\ NH_2 & \end{array}$$

(c)
$$\begin{array}{c} O \\ \| \\ C-OH \\ | \\ H-C-CH_2- \bigcirc \\ | \\ NH_2 \end{array}$$

(d)
$$\begin{array}{c} O \\ \| \\ C-OH \\ | \\ H-C-CH_3 \\ | \\ NH_2 \end{array}$$

21-5   PROPERTIES OF AMINO ACIDS

Optical Activity

All of the α-amino acids, with the exception of glycine, possess an asymmetric carbon (the α-carbon) and hence are optically active (see Figure 21-4). Only the L-isomers of the amino acids are synthesized and used by humans. In fact, all organisms in nature use only L-amino acids for protein synthesis. The few D-amino acids found in nature are used by some bacteria to form part of their cell wall.

Figure 21-4   Fischer projections for the D- and L-enantiomers of amino acids. The L-enantiomers of the amino acids are the predominant ones found in nature. D-Amino acids are found only in some bacteria.

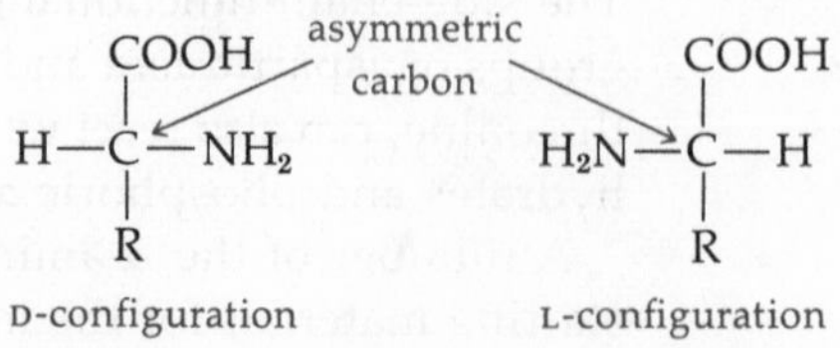

$$\begin{array}{ccc} & \text{asymmetric} & \\ \text{COOH} & \text{carbon} & \text{COOH} \\ | & & | \\ H-C-NH_2 & & H_2N-C-H \\ | & & | \\ R & & R \\ \text{D-configuration} & & \text{L-configuration} \end{array}$$

Acid-Base Properties

Every amino acid, regardless of its side chain, has an acidic carboxyl group and a basic amino group, so we shall consider first the acid-base properties of a typical neutral amino acid, such as alanine, which has no other acidic or basic functional groups. In the solid state, amino acids have saltlike properties because they have both a positively charged part and a negatively charged part. Such substances are called **zwitterions.** We can think of these zwitterions as being produced from the molecular form of the amino acids by an internal acid-base reaction:

(21-2)

$$\underset{\substack{\text{L-alanine}\\\text{(molecular form)}}}{\overset{\displaystyle\underset{\displaystyle CH_3}{\overset{\displaystyle O\quad OH}{\underset{|}{\overset{\diagdown\!/}{C}}}}}{H_2N-\underset{|}{\overset{|}{C}}-H}} \; \rightleftharpoons \; \underset{\substack{\text{L-alanine}\\\text{(zwitterion form)}}}{\overset{\displaystyle\underset{\displaystyle CH_3}{\overset{\displaystyle O\quad O^-}{\underset{|}{\overset{\diagdown\!/}{C}}}}}{H_3\overset{+}{N}-\underset{|}{\overset{|}{C}}-H}}$$

Note in reaction 21-2 that neither the molecular form of alanine nor the zwitterion form has a net electrical charge. In aqueous solution these two forms are in equilibrium, but this equilibrium overwhelmingly favors the zwitterion at any pH.

Alanine in solution is also involved in two other equilibria,

(21-3)

$$\underset{\text{positive form}}{H_3\overset{+}{N}-\underset{\underset{CH_3}{|}}{\overset{\overset{O\ \ OH}{\diagdown C /}}{C}}-H} + H_2O \rightleftharpoons H_3O^+ + \underset{\substack{\text{zwitterion}\\\text{form}}}{H_3\overset{+}{N}-\underset{\underset{CH_3}{|}}{\overset{\overset{O\ \ O^-}{\diagdown C /}}{C}}-H} + H_2O \rightleftharpoons \underset{\text{negative form}}{H_2N-\underset{\underset{CH_3}{|}}{\overset{\overset{O\ \ O^-}{\diagdown C /}}{C}}-H} + H_3O^+$$

At any given pH some of the alanine in solution exists in the positive ion form, some of it in the negative ion form, some in the zwitterion form, and some in the molecular form. If the solution pH is very high—that is, $[H_3O^+]$ is very low—both of the equilibria in reaction 21-3 are shifted far to the right and the negative ion form of alanine predominates. On the other hand, if the solution pH is very low—that is, $[H_3O^+]$ is very high—both equilibria in reaction 21-3 are shifted far to the left and the positive ion form of alanine predominates. At the pH of human cells and fluids (pH ~7), alanine exists primarily as the zwitterion.

In solutions that are moderately basic (pH ~8.5 to 10.5), no single form of alanine predominates. In this pH range there are roughly comparable amounts of the zwitterion and the negatively charged forms. Similarly, in moderately acidic solutions there are roughly comparable amounts of the zwitterion and the positively charged forms of alanine. Note that the equilibria in reaction 21-3 can also be viewed as the successive dissociation of the diprotic acid

$$H_3\overset{+}{N}-\underset{\underset{\displaystyle CH_3}{|}}{\overset{\overset{\displaystyle H\quad O}{|\quad\parallel}}{C}}-\overset{\parallel}{C}-OH$$

The other neutral hydrophilic or hydrophobic amino acids behave like alanine and are primarily zwitterions that have a net charge of zero at pH ~7.

Figure 21-5   At pH ~7, the predominant form of an acidic amino acid has a net negative charge.

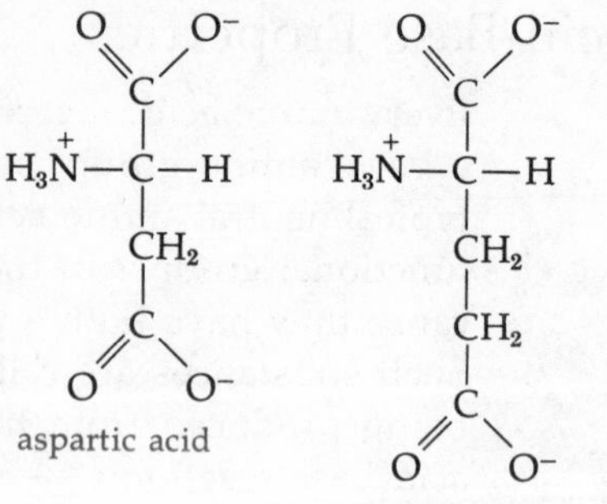

aspartic acid

glutamic acid

Figure 21-6   At pH ~7, the predominant forms of the basic amino acids lysine and arginine have a net positive charge.

lysine

arginine

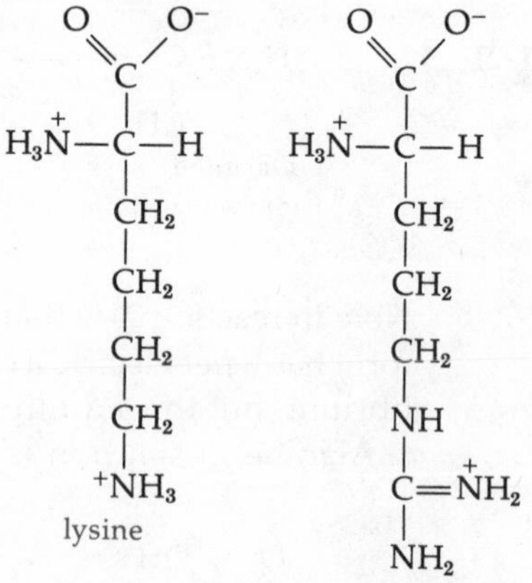

For the acidic and basic amino acids we must consider the acid-base behavior of their side-chain functional groups. The carboxyl group in the side chain of the acidic amino acids, aspartic acid and glutamic acid, is predominantly dissociated at neutral pH (see Figure 21-5). Two of the basic amino acids, lysine and arginine, exist mostly with a protonated nitrogen atom on their side chains at neutral pH and thus have a net positive charge (see Figure 21-6). The other basic amino acid, histidine, also contains a side-chain nitrogen that can serve as a proton acceptor.

(21-4)

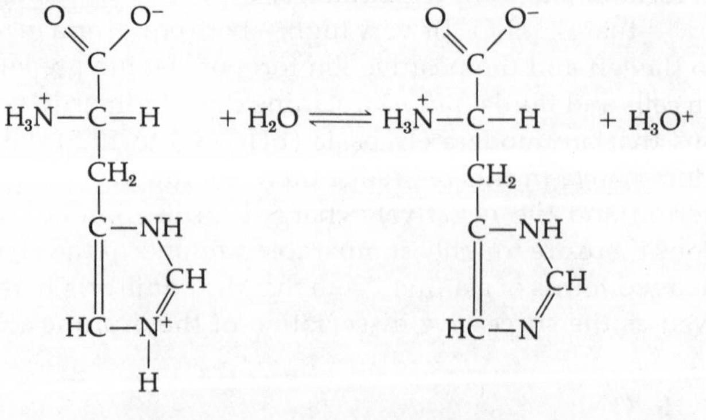

In an aqueous solution with a pH of about 6.5, the position of this equilibrium is such that about half of the histidine molecules have a protonated side-chain nitrogen and half do not. Both arginine and histidine have more than one nitrogen atom in their side chains, but they can accept only one proton per side chain. It is not necessary for us to consider why the particular nitrogen atoms shown in Figure 21-6 and reaction 21-4 are the ones that accept the protons.

Exercise 21-4

Draw structural formulas for the predominant forms of the following amino acids at pH values of 1, 7, and 14:

(a) Valine   (b) Aspartic acid   (c) Serine   (d) Lysine

21-6 AMINO ACID SEQUENCE

As we mentioned at the beginning of the chapter, the sequence of amino acids in a given peptide or protein is crucial to its function, in much the same way as the arrangement of words in a sentence determines its meaning. For example, the two sentences, "Ali beat Frazier" and "Frazier beat Ali" contain the same three words but have quite different meanings. Similarly, there are two different dipeptides with different properties composed of the amino acids glycine and alanine (see Figure 21-7).

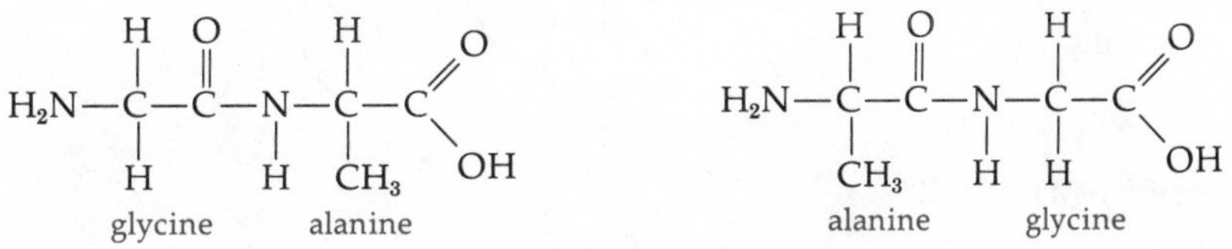

Figure 21-7   The two possible dipeptides composed of the amino acids glycine and alanine are gly-ala (a) and ala-gly (b).

Peptides and proteins usually have an amino acid with an uncombined α-amino group—called the **N-terminal amino acid**—at one end, and an amino acid with an uncombined α-carboxyl group—the **C-terminal amino acid**—on the other end. To avoid confusion, the structural formulas for peptides and proteins are always drawn with the N-terminal amino acid on the left and the C-terminal amino acid on the right, as we have done in Figure 21-7. The same convention is used when the structural formula for a protein or peptide is written in a condensed manner using the abbreviation for the names of the amino acids. Thus, for example, tyr-ala-ser is a condensed way of referring to the tripeptide with the structural formula

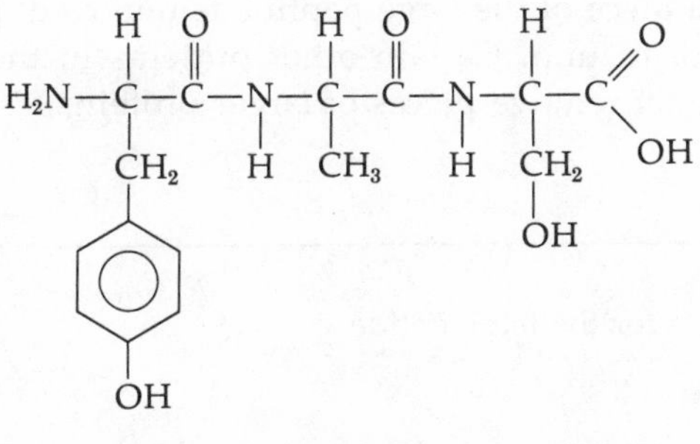

Notice that in this tripeptide the N-terminal amino acid is tyrosine and the C-terminal amino acid is serine.

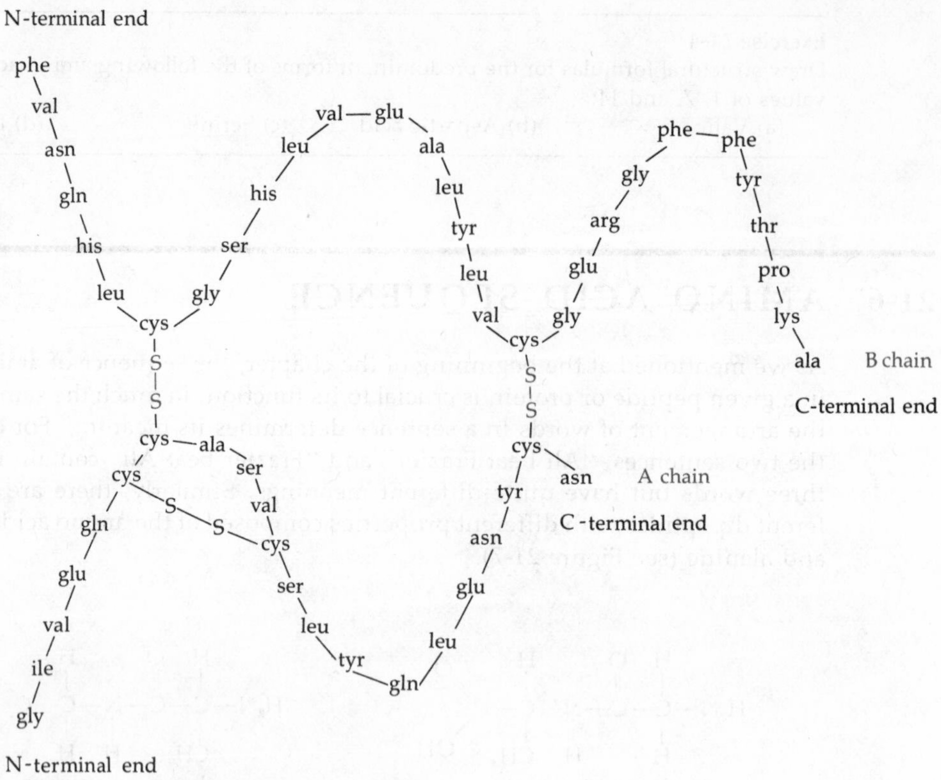

N-terminal end

phe
val
asn
gln
his
leu
cys

val — glu
leu   ala
his   leu
ser   tyr
gly   leu
       val
       cys

phe — phe
gly   tyr
arg   thr
glu   pro
gly   lys
cys   ala   B chain
               C-terminal end

S
S

cys — ala
cys   ser
gln   val
glu   cys
val   ser
ile   leu
gly   tyr — gln

cys
tyr
asn   A chain
asn
glu   C-terminal end
leu
gln

N-terminal end

Figure 21-8   A representation of the complete amino acid sequence of the poly-peptide hormone insulin (bovine). Note that the insulin molecule is composed of two peptide chains (A and B) held together by two disulfide bonds between cysteine side chains.

For very long complex sequences of amino acids, it is sometimes necessary to emphasize further which amino acids are at the ends. This is done by writing the labels "N-terminal end" and "C-terminal end" next to the corresponding ends of the chains. The amino acid sequence in the peptide hormone insulin is shown in Figure 21-8. Notice that insulin is a single molecule consisting of two polypeptides joined by two disulfide bonds between cysteine side chains. There is also one disulfide bond between two cysteine side chains of amino acids in the same chain. It is interesting to note that insulin is actually synthesized from a single large peptide called proinsulin. Proinsulin is made in the pancreas and then a piece of the large peptide is removed, leaving the two joined peptide chains of insulin. Certain other proteins in the body are also produced by reactions that remove pieces of larger proteins.

Exercise 21-5
Draw the structural formula for the tetrapeptide

cys — ala — asp — cys
          S — S

Optional

21-7 DETERMINATION OF AMINO ACID SEQUENCE

Until recently it was extremely difficult to determine the exact sequence of amino acids in peptides and proteins. Insulin (Figure 21-8), which has 51 amino acids, was the first large peptide whose amino acid sequence was determined. This was accomplished by Frederick Sanger and his colleagues in 1953, after several years of work. Basically, Sanger's technique involved, first, the hydrolysis of all the peptide bonds in insulin by strong acids, followed by separation and identification of the component amino acids, and finally, determination of the sequence of these amino acids.

When proteins are completely hydrolyzed, the constituent amino acids can be separated on the basis of their different chemical properties and identified. Today, a machine called an **amino acid analyzer** is routinely used for this purpose. This machine separates a mixture of amino acids according to their ability to interact with a highly charged insoluble matrix called an ion-exchange resin.

Determining the sequence of amino acids in a protein is more difficult. Sanger determined the sequence for insulin in a very laborious way. We can illustrate the principles he used by considering how the sequence of a tetrapeptide known to contain the amino acids tyrosine, alanine, valine, and serine could be determined. Twenty-four different tetrapeptides containing these four amino acids are possible. Suppose, however, that we react a sample of the tetrapeptide with an enzyme that can hydrolyze only those peptide bonds that contain the carbonyl group of tyrosine, and then determine that the products of this reaction are (1) the amino acid serine and (2) a tripeptide composed of alanine, valine, and tyrosine. Only two possible sequences are consistent with this information: val-ala-tyr-ser and ala-val-tyr-ser.

Now, suppose we take another sample of our tetrapeptide, react it with a reagent that can hydrolyze only the peptide bond involving the N-terminal amino acid, and determine that the products of this reaction are (1) a tripeptide composed of serine, tyrosine, and alanine, and (2) the amino acid valine.

The *only* sequence that is consistent with both experiments is val-ala-tyr-ser. Sanger performed a large number of these types of reactions on insulin, and after a great deal of effort was finally able to determine its amino acid sequence. Today, techniques are available whereby the amino acid sequence of small proteins can be determined relatively easily. Basically these methods involve the sequential hydrolysis and identification of amino acids, starting with the N-terminal amino acid. This new technique has even been automated, so that the sequence of small proteins such as insulin can now be determined in a matter of days, rather than years, as was the case for Sanger.

21-8 THE SHAPE OF PROTEINS

Proteins are very large molecules, and one could envision many possible shapes for a protein with a particular amino acid sequence. It is known, however, that every protein has its own definite, unique shape and that this shape must be maintained in order for the protein to function properly. Let us consider the factors that determine a protein's shape.

The Peptide Group

Thus far, we have considered peptides and proteins as strings of amino acids held together by peptide bonds (see Figures 21-2 and 21-3). Notice that these strings involve a repetitive pattern, called a **covalent backbone,** consisting of one nitrogen atom followed by two carbon atoms, and so on, which we can represent as

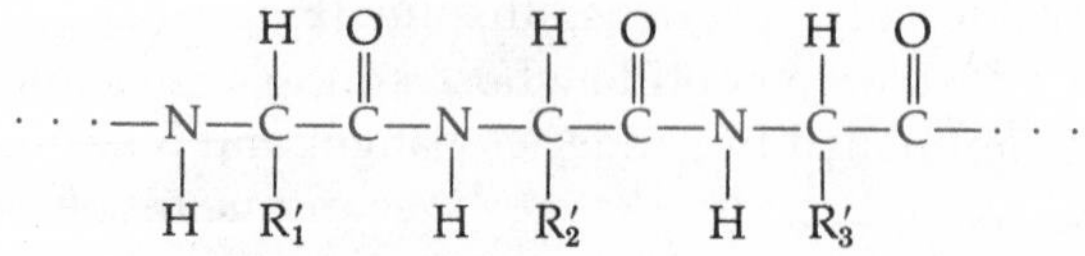

$$\cdots -\underset{\underset{H}{|}}{\overset{\overset{H}{|}}{N}}-\underset{\underset{R'_1}{|}}{\overset{\overset{O}{||}}{C}}-\overset{\overset{O}{||}}{C}-\underset{\underset{H}{|}}{\overset{\overset{H}{|}}{N}}-\underset{\underset{R'_2}{|}}{\overset{\overset{H}{|}}{C}}-\overset{\overset{O}{||}}{C}-\underset{\underset{H}{|}}{\overset{\overset{H}{|}}{N}}-\underset{\underset{R'_3}{|}}{\overset{\overset{H}{|}}{C}}-\overset{\overset{O}{||}}{C}-\cdots$$

We would expect the covalent backbone to be extremely flexible, since there is usually free rotation about single covalent bonds. Starting at the N-terminal end, however, the third bond in the covalent backbone and every third bond thereafter is an amide bond. In Chapter 18 we saw that (1) the carbon-nitrogen bond in amides is shorter than a typical carbon-nitrogen single bond; (2) there is *not* free rotation about amide bonds; and (3) amides are best represented by a composite picture of two resonance structures, one of which involves a C=N double bond. Resonance structures for a peptide group are shown in Figure 21-9. This representation of peptide groups agrees with their experimentally determined properties, including the fact that all six atoms in a peptide group lie in the same plane.

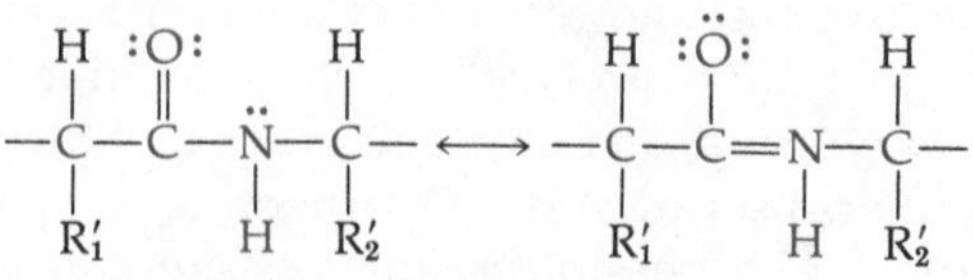

$$-\underset{\underset{R'_1}{|}}{\overset{\overset{H}{|}}{C}}-\overset{\overset{:O:}{||}}{C}-\underset{\underset{H}{|}}{\overset{\overset{H}{|}}{\ddot{N}}}-\underset{\underset{R'_2}{|}}{\overset{\overset{H}{|}}{C}}- \longleftrightarrow -\underset{\underset{R'_1}{|}}{\overset{\overset{H}{|}}{C}}-\overset{\overset{:\ddot{O}:}{|}}{C}=\underset{\underset{H}{|}}{N}-\underset{\underset{R'_2}{|}}{\overset{\overset{H}{|}}{C}}-$$

Figure 21-9   Resonance-structure representation for a peptide group. All six of the atoms in the peptide group (colored blue) lie in the same plane.

The rigid nature of peptide bonds restricts the possible shapes that a given protein can assume, but there is still free rotation of the group of atoms in one plane defined by one peptide bond relative to groups of atoms in other planes defined by other peptide bonds. The free rotation of these groups of atoms allows much flexibility in the covalent backbone of a protein (see Figure 21-10).

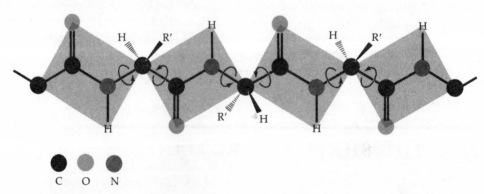

● C   ● O   ● N

Figure 21-10   Flexible and nonflexible bonds in a polypeptide. The shaded areas represent portions of the backbone containing a peptide group in which atoms all lie in the same plane (i.e., there is nonflexible bonding around peptide bonds). Between the shaded portions the backbone is free to rotate. Arrows indicate this flexibility.

Interactions Between Peptide Groups

A very important factor in determining the shapes of proteins is the ability of two peptide groups to form a strong hydrogen bond. This hydrogen bond can form between the hydrogen atom in one peptide group and the carbonyl oxygen in another peptide group (Figure 21-11). This hydrogen bond is exceptionally strong because there is a particularly large fractional negative charge on the oxygen atom of the peptide group and a particularly large fractional positive charge on the hydrogen atom of this group. Compare the resonance structures in Figure 21-11. In the structure on the right there are three unshared pairs of electrons on the oxygen atom, whereas there are only two in the structure on the left, and there are no unshared electron pairs on the nitrogen atom on the right. The resonance structure on the right implies that some of the valence electrons in the amide group are pulled toward the oxygen atom and away from the nitrogen atom and the nitrogen-hydrogen bond.

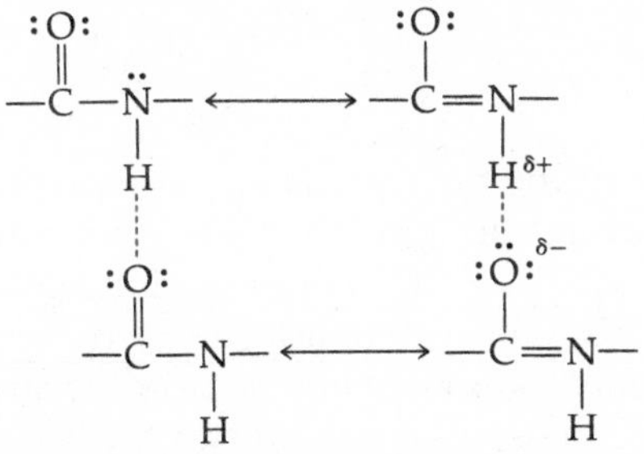

Figure 21-11   Hydrogen bonds (blue dashed lines) between peptide groups. These two resonance structures provide a composite representation that accounts for the particularly strong hydrogen bond between two peptide groups.

Interactions Between Side Chains

If the covalent backbone in a protein is folded or coiled, then the side-chain groups that extend out from the backbone can interact with one another. The shape of a particular protein is determined to a large degree by side-chain interactions. We can divide the attractive forces involved in side-chain interactions into four types: (1) hydrogen bonds, (2) disulfide bonds, (3) hydrophobic and hydrophilic interactions, and (4) salt bridges.

Hydrogen bonds can form between appropriate side-chain groups, and a disulfide bond can form between two cysteine side chains (Figure 21-3). Recall from Chapter 7 that hydrophilic parts of larger molecules interact strongly with water molecules and that hydrophobic parts of larger molecules tend to cluster together. Thus in an aqueous medium most proteins will tend to coil up like a ball of string so that the hydrophobic side chains extend inwards and the hydrophilic side chains are on the outside of the "ball," where they can interact with water molecules (see Figure 21-12).

Figure 21-12   The compact "glob" formed by a typical protein in an aqueous medium is schematically represented here. The interior, or hydrophobic core, of the protein is composed largely of nonpolar hydrophobic amino acids, whereas most hydrophilic side-chain groups are on the exterior, exposed to the aqueous environment.

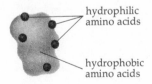

hydrophilic amino acids

hydrophobic amino acids

At the pH of body cells, the acidic side-chain groups of aspartic or glutamic acid are negatively charged, and there is a positive charge on the basic side-chain groups of lysine and arginine and on most of the side-chain groups of histidine. The attraction between a negatively charged acidic side chain and a positively charged basic side chain is called a **salt bridge** (see Figure 21-13).

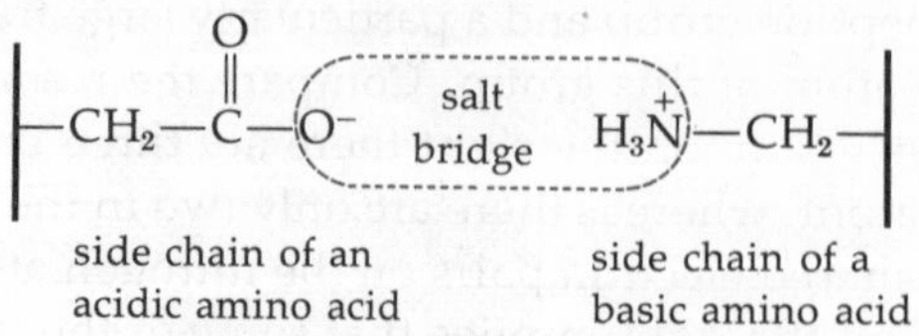

$$-CH_2-\overset{\overset{\textstyle O}{\|}}{C}-O^-\quad \underset{\text{bridge}}{\text{salt}}\quad H_3\overset{+}{N}-CH_2-$$

side chain of an   side chain of a
acidic amino acid   basic amino acid

Figure 21-13   A salt bridge between charged side chains of nearby acidic and basic amino acids.

Levels of Protein Structure

It is extremely difficult to determine the exact shape of a protein, although this has been accomplished for several proteins by using a technique called X-ray diffraction analysis. When discussing the complicated shapes of large protein molecules, it is useful to specify four different levels of protein structure.

The amino acid sequence of a protein is called its **primary structure.** The term *secondary structure* refers to the spatial relationship between amino acids fairly close together in the protein's amino acid sequence. To clarify this, let us represent a protein as a string with knots corresponding to amino acids (see Figure 21-14). We could visualize several possible types of secondary structure, including a linear chain, a zigzag chain, a twisted spiraling chain, as well as a chain without a repetitive pattern. A large chain could consist of portions with different types of secondary structure (see Figure 21-15). We shall study the common types of protein secondary structure in Section 21-9. The term *tertiary structure* refers to the overall shape of a protein, including the spatial relationship

Figure 21-14   Representations for possible types of protein secondary structure. The peptide backbone can be viewed as a string and the amino acid components as knots in the string.

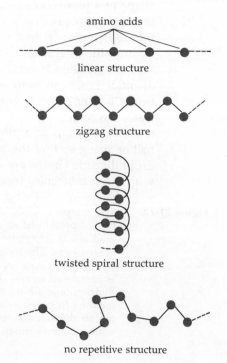

amino acids

linear structure

zigzag structure

twisted spiral structure

no repetitive structure

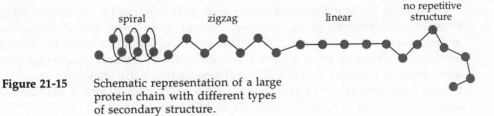

Figure 21-15   Schematic representation of a large protein chain with different types of secondary structure.

Figure 21-16   Schematic representation of the tertiary structure of a protein. Note that the amino acids labeled *a* and *b* in the figure are not nearby in the amino acid sequence but are close together in space because of the overall shape of this polypeptide.

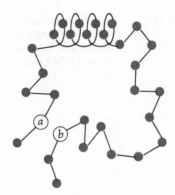

of amino acids that are far removed from each other in the amino acid sequence. This is represented schematically in Figure 21-16 and will be discussed further in Section 21-10.

Many proteins in the human body consist of more than one peptide chain. The spatial relationship of component peptide chains in such a protein is called the *quaternary structure* of the protein (see Section 21-11). For example, a protein composed of two peptide chains is analogous to two balls of string that have been glued together.

The precise shape of an individual protein is uniquely determined by its amino acid sequence. Dramatic support for this assertion was recently obtained. In 1969, R. B. Merrifield and his colleagues completed the laboratory synthesis of a protein with the same amino acid sequence as the naturally occurring enzyme, ribonuclease. Their synthetic protein curled up into exactly the same shape as that of natural ribonuclease.

Exercise 21-6
Using structural formulas, indicate side-chain interactions between the following pairs of amino acids at pH 7.0. If there is no interaction, write none.

(a) Glycine and serine
(b) Lysine and aspartic acid
(c) Glutamine and serine
(d) Tryptophan and phenylalanine

21-9 SECONDARY STRUCTURE

Secondary structure of a protein refers to the spatial relationship of amino acids that are fairly close together in the amino acid sequence. Several years ago, Linus Pauling showed that two types of protein secondary structure, called the α-helix and the β-pleated sheet, would allow a large number of hydrogen bonds to form between nearby peptide groups. He postulated that many proteins have an α-helix secondary structure or a β-pleated sheet secondary structure, and his prediction was later verified by experimental observations.

The **α-helix** is a spiral chain of amino acids held together by a hydrogen bond between each peptide group and the third peptide group farther along the amino acid chain (see Figure 21-17). The α-helix spiral might twist in either a left-handed direction or in a right-handed direction. Naturally occurring proteins contain L-amino acids, and a right-handed α-helix allows for more effective hydrogen bonding between L-amino acids than would a left-handed α-helix. Therefore natural α-helices are right-handed.

Figure 21-17   A schematic representation of the α-helix secondary structure. Hydrogen bonds are indicated by blue dashed lines.

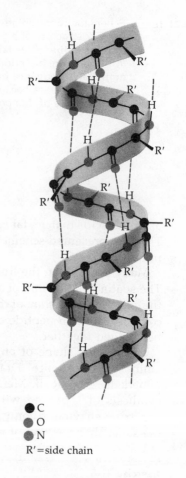

● C
● O
● N

R′=side chain

The **β-pleated sheet** is a zigzag type of secondary structure in which two or more chains of amino acids are held together by hydrogen bonds between peptide groups on adjacent chains (see Figure 21-18). A single chain of amino acids can also curl around so that a β-pleated sheet forms between parts of the same chain (see Figure 21-19).

Notice in Figures 21-17 and 21-18 that in both the α-helix and the β-pleated sheet secondary structures the amino acid side chains (R′) stick out from the covalent backbone and that adjacent side chains are fairly close together. The stability of an α-helix or a β-pleated sheet is enhanced when salt bridges can form between negatively charged (acidic) and positively charged (basic) side-chain groups. On the other hand, if a portion of a peptide contains large, bulky side chains or adjacent side chains with the same electrical charge, then the repulsion between side-chain groups in this portion of the chain is likely to prevent an α-helix, a β-pleated sheet, or any other characteristic repetitive secondary structure. It has been determined experimentally that some proteins are entirely α-helical. One example is the keratin in human hair, which consists of three

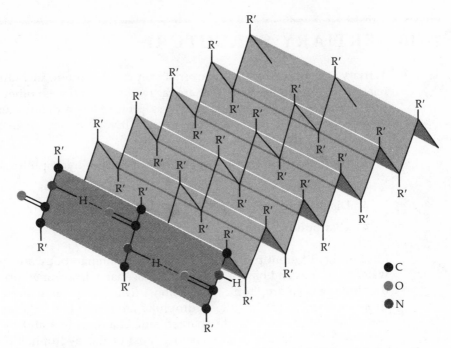

Figure 21-18   A schematic representation of the β-pleated sheet type of secondary structure. The hydrogen bonds (blue dashed lines) between the three adjacent peptide chains are shown only in the first plane.

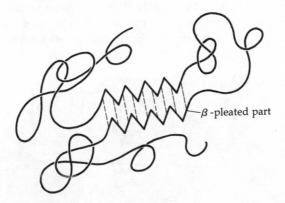

β -pleated part

Figure 21-19   Schematic representation of a β-pleated sheet formed between two parts of the same large protein. The blue dashed lines represent hydrogen bonds.

α-helices wound around each other as shown in Figure 21-20. Most proteins, however, contain only a part that is α-helical. For example, about 70% of the protein hemoglobin is helical, whereas only 7% of some other proteins is α-helical in shape.

Figure 21-20   The protein keratin, a component of hair, is a helix formed from three protein chains. Each component protein chain has an α-helical shape.

Exercise 21-7

Would you expect the peptide phe-tyr-ser-phe-tyr-glu-glu-glu-phe-tyr-lys-lys-lys to form an α-helix? Explain your answer.

21-10 TERTIARY STRUCTURE

Tertiary structure refers to the overall shape of a protein, including the spatial relationship of amino acids that are far removed from each other in the amino acid sequence (see Figure 21-16). Every protein has a definite unique overall (or tertiary) structure because of interactions involving side-chain groups. These interactions include:

1. Hydrogen bonding between pairs of neutral hydrophilic side chains
2. Salt bridges between oppositely charged side chains
3. Disulfide bonds between pairs of cysteines
4. Hydrophobic and hydrophilic interactions

Most of the different protein molecules in the human body are surrounded by water molecules and have a globular overall shape determined primarily by hydrophilic and hydrophobic interactions. The particular globular shape that these proteins adopt has the hydrophilic amino acids on the outside, where water molecules can hydrate charged side-chain groups and hydrogen bond with neutral hydrophilic side chains. Most of the hydrophobic amino acids, however, are packed very tightly together in the interior of the molecule, which is often called the **hydrophobic core.**

The tertiary structure of the protein myoglobin is represented in Figure 21-21. The myoglobin chain is packed so tightly that there is room for only four water molecules in its interior. Myoglobin is very similar to hemoglobin and is used to store oxygen in muscle cells. Notice the heme group tucked into a pocket on the

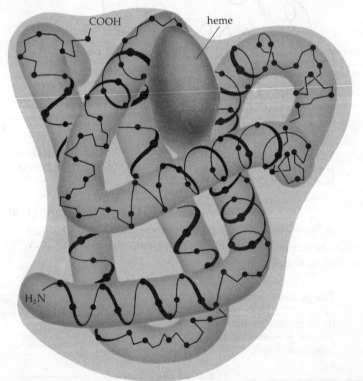

Figure 21-21   A schematic representation of the tertiary structure of whale myoglobin. The backbone is shown in black. The spaces between portions of the backbone are completely filled by side-chain groups, indicated here by pale blue shading. Tertiary structure is maintained by interactions between side-chain groups on different parts of the protein.

surface of myoglobin. This is where the oxygen molecule binds. Most other proteins, including enzymes, also have pockets where they can bind small molecules. We shall see that the existence and shape of such pockets is a crucial factor in determining how enzymes work (Chapter 23).

Exercise 21-8
Which of the following amino acids would you expect to find in the hydrophobic core of a protein?
(a) Lysine (b) Isoleucine (c) Glycine (d) Arginine (e) Histidine (f) Tryptophan

---

21-11 QUATERNARY STRUCTURE

Proteins that contain more than one polypeptide chain are called **oligomeric** proteins. The component polypeptide chains are called **subunits,** and they may all be the same or they may differ. Oligomeric proteins with either two or four subunits are the most common types, but proteins with three or many more subunits are known to exist.

Hemoglobin, for example, is an oligomeric protein composed of four subunits, all of which are quite similar in structure to myoglobin. In normal human adults, hemoglobin has two different subunits, called α and β. A single hemoglobin molecule consists of two α-type and two β-type subunits. Each subunit can bind oxygen. When oxygen binds to one subunit, the shapes of all the subunits change slightly. The functional significance of this slight change in shape will be discussed in the next chapter.

The subunits of an oligomeric protein usually interact considerably with one another, and an oligomeric protein is thus one grand protein molecule. Subunits are held together by hydrogen bonds, salt bridges between positively and negatively charged side-chain groups, hydrophilic and hydrophobic interactions, and, in some cases, disulfide bonds between cysteines on different subunits.

In the next chapter we shall see how oligomeric proteins can have certain functional advantages over single-subunit (monomeric) proteins. One advantage of making large proteins out of smaller subunits is analogous to the advantage a component stereo sound system has over a single-unit system. If the cell makes a defective component (subunit), it does not have to scrap the entire system. It may simply replace the defective subunit with a good one.

Isoenzymes, a special class of oligomeric proteins, are enzymes that catalyze the same reaction but have slightly different subunits. The best-known examples of isoenzymes are those of the enzyme lactic acid dehydrogenase (LDH). LDH has two types of subunits, called M and H, which differ to some extent in their amino acid sequence. These subunits associate to form a tetrameric (four-subunit) quaternary structure. There are five LDH isoenzymes: M_4, M_3H, M_2H_2, MH_3, and H_4. It is relatively easy to separate a mixture of LDH isoenzymes in the laboratory, because M and H subunits differ in the number of charged amino acid side-chain groups.

It has been found that heart muscle cells make almost entirely H_4-type LDH (H stands for heart), whereas cells that make up skeletal muscle make predominantly M_4-type LDH (M stands for muscle). White blood cells and other tissues make both types of subunits, so they have a mixture of LDH isoenzymes. This information is used clinically to aid in the diagnosis of certain diseases. For example, if a person's heart cells are damaged, the H_4 type of LDH will leak out of the damaged cells and the resulting increase in H_4 concentration in that person's serum can be detected.

21-12 PROTEIN DENATURATION

Under certain conditions it is possible to disrupt the unique shape of a protein without breaking any of the bonds that hold the covalent backbone together. Any process that drastically alters the shape of a protein but leaves the primary structure intact is called **denaturation.** A protein denatures when an increase in temperature or a change in the composition of the solution disrupts the hydrogen bonds, salt bridges, and/or hydrophobic interactions responsible for its secondary, tertiary, and quaternary structure. A change in solution pH or a change in salt concentration can cause a protein to denature. Substances that reduce disulfide bonds can also denature proteins.

A denatured protein loses its ability to function properly. For example, in order to measure the activity of blood serum enzymes in the clinical laboratory, it is imperative that the serum samples be protected from conditions that would denature and thus inactivate them.

Some proteins that have been denatured by a change in the composition of the solution or the breaking of disulfide bonds can be *renatured*—that is, refolded into their original shape. In such a case, we say that the denaturation is *reversible*. For example, the enzyme ribonuclease can be denatured by the addition of substances that disrupt hydrogen bonds, hydrophobic interactions, salt bridges, and disulfide bonds. Denatured ribonuclease cannot function as a catalyst. If the substances responsible for the denaturation are slowly removed, however, ribonuclease will refold into its original shape and regain its enzyme activity (see Figure 21-22). The phenomenon of reversible denaturation is additional evidence for the fact that the primary structure of a protein determines its overall shape.

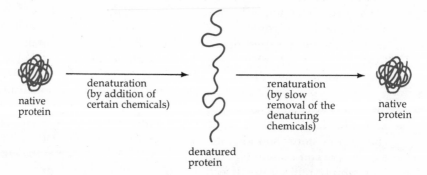

Figure 21-22   Schematic representation of the reversible denaturation of a protein. The shape and enzymatic activity of the renatured protein are the same as they are for the undenatured, native protein.

Some proteins cannot be reversibly denatured, and also, some methods of protein denaturation are not reversible (see Figure 21-23). A familiar example is cooking an egg. When an egg is cooked, the proteins in the egg are irreversibly denatured by the heat. When the egg cools, these proteins do not renature. Rather, large numbers of individual protein molecules clump together into an insoluble complex held together primarily by hydrophobic interactions. The irreversible denaturation of proteins that results in such insoluble complexes is called **coagulation.**

Partial denaturation and renaturation are involved in the process of hair curling. People commonly curl hair by using hot curlers or by a more involved process called permanent waving. We have seen that hair protein consists of three helices twisted around each other and held together by hydrogen bonds and disulfide bonds (see Figure 21-24). Hot curlers can disrupt some of these

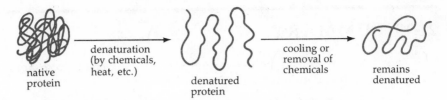

Figure 21-23   Schematic representation of irreversible protein denaturation.

hydrogen bonds (partial denaturation), and when the hair cools new hydrogen bonds form. Since the hair cools on the curlers, the new hydrogen bonds tend to stabilize the curled shape of the hair. However, hot, humid weather or a soapy shower can disrupt these hydrogen bonds and remove the curls.

In the permanent waving process, the disulfide bonds between the three interwoven helices of hair are broken by reducing agents in the wave treatment solution (Figure 21-24). The hair is then set on curlers and new disulfide bonds form, using oxygen in the air as the oxidizing agent; the new disulfide bonds hold the hair in a curled shape. These covalent disulfide bonds hold curls much more permanently than hydrogen bonds.

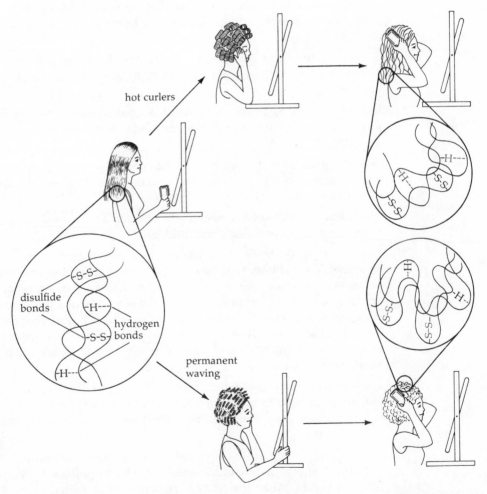

Figure 21-24   Curling hair. Hot curlers disrupt some hydrogen bonds. As the hair cools on the curlers, new hydrogen bonds form, maintaining the curled shape of the hair. However, by reducing and then re-forming disulfide bonds in the permanent waving process, more stable curls can be formed.

21-13 SUMMARY

1. Proteins are polymers composed of long chains of α-amino acids joined by peptide bonds.

2. A peptide bond is an amide bond formed between the carboxyl group of one amino acid and the amino group of another amino acid. Short chains of amino acids joined by peptide bonds are called peptides.

3. Each peptide or protein has a unique amino acid sequence, which is crucial to its overall shape and function. By convention, amino acid sequences are written with the N-terminal amino acid on the left and the C-terminal amino acid on the right.

4. The general formula for an α-amino acid is
$$H_2N-\overset{\overset{\displaystyle H}{|}}{\underset{\underset{\displaystyle R'}{|}}{C}}-\overset{\overset{\displaystyle O}{\|}}{C}-OH.$$
The 20 α-amino acids used to assemble proteins are classified according to the properties of their side-chain (R') groups.

5. The essential amino acids are those that are required for protein synthesis but that cannot be synthesized by the human body. They must be included in the diet.

6. Glycine is the only amino acid that is not optically active. Only the L-enantiomers of the other 19 amino acids are used in the synthesis of proteins.

7. In the solid state, amino acids exist as zwitterions. The predominant form of an amino acid in solution depends on the pH.

8. The unique overall shape of a protein is a result of the planar nature of its peptide groups, hydrogen bonds that occur between peptide groups, and various types of interactions among the side-chain groups of its component amino acids.

9. The attraction between a negatively charged and a positively charged side-chain group is called a salt bridge.

10. The shapes of proteins are discussed in terms of four levels of protein structure. The primary structure of a protein is its amino acid sequence. Secondary structure refers to the spatial relationship of amino acids that are fairly close together in the amino acid sequence of a protein. Tertiary structure refers to the overall shape of a protein, including the spatial relationship of amino acids that are far apart in the amino acid sequence. The spatial relationship of the subunits of an oligomeric protein is referred to as quaternary structure.

11. The α-helix and the β-pleated sheet are two types of secondary structure that are frequently found in proteins. These types of secondary structure are held together by hydrogen bonds between peptide groups of nearby amino acids.

12. Interactions among amino acid side-chain groups, including hydrogen bonding, salt bridges, disulfide bonds, and hydrophilic and hydrophobic interactions, stabilize the tertiary structure of proteins. These types of interactions also hold subunits together in oligomeric proteins.

13. Oligomeric enzymes that catalyze the same reaction but that have somewhat different subunits are called isoenzymes.

14. Any process that drastically alters the shape of a protein but leaves its primary structure intact is called denaturation. Under certain conditions the denaturation of some proteins is reversible.

15. Coagulation is the irreversible denaturation of a protein, which results in the formation of insoluble complexes.

PROBLEMS

1. Define the following terms: (a) zwitterion, (b) essential amino acid, (c) isoenzyme, (d) denaturation, (e) salt bridge.

2. Draw the general structural formula for α-amino acids.

3. What is the difference between a peptide and a protein?

4. The R group of the amino acid glutamic acid is $-CH_2-CH_2-\overset{\displaystyle O}{\overset{\|}{C}}-OH$. What is the net charge on the predominant form of glutamic acid at (a) pH = 1.0, (b) pH = 6.8, and (c) pH = 13.5?

5. What type of force is most important in stabilizing the α-helix form of a polypeptide chain?

6. The respective R groups of the amino acids glutamic acid, lysine, alanine, and serine are

$-CH_2-CH_2-\overset{\displaystyle O}{\overset{\|}{C}}-OH$, $-CH_2-CH_2-CH_2-CH_2-NH_2$, $-CH_3$, and $-CH_2OH$.

   (a) Draw the structural formula for the tetrapeptide ala-glu-lys-ser in its predominant ionic form at pH 7.0.
   (b) Identify the peptide bonds in ala-glu-lys-ser and indicate side-chain interactions if any occur.

7. Hydrophobic interactions are important in which levels of protein structure?

8. Which of the following changes is most likely to denature a globular protein?
   (a) A decrease in temperature from 37°C to −10°C
   (b) An increase in pH from 7.4 to 8.5
   (c) A decrease in pH from 7.4 to 6.5
   (d) An increase in temperature from 37°C to 85°C
   (e) None of the above

9. The predominant form of the amino acid histidine at pH 8.5 is

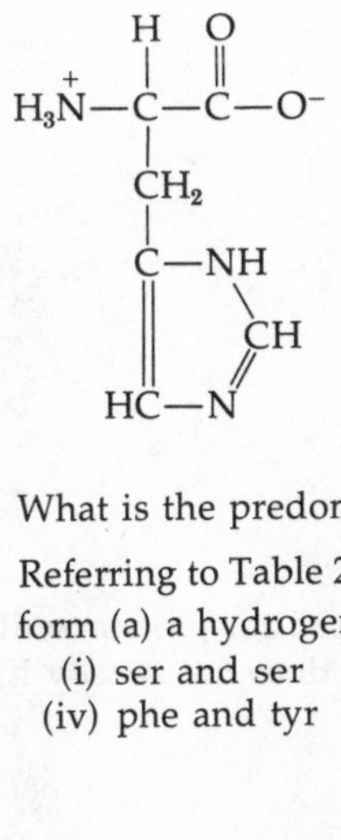

What is the predominant form of histidine at pH = 5.0?

10. Referring to Table 21-1, indicate which side chains of the following amino acid pairs will form (a) a hydrogen bond, (b) hydrophobic interactions, or (c) neither:
   (i) ser and ser   (ii) gln and thr   (iii) glu and phe
   (iv) phe and tyr   (v) tyr and thr

11. Define the terms (a) oligomer, (b) subunit, (c) coagulation, (d) hydrophobic core, and (e) β-pleated sheet, as they refer to proteins.

12. How are disulfide bonds involved in maintaining the shape of proteins?

13. Referring to Table 21-1, draw the structural formula for the predominant form of the tetrapeptide gly-ile-ser-val at pH 7.0.
    (a) Indicate which atoms in this tetrapeptide must lie in a common plane.
    (b) What is the net charge on this molecule at pH 1.0?

14. At what pH does the uncharged form of the amino acid arginine predominate?
    (a) pH = 1.0   (b) pH = 7.0
    (c) pH = 14.0   (d) None of these

SOLUTIONS TO EXERCISES

21-1 R'_1—R'_3—R'_2
      R'_2—R'_1—R'_3
      R'_2—R'_3—R'_1
      R'_3—R'_1—R'_2
      R'_3—R'_2—R'_1

21-2 (a) The structural formulas for the two acidic amino acids, aspartic acid and glutamic acid, are

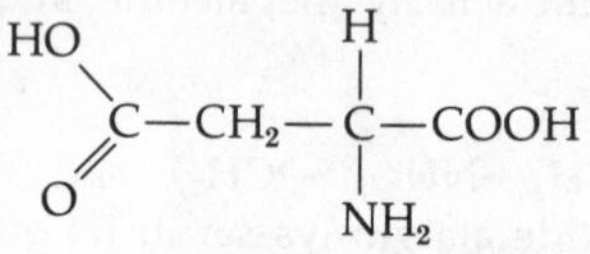

aspartic acid 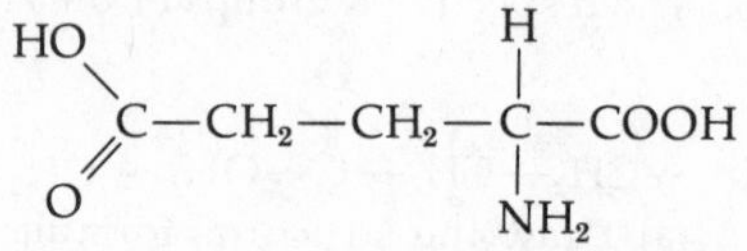   glutamic acid

(b) Structural formulas for the three basic amino acids, lysine, arginine, and histidine, are

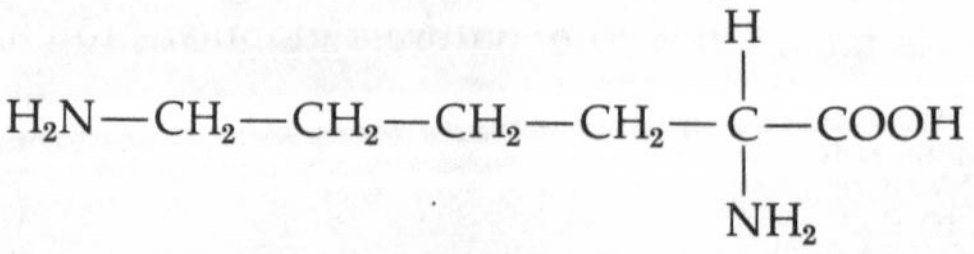

lysine

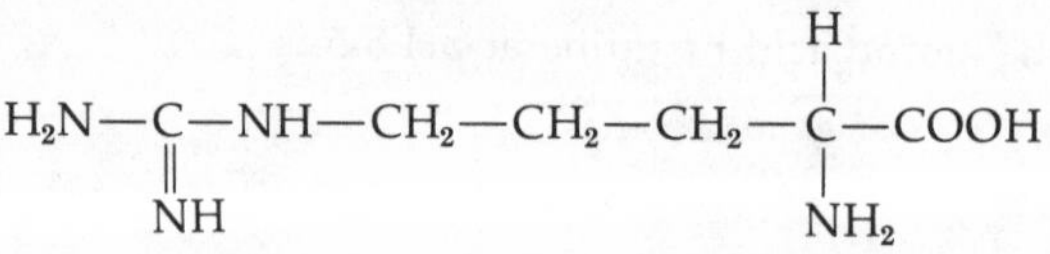

arginine

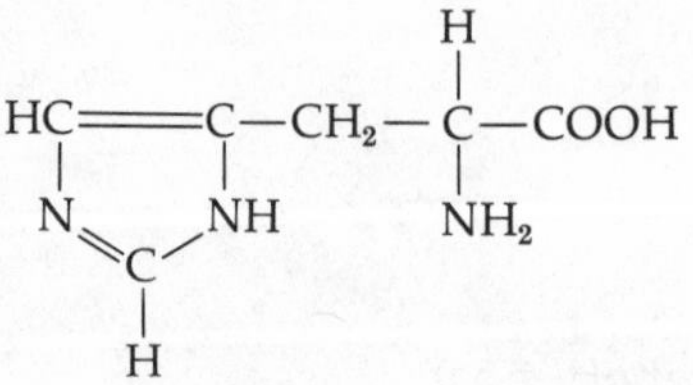

histidine

(c) The neutral amino acids (the other 15 amino acids used for protein synthesis) are shown in Table 21-1A and 21-1B. Note that this exercise does not specify hydrophilic or hydrophobic neutral amino acids, so either is correct.

The functional groups responsible for the classification of these compounds are as follows:

(a) Acidic: For both aspartic acid and glutamic acid, the side-chain carboxyl group.

(b) Basic: For lysine, the amino group ($-NH_2$) on the side chain.

For arginine, the $-N-C-NH_2$ group.
$$\overset{|}{\underset{H}{}}\ \overset{||}{\underset{NH}{}}$$

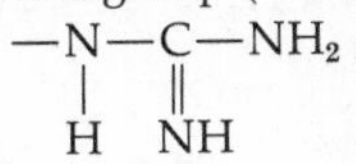

For histidine, the $-CH_2-C=CH$ group.

(with the imidazole ring: HN—C(H)=N attached)

(c) Neutral: The side chains of the various neutral amino acids contain several types of functional groups, including $-SH$, $-OH$, and so on, but none that act as proton donors or acceptors, so they are classified as neutral.

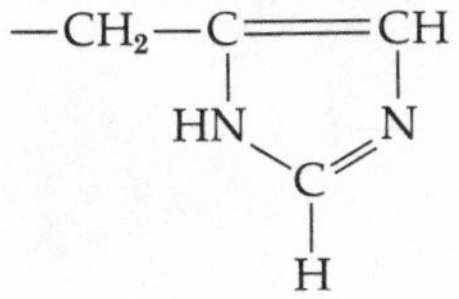

21-3 (a) Neutral hydrophilic
(b) Neutral hydrophilic
(c) Neutral hydrophobic
(d) Neutral hydrophobic

21-4

(a) Valine

pH = 1: $(H_3C)_2CH-CH(NH_3^+)-C(=O)-OH$

pH = 7: $(H_3C)_2CH-CH(NH_3^+)-C(=O)-O^-$

pH = 14: $(H_3C)_2CH-CH(NH_2)-C(=O)-O^-$

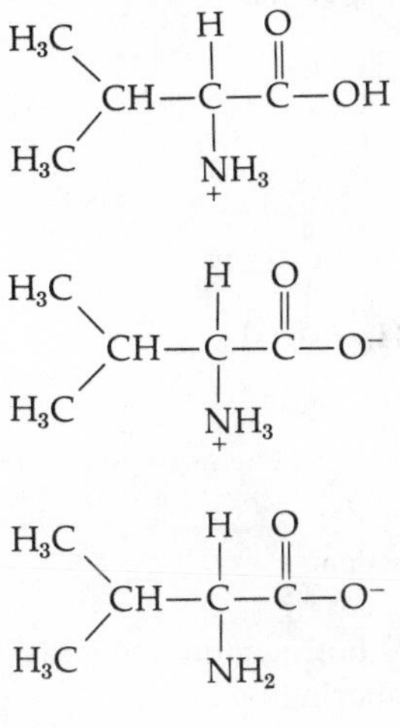

(b) Aspartic acid

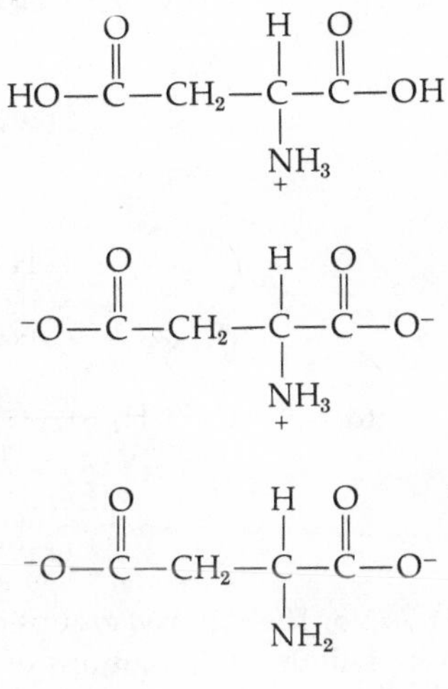

pH = 1: $HO-C(=O)-CH_2-CH(NH_3^+)-C(=O)-OH$

pH = 7: $^-O-C(=O)-CH_2-CH(NH_3^+)-C(=O)-O^-$

pH = 14: $^-O-C(=O)-CH_2-CH(NH_2)-C(=O)-O^-$

(c) Serine

pH = 1: $HO-CH_2-CH(NH_3^+)-C(=O)-OH$

pH = 7: $HO-CH_2-CH(NH_3^+)-C(=O)-O^-$

pH = 14: $HO-CH_2-CH(NH_2)-C(=O)-O^-$

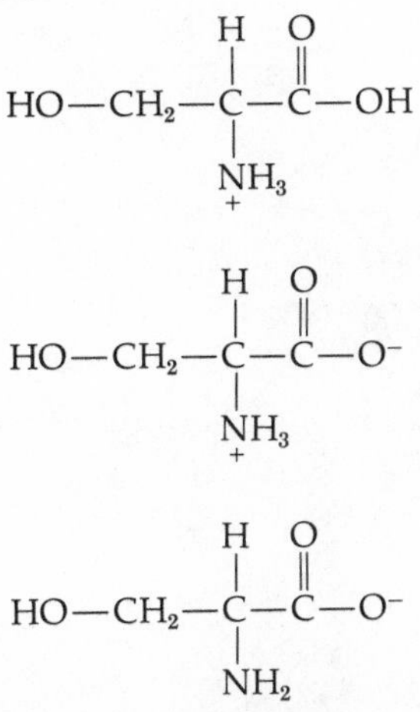

(d) Lysine

pH = 1: $H_3N^+-(CH_2)_4-CH(NH_3^+)-C(=O)-OH$

pH = 7: $H_3N^+-(CH_2)_4-CH(NH_3^+)-C(=O)-O^-$

pH = 14: $H_2N-(CH_2)_4-CH(NH_2)-C(=O)-O^-$

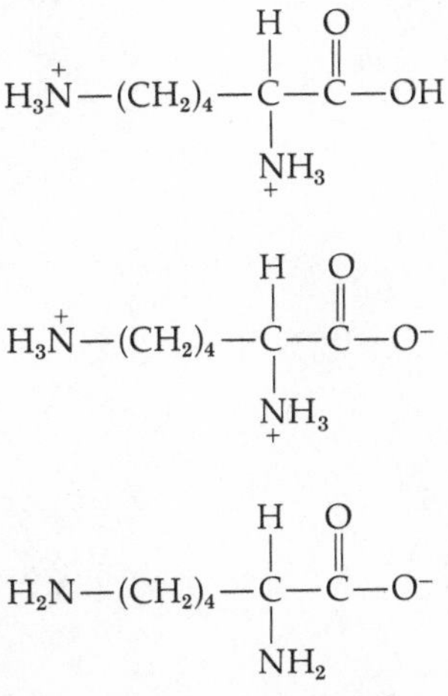

21-5 $H_2N-\overset{\overset{\displaystyle H}{|}}{\underset{\underset{\displaystyle CH_2}{|}}{C}}-\overset{\overset{\displaystyle O}{\|}}{C}-\overset{\overset{\displaystyle H}{|}}{N}-\overset{\overset{\displaystyle H}{|}}{\underset{\underset{\displaystyle CH_3}{|}}{C}}-\overset{\overset{\displaystyle O}{\|}}{C}-\overset{\overset{\displaystyle H}{|}}{N}-\overset{\overset{\displaystyle H}{|}}{\underset{\underset{\displaystyle CH_2}{|}}{C}}-\overset{\overset{\displaystyle O}{\|}}{C}-\overset{\overset{\displaystyle H}{|}}{N}-\overset{\overset{\displaystyle H}{|}}{\underset{\underset{\displaystyle CH_2}{|}}{C}}-\overset{\overset{\displaystyle O}{\|}}{C}-OH$

with the third residue side chain:
$C=O$
OH

and a disulfide bridge: $S—S$ connecting the first and fourth side chains.

21-6 (a) None

(b) $H-\overset{\overset{\displaystyle \overset{\overset{\displaystyle O}{\|}}{C}-O^-}{|}}{\underset{\underset{\displaystyle \overset{+}{NH_3}}{|}}{C}}-CH_2-CH_2-CH_2-CH_2-\boxed{\overset{+}{NH_3}}$ salt bridge $^-O-\overset{\overset{\displaystyle O}{\|}}{C}-CH_2-\overset{\overset{\displaystyle \overset{\overset{\displaystyle O}{\|}}{C}-O^-}{|}}{\underset{\underset{\displaystyle \overset{+}{NH_3}}{|}}{C}}-H$

(c) hydrogen bond

$H-\overset{\overset{\displaystyle \overset{\overset{\displaystyle O}{\|}}{C}-O^-}{|}}{\underset{\underset{\displaystyle \overset{+}{NH_3}}{|}}{C}}-CH_2-CH_2-\overset{\overset{\displaystyle O\cdots HO-CH_2-\overset{\overset{\displaystyle \overset{\overset{\displaystyle O}{\|}}{C}-O^-}{|}}{\underset{\underset{\displaystyle \overset{+}{NH_3}}{|}}{C}}-H}{\|}}{\underset{\underset{\displaystyle NH_2}{}}{C}}$

(d) $H-\overset{\overset{\displaystyle \overset{\overset{\displaystyle O}{\|}}{C}-O^-}{|}}{\underset{\underset{\displaystyle \overset{+}{H_3N}}{|}}{C}}-CH_2-$ [indole ring] $-CH_2-\overset{\overset{\displaystyle \overset{\overset{\displaystyle O}{\|}}{C}-O^-}{|}}{\underset{\underset{\displaystyle \overset{+}{NH_3}}{|}}{C}}-H$

hydrophobic interaction

21-7 No. The carboxyl groups of the three neighboring glutamic acids will repel each other, as will the amino groups of the three neighboring lysines.

21-8 (b) Isoleucine, (f) tryptophan, and possibly (c) glycine, because the side chain of glycine (—H) does not interact strongly with water.

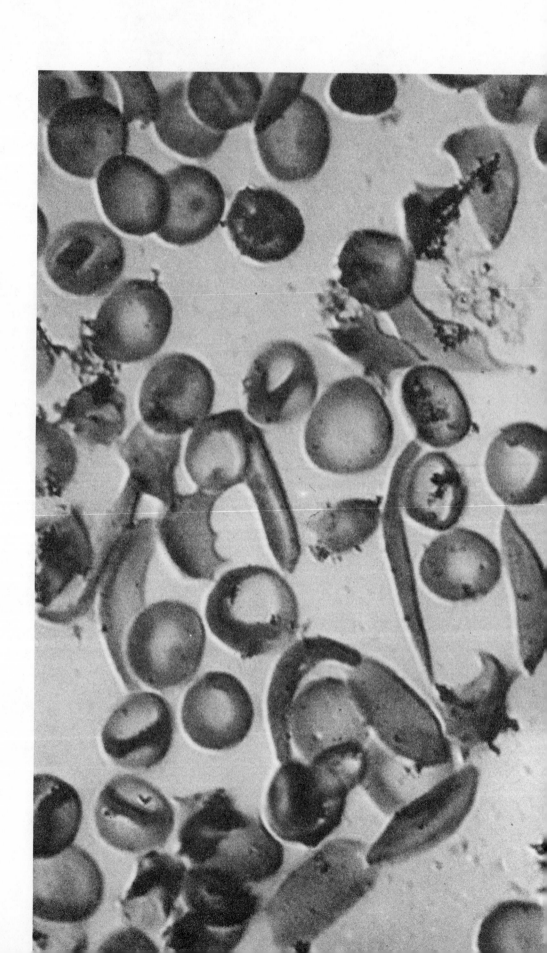

CHAPTER 22

Protein Function

22-1 INTRODUCTION

In the early years of this century, Henry Ford realized that a factory could operate much more efficiently if each worker assembled only a small portion of the total product. Of course, Ford was not the first person to realize the advantages of the assembly line, but he was extraordinarily successful at putting the idea into practice and so did as much as anyone to launch modern society into the age of specialization.

Long before Henry Ford promoted this industrial application of specialization, the proteins in living organisms had evolved their specialized functional roles. Each cell in the human body, for example, contains more than a thousand different proteins. Each protein has a specific, very limited role in the overall operation of the cell.

It would be impossible to study each protein individually; there are just too many. Instead, we shall concentrate on *functional categories* of proteins. The division of proteins into functional classes resembles the division of factory workers by classification. Many proteins function as *enzymes*, which work on cellular assembly lines called metabolic pathways. *Hormones* are chemical messengers and *carrier proteins* convey assembled products or nutrients between cells in different parts of the body. *Membrane transport proteins* work on the loading dock of the cellular factory, carrying molecules or ions into and out of cells. *Antibody molecules* guard the body against foreign invaders.

In the last chapter we saw that proteins are polymers of amino acids. Each different protein has a unique amino acid sequence that determines its overall structure. Its precise structure, in turn, determines its function in the cell. After we have classified proteins according to their roles in the cell, we shall look more closely at the structural and functional properties of protein hormones, transport proteins, and antibodies. Understanding how proteins function in the healthy individual will enable us to see how defects in protein function can lead to disease.

In the blood of a patient suffering from sickle cell disease, some of the red blood cells have a shape that is strikingly different from the normal doughnut-shaped red blood cell. The sickle shape is due entirely to a difference in a single amino acid in two of the four subunits of the hemoglobin molecule.

22-2   STUDY OBJECTIVES

After studying the material in this chapter, you should be able to:

1. Name the major functional classes of proteins and describe the role of each class in the cell and the whole body.

2. Describe how hormonal messages are amplified.

3. Differentiate among simple diffusion, passive transport, and active transport.

4. Define the terms prosthetic group and conjugated protein.

5. Describe specificity, saturation, and inhibition as they apply to membrane transport proteins.

6. Define the terms antigen, immunoglobulin, complement, agglutination, and vaccination.

7. Draw a diagram representing an IgG molecule and use the drawing to describe the structure and function of antibodies.

22-3   EVOLUTION OF PROTEIN FUNCTION

The complex and precise structure of each individual protein is essential to the survival of the living organism. During the course of evolution, the structures of individual proteins have been selected to optimize their specific functions. Even slight variations in the structure of a particular protein can lead to enormous consequences. For example, hemoglobin S, which occurs in people with sickle cell anemia, differs from normal hemoglobin A by only 1 amino acid out of 146 in half of the four hemoglobin subunits. The change of a single amino acid in the sequence of a protein is called an **amino acid substitution.** Such substitutions almost always lead to a functionally inferior or inactive protein, and thus to an organism with a poorer ability to survive and reproduce in a given environment.*

In our modern society, organisms with defects are not always selected against. Malfunctions such as diabetes or myopia (near-sightedness) can be compensated for. Given proper treatment, people with these and other disorders can survive and reproduce. In fact, the proportion of individuals with such disorders in the human population is actually increasing.

An amino acid substitution rarely leads to a functionally "better" protein. Yet these rare favorable changes are necessary, because eventually they lead to the evolution of organisms that are better adapted to their particular environment, organisms that have a better chance to survive and reproduce. These better adapted organisms are more likely to thrive and may even compete so successfully with those that do not possess the "better" protein that the latter organisms may become extinct.

---

* Interestingly, individuals possessing hemoglobin S have increased resistance to malaria. The **sickle cell trait,** in which a person has genes for, and synthesizes, both hemoglobin A and hemoglobin S, is quite common in individuals of African ancestry. Although most of the red blood cells of these individuals will sickle in laboratory tests, they have enough hemoglobin A to prevent this from occurring in their bodies. Individuals with **sickle cell disease** have genes for hemoglobin S only. Their red blood cells, which are devoid of any hemoglobin A, can sickle in their bodies — causing much pain and even death.

This process of evolution and selection has bestowed on modern organisms, including humans, a large number of functional proteins. Each of these proteins has evolved to the very specific structure it needs to function as it does. The general process of evolution of protein function has also made more highly evolved organisms different in appearance from their predecessors. The genetic information that each of us has inherited (such as height, a tendency for baldness, or the shape of our ears) is coded in our DNA. It is DNA that directs the synthesis of proteins in such a way as to give each of us our unique qualities.

Since virtually everything that happens in the body depends on the function of proteins, the study of human biochemistry can ultimately be viewed as the study of proteins and what they do. As we proceed to look at the functional roles and properties of proteins, we must always bear in mind that it is the highly evolved structure of a protein that determines its very specific function.

Exercise 22-1
In two subunits of hemoglobin S, the amino acid valine is substituted for a glutamic acid found in hemoglobin A. Hemoglobin S is much less soluble than hemoglobin A. How do you think a hypothetical hemoglobin, call it hemoglobin Z, with an aspartic acid substituted for this glutamic acid, would compare to hemoglobin A and hemoglobin S in solubility and function?

22-4 CLASSIFICATION OF PROTEINS

Proteins may be classified in two ways. They may be classified on the basis of their gross physical properties, such as solubility in water or coagulation by heat. Recall that coagulation is a special kind of irreversible denaturation in which protein precipitates out of solution, such as egg white solidifying upon cooking. Proteins may also be classified on the basis of their functional properties (that is, what they do).

Classification According to Physical Properties

The classification scheme based on physical properties was devised first, before the functions of most proteins were known. It is still a useful way to classify proteins because the physical properties of a protein are the result of its structure, as are its functional properties.

Table 22-1 shows the classification of proteins based on their physical properties. You will notice that **fibrous proteins** are important in the structure of hair, nails, and connective tissue. **Globular proteins,** on the other hand, include proteins with a wide variety of roles, such as enzymes and histones. For example, serum albumin is found in large quantities in human blood. One particular property, its high solubility, allows it to function as a carrier protein for molecules, such as fats, that would otherwise be insoluble in the aqueous medium of serum. Most other globular proteins are less soluble than albumin and are placed in the globulin subclass. Some globulins are antibodies, whose function is to bind and inactivate foreign substances and thus combat disease and infection. Most enzymes are also globulins. Enzymes work inside the cellular factories and are not usually found in the blood serum. Indeed, the presence of certain enzymes in serum is used to diagnose some diseases. For example, the injury of cardiac muscle cells caused by a myocardial infarction (heart attack) will cause these cells to release specific enzymes into the serum. Laboratory tests for these enzymes are used to detect the presence and severity of a heart attack.

Table 22-1 Classification of Proteins Based on Physical Properties

Class	Properties	Examples
Fibrous proteins	Insoluble in aqueous solutions; elongated molecules, often consisting of several coiled polypeptide chains	
Collagens	Can be converted into soluble gelatins by boiling; contain large amounts of hydroxy-proline and hydroxylysine but no cysteine or tryptophan	The major proteins of connective tissues
Elastins	Similar to collagens but cannot be converted to gelatins by boiling	Proteins of tendons and arteries
Keratins	Contain large amounts of cysteine	Hair, wool, nails (Hair is about 14% cysteine.)
Globular proteins	Soluble in aqueous solutions; spherical or ellipsoidal in shape	
Albumins	Readily soluble in pure water; coagulated by heat; function as carriers for hydrophobic molecules	Serum albumin, egg albumin
Globulins	Insoluble or only slightly soluble in pure water; very soluble in aqueous salt solutions; can be coagulated by heat	Enzymes and antibodies
Histones	Basic proteins; contain large amounts of arginine and lysine; soluble in pure water	Histones in chromatin
Protamines	Very basic proteins; contain large amounts of arginine, but no tryptophan or tyrosine	Found in sperm cell chromosomes

Other subclasses of proteins in Table 22-1 also have physical properties that are necessary for their particular function. For example, proteins known as histones contain large amounts of the basic amino acids arginine and lysine. This basic property allows histones to interact with the acidic phosphate portions of DNA molecules. The combination of histones and DNA in the cells of higher organisms such as humans is called chromatin.

Another practical use of this classification scheme can be found in the clinical laboratory, where the difference in solubility between albumins and globulins is used to determine their concentrations. The relative amount of albumins and globulins in serum is usually expressed as the ratio of albumin to globulin concentration, or the A/G ratio. Variations in the A/G ratio indicate several disease states, such as hypogammaglobulinemia, which means that a patient has less than (*hypo-*) the normal amount of serum gammaglobulins. Since most gammaglobulins are antibodies, a patient with a high A/G ratio may have an impaired resistance to infection, that is, a malfunctioning immune system.

Exercise 22-2
An individual is suspected of having a type of cancer called multiple myeloma, in which some of the cells that produce antibody molecules have become malignant. Would this individual's A/G ratio be lower or higher than normal? Explain your answer.

Classification According to Functional Properties

Classification of proteins based on function is relatively recent. We have already mentioned several of these functions, including transport, catalysis, and communication. Major functional classes of proteins are listed in Table 22-2. This

Table 22-2 Protein Classification by Function

Class	Properties	Examples
Catalytic proteins		
Enzymes	Catalyze chemical reactions	Lactate dehydrogenase (LDH), amylase, pyruvate dehydrogenase
Noncatalytic proteins		
Carrier proteins	Carry molecules or ions through the bloodstream	Hemoglobin, albumin
Receptor proteins	Bind hormones and neurotransmitters to cell membranes	The insulin receptor
Membrane transport proteins	Carry molecules across cell membranes	Na^+K^+-ATPase, which transports K^+ ions into cells and pumps Na^+ ions out of cells
Structural proteins	Form extracellular structures such as hair and nails	Collagen, keratin
Contractile proteins	Extend or contract to produce movement of muscles, cells, or subcellular parts	Myosin, tubulin
Protein hormones	Messenger molecules that direct the activities of various cells and organs	Insulin, adrenal corticotropic hormone (ACTH), growth hormone
Antibodies	Bind to foreign substances and activate their elimination from the body	Anti-Rh, Anti-A (antibodies to Rh factor and to blood group A)

method for classification of proteins is extremely useful. We want to know what proteins do and how they work. When a protein does not work, we want to know how this defect will affect the body as a whole. We shall spend the rest of this chapter and the next studying protein function.

Conjugated Proteins

Many proteins are self-sufficient, that is, they function perfectly well without the help of other molecules. Many other proteins require the assistance of a nonprotein molecule in order to function. For example, as we have previously mentioned, many enzymes require small molecules called coenzymes (such as NAD^+) in order to function properly as catalysts. As we shall see in Chapter 23, coenzymes are altered in enzyme-catalyzed reactions. In most enzyme-catalyzed reactions the altered coenzyme (such as NADH) quickly dissociates from the enzyme. On the other hand, several proteins, including some enzymes, have non-amino acid molecules or ions bound very tightly to them. We call such proteins **conjugated proteins,** and the auxiliary molecules bound to them are called **prosthetic groups.**

There are several types of prosthetic groups, ranging from metal ions to the complex heme group in hemoglobin. Heme is a complex ion containing a central iron ion bonded to a complicated nitrogen-containing organic molecule. Some examples of conjugated proteins and their prosthetic groups are listed in Table 22-3. Notice the wide variety of molecules that serve as prosthetic groups and the rather arbitrary distinction between glycoproteins and mucoproteins. Prosthetic groups are bound to their proteins very tightly, frequently via covalent bonds to certain amino acid side chains. They can also be bound via multiple charge interactions or by hydrogen bonds as in nucleoproteins. The role of prosthetic

Table 22-3 Conjugated Proteins

Type	Prosthetic Group	Properties	Example
Nucleoproteins	Nucleic acid (DNA, RNA)	Large, compact complexes	Chromatin, ribosomes
Mucoproteins*	Carbohydrate	More than 4% carbohydrate by weight	Human chorionic gonadotropin, a hormone used to test for pregnancy
Glycoproteins*	Carbohydrate	Less than 4% carbohydrate	Antibodies
Lipoproteins	Lipid	Water soluble	Serum lipoproteins
Proteolipids	Lipid	Not very water soluble, soluble in nonpolar solvents	Cell membranes
Hemoproteins	Heme group	Characteristic color	Hemoglobin, cytochrome c
Metalloproteins	Metal ion (Fe^{3+}, Zn^{2+}, Mg^{2+}, Mn^{2+})	Require a metal ion to function	Carbonic anhydrase

* The distinction between mucoproteins and glycoproteins is somewhat arbitrary.

groups in the function of a particular protein is not always completely understood. In hemoglobin, for example, oxygen actually binds to the heme group (see Figure 22-1), but the function(s) of the carbohydrate prosthetic groups found on antibodies and several enzymes is unclear.

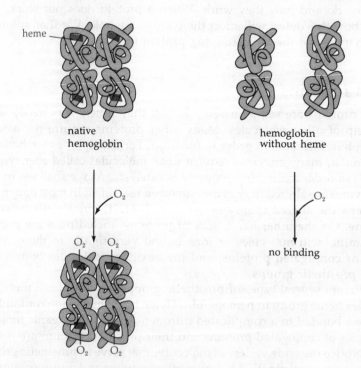

Figure 22-1   Hemoglobin has four subunits, each of which normally contains one heme prosthetic group. An O_2 molecule binds to each of the heme groups in native hemoglobin (left), but cannot bind to hemoglobin if the heme groups have been removed (right).

22-5 NONCATALYTIC PROTEINS

Proteins can be divided broadly into two functional categories: those that serve as catalysts—the enzymes—and those that do not. The noncatalytic proteins can be further classified according to their various functional roles. Carrier proteins, receptor proteins, membrane transport proteins, structural proteins, contractile proteins, hormones, and antibodies are all functional classes of noncatalytic proteins.

Carrier Proteins

Carrier proteins carry nutrients and water-insoluble vitamins to cells in need of them. Carrier proteins are the transportation workers of the human body. Albumin, hemoglobin, and transferrin are examples of carrier proteins. Albumin is relatively nonselective and transports a variety of fats through the bloodstream. Hemoglobin is more selective, but can still bind several different kinds of molecules. It binds oxygen tightly and carries it to cells that need it. Unfortunately, hemoglobin can also bind carbon monoxide (CO), and it does so more readily than it binds O_2. Therefore, exposure to too much CO leads to oxygen starvation of cells. The cells suffocate and the person dies. Transferrin is a carrier protein that is very selective. It carries iron from the intestines to the liver.

Receptor Proteins

The chemical messages between cells in the body, in the form of hormones and neurotransmitters, are received by specific **receptor proteins** located in cell membranes. After "capturing" the specific chemical messenger that it binds, a given receptor protein may (1) help escort the messenger across the cell membrane and into the cell, as is the case with steroid hormone receptors; or (2) activate or deactivate an enzyme located on the interior side of the cell membrane in order to convey the message to the interior of the cell. In this case the original messenger (a hormone) does not itself enter the cell. Receptors for the hormones adrenaline and insulin work in this fashion. The receptor protein may also (3) generate an electrical response. This occurs when the receptor proteins on nerve cells bind small molecules called **neurotransmitters,** which carry messages between nerve cells.

Membrane Transport Proteins

The **membrane transport proteins,** located in cell membranes, carry molecules and ions from the aqueous environment that surrounds the cell across the cell membrane, which contains a large amount of water-insoluble fats, and into the aqueous cytoplasm. (Some chemists refer to carrier proteins as transport proteins, but we shall not in order to avoid confusion.) Membrane transport proteins are analogous to workers on the loading dock of a factory. They handle shipping as well as receiving operations for the cellular factory. Sometimes transport across the cell membrane must be accomplished against a concentration gradient. This is hard work and requires energy input. As you will recall, molecules have a natural tendency to move from an area of high concentration to an area of low concentration. Molecules concentrated inside cells have a natural tendency to leak out. A great deal of the energy used daily by the body goes to reverse this tendency, moving molecules to a place where they are already more concentrated. We shall discuss transport across cell membranes in greater detail shortly.

Structural Proteins

Structural proteins form large polymers. Their functional properties, again, are due to their specific amino acid sequences. Hair and nails (keratins), the collagen of connective tissue, and the elastins of tendons and arteries are present in large amounts in the body. In fact, more than 30% of the total weight of protein in the body is collagen. Structural proteins have a broad range of properties. Keratins in nails are hard and brittle, elastins make tendons tough and elastic. Fibrin, the protein in blood clots, is formed from fibrinogen, a soluble globular protein normally present in unclotted blood. The process of blood clotting involves the conversion of fibrinogen into insoluble fibrin polymers by the enzyme thrombin (see Figure 22-2). The major difference between serum and plasma is the presence of fibrinogen in plasma. **Plasma** is obtained from blood that has been prevented from clotting and thus still contains fibrinogen. **Serum** is the fluid obtained from clotted blood and thus has no fibrogen.

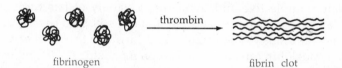

fibrinogen   fibrin clot

Figure 22-2   Certain stimuli, such as the injury   Thrombin in turn converts soluble
of a blood vessel, promote the   fibrinogen molecules into
conversion of prothrombin (an   insoluble fibrin clots.
inactive precursor) to thrombin.

Contractile Proteins

Contractile proteins function in the movement of cells and organs. A major example is muscle contraction. The contraction process in muscle cells is very complex and requires the association of several proteins into a network that contracts by the sliding of some proteins along a string of other proteins. Different contractile proteins are required for cells to divide, for sperm to swim, and so on. These contractile proteins, called *microtubules*, are long, hollow filaments that are built up from very large numbers of a subunit called tubulin. The spindle fibers of dividing cells, the tails of sperm cells, and the hairlike cilia on the cells lining the esophagus are examples of cellular components made from microtubules.

Hormones

Hormones are chemical messengers in the human body. Not all hormones are proteins. Some hormones are just modified amino acids, whereas others are lipid molecules. We shall discuss a number of these later in this text (see Chapters 26 and 28). However, several hormones are proteins or smaller polypeptides. Table 22-4 lists some of these polypeptide hormones and the messages they carry. By acting as chemical messengers, hormones allow cells in different parts of the body to communicate and thus to function in concert. They tell specific types of cells, called **target cells** (or target tissue), to speed up or slow down their assembly lines. One example is oxytocin:

$$
\begin{array}{c}
\text{tyr} \\
\text{ile} \diagdown \text{cys—NH}_2 \\
| \\
\text{S} \\
| \\
\text{S} \\
\text{gln} \diagdown \text{cys—pro—leu—gly—C—NH}_2 \\
\text{asn}
\end{array}
$$

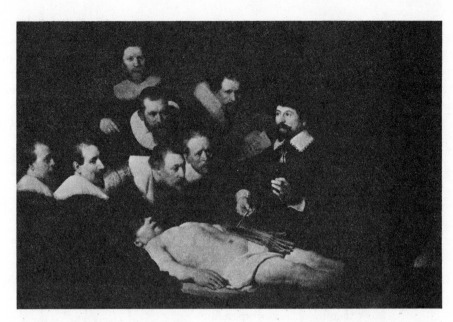

Tendons, which are composed to a large extent of elastic structural proteins, appear to be the subject of the anatomy lesson depicted in this Rembrandt painting.

The hormone oxytocin is released from the brain and travels through the bloodstream to its target tissue, the uterus. It tells the uterine muscle cells to contract, causing labor and childbirth. We can also use synthetic oxytocin or oxytocin obtained from another animal, such as a pig, to induce labor artificially.

Table 22-4 Some Polypeptide Hormones

Hormone	Producer Gland	Target Tissue	Effect
Luteotropin (prolactin)	Anterior pituitary	Mammary gland	Proliferation and initiation of milk secretion
		Corpus luteum of the ovary	Stimulation of progesterone secretion
Thyrotropin	Anterior pituitary	Thyroid	Formation and secretion of thyroid hormones
Somatotropin (growth hormone)	Anterior pituitary	Several	Several, including growth of bone and muscle
Follicle stimulating hormone	Anterior pituitary	Ovary	Secretion of estrogens and ovulation
		Testis	Development of seminiferous tubules, spermatogenesis
Oxytocin	Hypothalamus-posterior pituitary	Uterus	Contraction, parturition
		Mammary gland	Let-down of milk
Vasopressin	Hypothalamus-posterior pituitary	Arterioles	Regulates blood pressure
		Kidney tubules	Water resorption
Thymosin	Thymus	Lymphocytes	Maturation of T-lymphocytes
Insulin	Pancreas	General	Utilization of carbohydrate, increase in protein synthesis
Glucagon	Pancreas	Liver	Breakdown of glycogen to glucose
		Adipose tissue	Release of lipids

The pituitary gland, which sits at the base of the brain, makes a large number of polypeptide hormones. Pituitary hormones are released into the bloodstream upon orders from the brain. These orders are carried in the form of a small peptide, called a **releasing factor.** This multistep communication process between the brain and a target tissue is not wasteful. Each step in the process amplifies the signal. A minute quantity of a specific releasing factor from the brain causes the pituitary to release a large amount of a specific hormone. This specific pituitary hormone in turn stimulates an amplified response in its target tissue (see Figure 22-3).

Figure 22-3   The amplification that occurs in successive steps during the transmission of a hormonal message is represented by progressively larger arrows at each step. In this case, a stimulus causes the hypothalamus to release a tiny amount of a small peptide (a specific releasing factor), which in turn causes the pituitary gland to release a much larger amount of its corresponding specific hormone into the bloodstream. This hormone then binds to receptors on target cell membranes, causing the activation of a membrane-bound enzyme. Subsequent steps within the target cells result in an even greater amplification of the response.

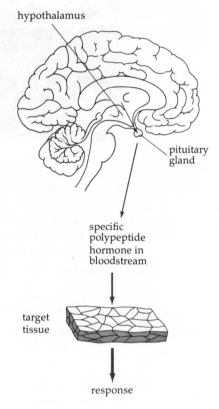

Hormones thus regulate the activities of cells throughout the body in response to several conscious and subconscious stimuli. Fright, for example, results in the brain's sending out the message: "Supply the muscle cells with more energy." The response to this hormonal message is the stimulation of enzymes in certain target cells. These enzymes break down stored glycogen polymers into their glucose components. In Chapter 25 we shall see that glucose is a very good source of energy.

Antibodies

Antibodies are the soldiers in the body's "defense department," which is called the immune system. Antibodies are fairly large proteins, containing more than 1400 amino acids each, and they are capable of specifically binding to certain molecules, called **antigens,** which are not normally found in an individual's body. The binding of antibody molecules to harmful antigens leads to their elimination from the body, as we shall see shortly.

Different individuals have different **blood types.** The most significant difference between blood types is the presence on an individual's red blood cell membranes of one of the three complex polysaccharides shown in Figure 22-4. These polysaccharides are the basis for the A, B, AB, and O blood groups. An individual with blood group A will have the group A polysaccharide on his or her red blood cell membranes but will not have the blood group B polysaccharide.

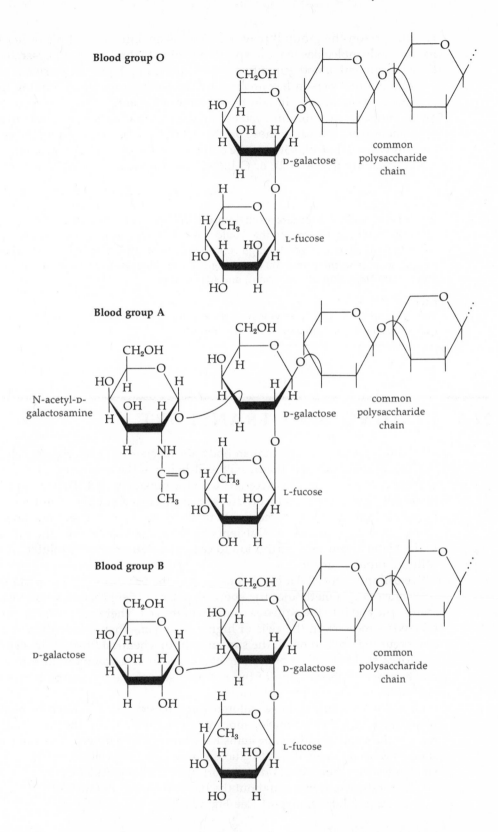

Figure 22-4   The complex polysaccharides that determine A, B, or O blood groups all consist of a polysaccharide chain, which is attached to the red blood cell membrane. Notice that the only difference between these polysaccharides is the presence of an additional galactose (group B) or N-acetylgalactosamine (group A) at the end of a common polysaccharide chain. These additional components are absent in the group O polysaccharide. For people with blood type AB, both group A and group B polysaccharide chains are present.

For this person the group B polysaccharide is an antigen, and his or her body has antibody molecules that can specifically bind to this group B polysaccharide. For a person with blood group B, however, the group B polysaccharide is not an antigen and that person has no antibody molecules capable of binding to this group B polysaccharide. In addition to producing antibodies to foreign blood group antigens, the immune system can make antibodies that will bind to many other molecules, including those on the cell walls of bacteria and on the surface coats of viruses. In fact, some of the antigens on the cell walls of certain bacteria are very similar to the A, B, or O blood group polysaccharides.

Exercise 22-3

A number of medical disorders result from abnormal hormone production. For each of the following, indicate a hormone that could possibly be involved: (a) dwarfism, (b) diabetes mellitus (characterized by high concentrations of glucose in blood and, when untreated, glucose in the urine), (c) diabetes insipidus (characterized by the elimination of abnormally large amounts of urine), and (d) sterility.

Exercise 22-4

The polypeptide hormone oxytocin is soluble in aqueous solution. Does oxytocin require a carrier protein to carry it to the uterus?

22-6   BINDING AND TRANSPORT

The cellular factories in the human body are specialized. Each type of cell has a specific responsibility for the overall well-being of the organism. The majority of these cells are not in direct contact with the outside world. They must rely on other cells and their protein products to provide the supply of nutrients and hormonal information they require to function properly. We just saw how hormones provide information. Nutrients must be transported from the intestines to the bloodstream for distribution to cells. If a nutrient is not soluble, it must utilize a protein carrier.

Even simple, unicellular organisms such as the bacterium *E. coli* in our intestines must have a method of transporting polar nutrients across a nonpolar cell membrane and into the aqueous confines of the cell. There are several aspects of transport processes in the cells of higher organisms that are identical to those that occur in *E. coli*. Indeed, the similarity of biochemical function in all organisms is a striking reflection of our evolution from a common ancestor. And much of our understanding of the biochemistry of the human body has been gained through the study of *E. coli*.*

Once food molecules have entered our digestive tract, enzymes begin to break them down into smaller molecular components. By the time the food reaches the small intestines, proteins have been broken down into amino acids, starch has been broken down to glucose, and so on. The cells in the lining of the small intestines then bind and transport these food molecules to the circulatory system for distribution throughout the body. We refer to transport across entire cells as **transcellular transport** (see Figure 22-5).

---

* This has led some biochemists to suggest that *all* biochemical events in humans are identical to those in *E. coli*, or: "The elephant is a large *E. coli*." This idea is true for many but not all of the biochemical events that occur in all organisms. The dissimilarities between humans and organisms as large and complex as mice have led to many difficulties in cancer research. Some treatments that kill tumors in mice have little or no effect on similar tumors in humans. Similarly, the drug thalidomide had no ill effects when tested on monkeys, but it produced disastrous birth defects in humans.

Figure 22-5   A schematic representation of transcellular transport. Molecules (indicated by blue circles and squares) are carried across the cells lining the small intestines by specific membrane transport proteins.

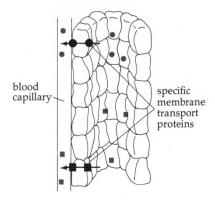

Very little is known about transcellular transport. Some molecules may even pass in between the intestinal cells. However, specific membrane transport proteins appear to be more frequently employed. In the case of amino acids, specific membrane transport proteins are required for different groups of amino acids. For example, the amino acids serine and threonine are transported by the same transport protein.

Once food molecules have been transported to the bloodstream, those that are not soluble in the aqueous plasma must be transported by carrier proteins to the cells. One example is apo-β-lipoprotein, which is made by the intestinal cells and apparently picks up its lipid passengers at or near the site of its synthesis.

Characteristics of Membrane Transport Proteins

Membrane transport proteins are often required to allow food molecules to enter "hungry" cells. Some molecules may pass through the membrane by a **simple diffusion process** (due to a concentration gradient) without the help of another molecule in the membrane (see Figure 22-6).

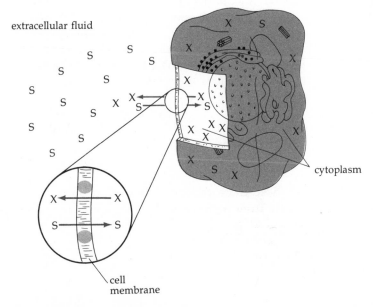

Figure 22-6   Schematic representation of simple diffusion of molecules across a cell membrane. The molecules simply move through the cell membrane from a region of high concentration toward one of lower concentration, as shown in a magnified view of a portion of the cell membrane. Note that the [S] is higher outside the cell, whereas [X] is higher inside the cell.

Very often, however, the transport of a molecule or ion requires the assistance of a membrane transport protein. There are three readily identifiable characteristics that distinguish protein-assisted transport across cell membranes from simple diffusion: **saturation, specificity,** and **inhibition.**

First, although the rate of diffusion increases with increased solute concentration, the rate of transport of solutes into or out of a cell is limited by the number and efficiency of its transport proteins in the cell membrane. Once all of these proteins are working as fast as they possibly can, an increase in solute concentration will not increase the rate of transport. When this happens, we say that the transport proteins are **saturated.** Second, diffusion is nonspecific, whereas transport is **specific.** The transport proteins are selective in what they will carry into and out of the cell. For example, amino acid transport proteins in humans will only carry L-amino acids; D-amino acids are discriminated against. Third, membrane transport proteins can be **inhibited** by molecules that are structurally similar to the molecule they transport. The transport protein is fooled and binds the inhibitor. The inhibited transport protein is then unable to transport its usual solute. Saturation, specificity, and inhibition are also important properties of enzymes, as we shall see later.

Exercise 22-5
Carrier proteins in the blood also show the characteristics of saturation, specificity, and inhibition. Use hemoglobin and O_2, N_2, and CO to describe these characteristics.

Passive Transport Versus Active Transport

The proteins that transport molecules across cell membranes can be further divided into two types, depending on whether or not the transport process requires energy.

When transport does not require energy it is called **passive transport** (see Figure 22-7). Passive transport is necessary to transfer polar molecules across the nonpolar cell membrane. A good example is the transport of glucose into human erythrocytes (red blood cells). The rate of entry of glucose into erythrocytes increases as the concentration of glucose goes up, but the rate reaches a maximum when the transport proteins are saturated. The glucose transport proteins are also specific. They transport D-sugars, such as D-glucose, D-galactose, and D-ribose. They will not bind and transport L-isomers of these sugars. There are hundreds of thousands of glucose transport protein molecules on each erythrocyte membrane, and they can be inhibited by molecules that are similar to glucose in structure. Therefore, the glucose transport protein satisfies all of the characteristics of protein-assisted transport (saturation, specificity, and inhibition).

In addition to these characteristics, **active transport** proteins use energy to transport molecules from regions where they are relatively dilute to regions where they are already concentrated. The energy is supplied by an exergonic chemical reaction, such as the hydrolysis of ATP, which we shall discuss in Chapter 25. For the time being, let us symbolize this reaction as follows:

$$ATP \rightarrow ADP + \text{phosphate ion} + \text{energy}$$

One of the best-known examples of active transport is the Na^+K^+-ATPase pump. This protein pumps K^+ into cells while simultaneously pumping out Na^+. The sodium ion concentration in blood plasma is already much higher than in cells, so energy must be expended to pump more Na^+ into plasma. Potassium

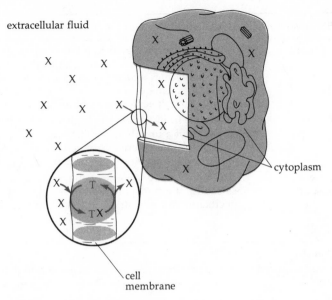

Figure 22-7   Schematic representation of passive transport of molecules via the membrane transport protein T. The molecules move from a region of high concentration to one of lower concentration, but they require the assistance of a membrane transport protein.

ions are more concentrated in cells than in plasma, so work is also needed to pump even more K^+ into cells. In both cases, the hydrolysis of ATP supplies the required energy (see Figure 22-8). Several other active transport proteins are known. They transport sugars, amino acids, and other molecules into or out of cells. Diffusion, active transport, and passive transport are compared in Table 22-5.

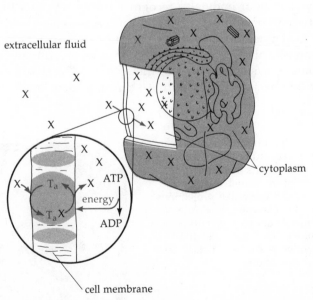

Figure 22-8   Active transport of the substance X into a cell where it is already concentrated is accomplished by a specific active transport protein, represented by T_a. The energy required to carry even more X into the cell is supplied by the hydrolysis of ATP.

Table 22-5 Movement of Molecules and Ions Across Cell Membranes

Diffusion:	Movement from a higher to a lower concentration without the assistance of a membrane transport protein. Only certain molecules and ions can readily diffuse across cell membranes. This process does not exhibit the characteristics of saturation, specificity, or inhibition.
Passive transport:	Movement from a higher to a lower concentration with the assistance of a membrane transport protein. This process exhibits the characteristics of saturation, specificity, and inhibition. No energy input is required.
Active transport:	Movement from a lower to a higher concentration with the assistance of a transport protein, exhibiting saturation, specificity, and inhibition. Energy input is required.

Exercise 22-6
Why is energy required to transport molecules from a lower to a higher concentration?

22-7 ANTIBODIES AND IMMUNITY

In order for the human body to survive and thrive in a world filled with disease-causing viruses and bacteria, it must be capable of defending itself against invasions by these agents. As we mentioned previously, this defense is the responsibility of the **immune system,** and antibody molecules can be considered to be the soldiers of this defense system. The antibodies are produced by a certain type of white blood cell called a **lymphocyte.** Lymphocytes circulate through the body in search of foreigners such as bacteria. When a foreigner is encountered it is usually inactivated and removed from the body.

Lymphocytes in your body also recognize cells from other humans and differences between your normal cells and cancerous cells. One consequence of this is that blood must be typed and matched to a recipient before a transfusion in order to prevent the destruction of the transfused red blood cells by the recipient's immune system. Likewise, transplanted kidneys, hearts, and other organs may be rejected if they do not share the same tissue type as the recipient's cells.

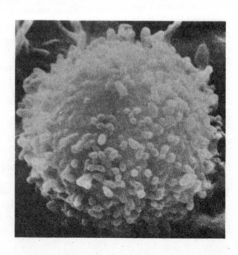

A human lymphocyte as it appears under the scanning electron microscope. These cells measure about 5×10^{-6} m in diameter and are covered with fingerlike projections called villi. The functional role of these membranous projections is not known. The molecules that bind antigens are too small to be seen in this photo.

To understand the chemical basis for the recognition and elimination of such foreigners by the immune system, we must first examine which kinds of molecules are recognized as foreign, that is, which are antigens. Then we must examine the structure of the molecules that lymphocytes frequently employ as soldiers—antibodies. We shall then study how antibodies are formed, how they work, and how we can stimulate the immune system to produce antibodies against deadly organisms in the process of immunization.

Antigens

In a healthy individual viruses, bacteria, mismatched red blood cells, and other foreign "invaders" are distinguished from native body cells by the immune system. The term **antigen** is used to describe those molecules that can be recognized as foreign by the immune system of a given individual. Any molecule, such as a protein, that is synthesized by the cells of an individual does not normally act as an antigen in that person's body, but it may be recognized as foreign (act as an antigen) if introduced into the body of a second individual. A very large number of molecules, including most foreign proteins and certain polysaccharides, can be recognized as antigens by the immune system of a normal person.

A very slight difference between a foreign molecule and one native to an individual may be sufficient to be recognizable by that person's immune system. For example, in Figure 22-4 we saw that the only difference between the A and B blood groups was the identity of the last sugar of a common polysaccharide chain. Similarly, a protein that differs by only a few amino acids from a protein native to an individual may be an antigen for that individual. For example, the hormone insulin, which is administered to persons with diabetes, is usually obtained from pigs or cows. The insulins obtained from these animals have slightly different amino acid sequences than human insulin, and some patients make antibodies against them. Table 22-6 lists various types of antigens that are important to human health.

Table 22-6 Some Common Types of Antigens

Type	Specific Examples	Importance
Viral	Polio virus coat protein	Used for polio vaccination
Bacterial	Bacillus of Calumette and Guerin (BCG)	Being tested for cancer immunotherapy
Blood group	B substance, Rh substance	Blood typing for transfusions
Histocompatibility (tissue antigens)	Histocompatibility antigen HLA-A7	Involved in organ transplantation
Fetal	Alphafetoprotein, carcinoembryonic antigen	Produced by fetuses and by some tumors, but not found in normal adults

Antibodies

Let us consider what happens when a foreign antigen enters the human body, as, for example, in a bacterial infection. The invading bacteria grow and reproduce rapidly in the warm, nutrient-rich body. Some of them also come into contact with lymphocytes, which constantly circulate through the body (see Figure 22-9). Attached to the membranes of some of these lymphocytes are antibody molecules specific for these bacterial antigens, allowing the lymphocytes to bind to the bacteria. This binding causes the lymphocytes to divide, producing many more cells that are capable of making antibodies against these specific bacterial antigens. In fact, an increase in the number of white blood cells in the circulatory system is used in the diagnosis of certain illnesses, such as acute infection of the appendix (appendicitis). Eventually the rate of antibody production exceeds the rate of bacterial multiplication and these antibody molecules bind to all of the bacteria and cause their inactivation and elimination from the body. This entire process may require several days, during which time the individual may feel quite ill.

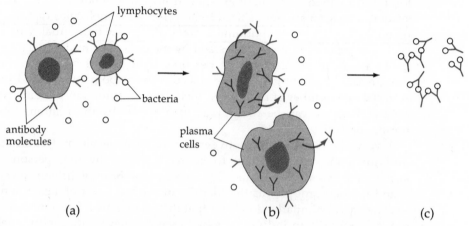

(a)   (b)   (c)

Figure 22-9   When the body is infected by bacteria, a few of the many circulating lymphocytes have antibody molecules on their membranes that specifically bind to antigens on the invading bacteria (a). These lymphocytes then divide several times—producing larger antibody-producing plasma cells and numerous identical lymphocytes. The plasma cells produce and secrete large amounts of antibodies directed against the bacterial antigens (b). These antibodies bind to the bacteria (c) and cause their elimination.

What special structural features enable antibody molecules to bind to antigens and cause their elimination from the body? Antibodies, which are also called **immunoglobulins,** are globular proteins. There are five types or classes of immunoglobulin in normal humans. The most prevalent one is immunoglobulin G, abbreviated IgG. IgG molecules are oligomers consisting of two large and two small polypeptide chains or subunits. The large chains are about twice as long as the small chains, and all four chains are held together by disulfide bonds and by hydrophilic and hydrophobic interactions (see Figure 22-10). Notice that an individual IgG molecule is symmetrical, so that it possesses two identical arms. At the tip of each arm is a binding site for the particular antigen with which it reacts. The binding sites of an antibody molecule are very specific for a particular antigen. The antigen is bound via hydrophilic and hydrophobic interactions with the side chains of individual amino acid components of the antibody.

Because one molecule of IgG can bind to two identical antigens, it is said to be **divalent.** This double-binding ability is crucial to the function of IgG. Consider the reaction between IgG antibody molecules that are specific for type B blood group antigen and type B red blood cells (see Figure 22-11). There are several type B antigen molecules on each type B red blood cell. Because the antibody molecules are divalent, some of them will bind to both an antigen on one red blood cell and to another antigen on a different red blood cell. When a large number of these divalent interactions occur, they cause the red blood cells to

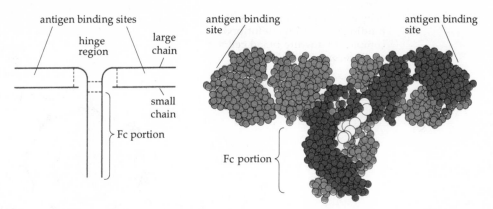

Figure 22-10   The structure of an IgG molecule. In the simplified schematic representation on the left, each component polypeptide chain is indicated by a solid line, and disulfide bonds by dotted blue lines. The figure on the right, a space-filling model of an actual human IgG molecule, shows the compact arrangement of the component polypeptide chains. The large chains are shaded in blue and the small chains in gray. The open circles represent a carbohydrate prosthetic group.

clump together or **agglutinate.** Agglutination of red blood cells can occur in the circulatory system of an individual given a transfusion of blood that is not matched to his or her blood type. This type of agglutination can produce severe clinical problems, and can lead to death. To prevent such problems, a few blood cells from any unit of blood considered for a transfusion are routinely mixed in a test tube with some serum from the patient who is to receive the blood. This procedure is called a **cross-match.** If agglutination occurs in the cross-match test, the unit of blood in question is not transfused into that patient.

Figure 22-11   Schematic representation of the agglutination of red blood cells. The binding of several antibody molecules to the antigens on individual red blood cells is accompanied by cross-linking of different cells by these divalent antibody molecules.

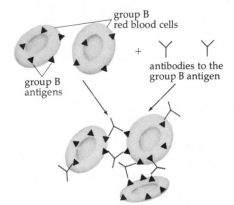

IgG and other immunoglobulins can also agglutinate invading bacteria. The human body can then easily eliminate the small clumps of bacteria.

An IgG molecule with a different amino acid sequence is required for each antigen that is to be recognized. The differences between these IgG molecules are restricted almost entirely to the antigen binding regions. The rest of the molecules are essentially identical, and they have common functions. The "hinge region" (see Figure 22-10) allows two antigens to be bound at varying angles. The Fc portion of an IgG molecule can function in a number of ways. It may allow an antibody and its antigen, an antigen-antibody complex, to be bound to different types of white blood cells via an **Fc receptor site** (see Figure 22-12). The binding of the complex to the Fc receptor site activates the destruction of certain types of antigens by mechanisms that are not well understood.

Figure 22-12   Schematic representation of the binding of an antigen-antibody complex to a white blood cell via an Fc receptor site.

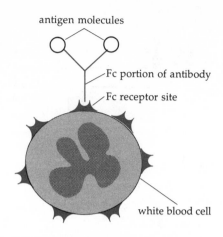

The Fc portions of antibodies are also involved in complement reactions. **Complement** is a term used to refer to a special group of proteins found in human blood. When several antibody molecules are bound to antigens on certain types of foreign cells, their Fc portions can bind to complement proteins (see Figure 22-13). The complement proteins then catalyze reactions that in turn cause the destruction of the foreign cell. Besides agglutination and complement reactions, there are a number of other ways in which foreign cells can be eliminated after binding by antibody molecules.

In addition to IgG, there are four other classes of immunoglobulins: IgA, IgM, IgD, and IgE. All are composed of small and large chains that combine to form molecules with more than one antigen binding site. IgA is found primarily in secretions (milk, saliva) and the intestinal tract, whereas IgG and IgM are found

Figure 22-13   When several antibody molecules attach to a foreign cell or organism (a), they trigger the binding of complement proteins to their Fc parts (b). The complement proteins then puncture the cell membrane (c), killing the cell.

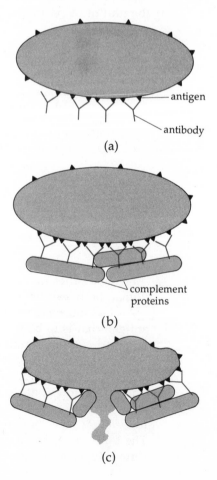

predominantly in blood. IgE antibodies are present in very low concentrations, but they play an important role in allergic reactions. Little is known of the IgD class of immunoglobulins. The properties of the five classes of immunoglobulins are summarized in Table 22-7. Note that an IgM molecule, for example, has 10 large and 10 small chains and can thus bind to 10 identical antigen molecules simultaneously.

Table 22-7 The Five Classes of Human Immunoglobulins

Immunoglobulin Class	IgG	IgA	IgM	IgD	IgE
Number of small chains per molecule	2	2 or 4	10	2	2
Number of large chains per molecule	2	2 or 4	10	2	2
Structural representation	T	T or /\ /\	(star shape)	T	T
Serum concentration (mg/ml)	12	1.8	1.0	0.03	0.0003
Functional role	Complement and agglutination reactions in serum	Some agglutination reactions	Complement and agglutination reactions in serum	Not known	Allergic reactions

Immunization

A cornerstone of modern medicine is the deliberate stimulation of lymphocytes in normal individuals to produce antibodies against disease-causing bacteria and viruses—a process called **immunization.** Basically, this process involves the injection of a specified dose of a bacteria or virus that has previously been killed or otherwise inactivated (to prevent illness). The antigens on the dead bacteria or virus stimulate lymphocytes to divide, just as the live infectious organisms would. The immunization results in the production of a large number of lymphocytes that produce antibody molecules. These antibodies do not remain in the bloodstream for very long. The larger number of lymphocytes, however, do remain in circulation and are quickly able to produce very large amounts of antibody should the individual be infected by that organism at a later time.

In 1798, Edward Jenner achieved the first successful immunization, which was made against the smallpox bacillus. Today we immunize individuals against a wide variety of diseases. However, we still do not know exactly how a foreign antigen can trigger a lymphocyte to divide and synthesize antibodies. There are many other aspects of the human immune system that are not yet well understood but that are actively being investigated. Experiments currently being conducted on the immune system offer great promise for significantly improving medical care within our lifetimes.

Exercise 22-7

In addition to group A and group B antigens, an individual's blood type is determined by a number of other antigens. Hemolytic disease of the newborn can result in the death of an infant whose blood type is Rh$^+$ (i.e., whose red blood cells have the Rh antigen), if the child's mother is Rh$^-$. We shall not discuss the detailed pathology of this disease, but with proper prenatal care most cases of hemolytic disease of the newborn can now be avoided. Suggest two laboratory tests that should be a part of good prenatal care in order to prevent this disease.

22-8   SUMMARY

1. The unique amino acid sequence of a protein determines its structure and its function.

2. The change of a single amino acid in a protein is called an amino acid substitution.

3. Proteins can be classified according to their physical properties or according to their functional properties.

4. There are two main functional classes of proteins: catalytic proteins or enzymes and noncatalytic proteins.

5. The subclasses of noncatalytic proteins are carrier proteins, receptor proteins, membrane transport proteins, structural proteins, contractile proteins, hormones, and antibodies.

6. A conjugated protein has an auxiliary molecule, called a prosthetic group, tightly bonded to it.

7. Membrane transport proteins exhibit three basic features: saturation, specificity, and inhibition.

8. There are three different ways in which molecules cross cell membranes: simple diffusion, passive transport, and active transport. Active transport requires energy.

9. The human immune system contains lymphocyte cells, which can bind to antigens and produce antibodies.

10. An antigen is a molecule that can be recognized as foreign by lymphocytes and antibodies.

11. An antibody or immunoglobulin is a protein, produced by lymphocytes, that possesses two or more binding sites for a particular antigen.

12. Agglutination and complement reactions are two ways in which antigens can be eliminated after their recognition by antibody molecules.

PROBLEMS

1. A substance X moves from the outside of a cell to the inside under conditions where the concentration of X inside the cell is greater than the concentration of X outside. This process is an example of which of the following: (a) simple diffusion, (b) passive transport, (c) active transport, (d) allosteric activation, (e) none of these?

2. Which of the following are found as prosthetic groups on at least some proteins: (a) carbohydrates, (b) lipids, (c) hemes, (d) metal ions, (e) none of these?

3. Which of the following classes of proteins is most soluble in pure water: (a) collagens, (b) albumins, (c) globulins, (d) keratins, (e) elastins? Explain your answer.

4. Thrombin is which of the following: (a) a protein hormone, (b) a membrane transport protein, (c) a coenzyme, (d) an enzyme, (e) a structural protein?

5. Which of the following are not characteristic properties of a membrane transport protein: (a) specificity, (b) inhibition, (c) coenzyme requirement, (d) saturation, (e) solubility?

6. Consider an IgG antibody molecule that binds to the blood group B antigen. The molecular weights of the large and small peptide chains in this antibody are 50,000 and 25,000, respectively. What is the molecular weight of the entire antibody molecule? How many antigen binding and Fc parts does this molecule contain?

SOLUTIONS TO EXERCISES

22-1 Hemoglobin Z, with the acidic glutamic acid side chains replaced by other acidic side chains in aspartic acid, should be as soluble as hemoglobin A, and probably would be indistinguishable from hemoglobin A in function.

22-2 An individual with multiple myeloma has malignant antibody-producing cells, which tremendously increases the concentration of antibodies—globulins—in the serum. A patient suffering from this malignancy should have a lower-than-normal A/G ratio.

22-3 (a) Lack of sufficient somatotropin (growth hormone)
  (b) Insufficient, or inactive, insulin
  (c) Insufficient, or inactive, vasopressin
  (d) Insufficient, or inactive, follicle stimulating hormone, or inappropriate amounts of one of the other sex hormones

22-4 Carrier proteins are used to transport relatively insoluble substances, such as oxygen and fats. Since oxytocin is soluble, it does not require a carrier protein.

22-5 A hemoglobin molecule is saturated when it has bound four O_2 molecules. Even with a large excess of O_2, no more oxygen can bind. Hemoglobin normally binds and transports O_2. It does not bind to N_2. CO, however, binds more readily to hemoglobin than does O_2 and can thus inhibit O_2 binding.

22-6 Molecules have a natural tendency to move from an area of high concentration to one of lower concentration. This is an example of the natural tendency toward disorder (increased entropy). Energy is required to overcome this tendency.

22-7 Good prenatal care should include determination of the mother's Rh group. If she is Rh^-, then tests should be performed to determine if she has antibodies directed against Rh^+ cells.

CHAPTER 23

Enzymes

23-1 INTRODUCTION

Using our factory analogy, enzymes are skilled workers who are employed on the cellular assembly lines. Two important facts should be remembered: Virtually all of the reactions that occur in living cells are catalyzed by enzymes, and all enzymes are proteins. Enzymes are the most numerous and diverse of all the functional types of protein. Enzymes possess the same basic structural features as all other proteins; that is, they are polymers of amino acids whose side chains interact to form a unique three-dimensional structure. Enzymes also share certain functional features with other proteins, such as membrane transport proteins, in that they show *specificity* for the molecules on which they work and also exhibit the properties of *saturation* and *inhibition*, as we shall see shortly.

Enzymes are very good catalysts. Many reactions that occur rapidly within living cells proceed at an incredibly slow rate in the absence of the required enzymes. Not only do enzymes increase the rates of reactions within cells, but they catalyze only those reactions that produce desired products.

Some enzyme-catalyzed reactions occur in every living cell. Others occur only in certain types of organisms or only in certain types of cells. Our appearance, as well as the way we function, is due to our unique assortment of enzymes. Major breakthroughs in the understanding of the biochemical workings of the human body have often resulted from the discovery of new enzymes. Increasingly, enzymes are being used in modern medicine. For example, certain diseases are diagnosed by measuring the levels of specific enzymes in a patient's serum. Other diagnostic tests make use of the tremendous specificity of enzymes to measure the quantity of smaller molecules in serum.

Remember that all catalysts, including enzymes, increase reaction rates but do not alter the position of equilibrium in the reactions they catalyze. Why, then, you might ask, don't all of the reactions in cells quickly reach a state of equilibrium? This does not happen because the products of one enzyme-catalyzed reaction are the starting materials for other reactions. Almost all of the molecules within each living cell are constantly being synthesized and used up. The reactions in cells reach a state of equilibrium only when the cells die.

Droplets secreted by the very specialized leaves of the Venus flytrap contain enzymes that digest the insect prey of this carnivorous plant.

In this chapter we shall examine the unique functional features of enzymes, including their specificity and the rates at which enzyme-catalyzed reactions proceed. Finally, we shall look at the way enzyme function is controlled by cells. This is very important. Using our factory analogy, cellular control of enzyme activity is like supervisors telling assembly line workers to speed up or slow down the production of a product so that the supply is kept in line with the demand.

23-2 STUDY OBJECTIVES

After studying the material in this chapter, you should be able to:

1. Describe the active site of an enzyme and the process of induced fit.
2. Discuss the specificity, saturation, and inhibition of enzymes.
3. Describe the simplest two-step mechanism, $E + S \rightleftharpoons ES$, followed by, $ES \rightarrow E + P$, for enzyme-catalyzed reactions.
4. Define molar activity and maximum rate (V_{max}).
5. Use the relationship among molar activity, maximum rate (V_{max}), and enzyme concentration to calculate any one of these quantities, given numerical values for the other two.
6. Explain the difference between a competitive and a noncompetitive enzyme inhibitor.
7. Explain how pH and temperature affect the rates of enzyme-catalyzed reactions.
8. Describe the function of coenzymes in some enzyme-catalyzed reactions, as well as the connection between coenzymes and vitamins.
9. Describe how an enzyme-catalyzed reaction can be regulated by the substrate concentration, rate of enzyme synthesis, allosterism, and chemical modification.
10. Describe how a sequence of enzyme-catalyzed reactions can be regulated by a process of feedback inhibition.
11. Describe the relationship between competitive inhibitors and the action of many drugs.

23-3 ENZYME CATALYSTS

Thousands of enzymes have been identified, and several have been purified and studied in great detail. We shall not attempt to describe all of the known enzymes. Instead we shall examine the common properties and characteristics of some typical enzymes and the reactions they catalyze. Again, we must keep in mind that enzymes are proteins. An enzyme's unique function is a result of its unique protein structure.

Catalytic Power

One common property of enzymes is their tremendous **catalytic power**—their ability to increase greatly the rate of a chemical reaction. As we discussed in Chapter 9, several factors influence the rate of any chemical reaction. For example, the rate of many chemical reactions doubles with each 10°C rise in temperature. The observed rate of a chemical reaction depends on the overall activation energy for that reaction. Recall that for a single step in a reaction mechanism, the activation energy (E_a) is the amount of energy required to form an activated complex in which the reactants are in close contact and in the correct orientation to each other to favor formation of products. Catalysts, as you will recall, alter the reaction pathway to one that has a lower overall E_a and thereby increase the reaction rate.

An example of the decrease in activation energy for an enzyme-catalyzed reaction is the reaction $2H_2O_2 \rightarrow 2H_2O + O_2$. Without the presence of an enzyme catalyst, this reaction has an activation energy in aqueous solution of 18 kcal/mole. The enzyme *catalase* decomposes hydrogen peroxide in an enzyme-catalyzed pathway with an overall E_a of only 2 kcal/mole. The activation energy for the reaction pathway catalyzed by catalase is only one-ninth (1/9) that of the uncatalyzed pathway. Therefore, the rate of the enzyme-catalyzed reaction is much faster than the rate for the uncatalyzed pathway.

In enzyme-catalyzed reactions, the reactant on which an enzyme works is called a **substrate.** The simplest general model, which is applicable to many but not all enzyme-catalyzed reactions, is the following two-step mechanism:

Step 1 $E + S \rightleftharpoons ES$

Step 2 $ES \rightarrow E + P$

where E, S, and P represent enzyme, substrate, and product, respectively, and ES represents an intermediate in which the enzyme and substrate are bound together. Notice that, according to this model, product(s) can be formed only if an enzyme and a substrate first bind together to form ES. The intermediate ES can either revert back to unbound enzyme and substrate or can break down to form product(s) and regenerate free enzyme.

There are a number of ways whereby an enzyme can lower the activation energy for a chemical reaction. First, it provides a surface on which the reactants can come together in close proximity. Second, it binds its substrates in the correct orientation for the reaction. Third, it may actually stretch some of the bonds in the reactants (substrates), making them easier to break. [Many of these properties are also common to nonenzyme catalysts. For example, platinum (a nonenzyme catalyst) is often used by chemists to provide a surface on which reactants can come close together.] The tremendous catalytic power of enzymes, however, results from the exact combination of effects necessary for an individual reaction. Enzymes align their substrates in the right orientation and under the right conditions to favor a rapid reaction.

Nomenclature for Enzymes

Most enzymes are named and classified according to the reactions they catalyze. Their names are their job descriptions in the cellular factory. For example, *urease* catalyzes a reaction in which urea is hydrolyzed to ammonia and carbon dioxide:

(23-1) $$H_2O + H_2N-\underset{\underset{\text{urea}}{O}}{\overset{\overset{\|}{C}}{}}-NH_2 \xrightarrow{\text{urease}} CO_2 + 2NH_3$$

In the early days of biochemistry, enzymes were often named by adding the suffix -*ase* to the name of the substrate. Thus the enzyme that hydrolyzes urea is called urease. An enzyme that hydrolyzes ATP would be an ATPase. However, we shall see that there are several enzymes that hydrolyze ATP, and, in order to be precise, names must be used that distinguish different enzymes that work on the same substrate.

Several enzymes were originally given nondescriptive names. Catalase is one example. The protein-splitting enzymes in our digestive tract were given the names trypsin, chymotrypsin, and pepsin. Even though the names of these enzymes do not indicate what reactions are catalyzed, nondescriptive names are still used for many common enzymes. As more enzymes were discovered, however, nondescriptive names became confusing. Therefore, an international commission on enzymes was convened in order to adopt a systematic enzyme nomenclature system. Each enzyme was given a precise name and a number. Although we shall not discuss this nomenclature system in great detail, a look at the major classes will be helpful.

Enzymes are divided into six major classes according to the type of reaction they catalyze. These classes are listed in Table 23-1, which illustrates the many types of reactions that can be catalyzed by enzymes. For example, the Enzyme Commission placed the enzymes that catalyze oxidation-reduction reactions in one class called **oxidoreductases.** One specific oxidoreductase is called glycerol dehydrogenase, which oxidizes glycerol to dihydroxyacetone (see Table 23-1) while reducing the coenzyme NAD^+. In Section 23-6 we shall discuss the role of hydrogen acceptors such as NAD^+, as well as acceptors for functional groups removed from substrates by enzymes. Another large class of enzymes, the **transferases,** transfer functional groups of atoms from one molecule to another. Glutamic-pyruvic transaminase is an example (Table 23-1). This enzyme catalyzes the exchange of an amine group on one molecule and a carbonyl oxygen of another molecule. Other classes of enzymes listed in Table 23-1 are hydrolases, lyases, isomerases, and ligases. **Hydrolases** break bonds with the addition of water. **Lyases** break bonds without the addition of water. Some lyases catalyze decarboxylation reactions. **Isomerases** interconvert isomers. **Ligases** join small molecules together into larger ones, with the energy for synthesis coming from hydrolysis of ATP (see Chapter 25). This latter reaction can be represented by $ATP + H_2O \rightarrow ADP + P_i + energy$, or $ATP + H_2O \rightarrow AMP + PP_i + energy$.

Table 23-1   Classification of Enzymes

Class 1. Oxidoreductases (catalyze oxidation-reduction reactions)
   Example: glycerol dehydrogenase

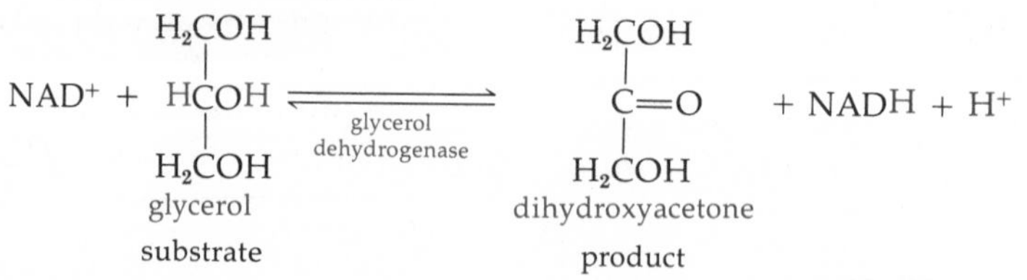

Class 2. Transferases (transfer groups of atoms)
   Example: glutamic-pyruvic transaminase

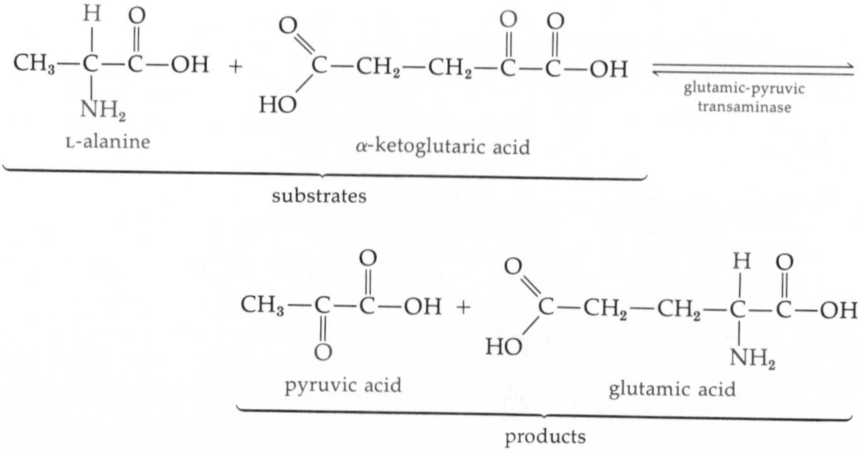

Class 3. Hydrolases (break bonds with addition of water)
   Example: carboxypeptidase A

$$H_2N\text{-gly-ala-phe-asp-gly-}\overset{\displaystyle O}{\overset{\|}{C}}\text{—OH} + H_2O \xrightarrow[\text{carboxypeptidase}]{}$$

(one of many substrates)

$$H_2N\text{-gly-ala-phe-asp—}\overset{\displaystyle O}{\overset{\|}{C}}\text{— OH} + H\text{—}\overset{\displaystyle H}{\overset{|}{N}}\text{—}CH_2\text{—}\overset{\displaystyle O}{\overset{\|}{C}}\text{—OH}$$

glycine

products

Class 4. Lyases (cleave bonds without addition of water)
   Example: pyruvate decarboxylase

$$H_3C\text{—}\overset{\displaystyle O}{\overset{\|}{C}}\text{—}C\overset{\displaystyle O}{\underset{\displaystyle OH}{}} \xrightarrow[\text{decarboxylase}]{\text{pyruvate}} H_3C\text{—}C\overset{\displaystyle H}{\underset{\displaystyle O}{}} + CO_2$$

pyruvic acid   acetaldehyde
substrate   products

Class 5. Isomerases (catalyze isomerization reactions)
   Example: triosephosphate isomerase

$$\begin{array}{l} H_2C\text{—OH} \\ | \\ C\text{=}O \\ | \\ H_2C\text{—O—}P\text{=}O \\ \qquad | \\ \qquad OH \end{array} \underset{\text{isomerase}}{\overset{\text{triosephosphate}}{\rightleftharpoons}} \begin{array}{l} H\quad O \\ \;\,C \\ | \\ H\text{—}C\text{—OH} \\ | \\ H_2C\text{—O—}P\text{=}O \\ \qquad | \\ \qquad OH \end{array}$$

dihydroxyacetone phosphate   glyceraldehyde 3-phosphate
substrate   product

Class 6. Ligases (join molecules via ATP hydrolysis)
   Example: pyruvate carboxylase

$$H_3C\text{—}\overset{\displaystyle O}{\overset{\|}{C}}\text{—}\overset{\displaystyle O}{\overset{\|}{C}}\text{—OH} + CO_2 + \begin{array}{c}ATP\\+\\H_2O\end{array} \xrightarrow[\text{carboxylase}]{\text{pyruvate}} HO\text{—}\overset{\displaystyle O}{\overset{\|}{C}}\text{—}CH_2\text{—}\overset{\displaystyle O}{\overset{\|}{C}}\text{—}\overset{\displaystyle O}{\overset{\|}{C}}\text{—OH} + \begin{array}{c}ADP\\+\\P_i\end{array}$$

pyruvic acid   oxaloacetic acid

substrates   products

Exercise 23-1

Several enzyme-catalyzed reactions follow. To which class does each of the enzymes E_a, E_b, E_c, and E_d belong?

(a)
$$\begin{array}{l} OH \\ | \\ C\text{=}O \\ | \\ CH_2 \\ | \\ CH_2 \\ | \\ C\text{=}O \\ | \\ S \\ | \\ R \end{array} + H_2O \xrightarrow{E_a} \begin{array}{l} OH \\ | \\ C\text{=}O \\ | \\ CH_2 \\ | \\ CH_2 \\ | \\ C\text{=}O \\ | \\ OH \end{array} + RSH$$

(b)
$$\begin{array}{l} S\text{—}R_1 \\ | \\ C\text{=}O \\ | \\ CH_3 \\ + \\ R_2 \\ | \\ C\text{=}O \\ | \\ OH \end{array} + \begin{array}{c}ATP\\+\\H_2O\end{array} \xrightarrow{E_b} \begin{array}{l} S\text{—}R_1 \\ | \\ C\text{=}O \\ | \\ CH_2 \\ | \\ C\text{=}O \\ | \\ OH \\ + \\ R_2H \end{array} + ADP + P_i$$

(c)

$$\underset{\substack{| \\ OH}}{\overset{\substack{OH \\ |}}{C=O}} \quad + FAD \xrightarrow{E_c} \quad + FADH_2$$

(d)

$$HOCH \xrightarrow{E_d} HCOH$$

Enzyme Specificity: The Active Site

Some enzymes will recognize only one kind of molecule as a substrate. Other enzymes are not so selective and catalyze reactions with a larger number of substrates. The selectivity an enzyme shows in choosing a substrate is called its **specificity.** Carboxypeptidase is an example of an enzyme with a broad range of specificity. It binds to the C-terminal amino acid of most proteins or peptides, and breaks the peptide bond holding this terminal amino acid to the rest of the protein chain (see Figure 23-1). Lipases are also fairly nonspecific and break ester bonds in a variety of lipids (see Chapter 27). Some enzymes, on the other hand, show absolute specificity. For example, lactic acid dehydrogenase catalyzes the oxidation of L-lactic acid to pyruvic acid. However, this enzyme will not bind D-lactic acid (see Figure 23-2).

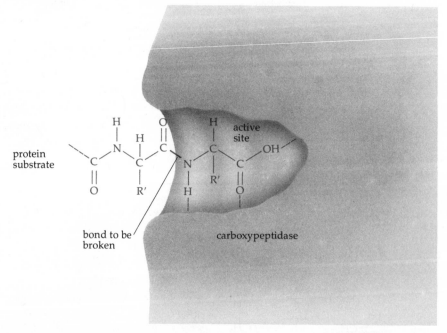

Figure 23-1   Schematic representation of the active site of the enzyme carboxypeptidase. The C-terminal ends of proteins, which are the substrates for this enzyme, bind at the active site. The active site is not absolutely specific, and can bind a variety of C-terminal amino acids.

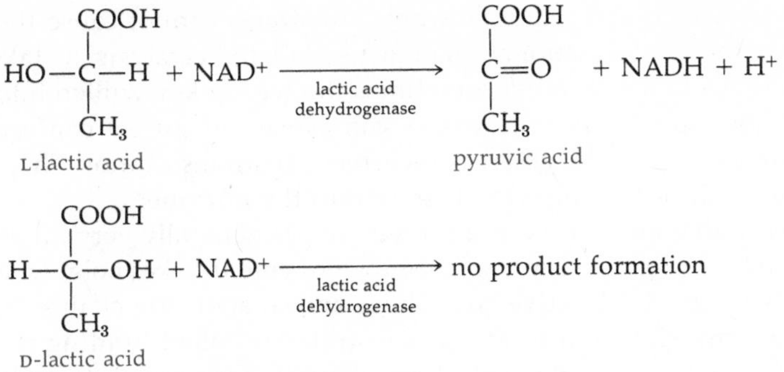

$$\underset{\text{L-lactic acid}}{\text{HO}-\overset{\displaystyle \overset{\text{COOH}}{|}}{\underset{\displaystyle \underset{\text{CH}_3}{|}}{\text{C}}}-\text{H}} + \text{NAD}^+ \xrightarrow[\text{dehydrogenase}]{\text{lactic acid}} \underset{\text{pyruvic acid}}{\overset{\displaystyle \overset{\text{COOH}}{|}}{\underset{\displaystyle \underset{\text{CH}_3}{|}}{\text{C}}}=\text{O}} + \text{NADH} + \text{H}^+$$

$$\underset{\text{D-lactic acid}}{\text{H}-\overset{\displaystyle \overset{\text{COOH}}{|}}{\underset{\displaystyle \underset{\text{CH}_3}{|}}{\text{C}}}-\text{OH}} + \text{NAD}^+ \xrightarrow[\text{dehydrogenase}]{\text{lactic acid}} \text{no product formation}$$

Figure 23-2   Lactic acid dehydrogenase is absolutely specific for L-lactic acid. D-Lactic acid is not a substrate for this enzyme.

This ability of an enzyme to discriminate between substrate and nonsubstrate is somewhat analogous to the fit of a key into a lock (see Figure 23-3). Not all of the parts of enzymes, or locks for that matter, are rigid. Recent experiments have shown that enzymes undergo a change in shape, called a **conformational change,** upon binding an allowable substrate. This is like the tumblers of a lock moving to accommodate the inserted key.

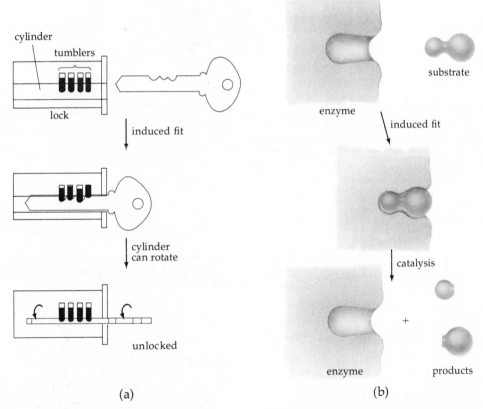

(a)   (b)

Figure 23-3   (a) The insertion of a key into its lock causes the tumblers of the lock to move into positions that allow the cylinder to rotate.

(b) The binding of a substrate to its enzyme induces an analogous change in the enzyme to a shape that allows for catalysis.

Molecules that are not allowable substrates cannot cause the enzyme to undergo the conformational change needed for catalysis to take place. This is analogous to the fact that sometimes the wrong key will go into a lock but then will not turn. Correct substrates can cause the correct conformational change needed for catalysis, whereas incorrect substrates cannot. This process is called **induced fit** of the correct substrate into the enzyme.

Recall that an enzyme is a protein and is generally very much larger than the substrate molecule. The region of the enzyme (the keyhole) where the substrate binds is called the **active site.** Those amino acid side chains that recognize the correct substrate and bind that substrate are called **binding residues.** They are like the tumblers of the lock. The substrate is recognized and bound by hydrophilic and hydrophobic interactions. Those amino acid side chains whose functional groups help to stretch bonds or otherwise participate in catalysis are called **catalytic residues.** Catalytic residues may also serve as proton donors or acceptors. They are like the cylinder of the lock, which turns to the unlocked position. The binding residues plus the catalytic residues constitute the active site (see Figure 23-4).

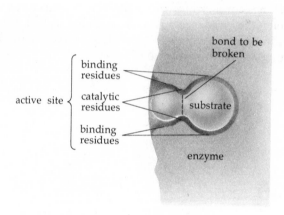

Figure 23-4   A simplified schematic representation of the active site of a typical enzyme. The actual shape of the active site of a given enzyme is unique for the reaction it catalyzes.

Exercise 23-2

Membrane transport and carrier proteins also have active sites. If the following were the amino acid sequence for the active site of Na^+K^+-ATPase protein, which amino acid side chain would you expect to bind the Na^+ and K^+ ions?

Exercise 23-3

The following amino acid sequence is part of the active center of an enzyme in its predominant form at pH 7. This enzyme interacts with its substrate by donating a proton. Which amino acid side chain can do this?

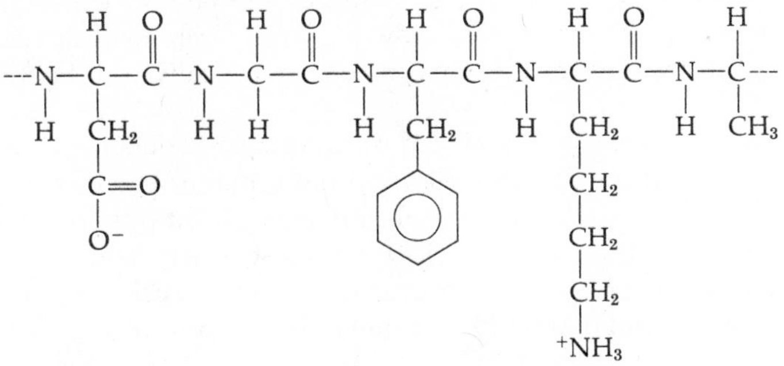

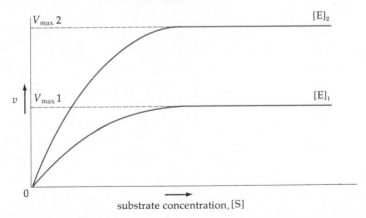

23-4 ENZYME ACTIVITY

Among the unique features of enzymes, there are three that are like those of transport proteins: *specificity*, *saturation*, and *inhibition*.

Enzyme Kinetics

Saturation of an enzyme with substrate means that all of the active sites of all of the available enzyme molecules are bound to substrate molecules. This is different from uncatalyzed reactions, where collisions between reactants and the formation of products continually increase as the concentration of the reactants increases. Figure 23-5 shows that the rate of an enzyme-catalyzed reaction, v, depends on the concentration of the substrate when relatively small amounts of substrate are present. In the presence of a relatively large amount of substrate, each of the enzyme molecules is saturated with substrate and the **maximum velocity** (V_{max}) is achieved. The value of V_{max} is directly proportional to the amount of enzyme present. If the amount of an enzyme is doubled, the V_{max} also doubles, all other things being equal.

| Figure 23-5 | The effect of substrate concentration on the rate of an enzyme-catalyzed reaction. The rate of product formation (v) increases with increasing substrate concentration until the enzyme is saturated, when it is working at its maximum velocity. Note that $[E]_2$ is twice $[E]_1$ and $V_{max}2$ is twice $V_{max}1$. |

Figure 23-5 illustrates the most general change that occurs in the rate of an enzyme-catalyzed reaction as the substrate concentration is increased. This relationship is typical for enzyme-catalyzed reactions that follow the two-step process described on page 555, and in certain other cases as well. Note that this figure shows the kind of data that can be obtained for experiments conducted in the laboratory. In the human body, enzymes rarely work at maximum velocity, because substrate concentrations in cells are usually well below the saturation level.

The rate of an enzyme-catalyzed reaction also depends on such factors as the temperature and the pH, as well as on the inherent efficiency of the enzyme in question. Each enzyme has a set of preferred conditions (pH, temperature, and so on) for maximum activity. Enzymes also differ widely in their inherent efficiency. The term used to express this inherent efficiency is the molar activity of an enzyme. **Molar activity** is defined as the ratio of V_{max} to [E]:

(23-2)
$$\text{Molar activity} = \frac{V_{max}}{[E]}$$

Thus, the molar activity of an enzyme can be determined by measuring V_{max} at a known concentration of enzyme. The molar activity is then obtained by dividing the V_{max} obtained by the enzyme concentration used. For example, if a 10^{-3} M solution of a particular enzyme exhibits a V_{max} of 1 M/min, then the molar activity of this enzyme is

(23-3)
$$\text{Molar activity} = \frac{1 \, M/\text{min}}{10^{-3} \, M} = 1000 \text{ min}^{-1}$$

Table 23-2 gives the molar activities of some specific enzymes. The large differences in molar activities between different enzymes is not surprising, since there is such a vast array of enzymes and reactions catalyzed. Remember that not all enzymes are specific for only one substrate. For an enzyme that is not specific for a single substrate, the values for V_{max} may be quite different for different substrates.

Table 23-2 Molar Activities of Some Enzymes

Enzyme	Molar Activity (min^{-1})
Carbonic anhydrase	36,000,000
Ketosteroid isomerase	17,100,000
Fumarase	1,200,000
β-Amylase	1,100,000
β-Galactosidase	12,500
Phosphoglucomutase	1,240
Succinate dehydrogenase	1,150
Aconitase	900

Exercise 23-4

A 1-liter solution that contains 10^{-5} mole of the enzyme catalase could theoretically produce 56 moles of H_2O from hydrogen peroxide, H_2O_2, per minute at substrate saturation. (The catalase reaction was discussed in Section 23-3.) What is the molar activity of the enzyme catalase?

Effect of pH

Many functional groups on the amino acid side chains of enzymes can lose protons or be protonated (see Chapter 21). The number of functional groups that are protonated depends on the pH. For most enzymes, salt bridges involving charged amino acid side chains are necessary to hold the enzyme together. In other words, the shape of an enzyme (its tertiary structure) depends on the pH. Now, since the activity of an enzyme depends critically on the shape of the enzyme, the activity of most enzymes varies with pH (see Figure 23-6). Trypsin is typical of most enzymes in the human body in that its activity varies substantially with pH, and it has a pH optimum very close to the actual pH of human cells. Some enzymes, however, are very active over a fairly broad range of pH values. Papain is an example of such an enzyme.

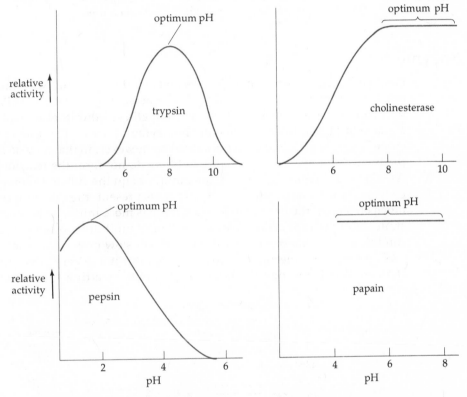

Figure 23-6   The effect of pH on the activity of some enzymes. Note the different pH optima of these enzymes and the broad pH optimum of papain. The behavior of most enzymes is similar to that of trypsin.

Effect of Temperature

The shape of an enzyme also depends on the temperature (Chapter 21). Very high temperatures are detrimental to enzymes. After all, they are proteins and can be denatured by heat. Unfolded, denatured enzymes are inactive (see Figure 23-7). Thus, as with noncatalyzed chemical reactions, the rate of an enzyme-catalyzed reaction first increases as the temperature increases, but then the rate decreases as the enzyme becomes denatured. Most enzymes in humans have an optimum temperature at which they are most active. This optimum temperature is close to 37°C (normal body temperature).

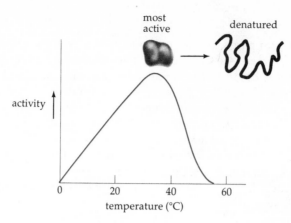

Figure 23-7   The activity of a typical enzyme   temperatures, the protein is
at different temperatures. At high   denatured.

Enzyme Inhibition

Frequently, the presence of molecules that are not substrates will slow down or stop an enzyme-catalyzed reaction. Such molecules are called **inhibitors,** and they work in various ways. In general, **competitive inhibitors** are molecules that are able to fit into the active site of an enzyme because they resemble the substrate. They are different from the substrate, however, in that they do not contain the bond that must be worked on, and they do not form the reaction products. As you can see from Figure 23-8, the inhibitor and the substrate compete for the enzyme. When a large amount of substrate is present, the inhibitor cannot have much effect. On the other hand, when there is more inhibitor around than substrate, most of the enzyme molecules will bind inhibitor molecules. Under these conditions, the rate of product formation will slow down drastically.

What kinds of molecules are competitive inhibitors? Very often, the product of an enzymatic reaction is similar enough to the starting substrate that it can

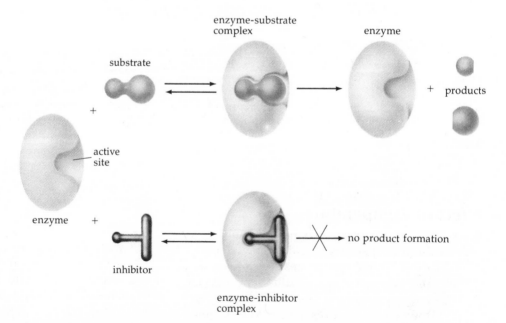

Figure 23-8   Schematic representation of com-   substrate for the enzyme. When
petitive inhibition of an enzyme.   inhibitor is bound, substrate
The inhibitor competes with the   cannot bind.

act as a competitive inhibitor and prevent further production of itself. Thus, the product-inhibited enzyme reaction slows down when the amount of product is too large.

Noncompetitive Enzyme Inhibition

There are other inhibitors that do not usually compete with the substrate for the active site of the enzyme, and usually are not similar to the substrate in structure. **Noncompetitive inhibitors** bind to the enzyme at some point other than the active site (see Figure 23-9). The substrate can still bind to the enzyme-inhibitor complex, but it cannot be converted to product. Since the substrate does not bind to the same site as a noncompetitive inhibitor, the presence of a large amount of substrate cannot overcome noncompetitive inhibition.

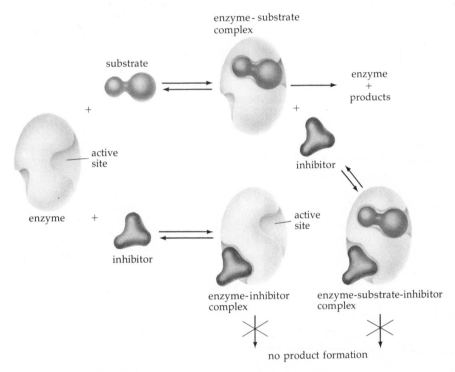

Figure 23-9   Noncompetitive inhibition of an enzyme. The inhibitor does not compete with the substrate for the active site. It can bind to either the free enzyme or the enzyme-substrate complex. When bound to the inhibitor, the enzyme is incapable of catalyzing the conversion of the substrate to products.

Exercise 23-5

The enzyme succinate dehydrogenase catalyzes the reaction

$$\underset{\text{succinic acid}}{\begin{array}{c} O \\ \| \\ C-OH \\ | \\ CH_2 \\ | \\ CH_2 \\ | \\ C-OH \\ \| \\ O \end{array}} + FAD \underset{\substack{\text{succinate} \\ \text{dehydrogenase}}}{\rightleftharpoons} \underset{\text{fumaric acid}}{\begin{array}{c} O \\ \| \\ C-OH \\ | \\ C-H \\ \| \\ H-C \\ | \\ C-OH \\ \| \\ O \end{array}} + FADH_2$$

Two of the following molecules are competitive inhibitors of succinate dehydrogenase. Which is not?

$$\underset{\text{malonic acid}}{O=\overset{\overset{\displaystyle OH}{|}}{C}-CH_2-\overset{\overset{\displaystyle OH}{|}}{C}=O} \qquad \underset{\text{oxaloacetic acid}}{O=\overset{\overset{\displaystyle OH}{|}}{C}-CH_2-\overset{\overset{\displaystyle O}{\|}}{C}-\overset{\overset{\displaystyle OH}{|}}{C}=O} \qquad \underset{\text{iodoacetamide}}{I-CH_2-\overset{\overset{\displaystyle O}{\|}}{C}-NH_2}$$

23-5 COENZYMES AND THE MECHANISMS OF ENZYME-CATALYZED REACTIONS

Chemical and physical studies of a number of typical enzymes have contributed significantly to our understanding of enzyme-catalyzed reactions. For example, the experimentally determined shape of the enzyme hexokinase is shown in Figure 23-10. The active site of this enzyme is at the bottom of a "pocket," and the binding of the substrate, glucose, to this enzyme induces a pronounced conformational change in the enzyme.

Studies on the substrate specificity of a large number of enzymes have shown that a molecule must possess certain characteristics in order to be a substrate for a particular enzyme: (1) A substrate must have a shape that allows it to fit into the active center; (2) it must also contain specific functional groups that can bind with the side chains of amino acids at the active center of the enzyme; and (3) in order for catalysis to occur, the substrate must also contain the type of chemical bond that the enzyme attacks. If a potential substrate molecule does *not* have this type of bond but can still bind at the active site, it will be a competitive inhibitor rather than a substrate. Thus enzyme catalysis requires the participation of both binding residues, which hold the substrate in place via combinations of salt bridges, hydrogen bonds, and hydrophobic interactions, and catalytic residues, which are involved in making and breaking bonds in the substrate (see Figure 23-4).

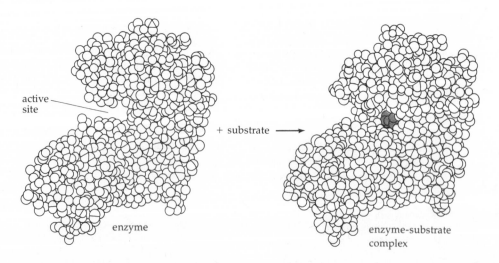

active site

+ substrate ⟶

enzyme

enzyme-substrate complex

Figure 23-10   The experimentally determined three-dimensional shape of the enzyme hexokinase isolated from yeast cells. The left figure shows the shape of the free enzyme. Note the deep pocket. The active site is at the bottom of this pocket.

Binding of substrate, glucose, induces hexokinase to undergo the conformational change necessary for it to catalyze its reaction. Note the molecule of glucose (in blue) bound at the active site in the right figure.

The nature of the catalytic residues (those amino acids required for bond making or breaking) depends on the reaction catalyzed. Some enzymes even form covalent bonds with their substrate as part of the reaction mechanism. For example, aldolase, one of the enzymes required for glucose oxidation, uses such a reaction pathway.

Many enzymes require small accessory molecules called **coenzymes** in order to function as catalysts. These accessory molecules supply groups of atoms to the substrate or accept groups of atoms from it. We have seen that the coenzyme NAD^+ accepts a hydride ion, $H:^-$. Some enzymes, called transaminases, remove amino groups from substrates. Since these enzymes must not be altered if they are to keep functioning as catalysts, transaminases require coenzymes that will accept these amino groups. Other coenzymes specialize in methyl groups, carboxyl groups, and so on. The number of different coenzymes in the human body is much less than the number of different enzymes, since many different enzymes can use the same coenzyme. Coenzymes are bound to enzymes at coenzyme binding sites in the same manner as substrates.

There is some similarity between coenzymes and prosthetic groups. However, prosthetic groups on proteins need not be altered in catalysis.

Let us consider one example of a reaction that requires a coenzyme:

(23-4)

alanine   pyridoxal phosphate

pyruvic acid   pyridoxamine phosphate

In reaction 23-4, catalyzed by the enzyme alanine transaminase, two bonds in the substrate alanine are broken (the bond between the α-carbon atom and a hydrogen atom and the bond between the α-carbon atom and the amine group) and two bonds are formed between the α-carbon atom and an oxygen atom to give the product, pyruvic acid. The enzyme does not accept the amine group from alanine, nor does it donate the oxygen atom needed to form pyruvic acid. Rather, these atoms are exchanged between the substrate alanine and a coenzyme called pyridoxal phosphate. In a subsequent enzyme-catalyzed reaction (not shown), the pyridoxamine phosphate produced in reaction 23-4 is converted back to pyridoxal phosphate. Pyridoxal phosphate is also involved in a number of other enzyme-catalyzed reactions in which amine groups are transferred.

Humans cannot synthesize some parts of coenzymes. Over the many years of evolution, we have lost the ability to do so. Those parts of coenzymes that we cannot synthesize must be included in our diet and are called **vitamins.** Vitamins were recognized as a necessary component of our diet long before we really knew why. If people did not have an adequate supply of them, they became ill. Now, we know why we need them. We shall see some of the specific reactions for which vitamins are required when we study the pathways for the synthesis and degradation of molecules in the body. We shall talk about nutritional requirements of the human body in Chapter 29. Structural formulas for some coenzymes, with their vitamin portions indicated, are shown in Appendix 3.

Unlike enzymes, coenzymes such as pyridoxal phosphate and NAD^+ are altered in enzyme-catalyzed reactions. However, they can be changed back to their original form by other reactions. Since coenzymes can be recycled, we only need small amounts of them to satisfy our needs.

Exercise 23-6

Figure 23-10 shows the enzyme hexokinase bound to its substrate, glucose. (a) List some amino acids that might function as binding residues at the active site of this enzyme. (b) Notice how small glucose is compared to hexokinase. Is the size of this enzyme wasteful? In other words, couldn't a lot smaller protein do the same job? Explain.

23-6 REGULATION OF ENZYME ACTIVITY

It is one thing to possess a collection of enzymes that are capable of making or breaking apart a wide variety of molecules. It is quite another thing to control the activity of these enzymes. We do not want to make too much or too little of any substance. Enzyme activity can be influenced by a number of factors. We have already seen that it depends on pH and temperature. For warm-blooded humans, a change in temperature is not an important means of controlling enzyme activity. At 37°C, almost all enzymes are working at or near their optimum temperature, and our body temperature does not vary significantly. The pH may be important in controlling the activity of an enzyme, but we do not yet know to what extent human metabolism makes use of this means of regulation. Enzyme activities are controlled in several other ways that are well understood.

Regulation of Substrate Concentration

One sure way to stop an enzyme from making too much product is to starve it, that is, to limit the amount of substrate available to it. If we use our factory analogy, this would be equivalent to stopping the assembly line. The worker (the enzyme) is there and ready, but has nothing on which to work. Substrate limitation can be achieved in a number of ways. Intracellular enzymes require that their substrates get into the cell. The rate of entry of substrates is, in turn, controlled by their ability to penetrate the cellular membrane or by their rate of transport into the cell. By controlling these factors, the rate of product formation can be controlled. For those enzymes that work farther down the assembly line, the availability of substrate depends on how fast the enzymes on the beginning of the assembly line are working (see Figure 23-11).

Figure 23-11   In this series of reactions, the concentrations of substrates B and C available to enzymes E_B and E_C depend on the rate of reaction A → B catalyzed by E_A. If E_A is inhibited, the formation of D will decrease.

$$A \xrightarrow[E_A]{} B \xrightarrow[E_B]{} C \xrightarrow[E_C]{} D$$

Regulation of Enzyme Concentration

Clearly, the rate of product formation is limited by the amount of enzyme present. If no enzyme is present, no product will form. Cells control the amounts of the various enzymes they contain by controlling their rate of synthesis. Synthesis of enzymes that are not needed can be shut off completely. However, even if synthesis of a particular enzyme is turned off completely, there will be a lot of it around for a while. Using our factory analogy, after a factory stops hiring one type of employee, there is usually a long time interval before the number of employees of that type decreases because some have quit or retired. (Actually, enzymes are "retired" quite forcibly—they are cut apart by other enzymes!) We shall see how enzyme synthesis is controlled in the next chapter.

Allosteric Regulation of Enzyme Activity

A faster way to stop enzyme workers in a cellular factory is to lay them off. This is accomplished by allosteric effects. **Allosteric enzymes** have more than one site (*allo-* = other, *-steric* = site) to which substrates or other small molecules can bind. The other sites may also be catalytic sites. Hemoglobin is an example of a multisite protein. Each of its four subunits can bind oxygen. The binding of an O_2 molecule by one of the four subunits makes binding of O_2 easier for the other subunits. This is because binding of O_2 by induced fit causes a conformational change in hemoglobin, as shown in Figure 23-12. The new conformation makes it easier for the other O_2 molecules to bind.

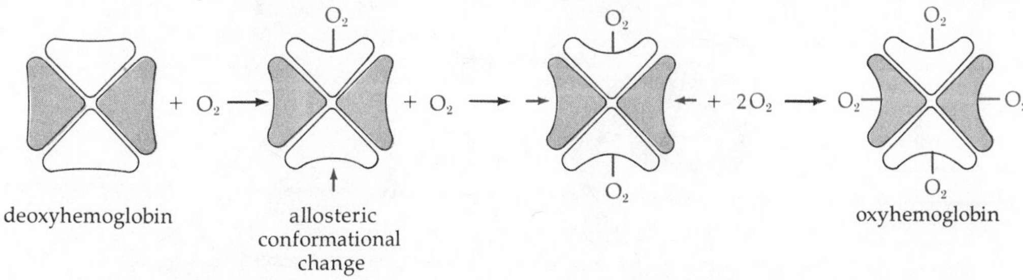

deoxyhemoglobin   allosteric conformational change   oxyhemoglobin

Figure 23-12   A model for the allosteric effect of oxygen binding to hemoglobin. The allosteric conformational change induced by the first molecule of O_2 to bind facilitates the binding of additional O_2 molecules. As indicated by the blue arrows, the subunits bound to O_2 exert pressure on the other subunits to change their shape as well.

Several allosteric proteins have binding sites for molecules that are not substrates. Many of these undergo an unfavorable conformational change when they bind this other molecule. In the unfavorable conformation it is more difficult for the substrate to bind to the active site. The other molecule is then called an **allosteric inhibitor** of the enzyme. End products of the assembly line on which the particular enzyme is working are frequently allosteric inhibitors. The enzyme that is inhibited is usually the first one on that assembly line. Thus when a product of the assembly line or pathway builds up, it stops production until it is used up. This type of control, called **feedback inhibition,** is illustrated in Figure 23-13. In this example, E_1D, the complex formed by the enzyme E_1 and the feedback inhibitor D, is in equilibrium with free enzyme and D, that is, $E_1D \rightleftharpoons E_1 + D$. When the concentration of D is large, most of the E_1 will be bound to D and inactivated; but when the concentration of D is small, most of the E_1 will be free to work. Thus, the formation of products, B, C, and D will slow down or speed up as the concentration of D increases or decreases.

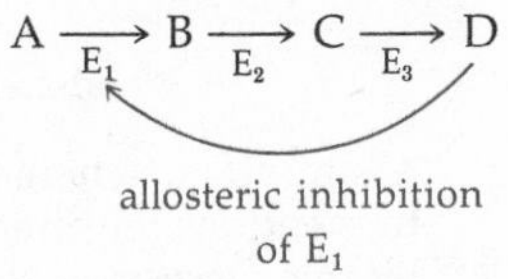

Figure 23-13   Feedback inhibition of an enzyme. The final product of this series of reactions, D, binds to an allosteric site on the first enzyme in the series, E_1. The activity of enzyme E_1 is thus dependent on the concentration of D, and large amounts of D shift the equilibrium to favor the inactive E_1D complex.

$$A \xrightarrow[E_1]{} B \xrightarrow[E_2]{} C \xrightarrow[E_3]{} D$$

allosteric inhibition of E_1

Regulation by Chemical Modification of Enzymes

Still another way of controlling enzyme activity is chemical modification. **Chemical modification** of an enzyme is achieved by the covalent attachment of a functional group to the enzyme. The attachment is catalyzed by yet another enzyme, and occurs at a specific site but usually not at the active site. For example, one of the serine residues on the enzyme glycogen phosphorylase can react to form a phosphate ester. This enzyme has a different shape when phosphate is attached to the serine residue, and is more active (has a higher molar activity).

Other enzymes can be modified by attachment of acetate or methyl or phosphate groups to them. The modified enzyme is either more or less active than the unmodified enzyme. Figure 23-14 shows how the control system works for glycogen phosphorylase. Here, a hormone reaches the cell and gives it the message to increase glucose production. The hormones glucagon and adrenaline (via their receptor proteins) are allosteric activators for a particular enzyme called adenyl cyclase, located in the cell membrane. These hormones change the con-

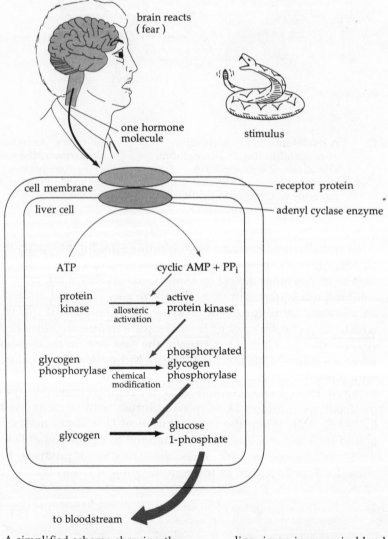

Figure 23-14   A simplified scheme showing the tremendous amplification achieved by the successive steps in the process whereby fright ultimately results, via the hormone adrenaline, in an increase in blood sugar. The amplification is represented by the increasing thickness of the blue arrows.

formation of adenyl cyclase to a highly active shape that can make a lot of cyclic AMP from ATP. The cyclic AMP then allosterically activates a second enzyme, a protein kinase, which results in the binding of phosphate groups to proteins, such as glycogen phosphorylase. Finally, the phosphorylated glycogen phosphorylase breaks glycogen down to glucose 1-phosphate. Thus, the activation of glycogen phosphorylase proceeds through several steps of allosteric activation and chemical modification. Each step also amplifies the message, so that about 3 million glucose molecules are produced for each messenger hormone molecule.

Exercise 23-7

Note that each successive step in the hormonal activation scheme in Figure 23-14 results in added amplification of the response. If each of the four steps within the cell resulted in only a 10-fold amplification, how many glucose 1-phosphate molecules would be produced as a result of the action of one hormone molecule at the cell membrane?

23-7 ON THE MECHANISM OF DRUG ACTION

Many drugs that are used to combat microbial infections or to fight cancer are competitive inhibitors of specific enzymes. For example, some enzymes that are vital to infectious organisms but are not found in healthy humans can be inhibited. Inhibitors of these enzymes can therefore kill the infectious organism but not affect human cells. Sulfanilamide was one of the first of these inhibitors to be identified. It is structurally similar to p-aminobenzoic acid, which bacteria use to make folic acid (see Figure 23-15).

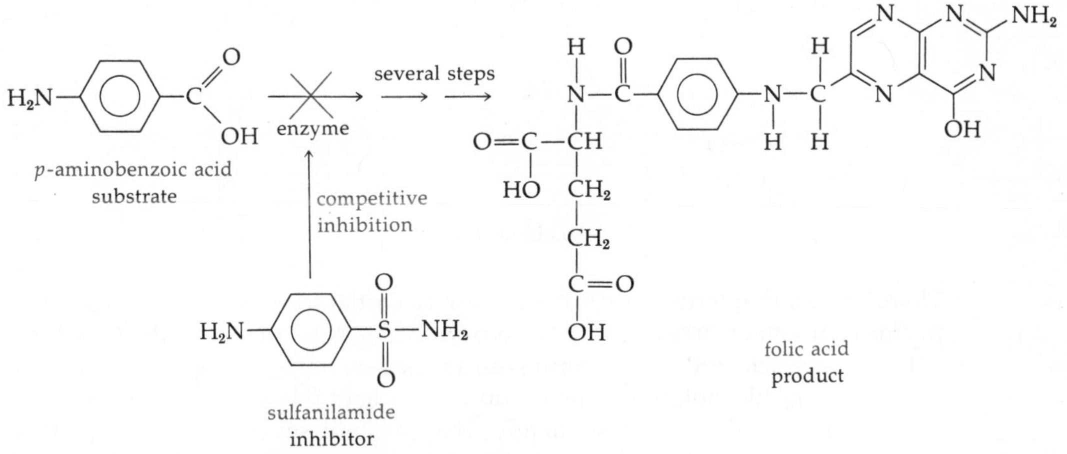

Figure 23-15   The drug sulfanilamide competitively inhibits the formation of folic acid from p-aminobenzoic acid in many microorganisms.

Humans cannot make folic acid. For us it is a vitamin. Thus, when we take sulfanilamide, bacteria die, but we are not affected. Several derivatives of sulfanilamide have now been synthesized, some of which are even better competitive inhibitors of folic acid synthesis. Together with sulfanilamide, these compounds are popularly referred to as the sulfa drugs, some of which have the following names and structures:

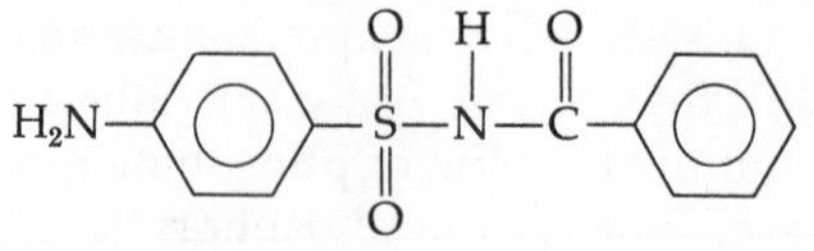

sulfabenzamide

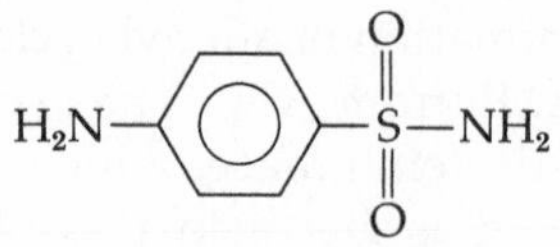

sulfanilamide

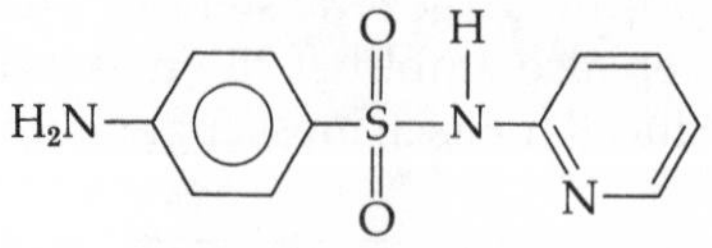

sulfapyridine

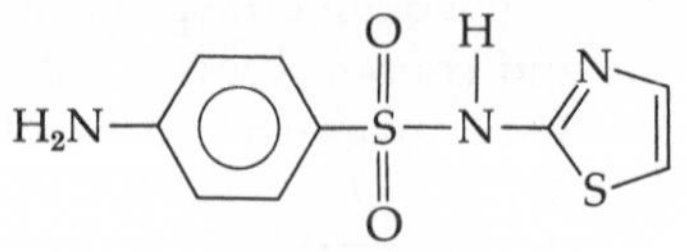

sulfathiazole

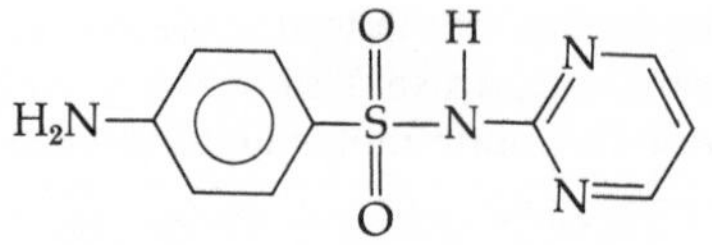

sulfadiazine

Potent competitive inhibitors of enzymes in important metabolic pathways, such as the sulfa drugs, are often referred to as **antimetabolites.** Other kinds of antimetabolites are used in the chemotherapy of cancer. They take advantage of the fact that cancer cells divide rapidly, and therefore need to make a lot of DNA, whereas normal cells do not. For example, aminopterin, an analog of folic acid, is a competitive inhibitor of purine biosynthesis.

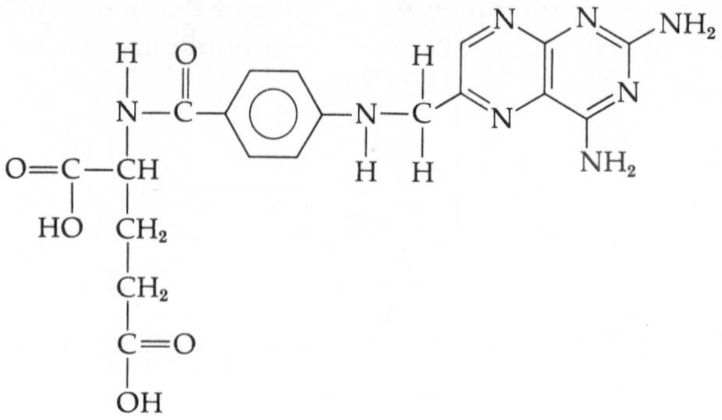

aminopterin

Therefore, aminopterin can be administered in order to starve cancer cells of the purine components needed for DNA biosynthesis. Note that the only difference between folic acid and aminopterin is an amine functional group in aminopterin in place of the alcohol functional group in folic acid (Figure 23-15). Since most normal cells divide very infrequently, they are not severely affected by this treatment. However, the cells of some tissues do divide frequently. Cells in the intestines and some other types of cells are also killed by aminopterin treatment. Since cancer cells are derived from normal cells, and thus contain the same enzymes, it is difficult to eliminate them selectively with antimetabolites. Since some human cancers now appear to be caused by viruses (Chapter 24), there may be a better chance of finding specific antimetabolites in these cases.

Other compounds used to treat microbial infections are the **antibiotics**—molecules made by one organism that are toxic to others. They are thought to act as antimetabolites, but the reactions they competitively inhibit are not always clearly understood. Penicillin was the first antibiotic to be discovered (accidentally, by Alexander Fleming in the 1920s); it interferes with cell wall biosynthesis in a variety of bacteria. The structural formulas of several antibiotics are shown at the top of the next page.

Penicillin G

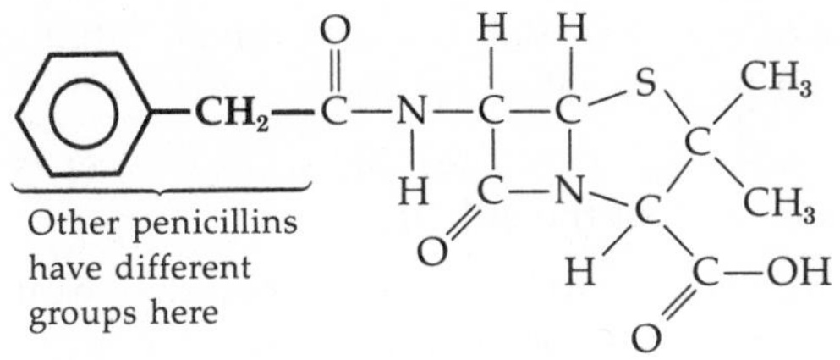

Other penicillins have different groups here

Penicillin F $CH_3-CH_2-CH=CH-CH_2-$

Penicillin K $CH_3-(CH_2)_6-$

Penicillin X $HO-\bigcirc-CH_2-$

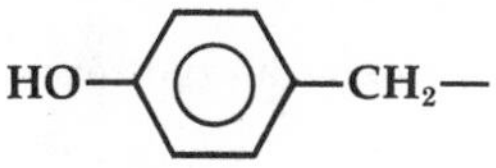

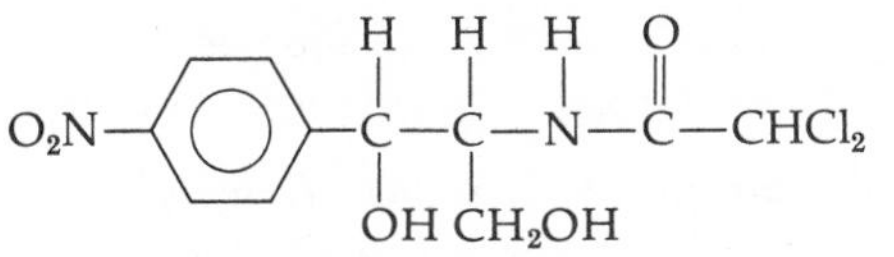

Chloramphenicol (chloromycetin)

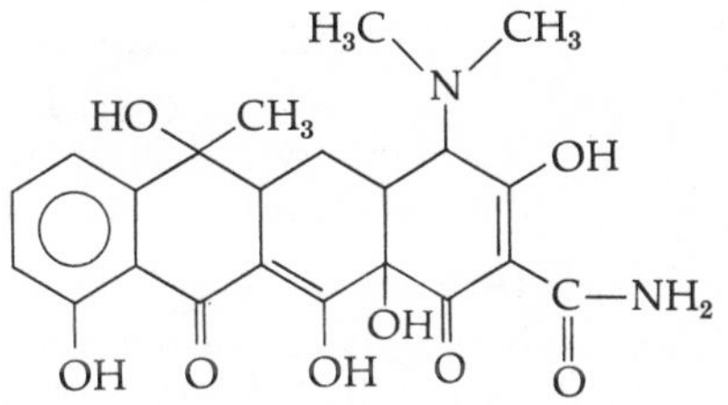

Tetracycline

Exercise 23-8

Would you expect aminopterin to be a useful drug for the treatment of bacterial infections? Explain your answer.

23-8 SUMMARY

1. Enzymes are proteins that catalyze chemical reactions. They all exhibit the basic functional features of specificity, saturation, and inhibition.

2. Enzymes are named and classified according to the reactions that they catalyze. There are six major classes of enzymes: oxidoreductases, transferases, hydrolases, lyases, isomerases, and ligases.

3. The region of an enzyme where the substrate binds is called the active site, which consists of binding residues and catalytic residues.

4. When a substrate binds to an enzyme, a conformational change in the enzyme occurs. This process is called induced fit.

5. One general mechanism for enzyme-catalyzed reaction is the two-step process $E + S \rightleftharpoons ES$ followed by $ES \rightarrow E + P$, where E, S, and ES are enzyme, substrate, and enzyme-substrate complex, respectively.

6. The molar activity of an enzyme is a measure of the catalytic power of the enzyme.

7. The rate of an enzyme-catalyzed reaction is influenced by temperature, pH of the reaction medium, enzyme concentration, substrate concentration, chemical modification of the enzyme, allosteric effects, and the presence of inhibitors.

8. Two general types of inhibitors are competitive inhibitors and non-competitive inhibitors.

9. A sequence of enzyme-catalyzed reactions is often regulated by allosteric enzymes in a process called feedback inhibition.

10. Coenzymes are small molecules that work with enzymes by supplying or accepting atoms from the substrate.

11. Vitamins are those parts of coenzymes that cannot be made by the human body and must be part of the diet.

12. Many drugs are competitive inhibitors of enzymes.

PROBLEMS

1. In a study of the following sequence of enzyme-catalyzed reactions,

$$S \xrightarrow{E_1} T \xrightarrow{E_2} U \xrightarrow{E_3} V \xrightarrow{E_4} W$$

it was observed that removing W from the reaction as soon as it was formed increased the rate of formation of T. Give a possible explanation for this observation.

2. The accompanying graph is for two different enzymes, A and B, which both produce NADH as one of their products. Use the graph to obtain approximate values for the molar activity of each enzyme. Which is more efficient in producing NADH?

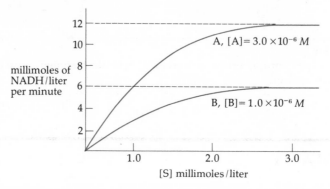

3. The molar activity of a certain enzyme, E, is 4800 min^{-1}. Calculate the rate of the reaction catalyzed by E when the concentration of E is 5.0×10^{-6} M and the substrate concentration is extremely large.

4. The enzyme amylase catalyzes the hydrolysis of starch to glucose. To which class of enzymes does amylase belong?

5. Explain what is meant by induced fit with regard to enzyme-catalyzed reactions.

6. The membrane transport protein Na$^+$K$^+$-ATPase, which is also an enzyme, has a binding site for ATP. The structural formula for the predominant form of ATP at a pH of about 7 is

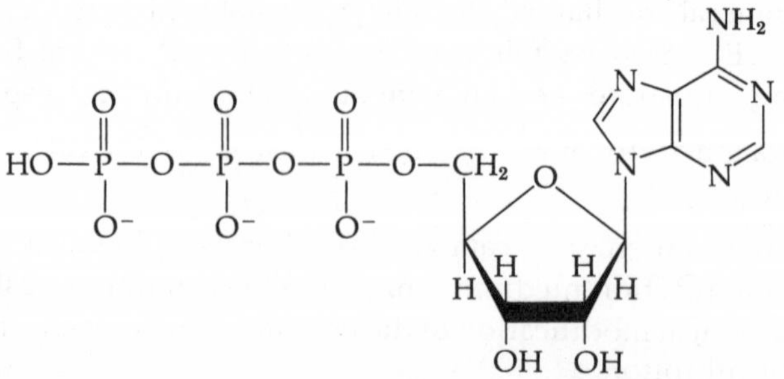

List some amino acids of Na$^+$K$^+$-ATPase whose side chains may be involved in binding ATP, and describe how each can interact with a portion of ATP.

SOLUTIONS TO EXERCISES

23-1 (a) A thioester bond is broken with the addition of water, so this enzyme is a hydrolase (Class 3).

(b) Two substrates are joined together via the exergonic hydrolysis of ATP, so this is a ligase reaction (Class 6).

(c) The $-CH_2-CH_2-$ portion of the substrate loses carbon-hydrogen bonds (an oxidation), while FAD is reduced. Therefore, this enzyme is classified as an oxidoreductase (Class 1).

(d) E_d catalyzes the interconversion of two isomers (Class 5).

23-2 The aspartic acid side chain contains a negatively charged carboxylate ion $\left(\begin{array}{c} O \\ \parallel \\ -C-O^- \end{array} \right)$, which will bind the positive Na^+ or K^+ ions.

23-3 At the pH of cells in the body ($\sim$7.4), the carboxyl group exists primarily as the negative ion and thus cannot donate a proton. However, the ammonium ion form of the amino acid lysine is a weak acid and can serve as a proton donor at this pH (see Chapter 18).

23-4 For catalase, V_{max} = 56 (moles/liter) min^{-1} at this enzyme concentration. Therefore,

$$\text{Molar activity of catalase} = \frac{V_{max}}{[E]} = \frac{56 \;\cancel{\text{(moles/liter)}}\, \text{min}^{-1}}{10^{-5} \;\cancel{\text{(mole/liter)}}} = 5.6 \times 10^6 \; \text{min}^{-1}$$

23-5 Malonic acid and oxaloacetic acid are similar to succinic acid in size and both possess two carboxyl groups. Iodoacetamide, however, is smaller and does not contain a carboxyl group and is not a competitive inhibitor of succinate dehydrogenase.

23-6 (a) Glucose, the substrate of this enzyme, contains several hydroxyl groups. We would therefore expect to find binding residues capable of forming hydrogen bonds with this substrate, possibly including serine, threonine, asparagine, glutamine, or tyrosine, as opposed to amino acids with hydrophobic side chains. (b) In addition to possessing the correct binding and catalytic residues, these residues must be correctly oriented in the active site of the enzyme. Hexokinase also undergoes a fairly substantial conformational change upon binding its substrate. The large sizes of enzymes are necessary in order for them to provide active sites that have very specific shapes as well as to undergo precise conformational changes.

23-7 There are four steps of amplification, so if we assume a 10-fold amplification per step, the overall amplification is $10 \times 10 \times 10 \times 10$ = 10,000 glucose 1-phosphate molecules.

23-8 Aminopterin is not useful for treating bacterial infections. Bacteria can make their own folic acid in the presence of aminopterin. A dose of aminopterin large enough to competitively inhibit the folic acid-dependent steps of DNA synthesis in bacteria would also severely hinder DNA synthesis in normal human cells.

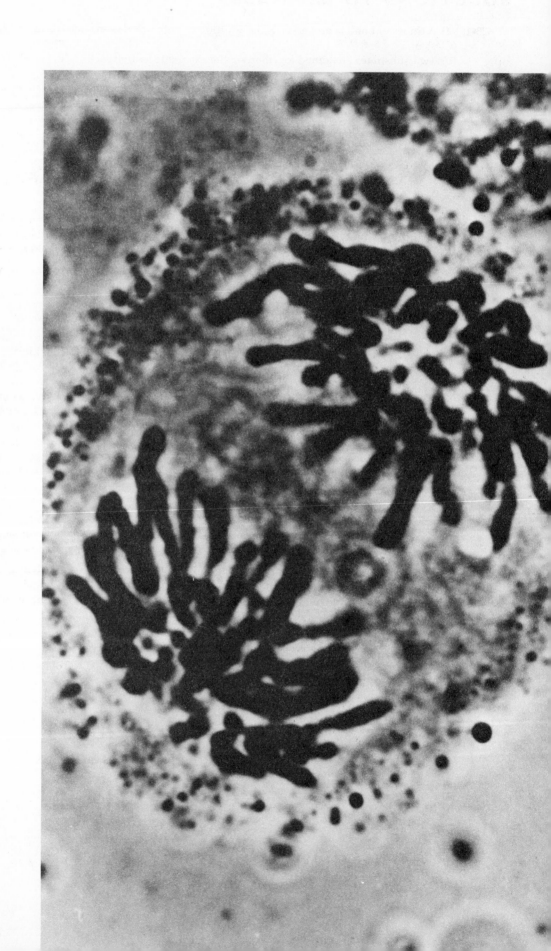

CHAPTER 24

Nucleic Acids and the Biosynthesis of Proteins

24-1   INTRODUCTION

In our analogy of cells in the human body as industrial factories, proteins play the roles of mail carriers, assembly-line workers, and so on. Obviously, someone needs to be in charge of an industrial factory, and there must be control of the cellular machinery as well. The task of directing the activities of the cellular factory begins with information stored in DNA molecules found in chromosomes located in the cell nucleus. Each chromosome consists of one DNA molecule and several proteins. There are 46 chromosomes (23 pairs) in all human cells except sperm and ova, which each have 23 unpaired chromosomes. The synthesis of a particular protein is directed by a definite portion of a DNA molecule called a gene. One DNA molecule contains hundreds of genes.

The total of all of the DNA present in all of the chromosomes in each cell is analogous to a set of master blueprints for the entire body. When a cell divides, each daughter cell must receive its own set of these master blueprints. Thus, prior to cell division, the DNA in a cell is reproduced. We shall see that the reproduction of DNA occurs in a process called *replication*, which produces an *exact* copy of a cell's DNA.

Using our analogy, managing a particular cellular factory is accomplished by determining which sections of the blueprint are to be read. We know quite a bit about how this blueprint is read, but we know relatively little about the process by which only certain portions are selected to be read by a given cell. As you study how cells read their DNA blueprints, you will see that the process of deciphering genetic information closely parallels the process of deciphering a code or a foreign language. Therefore, the steps in this process have been given names that suggest decoding, such as *transcription* and *translation*. The terms *code letters* and *codons* are also used in the description of protein synthesis. We shall also look at how the process of protein synthesis is controlled. Finally, we shall see how viruses invade our cells and use the cellular machinery to make their own proteins.

As a human cell divides, two sets of chromosomes (the oblong, dark objects in this micrograph) can be detected as they separate. Each chromosome consists of a single DNA molecule and several proteins.

24-2   STUDY OBJECTIVES

After careful study of this chapter, you should be able to:

1. Differentiate purine bases from pyrimidine bases, nucleosides from nucleotides, and DNA from RNA.

2. Discuss the three-dimensional structure of the nucleic acids and the relationship of their structure to their function.

3. Describe the role of hydrogen-bonded base pairs in the structure and function of nucleic acids.

4. Describe the process of DNA replication.

5. Compare and contrast transcription and replication.

6. Detail each of the steps in translation and its role in the overall process of protein synthesis.

7. Explain the necessity for regulating protein synthesis and the differences between induction and repression.

8. Describe what can happen when viruses infect cells.

24-3   GENES AND THEIR FUNCTION

A **gene** is a definite portion of a DNA molecule that codes for the synthesis of the specific sequence of amino acids in a particular protein. Since each cell in the human body (except sperm and ova) contains 23 pairs of chromosomes, each cell is, in theory, capable of producing all of the different proteins in the human body. Any *particular* cell, however, uses only some of its genes and synthesizes relatively few proteins. Which genes are used and which proteins are synthesized by a cell distinguishes one type of cell from another. As the human embryo matures, groups of cells become specialized as liver cells, brain cells, kidney cells, and so on. This process is called **differentiation.** We shall describe later what little is known about the complexities of differentiation.

Although we know relatively little about the mechanisms involved in differentiation, we do know a good deal about the everyday functioning of DNA molecules and their component genes. If we think of the DNA molecules in the nucleus as the master blueprints in the manager's office, then we realize that they must be kept safe for future reference. Therefore, in order to use the information contained in the DNA, the cell must make expendable copies of each portion of the blueprint (each gene) that it wants to use. Cells do not have Xerox machines, but they do have their own copying process, which we call **transcription.** The transcription process produces an expendable copy of a gene, called a **messenger RNA** molecule, that can be carried out into the factory proper. (As we shall see in Section 24-8, a messenger RNA molecule is not an exact copy of a DNA gene. A better, but still imperfect analogy of the relationship between messenger RNA and its DNA gene would be the relationship between a photographic negative and its positive print.)

Once outside of the nucleus, messenger RNA directs the assembly of the appropriate amino acids into the proper sequence to form the desired protein molecule. This process of protein assembly, called **translation,** takes place on ribosomes, specialized particles located in the cell's cytoplasm. The translation process involves several enzymes.

Basically, the management of cellular factories is accomplished according to the following model, reverently referred to as the **central dogma of molecular biology.**

DNA $\xrightarrow[\text{transcription}]{}$ messenger RNA $\xrightarrow[\text{translation}]{}$ protein

master   copy   product
blueprint

The arrows here represent information transfer, not simple chemical reactions. In order to understand the details of these processes, you must first become familiar with the structures and properties of the nucleic acid polymers DNA and RNA, and the components, building blocks, from which these nucleic acid polymers are assembled.

24-4 COMPONENTS OF NUCLEIC ACIDS

Nucleic acids, like proteins, are large polymers made up of a small number of different building blocks. The building blocks of nucleic acids are called **nucleotides.** Each nucleotide is in turn composed of three smaller parts: a phosphate group, a monosaccharide, and a nitrogen-containing base. The term **nucleoside** is used to refer to a nitrogenous base bound to a monosaccharide, and a nucleotide is a nucleoside phosphate (see Figure 24-1).

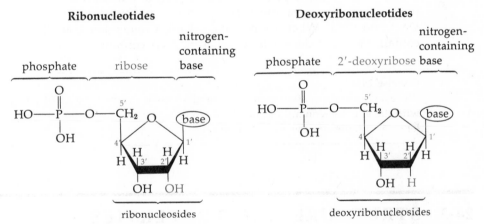

Figure 24-1   The components of nucleotides. Note the different sugar components in ribonucleotides and deoxyribonucleotides. Carbon atoms of the sugar component of nucleotides are numbered with primes (1', 2', etc.) to distinguish them from the atoms on the base component, which are numbered 1, 2, 3, and so forth.

There are two major types of nucleic acids: **deoxyribonucleic acid (DNA)** and **ribonucleic acid (RNA).** There are only two differences between the deoxyribonucleotide components of DNA and the ribonucleotide components of RNA. The first difference is in the sugar component. Both DNA and RNA have the sugar D-ribose, but in DNA there is no hydroxyl group on the number 2' carbon of ribose, so the prefix *deoxy-* is used to denote the absence of oxygen at this position. Second, each DNA and RNA nucleotide contains one of four possible bases. Three of them—adenine, guanine, and cytosine—are the same in both DNA and RNA. The fourth base is thymine in DNA and uracil in RNA. The structural formulas for these bases, as well as their names, are given in Figure 24-2. Adenine and guanine have two nitrogen-containing rings. They are called **purines** because of their similarity to the molecule purine. Likewise, cytosine, thymine, and uracil are called **pyrimidines,** because of their similarity to pyrimidine. Notice that thymine is structurally identical to uracil except that there is a methyl group on thymine and none on uracil.

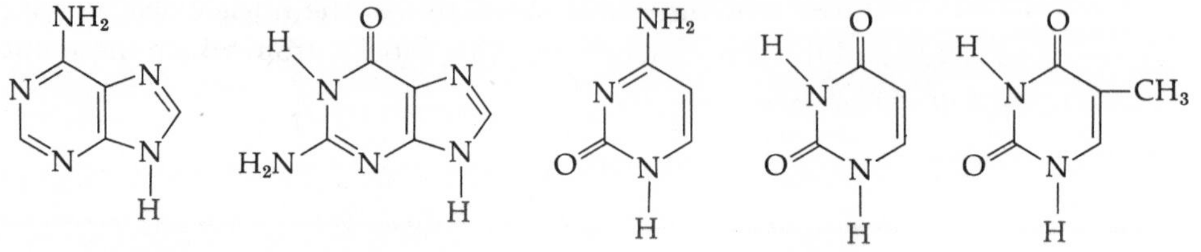

adenine (A)   guanine (G)   cytosine (C)   uracil (U)   thymine (T)

Figure 24-2   The nitrogenous bases of DNA and RNA. In a nucleoside, either ribose or deoxyribose replaces the hydrogen atom shown in blue. Note the similarity of adenine and guanine to the molecule purine, and of uracil, thymine, and cytosine to the molecule pyrimidine.

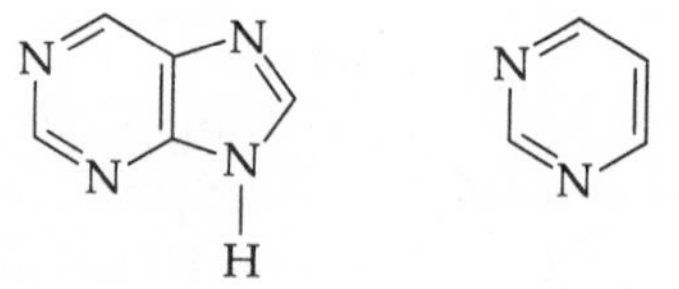

purine   pyrimidine

We can use the abbreviations A, G, U, T, and C to represent nucleotides with the bases adenine, guanine, uracil, thymine, and cytosine, respectively, in both DNA and RNA. When we do this, we must remember that DNA nucleotides contain the sugar deoxyribose, whereas RNA contains the sugar ribose.

Exercise 24-1

Write a chemical equation using structural formulas for the reaction of phosphoric acid with the adenine-containing nucleoside to form the corresponding nucleotide. Use the letter A to represent the adenine portion.

24-5   STRUCTURE AND PROPERTIES OF NUCLEOTIDE POLYMERS

In proteins, amino acids are joined together into long chains by peptide bonds formed between amino and carboxyl functional groups. How are nucleotides joined to form DNA and RNA? The bonds that hold these polymers together are ester linkages formed between the phosphate on the number 5' carbon of ribose in one nucleotide and the hydroxyl on the number 3' carbon of ribose in the next nucleotide (deoxyribose in the case of DNA). Thus, nucleic acids are said to have **3',5'-phosphate ester bridges** between their nucleotide components (see Figure 24-3). A nucleic acid molecule has a free phosphate on the ribose (or deoxyribose) located at one end and a free hydroxyl on the number 3' carbon of ribose (or deoxyribose) at the other end. This is similar to proteins, which have a free amino end and a free carboxyl end. When referring to nucleic acids, we use abbreviations to indicate the sequence of nucleotides. By convention, the nu-

Figure 24-3 The structure of a small piece of RNA. This piece of RNA can be represented by the simple notation $^{5'}ACGU^{3'}$.

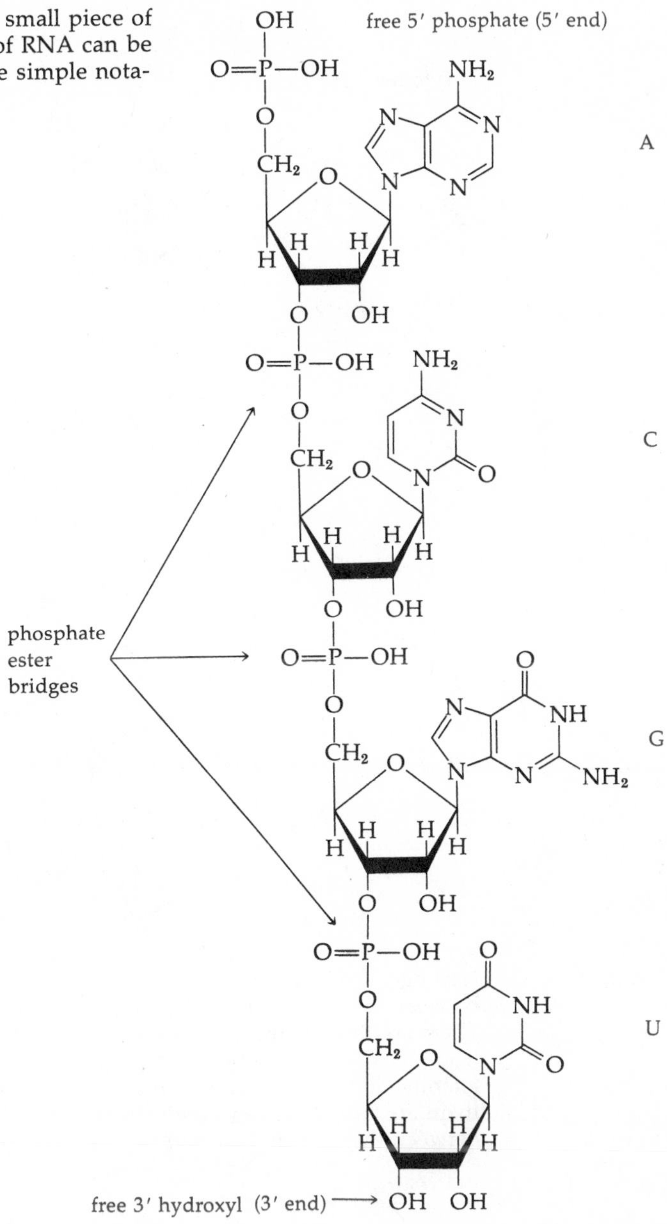

cleotide with the free phosphate group (the 5' end) is written on the left of the sequence and the nucleotide with the free hydroxyl on the number 3' carbon of the sugar (the 3' end) is written on the right. Thus, for example, we can represent the sequence of nucleotides in Figure 24-3 by $^{5'}ACGU^{3'}$.

The structure and function of a nucleic acid is determined by the particular sequence of nucleotides in that nucleic acid. We discussed in Chapters 21 and 22 how the sequence of amino acids in a protein determines the unique structure and function of that protein. What special feature of DNA allows it to function as a blueprint? The unique feature is that DNA consists of two long polynucleotide strands, and the base component of each nucleotide on one strand can form hydrogen bonds with only one specific nucleotide base on the other strand.

For example, a guanine base on one DNA strand can hydrogen bond only to a cytosine on another DNA strand. Two nucleotide units that are bound together by hydrogen bonds are called a **base pair.** Guanine-cytosine constitute one such base pair. We say that cytosine is the **complementary base** to guanine and that the complementary base of guanine is cytosine.

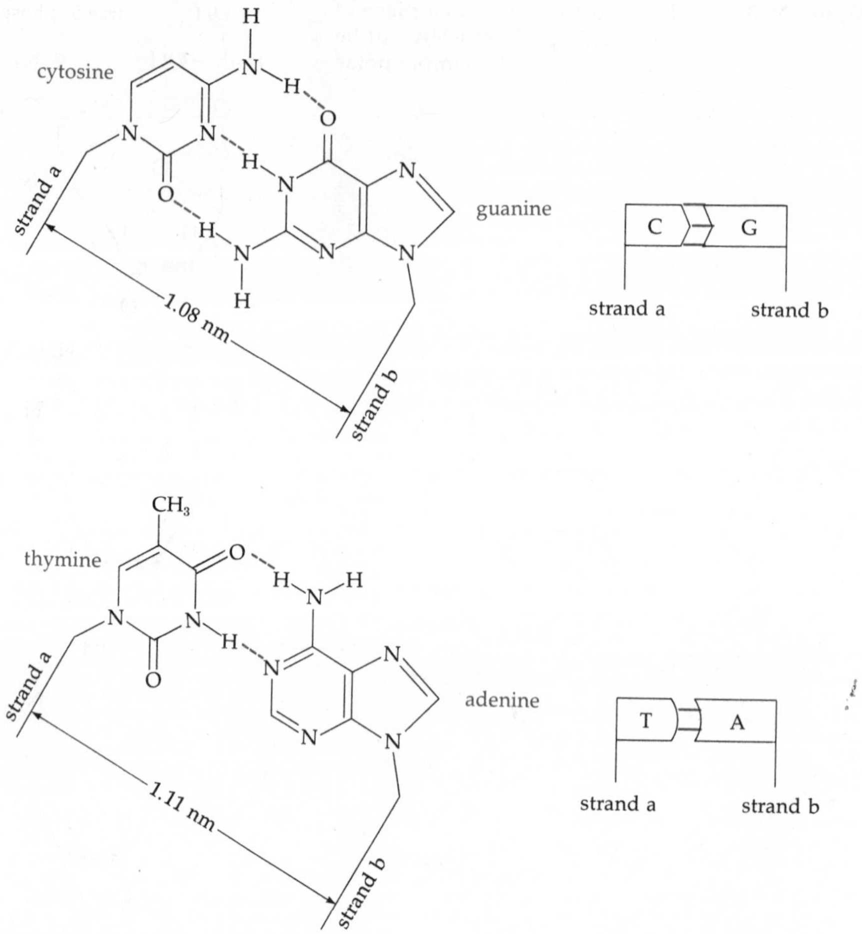

Figure 24-4   Hydrogen-bonded base pairs between two nucleic acid strands. There are three hydrogen bonds (shown as blue lines) between guanine and cytosine (top), and there are two hydrogen bonds between thymine and adenine (bottom). Adenine also forms a base pair with uracil. On the right are schematic representations for these base pairs in which the larger purine bases are represented by larger symbols than those used for pyrimidine bases.

Base pairing is illustrated in Figure 24-4. Note that A is capable of forming two hydrogen bonds with T or U, whereas G can form three hydrogen bonds with C, and G or C cannot form multiple hydrogen bonds with A, T, or U. For DNA, there are thus only two possible base pairs, A-T and G-C. Base pairing also occurs in processes involving RNA. For RNA, only the base pairs A-U and G-C are possible.

Note that the two long polynucleotide strands in DNA are held together by hydrogen-bonded base pairs. Now, hydrogen bonds are fairly weak bonds, so two or three hydrogen bonds could not hold these strands together very tightly. However, in one strand of DNA there are millions of nucleotide bases, so the forces holding the two DNA strands together can be quite strong. In order for this relatively strong bonding to occur, *each* nucleotide base on one strand must be able to hydrogen bond to its complementary base on the opposite strand. The sequences of two DNA strands arranged in this fashion are then said to be **complementary** to each other. For example, in complementary strands, wherever there is a G in one strand, it must be hydrogen bonded to a C in the other strand, and every T must be hydrogen bonded to an A. This is shown in Figure 24-5. Also notice in Figure 24-5 that if the direction of one strand is 3'-5', its

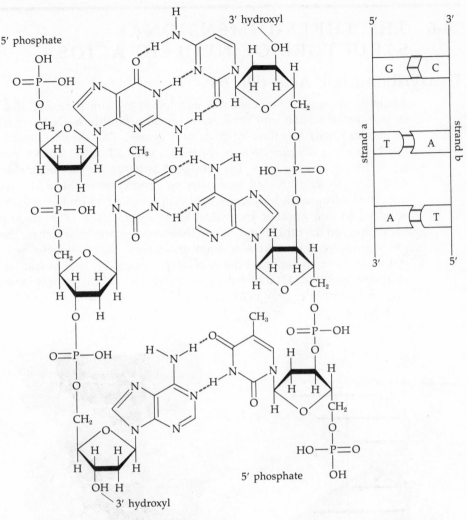

Figure 24-5   Complementary strands of DNA are represented by structural formulas on the left and schematically on the right. Hydrogen bonds are shown as blue lines. Notice that the sequence of bases in one DNA strand specifies the sequence of its complementary strand.

complementary strand must be in the opposite direction (5'-3'). This has been observed to be true for all double-stranded nucleic acids. Thus, <u>the sequence of one strand of DNA specifies the sequence and direction of its complementary strand. This is the crucial structural property of nucleic acids!</u> For example, if the sequence of a portion of one DNA strand is 5'CCA3', then the sequence of the corresponding portion of its complementary strand must be 5'TGG3', and these hydrogen-bonded strands can be represented by the notation

$$^{5'}CCA^{3'}$$
$$^{3'}GGT^{5'}$$

Exercise 24-2

Given the following abbreviated notation for part of one DNA strand,

$$^{5'}ATCGGATTCC^{3'}$$

show the sequence and direction of its complementary strand.

24-6 THE THREE-DIMENSIONAL STRUCTURE OF NUCLEIC ACIDS

Deoxyribonucleic Acid (DNA)

About 25 years ago, Francis Crick and James Watson proposed that the DNA strands twist around one another and form a **double helix** in which the two strands are held together by hydrogen bonds. The double helical model of DNA, shown in Figure 24-6, is consistent with all of the known properties of DNA. In this double helix, the nitrogenous bases are located inside the helix with the hydrogen-bonded base pairs perpendicular to the main axis. In addition, the hydrophobic base pairs, which are stacked up on top of each other, are attracted to one another by hydrophobic interactions. The sugar-phosphate backbones, on the other hand, are on the outside of the helix, where these hydrophilic groups can interact with water molecules. Thus a double helix structure stabilizes the association of the two DNA strands in three ways: (1) base-pair hydrogen bonding, (2) internal stacking of hydrophobic groups, and (3) exposing the hydrophilic groups to water.

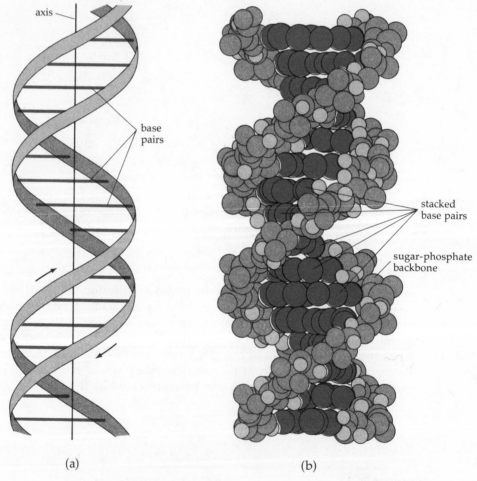

(a)   (b)

Figure 24-6   The double helix model of DNA. (a) The sugar-phosphate backbones of the two complementary strands are indicated as ribbons to which the hydrogen-bonded base pairs are attached. (b) A molecular model, with the sugar-phosphate backbone drawn in gray and the base pairs in blue.

Ribonucleic Acid (RNA)

Each RNA molecule consists of a single strand. There are three main types of RNA molecules: messenger RNA, ribosomal RNA, and transfer RNA.

Messenger RNA (mRNA) is the copy of one strand of the DNA gene that is used to specify the amino acid sequence of the corresponding protein. We do not believe that mRNA has an elaborate three-dimensional structure, but that it exists as a long filament, or a tape (see Figure 24-7a).

Ribosomal RNA (rRNA) molecules are components of ribosomes, the particles on which amino acids are bound together to form proteins. Ribosomes are structurally very complex. More is being learned about their structure and function every year. For our purposes, we can visualize ribosomes as composed of a large and a small subunit. Each subunit is made up of rRNA and several proteins (see Figure 24-7b). There are over 70 proteins in a human ribosome.

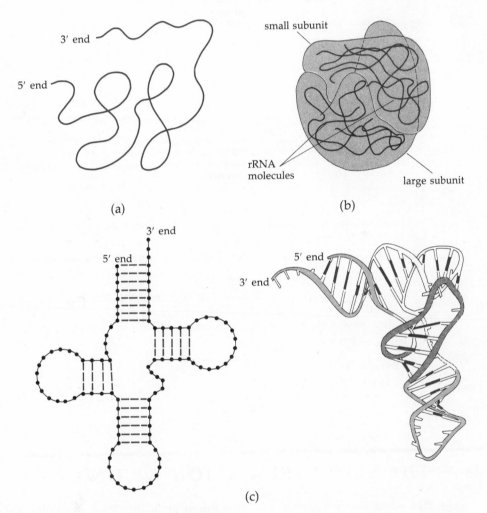

3' end	small subunit
5' end	rRNA molecules
(a)	large subunit
	(b)

3' end
5' end

5' end
3' end

(c)

Figure 24-7 The three-dimensional structure of RNA molecules. (a) Messenger RNA molecules probably exist as long strands. (b) A recently proposed model for the shape of ribosomes in *E. coli*. The presence of rRNA is represented by the blue lines. (c) Representations for a tRNA molecule. A tRNA can be represented as a cloverleaf (left), which twists up to form the three-dimensional shape represented on the right. The scales used in these four drawings are not the same. A tRNA molecule is actually much smaller than a ribosome.

Transfer RNA (tRNA) molecules are the carriers for the amino acids used in protein synthesis. Only one specific amino acid can bind to a particular tRNA carrier. There are more than 20 different tRNAs, at least one for each of the 20 different amino acids. The particular amino acid that can bind to a given tRNA molecule is indicated by a subscript. For example, a tRNA molecule that can bind glycine is represented by $tRNA_{gly}$.

A tRNA molecule is composed of only a single RNA strand, but it does have a characteristic three-dimensional structure. The general overall structure of these molecules is similar to a twisted-up cloverleaf. Base-pair formation between complementary bases on the same strand is responsible for this structure (see Figure 24-7c). Each tRNA has a different sequence of nucleotide components and thus has a slightly different overall shape.

Table 24-1 Size and Structure of Nucleic Acids

Nucleic Acid	Approximate Number of Nucleotides per Molecule	Structure
DNA*	2.5×10^8	Double helix
mRNA	300–9000	Long filament (single strand)
rRNA	100–4000	In complex ribosomes that contain two subunits and dozens of proteins
tRNA	75	Twisted cloverleaf

* There are 46 of these DNA molecules in the nucleus of every human cell except sperm and ova. Each DNA molecule consists of about 200 genes plus many segments that do not serve as blueprints for protein synthesis.

The structures and relative sizes of DNA and RNA molecules are compared in Table 24-1. Notice that DNA polymers are much larger than RNA polymers, since each DNA molecule carries the information for the sequence of several RNA molecules.

Exercise 24-3

Would you expect the two following nucleotide strands to form part of a double helix? Explain your answer.

Strand A: 5'ATCGCCG3'
Strand B: 3'ATCGCCG5'

24-7   THE SELF-REPLICATION OF DNA

The sequence of bases in the component genes of DNA molecules are coded blueprints for the amino acid sequences of proteins. If cells are to divide, and if parents are to pass hereditary information on to their children, there must be a mechanism for producing exact copies of DNA molecules. The process that accomplishes this is called **replication**. Replication is an example of the elegant simplicity of many biochemical processes. Basically, all that needs to be done is to separate the two complementary strands in the "parent" DNA helix and use each of these as a pattern or template for the synthesis of a new complementary strand (see Figure 24-8). The replication process is outlined in Figure 24-9.

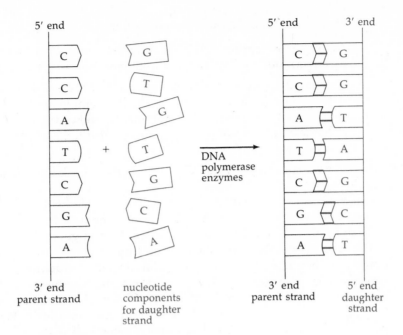

Figure 24-8   A parent DNA strand serves as a
template for the synthesis of a
daughter strand.

Figure 24-9   Replication of DNA involves (1)
unwinding of the double helix, (2)
polymerization of nucleotide components to form new "daughter"
strands, and (3) formation of two
new DNA helices, each composed
of one parent and one daughter
strand.

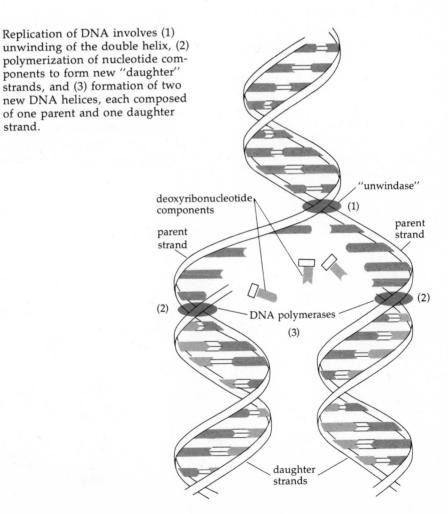

The steps in the replication process are as follows:

1. An enzyme (appropriately nicknamed *unwindase*) unravels the two parent strands by breaking the hydrogen bonds between base pairs.

2. The unpaired nucleotide bases on each of the single unwound parent strands form hydrogen bonds with new complementary nucleotides. A second enzyme, called **DNA polymerase,** takes these new complementary nucleotides and binds them together to form a new complementary daughter strand.

3. The parent and daughter strands then twist back up into double helices.

When the process is complete, there are two double helical DNAs, each composed of one parent and one daughter strand. The parent cell is now able to divide and give each daughter cell an identical set of genes.

The two sets of DNA genes that result from the replication process must be absolutely identical. Errors in replication are called **mutations.** Mutations can produce defective genes, which can direct the synthesis of defective proteins. Rarely, mutations are beneficial to the survival of the organism (Section 22-3). A defective gene often results in disease or death. When defective genes are inherited, the disease is called a **genetic disease.**

24-8   TRANSCRIPTION

The process of making an mRNA molecule corresponding to a DNA gene is called **transcription.** Transcription is very similar to replication. In transcription, the enzyme **RNA polymerase** binds to the DNA double helix at the beginning of a gene—it can recognize certain groups of nucleotides as starting points. The RNA polymerase then unravels a *portion* of the helix to expose the bases. RNA polymerase uses only one of the two DNA strands as a template—it "knows" which one to use. It then joins nucleotide components together into the corresponding mRNA molecule. It also "knows" when to stop the synthesis of

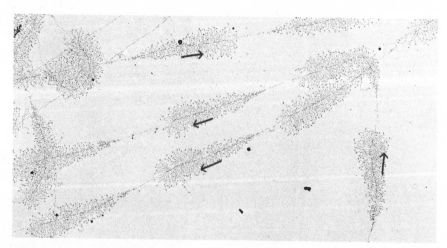

This electron micrograph shows large numbers of RNA molecules (the fine brushlike filaments) as they are being transcribed from DNA strands (the straighter lines). The arrows indicate the direction of movement of RNA polymerase molecules.

RNA—certain groups of nucleotides are recognized as stopping points. The RNA polymerase enzyme and the newly synthesized mRNA then detach from the gene, and the DNA re-forms its helical shape. As you can see, RNA polymerase is a versatile and complex enzyme. The steps in the transcription process are illustrated in Figure 24-10.

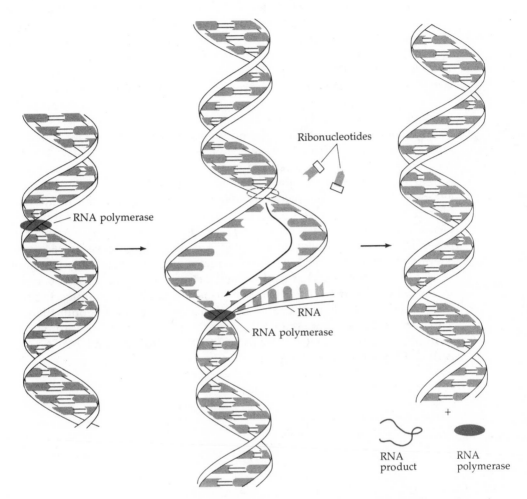

Figure 24-10   The process of RNA transcription begins with RNA polymerase binding to DNA at an appropriate point (left) and separating a portion of the helix. Using one DNA strand as a template, RNA polymerase then catalyzes the polymerization of ribonucleotides to form an RNA strand complementary to the DNA template (center). Transcription terminates when a stop signal on the DNA template is reached, and RNA polymerase and the newly synthesized RNA separate from the DNA template (right).

Exercise 24-4

Given the following sequence of nucleotides in part of a DNA gene, what is the sequence of nucleotides in the corresponding portion of the RNA molecule transcribed from this DNA strand?

DNA strand: · · ·$^{5'}$TCATGCA$^{3'}$· · ·

24-9 CRYPTOGRAPHY—ON CODES AND DECIPHERING THEM

The Genetic Code

Imagine yourself as an international spy. You have stolen a coded message that uses only four different *letters* (A, G, C, and U) and you know that, in some fashion, these letters are used to represent at most 20 different *words* (phe, gly, met, and so on). How can 20 words be formed with four letters? Obviously, there cannot be a one-to-one correlation between code letters and words, such as U = gly. There just aren't enough letters in the code to do it. What if each pair of two letters in the coded message corresponds to a word? Sixteen possible words could be formed in this manner, since there are $4 \times 4 = 16$ two-letter combinations of U, C, A, and G (such as UU, GA, CU, AG, etc.). There still aren't enough combinations to form 20 amino acid words. What about sets of three letters (GUG, AAU, GUC, GUU, etc.)? Now there are a total of $4 \times 4 \times 4 = 64$ possible coded words—more than enough. So you decide to try sets of three code letters and begin deciphering, that is, trying to match sets of three code letters to each word. An analogous coded message is contained in mRNA. The four different letters, A, G, C, and U, correspond to the four different nucleotide bases in RNA. The 20 words are the 20 different amino acids. A set of three bases in an mRNA molecule is called a **codon.** Thus there are 64 possible codons. After a great deal of experimental work, scientists have been able to crack the **genetic code.** That is, they have determined which codons correspond to each of the 20 amino acids. This information has been compiled to form the genetic code book, shown in Table 24-2. Since there are more codons than amino acids, in most cases more than one codon is used for the same amino acid. For example, $^5{}'CGU^3{}'$, $^5{}'CGC^3{}'$, $^5{}'CGA^3{}'$, $^5{}'CGG^3{}'$, $^5{}'AGA^3{}'$, and $^5{}'AGG^3{}'$ are all codons for the amino acid arginine.

Table 24-2 The Genetic Code Book

First Position (5' end)	Second Position				Third Position (3' end)
	U	C	A	G	
U	Phe	Ser	Tyr	Cys	U
	Phe	Ser	Tyr	Cys	C
	Leu	Ser	Term*	Term*	A
	Leu	Ser	Term*	Trp	G
C	Leu	Pro	His	Arg	U
	Leu	Pro	His	Arg	C
	Leu	Pro	Gln	Arg	A
	Leu	Pro	Gln	Arg	G
A	Ile	Thr	Asn	Ser	U
	Ile	Thr	Asn	Ser	C
	Ile	Thr	Lys	Arg	A
	Met**	Thr	Lys	Arg	G
G	Val	Ala	Asp	Gly	U
	Val	Ala	Asp	Gly	C
	Val	Ala	Glu	Gly	A
	Val	Ala	Glu	Gly	G

* Term = chain-terminating codon (stop signal)
** Met = chain-initiating codon (start signal)

Some codons are reserved to punctuate the message. The codons $^5{}'UAA^3{}'$, $^5{}'UAG^3{}'$, and $^5{}'UGA^3{}'$ are not codons for any amino acid. These three codons signal the termination of protein synthesis (the period at the end of the message). The codon $^5{}'AUG^3{}'$ codes for the amino acid methionine. This codon is also used to indicate the start of protein synthesis (the beginning of the message). All mRNA messages begin with the codon $^5{}'AUG^3{}'$.

Exercise 24-5
Without reference to Table 24-2, list all of the codons beginning with U.

Exercise 24-6
Using Table 24-2, write the sequence of amino acids corresponding to the mRNA sequence $^5{}'AUGCCCUGUAAUAGGCGAUAUUAG^3{}'$ (Note: Read the coded message from left to right.)

Machinery for Deciphering the Code

How do cells decipher the genetic code? The tRNA molecules do this job. You might consider tRNA molecules as interpreters that help translate an RNA language into a protein language. Every tRNA molecule has a set of three bases called an **anticodon** that can base pair with the three bases of a specific codon in mRNA. For example, the set of bases $^5{}'AGA^3{}'$ is the anticodon for the mRNA codon $^5{}'UCU^3{}'$, which specifies the amino acid serine. When an anticodon and a codon bind, the 3′ end of the anticodon binds to the 5′ end of the codon (as in the case of complementary strands in DNA). Thus, for example, the anticodon $^5{}'GAA^3{}'$ (rather than $^5{}'AAG^3{}'$) binds to the codon $^5{}'UUC^3{}'$ (see Figure 24-11). The codon $^5{}'UUC^3{}'$ corresponds to the amino acid phenylalanine according to the genetic code book.

Figure 24-11   Bases of an mRNA codon hydrogen bond with complementary bases of the anticodon of the appropriate tRNA molecule. Note that the 3′ end of an anticodon binds to the 5′ end of a codon.

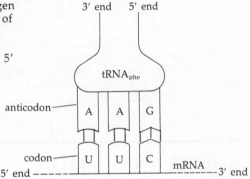

How does the tRNA molecule with the anticodon $^5{}'AGA^3{}'$ match up with the amino acid serine? This is accomplished by an enzyme called serine aminoacyl-tRNA synthetase. It binds serine—and no other amino acid—to the 3′ end of a tRNA$_{ser}$ molecule, which has the anticodon AGA. The hydrolysis of ATP supplies the energy required to join serine to its appropriate tRNA, forming serine-tRNA$_{ser}$. A schematic representation of this reaction is

$$\text{serine} + \text{tRNA}_{ser} + \text{ATP} \xrightarrow[\substack{\text{serine aminoacyl-}\\\text{tRNA synthetase}}]{} \text{serine-tRNA}_{ser} + \text{AMP} + \text{HO}-\overset{\overset{\displaystyle O}{\|}}{\underset{\underset{\displaystyle OH}{|}}{P}}-O-\overset{\overset{\displaystyle O}{\|}}{\underset{\underset{\displaystyle OH}{|}}{P}}-OH$$

Similarly, each of the tRNA molecules is bound to the correct amino acid by a specific aminoacyl-tRNA synthetase. An amino acid bound to its tRNA is called an **aminoacyl-tRNA.** The important structural features of one such aminoacyl-tRNA are illustrated in Figure 24-12.

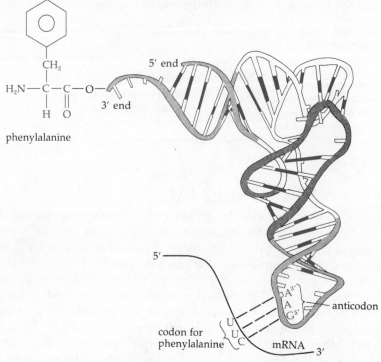

Figure 24-12   Schematic representation of the important features of phenyl-alanine-tRNA$_{phe}$. Note the binding of the anticodon portion of the tRNA to the appropriate codon on mRNA. The representations for the amino acid, for tRNA, and for mRNA are drawn to different scales.

Exercise 24-7

A tRNA molecule has the anticodon $5'GUA3'$. To what mRNA codon does this tRNA bind? What amino acid binds to this tRNA molecule?

24-10 TRANSLATION

With a good supply of aminoacyl-tRNAs and a messenger RNA template, the cell is ready to assemble the amino acids into a protein. This process, appropriately called **translation,** takes place on the ribosomes. Figure 24-13 shows the major steps of this process, which will be described here:

1. The 5' end of the mRNA molecule binds to the smaller subunit of the ribosome. Recall that all mRNA messages begin with the codon $5'AUG3'$, which corresponds to the amino acid methionine.* Methionyl-tRNA$_{met}$ recognizes this codon and binds to the mRNA.

---

* Methionine is always the N-terminal amino acid in newly made proteins. However, specific enzymes can cut an N-terminal peptide away from many newly synthesized proteins, so methionine is not the N-terminal amino acid of all "mature" proteins.

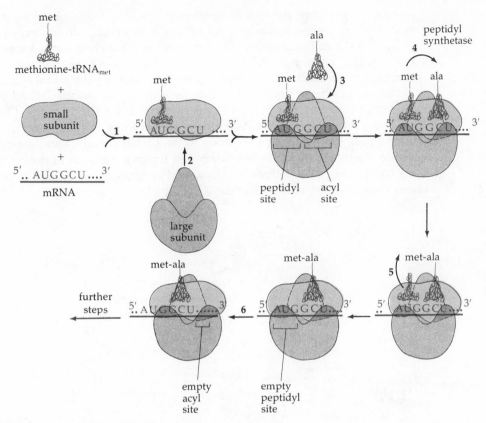

Figure 24-13   A schematic representation of the process of translation.

2. The larger ribosome subunit then binds to form a complete complex. The larger subunit has two active sites into which tRNA molecules can fit (think of them as keyholes), called the **peptidyl site** and the **acyl site.**

3. The next codon on the mRNA binds its aminoacyl-tRNA at the acyl site.

4. Now an enzyme, called **peptidyl synthetase,** located in the ribosome, removes the methionine from tRNA$_{met}$ and forms a peptide bond between methionine and the next amino acid (alanine, in this case), which remains attached to its tRNA at the acyl site.

5. Having lost its amino acid, tRNA$_{met}$ is released from the peptidyl site.

6. The ribosome slides along the mRNA tape by one codon's length, placing the newly formed dipeptide into the peptidyl site. The energy for this movement is supplied by hydrolysis of GTP (which can be represented by GTP + H$_2$O → GDP + P$_i$ + energy; see Chapter 25).

7. The process continues in this fashion, with the next aminoacyl-tRNA binding at the acyl site, followed by peptide bond formation and ribosome movement.

8. When a chain-terminating codon is finally encountered, the completed protein is detached from the last tRNA. The ribosome splits up into two subunits, which can go back and start the translation process over again.

In the past few years, several additional facts have been discovered about the process of translation. It is now known that a number of proteins that are not components of ribosomes are also involved in translation and that a second molecule of GTP must be hydrolyzed for every amino acid that is added to the growing peptide chain. Hence, the steps described here are somewhat simplified.

Granted, the entire process of synthesizing one protein chain is very involved, but the process has been clocked, and human cells can, for example, put together the 141 amino acids in one of the four peptide chains of hemoglobin in 3 minutes. And we don't hold the record. The bacterium *E. coli* can put together that many amino acids in 20 seconds! Figure 24-14 shows the actual synthesis of proteins in a bacterial cell. Notice that several ribosomes are bound to each mRNA molecule and that they follow each other down the message. One final word about protein assembly. As amino acids are being added to a growing protein chain (assembled from the amino to the carboxyl end), these amino acids begin to interact to form the secondary and tertiary structure of the protein. In other words, the peptide chain begins to fold up even as it is being synthesized, rather than after the entire assembly is completed.

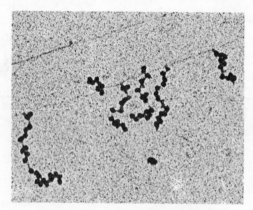

Figure 24-14   An electron micrograph showing transcription and translation in a bacterial cell. Along the lower DNA strand, several RNA polymerase molecules are at various stages in the transcription of mRNA molecules. As each mRNA is transcribed, it is used for the assembly of several identical protein molecules (not visible here) by several ribosomes. The mRNA molecules and the ribosomes are colored blue in the corresponding sketch on the right.

24-11 REGULATION OF PROTEIN SYNTHESIS

We have now seen how proteins are made. But since cells do not constantly need to synthesize all of the proteins coded for in their DNA blueprints, the process must be controlled. If a particular mRNA existed forever, we could never turn off the synthesis of the protein it codes for. The solution to this problem is simple. An enzyme in the cytoplasm, **ribonuclease,** chops up mRNA molecules into their component nucleotides. Thus a single mRNA molecule is usually able to supervise the assembly of only a small number of protein molecules before it is broken down.

Protein synthesis is controlled primarily by regulating the types and amounts of mRNA molecules produced in a cell. Thus the main regulation of protein synthesis is accomplished by controlling the process of transcription. In 1961, F. Jacob and J. Monod proposed a model that describes how transcription is regulated in *E. coli.* Two similar processes, called **induction** and **repression,** are involved in this model. Some of the same general mechanisms are used to control transcription in human cells, but the control process in human cells is very much more complicated than in *E. coli.*

Repression of Transcription

Genes that code for enzymes involved in a series of enzyme-catalyzed reactions are called **related genes.** For example, for the series of reactions

$$A \xrightarrow[E_1]{} B \xrightarrow[E_2]{} C \xrightarrow[E_3]{} D$$

the genes that code for the enzymes E_1, E_2, and E_3 are related genes. In *E. coli*, related genes are generally located next to each other on the chromosome, allowing them to be transcribed sequentially by the enzyme RNA polymerase. However, preceding each group of related genes is a set of nucleotides called an **operator site,** which can bind to a specific protein called a **repressor protein.** When the specific repressor protein in its active form is bound to its corresponding operator site, it prevents RNA polymerase from unwinding that portion of the DNA helix and transcribing the genes associated with that operator site (see Figure 24-15). The group of genes, together with its operator site, form a unit called an **operon.**

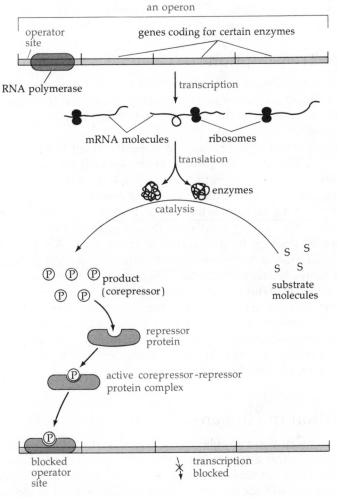

Figure 24-15   A model for the repression of genes. Enzyme molecules that are coded for by the genes in this operon catalyze the formation of a product P. Acting as a corepressor, P binds to a specific repressor protein, altering the conformation of the repressor protein to an active form, which then binds to its operator site on the DNA and prevents further transcription of the genes for the enzymes that produce P.

Now, the repressor protein itself can exist in two states, *active* and *inactive*. In its active state, the repressor binds to its operator site and prevents transcription. In its inactive state, it cannot bind to the operator site and transcription occurs. The repressor protein is normally inactive. However, when a small molecule called a **corepressor** binds to an allosteric site on a repressor protein, the repressor protein changes shape and becomes active. The active repressor protein stops transcription. The corepressor molecule is often the end product of the series of reactions catalyzed by the enzymes that are coded for by the genes in the operon. Do you now see how repression control operates? The more end product corepressor molecules are present, the more active repressor protein is formed. This results in a greater repression of mRNA formation and hence synthesis of less of the enzymes needed to form that end product (Figure 24-15). Thus, this process of repression prevents wasteful synthesis of unneeded enzymes.

The binding of the corepressor to the repressor protein is analogous to the binding and change of shape that occurs for allosteric enzymes. However, end product repression of enzyme synthesis is different from feedback inhibition of enzyme activity. Repression controls the amount of enzyme synthesized, whereas feedback inhibition controls the level of activity of enzymes.

Induction of Transcription

In induction, a different type of repressor protein binds to an operator site. However, in induction, the repressor protein is in its active state when *no* small molecule is bound to it. In its active state it binds to its operator and prevents transcription of appropriate genes.

However, when large amounts of a specific small molecule called an **inducer** are present, binding of the inducer at an allosteric site on the repressor protein inactivates the repressor protein, removing it from its operator site so that transcription and enzyme synthesis can then proceed (see Figure 24-16). Inducers are typically small molecules that can be broken down by enzymes produced by a cell. When the inducer inactivates its repressor protein, the inactive repressor protein can no longer bind at the operator site. Now, RNA polymerase can synthesize mRNA, and, via translation, the enzymes needed to break up the inducer are synthesized. This is illustrated in Figure 24-16. Of course, after a while, when the large supply of inducer has been used up by the newly synthesized enzymes, the repressor protein no longer has any inducer to bind to. It is therefore active again and turns off synthesis of any more of these enzymes. The overall effect of induction, like the overall effect of repression, is to stop the production of particular enzymes when they are not needed, and to allow the production of these enzymes when they are required.

Gene Regulation in Humans

How is transcription controlled in humans? Unfortunately, we do not have all the answers to this question yet. Some control is by induction and repression, but control mechanisms in humans are much more complicated than those in *E. coli* and are still the subject of intensive investigation. Some of these complications are the following: First, related genes are not always grouped together into an operon. For example, the genes for the α and β chains of hemoglobin are on different chromosomes. Second, the vast majority of genes in a given adult human cell are always kept turned off. Third, human cells undergo the process of **differentiation** as they mature from embryonic to specialized

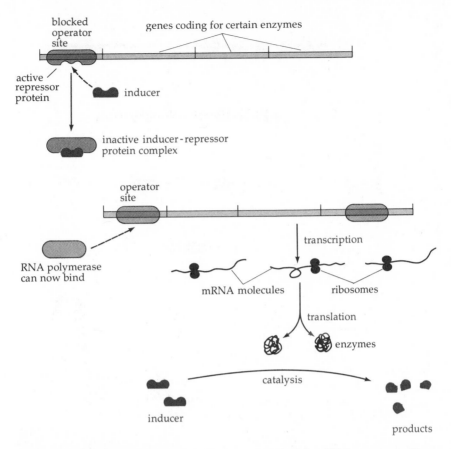

Figure 24-16 A model for the induction of genes. When the inducer is present, it binds to an allosteric site on its repressor protein. The repressor protein with its bound inducer now has a different conformation, and it "falls off" the operator site. This allows RNA polymerase to transcribe genes for enzymes catalyzing the conversion of inducer molecules to products.

adult cells; several genes must be switched on early in development, then turned off later. Fourth, hormones are used to help control transcription in humans. Fifth, the synthesis of at least some proteins in humans is regulated at the level of translation. For example, recent studies have shown that translation of the mRNA for hemoglobin does not occur if the heme prosthetic group is not available.

In the last few years, scientists have discovered that portions of some genes in higher organisms, including humans, are reorganized during differentiation, and that the messenger RNA molecules for some proteins are cut and spliced prior to translation. This reorganization of DNA segments during differentiation, called **recombination,** and the **splicing** of mRNA molecules are known to occur in the case of the DNA and RNA coding for antibody molecules (see Figure 24-17). The cutting and splicing of mRNA molecules to remove noncoding intravening segments, called **introns,** also occurs prior to the translation of several but not all other proteins. Introns have also been found in some newly synthesized tRNA molecules, and they must be removed in order for these tRNAs to function. The possible function of introns in the expression and regulation of human genes is currently being studied by a large number of scientists.

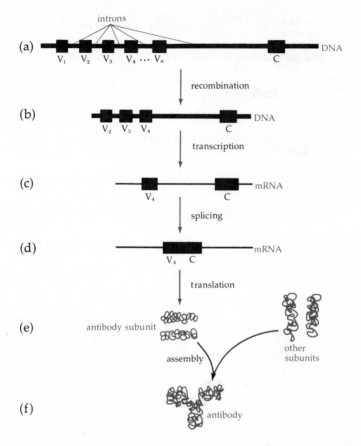

Figure 24-17   Steps involved in the expression of genes coding for antibody proteins. Part (a) represents a segment of DNA that contains genes coding for a variety of antibody subunits, as they are organized in embryonic cells. The solid rectangles represent individual genes coding for the antigen recognition portions, $V_1, \ldots, V_n$, of various antibody subunits, and a gene coding for C, a constant portion of these antibody subunits. The V and C genes are separated by introns. During differentiation, these DNA segments undergo recombination in some cells to allow for the various combinations of V and C genes found in the antibody-producing cells of adults. The recombinant DNA shown in (b), for example, can be transcribed to yield (c), a messenger RNA with a single V gene (V_4 in this case) separated from a C gene by an intron. This intron is spliced out to give (d), the mRNA that codes for the specific antibody subunit in (e). Assembly of this subunit with other subunits is followed by secretion from the cell of a functional antibody molecule (f).

Exercise 24-8

Consider the following sequence of enzyme-catalyzed reactions in which a molecule A is sequentially broken down to smaller molecules B, C, and D:

$$A \xrightarrow{E_1} B \xrightarrow{E_2} C \xrightarrow{E_3} D$$

Which of the molecules, A, B, C, D, might act as a corepressor molecule and which as an inducer?

24-12 VIRUSES

The ability to replicate DNA, transcribe DNA, and make proteins are major criteria that distinguish living organisms from nonliving things. Viruses do not contain DNA polymerase, RNA polymerase, or ribosomes. By themselves viruses are incapable of reproduction or growth. In fact, the simplest **viruses** are composed of just a small DNA molecule containing a few genes, and a protein coat to house the DNA (see Figure 24-18). However, when viruses infect (invade) living cells, they can use the machinery of the host cell to transcribe and replicate their own viral DNA. The mRNA molecules produced in this process are translated and viral proteins are assembled. Some of these proteins prevent normal function of the host cell, whereas the remainder are those needed for new envelopes to contain newly replicated viral DNA. Eventually, many new viruses are made and the host cell bursts, releasing the new viruses (see Figure 24-18).

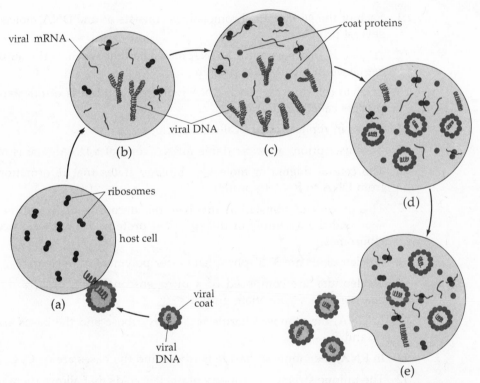

Figure 24-18 The life cycle of a DNA virus begins with a virus particle attaching to a host cell and injecting its viral DNA molecule inside (a). The viral DNA is then replicated and transcribed using the host cell's enzymes (b). Translation of viral mRNA on host ribosomes (c) yields viral coat proteins, which assemble spontaneously with new viral DNA molecules to form complete virus particles (d). Eventually the virus particles cause the host cell to burst (e), releasing the virus particles.

Other viruses contain RNA instead of DNA. They are called **RNA viruses.** The first thing an RNA virus must do when it infects a host cell is to make a master DNA blueprint from its RNA. This is accomplished with the aid of an enzyme called **reverse transcriptase:**

$$\text{Host deoxyribonucleotides} \xrightarrow[\text{reverse transcriptase}]{\text{viral RNA template}} \text{viral DNA}$$

Once viral DNA blueprints are made, they are used to make more viral RNA and protein coats for the daughter viruses.

Occasionally, a viral DNA will not replicate immediately upon infecting a host cell. Instead the viral DNA is inserted into the host cell's chromosome. The host cell's master blueprints now contain blueprints for this virus as well as for the normal host cell. Thus, the viral genes can be replicated along with the host's chromosome and given to daughter cells. At some later time, the viral DNA may be transcribed and new viruses made.

Very rarely a different process can occur. Viral DNA may cause rapid replication of the host DNA. This rare occurrence has been proposed to be one cause of human cancer. Although this has been shown to be a cause of cancer in many animals, it is unclear what role viruses play in human cancer.

24-13 SUMMARY

1. Each of the 46 human chromosomes consists of one DNA molecule and several proteins.

2. A gene is that portion of a DNA molecule that specifies the amino acid sequence in a particular protein.

3. Prior to cell division, each DNA molecule in a cell is duplicated in the process called replication.

4. Errors in replication are called mutations.

5. In transcription, an expendable mRNA copy of a DNA gene is made.

6. The central dogma of molecular biology states that information flows from DNA to RNA to protein.

7. The process of translation involves the decoding of the mRNA message and the assembly of the specified protein. This process occurs on ribosomes.

8. Nucleic acids are 3′,5′-phosphate ester polymers of nucleotides.

9. Nucleotides are composed of a nitrogen-containing base, a monosaccharide, and a phosphate group.

10. In DNA, the monosaccharide is 2′-deoxyribose and the bases are A, C, G, and T.

11. In RNA, the monosaccharide is ribose and the bases are A, C, G, and U.

12. The unique structural property of nucleic acids that allows them to function as blueprints is the ability of the bases to form specific base pairs by hydrogen bonding.

13. The allowable base pairs are A and T, C and G, and A and U.

14. Base pairing is responsible for the complementary relationship of the two DNA strands in the double helix.

15. There are three classes of RNA: mRNA, which are the copies of DNA genes; rRNA, which are components of ribosomes; and tRNA, which are used as carriers of amino acids in protein synthesis.

16. Codons on an mRNA molecule are sets of three nucleotides that specify one amino acid. They are recognized by the three complementary nucleotides of the anticodon on the tRNA for that amino acid.

17. In E. coli, protein synthesis is controlled by the induction or repression of mRNA transcription. Induction and repression involve the reversible binding of allosteric repressor proteins to DNA operator sites.

18. In humans, embryonic cells undergo the process of differentiation.

19. Viruses, which are particles containing a nucleic acid and proteins, cannot reproduce by themselves. Rather, they use their host cell's machinery for replication.

PROBLEMS

1. Draw the structural formula for one purine nucleotide triphosphate and the structural formula for one pyrimidine deoxyribonucleoside.

2. Draw simple diagrams representing the structures of DNA and the different types of RNA.

3. Translation is often divided into three operations, *initiation, elongation,* and *termination.* Which of the steps of protein synthesis shown in Figure 24-13 correspond to these operations?

4. Several enzymes in *E. coli* are required for the synthesis of the amino acid histidine from smaller molecules. Is the synthesis of these enzymes controlled by induction, or is it controlled by repression?

5. What differences in their substrates must be recognized by the enzymes DNA polymerase and RNA polymerase?

SOLUTIONS TO EXERCISES

24-1

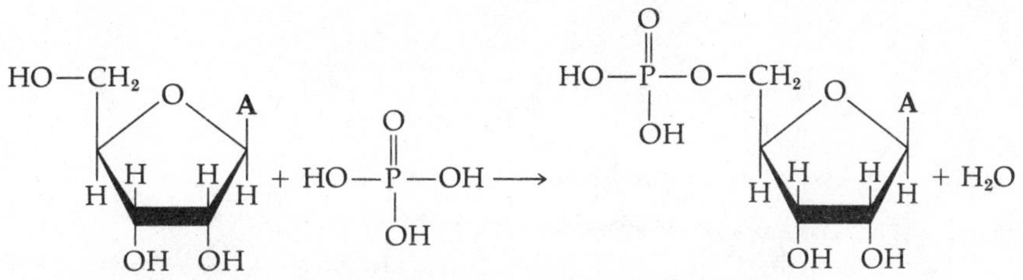

24-2 $3'$TAGCCTAAGG$^{5'}$. Note that the direction of this complementary strand is $3' \rightarrow 5'$.

24-3 No. These two DNA strands are going in opposite directions, however the sequences are *identical.* In order for a double helix to form, the two strands must be *complementary* to allow for hydrogen bonding of base pairs.

24-4 $3'$AGUACGU$^{5'}$

24-5 UUU, UUC, UUA, UUG, UAU, UAC, UAA, UAG, UCU, UCC, UCA, UCG, UGU, UGC, UGA, UGG

24-6 N-met-pro-cys-asn-arg-arg-tyr-C. The last codon, $^{5'}$UAG$^{3'}$, is a termination signal.

24-7 This anticodon will bind to the codon $^{5'}$UAC$^{3'}$, which specifies the amino acid tyrosine. Thus, this tRNA molecule binds to tyrosine.

24-8 The final product of this series of reactions, D, might serve as a corepressor, whereas the starting material, A, might be an inducer.

METABOLISM

Overview

In the last four chapters we looked at the structure and function of nucleic acids and proteins, the management and assembly-line workers in the cellular factories of the human body. We also looked at the complex machinery involved in the synthesis of these polymers from their monomeric components. In the following chapters we shall turn our attention to the cellular assembly lines themselves—we shall study the chemical reactions cells use to synthesize and break down molecules. The synthesis and breakdown of molecules occur in **metabolic pathways,** the collective name for cellular assembly lines. **Anabolic pathways** lead to the synthesis of molecules; **catabolic pathways** break apart molecules.

Most of these cellular assembly lines, or metabolic pathways, are used to assemble or take apart small molecules that are components of larger biomolecules. For example, some of these pathways make amino acids and nucleotides. Since these small molecules are generally used as *intermediates* in the manufacture of large biomolecules, pathways involving smaller molecules are frequently lumped together under the term **intermediary metabolism.**

Obviously, assembly lines need a source of energy to keep them moving. Cells need energy to fuel anabolic pathways and for other processes as well. This energy is usually supplied by the exergonic hydrolysis of a molecule called ATP. In Chapter 25 we shall discuss how the hydrolysis of ATP drives anabolic cellular assembly lines.

If cells hydrolyze ATP to get energy, then they need an adequate supply of energy to synthesize ATP. Where do they get it? We shall see that ATP is formed in conjunction with the catabolism of food molecules such as fats and sugars. The exergonic catabolism of these food molecules provides cells with the energy needed to form ATP. In order to understand how cells derive energy from the breakdown of sugars (carbohydrates) and fats (lipids), we shall need to understand some of the chemistry of these molecules. Carbohydrates have been discussed in Chapter 20. The chemistry of lipids will be discussed in Chapter 26, where we shall see that, in addition to serving as a source of energy, some fats are components of cell membranes and others are hormones and vitamins. We shall also see how carbohydrates and lipids are synthesized.

There are a great many cellular assembly lines, and we shall not be able to study all of them. In addition to pathways involving fats and sugars, we shall look at the pathways used to make and degrade amino acids (Chapter 28). However, we shall not study each and every step in all of these pathways. Rather, we shall concentrate on enzyme-catalyzed reactions that are similar in the synthesis of many amino acids and on the common features of amino acid catabolism. We shall see that amino acids are the body's major source of nitrogen. We shall also look at another pathway called the urea cycle. Urea is used by the human body as a vehicle for the excretion of excess nitrogen.

When we study a particular cellular assembly line, it is important to bear in mind that in a living cell many processes take place at the same time, and that all of the cellular assembly lines are interconnected. In fact, all of the metabolic pathways of intermediary metabolism are interrelated; they operate in an integrated manner, as we shall discuss in Chapter 29. When something goes wrong in one metabolic pathway, problems can arise in other connected pathways. Hence, cells use elaborate mechanisms to control their metabolic pathways. These control mechanisms prevent excess buildup or breakdown of any one type of molecule. We have discussed the basic features of these mechanisms: allosterism, hormones, chemical modification of enzymes, and competitive inhibition. In the following chapters we shall discuss a few important metabolic control points in some detail in order to gain an understanding of the principles of cellular control mechanisms.

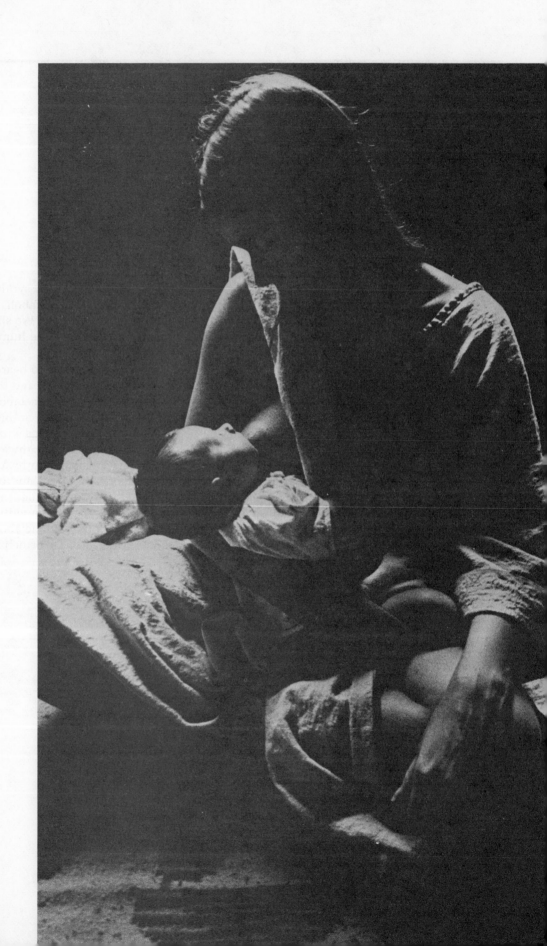

CHAPTER 25

Carbohydrate Metabolism and Bioenergetics

25-1 INTRODUCTION

Many people have a "sweet tooth." They crave sweets, even though an excess of sugar in the diet can lead to tooth decay, obesity, or worse. No one is more aware of the need for carefully controlling sugar intake than those persons who suffer from diabetes. More than 10 million people in the United States alone are afflicted with this disease. *Diabetes* is a disorder in which the metabolism of carbohydrates is not properly controlled. Understanding the pathology of diabetes requires an understanding of carbohydrate metabolism and how it is normally controlled in the human body.

Did you ever stop to wonder why people like sweet foods so much, especially when excess carbohydrate intake can cause medical problems? Our collective desire for sweets is probably due to the fact that our bodies actually need a somewhat generous (but not excessive) supply of carbohydrates—simple sugars or polymeric starch and glycogen—in order to supply the energy and carbon atoms needed to make nucleic acids, proteins, and membranes. We also need energy for movement, breathing, vision, active transport, maintaining body temperature, and so on. Carbohydrates are the major energy source for the human body. Even if we eat a lot of meat and our diet is therefore very rich in protein, the animals we eat lived off carbohydrates found in plants. Thus, directly or indirectly, the source of our biochemical energy is carbohydrates. In addition to energy, we need food molecules to supply the carbon, hydrogen, oxygen, and nitrogen that we need to make other biomolecules.

We do not use all of the atoms in our food to make biomolecules. The majority of the carbon atoms in the food we eat ends up in the carbon dioxide we exhale. Plants, in turn, require a supply of carbon dioxide so that they can make carbohydrates. Since green plants use solar energy in the process of photosynthesis to make carbohydrates from carbon dioxide and water, the ultimate source of energy for all plants and animals is the sun. Thus, all living things are dependent on one another for the atoms that are essential for life, and all rely on the sun

---

Nursing infants hydrolyze lactose to glucose and galactose. Galactosemic infants cannot properly metabolize galactose, and some of the galactose is converted to harmful by-products. The current therapy for galactosemia involves a lactose-free diet.

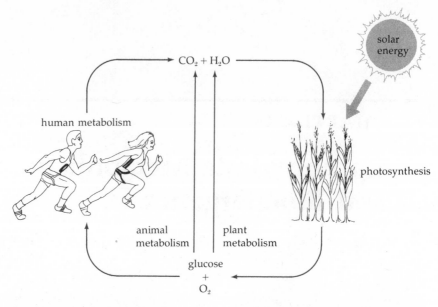

Figure 25-1   The flow of carbon atoms and energy in the biosphere. Photosynthetic organisms, such as green plants, use carbon dioxide, water, and solar energy to synthesize glucose, which is usually polymerized to cellulose and starch. All organisms rely on these carbohydrates for their energy needs, breaking glucose down to carbon dioxide and water, and using the energy released in this process.

to supply the necessary energy for life. This interdependence of the organisms that live on the earth for food and energy forms a network called the *biosphere*. The flow of energy and carbon in the biosphere is outlined in Figure 25-1. Other elements that are essential for life, such as nitrogen, are also circulated through the biosphere.

The overall reaction for the formation of 1 mole of glucose from carbon dioxide and water by the process of photosynthesis requires an input of at least 686 kcal of energy supplied by the sun:

$$\text{(25-1)} \qquad 6CO_2 + 6H_2O + 686 \text{ kcal} \longrightarrow \underset{\text{glucose}}{C_6H_{12}O_6} + 6O_2$$

Glucose therefore contains stored chemical energy. In this chapter we shall see how glucose is broken down in the human body in order to convert some of this stored chemical energy into more usable forms. We shall also see how some other carbohydrates—sucrose (table sugar) and the lactose in milk—are metabolized. Finally, we shall look at the control of carbohydrate metabolism and the imbalance in this control mechanism that is present in diabetics.

25-2   STUDY OBJECTIVES

After careful study of the material in this chapter, you should be able to:

1. Define the terms catabolism, anabolism, metabolism, and biosphere.

2. Differentiate endergonic from exergonic biochemical reactions.

3. Describe why ATP is called an energy carrier.

4. Recognize biochemical reactions that are energy coupled.

5. Show the steps in glycolysis that produce ATP from ADP, as well as those that produce $NADH + H^+$ from NAD^+.

6. List the different end products obtained from the glycolysis of glucose under aerobic and anaerobic conditions.

7. Explain what happens to the six carbon atoms in glucose as a result of either aerobic or anaerobic glycolysis.

8. Explain why the TCA cycle is called a cycle, where acetyl-S-CoA is fed into it, where carbon dioxide is released, and where ATP and reduced coenzymes are produced.

9. Recognize similarities in the enzymatic reactions of glycolysis and those of the TCA cycle.

10. Show the steps in the electron transport system and describe how energy can be made available for ATP synthesis via this system.

11. **(Optional)** Explain how the pentose phosphate pathway is used as a source of ribose 5-phosphate and the reduced coenzyme NADPH.

12. Write the overall reaction for gluconeogenesis, and describe the conditions that favor use of this pathway.

13. Explain the functions of the following three enzymes: lactase, sucrase, and hexose 1-phosphate uridylyltransferase.

14. Describe the disease galactosemia and explain its cause.

15. Describe the role of UDP-glucose in the synthesis of hexoses from glucose.

16. Describe in detail the glycogen synthesis/glycogen phosphorylase control point for glucose metabolism, including the roles of the hormones epinephrine (adrenaline), glucagon, and insulin.

17. Describe the major control point for glycolysis/gluconeogenesis.

18. Describe the various ways glucose is used in the human body and the general conditions under which each of these metabolic pathways is used.

19. Define the terms hyperglycemia, hypoglycemia, and hyperinsulinism.

20. Explain the role that insulin plays in diabetes and hypoglycemia.

21. Describe how a glucose tolerance test can be used to diagnose diabetes mellitus and hypoglycemia.

25-3 BIOENERGETICS

We burn food, and not just in the kitchen! In some areas crops are being grown for use as fuel to supplement our dwindling petroleum resources. This is an indirect use of solar energy. We can let plants capture the solar energy, then burn them to generate heat. In the body we burn food in a different way.

We use some energy to keep our bodies warm. But we need large amounts of energy in forms other than heat: We need a great deal of energy to synthesize complex biomolecules.

Energy-Coupled Reactions

Energy for biosynthesis is provided by the coupling of chemical reactions in the body. When a chemical reaction that releases energy, an **exergonic** reaction, supplies the energy to drive another reaction for which energy is required, an

endergonic reaction, we say that these two reactions are **energy coupled.** Recall that an exergonic reaction is one in which the free energy (G) of the products is less than the free energy of the reactants. Thus, for an exergonic reaction, the change in the free energy (ΔG) of the chemical system is negative. For an endergonic reaction, ΔG is positive. The concept of energy-coupled reactions is illustrated schematically in Figure 25-2.

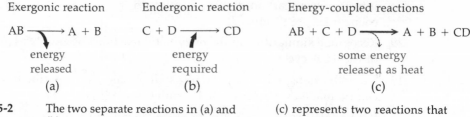

Exergonic reaction   Endergonic reaction   Energy-coupled reactions

AB ⟶ A + B   C + D ⟶ CD   AB + C + D ⟶ A + B + CD

energy   energy   some energy
released   required   released as heat

(a)   (b)   (c)

Figure 25-2   The two separate reactions in (a) and (b) are not energy coupled, whereas   (c) represents two reactions that are energy coupled.

The term **bioenergetics** is used to refer to energy changes and energy coupling in biochemical reactions. This coupling of two reactions is never perfect. Some of the energy released by the exergonic reaction is not used to drive the endergonic reaction, but rather is released in the form of heat.

Energy-coupled reactions occur together and are interdependent. A coupled reaction is somewhat analogous to the motion of two people on a seesaw (see Figure 25-3). When one person goes down, the other person is forced up. Similarly, in a coupled biochemical reaction, the decrease in free energy in the exergonic reaction supplies the energy needed for the endergonic reaction. Of course, the exergonic reaction must supply *at least* as much energy as is required to drive the endergonic reaction. The coupling of endergonic reactions to exergonic reactions occurs at the active sites of enzymes. Thus, enzymes are like the seesaw of our analogy.

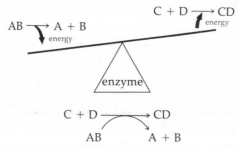

Figure 25-3   The interdependence of two energy-coupled reactions is somewhat like the motion of two people on a seesaw. The energy coupling of   two reactions is often represented by using a straight reaction arrow for one and a curved arrow for the other.

In the human body, carbohydrates, lipids, and proteins are broken down in a stepwise fashion. These reactions are exergonic, and the energy released when they occur is captured very efficiently by coupled endergonic reactions in which bonds are formed. Breaking foods down in a stepwise manner does not change the total amount of energy that is released. It does, however, permit us to capture some of the released energy in a usable form. The overall energy change in any chemical reaction still depends only on the reactants and products.

ATP

The energy required for essential endergonic reactions in the human body is not obtained directly from the stepwise breakdown of foods, but rather involves an energy-carrier intermediary as follows:

1. Many of the different exergonic reactions that occur in the stepwise breakdown of foods are coupled to the *same* endergonic reaction, in which a substance called adenosine diphosphate (ADP) is converted to one called adenosine triphosphate (ATP).

2. Hydrolysis of ATP to ADP releases energy.

3. This exergonic hydrolysis of ATP is the reaction that is directly coupled to most of the endergonic reactions in the human body.

Therefore, ATP acts as an energy carrier for the vast majority of endergonic biochemical reactions. The energy required for a few endergonic biochemical reactions is supplied directly by the hydrolysis of other substances. The hydrolysis reactions for these substances are similar to those for ATP.

Let us investigate some of the properties of ATP. Adenosine triphosphate, ATP, is very similar in structure to the RNA nucleotide containing the base adenine (Chapter 24). Recall that this RNA nucleotide has three component parts: the base adenine, the sugar ribose, and a phosphate group. ATP also has the base adenine and the sugar ribose. However, it has three phosphate groups bonded together by phosphoanhydride bonds (Chapter 17). The structural formula for ATP in its neutral form is shown in Figure 25-4. Notice that ATP has four acidic hydrogens and one basic amine group.

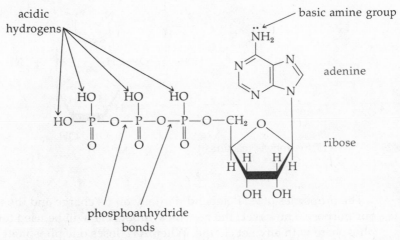

Figure 25-4   The structural formula for adenosine triphosphate, ATP.

We have seen that the actual charge on a molecule with many acidic and basic sites depends on the pH. The higher the pH, the more negatively charged the molecule is; and the lower the pH, the more positively charged the molecule is. At pH 7, which is approximately the pH of cells in the human body, about half of the ATP molecules have a net charge of -3 and about half have a net charge of -4. The structural formula for ATP with a net charge of -3 is shown on the left of the reactions in Figure 25-5.

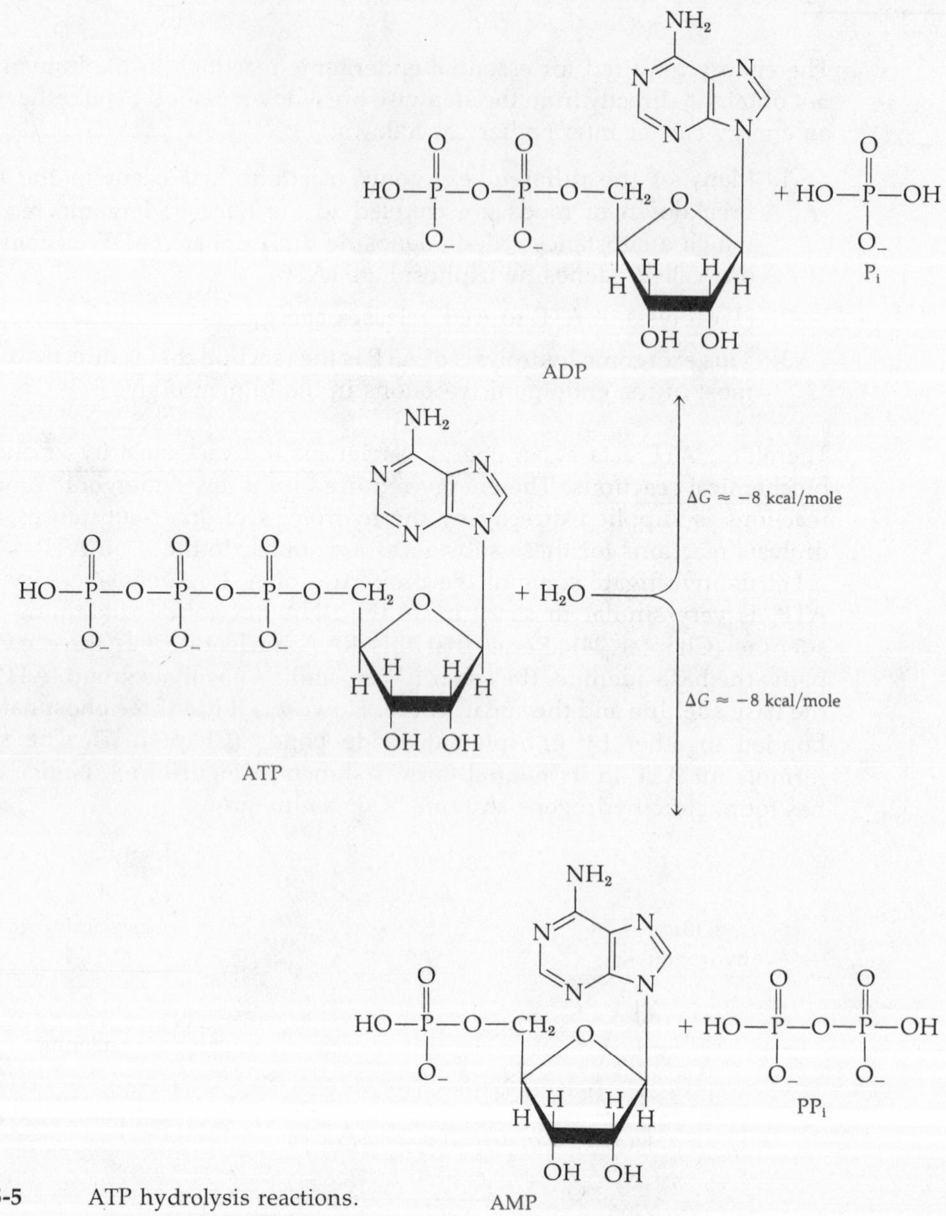

Figure 25-5   ATP hydrolysis reactions.

The properties of ATP depend partially on its charge and thus on the pH. For our purposes, however, the general symbol **ATP** will be used for adenosine triphosphate with any net charge. When ATP loses one phosphate group, a molecule of adenosine diphosphate, **ADP,** is produced. When ATP loses two phosphate groups, the product is adenosine monophosphate, **AMP.** We shall also use the general symbols P_i for inorganic phosphate (whether in the form H_3PO_4, $H_2PO_4^-$, HPO_4^{2-}, or PO_4^{3-}) and similarly, PP_i for pyrophosphate.

ATP can undergo two hydrolysis reactions (see Figure 25-5). Using our general symbols for phosphates, we can represent these reactions as

(25-2) $ATP + H_2O \longrightarrow ADP + P_i + energy$

and

(25-3) $ATP + H_2O \longrightarrow AMP + PP_i + energy$

In each of these reactions, a phosphoanhydride bond is broken and energy is released. Both of these reactions are therefore exergonic. Each reaction has a ΔG of about -8 kcal/mole. The precise value of ΔG depends on several factors, including the pH, because the charges on ATP, ADP, AMP, P_i, and PP_i depend on pH.

Catabolism

Chemical reactions in the body that break larger molecules (such as glucose or amino acids) into smaller ones (such as CO_2, H_2O, and NH_3) are called catabolic reactions, and the overall process is called **catabolism.** As we have noted, food molecules are broken down in a stepwise fashion in order to capture some of the free energy released in catabolism. Some catabolic reactions are coupled directly to the formation of ATP, but others are not. In the next section we shall consider the catabolism of glucose and the coupling of glucose breakdown to ATP formation. In later chapters we shall discuss how lipids (Chapter 27) and proteins (Chapter 28) can also be broken down in reactions that are coupled to ATP synthesis.

Exercise 25-1

Indicate which of the following reactions are endergonic and which are exergonic:

(a) $H_2O + CH_2=C-C\begin{smallmatrix}O\\\\OH\end{smallmatrix}$ with $O-P(=O)(OH)$ substituent $\longrightarrow H_3C-C-C\begin{smallmatrix}O\\\\OH\end{smallmatrix}$ with $=O$ $+ P_i$ $\Delta G = -12$ kcal/mole

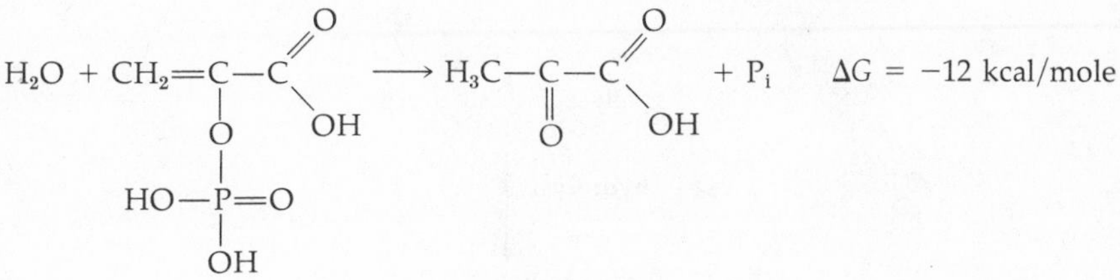

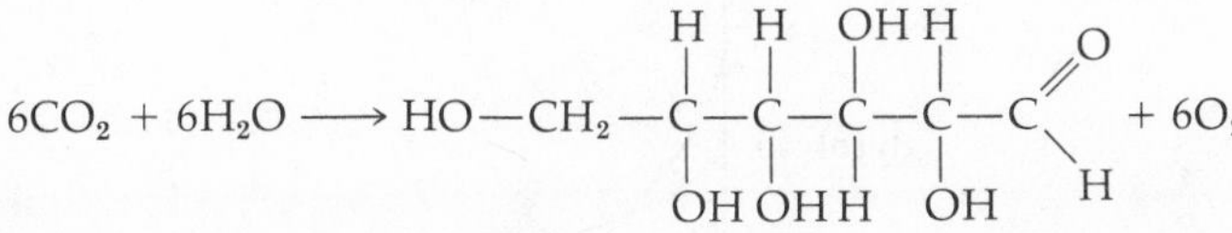

(b) $6CO_2 + 6H_2O \longrightarrow HO-CH_2-\overset{H}{\underset{OH}{C}}-\overset{H}{\underset{OH}{C}}-\overset{OH}{\underset{H}{C}}-\overset{H}{\underset{OH}{C}}-C\begin{smallmatrix}O\\\\H\end{smallmatrix} + 6O_2$

(c) $ADP + P_i \longrightarrow ATP + H_2O$

25-4 CARBOHYDRATE CATABOLISM AS A SOURCE OF ENERGY

The first catabolic pathway we shall study is the group of reactions that break down glucose to pyruvic acid. This series of reactions is called the glycolytic (sugar-breaking) pathway, or simply **glycolysis.** We shall look first at glucose because of its primary importance in our diet and its use in the body. We eat a lot of plant starch, animal starch (glycogen), and just plain sugar (sucrose). In the human body the concentration of glucose in the blood is carefully controlled so that enough will be available when needed. In fact, cells in the brain use only glucose to supply their energy needs and are extremely sensitive to changes in glucose concentration in the blood. We also store additional glucose in the polymeric form of glycogen.

We shall see that pyruvic acid, $H_3C-\overset{\overset{O}{\|}}{C}-\overset{\overset{O}{\|}}{C}-OH$, which is the end product of glycolysis, can suffer two different fates in the human body. If oxygen is not available, pyruvic acid is converted to lactic acid. On the other hand, if oxygen is available, pyruvic acid is further broken down to produce CO_2 and a substance called **acetyl-S-CoA,** $H_3C-\overset{\overset{O}{\|}}{C}-S-Coenzyme\ A.$

Acetyl-S-CoA plays a central role in metabolism. It is produced during the catabolism of lipids and proteins as well as carbohydrates. Acetyl-S-CoA is the input for the second catabolic pathway we shall study (Section 25-5), the *tricarboxylic acid pathway* or TCA cycle.

In Section 25-6 we shall look at another pathway linked to the TCA cycle and glycolysis. In this pathway, called the *electron transport system* (ETS), the reduced coenzymes produced during reactions in glycolysis and the TCA cycle are reoxidized.

The oxidation of coenzymes by the ETS is coupled to ATP synthesis from ADP and P_i. For these coenzymes to be oxidized, oxygen must be available. When cells need energy, and if enough oxygen is present, glucose is oxidized completely to CO_2 and water. Under these conditions, ATP is produced by the combined action of glycolysis, the TCA cycle, and the electron transport system (see Figure 25-6).

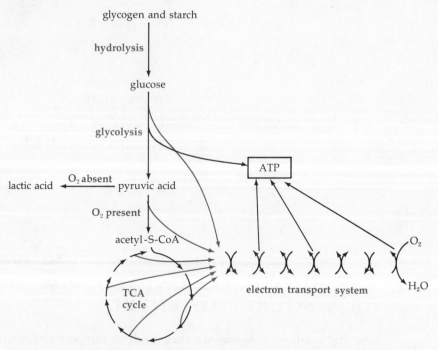

Figure 25-6   The relationship of glycolysis to the TCA cycle and the electron transport system. Hydrolysis of dietary polysaccharides in the gastrointestinal tract produces glucose, which is absorbed and subsequently catabolized in cells (via glycolysis) to pyruvic acid. When oxygen is not available, pyruvic acid is converted to lactic acid. When oxygen is available (it is needed for the electron transport system), pyruvic acid is further broken down via the TCA cycle. Some ATP is produced directly in glycolysis and the TCA cycle. Much more ATP is produced by the electron transport system, which reoxidizes coenzymes (indicated by blue arrows) that have been reduced in glycolysis and the TCA cycle.

Glucose in Polymeric Form

Before we begin our discussion of the catabolism of glucose, let us see what happens to food that contains glucose in polymeric form. The bonds that hold the glucose units together in both starch and glycogen are α-1,4- and α-1,6-glycosidic bonds, as we saw in Chapter 20. When we ingest starch or glycogen, these bonds are quickly broken by enzymes produced in the salivary glands and pancreas. In humans the major enzyme involved in this digestive process is α-amylase, which hydrolyzes α-1,4-glycosidic bonds (see Figure 25-7). Recall that α-amylase cannot break the β-1,4-glycosidic bonds in cellulose. Thus human beings cannot digest wood or vegetable cellulose, but a few animals can. For instance, organisms that can catalyze the hydrolysis of β-1,4-glycosidic bonds in cellulose live in the digestive systems of termites and ruminant animals.

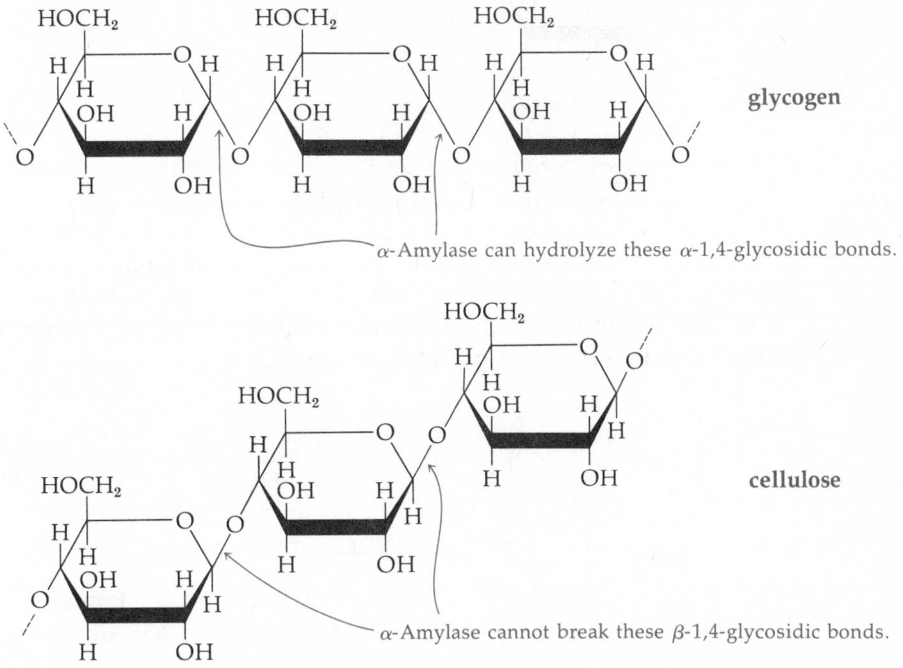

glycogen

α-Amylase can hydrolyze these α-1,4-glycosidic bonds.

cellulose

α-Amylase cannot break these β-1,4-glycosidic bonds.

Figure 25-7 α-Amylase is specific for α-1,4-glycosidic bonds.

Glycolysis

Glycolysis involves the breakdown of a glucose molecule to two molecules of pyruvic acid in a series of reactions coupled to the synthesis of two ATP molecules from ADP. Since the major purpose of glycolysis is the capture of energy, we shall keep close track of the energy-coupled reactions.

In the first step of glycolysis, energy is actually used to add a phosphate group to glucose, a process called **phosphorylation.** The energy for this reaction comes from the conversion of ATP to ADP:

Step 1:

(25-4)

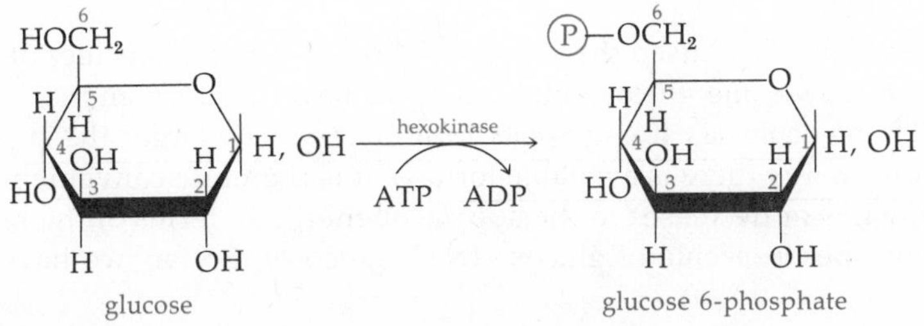

glucose   glucose 6-phosphate

Remember that the curved arrow indicates that these two reactions are energy coupled. The endergonic phosphorylation of glucose is coupled to the exergonic conversion of ATP to ADP. ATP actually donates the phosphate group in this reaction as well. The symbol (P) is a shorthand way of indicating a phosphate group in an organic molecule.

In the next step the aldose, glucose 6-phosphate, is converted to its structural isomer, the ketose, fructose 6-phosphate:

Step 2:

(25-5)

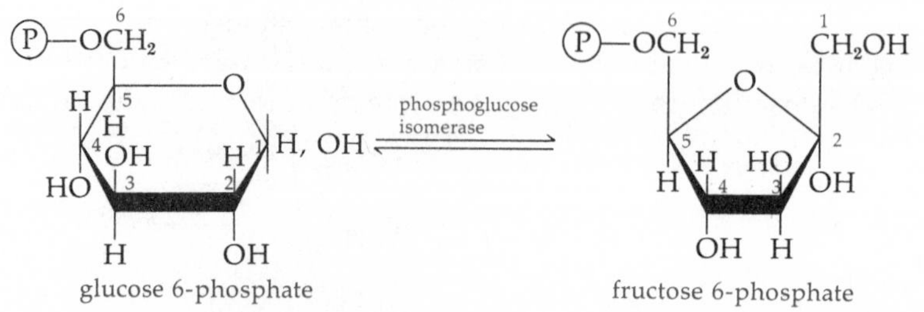

glucose 6-phosphate   fructose 6-phosphate

The double arrow used in this and later reactions indicates a reaction that is readily reversible. In this case, the enzyme phosphoglucose isomerase also converts fructose 6-phosphate back to glucose 6-phosphate. The net direction of the reaction depends on supply and demand. The larger the amount of glucose 6-phosphate, the greater the production of fructose 6-phosphate.

The third step in glycolysis is the phosphorylation of fructose 6-phosphate, which also requires energy supplied by ATP:

Step 3:

(25-6)

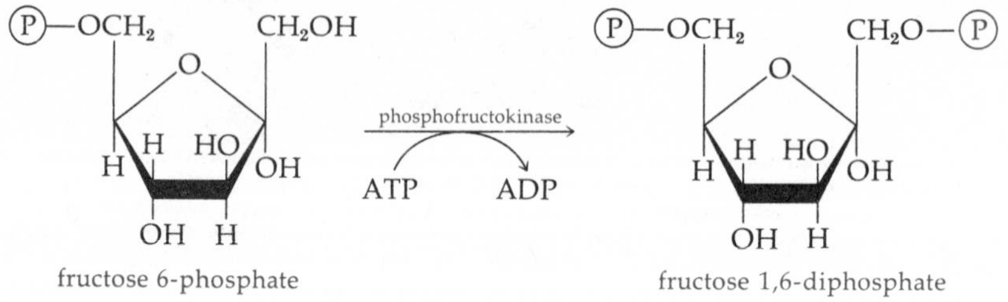

fructose 6-phosphate   fructose 1,6-diphosphate

In the fourth step of glycolysis, fructose 1,6-diphosphate is split into two three-carbon units by the enzyme aldolase:

Step 4:

(25-7)

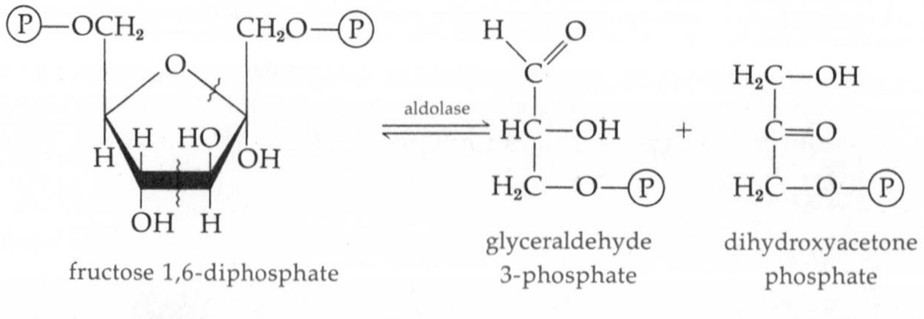

fructose 1,6-diphosphate   glyceraldehyde   dihydroxyacetone
                             3-phosphate   phosphate

So far we have used the energy contained in two molecules of ATP to partially break down one glucose molecule. We have used up energy. We can think of ATP molecules as storage places for energy. The larger the supply of ATP, the more energy there is available for use. It is therefore convenient to keep track of increases or decreases in the supply of energy in terms of the ratio of ATP molecules per molecule of glucose (ATP/glucose). So far, we have lost two molecules of ATP per molecule of glucose (one in step 1 and one in step 3), so our ATP/glucose ratio is -2.

The fifth step of glycolysis is a reaction that allows the dihydroxyacetone phosphate produced in step 4 to be converted to glyceraldehyde 3-phosphate, the other product of step 4. Because of step 5, only one subsequent pathway is required. Separate pathways for breaking down glyceraldehyde 3-phosphate and dihydroxyacetone phosphate are not needed.

Step 5:

(25-8)

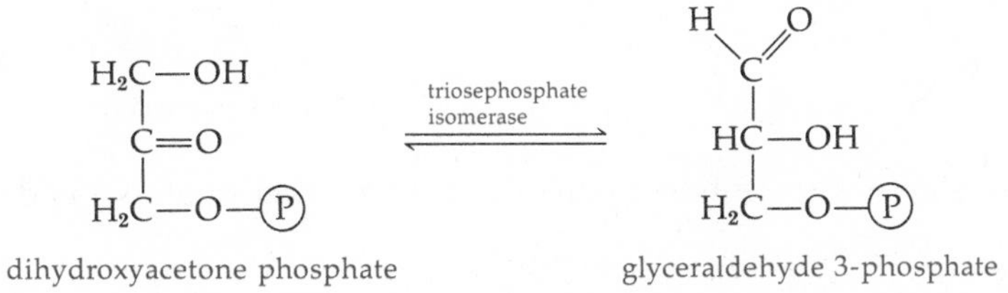

$$H_2C-OH \quad\quad C=O \quad\quad H_2C-O-\textcircled{P}$$
dihydroxyacetone phosphate

$$\xrightarrow[\text{isomerase}]{\text{triosephosphate}}$$

$$\underset{H}{\overset{O}{C}} \quad\quad HC-OH \quad\quad H_2C-O-\textcircled{P}$$
glyceraldehyde 3-phosphate

There is no change in the amount of ATP in step 5, and—like steps 2 and 4—this step is rapidly reversible. Keep in mind that after step 5, we have *two* molecules of glyceraldehyde 3-phosphate for every glucose molecule with which we started.

Step 6:

(25-9)

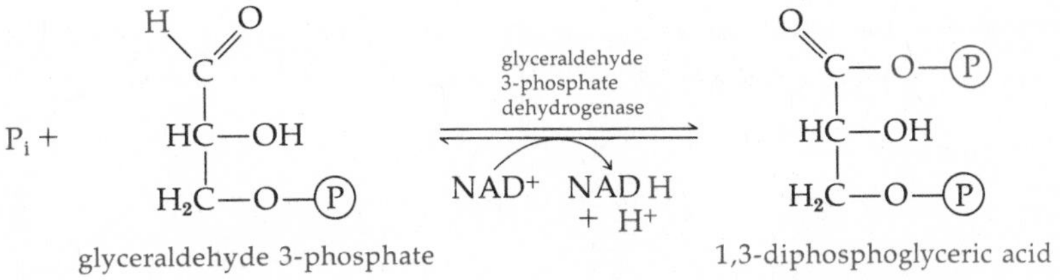

$$P_i + \quad \underset{H}{\overset{O}{C}} \quad HC-OH \quad H_2C-O-\textcircled{P}$$
glyceraldehyde 3-phosphate

$$\xrightarrow[\underset{NAD^+ \quad NADH}{\qquad\qquad +\ H^+}]{\substack{\text{glyceraldehyde}\\ \text{3-phosphate}\\ \text{dehydrogenase}}}$$

$$\underset{O}{\overset{O}{C}}-O-\textcircled{P} \quad HC-OH \quad H_2C-O-\textcircled{P}$$
1,3-diphosphoglyceric acid

In step 6 glyceraldehyde 3-phosphate is oxidized. This oxidation involves the conversion of an aldehyde to its corresponding acid, and the subsequent reaction of this acid to form an anhydride with phosphoric acid. The oxidation of glyceraldehyde 3-phosphate is coupled to the reduction of the coenzyme NAD^+ to NADH and H^+ (see Chapter 23). Notice that step 6 involves a combination oxidation-reduction reaction and the binding of a phosphate group to the acid product. The energy needed to form the phosphate anhydride comes from the energy released by the oxidation-reduction reaction.

Up to now in the glycolytic pathway, the energy resources of the cell have been drained. In the remaining steps of the pathway, energy reserves of the cell are replenished and increased.

Step 7:

(25-10)

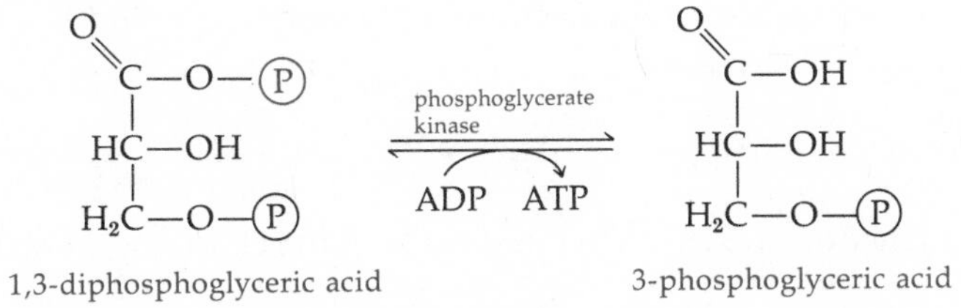

$$\underset{O}{\overset{O}{C}}-O-\textcircled{P} \quad HC-OH \quad H_2C-O-\textcircled{P}$$
1,3-diphosphoglyceric acid

$$\xrightarrow[\underset{ADP \quad ATP}{}]{\substack{\text{phosphoglycerate}\\ \text{kinase}}}$$

$$\underset{O}{\overset{O}{C}}-OH \quad HC-OH \quad H_2C-O-\textcircled{P}$$
3-phosphoglyceric acid

ATP is produced in step 7. For every molecule of glucose with which we started, there are now two molecules of 1,3-diphosphoglyceric acid. So step 7 gives us an ATP/glucose ratio of $+2$. Before step 7, the ATP/glucose ratio was -2. So now our *net* ATP/glucose ratio is zero.

We are now approaching the end of glycolysis. In step 8 another isomerization reaction takes place in which the phosphate group is moved to the next carbon on the chain:

Step 8:

(25-11)

$$
\underset{\text{3-phosphoglyceric acid}}{
\begin{array}{c}
\text{O} \\
\diagdown \\
\text{C—OH} \\
| \\
\text{HC—OH} \\
| \\
\text{H}_2\text{C—O—}\textcircled{P}
\end{array}
}
\quad \underset{\text{phosphoglyceromutase}}{\rightleftharpoons} \quad
\underset{\text{2-phosphoglyceric acid}}{
\begin{array}{c}
\text{O} \\
\diagdown \\
\text{C—OH} \\
| \\
\text{HC—O—}\textcircled{P} \\
| \\
\text{H}_2\text{C—OH}
\end{array}
}
$$

In step 9, the 2-phosphoglyceric acid formed in step 8 is dehydrated:

Step 9:

(25-12)

$$
\underset{\text{2-phosphoglyceric acid}}{
\begin{array}{c}
\text{O} \\
\diagdown \\
\text{C—OH} \\
| \\
\text{HC—O—}\textcircled{P} \\
| \\
\text{H}_2\text{C—OH}
\end{array}
}
\quad \underset{\text{enolase}}{\rightleftharpoons} \quad
\underset{\text{phosphoenolpyruvic acid}}{
\begin{array}{c}
\text{O} \\
\diagdown \\
\text{C—OH} \\
| \\
\text{C—O—}\textcircled{P} \\
\| \\
\text{H}_2\text{C}
\end{array}
}
\quad + \text{H}_2\text{O}
$$

In the tenth and last reaction of glycolysis, the phosphate group in the phosphoenolpyruvic acid is donated to ADP, and ATP is produced.

Step 10:

(25-13)

$$
\underset{\text{phosphoenolpyruvic acid}}{
\begin{array}{c}
\text{O} \\
\diagdown \\
\text{C—OH} \\
| \\
\text{C—O—}\textcircled{P} \\
\| \\
\text{H}_2\text{C}
\end{array}
}
\quad \xrightarrow[\text{ADP ATP}]{\text{pyruvate kinase}} \quad
\underset{\text{pyruvic acid}}{
\begin{array}{c}
\text{O} \\
\diagdown \\
\text{C—OH} \\
| \\
\text{C=O} \\
| \\
\text{CH}_3
\end{array}
}
$$

For the ten steps of glycolysis, the net ATP/glucose ratio is $+2$. The *overall* process of glycolysis can be symbolized as

(25-14)

$$
\underset{\text{glucose}}{\text{C}_6\text{H}_{12}\text{O}_6} \quad
\overset{\text{2NAD}^+ \quad \text{2NADH} + 2\text{H}^+}{\underset{2\text{ADP} + 2\text{P}_i \quad 2\text{ATP} + 2\text{H}_2\text{O}}{\rightleftharpoons}} \quad
\underset{\text{pyruvic acid}}{2\text{C}_3\text{H}_4\text{O}_3}
$$

We can consider the overall glycolytic process to be the sum of two separate reactions: (1) an oxidation-reduction reaction in which one molecule of glucose is oxidized—forming two molecules of pyruvic acid—and two molecules of the coenzyme NAD⁺ are reduced,

(25-15) $\text{C}_6\text{H}_{12}\text{O}_6 + 2\text{NAD}^+ \longrightarrow 2\text{C}_3\text{H}_4\text{O}_3 + 2\text{NADH} + 2\text{H}^+$

and (2) a reaction in which two molecules of ATP are synthesized,

(25-16) $2\text{ADP} + 2\text{P}_i \longrightarrow 2\text{ATP} + 2\text{H}_2\text{O}$

As we shall see, the NADH formed in glycolysis can be very useful. It can be used in the electron transport system to generate more ATP (see Section 25-6). The overall pathway of glycolysis is illustrated in Figure 25-8. What happens to the end product, pyruvic acid, depends on the availability of oxygen.

Figure 25-8   Glycolysis. The blue numbers indicate the number of moles of ATP per mole of glucose produced or consumed at the indicated points in the pathway.

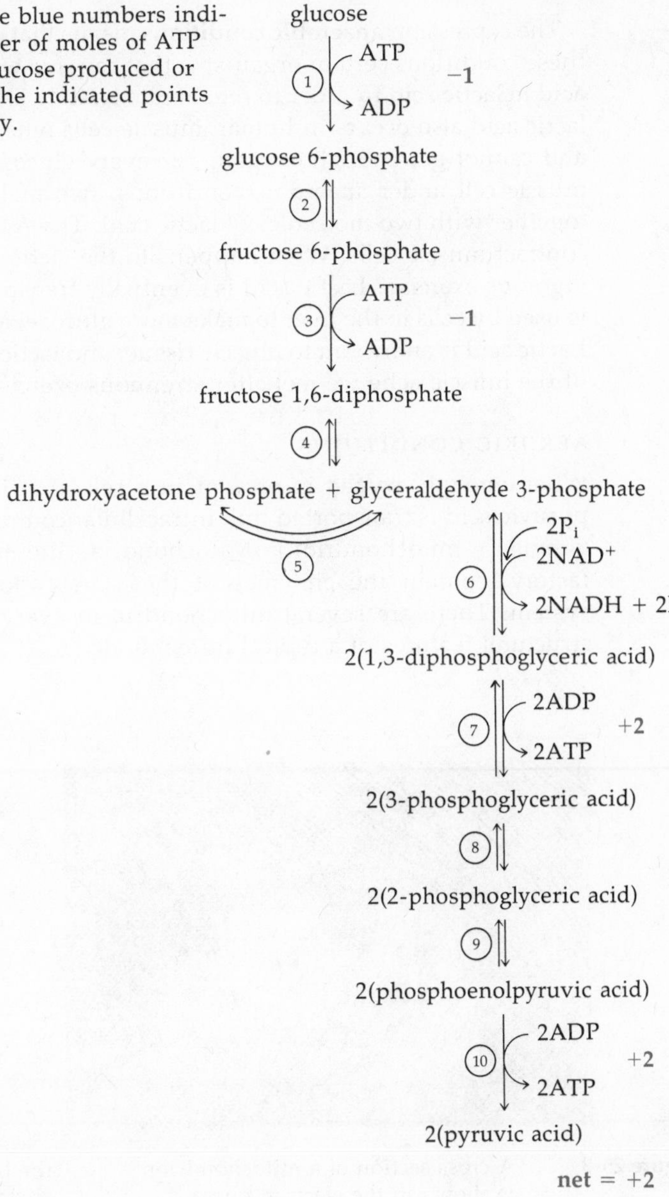

glucose

① ATP → ADP   −1

glucose 6-phosphate

②

fructose 6-phosphate

③ ATP → ADP   −1

fructose 1,6-diphosphate

④

dihydroxyacetone phosphate + glyceraldehyde 3-phosphate

⑤

⑥ 2P$_i$, 2NAD$^+$ → 2NADH + 2H$^+$

2(1,3-diphosphoglyceric acid)

⑦ 2ADP → 2ATP   +2

2(3-phosphoglyceric acid)

⑧

2(2-phosphoglyceric acid)

⑨

2(phosphoenolpyruvic acid)

⑩ 2ADP → 2ATP   +2

2(pyruvic acid)

net = +2

The Fate of Pyruvic Acid

Glycolysis and the production of pyruvic acid cannot continue for very long unless the coenzyme NAD$^+$, which was reduced to NADH and H$^+$ in step 6 of glycolysis, is oxidized back to NAD$^+$. There is a simple reason for this. The human body contains only a limited amount of NAD$^+$. Therefore, when NAD$^+$ is reduced to NADH, it must be reoxidized and reused.

ANAEROBIC CONDITIONS

In the absence of oxygen, NAD$^+$ is regenerated from NADH and H$^+$ in a reaction that is coupled to the reduction of pyruvic acid to lactic acid:

(25-17)

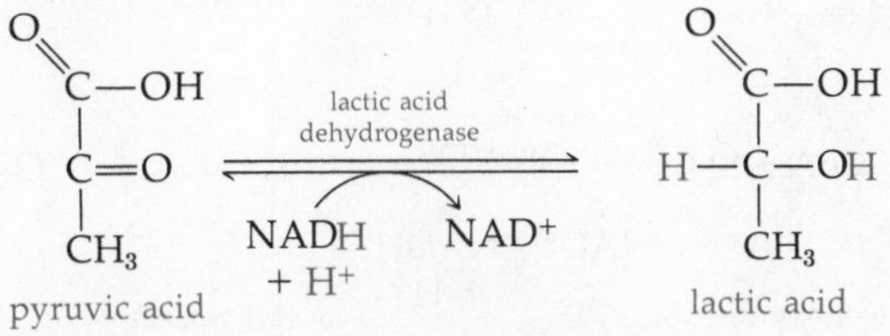

pyruvic acid $\xrightarrow[\text{NADH} + \text{H}^+]{\text{lactic acid dehydrogenase} \quad \text{NAD}^+}$ lactic acid

The expression **anaerobic conditions** means that oxygen is not available. Under these conditions certain organisms that can live without oxygen reduce pyruvic acid to lactic acid in order to regenerate NAD^+. The reduction of pyruvic acid to lactic acid also occurs in human muscle cells when they are working very hard and cannot get enough oxygen. For every glucose molecule metabolized by a muscle cell under anaerobic conditions, two molecules of ATP are produced, together with two molecules of lactic acid. The ATP is used to supply energy to contract muscle cells. What happens to the lactic acid that accumulates during vigorous exercise? Lactic acid is eventually transported out of muscle cells, and is used by cells in the liver to make more glucose, as we shall see in Section 25-9. Lactic acid is an irritant to muscle tissue, and lactic acid buildup is a major cause of the muscle ache we feel after strenuous exercise.

AEROBIC CONDITIONS

When enough oxygen is present in a cell, that is, under **aerobic conditions**, pyruvic acid is transported into intracellular compartments called **mitochondria** (singular: mitochondrion). Mitochondria, the powerhouses of the cellular factory, contain the enzymes of the TCA cycle and the electron transport system. There are several mitochondria in every cell. Figure 25-9 shows the structural features of a typical mitochondrion.

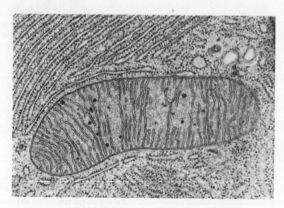 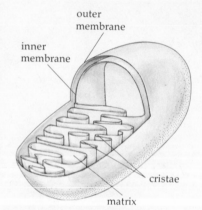

Figure 25-9   A cross section of a mitochondrion is shown in the electron micrograph on the left. The drawing on the right shows structural details of a mitochondrion, which has two membranes. The infoldings of the inner membrane, called cristae, contain the proteins of the electron transport system. The enzymes of the TCA cycle are located in the innermost matrix spaces of mitochondria.

Once inside the mitochondria, pyruvic acid is oxidized by a very complex enzyme, pyruvate dehydrogenase. This oxidation is coupled to the reduction of more NAD^+. Although this reaction produces even more NADH, under aerobic conditions NADH is readily reoxidized to NAD^+, as we shall see in Section 25-6. Pyruvate dehydrogenase catalyzes the reaction

(25-18)

pyruvic acid + CoA—SH $\xrightarrow[\text{pyruvate dehydrogenase}]{}$ acetyl-S-CoA + CO_2

The structural formula for coenzyme A was given in Chapter 18. We shall use the abbreviation CoA—SH for coenzyme A. Recall that —SH is the sulfhydryl functional group. Other coenzymes are also involved in the pyruvate dehydrogenase reaction, but they are reduced and then reoxidized during the reaction, so we need only consider the overall reaction (25-18).

Recall that two molecules of pyruvic acid are produced for every glucose with which we started. The molecule CH_3—$\overset{\displaystyle O}{\overset{\|}{C}}$—S—CoA is known as acetyl-S-CoA because CH_3—$\overset{\displaystyle O}{\overset{\|}{C}}$— is the **acetyl group.** Of the six carbon atoms in glucose, two are lost as CO_2 in reaction 25-18. The other four carbon atoms end up as part of two acetyl-S-CoA molecules. Acetyl-S-CoA is the entry point into the TCA cycle.

ANAEROBIC FERMENTATION

In some organisms, such as yeast, pyruvic acid can be converted to ethanol in a series of two reactions:

(25-19)

$$\underset{\text{pyruvic acid}}{\underset{\displaystyle CH_3}{\overset{\displaystyle O}{\underset{|}{\overset{\|}{C}}}}\underset{|}{\overset{|}{C}=O}}\quad \xrightarrow[\text{decarboxylase}]{\text{pyruvic acid}}\quad \underset{\text{acetaldehyde}}{\underset{\displaystyle CH_3}{\overset{\displaystyle H}{\underset{|}{\overset{|}{C}=O}}}} \quad +\; CO_2$$

(25-20)

$$\underset{\text{acetaldehyde}}{\underset{\displaystyle CH_3}{\overset{\displaystyle H}{\underset{|}{\overset{|}{C}=O}}}} \quad \xrightarrow[\underset{+\,H^+}{NADH \quad NAD^+}]{\text{alcohol}\atop\text{dehydrogenase}}\quad \underset{\text{ethanol}}{\underset{\displaystyle CH_3}{\overset{\displaystyle H}{\underset{|}{\overset{|}{H—C—OH}}}}}$$

These reactions also reoxidize the NADH produced in step 6 of glycolysis. The fate of pyruvic acid under different conditions is summarized in Figure 25-10.

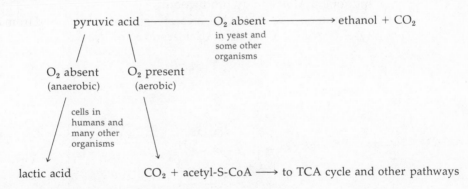

Figure 25-10   Possible fates of pyruvic acid.

Exercise 25-2
Write the overall reaction for glucose catabolism under anaerobic conditions.

Exercise 25-3
Why must active muscle cells convert pyruvic acid to lactic acid?

25-5 THE TRICARBOXYLIC ACID CYCLE

The tricarboxylic acid (TCA) cycle, also called the Krebs cycle or citric acid cycle, occurs in mitochondria under aerobic conditions. It is used to oxidize further the acetyl-S-CoA obtained from glycolysis and from other metabolic pathways in order to extract more energy (i.e., ATP). It is not a self-sufficient cycle. It is more like a gasoline engine: Acetyl-S-CoA is the fuel, and CO_2 and water are exhaust products. The molecules involved in the intermediate steps of the TCA cycle must be present in order for the cycle to operate. They are used in the cycle but are regenerated so that they can be used again. See Figure 25-11 for a comparison of the TCA cycle and the operation of a gasoline engine.

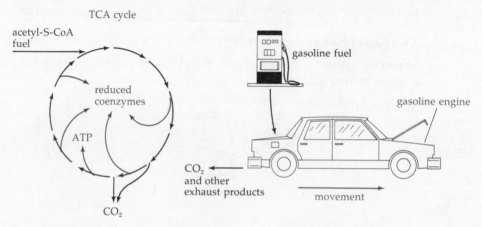

Figure 25-11   In the TCA cycle (left), breakdown of acetyl-S-CoA fuel is coupled to the synthesis of ATP and the reduction of coenzymes. The gasoline en- gine in a car (right) couples the breakdown of gasoline molecules to the movement of the car.

As we study the reactions of the TCA cycle (there are 10 steps in all), look for places where ATP or NADH are produced. In one step $FADH_2$, another reduced coenzyme, is produced from FAD. As we shall see, $FADH_2$ can also be used to make more ATP. After we have studied the individual reactions, we shall go back and total up the energy production.

In the first step of the TCA cycle, an acetyl group is transferred from a molecule of acetyl-S-CoA to a molecule of oxaloacetic acid:

Step 1:

(25-21)

acetyl-S-CoA   oxaloacetic acid   citric acid   coenzyme A

You can now see why the cycle is called the tricarboxylic acid or citric acid cycle. In this first step, two carbon atoms enter the TCA cycle via the acetyl group. Also notice that CoA—SH is regenerated in this step.

In the next two reactions of the TCA cycle, citric acid is converted to its structural isomer, isocitric acid. Both reactions are catalyzed by the same enzyme.

Step 2:

(25-22)

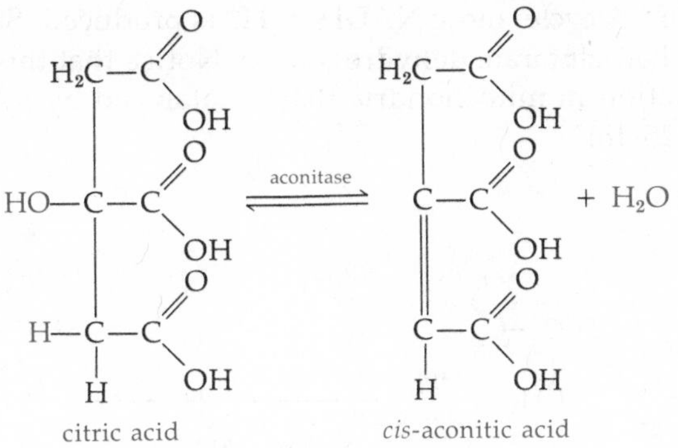

citric acid   *cis*-aconitic acid

Step 3:

(25-23)

cis-aconitic acid   isocitric acid

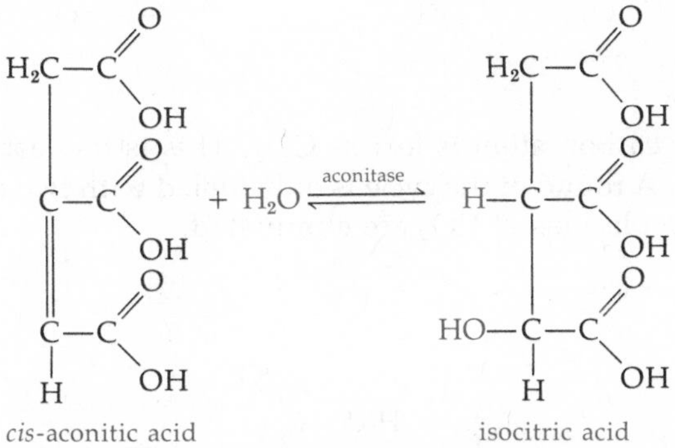

In the next two steps of the TCA cycle, isocitric acid is converted to α-ketoglutaric acid. Both steps 4 and 5 are catalyzed by the enzyme isocitrate dehydrogenase. In step 4, NAD^+ is reduced to $NADH + H^+$ in a reaction coupled to the oxidation of the secondary alcohol group of isocitric acid to form a carbonyl group. The product of the reaction is oxalosuccinic acid.

Step 4:

(25-24)

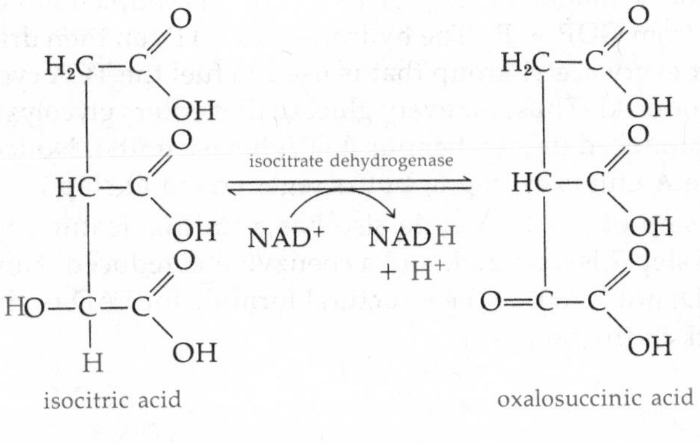

isocitric acid   oxalosuccinic acid

In step 5, a carboxyl group is removed in a decarboxylation reaction:

Step 5:

(25-25)

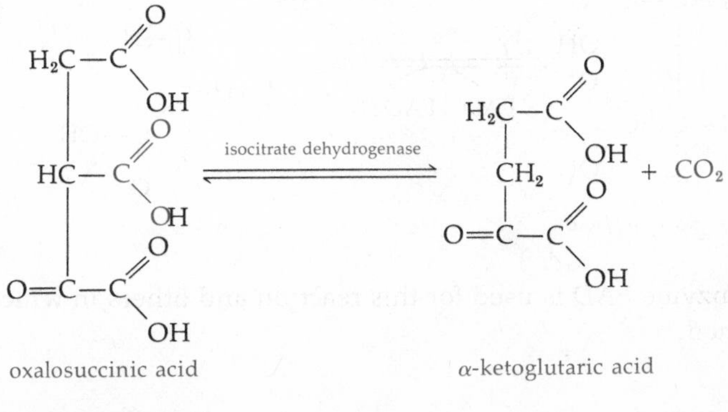

oxalosuccinic acid α-ketoglutaric acid

In step 6 of the TCA cycle, more NADH + H$^+$ is produced. Step 6 is catalyzed by the enzyme α-ketoglutarate dehydrogenase. Notice that this reaction is very similar to the reaction in mitochondria that is catalyzed by pyruvate dehydrogenase (reaction 25-18).

Step 6:

(25-26)

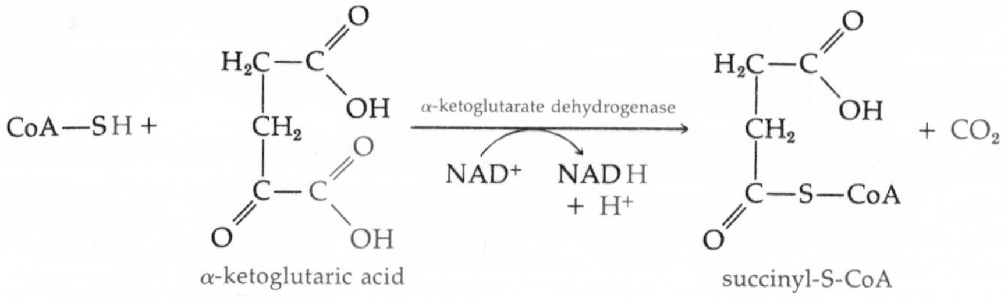

α-ketoglutaric acid → succinyl-S-CoA

In step 6, another carbon atom is lost as CO_2. This is the last CO_2 lost in one round of the cycle. A round of the cycle is only fueled with two carbon atoms, so *two and only two* molecules of CO_2 are eliminated.

Step 7:

(25-27)

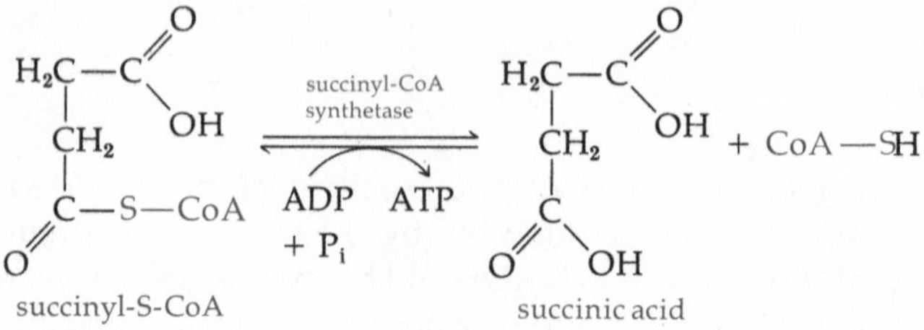

succinyl-S-CoA → succinic acid

In step 7 the highly exergonic hydrolysis of the succinyl-S-CoA thioester is coupled to the formation of ATP. (This reaction is coupled directly to the formation of GTP from GDP + P$_i$. The hydrolysis of GTP can then drive the formation of ATP.) For every acetyl group that is used to fuel the TCA cycle, one molecule of ATP is produced. Thus, for every glucose that enters glycolysis, *two* molecules of ATP are produced (remember our ATP/glucose units). Notice that a molecule of coenzyme A enters at step 6, but is regenerated at step 7.

The next step of the TCA cycle also has a unique feature. Succinic acid obtained from step 7 is oxidized, and a coenzyme is reduced. However, the coenzyme is FAD, not NAD$^+$. (The structural formula for FAD is given in Appendix 3 at the back of the book.)

Step 8:

(25-28)

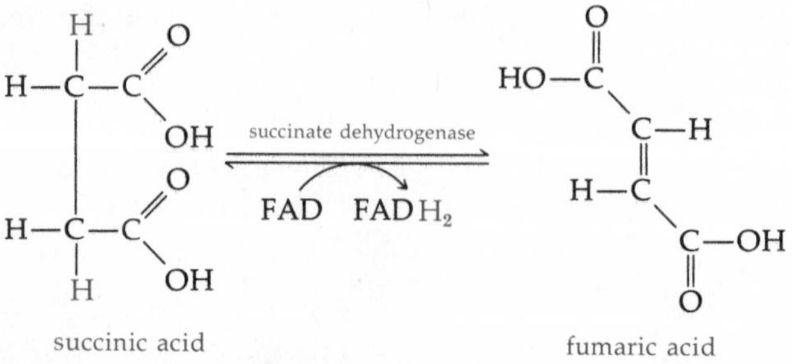

succinic acid → fumaric acid

The coenzyme FAD is used for this reaction and others in which double bonds are formed.

In the final two steps of the TCA cycle, water is added to the double bond in fumaric acid, and the resulting alcohol group is oxidized to a carbonyl group. This regenerates oxaloacetic acid for use in the next turn of the cycle.

Step 9:

(25-29)

fumaric acid $+ H_2O$ $\xrightleftharpoons{\text{fumarase}}$ L-malic acid

Step 10:

(25-30)

L-malic acid $\xrightleftharpoons{\text{L-malate dehydrogenase}}$ oxaloacetic acid

NAD^+ $NADH + H^+$

to step 1

This last step of the TCA cycle also produces more NADH. The TCA cycle is now complete. Oxaloacetic acid can pick up more acetyl group fuel and start the cycle all over again. A summary of the TCA cycle is shown in Figure 25-12.

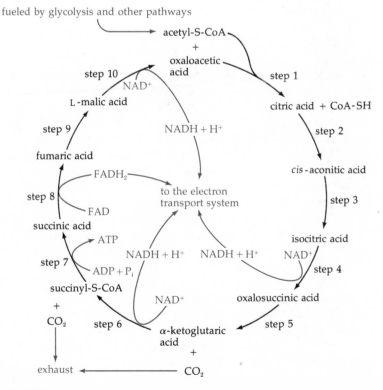

Figure 25-12   The TCA cycle.

What does our acetyl-group-burning TCA cycle engine give us? For every acetyl group we get one high-energy ATP molecule. We also get a lot of reduced coenzymes, 1FADH$_2$ + 3NADH + 3H$^+$. We shall see that reoxidation of these coenzymes in the electron transport system can be used to generate more ATP. Reoxidation of FADH$_2$ and NADH is also necessary in order to supply the TCA cycle with the FAD and NAD$^+$ coenzymes that it needs.

Exercise 25-4
Write a balanced overall reaction for the TCA cycle.

Exercise 25-5
What is the net ATP/glucose ratio generated in the TCA cycle? How many molecules of NADH and FADH$_2$ are obtained in the TCA cycle for every glucose molecule that entered glycolysis?

25-6   THE ELECTRON TRANSPORT SYSTEM; PHOSPHORYLATION OF ADP TO ATP

Some ATP is produced in a combination of glycolysis and the TCA cycle. But an ATP/glucose of +4 (or 32 kcal/mole of glucose) is all that is obtained. This is a fairly small amount compared to the 686 kcal that is available from the complete oxidation of 1 mole of glucose to CO$_2$ and H$_2$O. Also, several reduced coenzymes are produced in the TCA cycle and in glycolysis. They need to be reoxidized so that they can be used again. Reduced coenzymes such as NADH and FADH$_2$ are very strong reducing agents. For example, the oxidation of NADH back to NAD$^+$ is highly exergonic. It would be wasteful to release all of this energy without putting it to work. The **electron transport system**, ETS, is used to pass electrons and hydrogen from NADH (and FADH$_2$) to oxygen. The net overall reaction for the oxidation of NADH in the ETS is

(25-31) $$NADH + H^+ + \tfrac{1}{2}O_2 \longrightarrow NAD^+ + H_2O$$

We can think of this net overall reaction in the following manner: A hydrogen atom with two electrons, called a **hydride ion** (H:$^-$), leaves NADH, forming NAD$^+$. The hydride ion, H:$^-$, and the proton, H$^+$, combine with an atom of oxygen, $\tfrac{1}{2}O_2$, to form a molecule of water. It is in this sense that we speak of NADH passing electrons and hydrogen to oxygen in the ETS. To do this in one step would be a waste of energy. So, just as in glycolysis and the TCA cycle, this exergonic process is broken down into a series of steps with smaller energy changes. To make this point clear, consider the following: The overall free-energy change (ΔG) for reaction 25-31 is -53 kcal/mole. If we coupled this to the formation of one molecule of ATP, which requires 8 kcal/mole, we would be wasting a lot of energy. However, when the overall reaction 25-31 is broken down into a number of steps in the ETS, three molecules of ATP are formed for each molecule of NADH available.

In the ETS, electrons and hydrogen are passed along through a series of intermediates, or carriers. Each transfer to the next carrier is an exergonic oxidation-reduction reaction. Three of these transfers are sufficiently exergonic to be coupled to ATP formation.

All of the carriers in the ETS are found clustered together in the infolded inner membrane (cristae) of the mitochondria (see Figure 25-9). Each of the carriers in the ETS can exist in two states: oxidized and reduced. The major reactions, as we currently understand them, are as follows: The first carrier contains riboflavin (vitamin B_2) and is therefore a **flavoprotein.** It is represented by the symbol FP_1. This protein accepts the hydride ion from NADH along with a proton, H^+ (see Figure 25-13). Flavoprotein$_1$ is sometimes called NADH dehydrogenase. The next major electron carrier is **coenzyme Q,** abbreviated CoQ. CoQ accepts the hydrogen atoms from FP_1H_2 and is reduced to $CoQH_2$. The structure of CoQ is given in Appendix 3 at the back of the book. The remaining carriers in the ETS are called cytochromes. **Cytochromes** are proteins that contain a heme prosthetic group, as does the protein hemoglobin (Chapter 22). The major differences between individual cytochromes are in the protein portion of these molecules. All of the cytochromes are capable of undergoing the following oxidation-reduction reaction:

(25-32) $Fe^{2+} \rightleftharpoons Fe^{3+} + 1e^-$

Figure 25-13   Schematic representation of the major components in the transport of electrons from NADH to oxygen. The blue brackets represent oxidative phosphorylation sites, which are discussed on the following pages. These sites are the only ones sufficiently exergonic to drive ATP synthesis.

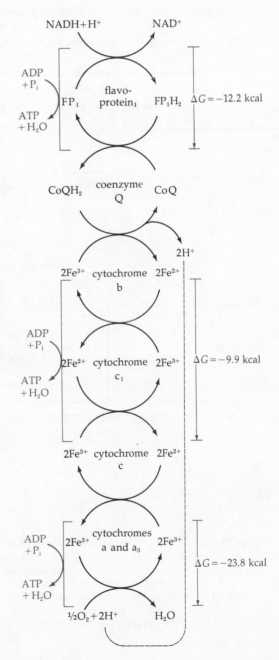

Notice that only one electron is involved in reaction 25-32. Thus, in order to pass the two electrons from $CoQH_2$ along to the cytochromes, two molecules of each cytochrome must be reduced at each step. Also notice that the hydrogen atoms from $CoQH_2$ are not passed along to the cytochromes. They are released as protons, $2H^+$, and are later used to form one molecule of water.

Since most of the ETS carriers are bound to the inner mitochondrial membrane, they have been difficult to study. Our information about the ETS has therefore been hard to acquire and is not complete. The important features of the ETS are represented in Figure 25-13, where the notation

$$\begin{array}{cc} A & B \\ & \times \\ C & D \end{array}$$

is used as a symbolic representation for the reaction $A + C \rightarrow B + D$. Note that the arrows go from reactants (arrow tails) to products (arrow heads). For example, the transfer of electrons from $CoQH_2$ to cytochrome b can be written as

(25-33) $CoQH_2 + 2(\text{cytochrome b} \cdot Fe^{3+}) \longrightarrow CoQ + 2(\text{cytochrome b} \cdot Fe^{2+}) + 2H^+$

Note the following important features of the ETS. *First*, NADH (and $FADH_2$) is reoxidized. The NAD^+ (and FAD) produced can be reused in the TCA cycle. We now see why there must be oxygen available to the cell (aerobic conditions) in order for the TCA cycle to work. The TCA cycle needs the ETS to supply it

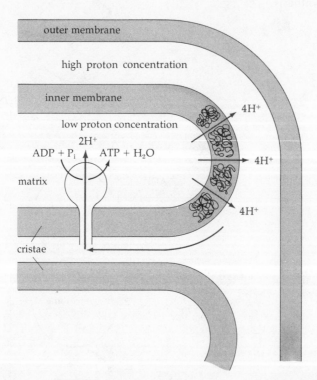

Figure 25-14   A model for oxidative phosphorylation. Peter Mitchell, who received the 1978 Nobel Prize in chemistry, has proposed that the coupling of electron transport to ATP synthesis occurs by protons being pumped out of the matrix space at three sites in the electron transport system. Recent experiments indicate that four protons are pumped out at each of these sites. According to this model, the diffusion of these protons back into the matrix drives oxidative phosphorylation.

with NAD^+ (and FAD), and the ETS cannot work without O_2. *Second*, you should see that each of the carriers in the ETS is first reduced and then reoxidized so that it can be used again. *Third,* notice that three portions of the ETS are exergonic enough to be coupled to ATP synthesis. The coupling of these three portions to ATP formation is called **oxidative phosphorylation,** and each portion of the ETS at which such coupling takes place is called an oxidative phosphorylation site. It has been proposed that the actual coupling mechanism involves the pumping of protons through the mitochondrial inner membrane (see Figure 25-14).

The electron transport system is also used to reoxidize $FADH_2$. However, $FADH_2$ cannot reduce FP_1. It enters the ETS at a different point. $FADH_2$ first reduces a different flavoprotein, FP_2, to FP_2H_2, and then FP_2H_2 reduces CoQ to $CoQH_2$ (see Figure 25-15). These steps ($FADH_2$ to CoQ) are not exergonic enough to allow for ATP synthesis, so that from the oxidation of one molecule of $FADH_2$ via the ETS, we get only two molecules of ATP.

Figure 25-15   The transfer of electrons from $FADH_2$ to a flavoprotein, FP_2, and then to coenzyme Q, is not sufficiently exergonic to drive ATP synthesis.

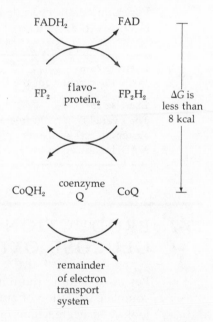

Hence, using the ETS, we can get three ATP molecules for every $NADH + H^+$ and two ATP molecules for every $FADH_2$ that were generated in the TCA cycle from the oxidation of acetyl-S-CoA. The NADH produced by the pyruvate dehydrogenase reaction (see reaction 25-18) in mitochondria also feeds the ETS and provides three ATPs per molecule of pyruvic acid.

NADH was also produced in step 6 of glycolysis. But step 6 of glycolysis takes place in cytoplasm, not in mitochondria. Can we generate any ATP in this case? In order for the NADH from glycolysis to be used in the ETS, it needs to enter the mitochondria. The NADH, however, cannot cross the mitochondrial membrane, so additional steps, involving additional molecules, are required by which NADH shuttles its electrons and hydrogen across to another molecule inside the mitochondria. In liver cells, the hydrogen atoms are shuttled to NAD^+ inside of the mitochondria. In muscle cells they are shuttled to a molecule of FAD. We do not need to examine this shuttle system in detail, but the overall reaction, shown in Figure 25-16, is important. Because $FADH_2$, not NADH, is the input to the ETS in muscle cells, only *two* ATP molecules can be made via the ETS for each NADH generated in glycolysis.

Figure 25-16   The shuttle of electrons across the mitochondrial membrane in a muscle cell is represented schematically. The process involves intermediates, X and XH_2, which are able to pass through the mitochondrial membrane.

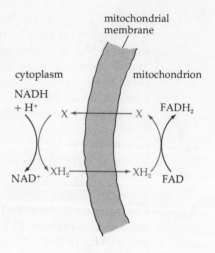

Several substances strongly inhibit, or block, the electron transport system. For example, the cyanide ion, CN^-, is poisonous because it blocks the transfer of electrons to oxygen.

Exercise 25-6
Write the reaction for the last step of the ETS.

Exercise 25-7
The overall change in free energy for the oxidation of NADH in the electron transport system is -53 kcal/mole. Why aren't six ATP molecules produced for every molecule of NADH oxidized in the ETS?

25-7 PRODUCTION OF ATP BY AEROBIC GLUCOSE OXIDATION

Let us determine the number of molecules of ATP that can be made by the complete oxidation of glucose. To do this, we shall take 1 mole of glucose and look at the net reactions of glycolysis, pyruvate dehydrogenase, the tricarboxylic acid cycle, and the electron transport system. We must keep track of the reduced coenzymes, NADH and $FADH_2$, because they are the inputs into the ETS.

A. Overall Reactions

Glycolysis (in the cytoplasm):

(25-34)   glucose + 2NAD$^+$ + 2ADP + 2P$_i$ $\longrightarrow$
2 pyruvic acid + 2ATP + 2H$_2$O + 2NADH + 2H$^+$

Pyruvate dehydrogenase (in mitochondria):

(25-35)   2(pyruvic acid + NAD$^+$ + CoA-SH $\longrightarrow$ acetyl-S-CoA + CO$_2$ + NADH + H$^+$)

The TCA cycle (in mitochondria):

(25-36)   2(acetyl-S-CoA + 3NAD$^+$ + FAD + 2H$_2$O + ADP + P$_i$ $\longrightarrow$
2CO$_2$ + 3NADH + 3H$^+$ + FADH$_2$ + ATP + CoA-SH)

B. Yield per Mole of Glucose

Yield

Glycolysis plus shuttle system
(in muscle cells): $2ATP + 2FADH_2$
Pyruvate dehydrogenase: $2NADH + 2H^+$
The TCA cycle: $2ATP + 2FADH_2 + 6NADH + 6H^+$
Subtotal: $\overline{4ATP + 4FADH_2 + 8NADH + 8H^+}$

Recall that in liver cells the shuttle system yields $2NADH^+$ in the mitochondria rather than $2FADH_2$ as in muscle cells. Thus, for liver cells, this subtotal would be $4ATP + 2FADH_2 + 10NADH + 10H^+$.

C. The Electron Transport System

Overall reactions:

(25-37) $8(NADH + H^+ + \tfrac{1}{2}O_2 + 3ADP + 3P_i \longrightarrow NAD^+ + 4H_2O + 3ATP)$

(25-38) $4(FADH_2 + \tfrac{1}{2}O_2 + 2ADP + 2P_i \longrightarrow FAD + 3H_2O + 2ATP)$

ATP production per mole of glucose:

$8(NADH + H^+) \longrightarrow 24ATP$
$4(FADH_2) \longrightarrow \underline{\ \ 8ATP}$
Subtotal: $32ATP$

D. ATP per Mole of Glucose

1. From part B: $4ATP + \underline{4FADH_2 + 8NADH + 8H^+}$

2. From part C: $\underline{32ATP} \longleftarrow$

3. Total $36ATP$ in muscle cells

Because of its different shuttle system, a total of 38 ATP/glucose are produced in liver cells.

The overall reaction for the complete oxidation of one molecule of glucose via glycolysis, the TCA cycle, and the ETS can now be derived.

First, take the sum of reaction 25-34 (glycolysis) + reaction 25-35 (pyruvate dehydrogenase) + reaction 25-36 (the TCA cycle), then (for muscle cells) adjust for the NADH shuttle into mitochondria (Figure 25-16) to obtain

(25-39) $Glucose + 8NAD^+ + 4FAD + 4ADP + 4P_i + 2H_2O \longrightarrow$
$6CO_2 + 4FADH_2 + 8NADH + 8H^+ + 4ATP$

Now, add to reaction 25-39 the ETS reactions 25-37 and 25-38 to get

(25-40) $Glucose + 36ADP + 36P_i + 6O_2 \longrightarrow 6CO_2 + 36ATP + 42H_2O$

Notice that in this overall reaction 25-40 the coenzymes are neither reactants nor products. That is because they are reoxidized and reused. The overall reaction 25-40 can be written as the sum of two reactions:

(25-41a)   Glucose $+ 6O_2 \longrightarrow 6CO_2 + 6H_2O$

and

(25-41b) $36ADP + 36P_i \longrightarrow 36ATP + 36H_2O$

Thus, in muscle cells under aerobic conditions, complete oxidation of glucose produces 36 moles of ATP per mole of glucose. If we use the value 8 kcal/mole for the endergonic reaction $ADP + P_i \rightarrow ATP + H_2O$, we have captured $8 \times 36 = 288$ kcal of energy in the form of ATP. The free-energy change for reaction 25-41a is about -686 kcal/mole. Thus we have captured 288/686 or about 42% of the energy released when 1 mole of glucose is oxidized. The rest of the energy is released in the form of heat. Hence the catabolism of glucose is not 100% efficient—no process is. However, 42% efficiency is not bad; it is much better than the energy efficiency of any machine ever made. For example, the average automobile has an operating efficiency of only about 3% under standard driving conditions.

Don't just memorize the number 36. It is important to know the reactions in which the ATP and reduced coenzymes are formed. We shall see that lactic acid and pyruvic acid can be made from molecules other than glucose, and we shall want to know how much ATP can be formed by the oxidation of these other molecules via the TCA cycle and the electron transport system.

Optional

25-8   THE PENTOSE PHOSPHATE PATHWAY

Glycolysis is the major pathway used by cells in the body to catabolize glucose. It is not, however, the only pathway. Another catabolic pathway, the **pentose phosphate pathway,** is used by cells under certain conditions. Skeletal and heart muscle cells hardly use it at all, but about 30% of the glucose in the liver is broken down by this pathway. Mammary gland cells use it even more.

Why should another pathway for glucose oxidation be needed? Glycolysis, operating in conjunction with the TCA cycle and the electron transport system, is highly efficient for capturing energy in the form of ATP. The pentose phosphate pathway is used for two other reasons: (1) production of the sugar ribose 5-phosphate, which is needed for the synthesis of nucleotides and nucleic acids; and (2) reduction of the coenzyme nicotinamide adenine dinucleotide phosphate, $NADP^+$, to form NADPH, which is structurally similar to NADH (see Appendix 3). NADPH is used to reduce molecules, regenerating $NADP^+$ in several anabolic biochemical pathways, as, for example, in the synthesis of lipids (see Chapter 27).

The starting material for the pentose phosphate pathway is glucose 6-phosphate, which is generated in step 1 of glycolysis (reaction 25-4). Glucose 6-phosphate can be converted into ribose 5-phosphate by several reaction steps. The overall reaction for this process is

(25-42)   Glucose 6-phosphate $+ 2NADP^+ + H_2O \longrightarrow$
$$\text{ribose 5-phosphate} + CO_2 + 2NADPH + 2H^+$$

By some additional steps, one molecule of glucose 6-phosphate can be completely degraded to CO_2. In this process, 12 molecules of NADPH are formed per molecule of glucose 6-phosphate. The overall reaction in this case is

(25-43)   Glucose 6-phosphate $+ 12NADP^+ + 7H_2O \longrightarrow 6CO_2 + 12NADPH + 12H^+ + P_i$

Thus, in the pentose phosphate pathway, glucose 6-phosphate can suffer two fates. The extent to which reactions 25-42 and 25-43 take place in a given cell depends on the requirements for ribose 5-phosphate and NADPH, respectively. If the need for NADPH is much larger than the need for ribose 5-phosphate, then reaction 25-43 will predominate. If neither NADPH nor ribose 5-phosphate is in great demand, almost all of the glucose 6-phosphate will be catabolized via glycolysis.

Exercise 25-8

During strenuous exercise, what would be the pathway used to metabolize glucose 6-phosphate? Explain your answer.

25-9   GLUCONEOGENESIS

Our major source of the glucose that we catabolize is dietary—coming primarily from plants, which make glucose in the process of photosynthesis. However, human beings are also capable of making glucose by a process (pathway) called **gluconeogenesis,** which means "making *new* glucose."

Gluconeogenesis occurs primarily in the liver, and is significant in the following situations (see Figure 25-17): (1) The lactic acid produced in active muscle cells as the product of anaerobic glycolysis can be used as a starting material for the synthesis of glucose, which is stored in the liver as glycogen for future use. (2) When the dietary sources of glucose are not sufficient, amino acids from proteins in the body can also be used as starting materials for gluconeogenesis.

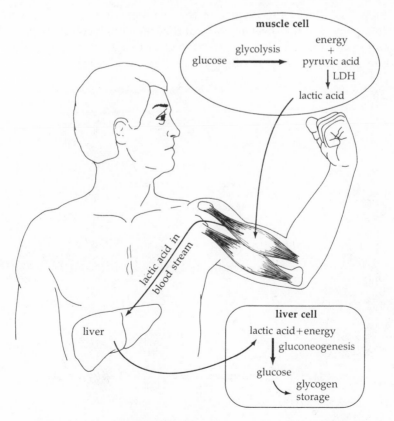

Figure 25-17   Lactic acid, which is produced in active muscle cells, can be converted to glucose via gluconeogenesis in liver cells.

As we have mentioned, lactic acid is produced from glucose in the process of anaerobic glycolysis. However, every step in gluconeogenesis cannot be the exact reverse of a step in the glycolysis pathway. At least some steps in the two pathways must be different. There is a very simple reason for this. The overall process of glycolysis is exergonic. Thus, gluconeogenesis must be an endergonic process. Now, since no biochemical process is 100% efficient, more ATP must be used to convert lactic acid into glucose than is produced when glucose is broken down to lactic acid in glycolysis. Some of the energy required for gluconeogenesis is supplied by the hydrolysis of ATP, and some by the hydrolysis of guanosine triphosphate, GTP, which is similar in structure to ATP, but contains the nitrogenous base guanine instead of the adenine found in ATP. The hydrolysis of GTP is similar to that of ATP:

(25-44) $GTP + H_2O \longrightarrow GDP + P_i$

The ΔG for this reaction is also about the same as for the hydrolysis of ATP, -8 kcal/mole. Thus, the energy supplied by hydrolysis of one GTP is equivalent to that supplied by one ATP.

Gluconeogenesis starts with the oxidation of lactic acid to pyruvic acid.

Step 1:

(25-45)

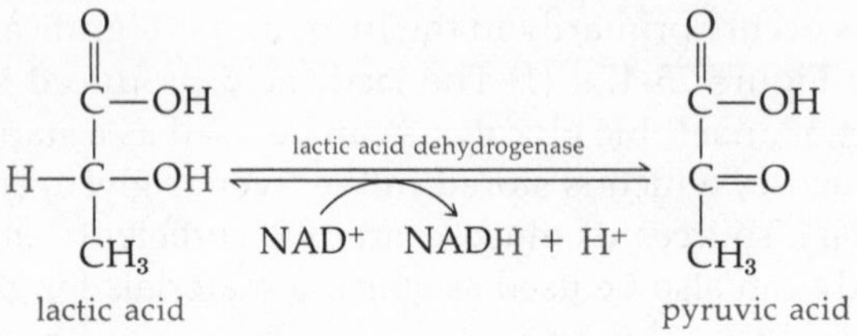

lactic acid   pyruvic acid

Step 1 is simply the reverse of the last step of anaerobic glucose metabolism (reaction 25-17). However, three of the steps in glycolysis (steps 1, 3, and 10) are not reversible. The next two steps in gluconeogenesis are designed to get around the irreversible step 10 of glycolysis (reaction 25-13).

Step 2:

(25-46)

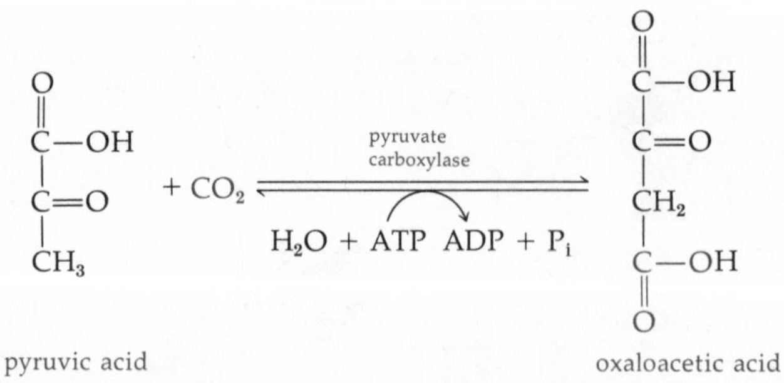

pyruvic acid   oxaloacetic acid

Step 3:

(25-47)

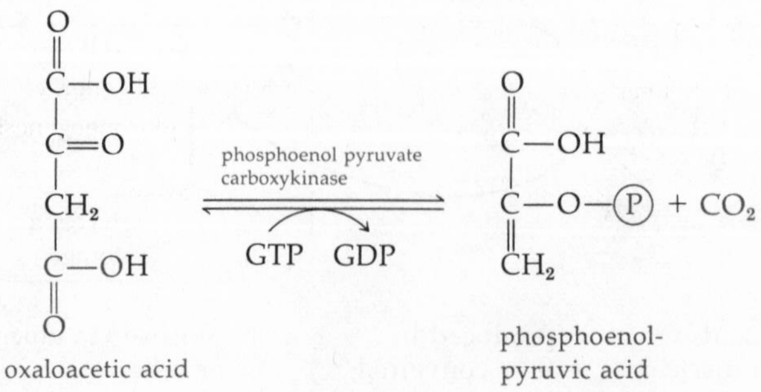

oxaloacetic acid   phosphoenol-
                               pyruvic acid

In these two steps, ATP and GTP are used to make the phosphoenolpyruvic acid molecule. The next several steps in gluconeogenesis use the reversible steps (9, 8, 7, 6, 5, and 4) of glycolysis (Figure 25-8) to produce fructose 1,6-diphosphate. The next irreversible step of glycolysis (step 3, reaction 25-6) is bypassed by the enzyme fructose 1,6-diphosphatase, which serves to catalyze the following reaction:

(25-48)

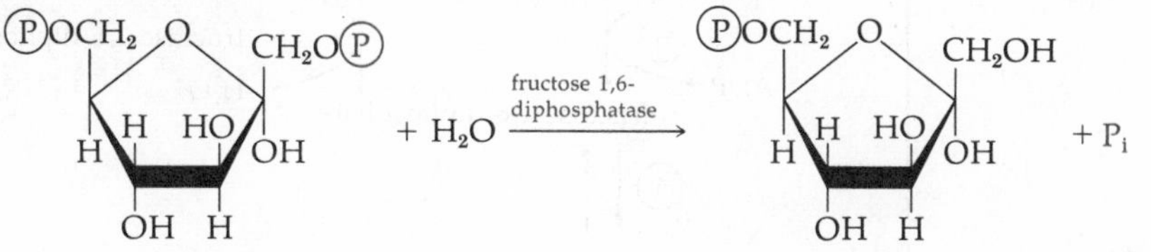

fructose 1,6-diphosphate   fructose 6-phosphate

The remaining irreversible step in glycolysis (step 1) is bypassed by the enzyme glucose 6-phosphatase in the reaction

(25-49)

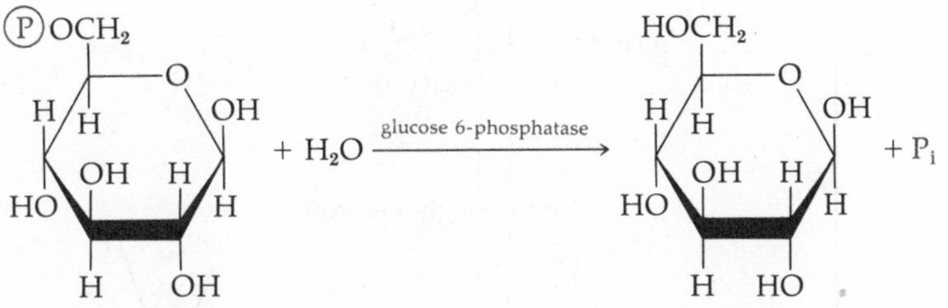

glucose 6-phosphate   glucose

Glycolysis and gluconeogenesis are compared in Figure 25-18 on page 636. The net overall reaction in gluconeogenesis is

(25-50)   2 Lactic acid + 4ATP + 2GTP + $6H_2O \longrightarrow$ glucose + 4ADP + 2GDP + $6P_i$

As you can see from reaction 25-50, the formation of one molecule of glucose from two molecules of lactic acid requires the hydrolysis of four molecules of ATP and two molecules of GTP, which is equivalent to a total of six ATP molecules.

We might ask the questions: Why isn't lactic acid just eliminated? Why does the human body utilize energy to convert lactic acid back to glucose? Isn't this a waste of energy? Actually, a lot of energy can be obtained from lactic acid, and eliminating lactic acid would be the wasteful procedure. Energy is used to convert lactic acid to glucose, but once formed, a molecule of glucose represents a potential source of an even larger amount of energy. Recall that a very large amount of ATP is produced when a molecule of glucose is completely degraded to CO_2 and H_2O.

Exercise 25-9

Compare the amount of energy (in terms of ATP) produced when 1 mole of glucose is completely catabolized to CO_2 and H_2O to the amount of ATP needed to form 1 mole of glucose from lactic acid.

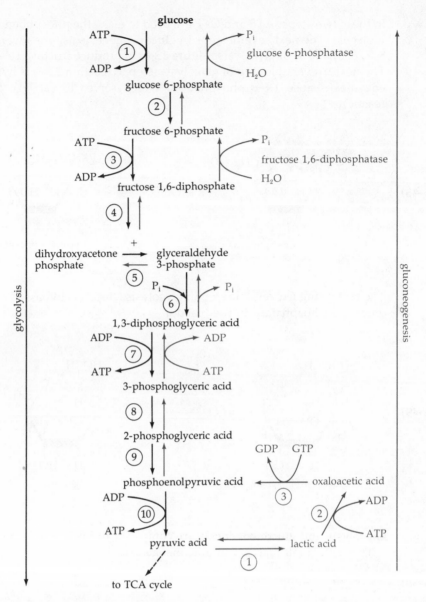

Figure 25-18   Reactions in glycolysis versus gluconeogenesis. Reactions of gluconeogenesis are shown in blue.

25-10 METABOLISM OF OTHER CARBOHYDRATES

Gluconeogenesis in humans and photosynthesis in plants lead to the formation of glucose. By now you are aware of the tremendous importance of glucose and glucose metabolism in the human body. However, in Chapter 20 we saw that glucose is only one of several six-carbon sugars. Fructose, which is part of the disaccharide sucrose, is another. Humans ingest about 100 g of sucrose each day. Are other hexoses important to the human body? If so, how are they metabolized? The answer to the first question is yes. Hexoses and their derivatives play several roles in the human body.

Functions of Hexoses in Humans

Large concentrations of free hexoses, other than glucose, are not found in human body cells. However, several hexoses are rather abundant as components of disaccharides and polysaccharides. For example, the sugar galactose is a component of lactose, the major carbohydrate in milk. Galactose is also a vital component of the carbohydrate prosthetic groups of several glycoproteins (see Chapter 22 for a review of prosthetic groups). Galactose, together with other hexoses and their chemical derivatives, are also components of polysaccharides found on cell surfaces. For example, the different ABO blood types of humans arise because of the presence of different sequences of sugars in certain polysaccharides on cell membranes.

Hexose Metabolism

In order for lactose and sucrose to be metabolized, these disaccharides must first be split into their monosaccharide components. Hydrolysis of lactose and sucrose is accomplished by two enzymes found in intestinal cells. These enzymes, lactase and sucrase, catalyze the following reactions:

(25-51)

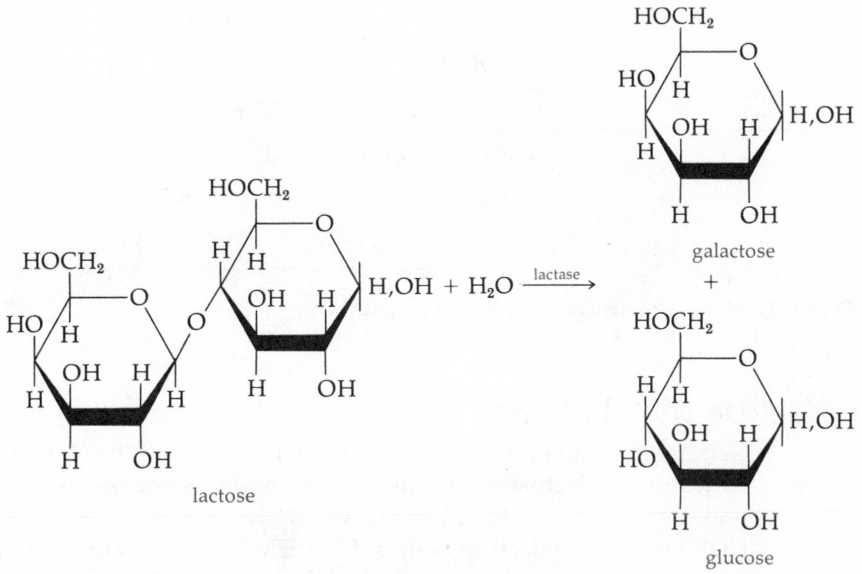

(25-52)

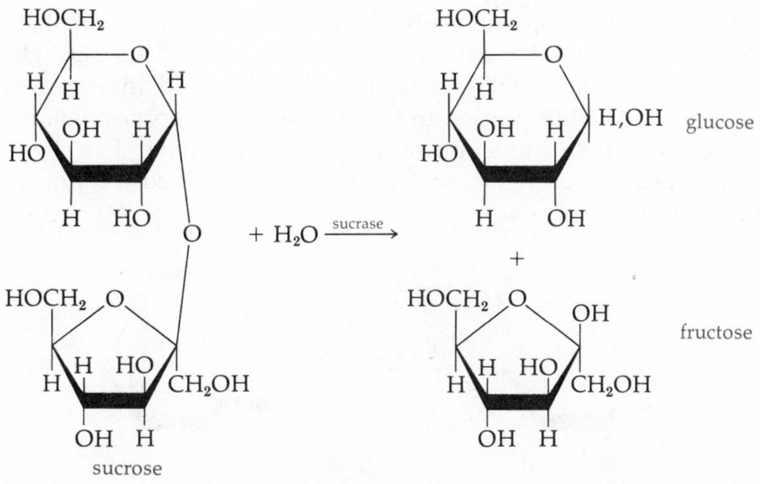

We have already discussed how the glucose portions of these two disaccharides are metabolized via glycolysis. What about galactose and fructose? Let us look at fructose first (because it is easier). Fructose can be fed into glycolysis via the fructose 1-phosphate pathway shown in Figure 25-19. Notice in this pathway that the fructose is first phosphorylated. The product of this reaction, fructose 1-phosphate, is then cleaved by a specific enzyme to give glyceraldehyde and dihydroxyacetone phosphate. Dihydroxyacetone phosphate can immediately enter the glycolytic pathway (step 5, Figure 25-8). The glyceraldehyde molecule is phosphorylated, so that it too can enter into glycolysis, instead of requiring another pathway.

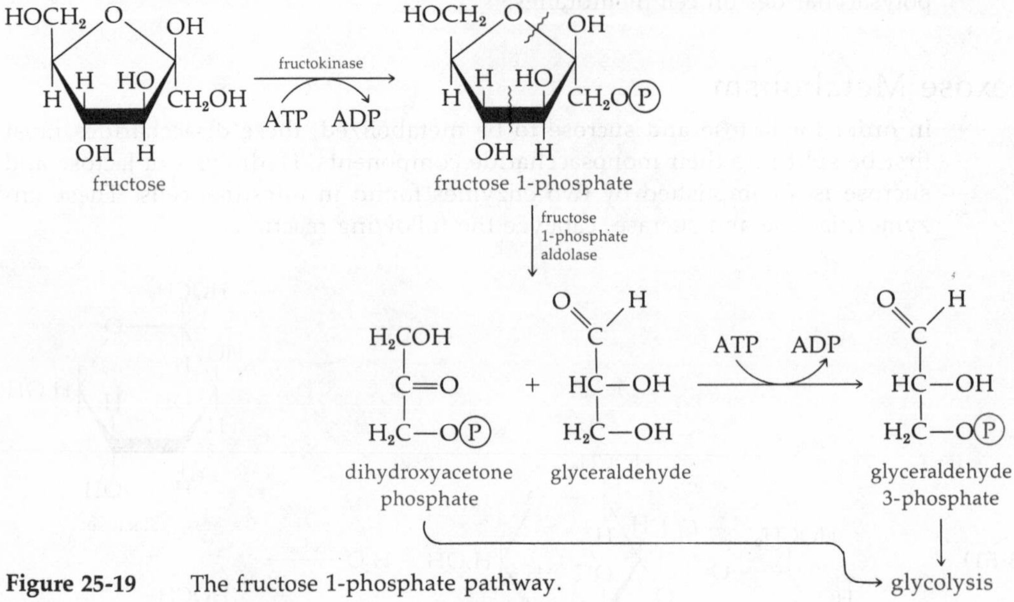

Figure 25-19   The fructose 1-phosphate pathway.

Galactose Metabolism

Galactose is converted to glucose in a series of reactions shown in Figure 25-20. Notice that galactose is first phosphorylated to galactose 1-phosphate. The second step of this pathway involves the transfer of a molecule of UMP from UDP-glucose to galactose 1-phosphate. UMP stands for uridine monophosphate, which is similar in structure to AMP but contains the base uracil instead of the base adenine (see Figure 24-2). UDP is likewise similar to ADP. (We shall see shortly where UDP-glucose comes from.) The products of the reaction catalyzed by hexose 1-phosphate uridylyltransferase are glucose 1-phosphate and UDP-galactose. (The UDP-galactose can be isomerized to UDP-glucose by an epimerase enzyme. UDP-glucose can then be used in several ways, including donating its UMP portion to another galactose 1-phosphate molecule.) The net reaction is the production of glucose 1-phosphate. The enzyme phosphoglucomutase converts glucose 1-phosphate into glucose 6-phosphate, which can enter into glycolysis.

(25-53)

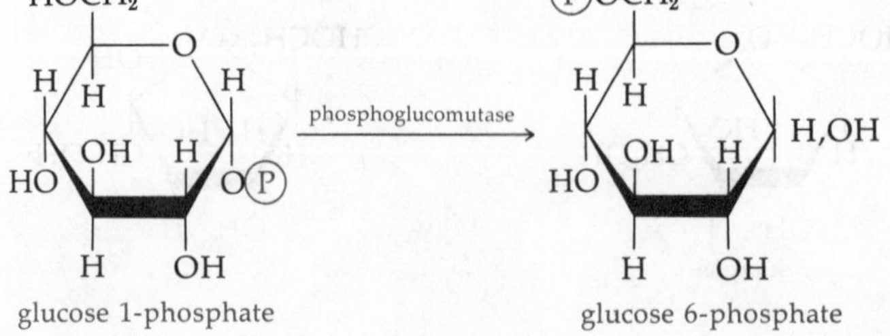

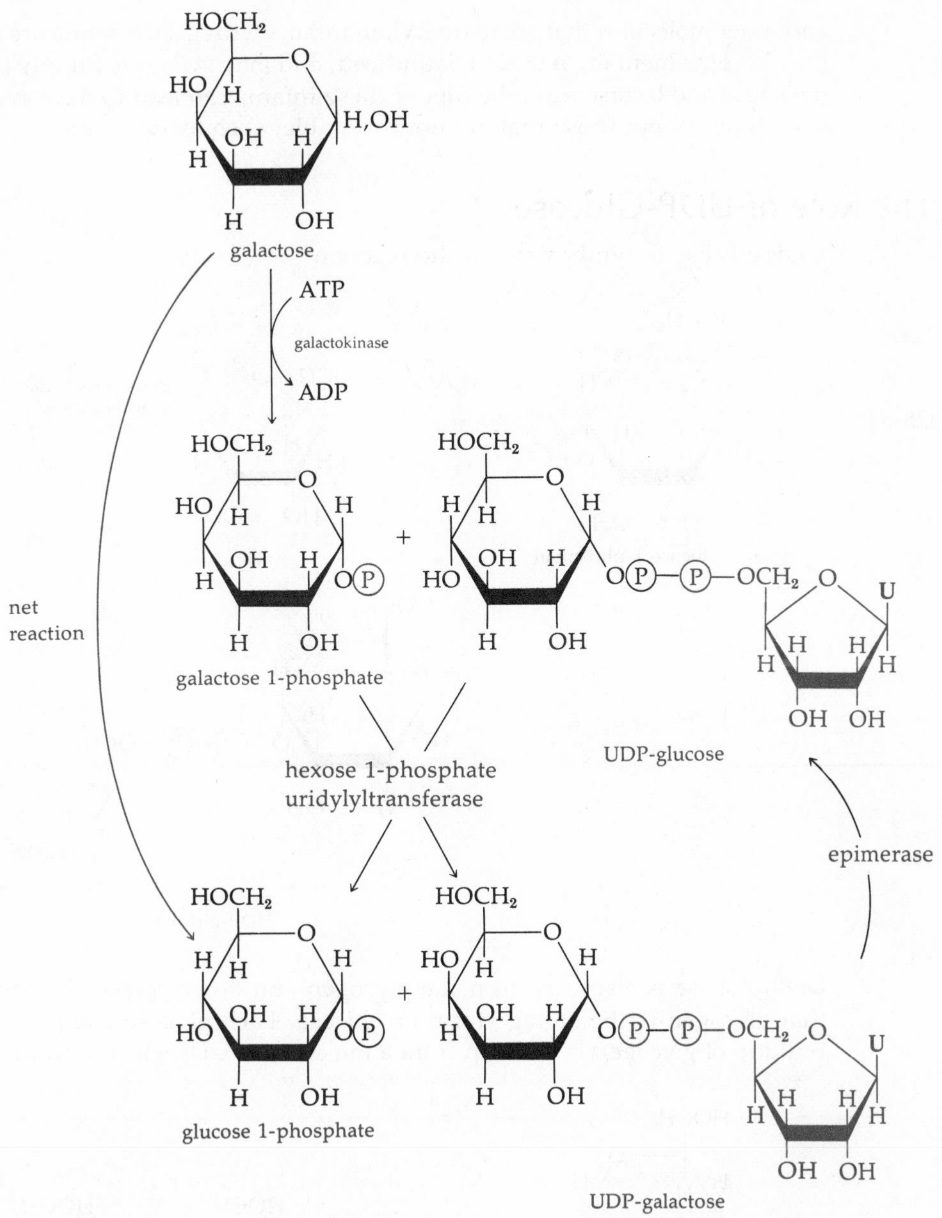

Figure 25-20   The conversion of galactose to glucose 1-phosphate. The UMP group that is transferred in the reaction catalyzed by hexose 1-phosphate uridylyltransferase is shown in blue.

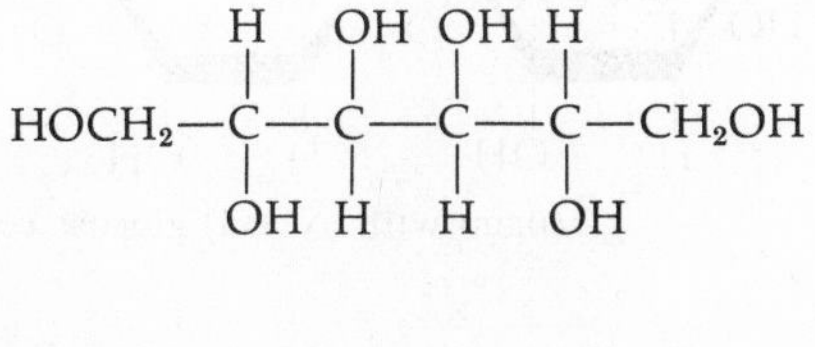

A crucial step in the conversion of galactose to glucose is the transfer of UMP. The enzyme hexose 1-phosphate uridylyltransferase, which catalyzes the transfer of UMP, is not present in some people because of a hereditary defect. The lack of this enzyme results in a disease called **galactosemia.** Infants suffering from galactosemia cannot metabolize milk properly, and large amounts of galactose 1-phosphate build up since it cannot be metabolized further. Some of the excess galactose 1-phosphate is converted to galactitol, and other molecules that are toxic. When infants with galactosemia are fed milk, they become mentally retarded, jaundiced, and may even die. Simply removing galactose and lactose from the diet of these infants can lead to recovery from all symptoms except those that are nonreversible, such as retardation.

The Role of UDP-Glucose

UDP-glucose is synthesized in the reaction

(25-54)

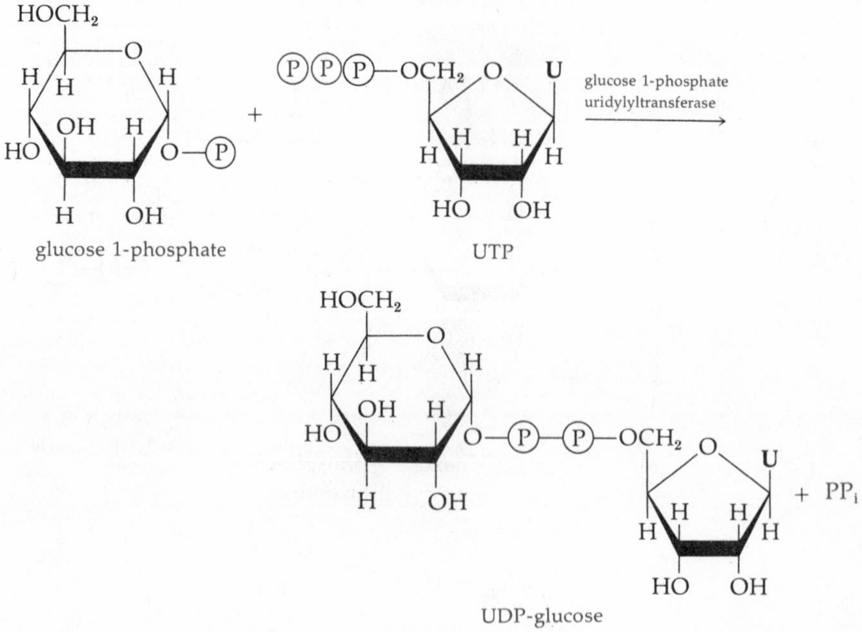

glucose 1-phosphate   UTP

UDP-glucose

UDP-glucose is also used to make glycogen and other polysaccharides. Recall that glycogen is the storage form of glucose. Each glucose component in the buildup of glycogen is donated from a molecule of UDP-glucose in the reaction

(25-55)

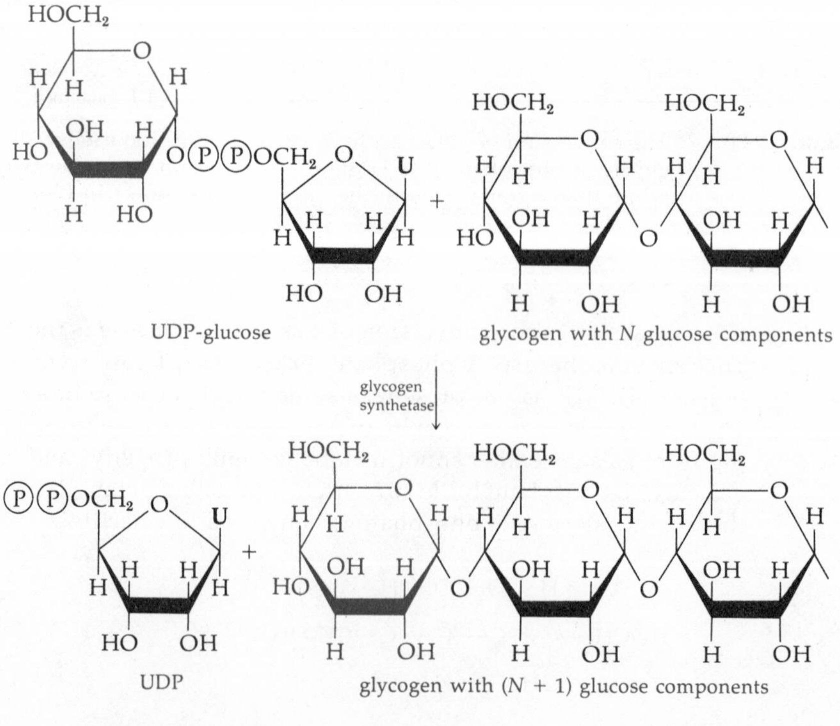

UDP-glucose   glycogen with N glucose components

glycogen
synthetase

UDP   glycogen with $(N + 1)$ glucose components

The glycogen synthetase reaction (25-55), as we shall see shortly, is a key control point in the metabolism of carbohydrates.

UDP-glucose and other UDP sugars serve as the precursors for all disaccharides and polysaccharides. For example, the disaccharide lactose is made from UDP-galactose and glucose by the enzyme lactose synthetase. Lactose synthetase is found only in mammary glands and catalyzes the reaction

(25-56) $$\text{Glucose + UDP-galactose} \xrightarrow{\text{lactose synthetase}} \text{lactose + UDP}$$

Several other hexoses and their derivatives are also produced in the human body. Several of these, such as mannose, mannosamine, glucosamine, and N-acetylgalactosamine, are important components of glycoprotein prosthetic groups and blood group substances.

Exercise 25-10
How many UDP-glucose molecules are needed to convert 100 galactose 1-phosphate molecules to 100 glucose 1-phosphate molecules? Why?

Exercise 25-11
How many UDP-glucose molecules are required in order to synthesize a glycogen molecule containing 100 glucose components? Why?

25-11 REGULATION OF CARBOHYDRATE METABOLISM

Human cells, as we have seen, have several assembly lines for carbohydrate metabolism. Glucose and its UDP and phosphate derivatives serve as the common connections for these pathways. Thus cells can make glucose via gluconeogenesis, break it down via glycolysis, polymerize it into glycogen, or convert it to other hexoses (see Figure 25-21). Operation of all of these assembly lines at

Figure 25-21   Pathways for glucose metabolism.

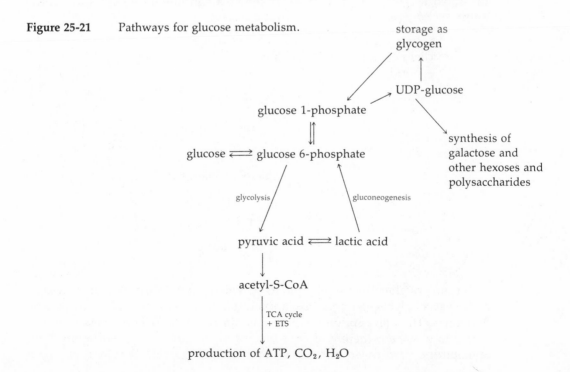

once would be counterproductive. For example, it is a waste of energy to run both the glycolytic and gluconeogenic pathways at the same time. Therefore cells have mechanisms to turn these assembly lines on and off as needed.

There is another important reason for carefully regulating carbohydrate metabolism. Certain cells in the human body, especially those in the brain, use the glycolytic pathway as their sole source of energy. These cells, whose operation is critically dependent on glucose, need a continuous and relatively constant supply of glucose.

A measure of the available glucose supply in the human body is the concentration of glucose in blood. The units usually used to express this concentration are milligrams of glucose/100 cc of blood, or mg/deciliter, or "milligrams %." The glucose concentration goes up after eating, but in normal healthy humans it quickly (within a few hours) returns to a level between 80 and 120 mg %. This range (80 to 120 mg %) is the normal "fasting" glucose level. (The normal range is somewhat dependent on the method used to analyze for glucose in blood. Subjects are required to fast 8 to 10 hours prior to testing.)

Too much or too little glucose in the blood is detrimental to cells, and can be fatal. **Hyperglycemia** is the term used for excessive amounts of glucose in the blood (above 120 mg %). **Hypoglycemia** refers to blood glucose concentrations less than 80 mg %. The glucose concentration in the blood is a sensitive measure of carbohydrate metabolism, and is used to diagnose and monitor certain diseases, most notably, diabetes.

Regulation of Glycogen Synthesis/Breakdown

You are already familiar with the types of mechanisms employed to control carbohydrate metabolism. They include allosteric and chemical modification of enzymes, and hormonal control. In referring back to Chapter 23 to review enzyme control mechanisms, you will notice that one of the processes discussed (Figure 23-14) was the control of glucose production by hydrolysis of glycogen. This process, together with the synthesis of glycogen from glucose, is a very important *control point* in the metabolism of carbohydrates because stored glycogen serves as a readily available source of glucose. Conversely, excess glucose can be stored as glycogen. Figure 25-22 shows a simplified view of this control point.

Figure 25-22   Reactions of glycogen synthesis and breakdown.

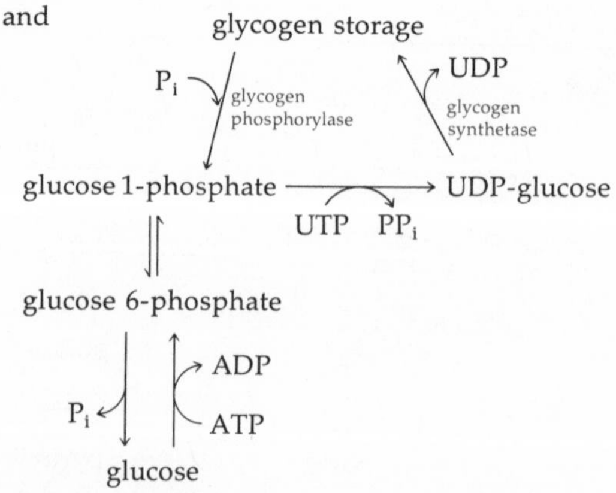

The entry of glucose into cells and its polymerization into glycogen, as well as hydrolysis of glycogen to glucose, are controlled in a fairly complicated way by hormones. The polypeptide hormone **insulin** increases polymerization of glucose into glycogen, thereby decreasing the blood glucose concentration. Both **epinephrine** (also called adrenaline) and the polypeptide hormone **glucagon**

increase hydrolysis of glycogen to glucose 1-phosphate, thereby increasing the blood glucose concentration. Epinephrine,

$$HO-\underset{HO}{\bigcirc}-\underset{OH}{\overset{H}{\underset{|}{C}}}-CH_2-\underset{H}{\overset{H}{\underset{|}{N}}}-CH_3$$

is a derivative of the amino acid tyrosine.

How do these hormones work? Epinephrine and glucagon directly influence the production of the compound **cyclic AMP** inside the cell. Cyclic AMP is frequently called "the second messenger" because it transmits hormonal messages to the proteins in the cell. Cyclic AMP is produced by the reaction

(25-56)

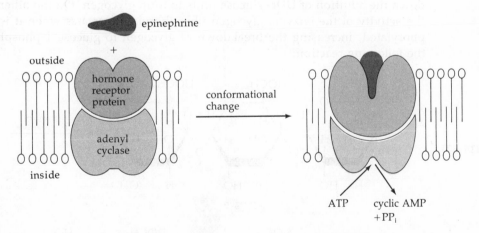

ATP $\xrightarrow{\text{adenyl cyclase}}$ cyclic AMP $+ PP_i$

The enzyme adenyl cyclase is located in cell membranes, right next to receptor proteins that can bind the hormones epinephrine or glucagon (Figure 25-23). The hormone receptor proteins on some cell membranes bind epinephrine, whereas the receptors on other cell membranes bind glucagon. These hormones induce a change in the shape of their respective receptor proteins when they bind to them. This, in turn, causes a change in the shape of the adenyl cyclase enzyme, converting it to a much more active form (see Figure 25-23). The result, for either epinephrine or glucagon binding, is increased production of cyclic AMP.

Figure 25-23   The binding of epinephrine to its receptor protein causes the allosteric activation of adenyl cyclase, which results in the increased production of cyclic AMP within the target cell. A number of other hormones function in a similar manner.

Cyclic AMP then binds to the allosteric site of an enzyme called a protein kinase, making it active. The active protein kinase is then capable of catalyzing reactions that result in the covalent bonding of phosphate groups to the enzymes glycogen synthetase and glycogen phosphorylase (see Figure 25-24). Attachment of phosphate groups to one of these enzymes, glycogen phosphorylase, involves an intermediate step that we need not consider.

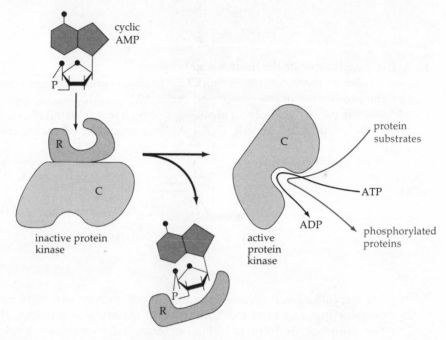

Figure 25-24   The enzyme protein kinase has two subunits, called the catalytic (C) and regulatory (R) subunits. When cyclic AMP binds to the regulatory subunit, R dissociates from C. The catalytic subunit then catalyzes the phosphorylation of certain proteins.

Glycogen synthetase catalyzes the synthesis of glycogen (reaction 25-55). Phosphorylation *lowers* the activity of glycogen synthetase, and thus slows down the addition of UDP-glucose units to form glycogen. On the other hand, the activity of the enzyme glycogen phosphorylase *increases* when it is phosphorylated, increasing the breakdown of glycogen to glucose 1-phosphate in the following reaction:

(25-57)

Thus, the net result of epinephrine or glucagon hormonal stimulation is an increased supply of glucose, accomplished by the allosteric activation and chemical modification of enzymes, and it can be quite rapid (see Figure 25-25). When you are frightened, there is a rapid rise in your blood glucose level caused by epinephrine. This extra glucose is then available for use by your muscle cells in whatever response (to "flee or fight") you may choose to make to the frightening stimulus.

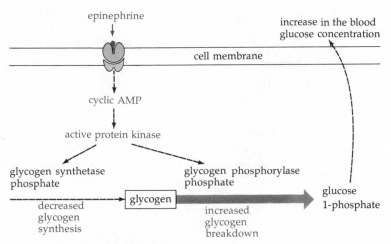

Figure 25-25   Steps leading to the increase in blood glucose concentration upon stimulation of a liver cell by the hormone epinephrine. The phosphorylated form of glycogen synthetase is less active than its unphosphorylated form, whereas phosphorylated glycogen phosphorylase is more active than its unphosphorylated form.

How does insulin influence the control of the concentration of glucose? Insulin has the opposite net effect from epinephrine and glucagon when it binds to its receptor protein on cell membranes. That is, it turns off glycogen phosphorylase, thus decreasing the breakdown of glycogen, while it turns on glycogen synthetase, leading to polymerization of glucose into glycogen. The precise molecular action of insulin is not yet understood. Insulin binding is known to increase glucose uptake by cells, thereby lowering the glucose concentration in the blood. In addition, glucose itself is known to bind to phosphorylated glycogen phosphorylase and promote its dephosphorylation by the enzyme phosphoprotein phosphatase, making it inactive. Phosphoprotein phosphatase also catalyzes the removal of the phosphate group from the phosphorylated form of glycogen synthetase, returning it to its more active, nonphosphorylated form. Research is actively being conducted in an attempt to arrive at a complete picture of the action of insulin.

Regulation of Glycolysis/Gluconeogenesis

The remaining control points in carbohydrate metabolism are less complicated than the control of glycogen metabolism. Let us look at the major control point of glycolysis and gluconeogenesis. Recall that in glycolysis, glucose is degraded to pyruvic acid and ATP is produced, whereas in gluconeogenesis, energy in the form of ATP is used to synthesize glucose.

You may remember that there are a few reactions in glycolysis that are irreversible (see Figure 25-18). In order for gluconeogenesis to occur, these steps must be bypassed. One of these irreversible reactions (step 3 of glycolysis) is

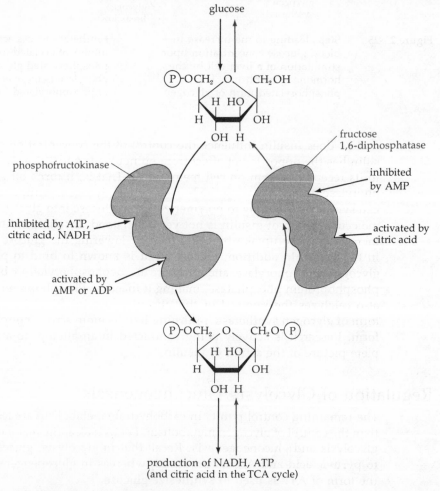

fructose 6-phosphate   fructose 1,6-diphosphate

phosphofructokinase

ATP ADP

In gluconeogenesis, this irreversible step is bypassed by reaction 25-48:

fructose 1,6-diphosphate   + H₂O $\xrightarrow{\text{fructose 1,6-diphosphatase}}$ fructose 6-phosphate   + P$_i$

This pair of reactions is the major control point for glycolysis/gluconeo-genesis. Phosphofructokinase and fructose 1,6-diphosphatase are allosteric enzymes that can be activated or inhibited by appropriate molecules (see Figure 25-26). ATP is an allosteric inhibitor of phosphofructokinase, as are citric

glucose

fructose
1,6-diphosphatase

phosphofructokinase

inhibited
by AMP

inhibited by ATP,
citric acid, NADH

activated by
citric acid

activated by
AMP or ADP

production of NADH, ATP
(and citric acid in the TCA cycle)

Figure 25-26   The allosteric control of glycolysis
(black arrows) and gluconeogenesis
(blue arrows).

acid and NADH. On the other hand, citric acid enhances the activity of fructose 1,6-diphosphatase. Conversely, AMP allosterically inhibits the phosphatase enzyme, whereas AMP and ADP enhance the activity of phosphofructokinase. These controls are logical. If the cell already has a lot of ATP, citric acid, and NADH, it does not need to produce any more. But if the cell has used up all of its ATP, then it will have a lot of AMP and ADP, which will signal the need for an increase in the rate of glycolysis and a shutdown of energy-using gluconeogenesis (see Figure 25-27)

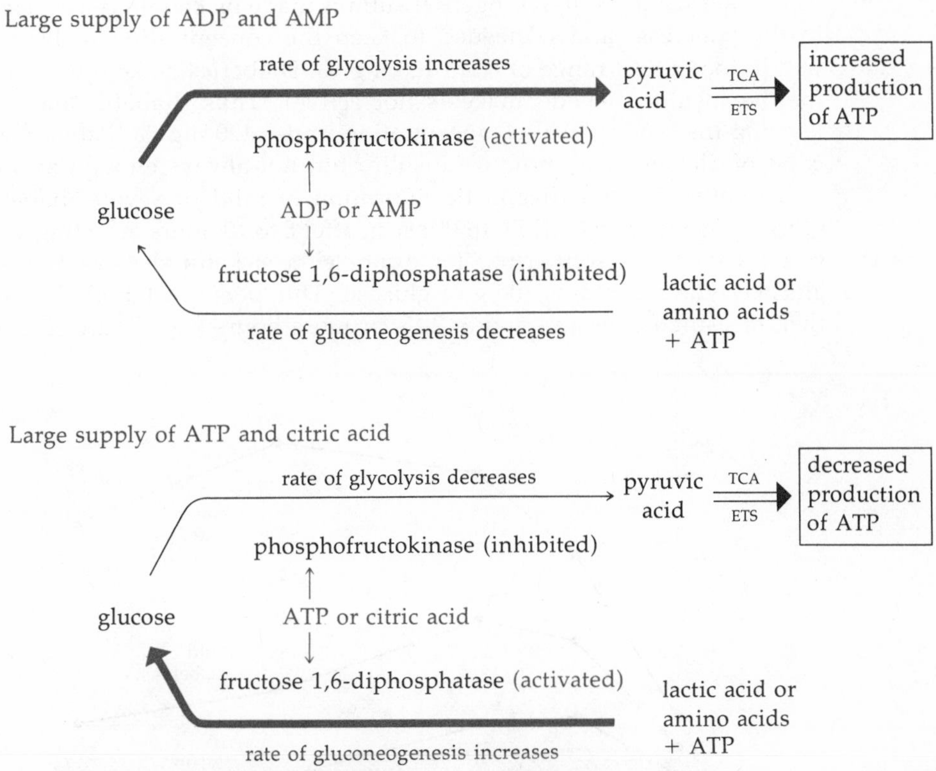

Figure 25-27   A high concentration of AMP or ADP signals the need for increased ATP production, thereby activating glycolysis and inhibiting gluco-neogenesis (top). A high concentration of ATP or critic acid, on the other hand, promotes gluconeogenesis (bottom).

There are several other control points in carbohydrate metabolism, but the two we have looked at appear to be the most important. Also, these two control points serve to illustrate the logical application of control mechanisms involved in cellular metabolism.

Exercise 25-12
When you are frightened, cells in your adrenal medulla produce adrenaline. What effect does this have on blood glucose levels? Does this help your body? How?

Exercise 25-13
Why does a large amount of ADP and AMP in a cell indicate that the cell has a decreased supply of ATP?

25-12 DIABETES

Diabetes mellitus is the disease that results from insufficient production of insulin and is commonly referred to simply as diabetes. The term comes from the Greek and means "sweet-tasting urine." Since an important symptom of this disease is the presence of glucose in the patient's urine, the taste of the urine was formerly used to diagnose the disease. Modern doctors and nurses are fortunate that more quantitative methods have been developed to determine the amount of glucose in the urine. We have seen that the polypeptide hormone insulin facilitates the transport of glucose into cells from the blood and promotes its polymerization into glycogen. Insulin is made by certain cells (called β-cells) in the pancreas, and is needed to keep the concentration of blood glucose within the critical range of 80 to 120 mg %. Diabetics make little or no insulin (or the insulin they do make is not active). Thus diabetics have difficulty keeping the blood glucose concentration under 120 mg %. Patients with mild cases of diabetes may produce insulin, but not always enough to do the job.

Currently the best diagnostic indication of mild or severe diabetes is the **glucose tolerance test** (GTT). In this test, after 8 to 10 hours of fasting, a person is given a doctored soft drink, for example, a sickeningly sweet syrup called glucola, which contains 100 g of glucose. This person's blood glucose level is then measured over a period of 2, 5, or more hours (see Figure 25-28).

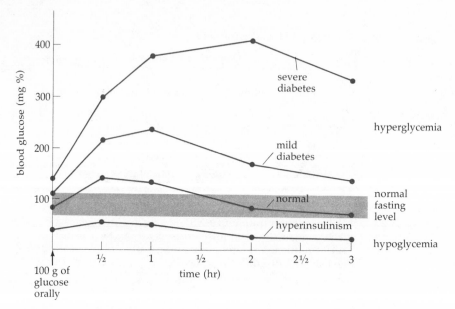

Figure 25-28   Typical results obtained in the glucose tolerance test. Each individual drinks 100 g of glucose at time zero, and the blood glucose concentration is then determined at given time intervals. The blood glucose concentration in normal fasting subjects is indicated by the blue shaded area.

Since glucose is rapidly absorbed into the bloodstream, a normal person will show a rapid increase in blood glucose concentration, peaking at about 150 mg % within 1 hr. This is followed by a rapid decrease in blood glucose concentration as insulin is produced. The blood glucose level of a healthy individual will return to the normal fasting range within 2 hr. People with mild diabetes will need a few hours longer to reduce the glucose concentration. The blood glucose level of severe diabetics may reach very high levels (see Figure

The incidence of diabetes is increasing in our population. Current therapy for severe diabetes involves daily injections of insulin obtained from animals—which is not exactly identical to human insulin. Researchers in California, however, have recently succeeded in splicing the human insulin gene into a bacterial cell, thus opening the way to large-scale production of human insulin by bacterial "slaves."

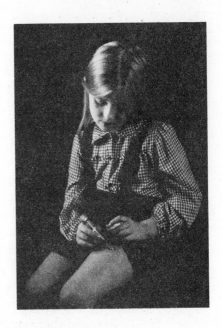

25-28) after ingestion of 100 g of glucose, and the blood glucose level will remain elevated for an extended period. When the blood glucose level reaches values of more than about 180 mg %, a decidedly hyperglycemic condition, some glucose is excreted by the kidneys and appears in the urine. Determination of glucose in urine is sometimes used to screen for diabetes. Testing for glucose in urine is easily done at home, and is routinely done by diabetics to monitor their condition.

Mild cases of diabetes can be controlled by diets that include only small amounts of carbohydrates. Severe diabetics require injections of insulin. There are more than 1 million insulin-dependent diabetics in the United States alone. Commercial insulin is usually obtained from the pancreas of animals. Synthetic insulin, or natural human insulin produced by genetic engineering, may soon be available. Uncontrolled diabetes can lead to a decrease in the pH of the blood, a condition called acidosis (Chapter 29), and can result in coma or even death. One major problem that contributes to these symptoms is the formation of large amounts of "ketone bodies" as a result of defective lipid metabolism (see Chapter 27).

Another clinical condition that results from the improper control of carbohydrate metabolism is hypoglycemia. In certain individuals, an increase in blood glucose level stimulates the pancreas to produce too much insulin. A condition in which the insulin level is above normal is called **hyperinsulinism.** What happens to an individual when the insulin level is too high? In this person, too much glucose is extracted from the blood, and the blood glucose level drops below normal, a hypoglycemic condition. Hypoglycemia is an extremely dangerous condition. When the blood glucose level drops below normal, brain cells quickly starve, which can lead to coma and even death. Hypoglycemic individuals can control their condition by a low-carbohydrate diet. Hypoglycemia can be detected by the glucose tolerance test. Diabetics who inject insulin must also be very careful not to inject too much insulin, which can lead to a hypoglycemic coma.

Exercise 25-14
Can you explain why diabetics frequently carry sugar cubes in addition to insulin?

25-13 SUMMARY

1. Carbohydrates are the primary source of usable energy for humans.

2. Catabolic pathways break down food molecules in a stepwise fashion, with energy-coupled steps leading to ATP synthesis.

3. The hydrolysis of ATP serves as the immediate source of energy used to drive most endergonic biochemical reactions.

4. In the glycolytic pathway, glucose is hydrolyzed to pyruvic acid, coupled to the net synthesis of 2 moles of ATP from ADP and 2 moles of reduced coenzyme NADH per mole of glucose.

5. Pyruvic acid can be reduced to lactic acid under anaerobic conditions, but under aerobic conditions it can be used to generate more ATP and reduced coenzymes via the TCA cycle.

6. The TCA cycle is fueled by acetyl-S-CoA and generates ATP and reduced coenzymes. CO_2 is the exhaust product of the cycle.

7. Reduced coenzymes from glycolysis, the pyruvate dehydrogenase reaction, and the TCA cycle are reoxidized and coupled to ATP synthesis in the electron transport system, with reduction of O_2 to H_2O.

8. In mitochondria, oxidation of 1 mole of NADH + H^+ generates 3 moles of ATP via the electron transport system, whereas oxidation of 1 mole of $FADH_2$ yields only 2 moles of ATP.

9. In the overall aerobic catabolism of glucose, about 42% of the liberated free energy is captured in ATP formation. The remainder is lost as heat. A net of 36 moles of ATP are formed in the complete aerobic catabolism of 1 mole of glucose in a muscle cell.

10. **(Optional)** The pentose phosphate pathway is an alternative pathway for glucose oxidation. It is a source of ribose 5-phosphate and the coenzyme NADPH.

11. The pathway for gluconeogenesis is not simply a reversal of glycolysis, but requires four different steps to circumvent the three irreversible steps of glycolysis.

12. Glucose can be converted to other hexoses, or incorporated into glycogen, via the formation of UDP-glucose.

13. Galactosemia is a disease resulting from the absence of the enzyme hexose 1-phosphate uridylyltransferase, which is required for the conversion of galactose to glucose.

14. The concentration of glucose in the blood is normally kept in the critical range of 80 to 120 mg %. This regulation is accomplished primarily by the complex hormonal control of glycogen formation and glycogen breakdown.

15. The key intermediate in controlling glycogen metabolism is cyclic AMP, which acts by allosteric activation of the enzyme protein kinase.

16. Glycolysis and gluconeogenesis are controlled by the allosteric activation and inhibition of the enzymes phosphofructokinase and fructose 1,6-diphosphatase.

17. Diabetes mellitus is a disease that results from the inability to synthesize sufficient amounts of active insulin. People with uncontrolled diabetes have high blood glucose levels, whereas those with hypoglycemia have inadequate blood glucose levels. Both of these conditions can result in death, but both can be detected by the glucose tolerance test.

PROBLEMS

1. Only two steps in glycolysis are sufficiently exergonic to drive the formation of ATP. Which steps are these?

2. Calculate the moles of ATP produced in muscle cells by the oxidation of the following:
   (a) 1 mole of glyceraldehyde 3-phosphate under aerobic conditions
   (b) 2 moles of fructose 1,6-diphosphate under anaerobic conditions
   (c) 1 mole of pyruvic acid under aerobic conditions
   (d) 1 mole of succinic acid under aerobic conditions

3. A glucose molecule contains six carbon atoms, but each acetyl group entering the TCA cycle contains only two. Account for the fate of all six carbon atoms of glucose under aerobic conditions.

4. Describe the similarities and differences between the pyruvate dehydrogenase reaction, the α-ketoglutarate dehydrogenase reaction, and the glyceraldehyde 3-phosphate dehydrogenase reaction.

5. Write a balanced equation for the aerobic oxidation of 1 mole of lactic acid in a liver cell. How many moles of ATP are produced in this process?

6. The fourth step of the TCA cycle involves the oxidation of a hydroxyl group on isocitric acid to form a carbonyl group. Explain why citric acid cannot serve as a substrate for such a reaction.

7. What effect does the ingestion of cyanide have on the TCA cycle?

8. NADH and $FADH_2$ donate electrons and protons to the ETS, but not with equal results. (a) Why are only two ATP molecules formed for each $FADH_2$ molecule in the ETS? (b) Is there a difference between these two coenzymes in the amount of water formed in the ETS (with oxidative phosphorylation)?

9. Explain the difference between the amount of ATP produced by the complete aerobic oxidation of 1 mole of glucose in muscle cells versus liver cells.

10. Draw structural formulas for the products of the following reactions:

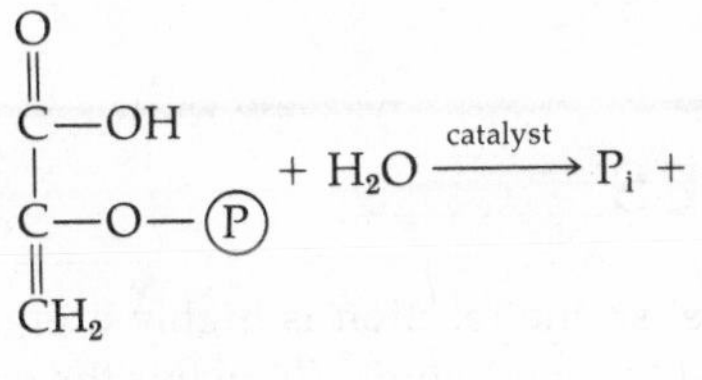

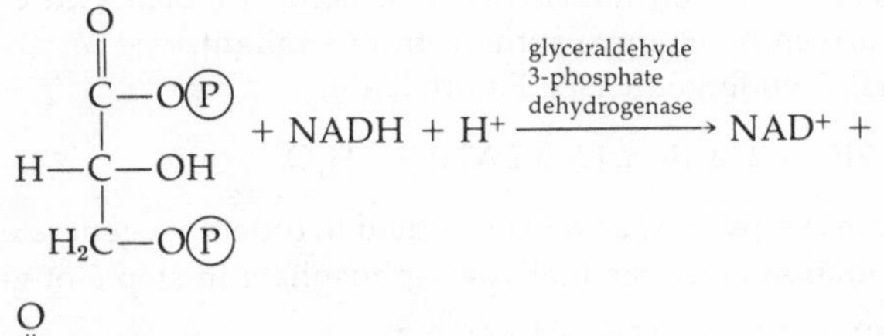

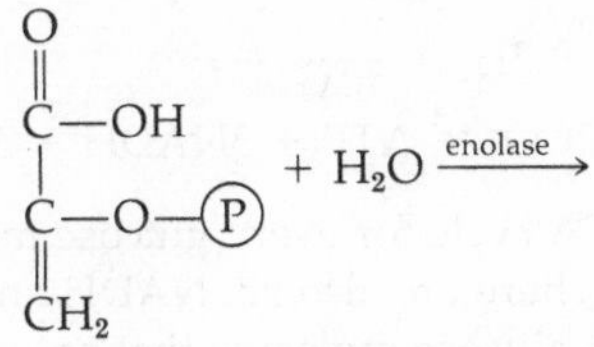

11. The total oxidation of 1 mole of glucose releases 686 kcal. Calculate the amount released per gram of glucose and compare this value to the number of kilocalories per gram of glucose that are captured in ATP when glucose is oxidized in muscle cells. (Use a value of 8.0 kcal/mole for ATP formation.)

12. **(Optional)** Is the operation of the pentose phosphate pathway required for the production of ribosomes? Explain your answer.

13. List any common intermediates in gluconeogenesis and the TCA cycle.

14. Do galactosemic children outgrow their disease? Explain.

15. List two metabolic pathways that involve UDP-galactose and two metabolic pathways that involve UDP-glucose.

16. Outline the reactions involved in the regulation of glycogen metabolism by epinephrine and show which reactions are controlled by allosteric modifiers and which are controlled by chemical modification.

17. In order to add one glucose unit to glycogen, how much energy (in ATP $\rightarrow$ ADP + P_i units) must be used (a) starting with glucose, and (b) starting with two molecules of lactic acid?

18. "Sugarholics," such as people who constantly drink soda pop and/or eat candy, should not stop their habit "cold turkey." Why?

19. Can you explain why many people feel very tired within a few hours after taking a morning break for coffee and donuts?

20. Free and phosphorylated "forms" of glucose are involved in several metabolic pathways. Outline the reactions that convert these different "forms" of glucose one to another.

21. Draw structural formulas for the products of the following reactions (assume that necessary coenzymes, ATP, etc., are present).

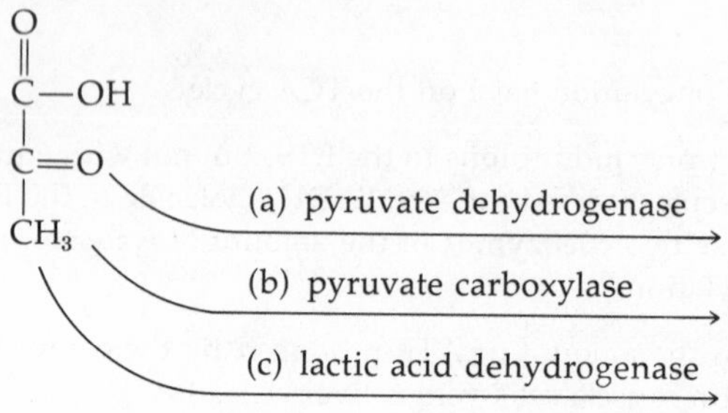

$$
\begin{array}{l}
\overset{\displaystyle O}{\overset{\displaystyle \|}{C}}\!-\!OH \\
\overset{\displaystyle |}{C}\!=\!O \\
\overset{\displaystyle |}{CH_3}
\end{array}
$$

(a) pyruvate dehydrogenase

(b) pyruvate carboxylase

(c) lactic acid dehydrogenase

SOLUTIONS TO EXERCISES

25-1 (a) ΔG for this reaction is -12 kcal/mole, so the reaction is highly exergonic.
(b) Synthesis of glucose from CO_2 and H_2O is endergonic. (In plants the endergonic synthesis of glucose is driven by energy captured from sunlight.)
(c) Formation of ATP is endergonic (see Figure 25-5).

25-2 Glucose + 2ADP + $2P_i$ $\rightarrow$ 2 lactic acid + 2ATP + $2H_2O$

25-3 Active muscle cells convert pyruvic acid to lactic acid in order to regenerate NAD^+, which is needed for the oxidation of glyceraldehyde 3-phosphate in step 6 of glycolysis.

25-4 Acetyl-S-CoA + ADP + P_i + $3NAD^+$ + FAD + $2H_2O$ $\xrightarrow[\text{TCA cycle}]{}$

CoA-SH + $2CO_2$ + ATP + 3NADH + $3H^+$ + $FADH_2$

25-5 Two ATP molecules are synthesized in the TCA cycle for every glucose molecule that is broken down to two pyruvic acid molecules. There are also six NADH and two $FADH_2$ molecules generated in the TCA cycle for each glucose molecule that entered glycolysis.

25-6 $2Fe^{2+} + 2H^+ + \frac{1}{2}O_2 \rightarrow 2Fe^{3+} + H_2O$

25-7 Only three steps are sufficiently exergonic to be coupled to ATP synthesis.

25-8 Glycolysis. To provide the energy for muscle contractions during strenuous exercise, cells in the body need to generate ATP, not NADPH or ribose.

25-9 Catabolism of 1 mole of glucose to CO_2 and H_2O in muscle yields 36 moles of ATP. Formation of 1 mole of glucose from lactic acid requires the hydrolysis of 6 moles of ATP. Thus the investment of 6 moles of ATP ensures the production of 36 moles of ATP at a later time.

25-10 Only one. The UMP group is donated to form UDP-galactose, which is then converted to UDP-glucose, which can donate its UMP group to the next galactose 1-phosphate molecule, and so on.

25-11 A hundred. Addition of each glucose unit to the growing glycogen chain requires the hydrolysis of one UDP. The UDP group is recycled.

25-12 Adrenaline (epinephrine) stimulates the breakdown of glycogen, resulting in a higher blood glucose level, which is then available for ATP production via glycolysis. The extra ATP may be needed to flee or fight.

25-13 The hydrolysis of ATP is the major exergonic reaction used to drive endergonic biochemical processes. Hydrolysis of ATP yields AMP + PP_i or ADP + P_i. Thus increased concentrations of ADP, AMP, or P_i indicate that a lot of ATP has been used and that more is needed.

25-14 If a diabetic accidentally injects too much insulin, he or she will begin to feel faint and may lapse into a hypoglycemic coma unless some glucose is quickly administered. Sugar cubes are a convenient source of glucose.

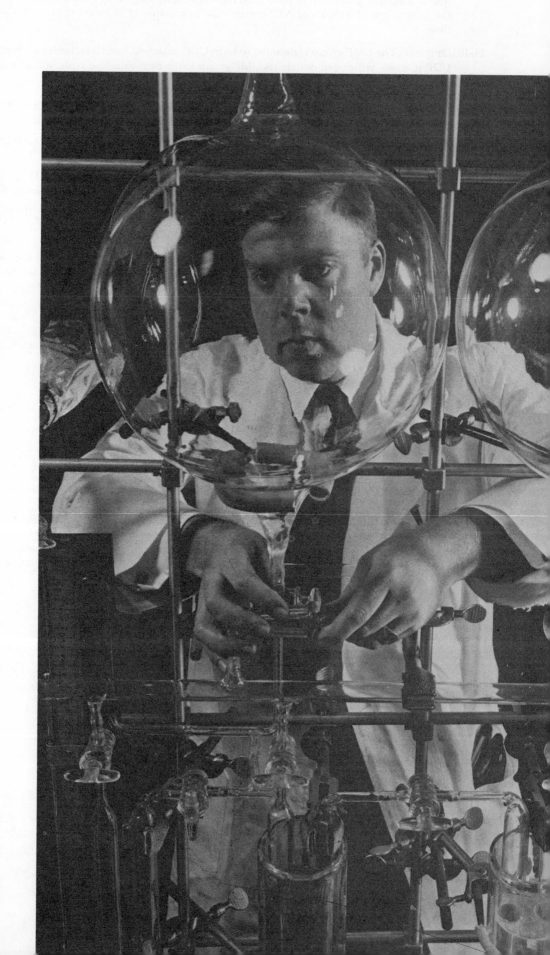

CHAPTER 26

Lipid Structure and Function

26-1 INTRODUCTION

Have you ever heard of "fat chemists"? They are not necessarily obese. Fat chemists are people who study fat molecules. Fat molecules aren't necessarily obese either. Fat molecules or, more generally, lipids, are biomolecules that are insoluble in water and other polar liquids, but soluble in nonpolar organic solvents such as benzene, chloroform, and ether. Many foods, including milk, cream sauces, and salad dressings, involve mixtures of lipids suspended in an aqueous medium. To remove these foods from dishes and cooking utensils, we use detergents to form emulsions of the lipids in water.

Several types of lipid molecules have important functions in the human body. Some, such as the phospholipids, are components of cell membranes. Others, called steroids, can serve as hormones. Still other lipids are vitamins. Some lipids are also a very energy-rich source of nourishment for humans. We shall see that catabolism of some simple lipids, called fatty acids, can be coupled to the production of large amounts of ATP. Because of this, the human body stores fatty acids and other lipids in adipose (fatty) tissue as a reserve supply of energy. Some humans have excessive reserves of lipids—they are obese. Obesity and other disorders, such as atherosclerosis, can result from improper nutrition and/or genetic defects in the metabolism of lipids.

Many lipids play very interesting roles in the human body, but it is not completely known how they function. For example, estrogen and testosterone, the major female and male sex hormones in humans, belong to a class of lipids called steroids. They carry very specific chemical messages to their target cells, but we are still not sure precisely how they work. In living cells, lipid molecules are often combined with other kinds of biomolecules. We saw in Chapter 22, for example, that they can combine with proteins to form lipoproteins and proteolipids. Lipids can also combine with carbohydrates to form molecules called glycolipids. These hybrid molecules also have unique functions, which in many cases are not very well understood.

In this chapter we shall study the structures and properties of lipid molecules. We shall also see how different classes of lipid molecules function in the human body. In the next chapter we shall study the metabolism of lipid molecules.

This experimental chemist is investigating new methods for producing synthetic rubber. Since rubber is a lipid, one could call such a scientist a "fat chemist."

26-2   STUDY OBJECTIVES

After careful study of the material in this chapter, you should be able to:

1. Define the terms fatty acid and essential fatty acid.

2. Describe the difference between a saturated fatty acid and an unsaturated fatty acid.

3. Describe the basic components of the 10 lipid classes.

4. Describe the major biological functions of the 10 classes of lipids.

5. Draw the structural formulas for glycerol and glycerol 3-phosphate.

6. Draw the structural formula for a monoglyceride, a diglyceride, or a triglyceride, given the structural formulas of its component parts.

7. Define the term saponification, and predict the products formed when a glyceride is saponified.

8. Describe a micelle and an emulsion, and show how a soap works.

9. Identify the hydrophilic and hydrophobic parts of a lipid, given its structural formula.

10. Draw a structural formula for an isoprene unit and the characteristic four-ring structure found in steroids.

11. Describe the relationship between fat-soluble vitamins and terpenes.

12. Describe a lipid bilayer and the mosaic model of cell membranes.

26-3   CLASSIFICATION OF LIPIDS

Lipids are a large group of naturally occurring organic molecules that are insoluble in water but soluble in nonpolar solvents. This is a very broad definition, based only on the relatively nonpolar nature of these molecules. Therefore you should not be surprised to discover that there are a variety of lipid molecules, each with a unique type of structure.

We have stressed the very close relationship between the structure of a biomolecule and its function in the human body. Lipids are divided into 10 classes, based on their structural features, and different classes of lipids have different functions in the cells of the human body. It is useful to consider the various roles of lipids in cells, using our analogy of cells as factories. For example, one class of lipid, the phosphoglycerides, are present in cell membranes and play a role comparable to the walls of a factory. We shall see that the structures of phosphoglycerides allow them to serve as excellent walls for cellular factories, separating the aqueous cytoplasm of the cell from its aqueous external surroundings. Some lipids serve as insulation against heat or cold for cellular factories, and for the entire body as well. Other lipids provide a source of energy to run cellular assembly lines.

The basic features of the 10 major classes of lipids are outlined in Table 26-1. Fatty acids comprise the first class. We shall define a **fatty acid** as a carboxylic acid with a long chain of carbon atoms (generally 14 to 22) extending from the carboxyl group. Fatty acids are components of glycerides, phosphoglycerides, sphingolipids, and waxes. Prostaglandins are also derivatives of fatty acids with a 20-carbon atom chain. Terpenes and steroids are classes of lipids that do not contain fatty acids.

Table 26-1 Classes of Lipids

Lipid Class	Fundamental Components
1. Fatty acids	Fatty acids
2. Glycerides	Fatty acids + glycerol
3. Phosphoglycerides	Fatty acids + glycerol + phosphoric acid
4. Sphingolipids	Fatty acids + sphingosine
5. Waxes	Fatty acids + long-chain alcohols
6. Prostaglandins	Derivatives of certain fatty acids
7. Terpenes	Contain isoprene units
8. Steroids	Characteristic four-ring structural unit
Hybrid Lipids	
9. Glycolipids	Lipids + carbohydrates
10. Lipoproteins	Lipids + proteins

Exercise 26-1

Is acetic acid, $CH_3-\overset{\overset{\displaystyle O}{\|}}{C}-OH$, a fatty acid? Is it a lipid? Explain your answer.

26-4 FATTY ACIDS

A fatty acid contains a long chain of carbon atoms. Most of the fatty acids found in the human body consist of a total of 14 to 22 carbon atoms. Most of them contain an even number of carbon atoms. The reason for an even number of carbon atoms will become evident in the next chapter, when we study the synthesis of fatty acids. Table 26-2 shows the condensed structural formulas for several fatty acids. The long carbon chains of these fatty acids may be **saturated** (contain only C—C single bonds), or they may be **unsaturated** (contain C=C double bonds).

Unsaturated fatty acids have lower melting points than saturated fatty acids of the same chain length. For example, compare the melting points of the two 16-carbon fatty acids in Table 26-2. Saturated palmitic acid melts at 63°C, but the double bond in palmitoleic acid lowers its melting point to −0.5°C.

Table 26-2 Some Common Fatty Acids

Name	Number of C Atoms	Condensed Structural Formula	Melting Point (°C)
Saturated			
Lauric	12	$CH_3(CH_2)_{10}COOH$	44
Myristic	14	$CH_3(CH_2)_{12}COOH$	54
Palmitic	16	$CH_3(CH_2)_{14}COOH$	63
Stearic	18	$CH_3(CH_2)_{16}COOH$	70
Arachidic	20	$CH_3(CH_2)_{18}COOH$	77
Lignoceric	22	$CH_3(CH_2)_{20}COOH$	86
Unsaturated			
Palmitoleic	16	$CH_3(CH_2)_5CH=CH(CH_2)_7COOH$	−0.5
Oleic	18	$CH_3(CH_2)_7CH=CH(CH_2)_7COOH$	13
Linoleic	18	$CH_3(CH_2)_4CH=CHCH_2CH=CH(CH_2)_7COOH$	−5
Linolenic	18	$CH_3CH_2CH=CHCH_2CH=CHCH_2CH=CH(CH_2)_7COOH$	−11
Arachidonic	20	$CH_3(CH_2)_4(CH=CHCH_2)_3CH=CH(CH_2)_3COOH$	−50

We shall see that cell membranes contain a mixture of saturated and un-saturated fatty acids and exhibit properties in between those of a liquid and a solid. Humans cannot synthesize some of the required unsaturated fatty acids containing more than one double bond. We require rather large amounts of linoleic and linolenic acids (Table 26-2) for use in making glycerides, phos-phoglycerides, and prostaglandins, and we must get these fatty acids in our diet. They are quite plentiful in plants, which serve as our source for these **essential fatty acids** (linoleic acid and linolenic acid). In countries such as the United States, the typical human diet contains a larger ratio of saturated to unsaturated fatty acids than it does in underdeveloped countries. This high dietary ratio favoring saturated fatty acids also favors the development of atherosclerosis, for reasons not yet well understood (see Section 26-6).

The long hydrocarbon chain of either saturated or unsaturated fatty acids is hydrophobic (Chapter 14), whereas the polar carboxyl group is hydrophilic (Chapter 17). Thus a fatty acid contains both hydrophobic and hydrophilic parts.

What happens when a fatty acid, say, oleic acid, is mixed with water? A solu-tion does not form, since the lipid oleic acid is not soluble in water. However, there is an attraction between the hydrophilic carboxyl groups on the oleic acid molecules and water molecules. Droplets of oleic acid form. Each of these drop-lets consists of a large number of oleic acid molecules, with their hydrophobic carbon chains close together and pointed toward the center of the droplet, and their hydrophilic carboxyl groups on the exterior of the droplet (see Figure 26-1). Droplets of this type are called **micelles.**

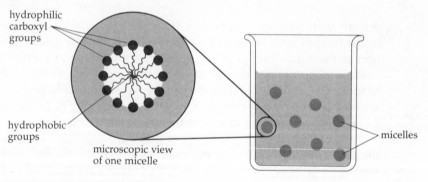

hydrophilic carboxyl groups

hydrophobic groups

microscopic view of one micelle

micelles

Figure 26-1   An emulsion of fatty acids in water. Emulsions are made up of micro-scopic droplets, called micelles, each consisting of many fatty acid mole-cules oriented so that their hydro-phobic portions are tightly packed at the center of the droplet and their hydrophilic parts are exposed to the aqueous environment.

A mixture of micelles in water is called an **emulsion.** It is also possible to layer oleic acid and other lipids on top of water, like an oil slick. In this case (see Fig-ure 26-2), the hydrophobic "tails" of the oleic acid molecules stick up into the air, whereas the hydrophilic carboxyl groups stick into the water. We shall see

Figure 26-2   A layer of oleic acid molecules on a water surface. Note that the hydro-philic "heads" of the fatty acid mol-ecules interact with the water, whereas the hydrophobic "tails" stick up into the air.

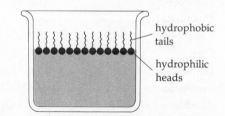

hydrophobic tails

hydrophilic heads

that such a layer of oleic acid on water is actually very similar in structure and properties to one-half of a cell membrane. For centuries people have made extensive use of the K^+ and Na^+ salts of fatty acids as soaps (see Figure 26-3). You should be able to explain how a fatty acid salt can function as a soap and emulsify hydrophobic greases and oils.

Figure 26-3   A soap molecule contains hydrophilic and hydrophobic parts.

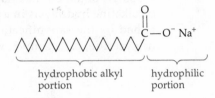

hydrophobic alkyl portion

hydrophilic portion

Exercise 26-2

What is the difference between a 20-carbon saturated fatty acid and a 20-carbon unsaturated fatty acid? Which has the lower melting point?

26-5 LIPIDS DERIVED FROM FATTY ACIDS

In the human body, fatty acids are not generally found as such, but almost always as parts of esters. Recall from Chapter 17 that an ester functional group is formed by the reaction of a carboxyl group on one molecule and an alcohol functional group on another molecule.

Glycerides

Glycerides, esters formed from fatty acids and glycerol, are one of the most common types of esters formed by fatty acids (see Figure 26-4). Since a molecule of glycerol has three alcohol groups, it is possible for one glycerol molecule to form one, two, or three fatty acid esters, called monoglycerides, diglycerides, and triglycerides, respectively.

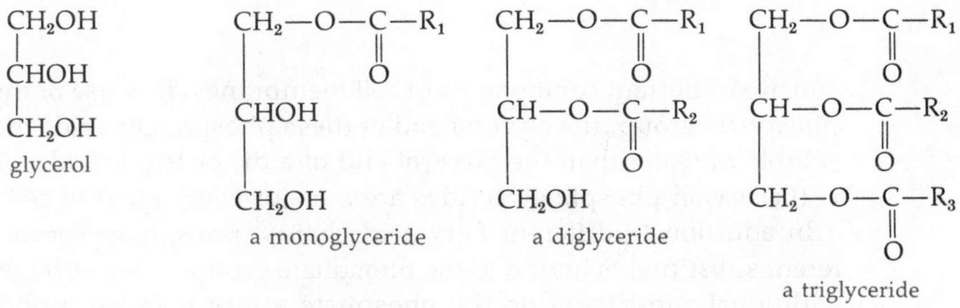

$$
\begin{array}{llll}
CH_2OH & CH_2-O-C-R_1 & CH_2-O-C-R_1 & CH_2-O-C-R_1 \\
| & \quad\quad\quad \| & \quad\quad\quad \| & \quad\quad\quad \| \\
CHOH & \quad\quad\quad O & \quad\quad\quad O & \quad\quad\quad O \\
| & | & | & | \\
CH_2OH & CHOH & CH-O-C-R_2 & CH-O-C-R_2 \\
\text{glycerol} & | & \quad\quad\quad \| & \quad\quad\quad \| \\
& CH_2OH & \quad\quad\quad O & \quad\quad\quad O \\
& \text{a monoglyceride} & | & | \\
& & CH_2OH & CH_2-O-C-R_3 \\
& & \text{a diglyceride} & \quad\quad\quad \| \\
& & & \quad\quad\quad O \\
& & & \text{a triglyceride}
\end{array}
$$

Figure 26-4   The structural formula for glycerol and general structural formulas for glycerides.

Obviously, the exact structure of a glyceride and its properties will depend on which fatty acids are in fact esterified to the glycerol molecule. Triglycerides are used as a means of food storage in animals. Much more reserve energy can be stored in triglyceride molecules than in glycogen. Later, we shall see that 1 g of triglycerides is capable of yielding more than double the energy of 1 g of carbohydrates. It is more convenient to store energy in the form of fat, since it requires less body weight to store a given amount of energy. Our triglyceride reserves, called **fat depots,** increase when we eat more food than is needed to supply us with the energy we use. We are all aware that it is very difficult to get rid of such fat depots once they have accumulated.

Since free fatty acids are not generally found in nature, you may wonder how soaps are obtained. In Chapter 17 we mentioned that ester bonds can be broken by alkaline hydrolysis in a process called **saponification.** Soaps are generally obtained by the saponification of triglycerides in animal fat. Saponification of a triglyceride, for example, will give a mixture of soaps plus a glycerol molecule:

(26-1)

$$
\begin{array}{c}
\underset{\text{a triglyceride}}{
\begin{array}{l}
CH_2-O-\overset{\displaystyle O}{\overset{\|}{C}}-R_1 \\[4pt]
CH-O-\overset{\displaystyle O}{\overset{\|}{C}}-R_2 \\[4pt]
CH_2-O-\overset{\displaystyle O}{\overset{\|}{C}}-R_3
\end{array}}
+ 3NaOH
\xrightarrow{\text{saponification}}
\underset{\text{glycerol}}{
\begin{array}{l}
H_2COH \\[4pt]
HCOH \\[4pt]
H_2COH
\end{array}}
+
\underset{\text{a mixture of soaps}}{
\begin{array}{l}
R_1-\overset{\displaystyle O}{\overset{\|}{C}}-O^-Na^+ \\[4pt]
R_2-\overset{\displaystyle O}{\overset{\|}{C}}-O^-Na^+ \\[4pt]
R_3-\overset{\displaystyle O}{\overset{\|}{C}}-O^-Na^+
\end{array}}
\end{array}
$$

Exercise 26-3

Draw the structural formula for a monoglyceride containing a 16-carbon saturated fatty acid.

Phosphoglycerides

Phosphoglycerides are esters that are formed from one or two fatty acids and glycerol 3-phosphate,

$$
\begin{array}{l}
CH_2OH \\[6pt]
HOCH \\[6pt]
H_2C-O-\overset{\displaystyle O}{\overset{\|}{P}}-OH \\[4pt]
OH
\end{array}
$$

and are important components of cell membranes. Because of the polarity of the phosphate group, the glycerol end of these phosphoglyceride molecules is more soluble in water than the glycerol end of a di- or triglyceride. Apparently, this is the reason phosphoglycerides are a major component of cell membranes.

In addition to different fatty acid chains, phosphoglycerides can have different substituents bound to the phosphate group. A phosphoglyceride with no additional substituent on the phosphate group is called a phosphatidic acid. Phosphatidic acids are not abundant in the body. They serve as the precursors for synthesis of other phosphoglycerides. The names and general structural formulas for the major phosphoglycerides are given in Table 26-3. Of these, phosphatidyl ethanolamine (or cephalin) and phosphatidyl choline (or lecithin) are those most commonly found in cell membranes.

Exercise 26-4

What is the difference between a phosphatidic acid and a cephalin? Identify the polar and nonpolar portions of these molecules.

Table 26-3 Phosphoglycerides

Name	General Structural Formula

Phosphatidic acid

$$\begin{array}{c} \quad\quad\quad\quad\quad\quad O \\ \quad\quad\quad\quad\quad\quad \| \\ H_2C-O-C-R_1 \\ O \quad\quad\quad | \\ \| \quad\quad\quad | \\ R_2-C-O-CH \\ \quad\quad\quad\quad | \quad\quad O \\ \quad\quad\quad\quad | \quad\quad \| \\ \quad\quad\quad H_2C-O-P-OH \\ \quad\quad\quad\quad\quad\quad\quad | \\ \quad\quad\quad\quad\quad\quad\quad OH \end{array}$$

Phosphatidyl ethanolamine (cephalin)

$$\begin{array}{c} \quad\quad\quad\quad\quad\quad O \\ \quad\quad\quad\quad\quad\quad \| \\ H_2C-O-C-R_1 \\ O \quad\quad\quad | \\ \| \quad\quad\quad | \\ R_2-C-O-CH \\ \quad\quad\quad\quad | \quad\quad O \\ \quad\quad\quad\quad | \quad\quad \| \\ \quad\quad\quad H_2C-O-P-O-CH_2-CH_2-NH_2 \\ \quad\quad\quad\quad\quad\quad\quad | \\ \quad\quad\quad\quad\quad\quad\quad OH \end{array}$$

phosphorylethanolamine

Phosphatidyl serine

$$\begin{array}{c} \quad\quad\quad\quad\quad\quad O \\ \quad\quad\quad\quad\quad\quad \| \\ H_2C-O-C-R_1 \quad\quad\quad\quad\quad O \\ O \quad\quad\quad | \quad\quad\quad\quad\quad\quad \| \\ \| \quad\quad\quad | \quad\quad\quad\quad\quad\quad C-OH \\ R_2-C-O-CH \\ \quad\quad\quad\quad | \quad\quad O \quad\quad\quad\quad | \\ \quad\quad\quad\quad | \quad\quad \| \quad\quad\quad\quad | \\ \quad\quad\quad H_2C-O-P-O-CH_2-C-NH_2 \\ \quad\quad\quad\quad\quad\quad | \quad\quad\quad\quad\quad\quad | \\ \quad\quad\quad\quad\quad\quad OH \quad\quad\quad\quad\quad H \end{array}$$

phosphorylserine

Phosphatidyl choline (lecithin)

$$\begin{array}{c} \quad\quad\quad\quad\quad\quad O \\ \quad\quad\quad\quad\quad\quad \| \\ H_2C-O-C-R_1 \\ O \quad\quad\quad | \\ \| \quad\quad\quad | \quad\quad\quad\quad\quad\quad\quad\quad CH_3 \\ R_2-C-O-CH \quad\quad\quad\quad\quad\quad\quad | \\ \quad\quad\quad\quad | \quad\quad O \quad\quad\quad\quad\quad\quad | \\ \quad\quad\quad\quad | \quad\quad \| \quad\quad\quad\quad\quad\quad + | \\ \quad\quad\quad H_2C-O-P-O-CH_2-CH_2-N-CH_3 \\ \quad\quad\quad\quad\quad\quad | \quad\quad\quad\quad\quad\quad\quad\quad | \\ \quad\quad\quad\quad\quad\quad OH \quad\quad\quad\quad\quad\quad\quad CH_3 \end{array}$$

phosphorylcholine

Sphingolipids

Sphingolipids contain a fatty acid joined by an amide bond to a molecule called sphingosine:

sphingosine

Sphingolipids are important components of membranes, especially in brain and nerve cells. The sphingolipid compound formed from sphingosine and a fatty acid is sometimes called a **ceramide.** The most abundant sphingolipids in human brain and nerve cells are called **sphingomyelins.** A sphingomyelin consists of a ceramide to which a polar molecule such as phosphorylcholine or phosphorylethanolamine (Table 26-3) is attached. The structure of one sphingo-myelin is given in Figure 26-5.

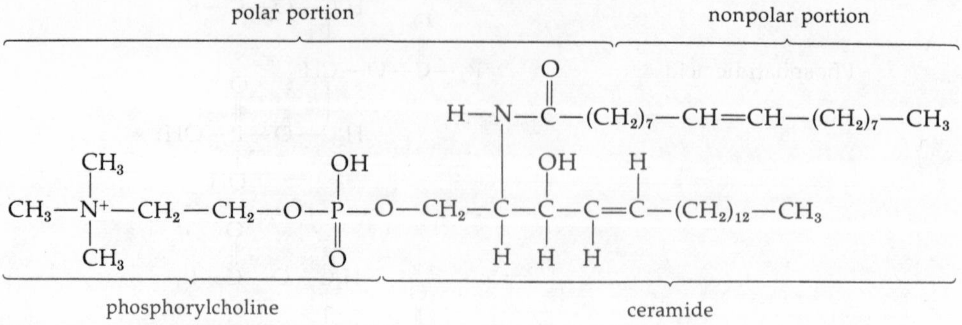

Figure 26-5 A sphingomyelin in which the ceramide consists of sphingosine (in blue) and oleic acid, joined by an amide bond.

Waxes

Waxes are esters of fatty acids and long-chain alcohols or steroid alcohols (see Figure 26-6 for an example). You should be familiar with some of the physical properties of waxes—for example, the flammability of candle wax. Waxes are soft and pliable when they are reasonably warm, but harden upon cooling. Waxes are used as protective coats by certain insects and by some plants. They are also found in the skin and fur of animals. The wax in your ears serves to trap potentially harmful entities that might enter the ear channel.

Figure 26-6 A wax.

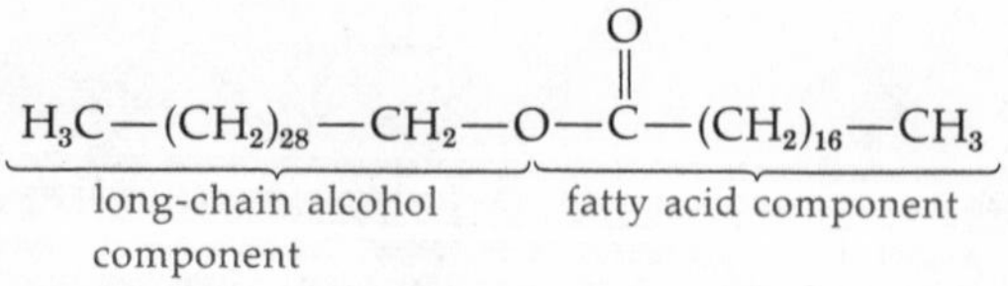

Prostaglandins

Another class of lipid molecules, the **prostaglandins,** are formed from polyun-saturated fatty acids containing 20 carbon atoms such as arachidonic acid (Table 26-2). These polyunsaturated fatty acids in turn must be synthesized from the essential fatty acids linoleic acid and linolenic acid. Two typical prostaglandins are prostaglandin E_1,

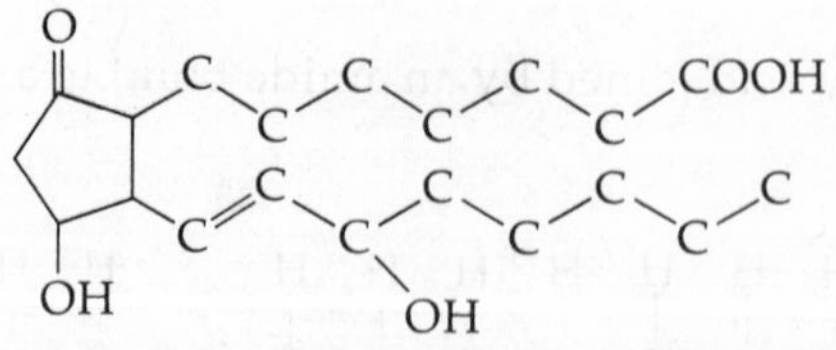

and prostaglandin $F_{1\alpha}$,

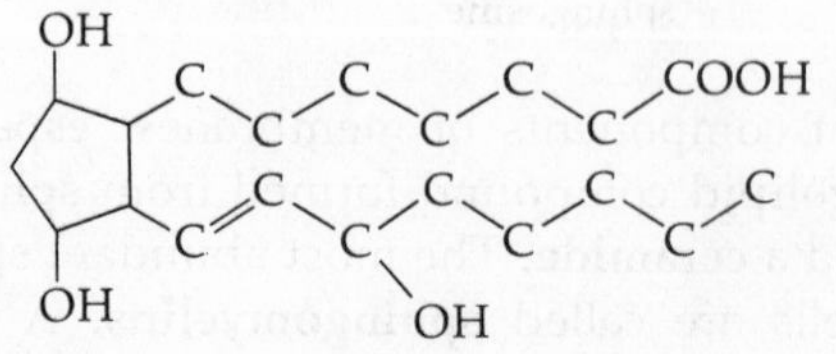

Prostaglandins have very interesting biological properties. They are generally thought to function as hormones and are currently the subject of intense investigations, which may have important medical applications. Some prostaglandins, for example, are known to be involved in the aggregation of platelets during blood clotting, whereas others may be useful for inducing relatively safe abortions.

26-6   TERPENES AND STEROIDS

Terpenes and steroids are quite different in structure from most of the lipids we have discussed so far. **Terpenes** are lipid molecules that can be considered to contain two or more isoprene units. An **isoprene unit** contains five carbon atoms. Its structure is given in Figure 26-7. Therefore, terpenes, by definition, must contain groups of carbon atoms in multiples of five, that is, 10, 15, 20, 25, and so on, carbon atoms.

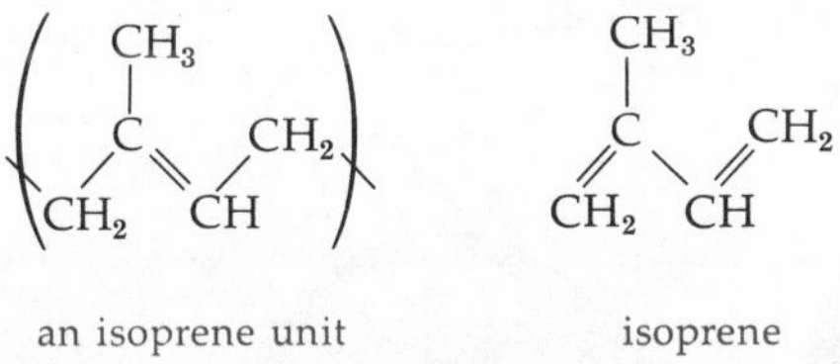

an isoprene unit   isoprene

Figure 26-7   The isoprene units found in terpenes are named after the structurally similar compound isoprene.

Many of the terpenes found in plants have odors and flavors that are familiar to you. For example, menthol and camphor are small terpenes, made from only two isoprene units. On the other hand, natural rubber is a polyterpene made from a very large number of isoprene units. In humans, the most important terpenes are vitamins A, E, and K, which we shall study in Section 26-7, and coenzyme Q, which is a component of the electron transport system (Chapter 25). The structural formula for coenzyme Q is given in Appendix 3. One terpene molecule with 30 carbon atoms, called **squalene,** has the following skeleton structural formula:

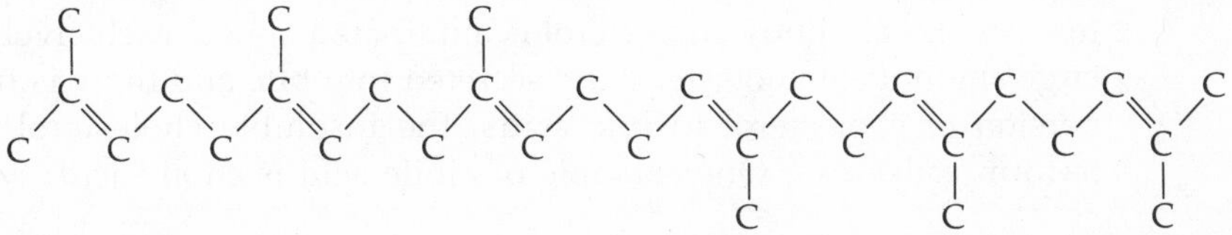

Squalene is used by the body to make steroid molecules.

Steroids are lipids containing a common structural element, which consists of 3 six-membered and 1 five-membered ring of carbon atoms, with the following skeleton structural formula:

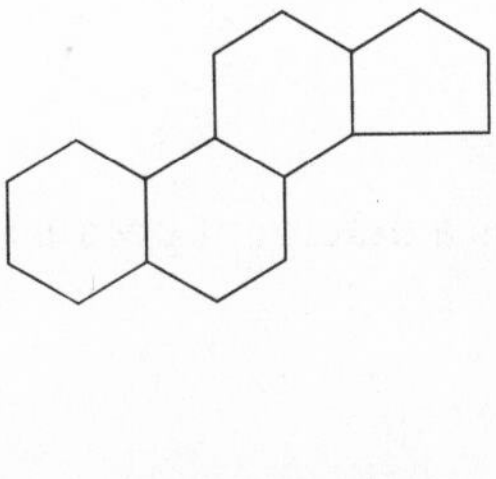

Steroids are all synthesized from acetyl-S-CoA, with squalene as an intermediate. The most abundant steroid molecule in humans is **cholesterol,**

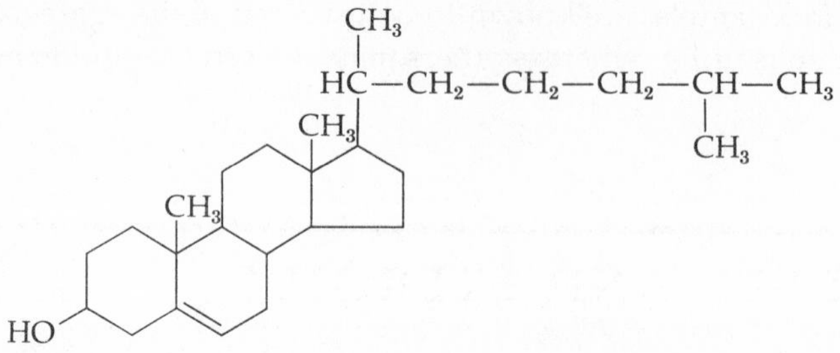

$$CH_3$$
$$HC-CH_2-CH_2-CH_2-CH-CH_3$$
$$CH_3$$
$$CH_3$$
$$CH_3$$
$$CH_3$$
$$HO$$

Cholesterol is a component of cell membranes and the starting material from which the human body manufactures various steroid hormones. Human beings synthesize cholesterol rapidly and do not need to obtain cholesterol in the diet. In fact, excess cholesterol can be deposited by overworked carrier proteins onto the walls of blood vessels. The disease that results from clogged vessels is called **atherosclerosis.** Restriction of dietary cholesterol is recommended to prevent atherosclerosis, but the solution is not that simple. Saturated fatty acids in the diet are also involved (see Section 26-4).

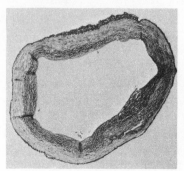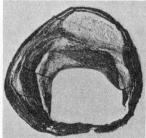

A cross-section of a normal artery is shown on the left. In atherosclerosis, cholesterol is deposited on artery walls and restricts blood flow (center). In severe cases, restricted blood flow may lead to the formation of deadly blood clots (right).

Other steroids function as vitamins and hormones. The structural formulas of several steroid hormones, as well as their functions, are given in Table 26-4.

Another group of steroids are the **bile acids,** which act as detergents in the intestines. Excess body cholesterol is eliminated almost exclusively in bile. When large amounts of cholesterol are secreted into bile and there is insufficient conversion of cholesterol to bile acids, the insoluble cholesterol may crystallize to form gallstones. One example of a bile acid is cholic acid,

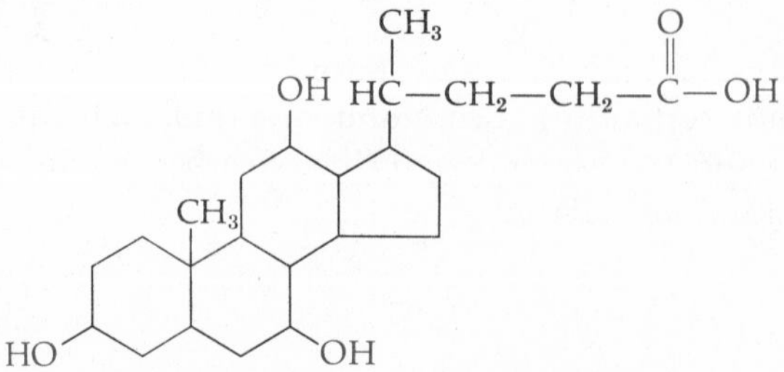

$$CH_3$$
$$O$$
$$OH \quad HC-CH_2-CH_2-\overset{\parallel}{C}-OH$$
$$CH_3$$
$$HO \qquad OH$$

Can you see how cholic acid can function as a detergent because of its polar and nonpolar components?

Table 26-4 Steroid Hormones

Name	Structural Formula*	Produced by	Target Tissue	Function
Corticosterone	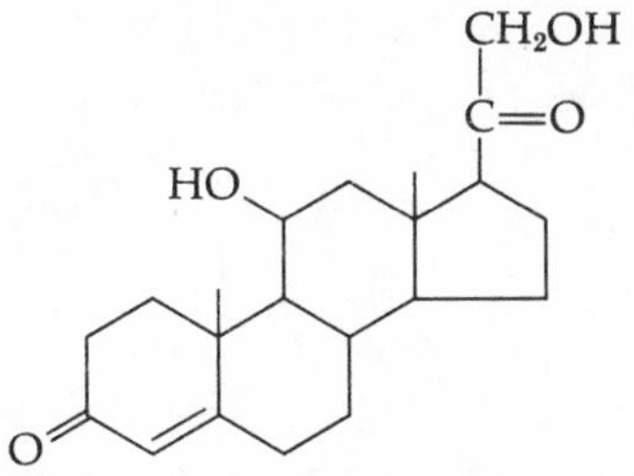	Adrenal gland	Liver and others	Increases glycogen synthesis and gluconeo-genesis
Aldosterone	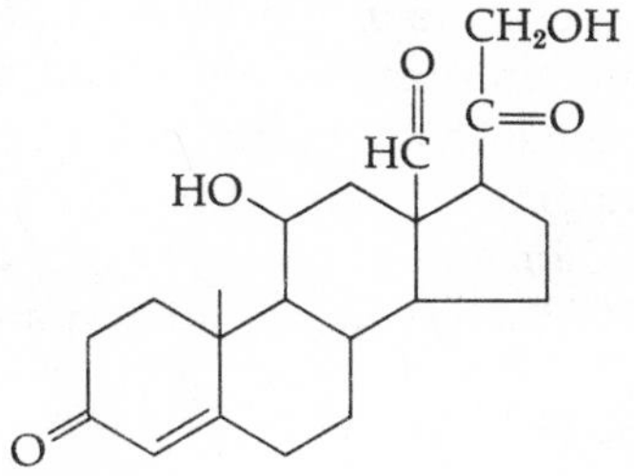	Adrenal gland	Primarily kidney	Na$^+$ and water retention
Estradiol	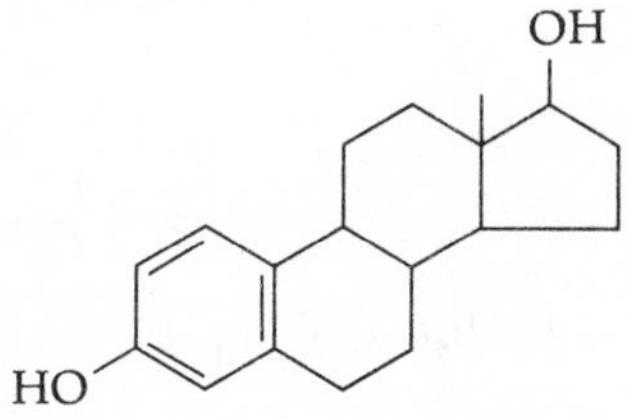	Ovary	Vagina, breast	Maturation and maintenance of reproductive capacity
Testosterone	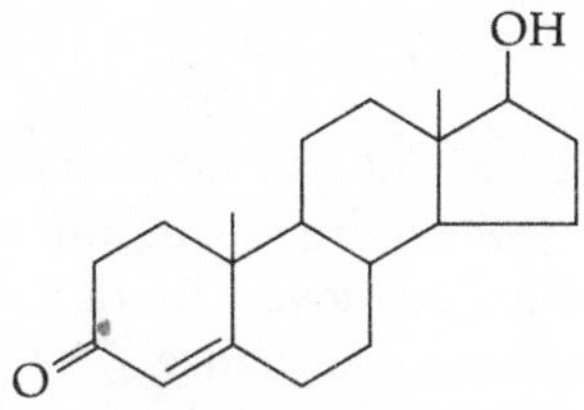	Testis and others	Seminal vesicles	Increased protein synthesis
Progesterone	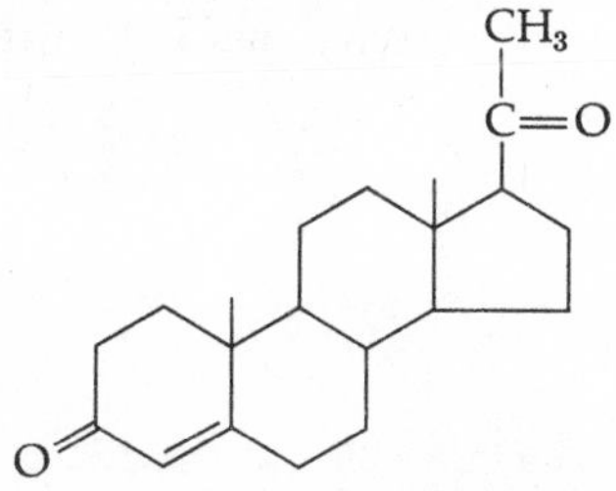	Ovary and others	Uterus	Keeps uterine muscle relaxed during pregnancy

* The short lines protruding from the steroid rings represent methyl groups.

26-7 FAT-SOLUBLE VITAMINS

Some steroids are vitamins. Recall that vitamins are compounds that are required in the human diet in small quantities. In Chapter 23 we mentioned that several coenzymes are derivatives of water-soluble vitamins. **Fat-soluble vitamins** are vitamins that are lipids and hence insoluble in water. The precise functions of a number of these vitamins are not clearly understood. All of the fat-soluble vitamins contain isoprene units.

Vitamin A

The structural formula of **vitamin A** is

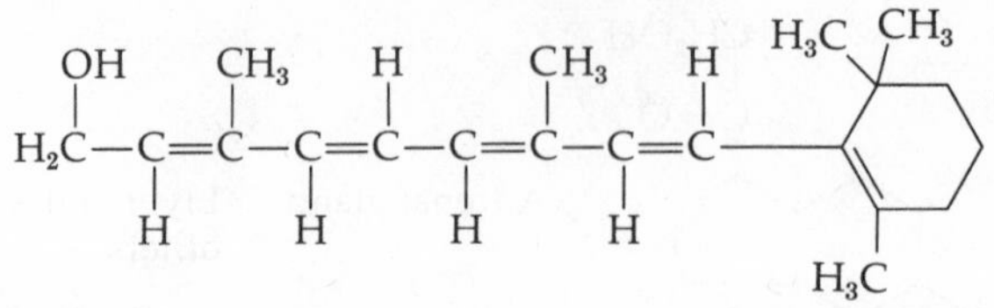

Vitamin A is also called **retinol,** and is needed by several types of cells in the human body, but most of all by retinal cells in the eyes. Absence of vitamin A causes "night blindness" and even permanent blindness in people whose nutrition is especially poor. Dietary vitamin A can be obtained from several vegetables and from fish liver oil. The human body needs about 1 mg of vitamin A per day and can store rather large amounts. It should be noted that excess vitamin A is toxic and can cause liver damage, fragile bones, and deformities in infants. One should keep this in mind when using vitamin supplements.

Vitamin E

The structural formula of **vitamin E** is

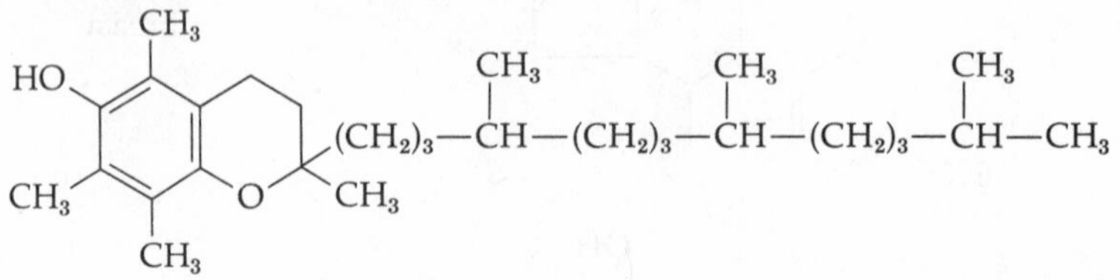

Vitamin E is also called **tocopherol.** Lack of vitamin E results in several symptoms in animals, including infertility. Interestingly, vitamin E has been suggested to be a natural human aphrodisiac. However, there is no concrete evidence that vitamin E affects human sexual performance. The dietary need for vitamin E may be due to its antioxidant activity, as it prevents the oxidation of unsaturated fatty acids. This antioxidant activity has led to the proposal that vitamin E retards aging.

Vitamin K

The structural formula of **vitamin K** is

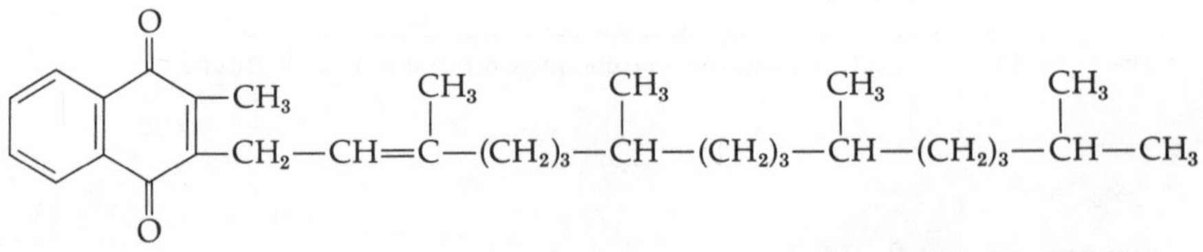

The only known effect of vitamin K deficiency is the inability of liver cells to synthesize an enzyme called **proconvertin.** Proconvertin is one of several enzymes required for blood clotting. A lack of vitamin K, therefore, can lead to defective blood clotting. We don't generally have to worry about getting enough vitamin K, however. Bacteria living in our intestines produce vitamin K which we absorb. It was recently discovered that vitamin K functions as a coenzyme in a number of reactions involving the formation of dicarboxylic acid side chains on certain proteins, including proconvertin.

The drug **Dicumarol,**

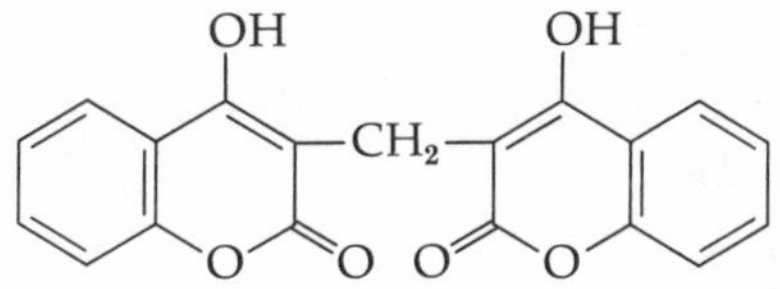

can act as a competitive inhibitor to the formation of proconvertin. Therefore, Dicumarol is used clinically to prevent the formation of blood clots in the blood vessels of patients suffering from phlebitis and other clotting disorders. A closely related compound, called warfarin, is the active ingredient in some rat poisons. Warfarin causes death by uncontrolled internal bleeding.

Vitamin D

There are several related lipids that are collectively called the **D vitamins.** The most important of these are vitamin D_2 (ergocalciferol),

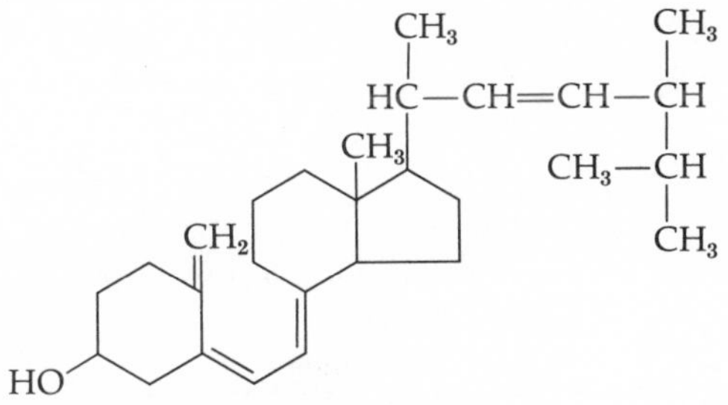

and vitamin D_3 (cholecalciferol),

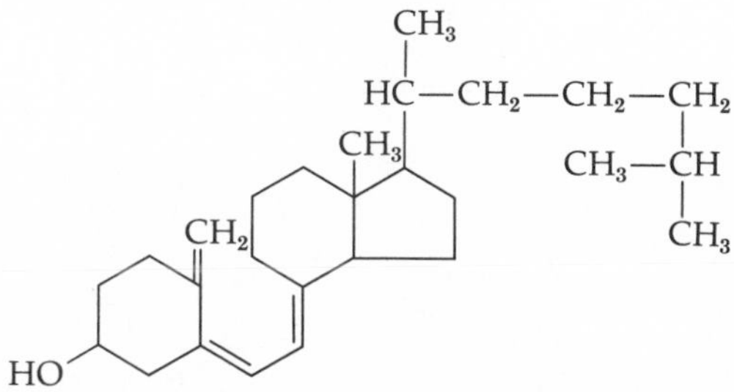

Vitamin D_3 is formed from a derivative of cholesterol by the action of sunlight on the skin. Thus, exposure to sufficient sunlight will assure an adequate supply of vitamin D_3 for humans. In fact, vitamin D-enriched milk is usually obtained merely by exposing milk to light.

What does vitamin D do in the body? Recent studies by Professor Hector De-Luca at the University of Wisconsin, and others, are providing an answer to this question. It appears that vitamin D is converted by kidney cells to a compound called 1,25-dihydroxycholecalciferol. This molecule appears to act as a hormone necessary for proper bone growth. When there is insufficient vitamin D, the kidney cannot make enough 1,25-dihydroxycholecalciferol, which in children can result in **rickets,** a disease in which bone growth is severely impaired.

Exercise 26-5
Dicumarol is used clinically as an anticoagulant. Why is it useful for this?

Exercise 26-6
Are fat-soluble vitamins terpenes or steroids? Explain.

26-8 HYBRID LIPIDS

The last two classes of lipid molecules listed in Table 26-1 are glycolipids and lipoproteins, which are hybrid molecules consisting of lipid molecules bound to carbohydrates and proteins, respectively.

Glycolipids

A **glycolipid** consists of a lipid component that is attached to a carbohydrate component by a glycosidic bond. For example, galactose attached to a diglyceride forms a galactosyldiacylglycerol:

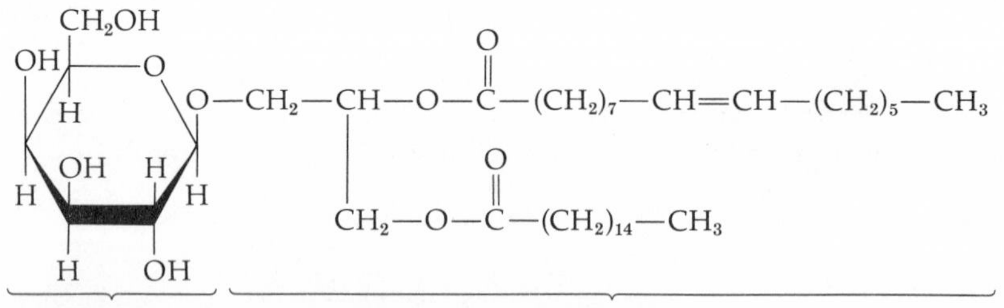

In another type of glycolipid, a sugar, such as glucose, is attached to a ceramide to form a glycosphingolipid molecule called a **cerebroside:**

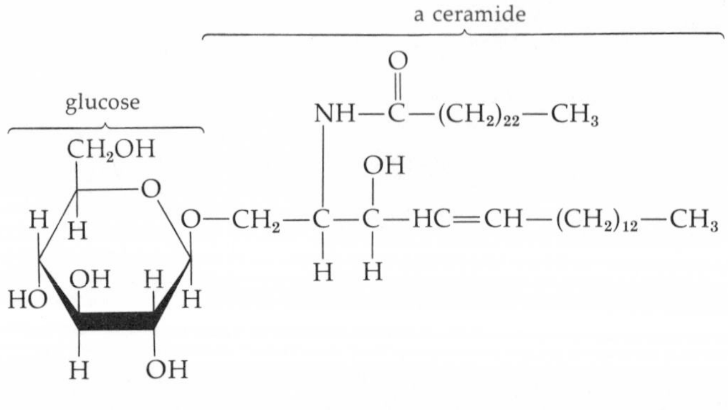

Cerebrosides, like sphingomyelins, are vital parts of nerve cell membranes. The functions of the sphingolipids and glycolipids in nerve cell membranes is being studied intensively. Several human diseases are now known to result from abnormal accumulation of these complex lipids (see Table 26-5). Recent research has shown that each of these diseases is due to the inherited absence of an enzyme needed to break down a sphingolipid or a glycolipid.

Table 26-5 Diseases Resulting from Abnormal Metabolism of Glycolipids and Sphingomyelins

Disease	Organ Affected	Lipid Accumulated
Tay-Sachs	Brain	A glycolipid called a GM_2 ganglioside
Fabry's	Several	A glycolipid consisting of the trisaccharide galactose-galactose-glucose and ceramide
Gaucher's	Spleen, liver	Cerebrosides containing glucose
Niemann-Pick	Several, particularly liver and spleen	Sphingomyelins

Lipoproteins

Lipoproteins, you may recall, are proteins that contain lipid prosthetic groups. Lipoproteins are held together mainly by hydrophobic interactions between nonpolar amino acid side chains and the lipid molecules. Lipoproteins are used as a means of transporting nonpolar lipids through the bloodstream. Thus, as shown in Figure 26-8, the protein component of a lipoprotein is a carrier protein for the lipid component.

Figure 26-8   A schematic representation of a lipoprotein. The carrier protein binds its lipid passenger by means of hydrophobic interactions.

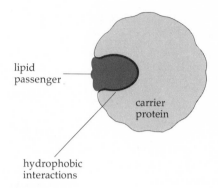

Another close association between lipids and proteins is found in cell membranes. We have already seen that proteins are needed in cell membranes to transport polar molecules across the nonpolar lipid membrane, and to serve as receptors for hormone molecules.

Exercise 26-7

Match each of the lipid classes listed on the left with the most appropriate function on the right.

Lipid	Function
(a) Prostaglandins	(1) Hormones and vitamins
(b) Glycerides	(2) Carry lipids in the blood
(c) Steroids	(3) Hormonelike activity
(d) Lipoproteins	(4) Food storage
(e) Phosphoglycerides	(5) Cell membranes

The biochemical cause of Tay-Sachs disease involves the absence of an enzyme that breaks down a lipid found in the brain. The lack of this enzyme leads to the harmful accumulation of fatty deposits within brain cells. Such fatty deposits appear at the bottom of this micrograph. A part of the nucleus of the brain cell is shown at the top.

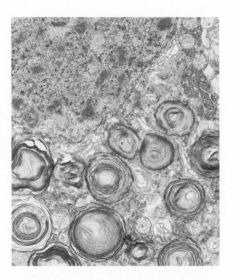

26-9   BIOLOGICAL MEMBRANES

We can consider cell membranes as the walls of cellular factories, and think of phosphoglycerides, cholesterol, and some other lipids as the bricks used to make the walls. A phosphoglyceride molecule, for example, is well suited for this purpose, since it has a polar "head" and two nonpolar, hydrophobic "tails" (see Figure 26-9).

$$
\begin{array}{l}
\quad\quad\;\; H \quad\quad\;\; O \\
\quad\quad\;\; | \quad\quad\quad\; \| \\
H-C-O-C-(CH_2)_{14}-CH_3 \\
\quad\quad\;\; | \quad\quad\quad\; O \\
\quad\quad\quad\quad\quad\quad\; \| \\
H-C-O-C-(CH_2)_7-CH=CH-(CH_2)_5-CH_3 \\
OH \quad\; | \\
| \quad\quad\;\; | \\
X-O-P-O-C-H \\
\| \quad\quad\;\; | \\
O \quad\quad\; H
\end{array}
$$

———————————————|—————————————————
        polar head   nonpolar tails

Figure 26-9   The polar and nonpolar regions of   represent a serine, ethanolamine,
a phosphoglyceride, where X can   or choline component.

Cell membranes are composed primarily of a lipid bilayer, with several protein components added in. The structure of a **phospholipid bilayer** is shown in Figure 26-10. Notice that the polar heads of the phosphoglycerides are on the aqueous outsides of the bilayer, and that the hydrophobic tails of the phosphoglycerides form the nonpolar interior of the bilayer. Because of this nonpolar interior, cell membranes can act as barriers to polar molecules and ions.

Figure 26-10   A phospholipid bilayer. The hydrophobic tails of the phosphoglycerides form a hydrophobic region separating two aqueous regions.

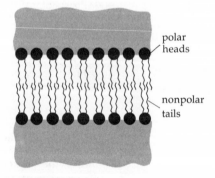

polar heads

nonpolar tails

Remember that the melting points of saturated fatty acids are above human body temperature, whereas those of unsaturated fatty acids are below (see Table 26-2). Cell membranes usually have just the right mixture of saturated and unsaturated fatty acid components to allow them to be semisolid. This gives cell membranes structural strength, but does not make them rigid or brittle.

We have seen that membrane transport proteins and hormone receptor proteins (Chapter 22) are also components of cell membranes. If we stretch our analogy a bit, we might consider them the windows and doors in the walls of the cellular factory. Thus, transport proteins, for example, help specific polar molecules to get into, and out of, cells. The complete structure of a cell membrane is currently viewed as consisting of a large number of these protein molecules embedded in a pliable wall of phosphoglycerides (see Figure 26-11). This is called the **mosaic model** of membrane structure. Notice in this illustration that the protein components are "floating" in a "sea" of phosphoglycerides. Since the phosphoglyceride bilayer is semisolid, the protein molecules can move around in the membrane. The portions of these proteins that are in the interior of the phosphoglyceride bilayer must be nonpolar. The proteins are actually held in the membrane by hydrophobic interactions between the fatty acid tails of the phosphoglycerides and the hydrophobic amino acid side chains of the proteins.

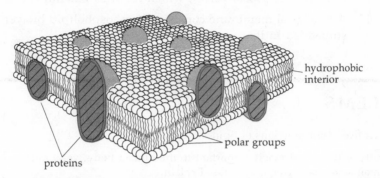

hydrophobic interior

polar groups

proteins

Figure 26-11   The mosaic model of membrane structure. The membrane consists of a phospholipid bilayer with   protein molecules penetrating into either side of, or extending through, the membrane.

Exercise 26-8
Why can't phosphoglyceride monolayers act as cell membranes?

26-10 SUMMARY

1. Lipids are insoluble in water but soluble in nonpolar organic solvents.

2. A fatty acid consists of a long nonpolar chain of carbon atoms, with a carboxyl group on one end.

3. Unsaturated fatty acids have lower melting points than saturated fatty acids.

4. Sodium and potassium salts of fatty acids are soaps.

5. The human body cannot synthesize some of the unsaturated fatty acids it requires. These fatty acids are called essential fatty acids because they must be obtained in the diet.

6. Glycerides are esters of glycerol and one, two, or three fatty acids. They serve as food reserves in the body.

7. Phosphoglycerides consist of two fatty acids that are esterified to glycerol 3-phosphate. Other polar groups may also be bonded to the phosphate group of phosphoglycerides.

8. Sphingolipids are esters of fatty acids and sphingosine.

9. Waxes are esters of fatty acids and long-chain alcohols.

10. Terpenes and steroids are lipid molecules made from isoprene units.

11. Cholesterol is the most abundant steroid in humans. It is a component of cell membranes. Other steroid molecules, such as estrogen and testosterone, are hormones.

12. Prostaglandins are lipids made from 20-carbon fatty acids and have hormonelike functions.

13. Glycolipids are lipid molecules that contain a carbohydrate portion attached to a lipid portion by a glycosidic bond.

14. Lipoproteins consist of lipids bound to their carrier proteins.

15. Vitamins A, E, K, and D are fat-soluble vitamins.

16. Dicumarol prevents clotting by acting as a competitive inhibitor to the formation of proconvertin, which requires vitamin K.

17. A biological membrane consists of a phospholipid bilayer with proteins embedded in it.

PROBLEMS

1. List five major functions of lipid molecules.

2. Fatty acids are extremely important to the human body, yet a normal human has only small amounts of free fatty acids. Explain.

3. Draw the structural formula for a triglyceride composed of glycerol and three lauric acid components.

4. Draw the structural formulas for the products obtained upon saponification of the triglyceride in Problem 3.

5. How does a soap consisting primarily of sodium palmitate enable you to remove grease, that is, hydrocarbons, from your hands and clothes?

6. Name two potential problems that could result from a lack of essential fatty acids in the diet.

7. Draw the characteristic structural component of steroids.

8. Why are molecules similar to vitamin K used in rat poison?

9. Siberian children have historically suffered form a high incidence of rickets. Explain a possible reason for this and suggest a solution that would not require dietary vitamin supplements.

10. The presence of phosphoglycerides containing unsaturated fatty acid components is an important part of the "fluid" mosaic model of cell membranes. Why wouldn't a membrane composed of phosphoglycerides with only saturated fatty acids function properly?

11. Describe the type of bonding found in lipoproteins.

SOLUTIONS TO EXERCISES

26-1 We do not consider acetic acid a fatty acid or a lipid because it does not have a long chain of carbon atoms and it is soluble in water and aqueous solutions.

26-2 The unsaturated fatty acid has at least one double bond, which results in a lower melting point than the saturated fatty acid.

26-3
$$
\begin{array}{l}
H_2C-O-\overset{\overset{\displaystyle O}{\displaystyle \|}}{C}-(CH_2)_{14}-CH_3 \\
H-\underset{|}{C}-OH \\
H_2C-OH
\end{array}
$$

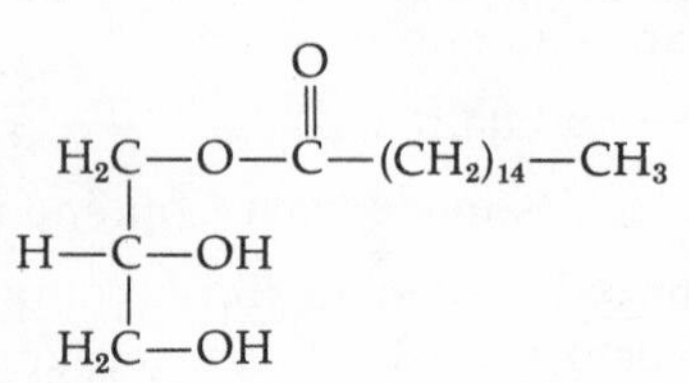

26-4 A phosphatidic acid can be considered to be the phosphate ester of a diglyceride, whereas a cephalin is a phosphatidic acid esterified to a molecule of ethanolamine. The phosphate and phosphorylethanolamine portions of these two types of molecules are polar, whereas the long fatty acid chains (indicated in Table 26-3 as R_1 and R_2) are not.

26-5 It has been found that Dicumarol can act as a competitive inhibitor of enzymes that require vitamin K. One such enzyme is needed to make proconvertin (it adds an additional carboxyl group to proconvertin to make it active). By reducing the rate of proconvertin production, the tendency to form blood clots is also reduced.

26-6 Some fat-soluble vitamins are terpenes (K, A, and E), whereas others (the D vitamins) are steroids.

26-7 (a) Prostaglandins: (3) Hormonelike activity
     (b) Glycerides: (4) Food storage
     (c) Steroids: (1) Hormones and vitamins
     (d) Lipoproteins: (2) Carry lipids in the blood
     (e) Phosphoglycerides: (5) Cell membranes

26-8 A phosphoglyceride monolayer has one polar side and one nonpolar side. The nonpolar side of such a monolayer would not be compatible with the aqueous environment inside or outside of a cell.

CHAPTER 27

Lipid Metabolism

27-1 INTRODUCTION

We have seen that lipids are a diverse group of compounds with correspondingly diverse biological functions. In this chapter we shall discuss the major pathways for the biosynthesis and catabolism of lipids in the human body. Since fatty acids are components of most classes of lipids, we shall pay particular attention to their metabolism.

We shall not study the detailed synthesis of each and every type of lipid molecule. Rather, we shall emphasize the basic features of the synthesis of fatty acids, glycerides, phosphoglycerides, and steroids. The starting material for the biosynthesis of all of these molecules is acetyl-S-CoA. Because they have this common feature, you should find the pathways for lipid biosynthesis fairly easy to follow.

What about pathways for the catabolism of lipids? In Chapter 26 we mentioned that triglycerides are the major food storage depot in the human body and are the source of a great deal of the energy needed by our cellular factories. We shall therefore study the major catabolic pathways for lipids and determine the amount of ATP that is synthesized in the process.

We shall see that the catabolism of a triglyceride involves its hydrolysis to glycerol and three fatty acids. The fatty acids are then degraded by a series of enzyme-catalyzed reactions, called the β-oxidation pathway, to give acetyl-S-CoA, whereas the glycerol is converted in a few steps to dihydroxyacetone phosphate. Since we have previously studied the catabolism of dihydroxyacetone phosphate via glycolysis and the catabolism of acetyl-S-CoA in the TCA cycle, we need only study β-oxidation and the steps in the conversion of glycerol to dihydroxyacetone phosphate in order to have a complete understanding of the catabolism of triglycerides.

Notice that acetyl-S-CoA is both the starting material for lipid biosynthesis and the common intermediate for lipid catabolism. Acetyl-S-CoA is also an intermediate in the catabolism of carbohydrates (Chapter 25) and some amino acids (Chapter 28). Since acetyl-S-CoA is involved in so many metabolic pathways, the human body must carefully regulate which of these pathways is used under a given set of conditions.

The flesh of many species of fish is rich in lipids. These fats contribute to the flavor and texture of such delicacies as smoked whitefish and salmon.

27-2   STUDY OBJECTIVES

After careful study of the material in this chapter, you should be able to:

1. Outline the relationship of lipid metabolism to carbohydrate metabolism and show the central role of acetyl-S-CoA in the metabolism of lipids and carbohydrates.

2. Write a balanced equation showing the action of a lipase enzyme on a triglyceride.

3. Write equations for the reactions involved in the conversion of glycerol to dihydroxyacetone phosphate.

4. Outline the steps of the β-oxidation pathway for fatty acids.

5. Calculate the amount of ATP produced by the complete catabolism of a fatty acid to CO_2 and H_2O, given its chemical formula.

6. Describe the need for the molecule carnitine in fatty acid catabolism.

7. Explain why storage of energy in the form of triglycerides is preferred to storage in the form of glycogen.

8. Describe the role of the acyl carrier portion of fatty acid synthetase, and the relationship between this prosthetic group and coenzyme A.

9. Describe how malonyl-S-CoA is synthesized and used for fatty acid synthesis.

10. Outline the fatty acid biosynthesis pathway.

11. Show how triglycerides are synthesized from fatty acids and glycerol 3-phosphate.

12. Show how phosphatidyl ethanolamine is synthesized in the human body.

13. List several controls for the synthesis and catabolism of fatty acids and triglycerides, and discuss how each works.

14. Define the term ketone bodies, explain their production in diabetics, and tell why they are harmful.

27-3   THE CENTRAL ROLE OF Acetyl-S-CoA

Before we proceed to study some of the individual pathways for lipid metabolism, it will be helpful to have an overall view of lipid metabolism and its interrelationship to carbohydrate metabolism (see Figure 27-1).

When the human body's need for energy cannot be met by the catabolism of glucose, stored triglycerides are catabolized. In this process, a triglyceride is first hydrolyzed to give fatty acids, plus glycerol. The glycerol is converted into dihydroxyacetone phosphate, which is an intermediate in the glycolytic pathway. The fatty acids are broken up into two-carbon-atom fragments which are used to form molecules of acetyl-S-CoA, the fuel for the TCA cycle. Since we have already studied glycolysis and the TCA cycle, we shall only need to study a few new enzyme-catalyzed reactions to understand glyceride catabolism. Phosphoglycerides are catabolized in an essentially identical manner.

When the human body has more than enough glucose to meet its energy requirements, glucose is converted to glycerides, which are stored for later use. The biosynthesis of the fatty acid components of triglycerides starts with acetyl-S-CoA produced, for example, from glucose by glycolysis. Fatty acids are made on a different assembly line from the one used for their catabolism.

Figure 27-1 An overview of lipid metabolism.

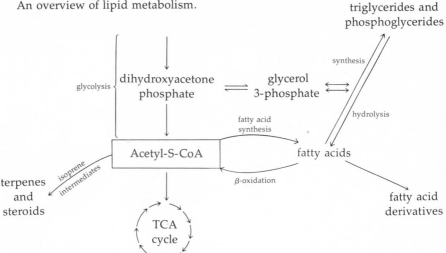

The glycerol or glycerol 3-phosphate portions of glycerides and phosphoglycerides are made from the dihydroxyacetone phosphate intermediate in the catabolism of glucose, in exactly the reverse of the process used to feed them into the glycolysis pathway.

How are needed steroids and terpenes synthesized by the human body? These molecules are built from isoprene units, which in turn are synthesized from acetyl-S-CoA molecules (see Figure 27-1). Therefore, acetyl-S-CoA is the common precursor for the synthesis of all lipids.

27-4 LIPID CATABOLISM

Triglycerides and phosphoglycerides are very good sources of energy for the human body. Triglycerides are stored in fat depots in the body in order to meet future nutritional needs. In fact, the body stores only enough glycogen to provide about one day's supply of energy, but enough triglycerides to supply energy for about one month! In a normal individual the stored triglycerides amount to about 10% of the total body weight. For an average 150-lb adult this amounts to about 15 lb. Storage of an equivalent amount of energy in the form of glycogen would require about 35 lb of glycogen.

We have already seen that both triglycerides and phosphoglycerides contain fatty acid molecules esterified to glycerol. The catabolism of these molecules begins with their hydrolysis to fatty acids and glycerol. This hydrolysis is accomplished by enzymes called lipases in the reaction

(27-1)

$$
3H_2O \; + \;
\begin{matrix}
R_1-\overset{\displaystyle O}{\overset{\|}{C}}-O-CH_2 \\[4pt]
R_2-\overset{\displaystyle O}{\overset{\|}{C}}-O-CH \\[4pt]
R_3-\overset{\displaystyle O}{\overset{\|}{C}}-O-CH_2
\end{matrix}
\;\xrightarrow{\text{lipases}}\;
\begin{matrix}
HO-CH_2 \\[4pt]
HO-CH \\[4pt]
HO-CH_2
\end{matrix}
\; + \;
\begin{matrix}
R_1-\overset{\displaystyle O}{\overset{\|}{C}}-OH \\[4pt]
R_2-\overset{\displaystyle O}{\overset{\|}{C}}-OH \\[4pt]
R_3-\overset{\displaystyle O}{\overset{\|}{C}}-OH
\end{matrix}
$$

a triglyceride   glycerol   3 fatty acids

One lipase is produced by the pancreas and released into the intestines, where it hydrolyzes the triglycerides in our diet. Another lipase is used to hydrolyze the triglycerides in our fat cells when this is required. Hormones, such as epinephrine and glucagon, activate this lipase when energy-rich fatty acids are needed (see Figure 27-2).

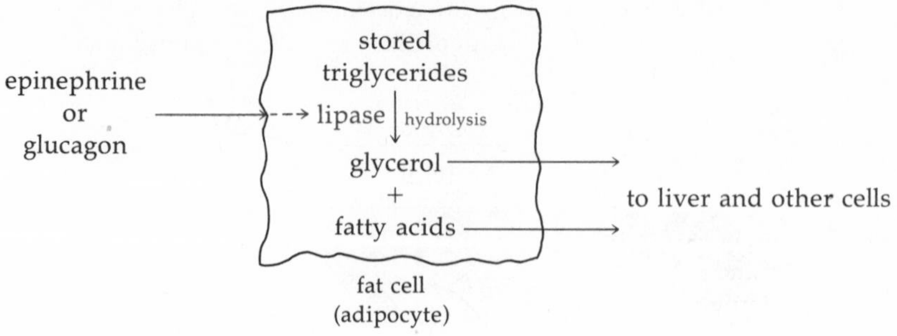

epinephrine
or
glucagon

stored
triglycerides
lipase | hydrolysis
glycerol
+
fatty acids

to liver and other cells

fat cell
(adipocyte)

Figure 27-2   The hormones epinephrine and glucagon activate the lipase enzyme in fat cells. The lipase, in turn, hydrolyzes triglycerides to give glycerol and fatty acids.

Glycerol

First, let us see what happens to the glycerol. Glycerol produced by the hydrolysis of one lipid can be used, if needed, as a component in the synthesis of other lipids. Glycerol can also be converted to dihydroxyacetone phosphate, which is then further catabolized via the glycolytic pathway. This conversion is accomplished in two steps: phosphorylation to glycerol 3-phosphate,

(27-2)

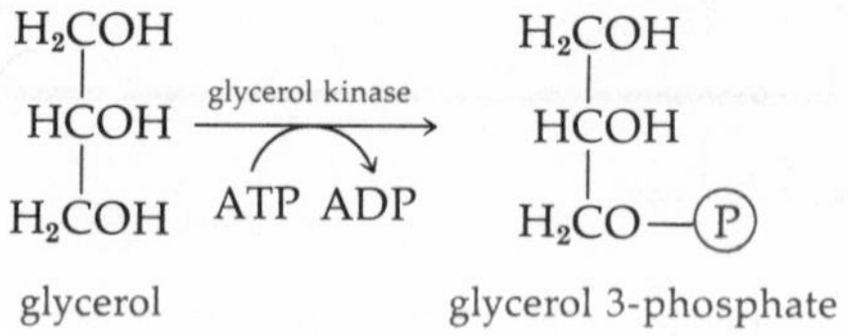

$$
\begin{array}{l}
\mathrm{H_2COH} \\
| \\
\mathrm{HCOH} \\
| \\
\mathrm{H_2COH}
\end{array}
\xrightarrow[\text{ATP ADP}]{\text{glycerol kinase}}
\begin{array}{l}
\mathrm{H_2COH} \\
| \\
\mathrm{HCOH} \\
| \\
\mathrm{H_2CO-\textcircled{P}}
\end{array}
$$

glycerol   glycerol 3-phosphate

and oxidation to dihydroxyacetone phosphate,

(27-3)

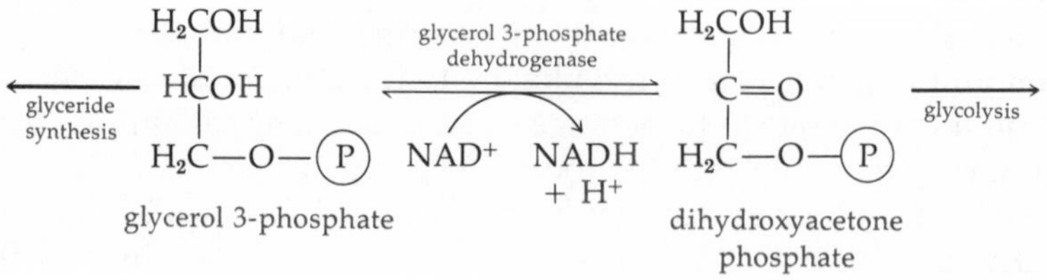

$$
\xrightleftharpoons[\substack{\text{glyceride}\\\text{synthesis}}]{}
\begin{array}{l}
\mathrm{H_2COH} \\
| \\
\mathrm{HCOH} \\
| \\
\mathrm{H_2C-O-\textcircled{P}}
\end{array}
\underset{\mathrm{NAD^+ \quad NADH \atop + \, H^+}}{\overset{\text{glycerol 3-phosphate} \atop \text{dehydrogenase}}{\rightleftharpoons}}
\begin{array}{l}
\mathrm{H_2COH} \\
| \\
\mathrm{C=O} \\
| \\
\mathrm{H_2C-O-\textcircled{P}}
\end{array}
\xrightarrow{\text{glycolysis}}
$$

glycerol 3-phosphate   dihydroxyacetone
phosphate

The reverse of reaction 27-3 is used to make glycerol 3-phosphate from dihydroxyacetone phosphate for use in lipid synthesis.

Fatty Acid Catabolism

The majority of fatty acids contain an even number of carbon atoms, and the pathway for their degradation involves their cleavage into two-carbon-atom fragments. This pathway is called the **β-oxidation pathway,** and the two-carbon-atom fragments produced are coupled to coenzyme A molecules.

The first step in the catabolism of fatty acids is called **activation.** In this step, the esterification of free fatty acids to coenzyme A molecules is catalyzed by an enzyme called acyl-S-CoA synthetase, thus producing acyl-S-CoA molecules—activated fatty acids. Recall from Chapter 17 that an acyl group is derived from a carboxylic acid by removing the —OH group.

Step 1:

(27-4)

$$R-CH_2-CH_2-\overset{\overset{\displaystyle O}{\|}}{C}-OH + CoA-SH \underset{\substack{\text{acyl-S-CoA synthetase}}}{\overset{}{\rightleftharpoons}} R-CH_2-CH_2-\overset{\overset{\displaystyle O}{\|}}{C}-S-CoA$$

a fatty acid   ATP   AMP + PP$_i$ an acyl-S-CoA

This activation step takes place in the cytoplasm of cells, and requires energy in the form of hydrolysis of 1 mole of ATP to AMP per mole of fatty acid. An additional mole of ATP is required to convert AMP to ADP:

(27-5) $$AMP + ATP \longrightarrow 2ADP$$

Thus, this first step requires an energy input equivalent to the hydrolysis of 2 moles of ATP to ADP per mole of fatty acid.

Further catabolism of fatty acids occurs inside the cell's mitochondria, and produces energy. Since an activated fatty acid does not easily permeate the mitochondrial membrane, it is bonded to a molecule of **carnitine** for transport across the membrane (see Figure 27-3). Once inside the mitochondria, an enzyme breaks the acyl carnitine molecule apart, producing an activated fatty acid and carnitine. The carnitine can then return to the cytoplasm for another fatty acid. Stepwise degradation of the activated fatty acid can now begin.

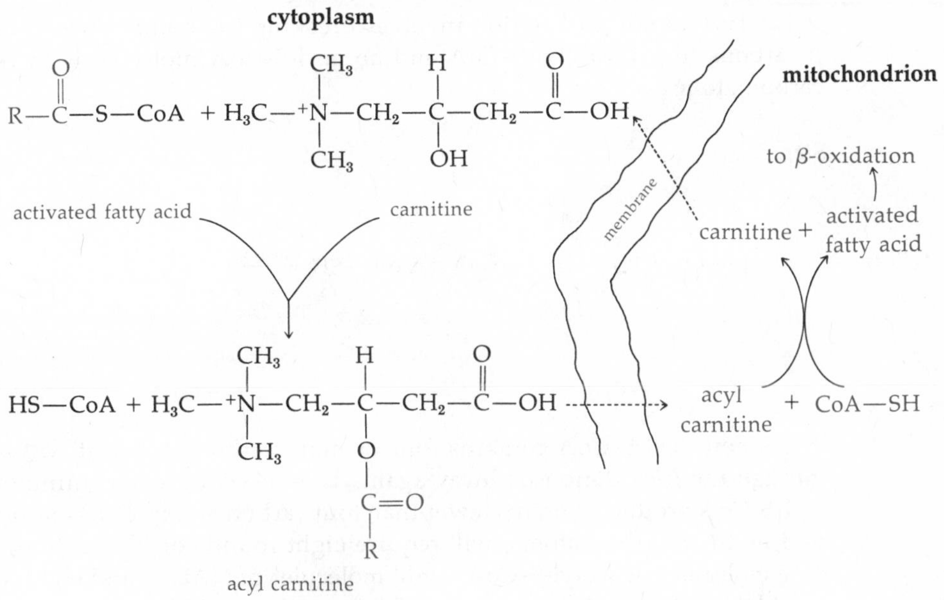

Figure 27-3   Carnitine serves as a carrier for the transport of activated fatty acids from the cytoplasm into mitochondria.

The next step of the β-oxidation pathway involves the oxidation of the single bond between the α and β carbons of the activated fatty acid:

Step 2:

(27-6)

$$R-\overset{\overset{\displaystyle H}{|}}{\underset{\underset{\displaystyle \beta}{}}{C}H}-\overset{\overset{\displaystyle H}{|}}{\underset{\underset{\displaystyle \alpha}{}}{C}H}-\overset{\overset{\displaystyle O}{\|}}{C}-S-CoA \underset{\substack{\text{FAD} \quad \text{FAD}H_2}}{\overset{\text{dehydrogenase}}{\longrightarrow}} R-\overset{\overset{\displaystyle H}{|}}{C}=\overset{\underset{\displaystyle H}{|}}{C}-\overset{\overset{\displaystyle O}{\|}}{C}-S-CoA$$

Notice that the oxidation of the saturated fatty acid is coupled to the reduction of an FAD molecule. The FADH$_2$ produced is used to generate ATP in the ETS.

In the third enzyme-catalyzed step of β-oxidation, a molecule of water is added to the double bond:

Step 3:

(27-7)

$$H_2O + R-\underset{\underset{H}{|}}{\overset{\overset{H}{|}}{C}}=C-\overset{\overset{O}{||}}{C}-S-CoA \xrightarrow{\text{hydratase}} R-\underset{\underset{H}{|}}{\overset{\overset{OH}{|}}{C}}-\underset{\underset{H}{|}}{\overset{\overset{H}{|}}{C}}-\overset{\overset{O}{||}}{C}-S-CoA$$

In this step, the hydroxyl group is always added to the β-carbon.

In the fourth step of this pathway, the alcohol group involving the β-carbon is oxidized to form a ketone, hence the name β-oxidation for this pathway. This step is coupled to the reduction of a NAD$^+$ coenzyme molecule.

Step 4:

(27-8)

$$R-\underset{\underset{H}{|}}{\overset{\overset{OH}{|}}{\underset{\beta}{C}}}-\underset{\alpha}{CH_2}-\overset{\overset{O}{||}}{C}-S-CoA \xrightarrow[\text{NAD}^+\ \text{NADH} \atop + \text{H}^+]{\text{dehydrogenase}} R-\overset{\overset{O}{||}}{C}-CH_2-\overset{\overset{O}{||}}{C}-S-CoA$$

The NADH produced is also used to generate ATP in the ETS.

The last step of β-oxidation involves breaking the bond between the α- and β-carbons to give acetyl-S-CoA and an acyl-S-CoA molecule with two fewer carbon atoms:

Step 5:

(27-9)

$$R-\underset{\beta}{\overset{\overset{O}{||}}{C}}-\underset{\alpha}{CH_2}-\overset{\overset{O}{||}}{C}-S-CoA + CoA-SH \xrightarrow{\text{thiolase}}$$

$$\underset{\text{an acyl-S-CoA}}{R-\overset{\overset{O}{||}}{C}-S-CoA} + \underset{\text{acetyl-S-CoA}}{CH_3-\overset{\overset{O}{||}}{C}-S-CoA}$$

If this new acyl-S-CoA contains four or more carbon atoms, it will be cycled through the β-oxidation pathway again. This recycling will continue until the acyl-S-CoA product contains fewer than four carbon atoms. For example, stearic acid, with 18 carbon atoms, will require eight rounds of β-oxidation, yielding nine molecules of acetyl-S-CoA, eight molecules of NADH, and eight molecules of FADH$_2$. This is analogous to breaking a string of 18 hot dogs eight times in order to get nine strings of two hot dogs each (see Figure 27-4).

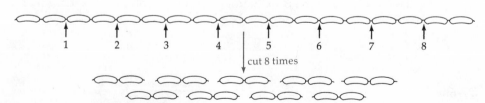

Figure 27-4   A string of 18 hot dogs must be broken eight times to get nine strings of two hot dogs each.

Therefore, for a fatty acid with 18 carbon atoms, the overall reaction for the β-oxidation of the initial activated fatty acid within the mitochondrion is

$$CH_3(CH_2)_{16}-\overset{\overset{\displaystyle O}{\|}}{C}-S-CoA + 8CoA-SH \longrightarrow 9CH_3-\overset{\overset{\displaystyle O}{\|}}{C}-S-CoA$$
$$+ 8H_2O + 8FAD + 8NAD^+ \qquad + 8FADH_2 + 8NADH + 8H^+$$

The molecules of acetyl-S-CoA produced by the β-oxidation of fatty acids can be further degraded by the TCA cycle (see Chapter 25). Remember that for each molecule of acetyl-S-CoA that enters the TCA cycle, 12 ATP molecules are formed. Thus, starting with the 18-carbon-atom stearic acid, eight rounds of β-oxidation produce nine molecules of acetyl-S-CoA. The oxidation of these acetyl-S-CoA molecules via the TCA cycle and the ETS results in the formation of 108 ATP molecules:

$$(9 \text{ acetyl-S-CoA}) \times \left(\frac{12 \text{ ATP}}{\text{acetyl-S-CoA}}\right) = 108 \text{ ATP}$$

In addition, eight rounds of β-oxidation produce eight molecules of $FADH_2$ and eight molecules of NADH. When the ETS is used to reoxidize these coenzymes, another 40 ATP molecules are formed:

$$\left(8 \text{ NADH} \times \frac{3 \text{ ATP}}{\text{NADH}}\right) + \left(8 \text{ FADH}_2 \times \frac{2 \text{ ATP}}{\text{FADH}_2}\right) = 40 \text{ ATP}$$

Therefore, the complete oxidation of an 18-carbon acyl-S-CoA molecule yields $40 + 108 = 148$ ATP molecules. Of course, the *net* ATP production from the complete oxidation of an 18-carbon-atom fatty acid is 146, since two ATP molecules are hydrolyzed to ADP in the activation step (reactions 27-4 and 27-5). A similar set of calculations are shown in Figure 27-5 for the 16-carbon fatty acid, palmitic acid.

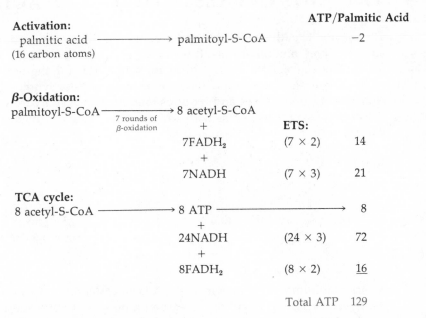

Figure 27-5   The complete oxidation of 1 mole of palmitic acid generates 129 moles of ATP via the β-oxidation pathway.

Complete oxidation of palmitic acid to CO_2 and H_2O involves β-oxidation, the TCA cycle, and the ETS, with the overall reaction

(27-10)
$$CH_3(CH_2)_{14}\overset{\overset{\displaystyle O}{\|}}{C}-OH + 23O_2 \longrightarrow 16CO_2 + 16H_2O + energy$$

How efficient is this process? The experimentally measured ΔG value for reaction 27-10 is -2340 kcal/mole. This is the maximum possible amount of usable energy that can be obtained from this reaction.

Now, in Figure 27-5 we calculated that 129 moles of ATP are produced by the complete oxidation of 1 mole of palmitic acid in the human body. Recall that for the hydrolysis of ATP, $\Delta G = -8$ kcal/mole (Chapter 25). The ΔG for the hydrolysis of the 129 moles of ATP produced by the oxidation of 1 mole of palmitic acid is thus -1032 kcal. Thus this biological oxidation of fatty acids is about 44% efficient, $(-1032 \text{ kcal}/-2340 \text{ kcal}) \times 100\% = 44\%$. This is comparable to the efficiency we determined for the oxidation of glucose in Chapter 25.

Note that in the formation of 129 moles of ATP from ADP + P_i, 129 moles of H_2O are produced in addition to the 16 moles of H_2O from reaction 27-10. Thus, the complete oxidation of 1 mole of palmitic acid, coupled to the formation of 129 moles of ATP, produces 145 moles of H_2O. This is a lot of water. For this reason a camel can store a large supply of water *and* energy in its fat hump.

The catabolism of unsaturated fatty acids and those fatty acids that contain an odd number of carbons require additional steps, which we need not consider.

Exercise 27-1

Compare the amount of ATP that can be formed by the complete oxidation of a six-carbon fatty acid to CO_2 and H_2O with the ATP formed by complete oxidation of a glucose molecule to CO_2 to H_2O.

27-5   THE BIOSYNTHESIS OF FATTY ACIDS

The metabolic pathway used to assemble fatty acids occurs in the cytoplasm of adipose, liver, and other cells. It is not simply the reverse of the pathway used for catabolism of fatty acids within mitochondria, for the same reason that gluconeogenesis is not simply the reverse of glycolysis. Synthesis of fatty acids is an endergonic process and requires energy from ATP hydrolysis.

Fatty acid synthesis begins with acetyl-S-CoA, most of which is first carboxylated to form malonyl-S-CoA:

Step 1:

(27-11)
$$H_3C-\overset{\overset{\displaystyle O}{\|}}{C}-S-CoA + CO_2 \xrightarrow[\substack{ATP \quad ADP \\ + H_2O \quad + P_i}]{\substack{acetyl\text{-}S\text{-}CoA \\ carboxylase \\ + biotin \\ (a\ coenzyme)}} HO-\overset{\overset{\displaystyle O}{\|}}{C}-CH_2-\overset{\overset{\displaystyle O}{\|}}{C}-S-CoA$$

acetyl-S-CoA   malonyl-S-CoA

Notice that formation of malonyl-S-CoA is an endergonic reaction and requires the hydrolysis of ATP. All of the carbon atoms in synthesized fatty acids are obtained from acetyl-S-CoA molecules. However, we shall see that only two carbon atoms are directly supplied by acetyl-S-CoA. The rest of the carbon atoms are supplied by malonyl-S-CoA molecules made from acetyl-S-CoA.

The remaining steps of fatty acid biosynthesis in humans take place on a very complex enzyme, called **fatty acid synthetase.** Many of the chemical reactions that are catalyzed by this enzyme are similar to the reverse of those used for fatty acid catabolism. Fatty acid synthetase has two prosthetic groups, called **acyl carrier portions** (ACP), which are identical to a portion of a coenzyme A molecule (see Figure 27-6).

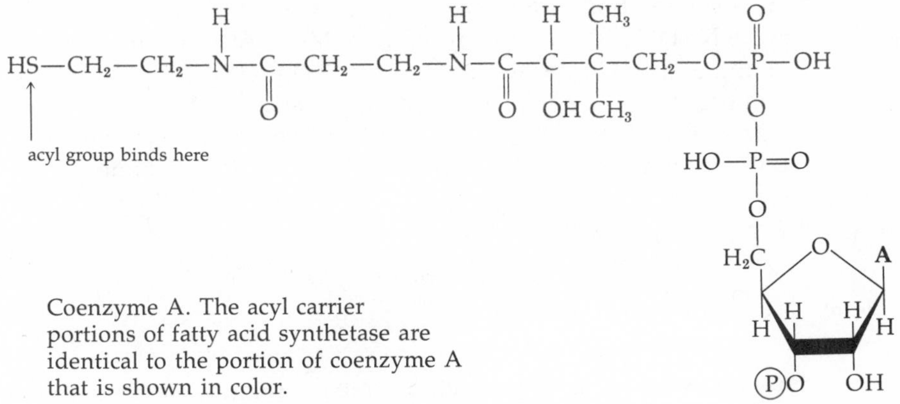

Figure 27-6   Coenzyme A. The acyl carrier portions of fatty acid synthetase are identical to the portion of coenzyme A that is shown in color.

Actual assembly of a fatty acid requires the covalent attachment of one acetyl group and one malonyl group to the ACP portions of fatty acid synthetase, as represented in the following reactions:

(27-12)

$$\underset{\text{acetyl-S-CoA}}{CoA-S-\overset{\overset{\displaystyle O}{\|}}{C}-CH_3} + \underset{\substack{\text{acyl carrier} \\ \text{portion}}}{ACP-SH} \rightleftharpoons \underset{\text{acetyl-S-ACP}}{ACP-S-\overset{\overset{\displaystyle O}{\|}}{C}-CH_3} + CoA-SH$$

(27-13)

$$\underset{\text{malonyl-S-CoA}}{CoA-S-\overset{\overset{\displaystyle O}{\|}}{C}-CH_2-\overset{\overset{\displaystyle O}{\|}}{C}-OH} + ACP-SH \rightleftharpoons \underset{\text{malonyl-S-ACP}}{ACP-S-\overset{\overset{\displaystyle O}{\|}}{C}-CH_2-\overset{\overset{\displaystyle O}{\|}}{C}-OH}$$

$$+ \ CoA-SH$$

The next step in the assembly of a fatty acid involves the transfer of the acetyl group from its ACP anchor to the malonyl group anchored to the other ACP portion of fatty acid synthetase. In this process, the recently added carboxyl group on malonyl-S-ACP is removed:

Step 2:

(27-14)

$$\underset{\text{malonyl-S-ACP}}{ACP-S-\overset{\overset{\displaystyle O}{\|}}{C}-CH_2-\overset{\overset{\displaystyle O}{\|}}{C}-OH} + \underset{\text{acetyl-S-ACP}}{ACP-S-\overset{\overset{\displaystyle O}{\|}}{C}-CH_3} \xrightarrow{\overset{\text{fatty acid}}{\text{synthetase}}}$$

$$\underset{\text{acetoacetyl-S-ACP}}{ACP-S-\overset{\overset{\displaystyle O}{\|}}{C}-CH_2-\overset{\overset{\displaystyle O}{\|}}{C}-CH_3} + CO_2 + ACP-SH$$

The product of reaction 27-14, acetoacetyl-S-ACP, remains anchored to fatty acid synthetase, which now catalyzes the reduction of the ketone functional group to an alcohol. Note that the carbonyl group bonded to the sulfur atom is not a ketone, but rather is part of a thioester.

Step 3:

(27-15)
$$
\underset{\text{acetoacetyl-S-ACP}}{\text{ACP}-\text{S}-\overset{\text{O}}{\overset{\|}{\text{C}}}-\text{CH}_2-\overset{\text{O}}{\overset{\|}{\text{C}}}-\text{CH}_3} \xrightarrow[\substack{\text{NADPH} \quad \text{NADP}^+ \\ + \text{H}^+}]{\substack{\text{fatty acid} \\ \text{synthetase}}} \text{ACP}-\text{S}-\overset{\text{O}}{\overset{\|}{\text{C}}}-\text{CH}_2-\overset{\overset{\text{OH}}{|}}{\underset{\overset{|}{\text{H}}}{\text{C}}}-\text{CH}_3
$$

Notice that this step is somewhat similar to the reverse of step 4 of fatty acid catabolism (reaction 27-8). However, a different enzyme is required and the coenzyme NADPH is oxidized in reaction 27-15. Also, in this case the substrate is bound to an ACP portion of fatty acid synthetase (not to CoA), and this reaction occurs in the cytoplasm (not in mitochondria).

   In step 4 of fatty acid synthesis, the hydroxyl group is removed in a dehydration reaction, forming an α, β-carbon-carbon double bond:

Step 4:

(27-16)
$$
\text{ACP}-\text{S}-\overset{\text{O}}{\overset{\|}{\text{C}}}-\overset{\overset{\text{H}}{|}}{\text{CH}}-\overset{\overset{\text{OH}}{|}}{\text{CH}}-\text{CH}_3 \xrightarrow{\substack{\text{fatty acid} \\ \text{synthetase}}} \text{ACP}-\text{S}-\overset{\text{O}}{\overset{\|}{\text{C}}}-\underset{\alpha}{\text{CH}}=\underset{\beta}{\text{CH}}-\text{CH}_3 + \text{H}_2\text{O}
$$

Again, this reaction is somewhat similar to the reverse of step 3 of fatty acid catabolism (reaction 27-7).

   In step 5 of fatty acid synthesis, the C=C double bond is reduced:

Step 5:

(27-17)
$$
\text{ACP}-\text{S}-\overset{\text{O}}{\overset{\|}{\text{C}}}-\text{CH}=\text{CH}-\text{CH}_3 \xrightarrow[\substack{\text{NADPH} \quad \text{NADP}^+ \\ + \text{H}^+}]{\substack{\text{fatty acid} \\ \text{synthetase}}} \text{ACP}-\text{S}-\overset{\text{O}}{\overset{\|}{\text{C}}}-\overset{\overset{\text{H}}{|}}{\text{CH}}-\overset{\overset{\text{H}}{|}}{\text{CH}}-\text{CH}_3
$$

Again, as in step 3, this reduction step is coupled to the oxidation of a molecule of NADPH. Step 5 completes one round of action of the fatty acid synthetase enzyme. Butyric acid, $\text{CH}_3-\text{CH}_2-\text{CH}_2-\text{COOH}$, a four-carbon carboxylic acid, could be obtained by the hydrolysis of this thioester. However, long-chain fatty acids are needed by the human body, so the product of step 5 is used for further rounds of fatty acid synthetase action. Each subsequent round begins with the binding of a malonyl group to the free ACP portion of the fatty acid synthetase (reaction 27-13). That is, additional two-carbon-atom pieces are added from additional malonyl-S-ACP molecules. The synthesis of a 16-carbon fatty acid thus requires seven rounds of fatty acid synthetase action. This process is summarized in Figure 27-7. The acyl group produced by seven rounds of fatty acid synthetase action (the palmitoyl group) is then transferred to a molecule of coenzyme A in the reaction

(27-18)
$$
\underset{\text{palmitoyl-S-ACP}}{\text{ACP}-\text{S}-\overset{\text{O}}{\overset{\|}{\text{C}}}-(\text{CH}_2)_{14}-\text{CH}_3} + \text{CoA}-\text{SH} \rightleftharpoons \underset{\text{palmitoyl-S-CoA}}{\text{CoA}-\text{S}-\overset{\text{O}}{\overset{\|}{\text{C}}}-(\text{CH}_2)_{14}-\text{CH}_3} + \text{ACP}-\text{SH}
$$

The hydrolysis of this palmitoyl-S-CoA thioester yields palmitic acid, the usual product of fatty acid synthetase action. The net overall reaction for the synthesis of palmitic acid from acetyl-S-CoA is

(27-19)
$$
8\text{CH}_3-\overset{\text{O}}{\overset{\|}{\text{C}}}-\text{S}-\text{CoA} + 7\text{ATP} + \text{H}_2\text{O} \longrightarrow \text{CH}_3(\text{CH}_2)_{14}\overset{\text{O}}{\overset{\|}{\text{C}}}-\text{OH} + 8\text{CoA}-\text{SH}
$$
$$
+ 14\text{NADPH} + 14\text{H}^+ \qquad\qquad\qquad\qquad + 7\text{ADP} + 7\text{P}_i + 14\text{NADP}^+
$$

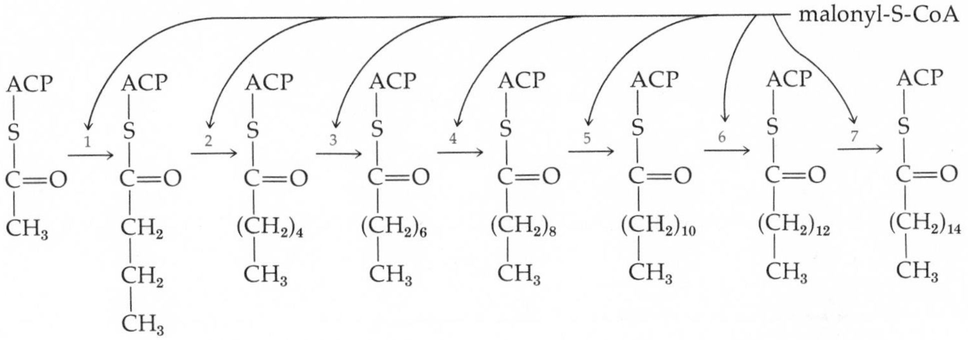

Figure 27-7   Synthesis of the 16-carbon-atom palmitic acid requires seven rounds of fatty acid synthesis. In each round, malonyl-S-CoA donates two carbon atoms.

Each round of the fatty acid synthesis pathway is fed by malonyl-S-CoA, which is immediately decarboxylated. Thus, only two carbon atoms are added per round, and you can now see why the majority of fatty acids found in nature have even-numbered chains of carbon atoms. The synthesis of unsaturated fatty acids begins with the synthesis of the corresponding saturated fatty acids. Double bonds are then introduced by a specific enzyme.

Conversion of Glucose into Lipids

Now let us consider why the human body converts glucose into lipids when the supply of glucose is in excess of the immediate need for energy. Take palmitic acid ($C_{16}H_{32}O_2$) as an example. To synthesize palmitic acid, eight acetyl-S-CoA molecules are needed. In glycolysis, one glucose molecule yields two molecules of acetyl-S-CoA (Chapter 25). Therefore, four glucose molecules are needed to make one molecule of palmitic acid. The overall reaction for this process is

(27-20) $$4C_6H_{12}O_6 + O_2 \longrightarrow C_{16}H_{32}O_2 + 8CO_2 + 8H_2O$$

Synthesis of palmitic acid from acetyl-S-CoA requires ATP hydrolysis, but palmitic acid is a rich potential source for ATP synthesis. Is the process of converting glucose molecules into molecules of palmitic acid worthwhile? In liver cells, complete catabolism of glucose to CO_2 and H_2O yields 38 ATP/glucose, so complete catabolism of four glucose molecules yields $4 \times 38 = 152$ ATP. Complete catabolism of palmitic acid yields 129 ATP, so the potential synthesis of 23 ATP molecules ($152 - 129$) is lost in converting glucose to palmitic acid. However, palmitic acid is a much more compact storage form for energy. One mole of palmitic acid weighs 256 g, whereas the molecular weight of glucose is 180. The potential supply of ATP per gram for each of these molecules is

(27-21) $$\frac{38 \text{ moles of ATP}}{1 \text{ mole of glucose}} \times \frac{1 \text{ mole of glucose}}{180 \text{ g}} = 0.21 \text{ mole of ATP/g of glucose}$$

(27-22) $$\frac{129 \text{ moles of ATP}}{1 \text{ mole of palmitic acid}} \times \frac{1 \text{ mole of palmitic acid}}{256 \text{ g}}$$
$$= 0.50 \text{ mole of ATP/g of palmitic acid}$$

More than twice as much ATP can be produced from the complete catabolism of 1 g of palmitic acid as from 1 g of glucose. If you used only glucose (or glycogen) to store energy, you would need to weigh a good deal more. And if you weighed more, you would need to use more energy to move around.

Exercise 27-2

How many molecules of glucose are required for the synthesis of one molecule of the 20-carbon-atom fatty acid, arachidic acid?

Exercise 27-3

(a) How many molecules of ATP are synthesized from the complete oxidation of arachidic acid? (b) Compare this amount to the amount of ATP produced by the complete oxidation in the liver of the number of glucose molecules you arrived at in Exercise 27-2.

27-6   THE BIOSYNTHESIS OF TRIGLYCERIDES AND PHOSPHOGLYCERIDES

In this section we shall see how newly synthesized fatty acids are used to form two major classes of lipid molecules, triglycerides and phosphoglycerides. The synthesis of both of these classes of lipids requires glycerol 3-phosphate in addition to fatty acyl-S-CoA molecules. Glycerol 3-phosphate can be formed from dihydroxyacetone phosphate by the reaction

(27-23)

$$
\begin{array}{ccc}
\text{H}_2\text{C}-\text{OH} & & \text{H}_2\text{COH} \\
| & \xrightleftharpoons[\text{dehydrogenase}]{\text{glycerol phosphate}} & | \\
\text{C}=\text{O} & & \text{HCOH} \\
| & & | \\
\text{H}_2\text{CO}-\textcircled{P} & \text{NADH}\quad \text{NAD}^+ & \text{H}_2\text{CO}-\textcircled{P} \\
& + \text{H}^+ & \\
\text{dihydroxyacetone} & & \text{glycerol} \\
\text{phosphate} & & \text{3-phosphate}
\end{array}
$$

Reaction 27-23 may look familiar, since it is exactly the reverse of reaction 27-3. Glycerol phosphate dehydrogenase is used for both the synthesis and catabolism of glycerol 3-phosphate.

The next step is the formation of a phosphatidic acid molecule by the esterification of two fatty acids to the alcohol groups on the glycerol 3-phosphate. This reaction is catalyzed by a transferase enzyme:

(27-24)

$$
\begin{array}{ll}
\text{H}_2\text{COH} & \quad\quad\quad\quad \overset{\text{O}}{\overset{\|}{\text{CoA}-\text{S}-\text{C}-\text{R}_1}} \\
| & \quad\quad\quad\quad \overset{\text{O}}{\overset{\|}{}} \\
\text{HCOH} & + \text{CoA}-\text{S}-\text{C}-\text{R}_2 \longrightarrow \\
| & \\
\text{H}_2\text{CO}-\textcircled{P} & \quad\quad \text{two fatty} \\
& \quad\quad \text{acyl-S-CoA's} \\
\text{glycerol} & \\
\text{3-phosphate} &
\end{array}
$$

$$
2\text{CoA}-\text{SH} +
\begin{array}{l}
\overset{\text{O}}{\overset{\|}{\text{H}_2\text{C}-\text{O}-\text{C}-\text{R}_1}} \\
\overset{\text{O}}{\overset{\|}{}} \\
\text{HC}-\text{O}-\text{C}-\text{R}_2 \\
| \\
\text{H}_2\text{C}-\text{O}-\textcircled{P} \\
\text{a phosphatidic acid}
\end{array}
$$

The formation of a triglyceride proceeds by removal of the phosphate group from a phosphatidic acid molecule,

(27-25)

$$
\text{H}_2\text{O} +
\begin{array}{l}
\overset{\text{O}}{\overset{\|}{\text{H}_2\text{C}-\text{O}-\text{C}-\text{R}_1}} \\
\overset{\text{O}}{\overset{\|}{}} \\
\text{HC}-\text{O}-\text{C}-\text{R}_2 \\
| \\
\text{H}_2\text{C}-\text{O}-\textcircled{P} \\
\text{a phosphatidic acid}
\end{array}
\longrightarrow
\begin{array}{l}
\overset{\text{O}}{\overset{\|}{\text{H}_2\text{C}-\text{O}-\text{C}-\text{R}_1}} \\
\overset{\text{O}}{\overset{\|}{}} \\
\text{HC}-\text{O}-\text{C}-\text{R}_2 + \text{P}_i \\
| \\
\text{H}_2\text{C}-\text{OH} \\
\text{a diglyceride}
\end{array}
$$

followed by the formation of a third fatty acid ester:

(27-26)

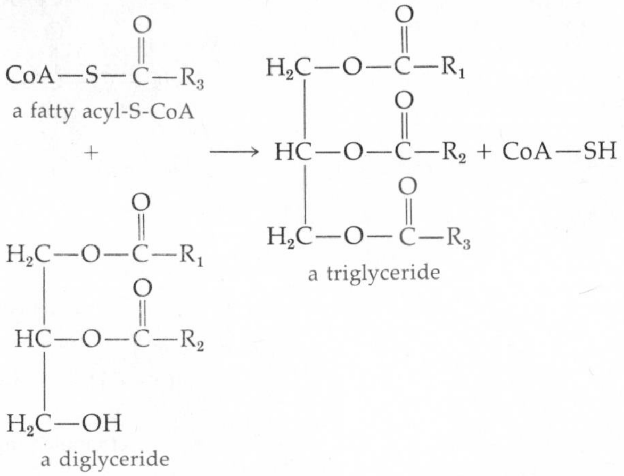

$$CoA-S-\overset{\overset{\displaystyle O}{\|}}{C}-R_3$$
a fatty acyl-S-CoA

+

$$H_2C-O-\overset{\overset{\displaystyle O}{\|}}{C}-R_1$$
$$HC-O-\overset{\overset{\displaystyle O}{\|}}{C}-R_2$$
$$H_2C-OH$$
a diglyceride

$\longrightarrow$

$$H_2C-O-\overset{\overset{\displaystyle O}{\|}}{C}-R_1$$
$$HC-O-\overset{\overset{\displaystyle O}{\|}}{C}-R_2 + CoA-SH$$
$$H_2C-O-\overset{\overset{\displaystyle O}{\|}}{C}-R_3$$
a triglyceride

In the synthesis of phosphoglycerides, human cells also use reaction 27-25 to produce diglyceride intermediates. We shall consider the synthesis of two common phosphoglycerides: phosphatidyl ethanolamine and phosphatidyl serine. The synthesis of phosphatidyl enthanolamine begins with the phosphorylation of ethanolamine to give phosphoethanolamine:

(27-27)

$$H_2N-CH_2-CH_2-OH \xrightarrow[\text{ATP \quad ADP}]{\text{ethanolamine kinase}} H_2N-CH_2-CH_2-O-\textcircled{P}$$
ethanolamine · phosphoethanolamine

Phosphoethanolamine is next activated by coupling it to a molecule of CMP in the reaction

(27-28)

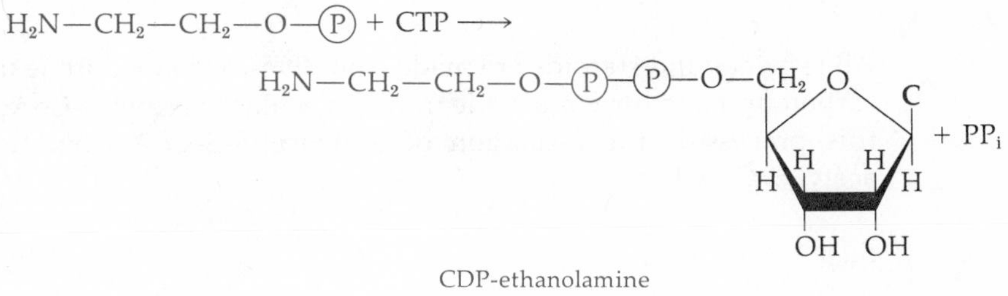

$$H_2N-CH_2-CH_2-O-\textcircled{P} + CTP \longrightarrow$$

$$H_2N-CH_2-CH_2-O-\textcircled{P}-\textcircled{P}-O-CH_2$$

(furanose ring with OH OH, H H H H, and C) $+ PP_i$

CDP-ethanolamine

CTP is similar in structure to ATP, except that it contains the base cytosine instead of adenine (Chapter 24).

Recall from Chapter 25 that the synthesis of polysaccharides also proceeds via similar nucleotide-activated intermediates. Phosphatidyl ethanolamine is then made in the following reaction:

(27-29)

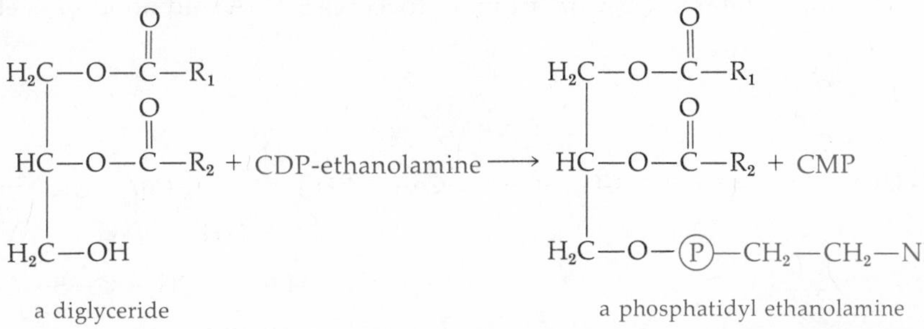

$$H_2C-O-\overset{\overset{\displaystyle O}{\|}}{C}-R_1$$
$$HC-O-\overset{\overset{\displaystyle O}{\|}}{C}-R_2 + CDP\text{-ethanolamine} \longrightarrow$$
$$H_2C-OH$$
a diglyceride

$$H_2C-O-\overset{\overset{\displaystyle O}{\|}}{C}-R_1$$
$$HC-O-\overset{\overset{\displaystyle O}{\|}}{C}-R_2 + CMP$$
$$H_2C-O-\textcircled{P}-CH_2-CH_2-NH_2$$
a phosphatidyl ethanolamine

Synthesis of phosphatidyl serine involves the exchange of the ethanolamine portion of a phosphatidyl ethanolamine molecule for the amino acid serine:

(27-30)

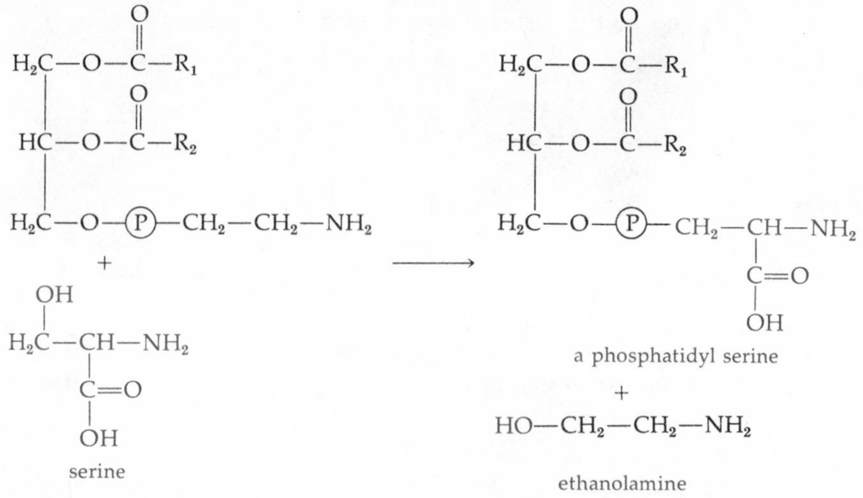

$$H_2C-O-\overset{\overset{O}{\|}}{C}-R_1$$
$$HC-O-\overset{\overset{O}{\|}}{C}-R_2$$
$$H_2C-O-\underset{}{\textcircled{P}}-CH_2-CH_2-NH_2$$

+

$$\overset{OH}{\underset{}{|}}$$
$$H_2C-CH-NH_2$$
$$\overset{}{\underset{}{|}}$$
$$C=O$$
$$\overset{}{\underset{}{|}}$$
$$OH$$

serine

$\longrightarrow$

$$H_2C-O-\overset{\overset{O}{\|}}{C}-R_1$$
$$HC-O-\overset{\overset{O}{\|}}{C}-R_2$$
$$H_2C-O-\underset{}{\textcircled{P}}-CH_2-CH-NH_2$$
$$\overset{}{\underset{}{|}}$$
$$C=O$$
$$\overset{}{\underset{}{|}}$$
$$OH$$

a phosphatidyl serine

+

$$HO-CH_2-CH_2-NH_2$$

ethanolamine

Exercise 27-4

What type of molecule is the last common intermediate in the synthesis of triglycerides and phosphoglycerides?

Exercise 27-5

What enzyme-catalyzed step in the synthesis of triglycerides is also used in their degradation?

27-7 BIOSYNTHESIS OF TERPENES AND STEROIDS

All terpenes and steroids are made from five-carbon isoprene units. These five-carbon units, in turn, are made from molecules of acetyl-S-CoA. The first step in this process is the formation of acetoacetyl-S-CoA from two molecules of acetyl-S-CoA:

Step 1:

(27-31)

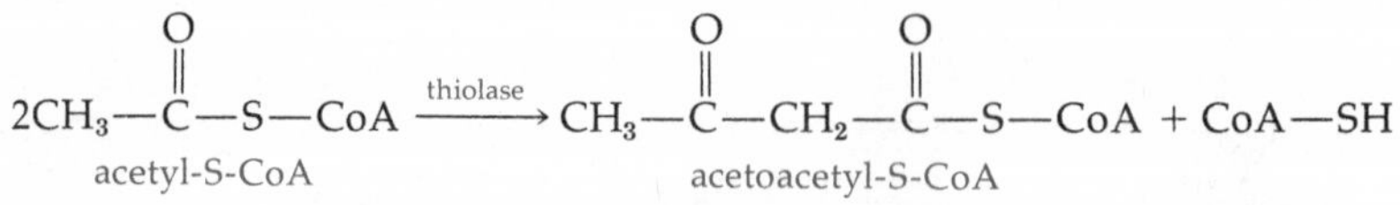

$$2CH_3-\overset{\overset{O}{\|}}{C}-S-CoA \xrightarrow{\text{thiolase}} CH_3-\overset{\overset{O}{\|}}{C}-CH_2-\overset{\overset{O}{\|}}{C}-S-CoA + CoA-SH$$

acetyl-S-CoA   acetoacetyl-S-CoA

Acetoacetyl-S-CoA is sometimes produced in excessively large amounts by people with diabetes, for reasons we shall discuss shortly. The next step in steroid biosynthesis is the formation of a six-carbon-atom unit called β-hydroxy-β-methylglutaryl-S-CoA from acetoacetyl-S-CoA and another acetyl-S-CoA:

Step 2:

(27-32)

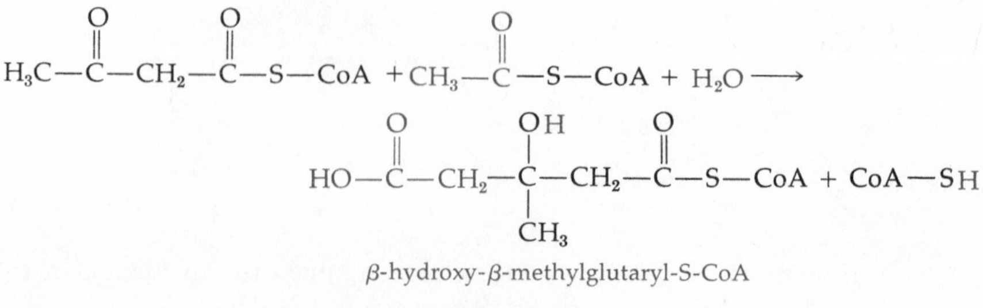

$$H_3C-\overset{\overset{O}{\|}}{C}-CH_2-\overset{\overset{O}{\|}}{C}-S-CoA + CH_3-\overset{\overset{O}{\|}}{C}-S-CoA + H_2O \longrightarrow$$

$$HO-\overset{\overset{O}{\|}}{C}-CH_2-\overset{\overset{OH}{|}}{\underset{\underset{CH_3}{|}}{C}}-CH_2-\overset{\overset{O}{\|}}{C}-S-CoA + CoA-SH$$

β-hydroxy-β-methylglutaryl-S-CoA

In five subsequent reactions of this pathway, β-hydroxy-β-methylglutaryl-S-CoA is converted to 3,3-dimethylallyl pyrophosphate, an isoprene derivative. The overall reaction for these five steps is

(27-33)

$$HO-\underset{\overset{\|}{O}}{C}-CH_2-\underset{\overset{|}{CH_3}}{\overset{|}{C}}-CH_2-\underset{\overset{\|}{O}}{C}-S-CoA \longrightarrow CH_3-\underset{\overset{|}{CH_3}}{C}=CH-CH_2-\text{\textcircled{P}}-\text{\textcircled{P}}$$

$$+ \text{ 3ATP} + \text{2NADPH} + \text{2H}^+ \qquad\qquad + \text{3ADP} + \text{P}_i + \text{2NADP}^+$$
$$+ \text{ CoA}-\text{SH} + CO_2 + 2H_2O$$

Note that a total of three acetyl-S-CoA molecules are required for the synthesis of one isoprene unit (shown in blue). Subsequent steps in this pathway involve joining isoprene units together to give terpenes with 10 carbon atoms, 15 carbon atoms, and so on. A 30-carbon-atom terpene called squalene is used as the precursor for cholesterol biosynthesis. In humans all other steroids are made from cholesterol.

Exercise 27-6
How many acetyl-S-CoA molecules are needed to synthesize one molecule of cholesterol?

27-8 THE REGULATION OF LIPID METABOLISM

Lipid molecules are all made from, and degraded to, acetyl-S-CoA molecules, which is also the fuel used in the TCA cycle (see Chapter 25). However, acetyl-S-CoA is not used to make fatty acids and other lipids unless there is an adequate supply to fuel the TCA cycle. In fact, when acetyl-S-CoA is needed, fats are broken down to supply it. Conversely, when there is more than enough acetyl-S-CoA present in a cell, it is used to make triglycerides for the fat depots. How are the synthesis and degradation of lipid molecules controlled in cells? There are several control mechanisms, each of which is very logical.

First, the synthesis of fatty acids depends upon the availability of NADPH, which is needed for the reduction steps in this biosynthetic pathway. NADPH can be generated by the pentose phosphate pathway (Chapter 25). Another pathway is also used to generate NADPH for fatty acid biosynthesis. In one of the reactions in this pathway, malic acid is oxidized to pyruvic acid, while NADP$^+$ is reduced to form NADPH + H$^+$:

(27-34)

$$HO-\underset{\overset{\|}{O}}{C}-CH_2-\underset{\overset{|}{OH}}{\overset{|}{\overset{H}{C}}}-\underset{\overset{\|}{O}}{C}-OH \xrightarrow[\underset{NADP^+ \quad NADPH}{\;}]{malic\ enzyme} H_3C-\underset{\overset{\|}{O}}{C}-\underset{\overset{\|}{O}}{C}-OH + CO_2$$
$$\underset{\text{malic acid}}{} \qquad\qquad + \text{H}^+ \qquad \underset{\text{pyruvic acid}}{}$$

When the cell's supply of acetyl-S-CoA exceeds that which is needed to run the TCA cycle, the excess acetyl-S-CoA activates malic enzyme to produce NADPH by reaction 27-34.

Other controls of fatty acid synthesis take the cell's energy supply into account. If the cell has a lot of ATP, the excess ATP can inactivate the TCA cycle enzyme, isocitrate dehydrogenase, by binding to an allosteric site. With that enzyme turned off, acetyl-S-CoA can be used for fatty acid synthesis instead of serving as fuel for the TCA cycle. A major control of fatty acid biosynthesis involves acetyl-S-CoA carboxylase. This enzyme, which synthesizes the malonyl-S-CoA required for fatty acid biosynthesis, is allosterically activated by the TCA cycle intermediates citric acid and isocitric acid. On the other hand, both acetyl-S-CoA carboxylase and fatty acid synthetase are inhibited by palmitoyl-S-CoA, the major product of the reactions catalyzed by fatty acid synthetase.

Several hormones also aid in the regulation of lipid metabolism. Epinephrine and glucagon stimulate the hydrolysis of triglycerides. Remember that epinephrine and glucagon also stimulate the breakdown of glycogen to glucose (see Chapter 25). Thus, these two hormones tell cells that energy is needed, and stimulate ATP formation from the catabolism of both fats and carbohydrates. Insulin, on the other hand, tells cells to store energy by synthesizing fats and glycogen. Controls for lipid metabolism are summarized in Figure 27-8.

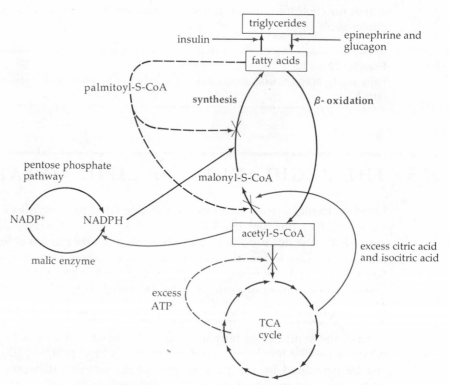

Figure 27-8   Regulation of lipid metabolism. Fatty acid synthesis is stimulated by citric acid, isocitric acid, acetyl-S-CoA, and by excess ATP (which inhibits operation of the TCA cycle). Fatty acid synthesis is inhibited by palmitoyl-S-CoA. Triglyceride synthesis is stimulated by insulin, whereas triglyceride hydrolysis is stimulated by epinephrine and glucagon.

In normal individuals, insulin stimulates cells to synthesize fatty acids and triglycerides when there is an excess supply of acetyl-S-CoA. However, in diabetic individuals, who have insufficient and/or inactive insulin, excess acetyl-S-CoA is not converted into fatty acids, but rather is used to make inordinate amounts of acetoacetyl-S-CoA (reaction 27-31).

Acetoacetyl-S-CoA is used to synthesize β-hydroxy-β-methylglutaryl-S-CoA (reaction 27-32), an intermediary used in terpene and steroid synthesis. In diabetics, however, there is a large excess of β-hydroxy-β-methylglutaryl-S-CoA, and most of it is converted to acetoacetic acid in the reaction

(27-35)

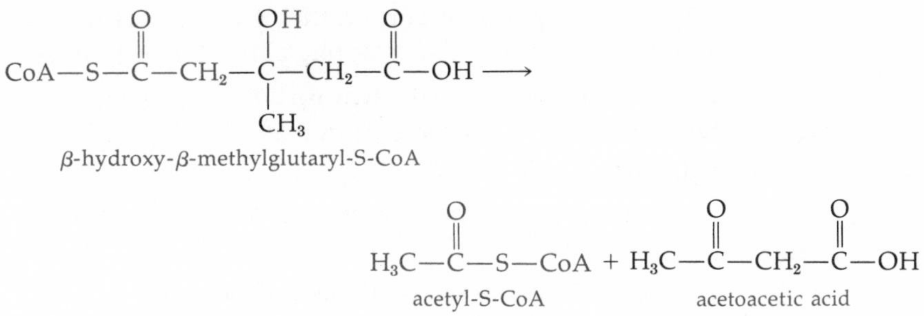

$$CoA—S—\overset{\overset{O}{\|}}{C}—CH_2—\overset{\overset{OH}{|}}{\underset{\underset{CH_3}{|}}{C}}—CH_2—\overset{\overset{O}{\|}}{C}—OH \longrightarrow$$

β-hydroxy-β-methylglutaryl-S-CoA

$$H_3C—\overset{\overset{O}{\|}}{C}—S—CoA + H_3C—\overset{\overset{O}{\|}}{C}—CH_2—\overset{\overset{O}{\|}}{C}—OH$$

acetyl-S-CoA   acetoacetic acid

Acetoacetic acid is also converted to either β-hydroxybutyric acid (reaction 27-36) or to acetone (reaction 27-37).

(27-36)

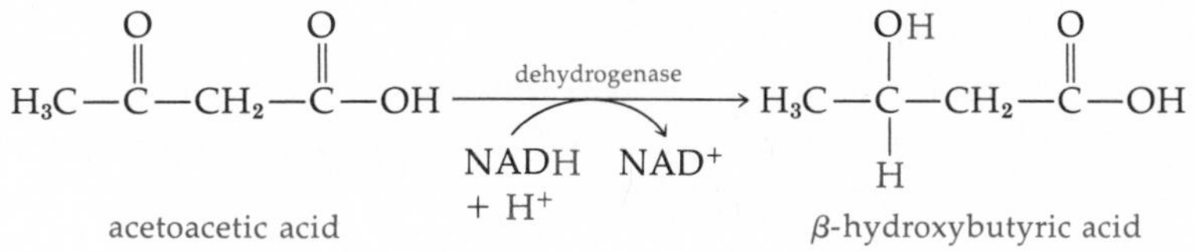

$$H_3C—\overset{\overset{O}{\|}}{C}—CH_2—\overset{\overset{O}{\|}}{C}—OH \xrightarrow[\underset{+ H^+}{NADH \quad NAD^+}]{dehydrogenase} H_3C—\overset{\overset{OH}{|}}{\underset{\underset{H}{|}}{C}}—CH_2—\overset{\overset{O}{\|}}{C}—OH$$

acetoacetic acid   β-hydroxybutyric acid

(27-37)

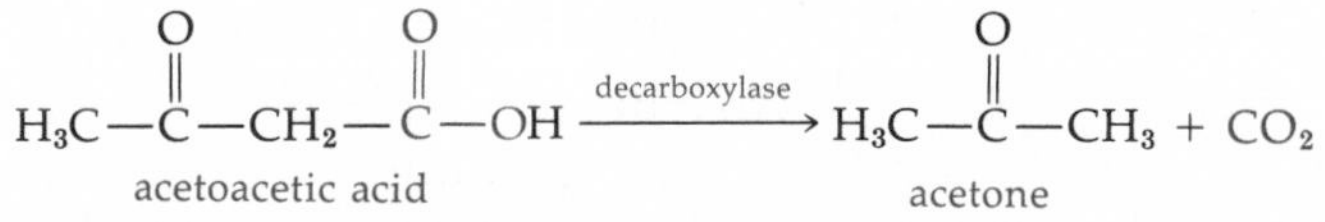

$$H_3C—\overset{\overset{O}{\|}}{C}—CH_2—\overset{\overset{O}{\|}}{C}—OH \xrightarrow{decarboxylase} H_3C—\overset{\overset{O}{\|}}{C}—CH_3 + CO_2$$

acetoacetic acid   acetone

In diabetics, abnormally large amounts of acetoacetic acid, β-hydroxybutyric acid, and acetone, collectively referred to as **ketone bodies,** can accumulate in the blood. Excessive amounts of acetoacetic acid and β-hydroxybutyric acid lower the blood pH and result in a condition called metabolic acidosis, a serious condition that can cause death (see Chapter 29). Acetone can be detected in the urine, and even on the breath, of severely affected diabetics.

Exercise 27-7

How can the supply of NADPH for fatty acid synthesis be increased?

27-9   SUMMARY

1. Acetyl-S-CoA is the starting material used for the synthesis of a large variety of lipids. Catabolism of several lipids also results in acetyl-S-CoA production.

2. The first step in the catabolism of triglycerides and phosphoglycerides is their hydrolysis by lipase enzymes.

3. Glycerol 3-phosphate is converted to dihydroxyacetone phosphate for catabolism in the glycolysis pathway. This compound can also be synthesized from dihydroxyacetone phosphate by the reverse of the same enzymatic reactions.

4. Catabolism of fatty acids begins in the cytoplasm, where they are activated by bonding to coenzyme A molecules. The activated fatty acids are then transported into mitochondria by carnitine molecules.

5. Degradation of fatty acids occurs in mitochondria via the β-oxidation pathway.

6. The β-oxidation pathway cleaves fatty acids into two-carbon-atom fragments, producing acetyl-S-CoA and reduced coenzymes.

7. Fatty acid synthesis occurs in the cytoplasm of human cells, and is not simply a reversal of the catabolic pathway for fatty acids.

8. In fatty acid synthesis, the two-carbon-atom fragments added to growing fatty acid molecules come from malonyl-S-CoA.

9. Fatty acid synthesis requires oxidation of the coenzyme NADPH, which is supplied by the pentose phosphate pathway and the malic enzyme reaction.

10. Triglycerides are made by (1) esterification of two fatty acids to a molecule of glycerol 3-phosphate to give a molecule of phosphatidic acid, (2) removal of the phosphate to give a diglyceride, and (3) esterification with a third fatty acid molecule.

11. Phosphoglycerides are made from diglycerides by attachment of phosphorylated molecules.

12. Fatty acid synthesis is controlled by (1) the available supply of NADPH, (2) the supply of ATP and TCA cycle intermediates, and (3) the concentration of palmitoyl-S-CoA.

13. Hydrolysis of triglycerides is stimulated by glucagon and epinephrine, whereas triglyceride synthesis is stimulated by insulin.

14. Acetoacetic acid, β-hydroxybutyric acid, and acetone, collectively referred to as ketone bodies, are produced in excessive amounts in diabetics. The acidic ketone bodies decrease the blood pH, resulting in metabolic acidosis, which can cause death.

PROBLEMS

1. Suppose that a particular triglyceride contains three hexanoic (six-carbon) acid esters. Outline the route whereby the entire molecule is catabolized to acetyl-S-CoA. Do all of the carbon atoms in this triglyceride become parts of acetyl groups? Explain.

2. Write a balanced equation for the action of pancreatic lipase on the triglyceride in Problem 1.

3. Calculate the total ATP production per mole by the complete aerobic catabolism in liver cells of the triglyceride in Problem 1.

4. Compare the structure and function of coenzyme A and the acyl carrier portion of fatty acid synthetase.

5. What is the role of phosphatidic acids in phosphoglyceride biosynthesis?

6. What is one similarity between the syntheses of phosphatidyl ethanolamine from ethanolamine and glycogen from glucose?

7. Outline the various controls used to regulate lipid metabolism and explain the role of each stimulator or inhibitor.

8. Draw structural formulas for the products of the following reactions:

(a)

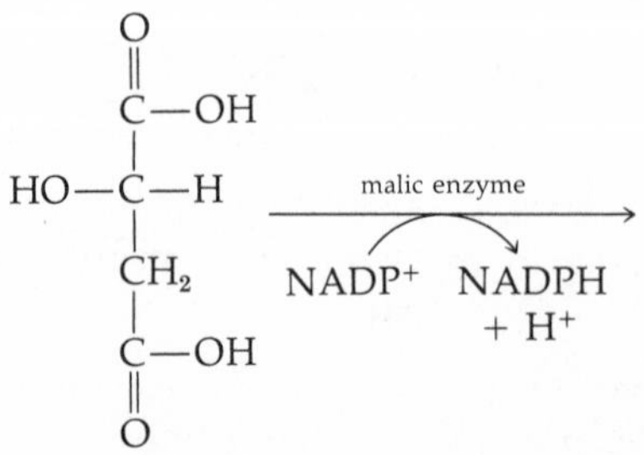

$$
\begin{array}{c}
\quad\quad O \\
\quad\quad \| \\
\quad\quad C-OH \\
\quad\quad | \\
HO-C-H \\
\quad\quad | \\
\quad\quad CH_2 \\
\quad\quad | \\
\quad\quad C-OH \\
\quad\quad \| \\
\quad\quad O
\end{array}
\xrightarrow[\text{NADP}^+ \quad \text{NADPH}]{\text{malic enzyme}}
$$

(b)

$$
\begin{array}{c}
\quad O \\
\quad \| \\
CH_3-C-S-CoA + CO_2 \xrightarrow{\text{acetyl-S-CoA carboxylase}}
\end{array}
$$

(c) $2[CH_3-\overset{\overset{\textstyle O}{\|}}{C}-S-CoA]\xrightarrow{\text{thiolase}} HS-CoA\ +$

(d) $\begin{matrix} H_2C-O-\overset{\overset{\textstyle O}{\|}}{C}-R_1 \\ | \quad\quad\ \ \overset{\overset{\textstyle O}{\|}}{} \\ HC-O-\overset{\overset{\textstyle O}{\|}}{C}-R_2\ +\ 3H_2O\ \xrightarrow{\text{lipase}} \\ | \quad\quad\ \ \overset{\overset{\textstyle O}{\|}}{} \\ H_2C-O-\overset{\overset{\textstyle O}{\|}}{C}-R_3 \end{matrix}$

SOLUTIONS TO EXERCISES

27-1 A six-carbon fatty acid: **ATP/mole**
      activation -2
      $2(\beta\text{-oxidation} + ETS)$ 10
      $3(TCA + ETS)$ $\underline{36}$
                                                           44 ATP/mole

      Glucose:
      glycolysis + ETS $+8$ (in liver)
      2(pyruvate dehydrogenase + ETS)   6
      $2(TCA + ETS)$ $\underline{24}$
                                                           38 ATP/mole

27-2 Since every glucose molecule can yield 2 molecules of acetyl-S-CoA, and 10 molecules of acetyl-S-CoA are required to make 1 molecule of arachidic acid, the total is

$$\frac{1\ \text{glucose}}{2\ \cancel{\text{acetyl-S-CoA}}} \times \frac{10\ \cancel{\text{acetyl-S-CoA}}}{1\ \text{arachidic acid}} = \frac{5\ \text{glucose}}{1\ \text{arachidic acid}}$$

27-3 (a) Arachidic acid: **ATP/mole**
      activation -2
      9 rounds (β-oxidation + ETS)   45
      10 acetyl-S-CoA (TCA + ETS) $\underline{120}$
                                                           163 ATP/mole

      (b) Glucose: 5 moles of glucose $\times$ 38 ATP/mole (in liver) = 190 moles of ATP/5 moles of glucose.

27-4 A diglyceride.

27-5 The glycerol phosphate dehydrogenase reaction,

$$\begin{matrix} H_2COH \\ | \\ C=O \\ | \\ H_2CO-\textcircled{P} \end{matrix} + NADH + H^+ \underset{\text{degradation}}{\overset{\text{synthesis}}{\rightleftharpoons}} \begin{matrix} H_2COH \\ | \\ HCOH \\ | \\ H_2CO-\textcircled{P} \end{matrix} + NAD^+$$

dihydroxyacetone   glycerol 3-phosphate
   phosphate

27-6 Three acetyl-S-CoA molecules are used to make each isoprene unit containing five carbon atoms (the sixth is lost as CO_2). Cholesterol is made from the terpene squalene, which consists of six isoprene units. Therefore, a total of $6 \times 3 = 18$ acetyl-S-CoA molecules are required for each cholesterol molecule.

27-7 The NADPH supply can be increased by activation of malic enzyme by acetyl-S-CoA.

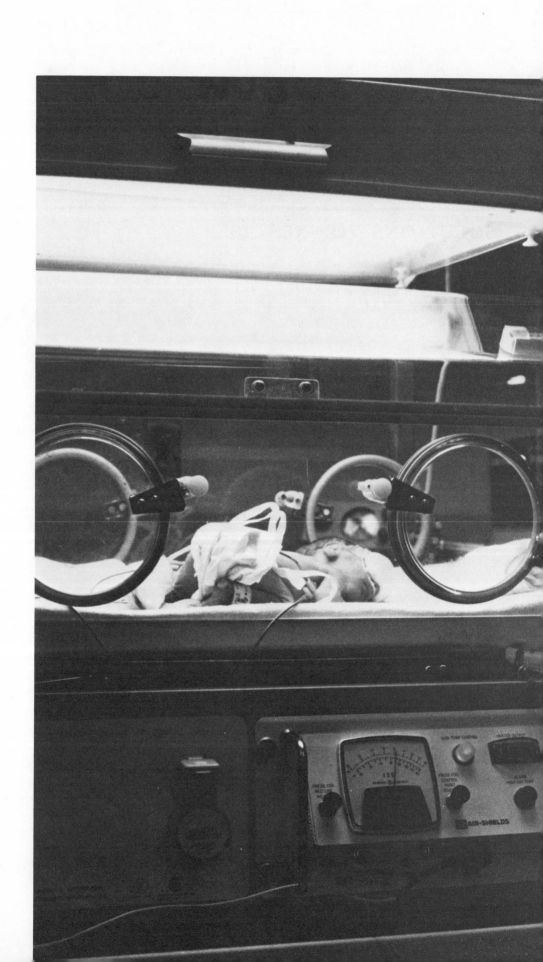

CHAPTER 28

Metabolism of Nitrogenous Compounds

28-1 INTRODUCTION

We have seen the importance of nitrogen-containing functional groups to the structure and function of a large number of biomolecules, including amino acids, proteins, nucleic acids, and some vitamins and lipids. When we discussed the biosynthesis of proteins and nucleic acids, we took for granted the presence in a cell of the requisite amino acids and nucleotides. Since there are 20 amino acids and 4 nucleotides, too much space would be required to discuss the synthesis and degradation of each in detail. However, these nitrogen-containing compounds are vital constituents of the cells in the human body, and we must now consider their metabolism. Fortunately, there are many common features in the metabolic pathways for these compounds. In this chapter we shall discuss the general features of the metabolism of amino acids and some other nitrogen-containing biomolecules. Since amino acids and proteins constitute the bulk of the nitrogen-containing compounds digested and synthesized by the human body, we shall concentrate our efforts on the study of their metabolism.

We shall discuss how the human body digests proteins in order to provide essential amino acids, how proteins are sometimes used as a source of energy, how proteins are constantly being turned over in the body, and how excess nitrogen is incorporated into urea for excretion in urine. In an optional section we shall describe a few of the pathways that lead to the synthesis of some interesting derivatives of amino acids, including some hormones and neurotransmitters. Finally, we shall briefly discuss the degradation of nucleotides. Since most of the reactions discussed in this chapter involve nitrogen-containing functional groups, you may wish to review the physical properties and reactivity of these functional groups, which we discussed in Chapter 18.

The jaundiced (yellow) color of some premature infants is due to a condition known as hyperbilirubinemia, in which the nitrogen-containing compound bilirubin—a degradation product of the heme prosthetic group—cannot be metabolized rapidly enough by the infant's immature liver cells. If not treated, hyperbilirubinemia can cause serious brain damage. This premature infant is being treated for hyperbilirubinemia by phototherapy, in which fluorescent light rays promote the conversion of bilirubin molecules to harmless products. Note that the infant's eyes must be protected from the strong light.

28-2   STUDY OBJECTIVES

After careful study of the material in this chapter, you should be able to:

1. Define the terms protein digestion, protein turnover, nitrogen balance, and amino acid pool.

2. Define the term proteolytic, and predict the products obtained by the action of proteolytic enzymes on peptides.

3. Outline the possible fates of amino acids in the amino acid pool.

4. Draw structural formulas for the products of transamination, oxidative deamination, and direct deamination reactions, given structural formulas for the reactants.

5. Draw structural formulas for the precursors and describe the enzyme-catalyzed reactions that lead to the synthesis of the amino acids alanine, glutamic acid, aspartic acid, glutamine, and asparagine, given structural formulas for these amino acids.

6. Define essential and nonessential amino acids.

7. Write the net overall reaction for the urea cycle, and explain the importance of carbamyl phosphate, glutamic acid, and aspartic acid in this cycle.

8. Describe how the aspartic acid used in the urea cycle is regenerated.

9. Describe the process of nucleic acid digestion.

10. Identify the source of the uric acid produced in the human body and explain its role in gout.

28-3   DIGESTION AND TURNOVER OF PROTEINS

Nitrogen is very abundant on earth. About 80% of air is molecular nitrogen (N_2). Nitrite (NO_2^-) and nitrate (NO_3^-) ions, as well as some ammonia (NH_3), are present in soil. The human body, however, cannot use these forms of nitrogen to synthesize useful nitrogen-containing biomolecules. Our existence depends on a variety of plants and bacteria. Working together, these plants and bacteria produce nitrogen-containing compounds that humans can use. When we ingest these compounds, some of them—including many vitamins—are quickly absorbed by cells in the intestinal tract and distributed by the circulatory system to the cells that need them. The major source of nitrogen in the human diet, however, is proteins, and proteins are too large to be absorbed by cells in the intestines.

Digestion of Dietary Protein

In order for the body to utilize ingested proteins, they must first be broken down into their constituent amino acids. This is accomplished in the digestive tract by a number of enzymes that hydrolyze the peptide bonds of the proteins. This hydrolysis process is aptly called **protein digestion.**

Besides the problem of absorption, there is another reason why the human body bothers to destroy all of the proteins that it eats, instead of immediately putting some of them (e.g., enzymes) to use. The proteins in our diet, unless we are cannibals, are not exactly identical in amino acid sequence to those synthesized in our body. The immune system (Chapter 22) would identify these proteins as foreign invaders and destroy them.

The digestion of proteins begins in the stomach, with the action of the enzyme **pepsin.** Pepsin is secreted into the stomach as a component of gastric juice, which is quite acidic (pH about 1–2). Pepsin is an unusual enzyme in that it is fairly resistant to denaturation by the acidic gastric juice. In fact, pepsin is maximally active at a pH of about 1–2. However, at this pH, ingested proteins are partially denatured and in their unfolded state they are hydrolyzed by pepsin into several smaller peptides. The reaction catalyzed by pepsin is illustrated in Figure 28-1. Pepsin predominately hydrolyzes peptide bonds that involve the aromatic amino acids phenylalanine, tyrosine, or tryptophan.

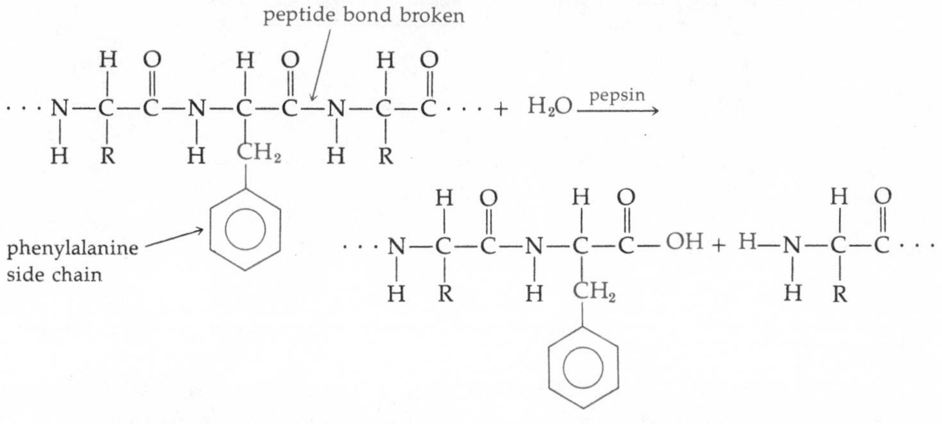

Figure 28-1   The hydrolysis by pepsin of a protein containing phenylalanine. Pepsin also hydrolyzes peptide bonds involving tyrosine, tryptophan, and some other amino acids.

From the stomach, the peptides that have been produced by the action of pepsin travel on to the small intestine, where their digestion continues. Several additional protein-hydrolyzing enzymes, including trypsin, chymotrypsin, carboxypeptidase, and aminopeptidase, are released into the small intestine in pancreatic juice. Protein-hydrolyzing enzymes are usually called **proteolytic** enzymes or proteases. The specificity of these proteolytic enzymes is shown in Table 28-1. Notice that trypsin and chymotrypsin hydrolyze peptide bonds at certain specific points within a peptide, whereas the enzymes carboxypeptidase and aminopeptidase "chew off" one amino acid at a time starting at different ends of a peptide. As a result of the combined action of all of these proteolytic

Table 28-1 Specificity of Intestinal Proteolytic Enzymes

Enzyme	Specificity*
Trypsin	R_2 is arginine or lysine
Chymotrypsin	R_2 is tyrosine, tryptophan, phenylalanine, or leucine
Carboxypeptidase	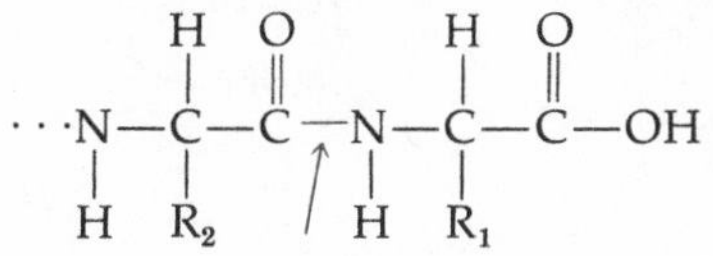 Hydrolyzes C-terminal amino acids
Aminopeptidase	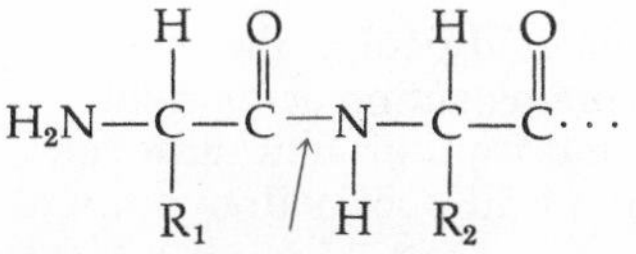 Hydrolyzes N-terminal amino acids

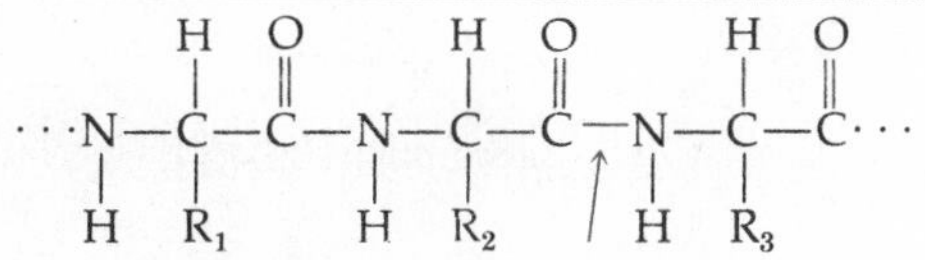

* Arrows indicate the peptide bond hydrolyzed.

enzymes, almost all ingested proteins are finally hydrolyzed into their component amino acids in the small intestine. These free amino acids are then transported across the cells of the intestinal walls into the bloodstream by means of a number of specific active transport proteins (see Figure 28-2).

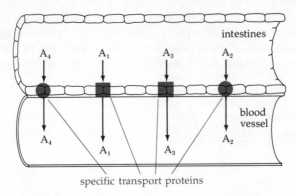

Figure 28-2   The absorption of amino acids from the intestines is accomplished by several specific active transport proteins.

Nitrogen Balance

Free amino acids in the bloodstream, obtained from ingested proteins and the turnover of body proteins (to be discussed shortly), form what is called the **amino acid pool** (see Figure 28-3). Cells in the body draw upon the amino acid pool for the synthesis of new proteins or, under certain conditions, as a source of energy. Excess amino acids and other small nitrogenous compounds are degraded and the nitrogen atoms excreted from the body in the compound urea (see Section 28-5). Healthy adults normally excrete about as much nitrogen per day as they receive in their diet, and are said to be in **nitrogen balance.** When suffering from certain diseases or when severely undernourished, however, an individual may need to catabolize body proteins for energy. This leads to the excretion of more nitrogen than is received in the diet, a condition referred to as **negative nitrogen balance.** On the other hand, healthy pregnant women and growing children need to ingest more nitrogen than they excrete. They require a state of **positive nitrogen balance.**

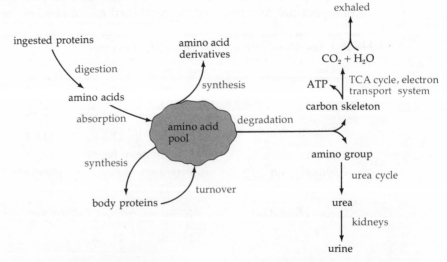

Figure 28-3   The amino acid pool in the bloodstream obtains amino acids from the diet and from protein turnover. The amino acid pool is drawn upon for the synthesis of proteins and amino acid derivatives, and on occasion to meet energy needs.

Turnover of Body Proteins

All of the various body proteins, including structural proteins, carrier proteins, enzymes, and so forth, are constantly being synthesized and degraded. This continual process of synthesis and breakdown of proteins is called **protein turnover.** Protein turnover provides new active proteins and removes proteins that have been around for a while and that may have become modified or damaged, and may also replace proteins that were not synthesized correctly. Some proteins, such as those in liver cells, are turned over fairly rapidly. About one-half of the proteins in the liver are turned over every 6 days. That is, during a period of 6 days, half of the proteins in the liver are degraded and replaced by newly synthesized ones. Proteins in other parts of the body are degraded and synthesized more slowly. For example, about half of the body's collagen is turned over every 3 years. The normal rate of turnover of various body proteins is generally related to their size, structure, and accessibility to proteolytic enzymes. The free amino acids produced by the degradation of body proteins enter the amino acid pool, where they can be used for synthesis of new proteins and other molecules or, when necessary, for energy production.

Exercise 28-1

List the products that would predominate if the peptide N-lys-phe-met-ser-ala-gly-tyr-C was "digested" by each of the following proteolytic enzymes:

(a) Pepsin   (b) Trypsin   (c) Chymotrypsin   (d) Pepsin, then trypsin

Exercise 28-2

Would you expect a healthy individual on a high-protein diet to be in a state of nitrogen balance? What about an individual on a 2-week fast?

28-4 ASPECTS OF AMINO ACID CATABOLISM

Usually a healthy individual is in nitrogen balance and will catabolize an amount of amino acids approximately equal to the amount ingested. However, when a person fasts or is starving or has diabetes, larger amounts of amino acids are catabolized to provide energy and the total amount of protein in the body becomes depleted. In the catabolism of amino acids, the amino groups of the individual amino acids are removed and then incorporated into the compound urea, which is excreted in urine (see Section 28-5). The remainder of each of the various amino acids, called the **carbon skeletons,** are converted into molecules that are used to produce glucose by the process of gluconeogenesis or are ultimately oxidized via the tricarboxylic acid cycle to produce CO_2 and reduced coenzymes. The reduced coenzymes are then used by the electron transport system to produce ATP from ADP (see Chapter 25). The overall process of amino acid catabolism, when used to supply energy for the body, is summarized in Figure 28-4.

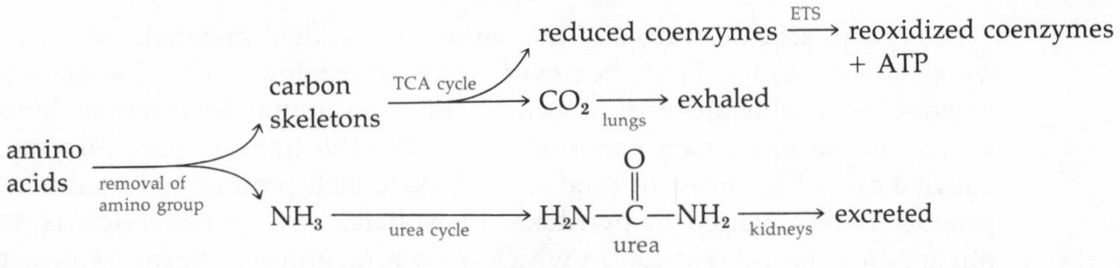

Figure 28-4   An overview of amino acid catabolism for energy production.

The first step in the catabolism of amino acids involves the removal of the amino group. Two general types of enzyme-catalyzed reactions are used for this purpose: transamination and oxidative deamination. Two of the amino acids, serine and threonine, lose their amino groups by a third process called direct deamination.

Transamination

Transamination reactions involve enzymes called **transaminases,** which catalyze the exchange of the amino group located on the α-carbon of particular amino acids for the carbonyl oxygen of the coenzyme **pyridoxal phosphate** (see Figure 28-5), which is bound to the transaminase enzyme.

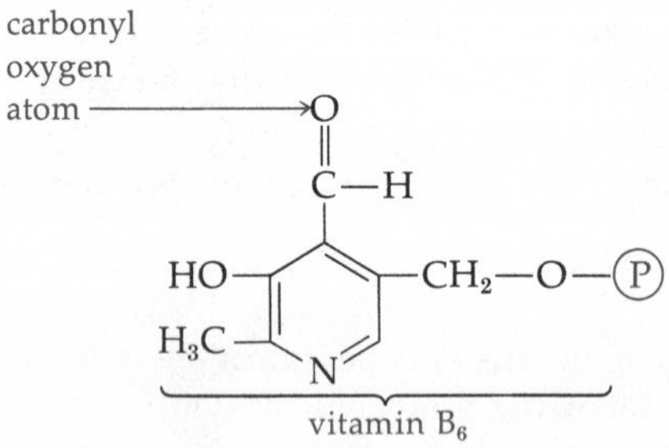

Figure 28-5   The coenzyme pyridoxal phosphate is a derivative of vitamin B_6.

The products of this reaction are the α-keto acid derivative of the reactant amino acid and a molecule of **pyridoxamine phosphate,** which remains attached to the transaminase:

(28-1)

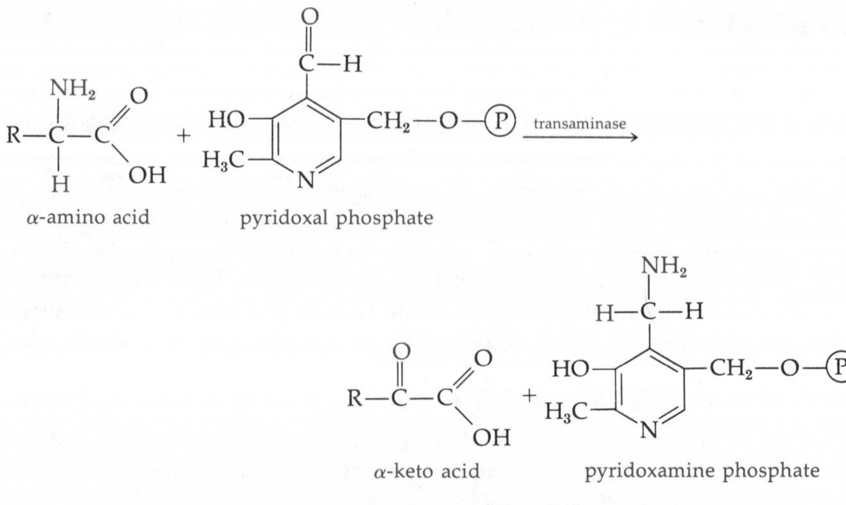

The α-keto acid derivative of the amino acid is then metabolized further, as we shall see shortly. First, however, we must examine what happens to the pyridoxamine phosphate produced in the transamination reaction. Since the pyridoxamine phosphate remains attached to the transaminase enzyme, this enzyme cannot continue to produce its α-keto acid product unless this second product is reconverted to pyridoxal phosphate. This reconversion is accomplished in a second reaction in which the amine group of pyridoxamine phosphate is exchanged for the α-keto oxygen of another α-keto acid, generally, that of α-ketoglutaric acid:

(28-2)

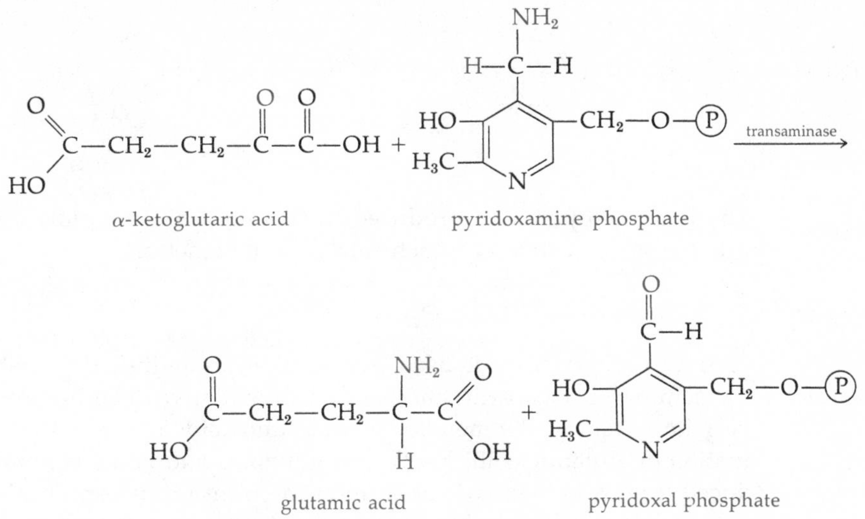

$$\underset{\text{HO}}{\overset{\text{O}}{\parallel}}\!\!C-CH_2-CH_2-\overset{\text{O}}{\overset{\parallel}{C}}-\overset{\text{O}}{\overset{\parallel}{C}}-OH \;+\; \text{pyridoxamine phosphate} \xrightarrow{\;\text{transaminase}\;}$$

α-ketoglutaric acid   pyridoxamine phosphate

$$C-CH_2-CH_2-\underset{H}{\overset{NH_2}{C}}-C \;+\; \text{pyridoxal phosphate}$$

glutamic acid   pyridoxal phosphate

Notice that reaction 28-2 is basically the reverse of reaction 28-1, except that a *particular* α-keto acid, α-ketoglutaric acid, is used as the donor of the carbonyl oxygen atom.

The combination of reactions 28-1 and 28-2 results in the net transfer of an amino group from a specific amino acid to α-ketoglutaric acid:

(28-3)   α-amino acid + α-ketoglutaric acid $\xrightarrow[\text{pyridoxal phosphate}]{\text{transaminase}}$ α-keto acid + glutamic acid

Notice in reaction 28-3 that the amino group becomes part of glutamic acid regardless of the particular reactant amino acid. For example, the enzyme alanine transaminase catalyzes the net conversion of the amino acid alanine to pyruvic acid:

(28-4)

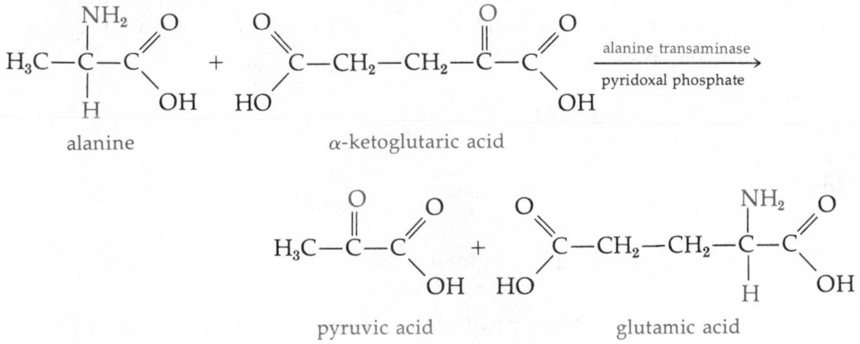

$$H_3C-\underset{H}{\overset{NH_2}{C}}-C \;+\; C-CH_2-CH_2-\overset{O}{\overset{\parallel}{C}}-C \xrightarrow[\text{pyridoxal phosphate}]{\text{alanine transaminase}}$$

alanine   α-ketoglutaric acid

$$H_3C-\overset{O}{\overset{\parallel}{C}}-C \;+\; C-CH_2-CH_2-\underset{H}{\overset{NH_2}{C}}-C$$

pyruvic acid   glutamic acid

Notice that reactions 28-3 and 28-4 each depict the sum of two reactions. Since the coenzyme pyridoxal phosphate used in the first reaction (28-1) is regenerated in the second reaction (28-2), we only need to indicate that this coenzyme is used when writing the overall reaction catalyzed by a transaminase.

Oxidative Deamination

The second general type of enzyme-catalyzed reaction that is used to remove amino groups from amino acids is called **oxidative deamination.** A single enzyme, amino acid oxidase, catalyzes a complex reaction that produces ammonia and hydrogen peroxide as products, together with the α-keto acid derivative of the reactant amino acid:

(28-5)

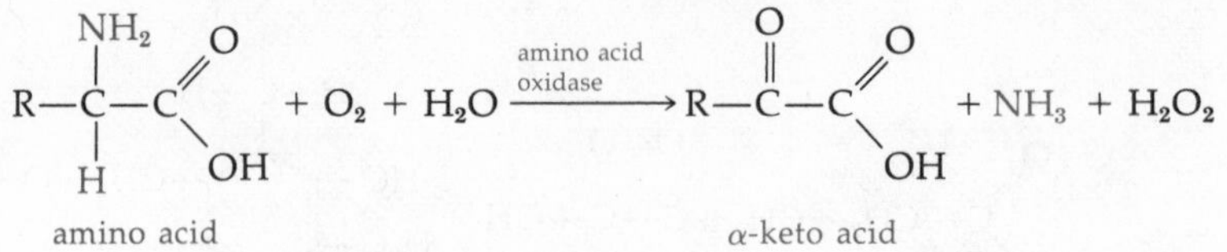

$$R-\underset{\underset{H}{|}}{\overset{\overset{NH_2}{|}}{C}}-C\overset{O}{\underset{OH}{\diagdown}} + O_2 + H_2O \xrightarrow{\text{amino acid oxidase}} R-\underset{}{\overset{\overset{O}{\|}}{C}}-C\overset{O}{\underset{OH}{\diagdown}} + NH_3 + H_2O_2$$

amino acid   α-keto acid

The hydrogen peroxide produced in this reaction is rapidly destroyed by another enzyme, catalase, which catalyzes the reaction

(28-6) $2H_2O_2 \xrightarrow{\text{catalase}} 2H_2O + O_2$

Amino acid oxidase is used far less often in amino acid catabolism than are transaminases. However, another type of oxidative deamination reaction that is very important in the metabolism of amino acids involves the oxidative deamination of glutamic acid. Recall that glutamic acid is the common product produced from α-ketoglutaric acid upon the transamination of other amino acids (reaction 28-3). The amino group of glutamic acid is eventually incorporated into urea for excretion, but it must first be removed from glutamic acid. The enzyme **glutamic acid dehydrogenase** catalyzes the reaction

(28-7)

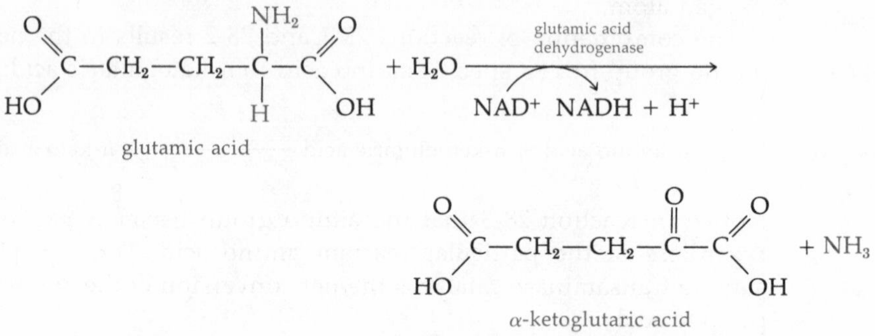

$$\underset{HO}{\overset{O}{\diagup}}C-CH_2-CH_2-\underset{\underset{H}{|}}{\overset{\overset{NH_2}{|}}{C}}-C\overset{O}{\underset{OH}{\diagdown}} + H_2O \xrightarrow[\text{NAD}^+ \quad \text{NADH} + H^+]{\text{glutamic acid dehydrogenase}}$$

glutamic acid

$$\underset{HO}{\overset{O}{\diagup}}C-CH_2-CH_2-\overset{\overset{O}{\|}}{C}-C\overset{O}{\underset{OH}{\diagdown}} + NH_3$$

α-ketoglutaric acid

Thus, by the *joint* action of transaminases and glutamic acid dehydrogenase, the amino groups from many amino acids can be converted to ammonia in a similar manner:

(28-8)

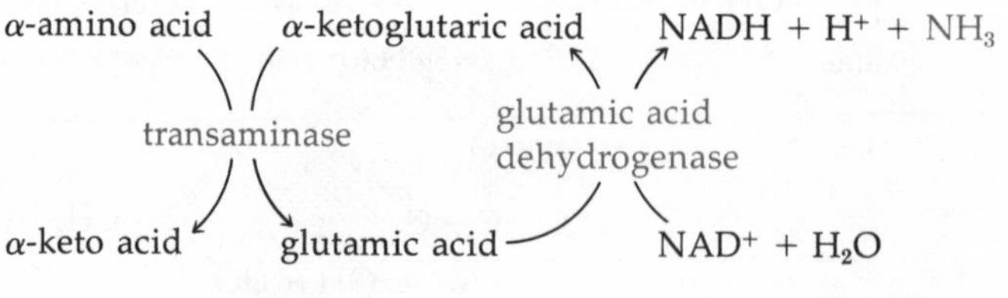

α-amino acid   α-ketoglutaric acid $NADH + H^+ + NH_3$

transaminase   glutamic acid dehydrogenase

α-keto acid   glutamic acid $NAD^+ + H_2O$

The ammonia thus produced is used to synthesize urea (Section 28-5).

Direct Deamination

The amino acids serine and threonine contain alcohol functional groups in their side chains. These amino acids can be directly deaminated by the enzyme **serine-threonine dehydratase.** This enzyme catalyzes the following reaction:

(28-9)

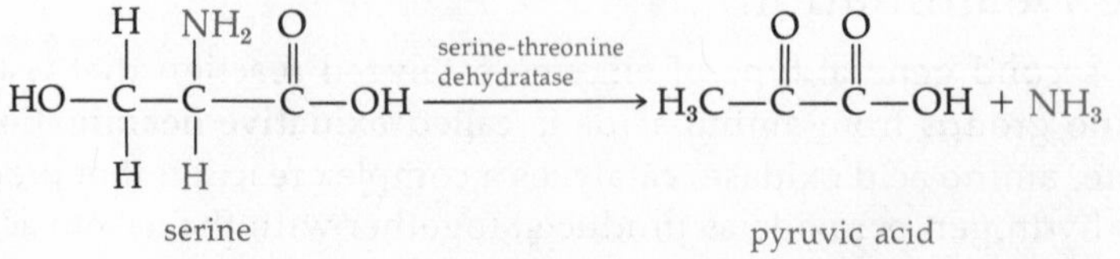

$$HO-\underset{\underset{H}{|}}{\overset{\overset{H}{|}}{C}}-\underset{\underset{H}{|}}{\overset{\overset{NH_2}{|}}{C}}-\overset{\overset{O}{\|}}{C}-OH \xrightarrow{\text{serine-threonine dehydratase}} H_3C-\overset{\overset{O}{\|}}{C}-\overset{\overset{O}{\|}}{C}-OH + NH_3$$

serine   pyruvic acid

Notice that serine-threonine dehydratase also removes the hydroxyl group from the side chain of serine, producing pyruvic acid. This enzyme also catalyzes the direct deamination of threonine. However, two other pathways for threonine catabolism also exist.

Fate of the α-Keto Acid Derivatives

After removal of the α-amino group, the α-keto acid derivatives of the various amino acids, containing the carbon skeletons of the original amino acids, are metabolized further. They may enter the gluconeogenic pathway or, when energy is required, be catabolized via the TCA cycle. We are already familiar with the metabolism of some of these α-keto acids. Pyruvic acid produced by the transamination of alanine (reaction 28-4) or by serine-threonine dehydratase (reaction 28-9) is converted to acetyl-S-CoA by pyruvic acid dehydrogenase (Chapter 25). α-Ketoglutaric acid, produced by glutamic acid dehydrogenase, is one of the intermediates in the TCA cycle (Chapter 25). The transamination of aspartic acid produces oxaloacetic acid, which is also an intermediate in the TCA cycle. The α-keto acids produced from the other amino acids are also funneled into the TCA cycle. For some of them, however, this requires a number of additional reaction steps. We shall not discuss these special pathways. Figure 28-6 summarizes the routes whereby the carbon skeletons of the various amino acids can be shuttled into the TCA cycle. The energy yield from the catabolism of 1 g of a typical protein is comparable to that obtained from the aerobic catabolism of 1 g of glucose.

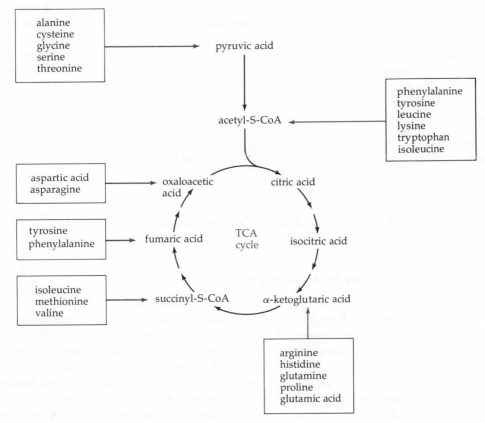

Figure 28-6   Pathways of entry of the carbon skeletons of amino acids into the TCA cycle.

Exercise 28-3
Draw the structural formulas for the products of the following reactions:

(a) 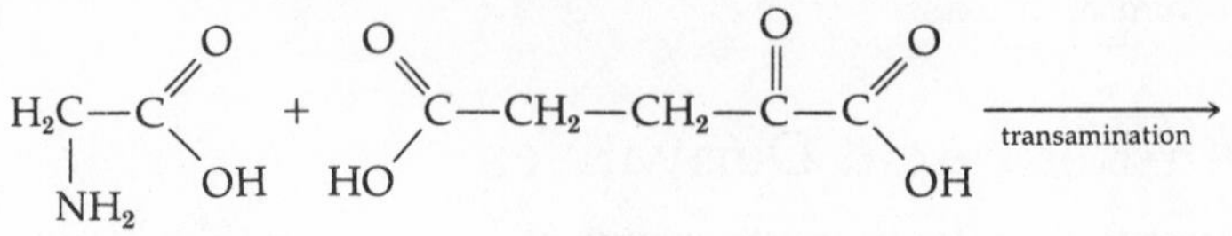 H_2C—C (with NH_2 below, O double bond above, OH) + C (O double bond, HO)—CH_2—CH_2—C(=O)—C(=O, OH) $\xrightarrow{\text{transamination}}$

(b) 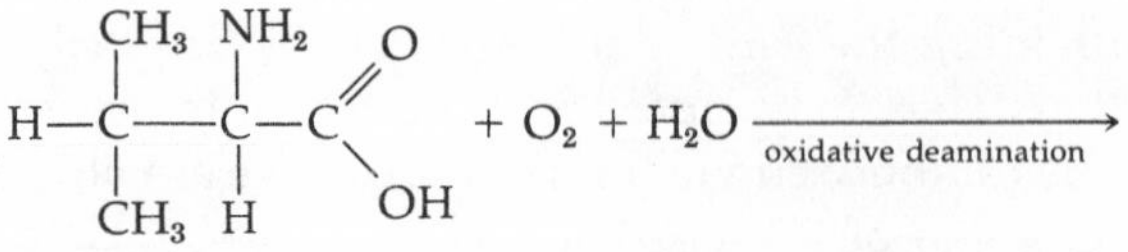 H—C(CH_3)(CH_3)—C(NH_2)(H)—C(=O, OH) + O_2 + H_2O $\xrightarrow{\text{oxidative deamination}}$

(c) H_2O_2 $\xrightarrow{\text{catalase}}$

(d) 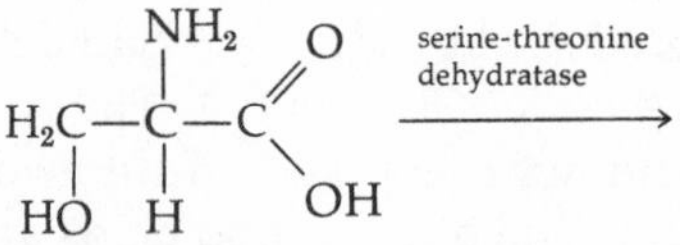 H_2C(HO)—C(NH_2)(H)—C(=O, OH) $\xrightarrow{\text{serine-threonine dehydratase}}$

28-5   THE UREA CYCLE

In order for a normal individual to be in a state of nitrogen balance, excess nitrogen generated in the catabolism of amino acids and other nitrogen-containing compounds must be excreted from the body. The excess nitrogen is incorporated into the nontoxic compound urea within liver cells. The urea is then excreted in urine.

Urea synthesis is accomplished in a cyclic series of reactions that is called the **urea cycle.** The overall net reaction of the urea cycle involves two molecules of ammonia, which originate from two amino acid molecules, and a molecule of carbon dioxide:

(28-10) $2NH_3 + CO_2 \longrightarrow H_2N-\overset{\overset{\displaystyle O}{\|}}{C}-NH_2 + H_2O$

Note that urea is the diamide of carbonic acid, $HO-\overset{\overset{\displaystyle O}{\|}}{C}-OH$.

Source of the Amino Groups

There are three sources for the amino groups used for the synthesis of urea: glutamic acid produced from α-ketoglutaric acid in the transamination of other amino acids (reaction 28-3), the direct deamination of serine or threonine, and the oxidative deamination of other amino acids. These two amino groups enter the urea cycle by different routes. One enters as free ammonia, which is produced either by direct deamination or by oxidative deamination reactions or from glutamic acid by the enzyme glutamic acid dehydrogenase (reaction 28-7). The second amino group enters the urea cycle as part of a molecule of aspartic acid, which is produced from oxaloacetic acid in a reaction catalyzed by the enzyme **aspartic acid transaminase.** Glutamic acid donates the amino group in this reaction.

(28-11)

$$HO-\overset{\overset{\displaystyle O}{\|}}{C}-CH_2-\overset{\overset{\displaystyle O}{\|}}{C}\diagdown_{OH} \quad + \quad HO-\overset{\overset{\displaystyle O}{\|}}{C}-CH_2-CH_2-\overset{\overset{\displaystyle NH_2}{|}}{\underset{\underset{\displaystyle H}{|}}{C}}-\overset{\overset{\displaystyle O}{\|}}{C}\diagdown_{OH} \xrightarrow{\text{aspartic acid transaminase}}$$

oxaloacetic acid   glutamic acid

$$HO-\overset{\overset{\displaystyle O}{\|}}{C}-CH_2-\overset{\overset{\displaystyle NH_2}{|}}{\underset{\underset{\displaystyle H}{|}}{C}}-\overset{\overset{\displaystyle O}{\|}}{C}\diagdown_{OH} \quad + \quad HO-\overset{\overset{\displaystyle O}{\|}}{C}-CH_2-CH_2-\overset{\overset{\displaystyle O}{\|}}{C}-\overset{\overset{\displaystyle O}{\|}}{C}\diagdown_{OH}$$

aspartic acid α-ketoglutaric acid

Exercise 28-4

The transamination of ingested amino acids leads to the formation of glutamic acid from α-ketoglutaric acid. What happens to the glutamic acid? What is the source of the α-ketoglutaric acid?

Carnivores obtain protein from both plant and animal tissue, whereas   herbivores obtain protein mainly from plant tissue.

Synthesis of Urea

The first step of the urea cycle is the synthesis of **carbamoyl phosphate,** which requires the hydrolysis of two molecules of ATP. The ammonia that is used in this step comes from one of the amino acids, and the CO_2 is a product of the TCA cycle. The carbamoyl phosphate is then used, together with ornithine, for the synthesis of **citrulline** in the second step of the urea cycle. In the third step of the cycle, the citrulline is joined to a molecule of aspartic acid (synthesized in reaction 28-11), which carries the second amino group that will end up in urea. This step requires the hydrolysis of a molecule of ATP to AMP and pyrophosphate, and results in the production of **argininosuccinic acid.** In the fourth step, argininosuccinic acid is cleaved to produce **arginine** and **fumaric acid.** If arginine is needed for protein synthesis, the urea cycle stops at this point. The fumaric acid produced in step 4 is used in the TCA cycle to produce more oxaloacetic acid, which can then be converted to aspartic acid by reaction 28-11.

The last step of the urea cycle (step 5) is the hydrolysis of arginine to give **urea** and **ornithine.** The urea is excreted and the ornithine is used in the next round of the cycle (in step 2). Figure 28-7 summarizes this cyclic process.

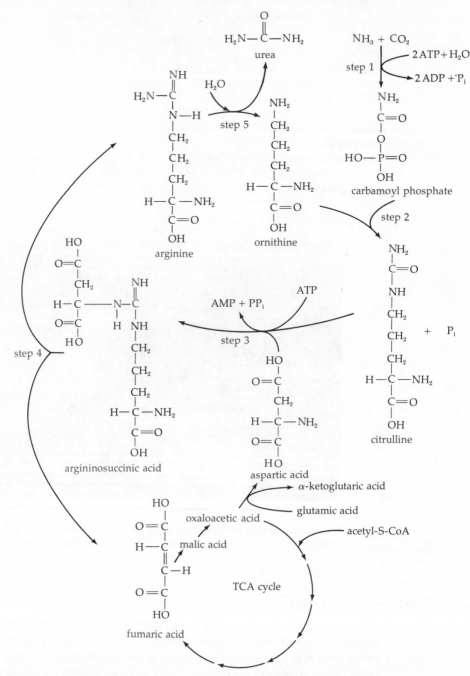

Figure 28-7   The urea cycle. Notice that the argininosuccinic acid is cleaved in step 4 to give arginine and fumaric acid. The arginine is then hydrolyzed to give urea in step 5, whereas the fumaric acid is converted to oxaloacetic acid in the TCA cycle.

Notice that there is no net consumption of aspartic acid nor production of fumaric acid in the urea cycle, since fumaric acid can be used to synthesize another molecule of aspartic acid. When we consider the energy (from ATP hydrolysis) that is required, the net overall reaction of the urea cycle can be written as:

(28-12)

$$2NH_3 + CO_2 \xrightarrow[\substack{urea\ cycle}]{\substack{3ATP + 2H_2O \quad\quad 2ADP + AMP + 2P_i + PP_i}} H_2N-\overset{\displaystyle O}{\overset{\|}{C}}-NH_2$$

urea

Recall that an additional molecule of ATP is required to convert the AMP produced in the urea cycle to ADP. Thus, the energy required for the operation of the entire urea cycle is equivalent to the hydrolysis of 4ATP to 4ADP + 4P_i.

Exercise 28-5

Two molecules of ATP are hydrolyzed in the first step of the urea cycle. What happens to the two phosphate groups that are removed from these two ATP molecules?

Exercise 28-6

Both of the amino groups in a molecule of urea may originate from glutamic acid molecules, but they are not transferred directly from glutamic acid to urea. How do these amino groups enter the urea cycle?

28-6 BIOSYNTHESIS OF AMINO ACIDS

Of the 20 α-amino acids required for the synthesis of proteins, the adult human body is able to synthesize only 12. Since it is not necessary for these 12 amino acids to be included in our diet, they are called **nonessential amino acids.** The 8 amino acids that we cannot synthesize are called **essential amino acids.** Table 28-2 lists the essential and nonessential amino acids for human adults. Nutritional requirements for amino acids and other compounds will be discussed further in Chapter 29.

Table 28-2 Essential and Nonessential Amino Acids for Human Adults

Nonessential	Essential
Glycine	Isoleucine
Alanine	Leucine
Serine	Lysine
Cysteine	Methionine
Proline	Phenylalanine
Glutamic acid	Threonine
Glutamine	Tryptophan
Aspartic acid	Valine
Asparagine	
Tyrosine	
Histidine*	
Arginine*	

* Although histidine and arginine do not appear to be essential for adults, they apparently are essential for the normal growth of children.

We have already discussed one mechanism used to synthesize amino acids, namely, transamination. (Remember that enzyme-catalyzed reactions are generally reversible.) Alanine, glutamic acid, and aspartic acid are produced directly by the respective transamination of pyruvic acid, α-ketoglutaric acid, and oxaloacetic acid.

(28-13)

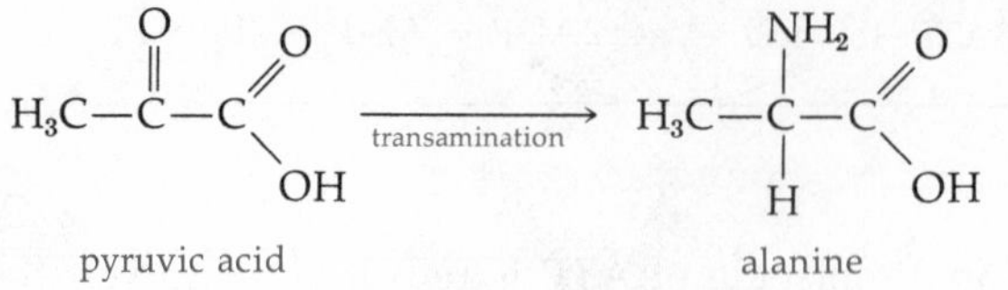

$$H_3C-\underset{O}{\overset{O}{||}}{C}-\underset{OH}{\overset{O}{||}}{C} \xrightarrow{\text{transamination}} H_3C-\underset{H}{\overset{NH_2}{|}}{C}-\underset{OH}{\overset{O}{||}}{C}$$

pyruvic acid   alanine

(28-14)

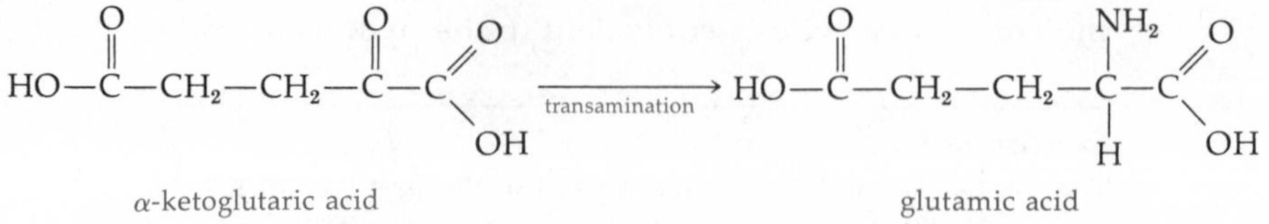

$$HO-\overset{O}{\overset{||}{C}}-CH_2-CH_2-\overset{O}{\overset{||}{C}}-\overset{O}{\overset{||}{C}}{-OH} \xrightarrow{\text{transamination}} HO-\overset{O}{\overset{||}{C}}-CH_2-CH_2-\overset{NH_2}{\underset{H}{|}}{C}-\overset{O}{\overset{||}{C}}{-OH}$$

α-ketoglutaric acid   glutamic acid

(28-15)

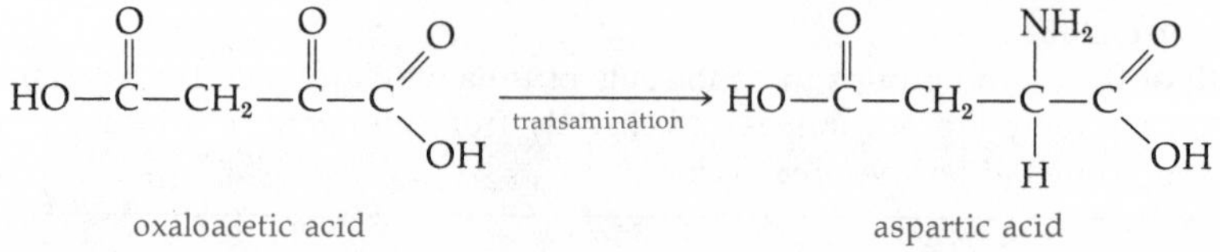

$$HO-\overset{O}{\overset{||}{C}}-CH_2-\overset{O}{\overset{||}{C}}-\overset{O}{\overset{||}{C}}{-OH} \xrightarrow{\text{transamination}} HO-\overset{O}{\overset{||}{C}}-CH_2-\overset{NH_2}{\underset{H}{|}}{C}-\overset{O}{\overset{||}{C}}{-OH}$$

oxaloacetic acid   aspartic acid

The amino acid glutamine is produced from glutamic acid and ammonia by the enzyme **glutamine synthetase,** in the following reaction:

(28-16)

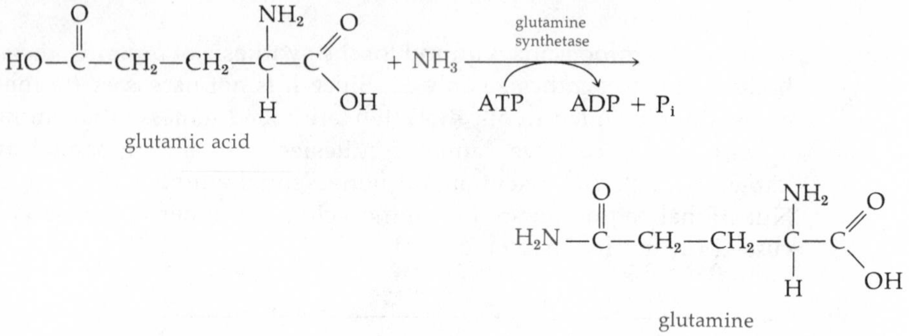

$$HO-\overset{O}{\overset{||}{C}}-CH_2-CH_2-\overset{NH_2}{\underset{H}{|}}{C}-\overset{O}{\overset{||}{C}}{-OH} + NH_3 \xrightarrow[\text{ATP} \quad \text{ADP} + P_i]{\text{glutamine synthetase}}$$

glutamic acid

$$H_2N-\overset{O}{\overset{||}{C}}-CH_2-CH_2-\overset{NH_2}{\underset{H}{|}}{C}-\overset{O}{\overset{||}{C}}{-OH}$$

glutamine

Asparagine is produced from aspartic acid in an identical manner by the enzyme **asparagine synthetase.**

The human body synthesizes tyrosine by the hydroxylation of phenylalanine, in a reaction catalyzed by the enzyme **phenylalanine hydroxylase:**

(28-17)

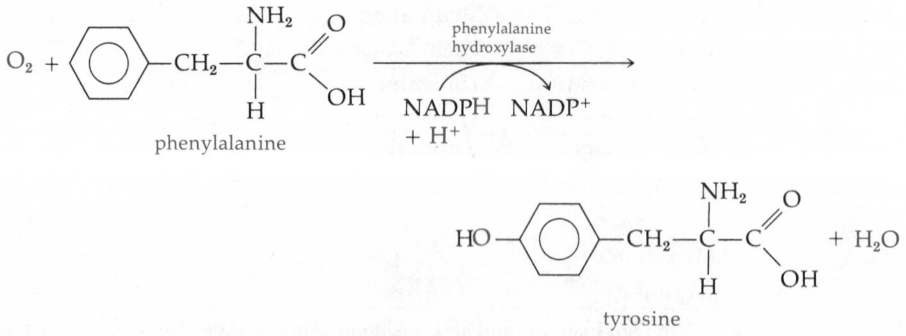

$$O_2 + \bigcirc-CH_2-\overset{NH_2}{\underset{H}{|}}{C}-\overset{O}{\overset{||}{C}}{-OH} \xrightarrow[\substack{\text{NADPH} \quad \text{NADP}^+ \\ + H^+}]{\substack{\text{phenylalanine} \\ \text{hydroxylase}}}$$

phenylalanine

$$HO-\bigcirc-CH_2-\overset{NH_2}{\underset{H}{|}}{C}-\overset{O}{\overset{||}{C}}{-OH} + H_2O$$

tyrosine

This reaction is also the first step in the synthesis of several other important compounds, including epinephrine and the neurotransmitter DOPA (see Section 28-7). In the disease **phenylketonuria** (PKU), the absence of this enzyme results in the buildup of phenylalanine, which causes mental retardation. The synthesis of the amino acids glycine, serine, cysteine, proline, and histidine require several specific steps (which we shall not discuss). Recall that the synthesis of arginine occurs as part of the urea cycle.

Exercise 28-7
Tyrosine is a nonessential amino acid. Can an adequate supply of tyrosine be synthesized by the human body without a source of protein for digestion or turnover?

Optional

28-7 BIOSYNTHESIS OF SOME IMPORTANT AMINO ACID DERIVATIVES

In addition to their use as components of proteins and occasionally as sources of energy for the human body, some amino acids are used as starting materials for the synthesis of several compounds involved in the regulation of metabolism. These amino acid derivatives include epinephrine (adrenaline), the thyroid hormones, and a variety of other small molecules with very specific functions. We shall discuss the synthesis and biological effects of a few of the more well-known amino acid derivatives.

Derivatives of Tyrosine

In the last section we saw that cells in the human body synthesize tyrosine by the hydroxylation of the essential amino acid phenylalanine. Tyrosine then serves as the starting material for the synthesis of a number of hormones and neurotransmitters. The biosynthetic pathway used for the synthesis of several of these compounds is shown in Figure 28-8. The first step in this pathway is the hydroxylation of tyrosine to form dihydroxyphenylalanine, or DOPA.

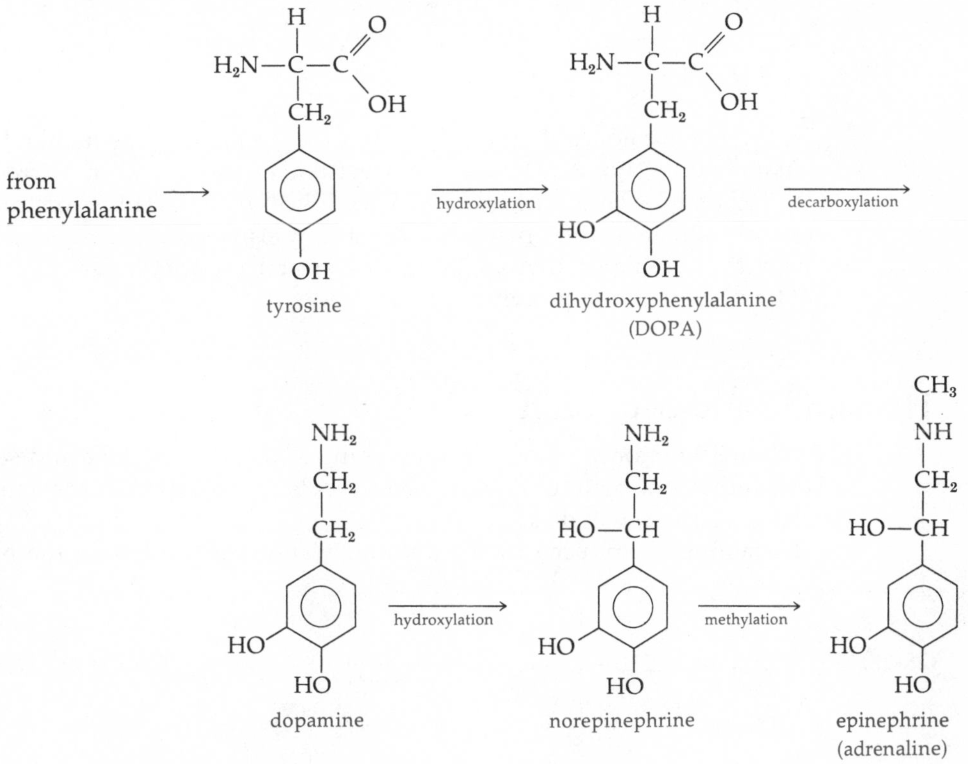

Figure 28-8 The synthesis of hormones and neurotransmitters from tyrosine.

DOPA is then decarboxylated to give the compound **dopamine,** which acts as a **neurotransmitter,** a substance that stimulates nerve cells. Dopamine works specifically at the junctions between nerve cells and muscle cells (also called sympathetic junctions). Dopamine is also the starting material for the further synthesis of norepinephrine (also called noradrenaline) and epinephrine (see Figure 28-8). We have already seen in Chapters 23 and 25 that epinephrine is a hormone used to regulate the synthesis and breakdown of glycogen. Both epinephrine and norepinephrine are also used as neurotransmitters in certain types of nerve junctions. In addition, norepinephrine causes constriction of blood vessels in peripheral areas of the body, helping to regulate blood pressure. Thus these tyrosine derivatives perform a wide range of extremely important regulatory functions in the human body.

Additional derivatives of tyrosine are made in the thyroid gland and are called **thyroid hormones.** Thyroid hormones are essential for maintenance of proper growth and metabolism of cells in the body. However, the exact mechanisms whereby this is accomplished are complex and not totally understood. The two primary thyroid hormones are **triiodothyronine,**

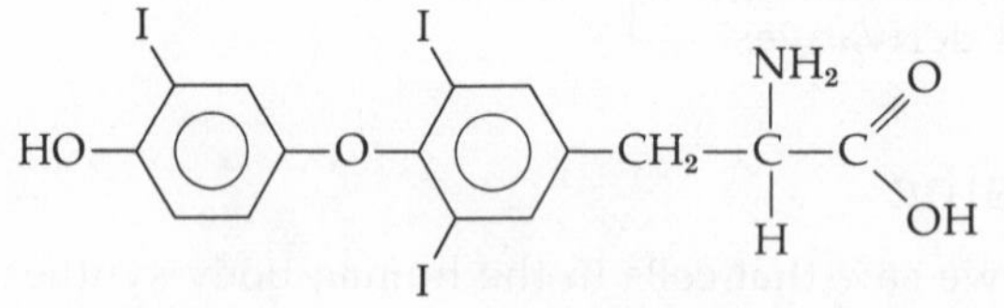

and tetraiodothyronine, or **thyroxine,**

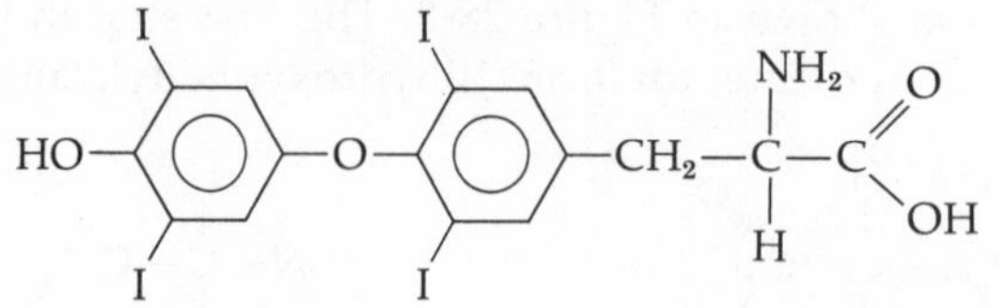

The thyroid hormones are the only molecules produced in the human body that contain iodine atoms. They are made from a protein called thyroglobulin, which contains a large amount of tyrosine. The tyrosine residues on this protein are first iodinated; then pairs of iodinated tyrosines are joined together; and finally the completed thyroid hormones are released upon the hydrolysis of peptide bonds in the protein.

Histamine and Serotonin

Two other hormones produced from amino acids are histamine and serotonin. The synthesis of both of these involves the decarboxylation of the amino acids used as starting materials.

Histamine is produced by the decarboxylation of histidine in the reaction

(28-18)

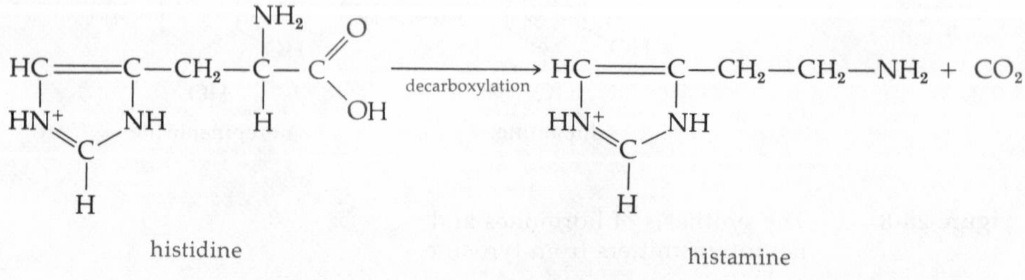

Histamine is synthesized predominately in skin and lung tissue, but is also made and stored in mast cells. This hormone causes the dilation of capillary blood vessels, which increases their permeability. Relatively large amounts of histamine are released into the bloodstream in traumatic shock and in allergic reactions. Several drugs called **antihistamines** are used to inhibit the adverse effects caused by an excessive release of histamine.

The synthesis of **serotonin** involves the hydroxylation of tryptophan followed by decarboxylation (see Figure 28-9). Serotonin has a variety of effects, including the constriction of blood vessels and the promotion of peristalsis in the intestinal system. It also serves as a neurotransmitter for certain types of nerve cells.

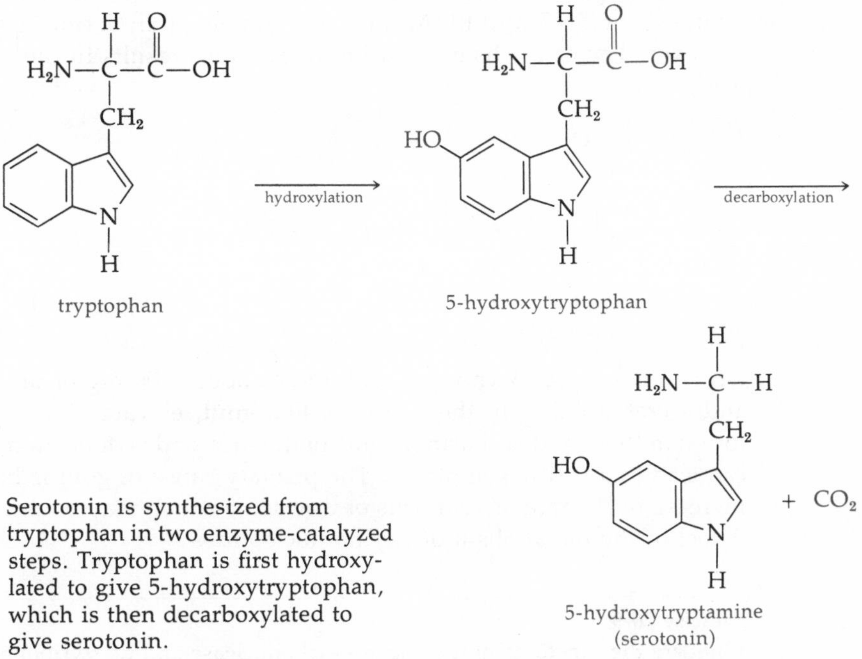

Figure 28-9   Serotonin is synthesized from tryptophan in two enzyme-catalyzed steps. Tryptophan is first hydroxylated to give 5-hydroxytryptophan, which is then decarboxylated to give serotonin.

Derivatives of several other amino acids also have important functions in the human body. The examples we have presented illustrate the vast differences in the functions of biomolecules that are caused by rather small molecular changes, such as the alteration or removal of only one functional group. These amino acid derivatives also illustrate the ability of cells in the human body to produce a large number of molecules with widely differing functions from only a relatively small number of precursors.

You will recall that various nucleotides also have a variety of functions. In addition to serving as component parts of nucleic acids, some nucleotides are used as sources of chemical energy, such as ATP. Others, such as cyclic AMP, are used as messengers, whereas still other nucleotide derivatives function as coenzymes.

Exercise 28-8
What type of reaction is common to the synthesis of serotonin, epinephrine, and histamine from their amino acid precursors? What other type of reaction is common to the synthesis of serotonin, norepinephrine, and DOPA?

28-8   METABOLISM OF NUCLEOTIDES

Cells in the human body are capable of synthesizing *all* of the nucleotides necessary for the assembly of DNA and RNA. These molecules are therefore not essential components of the human diet. However, nucleotides that are present in the human diet can be used for the synthesis of nucleic acids, which is more energy efficient than synthesizing these molecules from scratch. Nucleic acids in the diet are hydrolyzed to their component nucleotides during the course of digestion. This is accomplished by **ribonuclease** and **deoxyribonuclease,** two enzymes that are produced in the pancreas. The free nucleotides are then transported through the cells of the intestinal wall and distributed for use.

When larger amounts of nucleotides are ingested than are needed for biosynthesis of DNA and RNA, they are catabolized in various tissues, especially the liver. The catabolism of purine nucleotides results in the formation of uric acid,

uric acid

which is excreted. In normal individuals about 500 mg of uric acid is formed and excreted daily. In the disease called **gout,** elevated levels of uric acid are found in the blood and urine. Some of the uric acid is deposited in joints, which can lead to arthritic symptoms. The primary cause of gout is believed to be an increase in the rate of synthesis of uric acid, which may be due to an inherited defect in the metabolism of purine nucleotides.

Exercise 28-9
What are the functions of the enzymes ribonuclease and deoxyribonuclease?

Exercise 28-10
Cells in the human body turn over their proteins. Would you also predict substantial DNA turnover in human cells?

28-9   SUMMARY

1. The human body can only utilize nitrogen atoms that are already incorporated into biomolecules.

2. Digestion of proteins involves their hydrolysis into component amino acids by a number of proteolytic enzymes in the digestive system. These enzymes include pepsin, trypsin, chymotrypsin, carboxypeptidase, and aminopeptidase.

3. After hydrolysis of ingested proteins, the free amino acid products are transported into the bloodstream for distribution and use by the various cells in the body.

4. The amino acid pool is the collection of amino acids in the bloodstream that are available for synthesis or energy production by cells in the body.

5. Excess amino acids are catabolized and the nitrogen atoms are incorporated into urea for excretion.

6. Normal human beings, with the exception of pregnant women and growing children, ingest as much nitrogen per day as they excrete, and are said to be in nitrogen balance.

7. All of the proteins in the human body are constantly being synthesized and degraded. This process is called protein turnover.

8. When amino acids are catabolized as a source of energy, the nitrogen atoms are first removed for excretion in the form of urea, whereas the remainder of the molecules, called the carbon skeletons, are ultimately oxidized to CO_2.

9. The removal of amino groups from amino acids is accomplished primarily by transamination, and to a lesser extent by oxidative deamination and direct deamination.

10. Transamination, which is catalyzed by transaminase enzymes, involves the exchange of an amino group for a carbonyl oxygen atom, resulting in the production of an α-keto acid from an amino acid. Transaminase enzymes use the coenzyme pyridoxal phosphate, and generally they transfer amino groups to α-ketoglutaric acid to form glutamic acid.

11. Amino groups on glutamic acid are liberated as ammonia in a reaction that is catalyzed by the enzyme glutamic acid dehydrogenase.

12. The human body is capable of synthesizing 12 of the amino acids found in proteins. These are called the nonessential amino acids. The other 8 amino acids found in proteins must be included in the human diet and are called essential amino acids.

13. Some of the nonessential amino acids are synthesized by the transamination of α-keto acids. The amino acids glutamine and asparagine are synthesized from ammonia and glutamic acid and from ammonia and aspartic acid, respectively.

14. Urea, $NH_2{-}\overset{\overset{\displaystyle O}{\|}}{C}{-}NH_2$, is synthesized in the urea cycle from CO_2 and the α-amino groups from two amino acid molecules. One of the amino groups is used to synthesize carbamoyl phosphate. The second amino group enters the urea cycle via aspartic acid.

15. The net overall reaction of the urea cycle is

$$2NH_3 + CO_2 + 3ATP + 2H_2O \longrightarrow H_2N{-}\overset{\overset{\displaystyle O}{\|}}{C}{-}NH_2 + 2ADP + 2P_i + AMP + PP_i$$

16. The human body is capable of synthesizing all of the nucleotide components of DNA and RNA.

17. The digestion of ingested nucleic acids involves their hydrolysis into free nucleotides by the enzymes ribonuclease and deoxyribonuclease. These free nucleotides can then be used by the body for the synthesis of DNA and RNA.

18. Excess nucleotides from the diet are catabolized and their products excreted. The product of catabolism of purine nucleotides is uric acid. The disease gout involves the synthesis of abnormally large amounts of uric acid.

PROBLEMS

1. What is the difference between protein digestion and protein turnover?

2. The structural formulas for a few amino acids are shown below. Draw structural formulas for their precursors and briefly describe their enzymatic synthesis.

(a) 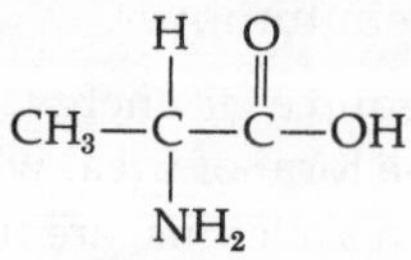 $CH_3-\overset{\overset{\displaystyle H}{|}}{\underset{\underset{\displaystyle NH_2}{|}}{C}}-\overset{\overset{\displaystyle O}{\|}}{C}-OH$

   alanine

(b) 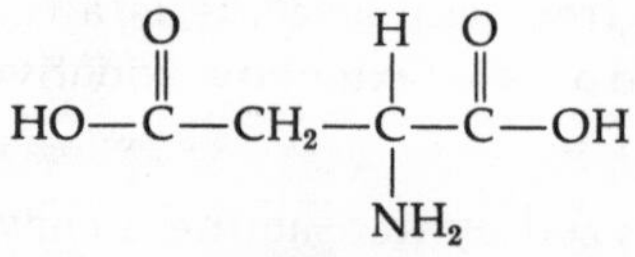 $HO-\overset{\overset{\displaystyle O}{\|}}{C}-CH_2-\overset{\overset{\displaystyle H}{|}}{\underset{\underset{\displaystyle NH_2}{|}}{C}}-\overset{\overset{\displaystyle O}{\|}}{C}-OH$

   aspartic acid

(c) 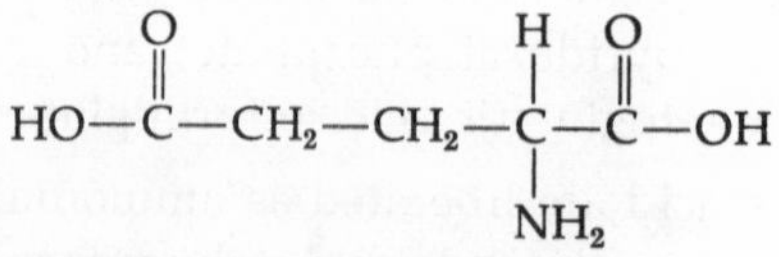 $HO-\overset{\overset{\displaystyle O}{\|}}{C}-CH_2-CH_2-\overset{\overset{\displaystyle H}{|}}{\underset{\underset{\displaystyle NH_2}{|}}{C}}-\overset{\overset{\displaystyle O}{\|}}{C}-OH$

   glutamic acid

(d) 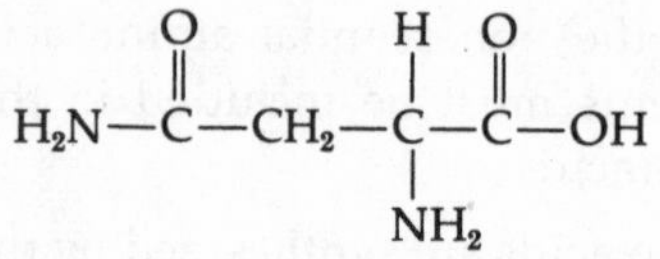 $H_2N-\overset{\overset{\displaystyle O}{\|}}{C}-CH_2-\overset{\overset{\displaystyle H}{|}}{\underset{\underset{\displaystyle NH_2}{|}}{C}}-\overset{\overset{\displaystyle O}{\|}}{C}-OH$

   asparagine

3. What is the function of aspartic acid in the urea cycle? How is this aspartic acid regenerated?

4. What is common in the intestinal fates of protein and nucleic acid polymers?

5. Draw the structural formulas for the products of the following reactions.

(a) 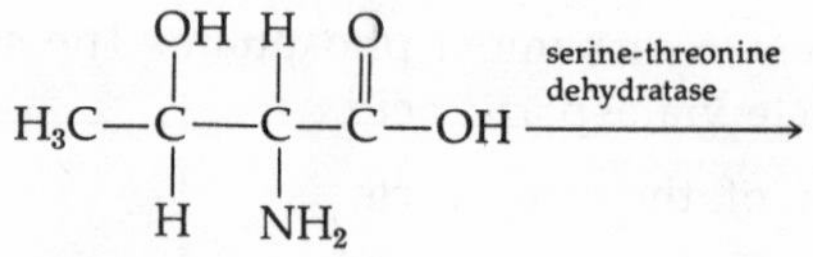 $H_3C-\overset{\overset{\displaystyle OH}{|}}{\underset{\underset{\displaystyle H}{|}}{C}}-\overset{\overset{\displaystyle H}{|}}{\underset{\underset{\displaystyle NH_2}{|}}{C}}-\overset{\overset{\displaystyle O}{\|}}{C}-OH \xrightarrow{\text{serine-threonine dehydratase}}$

(b) 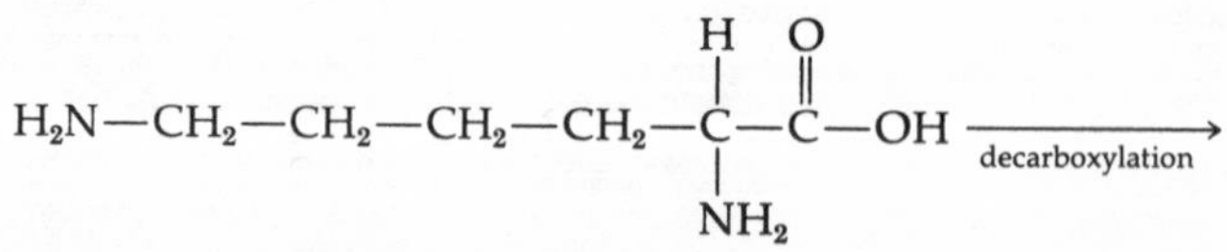 $H_2N-CH_2-CH_2-CH_2-CH_2-\overset{\overset{\displaystyle H}{|}}{\underset{\underset{\displaystyle NH_2}{|}}{C}}-\overset{\overset{\displaystyle O}{\|}}{C}-OH \xrightarrow{\text{decarboxylation}}$

(c) $\begin{matrix} H_3C \\ \diagdown \\ CH \\ \diagup \\ H_3C \end{matrix}-\overset{\overset{\displaystyle H}{|}}{\underset{\underset{\displaystyle NH_2}{|}}{C}}-\overset{\overset{\displaystyle O}{\|}}{C}-OH + HO-\overset{\overset{\displaystyle O}{\|}}{C}-CH_2-CH_2-\overset{\overset{\displaystyle O}{\|}}{C}-\overset{\overset{\displaystyle O}{\|}}{C}-OH \xrightarrow{\text{transamination}}$

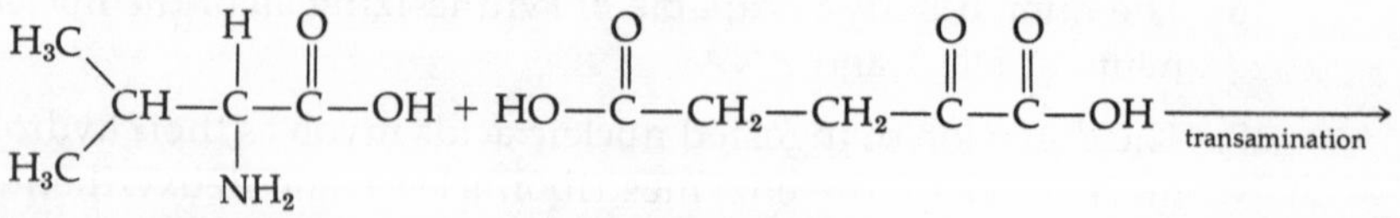

(d) 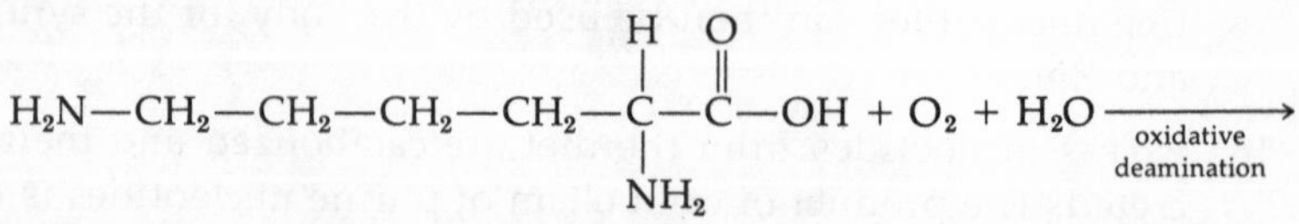 $H_2N-CH_2-CH_2-CH_2-CH_2-\overset{\overset{\displaystyle H}{|}}{\underset{\underset{\displaystyle NH_2}{|}}{C}}-\overset{\overset{\displaystyle O}{\|}}{C}-OH + O_2 + H_2O \xrightarrow{\text{oxidative deamination}}$

6. The transamination of alanine to pyruvic acid is accompanied by the conversion of pyridoxal phosphate to pyridoxamine phosphate. How can pyridoxal phosphate continue to function as a coenzyme when it is continually converted to pyridoxamine phosphate?

SOLUTIONS TO EXERCISES

28-1   trypsin   chymotrypsin

$$N—lys—phe—met—ser—ala—gly—tyr—C$$

pepsin

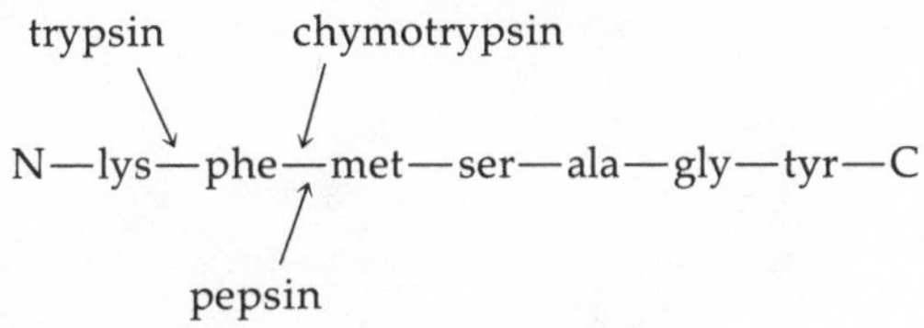

   (a)   N-lys-phe-C + N-met-ser-ala-gly-tyr-C
   (b)   lys + N-phe-met-ser-ala-gly-tyr-C
   (c)   same as (a)
   (d)   lys + phe + N-met-ser-ala-gly-tyr-C

28-2   A healthy individual on a high-protein diet will be in a state of nitrogen balance. Although more nitrogen is taken in, an equal amount is excreted. A fasting individual, however, will be in a state of negative nitrogen balance because body proteins will be used for energy production.

28-3   (a)

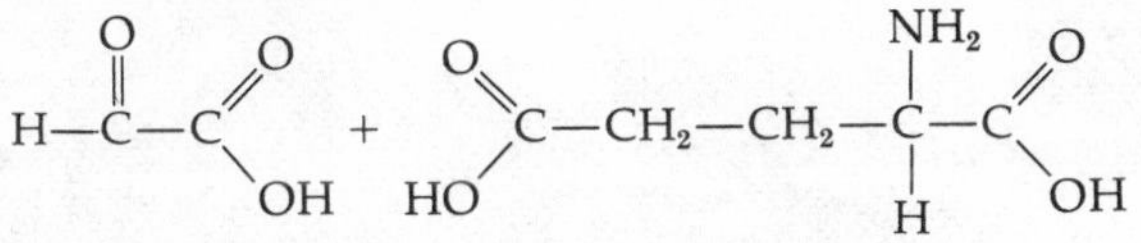

(structures)

   (b) 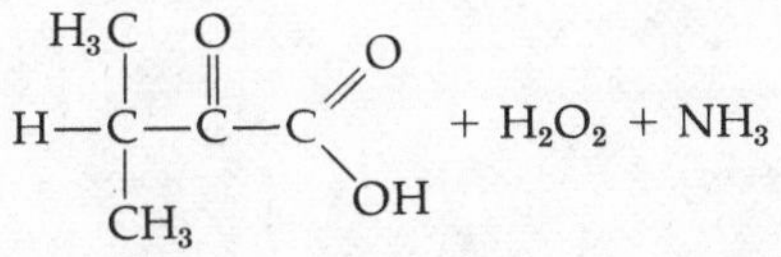 (structure) $+ H_2O_2 + NH_3$

   (c) $H_2O + \frac{1}{2}O_2$

   (d) 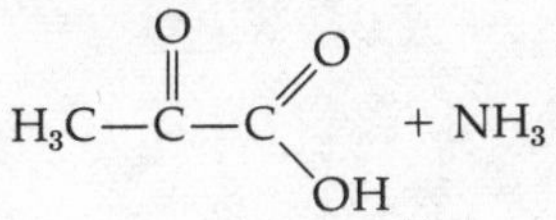 (structure) $+ NH_3$

28-4   The glutamic acid produced may then be used to donate an amino group to form urea. The amino group may be removed via reaction 28-7 as ammonia, or donated to oxaloacetic acid in reaction 28-11. The α-ketoglutaric acid is an intermediate in the TCA cycle.

28-5   One is immediately released as inorganic phosphate (P_i). The other is released, as P_i, in the second step of the urea cycle.

28-6   One enters the cycle as free ammonia in step 1 after release from glutamic acid. Alternatively, the amino group may come from an oxidase or deaminase reaction. The second amino group is transferred from glutamic acid to an oxaloacetic acid molecule by aspartic acid transaminase, and it enters the urea cycle as part of aspartic acid.

28-7   No. The synthesis of tyrosine requires a supply of phenylalanine, an essential amino acid.

28-8   **(Optional)** Decarboxylation of an amino acid or amino acid derivative is common to the synthesis of serotonin, epinephrine, and histamine. Hydroxylation is common to the synthesis of serotonin, norepinephrine, and DOPA.

28-9   These enzymes catalyze the hydrolysis of DNA and RNA into component nucleotides.

28-10   No. DNA must be protected against turnover in order to preserve the integrity of the cell's genetic information.

CHAPTER 29

The Integration and Regulation of Metabolism

29-1 INTRODUCTION

On several occasions we have explained biochemical concepts by using analogies between a living cell and a factory, and between a living organism and an industrial society. We are all aware of the interdependence of various industries in society. In this final chapter we shall examine the interdependence of the different metabolic pathways in the cells of the human body. No single metabolic pathway can function without regard to the needs of the whole body. Metabolic pathways are responsive to conditions of supply and demand. For example, when a sufficient supply of ATP is on hand to meet the body's needs, the oxidation of carbohydrates and fats in cells is retarded. We shall see that there are key control points for metabolic pathways which serve to maintain the body in proper working order under normal conditions.

Some biomolecules are common participants in a number of metabolic pathways. For example, acetyl-S-CoA is the starting material for fatty acid and steroid biosyntheses and, among other things, it is also the fuel for the TCA cycle. The fate of any given molecule of acetyl-S-CoA is determined by the demands of the various interconnected pathways.

In order for cellular factories to function smoothly to promote the well-being of the entire body, they must have an adequate supply of raw materials. Part of this chapter will therefore deal with the nutritional requirements of the cells in the human body. Even with proper nutrition, however, there are times when cells do not function properly. Fortunately, most of these disorders can now be detected and treated. The techniques for detection generally involve sampling a patient's blood or other body fluids, and determining what biomolecules are present in abnormally small or large amounts. The identity and quantity of these molecules is an indication of the nature and severity of the metabolic disorder. We have already seen, for example, that the concentration of glucose in the blood is abnormally high in untreated diabetics. Our discussion of diseases and their diagnosis is far from complete. It should serve, however, as a bridge between your study of basic chemical and biochemical concepts in this course and more specific studies of the human body in health and disease.

For centuries people have transformed the earth in order to cultivate crops that are rich sources of proteins, carbohydrates, and the other nutrients required for a balanced diet.

29-2 STUDY OBJECTIVES

After studying the material in this chapter, you should be able to:

1. Describe the interconnections of glycolysis and the TCA cycle, glycolysis and triglyceride metabolism, and the TCA cycle and the urea cycle.

2. Describe the central role of acetyl-S-CoA in metabolism.

3. Understand how the activities of pyruvate carboxylase and acetyl-S-CoA carboxylase are controlled, as well as the necessity for these controls.

4. Predict the effects of a diet that is unusually high in lipids, proteins, or carbohydrates on the operation of the metabolic pathways we have discussed.

5. Define nutrient and essential nutrient, and explain what is meant by the expression "recommended daily allowance" as applied to nutrition.

6. Explain why some nutrients are needed in much larger quantities than other nutrients.

7. List at least three factors that influence the daily nutritional needs of any individual.

8. Outline the major functions of extracellular fluids.

9. Describe what is meant by the terms electrolyte balance, pH balance, acidosis, and alkalosis.

29-3 METABOLIC INTERRELATIONSHIPS

Many different metabolic pathways operate in a typical cell at any given time. Exactly which metabolic pathways are functioning and to what extent depends on several factors, including the supply of nutrients, the need for energy, and the concentrations of the products of various pathways.

In the preceding chapters we have discussed several metabolic pathways, including glycolysis, gluconeogenesis, fatty acid synthesis and catabolism, and the TCA cycle. We also discussed some points of connection between specific pathways, and the involvement of acetyl-S-CoA in a number of pathways. The interrelationships of the major pathways we have studied are summarized schematically in Figure 29-1.

We have also discussed the different types of controls that regulate metabolic activity, including (1) allosteric enzymes, (2) hormones, (3) chemical modification of enzymes, and (4) the concentrations of substrates and products. In addition, you should recognize that different cells in the body are doing different things at any given time. For example, glycogen may be broken down to glucose in the liver, but the glucose is shipped to other cells (including nerve and muscle cells) for glycolysis. We shall now examine how the interconnections and controls of the various metabolic pathways work together for the well-being of the individual cell and the entire body.

Glycolysis and Gluconeogenesis

Recall that glycolysis is the major metabolic pathway for carbohydrate catabolism (Chapter 25). The catabolism (breakdown) of carbohydrates via glycolysis

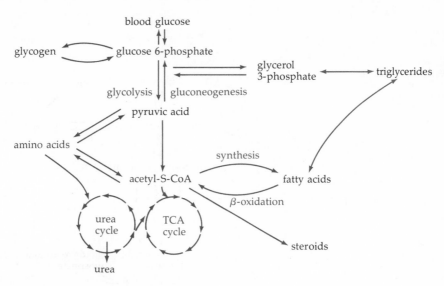

Figure 29-1   The interrelationships of some of the major metabolic pathways in human cells. (The electron transport system and pathways for nucleic acid and protein synthesis are not shown.)

and the β-oxidation of fatty acids are the major sources of metabolic energy in human cells. Glycolysis is interconnected with a number of metabolic pathways, including gluconeogenesis (glucose synthesis), glycogen synthesis and catabolism, the pentose phosphate pathway, the synthesis and catabolism of some amino acids, the synthesis and catabolism of glycerides and phosphoglycerides, and the TCA cycle (see Figure 29-2). All of the reactions that connect these pathways have already been presented in the preceding chapters, as were the three major control points indicated in Figure 29-2. Recall from Chapters 23 and 25 that the hormones epinephrine and glucagon turn off the formation of glycogen and initiate the breakdown of glycogen to glucose 1-phosphate, which is then catabolized via glycolysis in response to an energy demand on the body as a whole (control point 1 in Figure 29-2). The second major control point is responsive to the energy needs of the individual cells in the body. At this point, the rate of glycolysis is increased if the ATP supply is low, and it is decreased if the ATP supply is high. Recall that this occurs by means of the allosteric activation or inhibition of the enzymes fructose 1,6-diphosphatase and phosphofructokinase by ATP and other molecules.

The third control point indicated in Figure 29-2 involves the control of oxaloacetic acid synthesis by acetyl-S-CoA. Acetyl-S-CoA is produced by the catabolism of fatty acids, some amino acids, and glycolysis. When acetyl-S-CoA is produced in excessive amounts, it allosterically activates the enzyme pyruvate carboxylase, which makes oxaloacetic acid (see Figure 29-3). The oxaloacetic acid that is formed can be used in gluconeogenesis or in the synthesis of certain amino acids such as aspartic acid. Oxaloacetic acid is also used in the first step of the TCA cycle. The fate of a particular molecule of oxaloacetic acid will thus depend on other needs and controls. Sometimes more oxaloacetic acid is needed to combine with acetyl-S-CoA in the TCA cycle. At other times an excess of acetyl-S-CoA may indicate that sufficient fuel is already available for the TCA cycle and the oxaloacetic acid may then be used for gluconeogenesis. In the latter case, pyruvic acid is used to make glucose (via oxaloacetic acid) instead of being converted to more acetyl-S-CoA.

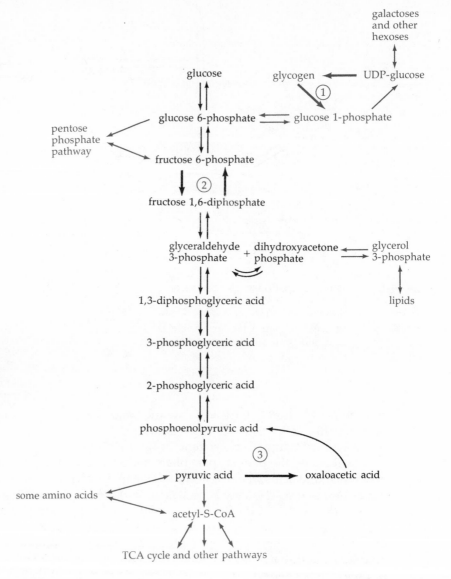

Figure 29-2   Connections between glycolysis and gluconeogenesis (shown in black) and some other metabolic pathways (shown in blue). The heavier arrows represent reactions involved in the three major control points that are discussed in the text.

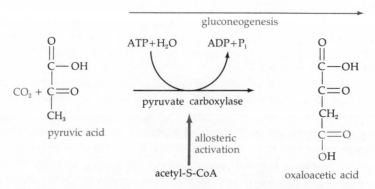

Figure 29-3   Pyruvate carboxylase is allosterically activated by acetyl-S-CoA, thereby increasing the rate of gluconeogenesis.

Lipid Metabolism

Recall that triglycerides are the major energy storage depot in the human body. Other lipids have a variety of functions (Chapter 26). The synthesis and catabolism of all lipids are connected to other metabolic pathways via acetyl-S-CoA (see Figure 29-4). Notice in this figure that metabolism of the glycerol portions of triglycerides is connected to glycolysis, as we explained in Chapter 27.

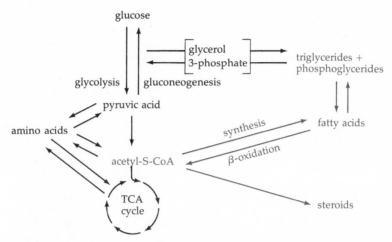

Figure 29-4   Pathways for the metabolism of lipids (shown in blue) are connected to other metabolic pathways by acetyl-S-CoA and glycerol 3-phosphate.

Fatty acids, you will recall, are components of most lipids. Fatty acids are synthesized from acetyl-S-CoA and can also be broken down via β-oxidation to acetyl-S-CoA. The synthesis and catabolism of fatty acids must be controlled, so that these two pathways do not continually run in a wasteful circle. A major control of fatty acid synthesis appears to be the rate of synthesis of malonyl-S-CoA from acetyl-S-CoA in step 1 of fatty acid synthesis (see Figure 29-5). The activity of this enzyme is allosterically increased by citric acid and is inhibited by palmitoyl-S-CoA. Recall that citric acid is an intermediate in the TCA cycle, and the concentration of citric acid will build up if the concentration of acetyl-S-CoA is large. Palmitoyl-S-CoA, the CoA thioester of the 16-carbon fatty acid palmitic acid, is the major product of fatty acid biosynthesis and acts as an end-product inhibitor of the first step in this pathway.

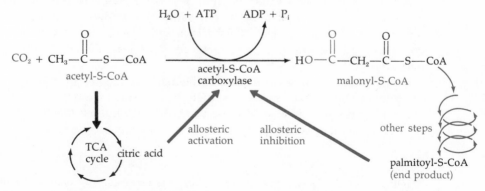

Figure 29-5   The first step in fatty acid biosynthesis, catalyzed by the enzyme acetyl-S-CoA carboxylase, is accelerated by excess citric acid and inhibited by excess palmitoyl-S-CoA.

The regulation of fatty acid catabolism is more complex. One major control of fatty acid catabolism is the activity of the lipase enzymes, which cleave triglycerides into glycerol and fatty acids. The hormone insulin slows down fatty acid catabolism by inhibiting lipase action in adipose (fat) cells when sufficient glucose is present to meet the body's energy needs.

The TCA and Urea Cycles

Two major connections to the TCA cycle are glycolysis and the β-oxidation of fatty acids, which provide acetyl-S-CoA—the fuel for this cycle. The interrelationships between the TCA cycle and other metabolic pathways are summarized in Figure 29-6.

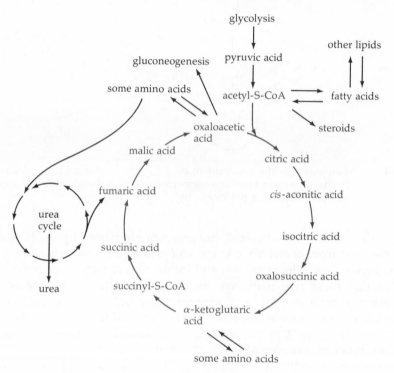

Figure 29-6   Connections between the TCA cycle
and other metabolic pathways.

In Chapter 28 we noted that the urea cycle is intimately connected to the TCA cycle. The TCA cycle is also interconnected to the synthesis and catabolism of a number of amino acids. The TCA cycle is additionally connected to gluconeogenesis via oxaloacetic acid, which is a common intermediate in both of these pathways.

We have touched briefly on a few of the major metabolic controls operating in the human body. Much more is known about these metabolic controls. However, there is still a great deal to be learned about human metabolism and its control. Considerable research is being done in this area.

Exercise 29-1

The final product of each round of the TCA cycle is a molecule of oxaloacetic acid, which *may* then combine with another molecule of acetyl-S-CoA to start the TCA cycle again. Explain why this cannot go on indefinitely without supplementing the supply of oxaloacetic acid. (Hint: See Figures 29-2 and 29-3.)

Exercise 29-2
People who consistently eat too much sugar tend to get fat. Under these conditions, explain how fatty acid synthesis is stimulated.

29-4 NUTRITIONAL REQUIREMENTS

In the preceding chapters we generally assumed that cells had sufficient supplies of all the raw materials needed to operate their metabolic pathways. Unfortunately, this is not always the case. In fact, poor nutrition is an ever-increasing problem around the world, not just in underdeveloped nations, but in parts of industrialized nations as well.

Good nutrition involves the regular ingestion of sufficient, but not overwhelming, amounts of all of the necessary nutrients. These **nutrients** include oxygen, water, such energy sources as carbohydrates and lipids, and a wide variety of other substances including proteins, vitamins, and minerals. The amounts of each type of nutrient that are required are shown in Figure 29-7. Notice the large variations in our requirements for different nutrients.

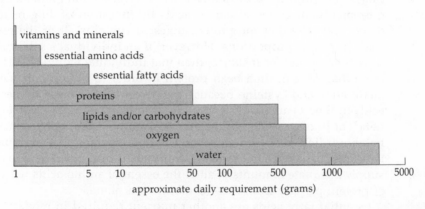

Figure 29-7   Approximate daily requirements (in grams) for the various nutrients required by the human body. Note that a sliding scale is used in this figure so that requirements for water (about 2800 g/day) and essential amino acids (about 2 g/day) can fit on the same graph. In spite of this device, the daily requirements for vitamins and minerals (ranging from 0.2 g/day down to 10^{-6} g/day) cannot be represented accurately on this scale.

Nutrients Required in Large Amounts (more than 100 g/day)

Water is needed in far greater amounts than any other nutrient. On the average, a total of almost 3 kg of water (from liquids and solid foods) is normally taken in daily by an adult, and approximately half of this is excreted in urine. The water in sweat and urine serves as a solvent for the elimination of some by-products of metabolism. The evaporation of water in perspiration also aids in temperature regulation.

The **oxygen** we inhale (about 0.6 to 0.8 kg/day, in about 2500 liters of air) is used almost exclusively as the final electron acceptor in the electron transport system, and is then eliminated in the form of water. Molecular oxygen also serves as a substrate for a few enzyme-catalyzed reactions.

Our daily requirements for **carbohydrates** and **lipids** reflect our daily energy needs. Adults with average energy needs require between 2000 and 3000 kcal* of energy per day. This corresponds to roughly 500 to 600 g of carbohydrates or 250 to 300 g of lipids per day. If a person's diet contains both carbohydrates and lipids, then these amounts are reduced proportionately; for example, 250 to 300 g of carbohydrates *plus* 125 to 150 g of lipids will satisfy the energy needs of the average adult. Note that lipids supply more calories per gram than carbohydrates (Chapter 27). You should keep in mind that these numbers represent *average* requirements. The actual amounts required can be quite different for persons with unusual energy needs. A lumberjack, for example, may require twice these amounts of carbohydrates and/or lipids in order to supply enough energy for the heavy labor required by that occupation.

Nutrients Required in Moderate Amounts (1 to 100 g/day)

Adults typically require about 50 g of **proteins** per day. In the last chapter we saw that the amino acids derived from digestion of proteins enter the body's amino acid pool, where they may be incorporated into newly synthesized proteins, converted (in some cases) to hormones, used for the synthesis of other nitrogen-containing biomolecules, or oxidized for energy production. The figure 50 g of protein per day is a very rough approximation. The actual requirement is for a certain amount of each of the **essential amino acids** (those that cannot be synthesized by human cells). A typical adult may satisfy his or her requirements for essential amino acids by ingestion of 40 g or less of protein daily, provided that the protein ingested contains all of the essential amino acids in proper proportions. However, if an individual's sole source of protein were lima beans, for instance, then that individual would need to ingest much more than 50 g of lima bean protein in order to obtain adequate amounts of methionine and cysteine because the percentage of these two essential amino acids in lima beans is quite low. Because many sources of dietary protein are deficient in one or more of the essential amino acids, nutritionists recommend **balanced diets** that contain several protein sources. For example, a diet containing equal amounts of protein from wheat and beans or rice and beans can supply adequate amounts of all of the essential amino acids with a total of 50 g of protein per day from these combinations alone.

Essential fatty acids are another nutrient required in moderate amounts, as was discussed in Chapter 26. Briefly, the average human requires about 5 g of these polyunsaturated fatty acids per day, an amount easily obtained in a normal diet. This requirement may not be met, however, when hospitalized patients are fed intravenously for extended periods of time.

Nutrients Required in Small Amounts (less than 1 g/day)

The nutrients in this group are the **vitamins** and several **minerals,** including the elements Fe, Zn, F, Mg, Mn, Cu, I, Mo, Se, Cr, Sn, and Co. Minimum daily requirements for vitamins and minerals have been established on the basis of the minimum amount needed to prevent certain deficiency diseases. For example, the lack of an adequate amount of vitamin C can cause a disease called **scurvy,** in which collagen synthesis is impaired. For some vitamins, the daily ingestion of larger amounts may be needed for peak operating efficiency and longevity. For example, it has been suggested that large daily doses of vitamin C may help to prevent colds and heart disease. However, excessive amounts of some vitamins, such as vitamin A, can be harmful. We understand the functions

* Recall that the kilocalorie is equivalent to the Calorie unit used by nutritionists.

of some vitamins and minerals, but the precise metabolic functions of others are not clearly understood. The very small amounts of vitamins and elements that are required indicates that they are not used up rapidly by the body. Most, in fact, are used as recyclable coenzymes or as prosthetic groups for proteins.

Variations in Required Daily Allowances

Published lists of recommended daily allowances should be viewed as only rough estimates of human nutritional needs. There are a number of factors that influence the daily nutritional requirements of any given individual, some of which have already been mentioned. These include:

1. The source of the nutrient, as in the case of dietary protein.

2. The health and physical state of the individual—growing children, pregnant and lactating women, and people recovering from an illness may need larger than average amounts of certain nutrients, whereas sedentary individuals need less fat or carbohydrate.

3. The overall diet, which influences the amounts of various components needed (for example, if a person is on a high-protein diet, he or she will need less carbohydrate or fat for energy).

4. A number of other factors, including symbiotic intestinal bacteria (which manufacture some vitamins for us), the composition of ingested foods, and the climate.

Exercise 29-3
On a diet in which all protein is obtained from lima beans, which of the following amino acids is not likely to be oxidized for energy production: (a) methionine, (b) glutamic acid, (c) alanine? Explain your answer.

Exercise 29-4
Figure 29-7 shows some very large differences in recommended daily allowances. Explain why hundreds of grams of carbohydrates or fat are required daily but only 5 g of essential fatty acids.

29-5 BODY FLUIDS

The transport of required nutrients to all of the cells in the body, the transmission of hormonal messages, and several other vital body functions are dependent upon the blood and other extracellular fluids. **Extracellular fluids** include blood plasma, spinal fluid, saliva, and other fluids that occur outside of body cells. About 20% of human body weight is due to extracellular fluids, and three-quarters of this extracellular fluid is found around and between individual cells. This fluid is called **interstitial fluid.** Blood plasma, the noncellular fluid portion of whole blood, accounts for most of the remaining 25% of the extracellular fluid. In previous chapters we have discussed several functions of blood, including the transport of nutrients such as glucose, lipids (bound to albumin or other carrier proteins), and oxygen (bound to hemoglobin in red blood cells), as well as the transport of hormones and metabolic waste products such as urea.

Blood plasma and interstitial fluid also function in the maintenance of the correct pH, ionic, and osmotic environment for body cells. The volumes of these extracellular fluids, and their pH and electrolyte concentrations, are therefore normally kept under precise control. Large variations in extracellular fluid composition cause related changes in body cells that are harmful and often deadly. Likewise, when body cells behave abnormally, there is likely to be a change in the composition of the extracellular fluids. Changes in the composition of body fluids, particularly that of plasma, since it is easy to obtain, are now used to diagnose certain diseases and metabolic disorders (see Table 29-1).

Table 29-1 Some Diagnostic Uses of Blood Plasma

Plasma Component	Normal Concentration	Abnormalities	
		Disorder	Concentrations
H^+ (measured as pH)	7.35–7.45	Acidosis	Less than 7.35
		Alkalosis	Greater than 7.45
CO_2 (measured as P_{CO_2})	35–40 mm Hg	Respiratory acidosis	Above 40 mm Hg
K^+	4.0–4.8 meq/liter	Renal disease	Less than 4 meq/liter
Cholesterol	150–250 mg/100 ml	Obstructive jaundice	250–500 mg/100 ml
		Biliary cirrhosis	Up to 1800 mg/100 ml
Glucose	80–120 mg/100 ml	Diabetes	150 mg/100 ml to 1000 mg/100 ml
Gammaglobulins	0.5–1.6 g/100 ml	Multiple myeloma	Above 2 g/100 ml
Glutamic oxaloacetic transaminase	9–25 Karmen units*	Myocardial infarction (heart attack)	50–400 Karmen units

* The activity of some enzymes is expressed using special units.

Notice that a wide variety of plasma components—from simple ions to complex proteins—are used for the diagnosis of a large number of disorders. During the treatment of many diseases the concentration of plasma components is also monitored regularly to assess the patient's progress and the effectiveness of the treatment.

pH Balance

Three interrelated mechanisms—**kidney function, respiration rate,** and **buffer actions**—are used to maintain blood pH within the normal limits of 7.35 to 7.45. The buffer systems include H_2CO_3/HCO_3^- (the major buffer), $H_2PO_4^-/HPO_4^{2-}$, and the buffering action of many proteins. Recall that carbonic acid, H_2CO_3, is in rapid equilibrium with dissolved CO_2 (reaction 29-1) and also with bicarbonate and H_3O^+ ions (reaction 29-2).

(29-1) $$CO_2(aq) + H_2O \xrightleftharpoons{\text{carbonic anhydrase}} H_2CO_3$$

(29-2) $$H_2CO_3 + H_2O \rightleftharpoons HCO_3^- + H_3O^+$$

Normally, on a short-term basis, the positions of these equilibria and hence the blood pH are controlled by the proper breathing rate. If the respiration rate is very low, however, a condition known as **hypoventilation,** not enough $CO_2(g)$ is expelled from the lungs and the partial pressure of $CO_2(g)$ in the lungs increases. As a consequence, the concentration of dissolved CO_2 in the blood plasma increases. Also, $[H_2CO_3]$, $[HCO_3^-]$, and $[H_3O^+]$ rise, as the positions of the equilibria for reactions 29-1 and 29-2 respond to the increase in $[CO_2(aq)]$. This condition, with its abnormally low blood pH, is referred to as **respiratory acidosis** (Table 29-2). Respiratory acidosis can be caused by asthma, pneumonia, drug overdoses, and other dysfunctions where insufficient CO_2 is exhaled.

Table 29-2 Alterations in Blood Plasma During Respiratory Acidosis and Alkalosis

	pH	$[H_3O^+]$	$[HCO_3^-]$	$[H_2CO_3]$
Respiratory acidosis	Below normal	Above normal	Above normal	Above normal
Respiratory alkalosis	Above normal	Below normal	Below normal	Below normal

On the other hand, prolonged **hyperventilation** (overbreathing) during, for example, anxiety, hysteria, crying, or mountain climbing, results in a decrease in the partial pressure of CO_2 in the lungs and a consequent decrease in $[CO_2(aq)]$. In this case $[H_2CO_3]$, $[HCO_3^-]$, and $[H_3O^+]$ decrease as the positions of the equilibria for reactions 29-1 and 29-2 respond to the decrease in $[CO_2(aq)]$. This condition, with its abnormally high blood pH, is referred to as **respiratory alkalosis** (see Table 29-2). The partial pressure of $CO_2(g)$ that can be in equilibrium with blood plasma is used as a measure of the concentration of carbon dioxide dissolved in the blood and also as a measure of $[H_2CO_3]$. Recall that the solubility of a gas is directly proportional to the partial pressure of the gas. The equilibrium constant expression for reaction 29-1 is

$$K_{eq} = \frac{[H_2CO_3]}{[CO_2(aq)]}$$

Therefore, $[H_2CO_3] = K_{eq}[CO_2(aq)]$, and $[H_2CO_3]$ is directly proportional to $[CO_2(aq)]$. In this relationship K_{eq} is actually very small (about 2.5×10^{-3}). Thus when $CO_2(g)$ dissolves in pure water, almost all of the dissolved carbon dioxide exists as $CO_2(aq)$, and very little of it is in the form of H_2CO_3.

The kidneys also help maintain the pH of blood plasma at its proper level. The kidneys can respond to a stress that tends to alter the pH of blood plasma from its normal value with more long-term effectiveness than can be achieved by an alteration of breathing rate.

For example, in an acidosis condition, where the blood plasma pH is below normal, the kidneys produce very acidic urine (pH as low as 4, compared to an average value of about 6). In this process,

1. $CO_2(aq)$ enters from the blood into the kidney cells, where it is used to produce H_3O^+ and HCO_3^-.

2. H_3O^+ ions go into the urine from the kidney cells, and an equal number of positively charged Na^+ ions enter the cells from the urine.

3. Na^+ ions and HCO_3^- ions enter the blood from the kidney cells (see Figure 29-8). The influx of HCO_3^- into the blood and $CO_2(aq)$ out of the blood shifts the equilibrium for reaction 29-2 to the left.

As we mentioned previously in this section, $[H_2CO_3]$ is directly proportional to $[CO_2(aq)]$, and according to equilibrium 29-2, $[H_3O^+]$ in blood plasma is related to the ratio $[H_2CO_3]/[HCO_3^-]$ by the expression

(29-3) $$[H_3O^+] = \frac{[H_2CO_3]}{[HCO_3^-]} \times (8.0 \times 10^{-7} \text{ mole/liter})$$

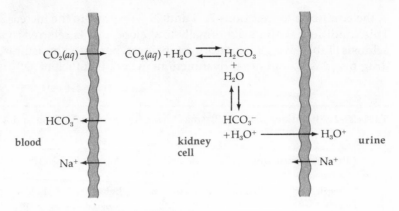

Figure 29-8   The response of kidney cells to a
condition of acidosis.

Thus, when $CO_2(aq)$ goes from the blood into the kidney cells, $[H_2CO_3]$ in the
blood decreases. This effect and the entry of HCO_3^- into the blood from the
kidney cells results in a decrease in the $[H_2CO_3]/[HCO_3^-]$ ratio and a decrease
in $[H_3O^+]$ by Eq. 29-3, thus helping to restore the blood pH to its normal level.

Electrolyte Balance

The kidneys also aid in the maintenance of correct concentrations of several
other electrolytes in the blood. Figure 29-9 shows the normal concentrations
of electrolytes in body fluids. Note that, whereas plasma and interstitial fluid
have similar compositions, the composition of the fluid within cells is quite
different. Also notice that each fluid is electrically neutral, containing equal
amounts of positive and negative equivalents. In this figure the concentration
unit milliequivalents per liter is used to put ions with different charges on an
equivalent basis.

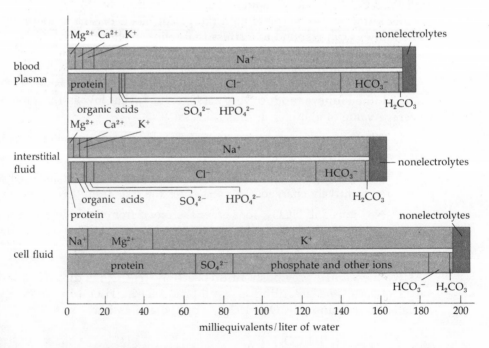

Figure 29-9   The normal concentrations of electrolytes in body fluids.

The term **electrolyte balance** is used to refer to the normal situation in which the concentrations of the electrolytes in body fluids are kept within close tolerances of their normal values in spite of the intake and removal of large volumes of water and solutes. Proper kidney function is primarily responsible for normal electrolyte balance in blood plasma and interstitial fluids. Specific membrane transport proteins are in turn responsible for the maintenance of the correct intracellular fluid composition. For example, the protein Na^+K^+-ATPase pumps K^+ ions into cells and Na^+ ions out of cells into the interstitial fluid.

For electrolyte balance, the overall input of water (in the form of ingested liquids, water in ingested solid foods, and water produced as a by-product of metabolic catabolism) must equal the overall output of water (eliminated in the urine and feces, from the lungs as water vapor, and through the skin as perspiration). There must also be a balance of the water within the body among blood plasma, interstitial fluid, and cell fluid. Recall that there will be a net flow of water across a semipermeable membrane separating two solutions unless the solutions have the same osmotic pressure. Also recall that the osmotic pressure of an aqueous fluid is directly proportional to the concentration of *all* the dissolved or dispersed particles in the fluid.

In arid desert climates, people must be especially careful to maintain sufficient quantities of water in their bodies.

The osmotic pressure of the body fluids (and thus the water flow within the human body) is controlled primarily by the kidneys, which regulate the volume of urine that is eliminated. For example, if you decrease the volume of water you ingest, then both the concentration of electrolytes in the body fluids and the osmotic pressure of these fluids increases. When the osmotic pressure of blood increases, the hormone **vasopressin** is released from the pituitary gland. This hormone triggers the reduction of urine output by the kidneys. The net result is that water is retained by the body, and this stabilizes the electrolyte concentrations and osmotic pressure of the body fluids.

Exercise 29-5
The acidic ketone bodies produced in diabetics often cause a condition called metabolic acidosis, which we shall study shortly. How would you expect kidney cells to respond to this condition?

29-6   METABOLIC DISORDERS

The malfunction of any one of the thousands of proteins in the human body will result in illness and, potentially, death. Other diseases are caused by pathogenic organisms and by external stresses that are placed on the body. An example of the latter case is **emphysema,** a progressive deterioration of the respiratory system caused by, among other things, cigarette smoke.

Let us consider a few common metabolic disorders in order to demonstrate that, because of the intricate interrelationships of metabolic pathways, a defect in one pathway can cause major alterations in the overall metabolic performance of the body. Such is the case in the disease **diabetes mellitus,** which might in the future afflict more than 20% of the population of the United States. Recall that diabetes results from an inability to produce a sufficient amount of active insulin and that the best diagnostic indication of diabetes is the glucose tolerance test (Section 25-12). Also recall that insulin normally functions in promoting glucose uptake by cells, stimulates the formation of glycogen, and inhibits the hydrolysis of triglycerides in adipose tissue. Lack of insulin, therefore, results in

1. Increased blood glucose levels

2. Increased fatty acid oxidation

3. Cessation of fatty acid synthesis

4. Increased gluconeogenesis

The abnormal fatty acid metabolism, items 2 and 3, results in an excessive production of acetyl-S-CoA (see Figure 29-10). Not all of this acetyl-S-CoA can be oxidized in the TCA cycle. The excess is converted to ketone bodies (see Chapter 27).

Recall that two of the ketone bodies are acids, so that an overproduction of ketone bodies causes a higher-than-normal $[H_3O^+]$ in blood and a lower-than-normal blood pH. This metabolic disorder is called **metabolic acidosis.** In an attempt to get rid of the increased amount of ketone bodies, the urine volume is increased. The urine volume can increase to such a point that a diabetic may become severely dehydrated. The normal concentration of ketone bodies in blood is less than 1 mg/100 ml, and the normal rate at which they are excreted in the urine is about 20 mg per day. If a person suffering from acute diabetes remains untreated, the concentration of ketone bodies in the blood may reach 90 mg/100 ml, and the rate at which they are excreted in the urine may go as high as 5000 mg per hour.

Since diabetics do not catabolize much glucose, ingested amino acids are also used for energy production and gluconeogenesis, with a concomitant increase in urea production. Thus, the lack of adequate amounts of a single hormone, insulin, can alter the normal functioning of virtually all of the metabolic pathways. This metabolic disruption results in easily detectable changes in the concentration of several components of extracellular fluids, including increased concentrations of glucose in the blood and urea in the urine, and possibly, a decrease in blood pH. These changes can be monitored during therapy.

Metabolic acidosis is a very serious condition. A decrease in blood pH interferes with the ability of hemoglobin to bind molecular oxygen. Recall that hemoglobin, HHb, and oxyhemoglobin, $HHbO_2$, are both weak acids. The effect of a decrease in blood pH on the binding of oxygen to hemoglobin is apparent when we consider the following equilibrium between hemoglobin, the oxyhemoglobin anion, molecular oxygen, and H^+:

(29-4) $$HHb + O_2 \rightleftharpoons HbO_2^- + H^+$$

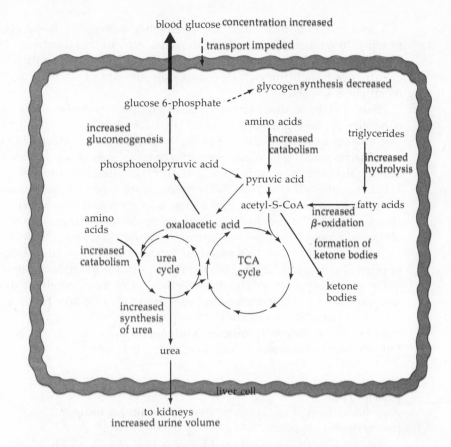

Figure 29-10 A schematic representation of the metabolic malfunctions in diabetes that result from the absence of sufficient active insulin.

A decrease in blood pH, that is, an increase in [H$^+$], will shift the position of equilibrium for reaction 29-4 to the left. Thus, a person suffering from metabolic acidosis has difficulty in providing the body cells with sufficient oxygen, and the person's respiration rate increases in an attempt to get more oxygen.

In an attempt to decrease [H$_3$O$^+$] and restore the blood pH to its normal value, the position of the H$_2$CO$_3$/HCO$_3^-$ buffer equilibrium,

(29-5) $$HCO_3^- + H^+ \rightleftharpoons H_2CO_3$$

is shifted to the right, thus decreasing [HCO$_3^-$] in blood plasma. Under these circumstances, [H$_2$CO$_3$] in blood does *not* increase, as it would appear from reaction 29-5, because the increased breathing rate causes an increase in the amount of CO$_2$(g) expelled by the lungs. The increased expulsion of CO$_2$(g) from the lungs more than offsets the shift in the equilibrium for reaction 29-5 to the right, and as a consequence, [H$_2$CO$_3$], in blood plasma *decreases*.

Table 29-3 compares blood pH, [HCO$_3^-$], and [H$_2$CO$_3$] (or P_{CO_2}) for metabolic acidosis with normal values. Metabolic acidosis can be caused by impairment of kidney function, starvation, and other disorders in addition to diabetes.

Table 29-3 Alterations in Blood Plasma During Metabolic Acidosis or Alkalosis

	pH	[H$_3$O$^+$]	[HCO$_3^-$]	[H$_2$CO$_3$]
Metabolic acidosis	Below normal	Above normal	Below normal	Below normal
Metabolic alkalosis	Above normal	Below normal	Above normal	Above normal

A less common condition than metabolic acidosis is **metabolic alkalosis,** in which the blood plasma pH is above normal. In this disorder, $[HCO_3^-]$ and $[H_2CO_3]$ (or P_{CO_2}) are above normal (see Table 29-3). An overdose of antacids or prolonged vomiting (which reduces the amount of stomach acid) can cause metabolic alkalosis.

Many metabolic disorders are not as complex as diabetes. In **galactosemia** (Chapter 25), for example, an enzyme that catalyzes the conversion of galactose to glucose is missing. Galactose is one-half of the milk disaccharide lactose. The other half is glucose. Clinical symptoms of galactosemia result from a buildup of abnormal amounts of galactose, not from generalized metabolic alterations. Likewise, **phenylketonuria** (PKU) results from a genetic deficiency of the enzyme that converts the essential amino acid phenylalanine to tyrosine. If tyrosine is provided in the diet there will be no difficulty in protein synthesis, but abnormal amounts of phenylalanine may build up unless the patient is placed on a diet that is low in this essential amino acid. The prime clinical symptom of phenylketonuria is an excessive concentration of phenylalanine in the blood.

These are only a few of the diseases that result from metabolic disorders. An understanding of their molecular basis has made effective treatment possible. The chemical and biochemical concepts you have learned in the preceding chapters will enable you to better understand not only normal body functions but also metabolic malfunctions and their treatment.

Exercise 29-6

A person suffering from untreated diabetes mellitus is likely to have which of the following clinical symptoms?

    (a) Below-normal urine pH and below-normal plasma $[HCO_3^-]$
    (b) Below-normal urine pH and above-normal plasma $[HCO_3^-]$
    (c) Above-normal urine pH and below-normal plasma $[HCO_3^-]$
    (d) Above-normal urine pH and above-normal plasma $[HCO_3^-]$
    (e) Normal urine pH and above-normal plasma $[HCO_3^-]$

29-7   SUMMARY

1. There are three major controls of carbohydrate metabolism. The first regulates the synthesis of glycogen and its degradation for use in glycolysis. The second controls glycolysis versus gluconeogenesis in response to the cell's energy needs. The third regulates gluconeogenesis in response to variations in the acetyl-S-CoA concentration.

2. Acetyl-S-CoA is a central intermediate connecting glycolysis and the metabolism of fatty acids and amino acids; it also serves as the entry point into the TCA cycle.

3. Fatty acid biosynthesis is stimulated by citric acid, which allosterically activates acetyl-S-CoA carboxylase; it is suppressed by palmitoyl-S-CoA, which allosterically inhibits this enzyme.

4. The TCA cycle and amino acid metabolism, as well as the TCA cycle and the urea cycle, share some common intermediates.

5. Humans require vastly different amounts of various nutrients, ranging from 10^{-6} to 10^3 g per day.

6. The actual nutritional requirements of the human body depend on the age, health, and level of activity of each individual. Recommended daily

allowances are average minimum requirements that have been suggested by nutritionists.

7. The extracellular fluids of the body function to (a) transport nutrients and wastes; (b) maintain the correct pH, ionic, and osmotic environment; and (c) transport hormones between cells.

8. Kidney function, respiration rate, and buffer actions are the mechanisms used to maintain blood pH within normal limits.

9. Hypoventilation can cause respiratory acidosis, whereas hyperventilation can cause respiratory alkalosis.

10. Electrolyte balance and water balance are regulated by the kidneys.

11. Metabolic disorders are diagnosed and monitored by determination of the concentrations of various components in several extracellular fluids (especially blood).

12. Diabetes is a metabolic disorder that involves the malfunction of several metabolic pathways.

13. Metabolic acidosis can be caused by diabetes, impaired kidney function, or starvation, whereas metabolic alkalosis can be caused by an antacid overdose or prolonged vomiting.

PROBLEMS

1. As you walk down a dark alley in the middle of the night, you hear a strange sound. You become frightened and start to run.
   (a) Your fear causes the release of epinephrine from cells in the adrenal gland. What effect will this have on glycogen and triglyceride metabolism?
   (b) In order to run, your muscle cells must use up large amounts of ATP. What effect will this rapid use of ATP have on glycolysis and gluconeogenesis in these muscle cells?
   (c) You continue to run for several miles. (You are really scared!) What is the major product of glucose catabolism in your muscle cells?
   (d) You stop running at last. What happens to the product of glycolysis from part (c)?

2. You are overweight and decide to go on a crash diet. There are several popular fad diets from which you can choose. What are some potential problems you might encounter with the following choices?
   (a) A diet totally devoid of unsaturated fats
   (b) A pure "liquid" protein diet
   (c) A cholesterol-free diet

3. As you climb a mountain, the partial pressure of oxygen in the atmosphere decreases and you may overbreathe to try to increase the partial pressure of oxygen in your lungs.
   (a) What effect, if any, would this overbreathing have on the partial pressure of CO_2 in your lungs? Explain your answer.
   (b) What effect, if any, might this overbreathing have on the pH of your blood plasma?

4. A person suffering from a stomach ulcer ingests an excessive amount of bicarbonate. Indicate by symbols whether the following are likely to have a value that is higher than normal (+), lower than normal (−), or normal (0).
   (a) Plasma pH
   (b) Plasma $[HCO_3^-]$
   (c) Plasma $[H_2CO_3]$
   (d) Respiration rate

5. Would you expect the hormone vasopressin to be released when you engage in prolonged strenuous exercise on a hot day?

6. Consider the two enzymes phosphofructokinase (PFK) and fructose 1,6-diphosphatase (FDP). When a person engages in vigorous exercise, which of the following will occur?
   (a) Both enzymes bind ATP at an allosteric site and the activity of both increases.
   (b) Both enzymes bind ADP at an allosteric site and the activity of PFK increases while the activity of FDP decreases.
   (c) FDP binds ATP at an allosteric site and its activity increases, while PFK binds ADP at an allosteric site and its activity also increases.
   (d) PFK binds ADP at an allosteric site and its activity decreases, while FDP binds ATP at an allosteric site and its activity increases.
   (e) None of the above.

7. Which of the following can result in a blood plasma pH that is above the normal value?
   (a) A woman panting during labor
   (b) Untreated diabetes
   (c) Overstrenuous exercise
   (d) A person nearly drowning
   (e) None of the above

SOLUTIONS TO EXERCISES

29-1 In addition to participating in the TCA cycle, oxaloacetic acid is used in gluconeogenesis and for the synthesis of aspartic acid. Thus, without replenishment, all of the oxaloacetic acid would eventually be drained off from the TCA cycle.

29-2 The large amounts of sugar you eat are catabolized and produce large amounts of acetyl-S-CoA, which allosterically activates acetyl-S-CoA carboxylase—the enzyme required for the first step of fatty acid biosynthesis.

29-3 (a) Methionine, since it is present in only relatively small amounts in lima bean protein.

29-4 Carbohydrate and/or fat must be catabolized to provide energy. Essential fatty acids, however, are generally not catabolized, but used by body cells for the synthesis of comparatively small amounts of more complex lipids.

29-5 Kidney cells will increase the acidity of the urine in an attempt to restore the plasma pH to normal (see Figure 29-8).

29-6 (a) The [HCO_3^-] in blood plasma is below normal (see Table 29-3), and kidney cells will acidify the urine in an effort to raise the plasma pH.

ESSENTIAL SKILLS

ESSENTIAL SKILLS 1

Exponential Numbers

The number 3×10^2 is an example of an **exponential number.** The factor 3 in this number is called the **coefficient** and the factor 10^2 is called the **exponential.** The superscript 2 in the exponential 10^2 indicates multiplication by 10 two times. The superscript is called the **power of ten** or the **exponent.** Therefore, the number $3 \times 10^2 = 3 \times 10 \times 10 = 3 \times 100 = 300$.

(1)

Exponential number

$$\overbrace{3 \times \underline{10^2}} \leftarrow \text{exponent}$$

coefficient   exponential

The preferred way to write an exponential number, called **standard scientific notation,** is with one digit before the decimal point in the coefficient. The number 9.82×10^5, for example, is written in standard scientific notation.

In general, the exponential 10^n, with n a positive integer, is equal to multiplication by 10 n times. Also, $10^0 = 1$. For example, the exponential number 1.65×10^5 is equal to $1.65 \times 10 \times 10 \times 10 \times 10 \times 10$ or 165,000. Notice that when 1.65 is multiplied by five powers of 10, the decimal point moves five places to the right. Thus:

(2)   Multiplying a number by 10 and moving the decimal point one place to the right are equivalent operations.

The number 10^{-2} is equal to $1/10^2$; in general, the exponential 10^{-n} is equal to $1/10^n$. Thus, for example, the exponential number 1.4×10^{-3} is equal to $1.4/10^3$ or $1.4/(10 \times 10 \times 10)$ or 0.0014. Notice that when 1.4 is divided by three powers of 10, the decimal point moves three places to the left. Thus:

(3)   Dividing a number by 10 and moving the decimal point one place to the left are equivalent operations.

Relationships (2) and (3) are the basis for the following two facts, which can be used to convert a decimal number into an exponential number in a very simple manner.

(4)   **Fact 1**   The value of an exponential number is unchanged when the decimal point is moved one place to the left, provided that the exponent is increased by one at the same time.

Moving the decimal point one place to the left is equivalent to dividing the number by 10, and increasing the exponent by one is equivalent to multiplying the number by 10. Dividing a number by 10, and at the same time multiplying it by 10, does not change the value of the number. For example, the implied position of the decimal point in the number 60315 is after the last digit. Therefore $60315 = 60315.0 \times 10^0$ (since $10^0 = 1$). If we use Fact 1 repeatedly, we obtain

$$
\begin{aligned}
60315 \times 10^0 &= 6031.5 \times 10^1 \\
&= 603.15 \times 10^3 \\
&= 60.315 \times 10^3 \\
&= 6.0315 \times 10^4 \quad \text{and so on}
\end{aligned}
$$

In standard scientific notation, the number 60315 is written as 6.0315×10^4.

(5)   **Fact 2**   The value of an exponential number is unchanged when the decimal point is moved one place to the right, provided that the exponent is decreased by one at the same time.

Moving the decimal point one place to the right is equivalent to multiplying the number by 10, and decreasing the exponent by one is equivalent to dividing the number by 10. Multiplying a number by 10, and at the same time dividing it by 10, does not change the value of the number. For example, the number $0.0023 = 0.0023 \times 10^0$, since $10^0 = 1$. If we use Fact 2 repeatedly, we obtain

$$
\begin{aligned}
0.0023 \times 10^0 &= 0.023 \times 10^{-1} \\
&= 0.23 \times 10^{-2} \\
&= 2.3 \times 10^{-3} \\
&= 23 \times 10^{-4} \quad \text{and so on}
\end{aligned}
$$

In standard scientific notation, the number 0.0023 is written as 2.3×10^{-3}.

When you want to write a number in exponential form or move the decimal point in a number, it is a good idea to keep in mind that:

1. If you decrease the coefficient of an exponential number, you must increase the exponent.

2. If you increase the coefficient, you must decrease the exponent.

Writing numbers in exponential form is particularly valuable for very large or very small numbers. For example, chemists often need to use the number 602200000000000000000000, which is called Avogadro's number. It is simpler to write Avogadro's number in standard scientific notation as 6.022×10^{23}.

Exercise 1
Write the following exponential numbers in decimal form:

(a) 5.23×10^{-4} (b) 7.90×10^5
(c) 61×10^3 (d) 0.93×10^{-2}

Exercise 2
Write the following numbers in standard scientific notation:

(a) 5060000   (b) 0.00000029   (c) 1350000000
(d) 0.000001   (e) 78×10^6 (f) 0.23×10^{-5}

ARITHMETIC OPERATIONS WITH EXPONENTIAL NUMBERS

Multiplication and Division

It is easy to multiply and divide exponential numbers using the following rules:

Multiplication Rule   To multiply two exponential numbers, multiply the coefficients and add the exponents.

For example,

$$(6.2 \times 10^2) \times (5.0 \times 10^3) = (6.2 \times 5.0) \times 10^{(2+3)}$$
$$= 31.0 \times 10^5$$
$$= 3.1 \times 10^6$$

and

$$(4.3 \times 10^4) \times (3.0 \times 10^{-6}) = (4.3 \times 3.0) \times 10^{(4-6)}$$
$$= 12.9 \times 10^{-2}$$
$$= 1.29 \times 10^{-1}$$

Division Rule   To divide two exponential numbers, divide the coefficient of the numerator by the coefficient of the denominator and subtract (algebraically) the exponent of the denominator from the exponent of the numerator.

For example,

$$\frac{8.4 \times 10^5}{2.0 \times 10^3} = \left(\frac{8.4}{2.0}\right) \times 10^{(5-3)} = 4.2 \times 10^2$$

and

$$\frac{2.0 \times 10^{-4}}{5.0 \times 10^{-6}} = \left(\frac{2.0}{5.0}\right) \times 10^{[-4-(-6)]} = 0.4 \times 10^2 = 4.0 \times 10^1$$

and

$$\frac{10^{-5}}{5 \times 10^{-3}} = \frac{1 \times 10^{-5}}{5 \times 10^{-3}} = \left(\frac{1}{5}\right) \times 10^{[-5-(-3)]} = 0.2 \times 10^{-2} = 2.0 \times 10^{-3}$$

Try working the above examples using the numbers in decimal form. Don't you agree that working with exponential numbers is easier?

Exercise 3

Express the answer to each of the following in standard scientific notation:

(a) $(3.0 \times 10^{10}) \times (2.5 \times 10^6) =$ (b) $(6.6 \times 10^{-27}) \times (3.0 \times 10^{10}) =$

(c) $(5.0 \times 10^{-8}) \times (2.0 \times 10^{-4}) =$ (d) $(6.0 \times 10^{10}) \div (3.0 \times 10^4) =$

(e) $(5.2 \times 10^{-8}) \div (2.1 \times 10^{-5}) =$ (f) $(1.8 \times 10^{-6}) \div (8.6 \times 10^{-10}) =$

(g) $\dfrac{(2.2 \times 10^{-6}) \times (4.3 \times 10^{-8})}{7.6 \times 10^{10}} =$ (h) $\dfrac{(5.1 \times 10^{-4}) \times (2.2 \times 10^9)}{4.8 \times 10^{-10}} =$

Addition and Subtraction

Addition and subtraction with exponential numbers is also quite easy using the following rule:

> **Addition and Subtraction Rule**   To add or subtract exponential numbers, first change the exponent of all but one of the numbers so that all the numbers have the same exponent (using Fact 1 and/or Fact 2). Then add or subtract the coefficients.

For example,

$$(5.30 \times 10^3) + (4.1 \times 10^2) = (5.30 \times 10^3) + (0.41 \times 10^3)$$
$$= (5.30 + 0.41) \times 10^3$$
$$= 5.71 \times 10^3$$

and

$$(6.4 \times 10^{-4}) - (1.33 \times 10^{-3}) = (0.64 \times 10^{-3}) - (1.33 \times 10^{-3})$$
$$= (0.64 - 1.33) \times 10^{-3}$$
$$= -0.69 \times 10^{-3}$$
$$= -6.9 \times 10^{-4}$$

Exercise 4

Express the answer to each of the following in standard scientific notation:

(a) $(9.61 \times 10^{-4}) + (6.2278 \times 10^{-2}) =$

(b) $(4.53 \times 10^3) - (1.628 \times 10^4) =$

(c) $(4.38 \times 10^3) + (7.2 \times 10^2) - (8.291 \times 10^4) =$

(d) $(2.28 \times 10^{-2}) - (5.3 \times 10^{-3}) + (4 \times 10^{-4}) =$

SOLUTIONS TO EXERCISES

1. (a) Moving the decimal point four places to the left and increasing the exponent by four gives 0.000523.

(b) Moving the decimal point five places to the right and decreasing the exponent by five gives 790,000.

(c) Moving the decimal point three places to the right and decreasing the exponent by three gives 61,000.

(d) Moving the decimal point two places to the left and increasing the exponent by two gives 0.0093.

2. (a) Moving the decimal point six places to the left and multiplying by 10^6 gives 5.06×10^6.

(b) Moving the decimal point seven places to the right and multiplying by 10^{-7} gives 2.9×10^{-7}.

(c) Moving the decimal point nine places to the left and multiplying by 10^9 gives 1.35×10^9.

(d) Moving the decimal point six places to the right and multiplying by 10^{-6} gives 1×10^{-6}.

(e) Moving the decimal point one place to the left and increasing the exponent by one gives 7.8×10^7.

(f) Moving the decimal point one place to the right and decreasing the exponent by one gives 2.3×10^{-6}.

3. (a) $(3.0 \times 10^{10}) \times (2.5 \times 10^6) = (3.0 \times 2.5) \times 10^{(10+6)} = 7.5 \times 10^{16}$

   (b) $(6.6 \times 10^{-27}) \times (3.0 \times 10^{10}) = (6.6 \times 3.0) \times 10^{(-27+10)} = 19.8 \times 10^{-17} = 1.98 \times 10^{-16}$

   (c) $(5.0 \times 10^{-8}) \times (2.0 \times 10^{-4}) = (5.0 \times 2.0) \times 10^{(-8-4)} = 10 \times 10^{-12} = 1.0 \times 10^{-11}$

   (d) $(6.0 \times 10^{10}) \div (3.0 \times 10^4) = \left(\dfrac{6.0}{3.0}\right) \times 10^{(10-4)} = 2.0 \times 10^6$

   (e) $(5.2 \times 10^{-8}) \div (2.1 \times 10^{-5}) = \left(\dfrac{5.2}{2.1}\right) \times 10^{[-8-(-5)]} = 2.5 \times 10^{-3}$

   (f) $(1.8 \times 10^{-6}) \div (8.6 \times 10^{-10}) = \left(\dfrac{1.8}{8.6}\right) \times 10^{[-6-(-10)]} = 0.21 \times 10^4 = 2.1 \times 10^3$

   (g) $\dfrac{(2.2 \times 10^{-6}) \times (4.3 \times 10^{-8})}{7.6 \times 10^{10}} = \left(\dfrac{2.2 \times 4.3}{7.6}\right) \times 10^{(-6-8-10)} = 1.2 \times 10^{-24}$

   (h) $\dfrac{(5.1 \times 10^{-4}) \times (2.2 \times 10^9)}{4.8 \times 10^{-10}} = \left(\dfrac{5.1 \times 2.2}{4.8}\right) \times 10^{[-4+9-(-10)]} = 2.3 \times 10^{15}$

4. (a) $(9.61 \times 10^{-4}) + (6.2278 \times 10^{-2}) = (9.61 \times 10^{-4}) + (622.78 \times 10^{-4})$
   $$= (9.61 + 622.78) \times 10^{-4}$$
   $$= 632.39 \times 10^{-4}$$
   $$= 6.3239 \times 10^{-2}$$

   (b) $(4.53 \times 10^3) - (1.628 \times 10^4) = (0.453 \times 10^4) - (1.628 \times 10^4)$
   $$= (0.453 - 1.628) \times 10^4$$
   $$= -1.175 \times 10^4$$

   (c) $(4.38 \times 10^3) + (7.2 \times 10^2) - (8.291 \times 10^4)$
   $$= (0.438 \times 10^4) \times (0.072 \times 10^4) - (8.291 \times 10^4)$$
   $$= (0.438 + 0.072 - 8.291) \times 10^4$$
   $$= -7.781 \times 10^4$$

   (d) $(2.28 \times 10^{-2}) - (5.3 \times 10^{-3}) + (4 \times 10^{-4})$
   $$= (2.28 \times 10^{-2}) - (0.53 \times 10^{-2}) + (0.04 \times 10^{-2})$$
   $$= (2.28 - 0.53 + 0.04) \times 10^{-2}$$
   $$= 1.79 \times 10^{-2}$$

ESSENTIAL SKILLS 2

Units

Every measured quantity is a multiple of some standard unit. For example, 5 meters refers to a length that is five times the standard unit "meter." Smaller and larger multiples of the standard unit are indicated by prefixes that are powers of ten (see Table 1-1, page 12). Both standard units and prefixes have abbreviations. When you write the abbreviation for a unit that includes a prefix, *no* extra space is included between the prefix and the standard unit. For example, the abbreviation for centimeter is written cm; never write c m.

ARITHMETIC OPERATIONS

Most physical quantities—density, specific heat, concentration, and so on—involve a combination of units. Arithmetic operations with units occur when you perform numerical calculations with the values of physical quantities. The general rule that applies when performing arithmetic operations with units is that they are treated like algebraic quantities. The following examples should make this clear.

EXAMPLE 1   The algebraic product of $(2a) \times (3a) \times (5a)$ is $30a^3$. Similarly, the volume of a box that is 2 cm long, 3 cm wide, and 5 cm high is

$$\begin{aligned} \text{Volume} &= (\text{length}) \times (\text{width}) \times (\text{height}) \\ &= (2 \text{ cm}) \times (3 \text{ cm}) \times (5 \text{ cm}) \\ &= 30 \text{ cm}^3 \end{aligned}$$

30 cm^3 is read "30 cubic centimeters."

EXAMPLE 2   The ratio $1/b$ can be written in the equivalent form b^{-1}, and therefore the ratio a/b can be written as ab^{-1}. Similarly, the density of iron, which is 7.85 g/ml (read as "7.85 grams per milliliter"), can be written as 7.85 g · ml^{-1} (read as "7.85 grams milliliters to the minus one"). Note that when you write a product of different units, such as g · ml^{-1}, a *dot* is included between the two different units.

EXAMPLE 3   The mass of a given volume of a substance can be calculated from the formula

Mass = (density) × (volume)

Thus, the mass of 5.00 ml of iron is

$$\text{Mass} = (7.85\ \text{g/ml}) \times (5.00\ \text{ml})$$

$$= (7.85 \times 5.00)\frac{\text{g}}{\cancel{\text{ml}}} \cdot \cancel{\text{ml}} = 39.3\ \text{g}$$

Note that we have used the general algebraic relationship that

$$\frac{a\cancel{b}}{\cancel{b}} = a$$

EXAMPLE 4   The volume of a given mass of a substance can be calculated from the formula

Volume = mass/density

Thus, the volume of 5.00 g of iron is

$$\text{Volume} = \frac{5.00\ \text{g}}{7.85\ \text{g/ml}} = 0.637\ \frac{\text{g}}{\text{g} \cdot \text{ml}^{-1}} = 0.637\ \frac{\cancel{\text{g}} \cdot \text{ml}}{\cancel{\text{g}}} = 0.637\ \text{ml}$$

Note that we have used the general algebraic relationship that

$$\frac{a}{a/b} = \frac{a}{ab^{-1}} = \frac{\cancel{a}b}{\cancel{a}} = b$$

Be careful. A very common mistake in working with an expression of this kind is the following:

$$\frac{\cancel{a}}{\cancel{a}/b} = \frac{1}{b} \qquad \text{incorrect}$$

This is *not* a true equation.

Exercise 1
Simplify the following expressions involving units:

(a) $\left(\dfrac{\text{cal}}{\text{g}}\right) \times (\text{g})$

(b) $\dfrac{(\text{ml})}{(\text{g/ml})}$

(c) $\dfrac{(\text{ml})}{(\text{ml/g})}$

(d) $\dfrac{(\text{cal})}{\left(\dfrac{\text{cal}}{\text{g} \cdot {}^{\circ}\text{C}}\right) \times (\text{g})}$

CONVERSION OF UNITS

Very often in scientific work, and in everyday experience as well, we need to convert the value of a quantity expressed in terms of one unit into an equivalent expression in terms of another unit. Isn't the common process of changing 50 cents into its equivalent, two quarters, an example of conversion of units?

There is a straightforward method for converting units that is "fail-safe" when it is used correctly. This general method can best be described by working through a specific example.

Suppose that we want to convert a length of 8.45 inches into centimeters. We begin with the equation relating the two units. A length of 1.00 in. is equivalent to a length of 2.54 cm. Therefore,

(1)
$$1.00 \text{ in.} = 2.54 \text{ cm}$$

If we divide both sides of Eq. (1) by 1.00 in., we obtain the following conversion factor, since 1.00 in./1.00 in. = 1:

(2)
$$1 = \frac{2.54 \text{ cm}}{1.00 \text{ in.}}$$

Similarly, if we divide both sides of Eq. (1) by 2.54 cm, we obtain the following conversion factor, since 2.54 cm/2.54 cm = 1:

(3)
$$1 = \frac{1.00 \text{ in.}}{2.54 \text{ cm}}$$

Now, multiplying the length 8.45 in., the length in our example, by 1 does not change its value. Thus we could multiply 8.45 in. by either the conversion factor in (2) or the conversion factor in (3) and not change its value, since both conversion factors have the value 1. The key point in converting units is that we multiply our original quantity by the conversion factor that will give us the *units* we want.

Therefore, to convert the length 8.45 in. into a length with centimeters as the units, we use the conversion factor in (2) because

(4)
$$(8.45 \text{ in.}) \times \frac{2.54 \text{ cm}}{1.00 \text{ in.}} = 21.5 \text{ cm}$$

If we use the conversion factor in (3), we do not obtain the units centimeters because

(5)
$$(8.45 \text{ in.}) \times \frac{1.00 \text{ in.}}{2.54 \text{ cm}} = 3.33 \frac{\text{in.}^2}{\text{cm}}$$

We would, however, use the conversion factor in (3) to convert the length 3.00 cm to a length with inches as the units, because

(6)
$$(3.00 \text{ cm}) \times \frac{1.00 \text{ in.}}{2.54 \text{ cm}} = 1.18 \text{ in.}$$

Notice that in Eq. (4) we used the conversion factor 2.54 cm/1.00 in. and in Eq. (6) we used the conversion factor 1.00 in./ 2.54 cm. The most important and simplifying feature of the method we have described for converting units is that we do not have to remember which of the two conversion factors to use in a particular situation: We simply use the conversion factor that gives us the correct units.

In general, we can convert units by considering that

(7)
$$\left(\begin{array}{c}\text{Quantity with}\\\text{given units}\end{array}\right) \times \left(\begin{array}{c}\text{appropriate}\\\text{conversion factor}\end{array}\right) = \left(\begin{array}{c}\text{quantity with}\\\text{units we want}\end{array}\right)$$

The general method we have described can be used whenever we want to convert units. It can also be used with the prefixes in Table 1-1. For example, suppose that we wish to convert the mass 0.523 kg to a mass with grams as units. From the relationship, 1.000 kg = 1000 g, we can obtain the two conversion factors,

$$1 = \frac{1000 \text{ g}}{1.000 \text{ kg}} \quad \text{and} \quad 1 = \frac{1.000 \text{ kg}}{1000 \text{ g}}$$

In this case, using the conversion factor 1000 g/1.000 kg gives us the units grams, so it is the correct factor to use.

(8)
$$(0.523 \text{ kg}) \times \left(\frac{1000 \text{ g}}{1.000 \text{ kg}}\right) = 523 \text{ g}$$

Exercise 2
Do the following conversions, using the information in Appendix 1:
- (a) 7.3 meters = _____ inches
- (b) 0.52 quart = _____ liter
- (c) 100 kilometers = _____ miles
- (d) 31 grains = _____ grams
- (e) 1.00 milliliter = _____ minims

SOLUTIONS TO EXERCISES

1. (a) $\dfrac{\text{cal}}{(\text{g})} \times (\text{g}) = \text{cal}$

   (b) $\dfrac{(\text{ml})}{(\text{g/ml})} = \dfrac{\text{ml}}{\text{g} \cdot (\text{ml})^{-1}} = \dfrac{\text{ml} \cdot \text{ml}}{\text{g}} = \dfrac{\text{ml}^2}{\text{g}}$

   (c) $\dfrac{(\text{ml})}{(\text{ml/g})} = \dfrac{\text{ml}}{\text{ml} \cdot (\text{g})^{-1}} = \dfrac{1}{\text{g}^{-1}} = \ = \text{g}$

   (d) $\dfrac{(\text{cal})}{\left(\dfrac{\text{cal}}{\text{g} \cdot {}^{\circ}\text{C}}\right) \times (\text{g})} = \dfrac{\text{cal}}{\text{cal} \cdot ({}^{\circ}\text{C})^{-1}} = \dfrac{1}{({}^{\circ}\text{C})^{-1}} = {}^{\circ}\text{C}$

2. (a) $(7.3 \text{ m}) \times \left(\dfrac{39.3 \text{ in.}}{1.00 \text{ m}}\right) = 287 \text{ in.}$

   (b) $(0.52 \text{ quart}) \times \left(\dfrac{1.00 \text{ liter}}{1.06 \text{ quarts}}\right) = 0.49 \text{ liter}$

   (c) $(100 \text{ km}) \times \left(\dfrac{1.00 \text{ mile}}{1.61 \text{ km}}\right) = 62.1 \text{ miles}$

   (d) $(31 \text{ grains}) \times \left(\dfrac{0.0648 \text{ g}}{1.00 \text{ grain}}\right) = 2.0 \text{ g}$

   (e) $(1.00 \text{ ml}) \times \left(\dfrac{1.00 \text{ minim}}{0.0616 \text{ ml}}\right) = 16.2 \text{ minims}$

ESSENTIAL SKILLS 3

Direct Proportionality and the Factor-Unit Method

DIRECT PROPORTIONALITY

The term "proportional to" is often used to describe a relationship between physical quantities. The most frequently encountered form of proportionality in introductory chemistry is **direct proportionality.** The mass of a sample of water, for example, is directly proportional to the volume of the sample. Let us consider this relationship between mass and volume. The mass of 2 ml of water is twice the mass of 1 ml (doubling the volume doubles the mass), the mass of 3 ml is three times the mass of 1 ml (tripling the volume triples the mass), the mass of $\frac{1}{2}$ ml is one-half the mass of 1 ml (halving the volume halves the mass), and so on. These facts about the mass and volume for samples of water can be summarized as follows: For samples of water (or samples of any other substance) at a fixed temperature, the *ratio* of the mass of the sample compared to the volume of the sample has a constant value. This statement can be put into the following mathematical form:

(1)
$$\frac{\text{Mass}}{\text{Volume}} = k$$

In Eq. (1), k is a constant, which we call the density. We can algebraically rearrange Eq. (1) to give

(2)
$$\text{Mass} = k \times \text{volume} \quad \text{or} \quad \text{Mass} = \text{density} \times \text{volume}$$

Equations (1) and (2) are examples of a direct-proportionality relationship. In general, for quantities A and B, when

(3)
$$A = k \times B \quad \text{or} \quad \frac{A}{B} = k$$

we say "A is **directly proportional** to B" or "A and B are **directly proportional** to one another," and we call the constant k the **proportionality constant,** that is, the amount of A per unit amount of B. For example, the density of a substance is its mass per unit volume.

THE FACTOR-UNIT METHOD

Given the volume of a sample of a substance, it is easy to determine the mass of the sample by using the substance's density. For example, let us determine the mass of 3.00 ml of iron given that the density of iron is 7.85 g/ml. Using Eq. (2), we obtain

(4) $$\text{Mass of 3.00 ml of iron} = \left(7.85\ \frac{g}{ml}\right) \times (3.00\ ml) = 23.6\ g$$

Now, it is quite useful to consider the density of a substance as a factor that relates the sample's mass to its volume. From this point of view, determining that the mass of 3.00 ml of iron is 23.6 g by using the density factor

$$7.85\ \frac{g}{ml} = \frac{7.85\ g}{1.00\ ml}$$

is analogous to converting 3.00 in. to 7.62 cm by using the conversion factor 2.54 cm/1.00 in.

There is also a simple factor that relates the volume of a sample to the sample's mass. For example, another way of saying "7.85 g of iron has a volume of 1.00 ml" is to say that "1.00 ml of iron has a mass of 7.85 g." Thus, the factor 1.00 ml/7.85 g can be used to determine the volume of a sample of iron given the sample's mass. The volume of 10.0 g of iron, for example, is

(5) $$\text{Volume of 10.0 g of iron} = 10.0\ g \times \frac{1.00\ ml}{7.85\ g} = 1.27\ ml$$

When you convert centimeters to inches or inches to centimeters, you do not have to remember which conversion factor to use,

$$\frac{1.00\ in.}{2.54\ cm} \quad \text{or} \quad \frac{2.54\ cm}{1.00\ in.}$$

Just use the conversion factor that gives you the correct units! Similarly, if you know the density of a substance, and you are given the volume of a sample of the substance, you can determine the mass of the sample by using the factor that gives you the correct units. We call this problem-solving method the **factor-unit method.**

Let us solve a problem using this factor-unit approach. The density of copper is 8.92 g/ml. What is the volume of 25.0 g of copper? The correct unit for volume is milliliters. If you multiply 25.0 g by the factor 1.00 ml/8.92 g, you will get an answer with the units ml. Therefore,

(6) $$\text{Volume of 25.0 g of copper} = (25.0\ g) \times \frac{1.00\ ml}{8.92\ g} = 2.80\ ml$$

The factor-unit method is applicable to problems involving the mass and the volume of a sample of a substance because these two quantities are directly proportional to one another. Similarly, the factor-unit method can be used for problems involving *any* two quantities that are *directly proportional* to one another. Thus, if two quantities A and B are directly proportional to one another,

(7)   (Amount of A) = (amount of B) × (appropriate factor)

If, for example, you are given a known amount of one quantity and you want to determine the amount of the other, then:

(8)   Quantity to be determined = (known quantity) × (appropriate factor)

Most of the quantitative problems you will face in an introductory chemistry course involve two quantities that are directly proportional to one another, and for this type of problem, the factor-unit method we have described is an easy way of obtaining the correct numerical answer. Consider, for example, the following problem: A certain drug is to be administered so that the mass of the drug given a person (the dosage) is *directly proportional* to that person's weight. Suppose that the correct drug dosage for a 120-lb person is 25 mg. What is the correct drug dosage for a 150-lb person?

We are told that the drug dosage is directly proportional to the person's weight, so the factor-unit method can be applied. We therefore use the information given in the problem to find the appropriate factor.

Since the correct drug dose for a 120-lb person is 25 mg, we have the factors

(A)   120 lb of person per 25 mg of drug   or $\left(\dfrac{120 \text{ lb of person}}{25 \text{ mg of drug}} \right)$

and

(B)   25 mg of drug per 120 lb of person   or $\left(\dfrac{25 \text{ mg of drug}}{120 \text{ lb of person}} \right)$

In our problem, since we want to determine the correct drug dosage for a 150-lb person, we want to multiply the quantity "150 lb of person" by the appropriate factor to get an answer with the unit "mg of drug." The appropriate factor is (B) above, so

(9)   Correct dose for 150-lb person = (150 ~~lb of person~~) × $\left(\dfrac{25 \text{ mg of drug}}{120 \text{ ~~lb of person~~}} \right)$

= 31 mg of drug

As a check on our work, note that a drug dose of 31 mg for a 150-lb person is a reasonably larger amount than the 25-mg dosage requirement for a 120-lb person.

Exercise 1

A certain drug is to be administered so that the dosage (the mass of the drug given a person) is directly proportional to the person's body surface area. The correct dosage for a person with a body surface area of 0.95 m² is 20 mg. What would be the correct dosage for a person with a body surface area of 0.60 m²?

Exercise 2

The mass of oxygen in a sample of pure water is directly proportional to the mass of the water sample. A 100-g sample of water contains 89 g of oxygen. What would be the mass of oxygen in a 72-g sample of water?

GENERALIZATIONS OF DIRECT PROPORTIONALITY

A quantity A can be directly proportional to more than one other quantity. For example, when

(10) $$A = k \times B \times C$$

we say that "A is directly proportional to *both* B and C." For example, if C remains the same and B is doubled, then A is doubled. If B remains the same and C is doubled, then A is also doubled. But if both B and C are doubled, then A is multiplied by 4 (quadrupled).

For example, the heat input to a body is related to the mass of the body and the temperature change by the following relationship (see Eq. 1-24, page 36):

(11) $$\text{Heat input} = (\text{specific heat}) \times (\text{mass}) \times (\text{temperature increase})$$

Thus the heat input is directly proportional to *both* the mass of the body and the temperature increase of the body. In this case the proportionality constant k is called the specific heat.

To raise the temperature of 1.0 g of water 1.0°C requires a heat input of 1.0 cal. The heat input required to raise the temperature of 2.0 g of water 2.0°C is four times larger, or 4.0 cal.

PROPORTIONALITY

As we mentioned previously, the statement "A is proportional to B" indicates a more general relationship between A and B than the statement "A is directly proportional to B." A few examples should indicate the usage of the term "proportional to."

EXAMPLE 1 The kinetic energy of a moving object is related to its mass (m) and its speed (s) (Section 1-5):

(12) $$\text{Kinetic energy} = \tfrac{1}{2} m s^2$$

We can express the content of Eq. (12) by saying that "the kinetic energy of a moving object is directly proportional to its mass and proportional to the square of its speed (or its speed squared)."

EXAMPLE 2 The pressure and volume of a fixed amount of a gas at a fixed temperature are related by Boyle's law (Section 5-3):

(13) $$\text{Boyle's law:} \quad \text{Pressure} = \frac{k}{\text{volume}}$$

The content of Boyle's law is expressed by the statement: "The pressure of a fixed amount of gas at a fixed temperature is inversely proportional to its volume." The relationship "inversely proportional to" reflects the fact that when the volume is doubled the pressure is halved, when the volume is tripled, the pressure drops to one-third its previous value, and so on.

EXAMPLE 3   Part of Coulomb's law (Section 1-9) is that the force between two electrical charges is inversely proportional to the square of the distance between them. This statement implies the following mathematical relationship between force and distance for electrical charges:

(14)
$$\text{Force} = \frac{k}{(\text{distance})^2}$$

Exercise 3

The rate of a certain chemical reaction depends on the concentrations of two substances, A and B. It is found that: (1) doubling the concentration of A and leaving the concentration of B the same doubles the rate; and that (2) doubling both the concentration of A and B causes the rate to become eight times larger.

(a) Is the rate of this reaction directly proportional to the concentration of A?
(b) Is the rate of this reaction directly proportional to the concentration of B?

SOLUTIONS TO EXERCISES

1. The factor-unit method can be applied because dosage and body surface area are directly proportional to one another. We have the factors $\dfrac{0.95 \text{ m}^2}{20 \text{ mg}}$ and $\dfrac{20 \text{ mg}}{0.95 \text{ m}^2}$ from the given data. The latter factor gives us the correct units:

$$\text{Dosage} = (0.60 \text{ m}^2) \times \left(\frac{20 \text{ mg}}{0.95 \text{ m}^2}\right) = 13 \text{ mg}$$

2. The factor-unit method can be applied because the mass of oxygen in water and the mass of water are directly proportional to one another. From the data given, we have the factors $\dfrac{100 \text{ g of water}}{89 \text{ g of oxygen}}$ and $\dfrac{89 \text{ g of oxygen}}{100 \text{ g of water}}$. The latter factor gives us the correct units:

$$\text{Mass of oxygen} = (72 \text{ g of water}) \times \left(\frac{89 \text{ g of oxygen}}{100 \text{ g of water}}\right) = 64 \text{ g}$$

3. (a) Yes. If the rate of the reaction is directly proportional to the concentration of A, then doubling the concentration of A and leaving the concentration of B the same should double the rate. This is the case.
    (b) No. If the rate of reaction were directly proportional to both the concentration of A and the concentration of B, then doubling both the concentration of A and B should cause the rate to become four times larger. This is not the case. Since the rate is directly proportional to the concentration of A [see part (a)], the rate is not directly proportional to the concentration of B.

ESSENTIAL SKILLS 4

Significant Figures: Rounding Off the Result of a Calculation

Significant figures are the actual numbers we obtain as a result of a measurement. When we compare two measurements of the same physical quantity, the measurement with more significant figures is the more precise measurement. Thus, for example, if the length of the same object is measured with two different instruments and values of 1.054 cm and 1.05 cm are obtained, then the value 1.054 cm, with four significant figures and an uncertainty of ±0.001 cm, is a more precise measurement than the value 1.05 cm, with only three significant figures and an uncertainty of ±0.01 cm (see Section 1-6).

DECIMAL POINT, ZEROS, AND SIGNIFICANT FIGURES

The number of significant figures in a measurement does *not* depend on the position of the decimal point. For example, 134.5 cm and 1.345 m are two equivalent ways of expressing the *same* measurement. Each measurement has four significant figures and the same uncertainty (±0.1 cm or, equivalently, ±0.001 m).

In determining the number of significant figures in a measured value, a zero can present a problem. A zero in a number can be either a significant figure (the result of a measurement) or may be used merely to indicate the position of the decimal point. For example, the zero in the length 1.054 cm is significant. It tells us that the measured number of tenths of a centimeter in this length is zero. On the other hand, the zeros in the length 0.0025 cm are not significant. Here the zeros just indicate the position of the decimal point.

Note that in the measurement 2.40 cal, the implied certainty is ±0.01 cal, whereas in the measurement 2.4 cal, the implied uncertainty is ±0.1 cal. Thus the zero in the measurement 2.40 cal is a significant figure. In the measurement 2400 cal, however, the function of the zeros is not clear. They may be significant figures, or they may be used merely to position the decimal point.

The difficulty with zero being used for two purposes is eliminated when numbers are written in exponential form. In an exponential number, the exponent (the power of 10) is used to indicate the position of the decimal point, so that for a number written in exponential form, a zero is just as significant as any other digit. For example, the measurement 2.400×10^3 cal has four significant figures and an uncertainty of $\pm 0.001 \times 10^3$ or 1 cal, whereas the measurement 2.4×10^3 cal has two significant figures and an uncertainty of $\pm 0.1 \times 10^3$ or 100 cal.

Exercise 1
Determine the number of significant figures in each of the following:

  (a) 2051   (b) 8.060×10^{-2} (c) 0.0052   (d) 9.1×10^3

ROUNDING OFF THE RESULT OF A CALCULATION

In scientific calculations, we perform arithmetic operations with measured values. We must know the precision (the uncertainty) in our calculated value if we are going to use our calculated value as the basis for an inference about a physical system. The uncertainty in a calculated value is related to the uncertainties in the measured values used in the calculation. We shall first consider the operations of multiplication and division and then the operations of addition and subtraction.

Multiplication and Division

Let us consider a specific example. Suppose that we measure the mass and volume of a sample of mercury and obtain the values 1.738 g and 0.128 ml. Using the equation, density = mass/volume, and with the aid of a small electronic calculator, we obtain a value of 13.578125 g/ml for the density of mercury. Common sense tells us that this calculated value, with an implied uncertainty of ± 0.000001 g/ml, is more precise than would be expected. The uncertainty in the mass measurement is ± 0.001 g, and the uncertainty in volume is ± 0.001 ml.

The mass of our sample could actually be 1.739 g; then the density would be 13.5859375 g/ml. Or the mass could be 1.737 g; then the density would be 13.570312 g/ml. Thus an uncertainty of ± 0.001 g in the mass measurement produces an uncertainty of almost ± 0.01 g/ml in the calculated value of the density (13.570312 − 13.578125 − 13.5859375). Similarly, the volume could actually be 0.129 ml; then the density would be 13.472868 g/ml. Or the volume could be 0.127 ml; the density would be 13.685039 g/ml. Therefore, an uncertainty of ± 0.001 ml results in an uncertainty of ± 0.1 g/ml in the calculated value of the density (13.472868 − 13.578125 − 13.685039). Both the uncertainty in the mass measurement and the uncertainty in the volume measurement cause an uncertainty in the calculated value of the density. But notice that in our example the uncertainty in volume causes the larger uncertainty.

Since the larger uncertainty in density is ± 0.1 g/ml, we should round off our calculated value so that the last digit retained is in the tenths place. When rounding off, we increase the last digit to be retained by one if the following digit is 5 or greater. Therefore, the rounded-off value for our calculated density of mercury is 13.6 g/ml, with an implied uncertainty of ± 0.1 g/ml.

You do not have to go through this procedure every time you do a multiplication or a division to determine the uncertainty in your calculated value. Fortunately, there is a much simpler method, which utilizes the concept of significant figures. Notice that in our example there are four significant figures in the measured value of the mass and three significant figures in the measured value of the volume. Extending the idea of significant figures to include calculated values (and rounding off appropriately), there are three significant figures in the calculated value of the density. The number of significant figures in the calculated value of the density is equal to the number of significant figures in the volume, which has fewer significant figures than the mass measurement.

A similar conclusion applies to any multiplication or division and is the basis for the following simple rule for determining the uncertainty in a calculated value obtained by multiplication or division.

> **Rule for Multiplication and Division**   When multiplying or dividing, round off the result so that the number of significant figures in the result is equal to the number of significant figures in the factor that has the *fewest* number of significant figures.

When applying this rule, we can either: (1) multiply or divide the given factors and then round off the result to the correct number of significant figures; or (2) first round off each factor to the correct number of significant figures and then proceed. The latter procedure is easier when multiplying or dividing without the aid of a calculator. Consider the following examples.

EXAMPLE 1

$$(2.12) \times (1.0517) = 2.229604 = 2.23$$
$$\underset{\text{round off}}{\underset{\uparrow}{}}$$

In this example there are three significant figures in the first factor and five in the second, so the result is rounded off to three significant figures. Equivalently,

$$(2.12) \times (1.05) = 2.226 = 2.23$$
$$\underset{\text{round off}}{\underset{\uparrow}{}}$$

EXAMPLE 2

$$\frac{(2.185) \times (4.21)}{(0.11)} = 83.6259 = 84 \qquad \text{or} \qquad \frac{(2.2) \times (4.2)}{(0.11)} = 84.0000 = 84$$
$$\qquad\qquad\underset{\text{round off}}{\underset{\uparrow}{}}\qquad\qquad\qquad\qquad\qquad\qquad\underset{\text{round off}}{\underset{\uparrow}{}}$$

In this example, the factor 0.11 has the fewest number of significant figures—two. Therefore, the result is rounded off to two significant figures.

Exercise 2

Perform the following operations and round off your answers to the correct number of significant figures.

(a) $(7.328) \times (5.23) =$

(b) $\dfrac{9.2648}{2.2} =$

(c) $(8.20 \times 10^2) \times (1.524 \times 10^{-8}) =$

(d) $\dfrac{(5.2607 \times 10^6) \times (1.2 \times 10^{-2})}{(2.76 \times 10^{-4})} =$

Addition and Subtraction

It is very simple to determine the uncertainty in a calculated value that is obtained by addition or subtraction by using the following rule:

Rule for Addition and Subtraction

Step 1 Determine which of the numbers has an uncertainty in the *largest* decimal place (the decimal place farthest to the *left*).

Step 2 Round off all the numbers to this decimal place.

Step 3 Add or subtract the rounded-off numbers.

For example, to add 35.12 cm, 1.572 cm, and 0.01664 cm, we proceed as follows:

Step 1 35.12 is the number with an uncertainty in the largest decimal place (the hundredths place).

Step 2 1.572 cm is rounded off to 1.57 cm, and 0.01664 cm is rounded off to 0.02 cm.

Step 3 35.12 cm + 1.57 cm + 0.02 cm = 36.71 cm.

Note the justification for this rule. In our example, 35.12 cm has an uncertainty in the hundredths place, so the result of adding 1.572 cm and 0.01664 cm to 35.12 cm will also be uncertain in the hundredths place.

$$
\begin{array}{l}
\quad\quad\text{uncertainty} \\
\quad\quad\quad\downarrow \\
35.1\overset{}{2} \\
1.572 \\
\underline{0.01664} \\
36.70864 \rightarrow 36.71 \\
\quad\quad\uparrow \\
\textbf{round off}
\end{array}
$$

Exercise 3

Perform the following operations and round off your answer to indicate the proper uncertainty in your result.

(a) $1.06 + 9.2387 - 8.132 =$

(b) $278.3 - 7.26 + 32.105 =$

(c) $(7.8 \times 10^3) + (5.6 \times 10^2) =$

(d) $(5.33 \times 10^{-4}) - (7.8 \times 10^{-3}) =$

(e) $(1.685 \times 10^5) + (4.457 \times 10^2) =$

SOLUTIONS TO EXERCISES

1. (a) four   (b) four   (c) two   (d) two

2. (a) $(7.328) \times (5.23) = 38.32544$
Round off to three significant figures: 38.3.

(b) $\dfrac{9.2648}{2.2} = 4.2112727$

Round off to two significant figures: 4.2.

(c) $(8.20 \times 10^2) \times (1.524 \times 10^{-8}) = 12.4968 \times 10^{-6}$
Round off to three significant figures: 1.25×10^{-5}.

(d) $\dfrac{(5.2607 \times 10^6) \times (1.2 \times 10^{-2})}{(2.76 \times 10^{-4})} = 2.2872608 \times 10^8$

Round off to two significant figures: 2.3×10^8.

3. (a) Uncertainty in hundredths place:
$1.06 + 9.24 - 8.13 = 2.17$

(b) Uncertainty in tenths place:
$278.3 - 7.3 + 32.1 = 303.1$

(c) $(7.8 \times 10^3) + (5.6 \times 10^2) = (7.8 \times 10^3) + (0.56 \times 10^3)$
Uncertainty in tenths place:
$(7.8 \times 10^3) + (0.6 \times 10^3) = 8.4 \times 10^3$

(d) $(5.33 \times 10^{-4}) - (7.8 \times 10^{-3}) = (5.33 \times 10^{-4}) - (78. \times 10^{-4})$
Uncertainty in units place:
$(5 \times 10^{-4}) - (78 \times 10^{-4}) = -73 \times 10^{-4} = -7.3 \times 10^{-3}$

(e) $(1.685 \times 10^5) + (4.457 \times 10^2) = (1.685 \times 10^5) + (0.004457 \times 10^5)$
Uncertainty in thousandths place:
$(1.685 \times 10^5) + (0.004 \times 10^5) = 1.689 \times 10^5$

ESSENTIAL SKILLS 5

Problem Solving

In scientific work it is frequently necessary to calculate the numerical value of a quantity from some given information. For example, consider the following problem: 2.50 g of water at 25.0°C loses 145 cal of heat. What is the new temperature of the water? (This problem is similar to Example 2 in Section 1-11). To do problems of this type you need to use some logical reasoning and perform some algebraic and arithmetic operations. We shall describe a general approach you can use to solve such problems. You must understand, however, that there is no simple set of formulas or rules you can learn that will allow you simply to plug numbers in a formula and arrive at the correct answer. There is only one way to gain facility doing problems, and that is by doing enough examples of them yourself.

Let us use the problem we stated above as an illustration of a general approach to problem solving. This approach involves the following steps:

1. Creation of a physical picture and a qualitative estimate of the answer

2. Determination of a method to solve the problem

3. Substitution of numerical values and checking of units

4. Completion of arithmetic operations and rounding off of the result

5. Verification that the answer is reasonable.

PROBLEM 25.0 g of water at 25.0°C loses 145 cal of heat. What is the new temperature of the water?

> *Step 1: Physical Picture and Qualitative Reasoning* Create a definite and precise picture in your mind of the system involved in the problem. Be sure that you understand what information is given, and what quantity you are trying to determine. A crude picture or diagram is extremely helpful. For example, you might draw something like this:

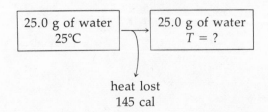

It is also a good idea to try and determine as much as possible about the answer to the problem using qualitative reasoning before you do any numerical calculations. This gives you a better "feel" for the problem, and will help you determine the correct method to use to solve the problem numerically. It is also a valuable way of checking your final answer.

In our problem, since the water *loses* heat, the new temperature of the water must be *less* than 25.0°C. Now, to decrease the temperature of 25 g of water by 1°C, 25 cal of heat must be lost, whereas 250 cal of heat must be lost for a 10°C decrease. Since 145 cal of heat is lost in our problem, the temperature decrease is between 1°C and 10°C, so the new temperature must be between 24°C and 15°C.

Step 2: Method   Consider the quantities involved in the problem and determine a method that you can use to go from the information given to the quantity you need to calculate. For certain simple problems, your method might involve only one or two steps, but you might need several steps for a more complicated problem. As we discussed in Essential Skills 3, when a problem involves two quantities that are directly proportional to one another, an easy way to obtain a numerical answer is to use the factor-unit method. For other problems it might be necessary or more convenient to (1) use what you have learned about the quantities involved in the problem to arrive at a relationship between them; and (2) algebraically rearrange this relationship to a more convenient form in which the quantity you want to determine is by itself on one side of the equation.

Our example problem involves the quantities heat lost, specific heat, mass, and temperature decrease. These quantities are connected by the following relationship (discussed in Section 1-11):

(1)   Heat lost = (specific heat) × (mass) × (temperature decrease)

Equation (1) can be algebraically rearranged to give the temperature decrease on one side of the equation and then the new temperature can be easily determined from the value of the temperature decrease and the original temperature.

If we divide both sides of Eq. (1) by the terms (specific heat) × (mass), we obtain

(2) $$\text{Temperature decrease} = \frac{\text{heat lost}}{(\text{specific heat}) \times (\text{mass})}$$

It is very important that (1) you use what you have learned about the quantities involved in the problem to arrive at a relationship between them [for example, Eq. (1)], and that (2) you do *not* search your mind for a formula [like Eq. (2)] that gives you the quantity you want to determine on one side of an equation and the quantities whose values you are given on the other side of the equation. For example, it is quite unlikely that you will arrive at Eq. (2) directly, because we do not think of a temperature decrease as a heat loss divided by the product of the specific heat and the mass. On the other hand, from your study of heat and temperature you should have in your memory a mental picture of the physical content of Eq. (1), that is, that the amount of heat lost is determined by the heat loss of 1 g of the substance for a temperature decrease of 1°C (the specific heat) multiplied by (1) the number of grams (the mass), and (2) how many °C the temperature is decreased. Thus, when you are presented with a problem involving the quantities heat loss, specific heat, mass, and temperature decrease, this mental picture and the associated Eq. (1) should come to your mind. Once you have Eq. (1), simple algebraic manipulation can be used to put it into a more convenient form, such as Eq. (2) in our example.

Step 3: Substitution of Numerical Values—Check on Units Substituting numerical values in Eq. (2) and using the fact that the specific heat of water is $1.00 \dfrac{cal}{g \cdot °C}$, we obtain

(3)
$$\text{Temperature decrease} = \frac{145 \text{ cal}}{1.00 \dfrac{cal}{g \cdot °C} \times 25.0 \text{ g}}$$

$$= \left(\frac{145}{1.00 \times 25.0}\right) \frac{\cancel{cal}}{\dfrac{\cancel{cal}}{\cancel{g} \cdot °C} \times \cancel{g}}$$

$$= \left(\frac{145}{1.00 \times 25.0}\right) °C$$

Notice that before you do any arithmetic operations you should check to see that your answer will have the correct units. Even if you obtain the correct units for your answer, it does not guarantee that you have not made a mistake. On the other hand, if the units are not correct, you can be certain that you have made some mistake in reasoning, algebra, or substitution. For example, in our problem, if you incorrectly rearranged Eq. (1), you might obtain the *incorrect* equation:

(4)
$$\text{Temperature decrease} = \frac{(\text{specific heat}) \times (\text{mass})}{\text{heat loss}} \qquad \text{incorrect}$$

If you substitute numerical values in this incorrect equation you do *not* get the correct units:

(5)
$$\text{Temperature decrease} = \frac{1.00 \dfrac{cal}{g \cdot °C} \times 25.0 \text{ g}}{(145 \text{ cal})}$$

$$= \left(\frac{1.00 \times 25.0}{145}\right) \frac{\dfrac{\cancel{cal}}{\cancel{g} \cdot °C} \times \cancel{g}}{\cancel{cal}}$$

$$= \left(\frac{1.00 \times 25.0}{145}\right) \frac{1}{°C} \qquad \text{incorrect}$$

Failure to get the correct units is a clear indication that you have made some error in your previous work.

Step 4 Arithmetic Operations and Rounding Off the Result

(6)
$$\text{Temperature decrease} = \left(\frac{145}{1.00 \times 25.0}\right) °C = 5.80°C$$

(7)
$$\text{New temperature} = 25.0°C - 5.80°C = 19.2°C$$

See Essential Skills 4 for a discussion of how to round off the result of an arithmetic operation so as to indicate correctly the uncertainty in the calculated value.

Step 5: Check   Think about your answer for a moment and assure yourself that it is (1) reasonable, and (2) in agreement with your previous qualitative expectations (step 1). Many students skip this step and thereby omit a valuable and simple way of checking their answer.

For our problem, a new temperature of 19.2°C is in agreement with our prior expectations (step 1) that the new temperature is less than 25°C and that the amount of the temperature decrease is between 1°C and 10°C.

Naturally, when you compare your initial qualitative considerations and your final numerical value, you must use some common sense. After you obtain a numerical answer you might realize that your initial qualitative expectation is in error. In any case, you should not be satisfied with your work on a problem until there is consistency between your qualitative reasoning and your numerical answer.

ESSENTIAL SKILLS 6

Percentage Composition and Percentage Change

PERCENTAGE COMPOSITION

We are all familiar with examples in which percentage composition is used to describe a collection of different things. For example, the statement that a certain class is 60% female means that, for this class,

(1)
$$\left(\frac{\text{Number of female students}}{\text{Total number of students}}\right) \times 100\% = 60\%$$

Thus, the percentage of female students is equal to the fraction of the class that is female times 100%.

Notice that we *cannot* determine the number of female students in the class from just the fact that it is 60% female. The class might be composed of 20 males and 30 females, 40 males and 60 females, 50 males and 75 females, and so on. If, however, we know the total number of students in the class, we *can* determine the number of female students in it. An algebraic rearrangement of Eq. (1) gives

(2)
$$\text{Number of female students} = \left(\frac{60\%}{100\%}\right) \times \text{total number of students}$$

For example, if the total number of students in the class is 125, then the class must have 75 female students. **Percent** literally means "parts per hundred." In other words, for a class that is 60% female, if the total number of students is 100, then the number of female students is 60.

In general, whenever the total amount of some quantity is composed of a part A, a part B, a part C, and so on, then the percentage composition is given by

(3)
$$\text{Percent } A = \left(\frac{\text{part } A}{\text{total amount}}\right) \times 100\%$$

$$\text{Percent } B = \left(\frac{\text{part } B}{\text{total amount}}\right) \times 100\%$$

$$\text{Percent } C = \left(\frac{\text{part } C}{\text{total amount}}\right) \times 100\% \quad \text{and so on}$$

The ratio, $\dfrac{\text{part } A}{\text{total amount}}$, is the fraction of the whole that is A. Thus, the percent A is determined by multiplying the fraction that is A by 100%, and similarly for B, C, and so on. The sum of the percentages of *all* the parts is 100%:

(4)
$$\text{Percent } A + \text{percent } B + \text{percent } C + \cdots = 100\%$$

For example, water is composed of the elements hydrogen and oxygen, and all samples of water have the same elemental composition. In a 150-g sample of water, the mass of hydrogen is 16.7 g. Thus, using Eq. (3), the percentage mass of hydrogen is

(5)
$$\text{Percent hydrogen} = \left(\frac{16.7\ \cancel{g}}{150\ \cancel{g}}\right) \times 100\% = 11\%$$

and, using Eq. (4),

(6)
$$\text{Percent oxygen} = 100\% - \text{percent hydrogen} = 100\% - 11\% = 89\%$$

The mass of oxygen in, for example, a 200-g sample of water is

(7)
$$\text{Mass of oxygen} = \left(\frac{\text{percent oxygen}}{100\%}\right) \times \text{mass of water}$$
$$= \left(\frac{89\%}{100\%}\right) \times 200\ g = 178\ g$$

Exercise 1
A class has 80 female and 35 male students. Calculate the percent female and the percent male in this class.

Exercise 2
There are a total of 60 students in a class and the class is 35% male. Calculate the number of male and the number of female students in this class.

Exercise 3
The substance carbon dioxide is composed of the elements carbon and oxygen, and all samples of carbon dioxide have the same elemental composition. A particular sample of carbon dioxide is composed of 18 g of carbon and 48 g of oxygen. Calculate the percent carbon and the percent oxygen in carbon dioxide.

Exercise 4
Use the result of Exercise 3 to calculate the mass of carbon and the mass of oxygen in a 150-g sample of carbon dioxide.

Exercise 5
A drug solution is prepared by dissolving 5.0 mg of the drug in water to form a solution with a mass of 250 g. (a) For this solution calculate the percent drug. (b) What amount of the drug solution contains 1.5 mg of the drug?

PERCENTAGE CHANGE

Percentage is also used when referring to *changes* in a quantity. In general, the percentage change in a quantity is determined by

(8)
$$\text{Percentage change} = \left(\frac{\text{amount of change}}{\text{original amount}}\right) \times 100\%$$

Often the percentage change in a quantity is more important than the absolute change. For example, a 10-lb weight loss for a 40-lb child (a 25% change) is more significant than a 10-lb loss for a 250-lb adult (only a 4% change). In our example, the percentage loss of weight for the child is determined as follows:

$$\text{Percentage change} = \left(\frac{10 \cancel{lb}}{40 \cancel{lb}}\right) \times 100\% = 25\%$$

The percentage loss of weight for the adult is:

$$\text{Percentage change} = \left(\frac{10 \cancel{lb}}{250 \cancel{lb}}\right) \times 100\% = 4\%$$

A 200% increase in a quantity is equivalent to tripling the quantity. For example, if a person's income increases from \$10,000 to \$30,000 (it triples), the percentage increase is (\$20,000/\$10,000) $\times$ 100% = 200%. Similarly, a 50% decrease in a quantity corresponds to reducing the quantity by a factor of one-half, or halving the original quantity.

Exercise 6
A patient was given 25 mg of a drug yesterday. If the drug dosage is to be decreased by 50% each day, what amount of drug should be administered to the patient today?

Exercise 7
A sample of iron is heated. Calculate the percentage change in temperature if the temperature of the iron (a) increases from 100 K to 150 K; (b) increases from 500 K to 550 K.

Exercise 8
Two people, A and B, each experience a 12% increase in weight. Calculate the new weight of each person if the initial weight of A was 120 lb and the initial weight of B was 190 lb.

SOLUTIONS TO EXERCISES

1. Percent female $= \left(\dfrac{80}{115}\right) \times 100\% = 70\%$

    Percent male $= \left(\dfrac{35}{115}\right) \times 100\% = 30\%$

    (Percent female + percent male = 70% + 30% = 100%.)

2. Number male $= \left(\dfrac{35\%}{100\%}\right) \times 60 = 21$

    Number female $= \left(\dfrac{65\%}{100\%}\right) \times 60 = 39$

    (Number male + number female = 21 + 39 = 60.)

3. Percent carbon $= \left(\dfrac{18 \cancel{g}}{66 \cancel{g}}\right) \times 100 = 27\%$

    Percent oxygen = 100% − 27% = 73%

4. Mass carbon $= \left(\dfrac{27\%}{100\%}\right) \times 150 \text{ g} = 41 \text{ g}$

    Mass oxygen = total mass − mass carbon = 150 g − 41 g = 109 g

5. (a) Percent drug $= \left(\dfrac{5.0 \times 10^{-3}\text{ g}}{250\text{ g}}\right) \times 100\% = 2 \times 10^{-3}\%$

    (b) Mass of drug $= \left(\dfrac{\text{percent drug}}{100\%}\right) \times (\text{mass of solution})$

    Therefore:

    Mass of solution $= \left(\dfrac{100\%}{\text{percent drug}}\right) \times (\text{mass of drug})$

    $\qquad\qquad\qquad = \left(\dfrac{100\%}{2 \times 10^{-3}\%}\right) \times (1.5\text{ mg}) = 75{,}000\text{ mg} = 75\text{ g}$

6. A 50% decrease in drug dosage means halving the drug dosage. Therefore, the new drug dosage should be 12.5 mg.

7. (a) Change in temperature $= 150\text{ K} - 100\text{ K} = 50\text{ K}$

    Percentage change $= \left(\dfrac{50\text{ K}}{100\text{ K}}\right) \times 100\% = 50\%$

    (b) Change in temperature $= 550\text{ K} - 500\text{ K} = 50\text{ K}$

    Percentage change $= \left(\dfrac{50\text{ K}}{500\text{ K}}\right) \times 100\% = 10\%$

8. Amount of change $= \left(\dfrac{\text{percentage change}}{100\%}\right) \times \text{original amount}$

    For person A:

    Amount of change $= \left(\dfrac{12\%}{100\%}\right) \times 120\text{ lb} = 14\text{ lb}$

    New weight $= 120\text{ lb} + 14\text{ lb} = 134\text{ lb}$

    For person B:

    Amount of change $= \left(\dfrac{12\%}{100\%}\right) \times 190\text{ lb} = 23\text{ lb}$

    New weight $= 190\text{ lb} + 23\text{ lb} = 213\text{ lb}$

ESSENTIAL SKILLS 7

Nomenclature for Some Simple Inorganic Compounds

Different rules are used for naming ionic compounds and covalent compounds that involve only nonmetallic elements. Binary covalent compounds that involve a metallic element are named as if they were ionic compounds. We shall discuss covalent compounds first, and limit our discussion to the simplest type—binary covalent compounds. We shall then discuss how simple ionic compounds are named.

BINARY COVALENT COMPOUNDS

Binary compounds contain two, and only two, elements. In the chemical formula, and in the name, for a binary covalent compound involving nonmetallic elements, the element with the lower electronegativity appears first. The *element name* is used for the element of lower electronegativity, and the stem of the element name, with the suffix *-ide* attached, is used for the element with the higher electronegativity. These stems with the suffix *-ide* attached, for several common elements, are oxide, sulfide, chloride, bromide, iodide, hydride, and phosphide. The number of atoms of each element in a molecule of a binary covalent compound is indicated in its name by the Greek prefixes: **mono** = one, **di** = two, **tri** = three, **tetra** = four, **penta** = five, **hexa** = six, and so on. When there is no risk of ambiguity, the prefix *mono-* is not used.

EXAMPLE 1 What is the name for the binary covalent compound with the molecular formula CCl_4?

As indicated by its position first in the molecular formula, C is the element with the lower electronegativity. Therefore, we use the element name carbon, and write it first in the name. The stem of the element name for Cl with the suffix *-ide*, or *-chloride*, is used for this element with the higher electronegativity. Since there are four chlorine atoms in a molecule of CCl_4, we use the prefix *tetra-* in front of chloride, or *tetrachloride*. The prefix *mono-* in front of carbon is not needed. Thus, the name for CCl_4 is **carbon tetrachloride.**

For ease in pronunciation, the "a" or "o" at the end of a Greek prefix is omitted when the prefix is used before an element name that begins with a vowel.

EXAMPLE 2 The name for N_2O_5 is **dinitrogen pentoxide.** Note that the "a" at the end of the Greek prefix *penta-* has been omitted

Some other examples of names for binary covalent compounds are

SF_6, sulfur hexafluoride

CO, carbon monoxide

CO_2, carbon dioxide

N_2O, dinitrogen monoxide

Several compounds have common names that are often, and in some cases always, used. The systematic name for H_2O is dihydrogen monoxide, but the common name **water** is always used for this substance. Another compound that is always referred to by its common name is **ammonia, NH_3**.

Exercise 1

Name the following compounds:

   (a) BF_3 (b) N_2O_4 (c) SO_2 (d) P_4S_3

Exericse 2

Write the molecular formula for each of the following:

   (a) Phosphorus pentachloride   (c) Oxygen difluoride

   (b) Diboron hexahydride   (d) Dinitrogen trioxide

IONIC COMPOUNDS

Ionic compounds contain positive and negative ions, and the name for an ionic compound consists of two parts: (1) the name for the positive ion, followed by (2) the name for the negative ion. By convention, any binary compound that involves a metallic element is named by using the rules for ionic compounds, even if the compound is not ionic. For example, $SnCl_4$, which is not an ionic compound but which contains the metallic element tin, is named by using the same rules as are used for ionic compounds.

Names for Positive Ions

The name for a *monatomic* positive ion, such as Na^+, Ca^{2+}, and so on, is just the name of the element from which the element is derived followed by the word *ion* or *cation*. Thus the monatomic positive ion Na^+ is called the **sodium ion** or the **sodium cation.** When an element forms two or more positive ions with different charges, such as Fe^{2+} and Fe^{3+}, the charge on the ion is indicated by a Roman numeral in parentheses after the name of the element. Thus, the names for Fe^{2+} and Fe^{3+} are iron(II) and iron(III), and the names for Sn^{2+} and Sn^{4+} are tin(II) and tin(IV).

An older way of referring to Fe^{2+} and Fe^{3+} is to use the names **ferrous** and **ferric,** respectively. These names are part of an older method of naming metal ions in which the Latin name for the metal is used with the suffix *-ous* for the ion with the lower charge, and the suffix *-ic* for the ion with the higher charge. This older method is still widely used but it is *not* the preferred method. For example, it is difficult to refer to the ions chromium(II), Cr^{2+}, chromium(III), Cr^{3+}, and chromium(VI), Cr^{6+}, using the two suffixes *-ous* and *-ic*.

The only common polyatomic cation, NH_4^+, is named the **ammonium ion.**

Names for Negative Ions

The name for a *monatomic* negative ion is formed from the name of the element from which the ion is derived by adding the suffix *-ide* to the stem of the name for the element. Thus, Cl^- is called the **chloride ion** or the **chloride anion,** and O^{2-} is called the **oxide ion** or the **oxide anion.**

There are many common polyatomic anions, such as HCO_3^-, OH^-, $H_2PO_4^-$, and so on, and the systematic naming of these ions is quite involved. You should memorize the names of the common polyatomic ions in Table 4-1.

The suffixes *-ate* and *-ite* are frequently used in the names of polyatomic anions. For example, NO_3^- is called the **nitrate ion,** and the name for NO_2^- is the **nitrite ion.** Anions that contain oxygen, such as NO_3^- and NO_2^-, are called **oxyanions.** When an element forms only two different oxyanions, the suffix *-ate* is used in the name of the oxyanion that contains the most oxygen, and the suffix *-ite* is used in the name of the other oxyanion with less oxygen.

Names for Ionic Compounds

Ionic compounds (and binary covalent compounds that contain a metallic element) are named by giving the name of the positive ion first, followed by the name for the negative ion. The positive ion is also written first in the chemical formula for an ionic compound.

EXAMPLE 3   What is the name for NH_4Cl?

NH_4Cl is the chemical formula for the ionic compound containing NH_4^+, the ammonium cation, and Cl^-, the chloride anion. Therefore, the name for NH_4Cl is **ammonium chloride.**

EXAMPLE 4   What is the name for MgF_2?

MgF_2 is the chemical formula for the ionic compound containing Mg^{2+}, the magnesium cation, and F^-, the fluoride anion. Since magnesium forms a $+2$ ion only, we do not indicate the charge with a Roman numeral. Therefore, the name for MgF_2 is **magnesium fluoride.**

EXAMPLE 5   What is the name for $Cu(NO_3)_2$?

$Cu(NO_3)_2$ is the chemical formula for the ionic compound containing Cu^{2+}, the copper(II) cation, and NO_3^-, the nitrate anion. There are two NO_3^- anions for each copper ion, thus, the charge on the copper ion must be $+2$ in order that the compound have a zero net electrical charge. Since copper forms ions with a charge of $+1$ in addition to ions with a charge of $+2$, we indicate the charge on the copper ion with a Roman numeral in the name. Therefore, the name for $Cu(NO_3)_2$ is **copper(II) nitrate.**

Note that in the names for ionic compounds, Greek prefixes are *not* used to indicate the number of atoms of each element in the chemical formula. Thus MgF_2 does *not* have the name magnesium difluoride. When you see, for example, the name "magnesium fluoride," you are expected to realize that: (1) the name refers to an ionic compound, (2) the charge on the magnesium ion is $+2$ (Mg^{2+}) and the charge on the fluoride ion is -1 (F^-), and thus that (3) the chemical formula for magnesium fluoride is MgF_2, because the compound must have zero net electrical charge (see Section 4-5). This is why Greek prefixes are unnecessary and therefore not used in the names of ionic compounds.

Exercise 3

Name the following compounds:

(a) $CaBr_2$	(b) $(NH_4)_2SO_4$	(c) $CuCl$	(d) SnF_2
(e) $FePO_4$	(f) $KHCO_3$	(g) MnO_2	(h) NaH_2PO_4

Exercise 4

Write chemical formulas for the following compounds:

    (a) potassium sulfide   (b) calcium nitrite

    (c) copper(II) chloride   (d) iron(II) nitrate

SOLUTIONS TO EXERCISES

1. (a) BF_3, boron trichloride

    (b) N_2O_4, dinitrogen tetroxide

    (c) SO_2, sulfur dioxide

    (d) P_4S_3, tetraphosphorus trisulfide

2. (a) Phosphorus pentachloride, PCl_5

    (b) Diboron hexahydride, B_2H_6 (also known as diborane)

    (c) Oxygen difluoride, OF_2

    (d) Dinitrogen trioxide, N_2O_3

3. (a) $CaBr_2$ contains Ca^{2+}, the calcium cation, and Br^-, the bromide anion. Therefore, the name for $CaBr_2$ is **calcium bromide.**

    (b) $(NH_4)_2SO_4$ contains NH_4^+, the ammonium cation, and SO_4^{2-}, the sulfate anion. Therefore, the name for $(NH_4)_2SO_4$ is **ammonium sulfate.**

    (c) $CuCl$ contains Cu^+, the copper(I) cation, and Cl^-, the chloride anion. Therefore, the name for $CuCl$ is **copper(I) chloride.**

    (d) SnF_2 contains Sn^{2+}, the tin(II) cation, and F^-, the fluoride anion. Therefore, the name for SnF_2 is **tin(II) fluoride.**

    (e) $FePO_4$ contains Fe^{3+}, the iron(III) cation, and PO_4^{3-}, the phosphate anion. Therefore, the name for $FePO_4$ is **iron(III) phosphate.**

    (f) $KHCO_3$ contains K^+, the potassium cation, and HCO_3^-, the monohydrogen carbonate anion. Therefore, the name for $KHCO_3$ is **potassium monohydrogen carbonate.**

    (g) MnO_2 contains Mn^{4+}, the manganese(IV) cation, and O^{2-}, the oxide anion. Therefore, the name for MnO_2 is **manganese(IV) oxide.**

    (h) NaH_2PO_4 contains Na^+, the sodium cation, and $H_2PO_4^-$, the dihydrogen phosphate anion. Therefore, the name for NaH_2PO_4 is **sodium dihydrogen phosphate.**

4. (a) The potassium cation is K^+, and the sulfide anion is S^{2-}. Two K^+ ions are required for every S^{2-} ion for zero net electrical charge. Therefore, the chemical formula for potassium sulfide is **K_2S**.

    (b) The calcium cation is Ca^{2+}, and the nitrite anion is NO_2^-. Two NO_2^- ions are required for every Ca^{2+} ion for zero net electrical charge. Therefore, the chemical formula for calcium nitrite is **$Ca(NO_2)_2$**.

    (c) The copper(II) cation is Cu^{2+}, and the chloride anion is Cl^-. Two Cl^- ions are required for every Cu^{2+} for zero net electrical charge. Therefore, the chemical formula for copper(II) chloride is **$CuCl_2$**.

    (d) The iron(II) cation is Fe^{2+}, and the nitrate anion is NO_3^-. Two NO_3^- ions are required for every Fe^{2+} ion. Therefore, the chemical formula for iron(II) nitrate is **$Fe(NO_3)_2$**.

ESSENTIAL SKILLS 8

Naming Organic Compounds

Common and IUPAC names for the various classes of organic compounds are described in detail in Chapters 13 through 18. The basic principles for naming organic compounds are summarized here.

COMMON NAMES

Common names for many organic compounds are formed by combining the name for the functional group with the names for the alkyl groups involved. Thus, CH_3—CH_2—OH is called ethyl alcohol, and CH_3—CH_2—O—CH_2—CH_3 is called diethyl ether. The common names for some of the more prevalent alkyl groups are given in Table 1.

Table 1 Common Alkyl Groups

Name	Structural Formula
Methyl	CH_3—
Ethyl	CH_3—CH_2—
n-Propyl	CH_3—CH_2—CH_2—
Isopropyl	CH_3—$\overset{\mid}{C}H$—CH_3
n-Butyl	CH_3—CH_2—CH_2—CH_2—
Isobutyl	CH_3—$\overset{\overset{\textstyle CH_3}{\mid}}{C}H$—$CH_2$—
t-Butyl	CH_3—$\overset{\overset{\textstyle CH_3}{\mid}}{\underset{\mid}{C}}$—$CH_3$

The common names for carboxylic acids, aldehydes, amides, and anhydrides, however, are generally not formed from the names of alkyl groups. Common names for simple compounds belonging to these classes are derived from a natural source of the carboxylic acid with that number of carbon atoms. The structural formulas and common names for a few simple carboxylic acids are given in Table 2. The Greek letters α, β, γ, and so on, are used as prefixes to the common names of carboxylic acids to specify the position of substituents, as in β-hydroxybutyric acid.

Table 2 Common Names for Some Carboxylic Acids

Common Name	Structural Formula
Formic acid	$\overset{\displaystyle O}{\overset{\displaystyle \|}{HC}}\!-\!OH$
Acetic acid	$CH_3\!-\!\overset{\displaystyle O}{\overset{\displaystyle \|}{C}}\!-\!OH$
Propionic acid	$CH_3\!-\!CH_2\!-\!\overset{\displaystyle O}{\overset{\displaystyle \|}{C}}\!-\!OH$
Butyric acid	$CH_3\!-\!CH_2\!-\!CH_2\!-\!\overset{\displaystyle O}{\overset{\displaystyle \|}{C}}\!-\!OH$
β-Hydroxybutyric acid	$CH_3\!-\!\overset{\displaystyle OH}{\overset{\displaystyle \|}{CH}}\!-\!CH_2\!-\!\overset{\displaystyle O}{\overset{\displaystyle \|}{C}}\!-\!OH$

Common Names for Acids and Bases

Recall that any carboxylic acid solution contains both carboxylic acid molecules and carboxylate ions that exist in equilibrium, and that the proportion of each form that is present depends on the pH. The higher the pH, the greater the proportion that exists as the carboxylate ion. For simplicity, however, we nearly always refer to these solutions as carboxylic acid solutions.

Similarly, bases are in equilibrium with their conjugate acid (the protonated cation). But, for simplicity, we refer to any equilibrium mixture that involves an amine and its conjugate acid by the name for the amine, even though in a sufficiently acidic solution the predominant form is the protonated cation.

Common Names for Complex Compounds

The common names for many other organic compounds, especially some very complex ones, are not formed in any systematic manner. Examples are glucose, toluene, and aspirin.

BASIC PRINCIPLES OF IUPAC NOMENCLATURE

The IUPAC name of an organic compound consists of three parts: a root, an ending, and one or more prefixes:

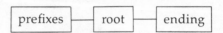

The **root** of a IUPAC name specifies the longest continuous chain of carbon atoms in a molecule of that compound, that is, the parent alkane molecule (see Table 3).

Table 3 Names for Unbranched Alkanes

Common Name	IUPAC Name	Structural Formula
Methane	Methane	CH_4
Ethane	Ethane	CH_3-CH_3
Propane	Propane	$CH_3-CH_2-CH_3$
n-Butane	Butane	$CH_3-CH_2-CH_2-CH_3$
n-Pentane	Pentane	$CH_3-CH_2-CH_2-CH_2-CH_3$
n-Hexane	Hexane	$CH_3-CH_2-CH_2-CH_2-CH_2-CH_3$
n-Heptane	Heptane	$CH_3-CH_2-CH_2-CH_2-CH_2-CH_2-CH_3$
n-Octane	Octane	$CH_3-CH_2-CH_2-CH_2-CH_2-CH_2-CH_2-CH_3$
n-Nonane	Nonane	$CH_3-CH_2-CH_2-CH_2-CH_2-CH_2-CH_2-CH_2-CH_3$
n-Decane	Decane	$CH_3-CH_2-CH_2-CH_2-CH_2-CH_2-CH_2-CH_2-CH_2-CH_3$

The **ending** of a IUPAC name specifies the class of organic compounds to which the molecule belongs or the major functional group of the molecule.

Prefixes are used to specify the position of functional groups and the identity and location of substituents attached to the longest carbon chain. Substituents are identified by name, and by a number that indicates the carbon atom of the longest chain to which they are attached, as follows:

1. The longest continuous chain must be numbered so that the positions of the substituents will have the lowest possible numbers.

2. The prefixes *di-*, *tri-*, and *tetra-* before the name of a substituent indicate two, three, or four of that substituent in the molecule.

3. When a molecule contains more than one substituent, the substituents are arranged alphabetically.

4. IUPAC names are written as a single word with numbers separated from one another by commas and numbers separated from letters by hyphens.

5. No punctuation or space is used between the name of a substituent and the root name.

Naming Alkanes

The common names for the alkanes containing an unbranched chain of one to ten carbon atoms are also given in Table 3. The common and IUPAC names of all alkanes end in -*ane*. The prefix *n-* (for normal) is not needed in the IUPAC name for an alkane. For example, the common name *n*-pentane specifies the compound with the formula $CH_3-CH_2-CH_2-CH_2-CH_3$, which must be

distinguished from its isomer isopentane $CH_3-CH_2-\overset{\displaystyle CH_3}{\underset{\displaystyle H}{\overset{\displaystyle |}{\underset{\displaystyle |}{C}}}}-CH_3$. The prefix

n- is not needed in the IUPAC name pentane, since the IUPAC name for

$CH_3-CH_2-\overset{\displaystyle CH_3}{\underset{\displaystyle H}{\overset{\displaystyle |}{\underset{\displaystyle |}{C}}}}-CH_3$ is 2-methylbutane, where the root of the IUPAC name,

but-, specifies that the longest carbon chain consists of four carbon atoms, and the prefix 2- specifies that the methyl substituent is bonded to the second carbon in that chain.

Table 4 Common and IUPAC Nomenclature for Some Classes of Organic Compounds

Class	General Formula	Ending	
		IUPAC	Common
Alkenes	$\begin{array}{cc} R_1' \\ \end{array} C = C \begin{array}{cc} R_3' \\ \end{array}$ R_2' R_4'	-ene	-ene
Alcohols	R—OH	-ol	alcohol
Ethers	R_1—O—R_2	—	ether
Aldehydes	$R'-\overset{\overset{\displaystyle O}{\|}}{C}-H$	-al	aldehyde
Ketones	$R_1-\overset{\overset{\displaystyle O}{\|}}{C}-R_2$	-one	ketone
Carboxylic acids	$R'-\overset{\overset{\displaystyle O}{\|}}{C}-OH$	-oic acid	-ic acid
Esters	$R_1'-\overset{\overset{\displaystyle O}{\|}}{C}-O-R_2$	-oate	-ate
Anhydrides	$R_1'-\overset{\overset{\displaystyle O}{\|}}{C}-O-\overset{\overset{\displaystyle O}{\|}}{C}-R_2'$	-oic anhydride	-ic anhydride
Amines	$R_3'-\overset{\overset{\displaystyle R_1'}{\|}}{N}-R_2'$	—	amine
Simple amides	$R'-\overset{\overset{\displaystyle O}{\|}}{C}-NH_2$	amide	amide
Thiols	R—SH	-thiol	mercaptan

Table 4 lists some of the major classes of organic compounds and the common and IUPAC endings used in their names. For some classes, common names are almost always used for simple compounds; the IUPAC endings for these compounds are not shown. Notice that the names for aldehydes, anhydrides, and amides are derived from the names for the corresponding carboxylic acids. Also note that esters are named in a manner analogous to carboxylic acid salts.

Drawing the Structural Formula for a Compound When Given Its IUPAC Name

1. Identify the root and ending of the IUPAC name.

2. Draw the longest continuous chain of carbon atoms (specified by the root).

3. Draw the functional group (specified by the ending) at the location on the chain specified by the prefix number for that functional group. (e.g., 2-pentanone has a ketone group at the number 2 carbon atom.)

	Example	
Structural Formula	IUPAC Name	Common Name
$CH_3—CH{=}CH_2$	Propene	Propylene
$CH_3—CH—CH_3$ $\quad\quad\mid$ $\quad\quad OH$	2-Propanol	Isopropyl alcohol
$CH_3—O—CH_3$	—	Dimethyl ether
$\quad\quad O$ $\quad\quad \|\|$ $CH_3—C—H$	Ethanal	Acetaldehyde
$\quad\quad\quad\quad O$ $\quad\quad\quad\quad \|\|$ $CH_3—CH_2—C—CH_2—CH_3$	3-Pentanone	Diethyl ketone
$\quad O$ $\quad \|\|$ $H—C—OH$	Methanoic acid	Formic acid
$\quad\quad O$ $\quad\quad \|\|$ $H_3C—C—O—CH_3$	Methyl ethanoate	Methyl acetate
$\quad\quad O\quad\quad O$ $\quad\quad \|\|\quad\quad \|\|$ $H_3C—C—O—C—CH_3$	Ethanoic anhydride	Acetic anhydride
$CH_3—CH_2—CH_2—NH_2$	—	n-Propylamine
$\quad\quad O$ $\quad\quad \|\|$ $CH_3—C—NH_2$	Ethanamide	Acetamide
$CH_3—CH_2—SH$	Ethanethiol	Ethyl mercaptan

4. Draw any substituents (specified by prefixes) at the locations specified by their prefix numbers.

Writing the IUPAC Name for a Compound When Given Its Structural Formula

1. Identify the longest continuous chain of carbon atoms and thus determine the root of the name.
2. Identify the class of compound and thus obtain the ending for the IUPAC name.
3. Identify the substituents and number the longest carbon chain so that the substituents have the lowest possible numbers.
4. Name the substituents as prefixes of the IUPAC name.

Practice in naming organic compounds can be gained by working the appropriate exercises and problems in Chapters 13 through 18.

ESSENTIAL SKILLS 9

Predicting the Products of Organic Reactions

As you study organic compounds and the chemical reactions they undergo, you will frequently find problems of the following type: "Draw the structural formula(s) for the product(s) of the following reaction, given the structural formula(s) for the reactant(s)." The problem may also state: "If no reaction occurs, write N.R." Some problems may specify the reaction conditions or indicate that an oxidizing agent or a reducing agent is involved in the reaction.

You can avoid some of the difficulties with this type of problem if you develop a systematic approach, such as the one illustrated in the example on page 747. You will find this approach extremely useful if: (1) you have gained the ability to recognize the functional groups in the structural formula for an organic compound; and (2) you know the general types of reactions that the class of compounds specified by a given functional group can undergo. For convenience, the general reactions of organic compounds that are presented in Chapters 13 through 18 are summarized in Table 1.

Table 1 General Types of Organic Reactions

Functional Group Class	General Reaction (In the problems in this text, it is assumed that the appropriate reaction conditions, catalyst, and so on, needed for the reaction to occur are present.)	
$\diagdown$C$=$C$\diagup$ Alkenes	Reduction (hydrogenation)	$R_1'-CH=CH-R_2' + H_2 \longrightarrow R_1'-CH_2-CH_2-R_2'$
	Halogenation	$R_1'-CH=CH-R_2' + HY \longrightarrow R_1'-CH_2-\overset{\overset{\displaystyle Y}{\displaystyle \vert}}{C}H-R_2'$ (Y = halogen atom)　　(Adds according to Markovnikov's rule)
		$R_1'-CH=CH-R_2' + Y_2 \longrightarrow R_1'-\underset{\underset{\displaystyle Y}{\displaystyle \vert}}{C}H-\underset{\underset{\displaystyle Y}{\displaystyle \vert}}{C}H-R_2'$
	Hydration	$R_1'-CH=CH-R_2' + H_2O \longrightarrow R_1'-\underset{\underset{\displaystyle H}{\displaystyle \vert}}{C}H-\underset{\underset{\displaystyle OH}{\displaystyle \vert}}{C}H-R_2'$ (Adds according to Markovnikov's rule)

Table 1 (Continued)

Functional Group Class	General Reaction

$-\overset{\displaystyle |}{\underset{\displaystyle |}{C}}-OH$

Alcohols

Dehydration

(1) $R'_1-\overset{\displaystyle H}{\underset{\displaystyle H}{\overset{\displaystyle |}{\underset{\displaystyle |}{C}}}}-\overset{\displaystyle OH}{\underset{\displaystyle H}{\overset{\displaystyle |}{\underset{\displaystyle |}{C}}}}-R'_2 \longrightarrow H_2O + R'_1-CH{=}CH-R'_2$

(2) $R-OH + R-OH \longrightarrow H_2O + R-O-R$

an ether

Oxidation

(1) $X + R'-CH_2-OH \longrightarrow XH_2 + R'-\overset{\displaystyle O}{\overset{\displaystyle ||}{C}}-H$

a primary alcohol   an aldehyde

(X is the general symbol for an oxidizing agent, and XH_2 represents a reducing agent.)

(2) $X + \quad R_1-\overset{\displaystyle OH}{\underset{\displaystyle H}{\overset{\displaystyle |}{\underset{\displaystyle |}{C}}}}-R_2 \longrightarrow XH_2 + R_1-\overset{\displaystyle O}{\overset{\displaystyle ||}{C}}-R_2$

a secondary alcohol   a ketone

(For the reactions of alcohols with aldehydes, hemiacetals, ketones, and hemiketals, see aldehydes and ketones.)

Ester formation

$R_1-OH + R'_2-\overset{\displaystyle O}{\overset{\displaystyle ||}{C}}-OH \longrightarrow R'_2-\overset{\displaystyle O}{\overset{\displaystyle ||}{C}}-O-R_1 + H_2O$

an ester

$\langle\!\!\bigcirc\!\!\rangle-OH$

Phenols

Ionization

$\langle\!\!\bigcirc\!\!\rangle-OH + H_2O \rightleftharpoons \langle\!\!\bigcirc\!\!\rangle-O^- + H_3O^+$

$-\overset{\displaystyle O}{\overset{\displaystyle ||}{C}}-H$

Aldehydes

Reduction

$R'-\overset{\displaystyle O}{\overset{\displaystyle ||}{C}}-H + XH_2 \longrightarrow R'-\overset{\displaystyle H}{\underset{\displaystyle H}{\overset{\displaystyle |}{\underset{\displaystyle |}{C}}}}-OH + X$

a primary alcohol

Oxidation

$R'-\overset{\displaystyle O}{\overset{\displaystyle ||}{C}}-H + X + H_2O \longrightarrow R'-\overset{\displaystyle O}{\overset{\displaystyle ||}{C}}-OH + XH_2$

a carboxylic acid

Hemiacetal formation

$R_1-OH + H-\overset{\displaystyle O}{\overset{\displaystyle ||}{C}}-R_2 \rightleftharpoons R_1-O-\overset{\displaystyle OH}{\underset{\displaystyle H}{\overset{\displaystyle |}{\underset{\displaystyle |}{C}}}}-R_2$

Acetal formation

$R_1-OH + R_1-O-\overset{\displaystyle OH}{\underset{\displaystyle H}{\overset{\displaystyle |}{\underset{\displaystyle |}{C}}}}-R_2 \xrightarrow{H^+} R_1-O-\overset{\displaystyle OR_1}{\underset{\displaystyle H}{\overset{\displaystyle |}{\underset{\displaystyle |}{C}}}}-R_2 + H_2O$

Aldol condensation reaction

$2\left[R-\overset{\displaystyle R'}{\underset{\displaystyle H}{\overset{\displaystyle |}{\underset{\displaystyle |}{C}}}}-\overset{\displaystyle O}{\overset{\displaystyle ||}{C}}-H\right] \xrightarrow{OH^-} R-\overset{\displaystyle R'}{\underset{\displaystyle H}{\overset{\displaystyle |}{\underset{\displaystyle |}{C}}}}-\overset{\displaystyle OH}{\underset{\displaystyle H}{\overset{\displaystyle |}{\underset{\displaystyle |}{C}}}}-\overset{\displaystyle R'}{\underset{\displaystyle R}{\overset{\displaystyle |}{\underset{\displaystyle |}{C}}}}-\overset{\displaystyle O}{\overset{\displaystyle ||}{C}}-H$

Table 1 (Continued)

Functional Group Class	General Reaction

Ketones — **Reduction**

$$XH_2 + R_1-\overset{O}{\underset{}{C}}-R_2 \longrightarrow R_1-\overset{OH}{\underset{H}{C}}-R_2$$

a secondary
alcohol

Hemiketal and ketal formation — As for aldehydes

Aldol condensation — As for aldehydes

Carboxylic acids

Ionization

$$R'-\overset{O}{C}-OH + H_2O \rightleftharpoons R'-\overset{O}{C}-O^- + H_3O^-$$

Salt formation

$$R'-\overset{O}{C}-OH + NaOH \rightleftharpoons R'-\overset{O}{C}-O^- Na^+ + H_2O$$

Reduction

$$R'-\overset{O}{C}-OH + XH_2 \longrightarrow R'-\overset{O}{C}-H + H_2O + X$$

Decarboxylation

$$R'-\overset{O}{C}-OH \longrightarrow R'H + CO_2$$

Condensation reactions

Anhydride formation

$$2R'-\overset{O}{C}-OH \longrightarrow R'-\overset{O}{C}-O-\overset{O}{C}-R' + H_2O$$

an anhydride

Ester formation

$$R'_1-\overset{O}{C}-OH + R_2-OH \longrightarrow R'_1-\overset{O}{C}-O-R_2 + H_2O$$

an ester

Amide formation

$$R'_1-\overset{O}{C}-OH + H-\overset{R'_2}{\underset{R'_4}{N}}-R'_3 \longrightarrow R'_1-\overset{O}{C}-\overset{R'_2}{\underset{R'_4}{N}}-R'_3 + H_2O$$

an amide

Thioester formation

$$R'_1-\overset{O}{C}-OH + R_2-SH \longrightarrow R'_1-\overset{O}{C}-S-R_2 + H_2O$$

a thioester

Phosphoester formation

$$R'_1-\overset{O}{C}-OH + HO-\overset{O}{\underset{OH}{P}}-OH \longrightarrow$$

$$R'_1-\overset{O}{C}-O-\overset{O}{\underset{OH}{P}}-OH + H_2O$$

a phosphoester

Note: The hydrolysis of anhydrides, esters, and any of the other products of condensation reactions involving carboxylic acids can be written as the reverse of the reactions in which they are formed.

Table 1 (Continued)

Functional Group Class	General Reaction	

$\begin{matrix} & R'_3 \\ & | \\ R'_1-N-R'_2 \\ \text{Amines} \end{matrix}$ Ionization $\begin{matrix} R'_3 \\ | \\ R'_1-N-R'_2 + H_2O \end{matrix} \rightleftharpoons \begin{matrix} R'_3 \\ | \\ R'_1-N^+-R'_2 + OH^- \\ | \\ H \end{matrix}$

Amide formation   See carboxylic acids
or hydrolysis

R—SH   Disulfide
Thiols   formation $R_1-SH + R_2-SH + X \longrightarrow R_1-S-S-R_2 + XH_2$
                                                                                          a disulfide

Thioester   See carboxylic acids
formation and
hydrolysis

EXAMPLE Draw the structural formulas for the products of the following reaction. (If no reaction occurs, write N.R.)

$$CH_3-CH_2-\overset{\overset{\textstyle O}{\|}}{C}-OH + CH_3-CH_2-OH \longrightarrow$$

Step 1 Identify the functional groups and determine the class of organic compounds to which each reactant belongs.

This example involves a carboxylic acid and an alcohol.

Step 2 Determine if a reaction can occur that involves a functional group on one reactant and a functional group on the other reactant. Write out the general reaction if one is possible (see Table 1).

In this example, the general reaction is

$$\underset{\text{carboxylic acid}}{R'_1-\overset{\overset{\textstyle O}{\|}}{C}-OH} + \underset{\text{alcohol}}{R_2-OH} \longrightarrow \underset{\text{ester}}{R'_1-\overset{\overset{\textstyle O}{\|}}{C}-O-R_2} + H_2O$$

Step 3 Write the structural formulas for the specific products of the reaction in question to conform with: (a) the general reaction in step 2, and (b) the structural formulas for the reactants.

In this example, $R'_1 = CH_3-CH_2-$ and $R_2 = CH_3-CH_2-$. Thus,

$$CH_3-CH_2-\overset{\overset{\textstyle O}{\|}}{C}-OH + CH_3-CH_2-OH \longrightarrow$$

$$CH_3-CH_2-\overset{\overset{\textstyle O}{\|}}{C}-O-CH_2-CH_3 + H_2O$$

APPENDIX 1

Conversion Factors

English and Metric Units

Mass	1.00 pound = 454 grams
	1.00 ounce = 28.4 grams
	1.00 kilogram = 2.20 pounds
Length	1.00 inch = 2.54 centimeters
	1.00 meter = 39.4 inches
	1.00 mile = 1.61 kilometers
Volume	1.00 pint = 0.473 liter
	1.00 liter = 1.06 quarts

Apothecary Units

Mass	1.00 grain = 0.0648 gram
	1.00 dram = 3.89 grams
	1.00 ounce (apothecary) = 31.1 grams
Volume	1.00 minim = 0.0616 milliliter
	1.00 fluid dram = 3.70 milliliters
	1.00 fluid ounce = 29.6 milliliters

APPENDIX 2

SI Units

Quantity	SI Unit (symbol)	Other Units
Mass	kilogram (kg)	
Length	meter (m)	
Volume	cubic meter (m³)	1 liter = 1 dm³ = 10^{-3} m³
Temperature	kelvin (K)	
Energy	joule (J)	1 cal = 4.184 J 1 electron volt (eV) = 1.602×10^{-19} J
Pressure	pascal (Pa)	1 atm = 1.013×10^5 Pa 1 torr = 1 mmHg = 133.3 Pa
Amount	mole (mol)	
Concentration	moles/cubic decimeter (mol/dm³)	1 mole/liter* = 1 mol/dm³

* The unit mole/liter is abbreviated M and is conventionally called the molarity. The use of the word "molarity" is not recommended in the SI system.

APPENDIX 3

Vitamins and Coenzymes

Vitamins are organic compounds that people must ingest in small amounts for proper body function. Vitamins are divided into two classes on the basis of

Coenzyme	Vitamin Precursor	Reaction	Group Transferred
Nicotinamide adenine dinucleotide (NAD$^+$)	Niacin (nicotinic acid)	Oxidation-reduction $$NAD^+ + SH_2 \rightleftharpoons NADH + S + H^+$$	H:$^-$ (hydride ion)
Nicotinamide adenine dinucleotide phosphate (NADP$^+$)	Niacin (nicotinic acid)	Oxidation-reduction $$NADP^+ + SH_2 \rightleftharpoons NADPH + S + H^+$$	H:$^-$ (hydride ion)
Thiamin pyrophosphate (TPP)	Thiamine (vitamin B$_1$)	Acyl group transfer	$$\overset{\displaystyle O}{\underset{\displaystyle \parallel}{R-C-}}$$

their solubility in water. The structural formulas for the fat-soluble (i.e., water-insoluble) vitamins A, D, K, and E are given in Chapter 27, together with what is known about their function.

The functions of the water-soluble vitamins are generally much better understood. All of the water-soluble vitamins are components of coenzymes. The structural formulas of these coenzymes are presented at appropriate places in the text, and are collected in the following table, which also summarizes the types of reactions in which these coenzymes are involved. The vitamin portions of these coenzymes are shown in blue.

Structural Formula

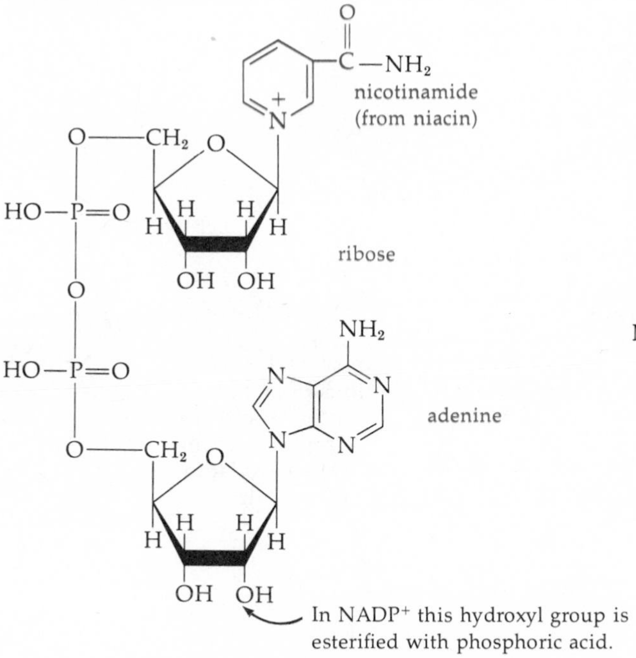

nicotinamide
(from niacin)

ribose

NAD+

adenine

In NAD+ this hydroxyl group is esterified with phosphoric acid.

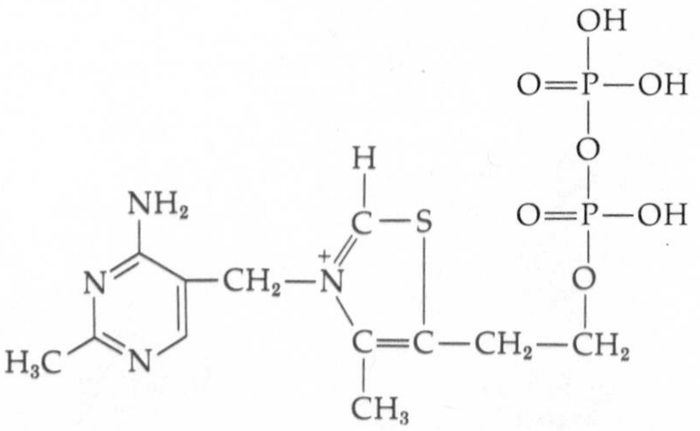

TPP

Coenzyme	Vitamin Precursor	Reaction	Group Transferred
Flavin mononucleotide (FMN)	Riboflavin (vitamin B_2)	Oxidation-reduction $$FMN + SH_2 \rightleftharpoons FMNH_2 + S$$	Hydrogen atoms
Flavin adenine dinucleotide (FAD)	Riboflavin	Oxidation-reduction $$FAD + SH_2 \rightleftharpoons FADH_2 + S$$	Hydrogen atoms
Pyridoxal phosphate	Pyridoxine (vitamin B_6)	In several reactions of amino acid metabolism: Transamination Decarboxylation Racemization	NH_3
Coenzyme A (CoA-SH)	Pantothenic acid	Biosynthesis of fatty acids and steroids, fatty acid oxidation	Acyl groups

Structural Formula

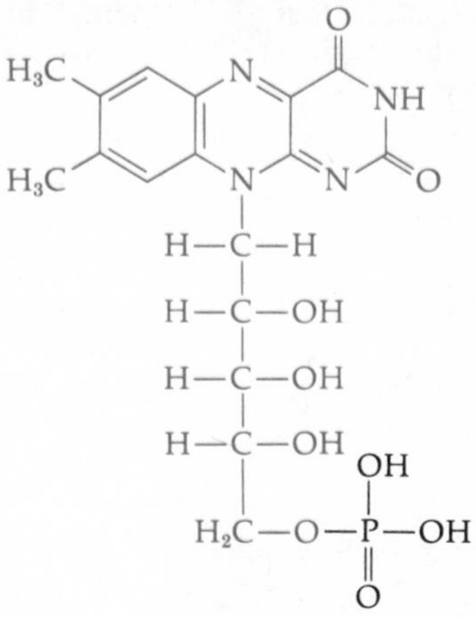

FMN

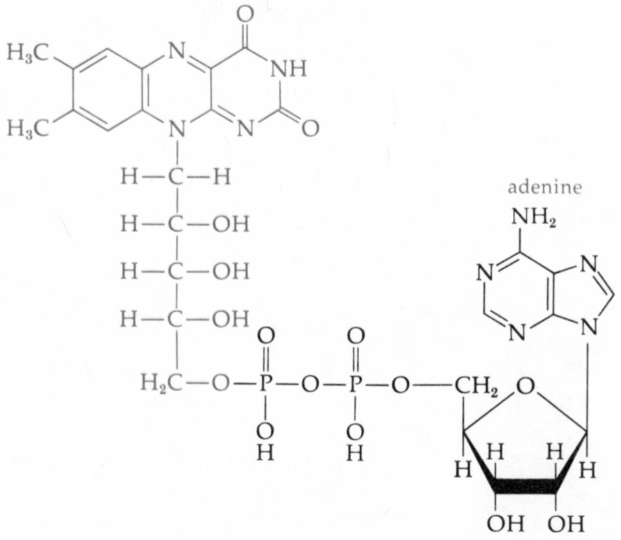

adenine

FAD

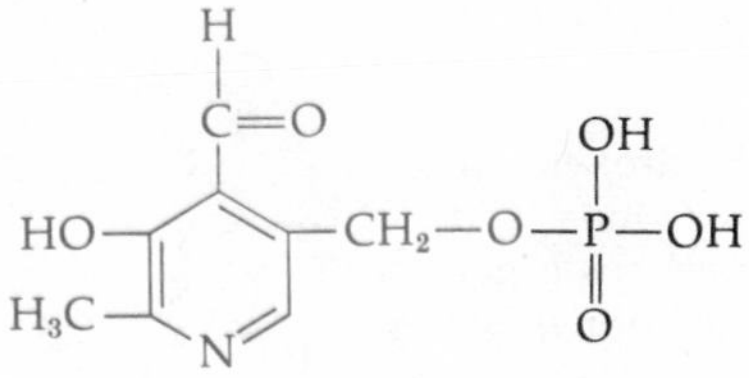

Pyridoxal phosphate

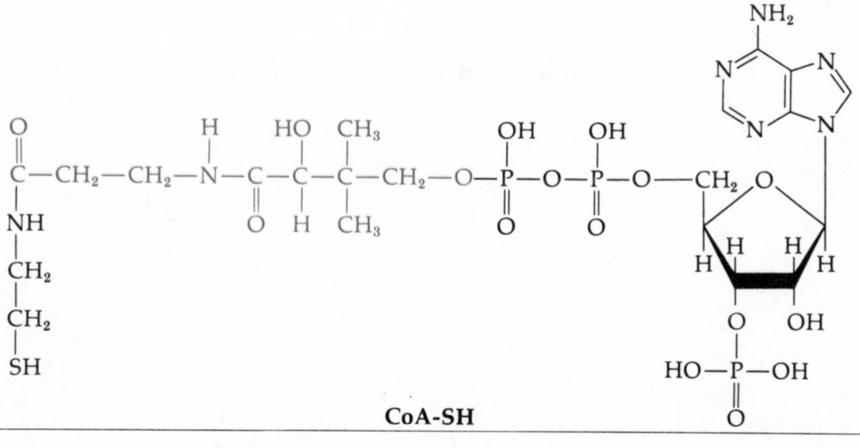

CoA-SH

Coenzyme	Vitamin Precursor	Reaction	Group Transferred
Tetrahydrofolic acid (THFA)	Folic acid	In reactions that transfer single carbon units	Methyl, formyl groups
Coenzyme Q (Ubiquinone)		Electron transport	Hydrogen atoms
Biotin*		In carboxylation reactions for the biosynthesis of purines, fatty acids, and urea	CO_2
Lipoic acid*		In generation of acyl groups, acyl group transfer, and electron transport	Acyl groups
Ascorbic acid* (Vitamin C)		Hydroxylation reactions (and perhaps other functions)	

* Biotin, lipoic acid, and ascorbic acid are both vitamins and coenzymes.

Structural Formula

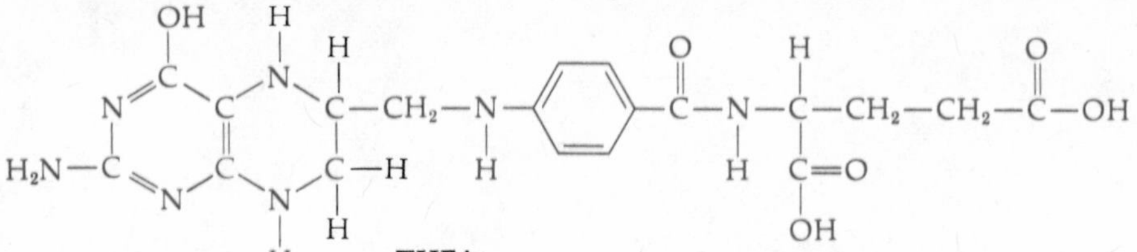

THFA

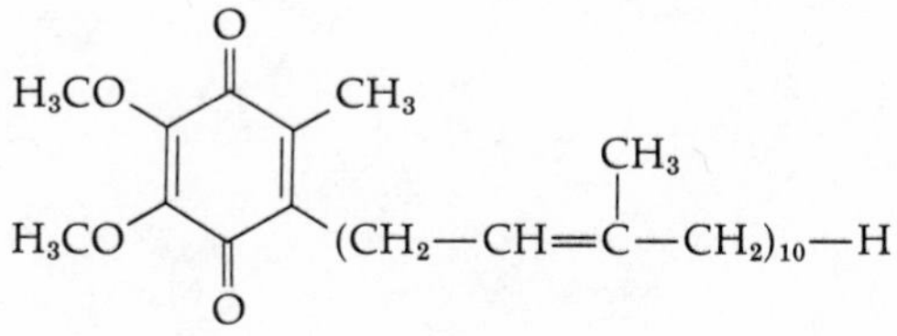

$$H_2N-C \quad N \quad OH \quad H \quad H \quad O \quad H \quad O$$

$$\text{THFA}$$

Coenzyme Q

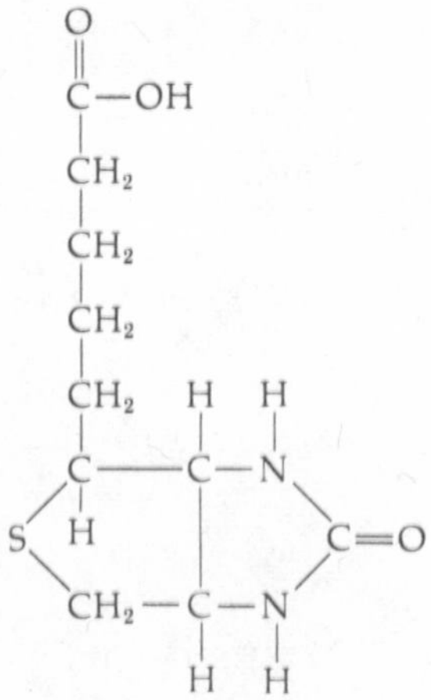

Biotin

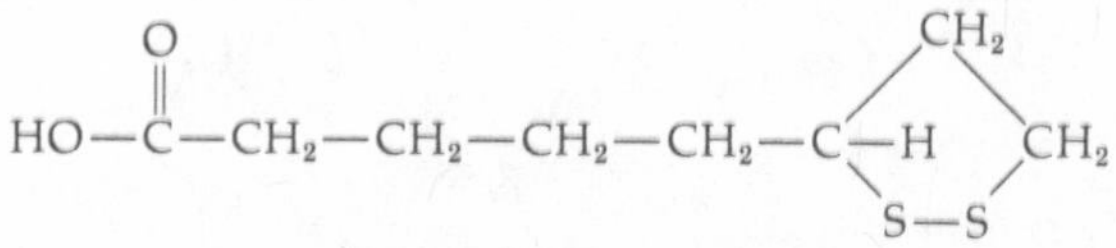

Lipoic acid

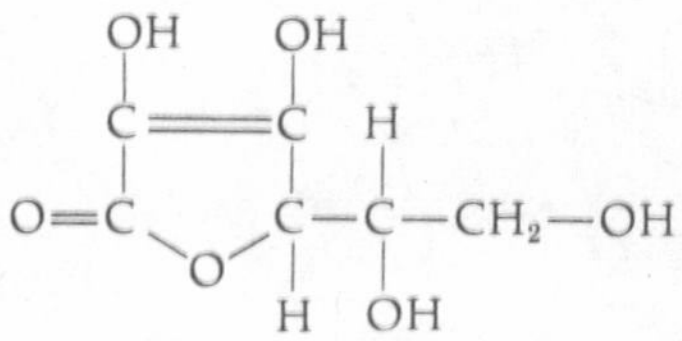

Ascorbic acid

Answers

CHAPTER 1

1. (a) 3.25×10^4 (b) 3.05×10^{-3} (c) 6.8×10^{-5} (d) 2.03×10^7

2. (a) 0.073   (b) 85,000   (c) 0.00209   (d) 62,800

3. 0.57 g

4. (a) 1.5×10^4 (b) 1.40×10^5 (c) 2.54×10^5 (d) 1.22×10^{-10}
   (e) 1.571×10^{-4} (f) 2.00   (g) 2.611×10^3 (h) 5.43×10^2

5. (a) 1.23×10^{-2} g   (b) 2.67×10^3 ml   (c) 71 lb   (d) 9.1×10^{-8} in.
   (e) 4.8 qt   (f) 5.2×10^2 μg   (g) 24 kcal

6. No. "1.1 g/ml" indicates that the true value could be between 1.0 and 1.2 g/ml.

7. 0.59 cm³

8. 2.9×10^{-2} g

9. Larger

10. 1.5 liters

11. Disappearance of the gray solid calcium metal. Appearance of the white solid calcium hydroxide. Bubbles of hydrogen gas in the water.

12. 7.50 ± 0.15 ml

13. ± 0.01 g

14. Kinetic energy has decreased.

15. Decreases

16. Potential energy increases and kinetic energy decreases.

17. $-196°C$, $-321°F$

18. 1074 K, 1474°F

19. 28.2°C

20. (a) 225 cal
    (b) 31 g

CHAPTER 2

1. Melting point between 180 K and 230 K
   Boiling point between 230 K and 280 K

2. 1.4×10^4 cal

3. Decreases

4. (a) Compound (b) Element (c) Solution (d) Heterogeneous mixture (e) Element

5. (a) $2C_3H_6 + 9O_2 \rightarrow 6CO_2 + 6H_2O$
   (b) $4Fe + 3O_2 \rightarrow 2Fe_2O_3$
   (c) $3NO_2 + H_2O \rightarrow 2HNO_3 + NO$
   (d) $4P + 5O_2 \rightarrow 2P_2O_5$
   (e) $4NH_3 + 5O_2 \rightarrow 4NO + 6H_2O$

6.

	No. of Protons	No. of Electrons	No. of Neutrons
^{33}S	16	16	17
$^{51}V^{3+}$	23	20	28
$^{18}O^{2-}$	8	10	10
$^{87}Sr^{2+}$	38	36	49
^{23}Na	11	11	12
$^{35}Cl^-$	17	18	18

7. $^{19}F^-$

8. $^{24}Mg^{2+}$

9. Same number of protons, different number of neutrons

10. (c)

11. $^{14}_7N + ^4_2He \rightarrow ^{17}_8O + ^1_1H$

12. 2 mCi after 18 hr, 1 mCi after 24 hr

13. 15 days

14. $^{90}_{37}Rb \rightarrow ^{90}_{38}Sr + ^{\ 0}_{-1}e$

15. Larger. The radioactivity used in diagnostic tests should be low enough so that cell damage is minimal.

CHAPTER 3

1. (a) Nonmetal (b) Metal (c) Metal (d) Metalloid (e) Nonmetal

2. Nonmetal

3. Sb_2O_5

4. SiF_4

5. Zn^{2+} and Br^-

6. Se

7. O

8. Larger. A Cl^- ion has one more electron than a Cl atom.

9. Smaller. A Na^+ ion has one less electron than a Na atom.

10. S

11. Smaller

12. The Bohr orbit with $n = 2$ is closer to the nucleus and has a lower energy than the Bohr orbit with $n = 3$. An energy input is required to take the electron from the $n = 2$ orbit and put it in the $n = 3$ orbit.

13. According to the Bohr model, the electron is definitely located in a circular orbit at a specific distance from the nucleus. For a quantum mechanical orbital, only the probability of finding the electron at a certain position from the nucleus is specified.

14. (a) Sr: 2 valence electrons in the fifth shell
    (b) P: 5 valence electrons in the third shell
    (c) I⁻: 8 valence electrons in the fifth shell
    (d) K: 1 valence electron in the fourth shell
    (e) Ga: 3 valence electrons in the fourth shell
    (f) B: 3 valence electrons in the second shell
    (g) C: 4 valence electrons in the second shell

15. (a) Ca^{2+} (b) Does not tend to form an ion (c) Se^{2-} (d) Li^+ (e) Does not tend to form an ion (f) H^+

16. (a) :N· (b) K· (c) :Br:⁻ (d) Mg^{2+} (e) :Se:

17. **(Optional)**
    (a) Ca: $1s^2 2s^2 2p^6 3s^2 3p^6 4s^2$
    (b) Al: $1s^2 2s^2 2p^6 3s^2 3p^1$
    (c) Cl: $1s^2 2s^2 2p^6 3s^2 3p^5$
    (d) Rb: $1s^2 2s^2 2p^6 3s^2 3p^6 3d^{10} 4s^2 4p^6 5s^1$

CHAPTER 4

1. (c)

2. (a), (d), (e), and (g) are ionic; (b), (c), and (f) are covalent.
    (a) K^+ and CO_3^{2-} (d) Ba^{2+} and OH^-
    (e) Na^+ and S^{2-} (g) NH_4^+ and SO_4^{2-}

3. (a) Dinitrogen trioxide   (b) Lead(II) sulfide
    (c) Copper(II) bromide   (d) Diarsenic pentoxide
    (e) Iron(III) sulfate   (f) Diphosphorus tetraiodide
    (g) Copper(II) monohydrogen carbonate   (h) Magnesium phosphate

4. (a) $SrBr_2$ (b) NH_4NO_3 (c) P_4Se_3 (d) FeI_2
    (e) $PbCO_3$ (f) OF_2 (g) K_2HPO_4

5. O

6. (a) Nonpolar covalent   (b) Polar covalent   (c) Ionic   (d) Nonpolar covalent
    (e) Polar covalent   (f) Polar covalent   (g) Ionic

7. (a) H—C≡N: (b) $\left[:\ddot{O}—H \right]^-$ (c)

(d)   (e)

$$\begin{array}{ccc} H & & H \\ & \diagdown & \diagup \\ & \ddot{N}—\ddot{N} & \\ & \diagup & \diagdown \\ H & & H \end{array}$$

(f) 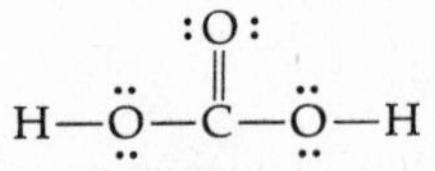 H—Ö—C—Ö—H with :O: double bonded to C

(g) 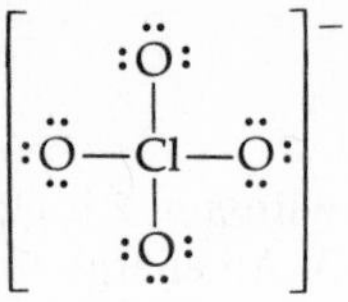 [:Ö—Cl—Ö: with :O: above and :O: below Cl]⁻

8. (a) 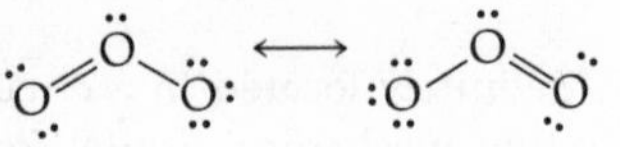 :O=O—Ö: ⟷ :Ö—O=O: (resonance structures with lone pairs)

(b) 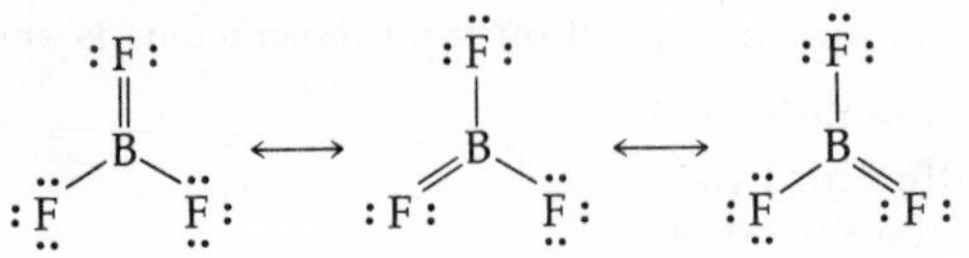 :F: B bonded to :F and F: (with B=F double bond) ⟷ :F: B=F ⟷ :F B=F: (three resonance structures)

(c) 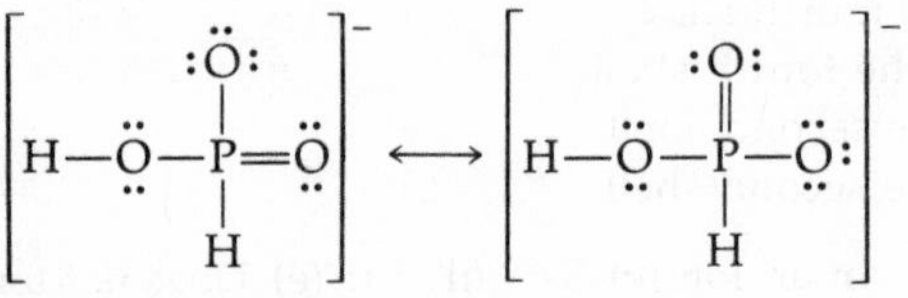 [H—Ö—P=Ö with H below P and :O: above]⁻ ⟷ [H—Ö—P—Ö: with H below P and :O: double bonded above]⁻

9. N₂

10. The carbon-oxygen bond energy is smaller and the bond length larger in ethyl alcohol.

11. **(Optional)**
   (a) Pyramidal   (b) Bent   (c) Tetrahedral   (d) Pyramidal
   (e) Tetrahedral   (f) Pyramidal   (g) Linear

CHAPTER 5

1. When the temperature decreases, the gas molecules move more slowly (on the average), so they hit the walls of the container less frequently and with less force.

2. As you move up from the earth's surface, the atmospheric pressure decreases. Therefore, the pressure inside the balloon must also decrease. To accomplish this, the volume of the balloon must increase.

3. 170 mm

4. 2.60 atm

5. 0.53 atm

6. 0.80 atm

7. 205°C

8. 2.0 atm

9. −113°C

10. 35 torr

11. CaF_2 and $NaNO_3$

12. As the temperature increases, proportionately more molecules in the liquid have sufficient kinetic energy to overcome the intermolecular forces of attraction in the liquid and go into the gas state. Thus, a higher vapor pressure is required to produce a balancing condensation rate.

13. (d) Liquid and vapor are in equilibrium at 25°C.

14. (a) Positive: vaporization is an endothermic process.
    (b) Positive: a gas is more disordered than a liquid.
    (c) Zero: the system is at equilibrium.

CHAPTER 6

1. Two, four, thirty-two

2. (a) 98 (b) 96 (c) 102 (d) 60 (e) 58 (f) 88

3. (a) 58 g (b) 9.7 g (c) 2.1×10^2 g (d) 0.45 g (e) 3.8×10^{-2} g

4. (a) 1.1×10^{-3} mole (b) 5.7×10^{-2} mole (c) 1.64 moles (d) 6.9×10^{-5} mole (e) 14 moles

5. (a) 6.2 moles of H_2O (b) 6.2 g of NH_3
   (c) Approximately the same number of molecules
   (d) 100 g of urea

6. (a) 3.0×10^{-22} g   (b) 1.1×10^{-22} g

7. (a) 9.6, 6.4 (b) 3.2 (c) 3.25, 9.75 (d) 1.7

8. (a) 5.0, 12.5 (b) 7.8, 6.2, 3.1 (c) 2.2 (d) 4.8

9. (a) 56.7, 73.3 (b) 6.4, 11.3, 3.4 (c) 61

10. (a) 275, 225 (b) 88, 128 (c) 20.6

11. (a) 13.3 kcal/g (methane), 11.9 kcal/g (acetylene)
    (b) 75 g (methane), 84 g (acetylene)

12. (a) Yes (b) 0.65 kcal
    (c) Maximum work for glycolysis is 0.261 kcal/g. Maximum work for oxidation of glucose to CO_2 and H_2O is 3.80 kcal/g (see Chapter 6, Example 12).

13. Yes. For this reaction ΔG is negative, so the reaction is a possible source of work.

14. Since ΔG is positive, the decomposition of Al_2O_3 is not a spontaneous process, and to accomplish the decomposition of Al_2O_3 by an electrical current, an input of electrical energy is required.

15. 34 liters

16. 1.4 g

CHAPTER 7

1. There are dipolar forces between polar CO molecules and no dipolar forces between non-polar N_2 molecules.

2. Lowest : CF_4, CCl_4, CBr_4 : Highest

3. (a) Polar (b) Nonpolar (c) Polar (d) Polar

4. A Cl atom is much larger than a N atom. Therefore, two NH_3 molecules can get closer together than two HCl molecules. In NH_3 there are three N—H bonds per molecule, whereas in HCl there is only one H—Cl bond.

5. No; yes

6. (a)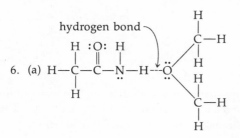

(c)

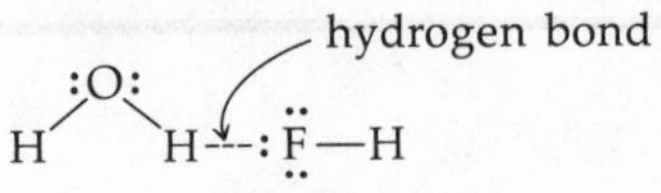

or

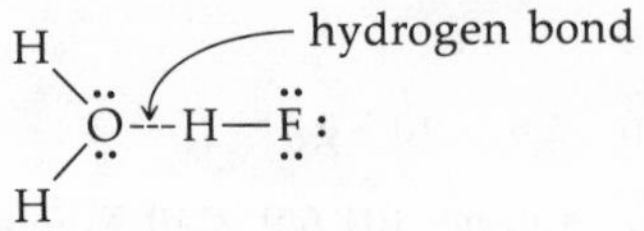

(d)

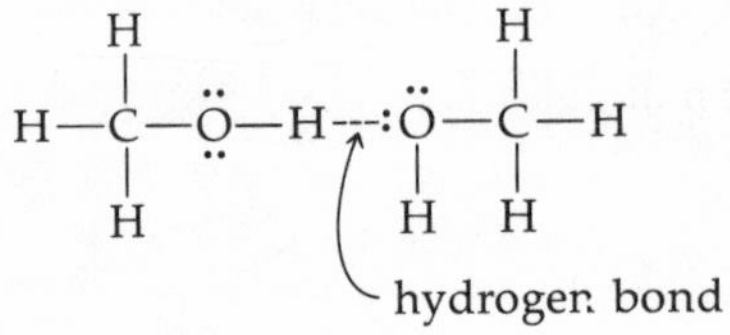

7. The hydrophilic parts of 1-hexanol and 1-butanol molecules are the same, but 1-hexanol has a larger hydrophobic part.

8. There are strong attractive interactions between positive ions and the oxygen atoms (which have large partial negative charges) of water molecules; and there are strong attractive interactions between negative ions and the hydrogen atoms (which have large partial positive charges) of water molecules.

CHAPTER 8

1. (a) 2.9×10^{-3} g/ml, 0.29%, 1.6×10^{-2} M
   (b) 1.7×10^{-3} g/ml, 0.17%, 2.9×10^{-2} M
   (c) 3.0×10^{-2} g/ml, 3.0%, 0.36 M
   (d) 1.4×10^{-2} g/ml, 1.4%, 0.19 M

2. (a) 2.0 g, 3.4×10^{-2} mole
   (b) 28 g, 0.33 mole
   (c) 17.5 g, 9.7×10^{-2} mole
   (d) 2.6 g, 3.5×10^{-2} mole

3. (a) 2000 ml (b) 390 ml (c) 80 ml (d) 160 ml

4. (a) 12.5 ml (b) 30 ml (c) 1.1×10^2 ml (d) 19 ml

5. (a) 0.56 M Na^+, 0.28 M CO_3^{2-}
      0.56 N Na^+, 0.56 N CO_3^{2-}
   (b) 0.52 M Mg^{2+}, 1.04 M Cl^-
      1.04 N Mg^{2+}, 1.04 N Cl^-
   (c) 0.12 M K^+, 0.12 M HCO_3^-
      0.12 N K^+, 0.12 N HCO_3^-
   (d) 0.28 M Na^+, 9.2×10^{-2} M PO_4^{3-}
      0.28 N Na^+, 0.28 N PO_4^{3-}

6. 7.2 g, 0.12 mole

7. 525 g

8. No

9. 5.3×10^{-3} g/liter

10. If it is a heterogeneous mixture, the dispersed particles will settle out under the influence of gravity upon standing. A solution and a colloidal dispersion can be distinguished by the absence (solution) or presence (colloidal dispersion) of the Tyndall effect.

11. Water will diffuse out of the red blood cells, and they will undergo crenation.

12. 5.5%

13. Solution A has the higher osmotic pressure. Solution B is hypotonic with respect to solution A.

14. (a) 0.1 M $CaCl_2$
    (b) Both solutions have approximately the same osmotic pressure.
    (c) 1.0 M NaCl

15. Water diffuses into the bag.

CHAPTER 9

1. 1.6×10^{-6} mole/liter · sec

2. (a) 3.8×10^{-4} mole (b) 3.8%

3. (a) The activation energy in the forward direction, $E_a(F)$, is greater than that in the reverse direction, $E_a(R)$.

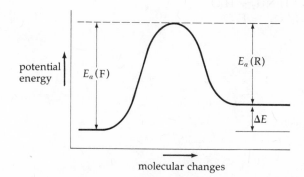

molecular changes

(b) The rate in the forward direction is less than that in the reverse direction.

4. At the lower temperature the rate for the souring reaction is slower.

5. (a) $2H_2 + 2NO \rightarrow 2H_2O + N_2$
   (b) N and O
   (c) No catalyst

6. (a) $K_{eq} = \dfrac{[O_2][NO_2]}{[O_3][NO]}$

   (b) $K_{eq} = \dfrac{[NO]^4}{[N_2O]^2[O_2]}$

   (c) $K_{eq} = [CO_2]$

   (d) $K_{eq} = \dfrac{[NO]^2[O_2]}{[NO_2]^2}$

7. Positive

8. (a) The reaction in container A.
   (b) We cannot tell. There is no information about the rates of the reactions.

9. (a) Shift to the left   (b) Shift to the right
   (c) Shift to the left   (d) No effect

CHAPTER 10

1. (a) $HI + NaOH \rightarrow NaI + H_2O$
   (b) $2HNO_3 + Ca(OH)_2 \rightarrow Ca(NO_3)_2 + 2H_2O$
   (c) $H_2SO_4 + 2KOH \rightarrow K_2SO_4 + 2H_2O$

2. (a) $HSO_3^- + H_2O \rightarrow H_3O^+ + SO_3^{2-}$
      Conjugate acid-base pairs: (HSO_3^-, SO_3^{2-}) and (H_3O^+, H_2O)
   (b) $HSO_3^- + H_2O \rightarrow H_2SO_3 + OH^-$
      Conjugate acid-base pairs: (H_2SO_3, HSO_3^-) and (H_2O, OH^-)
   (c) $HSO_3^- + OH^- \rightarrow SO_3^{2-} + H_2O$

3. (a) HPO_4^{2-} (b) H_3PO_4

4. Conjugate acid-base pairs $(H_3BO_3, H_2BO_3^-)$ and (H_3O^+, H_2O)

5. (a) Yes (b) Yes (c) CO_3^{2-} (d) CN^-

6. Less than

7. Larger

8. (a) $C_7H_5O_2^- + H_2O \rightleftharpoons HC_7H_5O_2 + OH^-$ (b) Stronger

9. (a) $HSO_4^- + OH^- \rightarrow SO_4^{2-} + H_2O$
   (b) $CO_3^{2+} + H_3O^+ \rightarrow HCO_3^- + H_2O$
   (c) No reaction
   (d) $CH_3NH_3^+ + OH^- \rightarrow CH_3NH_2 + H_2O$
   (e) $HPO_4^{2-} + H_3O^+ \rightarrow H_2PO_4^- + H_2O$

10. **(Optional)**

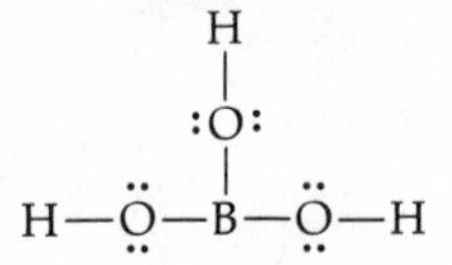

$$\begin{array}{c} H \\ | \\ :\ddot{O}: \\ | \\ H-\ddot{O}-B-\ddot{O}-H \end{array}$$

   (a) Weak acid   (b) Weaker

CHAPTER 11

1. (a) $4.3 \times 10^{-7} M$ (b) $1.7 \times 10^{-3} M$ (c) $1.4 \times 10^{-13} M$

2. (a) $2.2 \times 10^{-8} M$ (b) $1.5 \times 10^{-5} M$ (c) $4.8 \times 10^{-12} M$

3. (a) Acidic (b) Neutral (c) Basic (d) Basic

4. (a) Between 2 and 3 (b) Between 9 and 10 (c) Between 8 and 9

5. (a) Between $10^{-6} M$ and $10^{-7} M$ (b) Between $10^{-1} M$ and $10^{-2} M$
   (c) Between $10^{-8} M$ and $10^{-9} M$ (d) Between $10^{-11} M$ and $10^{-12} M$

6. (a) $HCO_3^- + OH^- \rightleftharpoons CO_3^{2-} + H_2O$ nearly complete
   (b) $H_3O^+ + OH^- \longrightarrow 2H_2O$ complete
   (c) $CN^- + H_3O^+ \rightleftharpoons HCN + H_2O$ nearly complete
   (d) $NH_4^+ + OH^- \rightleftharpoons NH_3 + H_2O$ nearly complete

7. $0.116 M$

8. $0.327 g$

9. (a) $9.3 \times 10^{-10} M$ (b) $6.0 \times 10^{-10} M$
   (c) $1.6 \times 10^{-9} M$ (d) $9.3 \times 10^{-10} M$

CHAPTER 12

1. (a) Hydrogen $(+1)$, nitrogen $(+5)$, oxygen (-2)
   (b) Carbon $(+4)$, oxygen (-2) (d) Iron $(+3)$, oxygen (-2)
   (c) Hydrogen $(+1)$, nitrogen (-3) (e) Sulfur $(+6)$, oxygen (-2)

2. (a) $Fe(s)$ is oxidized and is the reducing agent; $Cl_2(g)$ is reduced and is the oxidizing agent.
   (b) The sulfur in $Ag_2S(s)$ is oxidized and $Ag_2S(s)$ is the reducing agent. The nitrogen in $NO_3^-(aq)$ is reduced and $NO_3^-(aq)$ is the oxidizing agent.
   (c) Not an oxidation-reduction reaction.
   (d) $Br_2(l)$ is both oxidized [to $BrO_3^-(aq)$] and reduced [to $Br^-(aq)$]. Therefore, $Br_2(l)$ is both the oxidizing and reducing agent.

3. (a) Oxidation of carbon   (b) X is an oxidizing agent

4. **(Optional)** (a) Yes   (b) No   (c) Yes   (d) Yes

5. **(Optional)**
   (a) $I_2(s) + Cu(s) \rightarrow 2I^-(aq) + Cu^{2+}(aq)$ (b) No reaction
   (c) $3MnO_4^-(aq) + 4H^+(aq) + 5NO(g) \rightarrow 3Mn^{2+}(aq) + 2H_2O + 5NO_3^-(aq)$
   (d) $H_2O_2(aq) + 2Cl^-(aq) + 2H^+(aq) \rightarrow 2H_2O + Cl_2(g)$

6. (a) $PbSO_4(s)$ will precipitate   (b) No precipitate
   (c) $Fe(OH)_3(s)$ will precipitate   (d) $ZnCO_3(s)$ will precipitate

7. (a) $Pb^{2+}(aq) + SO_4^{2-}(aq) \rightarrow PbSO_4(s)$ (c) $Fe^{3+}(aq) + 3OH^-(aq) \rightarrow Fe(OH)_3(s)$
   (d) $Zn^{2+}(aq) + CO_3^{2-}(aq) \rightarrow ZnCO_3(s)$

CHAPTER 13

1. $CH_3-CH_2-CH_2-CH_2-$, n-butyl

$$CH_3-CH_2-\overset{|}{CH}-CH_3, sec\text{-butyl}$$

$$\begin{matrix} H_3C \\ \diagdown \\ CH-CH_2-, \text{ isobutyl} \\ \diagup \\ H_3C \end{matrix}$$

$$H_3C-\overset{\overset{\displaystyle CH_3}{|}}{\underset{\underset{\displaystyle CH_3}{|}}{C}}-, t\text{-butyl}$$

2. (a) $\begin{matrix} H_3C \\ \diagdown \\ CH-CH_3 \\ \diagup \\ H_3C \end{matrix}$ (b) and (c) $CH_3-CH_2-CH_2-CH_2-CH_2-CH_2-CH_2-CH_3$

   (d) $H_3C-\overset{\overset{\displaystyle CH_3}{|}}{\underset{\underset{\displaystyle CH_3}{|}}{C}}-CH_2-CH_2-CH_2-CH_2-CH_2-CH_2-CH_3$

   (e) $CH_3-CH_2-\overset{\overset{\displaystyle |}{CH}}{\underset{\underset{\displaystyle CH_3}{\underset{\displaystyle |}{CH_2}}}{|}}-CH_2-CH_3$

3. (a) and (c) are both pairs of positional isomers

4. (a) 2,2-Dimethylbutane   (b) 2,2,3,4-Tetramethylpentane
   (c) Butane   (d) Butane
   (e) 2,3-Dimethylpentane   (f) 2,3,5-Trimethylhexane

CHAPTER 14

1. (a) Cycloalkanes
   (c) Alkenes
   (e) Alkyl halides

   (b) Cycloalkanes
   (d) Aromatic hydrocarbons

2. (a) 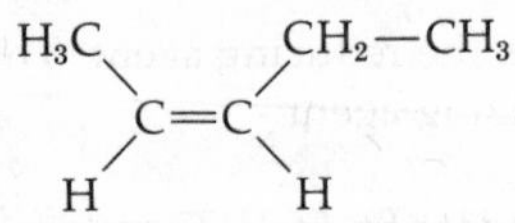
$$\underset{H}{\overset{H_3C}{\diagdown}}C=C\underset{H}{\overset{CH_2-CH_3}{\diagup}}$$

   (b)

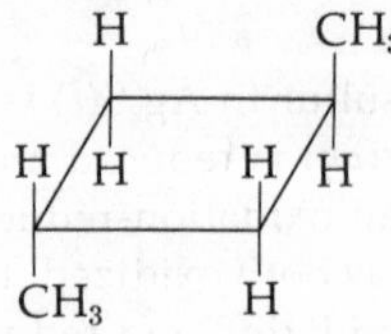

   (c) 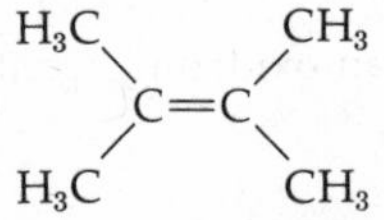
$$\underset{H_3C}{\overset{H_3C}{\diagdown}}C=C\underset{CH_3}{\overset{CH_3}{\diagup}}$$

   (d) 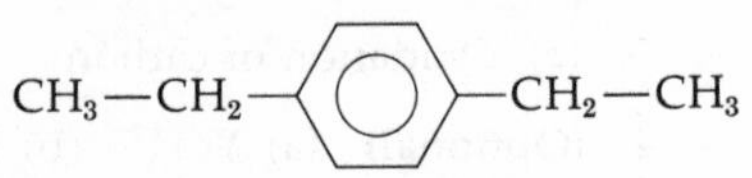 $CH_3-CH_2-\langle\bigcirc\rangle-CH_2-CH_3$

   (e)

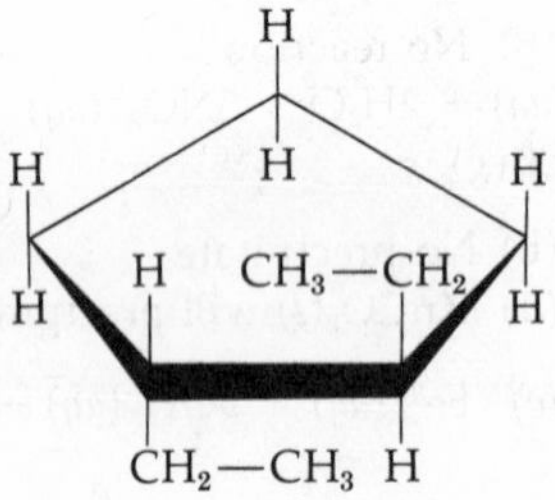

3. (a) 2,3-Dimethyl-2-pentene
   (c) Methylcyclohexane
   (e) *cis*-3,4-Dimethyl-2-pentene

   (b) *cis*-2-Pentene
   (d) 1,3-Diisopropylbenzene

4. $CH_2{=}CH-CH_2-CH_3$, 1-butene; $CH_3-CH{=}CH-CH_3$, 2-butene; $CH_2{=}CH-CH_3$,
$\overset{\qquad\qquad\qquad\qquad\qquad\qquad\qquad\qquad\qquad}{\underset{CH_3}{}}$

   2-methylpropene; $\triangleright-CH_3$, methylcyclopropane; and $\square$, cyclobutane are all
   structural isomers. There are two geometrical isomers for 2-butene: *cis*-2-butene,

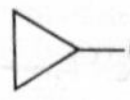

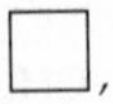

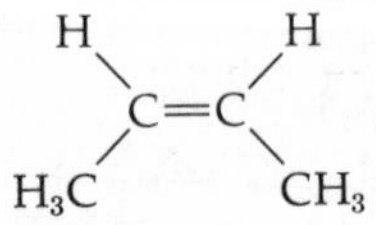

$$\underset{H_3C}{\overset{H}{\diagdown}}C=C\underset{CH_3}{\overset{H}{\diagup}}$$; and *trans*-2-butene, $$\underset{H_3C}{\overset{H}{\diagdown}}C=C\underset{H}{\overset{CH_3}{\diagup}}$$

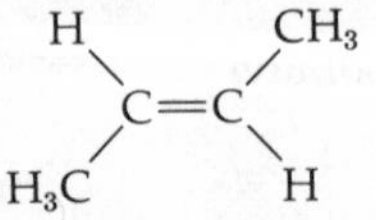

5. (a) 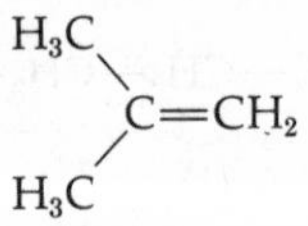
$$\underset{H_3C}{\overset{H_3C}{\diagdown}}C=CH_2$$

   (b) 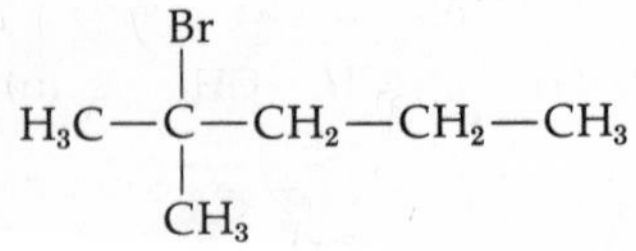 $H_3C-\overset{\overset{\displaystyle Br}{|}}{\underset{\underset{\displaystyle CH_3}{|}}{C}}-CH_2-CH_2-CH_3$

   (c)

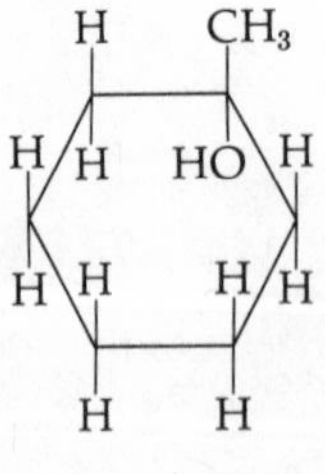

   (d) $CH_3-CH_2-CH_3$

6.

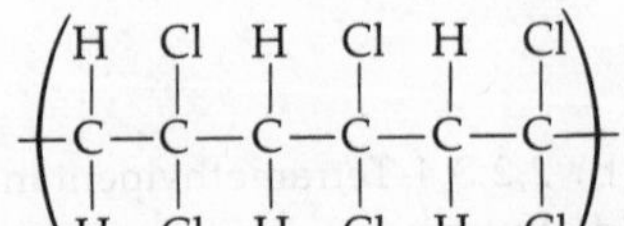

CHAPTER 15

1. (a) 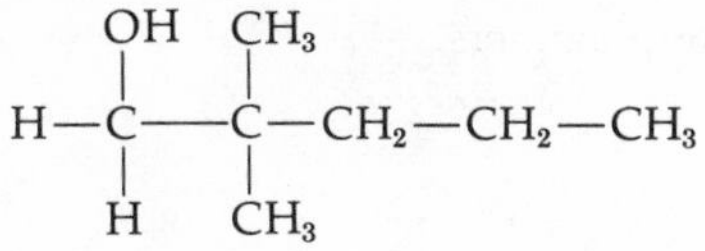
$$\underset{\underset{H}{|}}{\overset{\overset{OH}{|}}{H-C}}-\underset{\underset{CH_3}{|}}{\overset{\overset{CH_3}{|}}{C}}-CH_2-CH_2-CH_3$$

(b) 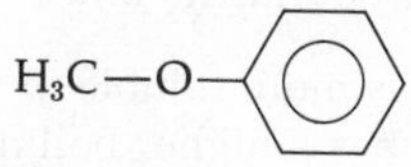 $H_3C-O-\bigcirc$

(c) $H_3C-O-CH_2-CH_3$

(d)

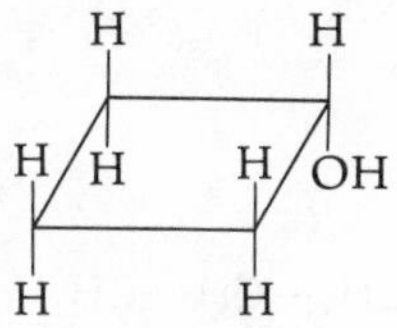

(e) 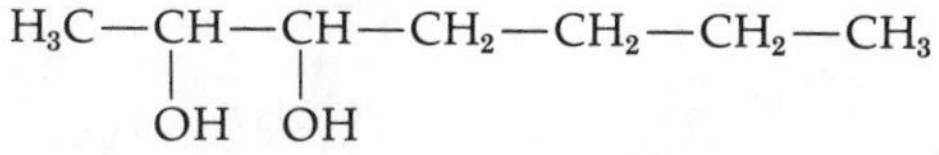 $H_3C-\underset{\underset{OH}{|}}{CH}-\underset{\underset{OH}{|}}{CH}-CH_2-CH_2-CH_2-CH_3$

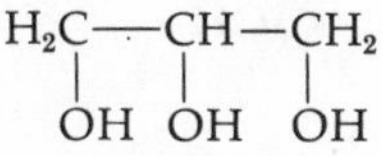

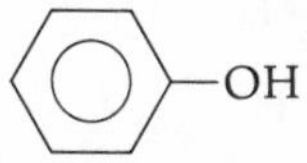

(f) $H_2C-\underset{\underset{OH}{|}}{CH}-\underset{\underset{OH}{|}}{CH_2}$
   $\;\;\;\;\underset{OH}{|}$

(g) $\bigcirc-OH$

(h) $H_3C-\underset{\underset{OH}{|}}{CH}-CH_3$

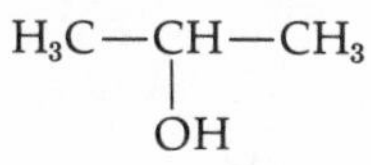

2. (a) 2,3,3-Trimethyl-2-butanol
    (b) 1,3-Cyclopentanediol
    (c) Methyl isopropyl ether

3. (a) is least soluble and (b) is most soluble; (c) has the lowest boiling point.

4. (a) 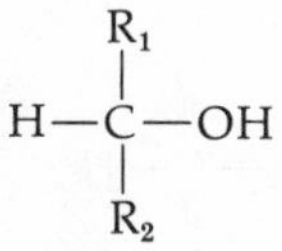 $H-\underset{\underset{R_2}{|}}{\overset{\overset{R_1}{|}}{C}}-OH$

(b) 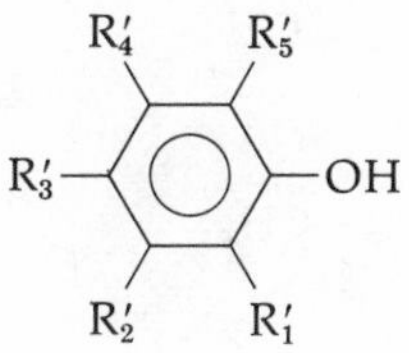 $R_3'-\bigcirc-OH$ with R_4', R_5' top and R_2', R_1' bottom

(c) R_1-O-R_2

5. (a) 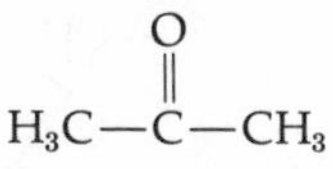 $H_3C-\overset{\overset{O}{||}}{C}-CH_3$

(b) $CH_3-CH=CH_2$

(c) N.R.

(d) N.R.

(e) 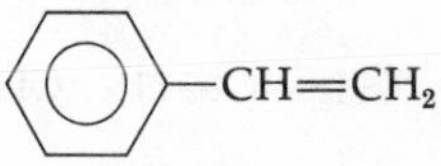 $\bigcirc-CH=CH_2$

(f) 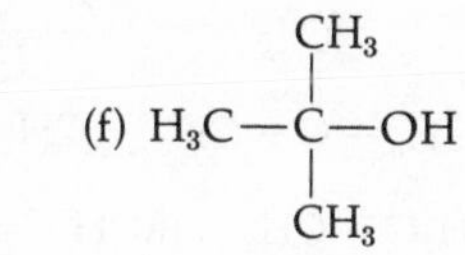 $H_3C-\underset{\underset{CH_3}{|}}{\overset{\overset{CH_3}{|}}{C}}-OH$

6. $H_3C-O-CH_2-CH_3$, $CH_3-CH_2-CH_2-OH$, and $CH_3-\underset{\underset{OH}{|}}{CH}-CH_3$

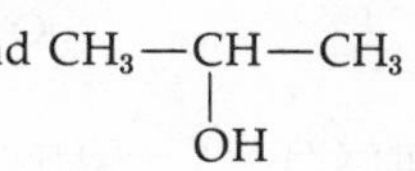

CHAPTER 16

1. (a) 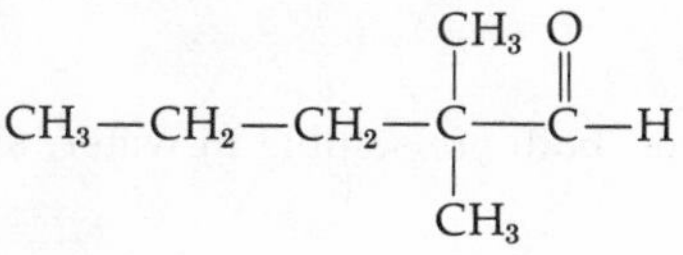 $CH_3-CH_2-CH_2-\underset{\underset{CH_3}{|}}{\overset{\overset{CH_3}{|}}{C}}-\overset{\overset{O}{||}}{C}-H$

(b) $CH_3-\underset{\underset{CH_3}{|}}{\overset{\overset{CH_3}{|}}{C}}-\overset{\overset{O}{||}}{C}-CH_2-CH_3$

(c) $HO-CH_2-CH_2-CH_2-\overset{\overset{O}{||}}{C}-H$

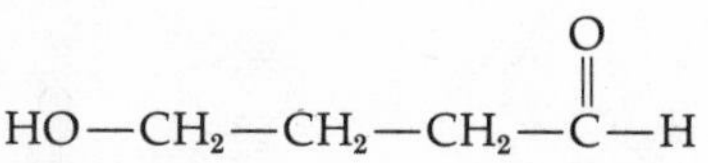

(d) 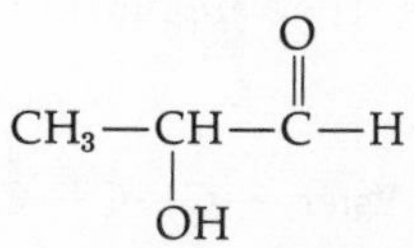 $CH_3-\underset{\underset{OH}{|}}{CH}-\overset{\overset{O}{||}}{C}-H$

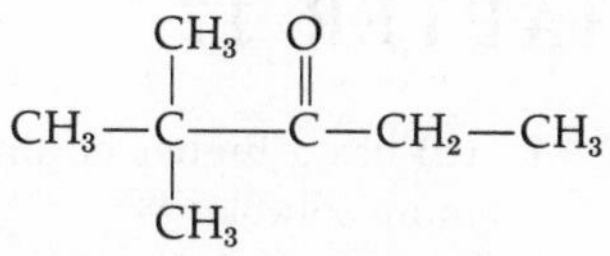

2. (a) 2-Methylpropanal

(b) 2,4-Dimethyl-3-pentanone

(c) 3-Hydroxybutanone

(d) Cyclopentanone

(e) Butanone

3. Compounds (a) and (e) are functional group isomers.

4. (c) is more soluble in water than (a)

(c) has a higher boiling point

5. (a) N.R.

(b) 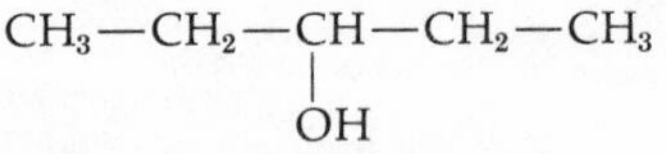 $CH_3-CH_2-\underset{\underset{OH}{|}}{CH}-CH_2-CH_3$ + X

(c) 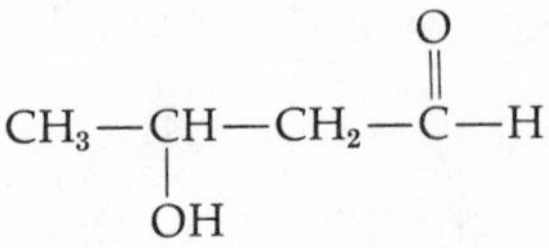 $CH_3-\underset{\underset{OH}{|}}{CH}-CH_2-\overset{\overset{O}{\|}}{C}-H$

(d) 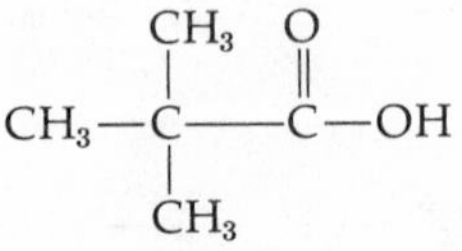 $CH_3-\underset{\underset{CH_3}{|}}{\overset{\overset{CH_3}{|}}{C}}-\overset{\overset{O}{\|}}{C}-OH$ + XH_2

(e) $CH_3-CH_2-CH_2-OH$ + X

(f) 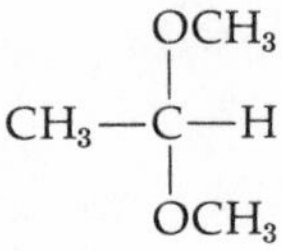 $CH_3-\underset{\underset{OCH_3}{|}}{\overset{\overset{OCH_3}{|}}{C}}-H$

(g) 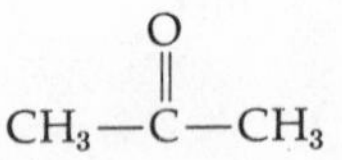 $CH_3-\overset{\overset{O}{\|}}{C}-CH_3$ + $2CH_3OH$

(h) N.R.

(i) 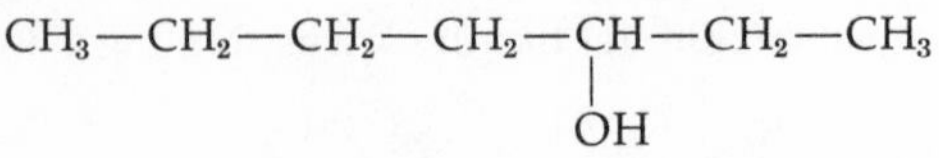 $CH_3-CH_2-CH_2-CH_2-\underset{\underset{OH}{|}}{CH}-CH_2-CH_3$ + X

6. (a) 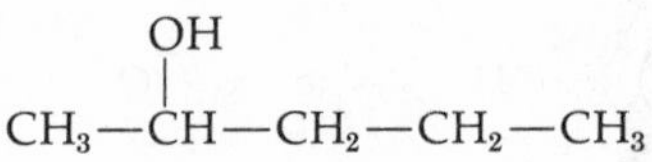 $CH_3-\underset{\underset{OH}{|}}{\overset{\overset{OH}{|}}{CH}}-CH_2-CH_2-CH_3$

(b) 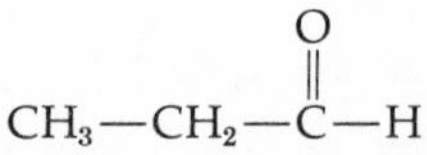 $CH_3-CH_2-\overset{\overset{O}{\|}}{C}-H$

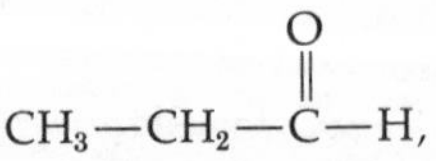

(c) and (d) 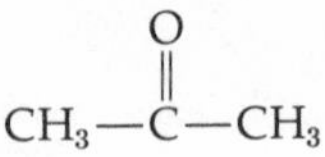 $CH_3-\overset{\overset{O}{\|}}{C}-CH_3$

(e) and (f) 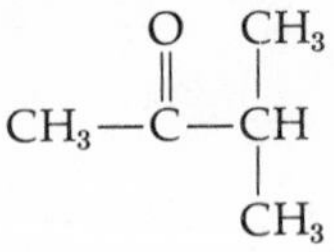 $CH_3-\overset{\overset{O}{\|}}{C}-\underset{\underset{CH_3}{|}}{\overset{\overset{CH_3}{|}}{CH}}$

7. 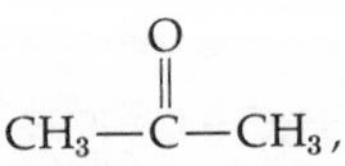 $CH_3-\overset{\overset{O}{\|}}{C}-CH_3$, $CH_3-CH_2-\overset{\overset{O}{\|}}{C}-H$, $CH_2{=}CH-CH_2OH$, $CH_2{=}CH-O-CH_3$,

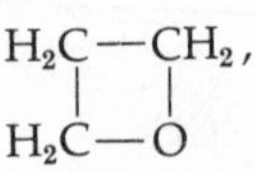

$\underset{\underset{H_2C-O}{|}}{\overset{\overset{H_2C-CH_2}{|}}{}}$, and $H_2C\underset{O}{\diagdown\diagup}CH-CH_3$. Note that a compound with the structural

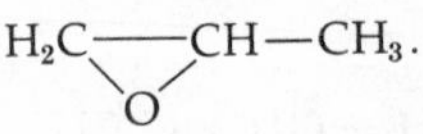

formula $CH_2{=}\underset{\underset{OH}{|}}{C}-CH_3$ cannot be isolated.

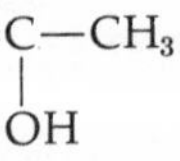

CHAPTER 17

1. (b) has a higher boiling point than (a), and both are soluble in water, although (b) is more soluble.

2. Yes. They are functional group isomers.

(a) Ester, $-\underset{|}{\overset{|}{C}}-O-\overset{\overset{O}{\|}}{C}-$

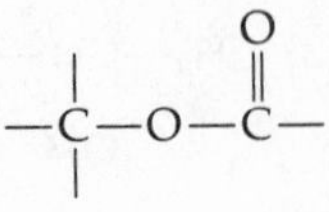

(b) Carboxylic acid, $-\overset{\overset{O}{\|}}{C}-OH$

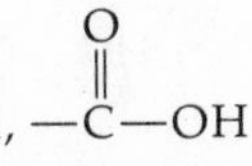

3. (a) 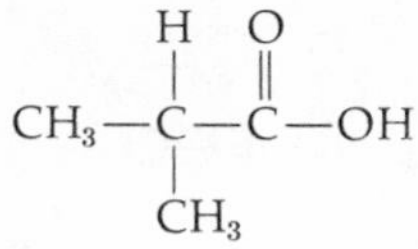 $CH_3-\overset{\overset{\displaystyle H}{|}}{\underset{\underset{\displaystyle CH_3}{|}}{C}}-\overset{\overset{\displaystyle O}{\|}}{C}-OH$ (b) $CH_3-CH_2-\overset{\overset{\displaystyle O}{\|}}{C}-O-CH_3$

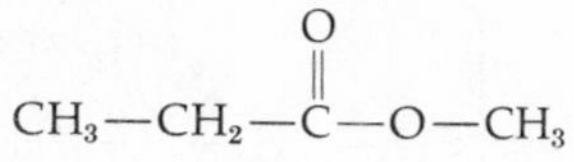

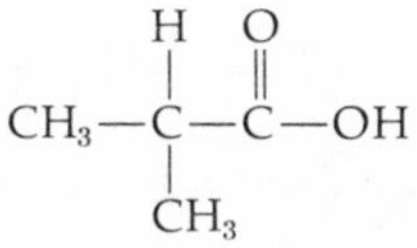

    (c) $CH_3-\overset{\overset{\displaystyle H}{|}}{\underset{\underset{\displaystyle CH_3}{|}}{C}}-\overset{\overset{\displaystyle O}{\|}}{C}-OH$ (d) 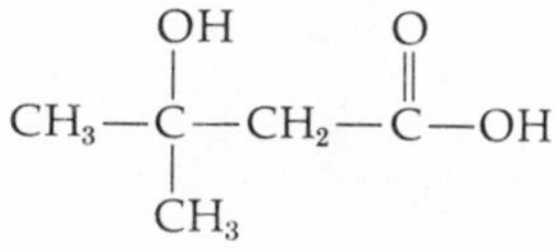 $CH_3-\overset{\overset{\displaystyle OH}{|}}{\underset{\underset{\displaystyle CH_3}{|}}{C}}-CH_2-\overset{\overset{\displaystyle O}{\|}}{C}-OH$

    (e) 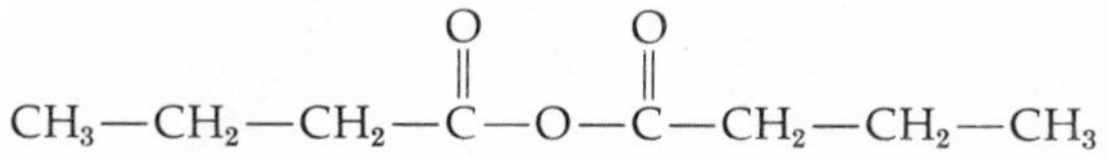 $CH_3-CH_2-CH_2-\overset{\overset{\displaystyle O}{\|}}{C}-O-\overset{\overset{\displaystyle O}{\|}}{C}-CH_2-CH_2-CH_3$

4. (a) α,α-Dimethylpropionic acid (b) Isopropyl α,α-dimethylpropionate
    (c) Ethyl propionate (d) Methyl benzoate
    (e) Ethyl phosphate

5. (a) 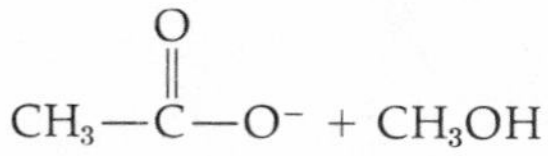 $CH_3-\overset{\overset{\displaystyle O}{\|}}{C}-O^- + CH_3OH$ (b) 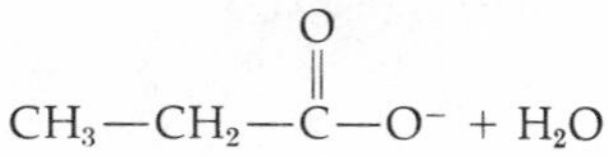 $CH_3-CH_2-\overset{\overset{\displaystyle O}{\|}}{C}-O^- + H_2O$

    (c) and (d) N.R. (e) CH_3-CH_3

    (f) 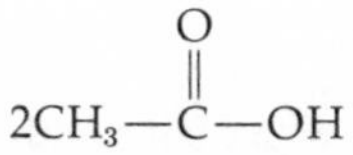 $2CH_3-\overset{\overset{\displaystyle O}{\|}}{C}-OH$

6. (a) is a functional group isomer of the reactant (propionic acid) in (b), (c), and (e).

7. N.R. in cold water. Ethyl acetate, $CH_3-\overset{\overset{\displaystyle O}{\|}}{C}-O-CH_2-CH_3$, is formed with a strong acid at a higher temperature.

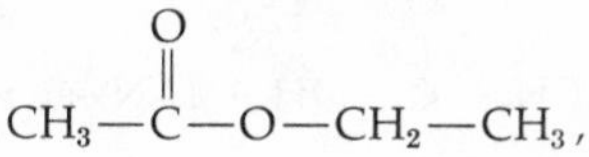

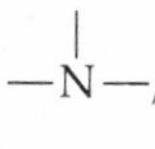

CHAPTER 18

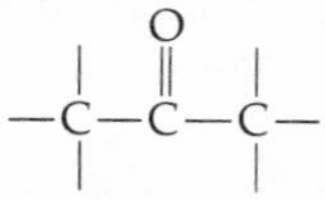

1. (a) Amide, $-\overset{\overset{\displaystyle O}{\|}}{C}-\underset{|}{N}-$ (b) Amine, $-\overset{|}{N}-$, and ketone, $-\overset{|}{\underset{|}{C}}-\overset{\overset{\displaystyle O}{\|}}{C}-\overset{|}{\underset{|}{C}}-$

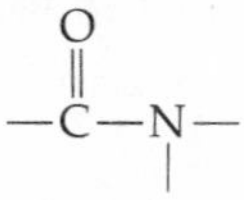

    (c) Amide, $-\overset{\overset{\displaystyle O}{\|}}{C}-\underset{|}{N}-$ (d) Disulfide, $-S-S-$ (e) Thioester, $-\overset{\overset{\displaystyle O}{\|}}{C}-S-\overset{|}{\underset{|}{C}}-$

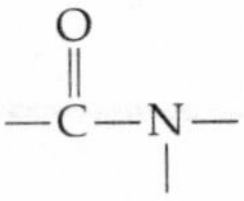

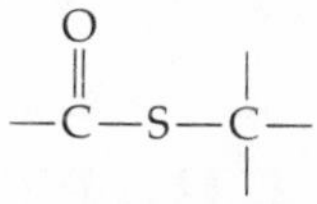

    (f) Ketone, $-\overset{|}{\underset{|}{C}}-\overset{\overset{\displaystyle O}{\|}}{C}-\overset{|}{\underset{|}{C}}-$, and sulfhydryl, $-SH$

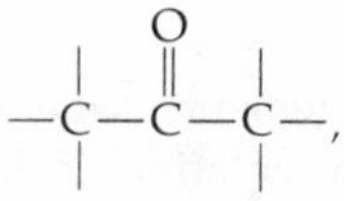

    (g) Tertiary ammonium ion, $-\overset{|}{\underset{|}{C}}-\overset{\overset{\displaystyle H}{|}}{\underset{\underset{\displaystyle \overset{|}{C}}{|}}{N^+}}-\overset{|}{\underset{|}{C}}-$

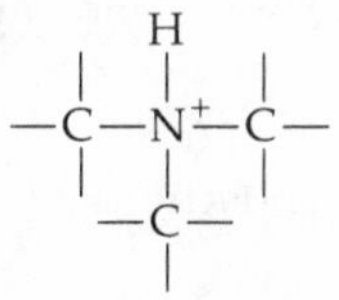

2. (a) N-ethylformamide (c) N-methylacetamide
    (d) Methyl disulfide (e) Ethyl thioacetate
    (g) Trimethylammonium chloride

3. (a) $\left[\begin{array}{c} \text{CH}_2-\text{CH}_3 \\ | \\ \text{CH}_3-\text{CH}_2-\overset{+}{\text{N}}-\text{CH}_2-\text{CH}_3 \\ | \\ \text{CH}_2-\text{CH}_3 \end{array} \right] \text{Br}^-$ (b) $\text{CH}_3-\overset{\overset{\text{O}}{||}}{\text{C}}-\text{S}-\text{CH}_2-\text{CH}_2-\text{CH}_3$

(c) $\text{CH}_3-\text{CH}_2-\text{CH}_2-\overset{\overset{\text{O}}{||}}{\text{C}}-\underset{\underset{\text{H}}{|}}{\text{N}}-\text{CH}_3$ (d) $\text{CH}_3-\underset{\underset{\text{H}}{|}}{\overset{\overset{\text{CH}_3}{|}}{\text{C}}}-\text{S}-\text{H}$

(e) $\text{CH}_3-\text{CH}_2-\text{CH}_2-\overset{\overset{\text{O}}{||}}{\text{C}}-\underset{\underset{\text{CH}_2-\text{CH}_3}{|}}{\text{N}}-\text{CH}_2-\text{CH}_3$

4. The ammonium salt should be more soluble in water than an N-substituted amide of comparable size.

5. (a) $\text{CH}_3-\overset{\overset{\text{O}}{||}}{\text{C}}-\text{S}-\underset{\underset{\text{CH}_3}{|}}{\overset{\overset{\text{CH}_3}{|}}{\text{C}}}-\text{H}$ (b) $\text{H}-\underset{\underset{\text{CH}_3}{|}}{\overset{\overset{\text{CH}_3}{|}}{\text{C}}}-\text{S}-\text{S}-\underset{\underset{\text{CH}_3}{|}}{\overset{\overset{\text{CH}_3}{|}}{\text{C}}}-\text{H}$

(c) Some $\text{H}_3\text{N}^+-\text{CH}_2-\text{CH}_3 + \text{OH}^-$, since ethylamine is a weak base, although most of the ethylamine will remain unprotonated.

(d) $\text{CH}_3-\underset{\underset{\text{CH}_3}{|}}{\overset{\overset{\text{CH}_3}{|}}{\text{C}}}-\text{CH}_2-\overset{\overset{\text{O}}{||}}{\text{C}}-\text{OH} + \text{H}_2\text{N}-\text{CH}_2-\text{CH}_3$

(e) $\text{CH}_3-\overset{\overset{\text{O}}{||}}{\text{C}}-\underset{\underset{\text{CH}_3}{|}}{\text{N}}-\text{CH}_2-\text{CH}_3$

(f) CH_3SH

6. Ethyl alcohol has a higher boiling point, because molecules of an alcohol form strong hydrogen bonds with each other, whereas thiols have a much weaker tendency to form hydrogen bonds.

CHAPTER 19

1. Structural isomers have different bonding arrangements, whereas stereoisomers have the same bonding arrangement but a different orientation of the bonds in space. The compounds ethanol, $\text{CH}_3\text{CH}_2\text{OH}$, and dimethyl ether, $\text{CH}_3-\text{O}-\text{CH}_3$, are structural isomers. The compounds D-lactic acid, $\text{H}-\underset{\underset{\text{CH}_3}{|}}{\overset{\overset{\text{COOH}}{|}}{}}-\text{OH}$, and L-lactic acid, $\text{HO}-\underset{\underset{\text{CH}_3}{|}}{\overset{\overset{\text{COOH}}{|}}{}}-\text{H}$, are stereoisomers.

2. (b)

3. Diastereomer pairs: I and III; I and IV; II and III; II and IV
   Enantiomer pairs: I and II; III and IV

4. I and II represent the same meso-type compound.
   I and III, and I and IV, are diastereomers.
   III and IV are enantiomers.

5. (d)

6. (a) and (b)

7. (a)

$$CH_3-\overset{*}{\underset{\underset{OH}{|}}{CH}}-\overset{\overset{O}{\|}}{C}-NH_2 \qquad HO-\overset{\overset{C-NH_2}{\overset{\|}{O}}}{\underset{CH_3}{|}}-H \qquad H-\overset{\overset{C-NH_2}{\overset{\|}{O}}}{\underset{CH_3}{|}}-OH$$

enantiomers

(b)

$$CH_3-\overset{\overset{Cl}{|}}{\underset{\underset{Cl}{|}}{C}}-\overset{\overset{CH_3}{|}}{\underset{\underset{H}{|}}{C}}-CH_3 \qquad \text{no asymmetric carbon}$$

(c)

$$CH_3-\overset{*}{\underset{\underset{OH}{|}}{CH}}-\overset{*}{\underset{\underset{OH}{|}}{CH}}-CH_2-CH_2-CH_3$$

$$
\begin{array}{cccc}
CH_3 & CH_3 & CH_3 & CH_3 \\
HO{-}{-}H & H{-}{-}OH & H{-}{-}OH & HO{-}{-}H \\
HO{-}{-}H & H{-}{-}OH & HO{-}{-}H & H{-}{-}OH \\
CH_2CH_2CH_3 & CH_2CH_2CH_3 & CH_2CH_2CH_3 & CH_2CH_2CH_3 \\
I & II & III & IV
\end{array}
$$

Enantiomer pairs: I and II; III and IV
Diastereomer pairs: I and III; I and IV; II and III; II and IV

(d)

$$CH_3-CH_2-\overset{*}{\underset{\underset{NH_2}{|}}{CH}}-\overset{\overset{O}{\|}}{C}-OH \qquad H_2N-\overset{\overset{COOH}{|}}{\underset{CH_2CH_3}{|}}-H \qquad H-\overset{\overset{COOH}{|}}{\underset{CH_2CH_3}{|}}-NH_2 \qquad \text{enantiomers}$$

8. (a)

$$
\begin{array}{cccc}
\triangle\!\!-CH_3 & H\!-\!\triangle\!-H & H_3C\!-\!\triangle\!-H & H\!-\!\triangle\!-CH_3 \\
| & | \quad | & | \quad | & | \quad | \\
CH_3 & CH_3 \;\; CH_3 & H \;\; CH_3 & CH_3 \;\; H \\
I & II & III & IV
\end{array}
$$

(b) Structural isomers: I and II; I and III; I and IV
   Stereoisomers: II and III; II and IV; III and IV
(c) III and IV are enantiomers.
   II and III, or II and IV, are diastereomers.

9. In linearly polarized light, the electric field is oriented in a single direction. Unpolarized light is a combination of electromagnetic waves with the electric field in all directions in the plane perpendicular to the direction of propagation.

10. You could measure the rotation of a solution of each isomer with a polarimeter. You could then determine which isomer is in which bottle from the results of this experimental test and the fact that D-glyceraldehyde is (+) and L-glyceraldehyde is (−).

CHAPTER 20

1. (a)

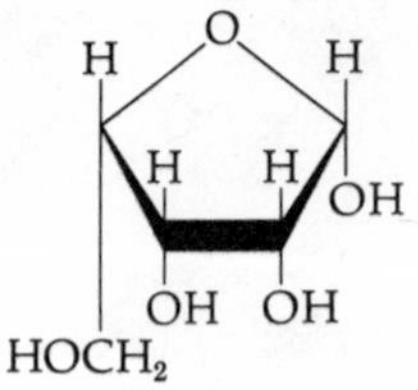

(b)

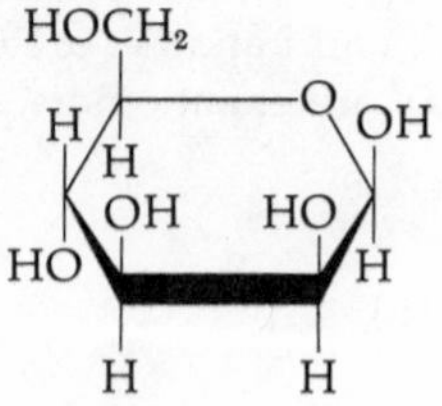

(c)

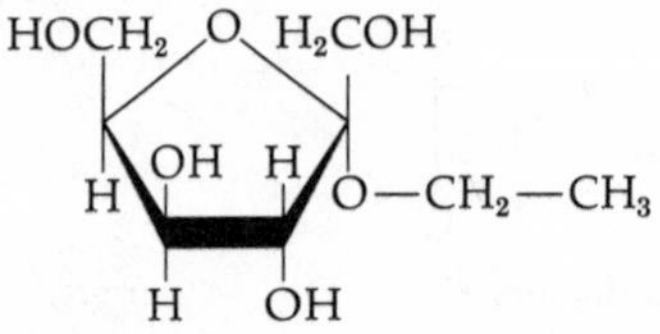

2. (a)

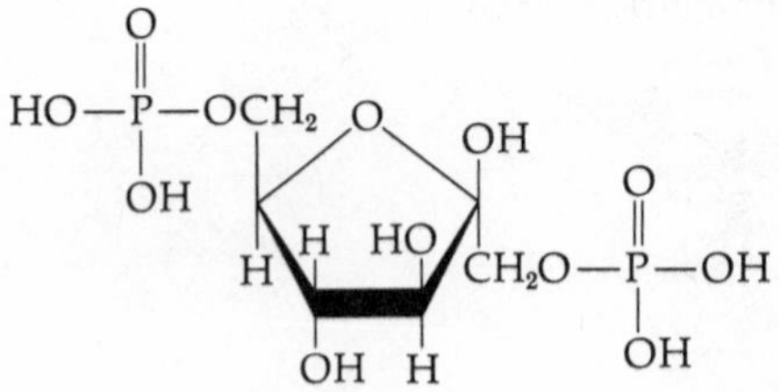

(b)

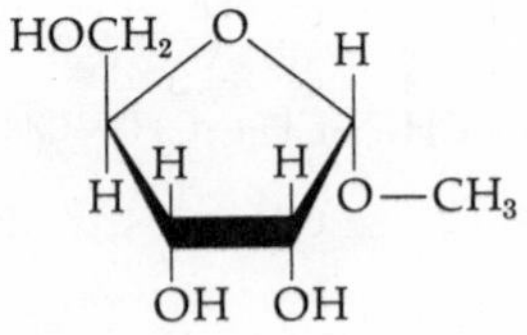

3.

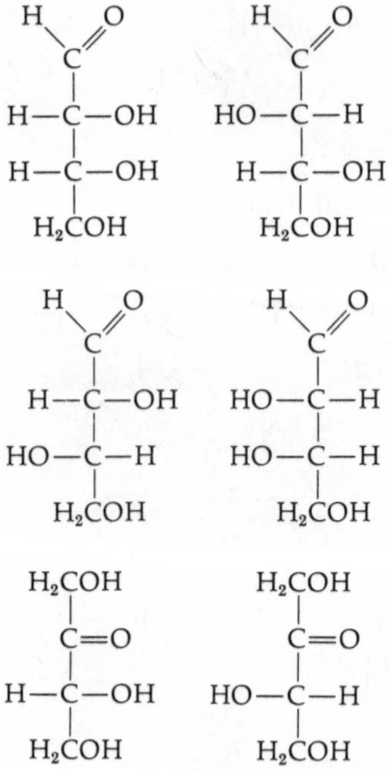

4. (a)

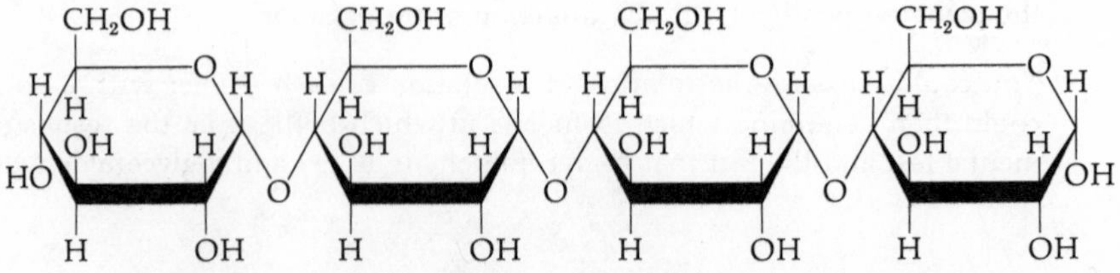

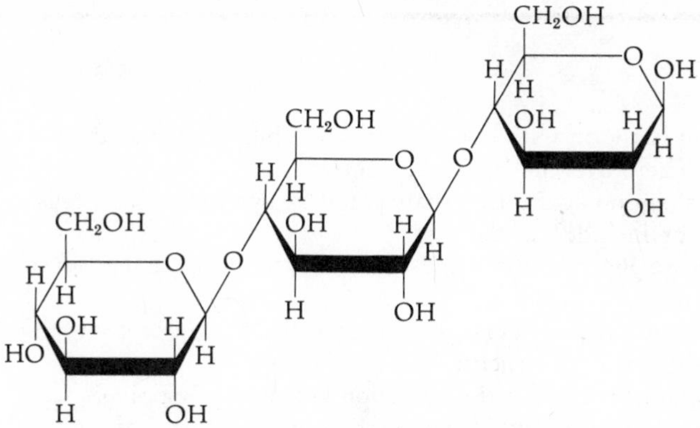

5. The carbon atom in the hemiacetal, hemiketal, acetal, or ketal form of a sugar that is the carbonyl carbon in the open-chain form of the sugar is called an anomeric carbon.

6. maltose:

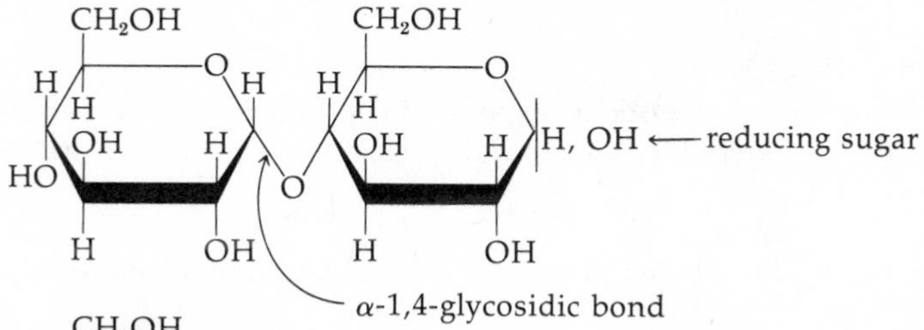

H, OH ⟵ reducing sugar

α-1,4-glycosidic bond

sucrose:

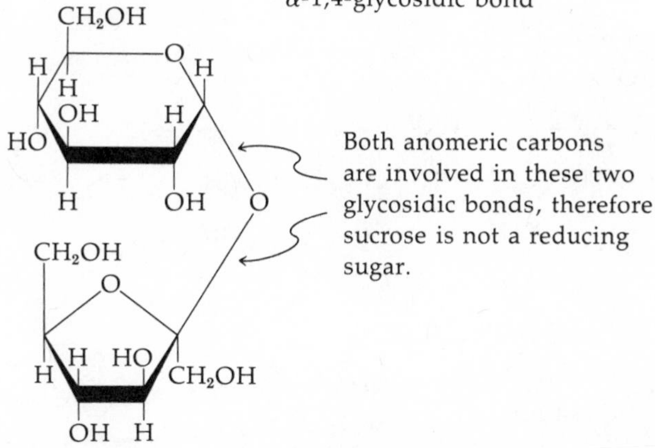

Both anomeric carbons are involved in these two glycosidic bonds, therefore sucrose is not a reducing sugar.

7. (a)

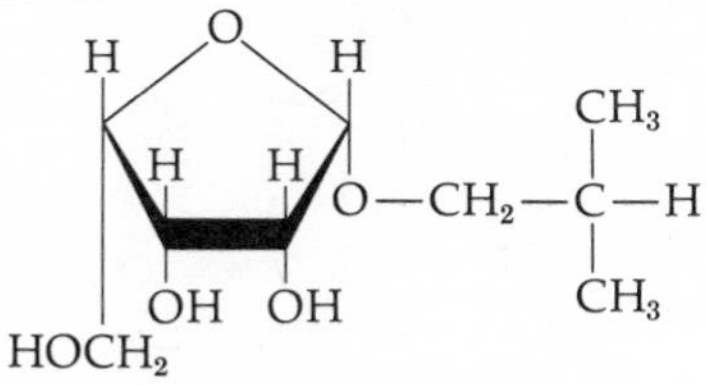

(b) N.R.

8.

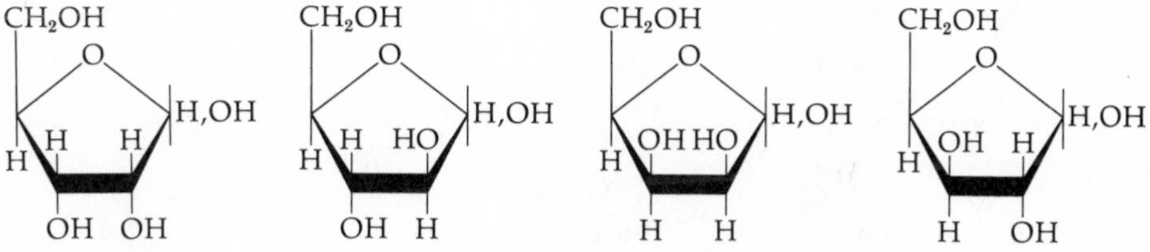

9. In addition to D-glyceraldehyde, they are

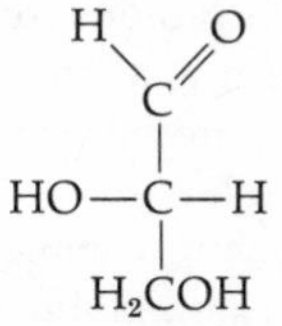

L-glyceraldehyde   dihydroxyacetone

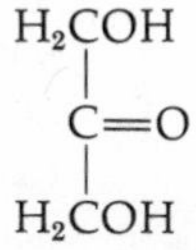

CHAPTER 21

1. (a) A zwitterion is an ion with both a positively charged part and a negatively charged part, but with a zero overall electrical charge.
(b) An essential amino acid is one that cannot be synthesized by cells in the body and must therefore be included in the diet.
(c) Isoenzymes are oligomeric enzymes that catalyze the same reaction but have slightly different subunits.
(d) The term denaturation refers to any process that drastically alters the shape of a protein but leaves its primary structure intact.
(e) A salt bridge in a protein is the attraction between a negatively charged amino acid side chain and a nearby positively charged amino acid side chain.

2. 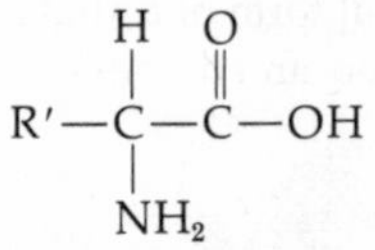
$$R'{-}\overset{\displaystyle H}{\underset{\displaystyle NH_2}{C}}{-}\overset{\displaystyle O}{C}{-}OH$$

3. Usually the term peptide refers to a string containing fewer than 100 amino acids joined together, whereas the term protein is used for a string with 100 or more amino acids.

4. (a) +1   (b) −1   (c) −2

5. Hydrogen bonds between every third peptide group along the amino acid chain

6.

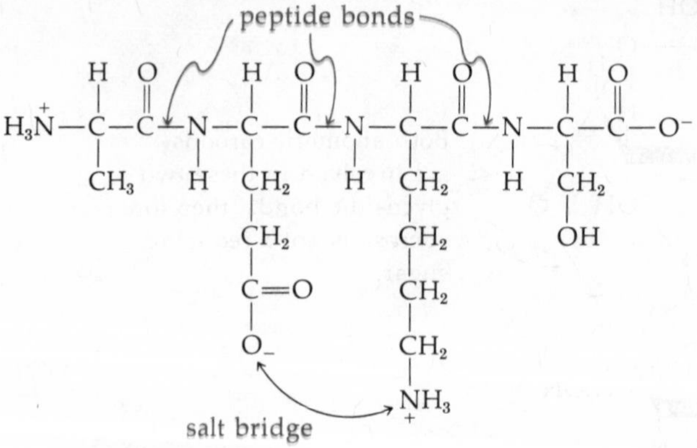

7. Tertiary and quaternary

8. (d) A drastic increase in temperature

9.

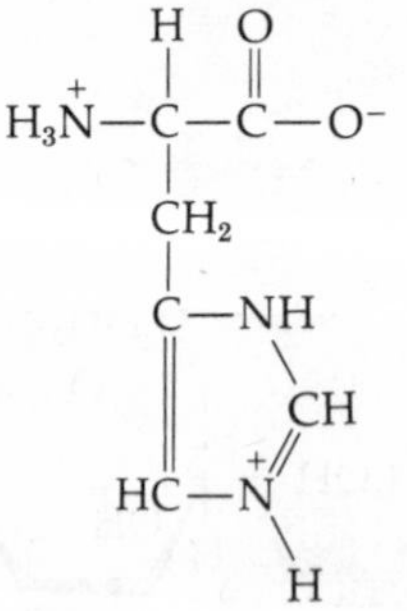

10. (i) a   (ii) a   (iii) c   (iv) b   (v) a

11. (a) An oligomer is a protein composed of subunits.
(b) A protein subunit is a single polypeptide chain component of an oligomer.
(c) Coagulation of a protein is its irreversible denaturation that results in the formation of an insoluble complex.
(d) A hydrophobic core consists of tightly packed side chains of hydrophobic amino acids in the interior of a globular protein.

(e) A β-pleated sheet is a zigzag secondary structural element involving hydrogen bonds between peptide groups of two or more chains of amino acids or two or more portions of the same chain.

12. Disulfide bonds are covalent bonds formed between cysteine side chains, which need not be near each other in the primary structure of a protein. They stabilize tertiary and sometimes quaternary structure.

13.

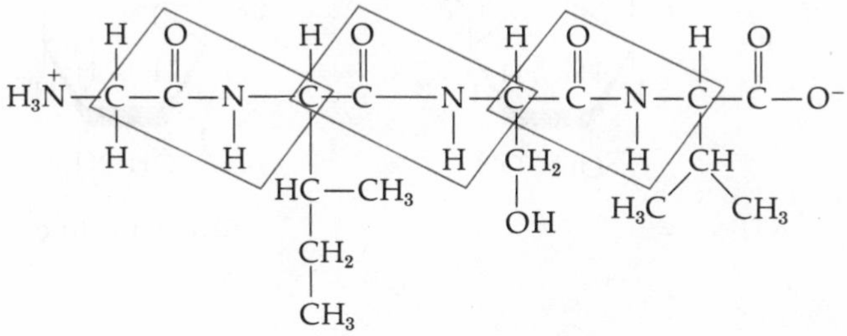

(a) The atoms enclosed by each of the three rectangles lie in a common plane.
(b) +1

14. (d) None of these

CHAPTER 22

1. (c) Active transport

2. (a), (b), (c), and (d)

3. (b) Albumins

4. (d) An enzyme

5. (c) Coenzyme requirement and (e) Water solubility

6. About 150,000; two antigen binding sites and one F_c part.

CHAPTER 23

1. W is an allosteric inhibitor of the enzyme E_1.

2. Molar activities: A = 4.0×10^3 min^{-1}, B = 6.0×10^3 min^{-1}. B is therefore more efficient.

3. 2.4×10^{-2} M/min

4. Hydrolases

5. The substrate, upon binding, induces the appropriate conformational change in the enzyme to allow for catalysis.

6. Lysine, arginine, and/or histidine can form ionic interactions with the negatively charged phosphate groups. Serine, threonine, tyrosine, or other polar amino acids can form hydrogen bonds with the hydroxyl groups on the ribose portion of ATP or the nitrogen atoms of the ring. Phenylalanine, tryptophan, and other amino acids with nonpolar side chains can form hydrophobic interactions with the nitrogen-containing ring of ATP.

CHAPTER 24

1.

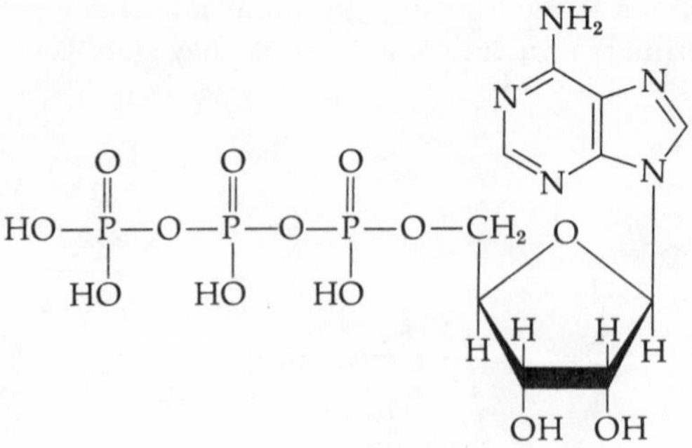

ATP

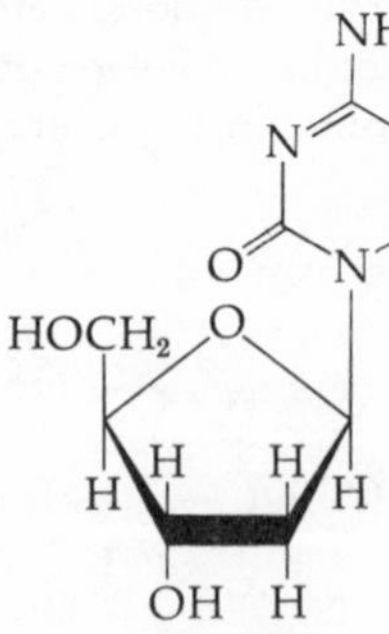

deoxycytidine

2. DNA,

rRNA is part of ribosomes,

tRNA,

mRNA (random strand),

3. Initiation: Step 1—Binding of met-tRNA$_{met}$ to mRNA and the smaller subunit
                 Step 2—Binding of the larger subunit to form the complete complex
   Elongation: Steps 3, 4, 5, 6, etc.
   Termination: Step 8 (not shown in Figure 24-13) involves release of the completed pro-
                tein when a chain-termination codon is encountered.

4. Repression. For drawing see Figure 24-15. The end product in this case is histidine. An
   accumulation of histidine leads to repression.

5. DNA polymerase must recognize and use deoxyribonucleotides involving adenine,
   guanine, cytidine, and thymine, whereas RNA polymerase uses ribonucleotides involving adenine, guanine, cytosine, and uracil.

CHAPTER 25

1. The two steps are step 7, catalyzed by phosphoglycerate kinase, and step 10, catalyzed
   by pyruvate kinase.

2. (a) Products, CO_2 and H_2O; 19 ATP   (b) Product, lactic acid; 8 ATP
   (c) Products, CO_2 and H_2O; 15 ATP   (d) Product, oxaloacetic acid; 5 ATP

3. All six end up in CO_2 molecules. The decarboxylation of two pyruvic acid molecules in
   the reaction catalyzed by pyruvate dehydrogenase accounts for the two carbon atoms
   that do not enter the TCA cycle as part of the two molecules of acetyl-S-CoA.

4. Both pyruvate dehydrogenase and α-ketoglutarate dehydrogenase catalyze the decarboxylation of substrates that are α-ketoacids and produce thioesters. The glyceraldehyde
   3-phosphate dehydrogenase reaction does not involve the decarboxylation of an α-
   ketoacid, but rather the formation of a phosphate ester. All three reactions are coupled to
   the formation of NADH from NAD$^+$.

5. $C_3H_6O_3 + 3O_2 \rightarrow 3CO_2 + 3H_2O$. Eighteen moles of ATP are produced.

6. The hydroxyl group of citric acid is attached to a tertiary carbon atom and cannot be oxidized (see Chapter 15).

7. Cyanide blocks the operation of the ETS. Thus NADH will build up, the supply of NAD$^+$ will become exhausted, and the TCA cycle will not continue to function.

8. (a) FADH$_2$ transfers its electrons to the ETS after the first site of coupling of the ETS with ATP synthesis.
   (b) No. Both form one molecule of water per coenzyme molecule.

9. The two NADH molecules produced in the cytoplasm during glycolysis are shuttled to form two FADH$_2$ molecules in muscle mitochondria, whereas they are shuttled to form two NADH molecules in liver mitochondria. An input of two FADH$_2$ molecules into the ETS yields four ATP molecules, whereas an input of two NADH molecules into the ETS yields six ATP molecules.

10. (a)
$$
\underset{\text{HO}}{\text{HO}}-\overset{\overset{\text{O}}{\|}}{\text{C}}-\overset{\overset{\text{OH}}{|}}{\text{C}}=\text{CH}_2
$$

   (b) P$_i$ +
$$
\text{H}-\overset{\overset{\overset{\text{O}}{\|}}{\text{C}-\text{H}}}{\underset{\underset{\text{H}_2\text{C}-\text{O}-\text{(P)}}{|}}{\overset{|}{\text{C}}}}-\text{OH}
$$

   (c)
$$
\text{H}-\overset{\overset{\overset{\text{O}}{\|}}{\text{C}-\text{H}}}{\underset{\underset{\text{H}_2\text{C}-\text{OH}}{|}}{\overset{|}{\text{C}}}}-\text{O}-\text{(P)}
$$

11. 3.8 kcal/g, 1.6 kcal/g in muscle cells

12. **(Optional)** Yes. A component of RNA nucleotides, ribose 5-phosphate, is produced.

13. Oxaloacetic acid.

14. No. The enzyme that galactosemic children cannot produce is also lacking when these children become adults.

15. Lactose catabolism and lactose synthesis involve UDP-galactose. Glycogen synthesis and lactose catabolism involve UDP-glucose.

16. Epinephrine $\xrightarrow{\text{allosteric}}$ adenyl cyclase $\longrightarrow$ cyclic AMP

   cyclic AMP $\xrightarrow{\text{allosteric}}$ protein kinase $\xrightarrow[\text{modification}]{\text{chemical}}$ glycogen synthetase and glycogen phosphorylase

17. (a) two (b) seven

18. Many sugarholics produce insulin at elevated levels. They may develop hypoglycemia if their pancreas continues to produce insulin at a high rate even though their sugar intake has decreased.

19. Overproduction of insulin in response to the ingested donuts (and possibly sugar in the coffee) may cause a temporary drop in the blood sugar level.

20.
glucose 1-phosphate

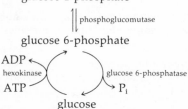

glucose 6-phosphate

21. (a) $CH_3-\overset{\overset{\displaystyle O}{\|}}{C}-S-CoA + NADH + H^+ + CO_2$

(b) $HO-\overset{\overset{\displaystyle O}{\|}}{C}-CH_2-\overset{\overset{\displaystyle O}{\|}}{C}-\overset{\overset{\displaystyle O}{\|}}{C}-OH + ADP + P_i$

(c) $CH_3-\overset{\overset{\displaystyle HO}{|}}{\underset{\underset{\displaystyle H}{|}}{C}}-\overset{\overset{\displaystyle O}{\|}}{C}-OH + NAD^+$

CHAPTER 26

1. Lipids can function as hormones, components of cell membranes, vitamins, and for energy storage and insulation.

2. They exist almost entirely as components of complex lipids rather than as free fatty acids.

3.
$$H_2C-O-\overset{\overset{\displaystyle O}{\|}}{C}-(CH_2)_{10}-CH_3$$
$$H-\overset{|}{C}-O-\overset{\overset{\displaystyle O}{\|}}{C}-(CH_2)_{10}-CH_3$$
$$H_2C-O-\overset{\overset{\displaystyle O}{\|}}{C}-(CH_2)_{10}-CH_3$$

4. $H-\overset{\overset{\displaystyle H_2C-OH}{|}}{\underset{\underset{\displaystyle H_2C-OH}{|}}{C}}-OH$ and $CH_3-(CH_2)_{10}-\overset{\overset{\displaystyle O}{\|}}{C}-O^-Na^+$

5. Micelles that can be washed away are formed with the long hydrophobic alkyl groups of the soap interacting with the hydrocarbon molecules, and the charged carboxylate groups of the soap interacting with water molecules.

6. (a) Abnormal membrane composition and therefore abnormal membrane function
   (b) Inability to manufacture prostaglandins

7.

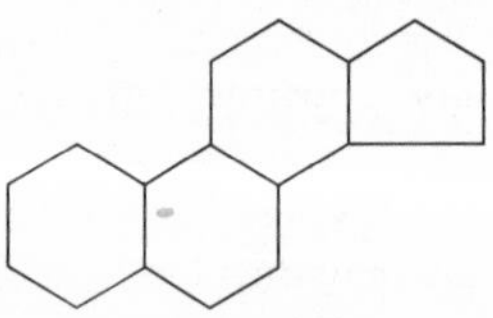

8. They competitively inhibit the formation of proconvertin. These rat poisons cause death as a result of uncontrolled internal bleeding.

9. Siberian children frequently do not receive adequate exposure to sunlight, which catalyzes the formation of vitamin D from precursors in human skin. Sunlamp treatments could be one solution to this problem.

10. Unsaturated fatty acid components lower the melting temperature of membranes, thus giving them more fluid characteristics at body temperature than membranes composed exclusively of higher-melting-point saturated fatty acids.

11. Lipoproteins are held together mainly by hydrophobic interactions between their non-polar lipid components and the hydrophobic side chains of some amino acids of the protein component.

CHAPTER 27

1. This triglyceride $\xrightarrow{\text{lipase}}$ glycerol   + 3 hexanoic acid
   (21 carbon atoms)

   glycerol $\downarrow$ glycerol kinase   3 hexanoyl-S-CoA $\downarrow$ activation

   glycerol 3-phosphate   3 hexanoyl-S-CoA

   $\downarrow$ glycerol 3-phosphate dehydrogenase $\downarrow$ β-oxidation

   dihydroxyacetone phosphate   9 acetyl-S-CoA

   $\downarrow$ glycolysis

   pyruvic acid

   $\downarrow$ pyruvate dehydrogenase

   CO_2 + acetyl-S-CoA

   Thus, 20 of the carbon atoms become part of acetyl groups, whereas one carbon atom from the catabolism of glycerol becomes part of a CO_2 molecule.

2.

3. 154 moles of ATP

4. The active "business end" of both coenzyme A and the acyl carrier portion of fatty acid synthetase are identical, ending with a sulfhydryl group, —SH, which combines with a carboxylic acid to form a thioester. The acyl carrier portion of fatty acid synthetase is used exclusively for fatty acid biosynthesis, whereas coenzyme A is involved in a variety of catabolic pathways.

5. They are intermediates that are converted to diglycerides and then to phosphoglycerides.

6. Both are synthesized from nucleoside diphosphate-activated precursors.

7. See Figure 27-8. Fatty acid synthesis from acetyl-S-CoA requires NADPH, which is provided by the pentose phosphate pathway and the pathway involving malic enzyme. Malic enzyme is allosterically activated by the precursor of fatty acids, acetyl-S-CoA. Other controls of fatty acid biosynthesis include allosteric inhibition of isocitrate dehydrogenase by excess ATP, which shunts acetyl-S-CoA to fatty acid synthesis instead of into the TCA cycle; allosteric activation of the enzyme acetyl-S-CoA carboxylase by excess TCA cycle intermediates; and inhibition of both acetyl-S-CoA carboxylase and fatty acid synthetase by excess palmitoyl-S-CoA. The synthesis of triglycerides from fatty acids is stimulated by insulin, whereas the hydrolysis of triglycerides to fatty acids is stimulated by epinephrine and glucagon.

8. (a) CO_2 + $CH_3-\overset{O}{\overset{||}{C}}-\overset{O}{\overset{||}{C}}-OH$

   (b) $HO-\overset{O}{\overset{||}{C}}-CH_2-\overset{O}{\overset{||}{C}}-S-CoA$

(c) $CH_3-\overset{O}{\overset{\|}{C}}-CH_2-\overset{O}{\overset{\|}{C}}-S-CoA$

(d) $H-\overset{H_2C-OH}{\underset{H_2C-OH}{\overset{|}{\underset{|}{C}}}-OH} + R_1-\overset{O}{\overset{\|}{C}}-OH + R_2-\overset{O}{\overset{\|}{C}}-OH + R_3-\overset{O}{\overset{\|}{C}}-OH$

CHAPTER 28

1. Protein digestion is the hydrolysis of ingested protein in the digestive tract, whereas protein turnover involves the hydrolysis of body proteins in cells and body fluids.

2. (a) Transamination of pyruvic acid, $CH_3-\overset{O}{\overset{\|}{C}}-\overset{O}{\overset{\|}{C}}-OH$, yields alanine.

   (b) Transamination of oxaloacetic acid, $HO-\overset{O}{\overset{\|}{C}}-CH_2-\overset{O}{\overset{\|}{C}}-\overset{O}{\overset{\|}{C}}-OH$, yields aspartic acid.

   (c) Transamination of α-ketoglutaric acid, $HO-\overset{O}{\overset{\|}{C}}-CH_2-CH_2-\overset{O}{\overset{\|}{C}}-\overset{O}{\overset{\|}{C}}-OH$, yields glutamic acid.

   (d) Asparagine is produced from aspartic acid, $HO-\overset{O}{\overset{\|}{C}}-CH_2-\overset{H}{\underset{NH_2}{\overset{|}{\underset{|}{C}}}}-\overset{O}{\overset{\|}{C}}-OH$, and ammo-

   nia by asparagine synthetase.

3. Aspartic acid provides one of the amino groups that will end up in urea. Aspartic acid is regenerated by the conversion of fumaric acid, produced in the urea cycle, to oxaloacetic acid by TCA cycle enzymes, followed by transamination of this oxaloacetic acid with glutamic acid serving as the amine donor.

4. Both protein and nucleic acid polymers are hydrolyzed to their respective monomeric components.

5. (a) $CH_3-CH_2-\overset{O}{\overset{\|}{C}}-\overset{O}{\overset{\|}{C}}-OH + NH_3$

   (b) $H_2N-CH_2-CH_2-CH_2-CH_2-CH_2-NH_2 + CO_2$

   (c) $\overset{H_3C}{\underset{H_3C}{}}{\diagdown}\!\!\diagup CH-\overset{O}{\overset{\|}{C}}-\overset{O}{\overset{\|}{C}}-OH + HO-\overset{O}{\overset{\|}{C}}-CH_2-CH_2-\overset{H}{\underset{NH_2}{\overset{|}{\underset{|}{C}}}}-\overset{O}{\overset{\|}{C}}-OH$

   (d) $H_2N-CH_2-CH_2-CH_2-CH_2-\overset{O}{\overset{\|}{C}}-\overset{O}{\overset{\|}{C}}-OH + NH_3 + H_2O_2$

6. The amino group of pyridoxamine phosphate is donated to an α-keto acid, which is usually α-ketoglutaric acid, thus regenerating pyridoxal phosphate.

CHAPTER 29

1. (a) Increased hydrolysis of triglycerides by lipases, increased breakdown of glycogen to glucose 1-phosphate, and decreased conversion of glucose to glycogen
   (b) Glycolysis will be accelerated, whereas gluconeogenesis will be inhibited.
   (c) Lactic acid
   (d) Gluconeogenesis in liver cells will convert most of it to glucose.

2. (a) Without a supply of essential fatty acids, you will lack some components of cell membranes and you will be unable to produce prostaglandins.
   (b) You will not be getting required vitamins, minerals, or essential fatty acids.
   (c) None. Humans synthesize cholesterol. It is not needed in the diet.

3. (a) It will decrease as you exhale rapidly.
   (b) It might increase.

4. (a) +
   (b) +
   (c) +
   (d) −

5. Yes

6. (b)

7. (a) and (c)

ILLUSTRATION ACKNOWLEDGMENTS

Chapter 1
p. 6   M.E. Warren/Photo Researchers
p. 11   Courtesy of the Science Museum, London
p. 13   Yvonne Freund/Photo Researchers
p. 18   CSTK/Sovfoto
p. 25   (clockwise from upper left) Stephen Feldman/Photo Researchers; Christa Armstrong/Photo Researchers; G. Gillette/Photo Researchers; J. Dermid/Bruce Coleman
p. 28   (left) Bruce Roberts/Photo Researchers; (right) Arthur Tress/Magnum
p. 29   Tom McHugh/Photo Researchers

Chapter 2
p. 42   Burk Uzzle/Magnum
p. 50   Brown Brothers
p. 51   Culver
p. 59   Brown Brothers
p. 63   Burk Uzzle/Magnum

Chapter 3
p. 68   Dennis Stock/Magnum
p. 72   Culver
p. 75   Sovfoto
p. 76   Earl R. Baker/USDA Soil Conservation Service
p. 82   Figure 3-7 (a) adapted from *Chemistry, Matter, and the Universe* by Richard E. Dickerson and Irving Geis, Benjamin-Cummings, Menlo Park, CA, 1976.
p. 85   Frank Meitz/FPG
p. 91   R.I. Nesmith & Associates/FPG

Chapter 4
p. 100   Wayne Miller/Magnum
p. 104   David Scharf/Peter Arnold
p. 122   Bruce Davidson/Magnum

Chapter 5
p. 132   Photo Researchers
p. 150   Figure 5-16 adapted from *Chemistry, Matter, and the Universe* by Richard E. Dickerson and Irving Geis, Benjamin-Cummings, Menlo Park, CA, 1976.
p. 152   (left) UPI; (right) J.P. Laffont/Sygma

Chapter 6
p. 164   Rene Burri/Magnum
p. 179   The Granger Collection
p. 184   Toni Schneiders/Bruce Coleman

Chapter 7
p. 192   Paul Fusco/Magnum

Chapter 8
p. 210   Photo Researchers
p. 214   Editorial Photocolor Archives
p. 227   Bruce Davidson/Magnum

Chapter 9
p. 238   FPG

Chapter 10
p. 264   Henri Cartier-Bresson/Magnum
p. 272   Michael Manheim/Photo Researchers

Chapter 11
p. 290   Kenneth R Ekkens

Chapter 12
p. 318   Doisneau/Photo Researchers
p. 327   Raimondo Borea
p. 336   Georg Gerster/Photo Researchers

Chapter 13
p. 344   Van Bucher/Photo Researchers

Chapter 14
p. 362   John Bryson/Photo Researchers
p. 366   Georg Gerster/Photo Researchers
p. 375   Michael Hayman/Photo Researchers

Chapter 15
p. 386   Novosti/Sovfoto

Chapter 16
p. 400   Photoworld/FPG

Chapter 17
p. 414   E.I. du Pont de Nemours & Co.

Chapter 18
p. 428   Bruce Coleman

Chapter 19
p. 448   Jeanloup Sieff/Photo Researchers

Chapter 20
p. 476   Henri Cartier-Bresson/Magnum
p. 487   Georg Gerster/Photo Researchers
p. 492   Henri Cartier-Bresson/Magnum

Chapter 21
p. 498   Carson Baldwin, Jr./FPG
p. 512   Figure 21-10 after Albert L. Lehninger, *Biochemistry*, 2nd ed., Worth Publishers, Inc., New York, 1975
p. 516   Figure 21-17 adapted from Helena Curtis, *Biology*, 2nd ed., Worth Publishers, Inc., New York, 1975
p. 518   Figure 21-21 adapted from Richard E. Dickerson in H. Neurath, ed., *The Proteins*, Academic Press, Inc., New York, 1964

Chapter 22
p. 528   Philip Harrington/Peter Arnold
p. 537   EPA
p. 544   Rockefeller University
p. 547   Figure 22-10 after E.W. Silverton, M.A. Navia, and D.R. Davies, *Proceedings of the National Academy of Sciences*, **74:** 5140, 1977

Chapter 23
p. 552   Bruce Coleman
p. 563   Figure 23-6 after Lehninger, *op. cit.*
p. 566   Figure 23-10 after C.M. Anderson, F.H. Zucker, and T. A. Steitz, *Science* **204:** 375, 1979. Copyright 1979 by American Association for the Advancement of Science

Chapter 24
p. 576   National Foundation of the March of Dimes
p. 582   Figure 24-4 after L. Pauling and R.B. Corey, *Archives of Biochemistry and Biophysics*, **65:** 164, 1956
p. 584   Figure 24-6 (a) from J.D. Watson and F.H.C. Crick, *Nature*, **171:** 737, 1953; 24-6 (b) from *DNA Synthesis* by Arthur Kornberg, W.H. Freeman and Company, copyright © 1974
p. 585   Figure 24-7 (b) reproduced, with permission, from C.G. Kurland, *Annual Review of Biochemistry*, Volume 46, © 1977 by Annual Reviews Inc.; 24-7 (c, left) adapted from S.H. Kim *et al.*, *Science*, **185:** 435, 1974, copyright 1974 by the American Association for the Advancement of Science; 24-7 (c, right) redrawn from S.H. Kim *et al.*, *Proceedings of the National Academy of Sciences*, **71:** 4970, 1974
p. 588   O.H. Miller, Jr., and Barbara R. Beatty, Biology Division, Oak Ridge National Laboratory
p. 592   Figure 24-12 adapted from Kim *et al.*, *op. cit.*
p. 593   Figure 24-13 adapted from Kurland, *op. cit.*
p. 594   Figure 24-14 from O.L. Miller, Jr., Barbara A. Hamkalo, and C.A. Thomas, Jr., *Science*, **169:** 392, 1970. Copyright 1970 by the American Association for the Advancement of Science

Chapter 25
p. 606   Victor Aleman/Photoworld
p. 620   Figure 25-9 adapted from Keith R. Porter in Helena Curtis, *Biology*, 3rd ed., Worth Publishers, Inc., New York, 1979
p. 628   Figure 25-14 adapted from "How Cells Make ATP" by Peter C. Hinkle and Richard E. McCarty. Copyright © 1978 by Scientific American Inc. All rights reserved
p. 649   Becton/Dickinson

Chapter 26
p. 654   Malak/Annan Photo Features/Photo Trends
p. 664   American Heart Association
p. 669   John S. O'Brien
p. 671   Figure 26-11 after S.J. Singer and G.L. Nicolson, *Science*, **175:** 720, 1972. Copyright 1972 by the American Association for the Advancement of Science

Chapter 27
p. 674   Burt Glinn/Magnum

Chapter 28
p. 694   Denis Callewaert
p. 705   George Rodger/Magnum

Chapter 29
p. 716   Burt Glinn/Magnum
p. 729   S. Treval/D.B./Bruce Coleman

Essential Skills
p. 738   Charles Harbutt/Magnum

INDEX

The page on which a term is defined, or on which the structural formula for a compound is given, is indicated in **boldface** type.